ENCYCLOPEDIA OF
OPERATIONS RESEARCH
AND
MANAGEMENT SCIENCE

ENCYCLOPEDIA OF
OPERATIONS RESEARCH & MANAGEMENT SCIENCE

Editors: Saul I. Gass and Carl M. Harris

ADVISORY BOARD

ENCYCLOPEDIA OF OPERATIONS RESEARCH AND MANAGEMENT SCIENCE

Edited by

Saul I. Gass
College of Business and Management
University of Maryland
College Park, Maryland, USA

and

Carl M. Harris
Department of Operations Research and Engineering
George Mason University
Fairfax, Virginia, USA

Kluwer Academic Publishers
Boston/Dordrecht/London

Distributors for North America:
Kluwer Academic Publishers
101 Philip Drive
Assinippi Park
Norwell, Massachusetts 02061 USA

Distributors for all other countries:
Kluwer Academic Publishers Group
Distribution Centre
Post Office Box 322
3300 AH Dordrecht, THE NETHERLANDS

Library of Congress Cataloging-in-Publication Data

A C.I.P. Catalogue record for this book is available
from the Library of Congress.

Printed on acid-free paper.

Printed in the United States of America

Table of Contents

CONTRIBUTORS

Leonard Adelman, George Mason University, Virginia. CHOICE THEORY

Susan Albin, Rutgers University, Piscataway, New Jersey. LITTLE'S LAW

Frank Alt, University of Maryland, College Park. QUALITY CONTROL

G. Anandalingam, University of Pennsylvania, Philadelphia. SIMULATED ANNEALING

J. Scott Armstrong, University of Pennsylvania, Philadelphia. FORECASTING

Jay E. Aronson, University of Georgia, Athens. CLUSTER ANALYSIS

Thomas E. Baker, Chesapeake Decision Sciences, New Providence, New Jersey. PETRO-CHEMICAL INDUSTRY

Osman Balci, VPI & SU, Blacksburg, Virginia. VERIFICATION, VALIDATION AND TESTING OF MODELS.

Stephen J. Balut, The Institute for Defense Analyses, Alexandria, Virginia. COST ANALYSIS.

Steve Bankes, The RAND Corporation, Santa Monica, California. EXPLORATORY MODELING

Arnold Barnett, MIT, Cambridge, Massachusetts. CRIME AND JUSTICE

David. J. Bartholomew, London School of Economics and Political Science, England. MANPOWER PLANNING

Frank M. Bass, University of Texas at Dallas. ADVERTISING

Isabel Beichl, National Institute of Standards and Technology, Gaithersburg, Maryland. COMPUTATIONAL GEOMETRY; VORONOI CONSTRUCTS

Peter C. Bell, University of Western Ontario, Canada. VISUALIZATION

Filmore Bender, MIAC, Nairobi, Kenya. AGRICULTURE AND THE FOOD INDUSTRY

Javier Bernal, National Institute of Standards and Technology, Gaithersburg, Maryland. COMPUTATIONAL GEOMETRY; VORONOI CONSTRUCTS

Gabriel R. Bitran, MIT, Cambridge, Massachusetts. PRODUCTION MANAGEMENT

John L. G. Board, London School of Economics and Political Science, England. PORTFOLIO THEORY: MEAN-VARIANCE

Lawrence Bodin, University of Maryland, College Park. VEHICLE ROUTING

Paul T. Boggs, National Institute of Standards and Technology, Gaithersburg, Maryland. INTERIOR POINT METHODS

Percy H. Brill, University of Windsor, Canada. LEVEL CROSSING METHODS

Robert G. Brown, Materials Management Systems, Vermont. EXPONENTIAL SMOOTHING

James R. Buck, University of Iowa, Iowa City. LEARNING

Dennis M. Buede, George Mason University, Virginia. DECISION MAKING; GROUP DECISION COMPUTER TECHNOLOGY

Laura I. Burke, Lehigh University, Bethlehem, Pennsylvania. NEURAL NETS

Richard M. Burton, Duke University, Durham, North Carolina. ORGANIZATION

Kenneth Chelst, Wayne State University, Detroit, Michigan. EMERGENCY SERVICES; URBAN SERVICES

Dilip Chhajed, University of Illinois, Champaign. FACILITY LOCATION

Clyde Chittister, Carnegie Mellon University, Pittsburgh. RISK ASSESSMENT; RISK MANAGEMENT

Nastaran Coleman, TRW, Fairfax, Virginia. BIN PACKING

Sue A. Conger, Southern Methodist University, Dallas, Texas. SYSTEMS ANALYSIS; YIELD MANAGEMENT

William W. Cooper, University of Texas at Austin. DATA ENVELOPMENT ANALYSIS

Richard W. Cottle, Stanford University, Stanford, California. COMPLEMENTARITY PROBLEMS

Sriram Dasu, University of California, Los Angeles. PRODUCTION MANAGEMENT

James A. Dewar, The RAND Corporation, Santa Monica. California. DELPHI METHOD

Ralph L. Disney, Texas A&M University, College Station. NETWORKS OF QUEUES

James S. Dyer, University of Texas at Austin. PREFERENCE THEORY

Joseph G. Ecker, RPI, Troy, New York. GEOMETRIC PROGRAMMING

Jonathan Eckstein, Rutgers University, New Brunswick, New Jersey. PARALLEL COMPUTING

Richard W. Eglese, Lancaster University, England. MATCHING

Jehoshua Eliashberg, University of Pennsylvania, Philadelphia. MARKETING

Joseph H. Engel, Bethesda, Maryland. ETHICS; LANCHESTER'S EQUATIONS

Stuart Eriksen, University of California, Irvine. DECISION TREES

Gerald W. Evans, University of Louisville, Kentucky. SPACE

Anthony V. Fiacco, George Washington University, Washington, DC. NONLINEAR PROGRAMMING

Martin Fischer, MITRE Corporation, Reston, Virginia. DICTIONARY

Peter Fishburn, AT&T Bell Laboratories, New Jersey. UTILITY THEORY

Gene H. Fisher, The RAND Corporation, Santa Monica, California. PUBLIC POLICY, RAND CORPORATION

Charles D. Flagle, Johns Hopkins University, Baltimore, Maryland. MEDICINE AND MEDICAL PRACTICE

Leonard Fortuin, Eindhoven University of Technology, The Netherlands. INDUSTRIAL APPLICATIONS AND OR

Richard L. Francis, University of Florida, Gainesville. FACILITY LOCATION

Linda Weiser Friedman, Baruch College of the City University of New York. METAMODELING

John A. Friel, The RAND Corporation, Santa Monica, California. DELPHI METHOD

Tomas Gal, Furnuniversitaet, Germany. PARAMETRIC PROGRAMMING

Roberto D. Galvão, Federal University of Rio De Janeiro. DEVELOPING COUNTRIES

Viviane Gascon, College militairè royal de St.-Jean, Québec. HEALTH CARE SYSTEMS

Saul I. Gass, University of Maryland, College Park. DICTIONARY; DOCUMENTATION; FUZZY SETS; GLOBAL MODELS; MODEL ACCREDITATION; MODEL EVALUATION

Denos C. Gazis, IBM Research, Yorktown Heights, New York. TRAFFIC ANALYSIS

Arthur M. Geoffrion, University of California, Los Angeles. California. STRUCTURED MODELING

Fred Glover, University of Colorado, Boulder. TABU SEARCH

Paul Gray, Claremont Graduate School, California. GEOGRAPHIC INFORMATION SYSTEMS

Harvey J. Greenberg, University of Colorado, Denver. ARTIFICIAL INTELLIGENCE

Irwin Greenberg, George Mason University, Virginia. DICTIONARY; REGRESSION ANALYSIS

Donald Gross, George Washington University, Washington, DC. DISTRIBUTION SELECTION FOR STOCHASTIC MODELING; RELIABILITY OF SYSTEMS; SIMULATION OF DISCRETE-EVENT STOCHASTIC SYSTEMS

Thomas R. Gulledge, George Mason University, Virginia. COST ANALYSIS

Robert W. Haessler, University of Michigan, Ann Arbor. CUTTING STOCK PROBLEMS

Yacov Y. Haimes, University of Virginia, Charlottesville. RISK ASSESSMENT; RISK MANAGEMENT

Leslie Hall, Johns Hopkins University, Baltimore, Maryland. COMPUTATIONAL COMPLEXITY

Carl M. Harris, George Mason University, Virginia. CENTER FOR NAVAL ANALYSES; MARKOV CHAINS; DICTIONARY; OPERATIONS RESEARCH OFFICE AND RESEARCH ANALYSIS CORPORATION; RISK ASSESSMENT

Dean S. Hartley, Oak Ridge National Laboratory, Tennessee. BATTLE MODELING

Arnoldo C. Hax, MIT, Cambridge, Massachusetts. CORPORATE STRATEGY; HIERARCHICAL PRODUCTION PLANNING

James C. Hearn, University of Georgia, Athens. HIGHER EDUCATION

Sidney W. Hess, Drexel University, Philadelphia, Pennsylvania. POLITICS

Daniel P. Heyman, Bell Communications Research, New Jersey. QUEUEING THEORY

Frederick S. Hillier, Stanford University, Stanford, California. LINEAR PROGRAMMING

James K. Ho, University of Illinois - Chicago. LARGE-SCALE SYSTEMS

Karla L. Hoffman, George Mason University, Virginia. COMBINATORIAL AND INTEGER OPTIMIZATION; TRAVELING SALESMAN PROBLEM

Clyde W. Holsapple, University of Kentucky, Lexington. EXPERT SYSTEMS

James P. Ignizio, University of Virginia, Charlottesville. NEURAL NETWORKS

Richard H. F. Jackson, National Institute of Standards and Technology, Gaithersburg, Maryland. FACTORABLE PROGRAMMING

Kamlesh Jain, University of Maryland, College Park. QUALITY CONTROL

John J. Jarvis, Georgia Institute of Technology, Atlanta. INDUSTRIAL ENGINEERING AND OPERATIONS RESEARCH

Jianmin Jia, University of Texas at Austin. PREFERENCE THEORY

Sharon A. Johnson, WPI, Worcester, Massachusetts. SPLINES

Albert Jones, National Institute of Standards and Technology, Gaithersburg, Maryland. JOB SHOP SCHEDULING

Gerald Kahan, McCormick and Company, Sparks, Maryland. AGRICULTURE AND THE FOOD INDUSTRY

Bharat K. Kaku, University of Maryland, College Park. FACILITIES LAYOUT

Gurumurthy Kalyanaram, University of Texas at Dallas. ADVERTISING

Edward H. Kaplan, Yale University, Connecticut. PROGRAM EVALUATION

Harry H. Kelejian, University of Maryland, College Park. ECONOMETRICS

L. Robin Keller, University of California, Irvine. DECISION TREES

Alan J. King, IBM, Yorktown Heights, New York. STOCHASTIC PROGRAMMING

Jack P.C. Kleijnen, Tilburg University, The Netherlands. MONTE-CARLO SAMPLING AND VARIANCE REDUCTION

Howard W. Kreiner, Center for Naval Analyses (retired), Virginia. FIELD ANALYSIS

Ramayya Krishnan, Carnegie Mellon University, Pittsburgh. MODEL MANAGEMENT

Roman Krzysztofowicz, University of Virginia Charlottesville. WATER RESOURCES

Shaul P. Ladany, Ben Gurion University, Israel. SPORTS

Richard C. Larson, MIT, Cambridge, Massachusetts. HYPERCUBE QUEUEING MODEL; QUEUE INFERENCE ENGINE

Kathryn Blackmond Laskey, George Mason University, Virginia. BAYESIAN DECISION THEORY, SUBJECTIVE PROBABILITY AND UTILITY

Eugene L. Lawler, University of California, Berkeley (deceased). COMBINATORICS

Pierre Ecuyer, Université de Montréal, Canada. RANDOM NUMBER GENERATORS

Reuven R. Levary, Saint Louis University, Missouri. CAPITAL BUDGETING

Bernard Levin, Rockville, Maryland. FIRE MODELS

Matthew J. Liberatore, Villanova University, Pennsylvania. AUTOMATION

Gary L. Lilien, Pennsylvania State University, University Park. MARKETING

Andrew G. Loerch, U.S. Army, Washington, DC. LEARNING CURVES

John R. Lough, University of Georgia, Athens. HIGHER EDUCATION

Timothy J. Lowe, University of Iowa, Iowa City. FACILITY LOCATION

William F. Lucas, Claremont Graduate School, California. GAME THEORY

Michael Magazine, University of Cincinnati, Ohio. SCHEDULING AND SEQUENCING

Thomas L. Magnanti, MIT, Cambridge, Massachusetts. NETWORK OPTIMIZATION

Nicolas S. Majluf, MIT, Cambridge, Massachusetts. CORPORATE STRATEGY

Andre Z. Manitius, George Mason University, Virginia, CONTROL THEORY

Michael D. Maltz, University of Illinois, Chicago. CRIME AND JUSTICE

William G. Marchal, University of Toledo, Ohio. OPERATIONS MANAGEMENT

Carl D. Martland, MIT, Cambridge, Massachusetts. RAIL FREIGHT OPERATIONS

Richard O. Mason, Southern Methodist University, Dallas, Texas. SYSTEMS ANALYSIS; YIELD MANAGEMENT

Christina M. Mastrangelo, University of Virginia, Charlottesville. TIME SERIES ANALYSIS

Zbigniew Michalewicz, University of North Carolina, Charlotte. GENETIC ALGORITHMS

Douglas Miller, George Mason University, Virginia. MARKOV PROCESSES

Hugh J. Miser, Farmington, Connecticut. PRACTICE OF OPERATIONS RESEARCH AND MANAGEMENT SCIENCE

Douglas C. Montgomery, Arizona State University, Tempe. TIME SERIES ANALYSIS

Richard C. Morey, Cornell University, Ithaca, New York. HOSPITALS

Heiner Müller-Merbach, Universität Kaiserslautern, Germany. INFORMATION SYSTEMS and DATABASE DESIGN

Frederic H. Murphy, Temple University, Philadelphia. ECONOMICS

Katta G. Murty, University of Michigan, Ann Arbor. QUADRATIC PROGRAMMING

Steven Nahmias, Santa Clara University, California. GANTT CHARTS

Stephen G. Nash, George Mason University, Virginia. CALCULUS OF VARIATIONS; NUMERICAL ANALYSIS

Marcel F. Neuts, University of Arizona, Tucson. MATRIX-ANALYTIC STOCHASTIC MODELS; PHASE-TYPE PROBABILITY DISTRIBUTIONS

Børge Obel, Odense University, Denmark. ORGANIZATION

Manfred Padberg, New York University, New York. COMBINATORIAL AND INTEGER OPTIMIZATION; TRAVELING SALESMAN PROBLEM

John C. Papageorgiou, University of Massachusetts, Boston. RESEARCH AND DEVELOPMENT

Chad Perry, Queensland Institute of Technology, Australia. CONSTRUCTION APPLICATIONS

William P. Pierskalla, University of California, Los Angeles. HEALTH CARE SYSTEMS

Donald R. Plane, Rollins College, Winter Park, Florida. SPREADSHEETS

Roman A. Polyak, George Mason University, Virginia. BARRIER AND DISTANCE FUNCTIONS

Israel Pressman, Baruch College of the City University of New York. METAMODELING

Ingmar R. Prucha, University of Maryland, College Park. ECONOMETRICS

Luis Rabelo, Ohio University, Athens. JOB SHOP SCHEDULING

John S. Ramberg, University of Arizona, Tucson. TOTAL QUALITY MANAGEMENT

Ramaswamy Ramesh, SUNY - Buffalo, New York. MULTIPLE CRITERIA DECISION MAKING

Graham K. Rand, Lancaster University, England. NETWORK PLANNING; PROJECT MANAGEMENT

Arnold Reisman, Case Western Reserve University, Cleveland, Ohio. LIBRARIES

Charles ReVelle, Johns Hopkins University, Maryland. ENVIRONMENTAL SYSTEM ANALYSIS; LOCATION ANALYSIS

George P. Richardson, SUNY - Albany, New York. SYSTEMS DYNAMICS

Meir J. Rosenblatt, Washington University, Missouri. MATERIAL HANDLING

Jonathan Rosenhead, London School of Economics and Political Science, England. PROBLEM STRUCTURING METHODS

Richard E. Rosenthal, Naval Postgraduate School, Monterey, California. ALGEBRAIC MODELING LANGUAGES FOR OPTIMIZATION

Michael H. Rothkopf, Rutgers University, New Brunswick, New Jersey. BIDDING MODELS

Reuven Y. Rubinstein, Technion Institute of Technology, Israel. MONTE-CARLO SAMPLING AND VARIANCE REDUCTION; SCORE FUNCTIONS

David M. Ryan, University of Auckland, New Zealand. AIRLINE INDUSTRY

Thomas L. Saaty, University of Pittsburgh, Pennsylvania. ANALYTIC HIERARCHY PROCESS

Andrew P. Sage, George Mason University, Virginia. CYBERNETICS

Rakesh K. Sarin, University of California, Los Angeles. MULTI-ATTRIBUTE UTILITY THEORY

Siegfried Schaible, University of California, Riverside. FRACTIONAL PROGRAMMING

Marc J. Schniederjans, University of Nebraska-Lincoln. GOAL PROGRAMMING

David A. Schum, George Mason University, Virginia. DECISION ANALYSIS

William Schwabe, The RAND Corporation, Santa Monica, California. GAMING

Alexander Shapiro, Georgia Institute of Technology, Atlanta. SCORE FUNCTIONS

Ramesh Sharda, Oklahoma State University, Stillwater. COMPUTER SCIENCE AND OPERATIONS RESEARCH

Bala Shetty, Texas A & M University, College Station. MULTI-COMMODITY NETWORK FLOWS

Douglas R. Shier, Clemson University, South Carolina. GRAPH THEORY

Edward A. Silver, University of Calgary, Canada. INVENTORY MODELING

Ariela Sofer, George Mason University, Virginia. DICTIONARY; UNCONSTRAINED OPTIMIZATION

Marius M. Solomon, Northeastern University, Boston, Massachusetts. LOGISTICS

Kathryn E. Stecke, University of Michigan, Ann Arbor. FLEXIBLE MANUFACTURING SYSTEMS

Ralph E. Steuer, University of Georgia, Athens. MULTIOBJECTIVE PROGRAMMING

William R. Stewart, Jr., College of William and Mary, Virginia. CHINESE POSTMAN PROBLEM

Lawrence D. Stone, Metron, Inc., Reston, Virginia. SEARCH THEORY

Todd Strauss, Yale University, Connecticut. PROGRAM EVALUATION

Susan Suchocki, Claremont Graduate School, California. GEOGRAPHIC INFORMATION SYSTEMS

Francis Sullivan, SRC, Bowie, Maryland. COMPUTATIONAL GEOMETRY; VORONOI CONSTRUCTS

Lakshmi Sundaram, University of Georgia, Athens. CLUSTER ANALYSIS

Edward A. Sykes, University of Virginia, Charlottesville. COMMUNICATIONS NETWORKS

Clayton J. Thomas: U. S. Air Force, Washington, DC. AIR FORCE OPERATIONS ANALYSIS; MILITARY OPERATIONS RESEARCH

Kaoru Tone, Saitama University, Japan. RETAILING

Alan Tucker, SUNY - Stony Brook. New York. MATRICES AND MATRIX ALGEBRA

Stanislav Uryasev, Brookhaven National Laboratory, New York. SCORE FUNCTIONS

Igor Ushakov, SOTAS, Rockville, Maryland. AVAILABILITY; MAINTENANCE; POINT STOCHASTIC PROCESSES; REDUNDANCY; RENEWAL PROCESSES; SAFETY

Paul van Beek, Agricultural University, Wageningen, The Netherlands. INDUSTRIAL APPLICATIONS

Luk Van Wassenhove, INSEAD, Fontainebleau, France. INDUSTRIAL APPLICATIONS

Andrew Vazsonyi, University of San Francisco, California. DECISION SUPPORT SYSTEMS

Eugene P. Visco, Office of Deputy Under Secretary of the Army, Washington, DC. OPERATIONS RESEARCH OFFICE AND RESEARCH ANALYSIS CORPORATION

Mark A. Vonderembse, University of Toledo, Ohio, OPERATIONS MANAGEMENT

Warren Walker, The RAND Corporation, Santa Monica, California. PUBLIC POLICY ANALYSIS; RAND CORPORATION

Pearl Wang, George Mason University, Virginia. BIN PACKING

Andrés Weintraub, Universidad de Chile, Chile. NATURAL RESOURCES

Andrew B. Whinston, University of Texas at Austin. EXPERT SYSTEMS

Chelsea C. White, III, University of Michigan, Ann Arbor. DYNAMIC PROGRAMMING; MARKOV DECISION PROCESSES

Yoram (Jerry) Wind, University of Pennsylvania, Philadelphia. MARKETING

Christoph Witzgall, National Institute of Standards and Technology, Gaithersburg, Maryland. COMPUTATIONAL GEOMETRY; VORONOI CONSTRUCTS

Norman Keith Womer, University of Mississippi, University, Mississippi. COST EFFECTIVENESS ANALYSIS

Carlos G. Wong-Martinez, Drexel University, Philadelphia, Pennsylvania. POLITICS

R. E. D. Woolsey, Colorado School of Mines, Golden. IMPLEMENTATION

Xiaomei Xu, Case Western Reserve, Cleveland, Ohio. LIBRARIES

Yuehwern Yih, Purdue University, Indiana. JOB SHOP SCHEDULING

Oliver S. Yu, SRI International, Menlo Park, California. ELECTRIC POWER SYSTEMS

Fatemeh (Mariam) Zahedi, University of Wisconsin, Milwaukee. GROUP DECISION MAKING

Stavros A. Zenios, University of Pennsylvania, Philadelphia. BANKING

William T. Ziemba, University of British Columbia, Vancouver, Canada. PORTFOLIO THEORY: MEAN VARIANCE

Stanley Zionts, SUNY - Buffalo, New York. MULTIPLE CRITERIA DECISION MAKING

Preface

The goal of the *Encyclopedia of Operations Research and Management Science* is to provide to decision makers and problem solvers in business, industry, government and academia a comprehensive overview of the wide range of ideas, methodologies, and synergistic forces that combine to form the preeminent decision-aiding fields of operations research and management science (OR/MS). To this end, we enlisted a distinguished international group of academics and practitioners to contribute articles on subjects for which they are renowned.

The editors, working with the *Encyclopedia's* Editorial Advisory Board, surveyed and divided OR/MS into specific topics that collectively encompass the foundations, applications, and emerging elements of this ever-changing field. We also wanted to establish the close associations that OR/MS has maintained with other scientific endeavors, with special emphasis on its symbiotic relationships to computer science, information processing, and mathematics. Based on our broad view of OR/MS, we commissioned 185 major expository articles and complemented them by numerous entries: descriptions, discussions, definitions, and abbreviations. The connections between topics are highlighted by an entry's final "See" statement, as appropriate. The reader will find that a few entries overlap somewhat in their coverage. We felt that such discussions by different authors would be of value, as they provide perspective and a diversity of viewpoints.

The *Encyclopedia's* intended audience is technically diverse and wide; it includes anyone concerned with the science, techniques, and ideas of how one makes decisions. As this audience encompasses many professions, educational backgrounds and skills, we were attentive to the form, format and scope of the articles. Thus, the articles are designed to serve as initial sources of information for all such readers, with special emphasis on the needs of students. Each article provides a background or history of the topic, describes relevant applications, overviews present and future trends, and lists seminal and current references. To allow for variety in exposition, the authors were instructed to present their material from both research and applied perspectives. In particular, the authors were allowed to use whatever mathematical notation they felt was "standard" for their topics. For space reasons, we attempted to limit each author to about the same size entry. However, we found this restriction to be an inhibiting one and, although some editorial constraints were imposed, the breadth and detail of an article's expository material and references were left to the author.

The *Encyclopedia* does include some historical material, especially on organizations that influenced the development of OR/MS. However, we do not provide major articles on the history of OR and MS. It was felt that these topics have been well covered in the readily available literature. For example, the following 1987 articles by Joseph F. McCloskey that appeared in *Operations Research*: "The Beginnings of Operations Research: 1934–1941," 35, 1, 143–152; "British Operational Research in World War II," 35, 3, 453–470; "U.S. Operations Research in World War II," 35, 6, 910–925; and the 1984 article by Harold Lardner that appeared in *Operations Research*: "The Origin of Operational Research," 32, 2, 465–475.

The *Encyclopedia* contains no entries that define the fields of operations research and management science. OR and MS are often equated to one another. If one defines them by the methodologies they employ, the equation would probably stand inspection. If one defines them by their historical developments and the classes of problems they encompass, the equation becomes fuzzy. The formalism OR grew out of the operational problems of the British and U.S. military efforts in World War II. It was augmented methodologically and computationally by the post-war developments of linear programming, game theory, dynamic programming, discrete-event simulation (among others), and the digital computer. A number of additional ideas and problem types from the pre-war years were incorporated into the field as well, including inventory and queueing theories, Markov modeling, and the basic methods of optimization. Early (1950s) practitioners of OR applied its philosophy and techniques to the solution of industrial and business operational problems with great success. It was soon recognized that whatever OR was as a scientific field, it could be used to study and solve the broader planning and strategic issues of management, organizations, financial analysis, and public policy. From this observation, MS began and flourished in a similar and somewhat overlapping manner to OR.

However, if one wants definitions of OR and MS, they are readily available. OR can be defined as: (1) Operations research is the application of the methods of science to complex problems arising in the direction and management of large systems of men, machine, materials and money in industry, business, government and defense; (2) Operations research is the science of deciding how to best design and operate man-machine systems; (3) Operations research is a scientific method for providing executive departments with a quantitative basis for decision. MS can be defined as: (1) Management science is the application of scientific methodology or principles to management decisions; (2) Management science is the use of quantitative methods for solving management and organizational decision problems. Together, OR and MS may be thought of as the science of operational processes, decision making, and management. However, to our minds, the definition of OR/MS is really given by the coverage of the material in this Encyclopedia.

We wish to thank all the contributors to the *Encyclopedia of Operations Research and Management Science* for their individual efforts and for their cooperation, support and patience given to us and the publisher. Our appreciation and thanks are also due the members of the Editorial Advisory Board for their help in the initial formulation of the *Encyclopedia* and their subsequent efforts in reviewing the articles. The editors and staff of Kluwer Academic Publishers were truly professional in their approach to the development and production of the *Encyclopedia*; our admiration and gratitude are due to all who were involved.

We want to emphasize that the *Encyclopedia of Operations Research and Management Science* is the responsibility of the editors. We made the final determination of the scope, topics and material. Any shortcoming (editorial, inclusion, omission, emphasis, factual) that the reader may perceive rests with us. Hence, we sincerely welcome comments and feedback on all aspects of the *Encyclopedia*. Please write to: Editor, *Encyclopedia of Operations Research and Management Science*, Kluwer Academic Publishers, 101 Philip Drive, Assinippi Park, Norwell, Massachusetts, 02061, USA.

Editors

Saul I. Gass
College of Business and Management
University of Maryland
College Park, Maryland

Carl M. Harris
School of Information Technology and Engineering
George Mason University
Fairfax, Virginia

A

A* ALGORITHM

A heuristic search procedure that selects a node in its search tree for expansion such that the selected node has minimum value of the sum of the cost to reach the node plus a heuristic cost value for that node, where the heuristic cost underestimates the true minimum cost of completion. See **Artificial intelligence**.

ACCEPTANCE SAMPLING

See **Statistical quality control**.

ACCOUNTING PRICES

See **Shadow prices**.

ACCREDITATION

See **Model accreditation**.

ACTIVE CONSTRAINT

A constraint in an optimization problem that is satisfied exactly by a solution. See **Inactive constraint**; **Slack variable**; **Surplus variable**.

ACTIVE SET METHODS

See **Quadratic programming**.

ACTIVITY

(1) A structural variable whose value (level) is to be computed in a linear programming problem. See **Structural variables**. (2) Project work items having specific beginning and completion points and durations. See **Network planning**; **Project management**.

ACTIVITY-ANALYSIS PROBLEM

A linear-programming problem of the form Maximize cx, subject to $Ax \leq b$, $x \geq 0$. The variables x_j of the vector x are quantities of products to be produced. The b_i coefficients of the resource vector b represent the amount of resource i that is available for production, the c_j coefficients represent the value (profit) of one unit of output x_j, and the coefficients a_{ij} of the technological matrix A represent the amount of resource i required to produce one unit of product j. The a_{ij} are termed technological or input-output coefficients. The objective function cx represents some measure of value of the total production.

ACTIVITY LEVEL

The value taken by a structural variable in an intermediate or final solution to a linear programming problem. See **Structural variables**.

ACYCLIC NETWORK

A network that contains no cycles. See **Network optimization**; **Graph Theory**.

ADJACENT

Nodes of a graph or network are adjacent if they are joined by an edge; edges are adjacent if they share a common node. See **Graph theory**; **network optimization**.

ADJACENT (NEIGHBORING) EXTREME POINTS

Two extreme points of a polyhedron that are connected by an edge of the polyhedron.

ADVERTISING

Gurumurthy Kalyanaram and Frank M. Bass

The University of Texas at Dallas

INTRODUCTION: Advertising research in the recent past has focused on three substantive areas – sales response to advertising, optimal advertising policy (constant spending or pulsing), and competitive reactions. The research has employed econometric, optimization and game theoretic analytical techniques to address the issues. The advent of enormous amounts of scanner panel data has led to some fruitful modeling at the individual household level. We will discuss contributions in each one of the three areas. For a thorough review of developments in optimal control advertising models, see Feichtinger, Hartl and Sethi (1993).

Mathematical programming has been a particularly useful technology. Since some early successful applications (Little and Lodish, 1969) of this technology for media planning, progress has been limited because of measurement problems relating to advertising response function. However, advances in research (Little, 1979; Eastlack and Rao, 1986) provide reasons for optimism in identifying the response function. Later, heuristic approaches (Rust and Eechambadi, 1989) were developed to estimate some of the media characteristics – reach and frequency.

SALES-ADVERTISING RELATIONSHIP: The first generally recognized model of importance was proposed by Vidale and Wolfe (1957). Building on a diffusion modeling framework, the Vidale-Wolfe model proposed that advertising directly persuades potential customers not currently buying from the firm, while those who are buying tend to forget (buy less) over time. Formally, the model is represented as follows:

$$x' = \rho u(1 - x) - kx, \qquad x(0) = x_o$$

where x is the market share and u is the level of advertising expenditure at time t, and k is the decay constant. The model suggests an exponential reach and decay phenomena with ρ and k rate parameters. In the next substantial step, Bass and Parsons (1969), following Bass (1969), developed a dynamic simultaneous equation model of sales and advertising and estimated this model on data for a frequently purchased consumer product. The empirical results from this analysis suggest that the advertising elasticity for the brand is small and the advertising expenditures are responsive to sales increases of other brands. A very interesting feature of this model is that it has good forecasting properties.

As far as estimation technology is concerned, there are three works that have provided insightful results (Bass and Clarke, 1972; Rao, 1986; Bass and Leone, 1986). Bass and Clarke (1972) showed that statistical models of sales-advertising relationship need not be limited to the Koyck (1954) model. For example, nonmonotonic lag distributions are more appropriate for monthly data. Bass and Leone (1986) further

examined the data interval issue. Rao (1986) has suggested that one should recognize the role of unobservable advertising expenditures in estimating the parameters of sales-advertising relationship associated with different data intervals.

With the evolution of a better appreciation of the advertising effects, the focus has shifted to models at the individual level. Blattberg and Jeuland (1981) postulated a micromodel which incorporated two well established advertising mechanisms, reach and decay. They assumed that the exposure of an individual to an advertisement can be characterized as a Bernoulli process, and the decay (forgetting) as an exponential process. These assumptions lead to a saw-tooth description of advertising effectiveness. The micro model is aggregated to derive a model of advertising effects on the firm's sales. The model, while fairly flexible and general, provides insightful interpretations. The advent and explosion of scanner panel data over the last decade has accelerated efforts to model at the individual level. The work by Pedrick and Zufryden (1991) is representative of this effort. They propose a nonstationary, integrative, stochastic model approach which melds brand choice, purchase incidence and exposure behavior components. The integrated model, calibrated on scanner panel data, provided good fit and fairly accurate forecasts.

OPTIMAL ADVERTISING POLICY: Researchers have been engaged in examining what might be the optimal advertising policy given a budget constraint. Some have argued that constant advertising or chattering (vacillate between two levels of spending with infinite frequency) is probably the optimal policy (Sasieni, 1971; Sethi 1973). However, others have found pulsing (Hahn and Hyun, 1991; Feinberg, 1992) to be optimal or better. Sasieni (1971) formulated the problem as follows:

$$\max \int_0^\infty [\pi x(t) - u(t)]e^{-rt}\, dt, \quad \text{with}$$

$$X' = g(x, u), \quad x(0) = X_0$$

where $x(t)$ and $u(t)$ refer to sales and advertising at time t and r is the discount rate. Letting g_u and g_x be the first partial derivatives and g_{uu} the second partial derivative and employing Bellman's approach in dynamic programming and the classical Poincare-Bendixson theorem for phase space of differential equations, Sasieni showed that the optimal advertising policy is constant spending when we assume that for a given sales and advertising (i) the sales response would be the same or more positive if advertising level were higher ($g_u \geq 0$), (ii) the sales response would be the same or more positive if sales were at a lower level ($g_x \leq 0$), and (iii) the sales response exhibits diminishing returns to increases in advertising level i.e. the response curve is concave ($g_{uu} \leq 0$). However, the optimal policy becomes chattering when the assumption of concavity of the response function is violated. Clearly, therefore, the shape of the response curve has become a matter of debate. There are many who find evidence for an S-shaped response curve (Little, 1979; Eastlack and Rao, 1986). A more definitive conclusion on the shape of the response curve would enhance the ability to model advertising better. This is, of course, a question for empirical examination.

Hahn and Hyun (1991) showed that when transaction costs above the ordinary media costs are included in Mahajan and Muller's model (1986), pulsing is the optimal policy. Feinberg (1992) introduced the concept of a filter and modified the Sasieni model to be:

$$x' = g(x, z), \quad z' = G(u - z),$$

where z characterizes the filter. The only way to produce something constant in the Sasieni formulation is to fluctuate very rapidly, that is, to chatter. Since chattering is impossible in principle and constant spending is impossible in practice, they are two unrealizable ends of a frequency spectrum and are, in a sense, perceptually equivalent. The introduction of a filter allows Feinberg to equate the two mathematically. The filter exponentially smooths out the input. If the input is constant or chattering advertising, the filter yields a constant output. However, the filter output is a pulsing policy for any nonconstant periodic input. Feinberg (1992) showed numerically that pulsing is a better policy than constant spending.

A goal-programming model for media planning is given in (Charnes *et al.*, 1968). The goals involve distribution of frequencies by demographic and other characteristics, as well as budget limitations.

COMPETITION: This issue has received considerable attention over the last decade. Two kinds of competitive models – differential game and hazard rate models – have emerged.

The differential game models have been built on the Vidale-Wolfe or Nerlove and Arrow models, and employed either open-loop (Rao, 1984) or closed-loop (Erickson, 1991) deterministic games to solve them. Rao (1984) casts the model in the standard format. The value to a firm is the discounted profit which is defined as sales minus the advertising cost. Sales and advertising response functions are assumed to be strictly concave and convex respectively. Further, sales at any time period are expressed as a geometric decay of last period's sales and this guarantees an upper bound for the value to the firm. The context is oligopolistic competition. In this setting, Rao demonstrates an industry open-loop Nash equilibrium. Most of the open-loop differential games confirm Rao's analysis. But in the recent past, there have been efforts to compare the open-loop and closed-loop solutions to differential games, and it is found that the closed-loop equilibrium strategies provide a better fit of the data i.e. actual spending levels in the market (Erickson, 1991).

The work by Bourguignon and Sethi (1981) is a good representation of the hazard rate models applied to study competition in the context of advertising. These models are useful in situations in which a firm must advertise to deal with the threat of entry by another firm. Bourguignon and Sethi characterize a special class of hazard rates given by $h(p, u) = (1 - F)^{n-1}$ where p and u represent price and advertising, $F(t)$ is the probability that the entry of firm has occurred in the time interval $[0, t)$, and n is a parameter depicting the nature of potential entrants. Employing Pontryagin's maximum principle, the researchers show that for certain conditions the optimal policy for a firm is to forbid the entry of any competitor by setting p and u aggressively.

The models reviewed suggest several conclusions relating to advertising. First, there are two mechanisms – growth and decay – in the advertising process. Second, the sales-advertising response curve is either concave or S-shaped. Third, pulsing is better (may be even optimal) advertising policy. Fourth, competition is better modeled using closed-loop equilibrium analysis.

See **Dynamic programming; Goal programming; Linear programming**.

References

[1] Bass, Frank M. (1969), "A Simultaneous Equation Regression Study of Advertising and Sales of Cigarettes," *Jl. Marketing Research*, 6, 291–300.
[2] Bass, Frank M. and D.G. Clarke (1972), "Testing Distributed Lag Models of Advertising Effect," *Jl. Marketing Research*, 9, 298–308.
[3] Bass, Frank M. and Robert P. Leone (1986), "Estimating Micro Relationships from Macro Data: A Comparative Study of Two Approximations of the Brand Loyal Model Under Temporal Aggregation," *Jl. Marketing Research*, 23, 291–297.

[4] Bass, Frank M. and L.J. Parsons (1969), "Simultaneous–Equation Regression Analysis of Sales and Advertising," *Applied Economics*, 1, 103–124.

[5] Blattberg, Robert C. and Abel P. Jeuland (1981), "A Micromodeling Approach to Investigate the Advertising–Sales Relationship," *Management Science*, 27, 988–1005.

[6] Bourguignon, F. and S.P. Sethi (1981), "Dynamic Optimal Pricing and (Possibly) Advertising in the Face of Various Kinds of Potential Entrants," *Jl. Economic Dynamics and Control*, 3, 119–140.

[7] Charnes, A., W.W. Cooper, J.K. DeVoe, D.B. Learner, and W. Reinecke (1968), "A Goal Programming Model for Media Planning," *Management Science*, 14, 8, B423–B430.

[8] Eastlack, J.O. and A. Rao (1986), "Modeling Response to Advertising and Pricing Changes for V8 Cocktail Vegetable Juice," *Marketing Science*, 5, 245–259.

[9] Erickson, G.M. (1991), "Empirical Analysis of Closed–Loop Duopoly Advertising Strategies," Working Paper, University of Washington, Seattle.

[10] Feinberg, Fred (1992), "Pulsing Policies For Aggregate Advertising Models," *Marketing Science*, 11, 221–234.

[11] Feichtinger, Gustav, Richard F. Hartl, and Suresh Sethi (1993), "Dynamic Optimal Control Models in Advertising: Recent Developments," *Management Science*, 40, 195–226.

[12] Hahn, M. and J.S. Hyun (1990), "Advertising Cost Interpretations and the Optimality of Pulsing," *Management Science*, 37, 157–169.

[13] Koyck, L.M. (1954), *Distributed Lags and Investment Analysis*, North Holland, Amsterdam.

[14] Little, J.D.C. (1979), "Aggregate Advertising Models, The State of the Art," *Operations Research*, 27, 629–667.

[15] Little, J.D.C. and Leonard M. Lodish (1969), "A Media Planning Calculus," *Operations Research*, 17, 1–35.

[16] Mahajan, V. and E. Muller (1986), "Advertising Pulsing Policies for Generating Awareness for New Products," *Marketing Science*, 5, 89–106.

[17] Pedrick, James H. and Fred S. Zufryden (1991), "Evaluating The Impact of Advertising Media Plans: A Model of Consumer Purchase Dynamics Using Single–Source Data," *Marketing Science*, 10, 111–130.

[18] Rao, Ram C. (1984), "Advertising Decisions in Oligopoly: An Industry Equilibrium Analysis," *Optimal Control Applications and Methods*, 5, 331–344.

[19] Rao, Ram C. (1986), "Estimating Continuous Time Advertising–Sales Models," *Marketing Science*, 5, 125–142.

[20] Rust, Roland, T. and Naras Eechambadi (1989), "Scheduling Network Television Programs: A Heuristic Audience Flow Approach to Maximizing Audience Share," *Jl. Advertising*, 18(2), 11–18.

[21] Sasieni, M.W. (1971), "Optimal Advertising Expenditures," *Management Science*, 18, 64–72.

[22] Sethi, S.P. (1973), "Optimal Control of the Vidale-Wolfe Advertising Model," *Operations Research*, 21, 998–1013.

[23] Vidale, M.L. and H.B. Wolfe (1957), "An Operations Research Study of Sales Response to Advertising," *Operations Research*, 5, 370–381.

AFFILIATED VALUES BIDDING MODEL

A bidding model in which a bidder, upon learning that a competitor's valuation for what is being sold is higher than previously thought, will raise (or at least not lower) the bidder's own valuation. Affiliated values models include common value models and independent private value models as limiting cases. See **Bidding**.

AFFINE TRANSFORMATION

A shifted linear transformation. An affine transformation on an n-dimensional vector space assigns to any point x the point $Ax + c$, where A is an $n \times n$ matrix and c is an n-dimensional vector.

AFFINE-SCALING ALGORITHM

An interior point method for linear programming based on affine transformations. In the primal affine-scaling algorithm, a problem in standard form is transformed so that the current solution estimate is mapped to the point $(1, 1, \ldots, 1)$. A movement is then made in the transformed space in the direction of the negative projected gradient. The inverse affine transformation is applied to the resulting point to obtain a new solution estimate in the original space. In the dual affine-scaling algorithm, similar ideas are used to solve the dual problem, with the affine transformations applied to the dual slack variables. See **Interior point methods**; **Nonlinear programming**.

AGENCY THEORY

See **Organization studies**.

AGRICULTURE AND THE FOOD INDUSTRY

Filmore Bender

University of Maryland, College Park

Gerald Kahan

McCormick and Company, Sparks, Maryland

It is often difficult to determine where the agricultural sector of an economy ends and the non-agricultural sector begins. For the purpose of this article, the agricultural sector of the economy is defined as production and supply of agricultural inputs, the production of agricultural goods on farms and ranches, the processing and transportation of those goods, as well as the wholesaling and retailing of finished products. Defined in this way, the agricultural sector of the economy in the United States represents approximately 24% of the gross national product.

As with the nonagricultural sector of the economy, operations research was first used to solve agricultural problems in the 1940s and 50s. *A Survey of Agricultural Economics Literature, Vol. 2: Quantitative Methods in Agricultural Economics, 1940s to 1970s* traces the devlopment of operations research in addressing problems of importance to agriculture (Judge *et al.*, 1977). This work ranged from quantifying production functions to the development of models; simulation structures and the use of linear programming and nonlinear optimization models to quantify or predict economic consequences; or alternatively the use of these tools to solve specific problems for specific firms.

Different components of the agricultural economy have embraced the tools of operations research with different levels of enthusiasm. Those sectors of agriculture that can exercise considerable control of inputs and environmental factors (e.g., feeder cattle, broilers, eggs, pork and dairy production) began adopting the tools of operations research during the late 1950s. By 1965, essentially all of the feed formulated for poultry in the United Sates was done using least-cost linear programming feed formulations. Simultaneously, during the 1960s, the beef industry began to adopt linear programming on a limited basis for least-cost feed formulation and for the development of optimal production and marketing strategies. The use of linear programming for least-cost dairy rations became standard practice during the late 1970s. Forestry, which is a branch of agriculture, uses multi-period linear programming models to determine optimal planting and harvesting schedules.

A number of interesting examples have been reported in the literature which describe how linear programming was used to solve a variety of agriculture problems. Upcraft *et al.* (1989) reported that the soil water deficit is the main decision variable that British farmers monitor in order to decide when to irrigate a particular field and how much water to use. The decision is generally based on the soil water deficit in the first strip to be irrigated within each field. A mixed linear program was constructed to model the short-term irrigation scheduling problem for a hosereel-raingun irrigation system. Optimal schedules were produced by quantifying the costs and benefits of irrigation, subject to the constraints of equipment, labor, and availability of water. The model is unique in producing whole farm-irrigation schedules, rather than individual field schedules for hosereel-raingun irrigation systems.

The efficient operation of a beef cattle feedlot is controlled by the price of the animals, purchase and selling weights, and the feeding system. The optimal feeding system involves feeding least-cost rations to animals at each stage in the production process. Glen (1980) reported the development of an optimal method for determining optimal feeding systems that meet the nutrient standards recommended by the US National Research Council. The approach involved using linear programming to determine the least-cost rations to produce specified liveweight gains in animals of known live weight. Dynamic programming was used to determine the optimal sequence of rations to feed to produce animals of specified live weight from known live weight at minimum cost, using least-cost rations from the linear programming model. Results from the dynamic programming model can be used to determine the optimal combination of purchase weight, selling weight, and feeding system. The linear programming model must be solved a large number of times to use the dynamic programming model.

In assessing feeding policy in livestock production, it is generally assumed that an optimal feeding policy will involve using least-cost rations throughout the production process. Glen (1987) showed that this assumption may not always be valid, particularly when the supply of some of the feedstuffs used for feeding the livestock is limited. A technique for testing the validity of this assumption was presented using a linear programming model of an integrated crop and intensive beef production enterprise in which some of the crops are used for livestock feeding. An interactive solution procedure was proposed for cases where this assumption was not valid. While the computational burden associated with the procedure for finding an improved solution is large, experience with realistic data suggests that the results from the linear programming model are likely to be optimal.

Intra-year milk supply patterns depend largely on the distribution of cow calving dates which are in turn influenced by climatic conditions. The most important and least-costly input to milk production is the fresh growth and high digestibility of grass in spring and early summer which often gives rise to a highly seasonal distribution of calving resulting in a seasonal milk supply pattern. However, milk for liquid consumption and production of perishable milk products must be geared to meet a constant consumer demand throughout the year, which necessitates a considerable amount of production outside the least-cost period. Killen and Keane (1978) reported that the development of a linear programming model which gives the distribution of calving dates, minimizes production costs, and meets consumer demand for milk and related products. In addition, the dual gives a set of seasonal prices which should be paid to producers, equitably compensating them for the costs they incur.

The agricultural sector deals with a biological system. By its very nature agriculture has elements that foster the use of operations research techniques and other elements that greatly impede the application of these tools. Because many agricultural production units are relatively small in size, they are unable to adopt operations research techniques in a cost effective manner. On the other hand, because agricultural firms are dispersed, those firms that either supply inputs to farms, or harvest and process agricultural products can make effective use of truck routing and other spatial optimization techniques.

Because of the savings in costs that can be achieved, as well as the increasing availability of computers and software, it is reasonable to expect increasing use of the tools of operations research in agriculture. In fact, as early as 1973, Beneke and Winterboer published *Linear Programming Applications to Agriculture*, a book devoted exclusively to the use of linear programming in agriculture.

In the food industry, linear programming is becoming increasingly common. Publications have reported the use of linear programming in formulating preblended meats (Rust, 1976); luncheon or sandwich meat (IBM, 1966; Wieske, 1981); a protein-enriched luncheon sausage (Nicklin, 1979); bologna (IBM, 1966); frankfurters (IBM, 1966); and a variety of sausage products (MacKenzie, 1964; IBM, 1966; Skinner and Debling, 1969). Ice cream is another food product which has been successfully formulated using linear programming (IBM, 1964; Dano, 1974; Singh and Kalra, 1979).

Cereal based food blends have been formulated using linear programming to insure adequate levels of good-quality protein. Since these blends are sometimes shipped to developing countries, linear programming has helped to ensure that the prominent grain of the country is present in the blend as a major ingredient. It is desirable to blend cereal grains since plant proteins are usually deficient in one or more of the essential amino acids. Inglett *et al.* (1969) used linear programming to bring the essential-amino-acid pattern of a cereal-based food as close as possible to the pattern found in a hen's egg. Cavins *et al.* (1972) used linear programming to formulate a least-cost cereal-based food. The protein quality was controlled by setting both lower and upper limits on each essential amino acid in terms of its percent of total essential amino acid content. Hsu *et al.* (1977 a,b) studied the blending of a wide range of plant and animal protein sources in formulations for bread, pasta, cookies, and extruded corn-meal snack and sausage. Constraints were used to restrict both the nutritional and functional properties.

A detailed description of the formulation of a low-cholesterol, low-fat beef stew using linear programming was given by Bender *et al.* (1976). The objective was to minimize cost while enforcing nutritional constraints and constraints based on the recommendations for fat-modified and low-cholesterol diets. These constraints were for a 100-g portion of stew and set an upper limit on cholesterol content; a lower limit on protein, vitamin A, thiamin, riboflavin, niacin, vitamin C, and iron; and both an upper and a lower limit on carbohydrate, fat, and calories.

Dano (1974) provided an excellent description of the application of linear programming to a beer-blending problem, and Wieske (1981) described the formulation of an optimal margarine product. Another application has been the formulation of mayonnaise (Bender *et al.*, 1982).

The feasibility of planning menus by computer was generally established in the early 1960s (Balintfy and Blackburn, 1964), as was the feasibility of computerized menu analysis (Brisbane, 1964). In these models, nutritional requirements were provided at lowest cost. Developing models which meet sensory objectives as well as nutritional requirements has proved to be a much more difficult problem.

See **Linear programming; Natural resources; Vehicle routing.**

References

[1] Balintfy, J.L. and Blackburn, C.R. (1964). "From New Orleans: A Significant Advance in Hospital Menu Planning by Computer." *Institutions Magazine*, 55(1), 54.

[2] Bender, F.E., Kahan, G. and Mylander, W.C. (1991). *Optimization for Profit*, Haworth Press, New York.

[3] Bender, F.E., Kramer, A. and Kahan, G. (1976). *Systems Analysis for the Food Industry*, AVI Publ., Westport, Connecticut.

[4] Bender, F.E., Kramer, A. and Kahan, G. (1982). "Linear Programming and its Application in the Food Industry." *Food Technology*, 36(7), 94.

[5] Beneke, R.R. and Winterboer, R.D. (1973). *Linear Programming Applications to Agriculture*, Iowa State Press, Ames.

[6] Brisbane, H.M. (1964). "Computing Menu Nutrients by Data Processing," *Jl. Amer. Dietetic Association*, 44, 453.

[7] Cavins, J.F., Inglett, G.E., and Wall, J.S. (1972). "Linear Programming Controls Amino Acid Balance in Food Formulation." *Food Technology*, 26(6), 46.

[8] Dano, S. (1974). *Linear Programming in Industry*, 4th ed., Springer-Verlag, New York.

[9] Glen, J.J. (1980). "A Mathematical Programming Approach to Beef Feedlot Optimization." *Management Science*, 26, 524–535.

[10] Hazell, P.B.R. (1986). *Mathematical Programming for Economic Analysis in Agriculture*, Macmillan, New York.

[11] Hsu, H.W., Satterlee, L.D. and Kendrick, J.G. (1977a). "Experimental Design: Computer Blending Predetermines Properties of Protein Foods, Part I." *Food Product Development*, 11(7), 52.

[12] Hsu, H.W., Satterlee, L.D. and Kendrick, J.G. (1977b). "Results and Discussion: Computer Blending Predetermines Properties of Protein Foods, Part II." *Food Product Development*, 11(8), 70.

[13] IBM (1964). *Linear Programming – Ice Cream Blending*. IBM Technical Publications Dept., White Plains, New York.

[14] IBM (1966). *Linear Programming – Meat Blending*. IBM Technical Publications Dept., White Plains, New York.

[15] Inglett, G.E., Cavins, J.F., Kwokek, W.F., and Wall, J.S. (1969). "Using a Computer to Optimize Cereal Based Food Composition." *Cereal Science Today*, 14(3), 69.

[16] Judge, G.G., Day, R., Johnson, S.R., Rausser, G., and Martin, L.R. (1977). *A Survey of Agricultural Economics Literature Vol. 2: Quantitative Methods in Agricultural Economics, 1940s to 1970s*, University of Minnesota Press, Minneapolis.

[17] Killen, L. and Keane, M. (1978). "A Linear Programming Model of Seasonality in Milk Production," *Jl. Operational Research Society*, 29, 625–631.

[18] Kreiner, H. W. (1994). "Operations Research in Agriculture: Thornthwaite's Classic Revisited," *Operations Research*, 42, 987–997.

[19] Love, R. R., Jr. and J. M. Hoey (1990). "Management Science Improves Fast-food Operations," *Interfaces*, 20(2), 21–29.

[20] MacKenzie, D.S. (1964). *Prepared Meat Product Manufacturing*. AMI Center for Continuing Education, Am. Meat Inst., Chicago.

[21] Nicklin, S.H. (1979). "The Use of Linear Programming in Food Product Formulations." *Food Technology in New Zealand*, 14(6), 2.

[22] Rust, R.E. (1976). "Sausage and Processed Meats Manufacturing." AMI Center for Continuing Education. Am. Meat Inst., Washington, DC.

[23] Singh, R.V. *et al.* (1979). "Least Cost Ice-Cream Mix Formulation: A Linear Programming Approach." *Agric. Situation in India*, 33(1), 7.

[24] Skinner, R.H. *et al.* (1969). "Food Industry Applications of Linear Programming." *Food Manufacturing*, 44(10), 35.

[25] Upcraft, M.J. *et al.* (1989). "A Mixed Linear Programme for Short-Term Irrigation Scheduling." *J. Operational Research Society*, 40, 923–931.

[26] Wieske, R. (1981). *Criteria of Food Acceptance*, 6th ed. Forster-Verlag, Zurich.

[27] Winston, W.L. (1987). *Operations Research: Applications and Algorithms*, Duxbury Press, Boston.

AHP

See **Analytic hierarchy process**.

AI

See **Artificial intelligence**.

AIR FORCE OPERATIONS ANALYSIS

Clayton J. Thomas

Air Force Studies and Analyses Agency, Washington, DC

INTRODUCTION: Air Force Operations Analysis (OA), as *military operations research* was often termed in the Air Force, began in the Army Air Force in World War II (see section below). After the war it was decided to continue operations analysis sections in the major commands, which led to procedures for "steady state" systems of analyst recruitment, training, rotation, etc. The Air Force became a separate service in 1947, and its AFR 20-7 regulated the OA program until 1971, when the OA office in Air Force Headquarters, which had been the focal point for implementing AFR 20-7, was merged into the Air Force Studies and Analyses office (which has had several titles and organizational settings since its creation in the mid-1960s). The latter office served informally as a focal point, and inter-office technical exchanges continued in the course of business and at meetings of professional societies. However, with the rapid changes of the defense establishment in recent years, the need for a more formal arrangement became clearer: in 1993 the Air Force created a Directorate of Modeling, Simulation, and Analysis with the Air Force Studies and Analyses Agency serving as its field operating agency.

In World War II, 245 analysts had been in the OA program at one time or another, the peak strength having been 175. With the war's end most of the analysts returned to universities, laboratories, or other civilian pursuits. Brothers (1951) reported that by January 1946 there were only a dozen left, about half of whom were finishing final reports, etc. As a stable program was established, numbers grew. By 1951 there were 70 assigned, with 95 authorized.

There was very rapid growth in the 1960s as the Office of the Secretary of Defense institutionalized military operations research as *systems analysis*, which increased the need for cost-effectiveness studies, etc. Also, the Air Force began to train significant numbers of uniformed analysts. The total number of Air Force analysts generally continued to increase, at a somewhat slower rate, through the mid-1980s.

In 1988 an Air Force personnel data base showed 476 civilian analysts in the operations research analyst career series. There were probably about as many uniformed analysts in roughly comparable military occupational series. With the end of the cold war in the late 1980s there began a general decrease in the size of the Defense Department, including military operations research. Air Force civilian analyst levels reported at the end of 1993 were about 20% lower than 1988 levels, and still declining.

WORLD WAR II AIR FORCE OA: Brothers (1951) noted that the 245 Air Force OA analysts (professional personnel, not including clerical and administrative staffs) of World War II were distributed over 26 OA sections, one with every combat air force plus several with other overseas Air Force headquarters and several with Air Force training establishments in the continental U.S. Studies were of many types:

"offensive ones dealing with bombing accuracy, weapons effectiveness, and target damage . . . defensive ones dealing with defensive formations of bombers, battle damage and losses of our aircraft, and air defense of our bases

. . . studies of cruise control procedures, maintenance facilities and procedures, accidents, in-flight feeding and comfort of crews, possibility of growing vegetables on South Pacific islands, and a host of others."

The first and largest of the OA sections was that at the Eighth Air Force. McArthur (1990) gave a detailed account of its work and much information about the analysts, with emphasis on the mathematicians. In its foreword, Miser noted:

"During the two and a half years of existence of the Eighth Air Force section, forty-eight persons with scientific and technical training were involved, representing more than a dozen specialties; mathematicians were the largest subgroup, with fifteen persons, thirteen of whom stayed with the section for six months or more . . . . it should be noted that the mathematicians were functioning, not just in a mathematical role, but as scientists, developing theories about actual phenomena and applying them to problems of operations, policy, and plans."

Brothers (1954) gave an account of the well-known improvement in bombing accuracy to which these analysts contributed. The commanding general had asked, "How can I put twice as many bombs on my targets?" In 1942 less than 15 percent of the bombs dropped fell within one thousand feet of the aiming point. The rate improved gradually, and within two years had reached 60 percent. Some of the analytical recommendations that played a part in this were the nearly simultaneous release of their bombs by all the bombardiers (instead of the practice of each bombardier aiming and releasing his own bombs), the salvoing of bombs instead of presetting them to release in a string, and the decrease in the number of aircraft per formation from a range of 18–36 to a range of 12–14.

The successful work of this first section made other Army Air Force commands aware of the OA concept and led to the establishment of the other OA sections. Those sections also had their successes, all of which led to the postwar continuation of OA in the Air Force.

POSTWAR AIR FORCE OA UNDER AFR 20-7: Brothers (1951) recalled that the Air Force, having decided to establish a peacetime OA program, also decided on the basis of wartime experience that it needed an analysis unit in the headquarters. The unit would have two functions: to furnish scientific assistance to the Air

Staff, and to serve as a focal point in the air force-wide OA organization. The Air Force AFR 20-7 established the OA Division in Headquarters, USAF, and authorized Air Force commanders to establish OA offices in their commands, getting needed help from the Headquarters OA office.

From the OA low point of January 1946, it had grown by mid-1951 to ten offices in field commands plus the headquarters office. The 95 authorized professional positions were mostly civilian (under Civil Service), as at that time there were few uniformed analysts available. The RAND corporation's work at that time emphasized problems of the far future, freeing the OA offices to work primarily on current and near-future problems. However, when analysts were needed in the Korean war, some came from RAND (and a smaller "think-tank" also), as well as from OA.

By the mid-1950s, the headquarters OA office had 25 professional positions divided among five "teams." Two of the teams were primarily concerned with implications of new types of weapons: one with atomic and nuclear weapons, and one with ballistic and cruise missiles. A third team dealt primarily with deriving information about combat operations from tests, exercises, etc. A fourth team integrated inputs from the previous three teams to use in assisting Air Staff planners. The fifth team maintained liaison with the existing field OA offices and helped commanders who wished to establish new field offices where they did not yet exist.

The field OA offices were organized according to the same general principles. There should be analysts available to study combat operations and related problems, as well as others with understanding of new technology and its implications for new weapons. Most of the growth in the OA program at that time came through the establishment of new offices, rather than the enlargement of existing offices.

It was only near the end of this period – the decade of the 1960s – that the situation began to change markedly, through the combination of two developments. One came fairly abruptly when the Kennedy administration institutionalized "systems analysis" (used to denote operations research on broad systems problems) in the Office of the Secretary of Defense, which greatly increased the demand for cost-effectiveness studies from the services. The other came throughout the decade as the increase in computer hardware and software capabilities led to great increases in the development, size, and use of computer simulation models.

The headquarters OA office was caught up in both of the above trends, which made it more difficult to devote as much effort as desired to the analysis of operations in Vietnam. Also, in the mid-1960s, a new and larger office of Studies and Analyses was formed from an office that had been set up in the late 1950s to operate what for that time was a large computer simulation model. It had been difficult to acquire the data and manpower to make effective use of that model, and the resources of that office became available to staff the new office created to meet the growing need for cost-effectiveness studies.

The newer office of Studies and Analyses and the smaller headquarters OA office (about 35 professionals at that time) both reported at high levels, required the same kind of competent analysts, used operations research techniques. These similarities suggested the merger of the smaller OA headquarters office into the larger office, and it was finally accomplished in the first six months of 1971.

THE 1970S AND 1980S: The Studies and Analyses office chose not to continue implementation of AFR 20-7. The immediate consequences were not striking. The field OA offices continued, though a few made slight changes in name. Most of the other trends noted above continued, or even accelerated. There was proliferation of computer simulation models and of their use in large studies. Air Force analysts no longer had the semi-annual OA technical symposia, but made increasing use of the multi-service classified symposia of the Military Operations Research Society.

The bulk of the studies dealt with future weapon systems and future force posture. The difference in emphasis between RAND and the in-house Air Force analytical offices that had prevailed in the 1950s diminished, to a large extent because of the impact of the institutionalization of systems analysis in the Defense Department (in which RAND "graduates" had played a significant role).

There continued to be very highly classified studies of "black" systems. There continued to be an effort to obtain, and thus to study, weapon systems exploiting the latest technology. The primary war in this period remained the "cold war," until, suddenly, it was "won."

ISSUES: The issues that now confront military operations research in the United States generally are important to all of the services. The major issues reflect one or more of the forces currently shaping future military operations research and analysis: the decrease in size of the defense establishment; the rich menu of technological options now available but not yet exploited, or even well understood; the still unsolved management problems of reducing undesirable duplication of models, simulations, and studies; the still challenging problem of formulating affordable programs of verification and validation of models and simulations, as prelude to determining suitable use of the models and simulations; etc.

Such issues have contributed to organizational changes in the Department of Defense. In the Air Force, the major recent organizational change affecting analysis has been the creation of the Directorate of Modeling, Simulation, and Analysis. In addition to speeding the application of new hardware and software technological options, it also promises to improve the management of studies, models, and simulations in the Air Force, and perhaps to recapture some of the better aspects of the former AFR 20-7.

See **Battle models; Military operations research; RAND Corporation.**

References

[1] Brothers, L.A. (1951). *Development of Operations Analysis.* Working Paper **17.1.4**, Operations Analysis Division, Directorate of Operations, Headquarters, United States Air Force, Washington, DC.
[2] Brothers, L.A. (1954). "Operations Analysis in the United States Air Force." *Opns. Res.* **2**, 1–16.
[3] McArthur, C.W. (1990). *Operations Analysis in the U.S. Army Eighth Air Force in World War II.* History of Mathematics Vol. **4**, American Mathematical Society.

AIRLINE INDUSTRY

David M. Ryan

University of Auckland, New Zealand

INTRODUCTION: The dramatic growth of the airline industry over the past thirty years into a highly competitive world-wide transport network has been accompanied by the extensive use of operations research and management science methodology in all areas of airline operations. All airlines make major investments in sophisticated aircraft and employ highly trained and skilled staff. Efficient utilization of such valuable resources is clearly an important objective in the management of a profitable airline.

In 1960, the airline industry recognized the potential benefits of OR/MS by setting up the Airline Group of the International Federation of Operations Research Societies (AGIFORS) as a special interest group. Since that time, annual AGIFORS symposia have been held and the proceedings of these meetings provide excellent documentation of the many applications and problems which have been addressed by the use of OR/MS techniques (Richter, 1989). A comprehensive discussion of the applications of OR/MS techniques in the airline industry is given by Teodorovic (1988) and a special issue of *Interfaces* edited by Cook (1989) presents six applications of OR/MS in the airline industry.

An extensive range of practical problems involving long-term planning, short-term planning and "day of operation" decision making have been considered and the full range of methods and techniques including forecasting, simulation, heuristics and optimization have been used to provide practical solutions and decision support systems. In particular, set partitioning and set covering optimization models have been widely applied in many airline scheduling problems and in recent years, linear optimization models generated from airline applications have stimulated much research into the development of interior point and improved simplex methods for solving such problems. The following broad application areas of OR/MS in the airline industry can be clearly identified and will be discussed in further detail:

- Flight Schedule Planning
- Fleet Assignment
- Crew Scheduling
- Yield Management

FLIGHT SCHEDULE PLANNING: The design of a flight schedule is probably the most important and fundamental task for any airline. The schedule which determines the frequency and departure times of flights between airports served by the airline is usually prepared and published many months before it is due to be operated. The preparation of the schedule must take into account forecast passenger demand, the operational limitations of both aircraft and crews and the access limitations imposed by airports either due to meteorological conditions, airport congestion, operational restricted hours or differential landing tariffs. Besides many constraints on the form of feasible flight schedules, there is also considerable variation in the choice of objective ranging from maximizing profit, maximizing passenger-kilometers, maximizing load factors, minimizing the number of aircraft and minimizing direct and indirect operating costs. A discussion of this problem was given by Soumis and Nagurney (1993).

Two particular forms of schedule also reflect the airlines' mode of operation as either a network or a "hub-and-spoke" operation. Airlines operating "hub-and-spoke" systems design schedules which bring many aircraft into a "hub" airport within a short space of time enabling passengers to transfer to another aircraft before all of the aircraft then depart (on the "spokes") from the "hub" over a short period of time. In both forms of operation, the airline schedule can be represented as a network problem in which one must determine "conserved flows" of aircraft between ports at times chosen to satisfy operational constraints and optimize a specified objective. Because of the enormous combinatorial complexity of the network model, many heuristic methods have been developed to assist airline schedule planners; but the problem continues to motivate the development of improved optimization methods.

FLEET ASSIGNMENT: Fleet assignment problems can be broadly classified into three major areas of

- Aircraft Allocation and Maintenance Scheduling
- Gate Allocation
- Schedules Disruption Recovery

The aircraft allocation problem involves first the allocation of an aircraft type to each flight to minimize operating costs and maximize revenue. This problem is discussed by Subramanian *et al.* (1994) who describe the successful solution of this allocation problem by linear optimization methods. They report expected savings for Delta Airlines of $100 million per year. In the second phase of aircraft allocation, sequences of flights are allocated to individual aircraft (so-called tail number allocation) taking into account maintenance scheduling.

The problem of gate allocation is particularly important in the "hub-and-spoke" airline systems in which many aircraft arrive at a hub within a short period of time. Often both crew members and passengers transfer to other aircraft before the next sequence of departures. Careful allocation of airport gates to incoming flights can minimize delays and disruptions during the transfer process.

Problems associated with schedules disruption recovery are particularly important in airline operations. Schedules disruptions occur through weather or mechanical failure and affect not only the aircraft patterns but also the crew patterns. Because of their "real-time" nature, the problems are especially difficult to formulate and to solve but the potential savings in improved decision making can be expected to motivate further research in this application area.

CREW SCHEDULING: Problems of crew scheduling have attracted considerable attention from OR/MS practitioners especially during the past decade. The problems can be broadly classified into three major areas of:
- Staff Planning
- Pairings Construction
- Rostering

Staff planning problems involve the forecasting of demand for crews in various ranks depending on future flight schedules and aircraft numbers. Crew members are often required to undergo significant retraining in moving from one crew rank and aircraft type to another and the "stovepipe" effect usually means that promotions to a senior rank must be preceded by a sequence of training programs and promotions starting with the most junior ranks. The solutions of staff planning problems also determine optimal sequencing of the training programs to provide suitably qualified staff in a "just-in-time" manner and at minimum cost.

The pairings or tours of duty (ToD) problems are amongst the most frequently studied airline problems. A pairing or ToD is a sequence of flights over one or more duty periods separated by rest periods which can be legally performed by a crew member. The sequences of flights must satisfy many constraints and conditions imposed by civil aviation regulations and union agreements. Typically ToDs originate and terminate at a crew base and depending on short-haul or long-haul airline operations can range from one day to many days in duration. The pairings problem then is to select from the set of all legal and feasible pairings, a minimal cost subset of pairings which cover all the scheduled flights. This problem is commonly formulated as a set partitioning or covering model and in recent years considerable progress has been made in the solution of such models (Barutt and Hull, 1990; Anbil *et al.*, 1991). Innovative methods including constraint branching (Foster and Ryan, 1980), branch and cut (Hoffman and Padberg, 1993) and column generation (Lavoie *et al.*, 1988; Barnhart *et al.*, 1991; Desrosiers *et al.*, 1993) have successfully solved problems with very large numbers of feasible pairings.

The rostering problem involves the allocation of pairings to crew members to build a legal and feasible roster for each crew member in a crew rank. Often such allocations are based on the so-called seniority preferential bidding (SPB) system in which pairings are allocated to crew members in decreasing order of seniority satisfying, whenever possible, each individual crew member's bids for certain types of work or rest periods. Heuristic algorithms of a greedy sequential type are most commonly used to solve this allocation problem but they usually result in an inequitable distribution of work and often some pairings (referred to as "open flying") remain unallocated. An alternative form of the rostering problem involves the equitable allocation of pairings to all crew members of a crew rank. Measures of equitability are usually based on the notion that all members of a crew rank should do approximately the same amount of work. Equitability rostering problems can again be formulated as specially structured and generalized set partitioning models (Ryan, 1992). Many alternative legal and feasible roster lines are generated for each crew member and the optimal solution of the model selects one line for each crew member to cover all pairings with the required number of crews and at a minimal total cost. The cost in this context can reflect an individual's preference for certain types of work. Column generation methods can again be used to reduce the need to generate many roster lines a priori for each crew member.

YIELD MANAGEMENT: Most airlines offer fare products at rates which reflect the nature of the travel demand. A limited number of tickets are made available to each fare class and the yield management problem attempts to maximize the total revenue generated by the mix of fare products being sold for each flight by adjusting the inventory levels of seats in each fare class to ensure that load factors are sufficiently high on each departing flight. Belobaba (1987) and Lee (1990) discuss this problem and develop probabilistic and statistical models to aid in decision making. This class of problems also includes consideration of standard airline practice of overbooking flights because of a predicted percentage of "no-show" passengers. Underlying the solution techniques for all of these problems are probabilistic models of passenger behavior.

See **Combinatorial and integer optimization; Linear programming; Yield management**.

References

[1] Anbil, R., E. Gelman, B. Patty and R. Tanga (1991). "Recent Advances in Crew Pairing Optimization at American Airlines," *Interfaces* 21, 62–74.

[2] Barnhart, C., E. Johnson, R. Anbil and L. Hatay (1991). *A Column Generation Technique for the Long-haul Crew Assignment Problem*, Industrial and Systems Engineering Reports Series, COC-91-01, Georgia Institute of Technology, Atlanta.

[3] Barutt, J. and T. Hull (1990). "Airline Crew Scheduling: Supercomputers and Algorithms," *SIAM News* 23(6), 1 & 20–22.

[4] Belobaba, P.P. (1987). "Airline Yield Management – An Overview of Seat Inventory Control," *Trans. Sci.* 21, 63–73.

[5] Cook, T.M., ed. (1989). "Airline Operations Research," special issue of *Interfaces* 19(4), 1–74.

[6] Desrosiers, J., Y. Dumas, M. Solomon and F. Soumis (1993). *The Airline Crew Pairing Construction Problem*, Working Paper, GERAD, Ecole des Hautes Etudes Commerciales, Montreal.

[7] Hoffman, K.L. and M. Padberg (1993). "Solving Airline Crew Scheduling Problems by Branch and Cut," *Mgmt. Sci.* 39, 657–682.

[8] Lavoie, S., M. Minoux and E. Odier (1988). "A New Approach of Crew Pairing Problems by Column Generation and Application to Air Transport," *Euro. Jl. Operational Res.* 35, 45–58.

[9] Lee, A.O. (1990). "Probabilistic and Statistical Models of the Airline Booking Process for Yield Management," PhD Dissertation in Civil Engineering, MIT, Cambridge, Mass.

[10] Richter, H. (1989). "Thirty Years of Airline Operations Research," *Interfaces* 19(4), 3–9.

[11] Ryan, D.M. and B.A. Foster (1981). "An integer programming approach to scheduling," in Wren, A., ed., *Computer Scheduling of Public Transport*. North-Holland, Amsterdam, 269–280.

[12] Ryan, D.M. (1992). "The Solution of Massive Generalised Set Partitioning Problems in Aircrew Scheduling," *Jl. Operational Res. Soc.* 43, 459–467.

[13] Soumis, F. and A. Nagurney (1993). "A Stochastic Multiclass Airline Network Equilibrium Model," *Operations Res.* 41, 710–720.

[14] Subramaniam, R., R.P. Scheff, Jr., J.D. Quillinan, D.S. Wiper and R.E. Marsten (1994). "Coldstart: Fleet Assignment at Delta Air Lines," *Interfaces* 24, 104–120.

[15] Teodorovic, D. (1988). *Airline Operations Research*. Gordon and Breech, New York.

ALGEBRAIC MODELING LANGUAGES FOR OPTIMIZATION

Richard E. Rosenthal

Naval Postgraduate School, Monterey, California

BUILDING OPTIMIZATION MODELS: Optimization models (linear, nonlinear and integer programs) have been used widely and with great success in industry, government and the military. As computers and algorithms for solving these models have become more and more powerful, and as a larger number of people in an ever-widening range of disciplines develop the expertise to pose important decision problems in the optimization modeling framework, there has been a growing awareness that the limiting factor in the application of this technology is often the modeler's ability to provide the necessary inputs to a computer algorithm and to make meaningful analysis of the output.

A complaint that has been made in the past about the viability of optimization modeling by some managers is: "By the time I receive the answer, I have forgotten the question." This complaint is not about computational limitations of the solution algorithms. It refers to the human time expended in converting a modeling idea into a form the computer can process. Before addressing the cause and cure of this bottleneck, it is valuable to review the steps involved in real-world optimization.

The process of building practical optimization models involves several interrelated steps. The first and most important is extensive communication with the client or owner of the decision problem to identify the problem ingredients and to ascertain the extent to which optimization is feasible within the managerial structure of the client organization and the cognitive limitations of the model user. The next step is to formulate the model's decision variables, constraints and objective function and to specify its data requirements. Further steps are computer implementation (generation and solution of instances of the model) and detailed analysis of results. These tasks must be followed by additional communication, which often results in model modifications and data refinements due to invalid assumptions, bad data, programming errors, and, most interestingly, the identification of previously unelucidated policies, constraints and preferences.

The faster the technical steps of model formulation, computer implementation and detailed analysis of results are performed, the smoother the overall process, and the greater the likelihood that the modeling effort will receive sufficient attention from the client in the communication phases. Without extensive client attention and feedback, the model, regardless of its technical brilliance, will never be adopted and supported.

ALGEBRAIC MODELING LANGUAGES: Algebraic modeling languages are highly specialized software packages which greatly reduce the time required for model formulation and analysis. They are based on the premise that modelers, using algebraic notation, can describe models to other algebraically literate people much more rapidly than they can convey the same information to computers using traditional (non-algebraic) methods. An astute diagnosis of this difficulty was furnished by Fourer (1983), who pointed out that the natural way for a modeler to think about and express models is in direct conflict with the input requirements of solution algorithms. Whereas the modeler's form is symbolic, general, concise, and understandable to other modelers, the solver's form is contrary in every respect: explicit, specific, extensive, and convenient for computation. An algebraic modeling language solves this problem by making the modeler's form acceptable computer input. (See Fourer, 1983, for historical comments on early modeling languages and for detailed discussion of their advantages over matrix generators, an earlier technology.)

The objects one can formally define with a modeling language are sets (indices), parameters (given or derived data), decision variables, objective functions, constraints, and various collections of the above. Sets may be entered as data or derived using standard set operations such as union, intersection, conditional selection, and Cartesian product. Parameters may be input in a variety of ways or derived using built-in mathematical functions. Primal and dual solutions from previous optimizations are accessible for use in these calculations. Variable and constraint definitions can be conditioned upon the derived sets.

EXAMPLE: As an example, consider the following constraint from a model for scheduling flight operations on an aircraft carrier, Rosenthal and Walsh (1996):

$$\sum_{jk} x_{ijk} + \sum_{j} (x_{i-1,j,SC} + x_{i-2,j,DC} + x_{i-3,j,TC})$$

$$\leq Max\_Ops\_i, \quad \forall i$$

The indices are i for flight operations cycles, j for aircraft types, and k for flight duration. Index k takes on three possible values: *SC DC TC*, meaning an aircraft stays airborne for a single-, double- or triple-cycle flight, respectively. The decision variable x_{ijk} is an integer variable representing the number of planes of type j launched in cycle i for a flight of duration k. The right-hand-side is a given parameter specifying the maximum number of launches and recoveries that may take place during cycle i. The purpose of the constraint is to enforce this limit, the first summation accounting for all the launches and the second summation accounting for all the recoveries, during cycle i.

Computer implementation of this constraint is straightforward using an algebraic modeling language such as AIMMS (Bisschop and Entriken, 1993); AMPL (Fourer *et al.*, 1993); GAMS (Bisschop and Meeraus, 1982); (Brooke *et al.*, 1992); LINGO (Cunningham and Schrage, 1994); or MPL (Maximal Software, 1993). The GAMS formulation is as follows:

FLIGHT_OPS(i) . .
  SUM((j,k), x(i,j,k))
  +SUM(j, x(i − 1,j,"SC") + x(i − 2,j,"DC") + x(i − 3,j,"TC"))
  = L = Max_Ops(i)

In order to accommodate the standard ASCII computer character set, the algebraic modeling language version of the constraint sacrifices some of the clarity of standard mathematical notation, such as the Greek Σ for summation, the use of subscripts for indices, and the symbol $\leq$. (The three terms in the second summation, representing recovery of planes launched one, two and three cycles earlier, could have been condensed into a double summation.)

ADVANTAGES AND LIMITATIONS: Once the modeler adjusts to above-noted typographical limitations, the advantages obtained are substantial:

- All the work involved in the encoding of particular instances of the optimization model for input to the solver is automated, sparing the modeler of responsibility for this extemely tedious task.
- Model formulation effort is independent of the scale of the problem. In our example, the flight operations constraint definition would require absolutely no modification even if the number of cycles quadrupled or the types of available aircraft changed or any other data alteration occurred. Indeed, very large-scale optimization models, with thousands of variables and constraints can be generated easily with an algebraic modeling language.
- Modeling languages permit separation of the model and the data, so that changes in input coefficients are easily handled by modifying a data table rather than changing any model equations.
- Because of the model/data separation and the model definition's lack of scale-dependency,

the modeler can build prototypes very rapidly. Prototypes supply early insights and help determine whether a full-scale modeling effort is justified before substantial resources are committed.

- It is just as easy to formulate nonlinear objective functions and constraints with an algebraic modeling language as it is to create linear programs. One simply expresses the nonlinear function in direct terms. For example, in a portfolio selection model where, x_i is a decision variable representing the proportion of the portfolio to be invested in security i, and v_{ij} is the covariance between securities i and j, the

function

$$\sum_{ij} v_{ij} x_i x_j$$

measures the variance of the portfolio. GAMS expresses this function as

SUM((i,j), v(i,j) ∗ x(i) ∗ x(j))

Modeling languages automatically generate instructions for the nonlinear programming solver on how to compute derivatives of the nonlinear functions analytically (not by numerical approximation). This is much easier and less error-prone than using the nonlinear programming solver directly, which would require writing FORTRAN subroutines to evaluate the nonlinear functions and their gradients.

- All the modeler's effort is concentrated on modeling-building itself, manipulating objects that are relevant to the problem context, rather than dealing with computational details and far-removed abstractions.
- Nearly all algebraic modeling languages offer the modeler a choice of solver. It is easy to switch between them and to change settings that may improve performance. For difficult optimization problems, such as nonlinear and integer programs, it is unlikely that the same solver will always work best. When a model is implemented with an algebraic modeling language, it can readily exploit improvements in solvers as they are released by their developers.
- Since algebraic modeling languages use a simple, standard set of characters, any model is readily portable without modification across a wide range of computing platforms and operating systems. Models can be easily shipped by

electronic mail. Algebraic modeling languages also have great facility for internal documentation, which helps not only with communication but also long-term maintenance.

- As noted, in practical applications of optimization, models often require modification as new aspects of the problem being modeled come to light. Another reason to revise a model is for computational performance, particularly with integer programs. The same model can often be formulated in several different ways in the sense that the different formulations, if solved to optimality, would give the same results. However, an important distinction between these seemingly equivalent formulations is that some are much easier to solve than others (Barnhart *et al.*, 1993; Schrage, 1991). Finding the more tractable formulations often calls for experimentation with competing formulations. Whether for this purpose or to respond to newly discovered problem features, modifying a model formulation with an algebraic modeling language takes much less time than traditional modeling tools.
- Some algebraic modeling languages contain advanced features that allow efficient implementation of very complex models and of advanced algorithms that involve iterative solution of multiple submodels. For example, it is very easy to implement Dantzig-Wolfe decomposition, Benders decomposition and other large-scale optimization algorithms, which iteratively solve different models, using output of the current model as input to the next (Lasdon, 1970). Another convenient use of this capability is on extremely difficult integer programs, where, due to large size or a nonlinear objective function, finding an optimal integer solution is impossible in a reasonable amount of time even after many simplification and reformulation tricks have been tried. An effective approach is to solve two easier optimization submodels in sequence. The first finds a continuous optimal solution. The second finds a feasible integer solution that is "optimally related" in some context-specific sense to the continuous solution. This heuristic approach is not recommended on a blanket basis, but it has been effective on a variety of real-world applications and is easy to implement with an algebraic modeling language (Rosenthal, 1994).

One should be aware that there are some circumstances when algebraic modeling languages may not be appropriate. These circumstances tend to cluster at two extremes. For casual users of optimization, whose models are not very complex, an algebraic modeling language may be an overly-specialized tool and perhaps of less value than spreadsheets, which have some (but not as much) embedded optimization modeling capability.

The advantages of modeling languages over spreadsheet optimizers are data/model separation, scale-independence, documentability, and ease of revision, all features noted above. Another advantage is dimension-independence, which means that it is extremely easy in a modeling language to define a variable with many indices or to convert a variable with one or two indices into a variable with three or more indices. In a spreadsheet, this conversion may be extremely difficult if not impossible.

Spreadsheet optimizers do have some advantages over modeling languages. They are currently more advanced in terms of integrating with database management systems, graphical user interfaces, and other components of corporate software systems. (Modeling languages are starting to improve in these areas.) Spreadsheets are also a more familiar and comfortable computing environment for most people, particularly those who started using computers in the personal-computer era.

The other extreme where algebraic modeling languages are not the best choice for implementation (although they can be of great value for prototyping) are problems of such great size or difficulty or time-criticality that they require special-purpose model generators and solvers. A common example is in the airline industry, where scheduling problems are often formulated as integer programs with so many columns they cannot be generated *a priori*. Between the extremes of simple models for which modeling languages may be too sophisticated, and exceptionally challenging models for which they may not be powerful enough, there is a vast middle range of optimization applications where algebraic modeling languages serve as an excellent implementation tool.

See **Structured modeling; Model management.**

References

[1] Barnhart, Cynthia, Ellis L. Johnson, George L. Nemhauser, Garbriele Sigismondi and Pamela Vance (1993). "Formulating a Mixed Integer Distribution Problem to Improve Solvability." *Operations Research* **41**, 1013–1019.

[2] Bisschop, Johannes and Robert Entriken (1993). *AIMMS: The Modeling System.* Paragon Decision Technology, Haarlem, The Netherlands.

[3] Bisschop, Johannes and Alexander Meeraus (1982). "On the Development of a General Alge-

braic Modeling System in a Strategic Planning Environment." *Mathematical Programming Study* **20**, 1–29.

[4] Brooke, Anthony, David Kendrick and Alexander Meeraus (1992). *GAMS: A User's Guide, Second Edition*. Boyd & Fraser – The Scientific Press Series, Danvers, Massachusetts.

[5] Cunningham, Kevin and Linus Schrage (1994). *LINGO User's Manual*, LINDO Systems, Chicago.

[6] Fourer, Robert (1983). "Modeling Languages vs. Matrix Generators for Linear Programming." *ACM Transactions on Mathematical Software* **9**, 143–183.

[7] Fourer, Robert, David M. Gay and Brian W. Kernighan (1993). *AMPL: A Modeling Language for Mathematical Programming*. Boyd & Fraser – The Scientific Press Series, Danvers, Massachusetts.

[8] Lasdon, Leon S. (1970). *Optimization Theory for Large Systems*. Macmillan, New York.

[9] Maximal Software (1993). *MPL Modeling System*, Arlington, Virginia.

[10] Rosenthal, Richard E. and William J. Walsh (1996). Optimizing Flight Operations for an Aircraft Carrier in Transit. *Operations Research* **44**, 2.

[11] Rosenthal, Richard E. (1994). *"Integerizing" Real-World Integer Programs*. Operations Research Dept., Naval Postgraduate School, Monterey, California.

[12] Schrage, Linus (1991). *LINDO: An Optimization Modeling System*, 4th ed., pp. 218–219. Boyd & Fraser – The Scientific Press Series, Danvers Massachusetts.

ALGORITHM

A computational procedure whose application yields a solution to an associated class of problems. See **Computational complexity**.

ALGORITHMIC COMPLEXITY

See **Computational complexity**.

ALTERNATE OPTIMA

Distinct solutions to the same optimization problem. See **Unique solution**.

ALTERNATE PATHS

In queueing networks, more than one arc connecting the same two nodes.

ANALYTIC COMBAT MODEL

A self-contained military model that directly computes its results from initial conditions, with no intermediate human interaction. See **Battle modeling**.

ANALYTIC HIERARCHY PROCESS

Thomas L. Saaty

University of Pittsburgh, Pennsylvania

The Analytic Hierarchy Process (AHP) is a general theory of measurement. It is used to derive ratio scales from both discrete and continuous paired comparisons in multilevel hierarchic structures. These comparisons may be taken from actual measurements or from a fundamental scale that reflects the relative strength of preferences and feelings. The AHP has a special concern with departure from consistency and the measurement of this departure, and with dependence within and between the groups of elements of its structure. It has found its widest applications in multicriteria decision making, in planning and resource allocation, and in conflict resolution (Saaty, 1990b; Saaty and Alexander, 1989). In its general form, the AHP is a nonlinear framework for carrying out both deductive and inductive thinking without use of the syllogism by taking several factors into consideration simultaneously and allowing for dependence and for feedback, and making numerical tradeoffs to arrive at a synthesis or conclusion (see Figures 1 and 2).

For a long time people have been concerned with the measurement of both physical and psychological events. By physical we mean the realm of what is fashionably known as the tangibles in so far as they constitute some kind of objective reality outside the individual conducting the measurement. By contrast, the psychological is the realm of the intangibles, comprising the subjective ideas, feelings, and beliefs of the individual and of society as a whole. The question is whether there is a coherent theory that can deal with both these worlds of reality without compromising either. The AHP is a method that can be used to establish measures in both the physical and social domains.

In using the AHP to model a problem, one needs a hierarchic or network structure to represent that problem, as well as pairwise comparisons to establish relations within the structure. In the discrete case these comparisons lead to dominance matrices and in the continuous case to kernels of Fredholm operators (Saaty and Vargas, 1993), from which ratio scales are derived in the form of principal eigenvectors, or eigenfunctions, as the case may be. These matrices, or kernels, are positive and reciprocal, for example,

$a_{ij} = 1/a_{ji}$. In particular, special effort has been made to characterize these matrices (Saaty, 1990b, 1993). Because of the need for a variety of judgments, there has also been considerable work done to deal with the process of synthesizing group judgments (Saaty, 1994). The axiomatic foundations of the AHP may be found in Saaty (1986a).

ABSOLUTE AND RELATIVE MEASUREMENT AND STRUCTURAL INFORMATION:

Cognitive psychologists have recognized for some time that there are two kinds of comparisons, absolute and relative. In absolute comparisons alternatives are compared with a standard in one's memory that has been developed through experience; in relative comparisons alternatives are compared in pairs according to a common attribute. The AHP has been used with both types of comparisons to derive ratio scales of measurement. We call such scales absolute and relative measurement scales. Relative measurement w_i, $i = 1, \ldots, n$, of each of n elements is a ratio scale of values assigned to that element and derived by comparing it in pairs with the others. In paired comparisons two elements i and j are compared with respect to a property they have in common. The smaller i is used as the unit and the larger j is estimated as a multiple of that unit in the form $(w_i/w_j)/1$ where the ratio w_i/w_j is taken from a fundamental scale of absolute values.

Absolute measurement (sometimes called scoring) is applied to rank the alternatives in terms of the criteria or else in terms of ratings (or intensities) of the criteria; for example: excellent, very good, good, average, below average, poor, and very poor; or A, B, C, D, E, F, and G. After setting priorities for the criteria (or subcriteria, if there are any), pairwise comparisons are also made between the ratings themselves to set priorities for them under each criterion and dividing their priorities each by the largest rated intensity (the ideal intensity). Finally, alternatives are scored by checking off their respective ratings under each criterion and summing these ratings for all the criteria. This produces a ratio scale score for the alternative. The scores thus obtained of the alternatives can in the end be normalized by dividing each one by their sum.

Absolute measurement has been used to rank cities in the United States according to nine criteria as judged by six different people (Saaty, 1986b). Another appropriate use for absolute measurement is that of schools admitting students (Saaty *et al.*, 1991). Most schools set their criteria for admission independently of the performance of the current crop of students seeking admission. Their priorities are then used to determine whether a given student meets the standard set for qualification. In that case absolute measurement should be used to determine which students qualify for admission.

THE FUNDAMENTAL SCALE:

Paired comparison judgments in the AHP are applied to pairs of homogeneous elements. The fundamental scale of values to represent the intensities of judgments is shown in Table 1. This scale has been validated for effectiveness, not only in many applications by a number of people, but also through theoretical comparisons with a large number of other scales (Saaty, 1990b).

There are many situations where elements are close or tied in measurement and the comparison must be made not to determine how many times one is larger than the other, but what fraction it is larger than the other. In other words there are comparisons to be made between 1 and 2, and what we want is to estimate verbally the values such as $1.1, 1.2, \ldots, 1.9$. There is no problem in making the comparisons by directly estimating the numbers. Our proposal is to continue the verbal scale to make these distinctions so that 1.1 is a "tad", 1.3 indicates moderately more, 1.5 strongly more, 1.7 very strongly more and 1.9 extremely more. This type of refinement can be used in any of the intervals from 1 to 9 and for further refinements if one needs them, for example, between 1.1 and 1.2 and so on.

COMMENTS ON COST/BENEFIT ANALYSIS:

Often, the alternatives from which a choice must be made in a choice-making situation have both costs and benefits associated with them. In this case it is useful to construct separate costs and benefits hierarchies, with the same alternatives on the bottom level of each. Thus one obtains both a costs-priority vector and a benefit-priority vector. The benefit/cost vector is obtained by taking the ratio of the benefit priority to the costs priority for each alternative, with the highest such ratio indicating the preferred alternative. In the case

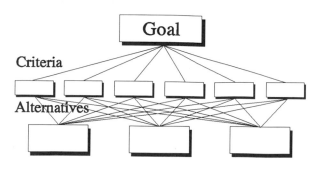

Fig. 1. A three level hierarchy.

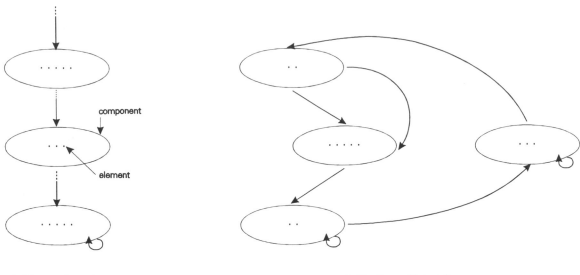

A Linear Hierarchy   A Nonlinear Network

(A) ——▶ (B)   means that A dominates B or that B depends on A

Fig. 2. Structural difference between a linear and a nonlinear network.

Table 1. The fundamental scale

Intensity of importance	Definition	Explanation
1	**Equal Importance**	**Two activities contribute equally to the objective**
2	Weak	
3	**Moderate importance**	**Experience and judgment slightly favor one activity over another**
4	Moderate plus	
5	**Strong importance**	**Experience and judgment strongly favor one activity over another**
6	Strong plus	
7	**Very strong or demonstrated importance**	**An activity is favored very strongly over another; its dominance demonstrated in practice**
8	Very, very strong	
9	**Extreme importance**	**The evidence favoring one activity over another is of the highest possible order of affirmation**
Reciprocals of above	**If activity i has one of the above nonzero numbers assigned to it when compared with activity j, then j has the reciprocal value when compared with i**	**A reasonable assumption**
Rationals	**Ratios arising from the scale**	**If consistency were to be forced by obtaining n numerical values to span the matrix**

where resources are allocated to several projects, such benefit-to-cost ratios or the corresponding marginal ratios prove to be very valuable.

For example, in evaluating three types of copying machines, one represents in the benefits hierarchy the good attributes one is looking for, and one represents in the costs hierarchy the pain and economic costs that one would incur in buying or maintaining the three types of machines. Note that the criteria for benefits and the criteria for costs need not be simply opposites of each other but may be totally different. Also note that each criterion may be regarded at a different threshold of intensity and that such thresholds may themselves be prioritized according to desirability, with each alternative evaluated only in terms of its highest priority threshold level. Similarly, three hierarchies can be used to assess a benefit/(cost × risk) outcome.

THE EIGENVECTOR SOLUTION FOR WEIGHTS AND CONSISTENCY:

There are an infinite number of ways to derive the vector of priorities from the matrix $A = (a_{ij})$. But emphasis on consistency leads to an eigenvalue formulation.

If a_{ij} represents the importance of alternative i over alternative j and a_{jk} represents the importance of alternative i over alternative j and a_{ik}, the importance of alternative i over alternative k, must equal $a_{ij}a_{jk}$ for the judgments to be consistent. If we do not have a scale at all, or do not have it conveniently as in the case of some measuring devices, we cannot give the precise values of $a_{ij} = w_i/w_j$ but only an estimate. Our problem becomes $A'w' = \lambda_{\max}w'$ where $\lambda_{\max}$ is the largest or principal eigenvalue of $A' = (a'_{ij})$, the perturbed value of $A = (a_{ij})$ with $a'_{ji} = 1/a'_{ij}$ forced. To simplify the notation we shall continue to write $Aw = \lambda_{\max}w$ where A is the matrix of pairwise comparisons.

The solution is obtained by raising the matrix to a sufficiently large power, then summing over the rows and normalizing to obtain the priority vector $w = (w_1, \ldots, w_n)$. The process is stopped when the difference between components of the priority vector obtained at the kth power and at the $(k + 1)$st power is less than some predetermined small value.

An easy way to get an approximation to the priorities is to normalize the geometric means of the rows. This result coincides with the eigenvector for $n \leq 3$. A second way to obtain an approximation is by normalizing the elements in each column of the judgment matrix and then averaging over each row.

We would like to caution the reader that for important applications one should use only the eigenvector derivation procedure because approximations can lead to rank reversal in spite of the closeness of the result to the eigenvector (Saaty and Vargas, 1992). It is easy to prove that for an arbitrary estimate x of the priority vector

$$\lim_{k \to \infty} \frac{1}{\lambda_{\max}^k} A^k x = cw$$

where c is a positive constant and w is the principal eigenvector of A. This may be interpreted roughly to say that if we begin with an estimate and operate on it successively by $A/\lambda_{\max}$ to get new estimates, the result converges to a constant multiple of the principal eigenvector.

A simple way to obtain the exact value (or an estimate) of $\lambda_{\max}$ when the exact value (or an estimate) of w is available in normalized form is to add the columns of A and multiply the resulting vector by the vector w. The resulting number is $\lambda_{\max}$ (or an estimate). This follows from

$$\sum_{j=1}^{n} a_{ij}w_j = \lambda_{\max}w_i$$

$$\sum_{i=1}^{n} \sum_{j=1}^{n} a_{ij}w_j = \sum_{j=1}^{n} \left(\sum_{i=1}^{n} a_{ij} \right)w_j$$

$$= \sum_{i=1}^{n} \lambda_{\max}w_i = \lambda_{\max}$$

The problem is now, how good is the principal eigenvector estimate w? Note that if we obtain $w = (w_1, \ldots, w_n)^T$, by solving this problem, the matrix whose entries are w_i/w_j is a consistent matrix which is our consistent estimate of the matrix A. The original matrix A itself need not be consistent. In fact, the entries of A need not even be transitive; that is, A_1 may be preferred to A_2 and A_2 to A_3 but A_3 may be preferred to A_1. What we would like is a measure of the error due to inconsistency. It turns out that A is consistent if and only if $\lambda_{\max} = n$ and that we always have $\lambda_{\max} \geq n$. This suggests using $\lambda_{\max} - n$ as an index of departure from consistency. But

$$\lambda_{\max} - n = -\sum_{i=2}^{n} \lambda_i; \quad \lambda_{\max} = \lambda_1$$

where λ_i, $i = 1, \ldots, n$ are the eigenvalues of A. We adopt the average value $(\lambda_{\max} - n)/(n - 1)$, which is the (negative) average of λ_i, $i = 2, \ldots, n$ (some of which may be complex conjugates).

It is interesting to note that $(\lambda_{\max} - n)/(n - 1)$ is the variance of the error incurred in estimating a_{ij}. This can be shown by writing

$$a_{ij} = (w_i/w_j)\varepsilon_{ij}, \quad \varepsilon_{ij} > 0 \text{ and } \varepsilon_{ij} = 1 + \delta_{ij}, \delta_{ij} > -1$$

and substituting in the expression for $\lambda_{\max}$. It is δ_{ij} that concerns us as the error component and its value $|\delta_{ij}| < 1$ for an unbiased estimator. The

Table 1a. Random consistency index

n	1	2	3	4	5	6	7	8	9	10
Random Consistency Index (R.I.)	0	0	.52	.89	1.11	1.25	1.35	1.40	1.45	1.49

measure of inconsistency can be used to successively improve the consistency of judgments.

The consistency index of a matrix of comparisons is given by $CI = (\lambda_{max} - n)/(n - 1)$. The consistency ratio (CR) is obtained by comparing the CI with the appropriate one of the following set of numbers each of which is an average random consistency index (RI) derived from a sample of size 500 of randomly generated reciprocal matrices using the scale $1/9, 1/8, \ldots, 1, \ldots, 8, 9$, Table 1a. A CR = CI/RI less than or equal to 0.10 is considered acceptable. If CR is larger than 0.10, the problem should be reanalyzed and the judgments revised.

The consistency index for an entire hierarchy is defined by

$$C_H = \sum_{j=1}^{j} \sum_{i=1}^{n_{ij}+1} w_{ij} \mu_{ij+1}$$

where $w_{ij} = 1$ for $j = 1$, and $n_{i_{j}+1}$ is the number of elements of the $(j + 1)$st level with respect to the ith criterion of the jth level.

Let $|C_k|$ be the number of elements of C_k, and let $w_{(k)(h)}$ be the priority of the impact of the hth component on the kth component, that is, $w_{(k)(h)} = w_{(k)}(C_h)$ or $w_{(k)}: C_h \rightarrow w_{(k)(h)}$.

If we label the components of a system along lines similar to those we followed for a hierarchy, and denote by w_{jk} the limiting priority of the jth element in the kth component, we have

$$C_S = \sum_{k=1}^{S} \sum_{j=1}^{n_k} w_{jk} \sum_{h=1}^{|C_k|} w_{(k)(h)} \mu_{k(j,h)}$$

where $\mu_k(j, h)$ is the consistency index of the pairwise comparison matrix of the elements in the kth component with respect to the jth element in the hth component.

HOW TO STRUCTURE A HIERARCHY:

Perhaps the most creative part of decision making that has a significant effect on the outcome is the structuring of the decision as a hierarchy. The basic principle to follow in creating this structure is always to see if one can answer the following question: "Can I compare the elements on a lower level in terms of some or all of the elements on the next higher level?"

A useful way to proceed is to come down from the goal as far as one can and then go up from the alternatives until the levels of the two processes are linked in such a way as to make comparison possible. Here are some suggestions for an elaborate design.

1. Identify overall goal. What are you trying to accomplish? What is the main question?
2. Identify subgoals of overall goal. If relevant, identify time horizons that affect the decision.
3. Identify criteria that must be satisfied to fulfill subgoals of the overall goal.
4. Identify subcriteria under each criterion. Note that criteria or subcriteria may be specified in terms of ranges of values of parameters or in terms of verbal intensities such as high, medium, low.
5. Identify actors involved.
6. Identify actor goals.
7. Identify actor policies.
8. Identify options or outcomes.
9. For yes-no decisions take the most preferred outcome and compare benefits and costs of making the decision with those of not making it.
10. Do benefit/cost analysis using marginal values. Because we are dealing with dominance hierarchies, ask which alternative yields the greatest benefit; for costs, which alternative costs the most.

The software program *Expert Choice* (1993) incorporates the AHP methodology and enables the analyst to structure the hierarchy and resolve the problem using relative or absolute measurements, as appropriate.

HIERARCHIC SYNTHESIS AND RANK:

Hierarchic synthesis is obtained by a process of weighting and adding down the hierarchy leading to a multilinear form. The hierarchic composition principle is a theorem in the AHP that is a particular case of network composition which deals with the cycles and loops of a network.

What happens to the synthesized ranks of alternatives when new ones are added or old ones deleted? The ranks cannot change under any single criterion, but they can under several criteria depending on whether one wants the ranks to remain the same or allow them to change. Many examples are given in the literature showing that allowing rank to change is natural. In 1990 Tversky *et al.* concluded that the "primary cause" of preference reversal is the "failure of procedure invariance." In the AHP there is no such methodological constraint.

In the distributive mode of the AHP, the principal eigenvector is normalized to yield a unique estimate of a ratio scale underlying the judgments. This mode allows rank to change and is useful when there is dependence on the number of alternatives present or on dominant new alternatives which may affect preference among old alternatives thus causing rank reversals (see phantom alternatives – Saaty, 1993). In the ideal mode of the AHP the normalized values of the alternatives for each criterion are divided by the value of the highest rated alternative. In this manner a newly added alternative that is dominated everywhere cannot cause reversal in the ranks of the existing alternatives (Saaty, 1994).

EXAMPLES: *Relative Measurement – Choosing the Best House*

When a family of average income is being advised on buying a house, the family identifies eight factors that they think they have to look for in a house. These factors fall into three categories: economic, geographic, and physical. Although one might begin by examining the relative importance of these clusters, the family feels they want to prioritize the relative importance of all the factors without working with clusters. The problem is to decide which of three candidate houses to choose. In applying the AHP, the first step is *decomposition*, or the structuring of the problem into a hierarchy (see Figure 3). On the first (or top) level is the overall goal of *Satisfaction with House*. On the second level are the eight factors or criteria that contribute to the goal, and on the third (or bottom) level are the three candidate houses that are to be evaluated in terms of the criteria on the second level. The definitions of the factor and the pictorial representation of the hierarchy follow.

The factors important to the individual family are:

1. *Size of House:* Storage space; size of rooms; number of rooms; total area of house.
2. *Transportation:* Convenience and proximity of bus service.

3. *Neighborhood:* Degree of traffic, security view, taxes, physical condition of surrounding buildings.
4. *Age of House:* Self-explanatory.
5. *Yard Space:* Includes front, back, and side space, and space shared with neighbors.
6. *Modern Facilities:* Dishwashers, garbage disposals, air conditioning, alarm system, and other such items.
7. *General Condition:* Extent to which repairs are needed; condition of walls, carpet, drapes, wiring; cleanliness.
8. *Financing:* Availability of assumable mortgage, seller financing, or bank financing.

The next step is *comparative judgment.* Arrange the elements on the second level into a matrix and elicit from the people buying the house judgments about the relative importance of the elements with respect to the overall goal, *Satisfaction with House*.

The questions to ask when comparing two criteria are of the following kind: of the two alternatives being compared, which is considered more important by the family buying the house and how much more important is it with respect to family satisfaction with the house, which is the overall goal?

The matrix of pairwise comparisons of the factors given by the home buyers in this case is shown in Table 2, along with the resulting vector of priorities. The judgments are entered using the Fundamental Scale, first verbally as indicated in the scale and then associating the corresponding number. The vector of priorities is the principal eigenvector of the matrix. This vector gives the relative priority of the factors measured on a ratio scale. That is, these priorities are unique to within a positive similarity transformation. However, if one insures that they add up to unity, they are always unique. In this case financing has the highest priority, with 33% of the influence.

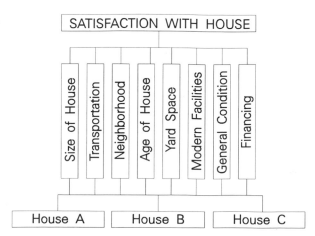

Fig. 3. Decomposition of the problem into a hierarchy.

Table 2. Pairwise comparison matrix for level 1

	1	2	3	4	5	6	7	8	Priority vector
1	1	5	3	7	6	6	1/3	1/4	.173
2	1/5	1	1/3	5	3	3	1/5	1/7	.054
3	1/3	3	1	6	3	4	6	1/5	.188
4	1/7	1/5	1/6	1	1/3	1/4	1/7	1/8	.018
5	1/6	1/3	1/3	3	1	1/2	1/5	1/6	.031
6	1/6	1/3	1/4	4	2	1	1/5	1/6	.036
7	3	5	1/6	7	5	5	1	1/2	.167
8	4	7	5	8	6	6	2	1	.333

$\lambda_{max} = 9.669$ C.I. = .238   C.R. = .169

In Table 2, instead of naming the criteria, we use the number previously associated with each.

We now move to the pairwise comparisons of the houses on the bottom level, comparing them pairwise with respect to how much better one is than the other in satisfying each criterion on the second level. Thus there are eight 3×3 matrices of judgments since there are eight elements on level two, and three houses to be pairwise compared for each element. The matrices (Table 3) contain the judgments of the family involved. In order to facilitate understanding of the judgments, a brief description of the houses is given below.

House A: This house is the largest of them all. It is located in a good neighborhood with little traffic and low taxes. Its yard space is comparably larger than that of houses B and C. However, its general condition is not very good and it needs cleaning and painting. Also, the financing is unsatisfactory because it would have to be financed through a bank at a high rate of interest.

House B: This house is a little smaller than House A and is not close to a bus route. The neighborhood gives one the feeling of insecurity because of traffic conditions. The yard space is fairly small and the house lacks the basic modern facilities. On the other hand, its general condition is very good. Also an assumable mortgage is obtainable, which means the financing is good with a rather low interest rate. There are several copies of B in the neighborhood.

House C: House C is very small and has few modern facilities. The neighborhood has high taxes, but is in good condition and seems secure. The yard space is bigger than that of House B, but is not comparable to House A's spacious surroundings. The general condition of the house is good, and it has a pretty carpet and drapes. The financing is better than for A but not better than for B.

Table 3 gives the matrices of the houses and their local priorities with respect to the elements on level two.

The next step is to synthesize the priorities. In order to establish the composite or global priorities of the houses we lay out in a matrix (Table 4) the local priorities of the houses with respect to each criterion and multiply each column of

Table 3. Pairwise comparison matrix for level 1

Size of house	A	B	C	Normalized priorities	Idealized priorities
A	1	6	8	.754	1.000
B	1/6	1	4	.181	0.240
C	1/8	1/4	1	.065	0.086
$\lambda_{max} = 3.136$		C.I. = .068		C.R. = .117	

Yard space	A	B	C	Normalized priorities	Idealized priorities
A	1	5	4	.674	1.000
B	1/5	1	1/3	.101	0.150
C	1/4	3	1	.226	0.335
$\lambda_{max} = 3.086$		C.I. = .043		C.R. = .074	

Transportation	A	B	C	Normalized priorities	Idealized priorities
A	1	7	1/5	.233	0.327
B	1/7	1	1/8	.055	0.007
C	5	8	1	.713	1.000
$\lambda_{max} = 3.247$		C.I. = .124		C.R. = .213	

Modern facilities	A	B	C	Normalized priorities	Idealized priorities
A	1	8	6	.747	1.000
B	1/8	1	1/5	.060	0.080
C	1/6	5	1	.193	0.258
$\lambda_{max} = 3.197$		C.I. = .099		C.R. = .170	

Neighborhood	A	B	C	Normalized priorities	Idealized priorities
A	1	8	6	.745	1.000
B	1/8	1	1/4	.065	0.086
C	1/6	4	1	.181	0.240
$\lambda_{max} = 3.130$		C.I. = .068		C.R. = .117	

General condition	A	B	C	Normalized priorities	Idealized priorities
A	1	1/2	1/2	.200	0.500
B	2	1	1	.400	1.000
C	2	1	1	.400	1.000
$\lambda_{max} = 3.000$		C.I. = .000		C.R. = .000	

Age of house	A	B	C	Normalized priorities	Idealized priorities
A	1	1	1	.333	1.000
B	1	1	1	.333	1.000
C	1	1	1	.333	1.000
$\lambda_{max} = 3.000$		C.I. = .000		C.R. = .000	

Financing	A	B	C	Normalized priorities	Idealized priorities
A	1	1/7	1/5	.072	0.111
B	7	1	3	.650	1.000
C	5	1/3	1	.278	0.428
$\lambda_{max} = 3.065$		C.I. = .032		C.R. = .056	

vectors by the priority of the corresponding criterion and add across each row, which results in the composite or global priority vector of the houses. Under the distributive mode, House A is preferred if for example copies of B matter. Under the ideal mode, House B is the preferred house if the family wanted the best house regardless of other houses and how many copies of it there are in the neighborhood. In a large number of situations with 10 criteria and 3 alternatives, the two modes gave the same best choice 92% of the time (Saaty, 1994).

Absolute Measurement – Evaluating Employees for Raises

Employees are evaluated for raises. The criteria are Dependability, Education, Experience,

and Quality. Each criterion is subdivided into intensities, standards, or subcriteria as shown in Figure 4. Priorities are set for the criteria by comparing them in pairs, and these priorities are then given in a matrix. The intensities are then pairwise compared according to priority with respect to their parent criterion (as in Table 5) and their priorities are divided by the largest intensity for each criterion (second column of priorities in Figure 4). Finally, each individual is rated in Table 6 by assigning the intensity rating that applies to him or her under each criterion. The scores of these subcriteria are weighted by the priority of that criterion and summed to derive a total ratio scale score for the individual. This approach can be used whenever it is possible to set priorities for intensities of criteria,

Table 4. Synthesis

				Distributive mode						
	1 (.173)	2 (.054)	3 (.188)	4 (.018)	5 (.031)	6 (.036)	7 (.167)	8 (.333)		
A	.754	.233	.754	.333	.674	.747	.200	.072		.396
B	.181	.055	.065	.333	.101	.060	.400	.650	=	.341
C	.065	.713	.181	.333	.226	.193	.400	.278		.263

				Ideal mode						
A	1.00	.327	1.00	1.00	1.00	1.00	.500	.111		.584
B	.240	.007	.086	1.00	.150	.080	1.00	1.00	=	.782
C	.086	1.00	.240	1.00	.335	.258	1.00	.428		.461

Table 5. Ranking intensities

	Outstanding	Above average	Average	Below average	Unsatisfactory	Priorities
Outstanding	1.0	2.0	3.0	4.0	5.0	0.419
Above average	1/2	1.0	2.0	3.0	4.0	0.263
Average	1/3	1/2	1.0	2.0	3.0	0.630
Below average	1/4	1/3	1/2	1.0	2.0	0.097
Unsatisfactory	1/5	1/4	1/3	1/2	1.0	0.062

Inconsistency ratio = 0.015

Table 6. Ranking alternatives

	Dependability .4347	Education .2774	Experience .1775	Quality .1123	Total
1. Adams, V	Outstanding	Bachelor	A little	Outstanding	0.646
2. Becker, L	Average	Bachelor	A little	Outstanding	0.379
3. Hayat, F	Average	Masters	A lot	Below average	0.418
4. Kesselman, S	Above average	H.S.	None	Above average	0.369
5. O'Shea, K	Average	Doctorate	A lot	Above average	0.605
6. Peters, T	Average	Doctorate	A lot	Average	0.583
7. Tobias, K	Above average	Bachelor	Average	Above average	0.456

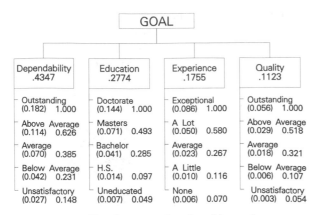

Fig. 4. Employee evaluation hierarchy.

which is usually possible when sufficient experience with a given operation has been accumulated.

APPLICATIONS IN INDUSTRY AND GOVERNMENT: The AHP has been applied in a variety of areas. It has been heavily used in the economics/management area in subjects including auditing, database selection, design, architecture, finance, macro-economic forecasting, marketing (consumer choice, product design and development, strategy), planning, portfolio selection, facility location, forecasting, resource allocation (budget, energy, health, project), sequential decisions, policy/strategy, transportation, water research, and performance analysis. In political problems, the AHP is used in such areas as arms control, conflicts and negotiation, political candidacy, security assessments, war games, and world influence. For social concerns, it is applied in education, behavior in competition, environmental issues, health, law, medicine (drug effectiveness, therapy selection), population dynamics (interregional migration patterns, population size), and public sector. Some technological applications include market selection, portfolio selection, and technology transfer. Additional applications are discussed in Golden *et al.* (1989) and Dyer and Forman (1989).

See **Decision analysis; Multi-attribute utility theory; Utility theory.**

References

[1] Dyer, R.F. and E.H. Forman (1989), *An Analytic Framework for Marketing Decisions: Text and Cases*, Prentice-Hall, Englewood Cliffs, New Jersey.

[2] Expert Choice Inc. (1993), *Expert Software*, 4922 Ellsworth Ave., Pittsburgh, PA 15213.

[3] Golden, B.L., P.T. Harker and E.A. Wasil (1989), *Applications of the Analytic Hierarchy Process*, Springer-Verlag, Berlin.

[4] Kinoshita, E. (1993), *The AHP Method and Application*, Sumisho Publishing Company, Tokyo.

[5] Saaty, T.L. (1986a), "Axiomatic Foundation of the Analytic Hierarchy Process," *Management Science* 32, 841–855.

[6] Saaty, T.L. (1986b), "Absolute and Relative Measurement with the AHP: The Most Livable Cities in the United States," *Socio-Economic Planning Sciences* 20, 327–331.

[7] Saaty, T.L. (1990a), *Decision Making for Leaders*, RWS Publications, 4922 Ellsworth Ave. Pittsburgh (first appeared in 1982, Wadsworth, Belmont, California).

[8] Saaty, T.L. (1990b), *The Analytic Hierarchy Process*, paperback edition, RWS Publications, Pittsburgh (first appeared in 1980, McGraw Hill, New York.)

[9] Saaty, T.L. (1993), "What is Relative Measurement? The Ratio Scale Phantom," *Mathematical and Computer Modelling* 17(4–5), 1-12.

[10] Saaty, T.L. (1994), *Fundamentals of Decision Making and Priority Theory*, RWS Publications, 4922 Ellworth Ave., Pittsburgh.

[11] Saaty, T.L. and J. Alexander (1989), *Conflict Resolution*, Praeger, New York.

[12] Saaty, T.L. and K.P. Kearns (1985), *Analytical Planning – The Organization of Systems*, International Series in Modern Applied Mathematics and Computer Science 7, Pergamon Press, Oxford.

[13] Saaty, T.L. and L.G. Vargas (1982), *The Logic of Priorities: Applications in Business, Energy, Health, Transportation*, Kluwer Academic, Norwell, Massachusetts.

[14] Saaty, T.L. and L.G. Vargas (1991), *Prediction, Projection and Forecasting*, Kluwer Academic, Norwell, Massachusetts.

[15] Saaty, T.L. and L.G. Vargas (1993), "A Model of Neural Impulse Firing and Synthesis," *Jl. Mathematical Psychology* 37, 200–219.

[16] Saaty, T.L., J.W. France and K.R. Valentine (1991), "Modeling the Graduate Business School Admissions Process," *Socio-Economic Planning Sciences* 25, 155–162.

[17] Tversky, A., P. Slovic and D. Kahneman (1990), "The Causes of Preference Reversal," *American Economic Review* 80, 204–215.

ANIMATION

See **Visualization.**

ANTICYCLING RULES

Rules that prevent simplex-type algorithms from cycling. See **Bland's anticycling rules; Cycling; Degeneracy.**

ANTITHETIC RANDOM VARIATES

See **Monte Carlo sampling and variance reduction**.

APPLIED PROBABILITY

The application of probability theory to the biological, physical, social, and engineering sciences. See **Stochastic models**.

ARC

An edge (link) connecting two nodes in a graph or network. The term arc usually means that it is directed. See **Digraph**.

ARCHIMEDEAN AXIOM

The property of real numbers that for any positive numbers a and b, there is a positive integer n such that $a < nb$.

ARIMA

Autoregressive Integrated Moving Averages. See **Time series**.

ARRIVAL PROCESS

A random point process or marked point process with marks denoting some aspects of the stream of customers arriving to a queue or some aspects of the queue itself at the times of arrival, with points representing the precise instants of arrivals. For example, in the marked point process $(\mathbf{X}^a, \mathbf{T}^a)$, the $\mathbf{X}^a$ process may represent the sequence of customer priority classes arriving to a queue, while the $\mathbf{T}^a$ process would be the sequence of actual arrival times.

ARRIVAL-POINT DISTRIBUTION

See **Customer distribution**.

ARROW DIAGRAM

A graphic use of arrows to represent component jobs of a project and the manner in which they are inter-connected. An arrow diagram is some-times also called a network diagram. See **Network planning**.

ARTIFICIAL INTELLIGENCE

Harvey J. Greenberg

University of Colorado at Denver

Both artificial intelligence (AI) and operations research (OR) have roots in the early years of computer science, both matured during the 1950s and 1960s, and both have undergone major changes in the last decade as a result of the explosive power and affordability of computers. Operations research is an inter-disciplinary approach to problem solving, generally using mathematical models to represent a system. Artificial intelligence involves making computers perform functions that are generally believed to require intelligence. Although the meaning of intelligence is subject to debate, one ingredient, which relates to OR, is being able to solve complex problems.

Two elements of problem-solving are *heuristics* and *reasoning*, and these AI topics have had the most in common with OR. Heuristic search is discussed first, noting similarities and differences between the OR and AI approaches. Then, one form of reasoning, namely logical inference, is described with its connection to combinatorial optimization. This is central for logic programming and the design of expert systems.

HEURISTIC SEARCH: A heuristic is a rule of thumb designed to solve a complex problem. Sometimes a heuristic, itself, is a completely defined procedure to obtain a solution. More commonly, a heuristic is a function used to guide a search strategy.

The first step in designing a heuristic search strategy is modeling. This defines not only a problem representation, but also a database, production rules and a control strategy. At any point during the search, the database contains subsets of candidate solutions, then the control strategy selects a production rule that generates a new candidate, which is tested. Examples of control strategies are depth-first and breadth-first search. Testing a candidate, such as a partial solution to a traveling salesman problem, can use a heuristic to guide the subsequent search for a tour. A heuristic can be simple, such as choosing the next link to be a nearest neighbor of an endpoint of the partial solution. It can be more computationally intensive, such as solving a linear programming relaxation, which provides a bound on the completion of the partial solution.

An AI paradigm for node selection in a search tree is the A* algorithm. This is a family of algorithms designed to find an optimal solution guided by a heuristic function. For each node, n, in the search tree, such as at a partial solution to a traveling salesman problem, the cost to reach that node is denoted by $g(n)$, and the estimated minimum cost to complete the solution is the heuristic function, denoted $h(n)$. The control strategy selects the node having the minimum value of $g + h$.

In this family of algorithms, the heuristic function is required to be *admissible*: $h(n) \leq h^*(n)$, where $h^*(n)$ is the actual minimum cost to complete the solution from node n. In words, the heuristic function is an optimistic estimate of completion; in particular, $h(n) = 0$ if n is a complete solution. Despite the mild condition of admissibility, A* always terminates with an optimal solution, if one exists. Further, because h is admissible, the search tree can be pruned below node n when $g(n) + h(n) \geq g(n')$ for some complete solution n' already found (in OR this is called *fathoming* a node).

A special case is breadth-first search, where $h \equiv 0$ and $g(n)$ is defined to be the depth of node n. Another special case is OR's branch-and-bound for general integer programming, where $g + h$ is the objective value of the linear relaxation at a node.

A heuristic can be improved to provide a better estimate of a node's cost. Using $h' \geq h$, however, can sometimes produce an anomaly, where the more informed heuristic, h', requires more nodes in the search tree (see Pearl, 1984). With a mild tie-breaking rule, such as left-most node, this anomaly cannot occur, so a more informed heuristic performs better in that any node that is pruned using h must also be pruned using h'. In AI there is interest in studying learning rules by which more informed heuristics can be derived automatically.

Unlike OR, AI has the dual mission of producing a solution and providing a laboratory tool to test explanatory theories of intelligent behavior. This dual role creates a difference in how AI approaches heuristic search. An OR design always seeks an optimal solution with a minimum of computational effort, and for complex problems must trade-off solution quality with computational effort. An AI design is additionally concerned with the meaning of the heuristic in relation to human reasoning, and the trade-off

between solution and computational effort can be different. Historically, OR has focused on computational efficiency by exploiting mathematical structures in a relatively narrow class of problems, while AI has focused on the use of logic over a broader class of problems. In recent years, however, OR and AI approaches have become more intermingled.

Both OR and AI researchers have been exploring new approaches to heuristic search: genetic algorithms, neural networks, simulated annealing, tabu search, and target analysis. These are inspired by metaphors of nature. Genetic algorithms are inspired by natural selection; neural networks by how the brain functions; simulated annealing by the second law of thermodynamics; tabu search by connections between intelligence and memory; and target analysis by learning from experiences.

COMPUTATIONAL LOGIC: One method of reasoning is by logical inference. The most elementary form uses propositional logic. We are given a set of propositions and logical expressions. The *satisfiability problem* is to find an assignment of truth values for the propositions such that the logical expressions are true. In logical inference, the given expressions are facts about how propositions relate to each other. For example, we can have a project selection problem, where proposition $P_j = TRUE$ means we select project j, and the truth of its negation means we do not select project j. Projects might be related by a simple precedence constraint, $P_i \rightarrow P_j$, which says that if project i is selected, we must also select project j (or the selection of project i must precede the selection of project j).

A feasible truth assignment exists when the facts, which comprise one form of a *knowledge base*, are consistent. Redundant facts occur when some logical expression is implied by the others. For example, if the knowledge base contains $P_i \rightarrow P_j$ and $P_j \rightarrow P_k$, then $P_i \rightarrow P_k$ is a redundant fact (from transitivity of implication). The knowledge base contains circular reasoning if it contains the implication $P_i \rightarrow P_i$ through a chain of implications. For example, the expressions, $P_i \rightarrow P_j$, $P_j \rightarrow P_k$ and $P_k \rightarrow P_i$, are circular. This means that projects i, j and k form an equivalence class: all are selected, or all are rejected. In managing a knowledge base, one wishes to know if it is consistent, non-redundant and non-circular, particularly when new facts are entered into it. In some cases, violations can mean an error in the rule entry. If, for example, precedence constraints are supposed to form a partial ordering, as in job scheduling, a circular chain implies that there is no feasible schedule. Similarly, a redundancy can be due to a subtle implication that the users of the rulebase should know to avoid a false perception of what an inference means.

The satisfiability problem can be represented by a system of linear inequalities with binary-valued variables. Let x_j be 1 or 0, according to whether P_j is true or false, respectively. A simple implication, $P_i \rightarrow P_j$, is true if, and only if, $x_j \geq x_i$. More complex logical expressions can also be represented by linear inequalities, but the effort to derive the inequalities is, itself, a difficult problem unless special forms are assumed.

One special form is a *Horn clause*. This is where the antecedent is a conjunction of propositions and the consequent is only one proposition: $P_1 \wedge P_2 \wedge \ldots P_n \wedge P_0$. This can be represented by $x_0 \geq x_1 + x_2 + \ldots + x_n - n + 1$. Horn clauses are common in logic programming, which is an AI approach to modeling problems and rules of inference based on deductive, logical reasoning.

To see if a particular proposition can be inferred from a set of facts, the logic programming problem becomes one of combinatorial optimization. Suppose we are given the truth value of P_i for i in I, where I is some index reference set over the propositions, and we want to know if P_j can be inferred from the knowledge base (where j is not in I). Then, we set $x_i = 1$ or 0, according to whether P_i is true or false, respectively, for i in I and extremize x_j subject to the logical constraints. The maximum of x_j is 1 if, and only if, P_j can be true, and the minimum of x_j is positive if, and only if, P_j must be true. Then, the facts imply P_j if P_j can be true (i.e., max $x_j = 1$) and P_j cannot be false (i.e., min $x > 0$). The facts are inconsistent if P_j cannot be true (i.e., max $x_j < 1$) and P_j cannot be false.

LOGIC PROGRAMMING AND EXPERT SYSTEMS:

Logic programming uses symbolic logic for problem representation and inferential reasoning. Its foundation is first-order predicate logic, which has greater expressive power than propositional logic, and its computational procedure is called an *inference engine*. One logic programming language is PROLOG, which was adopted by Japan as a standard for its Fifth Generation project.

Logic programming is being studied for its application to OR problems, like scheduling and routing. It is more generally used to build expert systems, designed to give advice about a particular problem. Knowledge representation in an expert system can include forms other than logical

expressions, and uncertainties can be represented by a variety of calculi.

Alternative logics arise naturally in both human reasoning and in AI. For example, we may accept the fact, *birds fly*, and deal with special cases, like penguins, without disturbing the main value of logical reasoning (key terms to investigate are non-monotonic and default logics).

FURTHER READING: Besides the many introductory texts now available, general background into AI terms and concepts can be found in Barr and Feigenbaum (1981) and Shapiro (1990). Traditional heuristic search is well covered by Pearl (1984); newer methods are surveyed by Glover and Greenberg (1989a). A wide variety of linkages between OR and AI appear in Glover and Greenberg (1989b). A particularly good reference for computational logic and its relation with OR is Chandru and Hooker (1992). Klir (1989) gives a good overview of different ways to represent uncertainty. An interesting account of the Japanese Fifth Generation project, which had major impacts on the reshaping of AI, is given by Feigenbaum and McCorduck (1983).

See **Combinatorial and integer optimization; Expert systems; Genetic algorithms; Heuristics; Horn clause; Inference engine; Neural networks; Simulated annealing; Tabu search.**

References

[1] A. Barr and E.A. Feigenbaum, eds. (1981). *The Handbook of Artificial Intelligence*, volume 1, Heuris Tech.

[2] V. Chandru and J.N. Hooker (1993). *Optimization Methods for Logical Inference*, John Wiley, New York.

[3] E.A. Feigenbaum and P. McCorduck (1983). *The Fifth Generation*, Addison-Wesley, Reading, Massachusetts.

[4] F. Glover and H.J. Greenberg (1989a). "New Approaches for Heuristic Search: A Bilateral Linkage with Artificial Intelligence," *European Jl. Operations Research* 39(2), 119–130.

[5] F. Glover and H.J. Greenberg, eds. (1989b.) *Annals of Operations Research* 21: *Linkages with Artificial Intelligence*.

[6] G. Klir (1989). "Is There More to Uncertainty Than Some Probability Theorists Might Have Us Believe?," *International Jl. General Systems* 15, 347–378.

[7] J. Pearl (1984). *Heuristics*, Addison-Wesley, Reading, Massachusetts.

[8] S.C. Shapiro, ed. (1990). *Encyclopedia of Artificial Intelligence*, volumes 1 & 2, John Wiley, New York.

ARTIFICIAL VARIABLES

A set of nonnegative variables added temporarily to a linear program to obtain an initial basic (artificial) feasible solution. If the original constraints are $Ax = b$, $x \geq 0$, then adding an artificial variable y_i to each equation yields the system $Ax + Iy = b$, $x \geq 0$, $y \geq 0$, where y is a column vector of artificial variables. Assuming the vector $b \geq 0$, this system has an obvious basic (artificial) feasible solution, with $y_i = b_i$ being the basic variables and the x_i the nonbasic variables. To obtain a basic solution to the original constraints, the artificial variables must be driven to zero. One way to do this is to solve an auxiliary linear program (known as Phase I) where the objective is to minimize the sum of the artificial variables. If the new system has no solution with all artificial variables equal to zero, then the original constraints are infeasible. See **Big M method**; **Phase I**; **Phase II**.

ASSIGNMENT PROBLEM

The problem of optimally assigning m individuals to m jobs, so that each individual is assigned to one job, and each job is filled by one individual. The problem can be formulated as a linear-programming problem with the objective function measuring the (linear) utility of the assignment as follows:

Maximize $\sum_i \sum_j c_{ij} x_{ij}$

subject to

$$\sum_i x_{ij} = 1 \quad j = 1, \ldots, m$$

$$\sum_j x_{ij} = 1 \quad i = 1, \ldots, m$$

$x_{ij} = 1$ if person i is assigned to job j

$x_{ij} = 0$ if person i is not assigned to job j

$c_{ij} =$ utility of person i assigned to job j.

The problem is a special form of the transportation problem and, as such, has an optimal solution in which each variable is either zero or one. The problem can be solved by the simplex method, but special assignment problem algorithms tend to be computationally more efficient. See **Hungarian method**; **Transportation problem**; **Transportation-simplex method**.

AUTOMATION

Matthew J. Liberatore

Villanova University, Villanova, Pennsylvania

INTRODUCTION: The word "automation" is claimed by some to be a contraction of the term "automatic operation." Automation has been defined by manufacturing engineers as "a technology concerned with the application of mechanical, electronic and computer-based systems to operate and control production" (Groover, 1987). The roots of automation can be traced to mechanization, which can be defined as "the transfer of skills and manual activities to machine operation. The primary difference between mechanization and automation is that automation includes feedback for controlling an automated system" (Odrey, 1992).

Automation is a dynamic technology that represents a continuous evolutionary process that began decades ago. D. S. Harder of Ford Motor Company described automation as "the automatic handling of parts between progressive production processes." Through the increased usage of computers, the scope of automation has expanded to include information collection and processing, process control, and communication links with the outside environment. Automated reasoning is rarely included explicitly into the definition of automation, although such subjects as machine learning, heuristics and knowledge-based systems are having an increasing impact on the scope of automation activities. Today we are moving towards building a high autonomy system which is "an intelligent, real-time system with self-determination capability for carrying out pre-defined objectives over an extended period of time in an uncertain environment" (Kim and Chung, 1991).

As a result, automation is defined here as a technology which minimizes various costs incurred by both routine physical labor and routine human intellectual reasoning, by designing and building machines which perform human-desired operations with minimal or no human intervention. These machines should be as self-activated, self-acting, self-determining, self-regulating, and self-reliant as is practical or necessary.

Nearly all human endeavors have been impacted by automation. Applications of automation in manufacturing include a variety of technologies, such as:

- automatic machine tools, including computer numerical controlled (CNC) tools which are controlled by software that is relatively easy to reprogram;
- automatic materials handling and storage systems, such as automatic storage/retrieval systems (AS/RS) and automatic guided vehicle (AGV) systems;
- industrial robots, which are reprogrammable multi-function manipulators, often arm-like in appearance, that can perform a variety of tasks;
- flexible manufacturing systems (FMS), which permit the manufacture of different items within a family of parts in small to moderate batches using a series of machining centers linked together by a common computer controller and materials handling devices;
- computer-aided design (CAD), which considers issues of design and geometric modeling, engineering analysis, computer kinetics (allows examination of the effects of moving parts on other parts of the structure), and drafting; and
- computer aided manufacturing (CAM), which is a general term addressing various forms of manufacturing automation such as those listed above. (Groover, 1987; Considine and Considine, 1989; Odrey, 1992; Kalpakjian, 1989).

Manufacturers have invested in automation to obtain important benefits such as: increased production capacity, improved inventory control, shorter lead times, lower unit cost of manufacture, reduced labor costs, improved product quality, reduced floor space, better worker safety, and increased flexibility to respond to changing market needs. The primary disadvantages of automation are the acquisition cost of the equipment and the elimination of jobs. Adoption and implementation of automation often requires worker retraining, investment in computer hardware and software, and the ability to handle start-up problems until the manufacturing process and the organization itself are stabilized.

THE ROLE OF OR/MS: OR/MS in manufacturing automation can be categorized according to the several phases that occur over the systems lifecycle (Singhal *et al.*, 1987):

1) *Technology Choice* – the evaluation, economic justification, and/or selection of different automation options or investments in automation. Many OR/MS models have been developed and applied to this decision problem. Some of these evaluate the basic design of the automation system to determine its costs and benefits, while others treat the sys-

tem output as exogenous. Many models include factors not captured by traditional net present value analysis since it may be difficult to quantify all of the important costs and benefits. As a result, a major issue in model development is the balancing of financial, non-financial and qualitative (including organizational and strategic) considerations. For this reason, multi-criteria approaches such as the analytic hierarchy process (AHP) or multi-attribute utility theory (MAUT) are often used to compare different automation options. Reviews of these models and related issues were given in Liberatore (1990) and Canada and Sullivan (1989).

2) *Physical Systems Design* – this category addresses both basic and detailed design issues. Important basic design issues included specifying the structure of the control system and the sizes and locations of the inventory buffers. Queueing network theory has been an important design tool (Buzzacott and Yao, 1986b). A review of various analytical models that can be applied to basic design of the physical systems was given in Buzzacott and Yao (1986a). Detailed design issues include layout, required machine accuracy and computer network reliability. Simulation is the most frequently used method for detailed design.

3) *Design of the Production Planning, Scheduling and Control System* – production planning includes decisions concerning when to produce which products, and when and which resources should be allocated to the production process; scheduling includes procedures for releasing, assigning and sequencing jobs; and control considers real-time control the workflow, and policies for equipment maintenance and repair, and quality monitoring. Simulation, mathematical programming and heuristics are often applied to these problems. A review of developments in machine scheduling including FMS was given in Blazewicz *et al.* (1988). Kusiak (1986) provided a framework for FMS planning and scheduling decision problems based on Materials Requirements Planning (MRP) Systems. Dhar (1991) reviewed the literature and presented a framework for FMS planning which emphasized the interrelationship between FMS planning and functional organization.

4) *Installation and Start-Up* – the OR/MS models developed for steady state systems operation are generally not applicable here. OR/MS modeling during this phase has been limited thus far to project management scheduling and control techniques.

5) *Steady-State Operation and Improvements* – monitoring systems performance, quality improvement studies, and so forth. Continual, planned learning efforts can lead to reduced costs, improved quality, and opportunities to improve product design and variety. Total Quality Management (TQM) and Business Process Redesign (BPR) can be applied to obtain some of these improvements.

Services as well as manufacturing have been impacted by automation. Some examples include automated communications systems, automated teller machines (ATMs), semi-automated mail processing, optical scanners, energy control systems in buildings and software code generators. OR/MS automation studies completed in the service sector include measuring the effect of ATM's on branch labor productivity, managing investment portfolios, trading fixed income securities, and processing loan applications. The continuing importance of improving productivity in the service sector will lead to additional expenditures in automation and OR/MS studies to improve the return from these investments.

Automation has also been applied to the primary administrative functions of the organization, such as accounting and finance. Here automation often involves computerizing the information flow in the office or business and transactions processing, but may address tactical planning issues as well. New technologies such as electronic imaging are having a major effect on work flow automation. Electronic imaging involves scanning and digitizing documents (e.g., routine reports, expense account reimbursements, purchase orders, etc.) allowing them to be stored into a database and retrieved. Justifying the investment in electronic imaging requires measuring its impact on workflow and overall system performance. Clearly, many of the OR/MS application areas that were described as part of the systems life cycle for manufacturing automation are equally applicable to new office automation technologies such as electronic imaging.

The advances in applied artificial intelligence (AI) are also having an important impact on automation. Automated technologies are becoming capable of learning from experience and making decisions with little or no human intervention to optimize operations and minimize costs. For example, in manufacturing automation, AI will become increasingly important as we move toward a true CIM (Computer Integrated Manufacturing) environment, where all of the organization's activities are linked together through computers. In particular, CIM addresses the automation of the information-processing activities in manufacturing: 1) business-related activities such as order entry, customer billing, etc.; 2) product design; 3) manufacturing planning; and 4) manufacturing control.

Increasingly, OR/MS and artificial intelligence approaches are being combined because of their synergistic effects. AI is influencing how we do the work of OR/MS, since it is helping us automate the selection, development and ongoing use of our existing tools and models. For example, AI can be applied to determine whether a mathematical programming model has a special structure which can be exploited by a more efficient algorithm. In addition, AI is helping us to develop new models in situations where the knowledge about operations is complex and qualitative. For example, expert systems, which are intelligent computer programs that have the capability to solve difficult problems using knowledge and inference, are increasingly being combined with OR/MS approaches in such areas as process planning and scheduling.

See **Analytic hierarchy process; Artificial intelligence; Flexible manufacturing; Multiattribute utility theory; Networks of queues; Operations management; Simulation of discrete-event stochastic systems.**

References

[1] Blazewicz, J., G. Finke, R. Haupt, and G. Schmidt (1988). "New Trends in Machine Scheduling," *European Jl. Operational Research*, 37, 303–317.

[2] Buzzacott, J. A. and D. D. Yao (1986a). "Flexible Manufacturing Systems: A Review of Analytical Models," *Management Science*, 32, 890–905.

[3] Buzzacott, J. A. and D. D. Yao (1986b). "On Queueing Network Models of Flexible Manufacturing Systems," *Queueing Systems*, 1, 5–27.

[4] Canada, J. R. and W. G. Sullivan (1989). *Economic and Multiattribute Evaluation of Advanced Manufacturing Technologies*. Prentice-Hall, Englewood Cliffs, New Jersey.

[5] Considine, D. M. and G. D. Considine, eds. (1989). *Standard Handbook of Industrial Automation*. Chapman and Hall, New York.

[6] Dhar, U.R. (1991). "Overview of Models and DSS in Planning and Scheduling of FMS," *International Jl. Production Economics*, 25, 121–127.

[7] Groover, M. (1987). *Automation, Production Systems and Computer-Aided Manufacturing* (2nd ed.). Prentice-Hall, Englewood Cliffs, New Jersey.

[8] Kalpakjian, S. (1989). *Manufacturing Engineering and Technology*. Addison-Wesley, Reading, Massachusetts.

[9] Kim, T. G. and M. Chung (1991). "Embedding Simulation Modeling in Development of High Autonomy Systems," *Proceedings of the Second Conference on AI, Simulation and Planning in High Autonomy Systems.*

[10] Kusiak, A. (1986). "Application of Operations Research Models and Technique in Flexible Manufacturing Systems," *European Jl. Operational Research*, 24, 336–345.

[11] Liberatore, M. J., ed. (1990). *Selection and Evaluation of Advanced Manufacturing Technologies.* Springer-Verlag, Berlin.

[12] Odrey, N. G. (1992). *Maynard's Industrial Engineering Handbook*, McGraw-Hill, New York.

[13] Singhal, K., C. H. Fine, J. R. Meredith, and R. Suri (1987). "Research and Models for Automated Manufacturing," *Interfaces*, 17(6), 5–14.

AVAILABILITY

Igor Ushakov

SOTAS, Rockville, Maryland

Availability is a property of a system requiring it to be ready for performing its required operation or task at time t. (It is often also referred to as *readiness*.) This is clearly related to the main system property, reliability. The main measure of availability is the so-called availability coefficient, $A(t)$, which is equal to the probability of finding the system in the operational state at the needed moment of time t.

If the process of the system's functioning is described in terms of an alternating sequence of lifetimes (i.e., times to failure) $\{X_i\}$ and repair times $\{Y_i\}$, then at any moment t, the availability coefficient can be determined as

$$A(t) = \Pr\{t \in X_i, i = 1, 2, \ldots\}.$$

For stationary processes, that is, where t goes to ∞, the stationary availability coefficient is defined as

$$A = \lim_{t \to \infty} A(t) = \frac{E[X]}{E[X] + E[Y]},$$

where $E[\,\cdot\,]$ is the expectation operator.

See **Reliability of stochastic systems.**

References

[1] Kozlov, B.A. and I.A. Ushakov (1970). *Reliability Handbook.* Holt, Rinehart and Winston, New York.

[2] Ushakov, I. A., ed. (1994). *Handbook of Reliability Engineering.* Wiley, New York.

AVERCH-JOHNSON HYPOTHESIS

See **Economics.**

B

BACKWARD CHAINING

An approach to reasoning in which an inference engine endeavors to find a value for an overall goal by recursively finding values for subgoals. At any point in the recursion, the effort of finding a value for the immediate goal involves examining rule conclusions to identify those rules that could possibly establish a value for that goal. An unknown variable in the premise of one of these candidate rules becomes a new subgoal for recursion purposes. See **Expert systems**.

BACKWARD KOLMOGOROV EQUATIONS

In a continuous-time Markov chain with state $X(t)$ at time t, define $p_{ij}(t)$ as the probability that $X(t + s) = j$, given that $X(s) = i$, $s, t \geq 0$, and r_{ij} as the transition rate out of state i to state j. Then Kolmogorov's backward equations say that, for all states i, j and times $t \geq 0$, the derivatives $dp_{ij}(t)/dt = \Sigma_{k \neq i} r_{ik} p_{kj}(t) - v_i p_{ij}(t)$, where v_i is the transition rate out of state i, $v_i = \Sigma_j r_{ij}$. See **Markov chains; Markov processes**.

BACKWARD-RECURRENCE TIME

Suppose events occur at times $T_1, T_2, \ldots$ such that the interevent times $T_k - T_{k-1}$ are mutually independent, positive random variables with a common cumulative distribution function. Choose an arbitrary time t. The backward recurrence time at t is the elapsed time since the most recent occurrence of an event prior to t.

BALANCE EQUATIONS

(1) In probability modeling, steady-state systems of equations for the state probabilities of a stochastic process found by equating transition rates. For Markov chains, such equations can be derived from the Kolmogorov differential equations or from the fact that the flow rate into a system state or level must equal the rate out of that state or level for steady state to be achieved. (2) In linear programming (usually referring to a production process model), constraints that express the equality of inflows and outflows of material.

BALKING

When customers arriving at a queueing system decide not to join the line and instead go away because they anticipate too long a wait.

BANKING

Stavros A. Zenios

University of Pennsylvania, Philadelphia & University of Cyprus

Operations research and management science techniques find applications in numerous and diverse areas of operation in a banking institution. Applications include, for example, the use of data-driven models to measure the operating efficiency of bank branches through *data envelopment analysis*, the use of image recognition techniques for cheque processing, the use of artificial neural networks for evaluating loan applications, and the use of facility location theory for opening new branches and placing automatic teller machines. A primary area of application, however, is that of risk control in developing broad asset/liability management strategies. Papers that summarize these areas are found in Zenios (1993) and Jarrow *et al.* (1994). This work can be classified into three categories: (1) pricing contingent cashflows, (2) portfolio immunization, and (3) portfolio diversification.

PRICING CONTINGENT CASHFLOWS: The fundamental pricing equation computes the price of a contingent cashflow as the *expected net present value* of the cashflows, discounted by an appropriate discount rate. In discrete time the pricing equation takes the form:

$$P_T = E_s \left\{ \sum_{t=0}^{T} \frac{C_{t+1}^S}{1 + r_t^S} \right\} \qquad (1)$$

where E denotes expectation over the set of scenarios indicated by index s, C_t^s denotes the cashflow received at period t under scenario s, r_t^s is the spot rate for the same period under the scenarios, and T denotes the maturity date. The vector (r_t) is known as the *term structure of interest rates*. For risk-free cashflows, the appropriate discount rate is the rate implied by the Treasury yield curve. At any given point in time, vector (r_t^{s0}) can be obtained using market data; this is the current term structure scenario. However, the temporal variation of the term structure is stochastic. This stochastic interest rate behavior, together with potential uncertainties in the level of the cashflows (i.e., the scenarios C_t^s) are the primary challenging issues behind the evaluation of Equation (1).

One major strand of research is devoted to the development of stochastic models for the term structure of interest rates. Cox, Ingersoll and Ross (1985) first described the interest rate dynamics via the (continuous) diffusion process:

$$dr = \kappa(\mu - r)\,dt + \sigma\sqrt{r}\,d\omega. \qquad (2)$$

Here, μ is the mean and σ the variance of the stochastic interest rate process, and $d\omega$ is the differential of a standard Wiener process. This model exhibits mean reversion with a *drift factor* $\kappa(\mu - r)$, and guarantees that interest rates remain positive. It is, however, a *single factor* model: the term structure of interest rates is represented by a single state variable, namely the spot rate, r.

A two-factor model for bond prices was developed by Brennan and Schwartz (1979). They consider two state variables, the spot rate r and a long-term (consol) rate L. The dynamics of these two variables are described by:

$$\begin{cases} dr = b_1(r,L,t)\,dt + a_1(r,L,t)\,d\omega_1 \\ dL = b_2(r,L,t)\,dt + a_2(r,L,t)\,d\omega_2 \end{cases} \qquad (3)$$

Here, the drift factors are denoted by the functions $b_1(r,L,t)$ and $b_2(r,L,t)$, and the variance terms are expressed by $a_1(r,L,t)$ and $a_2(r,L,t)$. The elements $d\omega_1$ and $d\omega_2$ are differentials of standard Wiener processes.

Despite the elegance of continuous-time models, since most practical applications deal with discrete time cashflows, there is increased interest in the development of discrete models. A popular choice of discrete models is based on binomial lattices. Such models typically assume that interest rates can move to one of two possible states, *up* or *down*, from period t to $t + 1$. The probability and magnitude of each step are calibrated using the Treasury yield curve and the volatility implied by the prices of traded option instruments. Ho and Lee (1986) and Black, Derman and Toy (1990) proposed some fundamental models. For example, the Black, Derman and Toy model describes the spot rates by the process:

$$r_t^\sigma = r_t^0(\kappa_t)^\sigma.$$

Here r_t denotes the spot rate that takes values r_t with possible states $\sigma = 0, 1, \ldots, t$; r_t^0 is the ground state; and κ_t is the volatility of the spot-rate in period t.

Models such as those described above generate the discount rates used in the pricing of riskless cashflows. For risky contingent cashflows (e.g., cashflows with credit, default, lapse, prepayment, and other such risks), the discount rates must be adjusted with a suitable *risk premium*. Such premiums can be computed from the observed market prices of actively traded securities with comparable risks through the use of *option adjusted analysis* (Babbel and Zenios, 1992).

Another important modeling issue in evaluating Equation (1) is the forecasting of the cash flow stream (C_t). Statistical analysis and econometric modeling can be used in this context, especially when dealing with the various complex securities that have emerged in the 1980s, like callable corporate bonds, mortgage and other asset-backed securities, and a range of insurance products. This kind of modeling is represented for insurance products by Asay, Bouyoucos and Marciano (1993), and for mortgage-backed securities by Richard and Roll (1989) and Kang and Zenios (1992).

PORTFOLIO IMMUNIZATION: This is a portfolio management strategy for locking in a fixed rate of return during a prespecified horizon. It assumes that all risk in the returns of the securities is systematic, that is, all risks are due to some common underlying factor(s). Portfolio immunization aims at eliminating this systematic risk. In the case of fixed-income securities, systematic risk is primarily due to changes in the term structure. Portfolio immunization traditionally deals with this type of risk.

The actuary F.M. Reddington (1952) was the first to introduce the notion of immunization, and also specified conditions for immunization. Portfolio immunization became a popular strategy in the 1970s at the aftermath of interest rate deregulation in the U.S. and the volatility of the fixed-income markets that followed. Fisher and Weil (1971) defined immunization as follows:

A portfolio of investments is immunized for a holding period if its value at the end of the holding period, regardless of the course of rates during the holding period, is at least as large as it would have been had the interest rate function remained constant throughout the holding period.

A portfolio of assets used to fund a stream of liabilities can be immunized if the following conditions are met: (1) The present value of the assets is equal to the present value of the liabilities, and (2) the *duration* of the assets is equal to the duration of the liabilities. The first condition guarantees that the target liabilities are funded if the interest rates remain constant throughout the target period. The second condition guarantees that assets and liabilities have identical sensitivities to parallel shifts of the interest rates. Hence, the target liabilities will be funded even if the term structure experiences parallel shifts. A general overview of portfolio immunization is given in Fabozzi (1991). Linear programming formulations are often used to structure immunized portfolios, as in Zenios (1993).

Briefly, let r_i be the yield of the ith security, and C_{it} be the cashflow of security i at time t. From the fundamental pricing Equation (1), we can obtain the price of the ith security by

$$P_i = \sum_{t=1}^{T} C_{it}(1 + r_i)^{-t}.$$

The sensitivity of the price – or *dollar duration* – of security i is obtained by differentiating with respect to cashflow yield, $(\partial P_i/\partial r_i)$,

$$k_i = -\sum_{t=1}^{T} t C_{it}(1 + r_i)^{-(t+1)}.$$

Given the present value P_L and dollar duration k_L of its liabilities, an immunized portfolio can be structured by solving the linear program:

$$
\begin{aligned}
\text{Maximize} \quad & \sum_i k_i r_i x_i \\
\text{s.t.} \quad & \sum_i P_i x_i = P_L \\
& \sum_i k_i x_i = k_L \\
& x_i \geq 0
\end{aligned}
$$

The objective function above maximizes an approximation to the portfolio yield, obtained as the dollar duration-weighted average yield of the individual securities in the portfolio.

Several variations exist on the theme of portfolio immunization. One extension is to structure a portfolio that matches not only present value and duration of assets with those of the liabilities, but that also matches *convexity*, that is,

second partial derivatives $(\partial^2 P_i/\partial r_i^2)$ as well. Another approach is to compute the sensitivity of the prices to more than one factor, than just to parallel shifts of interest rates. The precise form of these factors (i.e., parallel shifts, steepening of the term structure, or term structure inversions) can be obtained using factor analysis of market data. Factor analysis of the term structure was first proposed for the US market by Litterman and Scheinkman (1988). The use of linear programming for factor immunization was recently proposed by Dahl (1993).

PORTFOLIO DIVERSIFICATION: The principle of diversification – based on the adage "do not put all your eggs in one basket" – remains a universal strategy for portfolio management. It provides a systematic way for dealing with residual risk, assuming that residual risk is accurately represented by a function of the mean and variance in the return of the securities. It also assumes that investors have an (implied) utility function over the mean and variance of portfolio returns, favoring portfolios with higher means and lower variances. The efficient portfolios for an investor are those that achieve the highest expected return for a given level of variance or the smallest possible variance for a given level of return. Such portfolios are called *mean-variance efficient* portfolios. Mean-variance optimization models were proposed by Markowitz in the 1950s; Ingersoll (1987) gives an advanced textbook treatment.

Minimum variance portfolios, that is, portfolios with the lowest level of variance for a given target expected return, can be structured using nonlinear quadratic programming. Define:

Q as the covariance matrix $\{q_{ij}\}$ between securities i and j,

μ_i as the expected return of security i,

μ_p as the target expected return of the portfolio,

x_i as the fraction of the portfolio in security i.

Assuming that no short sales are allowed ($x_i \geq 0$ for all i), we can formulate the following program:

$$
\begin{aligned}
\text{Minimize} \quad & x^T Q x \\
\text{s.t.} \quad & \sum_i \mu_i x_i = \mu_p \\
& \sum_i x_i = 1 \\
& x_i \geq 0
\end{aligned}
$$

Other constraints, like limits on portfolio turnover, on minimum holdings, or limits of investments in different market segments, etc., can be captured with more complex formulations. These issues have been addressed by Perold (1984).

The major area of investigation in implementing minimum variance models in practice is in the estimation of the covariance matrix. Factor models that relate the returns and variances of individual securities to a set of common factors are widely used in practice (Elton and Gruber, 1984).

Mean-variance models have traditionally been used in managing portfolios of equities and for strategic asset allocation. By contrast, fixed-income portfolio management has traditionally been based on the principles of portfolio immunization. In the 1980s, however, we saw a convergence of portfolio management tools towards the ideas of portfolio diversification. More complex fixed income securities (e.g., corporate callable bonds, high-yield bonds, mortgages and other asset-backed securities) have very volatile returns. The notion of duration, as a measure of sensitivity, is extremely restrictive for such instruments. Mulvey and Zenios (1993) advocated the use of diversification models for fixed-income portfolios. They showed how pricing models can be developed to generate scenarios of holding period returns in order to calibrate the models, and illustrate that such models produce superior results than traditional portfolio immunization strategies.

Another recent development deals with the asymmetric returns of fixed-income securities, especially those with embedded options. Mean-variance models are valid if we assume a symmetric distribution of return. Furthermore, they penalize both upside and downside deviations from a target return. In the late 1980s we saw the development of more practical models for dealing with asymmetric returns and penalizing differentially upside from downside risk. (Such models, for example based on semi-variance, were investigated as far back as Markowitz's seminal work. They have now been refined to the point of becoming practical and started being used in practice.) Models in this later category include the *mean-absolute deviation* model of Konno and Yamazaki (1991) the *expected utility* optimization models of Grauer and Hakkanson (1985) and the *dynamic, multiperiod models* of Kallberg, White and Ziemba (1982), Mulvey and Vladimirou (1992) and Golub *et al.* (1994).

See **Data envelopment analysis; Facility location; Linear programming; Neural networks; Portfolio theory; Quadratic programming; Utility theory**.

References

[1] M.R. Asay, P.J. Bouyoucos, and A.M. Marciano (1993). "An economic approach to valuation of single premium deferred annuities." In S.A. Zenios, editor, *Financial Optimization*, 100–135. Cambridge University Press.

[2] D.F. Babbel and S.A. Zenios (1992). "Pitfalls in the analysis of option-adjusted spreads." *Financial Analysts Jl.*, July/August, 65–69.

[3] F. Black, E. Derman, and W. Toy (1990). "A one-factor model of interest rates and its application to treasury bond options." *Financial Analysts Jl.*, Jan/Feb, 33–39.

[4] M.J. Brennan and E.S. Schwartz (1979). "A continuous time approach to the pricing of bonds." *Jl. Banking and Finance*, 3, 133–155.

[5] John C. Cox, Jr. Jonathan E. Ingersoll, and Stephen A. Ross (1985). "A theory of the term structure of interest rates." *Econometrica*, 53, 385–407.

[6] E. Elton and M. Gruber (1984). *Modern Portfolio Theory and Investment Analysis.* John Wiley, New York.

[7] Frank J. Fabozzi, editor (1991). *The Handbook of Fixed-Income Securities.* Business One Erwin, Homewood, Illinois.

[8] Lawrence Fisher and Roman Weil (1971). "Coping with the risk of interest-rate fluctuations: returns to bondholders from naive and optimal strategies." *Jl. Business*, October, 408–431.

[9] B. Golub, M. Holmer, R. McKendall, L. Pohlman, and S.A. Zenios (1994). "Stochastic programming models for money management." *European Jl. Operational Research* (to appear).

[10] R.R. Grauer and N.H. Hakansson (1985). "Returns on levered actively managed long-run portfolios of stocks, bonds and bills." *Financial Analysts Jl.*, Sept., 24–43.

[11] H. Dahl (1993). "A flexible approach to interest-rate risk management." In S.A. Zenios, editor, *Financial Optimization*, 189–209. Cambridge University Press.

[12] Thomas S.Y. Ho and Sang-Bin Lee (1986). "Term structure movements and pricing interest rate-contingent claims." *Jl. Finance*, 41, 1011–1029.

[13] Ingersoll, J.E., Jr. (1987). *Theory of Financial Decision Making.* Studies in Financial Economics. Rowman & Littlefield.

[14] R. Jarrow, M. Maksimovic, and W. Ziemba, eds. (1994), *Handbooks in Operations Research and Management Science: Finance.* North Holland, Amsterdam.

[15] J.G. Kallberg, R.W. White, and W.T. Ziemba (1982). "Short term financial planning under uncertainty." *Management Science*, 28, 670–682.

[16] Pan Kang and Stavros A. Zenios (1992). "Complete prepayment models for mortgage backed securities." *Management Science*, 38, 1665–1685.

[17] H. Konno and H. Yamazaki (1991). "A mean-absolute deviation portfolio optimization model and its applications to the Tokyo stock market." *Management Science*, 37, 519–531.

[18] R. Litterman and J. Scheinkman (1988). "Common factors affecting bond returns." Technical report, Goldman, Sachs & Co., Financial Strategies Group, September.

[19] H. Markowitz (1952). "Portfolio selection." *Jl. Finance*, 7, 77–91.

[20] J.M. Mulvey and H. Vladimirou (1992). "Stochastic network programming for financial planning problems." *Management Science*, 38, 1643–1664.

[21] J.M. Mulvey and S.A. Zenios (1994). "Capturing the correlations of fixed-income instruments." *Management Science*, 40, 1329-1342.

[22] A.F. Perold (1984). "Large-scale portfolio optimization." *Management Science*, 30, 1143–1160.

[23] F.M. Reddington (1952). "Review of the principles of life-office valuations." *Journal of the Institute of Actuaries*, 78, 286–340.

[24] Scott F. Richard and Richard Roll (1989). "Prepayments on fixed-rate mortgage – backed securities." *Journal of Portfolio Management*, Spring, 73–82.

[25] S.A. Zenios, editor (1993). *Financial Optimization*. Cambridge University Press.

BAR CHART

See **Gantt chart; Quality control**.

BARRIER AND DISTANCE FUNCTIONS

Roman A. Polyak

George Mason University, Fairfax, Virginia

INTRODUCTION: In the mid-1950s and the early 1960s, Frisch (1955) and Carroll (1961) proposed the use of *classical barrier functions* (CBFs) for constrained optimization. Later, Huard (1967) introduced *interior distance functions* (IDFs) for the same purpose. Since then, the CBF and IDF have been extensively studied, with particularly major work in the area due to Fiacco and McCormick (1968) who developed the Sequential Unconstrained Minimization Technique (SUMT). Currently, methods based on barrier and distance functions make up a considerable part of modern optimization theory.

Interest in these functions was revived after N. Karmarkar (1984) developed his *projective scaling algorithm*, which is not only polynomial but shows the potential to rival the simplex method of linear programming for large-scale problems. Karmarkar's potential function is, in fact, an interior distance function.

In the late 1980s, Nesterov and Nemirovsky (1994) discovered that CBFs and IDFs possess "self-concordant" properties that guarantee the polynomial complexity of interior point methods (IPMs). In spite of the substantial progress which has been made on these two basic IPM tools, CBF and IDF still have inherent drawbacks: these functions, as well as their derivatives, do not exist at the solution; they grow infinitely large; and the condition number of their Hessians vanish when the approximation approaches the solution.

To eliminate the drawbacks, while still retaining the nice properties of the barrier and distance functions, Polyak (1986, 1991, 1992) introduced *modified barrier functions* (MBFs) and *modified interior distance functions* (MIDFs). Both the MBF and MIDF are particular cases of the *Nonlinear Rescaling Principle* (Polyak, 1986), that consists of transforming the objective function and/or the constraints into an equivalent problem and using the classical Lagrangian for both theoretical analysis and numerical methods.

MODIFIED BARRIER FUNCTIONS: Consider the constrained optimization problem

$$\text{minimize } f(x) \tag{1}$$

$$\text{subject to: } g_i(x) \geq 0, \quad i = 1, \ldots, m \tag{2}$$

For any $k > 0$, the problem (1)–(2) is equivalent to

$$\text{minimize } f(x) \tag{3}$$

$$\text{subject to: } k^{-1} \ln[kg_i(x) + 1] \geq 0, \quad i = 1, \ldots, m \tag{4}$$

where the constraints (2) are transformed by $\psi(t) = \ln(t + 1)$.

The classical Lagrangian for the equivalent problem (3)–(4),

$$F(x, \lambda, k) = f_0(x) - k^{-1} \sum \lambda_i \ln[(kg_i(x) + 1)], \tag{5}$$

is the MBF which corresponds to Frisch's (1955) CBF

$$F(x, k) = f_0(x) - k^{-1} \sum \ln g_i(x).$$

For any $k > 0$, the system (2) is equivalent to

$$k^{-1}[(kg_i(x) + 1)^{-1} - 1] \leq 0, \quad i = 1, \ldots, m, \tag{6}$$

where the constraints transformation is given by $v(t) = (t + 1)^{-1} - 1$.

The classical Lagrangian for the problem equivalent problem to (1) taken with (6), is the MBF

$$C(x, \lambda, k) = f_0(x) - k^{-1} \sum \lambda_i [(kg_i(x) + 1)^{-1} - 1], \tag{7}$$

which corresponds to Carroll's (1961) CBF

$$C(x, k) = f_0(x) - \sum g_i^{-1}(x).$$

The MBF's properties make them fundamentally different from the CBFs. The MBFs, as well as their derivatives, exist at the solution and for any Karush–Kuhn–Tucker pair (x^*, λ^*) and any $k > 0$, the following critical properties hold:

P1.　　　$F(x^*, \lambda^*, k) = C(x^*, \lambda^*, k) = f_0(x^*);$

P2.　$\nabla_x F(x^*, \lambda^*, k) = \nabla_x C(x^*, \lambda^*, k)$
$$= \nabla f_0(x^*) - \sum_1^m \lambda_i^* \nabla g_i(x^*) = 0;$$

P3.　$\nabla_{xx} F(x^*, \lambda^*, k) = \nabla_{xx} L(x^*, \lambda^*)$
$$+ k \nabla g^T(x^*) \Lambda^* \nabla g(x^*),$$

where $L(x, \lambda) = f(x) - \sum_1^m \lambda_i^* g_i(x)$ is the classical Lagrangian for (1), $\nabla g(x) = J[g(x)]$ is the Jacobian of the vector-function $g(x) = \{g_i(x), i = 1, \ldots, m\}$, and $\Lambda = \mathrm{diag}(\lambda_i)$.

The MBF's properties resemble that of Augmented Lagrangians (Bertsekas, 1982; Hestenes, 1969; Mangasarian, 1975; Powell, 1969; Rockefellar, 1974). One can consider the MBFs as Interior Augmented Lagrangians, although there are essential differences between MBFs and Augmented Lagrangians with quadratic kernel (Polyak, 1992). The MBFs' properties lead directly to the development of methods for their use.

Let $k > 0$, $\lambda^0 = e = (1, \ldots, 1) \in \mathscr{R}^m$ and $x^0 \in \Omega_k = \{x \mid g_i(x) \geq -k^{-1}, i = 1, \ldots, m\}$. We assume that $\ln t = -\infty$ for $t \leq 0$. The MBF method consists of generating two sequences $\{x^s\}$ and $\{\lambda^s\}$:

$$x^{s+1} \in \mathrm{argmin}\{F(x, \lambda^s, k) \mid x \in \mathscr{R}^n\} \qquad (8)$$

and

$$\lambda^{s+1} = \mathrm{diag}[k g_i(x^{s+1}) + 1]^{-1} \lambda^s. \qquad (9)$$

There is a fundamental difference between the MBF method and SUMT or other IPM that is based on CBF. It is that the MBF method converges to the primal-dual solution with *any* fixed $k > 0$ for any convex programming which has bounded sets of optimal primal and dual solutions (Jensen and Polyak, 1994). Moreover, if the second-order optimality conditions hold, then the primal-dual sequence converges with linear rate:

$$\max\{\|x^{s+1} - x^*\|, \|u^{s+1} - u^*\|\} \leq c k^{-1} \|u^s - u^*\|, \qquad (10)$$

where the constant $c > 0$ is independent of $k > 0$. Therefore, by increasing the barrier parameter $k > 0$, the ratio $q = ck^{-1}$ can be made as small as one wants (Polyak, 1992). In the case of linear programming, for any fixed barrier parameter, the MBF method produces such primal sequences that the objective function tends to its optimal value and constraint violations tend to zero with R-linear rate (Powell, 1992).

The numerical realization of the MBF method leads to the Newton MBF. The Newton method is used to find an approximation for x^s, followed by the Lagrange multiplier update. Because of the convergence of the MBF under the fixed barrier parameter $k > 0$, both the condition number of the MBF Hessian and the area where the Newton method is well defined are stable (Smale, 1986; Polyak, 1992). These properties contribute to both numerical stability and complexity, and they led to the discovery of the "hot" start phenomenon in constrained optimization (Melman and Polyak, 1994; Polyak, 1992). By hot start, we mean that, for a given accuracy of $\varepsilon > 0$, only $O(\ln \ln \varepsilon^{-1})$ Newton steps are needed from some point on for the Lagrange multiplier update if the constrained optimization problem is nondegenerate. Every Lagrange multiplier update shrinks the distance to the primal-dual solution by a factor of $0 < q_k = ck^{-1} < 1$, which can be made as small as we may want by choosing fixed but large enough $k > 0$.

MODIFIED INTERIOR DISTANCE FUNCTIONS: Another basic instrument in the IPM is the Interior Distance Function. Let $y \in \mathrm{int}\, \Omega = \{x \mid g_i(x) \geq 0, i = 1, \ldots, m\}$, and $\alpha = f(y)$. The best known IDFs,

$$F(x, \alpha) = -m \ln[\alpha - f(x)] - \sum \ln g_i(x) \qquad (11)$$

and

$$H(x, \alpha) = m[\alpha - f(x)]^{-1} + \sum g_i^{-1}(x) \qquad (12)$$

have been widely used in IPM developments.

The interior center method's step consists of finding a center of the Relaxation Feasible Set (RFS)

$$\Omega(y) = \{x \in \Omega \mid f(x) \leq f(y)\}$$

and updating the RFS by using the new objective function value. To find the center of the RFS $\Omega(y)$, one solves an unconstrained optimization problem

$$\hat{x} \equiv \hat{x}(\alpha) = \mathrm{argmin}\{F(x, \alpha) \mid x \in \mathscr{R}^n\}$$

assuming that $\ln t = -\infty$ for $t \leq 0$. By replacing y for $\hat{x}$ and α for $\hat{\alpha} = f(\hat{x})$, we complete a step of the Interior Center Method. The curve $\alpha \to \hat{x}(\alpha)$ has some key properties which are at the heart of the IPM complexity (Gonzaga, 1992; Renegar, 1988).

One of the most important elements of IPM is the way in which one changes the level value α from step to step. Starting with a point close enough to $\hat{x}(\alpha)$ ("warm" start) for a particular $\alpha = f(y)$, $y \in \mathrm{int}\, \Omega$, we perform one Newton step for the system $\nabla_x F(x, \alpha) = 0$, that is, we find

$$\hat{x} = \hat{x} - [\nabla_{xx} F(x,\alpha)]^{-1} \nabla_x F(x,\alpha),$$

followed by a "careful" update of α.

It turns out that there is a way for updating α, that is, replacing α for $\hat{\alpha}$, so that the new approximation $\hat{x}$ is again a "warm" start for the system $\nabla_x F(x,\hat{\alpha}) = 0$. The gap $\Delta(\alpha) = \alpha - f(x^*)$ shrinks by a factor $0 < q_n < 1$ which depends only on the size of the problem (Gonzaga, 1992; Renegar, 1988).

In contrast to the IDFs, the MIDFs have three tools to control the computational process: the center (or the objective function value); the barrier parameter; and the vector of Lagrange multipliers. This leads to the development of Modified Center Methods, which, for nondegenerate constrained optimization problems, converge to the primal-dual solution with linear rate, even when both the center and the barrier parameters are fixed.

Note that for any $k > 0$, the RFS

$$\Omega(y) = \{x \,|\, g_i(x) \geq 0, i = 1, \ldots, m; f(x) \leq f(y)\}$$

$$= \{x \,|\, k^{-1}[\ln(kg_i(x) + f(y) - f(x))$$

$$- \ln(f(y) - f(x))] \geq 0,$$

$$i = 1, \ldots, m; f(y) > f(x)\}.$$

Therefore, the problem

$$\text{minimize} \quad -\ln[(f(y) - f(x)] \tag{13}$$

$$\text{subject to:} \quad k^{-1}[\ln(kg_i(x) + f(y) - f(x)) - \ln(f(y)$$

$$- f(x))] \geq 0, \quad i = 1, \ldots, m \tag{14}$$

is equivalent to (1)–(2). The equivalent problem, (13)–(14), has been obtained by a monotone transformation of both the objective function (1) and the constraints (2). The classical Lagrangian for the equivalent problem (13)–(14),

$$F(x,y,\lambda,k) = \left(-1 + k^{-1} \sum \lambda_i\right) \ln[f(y) - f(x)]$$

$$- k^{-1} \sum \lambda_i \ln[kg_i(x) + f(y) - f(x)],$$

is the MIDF which corresponds to the $F(x,\alpha)$. The MIDF

$$H(x,y,\lambda,k) = \left(1 - k^{-1} \sum \lambda_i\right) [(f(y) - f(x)]^{-1}$$

$$+ k^{-1} \sum \lambda_i [kg_i(x) + f(y) - f(x)]^{-1}$$

corresponds to $H(x,\alpha)$. It is also a classical Lagrangian for an equivalent problem, which one obtains by using the inverse transformation for both the objective function and constraints. The MIDFs and their derivatives exist at the solu-

tion, and for any K-K-T pair (x^*,λ^*), properties similar to P1–P3 hold.

The MIDFs properties possess P1–P3 type which allows us to develop the Modified Center Method. Let $y \in \text{int } \Omega$, let $k > 0$ be large enough, define $e = (1, \ldots, 1) \in \mathcal{R}^m$, and assume that $\ln t = -\infty$ for $t \leq 0$. Then the Modified Center Method consists of finding two sequences, $\{x^s\}$ and $\{\lambda^s\}$:

$$x^{s+1} = \text{argmin}\{F(x,y,\lambda^s,k) \,|\, x \in \mathcal{R}^n \tag{15}$$

and

$$\lambda^{s+1} = \text{diag}\left\{\frac{f(y) - f(x^{s+1})}{kg_i(x^{s+1}) + f(y) - f(x^{s+1})}\right\} \lambda^s. \tag{16}$$

If the standard, second-order optimality conditions for (1)–(2) are fulfilled, then the primal-dual sequence $\{x^s,\lambda^s\}$ converges to the primal-dual solution, $\{x^*,\lambda^*\}$, at a linear rate of convergence when both the center $y \in \text{int } \Omega$ and the barrier parameter $k > 0$ are fixed. In such a case, the following estimation is appropriate, with c independent of k and y:

$$\max\{\|x^{s+1} - x^*\|, \|\lambda^{s+1} - \lambda^*\|\}$$

$$\leq ck^{-1}[f(y) - f(x^*)] \|\lambda^s - \lambda^*\|.$$

The primal unconstrained optimization of the MBF following the Lagrange multiplier update is equivalent to the unconstrained optimization of the shifted dual objective function augmented by a weighted classical barrier term (Jensen and Polyak, 1994; Teboulle, 1993). Thus, the MBF for the primal problem is equivalent to the IPM for the dual problem. This is very important numerically because it makes it possible to retain the smoothness of the initial functions and to use the Newton method for primal unconstrained minimization and to eliminate the combinatorial nature of the constrained optimization for both primal and dual problems.

Finally, the MBF method is equivalent to the Prox Method with entropy-like kernel for the dual problem (Teboulle, 1993).

See **Classical optimization; Computational complexity; Interior point methods; Nonlinear programming.**

References

[1] Bertsekas, D. (1982). *Constrained Optimization and Lagrange Multipliers Methods*, Academic Press, New York.

[2] Carroll, C. (1961). "The created response surface technique for optimizing nonlinear restrained systems," *Operations Research*, 9, 169–184.

[3] Fiacco, A.V. and McCormick, G.P. (1968). *Nonlinear Programming: Sequential Unconstrained Minimization Techniques*, John Wiley, New York.

[4] Frisch K. (1955). "The logarithmic potential method of convex programming," Technical Memorandum, May 13, University Institute of Economics, Oslo.

[5] Gonzaga, C. (1992). "Path Following Methods for Linear Programming," *SIAM Review*, 34, 167–224.

[6] Hestenes, M. (1969). "Multiplier and gradient methods," *Jl. Optimization Theory & Applications*, 4, 303–320.

[7] Huard, P. (1967). "Resolution of Mathematical Programming with Nonlinear Constraints by the Method of Centers" in *Nonlinear Programming*, J. Abadie, ed., North-Holland, Amsterdam.

[8] Jensen, D. and Polyak, R.P. (1994). "The convergence of the modified barrier method for convex programming," *IBM Jl. R&D*, 38, 307–321.

[9] Karmarkar, N. (1984). "A New Polynomial-time Algorithm for Linear Programming," *Combinatorica*, 4, 373–395.

[10] Mangasarian, O. (1975). "Unconstrained Lagrangians in Nonlinear Programming," *SIAM Jl. Control*, 13, 772–791.

[11] Melman, A. and Polyak R. (1994). "The Newton Modified Barrier Method for QP," ORE Report 94111, George Mason University, Fairfax, Virginia.

[12] Nesterov, Yu. and Nemirovskii, A. (1994). *Interior Point Polynomial Algorithms in Convex Programming*. SIAM Studies in Applied Mathematics, Philadelphia.

[13] Polyak, R. (1986). *Controlled processes in extremal and equilibrium problems*, VINITY, Moscow.

[14] Polyak, R. (1991). "Modified Interior Distance Functions and Centers Methods," Res. Report 17042, IBM T.J. Watson Research Center, New York.

[15] Polyak, R. (1992). "Modified Barrier Functions (Theory and Methods)," *Mathematical Programming*, 54, 177–222.

[16] Powell, M. (1969). "A Method for Nonlinear Constraints in Minimization Problems," in *Optimization*, R. Fletcher, ed., Academic Press, New York.

[17] Powell, M. (1992). "Some Convergence Properties of the Shifted Log-Barrier Method for Linear Programming," Numerical Analysis Report DAMTP 1992/NA7, Univ. of Cambridge.

[18] Renegar, J. (1988). "A polynomial-time algorithm, based on Newton's method for linear programming," *Mathematical Programming*, 40, 59–93.

[19] Rockafellar, R.T. (1974). "Augmented Lagrange Multiplier Functions and Duality in Nonlinear Programming," *SIAM Jl. Control & Optimization*, 12, 268–285.

[20] Smale, S. (1986). "Newton method estimates from data at one point" in *The Merging of Disciplines in Pure, Applied and Computational Mathematics*, R. Ewing, ed., Springer-Verlag, New York/Berlin.

[21] Teboulle, M. (1993). "Entropic proximal mappings with application to nonlinear programming," *Mathematics of Operations Research*, 17, 670–690.

BASIC FEASIBLE SOLUTION

A nonnegative basic solution to a set of $(m \times n)$ linear equations $Ax = b$, where $m \leq n$. The major importance of basic feasible solutions is that, for a linear-programming problem, they correspond to extreme points of the convex set of solutions. The simplex algorithm moves through a sequence of adjacent extreme points (basic feasible solutions). See **Adjacent extreme points; Basic solution**.

BASIC SOLUTION

For a set of $(m \times n)$ linear equations $Ax = b$ $(m \leq n)$, with rank m, a basic solution is a solution obtained by setting $(n - m)$ variables equal to zero and solving for the remaining m variables, provided that the column vectors associated with the m variables form a linearly independent set of vectors. The m variables are called basic variables, and the remaining $n - m$ variables that were set equal to zero are called nonbasic variables. The vectors associated with the basic variables form an $(m \times m)$ basis matrix B.

BASIC VARIABLES

The set of variables corresponding to the columns of a basis matrix in a linear system $Ax = b$. See **Basic solution; Basis**.

BASIS

A nonsingular square matrix B obtained by selecting linearly independent columns of a full row rank matrix A. The matrix B is then a basis matrix for the system $Ax = b$. The components of x associated with B are called the basic variables, and the remaining components are called the nonbasic variables. The term basis also refers to the set of indices of the basic variables. See **Basic variables**.

BASIS INVERSE

The inverse of a basis matrix. See **Basis**.

BASIS VECTOR

A column of a basis matrix. See **Basis**.

BATCH SHOPS

See **Production management**.

BATTLE MODELING

Dean S. Hartley, III

Oak Ridge National Laboratory

THE IDEAL BATTLE MODEL: The ideal battle model completely, accurately, quickly, and easily predicts the results of any postulated battle from the initial conditions. Several factors prevent the existence of an ideal battle model.

One factor is computational complexity. Medical planners could use such a battle model to determine the size of treatment facilities, the breakdown of physician skills needed, and the medical supply inventory requirements. It is reasonable to suppose a battle model would track individuals and their separate wounds for engagements of a dozen participants on a side; however, maintaining that level of detail for engagements of tens of thousands of people would be prohibitively expensive in time and hardware requirements. Thus, the requirement for complete predictions competes with the requirements for generality and speed of computation.

The second factor preventing the existence of an ideal battle model is the fact that we do not know enough about battle dynamics to model it accurately. Where we can model components accurately (e.g., firing disciplines for weapons and probabilities of kills given hits), we do not know how the components fit together (e.g., when do soldiers fire their weapons and how do conditions modify their ideal performance). Further, we do not know when, where, and why battles are joined or when and how they stop. Our ignorance is not absolute, but is relative to the accuracy we would like to achieve with our battle models.

A third factor also proceeds from our ignorance. We do not know which initial conditions are significant for determining battle results. In general, those battle models that deliver massive details about the model results require extremely large quantities of input data. Thus, perceived accuracy of results is a competitor of ease and rapidity of use.

BATTLE MODEL CLASSIFICATION: We cannot build the ideal battle model; however, we have built many individual battle models, each conceived to fulfill a particular set of objectives. These models of combat may be classified by their position along several dimensions; however, they all have one feature in common, and that is the object that is modeled is some aspect of combat. These dimensions are listed below with illustrative examples of positions along the dimension.

DOMAIN: land; air; naval; space; combinations.

SPAN (size of conflict): platoon battle; division combat; theater-level combat; global combat.

SCOPE (type of conflict): politico-military; special operations; low intensity conflict; urban warfare; conventional warfare; theater-level nuclear, chemical, and biological conflict; strategic nuclear conflict.

SCORING (adjudication topics and methodology): measures of merit: attrition, movement, tons of bombs dropped, supplies delivered, victory; methodologies: weapon weights (simple or complex, as in anti-potential potential), process simulations.

RANDOMNESS: deterministic or stochastic calculations.

COMBAT ACTIVITIES AND FORCE COMPOSITION (military assets and mission areas): small-arms; armor; aircraft; artillery; engineer; logistics; signal; command and control; intelligence; surface navy; submarine; electronic warfare; space assets; missiles.

LEVEL OF RESOLUTION OR DETAIL (smallest item modeled as a separate entity): bullet; soldier; tank; platoon; company; battalion; brigade; division; corps.

ENVIRONMENT: one-dimensional terrain (piston–model); two-dimensional terrain (including ocean or air), latitude–longitude or hexagonal grid-based; three-dimensional terrain; weather; day–night; smoke.

PURPOSE (design purpose or users' purpose): training; weapon system employment; force composition decisions; operations plans testing.

LEVEL OF TRAINING (training audience): individual skills; platoon leaders' skills; division staff skills; commanders' skills; combinations.

MODEL TREATMENT OF TIME: linear code with no time representation or algorithmically computed time (generally analytic combat models); time-stepped simulations;

event-driven simulations; expected value models; stochastic simulations.

HUMAN INTERACTION: data preparation and output interpretation; interruptable with modification and restart; computer-assisted human participation on one or more sides; continuous human participation on all sides.

SIDEDNESS: one-sided (e.g., strategic nuclear strike damage effects); two-sided; multisided; hard-coded identical properties for each side, hard-coded different properties for each side (e.g., U.S. vs Soviet tactics), or data-driven properties for each side.

COMPUTER INVOLVEMENT: none; moderate; complete.

SIZE COMPUTER REQUIRED: PC; mini-computer; mainframe; supercomputer; peripheral equipment required; large run-times, small run-times.

EXTERNAL INTERACTIONS (interfaces with parts of the real world): none; distributed processing; interfaces with weapon simulators; interfaces with real equipment; sand tables, scripting.

Battle modeling started the first time someone scratched a battle plan in the dirt and tried to conceive of the consequences. Sand tables, with miniature troops and landscaping, added discipline to the modeling process; however, the modeling remained essentially qualitative. Sand table models were used as war games, in which opposing players took turns moving the pieces and used rules to adjudicate the results of the moves. Modern war games include sand table games and computer adjudicated games.

ATTRITION LAWS: Lanchester (1916) introduced the concept of a quantitative model of attrition. (Osipov in Russia and Fiske in the U.S. introduced similar concepts at virtually the same time; however, most Western works refer to Lanchester's laws and Lanchestrian attrition.) Lanchester showed that one could express the value of concentration of forces precisely, using mathematics, and thus evaluate what forces would be needed for victory before a battle. Engel (1954) provided what many took to be proof that Lanchester's square law was correct.

Lanchester's simple concepts have been elaborated to the extent that Taylor (1983) required two volumes to discuss the many uses and implications of Lanchester theory. The computational power of computers has permitted this elaboration. First, heterogeneous Lanchester equations could be solved without undue manual labor. Once, heterogeneous equations were admitted, the coefficients could be represented as functions of other factors, such as weather, firing discipline, and distance to the target. Bonder and Farrel (in Taylor, 1983) introduced rigorous thinking into this area by observing direct fire activities and creating a mathematical model of those activities.

Dupuy (1985) argued that there are many important factors in combat that were not being included in the physics-based combat models. Morale, training, and leadership are at least as important as force sizes according to Dupuy. He proposed a model based on quantified judgments of these and other "soft" factors. His Quantified Judgment Model (QJM) stirred considerable controversy. Regardless of the merits of the QJM itself, the quantified judgments of soft factors is currently receiving more favorable reviews. The difference in public opinion at home during the Vietnam and Gulf wars and the impact on troop morale and the outcomes of the wars provides some justification for increased emphasis on soft factors.

Computers also made the computation of stochastic processes possible. The differential equations of Lanchester attrition were viewed as approximations to a random process model of the actual killing process that should be correct for large numbers. Stochastic duels addressed the results for small numbers. Ancker and Gafarian have made significant contributions in this area (see Ancker, 1994).

Helmbold has made contributions to both the theoretical and the practical aspects of battle modeling. His empirical studies of attrition (1961, 1964), breakpoints (1971), and movement (1990) injected the element of reality into the sometimes rarefied atmosphere of theoretical battle modeling. Hartley (1991) continued in this vein with results indicating that the best description of attrition (using a homogeneous approximation) is not the Lanchester square or linear law, but an intermediate form between the linear law and a logarithmic law.

With computer battle models also came a proliferation of structural types of models. Battle models involving anti-submarine warfare have a peculiar requirement of finding the enemy before the battle can be prosecuted. Search theory must be implemented in such models, just as it is used in actual battles or exercises (Shude, 1971). In some types of war, the proper allocation of resources or mix of strategies provides an easily defined variable (e.g., strategic nuclear targeting or allocation of combat air forces to mission types). Because game theory deals with optimal strategies considering both sides' options, it provided an obvious technique for addressing the problem and providing prescriptive models (Bracken, Falk and Miercort, 1974; HQ USAF/SAMA, 1974).

DIMENSION, DATA AND OUTPUT: In earlier times, land warfare models were one-dimensional: the forward edge of the battle area (FEBA) advanced or retreated. More sophisticated versions allowed one-dimensional structures for each sector (piston-models). More powerful computers now permit two-dimensional representations of the battlefield, using either *x,y* (or latitude, longitude) coordinates or (rectangular or hexagonal) grid structures. Some models are now three-dimensional, having terrain elevation and playing the effects of flying aircraft at different altitudes. (See, for example, the Research Evaluation and Systems Analysis (RESA) model (Naval Ocean Systems Center, 1992), which plays aircraft at different altitudes and submarines at different depths.)

Most large models have extremely large input and output data sets and require sophisticated database management systems to keep track of the data. These large output data sets also stress the human ability to understand the results. Sophisticated graphics are necessary adjuncts to most large models today. The graphics are required to define realistic scenarios and to understand the process and results of the model.

Advances in computer power have resulted in the capability for human interfaces that are qualitatively different from past capabilities. Such interfaces include real-time depictions of a battlefield from a human perspective and auditory and tactile interfaces. The first full-scale example of this kind of interface, called virtual reality, in a battle model was SIMNET. (The SIMNET Program Manager is at Fort Knox, which publishes SIMNET documentation – HQ US Army Armor School, 1987.) SIMNET is a network of tank and other vehicle simulators, each participating in a shared virtual battlefield.

Work is proceeding to tie virtual reality battle models to other, more conventional battle models. The success of connecting simulators has motivated recent work in connecting interactive training models. The connection of these battle models permits distributed processing and cost sharing among users.

The history of battle modeling has not been a smooth process of constant improvements. It has been beset with controversies in many areas. Some of the controversies have involved the standard resource allocation question: where do you spend the money? One of the first

of these concerned documentation. Early (1960s–70s) computer models were usually undocumented and, because of frequent modifications, had virtually indecipherable code. The need for proper documentation was obvious but the need for better (or at least more complex) models appeared overriding. While the readability of the documentation of models in the 1990s may be variable, most models are documented.

VERIFICATION, VALIDATION AND ACCREDITATION: One controversy probably began with the first model that produced a result someone did not like: is the model right? During the 1960s and early 1970s, we said there were two kinds of generals: those for whom computer printout was the gospel and those who would believe nothing produced by a computer. The problem in dealing with the first type was in conveying that there were caveats. All results had to be retyped manually to disguise their origin for the second type of general. Today's generals (and politicians) grew up with computers. They want to understand to what extent the results are believable. They require verification, validation, and accreditation. Although progress is being made, no one knows how to completely verify, validate, or accredit the general battle model.

There have also been technical controversies in battle modeling. Notable controversies have included the proper interpretation (and thus use) of the differences between the Lanchester linear and square laws, the connection between attrition and advance rates (if any), the value of force ratios, the connection between deterministic Lanchester formulations and stochastic attrition formulations and which should be used. There is a precept that states that a force ratio of 3–1, attacker-defender is required for a successful attack. Numerous studies have attacked this precept, yet it is still heard.

There are disagreements about the proper level of detail in deterministic models, despite agreement on the principle that what is appropriate depends on the uses to be made of a model. High resolution models of large span require tremendous quantities of data and run slowly. One camp advocates small, fast "roughly right" models as better than high resolution models. Another camp protests that such models will miss the critical points that differentiate the issues in question. The stochastic process camp protests that both the large, high resolution and the small, low resolution models are not grounded in the reality of stochastic battles, and cannot thus be even roughly right.

There have also been disagreements about the proper uses of models. At one time prescriptive battle models were popular (finding optimal strategies, where the definition of optimal varied with the model). Lately they have been out of favor. Complaints about the misuse of models have ranged from the use of models designed for other purposes and failing to understand the resulting mismatch of assumptions to charges of advocacy modeling. Advocacy modeling, in the pejorative sense, entails fiddling with input parameters until a combination is found that gives the desired result. Most large models have sufficient numbers of parameters with sufficiently tenuous connections to physical factors that plausible values can be found that generate almost any result.

An illuminating controversy involved the discovery that very simple deterministic battle models can exhibit chaos (Dewar *et al.*, 1991). The question of the impact of chaos on the more complex models that are actually used is obvious. Most issues are settled by point estimates. For example, suppose the impact of weapon X is being investigated. Model runs with 25% X, 50% X, 75% X, and 100% X are executed. The runs with 75% X and 100% X are found to have superior results. It is assumed that such results are valid for values between 75% and 100%. If the results are chaos driven, such an assumption is unwarranted. The question has not been finally answered; however, investigations with a complex model indicate that any uncertainty due to chaotic behavior in that model is no larger than a few percent. Because this is within the uncertainty that was already present in the model, the impact of possible chaotic behavior was claimed to be minimal.

Despite all controversy, battle modeling remains the only method of answering some questions and is widely used. Battle models are used to inform decisions on weapons' procurement issues, to test strategies and tactics, and to train personnel. Battle training models provide inexpensive tools for training commanders because the large numbers of combat personnel maneuver in the computer rather than on the ground. As military funding is reduced, this supplement to traditional training methods has become indispensable.

See **Accreditation; Game theory; Gaming; Lanchester equations; Military operations research; Operations Research Office and Research Analysis Corporation; RAND: Verification validation and testing of models.**

References

[1] Ancker, C.J., Jr. (1994). *An Axiom Set (Laws) for a Theory of Combat*, Technical Report, Systems Engineering, University of So. California, Los Angeles.

[2] Bracken, J., J.E. Falk, and F.A. Miercort (1974). *A Strategic Weapons Exchange Allocation Model*, Serial T-325. School of Engineering and Applied Science, The George Washington University, Washington, DC.

[3] Dewar, J.A., J.J. Gillogly, and M.L. Junessa (1991). *Non-Monotonicity, Chaos and Combat Models*, R-3995-RC, RAND, Santa Monica, California.

[4] Dupuy, T.N. (1985). *Numbers, Predictions & War.* Hero Books, Fairfax, Virginia.

[5] Engel, J. H. (1954). "A Verification of Lanchester's Law," *Operations Research* **2**, 163–171.

[6] Hartley, D.S., III (1991). *Predicting Combat Effects*, K/DSRD-412. Martin Marietta Energy Systems, Inc., Oak Ridge, Tennessee.

[7] Helmbold, R.L. (1961). *Historical Data and Lanchester's Theory of Combat*, AD 480 975, CORG-SP-128.

[8] Helmbold, R.L. (1964). *Historical Data and Lanchester's Theory of Combat, Part II*, AD 480 109, CORG-SP-190.

[9] Helmbold, R.L. (1971). *Decision in Battle*: *Breakpoint Hypotheses and Engagement Termination Data*, AD 729769. Defense Technical Information Center, Alexandria, Virginia.

[10] Helmbold, R.L. (1990). *Rates of Advance in Historical Land Combat Operations*, CAA-RP-90-1. Combat Analysis Agency, Bethesda, Maryland.

[11] HQ US Army Armor School (1987). *M-1 SIMNET Operator's Guide.* Fort Knox, Kentucky.

[12] HQ USAF/SAMA (1974). *A Computer Program for Measuring the Effectiveness of Tactical Fighter Forces (Documentation and Users Manual for TAC CONTENDER) SABER GRAND (CHARLIE).*

[13] Lanchester, F.W. (1916). "Mathematics in Warfare" in *Aircraft in Warfare*: *The Dawn of the Fourth Arm*, Constable and Company, London. (Reprinted in *The World of Mathematics*, ed. by R. Newman, Simon and Schuster, New York, 1956.)

[14] Naval Ocean Systems Center (1992). *RESA Users Guide Version* 5.5, *Vol* 1–8.

[15] Shudde, R.H. (1971). "Contact and Attack Problems" in *Selected Methods and Models in Military Operations Research*, ed. by P.W. Zehna. Military Operations Reserach Society, 125–146.

[16] Taylor, J.G. (1980). *Force-on-Force Attrition Modeling*. Military Applications Section, ORSA.

[17] Taylor, J.G. (1983). *Lanchester Models of Warfare, Volumes I and II*. Military Applications Section, ORSA.

BAYESIAN DECISION THEORY SUBJECTIVE PROBABILTITY AND UTILITY

Kathryn Blackmond Laskey

George Mason University, Fairfax, Virginia

In every field of human endeavor, individuals and organizations make decisions under conditions of uncertainty and ignorance. The consequences of a decision, and their value to the decision maker, often depend on events or quantities which are unknown to the decision maker at the time the choice must be made. Such problems of decision under uncertainty form the subject matter of Bayesian decision theory. Bayesian decision theory has been applied to problems in a broad variety of fields, including engineering, economics, business, public policy, and artificial intelligence.

A decision theoretic model for a problem of decision under uncertainty contains the following basic elements:

- A set of *options* from which the decision maker may choose;
- A set of *consequences* which may occur as a result of the decision;
- A *probability distribution* which quantifies the decision maker's beliefs about the consequences that may occur if each of the options is chosen; and
- A *utility function* which quantifies the decision maker's preferences among different consequences.

SUBJECTIVE PROBABILITY: Decision theory applies the probability calculus to quantify a decision maker's beliefs about uncertain events or quantities, and to update beliefs upon receipt of additional information. De Finetti (1974) showed that any decision maker who acts on degrees of beliefs not conforming to the probability calculus can be exploited by a series of gambles guaranteed to result in a net loss. Such a bet is called a *dutch book*. The Dutch Book Theorem and other related derivations of probability from axioms of rationality have been used to justify probability as a caclulus of rational degrees of belief (DeGroot, 1970; Pratt, *et al.*, 1965).

BAYES RULE: When a decision maker receives information bearing on an uncertain hypothesis, degrees of belief are updated by computing the conditional probability of the uncertain hypothesis given the new evidence. The equation expressing how beliefs change with new evidence has been attributed to the Reverend Thomas Bayes (1763) and is known as *Bayes Rule*. The odds-likelihood form of Bayes Rule is:

$$\frac{\Pr\{H_1|E\}}{\Pr\{H_2|E\}} = \frac{\Pr\{E|H_1\}\Pr\{H_1\}}{\Pr\{E|H_2\}\Pr\{H_2\}}.$$

In this equation, H_1 and H_2 refer to two uncertain hypotheses entertained by the decision maker and E refers to the new evidence or information received by the decision maker. Bayes rule quantifies how evidence is used to obtain the relative *posterior* probabilities $\Pr\{H_i|E\}$ of the hypotheses given the evidence. The ratio of posterior probabilities is determined by two factors. One is the ratio of *prior* probabilities $\Pr\{H_i\}$: all other things being equal, the stronger the prior belief in H_1 relative to H_2, the stronger the posterior belief in H_1 relative to H_2. The other is the *likelihood ratio*, or ratio of the probabilities $\Pr\{E|H_i\}$ of the evidence given each of the hypotheses. Again, all other things being equal, the better H_1 accounts for the evidence relative to H_2, the stronger the posterior belief in H_1 relative to H_2.

OTHER INTERPRETATIONS OF THE PROBABILITY CALCULUS: There has been considerable debate about how to interpret the concept of probability. The term *Bayesian*, after Bayes Rule, is used to refer to the *subjective* interpretation. A subjective probability distribution represents an individual's degrees of belief about the likelihood of uncertain outcomes. Alternative interpretations of probability include the classical, the logical, and the frequentist approaches (Fine, 1973). Much of standard statistical theory is based on the frequentist approach. Frequentists argue that probability models are appropriate only for repeatable phenomena exhibiting inherent randomness. For such phenomena, it is argued, there exist objectively correct probabilities intrinsic to the process producing the uncertain outcomes. Subjectivists apply probability theory to any outcomes about which a decision maker is uncertain. For subjectivists, no objectively correct probabilities need exist. Different decision makers are free to have different opinions about the probability of an outcome.

The only constraint subjective theory places on a probability distribution is that it be coherent, that is, that degrees of belief conform to the probability calculus. Within this constraint, a decision maker is free to choose any probability distribution to model his or her uncertainty about a problem. Its inherent subjectivity has been a persistent criticism of the subjectivist approach. This is often of little practical consequence for problems which can be said to exhibit inherent randomness. The subjectivist draws inferences about the posterior distribution of the unknown parameter while the frequentist draws inferences about the distribution of the data given different values of the unknown parameter. Nevertheless, it can be shown that when there are sufficient data to draw accurate inferences, the subjectivist and the frequentist will usually agree on the implications of the results. Thus, the major difference of practical import between the subjectivist and the frequentist is their attitudes toward problems for which there are too little data to estimate parameters accurately or for which the assumption of intrinsic objective frequencies is problematic. The frequentist maintains that probability models are inappropriate for such problems; the subjectivist argues that probabilities are appropriate and that it is legitimate for rational people to disagree until there are sufficient data to bring them to agreement.

UTILITY THEORY: Decision theory quantifies preferences by a utility function. It is assumed that the decision maker can assign a numerical utility to each possible consequence of each option being entertained. Consequences with higher utilities are preferred to consequences with lower utilities. When there is uncertainty, the decision maker selects the option for which the expected value of the utility function is the largest. For some problems it is customary to deal with losses, or negative utilities. Smaller losses are preferred to larger losses.

The concept of utility appears to have been first introduced by Daniel Bernoulli (1738) in his solution to a puzzle known as the St. Petersburg Paradox. Bernoulli considered the problem of what price to pay for the opportunity to play the following gamble. A fair coin (probability 0.5 of landing heads) is tossed repeatedly until the first head appears. If the first head appears on the nth toss, the decision maker receives a prize of 2^n units of currency. The decision maker's expected monetary prize is

$$2(0.5) + 2^2(0.5)^2 + 2^3(0.5)^3 + \ldots$$

which is infinite. A decision maker who maximized expected monetary value should be prepared to pay an arbitrarily large sum of money for the opportunity to play this gamble. As Bernoulli noted, most people would be willing to pay only a modest amount. Bernoulli suggested that the resolution to this apparent paradox was that a prize's worth to a decision maker was a nonlinear function of the monetary value of the prize. For example, replacing 2^n with $\log 2^n$ in the above equation yields a finite expected monetary prize.

Von Neumann and Morgenstern (1944) were the first to present a formal axiomatic development of utility theory. They defined the utility

of a consequence in terms of a comparison between two options, one sure and one uncertain. The sure option is the consequence itself; the uncertain option is a lottery between two standard reference prizes, one worth more and one worth less than the consequence in question. If the reference prizes are assigned utility one and zero then the utility of the consequence in question is defined as the probability at which the decision maker is indifferent between the two lotteries. Several similar axiom systems can be shown to lead to the maximization of expected utility as a principle of rational decision making (DeGroot, 1970; Pratt, *et al.*, 1965).

APPLYING DECISION THEORY: It has been observed that people systematically violate the axioms of expected utility theory in their everyday behavior. Some of these violations can be reversed by informing people of the implications of their stated preferences. In other cases, many people resist changes to their original judgments. Even when the decision maker regards expected utility theory as a norm of rational behavior, it cannot be assumed that unaided judgments will be consistent with the theory. The field of *decision analysis* applies theories and methods from decision theory and the psychology of human information processing to construct decision theoretic models for practical decision problems.

Interest has been growing in decision theoretic formulations of statistical problems. For example, to formulate an hypothesis testing problem, one defines a prior probability for the null and alternative hypotheses. One also defines losses associated with accepting a false alternative hypothesis and rejecting a true null hypothesis. The optimal decision rule is to accept or reject the hypothesis according to which decision yields the lower posterior expected loss given the observed sample. Similarly, decisions of whether to gather information and how large a sample to draw can be formulated as decision problems that consider both the cost of gathering information and the benefit of obtaining the information. Some problems that are quite complex when viewed from a frequentist perspective become straightforward when viewed from a Bayesian perspective. Examples include hierarchical models and problems of missing data (Rubin, 1984).

An area of application is the field of intelligent systems. Utility theory is being applied to planning and control of reasoning in expert systems. Diagnostic expert systems based on probability theory have achieved performance comparable to human decision makers (e.g., the Pathfinder system for diagnosing lymph node pathology, Heckerman, 1990). Perhaps the most important and challenging aspect of decision analysis is the creative process of model formulation. Decision theory takes options, consequences, and their interrelationships as given. Automated decision model generation is an open research area of great importance to application of decision theory to the field of intelligent systems (Laskey, *et al.*, 1993; Wellman, *et al.*, 1992).

See **Decision analysis; Decision problem; Decision trees; Expert systems; Utility theory.**

References

[1] Bayes, Thomas R. (1763). "An Essay Towards Solving a Problem in the Doctrine of Chances." *Philosophical Transactions of the Royal Society of London* 53, 370–418 (reprinted with biographical note by G. Barnard, 1958, in *Biometrika* 45, 293–315).

[2] Bernoulli, Daniel (1738). "Specimen Theoriae Novae de Mensura Sortis." *Commentarii Academiae Scientarium Imperalis Petropolitanae* 175–192 (translated in Sommer, L., 1984, *Econometrica* 22, 23–26).

[3] de Finetti, Bruno (1974). *Theory of Probability: A Critical Introductory Treatment*. John Wiley, New York.

[4] DeGroot, Morris H. (1970). *Optimal Statistical Decisions*. McGraw Hill, New York.

[5] Fine, Terrence L. (1973). *Theories of Probability*. Academic Press, New York.

[6] Heckerman, David (1990). "Probabilistic Similarity Networks." Ph.D. diss., Program in Medical Information Sciences, Stanford University, California.

[7] Laskey, Kathryn B., and Paul E. Lehner (1994). "Metareasoning and the Problem of Small Worlds." *IEEE Transactions on Systems, Man and Cybernetics* 24, 1643–1652.

[8] Pratt, John W., Howard Raiffa, and Ronald Schlaifer (1965). *The Foundations of Decision Under Uncertainty: An Elementary Exposition*. McGraw Hill, New York.

[9] Rubin, Donald B. (1984). "Bayesianly Justifiable and Relevant Frequency Calculations for the Applied Statistician." *Annals of Statistics* 12, 1151–1172.

[10] von Neumann, John and Oscar Morgenstern (1944). *Theory of Games and Economic Behaviour*. Princeton University Press, New Jersey.

[11] Wellman, Michael P., John S. Breese and R. P. Goldman (1992). "From Knowledge Bases to Decision Models." *The Knowledge Engineering Review* 7, 35–53.

BAYES RULE

When a decision maker receives data bearing on an uncertain event, the a priori probability of the event can be updated by computing the conditional probability of the uncertain hypothesis given the new evidence. The derivation of the revised or a posteriori probability can be easily derived from fundamental principles and its discovery has been attributed to the Reverend Thomas Bayes (1763). The result is therefore known as *Bayes rule* or *theorem*:

$$\Pr\{H_1|E\} = \frac{\Pr\{E|H_1\}\Pr\{H_1\}}{\Sigma_i \Pr\{E|H_i\}\Pr\{H_i\}}.$$

In this equation, H_1 refers to the specific, uncertain hypothesis entertained by the decision maker, the $\{H_i\}$ are the complete set of possible hypotheses, and E refers to the new evidence or information received.

BEALE TABLEAU

A modification of the simplex tableau arranged in an equation form such that the basic variables and the objective function value are expressed explicitly as functions of the nonbasic variables. This tableau is often used when solving integer-programming problems.

BENDER'S DECOMPOSITION METHOD

A procedure for solving integer-programming problems that have a few integer variables. These so-called complicating variables, when given specific values, enable the resulting problem to be readily solved as a linear-programming problem.

BEST-FIT DECREASING ALGORITHM

See **Bin packing**.

BIDDING MODELS

Michael H. Rothkopf

Rutgers University, New Brunswick, New Jersey

Bidding models have been constructed to help bidders decide how to bid, to help auction designers evaluate alternative rules and formats, and to help detect collusion. The models date back at least to 1956 when work that led to the first PhD in Operations Research was published by Friedman. Beginning in the late 1970s, there has been a surge of new theoretical work, much of it involving game theory models. The results of applying bidding models are known to be quite sensitive to the choice of modeling assumptions.

There are various types of auctions. Some bidding models have been constructed to analyze bidding in a particular type of auction, while others are intended to compare results across various types. There are four main types of auctions. One is standard sealed bidding in which the best bid wins and the sale price is the amount of that bid. A second, less common type is second-price or Vickrey sealed bidding, in which the high bid wins but the payment is the amount of the best losing bid (Vickrey, 1961). Oral progressive or English auctions in which bidders successively raise the bid until no bidder wishes to do so, at which point the sale takes place to the final bidder at the amount of his or her bid are another important auction type. The final type is the Dutch auction, in which the auctioneer or a mechanical "clock" progressively lowers the price until a bidder bids and wins at the amount of his or her bid. For each of these types, there is an essentially perfect analog for low–bid–wins auctions in which the bidders are sellers who get paid.

Often, bidding models deal with a single isolated auction. The earliest and most used bidding models are decision theory models of standard sealed bidding that help a bidder decide upon his bid. They assume, estimate statistically, or calculate a probability distribution, $F(x)$, for the best competitive bid, x, and then pick the bid, b, that maximizes expected profit, $E[b] = (v - b)F(b)$, from the auction relative to a known value, v, to the bidder of the item for sale. If v is a random variable, its expectation may be used in this calculation *provided* that it is independent of x. If $F(x)$ is calculated, it is frequently taken as the maximum of independent draws from the bid distributions, $F_j(x_j)$, of competitors: $\Pi_j F_j(x_j)$. Such models are known as independent *private values* models. In such models, the more competition a bidder faces, the more aggressively he or she should bid.

In many situations such as wildcat oil lease sales, a *common value* model is more appropriate than an independent private values model. In a common value model, the value of what is being sold is assumed to be the same to each of the bidders. Bidders are assumed to estimate this

common value making independent estimating errors. In such models, even when bidders' estimating errors are unbiased, the winning bidder is normally the bidder who has made the largest over estimate. Thus, the winner will, on the average, find what he or she has won to be worth less than he or she estimated it to be worth. This selection bias can be severe, especially when there are many competitors and when estimating accuracy is poor. Bidders must correct sufficiently for it or suffer the "winner's curse" (Capen, Clappand and Campbell, 1971). Making this correction leads bidders to bid *less* aggressively when the amount of competition is increased beyond some relatively low level (typically, two other bidders).

Decision theory models have also dealt with other situations such as unit price bidding and withdrawable bids. In unit price bidding, a single contract is awarded to the bidder whose set of offered unit prices gives the lowest total when summed after being multiplied by a corresponding set of bidder-supplied unit quantities. Actual quantities are uncertain, and payment is based upon them. Bidders have incentive to pick a set of unit prices that take advantage, to the maximum extent possible, of discrepancies between the way in which the bidder and the bid-taker view the situation. They may exploit differences in quantity estimates or in time of payment for different types of units. The problem of selecting an optimum set of unit prices has been modeled as a particularly easy to solve linear program and, for risk averse bidders, as a quadratic program.

The withdrawable bid models consider situations in which a bidder may, after the bids are opened, withdraw an otherwise winning bid, perhaps incurring a penalty for doing so. In some models, one or more bids may be withdrawn to win with another, less aggressive bid. In others, a rationally chosen bid may be withdrawn even if doing so incurs a penalty in order to lose so as to avoid what, given the now revealed bids of competitors, is likely to be a large winner's curse effect.

Many game theory models of auctions find Nash equilibrium bidding strategies for independent private values models in which there is a fixed set of, a priori, symmetric risk neutral bidders. Often, they compare results across auction types. Some results follow.

In an English auction, each bidder has a *dominant* strategy (i.e., optimal strategy no matter how rivals compete) in which he or she continues to compete up to his or her value. When all bidders follow such strategies, an efficient outcome is guaranteed: the item for sale is sold to the bidder valuing it most for a price at or marginally above the second-highest value attached to it by any bidder. A similar result holds for sealed second-price auctions. This time it is the optimal bids rather than the "strategies" that equal the bidders' values. As with oral auctions, the item goes to the highest valuer at the second-highest valuer's value. Dutch and standard sealed bidding call for identical strategies in which, in line with the decision theory approach for independent private values discussed above, the bidders select bids by balancing the risk of losing against the profitability of winning. Since the symmetric equilibrium bid function is increasing in the bidder's private valuation, the sale is made to the highest valuer.

Within this model, the seller's options to influence revenue are limited. The expected revenue to the seller is the same under any auction mechanisms in which (1) the asset always goes to the bidder with the highest value, and (2) every bidder would attain zero expected profit if his value were at its lowest possible level. In particular, this implies revenue equivalence for the four standard auction types.

For an important class of value functions called "affiliated values" that includes as limiting cases both independent private values and common values, English auctions produce at least as much expected revenue as Vickrey auctions which produce as much expected revenue as sealed bidding and Dutch auctions (Milgrom and Weber, 1982). However, these revenue ranking results are not robust. They require the assumptions of symmetry, risk neutrality and an exogenously fixed number of bidders (i.e., that bidders do not get to choose whether to incur the costs of bidding). Moreover, no general ranking of first-price and second-price expected revenue is robust to changes in these assumptions. Second-price procedures, both oral and sealed, continue to guarantee that in equilibrium the item sold always goes to a bidder who values it most highly, but first-price procedures no longer do so. A survey by McAfee and McMillan (1987) gives extensive results for game theory models of single isolated auctions with exogenously determined bidders.

To be *directly* useful for decision making in most situations, bidding models must go beyond the bounds of the single isolated auction. Bidding in simultaneous auctions subject to a constraint on exposure (i.e. the total of all bids) has been modeled both by dynamic programming and analytically. Sequential auctions have been modeled as well. In some of these models, the inter-relationships are internal to the bidding firm. In several such models, the optimal bid price can be characterized as the sum of direct costs, opportunity costs and a competitive advantage fee. In other sequential auction models, it is the effects of bids on the subsequent behavior of competitors that relates the auctions to each other. In one of these models, the appropriate bid becomes less aggressive as the discount factor between auctions and the anticipated magnitude of competitive response increase.

Concern about cheating may affect the appropriate choice of auction form. Several models consider cheating. They have shown that bidder conspiracies are stable, and hence may be more likely, in isolated oral and Vickrey auctions, and that fear of cheating by bid takers may be a factor in the rarity of Vickrey auctions. There has been statistical work on detecting patterns of rigged bids, but the most useful results are not in the public domain.

Further material on bidding can be obtained from: a description of auction use by Cassady (1967) and a sociological interpretation of oral auctions by Smith (1990); two edited volumes of papers by Amihud (1976) and Engelbrecht-Wiggans, Shubik and Stark (1983); two handbook chapters by Wilson (1992) and Rothkopf (1994); a survey paper by Engelbrecht-Wiggans (1980); and a critical essay by Rothkopf and Harstad (1994).

See **Decision analysis; Game theory.**

References

[1] Amihud, Y., ed. (1976), *Bidding and Auctioning for Procurement and Allocation*, New York University Press, New York.

[2] Capen, E., R. Clapp and W. Campbell (1971), "Competitive Bidding in High Risk Situations," *Jl. Petroleum Technology* 23, 641–653.

[3] Cassady, R., Jr. (1967), *Auctions and Auctioneering*, University of California Press, Berkeley.

[4] Engelbrecht-Wiggans, R. (1980), "Auctions and Bidding Models," *Management Science* 26, 119–142.

[5] Engelbrecht-Wiggans, R., M. Shubik and R.M. Stark, eds. (1983), *Auctions, Bidding, and Contracting: Uses and Theory*, New York University Press, New York.

[6] Friedman, L. (1956), "A Competitive Bidding Strategy, "*Operations Research* 4, 104–112.

[7] McAfee, R. P. and J. McMillan (1987), "Auctions and Bidding," *Jl. Economic Literature* 25, 699–738.

[8] Milgrom, P.R., and R.J. Weber (1982), "A Theory of Auctions and Competitive Bidding," *Econometrica* 50, 1089–1122.

[9] Rothkopf, M.H. (1994), "Models of Auctions and Competitive Bidding," *Handbooks in Operations Research, Vol. 7: Beyond the Profit Motive: Public Sector Applications and Methodology*, S. Pollock, A. Barnett and M.H. Rothkopf, eds., Elsevier Science Publishing, New York.

[10] Rothkopf, M.H. and R.M. Harstad (1994), "Modeling Competitive Bidding: A Critical Essay," *Management Science* 40, 364–384.

[11] Smith, C.W. (1990), *Auctions: The Social Construction of Value*, University of California Press, Berkeley.

[12] Wilson, R.B. (1992), "Strategic Analysis of Auctions," in *The Handbook of Game Theory, Volume 1*, R. Aumann and S. Hart, eds., North-Holland/ Elsevier Science Publishers, Amsterdam.

BIG-M METHOD

A method to drive artificial variables out of the basis in the simplex algorithm, by imposing a sufficiently large, finite penalty M for using these variables. See **Artificial variables**; **Phase I**; **Phase II**.

BILEVEL LINEAR PROGRAMMING

Bilevel linear programming (BLP) is a hierarchical, decentralized, multilevel mathematical programming problem in which the objective functions and constraints are linear. It can be stated in terms of upper and lower problems as follows:

Maximize $f_1(x,y) = c_1 x + d_1 y$
     x

where y solves:

Maximize $f_2(x,y) = c_2 x + d_2 y$
     y

subject to

$Ax + By \leq b$

$x, y \geq 0$

where c_1, c_2, d_1, d_2, and b are constant vectors, A and B are constant matrices; x and y are vectors of the decision variables of the upper and lower problems, respectively; f_1 and f_2 are the objective functions of the upper and lower problems, respectively. See **Linear programming**.

BINARY VARIABLE

A variable that is restricted to be equal to 0 or 1. Binary variables are often used to handle logical, nonlinear conditions associated with a problem whose constraining conditions are linear.

See **Combinatorial and integer optimization; Integer-programming problem**.

BIN-PACKING

Pearl Wang and Nastaran Coleman

George Mason University, Fairfax, Virginia

PROBLEM DEFINITION: The bin-packing problem is concerned with the determination of the minimum number of bins that are needed to pack a given set of input data items. The problem has numerous applications in operations research, computer science, and engineering, where the items and bins to be packed can be multi-dimensional. These applications include industrial manufacturing, stock cutting, military vehicle loading, television commercial scheduling, job scheduling on multiple processors, integrated circuit manufacturing and fault detection, location testing in linear circuits, and vehicle routing. Since the bin-packing problem is known to be NP-hard, it is of interest to find efficient heuristics that obtain near-optimal solutions to the problem (Garey and Johnson, 1981).

The classical one-dimensional bin-packing problem is defined as follows: Given a positive bin capacity C and a list of items $L = (p_1, p_2, \ldots, p_n)$, where p_i has size $s(p_i)$ satisfying $0 \leq s(p_i) \leq C$, determine the smallest integer m such that there is a partition $L = B_1 \cup B_2 \cup \ldots \cup B_m$ satisfying $\Sigma s(p_i) \leq C$, where $p_i \in B_j$, $1 \leq j \leq m$. The set B_j is usually viewed as the contents of a bin of capacity C. (In many cases, C is taken to be 1.)

Several versions of two-dimensional bin-packing problems have also been studied, where L is a set of rectangles p_i having heights h_i and widths w_i. In one type of problem, for example, the rectangles of L are to be packed into a single two-dimensional bin of width C and infinite height. The problem is to determine a minimum height packing of the pieces into this bin. In an alternative form of the problem, the rectangles of L must be packed into a minimum number of rectangular bins. A common version of the problem that has been studied concerns packing the list L into m unit squares with the objective being to minimize m.

Historically, heuristics for bin-packing problems were among the earliest algorithms studied in the literature. In the 1970s, it was shown that near-optimal solutions could be guaranteed for some commonly used one-dimensional packing techniques. Since then, many methods have been proposed for obtaining approximate solutions to both the one and two-dimensional problems for sequential and parallel models of computation. Higher-dimensional problems have been studied to a lesser degree. The performance of a given heuristic, as well as its packing behavior, are important considerations that have been analyzed by many researchers.

A detailed summary of many well-known bin-packing heuristics can be found in Coffman, *et al.* (1991) and Garey and Johnson (1981). A review of experimental studies performed on several bin packing algorithms appears in Dyckhoff (1990), and probabilistic analyses of several approaches for solving the problem are discussed in Coffman, *et al.* (1991). Algorithms for solving the problem on parallel models of computation can be found in Anderson *et al.* (1989) and Berkey (1990).

CLASSIFICATIONS OF BIN-PACKING ALGORITHMS:
Bin-packing algorithms have thus been proposed and analyzed for both sequential and parallel systems. Sequential heuristics can be classified as either on-line or off-line algorithms. On-line algorithms assign data items to bins in the same order as originally input, without utilizing any global knowledge of the data list. For example, the Next-Fit packing heuristic is an on-line algorithm. Off-line algorithms preprocess the data, usually by sorting. Well-known examples are the First-Fit Decreasing and Best-Fit Decreasing algorithms. Alternatively, Harmonic packing approaches partition the input data by size into subintervals, and pack the data using those subintervals. These techniques are described in more detail below.

Recently, approximation algorithms for solving the one-dimensional bin-packing problem on various models of parallel computation have been reported. It has been shown that several frequently used sequential bin packing strategies such as First-Fit Decreasing are P-Complete. Thus, it is unlikely that these heuristics can be parallelized into efficient algorithms for the theoretical Parallel Random Access Machine (PRAM) model of computation. However, other well-known sequential strategies such as Harmonic packing can be parallelized efficiently. Experimental studies of similar heuristics have been performed on Single-Instruction, Multiple-Data (SIMD) and Multiple-Instruction, Multiple-Data (MIMD) parallel computers.

PERFORMANCE STUDIES:
Several performance metrics have been formulated as a means to compare these different packing algorithms when executed on random data. Theoretical analyses typically include worst-case and average-case packing performance of the heuristics. The asymptotic worst-case performance is defined as the limiting ratio of an algorithm's worst instant packing to its optimal packing. For example, if $A(L)$ and $OPT(L)$ are the number of bins packed by an algorithm A and the optimal number of bins needed for a list L, respectively, then the asymptotic performance ratio can be defined as

$$R_A^\infty = \inf\{r \geq 1: \text{for some } N > 0, A(L)/OPT(L)$$
$$\leq r \text{ for all } L \text{ with } OPT(L) \geq N\}.$$

Two measures of average-case performance that have been studied are the expected values $E(R_N)$ and $E(U)$, where R_N is the ratio of the average number of bins packed by the algorithm to the average size of all data items, and U is the difference between these quantities. Further, an algorithm is often said to exhibit perfect packing if $E(R) = 1$, where $E(R)$ is the limiting distribution of $E(R_N)$, or when $E(U) = O(\sqrt{N})$.

These metrics are studied by theoretical analyses as well as simulation. The input data are usually assumed to come from a uniform distribution $[a,b]$. Recently, Coffman *et al.* (1991) used discrete uniform distributions to verify the aspects of average-case behavior that are lost in the passage to continuous approximation.

SOME ONE-DIMENSIONAL PACKING HEURISTICS:
The Next-Fit algorithm packs one-dimensional items into one-dimensional bins in the simplest fashion. The data items are processed one at a time, beginning with p_1, which is put into bin B_1. If item p_i is to be packed and B_j is the highest indexed non-empty bin, then p_i is placed into bin B_j if it fits into B_j, that is, $p_i + \text{size}(B_j) \leq C$. Otherwise, a new bin B_{j+1} is started and p_i is placed into it. In this manner, each successive piece is packed into the most recently used bin, and previously packed bins are not considered. Next-Fit is a fast on-line algorithm whose time complexity is $O(n)$. Its worst-case performance ratio is bounded by 2, and its average performance by 3/2. Variants of Next-Fit have been proposed and include Next-Fit-Decreasing, Next-1-Fit, and Next-K-fit. The basic approach is also used to obtain level-oriented heuristics for solving two-dimensional bin packing problems.

The First-Fit heuristic packs each successive data item p_i into the lowest indexed bin B_j into which it fits. When this is not possible, a new bin is created. Thus, it is necessary to maintain a list of all partially filled bins. For the worst-case, average case, and lower bound performance of First-Fit, it has been shown that the number of bins used by this algorithm is 1.7 $OPT(L) \pm 2$. The time complexity of First-Fit is $O(n \log n)$. If the items are initially sorted in non-increasing order before packing proceeds, the heuristic is referred to as First-Fit Decreasing, and the performance bound decreases to 11/9. Other algorithms that are based on this approach include Best-Fit (where the "best" bin is chosen if there is more than one possibility), Best-Fit Decreasing, Worst-Fit, Almost Worst-Fit, Revised First-Fit, and Modified First-Fit Decreasing. When the data items are drawn from a uniform distribution, then $E(A(L)) - n/2 = O(n)$ for the First-Fit Decreasing and Best-Fit Decreasing algorithms.

The Harmonic packing algorithm begins by partitioning the unit interval into the set of intervals $I_k = (1/(k+1), 1/k]$, $1 \le k < m$ and $I_m = (0, 1/m]$. The bins are divided into m categories and an I_k-bin packs atmost $k I_k$ data. The packing of each I_k piece into an I_k-bin is done using the Next-Fit Algorithm. At any given time, an active list of all unfilled I_k-bins is kept. The Harmonic algorithm has a worst-case performance bound of 1.67 contain some variants of this algorithm (Gambosi *et al.*, 1989; Rhee and Talagrand, 1988).

SOME MULTI-DIMENSIONAL PACKINGS HEURISTICS:

The Bottom-Left (BL) approach is a well-known heuristic that can be used to pack rectangles into a single two-dimensional bin that has infinite height. In this approach, rectangles are successively placed into the bottom-most, left-most position in the bin into which they fit without overlapping the rectangles that have already been packed. If the items are preordered by non-increasing width, then the worst case bound for this heuristic indicates that the height of the packing does not exceed twice the height of an optimal packing. The algorithm can be implemented in $O(n^2)$ time.

Level-oriented packings can be obtained if the rectangles to be packed are first ordered by non-increasing height. The packing is constructed as a sequence of levels, whose heights are defined by the heights of the first rectangles placed in the respective levels. The Next-Fit or First-Fit approaches can be used to define and fill these levels of the bin. The asymptotic performance bounds of the Next-Fit Decreasing Height and First-Fit Decreasing Height heuristics are 2 and 1.7, respectively. Similar approaches in which the heights of the levels are preset by a parameter yield a variety of shelf heuristics, where

these levels can be packed in a similar fashion. Next-Fit Shelf and First-Fit Shelf are examples of these heuristics. Their corresponding execution times are $O(n)$ and $O(n^2)$. If the parameter is defined by r, then these methods have asymptotic performance bounds of $2/r$ and $1.7/r$, respectively.

Alternative methods may divide the set of items being packed into sublists that are used to obtain split packings. In this case, the infinite height bin is also divided into subregions where one-dimensional heuristics are used to pack the rectangles. These techniques include Split-Fit, Mixed-Fit, and Up-Down which require $O(n \log n)$ time. Performance ratios of 2, 1.33, and 1.25, respectively, have been proven for these approaches. Other similar methods are summarized in Coffman *et al.* (1984); Coffman and Lueker (1991).

One particular heuristic (Coffman *et al.*, 1988) that uses this approach addresses the problem of packing squares into a two-dimensional strip. The squares whose widths are greater than 1/2 are first stacked along the left edge of the strip in order of decreasing width. Starting at the height, $H_{1/2}$, where the sum of the sizes of packed squares exceeds 1/2, the remaining squares are stacked along the right edge of the strip in order of decreasing width. This stack is then slid downward until it either rests on the bottom of the strip, or a square in the right stack comes in contact with a square in the left stack, which ever occurs first. Finally, all the squares lying entirely above $H_{1/2}$ are repacked into two stacks, one against the left edge of the strip and the other against the right edge. This is done in decreasing order of size, placing each successive square on the shorter of the two stacks already created. It can be shown for this algorithm, that $E(A(L)) = E(OPT(L)) + O(1)$.

When multi-dimensional objects are to be packed into a minimum number of multi-dimensional bins, the vector packing approach can be used. This technique is a direct generalization of the one-dimensional problem. For example, if rectangles are to be packed into square bins, then the only types of packings that are permitted are those where the rectangles are diagonally placed corner-to-corner across the bins. In general, if a vector packing algorithm is such that no two nonempty bins can be combined into a single bin, then the ratio of the number of bins packed to the optimal solution does not exceed $d + 1$, where d is the number of dimensions. Extensions of the First-Fit and First-Fit Decreasing heuristics to this multi-dimensional case have yielded approaches whose asymptotic worst case ratio is $d + 7/10$ and $d + 1/3$, respectively.

PARALLEL ALGORITHMS: Heuristics have also been proposed which obtain approximate solutions to the one-dimensional bin packing problem on various models of parallel computation. For the shared-memory Exclusive-Read Exclusive-Write PRAM model of computation, a heuristic based on First-Fit Decreasing has been proposed which runs in $O(\log n)$ time on $n \log n$ processors (Anderson, 1989). This approach divides the data items into two groups. Items in the first group are partitioned into sublists that are packed into "runs" of bins. The bins are then filled using items in the second group. The algorithm relies on parallel prefix, merging, and parenthesis matching operations, and has a worst-case performance bound of 11/9.

Practical one-dimensional bin packing algorithms (including parallelizations of the Harmonic algorithm) have also been proposed and implemented on parallel architectures such as systolic arrays, SIMD arrays, and MIMD hypercubes. Quantitative studies and theoretical analyses have been performed on some of these approaches. The Systolic packing algorithm, for example, has a worst-case performance bound of 1.5 and executes in $O(n)$ time. Similar results are reported in Berkey (1990).

See **Combinatorics; Computational complexity; Parallel computing.**

References

[1] Anderson, R.J., Mayr, E.W. and Warmuth, M.K. (1989), "Parallel Approximation Algorithms for Bin Packing," *Information and Computation*, 82, 262–271.

[2] Berkey, J.O. (1990), "The Design and Analysis of Parallel Algorithms for the One-Dimensional Bin Packing Problem," Ph.D. Dissertation, School of Information and Technology, George Mason University, Fairfax, Virginia.

[3] Coffman, E.G., Jr., Courcoubetis, C.A., Garey, M.R., Johnson, D.S., McGeoch, L.A., Shor, P.A., Weber, R.R., and Yannakakis, M. (1991), "Average-Case Performance of Bin Packing Algorithms under Discrete Uniform Distributions," *Proceedings of the 23rd ACM Symposium on the Theory of Computing*, 230–241, ACM Press, NewYork.

[4] Coffman Jr., E.G., Garey, M.R. and Johnson, D.S. (1984), "Approximation Algorithms for Bin-Packing – An Updated Survey," in *Algorithm Design for Computer System Design*, Ausiello, G., Lucertini, M. and Serafini, P., eds., Springer, NewYork.

[5] Coffman, Jr., E.G. and Lueker, G.S. (1991), *Probabilistic Analysis of Packing and Partitioning Algorithms*, John Wiley, New York.

[6] Coffman Jr., E. G. and Lueker, G. S. and Rinnooy Kan, A. H. G. (1988), "Asymptotic Methods in the Probabilistic Analysis of Sequencing and Packing Heuristics," *Management Science*, 34, 266-291.

[7] Dyckhoff, H. (1990), "Typology of Cutting and Packing Problems," *European Jl. Operational Research*, 44, 145-159.

[8] Gambosi, G., Postiglione, A. and Talamo, M. (1989), "On the On-line Bin-packing Problem," IASI Report R.263.

[9] Garey, M.R. and Johnson, D.S.(1981), "Approximation Algorithms for Bin Packing Problems: A Survey," in *Analysis and Design of Algorithms in Combinatorial Optimization*, Ausiello, G. and Lucertini, M., eds., 147–172, Springer-Verlag, New York.

[10] Rhee, W.T. and Talagrand, M. (1988), "Some Distributions that Allow Perfect Packing," *JACM*, 35, 564–573.

BIPARTITE GRAPH

A graph or network whose nodes can be partitioned into two subsets such that its edges connect a node in each partition. See **Assignment problem; Graph theory; Network optimization; Transportation problem.**

BIRTH-DEATH PROCESS

A stochastic counting process that satisfies the following is called a birth-death process: (1) changes from state n (sometimes written more generally as state E_n) may only be to states $n + 1$ or $n - 1$ (i.e., changes can only be ± 1 unit); (2) the probability of a birth (death) occurring in the "small" interval of time, $(t, t + dt)$, given that the process was in state n at the start of the interval, is $\lambda_n dt + o(dt)[\mu_n dt + o(dt)]$, where $o(dt)$ is a function going to 0 faster than dt. Such processes are in fact Markov chains in continuous time. The system size of an M/M/1 queueing system is an example of a birth-death process where $\lambda_n = \lambda$ $(n = 0,1,2,\ldots)$ and $\mu_n = \mu$ $(n = 1,2,\ldots)$. See **Markov chains; Markov processes.**

BLAND'S ANTICYCLING RULES

A set of pivot rules, the application of which to linear-programming (degenerate) problems, prevents cycling in the simplex algorithm. Their basic principle is that whenever there is more than one eligible candidate in selection of the variable entering the basis, or the variable leaving the basis, the candidate with the smallest index is chosen. See **Anticycling rules; Cycling; Degeneracy**.

BLENDING PROBLEM

The linear-programming problem of blending raw materials, for example, crude oils, meats, to produce one or more final products, for example, fuels, sausages, so that the total cost of production is minimized. The problem is subject to restrictions on material availability, blending requirements, quality restrictions, etc. See **Activity-analysis problem**.

BLOCK-ANGULAR SYSTEM

A linear system of equations for which its matrix of coefficients A can be decomposed into k separate blocks of coefficients A_i, where each A_i represents the coefficients of a different set of equations. This structure typically represents a system consisting of k subsystems, whose activities are almost autonomous, except for a few top-level system constraints whose variables couple the k blocks of the subsystems. Such systems can also have a few variables external to the blocks that couple the blocks. See **Dantzig-Wolfe decomposition algorithm; Large-scale systems; Weakly-coupled systems**.

BLOCK PIVOTING

The process of entering several nonbasic variables simultaneously into the basis in the simplex algorithm. See **Simplex algorithm**.

BLOCK-TRIANGULAR MATRIX

A matrix which is lower (upper) triangular except for a number of blocks along the diagonal. See **Triangular matrix**.

BOOTSTRAPPING

In forecasting, the term bootstrapping refers to models that have been developed by regressing an individual's (or group's) forecasts against the inputs that the individual used to make the forecasts. See **Forecasting; Regression analysis**.

BOUNDED RATIONALITY

The concept that a decision maker lacks both the knowledge and computational skill required to make choices in a manner compatible with economic notions of rational behavior. See **Choice theory; Decision analysis; Organization studies; Satisficing**.

BOUNDED VARIABLE

A variable x_j in a linear-programming problem that is required to satisfy a constraint of the form $0 \leq x_j \leq b, -b \leq x_j \leq 0$, or $b_1 \leq x_j \leq b_2$, where b is some positive constant and $b_1 \leq b_2$.

BRANCH

To move and analyze a new computational path (i.e., branch) based on the results obtained from a previous path. See **Branch and bound**.

BRANCH AND BOUND

A method for solving an optimization problem, by successively partitioning (branching) the set of feasible points to smaller subsets, and solving the problem over each subset. The resulting problems are called subproblems or nodes in the enumeration tree. The idea in branch and bound is that the optimal solution to the problem is the best among the optimal solutions to the subproblems. To reduce the number of subproblems solved, best-case bounds are computed by solving relaxed problems defined at the nodes. If the best-case bound on a solution to a subproblem is worse than the best available solution, the subproblem is eliminated from consideration (fathomed). Branch and bound techniques are frequently used to solve integer-programming problems, as well as in global optimization. See **Combinatorial and integer optimization; Integer-programming problem**.

BROWNIAN MOTION

A one-dimensional Brownian motion $\{B(t), 0 \leq t\}$ is a continuous-time, Markovian, real-valued stochastic process having continuous sample paths; its distribution is Gaussian with mean function $E[B(t)] = \mu t$ and covariance function $Cov[B(s),B(t)] = \sigma^2 min(s,t)$. An n-dimensional Brownian motion is a stochastic process on $\mathbb{R}^n$

whose *n* components are independent one-dimensional Brownian motions. See **Markov processes**.

BTRAN

The procedure for computing the dual variables in a simplex iteration, when the *LU* factors of the basis matrix are given in product form. The name BTRAN (backward transformation) derives from the fact that the eta file is scanned backwards in the solution process. See **Eta file**.

BUFFER

The queue or the waiting "room." The term is most often used in the context of tandem or series queues. See **Queueing theory**.

BULK QUEUES

Arrivals to a queueing system may consist of more than one customer at a time, and/or service might process more than one customer simultaneously. See **Queueing theory**.

BURKE'S THEOREM

The steady-state departure process of a stable M/M/c queueing system is a Poisson process with the same rate as the arrival process, irrespective of the service rate. See **Queueing theory**.

BUSY PERIOD

A time interval that starts when all the servers of a queueing system become busy and ends when at least one server becomes free. See **Queueing theory**.

CALCULUS OF VARIATIONS

Stephen G. Nash

George Mason University, Fairfax, Virginia

INTRODUCTION: The calculus of variations is the grandparent of mathematical programming. From it we have inherited such concepts as duality and Lagrange multipliers. Many central ideas in optimization were first developed for the calculus of variations, then specialized to nonlinear programming, all of this happening years before linear programming came along.

The calculus of variations solves optimization problems whose parameters are not simple variables, but rather functions. For example, how should the shape of an automobile hood be chosen so as to minimize air resistance? Or, what path does a ray of light follow in an irregular medium? The calculus of variations is closely related to optimal control theory, where a set of "controls" are used to achieve a certain goal in an optimal way. For example, the pilot of an aircraft might wish to use the throttle and flaps to achieve a particular cruising altitude and velocity in a minimum amount of time or using a minimum amount of fuel. We are surrounded by devices designed using optimal control – in cars, elevators, heating systems, stereos, etc.

BRACHISTOCHRONE PROBLEM: The calculus of variations was inspired by problems in mechanics, especially the study of three-dimensional motion. It was used in the 18th and 19th centuries to derive many important laws of physics. This was done using the Principle of Least Action. Action is defined to be the integral of the product of mass, velocity, and distance. The Principle of Least Action asserts that nature acts so as to minimize this integral. To apply the principle, the formula for the action integral would be specialized to the setting under study, and then the calculus of variations would be used to optimize the integral. This general approach was used to derive important equations in mechanics, fluid dynamics, and other fields.

The most famous problem in the calculus of variations was posed in 1696 by John Bernoulli. It is called the Brachistochrone ("least time") problem, and asks what path a pellet should follow to drop between two points in the shortest amount of time, with gravity the only force acting on the pellet. The solution to the Brachistochrone problem can be found by solving

$$\underset{y(t)}{minimize} \ \frac{1}{\sqrt{2g}} \int_{t_1}^{t_2} \sqrt{\frac{1 + y'(t)^2}{y(t)}} \, dt$$

where g is the gravitational constant. If this were a finite-dimensional problem then it could be solved by setting the derivative of the objective function equal to zero, but seventeenth-century mathematics did not know how to take a derivative with respect to a function.

The Brachistochrone problem was solved at the time by Newton and others, but the general techniques that inspired the name "calculus of variations" were not developed until several decades later. The first major results were obtained by Euler in the 1740s. He considered various problems of the general form

$$\underset{y(t)}{minimize} \ \int_{t_2}^{t_2} f(t, y(t), y'(t)) \, dt.$$

The Brachistrochrone problem is of this form. Euler solved these problems by discretizing the solution $y(t)$ – approximating the solution by its values at finitely many points. This gave a finite-dimensional problem that could be solved using the techniques of calculus. Euler then took the limit of the approximate solutions as the number of discretization points tended to infinity. This approach was difficult and restrictive, because it had to be adapted to the specifics of the problem being solved, and because there were restrictions on the types of problems for which it was successful.

Far more influential was the approach of Lagrange. He suggested that the solution be perturbed or "varied" from $y(t)$ to $\in y(t) + z(t)$, where $\in$ is a small number and $z(t)$ is some arbitrary function that satisfies $z(t_1) = z(t_2) = 0$. For the Brachistochrone problem this latter condition ensures that the perturbed function still represents a path between the two points.

If $y(t)$ is a solution to the problem

$$\underset{y(t)}{minimize} \int_{t_1}^{t_2} f(t, y(t), y'(t)) \, dt,$$

then $\epsilon = 0$ will be a solution to

$$\underset{\epsilon}{minimize} \int_{t_1}^{t_2} f(t, y(t) + \epsilon z(t), y'(t) + \epsilon z'(t)) \, dt$$

This observation allowed Lagrange to convert the original infinite-dimensional problem to a one-dimensional problem that could be analyzed using ordinary calculus. Setting the derivative of the integral with respect to ϵ equal to zero at the point $\epsilon = 0$ leads to the equation

$$\frac{d}{dt} \frac{\partial f}{\partial y'} - \frac{\partial f}{\partial y} = 0.$$

This final condition is a first-order optimality condition for an unconstrained calculus-of-variations problem. It was first discovered by Euler, but the derivation here is due to Lagrange.

The name "calculus of variations" was chosen by Euler and was inspired by Lagrange's approach in "varying" the function $y(t)$. The optimality condition is stated as "the first variation must equal zero" by analogy with the condition $f'(x) = 0$ for a one-variable optimization problem. Euler was so impressed with Lagrange's work that he held back his own papers on the topic so that Lagrange could publish first, a magnanimous gesture by the renowned Euler to the then young and unknown Lagrange.

There are additional first-order optimality conditions for calculus of variations problems. The theory is more complicated than for finite-dimensional optimization, and the necessary and sufficient conditions for an optimal solution were not fully understood until the 1870s, when Weierstrass studied this topic. A discussion of this theory can be found in Gregory and Lin (1992).

MULTIPLIERS: Constraints can be added to problems in the calculus of variations just as in other optimization problems. A constraint might represent the principle of conservation of energy, or perhaps that the motion was restricted in some way, for example that a planet was traveling in a particular orbit around the sun.

Both Euler and Lagrange considered problems of this type, and both were led to the concept of a multiplier. In the calculus of variations the multiplier might be a scalar (as it is in finite-dimensional problems) or, depending on the particular form of the constraint, it might be a function of the independent variable t. They have come to be called "Lagrange multipliers"; but, as with the optimality condition, Euler discovered them first.

In his book *Mécanique Analytique*, Lagrange includes an interpretation of the multiplier terms. He writes that they can be considered as representing the moments of forces acting on the moving particle, and serving to keep the constraints satisfied. This point of view is the basis for duality theory, although Lagrange does not seem to have followed up on this idea.

DUALITY: Duality theory did not become fully developed until early in this century, with many of the important steps coming from the calculus of variations. At first there were only isolated examples of duality. That is, someone would notice that a pair of problems – one a maximization problem, one a minimization problem – would have optimal solutions that were related to each other. An early example of this type was published in 1755, and is described in Kuhn (1991). In the nineteenth century various other examples were noticed, such as the relationship between currents and voltages in an electrical circuit. Gradually it was understood that duality was not an accidental phenomenon peculiar to these examples but rather a general principle that applied to wide classes of optimization problems. By the 1920s techniques had been developed for obtaining upper and lower bounds on the solutions to optimization problems by finding approximate solutions to the primal and dual problems. Duality as a general idea is described in the book by Courant and Hilbert (1953).

Euler and Lagrange only considered problems with equality constraints, but later authors allowed inequality constraints as well. When specialized to finite-dimensional problems, the optimality condition is referred to as the Karush-Kuhn-Tucker condition. Kuhn and Tucker derived this result in a 1951 paper. It was later discovered that Karush had proven the same result in his Master's thesis (1939) at the University of Chicago under the supervision of Bliss. There are two aspects to the result: its treatment of inequality constraints, and the assumption or "constraint qualification" that was used to prove it. The first idea can be traced to Weierstrass and the second to Mayer (1886), and both are outgrowths of the calculus of variations.

In the 1870s Weierstrass studied the calculus of variations and presented the results of his investigations in lectures. Weierstrass did not publish his work and it only became widely known years later through the writings of those in attendance. According to Bolza (1904), Weierstrass converted the inequality constraint

$$g(y) \leq 0$$

to an equivalent equality constraint

$$g(y) + s^2 = 0$$

using a squared "slack variable" s. This technique is described in many sources from 1900 onward. Bolza later became a professor at the University of Chicago, establishing a connection from Weierstrass to Bliss to Karush. Karush used this technique in his thesis.

The constraint qualification used by Karush, Kuhn and Tucker relates feasible arcs (paths of feasible points leading to the solution) and the gradients of the constraints at the solution. This same condition was used by Mayer (1886), although applied to a calculus of variations problem with equality constraints, and then in a chain of papers by various authors (including Bliss) leading to Karush's thesis. In these papers it is called a "normality" condition, and it is equivalent to requiring that the matrix of constraint gradients at the solution be of full rank. The implicit function theorem can be used to relate this to the condition on feasible arcs, an observation that is explicit in Mayer's work.

The calculus of variations has influenced many areas of applied mathematics. It is a technical tool for solving optimization problems whose parameters are functions, and in this way it continues to be used in optimal control. It was the setting for the development of the most important concepts in optimization, such as duality and the treatment of constraints. And, when coupled with the Principle of Least Action, it was the vehicle for deriving the fundamental laws of physics.

See **Control theory; Lagrange multipliers; Linear programming; Nonlinear programming.**

References

[1] G.A. Bliss (1925), *Calculus of Variations*, Open Court, Chicago.

[2] O. Bolza (1904), *Lectures on the Calculus of Variations*, University of Chicago Press, Chicago.

[3] R. Courant and D. Hilbert (1953), *Methods of Mathematical Physics, Volume I*, Interscience, New York, 1953.

[4] H.H. Goldstine (1980), *A History of the Calculus of Variations from the 17th through the 19th Century*, Springer-Verlag, New York.

[5] J. Gregory and C. Lin (1992), *Constrained Optimization in the Calculus of Variations and Optimal Control Theory*, Van Nostrand Reinhold, New York.

[6] M.R. Hestenes (1966), *Calculus of Variations and Optimal Control Theory*, John Wiley, New York.

[7] H.W. Kuhn (1991), *Nonlinear Programming: A Historical Note*, in *History of Mathematical Programming*, J.K. Lenstra, A.H.G. Rinnooy Kan, and A. Schrijver, eds., North-Holland (Amsterdam), 82–96.

[8] J.L. Lagrange (1888–89), *Oeuvresde Lagrange, Volumes XI and XII*, Gauthier-Villars, Paris.

[9] A. Mayer (1886), "Begrndung der Lagrange'schen Multiplicatorenmethode in der Variationsrechnung," *Mathematische Annalen*, 26, 74–82.

CALL PRIORITIES

A strategy for handling calls with varying degrees of urgency. Many emergency services have instituted formal procedures for responding differently (e.g., with and without flashing lights and sirens) to calls depending upon priority level. See **Emergency services**.

CANDIDATE RULES

A group of rules that the inference engine has determined to be of immediate relevance at the present juncture in a reasoning process. These rules will be considered according to a particular selection order and subject to a prescribed degree of rigor. See **Artificial intelligence; Expert systems**.

CAPACITATED TRANSPORTATION PROBLEM

A version of the transportation problem in which upper bounds are imposed on some or all of the flows between origins and destinations. See **Transportation problem**.

CAPITAL BUDGETING

Reuven R. Levary

Saint Louis University, Missouri

The desired end result of the capital budgeting process is the selection of an optimal portfolio of investments from a set of alternative investment proposals. An *optimal portfolio of investments* is defined as the set of investments that makes the greatest possible contribution to the achievement of the organization's goals, given the organization's constraints. The constraints faced by a corporation in the capital budgeting process can include limited supplies of capital or other resources as well as dependencies between investment proposals. A dependency occurs if two projects are mutually exclusive, acceptance of one requires rejection of the other, or if one project can be accepted only if another is accepted. Assuming that the organizational goals and constraints can be formulated as linear functions, the optimal set of capital investments can be found using linear programming (LP).

CAPITAL BUDGETING UNDER CAPITAL RATIONING:

Capital rationing is a constrained capital budgeting problem in which the amount of capital available for investment is limited. *Capital Budgeting Under Pure Capital Rationing, with No Lending or Borrowing Allowed:* Consider a firm that has an opportunity to invest in several independent projects. It is assumed that both the future cash flows associated with each project and the firm's future cost of capital can be forecast. These forecasts enable calculation of the net present value for each project, assuming that the firm expects to be affiliated with the projects for a period of N years. It is also assumed that the firm has a given fixed budget for funding the projects for each of the N years, with both the budget and the cost of capital in future periods being unaffected by investments made in previous periods. Finally, it is assumed that any portion of the budget not used in one year cannot be carried over to future years.

The basic model for capital budgeting under pure capital rationing is as follows:

$$\text{maximize} \sum_{i=1}^{M} P_i x_i \tag{1}$$

$$\text{subject to} -\sum_{i=1}^{M} f_{it} x_i \le b_t \text{ for } t = 1, 2, \ldots, N \tag{2}$$

$$0 \le x_i \le 1 \text{ for } i = 1, 2, \ldots, M \tag{3}$$

where P_i is the net present value for the ith project (calculated based on forecasts of future cash flows), f_{it} is the expected cash flow for project i during year t (cash flow is defined to be positive if it is inflow and negative if it is outflow), b_t is the available budget for year t, M is the number of alternative projects and x_i is the fraction of project i to be funded.

The objective function (1) represents the total expected net present value of the investment proposals that should be funded. Constraints (2) represent restrictions on the available yearly budget. Constraints (3) ensure that no more than one project of a given type will be included in the optimal portfolio. By adding the constraint that x_i be integer for $i = 1, 2, \ldots, M$, the problem becomes an integer program. In this case, no fractional projects will be allowed; a project is either accepted or rejected. Constraints on scarce resources, mutually exclusive projects, and contingent projects can easily be added to the above model when necessary.

Capital Budgeting Where Borrowing and Lending are Allowed: In this model, the amount available for lending in a given year is the "left-over" money for that year. This amount can be carried over to the next year at a given rate of interest r. Consider the case when the interest rate for borrowing, or cost of funds, depends on the amounts borrowed. The cost of borrowing is assumed to have the shape of a step function; that is, the larger the amount borrowed, with limits, the higher the interest rate. Let r_j be the interest rate that applies to borrowing an amount greater than C_{j-1} and less than or equal to C_j. A firm will borrow at interest rate r_j if it exhausts the limits placed on its borrowing at lower interest rates.

If the firm expects to be affiliated with the proposed projects for N years, then the objective is to maximize the total related cash flows at the end of the Nth year, that is, the horizon. Let and be, respectively, the amount lent and the amount borrowed (at interest rate r_j) in year t. Also, let f_{it} be the cash flow in year t resulting from approval of project i. All flows in this model are current values, that is, not present values. Revenues and expenditures are defined, respectively, to be positive and negative cash flows. A given project can generate cash flows after the Nth year as well. Let $\hat{f}_i$ be the present value of total cash flows at the horizon (i.e., year N) that are expected to be generated by project i at years following year N. These flows are discounted to year N, assuming an interest rate equivalent to the firm's weighted average cost of capital. The model is formulated as follows:

$$\text{maximize} \sum_{i=1}^{M} \hat{f}_i x_i + \alpha_N - \sum_{j=1}^{m} \beta_{jN} \tag{4}$$

$$\text{subject to} \sum_{i=1}^{M} f_{i1} x_i + \alpha_i - \sum_{j=1}^{m} \beta_{j1} \le b_1 \tag{5}$$

$$-\sum_{i=1}^{M} f_{it} x_i - (1+r)\alpha_{t-1} + \alpha_t + \sum_{j=1}^{m} (1+r_j)\beta_{j,t-1}$$

$$-\sum_{j=1}^{m} \beta_{jt} \le b_t \quad \forall t = 2, 3, \ldots, N \tag{6}$$

$$\beta_{jt} \le C_{jt} \quad \forall t = 1, 2, \ldots, N; j = 1, 2, \ldots, m \qquad (7)$$

$$0 \le x_i \le 1 \quad \forall i = 1, 2, \ldots, M \qquad (8)$$

$$\alpha_t, \beta_{jt} \ge 0 \quad \forall t = 1, 2, \ldots, N; j = 1, 2, \ldots, m \qquad (9)$$

where m represents the number of different interest rates in the supply of funds schedule. The limit on borrowing during year t, at interest rate r_j is denoted by C_{jt}. Objective function (4) represents the total flows resulting from the proposed projects at the end of the Nth year. The first component $\Sigma_{i=1}^{M} \hat{f}_i x_i$ of the objective function represents the present value at the horizon of the cash flows expected to be generated by the projects in years following the horizon year N. The second component $\alpha_N - \Sigma_{j=1}^{m} \beta_{iN}$ is the amount lent minus the amount borrowed during the horizon year N. Inequality (5) and inequalities (6) represent the constraints on the available budget for a given year. The limits on borrowing are represented by constraints (7). This model can be extended by adding constraints on scarce resources and by incorporating mutually exclusive and contingent projects when applicable.

FRACTIONAL PROJECTS:

All LP models can result in an optimal portfolio of projects composed of fractional projects. Weingartner (1967) showed that the number of fractional projects in the optimal solution set of the basic LP model (described by relations (1)–(3)) cannot exceed the number of time periods for which constraints are imposed. Additional constraints such as mutual exclusion, contingency, and scarce resources can increase the maximum number of fractional projects. Each additional constraint increases the maximum number of fractional projects by one. Weingartner (1967) also showed that the number of fractional projects in the optimal solution of the model where borrowing and lending are allowed is no larger than the number of time periods during which the firm does not lend or borrow money.

Because solutions to LP models can include fractional projects, these models are only an approximation of the exact solution. The exact solution can be obtained by applying integer programming solution procedures. The fractions of mutually exclusive projects, which can be the solution of an LP model, may have a useful interpretation. Fractional projects may suggest the possibility of a partnership. For example, one might interpret the decision to fund the expenses of building a fraction of a shopping center to mean that it would be beneficial for the company to engage in a partnership arrangement.

DUAL LINEAR PROGRAMMING AND CAPITAL BUDGETING:

Consider the basic model for capital budgeting under pure capital rationing formulated by relations (1)–(3). To evaluate the profitability of various projects, a discount factor must be incorporated into the capital budgeting analysis. Define d_t as the discount factor for period t: $d_t = (1 + r_t)^{-1}$ where r_t is the interest rate at period t. The net present value for project i is:

$$P_i = \sum_{t=1}^{N} f_{it} d_t \qquad (10)$$

Substitution of Equation (10) into (1) results in the following formulation, called Problem P:

$$\text{maximize } Z = \sum_{i=1}^{M} \sum_{t=1}^{N} f_{it} d_t X_i \quad \text{subject to (2) and (3).}$$

$$\textbf{(P)}$$

Let y_t be the dual variable associated with the budget constraint for year t. The value of y_t at the optimal solution, y_t^*, represents the increase in the total combined net present value of the projects that results from an addition of $1 to the budget for year t.

Assume that dollars are added to the budget in period t. This results in an increase of the net present value (the objective function) by $v y_t^*$. The net present value of v is $v d_t$. This implies that the discount factor d_t should be equal to the dual variable y_t^* at the optimal solution (Baumol and Quandt, 1965). Problem **P** is called consistent if its optimal solution has the property $d_t = y_t^* \, \forall t$. A solution to a capital budgeting problem under pure capital rationing where dual variables do not equal the discount factor is not optimal. Therefore, such a problem is inconsistent.

An analysis of consistent solutions helps clarify the relationship between discount factors and dual variables, as well as the choice of an objective function. Several properties of consistent solutions were summarized by Freeland and Rosenblatt (1978) and are:

1. The value of the objective function of Problem **P** equals zero if there are no upper bounds on the decision variables (that is, in the case when the X_i are not restricted to be less than one).

2. When the value of the objective function is zero, the only way to obtain a consistent solution is by having all discount factors equal zero. This is a meaningless situation.

3. To ensure a meaningful consistent solution, the decision variables must have upper bounds. Furthermore, some projects must be fully accepted.

4. For a consistent solution to be meaningful, the optimal value of the objective function must

be positive and the budget vector must include both positive and negative components.

5. If unused funds cannot be carried forward, the discount factor in period t may exceed the discount factor in period $t + 1$.

FINDING THE "RIGHT" DISCOUNT FACTORS: Because different optimal solutions to Problem P are obtained for various values of the discount factor, it is necessary to find the "right" discount factor for the pure capital rationing case before Problem P is solved. Freeland and Rosenblatt (1978) reported that most of the proposed iterative procedures for finding the "right" discount factors described in the literature do not work properly. Problems involved in finding the "right" discount factors are avoided by using horizon models, such as (4)–(9).

The horizon value of the model where borrowing and lending are allowed is $\alpha_N - \Sigma_{j=1}^m \beta_{jN}$ [see relation (4)] when there are no cash flows beyond the horizon. In this case, no discount rate is used in maximizing the horizon value and therefore the problem of finding the "right" discount factor is irrelevant. In the case where there are cash flows beyond the horizon, management must estimate the respective discount rates using financial and economic forecasting. The calculation of these estimates is external to the LP models used in capital budgeting decisions, and therefore is not linked to the solution procedure of the LP model.

ALTERNATIVE CAPITAL BUDGETING MODELS: Some capital budgeting problems have multiple objectives. Such problems can be formulated as goal programming problems. In many cases, the values of variables affecting the cash flows of the projects are not known with certainty. Such variables include future interest rates, length of useful economic lives, and salvage values. Computer simulation can be used to handle the uncertainty surrounding capital budgeting decisions (Levary and Seitz, 1990). Simulation can also be used to analyze the risk consequences of various capital budgeting alternatives. Decision tree analysis is a widely used method for analyzing risk associated with a single investment alternative (Levary and Seitz, 1990). Expected return on investments can be adjusted for risk using the capital asset pricing model (CAPM). CAPM was generalized by Richard (1979) to include environmental uncertainty.

Relationships among investments contribute to portfolio risk and can be measured by covariances. Quadratic programming models for capital budgeting can be used in situations where the covariances between returns of various projects can be estimated. Various characteristics of a specific capital budgeting problem, like tax consequences, can be modeled using mathematical programming.

See **Combinatorial and integer optimization; Goal programming; Linear programming; Mean-value portfolio analysis.**

References

[1] Baumol, W.J. and R.E. Quandt (1965). Investment and Discount Rates Under Capital Rationing – A Programming Approach. *The Economic Journal* 75, 298, 317–329.
[2] Freeland, J.R. and M.J. Rosenblatt (1978). An Analysis of Linear Programming Formulations for the Capital Rationing Problems. *The Engineering Economist.* Fall, 49–61.
[3] Levary, R.R. and N.E. Seitz (1990). *Quantitative Methods for Capital Budgeting.* South-Western Publishing, Cincinnati.
[4] Richard, S.F. (1979). "A Generalized Capital Asset Pricing Model." *Studies in the Management Sciences*, 11, North Holland, Amsterdam, 215–232.
[5] Weingartner, H.M. (1967). *Mathematical Programming and the Analysis of Capital Budgeting Problems.* Markham Publishing, Chicago.

CASE

Computer-aided software systems engineering. See **Systems analysis.**

CDF

Cumulative distribution function.

CENTER FOR NAVAL ANALYSES

Carl M. Harris

George Mason University, Fairfax, Virginia

In the pre-World-War-II year of 1940, many scientists believed that organizing the nation's scientific research would strengthen national defense. As a result, the National Defense Research Committee (NDRC) was established by Presidential Executive Order.

The NDRC was placed under the direction of the newly created Office of Scientific Research and Development (OSRD), which reported directly to the president. NDRC's contact with British researchers indicated that studying actual operations was an essential part of any assessment process. Because the need for operations research was particularly pressing in the area of antisubmarine warfare (ASW), the Navy created the Antisubmarine Warfare Operations Research Group (ASWORG). In 1942, comprising at first fewer than a dozen scientists, it was the first civilian group engaged in military operations research in the country. The Center for Naval Analyses (CNA) traces its origins to ASWORG.

Today, CNA analysts provide the Navy and Marine Corps with objective studies of a wide variety of operations, systems and programs. Such studies range from the support of training and testing activities to the evaluation of new technologies and alternative force structures for top-level decision-makers. The following short history of CNA recounts the highlights of its evolution and contribution to national security.

WORLD WAR II: During the 1940s, the United States was preoccupied first with the war in Europe and then with the war in the Pacific. As soon as the United States entered the war, German submarines began to patrol the U.S. East Coast and Atlantic shipping lanes in earnest. The Navy's immediate focus was on the U-boat threat and the Battle of the Atlantic.

In Britain, Professor P.M.S. Blackett had demonstrated the value of operations research in solving military problems. Captain Wilder Baker, leader of the newly formed U.S. Navy Antisubmarine Warfare Unit in Boston, was inspired by Blackett's paper, "Scientists at the Operation Level." Baker believed that a cadre of civilian scientists could also help the U.S. Navy. He asked Professor Philip M. Morse of MIT to head such a group. ASWORG was formed in April 1942 with a mission to help defeat the German U-boats. The contract for ASWORG was administered by Columbia University, which already had an existing contract with the NDRC that focused on anti-submarine warfare.

ASWORG set a major precedent when it required its analysts to gather field data first hand. Sending civilian experts to military commands was a delicate matter. In June 1942, the field program began when an ASWORG analyst assisted the Gulf Sea Frontier Headquarters in Miami. Shortly afterward, several analysts were assigned to the Eastern Sea Frontier in New York. The field analysts quickly became accepted; most of ASWORG's noteworthy work was achieved in the field.

In June 1942, ASWORG was assigned to the Headquarters of Commander in Chief, U.S. Fleet (CominCh). Admiral Ernest J. King was both

CominCh and the Chief of Naval Operations (CNO). The Tenth Fleet was formed in 1943 to consolidate U.S. ASW operations. In July 1943, ASWORG became part of the Tenth Fleet.

In October 1944, because of the decrease in enemy submarine activity and the increase in operations research requirements on subjects other than ASW, ASWORG was transferred from the Tenth Fleet to the Readiness Division of the Headquarters of CominCh. It was also renamed the Operations Research Group (ORG) as its analysis efforts had become more diversified.

By the end of the war, ORG had about 80 scientists whose scope of study was all forms of naval warfare. During most of World War II, about 40 per cent of the group was assigned to various operating commands. These field analysts developed immediate, practical answers to tactical and force allocation questions important to their commands. Concurrently, they fed back practical experiences and understanding to the central Washington group, a practice still continued a half century later.

Among its many World War II contributions, ORG devised more effective escort screening plans; determined the optimum size of convoys; developed ASW tactics, such as optimum patterns and altitudes for flying AWS patrol aircraft; developed counter measures to German acoustic torpedoes and snorkeling U-boats; and contributed to the use of airborne radar.

POST-WAR PERIOD: In August 1945, Admiral King, in a letter to Secretary of the Navy James V. Forrestal, recommended and requested that ORG be allowed to continue into peacetime at about 25% of its wartime size. Secretary Forrestal gave his approval shortly thereafter.

Both Admiral King and Secretary Forrestal concluded that much of ORG's unique value was due to its ability to provide an independent, scientific viewpoint to a broad range of Navy problems. Consequently, in extending the service of ORG into peacetime, it was decided that its character could best be preserved by perpetuating the wartime arrangement through a contract with an academic institution. Such a contract was entered into with MIT in November 1945. At that time, ORG was renamed the Operations Evaluation Group (OEG). OEG was to assist the Navy and its research laboratories in analyzing and evaluating new equipment, tactical doctrine and strategic warfare.

After the war, OEG published several comprehensive reports on important naval operations, which included many new methodologies. Although some were originally classified secret, they later appeared in Morse and Kimball's *Methods of Operations Research*, Bernard Koopman's *Search and Screening*, and Charles Sternhell and Alan Thorndike's *Antisubmarine Warfare in World War II*. Taken together, these reports provided a record of vital lessons learned in World War II, as well as important operations research methods. With the Korean War and the intensification of the Cold War, the role of analysis in defense planning expanded in the 1950s. Once the Soviets had detonated their first thermonuclear device, the United States had to revise its thinking on many critical defense issues. As the consequences of nuclear war loomed and the cost of military preparedness escalated, the government, more than ever, needed reliable scientific information on which to base its strategic decision-making.

Before the Korean War, OEG began a slow but steady buildup. By 1950, the research staff had grown to about 40. As the war began, OEG received requests for analysts from combat commands. These analysts collected data, solved tactical problems and recommended improvements in procedures, improvements that were sometimes used immediately. OEG expended its major efforts on such specific tactical problems as: selection of weapons for naval air attack on tactical targets; scheduling of close air support; analysis of air-to-air combat; naval gunfire in shore bombardment; blockade tactics; and interdiction of land transportation. By the end of the war, OEG had 60 research staff members.

After the war, OEG continued to grow, albeit slowly. Analysts participated with naval forces in all post-Korean crises. The most important changes in the nature of the group's post-Korean activity were the results of major technological advances, particularly in the filed of atomic energy and guided missiles. Issues were broadened to include the possible enemy use of nuclear weapons and the effect of U.S. policies and weapon system choices on the nature of wars the United States would have to be prepared to fight. During this period, the Navy also established the Long-Range Studies Project of MIT; it was later renamed the Institute for Naval Studies (INS).

DEFENSE MANAGEMENT: By the 1960s, advances in weapons technology were causing defense costs to rise dramatically, and the increasing tempo of the Vietnam War later in the decade would cause the defense budget to balloon still further. The swearing in of Secretary of Defense Robert S. McNamara in 1961 marked the beginning of a new philosophy of defense management. Emphasis began to be placed on cost as well as effectiveness. McNamara believed that integrated systems analysis throughout the defense establishment was required to achieve a balanced, affordable military structure.

In 1961, MIT established an Economics Division within OEG because the cost of weapon systems was becoming a dominant factor in military decision-making. Until 1961, the Marine Corps had only one OEG analyst. By the early 1960s, however, Marine Corps requirements for operations research had increased substantially. The Marine Corps Section of OEG was established in December 1961.

By 1962, the Secretary of the Navy wanted to consolidate the study efforts of OEG and INS and began to look for a contractor. MIT, which had managed OEG since 1945, declined an invitation to manage this proposed new enterprise. The Navy then selected the Franklin Institute to administer the contract for the new organization. In August 1962, OEG and INS were brought under the common management of a new entity, the Center for Naval Analyses (CNA).

CENTER FOR NAVAL ANALYSES: Shortly after CNA was formed, OEG (now as a division) again became involved in an actual naval operation. In October 1962, it helped the Office of the Chief of Naval Operations (OPNAV) develop plans for the naval quarantine of Cuba and assessed the effectiveness of surveillance operations.

As combat escalated in Southeast Asia, so did the number of CNA field representatives providing direct support to the naval operating forces. CNA participated in the study of many operations, such as interdiction campaigns in North Vietnam and infiltration rates in South Vietnam. Also, a large data base on war-related activities was being developed and maintained in CNA's Washington office. In August 1967, management of the CNA contract transferred from the Franklin Institute to the University of Rochester.

Because the war in Vietnam was escalating, the Navy needed more combat analysis. As a result, the Southeast Asia Combat Analysis Group (SEACAG) was established within OPNAV. Shortly thereafter, the Southeast Asia Combat Analysis Division (SEACAD) was established within OEG. SEACAD's role was to support SEACAG and to increase the amount of war-related analysis that CNA was performing. CNA analyzed various operations of the Southeast Asian conflict, including combat aircraft losses, interdiction, strike warfare and carrier defense, surveillance and naval gunfire support.

In the 1970s, as the war in Vietnam wound down, military budgets, forces and equipment began to deteriorate. To maintain effectiveness in the face of reduced budgets, the Navy increased its emphasis on analysis. As new systems became available, the Navy needed to determine how best to exploit their capabilities. With old systems that were already deployed, the Navy needed to develop tactics that overcame technical shortcomings.

MILITARY BUILDUP: The 1980s witnessed a major buildup of U.S. forces in response to the growth of Soviet military power during the 1970s. For the Navy, this meant not only more ships and aircraft but also more emphasis on a maritime strategy and on specific concepts of operations for employing the Fleet in a global war. These efforts matured by 1987, just as Gorbachev unleashed the forces that would lead to the razing of the Berlin Wall and, ultimately, the demise of the Soviet Union.

In 1982, CNA began a major study of concepts of operations for employing the Atlantic Fleet in a global war. This work involved issues ranging from Soviet objectives and intentions in a war to actions the Navy could take to counter Soviet strategy, as well as theater-level tactics that would be executable in the face of a concerted Soviet threat. The results of this work were put into practice in 1984 by Commander, Second Fleet, who also added important tactical innovations. The resulting interaction and cooperation of Washington and the Fleet (and of CNA-Washington and the field analysts) set the tone for similar efforts at other fleet commands.

By December 1982, differences concerning the management of CNA had arisen between the Department of the Navy and the University of Rochester. The Secretary of the Navy decided to open the CNA contract to competition, and several universities and nonprofit research organizations responded. In August 1983, the Navy announced that the Hudson Institute had been awarded the contract for the management of CNA, effective October 1983.

NEW WORLD ORDER: The 1990s ushered in an entirely new security environment. In light of the collapse of the Soviet Union and the new emphasis on Third World threats, the Navy and Marine Corps are re-evaluating their structure. Unlike the threat of the Cold War era, these new threats are smaller and more diffuse. They require smaller units that can operate jointly in distant areas where the United States often has a limited number of forces and restricted access to bases. Developing these types of forces and operations is a continuing theme for defense planning in the 1990s.

During the 1980s, some significant events had solidified CNA's stature in the analytical field. Demands for CNA's analytical assistance had grown, particularly from senior Navy and Marine Corps leaders. CNA had become more involved in critical issues and issues of concern to top-level decision makers, and CNA's staff had increased in size and quality to meet those growing demands.

Organizationally, CNA had changed often over the years to meet the demands of a changing world and a changing military environment. In the spring of 1990, CNA's management, the Board of Overseers, the Navy, and the Hudson Institute all agreed that CNA could function as an independent organization.

On October 1, 1990, CNA became independent and began operating under a direct contract with the Department of the Navy, ready to help the Navy and Marine Corps cope with the impending changes in national security policy, defense strategy, defense budgets and defense management practices.

After Iraq annexed Kuwait in August 1990, the CNO asked CNA to track and document the events in the Middle East, to analyze activities, and to develop a lessons-learned data base. CNA had up to 20 field representatives providing support to various naval commands in the Middle East, including Commander, U.S. Naval Central Command.

After the war, CNA was designated the Navy's lead agency for Desert Shield/Storm data collection and analysis. The Navy believed that future force composition, systems design and budget decisions would be shaped by events of the war and the subsequent analysis. CNA led the reconstruction of Desert Shield/Storm and provided the Navy with a 14-volume report. In addition, CNA is continuing its analysis of the war and is archiving all the fleet data for the National Archives.

During Desert Storm, the value of concepts that CNA had analyzed for the Navy and Marine Corps – the Tomahawk cruise missile, the air-cushioned landing craft (LCAC), the maritime prepositioning – became evident. The Tomahawk land-attack missile was one of the high-tech "stars" of the war; the LCAC played an important role in creating fear of an amphibious assault; and maritime prepositioning allowed two brigades of Marines to deploy to the Gulf in record time.

In the 1990s, CNA's most important task is to help the Navy and Marine Corps make the transition to a post-Cold War security environment. To do this, CNA's research program plan is to emphasize areas of immediate importance of this transition: the new security environment, littoral operations, communications, warfare area adjustments, training and education, investment alternatives, force structure, and economies and efficiencies. Across the entire research program, special emphasis will be given to joint operations and the naval infrastructure.

See **Field analysis; Military operations research; RAND Corporation; Operations Research Office and Research Analysis Corporation.**

References

[1] Center for Naval Analyses (1993), "Victory at Sea: A Brief History of the Center for Naval Analyses," *OR/MS Today*, 20(2), 46–51.

[2] Kreiner, H.W. (1992), *Fields of Operations Research*, Operations Research Society of America, Baltimore.

[3] Morse, P.M. and G.E. Kimball (1946), "Methods of Operations Research," OEG Report 54, Operations Evaluation Group (CNA), U.S. Department of the Navy, Washington, DC.

CERTAINTY EQUIVALENT

A certainty equivalent of Lottery L is an amount x' such that the decision maker is indifferent between L and the amount x' guaranteed for certain. See **Decision analysis; Lottery; Utility theory.**

CERTAINTY FACTOR

A numeric measure of the degree of certainty about the goodness, correctness, or likelihood of a variable value, an expression (e.g., premise) value, or conclusion. See **Expert systems.**

CHAIN

A chain in a network is a sequence of arcs connecting a designated initial node to a designated terminal node such that the direction (orientation) of flow in the arcs is from the initial node to the terminal node. See **Cycle; Path; Markov chain.**

CHANCE-CONSTRAINED PROGRAMMING

A mathematical programming problem in which the parameters of the problem are random variables and for which a solution must satisfy the constraints of the problem in a probabilistic sense. Here the usual linear-programming constraints are given as probability statements of the form $\Pr\{\Sigma_{j=1}^{n} a_{ij} x_j \le b_i\} \ge \alpha_i$ for $i = 1, \ldots, m$, where the α_i are given constants between zero and one. Some forms of the chance-constrained programming problem can be transformed to an equivalent linear-programming problem. See **Linear programming; Stochastic programming.**

CHANCE CONSTRAINT

A constraint that restricts the probability of a certain event to a prespecified range of values. Under certain conditions, chance constraints can be incorporated into mathematical-programming problems. See **Chance-constrained programming; Linear programming; Stochastic programming.**

CHAOS

A mathematical term describing a situation in which arbitrarily small variations in independent variable values can produce large variations in the dependent variable. The term is most typically used to characterize the behavior of deterministic, nonlinear, differentiable dynamic systems. The term is sometimes used to describe situations in which true mathematical chaos is not present, but where the results are similarly disturbing. The disturbing effect in battle modeling, for example, is the apparent loss of deterministic behavior.

CHAPMAN-KOLMOGOROV EQUATIONS

In a parameter-homogeneous Markov chain $\{X(t)\}$ with state space S, define $p_{ij}(t)$ as the probability that $X(t+s) = j$, given that $X(s) = i$ for $s, t \ge 0$. Then, for all states i, j and index parameters $s, t \ge 0$,

$$p_{ij}(t+s) = \sum_{k \in S} p_{ik}(t) p_{kj}(s)$$

are the Chapman-Kolmogorov equations. There is a comparable definition when the state space is instead continuous. See **Markov chains; Markov processes.**

CHINESE POSTMAN PROBLEM

William R. Stewart, Jr.

College of William and Mary, Williamsburg, Virginia

INTRODUCTION: The Chinese Postman Problem acquired its name from the context in which it was first popularly presented. The Chinese mathematician Mei-Ko Kwan (1962) addressed the question of how, given a postal zone with a number of streets that must be served by a postal carrier (postman), does one develop a tour or route that covers every street in the zone and brings the postman back to his point of origin having traveled the minimum possible distance. Researchers who have followed on Kwan's initial work have since referred to this problem as the Chinese Postman Problem or CPP. In general, any problem that requires that all of the edges of a graph (streets, etc.) be traversed (served) at least once while traveling the shortest total distance overall is a CPP. Like its cousin the traveling salesman problem, which seeks a route of minimum cost that visits every vertex of a graph exactly once before returning to the vertex of origin, the CPP has many real world manifestations, not the least of which is the scheduling of letter carriers. Such problems as street sweeping, snow plowing, garbage collection, meter reading and the inspection of pipes or cables can and have all been treated as CPPs.

In the following discussion, the terms tour and cycle will be used interchangeably to refer to a route on a graph that begins and ends at the same vertex and that traverses all of the edges of that graph *at least once*, and, unless otherwise noted, the edges are assumed to be undirected (i.e. they may be traversed in either direction).

CYCLES AND TOURS: The CPP and its many variants have their roots in the origins of mathematical graph theory. The problem of finding a cycle (tour/route) on a graph which traverses all of the edges of that graph and returns to its starting point dates back to the mathematician Leonid Euler and his analysis in 1736 of a popular puzzle of that time. Euler's problem of traversing all of the bridges of Königsberg and returning to his starting point without retracing his steps is equivalent to asking if there is a tour of the graph shown in Figure 1 that traverses all of the edges *exactly once*. Euler showed that such a cycle exists in a graph if and only if each vertex in the graph has an even number of edges connecting to it or in mathematical terms each vertex is of even cardinality. This follows logically from the observation that, in a tour that traverses all of the edges *exactly once*, each vertex must be exited the same number of times it is entered. Tours that traverse each edge of a graph *exactly once* are termed Euler cycles or tours, and graphs that contain an Euler cycle are appropriately called Eulerian. When costs are assigned to each of the edges and edges may be repeated, the problem of finding a *minimum cost tour* is a CPP.

When a graph is Eulerian, the cost of a tour is just the sum of the costs of all of the edges in the graph, and the solution to the CPP is any Eulerian tour, of which there are usually many. In general, an Eulerian tour can easily be found when one exists. When a graph has more than one odd cardinality vertex (exactly one such vertex is impossible), the CPP is the problem of finding which of the edges must be traversed more than once in order to produce a minimum cost tour. The graph shown in Figure 1 has four vertices with odd cardinality, and a tour of this graph requires that one or more of the edges be crossed *more than once*. Figure 2 shows hypothetical costs on each edge, and the dashed lines indicate the edges that must be traversed twice in order to achieve a minimal cost tour. This tour will have a total cost of 23, the cost of crossing each edge once plus the cost of crossing edges (a,b) and (c,d) a second time each.

In mathematical terms, the CPP can be stated as follows: given a graph $G = \{V,E\}$, where V is a set of n vertices, E is a set of edges connecting these vertices, and each edge (i,j) connecting vertices i and j has a nonnegative cost, c_{ij}, find x_{ij}, the number of times that edge (i,j) is to be traversed from i to j so that the total cost of traversing all of the edges in E at least once is a minimum. The sum of x_{ij} and x_{ji} is the number of times that the edge between vertices i and j must be traversed in an optimal tour.

CPP Formulation:

$$\text{Minimize} \quad \sum_i \sum_j c_{ij} x_{ij} \tag{1}$$

$$\text{Subject to} \quad \sum_i x_{ik} - \sum_j x_{kj} = 0, \qquad \text{for } k = 1, \ldots, n, \tag{2}$$

$$x_{ij} + x_{ji} \geq 1, \qquad \text{for all } (i,j) \text{ and } (j,i) \in E, \tag{3}$$

$$x_{ij} \geq 0, \text{ and integer}, \qquad \text{for all } (i,j) \in E. \tag{4}$$

For ease of exposition, this formulation assumes that there is a maximum of one edge between any two vertices. As can be seen in the illustration in Figures 1 and 2, this may not always be the case. However, cases where there are multiple edges between the same pair of vertices do not complicate the treatment since those cases can easily be transformed into the form shown in (1)–(4).

As pointed out by Edmonds and Johnson (1973) and Christofides (1973), when there are odd cardinality vertices in the graph, the CPP reduces to the problem of finding a minimum cost matching among the odd cardinality vertices. A minimum cost matching on a graph is a pairing of the vertices on that graph such that each vertex is paired with exactly one other vertex and the total cost of the edges connecting the pairs is a minimum. When no edge exists between a pair of vertices, the cost of pairing them is the cost of the shortest path running between the pair. Replicating the edges that connect each pair of odd cardinality vertices in the minimum matching produces an Eulerian graph (i.e. all vertices now have even cardinality) where the total cost of all the edges, the edges in the original graph plus the edges that have been replicated as a result of the matching, is the cost of the optimal tour on the original graph.

To illustrate the general solution process, Figure 3 presents a graph with four odd cardinality

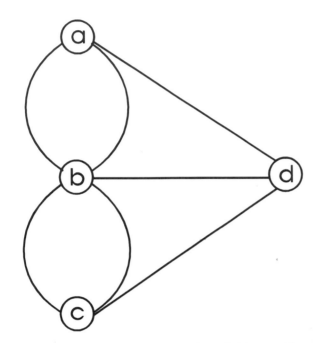

Fig. 1. A graph of Euler's Konigsberg bridge problem.

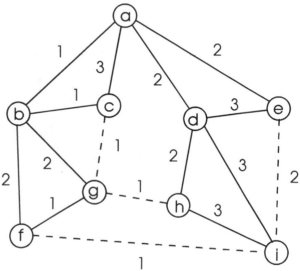

Fig. 3. The graph G.

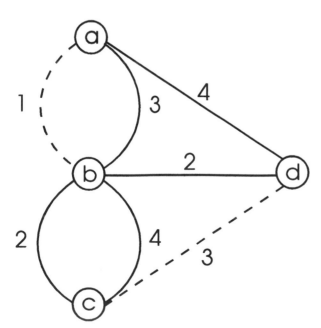

Fig. 2. The Königsberg bridge problem with a cost on each edge.

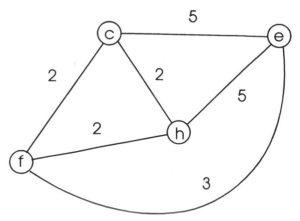

Fig. 4. The graph G'.

vertices (*c,e,f, & h*). None of the four vertices is directly connected to another of the four. To find the required minimum cost matching requires the construction of the graph *G'*, shown in Figure 4, which consists of the four odd cardinality vertices connected by edges whose costs are the cost of the shortest path between each pair on the original graph. The problem is then to find a minimum cost matching on the graph *G'*. This matching will determine which edges must be traversed twice to achieve a minimum cost tour on *G*. A quick inspection of *G'* shows that the edges (*c,h*) and (*e,f*) constitute a minimal matching on *G'*. The paths (*c*–*g*–*h*) and (*e*–*i*–*f*) on graph *G* in Figure 3 correspond to this matching, and the edges along these paths will be traversed twice each in an optimal tour and are shown as dashed lines in Figure 3.

Solving the CPP requires two operations, both of which can be performed in polynomial time. A matching of the odd cardinality vertices must be found, and the corresponding edges replicated which results in an Eulerian graph. An Eulerian tour of this expanded graph must then be found. The complexity of the CPP is dominated by the complexity of solving the minimum cost matching problem, which can be solved in at most $O(n^3)$ time. Variations of the basic CPP, briefly described below, are generally not as tractable.

VARIATIONS OF THE CHINESE POSTMAN PROBLEM: The CPP has many variations that can and do occur on a regular basis. In the CPP, the edges are undirected and they may be traversed in either direction. The most obvious variation of the CPP is the directed postman problem where each of the edges has a direction associated with it. This is often encountered when an edge represents a one way street in a routing problem, or an edge must be traversed twice, once in each direction, as might occur in routing a street sweeper. In this latter case, each street would be represented in the graph by two edges, one in each direction. Like the CPP, the directed postman problem can be solved in polynomial time, and in a sense it is even easier than the CPP since it requires a network flow algorithm rather than a matching algorithm.

When the graph contains a mixture of both directed and undirected edges, the problem of finding a minimum cost tour is called the mixed postman problem. The mixed postman problem has been shown to be NP-hard. The rural postman problem is a variation of the CPP where a subset of the edges in the graph must be traversed. The rural postman problem has been shown to be equivalent to a traveling salesman problem, and

as such it is also an NP-hard problem. Finally, the capacitated chinese postman problem recognizes that each edge may have a nonzero demand for service and that the server (postman) may have a finite capacity for supplying service. In the general case, multiple servers must be assigned to routes such that the demands on all of the edges are met and no server is assigned a route that exceeds his capacity. This then is the problem of partitioning the edges of the graph into subsets and assigning a server (postman) to each subset in such a way that all capacity constraints are met and the total distance covered by all of the servers is a minimum. As with the directed and rural postman problems, the capacitated postman problem has been shown to be NP-hard.

See **Combinatorics; Computational complexity; Graph theory; Matching; Networks, Traveling salesman problem; Vehicle routing.**

References

[1] Christofides, N. (1973), "The Optimum Traversal of a Graph," *Omega*, 1, 719–732.

[2] Edmonds, J. and E. Johnson (1973), "Matching, Euler Tours, and the Chinese Postman," *Mathematical Programming*, 5, 88–124.

[3] Kwan, M.-K. (1962), "Graphic Programming Using Odd or Even Points," *Chinese Mathematics*, 1, 273–277.

CHOICE STRATEGIES

The different approaches people use to combine deterministic information in their mind; sometimes referred to as combination rules. See **Choice theory; Decision analysis; Decision making.**

CHOICE THEORY

Leonard Adelman

George Mason University, Fairfax, Virginia

There is no one descriptive theory of human choice. Instead, there are different theoretically and empirically-based approaches for describing choice behavior. This piece briefly overviews three approaches: bounded rationality, prospect theory, and choice strategies. These approaches are descriptive in the sense that they describe certain aspects of how people actually make choices. They contrast with prescriptive approaches, such as the rational economic model (or expected utility theory), which prescribe how one should make decisions, but do not necessarily describe choice behavior. An overriding principle is that how people make choices is highly dependent on characteristics of the choice problem.

BOUNDED RATIONALITY: The concept of bounded rationality is attributed to Nobel laureate H. Simon (1955, 1979), who argued that humans lack both the knowledge and computational skill required to make choices in a manner compatible with economic notions of rational behavior (Hogarth, 1987). The rational economic model's requirements are illustrated by the concept of a payoff matrix, an example of which is presented in Table 1.

Table 1. The rational economic model's decision making requirements as a payoff matrix

	States of the World			
Alternatives	$S_1 \ (p_1)$	$S_2 \ (p_2)$	.....	$S_k \ (p_k)$
A	a_1	a_2	.....	a_k
B	b_1	b_2	.....	b_k
.	.	.	.....	.
N	n_1	n_2	.....	n_k

The rows of the matrix represent all the different alternatives available to the decision maker for solving a choice problem. The columns represent all of the different states of the world, as defined by future events, that could affect the attractiveness of the alternatives. The $p_1, \ldots, p_k$ values represent the probabilities for each state of the world. The cell entries in the matrix indicate the value (or "utility") of the outcome or "payoff" for each combination of alternatives and states of the world. Each outcome represents a net payoff from the alternative after combining the perceived advantages and disadvantages on multiple criteria of varying importance to the decision maker. Finally, the rational decision maker is assumed to select the alternative that maximizes expected utility, which is calculated for each alternative by multiplying the values for the outcomes by the probabilities for the future states, and then summing the products.

Numerous studies have shown that, unaided, people do not employ the above decision matrix due to the complex, dynamic nature of the envi-

ronment and to basic human information acquisition and processing limitations. Therefore, how does unaided human choice remain purposeful and "reasonable"? Simon suggested that people employ three simplification strategies, which result in a "bounded rationality." First, people simplify the problem by only considering a small number of alternatives and states of the world at a time. Second, people simplify the problem by setting aspiration (or acceptability) levels on the outcomes. And, third, people choose the first alternative that satisfies the aspiration levels. In other words, people do not optimize (i.e., choose the best of all possible alternatives), but satisfice (i.e., choose the first satisfactory alternative). In this way, people can reduce information acquisition and processing demands and act in a purposeful, reasonable manner.

PROSPECT THEORY: Like Simon's bounded rationality, prospect theory is juxtaposed against expected utility theory. For example, this prospect (or choice) is taken from Kahneman and Tversky (1979):

Choice A: ($4000 with $p = .8$; $0 with $p = .2$) or
Choice B: ($3000 for sure; that is, $p = 1.0$).

The majority of participants will select Choice B. Yet, Choice A has the greater expected value; that is, $\$4000 \times .8 = 3200$. Now, consider the following prospect:

Choice C: ($-\$4000$ with $p = .8$; $0 with $p = .2$) or
Choice D: ($-\$3000$ for sure; that is, $p = 1.0$).

The only change in the second prospect is that the sign has been reversed so that one is now considering losses, not gains. However, in this case, the majority of the subjects picked Choice C. That is, they would now be willing to take a gamble of losing $4000 with a probability of .8, which has an expected value of losing $3200, instead of taking a sure loss of $3000. Again, they selected the choice with the lower expected value. In addition, they switched from the "sure thing" to preferring the gamble.

What Kahneman and Tversky (1979, 1981) have shown is that the way the choice problem is presented (or "framed") significantly affects how people evaluate it, such that information that should result in the same choice from the perspective of expected utility theory actually results in different choices. In particular, people perceive outcomes as gains or losses from a reference point rather than from final states (e.g., of wealth), as assumed by economic-based models of choice. The current position is usually considered as the reference point. However, the location of the reference point and, in turn, the coding of outcomes as either gains or losses, can be affected by how the choices are "framed." This framing is particularly important for choice because, as the example presented above indicates, people tend to be risk adverse when considering gains and risk seeking when considering losses, particularly if one of the prospects is certain. Moreover, the value function is steeper for losses than for gains, consistent with the observation that "losses loom much larger than gains." For these reasons, many people are willing to gamble to avoid a "sure loss," but unwilling to gamble when they have a "sure gain," even when both choices have a lower expected value than another choice.

CHOICE STRATEGIES: Substantial research has focused on describing the different strategies people use to combine information when facing a choice. In contrast to bounded rationality and prospect theory, these strategies are used when people (a) have information on a number of different dimensions (or attributes) describing the alternatives, and (b) do not consider probabilities, either in terms of different states of nature or the reliability (or accuracy) of the information. A representative type of problem is making a purchase decision, such as choosing a car.

The literature makes a distinction between two classes of choice strategies: compensatory and noncompensatory (e.g., Beach, 1990; Hogarth, 1987). Compensatory strategies are used when one trades off (e.g., via relative importance weights) a low value on one attribute for a high value on another. For example, when choosing among cars, one may trade off gas mileage for comfort. Non-compensatory strategies do not employ trade-offs but, rather, employ thresholds (or cut-offs) that need to be achieved for choice of an alternative. For example, one eliminates all cars that do not get at least 25 miles per gallon, regardless of comfort. Some of the strategies identified in the literature are defined below. The literature cites three different types of compensatory models:

1. *Linear, additive strategy* – the value of an alternative is equal to the sum of the products, overall the dimensions, of the relative weight times the scale value for the dimension.
2. *Additive difference strategy* – the decision maker evaluates the differences between the alternatives on a dimension by dimension basis, and then sums the weighted differences in order to identify the alternative with the highest value overall.

3. *Ideal point strategy* – is similar to the additive difference model, except the decision maker compares the alternatives against an ideal alternative instead of each other.

The literature cites four different types of noncompensatory strategies:

4. *Dominance strategy* – select the alternative that is at least as attractive as the other alternatives on all the dimensions, but is better than them on at least one dimension. Although the dominance strategy is easier for an unaided decision maker to use, all three compensatory strategies will identify the dominant alternative too. Moreover, the compensatory strategies can be used if there is no dominant alternative; the dominance strategy cannot.
5. *Conjunctive strategy* – select the alternative that best passes some critical threshold on all dimensions. This is the satisficing strategy when one selects the first option that passes a threshold on all dimensions. A variant of the conjunctive strategy can be used to reduce the set of alternatives by eliminating all alternatives that fail to pass a threshold on all dimensions.
6. *Lexicographic strategy* – select the alternative that is best on the most important dimension. If two or more alternatives are tied, select among them by choosing the alternative that is best on the second most important dimension, and so on.
7. *Elimination by aspects* – sequentially identify different dimensions, either according to their importance or some more probabilistic scheme. Eliminate all alternatives that fail to pass the threshold or aspect for each dimension until only one alternative is left.

Research has shown that people often use multiple strategies when considering choice alternatives (e.g., Hogarth, 1987). Typically, they use noncompensatory strategies to reduce the number of alternatives and dimensions under consideration. To use a job selection example, a person might first eliminate all alternatives that fail to pass a specific threshold on job security, which may no longer be as important when considering the reduced set of alternatives. Then, after the set of alternatives and dimensions have been reduced to a smaller, more manageable set, people often employ a compensatory strategy where they weigh the strengths and weaknesses of the remaining alternatives in order to select the one which best satisfies their values.

SUMMARY: We emphasize that there is no one descriptive theory of human choice. Instead, there are different theoretically and empirically-based approaches for describing choice behavior. This paper briefly overviewed three of them: bounded rationality, prospect theory, and choice strategies. These approaches were contrasted with prescriptive approaches, such as decision analysis or other economic-based theories of choice, which prescribe how one should make decisions, but do not necessarily describe choice behavior.

See **Decision analysis; Decision making; Preference theory; Utility theory.**

References

[1] Beach, L.R. (1990). *Image Theory: Decision Making in Personal and Organizational Contexts.* Wiley, New York.

[2] Hogarth, R.M. (1987). *Judgment and Choice.* Wiley, New York.

[3] Kahneman, D. and A. Tversky (1979). "Prospect Theory: An Analysis of Decision Making Under Risk," *Econometrica,* 47, 263–289.

[4] Simon, H.A. (1955). "A Behavioral Model of Rational Choice," *Quarterly Jl. Economics,* 69, 99–118.

[5] Simon, H.A. (1979). "Rational Decision Making in Business Organizations," *American Economic Review,* 69, 493–513.

[6] Tversky, A. and D. Kahneman (1981). "The Framing of Decisions and the Psychology of Choice," *Science,* 211, 453–458.

CHROMATIC NUMBER

In a graph, the minimum number of colors needed to ensure that adjacent nodes receive different colors. See **Graph theory.**

CHROMOSOME

In genetic algorithms, a chromosome represents a potential solution to the problem at hand.

CIRCLING

See **Cycling.**

CIM

Computer integrated manufacturing. See **Automation; Flexible manufacturing.**

CLASSICAL OPTIMIZATION

See **Unconstrained optimization.**

CLOSED NETWORK

A queueing network in which there is neither entrance nor exit but only a fixed number of customers endlessly circulating. See **Networks of queues.**

CLUSTER ANALYSIS

Jay E. Aronson and Lakshmi Sundaram

The University of Georgia, Athens, Georgia

INTRODUCTION: Cluster analysis is a generic term for various procedures that are used objectively to group entities based on their similarities and differences. In applying these procedures, the objective is to group the entities (elements, items, objects, etc.) into mutually exclusive clusters so that elements within each cluster are relatively homogeneous in nature while the clusters themselves are distinct. The key purposes of cluster analysis are reduction of data, data exploration, determination of natural groups, prediction based on groups, classification, model fitting, generation and testing of hypotheses (Everitt, 1980; Aldenderfer and Blashfield, 1984; Lorr, 1983).

Due to the importance of clustering in different disciplines such as psychology, zoology, botany, sociology, artificial intelligence and information retrieval, a variety of names such as Q-analysis, typology, grouping, clumping, classification, numerical taxonomy and unsupervised pattern recognition (apart from cluster analysis) have been used to refer to such techniques (Everitt, 1980). In fact, Jain and Dubes (1988) quoted, "*I.J. Good (1977) has suggested the new name botryology for the discipline of cluster analysis, from the greek word for a cluster of grapes.*"

Though clustering techniques have been in existence for many years, profuse work in this area has been accomplished only in the past two decades. The primary stimuli for this were the founding of the Classification Society in 1970 and the publication of the *Principles of Numerical Taxonomy* by Sokal and Sheath in 1963 (Lorr, 1983). Other reasons for the rapid growth in cluster analysis literature are: the basic importance of classification as a scientific procedure and prolific developments in high-speed computers. The complexity of clustering methods are known to increase tremendously with increase in problem sizes. With the availability of sophisticated computing power, handling of large, practical problems is of less concern now.

APPLICATIONS OF CLUSTER ANALYSIS: Clustering methods are applied in a variety of fields such as psychology, biology, medicine, economics, marketing research, pattern recognition, weather prediction, information systems design and flexible manufacturing systems. Some interesting cluster analyses include analyzing large engineering records collections (Homayoun, 1984), measuring welfare and quality of life across countries (Hirschberg *et al.*, 1991), management of cutting tools in flexible manufacturing systems (DeSouza and Bell, 1990), clustering as a quality management tool (Spisak, 1992), identifying the structure and content of human decision making (Allison *et al.*, 1992), mapping consumers' cognitive structures (Hodgkinson *et al.*, 1991), information systems design (Aronson and Klein, 1989; Karimi, 1986; Klein and Aronson, 1991), vehicle routing, production scheduling and sampling (Romesburg, 1984) and income tax bracket determination (Mulvey and Crowder, 1979). Punj and Stewart (1983) provide a good description of the applications of cluster analysis including some details on the various clustering packages and programs that are available.

CLUSTERING TECHNIQUES: Authors such as Everitt (1980), Cormack (1971), Aldenderfer and Blashfield (1984), Hartigan (1975), and Anderberg (1973) have provided good reviews on existing clustering methods. Nevertheless, there has been no unique classification of the various clustering methods. In fact, this is one of the pitfalls of cluster analysis. It is due to the excellent work by Cormack, also lauded by Punj and Stewart (1983) and Everitt (1980), that the following five categories have been accepted as a basis:
1. Hierarchical methods,
2. Optimization techniques,
3. Density search techniques,
4. Clumping methods, and
5. Other techniques.
1. *Hierarchical methods:* Hierarchical procedures are tree-like structures in which elements are first separated into broad classes. These classes are further subdivided into smaller classes and so on until the terminal classes are not further subdivisible. These methods are most frequently used in the biological sciences. The hierarchical methods are basically of two types – *agglomerative and divisive.*

The agglomerative methods begin by making each item its own cluster. In subsequent iterations two or more closest clusters are combined to form a new, aggregate cluster. Eventually, all items are grouped into one large cluster. Hence, these methods are sometimes referred to as build-up methods (Hair *et al.*, 1987).

In contrast to agglomerative methods are the divisive methods which proceed in the opposite direction. That is, beginning with one large cluster, groups of items that are most dissimilar are removed and turned into smaller clusters. The process continues until each item becomes a cluster by itself. Cormack (1971), Everitt (1980), Aldenderfer and Blashfield (1984) and Hair *et al.* (1987) have provided comprehensive descriptions on the various agglomerative and divisive procedures.

2. *Optimization techniques:* These methods allow relocation of items during the clustering process, improving from an initial solution to optimality. The number of clusters needs to be decided *a priori*, although some methods allow for changes during analysis. Differences in optimization techniques exist due to the different methods used to obtain a starting solution and different clustering criteria used for obtaining the optimal solution (Everitt, 1980).

Since most of the optimization techniques are based on well-established statistical concepts, very few mathematical programming approaches have been developed to solve these problems. Mulvey and Crowder (1979) use a subgradient method coupled with a simple search procedure for solving the clustering problem. However, their method does not yield an exact optimum solution. Though heuristics seem to be efficient at large, the need to obtain optimal solution for problems such as designing effective information systems (Klein *et al.*, 1988) make heuristics less attractive. Klein and Aronson (1991) have developed a mixed-integer programming model to find an optimal solution for clustering problems. Their method is based on the implicit enumeration method due of Balas (1965). Extensions including precedence and group size limits may be found in Aronson and Klein (1989). Earlier, Gower and Ross (1969) and Rohlf (1974) showed that there is a direct relationship between some common cluster formulations and certain types of well-known graph theoretic problems, primarily the minimum spanning tree. A further expansion on the use of graph theoretic techniques in cluster analysis may be found in Matula (1977).

3. *Density search techniques:* This concept, proposed by Gengerelli (1963), depicts the items as points in a metric space. Parts of the space where the distribution of points is very dense but separated by parts of low density suggest natural clusters. Everitt (1980) describes the different types of density search techniques.

4. *Clumping techniques:* These techniques are most popular in language studies where words that tend to have several meanings, when classified based on their meaning, belong to several groups. Thus, in general, clumping techniques allows for overlapping clusters. This terminology was introduced by Jones, Needham, and co-workers at the Cambridge Language Research Unit (Everitt, 1980). This method attempts to partition entities into two groups based on the similarity matrix from the original data. Needham's (1967) criteria is to minimize the cohesion function between the two groups. Other clumping procedures are also discussed in Rohlf (1974) and Everitt (1980).

5. *Other techniques:* This comprises all clustering techniques that do not fall into the above four categories. For example, description of the inverse 'Q' factor analysis that is commonly used in behavioral sciences can be found in Cattell (1952). Gower (1966) provided a good review of the properties of various 'R' and 'Q' factor analysis techniques. The 'R' factor analysis is a type of 'Q' factor analysis that utilizes the correlations between variables. Everitt (1980) and Aldenderfer and Blashfield (1984) included a good summary of various other clustering methods.

ISSUES OF CONCERN: Though initially the concepts of cluster analysis seem to be intuitive, one can encounter a host of problems while performing an actual analysis. Some of the problems include selection of data units and variables, knowing exactly what to cluster, distance or similarity measures, transformation of measures, clustering criterion, the clustering method to use, the number of clusters and interpretation of the results (Anderberg, 1973). Authors such as Aldenderfer and Blashfield (1984), Everitt (1980), Hair *et al.* (1987) and Anderberg (1973) have discussed some of the issues in great detail. We next briefly discuss some of the more critical issues.

Measurement of distance or similarity matrix – The relationship between elements are represented by using either a similarity or distance measure. While similarity measures (indicating cohesion) take values between 0 and 1, distance measures can be any positive value. The output of any clustering method depends on the type of input measure used. One of the most commonly used measures is the Euclidean distance. This concept can be easily generalized for additional variables (Hair *et al.*, 1987).

Another measure which allows for correlations between variables was originally proposed by Mahalanobis in 1936 (Everitt, 1980). This is similar to Euclidean distance measure using standardized variables when the correlations are zero. The Mahalanobis distance measure has been used by McRae (1971). Everitt (1980), Hair *et al.* (1987) and Hartigan (1975) have provided some discussions on other types of distance measurements. The clustering model for computer-assisted organization presented by Klein and Aronson (1991) accounts for total interactions. The need to consider all interactions among items in each cluster led to the formulation of a mixed-integer model for optimal clustering based on scaled distance (Klein and Aronson, 1991).

Which clustering method to use? The problem of choosing an appropriate clustering method generally arises after one has determined the variables, distance measure and criterion for clustering. A number of software packages and programs are available for clustering. Punj and Stewart (1983) and Anderberg (1973) have identified various programs available for clustering. They not only cite their original source but also provide an empirical comparison of their performance. For selecting the best clustering method one should be aware of the performance characteristics of the various methods (Hair *et al.*, 1987).

Appropriate number of clusters – One of the practical issues of concern in clustering is choosing the number of clusters. Some algorithms find the best fitting structure for a given number of clusters while others, like the hierarchical methods, provide configurations from the number of entities to one large cluster, that is, the entire data set as one cluster. However, if the number of clusters cannot be predetermined, a range of clusters can be selected, solving the problem for each of those cluster sizes and then selecting the best alternative (Hair *et al.*, 1987).

See **Combinatorial and integer optimization; Decision making; Graph theory; Information systems and database design; Minimum spanning tree problem; Vehicle routing.**

References

[1] Aldenderfer, M.S. and R.K. Blashfield (1984). *Cluster Analysis.* Sage Publications, California.

[2] Allison, S.T., A.M.R. Jordan, and C.E. Yeatts (1992). "A Cluster-analytic Approach Toward Identifying the Structure and Content of Human Decision Making." *Human Relations,* **45**, 49–73.

[3] Anderberg, M.R. (1973). *Cluster Analysis for Applications.* Academic Press, New York.

[4] Aronson, J.E. and G. Klein (1989). "A Clustering Algorithm for Computer-Assisted Process Organization." *Decision Sciences,* **20**, 730–745.

[5] Balas, E. (1965). "An Additive Algorithm for Solving Linear Programs with Zero-One Variables." *Operations Research*, **13**, 517–546.

[6] Cattell, R.B. (1952). *Factor Analysis: An Introduction and Manual for the Psychologist and Social Scientist*. Harper, New York.

[7] Cormack, R.M. (1971). "A Review of Classification." *Jl. Royal Statistical Society* (Series A), **134**, 321–367.

[8] DeSouza, R.B.R. and R. Bell (1991). "A Tool Cluster Based Strategy for the Management of Cutting Tools in Flexible Manufacturing Systems." *Jl. Operations Management*, **10**, 73–91.

[9] Everitt, B. (1980). *Cluster Analysis* (2nd ed.), Halsted Press, New York.

[10] Gengerelli, J.A. (1963). "A Method for Detecting Subgroups in a Population and Specifying their Membership." *Jl. Psychology*, **5**, 456–468.

[11] Gower, J.C. (1966). "Some Distance Properties of Latent Root and Vector Methods Used in Multivariate Analysis." *Biometrika*, **53**, 325–338.

[12] Gower, J.C. and G.J.S. Ross (1969). "Minimum Spanning Trees and Single Linkage Cluster Analysis." *Appl. Statist.*, **18**, 54–64.

[13] Hair, J.F. Jr., R.E. Anderson, and R.L. Tatham (1987). *Multivariate Data Analysis* (2nd ed.), Macmillan, New York.

[14] Hartigan, J.A. (1975). *Clustering Algorithms*. John Wiley, New York.

[15] Hirschberg, J.G., E. Maasoumi, and D.J. Slottje (1991). "Cluster Analysis for Measuring Welfare and Quality of Life Across Countries." *Jl. Econometrics*, **50**, 131–150.

[16] Hodgkinson, G.P., J. Padmore, and A.E. Tomes (1991). "Mapping Consumers' Cognitive Structures: A Comparison of Similarity Trees with Multidimensional Scaling and Cluster Analysis." *European Jl. Marketing*, **25**, 41–60.

[17] Homayoun, A.S. (1984). "The Use of Cluster Analysis in Analyzing Large Engineering Records Collection." *Records Management Quarterly*, October, 22–25.

[18] Jain, A.K. and R.C. Dubes (1988). *Algorithms for Clustering Data*. Prentice Hall, Englewood Cliffs, New Jersey.

[19] Karimi, J. (1986). "An Automated Software Design Methodology Using CAPO." *Jl. Management Information Systems*, **3**, 71–100.

[20] Klein, G. and J.E. Aronson (1991). "Optimal Clustering: A Model and Method." *Naval Research Logistics*, **38**, 447–461.

[21] Klein, G., P.O. Beck, and B.R. Konsynski (1988). "Computer Aided Process Structuring via Mixed Integer Programming." *Decision Sciences*, **19**, 750–761.

[22] Lorr, M. (1983). *Cluster Analysis for Social Scientists*. Jossey-Bass Publishers, California.

[23] Matula, D.W. (1977). "Graph Theoretic Techniques for Cluster Analysis Algorithms," in *Classification and Clustering*, J. Van Ryzin, ed., Academic Press, New York.

[24] McRae, D.J. (1971). "MICKA, A FORTRAN IV Iterative K-means Cluster Analysis Program." *Behavioural Science*, **16**, 423–424.

[25] Mulvey, J. and H. Crowder (1979). "Cluster Analysis: An Application of Lagrangian Relaxation." *Management Science*, **25**, 329–340.

[26] Needham, R.M. (1967)."Automatic Classification in Linguistics." *The Statistician*, **17**, 45–54.

[27] Punj, G. and D.W. Stewart (1983). "Cluster Analysis in Marketing Research: Review and Suggestions for Application." *Jl. Marketing Research*, **20**, 134–148.

[28] Rohlf, F.J. (1974). "Graphs Implied by the Jardine-Sibson Overlapping Clustering Methods." *Jl. American Statistical Association*, **69**, 705–710.

[29] Romesburg, H. (1984). *Cluster Analysis for Researchers*. Lifetime Learning Publications, Belmont, California.

[30] Sneath, P.H.A. and Sokol, R.R. (1973). *Principles of Numerical Taxonomy*. W.H. Freeman, San Francisco.

[31] Spisak, A.W. (1992). "Cluster Analysis as a Quality Management Tool." *Quality Progress*, **25**, 33–38.

COBB-DOUGLAS PRODUCTION FUNCTION

See **Economics**.

COEA

Cost and operational effectivness analysis. See **Cost analysis**.

COEFFICIENT OF VARIATION

The ratio of the standard deviation to the mean of a random variable.

COGNITIVE MAPPING

A graphical notation for capturing concepts in use by decision makers for understanding a problematic situation. Concepts are fixed by reference to polar opposites, and directed arcs indicate perceived causal relationships. See **Problem structuring methods**.

COHERENT SYSTEM

See **System reliability**.

COLUMN GENERATION

A technique that permits solution of very large linear-programming problems by generating the columns of the constraint matrix only when they are needed. It is typically employed when the constraint matrix is too large to be stored, or when it is only known implicitly. Column generation, as imbedded in the revised simplex method, has been used to solve the trim problem and other such problems in which the columns are formed from combinatorial considerations. See **Trim problem**.

COLUMN VECTOR

One column of a matrix or a matrix consisting of a single column. See **Matrices and matrix algebra**.

COMBAT MODEL

A model whose object is military combat or some aspect of combat. Three associated terms are often used as synonyms, but are frequently used to differentiate three common aspects of combat modeling: "combat model," "combat simulation," and "war game." When "combat model" is used as a discriminator it often is used to mean that the model in question is an analytic combat model. See **Analytic combat model**; **Battle modeling**.

COMBAT SIMULATION

A type of model whose object is military combat or some aspect of combat. Combat simulation is used as a discriminator to emphasize the time or process aspect of the model in question. See **Battle modeling**.

COMBINATORIAL AND INTEGER OPTIMIZATION

Karla L. Hoffman

George Mason University, Fairfax, Virginia

and

Manfred Padberg

New York University, New York, NY

INTRODUCTION: Combinatorial optimization problems are concerned with the efficient allocation of limited resources to meet desired objectives when the values of some or all of the variables are restricted to be integral. Constraints on basic resources, such as labor, supplies, or capital restrict the possible alternatives that are considered feasible. Still, in most such problems, there are many possible alternatives to consider and one overall goal determines which of these alternatives is best. For example, most airlines need to determine crew schedules which minimize the total operating cost; automotive manufacturers may want to determine the design of a fleet of cars which will maximize their share of the market; a flexible manufacturing facility needs to schedule the production for a plant without having much advance notice as to what parts will need to be produced that day. In today's changing and competitive industrial environment the difference between using a quickly derived "solution" and using sophisticated mathematical models to find an *optimal* solution can determine whether or not a company survives.

The versatility of the combinatorial optimization model stems from the fact that in many practical problems, activities and resources, such as machines, airplanes and people, are indivisible. Also, many problems have only a finite number of alternative choices and consequently can appropriately be formulated as combinatorial optimization problems – the word combinatorial referring to the fact that only a finite number of alternative feasible solutions exists. Combinatorial optimization models are often referred to as *integer programming* models where programming refers to "planning" so that these are models used in planning where some or all of the decisions can take on only a finite number of alternative possibilities.

Combinatorial optimization is the process of finding one or more best (optimal) solutions in a well defined discrete problem space. Such problems occur in almost all fields of management (e.g. finance, marketing, production, scheduling, inventory control, facility location and layout, data-base management), as well as in many engineering disciplines (e.g. optimal design of waterways or bridges, VLSI-circuitry design and testing, the layout of circuits to minimize the area dedicated to wires, design and analysis of data networks, solid-waste management, determination of ground states of spin-glasses, determination of minimum energy states for alloy construction, energy resource-planning models, logistics of electrical power generation and transport, the scheduling of lines in flexible manufacturing facilities, and problems in crystallography). A survey of related applications of combinatorial optimization is given in Grötschel (1992).

We assume throughout this discussion that both the function to be optimized and the functional form of the constraints restricting the possible solutions are linear functions. Although some research has centered on approaches to problems where some or all of the functions are nonlinear, most of the research to date covers only the linear case. A survey of nonlinear integer programming approaches is given in Cooper and Farhangian (1985).

The general linear *integer* model is

$$\max \sum_{j \in B} c_j x_j + \sum_{j \in I} c_j x_j + \sum_{j \in C} c_j x_j$$

subject to:

$$\sum_{j \in B} a_{ij} x_j + \sum_{j \in I} a_{ij} x_j + \sum_{j \in C} a_{ij} x_j \sim b_i \ (i = 1, \dots, m)$$

$$l_j \le x_j \le u_j \quad (j \in I \cup C)$$

$$x_j \in \{0, 1\} \quad (j \in B)$$

$$x_j \in integers \quad (j \in I)$$

$$x_j \in reals \quad (j \in C)$$

where B is the set of zero-one variables, I is the set of integer variables, C is the set of continuous variables, and the $\sim$ symbol in the first set of constraints denotes the fact that the constraints $i = 1, \dots, m$ can be either $\le$, $\ge$, or $=$. The data l_J and u_j are the lower and upper bound values, respectively, for variable x_j. As we are discussing the integer case, there must be some variable in $B \cup I$. If $C = I = \phi$, then the problem is referred to as a *pure 0-1 linear-programming problem*; if $C = \phi$, the problem is called a *pure integer (linear) programming problem*. Otherwise, the problem is a *mixed integer (linear) programming problem*. Throughout this discussion, we will call the set of points satisfying all constraints S, and the set of points satisfying all but the integrality restrictions P.

SOME APPLICATIONS OF COMBINATORIAL OPTIMIZATION: We describe some classical combinatorial optimization models to provide both an overview of the diversity and versatility of this field and to show that the solution of large real-world instances of such problems requires the solution method exploit the specialized mathematical structure of the specific application.

Knapsack problems: Suppose one wants to fill a knapsack that can hold a total weight of W with some combination of items from a possible n items each with weight w_i and value v_i so that

the value of the items packed into the knapsack is maximized. This problem has a single linear constraint (that the weight of the items in the knapsack not exceed W), a linear objective function which sums the values of the items in the knapsack, and the added restriction that each item either be in the knapsack or not – a fractional amount of the item is not possible. For solution approaches specific to the knapsack problem, see Martello and Toth (1990).

Although this problem might seem almost too simple to have much applicability, the knapsack problem is important to cryptographers and to those interested in protecting computer files, electronic transfers of funds and electronic mail. These applications use a "key" to allow entry into secure information. Often the keys are designed based on linear combinations of some collection of data items which must equal a certain value. This problem is also structurally important in that most integer programming problems are generalizations of this problem (i.e. there are many knapsack constraints which together compose the problem). Approaches for the solution of multiple knapsack problems are often based on examining each constraint separately.

An important example of a multiple knapsack problem is the *capital budgeting problem*. This problem is one of finding a subset of the thousands of capital projects under consideration that yields the greatest return on investment, while satisfying specified financial, regulatory and project relationship requirements (Markowitz and Manne, 1957; Weingartner, 1963).

Network and graph problems: Many optimization problems can be represented by a network where a network (or graph) is defined by nodes and by arcs connecting those nodes. Many practical problems arise around physical networks such as city streets, highways, rail systems, communication networks, and integrated circuits. In addition, there are many problems which can be modeled as networks even when there is no underlying physical network. For example, one can think of *the assignment problem* where one wishes to assign a set of persons to some set of jobs in a way that minimizes the cost of the assignment. Here one set of nodes represents the people to be assigned, another set of nodes represents the possible jobs, and there is an arc connecting a person to a job if that person is capable of performing that job.

Space-time networks are often used in scheduling applications. Here one wishes to meet specific demands at different points in time. To model this problem, different nodes represent the same entity at different points in time. An example of the many scheduling problems that can be represented as a space-time network is the airline *fleet assignment problem*, which requires that one assign specific planes to pre-scheduled flights at minimum cost. Each flight must have one and only one plane assigned to it, and a plane can be assigned to a flight only if it is large enough to service that flight and only if it is on the ground at the appropriate airport, serviced and ready to depart when the flight is scheduled to take off. The nodes represent specific airports at various points in time and the arcs represent the flows of airplanes of a variety of types into and out of each airport. There are layover arcs that permit a plane to stay on the ground from one time period to the next, service arcs which force a plane to be out of duty for a specified amount of time, and connecting arcs which allow a plane to fly from one airport to another without passengers. A general survey of network applications is given in Ahuja *et al.* (1993) and solution procedures in Ahuja *et al.* (1992).

In addition, there are many graph-theoretic problems which examine the properties of the underlying graph or network. Such problems include the *Chinese postman problem* where one wishes to find a path (a connected sequence of edges) through the graph that starts and ends at the same node, that covers every edge of the graph at least once, and that has the shortest length possible. If one adds the restriction that each node must be visited exactly one time and drops the requirement that each edge be traversed, the problem becomes the notoriously difficult *traveling salesman problem*. Other graph problems include the *vertex coloring problem*, the object of which is to determine the minimum number of colors needed to color each vertex of the graph in order that no pair of adjacent nodes (nodes connected by an edge) share the same color; the *edge coloring problem*, whose object is to find a minimum total weight collection of edges such that each node is incident to at least one edge; the *maximum clique problem*, whose object is to find the largest subgraph of the original graph such that every node is connected to every other node in the subgraph; and the *minimum cut problem*, whose object is to find a minimum weight collection of edges which (if removed) would disconnect a set of nodes *s* from a set of nodes *t*.

Although these combinatorial optimization problems on graphs might appear, at first glance, to be interesting mathematically but have little application to the decision making in management or engineering, their domain of applicabil-

ity is extraordinarily broad. The traveling salesman problem has applications in routing and scheduling, in large scale circuitry design and in strategic defense. The four-color problem (Can a map be colored in four colors or less?) is a special case of the vertex coloring problem. Both the clique problem and the minimum cut problem have important implications for the reliability of large systems.

Approximating nonlinear functions by piecewise linear functions: The versatility of the integer programming model might best be exemplified by the fact that many nonlinear programming problems can be modeled as mixed-integer linear programs. The "trick" in this case, is to find a piecewise-linear approximation to each nonlinear function. The simplest example of such a transformation is the *fixed charge problem* where the cost function has both a fixed charge for initiating the activity as well as marginal costs associated with activities. One example of a fixed-charge problem is the *facility location problem* where one wishes to locate facilities such that the combined cost of building the facility (a one time fixed cost) and the cost of production and shipping to customers (marginal costs based on the amount shipped and produced) is minimized. The fact that nothing can be produced in the facility unless the facility exists, creates a discontinuity in the cost function at zero. This function can be transformed to a linear function by the introduction of additional variables that take on only the values zero or one. Similar transformations allow one to model separable nonlinear functions as integer (linear) programming problems.

Scheduling problems which are rule-based: There are many problems where it is impossible to write down all of the restrictions in a mathematically "clean" way. Such problems often arise in scheduling where there are a myriad of labor restrictions, corporate scheduling preferences and other rules related to what constitutes a "feasible schedule." Such problems can be solved by generating all, or some reasonable large subset of the feasible schedules for each worker. One associates a matrix with such problems whose rows correspond to the tasks considered and whose columns correspond to individual workers, teams or crews. A column of the matrix has an entry of one in those rows that correspond to tasks that the worker will be assigned and a zero otherwise. Each "feasible" schedule defines one column of the constraint matrix and associated with each such schedule is a value. Thus the matrix of constraints consists of all zeroes and ones and the sense of the inequality indicates whether that job must be covered by exactly a specified number of people (called *set partitioning*), that it must be covered by at least a specific number (called *set covering*) or that it must be covered by not more than a specified number (called *set packing*). The optimization problem is then the problem of finding the best collection of columns which satisfy these restrictions. Surveys on set partitioning, covering and packing, are given in Balas and Padberg (1976) and Padberg (1979).

FORMULATION CONSIDERATIONS: The versatility of the integer programming formulation, as illustrated by the above examples, should provide sufficient explanation for the high activity in the field of combinatorial optimization which is concerned with developing solution procedures for such problems. Since there are often different ways of mathematically representing the same problem, and since obtaining an optimal solution to a large integer programming problem in a reasonable amount of computer time may well depend on the way it is "formulated," much recent research has been directed toward the reformulation of integer programming problems. In this regard, it is sometimes advantageous to increase (rather than decrease) the number of integer variables, the number of constraints or both. More will be said about this in the section on solution techniques for integer programming. Discussions of alternative formulation approaches are given in Guignard and Spielberg (1981) and Williams (1985) and a description of approaches to "automatic" reformulation or preprocessing are described in Brearley, Mitra and Williams (1975) and Hoffman and Padberg (1991). We now give a short description of solution approaches to these problems.

SOLUTION TECHNIQUES FOR INTEGER PROGRAMMING: *Solving* combinatorial optimization problems, i.e. finding an *optimal* solution to such problems, can be a difficult task. The difficulty arises from the fact that unlike linear programming, for example, whose feasible region is a convex set, in combinatorial problems, one must search a lattice of feasible points or, in the mixed-integer case, a set of disjoint halflines or line segments to find an optimal solution. Thus, unlike linear programming where, due to the convexity of the problem, we can exploit the fact that any local solution is a global optimum, integer programming problems have many local optima and finding a *global* optimum to the problem requires one to *prove* that a particular solution dominates all feasible points by arguments other than the calculus-based derivative approaches of convex programming.

There are, at least, three different approaches for solving integer programming problems, although they are frequently combined into "hybrid" solution procedures in computational practice. They are

● enumerative techniques
● relaxation and decomposition techniques
● cutting-plane approaches based on polyhedral combinatorics

Enumerative approaches: The simplest approach to solving a *pure* integer programming problem is to enumerate all finitely many possibilities. However, due to the "combinatorial explosion" resulting from the parameter "size," only the smallest instances could be solved by such an approach. Sometimes one can *implicitly* eliminate many possibilities by domination or feasibility arguments. Besides straight-forward or implicit enumeration, the most commonly used enumerative approach is called *branch and bound*, where the "branching" refers to the enumeration part of the solution technique and bounding refers to the fathoming of possible solutions by comparison to a known upper or lower bound on the solution value (Land and Doig, 1960). To obtain an upper bound on the problem (we presume a maximization problem), the problem is relaxed in a way which makes the solution to the relaxed problem relatively easy to solve.

All commercial branch-and-bound codes relax the problem by dropping the integrality conditions and solve the resultant continuous linear programming problem over the set P. If the solution to the relaxed linear programming problem satisfies the integrality restrictions, the solution obtained is optimal. If the linear program is infeasible, then so is the integer program. Otherwise, at least one of the integer variables is fractional in the linear programming solution. One chooses one or more such fractional variables and "branches" to create two or more subproblems which exclude the prior solution but do not eliminate any feasible integer solutions. These new problems constitute "nodes" on a branching tree, and a linear programming problem is solved for each node created. Nodes can be fathomed if the solution to the subproblem is infeasible, satisfies all of the integrality restrictions, or has an objective function value worse than a known integer solution. A variety of strategies that have been used within the general branch-and-bound framework is described in Johnson and Powell (1978).

Lagrangian Relaxation and Decomposition Methods: Relaxing the integrality restriction is not the only approach to relaxing the problem. An alternative approach to the solution to integer programming problems is to take a set of "complicating" constraints into the objective function in a Lagrangian fashion (with fixed multipliers that are changed iteratively). This approach is known as *Lagrangian relaxation.* By removing the complicating constraints from the constraint set, the resulting sub-problem is frequently considerably easier to solve. The latter is a necessity for the approach to work because the subproblems must be solved repetitively until optimal values for the multipliers are found. The bound found by Lagrangian relaxation can be tighter than that found by linear programming, but only at the expense of solving subproblems in *integers*, i.e., only if the subproblems do not have the *Integrality Property.* (A problem has the integrality property if the solution to the Lagrangian problem is unchanged when the integrality restriction is removed.) Lagrangian relaxation requires that one understand the structure of the problem being solved in order to then relax the constraints that are "complicating" (Fisher, 1981). A related approach which attempts to strengthen the bounds of Lagrangian relaxation, is called *Lagrangian decomposition* (Guinard and Kim, 1987). This approach consists of isolating sets of constraints so as to obtain separate, easy problems to solve over each of the subsets. The dimension of the problem is increased by creating linking variables which link the subsets. All Lagrangian approaches are problem dependent and no underlying general theory – applicable to say, an arbitrary zero-one problem – has evolved.

Most Lagrangian-based strategies provide approaches which deal with special row structures. Other problems may possess special column structure, such that when some subset of the variables are assigned specific values, the problem reduces to one that is easy to solve. Benders' decomposition algorithm fixes the complicating variables, and solves the resulting problem iteratively (Benders, 1962). Based on the problem's associated dual, the algorithm must then find a cutting plane (i.e. a linear inequality) which "cuts off" the current solution point but no integer feasible points. This cut is added to the collection of inequalities and the problem is re-solved.

Since each of the decomposition approaches described above provide a bound on the integer solution, they can be incorporated into a branch and bound algorithm, instead of the more commonly used linear programming relaxation. However, these algorithms are special-purpose algorithms in that they exploit the "constraint pattern" or special structure of the problem.

Cutting Plane algorithms based on polyhedral combinatorics: Significant computational advances in exact optimization have taken place. Both the size and the complexity of the problems solved have been increased considerably when *polyhedral theory*, developed over the past twenty five years, was applied to numerical problem solving. The underlying idea of polyhedral combinatorics is to replace the constraint set of an integer programming problem by an alternative convexification of the feasible points and extreme rays of the problem.

Many years ago, H. Weyl (1935) established the fact that a convex polyhedron can alternatively be defined as the intersection of finitely many halfspaces *or* as the convex hull plus the conical hull of some finite number of vectors or points. If the data of the original problem formulation are *rational* numbers, then Weyl's theorem implies the existence of a finite system of linear inequalities whose solution set coincides with the *convex hull* of the mixed-integer points in S which we denote $conv(S)$. Thus, if we can list the set of linear inequalities that completely define the *convexification of S*, then we can solve the integer programming problem by linear programming. Gomory (1958) derived a "cutting plane" algorithm for integer programming problems which can be viewed as a *constructive* proof of Weyl's theorem, in this context.

Although Gomory's algorithm converges to an optimal solution in finite number of steps, the convergence to an optimum is extraordinarily slow due to the fact that these algebraically-derived cuts are "weak" in the sense that they frequently do not even define supporting hyperplanes to the convex hull of feasible points. Since one is interested in a linear constraint set for $conv(S)$ which is as small as possible, one is led to consider *minimal* systems of linear inequalities such that each inequality defines a *facet* of the polyhedron $conv(S)$. When viewed as cutting planes for the original problem then the linear inequalities that define facets of the polyhedron $conv(S)$ are "best possible" cuts – they cannot be made "stronger" in any sense of the word without losing some feasible integer or mixed-integer solutions to the problem. Considerable research activity has focused on identifying part (or all) of those linear inequalities for specific combinatorial optimization problems – problem-dependent implementations, of course, that are however derived from an underlying *general* theme due to Weyl's theorem which applies generally. Since for most interesting integer-programming problems the minimal number of inequalities necessary to describe this polyhedron is exponential in the number of variables, one is led to wonder whether such an approach could ever be computationally practical. It is therefore all the more remarkable that the implementation of cutting plane algorithms based on polyhedral theory has been successful in solving problems of sizes previously believed intractable. The numerical success of the approach can be explained, in part, by the fact that we are interested in *proving* optimality of a *single extreme point* of $conv(S)$. We therefore do not require the *complete* description of $conv(S)$ but rather only a partial description of $conv(S)$ in the *neighborhood* of the optimal solution.

Thus, a general cutting plane approach relaxes in a first step the integrality restrictions on the variables and solves the resulting linear program over the set P. If the linear program is unbounded or infeasible, so is the integer program. If the solution to the linear program is integer, then one has solved the integer program. If not, then one solves a *facet-identification problem* whose objective is to find a linear inequality that "cuts off" the fractional linear programming solution while assuring that all feasible integer points satisfy the inequality – i.e. an inequality that "separates" the fractional point from the polyhedron $conv(S)$. The algorithm continues until: 1) an integer solution is found (we have successfully solved the problem); 2) the linear program is infeasible and therefore the integer problem is infeasible; or 3) no cut is identified by the facet-identification procedures either because a full description of the facial structure is not known or because the facet-identification procedures are inexact, i.e. one is unable to *algorithmically* generate cuts of a known form. If we terminate the cutting plane procedure because of the third possibility, then, in general, the process has "tightened" the linear programming formulation so that the resulting linear programming solution value is much closer to the integer solution value. We next explain how much of the research and development of integer programming methods can be incorporated into a super-algorithm which uses all that is known about the problem. This method is called "branch-and-cut" (Padberg and Rinaldi, 1991; Hoffman and Padberg, 1985, 1991, 1992).

The major components of this algorithm consist of automatic reformulation procedures, heuristics which provide "good" feasible integer solutions, and cutting plane procedures which tighten the linear programming relaxation to the combinatorial problem under consideration – all of which is embedded into a tree-search framework as in the branch-and-bound approach to integer programming. Whenever possible, the procedure permanently fixes variables (by reduced cost implications and logical implications) and does comparable conditional fixing throughout the search-tree. These four components are combined so as to guarantee optimality of the solution obtained at the end of the calculation. However, the algorithm may also be stopped *early* to produce suboptimal solutions along with a bound on the remaining error. The cutting planes generated by the algorithm are facets of the convex hull of feasible integer solutions or good polyhedral approximations thereof and as such they are the "tightest cuts" possible. Lifting procedures assure that the cuts generated are valid throughout the search tree which aids the search process considerably and is a substantial difference to traditional (Gomory) cutting-plane approaches.

Mounting empirical evidence indicates that both pure and mixed integer programming problems can be solved to *proven* optimality in economically feasible computation times by methods based on the polyhedral structure of integer programs. For applications which use this branch-and-cut approach, see, among others, Barahona *et al.* (1988), Chopra *et al.* (1991), Grötschel *et al.* (1989), Magnanti and Vachani (1989), Pochet and Wolsey (1991), and Van Roy and Wolsey (1987). A direct outcome of these research efforts is that similar preprocessing and constraint generation procedures can be found in commercial software packages for combinatorial problems.

The computational successes for difficult combinatorial optimization problems reflects the intense effort devoted to developing the underlying polyhedral structure of these problems. Thus, to use this approach, one must be able to both *identify* specific mathematical structures inherent in the problem and then study the polyhedron associated with that structure. As more structures are understood, and can be detected automatically, we will see larger classes of problems solved by these methods. These codes will certainly be complex, but they are likely to lead to methods for solving *to optimality* – with reasonable computational effort – many of the difficult combinatorial problems for which only heuristic "guesses" are known today. Different from decomposition methods, the related computational successes are based on a *mathematical* understanding of the problem and not just the "structure," that is, the particular constraint "pattern" itself.

We briefly note some topics related to combinatorial and integer programming. One such topic is the complexity of integer programming problems (Garey and Johnson, 1979). Another topic is that of heuristics solution approaches – i.e. techniques for obtaining "good" but not necessarily optimal solutions to integer programming problems quickly and, in general, without any guarantee as to their "closeness" to an optimal solution. Heuristics are, however, important for a variety of reasons. They may provide the only usable solution to very difficult optimization problems for which the current exact algorithms are incapable of providing an optimal solution in reasonable times; when heuristics are used within an exact algorithm, they provide a bound to fix variables and to fathom branches on a search-tree. Recent research into heuristic algorithms has applied techniques from the physical sciences to the approximate solution of combinatorial problems. For surveys of research in simulated annealing (based on the physical properties of heat), genetic algorithms (based on properties of natural mutation) and neural networks (models of brain function), see Hansen (1986), Mühlenbein (1992), and Beyer and Ogier (1991), respectively. Glover and Laguna (1992) have generalized some of the attributes of these methods into a method called *tabu-search*. Worst-case and probabilistic analysis of heuristics are discussed in Cornuejols, *et al.* (1980), Rinnooy Kan (1986) and Karp (1976). Text books on integer programming and related topics include Grötschel, Lovasz and Schrijver (1988), Nemhauser and Wolsey (1988), Parker and Rardin (1988) and Schrijver (1986).

See **Assignment problem; Branch and bound; Bender's decomposition; Bin packing; Capital budgeting; Chinese postman problem; Combinatorial explosion; Combinatorics; Facility location; Fathom; Global optimum; Lagrangian function; Linear programming; Local optimum; Networks; Packing problem; Relaxed problem; Set-covering problem; Set-partitioning problem; Traveling salesman problem.**

References

[1] R.K. Ahuja, T.L. Magnanti, J. Orlin (1992). *Network Flows: Theory, Algorithms and Applications*. Prentice-Hall, New Jersey.

[2] R.K. Ahuja, T.L. Magnanti, J.B. Orlin and M.R. Reddy (1993). "Applications of Network Optimization," *Operations Research* **41**.

[3] E. Balas and M. Padberg (1976). "Set Partitioning: A Survey," *SIAM Review*, **18**, 710–760.

[4] M. Barahona, M. Grötschel, M., G. Jünger, and G. Reinelt (1988). "An Application of Combinatorial Optimization to Statistical Physics and Circuit Layout Design," *Operations Research*, **18**, 493–513.

[5] J.F. Benders (1962). "Partitioning procedures for solving mixed-variables programming problems," *Numerische Mathematik*, **4**, 238–252.

[6] D. Beyer and R. Ogier (1991). "Tabu learning: a neural network search method for solving nonconvex optimization problems," *Proceedings of the International Joint Conference on Neural Networks*. IEEE and INNS, Singapore.

[7] A.L. Brearly, G. Mitra and H.P. Williams (1975). "Analysis of mathematical programming problems prior to applying the simplex method," *Mathematical Programming*, **8**, 54–83.

[8] S. Chopra, E. Gorres and M.R. Rao (1992). "Solving the Steiner Tree Problem on a Graph Using Branch and Cut," *ORSA Journal on Computing*, **4**, 320–335.

[9] M.W. Cooper and K. Farhangian (1985). "Multicriteria Optimization for Nonlinear Integer-variable Problems," *Large Scale Systems*, **9**, 73–78.

[10] G. Cornuejols, G.L. Nemhauser, and L.A. Wolsey (1980). "Worst Case and Probabilistic Analysis of Algorithms for a Location Problem," *Operations Research*, **28**, 847–858.

[11] M.L. Fisher (1981). "The Lagrangian Method for Solving Integer Programming Problems," *Management Science*, **27**, 1–18.

[12] M.R. Garey and D.S. Johnson (1979). *Computers and Intractibility: A Guide to the Theory of NP-Completeness*. W.H. Freeman, San Fransisco, California.

[13] F. Glover and M. Laguna (1992). "Tabu Search," a chapter in *Modern Heuristic Techniques for Combinatorial Optimization*.

[14] R.E. Gomory (1958). "Outline of an Algorithm for Integer Solution to Linear Program," *Bulletin American Mathematical Society*, **64**, 275–278.

[15] R.E. Gomory (1960). "Solving Linear Programming Problems in Integers," *Combinatorial Analysis* (R.E. Bellman and M. Hall, Jr. eds, American Mathematical Society), 211–216.

[16] M. Grötschel (1992). "Discrete mathematics in manufacturing," Preprint SC92–3, ZIB.

[17] M. Grötschel, L. Lovasz, and A. Schrijver (1988). *Geometric Algorithms and Combinatorial Optimization*, Springer, Berlin.

[18] M. Grötschel, C.L. Monma, and M. Stoer (1989). "Computational results with a cutting plane algorithm for designing communication networks with low-connectivity constraint," Report No. 187, Schwerpunktprogramm der Deutschen Forschungsgemeinschaft, Universität Augsburg.

[19] M. Guignard and K. Spielberg (1981). "Logical Reduction Methods in Zero-one Programming: Minimal Preferred Inequalities," *Operations Research*, **29**, 49–74.

[20] M. Guignard and S. Kim (1987). "Lagrangian decomposition: a model yielding stronger Lagrangian bounds," *Mathematical Programming*, **39**, 215–228.

[21] P. Hansen (1986). "The steepest ascent mildest descent heuristic for combinatorial programming," *Proceedings of Congress on Numerical Methods in Combinatorial Optimization*, Capri, Italy.

[22] K.L. Hoffman and M. Padberg (1985). "LP-based Combinatorial Problem Solving," *Annals Operations Research*, **4**, 145–194.

[23] K.L. Hoffman and M. Padberg (1991). "Improving the LP-representation of Zero-one Linear Programs for Branch-and-Cut," *ORSA Journal Computing*, **3**, 121–134.

[24] K.L. Hoffman and M. Padberg (1993). "Solving Airline Crew Scheduling Problems by Branch-and-Cut," *Management Science*, **39**, 657–682.

[25] E.L. Johnson and S. Powell (1978). "Integer programming codes," *Design and Implementation of Optimization Software* (ed. H.J. Greenberg), NATO Advanced Study Institute Series, Sijthoff & Noordhoff, 225–248.

[26] R.M. Karp (1976). "Probabilistic analysis of partitioning algorithms for the traveling salesman problem," in *Algorithms and Complexity: New Directions and Recent Results* (J.F. Traub, ed.) Academic Press, New York, 1–19.

[27] A.H. Land and A.G. Doig (1960). "An automatic method for solving discrete programming problems" *Econometrica*, **28**, 97–520.

[28] T.L. Magnanti and R. Vachani (1990). "A Strong Cutting Plane Algorithm for Production Scheduling with Changeover Costs," *Operations Research*, **38**, 456–473.

[29] S. Martello and P. Toth (1990). *Knapsack Problems*, John Wiley, New York.

[30] H. Markowitz and A. Manne (1957). "On the solution of discrete programming problems," *Econometrica*, **25**, 84–110.

[31] H. Mühlenbein (1992). "Parallel genetic algorithms in combinatorial optimization," *Computer Scienceand Operations Research* (ed. by Osman Blaci), Pergamon Press, New York.

[32] G.L. Nemhauser and L.A. Wolsey (1988). *Integer and Combinatorial Optimization*, John Wiley, New York.

[33] M. Padberg (1979). "Covering, packing and knapsack problems," *Mathematical Programming*, **47**, 19–46.

[34] M. Padberg and G. Rinaldi (1991). "A Branch-and-Cut Algorithm for the Resolution of Large-scale Symmetric Traveling Salesman Problems," *SIAM Review*, **33**, 60–100.

[35] R.G. Parker and R.L. Rardin (1988). *Discrete Optimization*, Academic Press, San Diego.

[36] Y. Pochet, and L.A. Wolsey (1991). "Solving Multi-item Lot Sizing Problems Using Strong Cutting Planes," *Management Science*, **37**, 53–67.

[37] A.H.G. Rinooy Kan (1986). "An introduction to the analysis of approximation algorithms," *Discrete Applied Mathematics*, **14**, 111–134.

[38] A. Schrijver (1984). *Linear and Integer Programming*, Wiley, New York.

[39] T.J. VanRoy and L.A. Wolsey (1987). "Solving Mixed Integer Programming Problems Using Automatic Reformulation," *Operations Research*, **35**, 45–57.

[40] H. Weingartner (1963). *Mathematical Programming and the Analysis of Capital Budgeting Problems*, Prentice Hall, Englewood Cliffs, New Jersey.

[41] H. Weyl (1935). "Elementare theorie der konvexen polyheder," *Comm. Math. Helv*, **7**, 290 (Translated in *Contributions to the Theory of Games*, **1**, 3, 1950).

[42] H.P. Williams (1985). *Model Building in Mathematical Programming*, 2nd ed. Wiley, New York.

COMBINATORIAL EXPLOSION

The phenomenon associated with optimization problems whose computational difficulty increases exponentially with the size of the problem. One common paradigm is the traveling salesman problem. See **Combinatorics; Combinatorial and integer optimization; Curse of dimensionality; Traveling salesman problem**.

COMMON RANDOM VARIATES

See **Monte Carlo sampling and variance reduction**.

COMBINATORICS

E. L. Lawler

University of California at Berkeley

Combinatorics is the branch of mathematics that deals with arrangements of objects, usually finite in number. The term *arrangement* encompasses, among other possibilities, selection, grouping, combination, ordering or placement, subject to various constraints.

Elementary combinatorial theory concerns permutations and combinations. For example, the number of permutations or orderings of n objects is $n! = n(n-1)\ldots(2)(1)$, and the number of combinations of n objects taken k at a time is given by the binomial coefficient $\binom{n}{k} = n!/k!(n-k)!$. In order to compute the probability of throwing a 7 with two dice, or of drawing an inside straight at poker, one must be able to count permutations and combinations, as well as other types of arrangements. Indeed, combinatorics is said to have originated with investigations of games of chance. Combinatorial counting theory is the foundation of discrete probability theory as we know it today.

Experimental design provides the motivation for another classic area of combinatorial theory. Suppose five products are to be tested by five experimental subjects over a period of five days, with each subject testing one product per day. Labeling the subjects *A,B,C,D,E*, the products *1,2,3,4,5*, and the days *M,Tu,W,Th,F*, one way to schedule the tests is shown below:

	M	Tu	W	Th	F
1	A	B	C	D	E
2	B	C	D	E	A
3	C	D	E	A	B
4	D	E	A	B	C
5	E	A	B	C	D

A square array of symbols, with each symbol occurring in each row exactly once and in each column exactly once, is called a *Latin square*.

Now suppose each of the tests is to be performed by a subject in the presence of an observer. In order to reduce the effects of bias due to subject-observer interactions, we should like the Latin square representing the schedule for the subjects to be *combinatorially orthogonal* to the Latin square for the observers. This means that when the two Latin squares are superimposed, each of the 25 possible subject-observer pairs appears exactly once in the resulting array, called a *Graeco-Latin square*. Labeling the observers *a,b,c,d,e*, a *5 × 5* Graeco-Latin for our experiment is as shown below.

Aa	Bb	Cc	Dd	Ee
Bc	Cd	De	Ea	Ab
Ce	Da	Eb	Ac	Bd
Db	Ec	Ad	Be	Ca
Ed	Ae	Ba	Cb	Dc

Leonhard Euler observed that no *2 × 2* Graeco-Latin square exists and found he was able to construct examples of *n × n* Graeco-Latin squares for *n* up to 5, but had trouble with *6*. In 1782 Euler conjectured the nonexistence of such an arrangment for any *n = 4k + 2*, where *k* is an integer. About 1900, Euler's conjecture was comfirmed, by systematic examination of cases, for *n = 6*. However, his more general conjecture remained unsettled until 1959 when Bose, Shirkhande and

Parker exhibited a *22 × 22* Graeco-Latin square. Shortly after, these same investigators ("Euler's Spoilers") demolished what remained of Euler's conjecture by establishing that Graeco-Latin squares do exist for all *n* other than *2* and *6*. Their work made use of results of number theory, a branch of mathematics with which combinatorics exists in happy symbiosis.

Another investigation of Euler turned out to have considerable importance for combinatorial mathematics. In the old city of Königsberg in Eastern Prussia the River Pregel divided into two branches surrounding an island. The river was spanned by seven bridges. It is said that the people of Königsberg entertained themselves by trying to find a route around the city that would cross each of the bridges exactly once. In 1736, Euler provided a definitive answer to the Königsberg bridge problem, and any related instances: "If there are no more than two areas to which an odd number of bridges lead, then such a journey is not possible. If, however, the number of bridges is odd for exactly two areas, then the journey is possible if it starts in either of these areas. If, finally, there are no areas to which an odd number of bridges leads, then the required journey can be accomplished from any area." This result has been viewed as the oldest theorem of what is now known as graph theory.

With the advent of digital computers and operations research, the emphasis of combinatorics shifted from problems of counting and existence of arrangements to problems of optimization. Modern combinatorics may be said to have come of age with the development of network flow theory by Lester Ford and Ray Fulkerson in the 1950s. This remarkable theory enables a great variety of practical optimization problems to be solved by efficient algorithms. More theoretically, a number of elegant duality results follow directly from Ford and Fulkerson's Max-Flow Min-Cut Theorem. As one example, consider the König-Egervary Theorem, which can be stated as follows: Let us call a subset of elements of a matrix *independent* if no two of the elements lie in the same row or the same column. Then the maximum size of an independent set of 1s is equal to the minimum number or rows and columns containing all the 1s in the matrix.

In the 1960s Jack Edmonds generalized many of the results of Ford and Fulkerson by exploiting the concept of a *matroid*, a combinatorial structure abstracting the notion of linear independence. Edmonds also developed a general theory of matching in graphs, where a *matching* is a subset of edges, no two of which are incident to the same vertex. He also proved a generalization of the König-Egervary Theorem, which may be viewed as a duality theorem for matchings in the special case of bipartite graphs.

Edmonds (1965) further observed that the running time of his general matching algorithm was bounded by a polynomial in the size of the graph it is applied to, and made an eloquent argument for the goodness of polynomial-time bounded algorithms. The significance of polynomial time bounds came to be more fully appreciated with the development of NP-completeness theory by Stephen Cook, Richard Karp and Leonid Levin in 1973. The theory of NP-completeness has been an essential tool for researchers in combinatorial optimization ever since.

Algorithms arising from network flow theory, matroid optimization theory, matching theory, or similar theories, may all be viewed as special-purpose linear programming algorithms. Combinatorial duality results, including the Max-Flow Min-Cut Theorem and the König-Egervary Theorem, are most often special cases of linear programming duality. The term applied to the general paradigm of formulating and solving combinatorial problems by linear programming techniques is *polyhedral combinatorics*.

More often than not, combinatorial optimization problems that arise in the real world are too idiosyncratic and complicated to be fully tamed by polyhedral techniques alone.

For these problems, it is usually necessary to engage in some form of enumeration of cases if one seeks to find a provably optimal solution. The Traveling Salesman Problem (TSP) is prototypical of a difficult (NP-complete) problem with a real-world flavor. In this problem, one is asked to find a shortest closed tour of n cities (visiting each city exactly once, and ending at the starting point), given an $n \times n$ matrix of intercity distances. The number of possible tours is, of course, finite: $(n-1)!$. But for any interesting value of n, say *100* or *1,000*, the number of tours is so astronomically large as to be effectively infinite. An exhaustive enumeration of even a tiny fraction of the tours is out of the question. Hence if the TSP is to be solved by enumeration, the enumeration must be very artfully limited.

The TSP has served as a testbed for algorithmic research. Indeed, the approaches that have been applied to the TSP are representative of the full range of techniques of combinatorial optimization. These include polyhedral and integer linear programming, Lagrangian relaxation, nondifferentiable optimization, heuristic and approximation algorithms, branch-and-bound, dynamic programming, neighborhood search, and simulated annealing. With much effort by many investigators, it is today possible to find optimal, or

provably near-optimal, solutions to instances of the TSP with hundreds, even thousands of cities.

Combinatorial optimization has assumed great practical importance, in such diverse problem areas as machine scheduling and production planning, vehicle routing, plant location, network design, VLSI design, among many others. The practical and theoretical importance of this field can only be expected to grow in the future.

See **Chinese postman problem; Combinatorial and integer optimization; Computational complexity; Graph theory; Matching; Networks; Traveling salesman problem; Vehicle routing.**

References

[1] N.L. Biggs, E.K. Lloyd, and R.J. Wilson (1976), *Graph Theory: 1736–1936*, Oxford Univ. Press.

[2] J. Edmonds (1965), "Paths, Trees, and Flowers," *Canad. Jl. Math.*, 17, 449–467.

[3] M.R. Garey and D.S. Johnson (1979), *Computers and Intractability: A Guide to NP-Completeness*, W.H. Freeman, San Franciso.

[4] R.L. Graham, B.L. Rothchild, J.H. Spencer (1980), *Ramsey Theory*, John Wiley, New York.

[5] E.L. Lawler (1976), *Combinatorial Optimization: Networks and Matroids*, Holt, Rinehart and Winston, New York.

[6] E.L. Lawler, J.K. Lenstra, A.H.G. Rinnooy Kan, and D.B. Shmoys, eds. (1985), *The Traveling Salesman Problem: A Guided Tour of Combinatorial Optimization*, John Wiley, New York.

[7] L. Lovasz (1979), *Combinatorial Problems and Exercises*, North Holland, Amsterdam.

[8] L. Lovasz and M.D. Plummer (1986), *Matching Theory*, North Holland, Amsterdam.

[9] G.L. Nemhauser and L.A. Wolsey (1988), *Integer Programming and Combinatorial Optimization*, John Wiley, New York.

[10] A. Schrijver (1986), *Theory of Linear and Integer Programming*, John Wiley, New York.

[11] R.J. Wilson and J.J. Watkins (1990), *Graphs: An Introductory Approach*, John Wiley, New York.

COMMON RANDOM VARIATES

See **Monte Carlo sampling and variance reduction.**

COMMON VALUE BIDDING MODEL

A bidding model in which the value of what is being auctioned, while unknown at the time of the auction, is known to be the same for all bidders. In such a model, bidders must correct for the selection bias, often called the "winner's curse," caused by the fact that winning bidder is likely to have been the one who most overestimated the value. See **Bidding**.

COMMUNICATIONS NETWORKS

Edward A. Sykes

Make Systems Laboratories, Carey, North Carolina

INTRODUCTION: Communications networks are systems of electronic and optical devices which support information exchange among their subscribers. Examples of communications networks are abundant in everyday life: telephone networks, broadcast and cable television networks, and computer communications networks such as the Internet. The impacts of communications networking on the individual, society and the planet are staggering, rivaling that of the tall ship and the automobile. In just under two centuries, humanity has been transformed from myriad villages and towns isolated in obscure corners of the continents to one "global information village." This transformation is no more evident than in the fact that the very boundaries between information transfer and information processing are increasingly hard to define. The integration of communications networks, computing technology, and end-user devices (e.g., the telephone, television, personal computer) is increasingly being referred to simply as the "information infrastructure."

OR and MS have been major players in the development, deployment and management of information technologies and infrastructure. Applications of OR/MS in modeling, analysis and design of communications networks are among the oldest of the fields, dating from the late nineteenth and early twentieth century. Among the most notable of all work in OR/MS history is queueing modeling of telephony by A. K. Erlang. Modeling, analysis and design of communications networks, moreover, is an area rich in applications of more generic OR/MS work. Communications networks are, fundamentally, networks and thus, almost all generic discussion of networks applies. Analogous remarks are appropriate: in communications network modeling and analysis for topics such as queueing and queueing networks, simulation, and network reliability; and in communications network design for topics such as facility location, topological design and optimization, capacity optimization and allocation. Finally, communications networking problems have a great deal of commonality with problems arising in other domains, for example, modeling, analysis and design of transportation systems, water resource distribution systems, etc.

A discussion of the wealth of communications networking issues arising in the application of OR/MS techniques would be quite extensive. Here we focus on several classes of modeling, analysis and design problems arising in a variety of modern communications technologies.

BASIC STRUCTURE AND CONCEPTS: A typical communications network comprises a set of *subscribers* that offer subscriber-to-subscriber *traffic requirements* to be supported on the given network *architecture*. For example, a typical household (subscriber) makes telephone calls (traffic requirements) to be supported on a voice network switching fabric (architecture). In most communications architectures, a hierarchy of communications devices exist to support traffic, but the most basic of these are *customer premises equipment*, *local access equipment* and *switching equipment*. Customer premises equipment is associated directly or indirectly with the generation of traffic requirements. Local access equipment provides a means of connecting the subscriber to the network, that is, the interface between the subscriber and the network necessary for traffic to enter the network and be routed over it. Switching equipment routes the traffic from its source subscriber to its destination subscriber.

All three types of equipment are determined by the nature of the traffic requirements and their associated technology and architecture. In a voice (i.e., telephone) network, the customer premises equipment is generally just a telephone – in this case, the "subscriber" is the household whose aggregate traffic (telephone calls) enters and leaves the network at the telephone. The local access equipment in this case is owned and provided by the local telephone company. Although there typically is switching in the local access in this case (for local calls), for purposes of the discussion here, the long haul switching equipment is owned and provided by a common carrier such as AT&T. Analogous examples can be provided for data communications networks, videotele conferencing networks, etc.

Communications networks differ on the manner in which they carry traffic requirements. Most voice networks set-up calls from source to destination in a circuit switched manner, that is, dedicating capacity along the entire path of the call. Most data networks segment information into streams of packets which are routed independent from one another from source to destination and reassembled into the original information at the destination. Many variations and hybrids of these basic approaches exist and the evolution of technology is becoming increasingly toward supporting traffic sources with differing traffic characteristics and differing service requirements differently. For example, voice traffic is error tolerant (one can tolerate a little static on the line) but delay sensitive (one cannot tolerate long delays between the time a word is spoken and the time it is received at the destination). Some data traffic (e.g. file transfer) is typically error intolerant but delay insensitive. Consideration of these kinds of issues is addressed in network modeling and simulation.

A common thread among most network modeling, analysis and design conceptualizations is the view of a network as a graph comprising nodes and links. A node is used to abstractly represent a device location (e.g., a subscriber location or a switching location). A link is used to represent connections between subscribers and switches and between switches. A link typically has a capacity for supporting traffic. One can view a link as analogous to a pipe and the capacity of the link as analogous to the diameter of the pipe, but with one caveat. A communications link of a given capacity typically supports traffic at that capacity in both directions, that is, it is more properly viewed as two pipes of equal capacity in parallel, each flowing in a direction opposite the other. Design of communications networks typically addresses selecting the number of and the placement of backbone (central) nodes, selecting and sizing the links between subscribers and backbone nodes, and selecting and sizing the links between pairs of backbone nodes.

MODELING: Communications networks are large scale systems with enormous complexity. As with most such systems, modeling relies heavily on computer-based techniques and the nature of the models developed depends strongly on the questions the model is intended to answer. For example, a simulation may be used to answer detailed questions regarding the interaction of communications devices or protocols. Often these studies address questions as to the feasibility of a given device or protocol to support certain types of traffic requirements with acceptable performance. Such models can be used to design the devices or protocols as well. Simula-

tion of communications systems typically models the generation, transfer, and disposition of each unit of information (e.g., call, packet, cell), the protocol decisions as the system operates and the physical behavior of the devices that make up the network. As with any simulation, various aspects of the system may be ignored or aggregated to improve the computational speed of the simulation.

An alternative to simulation approaches is analytical modeling (Kleinrock, 1976), which typically implicitly aggregates traffic units into flows whose characteristics are captured using statistical or probabilistic models. The advantage of analytical modeling is that the behavior of a network can be predicted by a system of equations more quickly computed than the operation of the network can be simulated. The disadvantage is in the aggregation and averaging of detail, effectively capturing the behavior of the network on average rather than accurately depicting a realization of performance over time. Most analytical models of communications systems employ individual and network queueing models. Information units (calls, packets, cells) are the customers in these queueing systems and communications devices (switches, links, etc.) are the servers.

Hybrid simulation/analytical modeling is a third and increasingly popular approach to communications network modeling (Sage and Sykes, 1994). The tenets of this approach are to use simulation techniques in capturing key protocol decisions in traffic admission, routing, congestion control, and resource allocation, but to use analytical techniques for modeling the behavior of the traffic itself, thus avoiding the computational complexity incurred if each packet or cell were to be simulated individually. Hybrid simulation/analytical models of communications networks also have been described as "flow-based simulations," in which the paths that traffic flows take are simulated while the flows themselves are modeled analytically.

Selection of modeling approach depends strongly on the purpose to which the model is applied. For purposes of protocol or device design, where many replications of realizations of performance are required to observe the entity under a wide variety of operational conditions and circumstances, simulation approaches dominate. For analysis and design purposes, where often the intent is to assess the quality of the design or to compare alternative designs, models which provide average behavior over many potential realizations of performance are useful. Performance can be computed over multiple simulation replications, but analytical tools or simulation analytical hybrids which compute those averages directly and more efficiently are dominant.

ANALYSIS: Network analysis is the application of one or more network models to characterize a communications network. In many communications network design contexts, the central step of the design process is to characterize a design on a number of categories of measures: cost, topological properties, and performance being the major ones. For each of these categories of measures, models which compute specific measures of interest can be applied, with the aggregate network analysis being produced in summary from the results of the individual models. Cost measures can include one-time (e.g., device purchase) and recurring costs (e.g., link leasing), often commensurated to the same units. Topological measures are generally technology independent characterizations of the network structure along gross lines (e.g., measures summarizing path availability and diversity, path lengths in number of links or hops from source to destination, etc.). Performance measures are generally technology dependent characterizations of the ability of the network to support the offered traffic and the quality of that support.

DESIGN: A common paradigm for design of communications networks is one in which the design process is broken down into two phases: *access area design* and *backbone design* (Boorstyn and Frank, 1977). Access area design determines the number and location of backbone nodes and homes (i.e., provides a link from) each subscriber to a backbone node. Backbone design determines the interconnections among (links between) backbone nodes. The process is depicted in Figure 1. Figure 1(a) represents the starting point, where the subscriber locations (black circles) and candidate backbone node locations (grey squares) are given. Figure 1(b) represents the completion of the access area design phase, where the black squares are the selected backbone nodes and the lines from the subscribers to the backbone nodes are the homings (implicit in the homings is the assumption that the communications links from each subscriber to its switch is of type and capacity to support the subscriber's offered traffic requirements). The output of the access area design phase is the input to the backbone design phase: the number and location of backbone nodes and the aggregate traffic requirements among the backbone nodes. The aggregate traffic is computed based on the homings. In the backbone phase, the interconnections among backbone nodes are designed to support the backbone traffic with adequate performance, to meet other constraints, and typically to minimize cost. Figure 1(c) depicts a backbone design and Figure 1(d) depicts the final overall solution.

It is notable that solution of the global design problem (including all access and backbone components) is precluded by the computational complexity of the design problem for all but a few special cases which will be ignored here. It also is notable that the structure of the decomposition of the global problem into access area and backbone design phases can lead to gross suboptimalities in the overall solution. To illustrate this assertion, consider a global design problem in which the total cost of the network includes three components:

● homing link costs, the sum of the costs of links homing subscribers to nodes, which can vary for each subscriber-node pair;

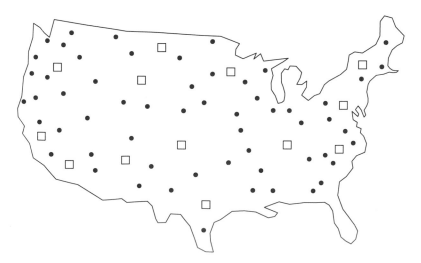

Fig. 1(a). Network design – starting point.

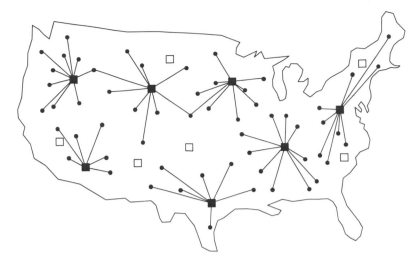

Fig. 1(b). Network design – access area design.

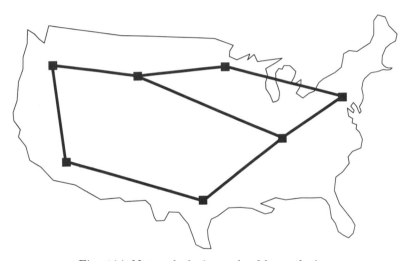

Fig. 1(c). Network design – backbone design.

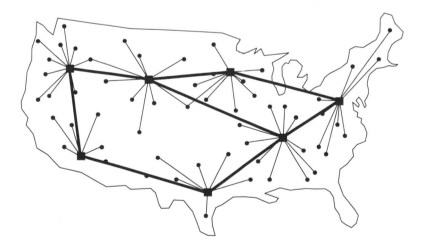

Fig. 1(d). Network design – integrated solution.

- backbone node costs, typically the cost associated with purchasing each node selected as part of the backbone, which typically is uniform over all candidate nodes; and
- backbone link costs, the sum of the costs of links between backbone nodes, which can vary for each subscriber-node pair.

Under fairly general assumptions, the following relationships hold as the number of nodes selected for the backbone increases:

- the access area homing costs tend to decrease (because the access links tend to decrease in length and hence cost);
- the node activation costs increase linearly (directly with the number of nodes selected); and
- the backbone link costs tend to increase (as the number of backbone nodes increases, more backbone links are required).

Thus, if the access area design phase optimizes solely on the basis of homing and node activation costs, it tends to select too many nodes. Two remedies to this pathology are commonly employed: (i) using some estimate of the backbone cost in the access area design problem; or (ii) iterating on the number of backbone nodes selected (i.e., fixing the number of activated nodes to given number, solving the access area and backbone problems in sequence for that number, and computing the total solution cost, but doing so over a wide range of numbers of nodes and selecting the best total cost solution obtained). Neither remedy guarantees global optimality, but both approaches can improve solution costs substantially.

Access area design problems often are formulated as 0-1 integer programming problems (Fischer *et al.*, 1993) that are strongly related to discrete location problems and/or facility location problems generally (Mirchandani and Francis, 1990). In many cases, these integer programs are too large to be solved directly, so a variety of solution approaches are used; e.g. linear programming relaxation methods, Lagrangian relaxation methods, cutting plane and column generation methods, etc. (Ahuja *et al.*, 1993). Alternatively, heuristic algorithms can be used to solve the access problem, and perhaps, more generally, clustering techniques can be used as a solution approach. The basic access area design problem can be stated as follows:

Given: Subscriber-to-subscriber traffic requirements
Candidate node locations

Minimize: Sum of costs of homing + Sum of node
each subscriber   activation
to a backbone node   costs

Over: Node activations
Subscriber homings

Subject To: Node port constraints (limit on the number of subscriber, than can be homed to a node)

Node traffic constraints (limit on the total amount of subscriber traffic that can be homed to a node)

Each subscriber must be homed to a node (occasionally subscribers must be homed to more than one node)

(Optionally, a constraint fixing the number of node activations)

Backbone design problems can be formulated as 0-1 or general integer programming problems (Gavish, 1986), however, it is difficult if not impossible to accurately capture or predict network performance in that context. Moreover, many of the critical aspects of the backbone problem that can be captured in the integer programming formulation (e.g., topological constraints) can also cause a combinatoric explosion in its solution time. Nonetheless, OR/MS literature is replete with many IP backbone design formulations. In these cases, the solution techniques again typically rely on LP relaxation or Lagrangian relaxation approaches.

An alternative to the mathematical programming approach to backbone design is commonly employed in interactive software based tools for solving design problems (Stiffler and Sykes, 1990; Monma and Shallcross, 1989). This iterative approach:

- starts the design process with an initial design;
- analyzes the design using a series of models assessing measures of various aspects of cost, topological properties, performance, and physical constraints on feasibility;
- makes an assessment as to whether the design is satisfactory, stopping if so; and if not
- improves one or more design deficiencies and returns to the analysis step.

This iterative paradigm for backbone design has been used extensively and successfully for design of communications networks with a wide range of architectures (e.g., voice, packet data, multiplexer, asynchronous transfer mode). It also can capture directly a broader set of design objectives and constraints than mathematical programming methods, as well as be implemented in ways which more accurately predict network performance. All of this is possible through the embedding of the comprehensive network analysis at the core of the process, along with the decomposition of the optimization process into smaller steps aimed at initial design generation and design improvement. Unlike mathematical programming approaches, which often can be solved to optimality or at least provide bounds from optimality for the solutions they produce, iterative approaches typically cannot guarantee nor bound optimality.

A typical backbone design problem can be stated as follows:

Given: Backbone node-to-node traffic requirements
Node locations
Link availability and costing

Minimize: Sum of link costs

Over: Link Placement

Subject To: *Topological Constraints:*

Node connectivity (lower bound on the number of node-disjoint paths available between each node pair)

Diameter (upper bound on the minimum number of links a node pair must traverse in order to communicate)

Node port degree (upper bound/physical limit on the number of links that can be incident to each node)

Performance Constraints:

Constraints on throughput, utilization, delay, blocking, etc., as appropriate for a given network architecture.

REMARKS: For an overview of telecommunications systems and their design, see Schwartz (1987), Tanenbaum (1987), Bertsekas and Gallager (1987) or Kershenbaum (1993). For a classical introduction to data communications networking, see Kleinrock (1976), which contains extensive modeling and optimization discussions. See Partridge (1994) for a discussion of the emerging technologies of communications networking.

See **Combinatorial and integer optimization; Networks; Networks of queue; Queueing theory.**

References

[1] Ahuja, R.K., T.L. Magnanti and J.B. Orlin (1993). *Network Flows*. Prentice Hall, Englewood Cliffs, New Jersey.
[2] Bersekas, D. and R. Gallager (1987). *Data Networks*. Prentice Hall, Englewood Cliffs, New Jersey.
[3] Boorstyn, R.R. and H. Frank (1977). "Large Scale Network Topological Optimization," *IEEE Transactions on Communications*, COM-25, 29–47.
[4] Fischer, M.J., G.W. Swinsky, D.P. Garland, and L.E. Stanfel (1993). "A Methodology for Designing Large Private Line Transmission Networks with Multiple Facilities," *Telecommunication Systems*, 1, 243–261.
[5] Gavish, B. (1986). "A General Model for the Topological Design of Communications Networks," *Proceedings GLOBCOM '86*, 1584–1588.
[6] Kershenbaum, A. (1993). *Telecommunications Network Design Algorithms*. McGraw-Hill, New York.
[7] Kleinrock, L. (1975). *Queueing Systems Volume I: Theory*. John Wiley, New York.
[8] Kleinrock, L. (1976). *Queueing Systems Volume II: Computer Applications*. John Wiley, New York.
[9] Mirchandani, P.B. and R.L. Francis, editors (1990). *Discrete Location Theory*. John Wiley, New York.
[10] Monma, C.L. and D.F. Shallcross (1989). "Methods for Designing Communications Networks with Certain Two-Connected Survivability Constraints," *Operations Research*, 37, 531–541.
[11] Partridge, C. (1994). *Gigabit Networking*. Addison-Wesley, Reading, Massachusetts.
[12] Sage, K.M. and E.A. Sykes (1994). "Evaluation of Routing-Related Performance for Large Scale Packet-Switched Networks with Distributed, Adaptive Routing Policies," *Information and Decision Technologies*, 19, 545–562.
[13] Schwartz, M. (1987). *Telecommunication Networks, Protocols, Modeling and Analysis*. Addison-Wesley, Reading, Massachusetts.
[14] Stiffler, J.A. and E.A. Sykes (1990). "An AI/OR Hybrid Expert System for Data Network Design," *Proceedings of the 1990 IEEE International Conference on Systems, Man and Cybernetics*, 307–313.
[15] Tanenbaum, A.S. (1988). *Computer Networks*. Prentice Hall, Englewood Cliffs, New Jersey.

COMMUNITY OPERATIONS RESEARCH

An application area of OR/MS in which the client is a community group. Such organizations are defined by 1) existence to protect or advance the interests of their members; 2) absence of managerial hierarchy; 3) possession of scant physical or financial resources; and 4) operation through consensus or democratic forms. See **Problem structuring methods.**

COMPLEMENTARITY CONDITION

A relation between two nonnegative vectors in which, whenever a given component of one of the vectors is positive, the corresponding component of the other vector must be zero. For example, two nonnegative n-dimensional vectors X and Y satisfy a complementarity condition if their ith components are such that $x_i y_i = 0$, $i = 1, \ldots, n$. See **Complementarity problem; Complementary slackness theorem.**

COMPLEMENTARY PIVOT ALGORITHM

See **Quadratic programming**.

COMPLEMENTARITY PROBLEMS

Richard W. Cottle

Stanford University, Stanford, California

DEFINITION: In its most elementary form, a complementarity problem is an inequality system stated in terms of a mapping $f: R^n \to R^n$. Given f, one seeks a vector $x \in R^n$ such that

$$x_i \geq 0, \quad f_i(x) \geq 0, \quad x_i f_i(x) = 0 \quad (i = 1, \ldots, n). \quad (1)$$

When the mapping f is affine, say of the form $f(x) = q + Mx$, the problem (1) is called a *linear complementarity problem*, denoted LCP(q,M) or sometimes just (q,M). Otherwise, it is called a *nonlinear complementarity problem* and is denoted CP(f).

If $\bar{x}$ is a solution to (1) satisfying the additional *nondegeneracy condition* $\bar{x}_i + f_i(\bar{x}) > 0$, $i = 1, \ldots, n$, the indices i for which $\bar{x}_i > 0$ or $f_i(\bar{x}) > 0$ form complementary subsets of $\{1, \ldots, n\}$. This is believed to be the origin of the term *complementary slackness* as used in linear and nonlinear programming. It was this terminology that inspired the name *complementarity problem*.

SOURCES OF COMPLEMENTARITY PROBLEMS: The complementarity problem is intimately linked to the Karush-Kuhn-Tucker necessary conditions of local optimality found in mathematical programming theory. This connection was brought out by Cottle (1964, 1966) and later by Cottle and Dantzig (1968). Finding solutions to such systems was one of the original motivations for studying the subject. Another was the finding of equilibrium points in bimatrix and polymatrix games. This kind of application was emphasized by Howson (1963), and Lemke and Howson (1964). These early contributions also included essentially the first algorithms for these types of problems. There are numerous applications of the linear and nonlinear complementarity problems in computer science, economics, various engineering disciplines, finance, game theory, and mathematics. Descriptions of and references to these applications can be found in the books by Murty (1988), Cottle, Pang and Stone (1992), and Isac (1992).

EQUIVALENT FORMULATIONS: The problem CP(f) can be formulated in several equivalent ways. An obvious one calls for a solution (x,y) to the system

$$y - f(x) = 0, \quad x \geq 0, \quad y \geq 0, \quad x^T y = 0. \quad (2)$$

Another is to find a zero x of the mapping

$$g(x) = \min\{x, f(x)\} \quad (3)$$

where the symbol $\min\{a,b\}$ denotes the componentwise minimum of the two n-vectors a and b. Another equivalent formulation asks for a fixed point of the mapping

$$h(x) = x - g(x),$$

that is, a vector $x \in R^n$ such that $x = h(x)$.

The formulation of CP(f) given in (3) is related to the (often nonconvex) optimization problem:

$$\begin{aligned} \text{minimize} \quad & x^T f(x) \\ \text{subject to} \quad & f(x) \geq 0 \\ & x \geq 0 \end{aligned} \quad (4)$$

In such a problem, the objective is bounded below by zero, thus any feasible solution of (4) for which the objective function $x^T f(x) = 0$ must be a global minimum as well as a solution of CP(f). As it happens, there are circumstances (for instance, the monotonicity of the mapping f) under which all the local minima for the mathematical programming problem (4) must in fact be solutions of (3). A discussion of nonlinear programming attacks on the nonlinear complementarity problem is given in Mangasarian and Solodov (1993) and the references therein.

Also noteworthy is a result of Eaves and Lemke (1981) showing that the LCP is equivalent to solving a piecewise linear system of equations $y = \varphi(x)$ where the mapping $\varphi: R^n \to R^n$ is piecewise linear. In particular, LCP(q,M) is equivalent to finding a vector u such that

$$q + Mu^+ - u^- = 0$$

where, for $i = 1, \ldots, n$, $u_i^+ = \max\{0,u_i\}$ and $u_i^- = -\min\{0,u_i\}$.

THE LINEAR COMPLEMENTARITY PROBLEM: The LCP has quite an extensive literature, far more so than the CP. This is most likely attributable to the LCP's relatively greater accessibility. Within this field of study, there are several main directions: the existence and uniqueness (or number of) solutions, mathematical properties of the problem, generalizations of the problem, algorithms, applications, and implementations.

Much of the theory of the linear complementarity problem is strongly linked in various ways to matrix classes. For instance, one of the earliest theorems on the existence of solutions is due to Samelson, Thrall and Wesler (1958). Motivated by a problem in structural mechanics, they showed that the LCP(q,M) has a unique solution for every $q \in R^n$ if and only if the matrix M has positive principal minors. (That is, the determinant of every principal submatrix of M is positive.) The class of such matrices has come to be known as **P**, and its members are called **P**-matrices. (It is noteworthy that the Samelson-Thrall-Wesler theorem characterizes a class of matrices in terms of the LCP.) The class **P** includes all positive definite (**PD**) matrices, that is, those square matrices M for which $x^T M x > 0$ for all $x \neq 0$. In the context of the LCP, the term **PD** does not require symmetry. An analogous definition (and usage) holds for positive semi-definite (**PSD**) matrices, namely, M is **PSD** if $x^T M x \geq 0$ for all x. Some authors refer to such matrices as *monotone* because of their connection with monotone mappings. **PSD**-matrices have the property that associated LCPs (q,M) are solvable whenever they are feasible, whereas LCPs (q,M) in which $M \in$ **PD** are always feasible and (since **PD** $\subset$ **PSD**) are always solvable. Murty (1968, 1972) gave this distinction a more general matrix form. He defined Q as the class of all square matrices for which LCP(q,M) has a solution for all q and Q_0 as the class of all square matrices for which LCP(q,M) has a solution whenever it is feasible. Although the goal of usefully characterizing the classes q and Q_0 has not yet been realized, much is known about some of their special subclasses. Indeed, there are now literally dozens of matrix classes for which LCP existence theorems have been established (Murty, 1988; Cottle, Pang and Stone, 1992; Isac, 1992).

ALGORITHMS FOR SOLVING LCPS: The algorithms for solving linear complementarity problems are of two major types: pivoting (or, direct) and iterative (or, indirect). Algorithms of the former type and finite procedures that attempt to transform the problem (q,M) to an equivalent system of the form (q',M') in which $q' \geq 0$. Doing this is not always possible; it depends on the problem data, usually on the matrix class (such as **P**, **PSD**, etc.) to which M belongs. When this approach works, it amounts to carrying out a *principal pivotal transformation* on the system of equations

$$w = q + Mz.$$

To such a transformation, there corresponds an index set α (with complementary index set $\bar{\alpha} = \{1, \ldots, n\} \setminus \alpha$) such that the *principal submatrix* $M_{\alpha\alpha}$ is nonsingular. When this (block pivot) operation is carried out, the system

$$w_\alpha = q_\alpha + M_{\alpha\alpha} z_a + M_{\alpha\bar{\alpha}} z_{\bar{\alpha}}$$
$$w_{\bar{\alpha}} q_{\bar{\alpha}} + M_{\bar{\alpha}\alpha} z_\alpha + M_{\bar{\alpha}\bar{\alpha}} z_{\bar{\alpha}}$$

becomes

$$z_\alpha = q'_\alpha + M'_{\alpha\alpha} w_a + M'_{\alpha\bar{\alpha}} z_{\bar{\alpha}}$$
$$w_{\bar{\alpha}} = q'_{\bar{\alpha}} + M'_{\bar{\alpha}\alpha} w_\alpha + M'_{\bar{\alpha}\bar{\alpha}} z_{\bar{\alpha}}$$

where

$$q'_\alpha = -M_{\alpha\alpha}^{-1} q_\alpha \quad M'_{\alpha\alpha} = M_{\alpha\alpha}^{-1} \quad M'_{\alpha\bar{\alpha}} = -M_{\alpha\alpha}^{-1} M_{\alpha\bar{\alpha}}$$
$$q'_{\bar{\alpha}} = q_{\bar{\alpha}} - M_{\bar{\alpha}\alpha} M_{\alpha\alpha}^{-1} q_\alpha \quad M'_{\bar{\alpha}\alpha} = M_{\bar{\alpha}\alpha} M_{\alpha\alpha}^{-1}$$
$$M'_{\bar{\alpha}\bar{\alpha}} = M_{\bar{\alpha}\bar{\alpha}} - M_{\bar{\alpha}\alpha} M_{\bar{\alpha}\alpha}^{-1} M_{\alpha\bar{\alpha}}$$

There are two main pivoting algorithms used in processing LCPs. The more robust of the two is due to Lemke (1965). Lemke's method embeds the LCP (q,M) in a problem having an extra "artificial" nonbasic (independent) variable z_0 with coefficients specially chosen so that when z_0 is sufficiently large, all the basic variables become nonnegative. At the least positive value of z_0 for which this is so, there will (in the nondegenerate case) be (exactly) one basic variable whose value is zero. That variable is exchanged with z_0. Thereafter the method executes a sequence of (almost complementary) simple pivots. In each case, the variable *becoming* basic is the complement of the variable that became nonbasic in the previous exchange. The method terminates if either z_0 decreases to zero – in which case the problem is solved – or else there is no basic variable whose value decreases as the incoming nonbasic variable is increased. The latter outcome is called *termination on a secondary ray*. For certain matrix classes, termination on a secondary ray is an indication that the given LCP has no solution. Eaves (1971) was among the first to study Lemke's method from this point of view.

The other pivoting algorithm for the LCP is called the Principal Pivoting Method (Cottle and Dantzig, 1968; and Cottle, Pang and Stone, 1992). The algorithm has two versions: symmetric and asymmetric. The former executes a sequence of principal (block) pivots of order 1 or 2, whereas the latter does sequences of almost complementary pivots, each of which results in a block principal pivot or order potentially larger than 2. The class of problems to which the Principal Pivoting Method applies is more restrictive. Cottle, Pang and Stone (1992) provide an up-to-date treatment of this algorithm.

Iterative methods are often favored for the solution of very large linear complementarity problems. In such problems, the matrix M tends to be sparse (i.e., to have a small percentage of nonzero elements) and frequently structured. Since iterative methods do not modify the problem data, these features of large-scale problems can be used to advantage. Ordinarily, however, an iterative method does not terminate finitely; instead, it generates a convergent sequence of trial solutions. As is to be expected, the applicability of algorithms in this family depends on the matrix class to which M belongs. Details on several algorithms of this type are presented in the book by Kojima, Megiddo, Noma and Yoshise (1991), as well as the one by Cottle, Pang and Stone (1992).

SOME GENERALIZATIONS: The linear and non-linear complementarity problems have been generalized in numerous ways. One of the earliest generalizations was given by Habetler and Price (1971) and Karamardian (1971) who defined the problem CP(K,f) as that of finding a vector x in the closed convex cone K such that $f(x) \in K^*$ (the dual cone) and $x^T f(x) = 0$. Through this formulation, a connection can be made between complementarity problems and variational inequality problems, that is, problems VI(X,f) wherein one seeks a vector $x^* \in X$ (a nonempty subset of R^n) such that $f(x^*)^T(y - x^*) \geq 0$ for all $y \in X$. Karamardian (1971) established that when X is a closed convex cone, say K, with dual cone K^*, then CP(K,f) and VI(X,f) have exactly the same solutions (if any).

Robinson (1979) has considered the generalized complementarity problem CP(K,f) defined above as an instance of a *generalized equation*, namely to find a vector $x \in R^n$ such that

$$0 \in f(x) + \partial \psi_K(x)$$

where ψ_K is the indicator function of the closed convex cone K and ∂ denotes the subdifferential operator as used in convex analysis.

Among the diverse generalizations of the linear complementarity problem, the earliest appears in Samelson, Thrall and Wesler (1958). There, for given $n \times n$ matrices A and B and n-vector c, the authors considered the problem of the finding n-vectors x and y such that

$$Ax + By = c, \quad x \geq 0, y \geq 0 \quad and \quad x^T y = 0.$$

A different generalization was introduced by Cottle and Dantzig (1970). In this sort of problem, one has an affine mapping $f(x) = q + Nx$ where N is of order $\Sigma_{j=1}^k p_j \times n$ partitioned into k blocks; the vectors q and $y = f(x)$ are partitioned

conformably. Thus,

$$y^j = q^j + N^j x \quad for \quad j = 1, \ldots, k.$$

The problem is to find a solution of the system

$$y = q + Nx, \quad x \geq 0, \quad y \geq 0,$$

$$and \quad x_j \prod_{i=1}^{p_j} y_i^j = 0 \quad (j = 1, \ldots, k).$$

Several authors have further investigated this *vertical generalization* while others have studied some analogous *horizontal generalizations*.

See **Game theory; Matrices and matrix algebra; Nonlinear programming; Quadratic programming.**

References

[1] Cottle, R.W. (1964), *Nonlinear Programs with Positively Bounded Jacobians*, Ph.D. Thesis, Department of Mathematics, University of California, Berkeley. [See also, Technical Report ORC 64-12 (RR), Operations Research Center, University of California, Berkeley.]

[2] Cottle, R.W. (1966), "Nonlinear Programs with Positively Bounded Jacobians," *SIAM Jl. Applied Mathematics* 14, 147–158.

[3] Cottle, R.W. and G.B. Dantzig (1968), "Complementary Pivot Theory of Mathematical Programming," *Linear Algebra and its Applications* 1, 103–125.

[4] Cottle, R.W. and G.B. Dantzig (1970), "A Generalization of the Linear Complementarity Problem," *Jl. Combinatorial Theory* 8, 79–90.

[5] Cottle, R.W., J.S. Pang and R.E. Stone (1992), *The Linear Complementarity Problem*, Academic Press, Boston.

[6] Eaves, B.C. (1971), "The Linear Complementarity Problem," *Management Science* 17, 612–634.

[7] Eaves, B.C. and C.E. Lemke (1981), "Equivalence of LCP and PLS," *Mathematics of Operations Research* 6, 475–484.

[8] Habetler, G.J. and A.J. Price (1971), "Existence Theory for Generalized Nonlinear Complementarity Problems," *Jl. Optimization Theory and Applications* 7, 223–239.

[9] Harker, P.T. and J.S. Pang (1990), "Finite-Dimensional Variational Inequality and Nonlinear Complementarity Problems: A Survey of Theory, Algorithms and Applications," *Mathematical Programming*, Series B 48, 161–220.

[10] Howson, J.T., Jr. (1963), *Orthogonality in Linear Systems*, Ph.D. Thesis, Department of Mathematics, Rensselaer Institute of Technology, Troy, New York.

[11] Isac, George (1992), *Complementarity Problems*, Lecture Notes in Mathematics 1528, Springer-Verlag, Berlin.

[12] Karamardian, S. (1971), "Generalized Complementarity Problem," *Jl. Optimization Theory and Applications* 8, 161–168.

[13] Kojima, M., N. Megiddo, T. Noma and A. Yoshise (1991), *A Unified Approach to Interior Point Algorithms for Linear Complementarity Problems*, Lecture Notes in Computer Science 538, Springer-Verlag, Berlin.

[14] Lemke, C.E. (1965), "Bimatrix Equilibrium Points and Mathematical Programming," *Management Science* 11, 681–689.

[15] Lemke, C.E. and J.T. Howson, Jr. (1964), "Equilibrium Points of Bimatrix Games," *SIAM Jl. Applied Mathematics* 12, 413–423.

[16] Murty, K.G. (1968), *On the Number of Solutions to the Complementarity Problem and Spanning Properties of Complementary Cones*, Ph.D. Thesis, Department of Industrial Engineering and Operations Research, University of California, Berkeley.

[17] Mangasarian, O.L. and M.V. Solodov (1993), "Nonlinear Complementarity as Unconstrained and Constrained Minimization," *Mathematical Programming, Series B* 62, 277–297.

[18] Murty, K.G. (1972), "On the Number of Solutions to the Complementarity Problem and Spanning Properties of Complementary Cones," *Linear Algebra and Its Applications* 5, 65–108.

[19] Murty, K.G. (1988), *Linear Complementarity, Linear and Nonlinear Programming*, Heldermann-Verlag, Berlin.

[20] Robinson, S.M. (1979), "Generalized Equations and Their Solutions, Part I: Basic Theory," *Mathematical Programming, Study* 10, 128–141.

[21] Samelson, H., R.M. Thrall and O. Wesler (1958), "A Partition Theorem for Euclidean n-Space," *Proceedings American Mathematical Society* 9, 805–807.

COMPLEMENTARY SLACKNESS THEOREM

For the symmetric form of the primal and dual problems the following theorem holds: For optimal feasible solutions of the primal and dual (symmetric) systems, whenever inequality occurs in the kth relation of either system (the corresponding slack variable is positive), then the kth variable of its dual is zero; if the kth variable is positive in either system, the kth relation of its dual is equality (the corresponding slack variable is zero). Feasible solutions to the primal and dual problems that satisfy the complementary slackness conditions are also optimal solutions. A similar theorem holds for the unsymmetric primal-dual problems: For optimal feasible solutions of the primal and dual (unsymmetric) systems, whenever the kth relation of the dual is an inequality, then the kth variable of the primal is zero; if the kth variable of the primal is positive, then the kth relation of the dual is equality. This theorem just states the optimality conditions of the simplex method. See **Complementarity condition**; **Complementarity problem**; **Symmetric primal-dual problems**; **Unsymmetric primal-dual problems**.

COMPUTATIONAL COMPLEXITY

Leslie Hall

The Johns Hopkins University, Baltimore, *Maryland*

The term "computational complexity" has two usages which must be distinguished. On the one hand, it refers to an *algorithm* for solving instances of a *problem*: broadly stated, the computational complexity of an algorithm is a measure of how many steps the algorithm will require in the worst case for an instance or input of a given size. The number of steps is measured as a function of that size.

The term's second, more important usage, is in reference to a problem itself. The theory of computational complexity involves classifying problems according to their inherent tractability or intractability – that is, whether they are "easy" or "hard" to solve. This classification scheme includes the well-known classes P and NP; the terms "NP-complete" and "NP-hard" are related to the class NP.

ALGORITHMS AND COMPLEXITY: In order to understand what is meant by the complexity of an algorithm, we must define algorithms, problems, and problem instances. Moreover, we must understand how one measures the size of a problem instance, and what constitutes a "step" in an algorithm. A *problem* is an abstract description coupled with a question requiring an answer; for example, the Traveling Salesman Problem (TSP) is: "Given a graph with nodes and edges and costs associated with the edges, what is a least-cost closed walk (or *tour*) containing each of the nodes exactly once?" An *instance* of a problem, on the other hand, includes an exact specification of the data: for example, "The graph contains nodes 1, 2, 3, 4, 5, and 6, and edges (1,2) with cost 10, (1,3) with cost 14, . . ." and so on. Stated more mathematically, a problem can be thought of as a function p that maps an instance x to an output $p(x)$ (an answer).

An *algorithm* for a problem is a set of instructions guaranteed to find the correct solution to any instance in a finite number of steps; in other words, for a problem p an algorithm is a finite procedure for computing $p(x)$ for any given input x. Computer scientists model algorithms by a mathematical construct called a *Turing machine*, but we will consider a more concrete model here. In a simple model of a computing device, a "step" consists of one of the following operations: addition, subtraction, multiplication, finite-precision division, and comparison of two numbers. Thus if an algorithm requires one hundred additions and 220 comparisons for some instance, we say that the algorithm requires 320 steps on that instance. In order to make this number meaningful, we would like to express it as a function of the size of the corresponding instance, but determining the exact function would be impractical. Instead, since we are really concerned with how long the algorithm will take in the worst case, we formulate a simple function of the input size that is a reasonably tight upper bound on the actual number of steps. Such a function is called the *complexity* or *running time* of the algorithm.

Technically, the *size* of an instance is the number of bits required to encode it. It is measured in terms of the inherent dimensions of the instance (such as the number of nodes and edges in a graph), plus the number of bits required to encode the numerical information in the instance (such as the edge costs). Since numerical data are encoded in binary, an integer C requires about $log_2|C|$ bits to encode and so contributes logarithmically to the size of the instance. The running time of the algorithm is then expressed as a function of these parameters, rather than the precise input size; for example, for the TSP, an algorithm's running time might be expressed as a function of the number of nodes, the number of edges, and the maximum number of bits required to encode any edge cost.

As we have seen, the complexity of an algorithm is only a rough estimate of the number of steps that will be required on an instance. In general – and particularly in analyzing the inherent tractability of a problem – we are interested in an asymptotic analysis: how does the running time grow as the size of the instance gets very large? For these reasons, it is useful to introduce *Big-O notation*. For two functions $f(t)$ and $g(t)$ of a nonnegative parameter t, we say that $f(t) = O(g(t))$ if there is a constant $c > 0$ such that, for all sufficiently large t, $f(t) \leq cg(t)$. The function $cg(t)$ is thus an asymptotic upper bound on f. For example, $100(t^2 + t) = O(t^2)$,

since by taking $c = 101$ the relation follows for $t \geq 100$; however, $0.0001t^3$ is not $O(t^2)$. Notice that it is possible for $f(t) = O(g(t))$ and $g(t) = O(f(t))$ simultaneously.

We say that an algorithm runs in *polynomial time* (is a polynomial-time algorithm) if the running time $f(t) = O(P(t))$, where $P(t)$ is a polynomial function of t. Polynomial-time algorithms are generally (and formally) considered efficient, and problems for which polynomial time algorithms exist are considered "easy." For the remainder of this article, when we use the term "polynomial," we mean as a function of the input size.

THE CLASSES P AND NP: In order to establish a formal setting for discussing the relative tractability of problems, computer scientists first define a large class of problems called *recognition* (or decision) *problems*. This class comprises precisely those problems whose associated question requires the answer "yes" or "no." For example, consider the problem of determining whether an undirected graph is connected (that is, whether there is a path between every pair of nodes in the graph). This problem's input is a graph G consisting of nodes and edges, and its question is, "Is G connected?" Notice that most optimization problems are not recognition problems, but most have recognition counterparts. For example, a recognition version of the TSP has as input both a graph G, with costs on the edges, *and* a number K; and the associated question is, "Does G contain a traveling salesman tour of length less than or equal to K?" In general, an optimization problem is not much harder to solve than its recognition counterpart; one can usually embed the recognition algorithm in a binary search over the possible objective function values to solve the optimization problem with a polynomial number of calls to the embedded algorithm.

The class P is defined as the set of recognition problems for which there exists a polynomial-time algorithm; "P" stands for "polynomial time." Thus, P comprises those problems that are formally considered "easy." The larger problem class NP contains the class P. The term "NP" stands for "nondeterministic polynomial" and refers to a different, hypothetical model of computation, which can solve the problems in NP in polynomial time (for further explanation, see references).

The class NP consists of all recognition problems with the following property: for any "yes"-instance of the problem there exists a polynomial-length "certificate" or proof of this fact that can be verified in polynomial time. The easiest way to understand this idea is by considering the position of an omniscient being (say, Merlin) who is trying to convince a mere mortal that some instance is a yes-instance. Suppose the problem is the recognition version of the TSP, and the instance is a graph G and the number $K = 100$. Merlin knows that the instance does contain a tour with length at most 100, and to convince the mortal of this fact, he simply hands her a list of the edges of this tour. This list is the certificate: it is polynomial in length, and the mortal can easily verify, in polynomial time, that the edges do in fact form a tour with length at most 100.

There is an inherent asymmetry between "yes" and "no" in the definition of NP. For example, there is no obvious, succinct way for Merlin to convince a mortal that a particular instance does NOT contain a tour with length at most 100. In fact, by reversing the roles played by "yes" and "no" we obtain a problem class known as *Co-NP*. In particular, for every recognition problem in NP there is an associated recognition problem in *Co-NP* obtained by framing the NP question in the negative (e.g., "Do *all* traveling salesman tours in G have length *greater* than K?"). Many recognition problems are believed to lie outside both of the classes NP and *Co-NP*, because they seem to possess no appropriate "certificate;" an example would be the problem consisting of a graph G and two numbers K and L, with the question, "Is the number of distinct traveling salesman tours in G with length at most K exactly equal to L?"

NP-COMPLETE PROBLEMS: To date, no one has found a polynomial-time algorithm for the TSP. On the other hand, no one has been able to prove that no polynomial-time algorithm exists for the TSP. How, then, can we argue persuasively that the TSP, and many problems in NP, are "intractable"? Instead, we offer an argument that is slightly weaker but also compelling. We show that the recognition version of the TSP, and scores of other NP problems, are the *hardest* problems in the class NP in the following sense: if there is a polynomial-time algorithm for *any one* of these problems, then there is a polynomial-time algorithm for *every* problem in NP. Observe that this is a very strong statement, since NP includes a large number of problems that appear to be extremely difficult to solve, both in theory and in practice! Problems in NP with this property are called *NP-complete*. Otherwise stated, it seems

highly unlikely that a polynomial algorithm will be found for *any* NP-complete problem, since such an algorithm would actually provide polynomial time algorithms for *every* problem in NP!

The class NP and the notion of "complete" problems for NP were first introduced by Cook (1971). In that paper, he demonstrated that a particular recognition problem from logic, SATISFIABILITY, was NP-complete, by showing directly how every other problem in NP could be encoded as an appropriate special case of SATISFIABILITY. Once the first NP-complete problem had been established, however, it became easy to show that others were NP-complete. To do so requires simply providing a *polynomial transformation* from a known NP-complete problem to the candidate problem. Essentially, one needs to show that the known "hard" problem, such as SATISFIABILITY, is a special case of the new problem; thus, if the new problem has a polynomial-time algorithm, then the known hard problem has one as well.

RELATED TERMS: The term *NP-hard* refers to any problem that is at least as hard as any problem in NP. Thus, the NP-complete problems are precisely the intersection of the class of NP-hard problems with the class NP. In particular, optimization problems whose recognition versions are NP-complete (such as the TSP) are NP-hard, since solving the optimization version is at least as hard as solving the recognition version.

The *polynomial hierarchy* refers to a vast array of problem classes both beyond NP and *Co-NP* and within. There is an analogous set of definitions which focuses on the space required by an algorithm rather than the time, and these time and space definitions correspond in a natural way. There are parallel complexity classes, based on allowing a polynomial number of processors, and there are classes corresponding to randomized algorithms, those that allow certain decisions in the algorithm to be made based on the outcome of a "coin toss." There are also complexity classes that capture the notions of optimization and approximability. The most famous open question concerning the polynomial hierarchy is whether the classes P and NP are the same, that is, $P = ? NP$. If a polynomial algorithm were discovered for any NP-complete problem, then all of NP would "collapse" to P; indeed, most of the polynomial hierarchy would disappear.

In algorithmic complexity, two other terms are heard frequently: *strongly polynomial* and *pseudo-polynomial.* A strongly polynomial-time algorithm is one whose running time is bounded polynomially by a function *only* of the inherent dimensions of the problem and *independent* of the sizes of the numerical data. For example, most sorting algorithms are strongly polynomial, since they normally require a number of comparisons polynomial in the number of entries and do not depend on the actual values being sorted; an algorithm for a network problem would be strongly polynomial if its running time depended only on the numbers of nodes and arcs in the network, and not on the sizes of the costs or capacities.

A *pseudo-polynomial-time* algorithm is one that runs in time polynomial in the dimension of the problem and the *magnitudes* of the data involved (provided these are given as integers), rather than the base-two logarithms of their magnitudes; such algorithms are technically exponential functions of their input size, and are therefore not considered polynomial. Indeed, some *NP*-complete and *NP*-hard problems are pseudo-polynomially solvable (sometimes these are called *weakly NP*-hard or -complete); for example, the *NP*-hard knapsack problem can be solved by a dynamic programming algorithm requiring a number of steps polynomial in the size of the knapsack and the number of items (assuming that all data are scaled to be integers). This algorithm is exponential-time since the input sizes of the objects and knapsack are logarithmic in their magnitudes. However, as Garey and Johnson (1979) observe, "A pseudo-polynomial time algorithm ... will display 'exponential behavior' only when confronted with instances containing 'exponentially large' numbers, [which] might be rare for the application we are interested in. If so, this type of algorithm might serve our purposes almost as well as a polynomial time algorithm." The related term *strongly NP-complete* (or unary *NP*-complete) refers to those problems that remain *NP*-complete even if the data are encoded in unary (or even if the data are "small" relative to the overall input size); consequently, if a problem is strongly *NP*-complete then it cannot have a pseudo-polynomial-time algorithm unless $P = NP$.

See **Combinatorial and integer optimization; Combinatorics.**

References

[1] Cook, S.A. (1971). "The complexity of theorem-proving procedures," *Proc. 3rd Annual ACM Symp. Theory of Computing*, 151–158.

[2] Garey, M.R. and D.S. Johnson (1979). *Computers and Intractability: A Guide to the Theory of NP-Completeness.* W.H. Freeman, New York.

[3] Karp, R.M. (1975). "On the computational complexity of combinatorial problems," *Networks* **5**, 45–68.

[4] Papadimitriou, C.H. (1985). "Computational complexity," in E.L. Lawler, J.K. Lenstra, A.H.G. Rinnooy Kan, and D.B. Shmoys, eds., *The Traveling Salesman Problem: A Guided Tour of Combinatorial Optimization.* Wiley, Chichester.

[5] Papadimitriou, C.H. (1993). *Computational Complexity.* Addison-Wesley, Redwood City, California.

[6] Papadimitriou, C.H. and K. Steiglitz (1982). *Combinatorial Optimization: Algorithms and Complexity.* Prentice-Hall, Englewood Cliffs, New Jersey [Chapters 8 (pp. 156-192), 15, and 16 (pp. 342–405)].

[7] Shmoys, D.B. and E. Tardos (1989). "Computational complexity of combinatorial problems," in L. Lovasz, R.L. Graham, and M. Groetschel, eds., *Handbook of Combinatorics.* North-Holland, Amsterdam.

[8] Stockmeyer, L.J. (1990). "Complexity theory," in E.G. Coffman, Jr., J.K. Lenstra, and A.H.G. Rinnooy Kan, eds., *Handbooks in Operations Research and Management Science; Volume 3: Computation*, Chapter 8. North Holland, Amsterdam.

COMPUTATIONAL GEOMETRY

Isabel Beichl, Javier Bernal and Christoph Witzgall

National Institute of Standards & Technology, Gaithersburg, Maryland

Francis Sullivan

Supercomputing Research Center, Bowie, Maryland

INTRODUCTION: Computational geometry is the discipline of exploring algorithms and data structures for computing geometric objects and their – often extremal – attributes. The objects are predominantly finite collections of points, flats, hyperplanes – "arrangements" –, or polyhedra, all in finite dimensions. The algorithms are typically finite, their *complexity* playing a central role. Emphasis is on problems in low dimensions, exploiting special properties of the plane and 3-space.

A young field – its name coined in the early 1970s – it has since witnessed explosive growth, stimulated in part by the largely parallel development of *computer graphics*, *pattern recognition*, *cluster analysis*, and modern industry's reliance on computer-aided design (CAD) and robotics (Forrest, 1971; Graham and Yao, 1990; Lee and Preparata, 1984). It plays a key role in the emerging fields of automated cartography and computational metrology.

For general texts, consult Preparata and Shamos (1985), O'Rourke (1987), Edelsbrunner (1987), and Agarwal (1991). Pertinent geometrical concepts are presented in Grünbaum (1967).

There are strong connections to operations research, whose classical problems such as finding a *minimum spanning tree*, a maximum-length *matching*, or a *Steiner tree* become problems in computational geometry when posed in Euclidean or related normed linear spaces. The Euclidean *traveling salesman problem* remains *NP*-complete (Papadimitriou, 1977). *Facility location*, and *shortest paths* in the presence of obstacles, are other examples. *Polyhedra* and their extremal properties, typical topics of computational geometry, also lie at the foundation of *linear programming*. Its complexity, particularly in lower dimensions, attracted early computational geometric research, heralding the achievement of linear complexity for arbitrary fixed dimension (Megiddo, 1982, 1984; Clarkson, 1986).

PROBLEMS: A fundamental problem is to determine the *convex hull* conv(S) of a set S of n points in d-dimensional Cartesian space $\Re^d$. This problem has a weak and a strong formulation. Its weak formulation requires only the identification of the extreme points of conv(S). In operations research terms, that problem is well known as (the dual of) identifying redundant constraints in a system of linear inequalities. The strong formulation requires, in addition, characterization of the facets of the polytope conv(S). For dimension $d > 3$, the optimal complexity of the strong convex hull problem in $\Re^d$ is $O(n^{\lfloor d/2 \rfloor})$ (Chazelle, 1991).

Early $O(n \log n)$ methods for delineating convex hulls in the plane – vertices and edges of the convex hull of a simple polygon can be found in linear time – were based on *divide-and-conquer* (Graham and Yao, 1983) and (Preparata and Hong, 1977). In this widely used recursive strategy, a problem is divided into subproblems whose solutions, having been obtained by further subdivision, are then combined to yield the solution to the original problem. Divide-and-conquer heuristics find applications in Euclidean optimization problems such as optimum-length matching (Reingold and Supowit, 1983).

The following *bridge problem* is, in fact, a linear program: given two sets S_1 and S_2 of planar points separated by a line, find two points $p_1 \in S_1$ and $p_2 \in S_2$ such that the line segment $[p_1, p_2]$ is an upper edge of the convex hull conv($S_1 \cup S_2$), bridging the gap between the two sets. Or, through which edge does a given directed line leave the – not yet delineated – convex hull of n points in the plane? As a linear program of fixed dimension 2, the bridge problem can be solved in linear time. Kirkpatrick and Seidel (1986) have used it along with a divide-and-conquer paradigm to devise an $O(n \log m)$ algorithm for the planar convex hull of n points, m of which are extreme.

When implementing a divide-and-conquer strategy, one typically wishes to divide a set of points $S \subset \Re^d$ by a straight line into two parts of essentially equal cardinality, that is, to execute a *ham-sandwich cut*. This can be achieved by finding the median of, say, the first coordinates of the points in S. It is a fundamental result of the theory of algorithms that the median of a finite set of numbers can be found in linear time. The bridge problem is equivalent to a double ham-sandwich cut of a planar set: given a first cut, find a second line quartering the set. Three-way cuts in three dimensions and results about higher dimensions were reported in Dobkin and Edelsbrunner (1984).

The Euclidean *post office problem* is a prototype for a class of proximity search problems encountered, for instance, in the implementation of *Expert Systems*. "Sites" p_i of n "post offices" in $\Re^d$ are given, and the task is to provide suitable preprocessing for efficiently identifying a post office closest to any client location.

Associated with this problem is the division of space into "postal" regions, that is, sets of locations $V_i \subset \Re^d$ closer to "postal" site p_i than to any other site p_j. Each such region V_i around site p_i is a convex polyhedron, whose facets are determined by perpendicular bisectors, that is, (hyper)planes or lines of equal distance from two distinct sites. Those polyhedra form a polyhedral complex covering $\Re^d$ known as a *Voronoi diagram*. The Voronoi diagram and its dual, the *Delaunay triangulation*, are important related concepts in computational geometry.

Once a Delaunay triangulation of a planar set of n sites has been established – an $O(n \log n)$ procedure – a pair of *nearest points* among these sites can be found in linear time. The use of Delaunay triangulations for computational geometric problems was pioneered by Shamos and Hoey (1975).

The problem of efficiently finding for an arbitrary query point p a Voronoi cell V_i such that p is an instance of *point location* in subdivisions. Practical algorithms for locating a given point in a subdivision of the plane generated by n line segments in time $O(\log n)$ requiring preprocessing of order $O(n \log n)$ and storage of size $O(n \log n)$ or $O(n)$, respectively, have been proposed (Preparata, 1990). For point location in planar Voronoi diagrams, Edelsbrunner and Maurer (1985) utilized acyclic graphs and *packing*. A probabilistic approach to the post office problem is given in Clarkson (1985).

Whether a given point lies in a certain simple polygon can be decided by an $O(n)$ process of examining the boundary intersections of an arbitrary ray emanating from the point in question. For convex polygons, an $O(n)$ preprocessing procedure permits subsequent *point inclusion* queries to be answered in $O(\log n)$ time (Bentley and Carruthers, 1980).

Let $h_e(x)$ be the truth function expressing point inclusion in the halfplane to the left of a directed line segment e. Muhidinov and Nazirov (1978) have shown that a polygonal set can be characterized by a Boolean expression of n such functions, one for each edge e of the polygonal set, where each such function occurs only once in the expression. This Boolean expression transforms readily to an algebraic expression for the characteristic function of the polygon. For 3-dimensional polyhedral bodies, Dobkin, Guibas, Hershberger, and Snoeyink (1988) investigated the existence and determination of analogous *constructive solid geometry* (CSG) representations (they may require repeats of halfspace truth functions). In general, CSG representations use Boolean operations to combine primitive shapes, and are at the root of some commercial CAD/CAM and display systems. For a survey of methods for representing solid objects see Requicha (1980).

Given a family of polygons, a natural generalization of point inclusion is to ask how many of those polygons include a query point. This and similar intersection-related problems are subsumed under the term *stabbing*. The classical 1-dimensional stabbing problem involves n intervals. Here the "stabbing number" can be found in $O(\log n)$ time and $O(n)$ space after suitable preprocessing. Similar results hold for special classes of polygons such as rectangles (Edelsbrunner, 1983).

Sweep-techniques rival divide-and-conquer in popularity. *Plane-sweep* or *line-sweep*, for instance, conceptually moves a vertical line from left to right over the plane, registering objects as it passes them. Plane-sweep permits to decide in $O(n \log n)$ time (optimal complexity) whether n line segments in the plane have at least one intersection (Shamos and Hoey, 1976).

Important special cases of the above intersection problem are testing for (self-)intersection of paths and polygons. *Polygon simplicity* can be tested for in linear time by trying to "triangulate" the polygon.

Polygon triangulation, more precisely, decomposing the interior of a simple polygon into triangles whose vertices are also vertices of the polygon, is a celebrated problem of computational geometry. In a seminal paper, Garey, Johnson, Preparata, and Tarjan (1978) proposed an $O(n \log n)$ algorithm for triangulating a simple polygon of n vertices. They used a plane sweep approach for decomposing the polygon into "monotone polygons," which can each be triangulated in linear time. A related idea is to provide a "trapezoidization" of the polygon, from which a triangulation can be obtained in linear time. Chazelle (1990) introduced the concept of a "visibility map," a tree structure which might be considered a local trapezoidization of the polygon, and based on it an $O(n)$ triangulation algorithm for simple polygons. In 3-space, an analogous "tetrahedralization" (without additional "Steiner" points for vertices) for nonconvex polyhedral bodies may not exist. Moreover, the problem of deciding such existence is *NP*-complete (Ruppert and Seidel, 1989).

For algorithms that depend on sequential examination of objects, *bucketing* may improve performance by providing advantageous sequencing (Devroye, 1986). The idea is to partition an area into a regular pattern of simple shapes such as rectangles to be traversed in a specified sequence. The problem at hand is then addressed locally within buckets or bins followed by adjustments between subsequent or neighboring buckets. Bucketing-based algorithms have provided practical solutions to Euclidean optimization problems, such as shortest paths, optimum-length matching, and a Euclidean version of the *Chinese postman problem*: minimizing the pen movement of a plotter (Asano, Edahiro, Imai, and Iri, 1985). The techniques of *quadtrees and octrees* might be considered as hierarchical approaches to bucketing, and are often the methods of choice for image processing and spatial data analysis including surface representation (Samet, 1990, 1990a).

The position of bodies and parts of bodies, relative to each other in space, determines visibility from given vantage points, shadows cast upon each other, and impediments to motion.

Hidden line and *hidden surface* algorithms are essential in computer graphics, as are procedures for shadow generation and shading (Sutherland, Sproull, and Shumacker, 1974) and (Atherton, Weiler, and Greenberg, 1978). Franklin (1980) used bucketing techniques for an exact hidden surface algorithm.

Lozano-Pérez and Wesley (1979) used the concept of a *visibility graph* for planning collision-free paths: given a collection of mutually disjoint polyhedral objects, the node set of the above graph is the set of all vertices of those polyhedral objects, and two such nodes are connected if the two corresponding vertices are visible from each other.

The *piano movers problem* captures the essence of "motion planning" (Schwartz and Sharir, 1983, 1989). Here a 2-dimensional polygonal figure, or a line segment ("ladder"), is to be moved, both translating and rotating, amidst polygonal barriers.

Geometric objects encountered in many areas such as Computer-Aided Design (CAD) are fundamentally nonlinear (Dobkin and Souvaine, 1990). The major thrust is generation of classes of curves and surfaces with which to interpolate, approximate, or generally speaking, represent data sets and object boundaries (Barnhill, 1977; Bartels, Beatty, and Barski, 1987; Farin, 1988). A classical approach – building on the concepts of *splines* and *finite elements* – has been to use piecewise polynomial functions over polyhedral tilings such as triangulations. Examples are the "TIN(= triangulated irregular network)" approach popular in terrain modeling, C^1 functions over triangulations, and the arduous solution of the corresponding C^2 problem (Heller, 1990; Lawson, 1977; and Alfeld and Barnhill, 1984).

Bézier curves and surfaces involve an elegant concept: the use of "control points" to define elements of curves and surfaces, permitting intuition-guided manipulation important in CAD (Forrest, 1972). In general, polynomials are increasingly supplanted by rational functions, which suffer fewer oscillations per numbers of coefficients (Tiller, 1983). All these techniques culminate in "NURBS (= non-uniform rational B-splines)", which are recommended for curve and surface representation in most industrial applications.

In geometric calculations, round-off errors due to floating-point arithmetic may cause major problems (Fortune and Milenkovic, 1991). When testing, for instance, whether given points are collinear, a tolerance level "*eps*" is often specified, below which deviations from a collinear-ity criterion are ignored. Points p_1, p_2, p_3 and p_2, p_3, p_4, but not p_1, p_2, p_4 may thus be found collinear. Such and similar inconsistencies may cause a computation to abort. "Robust" algorithms are constructed so as to avoid breakdown due to inconsistencies caused by round-off (Guibas, Salesin, and Stolfi, 1989; Beichl and Sullivan, 1990). Alternatively, various forms of "exact arithmetic" are increasingly employed (Fortune and Van Wyck, 1993; Yap, 1993). Inconsistencies occur typically whenever an inequality criterion is satisfied as an equality. An example is the degeneracy behavior of the *Simplex Method* of linear programming. Lexicographic perturbation methods can be employed to make consistent selections of subsequent feasible bases and thus assure convergence. Similar consistent tie breaking, coupled with exact arithmetic, is the aim of the *simulation of simplicity* approach proposed by Edelsbrunner and Mücke (1988) in a more general computational context.

See **Cluster analysis; Convex hull; Facility location; Minimum spanning tree; Traveling salesman problem; Voronoi/Delaunay constructs.**

References

[1] P.K. Agarwal (1991), *Intersection and Decomposition Algorithms for Planar Arrangements*, Cambridge University Press, New York.

[2] P. Alfeld and R.E. Barnhill (1984), "A Transfinite C^2 Interpolant over Triangles," *Rocky Mountain Journal of Mathematics*, 14, 17–39.

[3] T. Asano, M. Edahiro, H. Imai, and M. Iri (1985), "Practical Use of Bucketing Techniques in Computational Geometry," in *Computational Geometry*, G.T. Toussaint (ed), North Holland, New York.

[4] P. Atherton, K. Weiler, and D.P. Greenberg (1978), "Polygon Shadow Generation," *Comput. Graph.*, 12, 275–281.

[5] R.E. Barnhill (1977), "Representation and Approximation of Surfaces," in *Mathematical Software III*, J.R. Rice (ed), Academic Press, New York.

[6] R.H. Bartels, J.C. Beatty, B.A. Barski (1987), *An Introduction to Splines for Use in Computer Graphics*, Morgan Kaufmann, Los Altos, CA.

[7] I. Beichl and F. Sullivan (1990), "A Robust Parallel Triangulation and Shelling Algorithm," *Proc. 2nd Canad. Conf. Comput. Geom.*, 107–111.

[8] J.L. Bentley and W. Carruthers (1980), "Algorithms for Testing the Inclusion of Points in Polygons," *Proc. 18th Allerton Conf. Commun. Control Comput.*, 11–19.

[9] J.L. Bentley, B.W. Weide, and A.C. Yao (1980), "Optimal Expected-Time Algorithms for Closest

Point Problems," *ACM Trans. Math. Software*, 6, 563–580.

[10] B. Chazelle (1990), "Triangulating the Simple Polygon in Linear Time," *Proc. 31st Annu. IEEE Sympos. Found. Comput. Sci.*, 220–230.

[11] B. Chazelle (1991), "An Optimal Convex Hull Algorithm and New Results on Cuttings," *Proc. 32nd Annu. IEEE Sympos. Found. Comput. Sci.*, 29–38.

[12] K.L. Clarkson (1985), "A Probabilistic Algorithm for the Post Office Problem," *Proc. 17th Annu. ACM Sympos. Theory Comput.*, 175–184.

[13] K.L. Clarkson (1986), "Linear Programming in $0(n3^{d_2})$ Time," *Inform. Process. Lett.*, 22, 21–24.

[14] L. Devroye (1986), *Lecture Notes on Bucket Algorithms*, Birkhuser Verlag, Boston, Massachusetts.

[15] D.P. Dobkin and H. Edelsbrunner (1984), "Ham-sandwich Theorems Applied to Intersection Problems," *Proc. 10th Internat. Workshop Graph-Theoret. Concepts Comput. Sci.* (WG 84), 88–99.

[16] D.P. Dobkin and D.L. Souvaine (1990), "Computational Geometry in a Curved World," *Algorithmica*, 5, 421–457.

[17] D. Dobkin, L. Guibas, J. Hershberger, and J. Snoeyink (1988), "An Efficient Algorithm for Finding the CSG Representation of a Simple Polygon," *Computer Graphics*, 22, 31–40.

[18] H. Edelsbrunner (1983), "A New Approach to Rectangle Intersections, Parts I and II," *Internat. Jl. Comput. Math.*, 13, 209–219, 221–229.

[19] H. Edelsbrunner (1987), *Algorithms in Combinatorial Geometry*, Springer Verlag, New York.

[20] H. Edelsbrunner and H.A. Maurer (1985), "Finding Extreme Points in Three Dimensions and Solving the Post-Office Problem in the Plane," *Inform. Process. Lett.*, 21, 39–47.

[21] H. Edelsbrunner and E.P. Mcke (1988), "Simulation of Simplicity: a Technique to Cope with Degenerate Algorithms," *Proc. 4th Annu. ACM Sympos. Comput. Geom.*, 118–133.

[22] G. Farin (1988), *Curves and Surfaces for Computer Aided Geometric Design*, Academic Press, New York.

[23] A.R. Forrest (1971), "Computational Geometry," *Proc. Roy. Soc. Lond. Ser. A*, 321, 187–195.

[24] A.R. Forrest (1972), "Interactive Interpolation and Approximation by Bzier Polynomials," *The Computer Journal*, 15, 71–79.

[25] W.R. Franklin (1980), "A Linear Time Exact Hidden Surface Algorithm," *Proc. SIGGRAPH '80, Comput. Graph.*, 14, 117–123.

[26] S. Fortune and V. Milenkovic (1991), "Numerical Stability of Algorithms for Line Arrangements," *Proc. 7th Annu. ACM Sympos. Comput. Geom.*, 334–341.

[27] S. Fortune and C. Van Wyck (1993), "Efficient Exact Arithmetic for Computational Geometry," *ACM Symposium on Computational Geometry*, 9, 163–172.

[28] M.R. Garey, D.S. Johnson, F.P. Preparata, and

R.E. Tarjan (1978), "Triangulating a Simple Polygon," *Inform. Process. Lett.*, 7, 175–179.

[29] R.L. Graham and F.F. Yao (1983), "Finding the Convex Hull of a Simple Polygon," *Jl. Algorithms*, 4, 324–331.

[30] R. Graham and F. Yao (1990), "A Whirlwind Tour of Computational Geometry," *Amer. Math. Monthly*, 97, 687–701.

[31] B. Grünbaum (1967), *Convex Polytopes*, Wiley Interscience, New York.

[32] L.J. Guibas, D. Salesin, and J. Stolfi (1989), "Epsilon Geometry: Building Robust Algorithms from Imprecise Computations," *Proc. 5th Annu. ACM Sympos. Comput. Geom.*, 208–217.

[33] M. Heller (1990), "Triangulation Algorithms for Adaptive Terrain Modeling," *4th Symposium on Spatial Data Handling*, 163–174.

[34] D. Kirkpatrick (1983), "Optimal Search in Planar Subdivisions," *SIAM Jl. Comput.*, 12, 28–35.

[35] D.G. Kirkpatrick and R. Seidel (1986), "The Ultimate Planar Convex Hull Algorithm?," *SIAM Jl. Comput.*, 15, 287–299.

[36] C.L. Lawson (1977), "Software for C^1 Surface Interpolation," in *Mathematical Software III*, J.R. Rice (ed), Academic Press, New York.

[37] D.T. Lee and F.P. Preparata (1984), "Computational Geometry – A Survey," *IEEE Transactions on Computers*, c-33, 1072–1101.

[38] T. Lozano-Prez and M.A. Wesley (1979), "An Algorithm for Planning Collision-Free Paths Among Polyhedral Obstacles," *Commun. ACM*, 22, 560–570.

[39] N. Megiddo (1982), "Linear-Time Algorithms for Linear Programming in R^3 and Related Problems," *Proc. 23rd Annu. IEEE Sympos. Found. Comput. Sci.*, 329–338.

[40] N. Megiddo (1984), "Linear Programming in Linear Time When the Dimension is Fixed," *Jl. ACM*, 31, 114–127.

[41] N. Muhidinov and S. Nazirov (1978), "Computerized Recognition of Closed Plane Domains," *Voprosy Vychisl. i Prikl. Mat. (Tashkent)*, 53, 96–107, 182.

[42] E.M. Reingold and K.J. Supowit (1983), "Probabilistic Analysis of Divide-and-Conquer Heuristics for Minimum Weighted Euclidean Matching," *Networks*, 13, 49–66.

[43] J. O'Rourke (1987), *Art Gallery Theorems and Algorithms*, Oxford University Press, New York.

[44] C.H. Papadimitriou (1977), "The Euclidean Traveling Salesman Problem is *NP*-Complete," *Theoret. Comput. Sci.*, 4, 237–244.

[45] F.P. Preparata (1990), "Planar Point Location Revisited," *Internat. Jl. Found. Comput. Science* 24, 1, 71–86.

[46] F.P. Preparata and S.J. Hong (1977), "Convex Hulls of Finite Sets of Points in Two and Three Dimensions," *Commun. ACM*, 20, 87–93.

[47] F.P. Preparata and M.I Shamos (1985), *Computational Geometry: An Introduction*, Springer Verlag, New York.

[48] A.A.G. Requicha (1980), "Representations for Rigid Solids: Theory, Methods, and Systems," *ACM Comput. Surveys*, 12, 437–464.

[49] J. Ruppert and R. Seidel (1989), "On the Difficulty of Tetrahedralizing 3-dimensional Non-convex Polyhedra," *Proc. 5-th Annu. ACM Sympos. Comput. Geom.*, 380–392.

[50] J.T. Schwartz and M. Sharir (1983), "On the 'Piano Movers' Problem, I: The Case of a Two-dimensional Rigid Polygonal Body Moving Amidst Polygonal Barriers," *Commun. Pure Appl. Math.*, 36, 345–398.

[51] J.T. Schwartz and M. Sharir (1989), "A Survey of Motion Planning and Related Geometric Algorithms," in *Geometric Reasoning*, D. Kapur and J. Mundy (eds.), 157-169, MIT Press, Cambridge, Massachusetts.

[52] H. Samet (1990), *The Design and Analysis of Spatial Data Structures*, Addison Wesley, Reading, Massachusetts.

[53] H Samet (1990), *Applications of Spatial Data Structures: Computer Graphics, Image Processing and GIS*, Addison Wesley, Reading, Massachusetts.

[54] M.I. Shamos and D. Hoey (1975), "Closest-Point Problems," *Proc. 16th Annu. IEEE Sympos. Found. Comput. Sci.*, 151–162.

[55] M.I. Shamos and D. Hoey (1976), "Geometric Intersection Problems," *Proc. 17th Annu. IEEE Sympos. Found. Comput. Sci.*, 208–215.

[56] I.E. Sutherland, R.F. Sproull, and R.A. Shumacker (1974), "A Characterization of Ten Hidden Surface Algorithms," *ACM Comput. Surv.*, 6, 1–55.

[57] W. Tiller (1983), "Rational B-splines for Curve and Surface Representation," *IEEE Computer Graphics and Applications*, 3, (6), 61–69.

[58] C. Yap (1993), "Towards Exact Geometric Computation," *Proc. 5th Canadian Conference on Computational Geometry*, 405–419.

COMPUTATIONAL PROBABILITY

Broadly defined, computational probability is the computer-based analysis of stochastic models with a special focus on algorithmic development and computational efficacy. The computer and information revolution has made it easy for stochastic modelers to build more realistic models even if they are large and seemingly complex. Computational probability is not just concerned with questions raised by the numerical computation of existing analytic solutions and the exploitation of standard probabilistic properties. It is the additional concern of the probabilist, however, to ensure that the solutions obtained are in the best and most natural form for numerical computation. Before the advent of modern computing, much effort was directed at obtaining insight into the behavior of formal models, while avoiding computation. On the other hand, the early difficulty of computation has allowed the development of a large number of formal solutions from which limited qualitative conclusions may be drawn, and whose appropriateness for algorithmic implementation has not been seriously considered. Ease of computation has now made it feasible to have the best of all worlds: computation is now possible for classical models heretofore not completely solved, while complex algorithms can be developed for providing often needed insights on stochastic behavior. See **Applied probability**; **Computer science and operations research**; **Matrix-analytic stochastic models**; **Phase-type probability distributions**; **Simulation of stochastic systems**; **Stochastic model**.

COMPUTER SCIENCE AND OPERATIONS RESEARCH

Ramesh Sharda

Oklahoma State University, Stillwater, Oklahoma

INTRODUCTION: Operations research (OR) and computer science (CS) have evolved together. One of the first business applications for computers was to solve OR problems for the petroleum industry when linear programming (LP) was used to determine the optimum blends of gasoline. OR problems have constantly challenged the limits of computer technology and, in turn, have also taken advantage of developments in hardware and software. Here we summarize some of the interfaces between OR and CS; it is not intended to be exhaustive. Rather, we provide an overview of where OR and CS have benefitted from each other.

THREE FACETS OF OR/CS INTERFACES – COMMON PROBLEMS: Figure 1 illustrates the relationship between OR and CS. The shaded area represents the set of problems that both OR and CS have attempted to solve. These problems are generally known as combinatorial problems. A specific problem, the traveling salesman problem (TSP), has been the subject of much research in both OR and CS. Various OR approaches, such as branch-and-bound and heuristics, have focused on solving the TSP. Computer scientists have also attempted to solve TSP using heuristics and tree search algorithms, which are very similar to the branch-and-bound procedures used by OR analysts. The artificial intelligence and

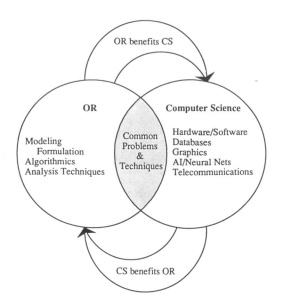

Fig. 1. OR-CS interfaces.

neural network communities have also focused on solving the TSP. Other heuristics approaches, such as genetic algorithms, simulated annealing, and tabu search, are being employed by both OR and CS specialists to solve combinatorial optimization problems. Computer scientists are using logic programming to solve routing and scheduling problems. These problems are also combinatorial in nature, and have been the focus of much research in the OR community.

IMPACTS OF CS ADVANCES ON OR: We now focus on how OR has benefitted from the advances in CS. In a broad sense, we can divide the research and practice of OR into two subfields: modeling and algorithmics. Modeling refers to preparation and delivery of a model, its results, and the analyses to a decision maker. Algorithmics refers to the actual solution procedures. Both modeling and algorithmics have benefitted from advances in CS.

Modeling – The modeling process starts with preparation of a model from the user form (possibly a verbal description of the problem) to the algorithmic form (a file that an algorithm is able to read as input). Various modeling languages have been developed to aid the analyst in model formulation. These languages (Fourer, 1983) have been influenced by and taken advantage of developments in database technology, specifically the relational database model. Papers by Choobineh (1991) and Geoffrion (1992) demonstrate the use of database developments in modeling. Model management environments employ the developments in database modeling to manage groups of models (Krishnan,

1993). Formulation aids employing knowledge-based tools from artificial intelligence are also being developed. Collins and Baker (1989) outline some attempts of combining OR and expert systems.

Computer science is also playing a role in the post-solution phase of modeling. Diagnosis of solution results is exemplified by systems such as ANALYZE (Greenberg, 1992). This system employs graph theory and natural language discourse, two developments from CS. Analysis of multiple scenarios is possible through the use of scenario manager in Microsoft Excel. Recent research by Sharda and Steiger (1994) focuses on the use of neural networks to gain insight from results of multiple scenarios.

Perhaps the most visible impact of computer science developments is in the user interfaces. Spreadsheets have become the ubiquitous paradigm for managing models and associated data as well as tools for delivery and presentation of results. Several spreadsheets include linear and nonlinear programming algorithms as a part of the standard function set. The spreadsheets are changing the way OR analysts prepare, manage, and deliver the models.

Many graphical user interfaces are being developed to aid in formulation. These include systems such as LPFORMS (Ma, Murphy and Stohr, 1989). Jones (1994) presents an excellent overview of the use of graphics and visualization technologies in modeling and solutions.

Finally, advances in telecommunications are becoming critical in timely delivery of OR models and solutions. Real time data access is a key in successful implementation of many models, and that is possible only because of advances in telecommunications.

Algorithmics – Algorithmic aspects of OR have benefitted from advances in both hardware and software in computing. Improvements in hardware have allowed us to solve realistic, large OR problems on a desktop computer (Nemhauser, 1994; Sharda, 1993). These developments include cache memory and superscalar computation. Since the LP matrix computations involve working with sparse matrices, use of cache memory allows faster access and manipulation of matrix elements. Similarly, superscalar architectures as well as vectorization facilities of the new computers allow vectorized calculations. LP codes such as OB1(now incorporated within CPLEX) and OSL are examples of codes that have exploited the recent developments in computer architecture quite well.

Developments in parallel processing have led to a new class of algorithms to solve various OR problems faster. Capabilities such as pipelining, vectorization, and superscalar computations are available in today's workstations and have been employed in the latest implementations of simplex as well as interior point methods. Algorithms are being developed to exploit multiple as well as massively parallel processors. A summary of related issues is provided by Zenios (1989) and Eckstein (1993).

Developments in graphics have also helped in visualizing algorithmic progress, which leads to optimization of algorithmic parameters (Jones, 1994). For example, OSL/6000 offers a graphical view of the objective function value achieved at any particular iteration of the LP algorithms.

Software advances in CS have also had an impact on OR algorithms. For example, new data structures developed in CS are routinely used in OR algorithms. As any serious OR algorithm developer knows very well, learning about sparse matrix approaches such as linked lists, arrays, orthogonal lists, etc. is a key in implementing an algorithm. As an example, the paper by Adler *et al.* (1989) focuses on the data structures employed in their implementation of interior point methods.

Another example of the impact of CS on OR is the field of computational probability. Researchers are working to develop numerical techniques to solve systems of stochastic equations (Albin and Harris, 1987).

Simulation research and practice has also been a beneficiary of CS advances. One example is the use of artificial intelligence techniques in design and interpretation of simulations. Advances in parallel processing have led to active research in parallel simulation to speed up the computations (Fujimoto, 1993).

CS BENEFITS FROM OR: Just as OR has benefited from CS advances, CS practitioners and researchers have employed the developments in OR. OR algorithms have proven to be very useful in computer performance modeling. An example given by Greenberg (1988) includes the use of random walk theory to analyze various storage allocation approaches, an important issue in operating systems.

As mentioned earlier, the developments in databases and data structures have helped OR in modeling and algorithmics. But OR has been a key player in designing distributed databases. OR models and their solutions are important in designs of such databases. Information storage and retrieval research has also been the beneficiary of OR algorithms for query optimization.

Kraft (1985) provides a good survey of this interface between OR and CS.

OR approaches (e.g. mixed-integer programming) are also used in artificial intelligence. Specific examples include use of MIP in automated theorem proving.

Another significant impact of OR in CS is in the telecommunications area. Various telecommunications problems are the targets of much OR research: network design and routing, location analysis, etc. Decisions in telecommunications networks are based on OR approaches using queueing theory, Markov analysis, simulation, and MIP models.

SUMMARY: Our purpose here has been to point to some links between OR and CS to show the vibrancy of this interface. As an article in *Computer-World* (Betts, 1993) noted, OR/MS needs corporate data for its algorithms and needs the algorithms used in strategic information systems to make a real impact. On the other hand, information systems (IS) groups need OR to build smart applications. Betts calls the individuals with significant OR/MS and CS/IS skills the new "Efficiency Einsteins," a term that indeed appropriately describes the individuals trained in this interface.

See **Artificial intelligence; Combinatorial and integer optimization; Information systems/ database design; Linear programming; Algebraic modeling languages; Nonlinear programming; Parallel computing; Structured modeling; Vehicle routing; Visualization.**

References

[1] Adler, I., N. Karmarkar, M.C.G. Resende, and G. Beiga (1989). "Data Structures and Programming Techniques for the Implementation of Karmarkar's Algorithm." *ORSA Jl. Computing*, 1, 84–106.

[2] Albin, S.L. and C.M. Harris (1987). "Statistical and Computational Problems in Probability Modeling." *Annals of Operations Research*, 8/9.

[3] Betts, M. (1993). "Efficiency Einsteins." *Computer-World*, March 22, 63–65.

[4] Choobineh, J. (1991). "SQLMP: A Data Sublanguage for Representation and Formulation of Linear Mathematical Models." *ORSA Jl. Computing*, 3, 358–375.

[5] Collins, D.E. and T.E. Baker (1989). "Using OR to Add Value in Manufacturing." *OR/MS Today*, 16, 6, 22–26.

[6] Eckstein, J. (1993). "Large-Scale Parallel Computing, Optimization, and Operations Research: A Survey." *ORSA/CSTS Newsletter*, 14, 2, 11–12, 25–28.

[7] Fourer, R. (1983). "Modeling Languages Versus Matrix Generators for Linear Programming." *ACM Trans. Math. Software*, 9, 143–183.

[8] Fujimoto, R. M. (1993). "Parallel Discrete Event Simulation: Will the Field Survive?," *ORSA Jl. Computing*, 5, 213–230.

[9] Geoffrion, A.M. (1992). "The SML Language for Structured Modeling: Levels 1 & 2." *Operations Research*, 40, 38–75.

[10] Greenberg, H.J. (1988). "Interfaces Between Operations Research and Computer Science." *OR/MS Today*, 15, 5.

[11] Greenberg, H.J. (1992). "Intelligent Analysis Support for Linear Programs." *Computers and Chemical Engineering*, 16, 659–674.

[12] Jones, C.V. (1994). "Visualization and Mathematical Programming." *ORSA Jl. Computing*, 6 (to appear).

[13] Kraft, D.H. (1985). "Advances in Information Retrieval: Where Is That /*&% Record?" *Advances in Computers*, 24, 277–318.

[14] Krishnan, R. (1993). "Model Management: Survey, Future Research Directions, and a Bibliography." *ORSA/CSTS Newsletter*, 14, 1, 1–16.

[15] Lustig, I. J., R. E. Marsten, and D. F. Shanno (1994). "Interior Point Methods for Linear Programming: Computational State of the Art." *ORSA Jl. Computing*, 6, 1–14.

[16] Ma, P.-C., F.H. Murphy, and E.A. Stohr (1989). "A Graphics Interface for Linear Programming." *Communications of the ACM*, 32, 996–1012.

[17] Nemhauser, G.L. (1994). "The Age of Optimization: Solving Large Scale Real-World Problems." *Operations Research*, 42, 5–13.

[18] Sharda, R. (1993). *Linear and Discrete Optimization Modeling and Optimization Software: An Industry Resource Guide*, Lionheart Publishing, Atlanta, Georgia.

[19] Sharda, R. and D. Steiger (1994). "Enhancing Model Analysis Using Neural Networks," presented at the ORSA/CSTS Meeting, Williamsburg, Virginia, January.

[20] Zenios, S. (1989). "Parallel Numerical Optimization: Current Status and an Annotated Bibliography." *ORSA Jl. Computing*, 1, 20–43.

CONCAVE FUNCTION

A function that is never below its linear interpolation. Mathematically, a function $f(x)$ is concave over a convex set S, if for any two points, x_1 and x_2 in S and for any $0 \leq \alpha \leq 1$, $f[\alpha x_1 + (1 - \alpha)x_2] \geq \alpha f(x_1) + (1 - \alpha)f(x_2)$.

CONCLUSION

A portion of a rule composed of series of one or more actions that the inference engine can carry out if a rule's premise can be established to be true. See **Artificial intelligence**; **Expert systems**.

CONDITION NUMBER

See **Numerical analysis**.

CONE

A set which contains the ray generated by any of its points. Mathematically, a set S is a cone if the point x in S implies that αx is in S for all $\alpha \geq 0$.

CONGESTION SYSTEM

Often used to be synonymous with queueing system because congestion refers to the inability of arriving customers to get immediate service, which is the reason behind doing queueing analyses. See **Queueing theory**.

CONJUGATE GRADIENT METHOD

See **Quadratic programming**.

CONNECTED GRAPH

A graph (or network) in which any two distinct nodes are connected by a path.

CONSERVATION OF FLOW

(1) A set of flow-balance equations governing the flow of a commodity in a network that state that the difference between the amount of flow entering and leaving a node equals the supply or demand of the commodity at the node. See **Network optimization**. (2) A set of equations that state that the limiting rates that units enter and leave a state or entity of a queueing system or related random process must be equal. The entities may be service facilities (stages), where the limiting number of units coming in must equal the limiting departing; balance at a state might mean, for example, that the rate at which a queueing system goes up to n customers equals the rate at which it goes down to n from above. See **Queueing theory**.

CONSTRAINED OPTIMIZATION PROBLEM

A problem in which a function $f(X)$ is to be optimized (minimized or maximized), where the possible solutions X lie in a defined solution subspace S. S is usually determined by a set of linear and/or nonlinear constraints.

CONSTRAINT

An equation or inequality relating the variables in an optimization problem; a restriction on the permissible values of the decision variables of a given problem.

CONSTRAINT QUALIFICATION

A condition imposed on the constraints of an optimization problem so that local minimum points will satisfy the Karush-Kuhn-Tucker conditions. See **Nonlinear programming**.

CONSTRUCTION APPLICATIONS

C. Perry

Queensland University of Technology,
Brisbane, Australia

Due to their size and complexity, most construction projects would appear to offer a wide potential for OR/MS applications. For example, the standard critical path models of PERT, CPM and precedence diagrams are particularly successful in construction. However, apart from these models, we find that OR/MS are not often used in construction. Schelle (1990, p. 111) summarizes, "In project management the large number of publications about operations research topics contrast to the small number of real applications."

Here, we review three major areas of construction where OR/MS applications could occur – job estimation and tendering, project planning, and project management and control. Factors inhibiting the application of OR/MS in construction projects are discussed and possible future developments are canvassed.

JOB ESTIMATION AND TENDERING: Some OR/MS models have been applied to job estimation. Job estimation requires trade-offs between time and cost. Early OR/MS work assumed direct costs for each activity increased linearly with time, and therefore, used linear programming. But construction usually does not fit this assumption. Dynamic programming and integer linear programming have also been used, but the large number of variables and constraints of construction projects made them unworkable. Models based on heuristic and nonlinear curves have been found to be almost as accurate and more friendly for construction managers, and have been tried (Cusack, 1984). In addition, the Line of Balance (LOB) model, originally developed for the U.S. Navy, is used to make trade-offs between alternative schedules, and a modified LOB model called Time Chainage is used in the U.K. for estimating schedules for construction of roads, bridges and other civil engineering projects (Wager and Pittard, 1991).

Allied to job estimation is *tendering*, which must consider competitors' likely actions along with the bidder's decisions. It is a relatively more open and therefore more difficult system to model. Hence, although ARIMA and regression, plus other statistical and simulation models, have been developed to assist tendering, they have rarely been applied.

If tendering is considered from the selector's point of view, rather than a bidder's point of view, variables are not so uncertain because the selector will have certain information about all the bids. Nevertheless, the complexity of construction projects again makes application of conventional OR/MS models difficult, especially as prior knowledge about bidders is an important choice factor. A hybrid model using linear programming, multiattribute utility, regression and expert systems seems appropriate here (Russell, 1992).

PROJECT PLANNING: While preparing a tender, construction managers must start to *plan the project* in more detail. The critical path models, integrated with cost control and reporting models, are widely used in construction for this purpose (Wager and Pittard, 1991). Their application in complex construction projects has suggested theoretical extensions, for example, incorporating the stochastic relationship of cost with time. One such extension for the complex construction industry is a suite of PC programs, Construction Project Simulator (CPS), which incorporates productivity variability and external interferences to the construction process on site. It then produces barcharts, cost and resource schedules like the critical path models (Bennett and Ormerod, 1984). However, most of these extensions have unrealistic data requirements and are rarely applied even if they are tried.

Modeling could be especially useful in planning tunnel construction projects. For example, Touran and Toushiyuki (1987) demonstrated a simulation model for tunnel construction and design. But model use is limited to very large projects.

Project planning usually involves more than cost minimization with constraints, for example, environmental considerations. Some OR/MS multi-objective models have provided assistance here. For example, Scott (1987) applied multi-objective valuation to roads construction, using a step-by-step procedure to evaluate all objectives, without having to assume all quantified data as being equally accurate and reliable.

MANAGEMENT AND CONTROL: After a project is planned, it must be *managed and controlled*. Linked with the project plan are straightforward accounting models. With increasing use of "real time" reporting, they allow closer management of costs. It is in this relatively stable field of managing and controlling the project after it has begun that conventional OR/MS models offer most promise, that is, at a tactical and relatively deterministic and repetitive level. For example, standard cost-minimization models could be applied to the management of construction equipment, to location and stocking of spare parts warehouses, and to selecting material handling methods. In one of few actual OR/MS applications, Perry and Iliffe (1983) used a transhipment model to manage movement of sand during an airport construction project. Two other possible areas in where OR/MS models might be applied are multiple projects (where several projects are designed and built somewhat concurrently to minimize costs), and marketing.

In summary, although potential applications of OR/MS in construction appear at first glance to be plentiful, progress with actual OR/MS applications is slow. One reason for this is that risks in using unproven OR/MS models are high in commercial operations where claims resulting from mistakes can be taken to court. Moreover, each construction appears to be "one-off," that is, the building is more or less different than previous ones of the constructor: at a different site with different subsurface conditions; involving different organizations and individuals with different goals; different weather; different material, labor requirements and shortages; different errors in estimates of time and cost; and different levels of interference from outside. Given this lack of standardization, OR/MS

modeling has tended to move towards more general simulation models (which have large data requirements) or heuristic models. Still, OR/MS applications are few and although "computers are installed extensively throughout...consultants and construction site offices...their role appears to make the former manual processes more efficient rather than exploit the increased potential brought by the machine" (Brandon, 1990, p. 285).

What does the future hold for OR/MS applications in the construction industry? A probable development is their increasing use in conjunction with user-friendly software on PCs. Research in the construction industry suggests that the key to successful implementation of research is a powerful intermediary like construction managers. Developments in PC-based software such as simulations and expert systems, which assist rather than replace the experience-based knowledge of people like site managers, offer promise of more OR/MS applications, especially in the complex and expensive field of contractual disputes. These possibilities will be enhanced by interactive, three-dimensional graphical interfaces. In particular, expert systems should be used more frequently because they incorporate the existing knowledge of construction managers.

See **CPM; Expert systems; PERT; Project management.**

References

[1] Bennett, J. and Ormerod, R.N. (1984), "Simulation applied to construction projects," *Construction Management and Economics*, 2, 225–263.

[2] Brandon, P.S. (1990), "The development of an expert system for the strategic planning of construction projects," *Construction Management and Economics*, 8, 285–300.

[3] Cusack, M.M. (1985), "A simplified approach to the planning and control of cost and project duration," *Construction Management and Economics*, 3, 183–198.

[4] Perry, C. and Iliffe, M. (1983), "Earthmoving on construction sites," *Interfaces*, 13(1), 79–84.

[5] Russell, J.S. (1992), "Decision models for analysis and evaluation of construction contractors," *Construction Management and Economics*, 10, 185–202.

[6] Schelle, H. (1990), "Operations research and project management past, present and future," in Reschke, H. and Schelle, H., eds., *Dimensions of Project Management*, Springer-Verlag, Berlin.

[7] Scott, D. (1987), "Multi-objective economic evaluation of minor roading projects," *Construction Management and Economics*, 5, 169–181.

[8] Slowinski, R. and Weglarz, R., eds. (1989), *Advances in Project Scheduling Studies in Production and Engineering Economics* 9, Amsterdam.

[9] Touran A. and Toshyuki, A. (1987), "Simulation of tunnelling operations," *Construction Engineering and Management*, 113, 554–568.

[10] Wager, D.M. and Pittard, S.J. (1991), *Using Computers in Project Management*, Construction Industry Computing Association, Cambridge, England.

CONTINUOUS-TIME MARKOV CHAIN (CTMC)

A Markov process with a continuous parameter but countable state space. The stochastic process $\{X(t)\}$ has the property that, for all $s,t \geq 0$ and nonnegative integers i,j, and $x(u), 0 \leq u < \infty$,

$$Pr\{X(t+s) = j | X(s) = i, \ X(u) = x(u), 0 \leq u < s\}$$
$$= Pr\{X(t+s) = j | X(s) = i\}.$$

See **Markov chains**; **Markov processes**.

CONTROL CHARTS

See **Quality control**.

CONTROLLABLE VARIABLES

In a decision problem, variables whose values are determined by the decision process and/or decision maker. Such variables are also called decision variables. See **Decision maker; Decision problem; Mathematical model**.

CONTROL THEORY

Andre Z. Manitius

George Mason University, Fairfax, Virginia

INTRODUCTION: Although the use of control theory is normally associated with applications in electrical and mechanical engineering, it shares much of its mathematical foundations with operations research and management science. These foundations include differential and difference equations, stochastic processes, optimization, calculus of variations, and others.

In application, control theory is concerned with steering dynamical systems to achieve desired results. Both types of systems to be controlled and the goals of control include a wide variety of cases. Control theory is strongly related to control systems engineering, which is fundamental to many advanced technologies. In a broader sense, control theoretic concepts are applicable not just to technological systems, but also to dynamical systems encountered in biomedical, economic and social sciences. Control theory has also had a fundamental impact on many areas of applied mathematics and continues to be a rich source of research problems.

Systems to be controlled may be of various forms: they could be mechanical, electrical, chemical, thermal or other systems that exhibit dynamical behavior. Control of such systems requires that the system dynamics be well understood. This is usually accomplished by formulating and analyzing a mathematical model of the system. Physical properties of the system play an important role in establishing the mathematical model. However, once the model is established, the control theoretic considerations are independent of the exact physical nature of the system. Since different physical systems often have similar mathematical models, similar control principles are applicable to them. For example, a mechanical system of interconnected masses and springs is described by the same mathematical model as an electrical circuit of interconnected capacitors and inductors. From the control theoretic point of view, the two systems can be treated in the same way.

The control of a system is usually accomplished by providing an input signal which affects the system behavior Figure 1. Physically, the input signal often changes the energy flow in the system, much like the pilot's commands change the thrust of the engines in the aircraft. The conversion of input signals into physical variables, such as the energy of the mass flow, is done by devices called actuators. System response is measured by various instruments, called sensors. The measurements, called output signals, are fed to a controller, which usually means a control computer. The controller determines the successive values of the input signals that are then passed on to the actuators. While the control computer hardware is the physical location where the control decisions are being made, the essence of the control is a control algorithm imbedded in the computer software. The development of control algorithms is often based on sophisticated mathematical theory of control and on specific models of systems under control.

One of the key difficulties of control is the uncertainty about the system model and system outputs. The uncertainty has several origins. Mathematical models of systems under control

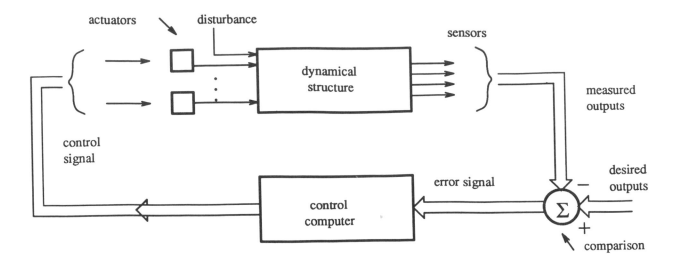

Fig. 1. Closed loop multivariate system.

are based on many simplifying assumptions and thus contain errors due to approximations. Properties or parameters of the system may change in unpredictable ways. Systems may be subject to unknown external inputs, such as, gusts of winds acting on the aircraft. Output signals provided by sensors contain sensor noise or communication channel noise. By its very nature, the control problem formulation usually includes uncertain parameters and signals. The task of control theory is to provide solutions which guarantee, whenever possible, good system performance in spite of the uncertainties.

HISTORICAL DEVELOPMENT: The first systematic study of feedback control of steam engines by J.C. Maxwell appeared in 1868. In 1893, A.M. Lyapunov published a first paper on the stability of motion, but his work made an impact on the control theory literature only 55 years later. When the first electronic amplifiers appeared in the long-distance telephone lines after World War I, high-gain feedback coupled with high-order dynamics of amplifiers led to stability problems. In 1932, H. Nyquist provided a method of feedback stability analysis based on the frequency response. In the late 1930s, devices for controlling aircraft were introduced. World War II gave a big boost to the field of feedback control. Norbert Wiener's theory of filtering of stochastic processes, combined with the servomechanism theory, provided a unified framework for the design of control mechanisms in aircraft and ships and became what is known as "classical control theory."

In the late 1950s and in the 1960s, an extensive development of control theory took place, coinciding with manned space flight and other aerospace applications, and with the advent of computers. Bellman's principle of optimality embedded in Dynamic Programming, Pontryagin's Maximum Principle of Optimal Control, and the Kalman Filter, were invented between 1956 and 1960. State-space methods of analysis, based on differential equations and matrix computations, have become the main tools of what was then named "modern control theory." Control theory played a crucial role in the success of the Apollo moon-landing project in 1969. In the 1970s, substantial progress was made in the control of systems governed by partial differential equations, adaptive control and nonlinear control. The applications of control theory became very diverse, including complex material processing, bio-medical problems, and economic studies. In the 1980s, robust control theory was formulated and reached a significant level of maturity. Robust control theory has by now provided a synthesis of the "classical" and the "modern" (state-space based) control theory.

In general, current research in control aims at studying the limits of performance of feedback control systems in some advanced applications. Computational tools of control have been coded in MATLAB software system and in similar software. During the last decade, control hardware has been revolutionized by microprocessors and new sensor and actuator technologies, such as "smart materials." Some tools of the intelligent control approach have been applied to on-board guidance and navigation systems. Anti-lock brake systems, computerized car engine control, and geographic positioning systems are a few contemporary examples of systems where the principles and tools of control theory are at work.

MATHEMATICAL MODELS FOR CONTROL AND THE IDENTIFICATION ISSUE:

Currently, the most commonly used mathematical control is the linear state-space model. This is a system of first-order, time-invariant, linear differential equations with inputs and outputs. Such a linear system can be written as:

$$\begin{cases} \dfrac{d}{dt} x(t) = \boldsymbol{A}x(t) + \boldsymbol{B}u(t) \\ \quad y(t) = \boldsymbol{C}x(t) + \boldsymbol{D}u(t) \end{cases}$$

where $\boldsymbol{x}(t) =$ state vector, $\boldsymbol{u}(t) =$ control, $\boldsymbol{y}(t) =$ output, and $\boldsymbol{A}, \boldsymbol{B}, \boldsymbol{C}, \boldsymbol{D}$ are matrices of appropriate dimensions.

In practical applications, engineers often use scalar or matrix transfer functions. These are rational functions of the complex variable s arising in the Laplace transform or the variable z from the Z-transform, the latter being used for discrete-time systems. There are close relationships between state-space and transfer-function models.

In the last 20 years, many other systems have been analyzed in the control theory literature, such as nonlinear ordinary differential systems, differential equations with delay, integro-differential equations, linear and nonlinear partial differential equations, stochastic differential equations, both ordinary and partial, semigroup theory, discrete-event systems, queueing systems, Markov chains, Petri nets, neural network models, and others. In many cases, research on those systems has resulted in precise mathematical conditions under which the main paradigms of linear system theory extend to those systems.

Given an existing physical system, one of the most challenging tasks is the determination of the mathematical model for control. This is usually done in one of two ways: either the model equations are derived from physical laws and the few unknown parameters are estimated from input and output data, or a general model form is assumed with all the parameters unknown, which then requires a more extensive parameter estimation and model validation procedure. In either case, the overall step of model determination from experimental data is called system identification. Well developed methods and computer algorithms exist to assist the control designer in this task.

THE MAIN IDEAS:

Feedback is a scheme in which the control of the system is based on a concurrent measurement of the system's output. Usually, the system output is being compared to a given desired value of the output and the control is adjusted so as to steer the system output closer to the desired value. Feedback creates a directed loop linking the output to the input.

Complicated systems may have many feedback loops, either nested or intersecting one another. Feedback results in a change of a system's internal dynamics and system's input-output characteristics. A system with a properly designed feedback is capable of responding correctly to input commands even in the presence of uncertainties about the model and the external perturbations. An effective feedback reduces the effects of uncertainties, regardless of their origin. Feedback is also being used to improve stability margins, eliminate or attenuate some undesirable nonlinearities, or to shape system's bandwidth. Some systems cannot even function in a stable way without feedback. An example is the modern "fly-by-wire" fighter jet in which the feedback control loop keeps the aircraft in a stable flight envelope. The mechanism of feedback is well understood in case of linear systems. However, feedback mechanisms in nonlinear systems, especially those with many degrees of freedom, remain the subject of continued investigations. In a broader sense, the concept of feedback may be used to interpret various closed-loop interactions taking place in dynamical systems in physics, biology, economics, etc. (e.g., see Franklin, Powell and Emami-Naeni, 1994; SIAM, 1988).

OPTIMAL CONTROL:

In many cases, the goal of control may be mathematically formulated as the optimization of a certain performance measure. The tools of optimization theory and calculus of variations have been applied to derive certain optimal control principles. For example, one of the fundamental results valid for a broad class of linear systems with a quadratic performance measure says that the optimal control is accomplished by a linear feedback based on the measurement of the internal "state vector" of the system. Parameters of that linear feedback are obtained by solving a quadratic equation called the Riccati equation. Another fundamental result says that the control of linear systems with bounded control function and the transition time as a performance measure is accomplished by using only the extremum values of control (a "bang-bang control"). Solution of optimal control problems often requires iterative numerical computations to find a control that yields the best performance.

ROBUST CONTROL: Control methods have been developed to design feedback that minimizes the effect of uncertainty. Systems of this type are called robust. For example, one can design a feedback which minimizes the norm of the transfer function from unwanted disturbances to the output. Another design of that type makes the feedback system maximally insensitive to parameter variations. One of the key ideas in robust control is the use of norms in the Hardy function space H_∞, for both signals and operators (transfer functions). A close connection between the minimum H_∞ norm solutions and the solutions of certain systems of matrix Riccati equations has been discovered.

Robust control theory is well understood for linear time invariant systems, and some results have been obtained for nonlinear systems. A link has been discovered between the game-theoretic approach to control problems with uncertainty and the linear and nonlinear robust control.

STOCHASTIC CONTROL: *Stochastic control theory* involves the study of control and recursive estimation problems in which the uncertainty is modeled by random processes. One of the most significant achievements of the linear theory was the discovery of Kalman filtering algorithms and the separation principle of the optimal stochastic control. The principle states that under certain conditions the solution of the optimal stochastic control problem combines the optimal deterministic state feedback and the optimal filter estimate of the state vector, which are obtained separately from each other.

For nonlinear systems, Markov diffusions have become the tool of analysis. Stochastic optimal control conditions lead to certain nonlinear second order partial differential equations which may have no smooth solutions satisfying appropriate initial and boundary conditions. "Weak solutions" and "viscosity solutions" have recently been used to describe solutions to such optimal control problems.

ADAPTIVE CONTROL: One possible remedy against the uncertainty about the system and external signals is the use of adaptive feedback mechanism. During the system operation under a regular feedback, input and output signals can be processed to produce increasingly accurate estimates of system parameters which in turn can be used to adjust the regular feedback loop.

Alternatively, the step of estimating the original system can be bypassed in favor of a direct tuning of the feedback controller to minimize the error. The control system built this way contains two feedback loops, one regular but with adjustable parameters, and one that provides the adjustment mechanism. Adaptive systems are inherently nonlinear.

The main theoretical issue is the question of stability of the adaptive feedback loop; stable adaptive feedback laws for certain classes of nonlinear systems have been discovered. In contrast, bursting phenomena, oscillations and chaos have also been found in certain simple adaptive systems. Current research efforts are directed at finding robust adaptive control laws and at solving stochastic adaptive control problems for systems governed by some partial differential equations.

INTELLIGENT CONTROL: The term intelligent control is meant to describe control which includes decision making in uncertain environments, learning, self-organization, evolution of the control laws based on adaptation to new data, and to changes in the environment. An intelligent controller may deal with situations that require deciding which variables should be controlled, which models should be used, and which control strategy should be applied at any particular stage of operation. In some situations, no precise mathematical model of the system may exist, with the only information about the process being descriptive.

Intelligent control is a blend of control theory with artificial intelligence. In contrast to mathematical control theory, which uses precisely formulated models and control laws, intelligent control relies in many cases on heuristic models and rules. It is an area under development and few established paradigms exist. Current tools of intelligent control includes expert systems, fuzzy set theory and fuzzy control algorithms, and artificial neural networks. Examples of systems where the intelligent control may become effective are autonomous robots and vehicles, flexible manufacturing systems, and traffic control systems.

RESEARCH CHALLENGES AND DIRECTIONS: Among the main control theory challenges are: feedback control laws for nonlinear systems with many degrees of freedom, including systems governed by nonlinear partial differential equations (e.g. control of fluid flow); adaptive and robust control of such systems; control of systems based on incomplete models with learning and intelligent decision making; and feedback mechanisms based on vision and other non-traditional sensory data (SIAM, 1988).

See **Artificial intelligence; Calculus of variations; Dynamic programming; Neural networks and algorithms.**

References

[1] K. Astrom and B. Wittenmark (1989), *Adaptive Control*, Addison-Wesley, Reading, Massachusetts.

[2] B.D.O. Anderson and J.B. Moore (1990), *Optimal Control*, Prentice Hall, Englewood Cliffs, New Jersey.

[3] W. Fleming and M. Soner (1994), *Controlled Markov Processes and Viscosity Solutions*, Springer Verlag, New York.

[4] G. F. Franklin, J.D. Powell, and A. Emami-Naeni (1994), *Feedback Control of Dynamic Systems*, 3rd ed., Addison-Wesley, Reading, Massachusetts.

[5] M. Green and D.J.N. Limebeer (1995), *Linear Robust Control*, Prentice Hall, Englewood Cliffs, New Jersey.

[6] C.F. Lin (1994), *Advanced Control Systems Design*, Prentice Hall, Englewood Cliffs, New Jersey.

[7] SIAM (1988), *Future Directions in Control Theory: A Mathematical Perspective*, SIAM Reports on Issues in the Mathematical Sciences, Philadelphia.

[8] E.D. Sontag (1990), *Mathematical Control Theory: Deterministic Finite Dimensional Systems*, Springer Verlag, New York.

[9] A. Stoorvogel (1992), *The H_∞ Control Problem*, Prentice Hall International, London.

[10] J. Zabczyk (1992), *Mathematical Control Theory: An Introduction*, Birkhauser, Boston.

CONVEX COMBINATION

A weighted average of points (vectors). A convex combination of the points $x_1, \ldots, x_k$ is a point of the form $x = \alpha_1 x_1 + \ldots + \alpha_k x_k$, where $\alpha_1 \geq 0, \ldots, \alpha_k \geq 0$, and $\alpha_1 + \ldots + \alpha_k = 1$.

CONVEX CONE

A cone that is also a convex set.

CONVEX FUNCTION

A function that is never above its linear interpolation. Mathematically, a function $f(x)$ is convex over a convex set S, if for any two points x_1 and x_2 in S and for any $0 \leq \alpha \leq 1$, $f[\alpha x_1 + (1 - \alpha)x_2] \leq \alpha f(x_1) + (1 - \alpha)f(x_2)$.

CONVEX HULL

The smallest convex set containing a given set of points S. The convex hull of a given set S is the intersection of all convex sets containing S. The convex hull of a given set of points S is the set of all convex combinations of sets of points from S. If the set S is a finite set of points in a finite-dimensional space, then the convex hull is a polyhedron. See **Convex set; Polyhedron.**

CONVEX POLYHEDRON

See **Polyhedron.**

CONVEX-PROGRAMMING PROBLEM

A minimization programming problem with convex objective function and convex inequality constraints. It is typically written as

Minimize $f(x)$

subject to

$g_i(x) \leq 0$, $i = 1, \ldots, m$,

where the functions $f(x)$ and the $g_i(x)$ are convex functions defined on Euclidean n-space. See **Mathematical-programming problem; Nonlinear programming.**

CONVEX SET

A set of points that contains the line segment connecting any two of its points. Mathematically, the set S is convex if for all $0 \leq \alpha \leq 1$ and for all x_1 and x_2 in S, the point $\alpha x_1 + (1 - \alpha)x_2$ is also in S.

CONVEXITY ROWS

The constraints in the decomposition algorithm master problem that require solutions to be convex combinations of the extreme points of the subproblems. See **Dantzig-Wolfe decomposition algorithm.**

CORNER POINT

See **Extreme point.**

CORPORATE STRATEGY

Arnoldo C. Hax and Nicolas S. Majluf

*Massachusetts Institute of Technology,
Cambridge, Massachusetts*

STRATEGIC TASKS AT THE CORPORATE LEVEL: In a formal strategic planning process, we distinguish three perspectives – corporate, business, and functional. These perspectives are different both in term of the nature of the decisions they address, as well as the organizational units and managers involved in formulating and implementing the corresponding action programs generated by the strategy formation process.

At the corporate level we deal with the tasks that can not be delegated downward in the organization, because they need the broadest possible scope – involving the whole firm – to be properly addressed. At the business level we face those decisions that are critical to establish a sustainable competitive advantage, leading toward superior economic returns in the industry where the business competes. At the functional level we attempt to develop and nurture the core competencies of the firm – the capabilities that are the sources of the competitive advantages.

Here, we deal exclusively with corporate strategic tasks (Hax and Majluf, 1991). There are three different imperatives – leadership, economic, and managerial – that are useful to characterize these tasks, depending on whether we are concerned with shaping the vision of the firm, extracting the highest profitability levels, or assuring proper coordination and managerial capabilities.

THE LEADERSHIP IMPERATIVE: This imperative is commonly associated with the person of the CEO, who is expected to define a vision for the firm, and communicate it in a way that generates contagious enthusiasm.

The CEO's vision provides a sense of purpose to the organization, poses a significant but yet attainable challenge, and draws the basic direction to the pursuit of that challenge. Successful organizations invariably seem to have competent leaders who are able to define and transmit a creative vision, that generates a spirit of success. In other words, success breeds success.

Hamel and Prahalad (1989) argue that the vision of the firm should carry with it an "obsession" that they refer to as "Strategic Intent." It implies a sizable stretch for the organization that requires leveraging resources to reach seemingly unattainable goals.

Much has been written and said about leadership including the controversy on "nature or nurture" – whether leaders are born or made – and on the existence of common characteristics to describe successful leaders (Schein, 1992; Kotter, 1988). We do not review this literature here, since we concentrate on the economic and managerial imperatives of the corporate strategic tasks. Nonetheless, the set of corporate tasks that deal with the economic and managerial imperatives are the critical instruments to imprint the vision of the firm. The leadership capabilities are expressed and made tangible through the tasks that are discussed herein (Pfeffer, 1992).

THE ECONOMIC IMPERATIVE: This imperative is concerned with creating value at the corporate level. The acid test is whether the businesses of the firm are benefiting from being together, or if they would be better off as separate and autonomous units. From this point of view, the essence of corporate strategy is to assure that the value of the whole firm is bigger than the sum of the contributions of its businesses as independent units.

The economic imperative involves three central issues: the definition of the businesses of the firm; the identification and exploitation of interrelationships across those businesses, and the coordination of the business activities that allow sharing assets and skills (Porter, 1987; Pearson, 1989).

There are eight corporate tasks that we associate with the economic imperative of corporate strategy. The first one is the Environmental Scan at the Corporate Level, which allow us to start the reflection of the firm's competitive position by a thorough understanding of the external forces that it is facing. One of the principal objectives of strategy is to seek a proper alignment between the firm and its environment. Therefore, it seems logical to start the corporate strategic planning process with a rigorous examination of the external environment.

The seven additional tasks imply critical strategic decisions seeking the attainment of corporate competitive advantages. They are mission of the firm, business segmentation, horizontal strategy, vertical integration, corporate philosophy, strategic posture of the firm, and portfolio management. We comment now on the essence of these tasks.

1. *Environmental Scan at the corporate level: Understanding the external forces impacting the firm*

– The Environmental Scan provides an assessment of the distinct business opportunities offered by the geographical regions in which the firm operates. It also examines the general trends of the various industrial sectors related to the portfolio of businesses of the corporation. Finally, it describes the favorable and unfavorable impacts to the firm from technological trends, supply of human resources; and political, social, and legal factors. The output of the Environmental Scan is the identification of key opportunities and threats resulting from the impact of external factors.

2. *The mission of the firm: choosing competitive domains and the way to compete* – The mission of the firm defines the business scope – products, markets, and geographical locations – as well as the unique competencies that determine its capabilities. The level of aggregation used to express this mission statement is very broad, because we need to encompass all the critical activities and capabilities of the corporation.

The mission of the firm defines the overall portfolio of businesses. It selects the businesses in which the firm will enter or exit, as well as the discretionary allocation of tangible and intangible resources assigned to them. The selection of a business scope at the level of the firm is often very hard to reverse without incurring significant or prohibitive costs. The development of unique competencies shape the *corporate advantage*, namely, the capabilities that will be transferred across the portfolio of businesses.

The mission of the firm involves two of the most essential decisions of corporate strategy: selecting the businesses of the firm, and integrating the business strategies to create additional economic value. Mistakes in these two categories of decisions could be painful, because the stakes that are assigned to the resulting bets are very high indeed.

3. *Business segmentation: selecting planning and organizational focuses* – The mission of the firm defines its business scope, namely the products and services it generates, the markets it serves, and the geographical locations in which it operates. The business segmentation defines the perspectives or dimensions that will be used to group these activities in a way that will be managed most effectively. It adds planning and organizational focuses which are central for both the strategic analysis and the implementation of the business strategies. This concept is of great importance in the conduct of a formal strategic planning process, since the resulting businesses are the most relevant units of analysis in that process.

4. *Horizontal strategy: pursuing synergistic linkages across business units* – One could argue that horizontal strategies are the primary sources for corporate advantage of a diversified firm. It is through the detection and realization of the existing synergy across the various businesses that significant additional economic value can be created. The value chain is the basic framework that is used to detect opportunities for sharing resources and activities across businesses (Porter, 1985). The resulting degree of linkages among businesses determines their relative autonomy and independence.

The mission of the firm defines the business scope; business segmentation organizes the businesses into planning and managerial units; horizontal strategies determines their degree of interdependence. Consequently, these tasks are highly linked. Moreover, the mission of the firm also defines the current and future corporate core competencies, which are the basis that supports the relationship among the various businesses, and the role to be played by horizontal strategy.

5. *Vertical integration: defining the boundaries of the firm* – Vertical integration determines the breadth of the value chain, as well as the intensity of each of the activities performed internally by the firm. It specifies the firm's boundaries, and establishes the relationship of the firm with its primary outside constituencies – suppliers, distributors, and customers.

The major benefits of vertical integration are realized through: cost reductions from economies of scale and scope; creation of defensive market power against suppliers and clients; and creation of offensive market power to profit from new business opportunities. The main deterrents of vertical integration are: diseconomies of scale from increases in overhead and capital investments; loss of flexibility; and administrative penalties stemming from more complex managerial activities (Stuckey and White, 1993; Harrigan, 1985; Walker, 1988; Teece, 1987).

6. *Corporate philosophy: defining the relationship between the firm and its stakeholders* – The corporate philosophy provides a unifying theme and a statement of basic principles for the organization. First, it addresses the relationship between the firm and its employees, customers, suppliers, communities, and shareholders. Second, it specifies broad objectives for the firm's growth and profitability. Third, it defines the basic corporate policies; and finally, it comments on issues of ethics, beliefs, and rules of personal and corporate conduct.

The corporate philosophy is the task that is most closely related to the leadership imperative, insofar as bringing a capability to articulate key elements of the CEO's vision.

7. *Strategic posture of the firm: identifying the strategic thrusts, and corporate performance objectives* – The strategic posture of the firm is a set of pragmatic requirements developed at the corporate level to guide the formulation of corporate, business, and functional strategies. The strategic thrusts characterize the strategic agenda of the firm. They identify all of the key strategic issues, and signal the organizational units responsible to respond to them. The corporate performance objectives define the key indicators used to evaluate the managerial results, and assign numerical targets as an expression of the strategic intent of the firm. The strategic posture captures the outputs of all of the previous tasks and uses them as challenges to be recognized and dealt with in terms of action-driven issues.

8. *Portfolio management: assigning priorities for resource allocation and identifying opportunities for diversification and divestment* – Portfolio management and resource allocation have always been recognized as responsibilities that reside squarely at the corporate level. We already have commented that the development of core competencies shared by the various businesses of the firm constitute a critical source of corporate advantage. Those competencies are borne from resources that the firm should be able to nurture and deploy effectively, including: physical assets, like plant and equipment; intangible assets, like highly-recognized brands; and capabilities, like skills associated with product design and development.

The heart of an effective resource allocation process is the capacity to create economic value. Sometimes, this value emerges from internal activities of the firm, other times it is acquired from external sources through mergers, acquisitions, joint ventures, and other forms of alliances. Even, on occasions, value can be created by divesting businesses that are not earning their cost of capital – i.e. they are destroying instead of adding value to the firm. Portfolio management deals with all of these critical issues.

In the 1980s, most developed economies faced periods of stagnation which have forced firms to implement drastic restructuring policies. Restructuring leads to the realignment of physical assets – including divestment – human resources, and organizational boundaries of the various businesses with the intent of reshaping their structure and performance. Restructuring decisions are also part of portfolio management (Donaldson, 1994).

THE MANAGERIAL IMPERATIVE: This imperative is the major determinant for a successful implementation of corporate strategy. It involves two additional important corporate tasks: the design of the firm managerial infrastructure, and the management of its key personnel.

9. *Managerial infrastructure: designing and adjusting the organizational structure, managerial processes, and systems in consonance with the culture of the firm to facilitate the implementation of strategy* – Organizational structure and administrative systems constitute the managerial infrastructure of the firm. An effective managerial infrastructure is critical for the successful implementation of the strategies of the firm. Its ultimate objective is the development of corporate values, managerial capabilities, organizational responsibilities, and managerial processes to create a self-sustaining set of rules that allow the decentralization of the activities of the firm.

The term organizational architecture is commonly used to designate the design efforts that produce an alignment between the environment, the organizational resources, the culture of the firm, and its strategy (Nadler *et al.*, 1992).

10. *Human resources management of key personnel: selection, development, appraisal, rewards, and promotion* – Regardless how large a corporation is, it will be always managed by a few key individuals. Percy Barnevik, the CEO of Asea Brown-Boveri, a successful global company, stated that one of ABB's biggest priorities and crucial bottlenecks is to create global managers. However, he immediately added that a global company does not need thousands of them. At ABB, five hundred out of a total of fifteen thousand managers are enough to make ABB work well (Taylor, 1991).

Tom MacAvoy, the former President of Corning Glass-Works, used to talk in a rather colorful way about the need for "one hundred centurions" to run an organization. These are huge corporations, with operations in over one hundred countries. When it comes to identifying the key personnel they need, the numbers are surprisingly small; yet, the process of identifying, developing, promoting, rewarding, and retaining them, is one of the toughest challenges that an organization faces.

THE FUNDAMENTAL ELEMENTS IN THE DEFINITION OF CORPORATE STRATEGY: We can organize the corporate strategic tasks in a strategic planning framework that we label "The Fundamental Elements in the Definition of Corporate Strategy: The Ten Tasks" (Figure 1).

The first element of the framework – The Central Focus of Corporate Strategy – consists in identifying the entity that is going to be part of the corporate strategic analysis. As opposed to the case of business strategy, where the unit of analysis is the Strategic Business Unit (SBU), corporate strategy can be applied at different levels in a large diversified organization. The amplest possible scope is the firm as a whole. However, there are circumstances under which we want to narrow the scope of the analysis to a sector, group or division of a given organization. These entities should encompass a number of

different business units to be the subject of a meaningful corporate strategic analysis.

Next, there are two important set of issues that we label Corporate Environmental Scan and Corporate Internal Scrutiny. Before we address the set of tasks associated with these issues, we need to define the time frame to be used. There is an underlying time frame which has to be spelled out at the beginning of the planning process. Throughout the corporate strategic analysis, we are contrasting existing conditions with future ones.

In the case of the Environmental Scan, there are two different treatments of the future. When we are dealing with completely uncontrollable factors, we need to forecast their most likely

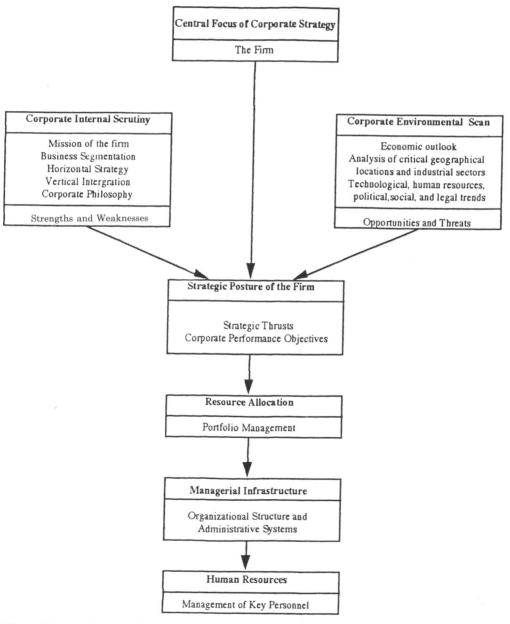

Fig. 1. The fundamental elements in the definition of corporate strategy: the ten tasks.

trends to be able to understand their potential impacts. However, there are cases in which we would like to influence future events, when we can exercise some degree of control that will allow us to shape the future in our advantage.

By contrast, in all of the tasks that are part of the Internal Scrutiny, the future represents a state that we are aiming at through a set of controllable decisions.

The Corporate Environmental Scan should be conducted first in the planning process, because it serves to frame the impacts resulting from the external environment. It has also the important role of transferring a common set of assumptions to the various businesses and functional managers of the firm, to serve as inputs in their own strategic planning efforts. It gives a sense of uniformity to the strategic planning thinking across all the key organizational units of the firm. This task culminates with the recognition of opportunities – the favorable impacts of the external environment which we would like to seize – and threats – the unfavorable impacts which we would like to neutralize.

The Corporate Internal Scrutiny captures the key actions and decisions the corporation has to address to gain a competitive position that is in line with the challenges generated by the external environment, and conducive to the development of a sustainable corporate advantage. As we have indicated, this advantage is transferable to the various business units of the firm, and enhances its resources and capabilities. The tasks which are part of the Internal Scrutiny in our framework are:

- Mission of the Firm
- Business Segmentation
- Horizontal Strategy
- Vertical Integration
- Corporate Philosophy

In all of these decisions we contrast the current state with a desirable future one, and we proceed to define the challenges those changes generate for the formulation of corporate strategy. The Internal Scrutiny concludes with an overall statement of corporate strengths – that the firm wishes to maintain and reinforce – as well as a statement of corporate weaknesses – that the firm wishes to correct or eliminate.

The Corporate Environmental Scan and the Corporate Internal Scrutiny provide the basic inputs that will define the Strategic Posture of the firm. This task serves as a synthesis of the analysis conducted so far, and captures the strategic agenda of the firm. The strategic thrusts are a powerful expression of all of the issues that, from the perspective of the firm, need

to be addressed to come out with an integrative strategy. The Corporate Performance Objectives define the key indicators that will be used to detect the operational and strategic effectiveness of the firm. The Strategic Posture is the essence of the formulation of the corporate strategy, and as such, it is a task that should receive the utmost attention. When properly conducted, the firm is able to frame the activities, responsibilities and performance measurements that are critical for its superior strategic position.

The subsequent task – Resource Allocation and Portfolio Management – permits to backup the strategic actions implicit in the Strategic Posture of the firm with the necessary resources needed for their deployment. We are entering now into the realm of strategy implementation. These implementation efforts are going to be strongly reinforced by the remaining two corporate tasks: Managerial Infrastructure, and Human Resources Management of Key Personnel.

See **Organization studies.**

References

[1] G. Donaldson (1994), *Corporate Restructuring, Managing the Change Process from Within*, Harvard Business School Press, Boston.

[2] G. Hamel and C.K. Prahalad (1989), "Strategic Intent," *Harvard Business Review*, 67 (3), 63–76.

[3] K.R. Harrigan (1985), *Strategic Flexibility: A Management Guide for Changing Times*, Lexington Books, Lexington, Massachusetts.

[4] A.C. Hax and N.S. Majluf (1991), *The Strategy Concept and Process: A Pragmatic Approach*, Prentice Hall, Englewood Cliffs, New Jersey.

[5] J.P. Kotter (1988), *The Leadership Factor*, Free Press, New York.

[6] D.A. Nadler, M.S. Gerstein, R.B. Shaw, and Associates (1992), *Organizational Architecture: Designs for Changing Organizations*, Jossey-Bass, San Francisco.

[7] A.E. Pearson (1989), "Six Basics for General Managers," *Harvard Business Review* 67 (4), 94–101.

[8] J. Pfeffer (1992), *Managing with Power: Politics and Influence in Organizations*, Harvard Business School Press, Boston.

[9] M.E. Porter (1985), *Competitive Advantage*, Free Press, New York.

[10] M.E. Porter (1987), "From Competitive Advantage to Corporate Strategy," *Harvard Business Review*, 65 (3), 43–59.

[11] E.E. Schein (1992), *Organizational Culture and Leadership*, 2nd ed., Jossey-Bass, San Francisco.

[12] J. Stuckey and D. White (1993), "When and When Not to Vertically Integrate," *Sloan Management Review*, (Spring), 34 (3), 71–83.

[13] W. Taylor (1991), "The Logic of Global Business: An Interview with ABB's Percy Barnevik," *Harvard Business Review*, 69 (2), 90–105.

[14] D.J. Teece (1987), "Profiting from Technological Innovations: Implications for Integration, Collaboration, Licensing, and Public Policy," in *The Competitive Challenge*: *Strategies for Industrial Innovations and Renewal*, D.J. Teece, ed., Ballinger Publishing, Cambridge, Massachusetts.

[15] G. Walker (1988), "Strategic Sourcing, Vertical Integration and Transaction Costs," *Interfaces*, 19 (3), 62–73.

COST ANALYSIS

Stephen J. Balut

The Institute for Defense Analyses, Alexandria, Virginia

Thomas R. Gulledge

George Mason University, Fairfax, Virginia

Cost analysis is the process of estimating the individual and comparative costs of alternative ways of accomplishing an objective. The goal is not to forecast precisely accurate costs, but rather to reveal the extent to which one alternative costs more or less than another. A cost analysis is often conducted in conjunction with an effectiveness analysis to aid in the selection of one alternative over others.

EVOLUTION: Cost analysis emerged as part of a broader initiative in the late 1940s and early 1950s to apply economic principles to the decision making process of the Department of Defense (DoD). A confluence of events following World War II resulted in a dramatic and enduring change in the way resource allocation decisions were made in public organizations. The development and evolution of cost-effectiveness analysis and cost analysis occurred nearly simultaneously and are therefore closely related. Both types of analysis make use of operations research methods.

Operations research was invented and applied mainly by civilian scientists in support of the war effort. From its inception, operations research sought to "use scientific methods to get the most out of available resources" (Quade, 1971). Immediately following the war, many of these scientists were retained by the Military Departments to apply newly developed quantitative methods to aid defense decisions. The forerunners of the RAND Corporation, the Institute for Defense Analyses (IDA), and the Center for Naval Analyses (CNA) were formed during this period.

After the war, separation of military responsibilities between the U.S. Armed Services broke down as a consequence of the rapid development of military technology and the different character of the military threat (Smale, 1967). The Services began competing for missions and disputes were settled via approval of budgets for new weapon systems. Competing systems were considered on the basis of cost-effectiveness. When equally effective weapon systems were compared, those estimated to cost the least won funding approvals. The analytical procedure applied to such decisions was first named "weapon systems analysis," a term later shortened to "systems analysis." The first documented systems analysis was accomplished in 1949 by the RAND Corporation and compared the B-52 to a turbo-prop bomber. The use of "dollar costs" as a proxy for real costs changed the basic systems analysis question from "Which weapon system is best for the job?" to "Given a fixed budget, which weapon system is most cost effective?" (Smale, 1967; Novick, 1988).

The birth of cost analysis as a separate activity occurred in the early 1950s and is attributed to Novick (1988), a cost analyst with the RAND Corporation. Novick pioneered weapon system cost analysis and is referred to as the "father of cost analysis." Novick and his group at RAND are attributed with development of the fundamental building blocks of cost analysis. These include separation of total costs into cost elements, separation of one-time and recurring costs, development of cost estimating relationships, and development of conceptual costs or order-of-magnitude estimates used to compare future system proposals. Novick's group went on to invent parametric cost estimating, incremental costing, and "Total Force Costing" (Novick, 1988; Hough, 1989).

In the early 1960s, the Department of Defense established and implemented a centralized resource allocation process called the Planning, Programming and Budgeting System (PPBS). Under this system, future defense resources were allocated to missions in a systematic, rational manner using cost-effectiveness as the decision criterion. In 1961, a Systems Analysis office was established within the Office of the Secretary of Defense (OSD) to help implement this new resource allocation procedure. In 1965, a Cost Analysis Division was established within the office of the Assistant Secretary of Defense, Systems Analysis. With this act, cost analysis gained a primary role in the examination of alternative force structures at the OSD level. Also in 1965, the PPBS system was extended to all federal agencies by U.S. President Lyndon Johnson.

The next few decades brought initiatives that strengthened the cost analysis capabilities of the DoD. The military departments established cost analysis offices at headquarters and major commands and staffed them, at least in part, with people trained and experienced in the methods of operations research. The DoD initiated systematic collection of cost information from defense contractors to provide defense cost analysts with records of cost experiences on major weapon system acquisitions. These records formed the bases of estimates of the costs of proposed systems at acquisition milestone decision points, strengthened the DoD's position during contract negotiations, and provided for DoD tracking of negotiated costs. In 1971, U.S. Deputy Secretary of Defense Packard instituted defense acquisition reforms that included establishment of the DoD Cost Analysis Improvement Group (Hough, 1989), the requirement for independent parametric estimates for new systems acquisitions, formalization of cost analysis reviews at milestone decision points, and requirements for the military departments to improve their cost-estimating capabilities. As part of the Packard reform, cost was elevated to a principal design parameter with implementation of the "Design to Cost" initiative (Hough, 1989). Ten years later, in 1981, U.S. Deputy Secretary of Defense Carlucci placed further demands on the DoD's cost analysis capabilities. Mr. Carlucci instituted the practice of "Multi-Year Procurement" based on benefit/risk analyses, "Budget to Most Likely or Expected Cost," budgeting more realistically for inflation, the use of economic production rates, the requirement to forecast business base at defense contractors' plants, increased efforts to quantify cost risk and uncertainty, and provision of greater incentives on design-to-cost goals by tying award fees to actual costs achieved in production.

Throughout the 1970s and 1980s, the practice of cost analysis continued to expand mainly in the public sector. Government cost analysis organizations grew in size by drawing people skilled in engineering, economics, operations research, accounting, mathematics, statistics, business, and related fields. Several focused educational programs were initiated to support this budding profession at military universities, including the Air Force Institute of Technology, the Naval Postgraduate School, and the Defense Systems Management College.

The 1990s brought a surge of activity in cost analysis with institutionalization of a Cost and Operational Effectiveness Analysis (COEA) as an integral part of the defense acquisition process. COEAs are now required to be conducted and presented to the Defense Acquisition Executive at each major milestone in the acquisition of a major weapon system.

METHODS: Cost analysis is a sequential process: first identification, then measurement, and finally evaluation of alternatives. This involves the structuring and analysis of resource alternatives in a full planning context. In the case of defense, the size of the U.S. defense budget limits the dollars available to provide for the national defense. Monies spent on one mission/capability/weapon system are not available to spend on another. "Therefore, properly constructed cost estimates and cost analyses are essential because an accurate assessment of the cost of individual programs is the first necessary step towards understanding the comparative benefits of alternative programs and capabilities" (Smale, 1967).

"Economic costs" are benefits lost and are often referred to as "alternative costs" or "opportunity costs" (Fisher, 1970). An estimate of the economic cost of one choice, decision or alternative, within this context, is an estimate of the benefits that could otherwise have been obtained by choosing the best of the remaining alternatives. When constructed in this way, costs have the same dimension as benefits, and direct comparison is possible.

The following cost analysis concepts are briefly described here: the Work Breakdown Structure (WBS), estimating relationships, and cost progress curves. The treatment is not comprehensive in any sense and is provided to give those completely unfamiliar with the methods of cost analysis an idea of what is involved.

Work Breakdown Structure – Cost analysts break complex systems down into pieces before attempting to estimate their costs. A notion fundamental to this process is the Work Breakdown Structure (WBS) (U.S. Air Force Material Command, 1993). The basic concept of a WBS is to represent an aircraft system, for example, as a hierarchical tree composed of hardware, software, facilities, data, services, and other work tasks. This tree completely defines the product and the work to be accomplished. It relates elements of work to each other and to the end product. Cost analysts usually estimate total systems costs as the sum of the costs of the individual elements of the WBS.

Estimating Relationships – Another tool that is fundamental to cost analysis is the estimating relationship (ER). In a broad sense, "estimating relationships are 'transformation devices' which permit cost analysts to go from basic inputs (for example, descriptive information for some future weapon system) to estimates of the cost of output-oriented packages of military capability" (Fisher, 1970). More specifically, ERs are analytic devices that relate various categories of cost (e.g., dollars or physical units) to explanatory variables referred to as "cost drivers." While taking many different forms, ERs are usually mathematical functions derived from empirical data using statistical analyses.

Cost Progress Curves – The basic notion of a learning curve is that, as a work procedure (e.g. sequence of steps/activities) is repeated, the person performing the procedure normally becomes better or more efficient at performing the procedure. The reduction in time or cost to perform the procedure is commonly attributed to "learning." Cost analysts, who are more interested in reductions in cost, refer to this phenomenon as "cost progress" rather than learning.

The theory of cost progress curves states that as the total quantity of units (e.g. aircraft, wings, or fuselages) produced doubles, the cost per unit declines by some constant percentage. Wright (1936) empirically demonstrated the principle (Asher, 1956). The standard mathematical model is a power function that relates manufacturing labor hours required to produce a particular unit to the cumulative number of units produced. The functional form is simply:

$$C = aQ^b$$

where C is the number of hours required to produce unit Q, a is the labor hours required to produce the first unit, and b is a parameter that measures the amount of "cost progress" reflected in the data used to estimate the model parameters. The form is a hyperbolic function that is linear in logarithmic space. The characteristic of linearity in logarithmic space and the ease of application account for the general acceptance and popularity of the cost progress curve among cost analysts. The cost progress curve is applied widely by defense cost analysts when estimating the costs of alternative force sizes and compositions.

PROFESSIONAL ORGANIZATIONS: As cost analysis evolved over the past few decades, a number of professional organizations were formed to further advance cost analysis and related professional activities. The Cost-Effectiveness Technical Section of the Operations Research Society of America (now the Institute for Operations Research and the Management Sciences – INFORMS) was formed in 1956 to provide for the exchange of experiences in conducting such analyses. This organization has since changed its name to the Military Application Section (MAS) of INFORMS.

The National Estimating Society (NES) was formed in 1978. This organization's focus was on cost estimating from the perspective of the private sector. The formation of the Institute of Cost Analysis (ICA) in 1981 was referred to as the most significant event of the decade for DoD cost analysts (Hough, 1989). ICA was dedicated to the furtherance of cost analysis in the public and private sectors. Both ICA and NES established programs under which the technical competence of members were certified, leading to a designation of "Certified Cost Analyst," or "Certified Cost Estimator." ICA and NES subsequently merged to form the Society of Cost Estimating and Analysis (SCEA). SCEA continues the certification process by conferring the "Certified Cost Estimator/Analyst" designation to those who pass a qualifying examination.

See **Center for Naval Analyses; Cost effectiveness analysis; RAND Corporation.**

References

[1] Asher, H. (1956), "Cost-Quantity Relationships in the Airframe Industry," R-291, The RAND Corporation, Santa Monica, California.

[2] Fisher, G.H. (1970), "Cost Considerations in Systems Analysis," R-490-ASD, The RAND Corporation, Santa Monica, California.

[3] Hough, P.G. (1989), "Birth of a Profession: Four Decades of Military Cost Analysis," The RAND Corporation, Santa Monica, California.

[4] Novick, D. (1988), "Beginnings of Military Cost Analysis: 1950–1961," P-7425, The RAND Corporation, Santa Monica, California.

[5] Smale, G.F. (1967), "A Commentary on Defense Management," Industrial College of the Armed Forces, Washington D. C.

[6] Quade, E.S. (1971), "A History of Cost-Effectiveness Analysis," Paper P-4557, The RAND Corporation, Santa Monica, California.

[7] U.S. Air Force Materiel Command (1993), "Work Breakdown Structures for Defense Material Items," Military Standard 881B.

[8] Wright, T.P. (1936), "Factors Affecting the Cost of Airplanes," *Jl. Aeronautical Sciences*, 3, 122–128.

COST COEFFICIENT

In a linear-programming problem, the generic name given to the objective function coefficients.

COST EFFECTIVENESS ANALYSIS

Norman Keith Womer

University of Mississippi

INTRODUCTION: Cost effectiveness analysis is a practical way of assessing the usefulness of public projects. The history of the subject can be traced to Dupuit's classic 1844 paper, "On the Measurement of the Utility of Public Works." The technique has been a mainstay of the Army Corps of Engineers since 1902. Recent variations of the technique have been labeled cost effectiveness analysis, cost benefit analysis, systems analysis, or merely analysis. It has been extensively applied to projects in defense, transportation, irrigation, waterways, and housing.

Cost effectiveness analysis is the process of using theory, data and models to examine a problem's relevant objectives and alternative means of achieving them. It is used to compare the costs, benefits, and risks of alternative solutions to a problem and to assist decision makers in choosing among them. The differences between cost effectiveness analysis and the discipline of operations research itself are subtle and, in some treatments, merely a "matter of emphasis." (See the discussion in Quade, 1971.) The convention adopted here is that operations research is a body of knowledge that includes all of the tools and methods that might be used in any study. *Cost effective analysis* (CEA) is a particular application of models and methods to a choice problem.

Sometimes CEA is portrayed as the combination of the difficult problem of measuring effectiveness with the rather mundane problem of cost estimation. In fact, cost measurement is an important issue. Cost effectiveness analysis provides a tool for effective resource allocation only when all the resource implications associated with each alternative – both direct and indirect – are included in the analysis. The opportunity cost of some proposed allocation of resources is the value of those resources in their best alternative use. The very concept of opportunity cost therefore requires knowledge of the goals and objectives, measures of effectiveness, the other alternatives and constraints that the organization. That is, to employ this basic concept of cost, a careful analysis of the problem must be accomplished.

Therefore, CEA must focus on the process of modelling both cost and effectiveness to develop relevant measures that shed light on the problem under study. Ultimately, CEA consists of methods for evaluation vectors of measure. In the process, CEA must grapple with issues like the scale of operations, risk, uncertainty, timing, and actions of other players.

THE ROLE OF MODELS: Figure 1, adapted from Quade (1970), portrays the elements of CEA. Models are used in CEA to aid in the evaluation of alternatives. These models often take the form of equations which relate the physical description of alternative systems to various impacts on their production and use. The models may concern the acquisition of the systems, their operation, or various circumstances associated with applying the system in an environment.

There are many assumptions in any analysis. One important class of assumptions that is often left out concerns the behavior of key players in the process. Traditionally, CEA's have been based on rather mechanical models that relate a system's physical characteristics (e.g., weight and speed) to production cost. Any reference to behavior has often been confined to vague statements about efficiency. In fact, costs and benefits result only from actions. Thus, the motivation to act is an important part of modeling costs and benefits. Unfortunately, these behavioral assumptions are often not stated explicitly. Instead, they are frequently imbedded in detailed computer simulations that attempt to emulate the simultaneous operation of complex systems in realistic environments.

INCOMMENSURABLE IMPACTS: The output of a suite of models may be a rather long list of measured system impacts. Some of the system impacts are measurable in units of effectiveness

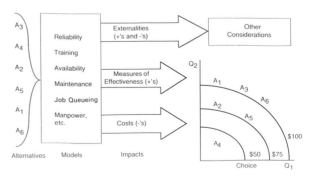

Fig. 1. The elements of cost effectiveness analysis.

or costs while others are external to our frame of reference. Generally, each of the impacts will be measurable in units that are unique to that impact, for example, number of lives lost, replacement cost of lost equipment, number of minutes of error free transmission accomplished, etc. Choice requires not only the objective consideration of the measurable impacts, but also the consideration of the often unmeasurable externalities. As a result, it is important that the analyst carefully report both impact measure and their accuracy and those impacts that remain unmeasured. Choice also requires the explicit use of a criterion which evaluates the impacts and their relation to the choice problem at hand.

THE ANALYST AND THE DECISION MAKER:
In doing analysis, the first and most important issue to understand is the decision maker's problem. Answering the question "What is the problem?" often requires understanding both the organization for whom the analysis is performed and the physical system or structural change that is under study. The problem may be stated in different forms at different points in time and at different levels in the organization. Thus, understanding the problem requires understanding the objectives of the entire organization.

For example, consider the problem of analyzing the cost of a mission currently assigned to an aircraft system. What is the problem? Some candidates are:
- Should the existing system be replaced?
- What design should be chosen?
- Who should produce the system?
- How should the mission be performed?
- Is the mission affordable?

Often analysis is done with reference to one of these problems and then later the same study is applied to a different problem. Clearly, the alternatives, the risks, the objectives and the cost are not independent of the problem being addressed.

Whose problem is this? It is the analyst who must choose the techniques, collect the data, model the processes and measure the costs and outputs. It is the analyst who must justify the choices made in the particular context of the problem being addressed. Thus, it is the analyst who must justify the choices made in the particular context of the problem being addressed. Thus, it is the analyst who must be able to answer the question, "What is the Problem?"

If the role of the analyst is so large, what is left for the decision maker? The decision maker must also completely understand the problem and

judge the value of the analysis. The decision maker must examine the completeness of the alternatives, evaluate the assumptions, examine the measurement of the impacts, and determine if risks are adequately addressed. All of these tasks are important, but the most important task of the decision maker is the task of evaluating the relative importance of the various positive and negative impacts. This includes not only the impacts that are internal to the organization, but also the externalities. Evaluating the impacts also means dealing with their risks and uncertainties. The decision maker's values also include his or her attitudes toward risk. It is in this effort that the decision maker's role is uniquely different from that of the analyst. Once the impacts have been evaluated, then choice is merely a matter of adding them up and comparing the weighted impacts of each of the alternatives.

CRITERIA: *Cost benefit ratios* – Cost effectiveness analysis often is implemented by classifying each impact of a system as either a cost or a benefit. Common units are then found for costs and for benefits and the discounted present value of each is calculated. Alternatives are compared by the ratio of these two measures.

Using a cost benefit ratio to choose among alternatives presents several problems. Often, this approach leaves out relevant measures (treating them as the externalities depicted in Figure 1) because those impacts cannot be evaluated in units that are comparable to the main impacts. Choosing units for the main impacts involves subjective decisions that trade-off relative values of measures of merit. For example, lives lost must be compared to visual pollution or environmental impact must be valued relative to economic loss. The person who determines common units for such diverse measures of merit is no longer playing the role of an analyst. That person is acting as the decision maker.

The alternative is to leave the various measures of merit uncombined. But a major problem occurs with ratio analysis when the analysis must consider multiple inputs and multiple outputs. Several ratios may be constructed but then it is not clear how these multiple ratios should be combined to determine the overall value of an alternative. Cost benefit ratios provide the decision maker with little guidance on how to proceed in this case.

Another problem with ratio analysis is the constant returns to scale assumption that is implicit in calculating a cost benefit ratio. By displaying the results in ratio form the analyst implies that if the system is expanded or contracted, the costs and the benefits both change proportionately. Unfortunately, the world is replete with examples of alternatives that violate such proportionality rules. Finally, ratio analysis does not lend itself to explicit treatments of risk and uncertainty.

Production functions – The production function approach to CEA can deal with variable returns to scale and with other nonlinearities in technology. Numerous estimates of costs and benefits for various alternatives at different scale levels are used to fit a nonlinear production function by regression. This technique can deal with several measures of input and can therefore overcome some of the difficulties of the cost benefit ratio. Production functions can also incorporate risk described with random variables. But, the multiple regression production function also has some drawbacks. First, the use of regression tends to measure efficiency relative to average performance instead of best performance. That is, all the observations are pooled to fit the production function, a measure of average efficiency, then each alternative is compared to that average measure. Also, multiple regression requires that a single indicator for output be used. Thus, multiple outputs must be combined into a single effectiveness indicator, similar to ratio analysis. This type of problem is especially severe in non-profit and governmental organizations where prices for outputs are unavailable or incomplete. Charnes and Cooper (1985) also criticized regression's lack of ability "in identifying the underlying sources and amounts of inefficiencies."

Data envelopment analysis – A recently developed technique, Data Envelopment Analysis provides an efficiency measure that offers some aid for the criterion problem. This linear programming based measure has its origin in linear production theory. Golany (1988) pointed out the "DEA is quickly emerging as the leading method for efficient evaluation, in terms of both the number of research papers published and the number of applications to real world problems."

DEA is a procedure that has been designed specifically to measure relative efficiency in situations in which there are multiple measures of merit and there is no obvious objective way of aggregating measures of merit into a meaningful index of productive efficiency. Compared to regression, which averages the aggregate impact of a system, DEA is an extremal method. DEA calculates the efficiency of each alternative by comparing (via mathematical programming models) an alternative's measures of merit with the measures of merit of the other alternatives. Each

alternative's measures of merit are weighed as favorably as possible. If the alternative is inefficient, DEA indicates which of its measures of merit imply its inefficiency. Also, DEA does not require the parametric specification of a production function; it derives an estimate of the production function directly from the observed data on elements of cost and effectiveness that are model outputs. DEA has been used to measure the productivity and efficiency of many organizations. It has been particularly useful for public sector organizations where market prices of outputs are not available. DEA has the potential to be extremely helpful in developing criteria in cost effectiveness analyses.

EXAMPLES: Cost effectiveness analyses have been conducted in support of (and in opposition to) numerous significant national decisions. For example, the study of alternative delivery systems that resulted in the choice of the space shuttle, the series of studies on the anti-ballistic missile, and the studies for and against the break up of AT&T are classic studies that illustrate both the power and the fragility of this important concept.

See **Cost analysis; Data envelopment analysis; Measure of effectiveness; Multicriteria decision making; Opportunity cost.**

References

[1] Charnes, A. and W.W. Cooper (1985), "Preface to Topics in Data Envelopment Analysis," *Annals of Operations Research*, 2, 59–94.

[2] Sueyoshi, T. (1988), "A Goal Programming/Constrained Regression Review of the Bell System Breakup," *Management Science*, 34, 1–26.

[3] Dupuit, J. (1844), "De la mesure de l'utilité des travaux publics," Reprinted in *Jules Dupuit, De l'utilité et de sa mesure*, Torino, la Roforma sociate, 1933.

[4] Evans, D.S. and J.J. Heckman (1983), "Natural Monopoly," in *Breaking Up Bell*, D.S. Evans, ed., North Holland, New York, 127–156.

[5] Evans, D.S. and J.J. Heckman (1988), "Natural Monopoly and the Bell System: Response to Charnes, Cooper and Sueyoshi," *Management Science*, 34, 27–38.

[6] Gregory, W.H. (1973), "NASA Analyzes Shuttle Economics," *Aviation Week and Space Technology*, September.

[7] Golany, B. (1988), "An Interactive MOLP Procedure for the Extension of DEA to Effectiveness Analysis," *Jl. Operational Research Society*, 39, 725–734.

[8] Heiss, K.P. and O. Morgenstern (1971), "Factors for a Decision on a New Reusable Space Trans-

portation System," memorandum to Dr. James C. Fletcher, Administrator, NASA, Mathematica Corp., Princeton, New Jersey.

[9] Operations Research Society of America, "Guidelines for the Practice of Operations Research," *Operations Research*, 19, 1123–1258.

[10] Quade, E.S. (1964), *Analysis of Military Decisions*, United States Air Force Project Rand, R-387-PR, Santa Monica, California, p. 382.

[11] Quade, E.S. (1971), "A History of Cost-Effectiveness," United States Air Force Project Rand, P-4557, Santa Monica, California.

[12] Sueyoshi, T. (1991), "Estimation of Stochastic Frontier Cost Function Using Data Envelopment Analysis: An Application to the AT&T Divestiture," *Jl. Operational Research Society*, 42, 463–477.

COST RANGE

See **Ranging; Sensitivity analysis**.

COST ROW

The row in a simplex tableau that contains the reduced costs of the associated feasible basis. See **Simplex method**.

COST SLOPE

The rate of cost change per unit of time duration of a project's work item. See **Network planning**.

COST VECTOR

In a linear-programming problem, a row vector c whose components are the objective function coefficients of the problem. See **Cost coefficient**.

COV

See **Coefficient of variation**.

COVERING PROBLEM

See **Set-covering problem**.

COXIAN DISTRIBUTION

A probability distribution whose Laplace-Stieltjes transform may be written as the quotient of two polynomials (i.e., a rational function). All Coxian distributions have a phase-type formulation which may include fictitious stages. See **Queueing theory**.

CPM

Critical path method. See **Critical path method; Network planning; Research and development**.

CPP

See **Chinese postman problem**.

CRAMER'S RULE

A formula for calculating the solution of a nonsingular system of linear equations. Cramer's rule states that the solution of the $(n \times n)$ nonsingular linear system $Ax = b$ is $x_i = det\ A_i(b)/det\ A$, $i = 1, \ldots, n$, where $det\ A$ is the determinant of A, and $det\ A_i(b)$ is the determinant of the matrix obtained by replacing the ith column of A by the right-hand-side vector b. This rule is inefficient for numerical computation and its main use is in theoretical analysis. See **Matrices and matrix algebra**.

CRASH COST

The estimated cost for a job (project) based on its crash time. See **Network planning**.

CRASH TIME

The minimal time in which a job may be completed by expediting the work. See **Network planning**.

CREW SCHEDULING

The determination of the temporal and spacial succession of the activities of staff personnel, as, for example, in an airline, train, factory, etc. Such problems are often modeled as mathematical programs.

CRIME AND JUSTICE

Arnold Barnett

*Massachusetts Institute of Technology,
Cambridge*

Michael D. Maltz

University of Illinois at Chicago

INTRODUCTION: Few OR/MS professionals have done research into crime and justice. And few criminal justice researchers have formal backgrounds in OR/MS. Yet, exerting an influence all out of proportion to their numbers, OR/MS scholars have transformed the way many decision-makers think about problems of crime and punishment.

The OR/MS contribution pervades quantitative discussions about crime and justice systems. It has generated a more precise and transparent description of the crime problem than had hitherto been available. It has achieved uneven but sometimes magnificent successes in both identifying and implementing crime-reduction strategies. And it has enhanced the scientific rigor with which criminal justice policy experiments are analyzed and interpreted.

It is not commonly known that some of the most frequently used tools of OR/MS were developed because of crime and justice problems. In the early 19th century, France began to amass statistics on the operation of the criminal justice system (Daston, 1988), and the richness of these data led statisticians to devise new techniques to analyze them. Stigler (1986) describes how Simeon Denis Poisson developed the statistical distribution that bears his name – arguably the "union label" of the OR/MS profession – while modeling conviction rates in French courtrooms. Similarly, Hacking (1990) shows how Poisson developed the law of large numbers by modeling the reliability of jurors in criminal trials.

The modern application of operations research to crime and justice began in the mid-1960s, when operations researchers and systems analysts on a Presidential Crime Commission (STTF, 1967) directed their talents to this area. Since then, the application of OR/MS ideas in this area has burgeoned, as detailed in a recent survey article (Maltz, 1994). We here illustrate some of the more salient roles played by OR/MS in this field. In particular, we discuss how OR/MS has been used in analyzing crime statistics, offender behavior, and criminal justice system dynamics. It also described how queueing models and optimization techniques have been applied in criminal justice contexts, and how OR/MS has caused (some) criminologists to rethink some of their conclusions.

HOMICIDE: In discussing crime, it is natural to start with the most serious offense–murder. Led by the FBI, those assessing homicide patterns had thought it sufficient to consider annual murder rates, expressed in killings per 100,000 citizens per year. The calculated rates had a reassuring quality about them: if 50 per 100,000 citizens were murdered last year, then the other 99,950 were not murdered. Thus, after Detroit had precisely that murder rate in 1973, *The New York Times* reported that "If you live in Detroit, the odds are 2000 to 1 (i.e., 99,950 to 50) that you will not be killed by one of your fellow citizens. Optimists searching for perspective in the city's murder statistics insist that these odds are pretty good."

But some OR/MS scholars raised a question: why measure homicide risk per year as opposed to (say) per day, per month, or per decade? Given that an urban resident is in danger of being murdered throughout his life, that would seem the natural time frame over which to measure the risk. And at an annual risk of 1 in 2000, a person with a natural lifespan of 70 years would face a lifetime murder risk of 1 in 28 (!!). Refinements of this raw calculation leave its result virtually unchanged. (Barnett, Kleitman and Larson, 1975; Barnett, Essenfeld and Kleitman, 1980; Barnett and Schwartz, 1989.)

The idea of estimating lifetime risk of murder has come into general use: detailed projections appeared in the 1981 FBI *Uniform Crime Reports*, and such forecasts have since been incorporated into the actuarial projections of the (U.S.) Centers for Disease Control. There is now widespread awareness that homicide is not a tragic, rare phenomenon, but instead a critical public health problem.

OFFENDER BEHAVIOR: From the standpoint of public policy, it makes a great deal of difference whether existing crimes are committed by relatively few individuals who all offend frequently or by a large number who all offend rarely. OR/MS researchers have taken part in efforts to estimate the total number of offenders, some of whom may never be apprehended for their crimes (Greene and Stollmack, 1981; Greene, 1984). Of course, the offender population is highly diverse in terms of both frequency of criminal activity and types of crime committed

(Chaiken and Chaiken, 1982). A major OR/MS contribution to criminal justice has been in creating succinct models that can characterize both individual criminal behavior and the variation of that behavior across offenders.

Most offenders do not commit crimes according to some deterministic schedule. The exact nature of their crime-generation process is by and large unknown, but it is generally safe to say that the aggregate crime commission by a group of offenders can be modeled by the Poisson distribution. This distribution plays the same role in aggregating point processes that the normal distribution plays in aggregating continuous processes.

In highly influential work, Shinnar and Shinnar (1975) proposed a simple but insightful model of the crime and punishment process. The authors assumed that an active offender commits crimes at a Poisson rate λ per year over a career of length T years. If not arrested, he would commit on average λT crimes over his career. But things change if, like Shinnar and Shinnar, we assume that the offender's probability of arrest for each crime is q, that his probability of imprisonment given arrest is Q, and that the average sentence length per prison term is S. If career length T is long relative to sentence length S, then steady-state arguments imply that, under the revised scenario, the offender is free on average for only $1/\lambda qQ$ years between successive imprisonments, Figure 1. Thus, because of detention, he is free and active only for fraction $(1/\lambda qQ)/(1/\lambda qQ + S)$ of his career rather than for all of it. It follows that incapacitation has reduced his total number of offenses by the proportion $S/(S + 1/\lambda qQ)$ compared the number in a world free of punishment.

There are some gross simplifications in this model (for example, the career length L is assumed independent of the punishment policy in place). But it encapsulates in one equation the effects of all primary elements of the criminal justice system: the offender (via crime commission rate λ), the police (arrest probability q), the courts (chance of imprisonment given arrest Q), and the correctional system (sentence length S). The model also provided guidance to those exploring empirical data (e.g., offenders' arrest, sentencing, and conviction records) about which quantities were especially worth trying to estimate.

OR/MS professionals like Blumstein and his colleagues worked to flesh out the description of the individual criminal career (Blumstein, Cohen, Roth and Visher, 1986). The estimated key parameters like the proportion of citizens who participated at some time in criminal behavior, the frequency of crime-commission during the career, the degree to which offenders specialize by crime-type, and the duration of the criminal career. A simple model of the career might summarize it with four parameters: P, the fraction of individuals in a birth cohort who initiate criminal careers; A, the age of onset for the career; λ, the average annual crime commission rate while free and active, and ρ, the annual probability that the career ends. A macroscopic model could reflect diversity among offenders by assigning a population-wide distribution to each of these parameters. (Other distributions about offense type would fill out the description; for example, Chaiken and Chaiken, 1982.) Interestingly, offenders who differ greatly on some parameters may be quite similar on others. In a cohort of London multiple-offenders, for example, individuals appear to differ far more in their λ-values than their ρ-values (Barnett, Blumstein, and Farrington, 1987). Thus, their career lengths may diverge far less than do intensities of activity during their careers.

As is shown in Figure 1, many offenders continue to commit offenses despite their having

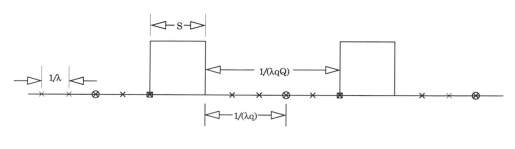

×	crime, at rate λ/year
⊗	arrest, at rate λq/year
⊠	arrest, conviction, and incarceration at rate λqQ/year
S	sentence length

Fig. 1. A (deterministic) criminal career.

been in "correctional" institutions. But not all do, and the extent of recidivism (commission of additional offenses) is an important concern in criminal justice research. OR/MS researchers have devised, calibrated, and tested probabilistic models that assess the likelihood that given offenders with given past records will again commit crimes within particular future time periods (Stollmack and Harris, 1974; Harris, Kaylan and Maltz, 1981; Maltz, 1984; Ellerman, Sullo, and Tien, 1992). These flexible and mathematically-rich techniques allow frequent updating of the prognoses for particular individuals.

THE CRIMINAL JUSTICE SYSTEM: With mathematical models, OR/MS professionals described the idea that the criminal justice system – composed of police, courts, and corrections – is, in fact, a system (STTF, 1967), within which policy shifts in one component generally have consequences for the others. An increase in arrests aimed at reducing crime, for example, can first clog the courts and then overcrowd jails and prisons which, in turn, may be required to reduce surging inmate populations by instituting early release programs for those incarcerated. One of the earliest models to incorporate such "feedback" effects was JUSSIM (Belkin, Blumstein, Glass and Lettre, 1972); subsequent efforts include Cassidy (1985) and Morgan (1985). JUSSIM has since been updated and software written for personal computers by the US Department of Justice to permit its widespread use (ILJ, 1991).

While it is hoped that the criminal justice system provides a fair and cost-effective way to reduce crime, there is a national debate about whether this goal is being achieved. The aims of criminal justice system (CJS) activity are to *deter* potential offenders from committing crime, to *incapacitate* those who have been convicted by imprisoning them, and to *rehabilitate* past offenders so that they are harmless in the future. Of course, the system might induce undesirable changes in criminal behavior, such as *brutalization* under which an offender released from prison is more violent than ever before. Statistical investigations by OR/MS researchers have tried to estimate various net effects of the CJS on crime levels (Blumstein, Cohen, and Nagin, 1978; Blumstein, Cohen, Roth and Visher, 1986), as well as to assess the realism of specific attempts to estimate such effects from aggregate data (Barnett, 1981).

QUEUEING MODELS: While everyone recognizes that crime rates vary from neighborhood to

neighborhood and by time of day, OR/MS analysts have built probabilistic models that allow exploration of the consequences of such heterogeneity, especially with respect to practical issues of police deployment and staffing of 911 emergency centers (Larson, 1972; Chelst, 1978). From such models and OR/MS insights into queueing theory have come an unpleasant realization: randomness in the arrival times of calls for service can cause surprisingly large delays in responding to them. Getting six calls randomly distributed over a one-hour period, for example, can yield much slower responses than getting six calls spaced exactly ten minutes apart.

Queueing theory has been applied and extended in developing improved allocation methods for police patrol resources. Such formulations as the hypercube queueing model (Larson, 1974; Larson and Odoni, 1981; Larson and Rich, 1987) and RAND's Patrol Car Allocation Model (Chaiken and Dormont, 1978) have depicted with great accuracy how particular police response strategies affect mean response times, workload imbalance across officers, and a host of other performance measures. The models, which are used by many American cities to set police dispatching strategies, allow the user to vary the number of patrol cars and the deployment rules, and then to observe on a computer screen the performance statistics under each scenario. Recent OR/MS advances allow the user to set priorities in responding to calls for service and to analyze sending multiple vehicles to incidents (Green and Kolesar, 1984). A review of this work is provided in Swersey (1994).

OPTIMIZATION: Optimization, one of the strongest OR/MS specialties, has played a relatively small role in the profession's contribution to criminal justice. For example, limited success has attended OR/MS efforts to suggest "optimal" punishment policies. Under particular assumptions about crime-commission processes and their sensitivity to the sentencing strategy in place, Blumstein and Nagin (1978) and Barnett and Lofaso (1986) have worked out optimal allocations of prison space. But the verification of such assumptions – let alone the estimation of key model parameters – has not gone far enough that such models are taken very seriously. Associated attempts to estimate how prison populations vary with changes in demography and sentencing policy have yielded prison population forecasts that, to put it delicately, do not immediately demonstrate the practicality of the models (Blumstein, Cohen, and Miller, 1980; Barnett, 1987).

Perhaps the most famous OR/MS proposal for optimal prison sentencing was Greenwood's "selective incapacitation" scheme in which heavy sentences would be imposed on offenders with at least four of seven high-risk characteristics (Greenwood, 1982; Chaiken and Rolph, 1980). But data analyses revealed difficulties with implementing such policies (Chaiken and Chaiken, 1982; Greenwood and Turner, 1987), including a high rate of "false positives" (people incarcerated to prevent projected future crimes that would never have occurred were they free). These false positives raise controversies about "sentencing by conjecture" and yield smaller crime-reduction benefits in practice than the strategy can achieve in theory.

Caulkins (1993a,b) has used OR/MS techniques to evaluate strategies for cutting illicit drug use. He argued that "zero tolerance" sentencing policies – which mete out drastic punishment for the sale or possession of even minute quantities of narcotic – might stimulate rather than curb consumption, in part because they could engender an "in for a penny, in for a pound" mentality among pushers and users. He also characterized conditions under which periodic crackdowns at major drug selling sites would reduce the size of the market rather than geographically displace its activities. The general points Caulkins makes could be advanced without mathematics; what Caulkins has done, however, is to show that if one accepts certain plausible assumptions about the behavior of drug offenders, then advocating particular policies may be simply illogical.

A SENSE OF AMBIGUITY: Sometimes OR/MS people have contributed to criminal justice research less by what they said than by what they didn't say. OR/MS scholars approach data with a "sense" of ambiguity: an awareness that a particular empirical pattern is often consistent with a broad range of possibilities. Thus, they have usefully called out "not so fast!" when the most obvious interpretation of certain data was being treated as the only viable one. Four examples of such rescue activities are described below.

One case concerns the Kansas City Preventive Patrol Experiment conducted in the early 1970s. When not responding to calls to service, patrol cars drive randomly through their districts; in theory, such "preventive" patrol reduce crimes because would-be offenders realize that, even if their victims cannot contact the police, a patrol car might reach the crime scene purely by chance. That theory was called into doubt after Kansas City, in a prearranged experiment, acted to increase preventive patrol sharply in some "beats" and virtually eliminate it in others. When neither beat-by-beat crime rates nor citizen perceptions about police presence changed visibly during the (unannounced) experiment, some people saw preventive patrol as having lost any rationale.

Larson (1975) demonstrated with detailed calculations, however, that actual conditions during the Kansas City experiment were quite different from the anticipated ones. Patrol cars from high-activity beats were spending much of their time responding to calls for service from low-activity ones, which had been deprived of all local police vehicles. The upshot was that there was a great deal of police-car movement – often with sirens screaming – in the districts supposedly *without* preventive patrol, and surprisingly little increase in patrol in the districts supposedly saturated with it. Perhaps, Larson argued, the reason crime rates and citizen perceptions did not change was that police activity itself had not meaningfully changed.

A second example concerned the relationship between arrests and age well-known to criminologists. The graph of arrests vs. age is unimodal, reaching a peak in the late teens and then dropping off steadily and sharply. Given this curve, some people argued that it was not cost-effective to give long sentences to offenders convicted at age 30; such offenders, it was contended, were already far less active than at their "primes" and were unlikely to do much harm even if left on the streets.

But Blumstein, Cohen and Hsieh (1982) pointed out that such an analysis was vulnerable to a variant of the well-known ecological fallacy: Even if arrests in the *aggregate* were dropping rapidly with age, it did not follow that *individual* offenders exhibited this pattern. Having studied longitudinal data about individual offenders, they found that the drop in arrests with age reflected not less activity per year among active offenders, but rather a growing fraction of offenders who had retired from criminal activity. Statistically, an individual convicted at 30 (and thus presumably still active at that age) would be expected, if allowed to go free, to commit as many crimes over the next several years as someone several years his junior.

While Americans were debating in the mid-1970s whether to restore the death penalty, several economists came forth with historically-based analyses that purported to weed out extraneous factors and estimate how each execution affects the overall homicide level. The model, whose findings were cited by the US Supreme

Court, purported to show that each execution deterred eight homicides. But Barnett (1981) wondered whether the econometric models being used had sufficient explanatory power to fulfill their ambitious goals. Arguing that homicide levels were subject to roughly Poisson-level statistical noise, he proposed a test of how well the econometric models could "forecast" state-by-state homicide levels in the very data sets used to calibrate them. The test results indicated that the predictions from all the models suffered large systematic errors of unknown cause and that, indeed, the errors were far larger than any reasonable estimate of the size of the effect the models sought to measure. Thus, Barnett concluded, the analyses were not sensitive enough to answer the question that motivated them.

A final example arises from a study that concluded that juvenile detention acts to reduce delinquency. The study found that an average Illinois youth sent to a reformatory, though not "cured" of criminal activity by his stay there, was arrested far fewer times per month *after* his release than just *prior* to his detention. The decline was interpreted as the post-release "suppression effect" on incarceration: getting tough works.

Maltz and Pollock (1980), however, saw another possibility, tied to the phenomenon called regression-to-the-mean. Even if a youth commits crime at a steady frequency and has an unchanging probability of arrest per offense, varying luck in the arrest "lottery" will cause his observed arrest rate to fluctuate from month to month. But the authorities are especially likely to send him to a reformatory after an upsurge of arrests – i.e. at a *peak* of the fluctuating pattern. Thus, even if the reformatory has no effect on his underlying pattern of criminal behavior, his post-detention arrests would likely fall in frequency compared to his "unluckily" high pre-detention levels. Tierney (1983) proposed a revision in their analysis that modified its result, but the work still showed that the suppression effect was quite possibly just an illusion.

These four examples show one of the primary assets of OR/MS thinking as applied to a field so "data rich" as criminal justice. Although data may exhibit certain *aggregate* patterns, these patterns need not illuminate what is happening at the more *detailed* level that is, quite often, the appropriate focus of policy analysis. OR/MS analysts never forget the importance of studying a problem's "molecular structure."

FINAL REMARKS: Given that the U.S. crime problem as of this writing (1994) is about as bad as it has ever been, it is not easy to quantify the overall OR/MS contribution to public safety. OR/MS research has brought about a deeper understanding of the crime problem and how it affects and is affected by the criminal justice system. Crime, however, has such deep psychological, cultural, economic, and social roots that there are limits to what mathematical models can be expected to accomplish on their own.

See **Emergency services; Hypercube model; Program evaluation; Public policy analysis; Queueing theory.**

References

[1]   Barnett, A. (1981). "The Deterrent Effect of Capital Punishment: A Test of Some Recent Studies." *Operations Research* 29, 346–370.

[2]   Barnett, A. (1987). "Prison Populations: A Projection Model." *Operations Research* 35, 18–34.

[3]   Barnett, A., A. Blumstein, and D.P. Farrington (1987). "Probabilistic Models of Youthful Criminal Careers." *Criminology* 30, 83–108.

[4]   Barnett, A., E. Essenfeld, and D.J. Kleitman (1980). "Urban Homicide: Some Recent Developments," *Jl. Criminal Justice* 8, 379–385.

[5]   Barnett, A., D.J. Kleitman, and R.C. Larson. 1975. "On Urban Homicide: A Statistical Analysis." *Jl. Criminal Justice* 3, 85–110.

[6]   Barnett, A., and A.J. Lofaso (1986). "On the Optimal Allocation of Prison Space." In A.J. Swersey and E. Ignall, eds. *Delivery of Urban Services.* TIMS Series in the Management Sciences, 22, 249–268, Elsevier-North Holland, Amsterdam.

[7]   Barnett A., and E. Schwartz (1989). "Urban Homicide: Still the Same." *Jl. Quantitative Criminology* 5, 83–100.

[8]   Belkin, J., A. Blumstein, W. Glass and M. Lettre (1972). "JUSSIM: An Interactive Computer Program and Its Uses in Criminal Justice Planning." In G. Cooper, ed., *Proceedings of International Symposium on Criminal Justice Information and Statistics Systems.* Project SEARCH, Sacramento, 467–477.

[9]   Blumstein, A., J. Cohen and P. Hsieh (1982). *The Durations of Adult Criminal Careers.* Final Report to National Institute of Justice, Carnegie-Mellon University, Pittsburgh.

[10] Blumstein, A., J. Cohen, and H. Miller (1980). "Demographically Disaggregated Projections of Prison Populations." *Jl. Criminal Justice* 8, 1–25.

[11] Blumstein, A., J. Cohen, and D. Nagin, eds. (1978). *Deterrence and Incapacitation: Estimating the Effects of Criminal Sanctions on Crime Rates.* National Academy of Sciences, Washington, D.C.

[12] Blumstein, A., J. Cohen, J.A. Roth, and C. Visher, eds. (1986). *Criminal Careers and "Career Criminals."* Vols. I and II, National Academy of Sciences, Washington, D.C.

[13] Blumstein, A. and D. Nagin (1978). "On the Optimum Use of Incarceration for Crime Control." *Operations Research* 26, 383–405.

[14] Cassidy, R.G. (1985). "Modelling a Criminal Justice System." In D.P. Farrington and R. Tarling, eds. *Prediction in Criminology*. State University of New York Press, Albany.

[15] Caulkins, J. (1993a). "Zero-Tolerance Policies: Do They Inhibit or Stimulate Illicit Drug Consumption?" *Management Science* 39, 458–476.

[16] Caulkins, J. (1993b). "Local Drug Markets' Response to Focused Police Enforcement." *Operations Research* 41, 843–863.

[17] Chaiken, J.M. and M.R. Chaiken (1982). *Varieties of Criminal Behavior*. Report R-2814-NIJ, The Rand Corporation, Santa Monica, California.

[18] Chaiken, J.M., and P. Dormont (1978). "A Patrol Car Allocation Model: Background, Capabilities, and Algorithms." *Management Science* 24, 1280–1300.

[19] Chaiken, J.M. and J. Rolph (1980). "Selective Incapacitation Strategies Based on Estimated Crime Rates." *Operations Research* 28, 1259–1274.

[20] Chelst, K. (1978). "An Algorithm for Deploying a Crime-Directed (Tactical) Patrol Force." *Management Science* 24, 1314–1327.

[21] Daston, L. (1988). *Classical Probability in the Enlightenment*. Princeton University Press, New Jersey.

[22] Ellerman, P., P. Sullo, and J.M. Tien (1992). "An Alternative Approach to Modeling Recidivism Using Quantile Residual Life Functions." *Operations Research* 40, 485–504.

[23] Green, L. and P. Kolesar (1984). "A Comparison of Multiple Dispatch and M/M/C Priority Queueing Models of Police Patrol." *Management Science* 30, 665–670.

[24] Greene, M.A. (1984). "Estimating the Size of the Criminal Population Using an Open Population Approach." *Proceedings of the American Statistical Association, Survey Methods Research Section*, 8–13.

[25] Greene, M.A. and S. Stollmack (1981). "Estimating the Number of Criminals." In J.A. Fox, ed., *Models in Quantitative Criminology*. Academic Press, New York, 1–24.

[26] Greenwood, P.W. (with A. Abrahamse) (1981). *Selective Incapacitation*. Report R-2815-NIJ, The Rand Corporation, Santa Monica, California.

[27] Greenwood P.W., and S. Turner (1987). *Selective Incapacitation Revisited: Why the High-Rate Offenders are Hard to Predict*. Report R-3397-NIJ, The Rand Corporation, Santa Monica, California.

[28] Hacking, I. (1990). *The Taming of Chance*. Cambridge University Press, England.

[29] Harris, C.M., A.R. Kaylan, and M.D. Maltz (1981). "Recent Advances in the Statistics of Recidivism Measurement." In J.A. Fox, ed., *Models of Quantitative Criminology*. Academic Press, New York, 61–79.

[30] Institute for Law and Justice (1991). "CJSSIM: Criminal Justice System Simulation Model: Software and User Manual." Institute for Law and Justice, Alexandria, Virginia.

[31] Larson, R.C. (1972). *Urban Police Patrol Analysis*. MIT Press, Cambridge, Massachusetts.

[32] Larson, R.C. (1974). "A Hypercube Queueing Model for Facility Location and Redistricting in Urban Emergency Services." *Jl. Computers & Operations Research* 1, 67–95.

[33] Larson, R.C. (1975). "What Happened to Patrol Operations in Kansas City?: A Review of the Kansas City Preventive Patrol Experiment." *Jl. Criminal Justice* 3, 267–297.

[34] Larson, R.C. and A.R. Odoni (1981). "The Hypercube Queueing Model." In *Urban Operations Research*, Prentice-Hall, Englewood Cliffs, New Jersey, 292–335.

[35] Larson, R.C. and T. Rich (1987). "Travel Time Analysis of New York City Police Patrol Cars." *Interfaces* 17, 15–20.

[36] Maltz, M.D. (1984). *Recidivism*. Academic Press, Orlando, Florida.

[37] Maltz, M.D. (1994). "Operations Research in Studying Crime and Justice: Its History and Accomplishments." In S.M. Pollock, A. Barnett, and M. Rothkopf, eds. *Operations Research and Public Systems*. Elsevier, Amsterdam.

[38] Maltz, M.D. and S.M. Pollock (1980). "Artificial Inflation of a Delinquency Rate by a Selection Artifact." *Operations Research* 28, 547–559.

[39] Morgan, P.M. (1985). *Modelling the Criminal Justice System*. Home Office Research and Planning Unit Paper 35, Home Office, London.

[40] STIF: Science and Technology Task Force (1967). *Task Force Report: Science and Technology*. President's Commission on Law Enforcement and the Administration of Justice, US Government Printing Office, Washington, D.C.

[41] Shinnar, R. and S. Shinnar (1975). "The Effects of the Criminal Justice System on the Control of Crime: A Quantitative Approach." *Law and Society Review* 9, 581–611.

[42] Stollmack, S. and C. Harris (1974). "Failure-rate Analysis Applied to Recidivism Data." *Operations Research* 22, 1192–1205.

[43] Stigler, S.M. (1986). *The History of Statistics: The Measurement of Uncertainty before 1900*. The Belknap Press of Harvard University Press, Cambridge, Massachusetts.

[44] Swersey, A.J. (1994). "The Deployment of Police, Fire, and Emergency Medical Units." In S.M. Pollock, A. Barnett, and M. Rothkopf, eds. *Operations Research and Public Systems*. Elsevier, Amsterdam.

[45] Tierney, L. (1983). "A Selection Artifact in Delinquency Data Revisited." *Operations Research* 31, 852–865.

CRITERION CONE

See **Multi-objective programming**.

CRITERION SPACE

See **Multi-objective programming**.

CRITERION VECTOR

See **Multi-objective programming**.

CRITICAL ACTIVITY

A project work item on the critical path having zero float time. See **Critical path**; **Critical path method**; **Network planning**.

CRITICAL PATH

The longest continuous path of activities through a project network from beginning to end. The total time elapsed on the critical path is the shortest duration of the project. The critical path will have zero float time, if a date for completion has not been specified. Any delay of activities on the critical path will cause a corresponding delay in the completion of the project. It is possible to have more than one critical path. See **Network planning**.

CRITICAL PATH METHOD (CPM)

A project planning technique that is used for developing strategy and schedules for an undertaking using a single-time estimate for each activity of which the project is comprised. In its basic form, it is concerned with determining the critical path, that is, the longest sequence of activities through the project network from beginning to end. See **Project evaluation and review technique**; **Network planning**; **Project management**.

CROSSOVER

A genetic-algorithm operator which exchanges corresponding genetic material from two parent chromosomes (i.e., solutions), allowing genes on different parents to be combined in their offspring. See **Genetic algorithms**.

CURSE OF DIMENSIONALITY

The situation that arises in such areas as dynamic programming, control theory, integer pro-

gramming, and, in general, time-dependent problems in which the number of states and/or data storage requirements increases exponentially with small increases in the problems' parameters or dimensions; sometimes referred to as *combinatorial explosion*. See **Control theory**; **Combinatorial and integer programming**; **Dynamic programming**.

CUSTOMER DISTRIBUTION

The probability distribution of the state of the process that customers observe upon arrival to a queueing system. In general, it is not the same as the distribution seen by a random outside observer; but the two distributions are the same for queueing systems with Poisson arrivals. Since customers entering a queue must also exit, the probability distribution seen by arriving customers who are accepted is the same as that for the number of customers left behind by the departures. See **Outside observer distribution**; **Queueing theory**.

CUT

A set of arcs in a graph (network) whose removal eliminates all paths joining a node s (source node) to a node t (sink node). See **Graph theory**.

CUT SET

A minimal set of edges whose removal disconnects a graph. See **Cut**; **Graph theory**.

CUTTING STOCK PROBLEMS

Robert W. Haessler

University of Michigan, Ann Arbor

INTRODUCTION: Solid materials such as aluminum, steel, glass, wood, leather, paper and plastic film are generally produced in larger sizes than required by the customers for these materials. As a result, the producers or primary converters must determine how to cut the production units of these materials to obtain the sizes required by their customers. This is known as a cutting stock problem. It can occur in one, two or three dimensions depending on the material. The production units may be identical, may consist of a few different sizes, or may be unique.

They may be of consistent quality throughout or may contain defects. The production units may be regular (rectangular) or irregular. The ordered sizes may be regular or irregular. They may all have the same quality requirements or some may have different requirements. They may have identical or different timing requirements which impact inventory.

Some examples follow:

- cutting rolls of paper from production reels of the same diameter.
- cutting rectangular pieces of glass from rectangular production sheets.
- cutting irregular pieces of steel from rectangular plates.
- cutting rectangular pieces of leather from irregular hides.
- cutting dimensional lumber from logs of various size.

There are two other classes of problems which are closely related to the cutting problems described above. The first is the layout problem. An example of this would be the problem of determining the smallest rectangle which will contain a given set of smaller rectangles without overlap. Solving this problem is essentially the same as being able to generate a cutting pattern in the discussion of cutting stock problems which follows. The second type of problem which in many cases can be solved by the same techniques as cutting stock problems is the (bin) packing problem. A one-dimensional example of this would be to determine the minimum number of containers required to ship a set of discrete items where weight and not floor space or volume is the determinant of what can be placed in the container. If floor space or volume is the key determinant then the problem is equivalent to a two or three-dimensional cutting stock problem in which guillotine cuts are not required. Even though the following discussions focus on cutting stock problems it is also applicable to solving both packing and layout problems.

Although cutting stock problems are relatively easy to formulate, many of them especially those with irregular shapes, are difficult to solve and there are no efficient procedures available in the literature. The major difficulty has to do with the generation of feasible low trim loss cutting pattern. As will be seen later, this ranges from being simple in one-dimension to complex in two-dimensions even with regular shapes.

The first known formulation of a cutting stock problem was given in 1939 by the Russian economist Kantorovich (1960). The first and most significant advance in solving cutting problems was the seminal work of Gilmore and Go-mory (1961, 1963) in which they described their delayed pattern generation technique for solving the one-dimensional trim loss minimization problem using linear programming. Since that time there has been an explosion of interest in this application area. Sweeney and Paternoster (1991) have identified more than 500 papers which deal with cutting stock and related problems and applications. The primary reasons for this activity are that cutting stock problems occur in a wide variety of industries, there is a large economic incentive to find more effective solution procedures, and it is easy to compare alternative solution procedures and to identify the potential benefits of using a proposed procedure.

Cutting stock problems are introduced with a discussion of the one-dimensional problem and the techniques available for solving it. The paper concludes with an extension to the regular two-dimensional problem.

ONE-DIMENSIONAL PROBLEMS: An example of a one-dimensional cutting stock problem is the trim loss minimization problem which occurs in the paper industry. In this problem, known quantities of rolls of various widths and the same diameter are to be slit from stock rolls of some standard width and diameter. The objective is to identify slitting patterns and their associated usage levels which satisfy the requirements for ordered rolls at the least possible total cost for scrap and other controllable factors. The basic cutting pattern feasibility restriction in this problem is that the sum of the roll widths slit from each stock roll must not exceed the usable width of the stock roll.

Let R_i be the nominal order requirements for rolls of width W_i, $i = 1, \ldots, n$, to be cut from stock rolls of usable width UW. RL_i and RU_i are the lower and upper bounds on the order requirement, for customer order i reflecting the general industry practice of allowing overruns or underruns within specified limits. Depending on the situation, R_i may be equal to RL_i and/or RU_i. All orders are for rolls of the same diameter. This problem can be formulated as follows, with X_j as the number of stock rolls to be slit using pattern j and T_j as the trim loss incurred by pattern j:

$$MIN \sum_j T_j X_j \tag{1}$$

$$s.t. \quad RL_i \leq \sum_j A_{ij} X_j \leq RU_i \quad \text{for all } i, \tag{2}$$

$$T_j = UW - \sum_i A_{ij} W_i \quad \text{for all } j, \tag{3}$$

$$X_j \text{ integer}, \ \geq 0. \tag{4}$$

where A_{ij} is the number of rolls of width W_i to be slit from each stock roll that is processed using pattern j. In order for the elements A_{ij}, $i = 1, \ldots, n$, to constitute a feasible cutting pattern, the following restrictions must be satisfied:

$$\sum_i A_{ij} W_i \leq UW \qquad (5)$$

$$A_{ij} \; integer, \; \geq 0. \qquad (6)$$

Note that the objective in this example is simply to minimize trim loss. In most industrial applications, it is necessary to consider other factors in addition to trim loss. For example, there may be a cost associated with pattern changes and, therefore, controlling the number of patterns used to satisfy the order requirements would be an important consideration.

Because optimal solutions to integer cutting stock problems can be found only for values of n smaller than typically found in practice, heuristic procedures represent the only feasible approach to solving this type of problem. Two types of heuristic procedures have been widely used to solve one-dimensional cutting stock problems. One approach uses the solution to a linear programming (LP) relaxation of the integer problem above as its starting point. The LP solution is then modified in some way to provide an integer solution to the problem. The second approach is to generate cutting patterns sequentially to satisfy some portion of the remaining requirements. This sequential heuristic procedure (SHP) terminates when all order requirements are satisfied.

LINEAR PROGRAMMING SOLUTIONS: Almost all LP based procedures for solving cutting stock problems can be traced back to Gilmore and Gomory (1961, 1963). They described how the next pattern to enter the LP basis could be found by solving an associated knapsack problem. This made it possible to solve the trim loss minimization problem by linear programming without first enumerating every feasible slitting pattern. This is extremely important because a large number of feasible patterns may exist when narrow widths are to be slit from a wide stock roll. Pierce (1964) showed that in such situations the number of slitting patterns can easily run into the millions. Because only a small fraction of all possible slitting patterns need to be considered in finding the minimum trim loss solution, the delayed pattern generation technique developed by Gilmore and Gomory made it possible to solve trim loss minimization problems in much less time than would be required if all the slitting patterns were input to a general-purpose linear programming algorithm.

A common LP relaxation of the integer programming problem given in (1) through (3) can be stated as follows:

$$MAX \sum_j X_j \qquad (7)$$

$$s.t. \sum_j A_{ij} X_j \geq R_i \quad for \; all \; i \qquad (8)$$

$$x_j \geq 0. \qquad (9)$$

Let U_i be the dual variable associated with constraint i. Then the dual of this problem can be stated as:

$$MAX \sum_i R_i U_i \qquad (10)$$

$$s.t. \sum_i A_{ij} U_i \leq 1 \qquad (11)$$

$$U_i \geq 0. \qquad (12)$$

The dual constraints in (11) provide the means for determining if the optimal LP solution has been obtained or if there exists a pattern which will improve the LP solution because the dual problem is still feasible.

The next pattern $A = (A_l, \ldots, A_n)$ to enter the basis, if one exists, can be found by solving the following knapsack problem:

$$Z = MAX \sum_i U_i A_i, \qquad (13)$$

$$s.t. \sum_i W_i A_i \leq UW, \qquad (14)$$

$$A_i \; integer, \; \geq 0. \qquad (15)$$

If $Z \leq 1$, the current solution is optimal. If $Z > 1$, then A can be used to improve the LP solution.

Once found, the LP solution can be modified in a number of ways to obtain integer values for the X_j which satisfy the order requirements. One common approach is to round the LP solution down to integer values, then increase the values of X_j by unit amounts for any patterns whose usage can be increased without exceeding RU_i. Finally, new patterns can be generated for any rolls still needed using the sequential heuristic described in the next section.

SEQUENTIAL HEURISTIC PROCEDURES: With an SHP, a solution is constructed one pattern at a time until all the order requirements are satisfied. The first documented SHP capable of finding better solutions than those found manually by schedulers was described by Haessler (1971). The key to success with this type of procedure is to make intelligent choices as to the patterns which are selected early in the SHP. The patterns selected initially should have low trim loss, high usage and leave a set of requirements for future patterns which will combine well without excessive side trim.

The following procedure is capable of making effective pattern choices in a variety of situations:

1. Compute descriptors of the order requirements yet to be scheduled. Typical descriptors would be the number of stock rolls still to be slit and the average number of ordered rolls to be cut from each stock roll.

2. Set goals for the next pattern to be entered into the solution. Goals should be established for trim loss, pattern usage, and number of ordered rolls in the pattern.

3. Search exhaustively for a pattern that meets those goals.

4. If a pattern is found, add this pattern to the solution at the maximum possible level without exceeding R_i, for all i. Reduce the order requirements and return to 1.

5. If no pattern is found, reduce the goal for the usage level of the next pattern and return to 3.

The pattern usage goal provides an upper bound on the number of times a size can appear in a pattern. For example, if some ordered width has an unmet requirement of 10 rolls and the pattern usage goal is 4, that width may not appear more than twice in a pattern. If after exhaustive search no pattern satisfies the goals set, then at least one goal, most commonly pattern usage, must be relaxed. This increases the number of patterns to be considered. If the pattern usage goal is changed to 3 in the above example, then the width can appear in the pattern three times. Termination can be guaranteed by selecting the pattern with the lowest trim loss at the usage level of one.

The primary advantage of this SHP is its ability to control factors other than trim loss and to eliminate rounding problems by working only with integer values. For example, if there is a cost associated with a pattern change, a sequential heuristic procedure which searches for high usage patterns may give a solution which has less than one-half the number of patterns required by an LP solution to the same problem. The major disadvantage of an SHP is that it may generate a solution which has greatly increased trim loss because of what might be called ending conditions. For example, if care is not taken as each pattern is accepted and the requirements reduced, the widths remaining at some point in the process may not have an acceptable trim loss solution. Such would be the case if only 34-inch rolls are left to be slit from 100-inch stock rolls.

RECTANGULAR TWO-DIMENSIONAL PROBLEMS:

The formulation of a higher dimensional cutting stock problem is exactly the same as that of the one-dimensional problem given in (1) through (3). The only added complexity comes in trying to define and generate feasible cutting patterns. The simplest two-dimensional case is one in which both the stock and ordered sizes are rectangular. Most of the important issues regarding cutting patterns for rectangular two-dimensional problems can be seen in the examples shown in Figure 1.

One important issue not covered in Figure 1 is a limit on the number of times an ordered size can appear in a pattern. This generally is a function of the maximum quantity of pieces, RU_i, required for order i. If R_i is small, it is just as important for the two-dimensional case as the one dimensional case that the number of times size i appears in a pattern should be limited. This becomes less important as R_i becomes larger and as the difference between RU_i and RL_i becomes larger.

The cutting pattern shown in Figure 1(a) is an example of two-stage guillotine cuts. The first cut can be in either the horizontal or vertical direction. A second cut perpendicular to the first, yields a finished piece. Figure 1(b) is similar except a third cut can be made to trim the pieces down to the correct dimension. Figure 1(c) shows the situation in which the third cut can create 2 ordered pieces.

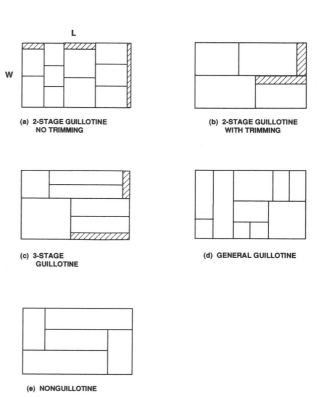

Fig. 1. Sample cutting patterns.

For simple staged cutting such as shown in Figure 1(a,b,c), Gilmore and Gomory (1965) showed how cutting patterns can be generated by solving two one-dimensional knapsack problems. To simplify the discussion, assume that the orientation of each ordered piece is fixed relative to stock piece and the first guillotine cut on the stock pieces must be along the length (larger dimension) of the stock piece. For each ordered width W_k, find the contents of a strip of width W_k and length L which gives the maximum contribution to dual infeasibility:

$$Z_k = MAX \sum_{i \in I_k} U_i A_{ik} \tag{16}$$

$$s.t. \sum_{i \in I_k} L_i A_{ik} \leq L \tag{17}$$

$$A_{ik} \ integer, \ \leq 0 \tag{18}$$

$$I_k = \{i \,|\, W_i \leq M_k\}. \tag{19}$$

Next find the combination of strips which solve the problem

$$Z = MAX \sum_k Z_k A_k \tag{20}$$

$$\sum_k W_k A_k \leq W \tag{21}$$

$$A_k \ integer, \ \leq 0. \tag{22}$$

Any pattern for which Z is greater than one will yield an improvement in the LP solution.

The major difficulty with this approach is the inability to limit the number of times an ordered size appears in a pattern. It is easy to restrict the number of times a size appears in a strip and to restrict the number of strips in a pattern. The problem is that small ordered sizes with small quantities may end up as filler in a large number of different strips. This makes the two-stage approach to developing patterns ineffective when the number of times a size appears in a pattern must be limited.

Wang (1983) developed an alternative approach to generating general guillotine cutting patterns which limits the number of times a size appears in a pattern. She combined rectangles in a horizontal and vertical build process as shown in Figure 2 where O_i is an ordered rectangle of width W_i and length L_i.

She used an acceptable value for trim loss, B, rather than the shadow price of the ordered sizes to drive her procedure which is as follows:

Step 1(a). Choose a value for B the maximum acceptable trim waste.
    (b). Define $L^{(O)} = F^{(O)} = \{O_1, O_2, \ldots, O_n\}$, and set $K = 1$.
Step 2(a). Compute $F^{(K)}$ which is the set of all rectangles T satisfying

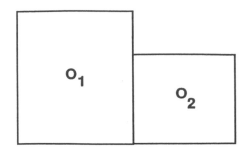

(a)   **Horizontal build of O_1 and O_2**

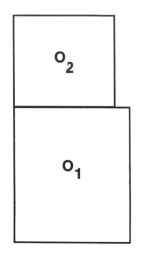

(b)   **Vertical build of O_1 and O_2**

Fig. 2. Guillotine cutting patterns.

      (i) T is formed by a horizontal or vertical build of two rectangles from $L^{(K-1)}$,
      (ii) the amount of trim waste in T does not exceed B, and
      (iii) those rectangles O_i, appearing in T do not violate the constraints on the number of times a size can appear in a pattern.
    (b). Set $L^{(K)} = L^{(K-1)} \ U \ F^{(K)}$. Remove any equivalent (same component rectangles) rectangle patterns from $L^{(K)}$.
Step 3.   If $F^{(K)}$ is non-empty, set $K = K + 1$ and go to Step 2;
otherwise, set $M = K - 1$, and choose the rectangle in $L^{(M)}$ which has the smallest total trim waste when placed in the stock rectangle.

CONCLUSION: It is clear that moving from one to two-dimensions causes significant difficulty in the pattern generating process. This is all the more alarming in light of the fact that only rectangular shapes were considered.

This clearly suggests that there is much more research needed on procedures for solving two-dimensional cutting stock problems. An alternative worth considering, especially in those cases where there are many different ordered sizes with small order quantities, might be to first select a subset of orders to consider by solving a one-dimensional knapsack problem as in (13) through (15) based on area and then see if the resulting solution can be put together into a feasible two-dimensional pattern. Wang's algorithm seems to be ideal for this purpose inasmuch as the trim loss in the pattern would be known.

A candidate set of items to be included in the next pattern could be found by solving the following problem:

$$Z = MAX \sum_i U_i A_i \qquad (23)$$

$$\sum_i AR_i A_i \leq UAR \quad \text{for all } i \qquad (24)$$

$$A_i \leq b_i \qquad (25)$$

$$A_i \text{ integer}, \geq 0 \qquad (26)$$

where AR_i is the area of ordered rectangle i; UAR is the usable area of the stock rectangle; and b_i is the upper limit on the number of times order i can be included in the pattern.

This candidate pattern $(A_1, \ldots, A_n)$ could then be tested for feasibility using Wang's procedure. If the AR_i are small, the chances are that there will be little trim loss in the candidate patterns generated. This may require that UAR be reduced to force some trim loss to make it more likely that feasible patterns are found.

See **Bin packing; Combinatorial and integer optimization; Linear programming.**

References

[1] Gilmore, P.C. and R. E. Gomory (1961), "A Linear Programming Approach to the Cutting Stock Problem," *Operations Research*, 9, 848–859.

[2] Gilmore, P.C. and R. E. Gomory (1963), "A Linear Programming Approach to the Cutting Stock Problem, Part II," *Operations Research*, 11, 863–888.

[3] Gilmore, P.C. and R.E. Gomory (1965) "Multistage Cutting Stock Problems of Two and More Dimensions," *Operations Research*, 13, 94–120.

[4] Gilmore, P. C. and R. E. Gomory (1966), "The Theory and Computation of Knapsack Functions," *Operations Research*, 14, 1045–1074.

[5] Haessler, R. W. (1971) "A Heuristic Programming Solution to a Nonlinear Cutting Stock Problem," *Management Science*, 17, 793–802.

[6] Kantorovich, L.V. (1960), "Mathematical Methods of Organizing and Planning Production," reprinted in *Management Science*, 6, 366–422.

[7] Paull, A.E. (1956), "Linear Programming: A Key to Optimum Newsprint Production," *Paper Magazine of Canada*, 57, 85–90.

[8] Pierce, J. F. (1964), *Some Large Scale Production Problems in the Paper Industry*, Prentice-Hall, Inc. Englewood Cliffs, New Jersey.

[9] Sweeney, P.E. and E.R. Paternoster (1991), "Cutting and Packing Problems: An Updated Literature Review," working paper No. 654, University of Michigan, School of Business.

[10] Wang, P.Y. (1983), "Two Algorithms for Constrained Two-Dimensional Cutting Stock Problems," *Operations Research*, 31, 573–586.

CV

See **Coefficient of variation.**

CYBERNETICS

Andrew P. Sage

George Mason University, Fairfax, Virginia

INTRODUCTION AND EARLY HISTORY: Cybernetics is a term that is occasionally used in the systems engineering and OR/MS literature to denote the study of control and communication in, and, in particular between humans, machines, organizations, and society. The word cybernetics comes from the Greek word *Kybernetes*, which means "controller," or "governor," or "steersman." The first modern use of the term was due to Professor Norbert Wiener, an MIT professor of mathematics, who made many early and seminal contributions to mathematical system theory (Wiener, 1949). The first book formally on this subject was titled *Cybernetics* and published in 1948 (Wiener, 1948). In this book, Wiener defined the term as "*control and communication in the animal and the machine.*" This emphasized the concept of feedback control as a construct presumably of value in the study of neural and physiological relations in the biological and physical sciences. In the historical evolution of cybernetics, major concern was initially devoted to the study of feedback control and servomechanisms, studies which later evolved into the area of control systems or control engineering (Singh, 1987). Cybernetic concerns also have involved analog and digital computer development, especially computer efforts that were presumed to be models of the human brain (Mc-Culloch, 1965) and the combination of computer and control systems for purposes of automation and remote control (Ashby, 1952, 1956; George, 1971; Lerner, 1976).

There were a number of other early influences on cybernetics, including artificial intelligence (McCulloch, 1974). It was the initial presumed resemblance, at a neural or physiological level, between physical control systems and the central nervous system and human brain that concerned Wiener. He and close associates, Warren McCulloch, Arturo Rosenblueth, and Walter Pitts, were the initial seminal thinkers in this new field of cybernetics. Soon, it became clear that it was fruitless to study control independent of information flow; cybernetics thus took on an identification with the study of communications and control in humans and machines. An influence in the early notions of cybernetics was the thought that physical systems could be made to perform better by, somehow, enabling them to emulate human systems at the physiological or neural level. Thus, early efforts in what is now known as neural networks began as cybernetic studies.

Another early concept explored in cybernetics was that of homeostasis, which has come to be known as the process by which systems maintain their level of "organization" in the face of disturbances, often occurring over time, and generally of a very large scale (Ashby, 1952). Cybernetics soon became concerned with purposive organizational systems, or viable systems, as contrasted with systems that are static over time and purpose (Beer, 1979). Further, organizations operate in the face of incomplete and redundant information by establishing useful patterns of communications (Beer, 1979). Thus, organizations can potentially be modeled and have been modeled as cybernetic systems (Steinbrunner, 1974).

Cybernetics has often been viewed as a way of looking at systems, or a philosophical perspective concerning inquiry, as contrasted with a very specific method. An excellent collection of Norbert Wiener's original papers on cybernetics studies is contained in Volume IV of an edited anthology (Masani, 1985). Fundamental to any cybernetic study is the notion of modeling, and, in particular, the interpretation of the results of a modeling effort as theories that have normative or predictive value. Today, there is little explicit or implicit agreement concerning a precise definition for cybernetics. Some users of the term cybernetics infer that the word implies a study of control systems. Some uses refer to modeling only at the neural and physiological level. Some refer to cognitive ergonomic model-

ing without necessary consideration of, or connection to, neuronal level elements. Other uses of the word are so general that cybernetics might seem to infer either nothing, or everything. Automation, robotics, artificial intelligence, information theory, bionics, automata theory, pattern recognition and image analysis, control theory, communications, human and behavioral factors, and other topics have all, at one time or another, been assumed to be a portion of cybernetics.

DEFINITION OF CYBERNETICS: The notion of the physiological aspects of human nervous system as playing a necessarily critical role in modern cybernetics has all but vanished today, except in very specialized classic works. This does not suggest that interest in neural type studies has vanished as there is, indeed, much interest today in neural networks and related subjects (Freeman and Skapura, 1991; Zurada, 1992). A much more cognitive perspective is prevalent today, at least in many systems engineering views of cybernetics. In this article, *cybernetics is defined as the study of the communication and control processes associated with human-machine interaction in systems that are intended to support accomplishment of purposeful tasks.* While this is not a universally accepted definition of cybernetics, it is a useful one for many systems engineering studies involving human-system interaction through communications and control (Sage, 1992).

A COGNITIVE ERGONOMICS VIEW: A purpose of this article is to discuss cybernetics, and the design of cybernetic based support systems, for such purposes as knowledge support to humans. We are especially concerned with the human-system interactions that occur in such an effort. Thus our discussions are particularly relevant to knowledge based system design concerns relative to human-machine cybernetic problem solving tasks, such as fault detection, diagnosis and correction. These are very important concerns for a large number of decision support systems engineering applications that require fundamentally cognitive support to humans in supervisory control tasks (Sage, 1991; Sheridan, 1992). Advances in technology involving computers, automation, robotics, and many other recent innovations, together with the desire to improve productivity and the human condition, render physiological skills that involve strength and motor abilities relatively less important than they have been. In many instances, these physical tasks are now accomplished by robots. These changes diminish the need for physical abilities, and increase requirements for cognitive abilities that are related to human supervisory control of systems.

The need for humans to monitor and maintain the conditions necessary for satisfactory operation of systems, and to cope with poorly structured and imprecise knowledge is greater than ever. Ultimately, these primarily cognitive efforts, which involve a great variety of human problem solving activities, are often translated into physical control signals, which control or manipulate some physical process. As a consequence of this, there are a number of human interface issues that naturally occur between the human and the machines over which the human must exercise control.

A number of advances in information technology provide computer, control, and communication systems that enable a significant increase in the amount of information that is available for judgment and decision-making tasks at the problem solving level. However, even the highest quality information will generally be associated with considerable uncertainty, imprecision, and other forms of knowledge imperfection. Computer, control, and communications technology can assist in human problem solving tasks by enhancing the quality of the information and knowledge that is available for decision making (Sheridan and Ferrell, 1974). A system which assists in this function is called an expert system by the artificial intelligence community (Barr, Cohen, and Feigenbaum, 1981, 1982; Shapiro, 1987) or a decision support system (Sage, 1991) or executive support system by the systems engineering, management science and decision analysis communities (Rockart and DeLong, 1988). The term knowledge support system is often used to denote the integrated use of expert system and decision support system technologies. Through use of knowledge based systems, we create the need for another human-system interface; one between the human and the computer, and the need for proper system design to ensure appropriate interaction between the human and the computer (Mayhew, 1992). This contemporary use of information technology is expected to lead to major organizational transformations in the future (Harrington, 1991; Scott Morton, 1991; Davenport, 1993).

CYBERNETICS AND SYSTEMS MANAGEMENT: A human-machine cybernetic system may be defined as a functional synthesis of a human system and a technological system or machine. Human-machine systems are predominantly characterized by the interaction and functional interdependence between these two elements. The introduction of communication and control concerns results in a *cybernetic* system. All kinds of technological systems, regardless of their degree of complexity, may be viewed as parts of a human-machine cybernetic system: industrial plants, vehicles, manipulators, prostheses, computers or management information systems. A human-machine system may, of course, be a subsystem that is incorporated within another system. We may, for example, incorporate a decision support system as part of a larger enterprise management, process control, or computer-aided design system which also involves human interaction. This use of the term *human-machine cybernetic system* corresponds, therefore, to a specific way of looking at technological systems through the integration of technological systems and human-enterprise systems, generally through a systems management or systems engineering process.

The overall purpose of any *human-machine cybernetic system* is to provide a certain function, product, or service with reasonable cost under constraint conditions and disturbances. This concept involves and influences the human, the machine, and the processes through which they function as an integrated whole. Figure 1 presents a simple information technology based conceptualization of a human-machine cybernetic system. The primary "inputs" to a human-machine cybernetic system are a set of purposeful performance objectives that are typically translated into a set of expected values of perfor-

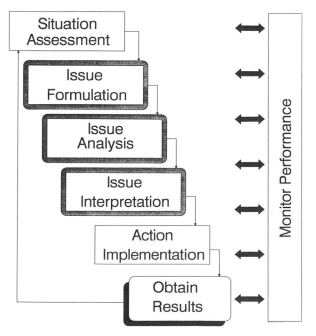

Fig. 1. Activities involved in task performance.

mance, costs, reliability, and safety. Also, the design must be such that an acceptable level of workload and job satisfaction is maintained. It is on the basis of these that the human is able to perform the following activities:

1. Identify task requirements, such as to enable determination of the issues to be examined further and the issues to be not considered;

2. Identify a set of hypotheses or alternative courses of actions which may resolve the identified issues to be resolved;

3. Identify the probable impacts of the alternative courses of action;

4. Interpret these impacts in terms of the objectives or "inputs" to the task;

5. Select an alternative for implementation and implement the resulting control;

6. Monitor performance such as to enable determination of how well the integrated combination of human and system are performing (Sage, 1992).

Many researchers have described activities of this sort in a number of frameworks that include behavioral psychology, organizational management, human factors, systems engineering, operations research and management science.

Many questions can be raised concerning the use of information for judgment and choice activities, as well as activities that lead to the physical control of an automated process. Any and all of these questions can arise in different application areas. These questions relate to the control of technological systems. They concern the degree of automation with respect to flexible task allocation. They also concern the design and use of computer-generated displays. Further, they relate to all kinds of human-computer interaction concerns, as well as management tasks at different organizational levels: strategic, tactical, and operational. For example, computer-based support systems to aid human performance now invade more and more areas of systems design, operation, maintenance, and management. The importance of augmenting hardware and micro-level programming aspects of system design to architectural and software systems management considerations is great. The integrated consideration of systems engineering and systems management for software productivity is expressed by the term *software systems engineering* (Sage and Palmer, 1990).

Human tasks in human-machine cybernetic systems can be condensed into three primary categories: (1) controlling (physiological); (2) communicating (cognitive), and (3) problem solving (cognitive) (Johannsen, Rijnsdorp, and Sage, 1983). In addition, there exists a monitoring or

feedback portion of the effort that enables learning over time (Johannsen, Rijnsdorp, and Sage, 1983). Ideally, but not always, we learn well. There needs to be meta-level learning, or learning how to learn if improvements are to truly be lasting, as contrasted with only specific task performance learning. Associated with the rendering of a single judgment and the associated control implementation, the human monitors the result of the effect of these activities. The effect of present and past monitoring is to provide an experiential base for present problem conceptualization. In our categorization above, activities 1 through 4 may be viewed as problem (finding and) solving, activity 5 involves implementation or controlling, and activity 6 involves communications or monitoring and feedback in which responses to the question "How good is the process performance?" enables improvement and learning through iteration. Of course, the notion of information flow and communication is involved in all of these activities.

These three human task categories are fairly general. Figure 2 shows an attempt to integrate them into a schematic block diagram. Controlling should be understood in a much broader sense than in many control theory studies. Controlling in this narrower sense includes open-loop vs. closed-loop and continuous vs. intermittent controlling, as well as discrete tasks such as reaching, switching and typing. It is only through these physiological aspects of con-

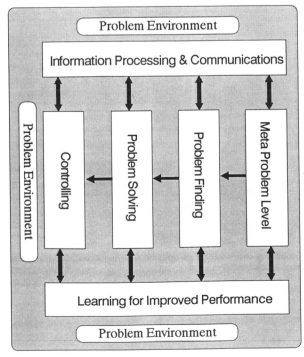

Fig. 2. Cognitive and physical efforts involved in system operation.

trolling that outputs of the human-machine cybernetic system can be produced, as shown in Figure 2. Controlling, in the sense of the cognitive ergonomic concerns that support human information processing and associated judgment and choice, is also now included. Although human functions on a cognitive level can and do play a role in control implementation, their major importance occurs in problem solving activities. Tasks such as fault detection, fault diagnosis, fault compensation or managing, and planning are particularly important in "problem solving." Fault detection concerns the identification of a potential difficulty concerning the operation of a system. Fault diagnosis is concerned with identification of a set of hypotheses concerning the likely cause of a system malfunction, and the evaluation and selection of a most likely cause. It is primarily a cognitive activity. Fault compensation or managing is concerned with solving problems in actual failure situations. This may occur through the use of rules that are based on past experience, and the updating of certain rules based on the results of their present application. It is accomplished with the objective of returning the overall system to a good operating state. Fault compensation or managing involves both cognitive and physiological activities. Planning is a cognitive activity concerned with solving possible future problems in the sense of mentally generating a sequence of appropriate alternatives. Appropriate planning involves the use of knowledge perspectives, knowledge principles, and knowledge practices (Sage, 1992). They are based on experiential familiarity with analogous situations and are often expressed in the form of and through the use of skills, rules, and formal knowledge based reasoning efforts (Rasmussen, 1986). Human error issues are of particular importance, especially those concerned with the design of systems that cope with human error through avoidance and amelioration efforts (Reason, 1990).

THE DESIGN OF CYBERNETIC SYSTEMS: All of this has major implications with respect to the design of systems for the human user, and for associated cybernetic systems as well. It requires, for appropriate system design, an understanding of human performance in problem solving and decision-making tasks. This understanding has to be at a descriptive level, such that we can predict what humans will likely do in particular situations. It has to be at a normative level, such that we can understand what would be best performance under restrictive axiomatic conditions that will generally not exist

in practice. Also, this understanding has to be at a prescriptive level, such that we can aid actual real-world humans in various real-world cognitive tasks.

Technological advances have changed, and will continue to change, the specific design requirements for human-machine cybernetic systems needed in any given application area. This is especially true due to the many advances made possible through modern information technologies, for industrial plants with integrated automated manufacturing capabilities, and for aids to cognitive activities in strategic planning, design, or operational activities. Office automation systems and information systems for observation, planning, executive support, management, and command and control tasks in business, defense, and medicine are similarly influenced by efforts in human-machine cybernetic systems. These involve not only the operation of technological and management oriented information systems by highly skilled and knowledgeable personnel, but also systems that are intended for use by the less skilled. A major use for new generation human-machine cybernetic systems is to provide computer assistance for the maintenance of existing systems and for the design of new technological systems.

A problem common to all these applications is the design of appropriate human-computer interaction subsystems. The possibly adaptive task allocation between human and computer, the dialog design including the use of natural language, and other software systems engineering aspects, are especially important contemporary topics of research and development. With more computerization and higher degrees of automation, much greater attention must be paid to system design for human interaction. Contemporary advanced technologies potentially allow a more flexible work organization with higher user acceptance and job satisfaction. However, this potential advantage can only be achieved if the behavioral implications are seriously considered by the designers of computerized human-machine cybernetic systems. Paradoxically, more sophisticated automation systems require greater human knowledge and a more sophisticated quality of human skills. These are needed such that it becomes possible for humans, at least in a supervisory capacity, to compensate for inevitable physical system limitations that are due to physical equipment faults, human errors, and changes in the environment and culture extant. These will always occur in automated systems. The potential for catastrophic behavior increases along with the degree of automation.

Yet, automation through the use of cybernetic systems is an increasing need. This is but one of many contemporary paradoxes affecting modern information technology, society, and cybernetic systems.

See **Control theory; Mathematical models; Neural networks and algorithms.**

References

[1] Ashby, W.R. (1952), *Design for a Brain*, Chapman and Hall, London.

[2] Ashby, W.R. (1956), *An Introduction to Cybernetics*, Chapman and Hall, London.

[3] Barr, A., Cohen, P.R., and Feigenbaum, E.A. (eds.) (1981, 1982), *Handbook of Artificial Intelligence, Vol. I, II, and III*, William Kaufman.

[4] Beer, S. (1979), *The Heart of Enterprise*, John Wiley, Chichester, UK.

[5] Davenport, T.H. (1993), *Process Innovation: Reengineering Work through Information Technology*, Harvard Business School Press, Boston, Massachusetts.

[6] Freeman, J.A., and Skapura, D. (1991), *Neural Networks: Algorithms, Applications and Programming Techniques*, Addison-Wesley Publishing Co., Reading, Massachusetts.

[7] George, F.H. (1971), *Cybernetics*, St. Paul's House, Middlegreen, Slough, UK.

[8] Harrington, H.J. (1991), *Business Process Improvement: The Breakthrough Strategy for Total Quality, Productivity, and Competitiveness*, McGraw-Hill, New York.

[9] Johannsen, G., Rijnsdorp, J.E., and Sage, A.P. (1983), "Human Interface Concerns in Support System Design," *Automatica*, 19(6), 1–9.

[10] Lerner, A.Y. (1976), *Fundamentals of Cybernetics*, Plenum, New York.

[11] Masani, P. (ed) (1985), *Norbert Wiener: Collected Works Volume IV – Cybernetics, Science and Society; Ethics, Aesthetics, and Literary Criticism; Book Reviews and Obituaries*, MIT Press, Cambridge, Massachusetts.

[12] Mayhew, D.J. (1992), *Principles and Guidelines in Software User Interface Design*, Prentice Hall, Englewood Cliffs, New Jersey.

[13] Rasmussen, J. (1986), *Information Processing and Human Machine Interaction: An approach to Cognitive Engineering*, North Holland Elsevier, Amsterdam.

[14] Reason, J. (1990), *Human Error*, Cambridge University Press, Cambridge, UK.

[15] Rockart, J.F., and DeLong, D.W. (1988), *Executive Support Systems: The Emergence of Top Management Computer Use*, Dow Jones-Irwin, Homewood, Illinois.

[16] Sage, A.P. (ed) (1990), *Concise Encyclopedia of Information Processing in Systems and Organizations*, Pergamon Press, Oxford.

[17] Sage, A.P. (ed) (1987), *System Design for Human Interaction*, IEEE Press, New York.

[18] Sage, A.P. and Palmer, J.D. (1990), *Software Systems Engineering*, John Wiley, New York.

[19] Sage, A.P. (1991), *Decision Support Systems Engineering*, John Wiley, New York.

[20] Sage, A.P. (1992), *Systems Engineering*, John Wiley, New York.

[21] Scott Morton, M.S. (ed) (1991), *The Corporation of the 1990s: Information Technology and Organizational Transformation*, Oxford University Press, New York.

[22] Shapiro, S.C. (ed) (1987), *Encyclopedia of Artificial Intelligence*, John Wiley, New York.

[23] Sheridan, T.B., and W.R. Ferrell (1974), *Man-Machine Systems: Information, Control, and Decision Models of Human Performance*, MIT Press, Cambridge, Massachusetts.

[24] Sheridan, T.B. (1992), *Telerobotics, Automation, and Human Supervisory Control*, MIT Press, Cambridge, Massachusetts.

[25] Singh, M.G., (ed) (1990), *Systems and Control Encyclopedia*, Pergamon Press, Oxford, UK.

[26] Steinbruner, J.D. (1974), *The Cybernetic Theory of Decision*, Princeton University Press, New Jersey.

[27] Wiener, N. (1948), *Cybernetics, or Control and Communication in the Animal and the Machine*, John Wiley, New York.

[28] Wiener, N. (1949), *Extrapolation, Interpolation and Smoothing of Stationary Time Series with Engineering Applications*, MIT Press, Cambridge, Massachusetts.

[29] Zurada, J. (1992), *Introduction to Artificial Neural Systems*, West Publishing, St. Paul, Minnesota.

CYCLE

A path in a graph (network) joining a node to itself. See **Chain; Path**.

CYCLIC QUEUEING NETWORK

A closed network of queues in which customer routing is serial. See **Networks of queues**.

CYCLIC SERVICE DISCIPLINE

When a congestion system with several different locations (service centers) of customers are served by a single service "facility." For a given period of time determined by an a priori rule, the service process only works on customers from (at) a given location and then switches to the next group when the period is over. See **Queueing theory**.

CYCLING

A situation where the simplex algorithm cycles (circles) repeatedly through some sequence of bases and corresponding basic feasible solutions. This can occur at a degenerate extreme point solution where several bases correspond to the same extreme point. See **Anticycling rules**; **Degeneracy**.

DANTZIG-WOLFE DECOMPOSITION ALGORITHM

A variant of the simplex method designed to solve block-angular linear programs in which the blocks define subproblems. The problem is transformed into one that finds a solution in terms of convex combinations of the extreme points of the subproblems. See **Block-angular system**.

DATABASE DESIGN

See **Information systems and database design**.

DATA ENVELOPMENT ANALYSIS

William W. Cooper

The University of Texas at Austin

INTRODUCTION: DEA (Data Envelopment Analysis) is a "data oriented approach" for evaluating the performance of a collection of entities called DMUs (Decision Making Units) which are regarded as responsible for converting inputs into outputs. Examples have included hospitals and U.S. Air Force Wings, or their subdivisions, such as surgical units and squadrons. The definition of a DMU is generic and flexible. Uses that have been accommodated include (i) discrete periods of production in a plant producing semiconductors and (ii) marketing regions to which advertising and other sales activities have been directed. Inputs as well as outputs may be multiple and each may be measured in different units.

A variety of models have been developed for implementing the concepts of DEA. To start, we use the following dual pair of linear programming models

$$\min \quad h_0 = \theta_0 - \varepsilon\left(\sum_{i=1}^{m} s_i^- + \sum_{r=1}^{s} s_r^+\right) \tag{1a}$$

subject to

$$0 = \theta_0 x_{i0} - \sum_{j=1}^{n} x_{ij}\lambda_j - s_i^-$$

$$y_{r0} = \sum_{j=1}^{n} y_{rj}\lambda_j - s_r^+$$

$$0 \leq \lambda_j, s_r^+, s_i^-$$

and

$$\max \quad y_0 = \sum_{r=1}^{s} \mu_r y_{r0} \tag{1b}$$

subject to

$$1 = \sum_{i=1}^{m} v_i x_{i0}$$

$$0 \geq \sum_{r=1}^{s} \mu_r y_{rj} - \sum_{i=1}^{m} v_i x_{ij}$$

$$\varepsilon \leq \mu_r, v_i$$

where x_{ij} = amount of input i used by DMU$_j$ and y_{rj} = amount of output r produced by DMU$_j$, with $i = 1, \ldots, m; r = 1, \ldots, s; j = 1, \ldots, n$. For convenience we assume that all inputs and outputs are positive. (This condition may be relaxed, Charnes, Cooper and Thrall, 1991.)

EFFICIENCY: We have here changed the orientation of linear programming from *ex-ante* uses, for *planning*, and apply it *ex-post*, to choices already made, for purposes of evaluation and *control*. To evaluate the performance of any DMU, (1) is applied to the input-output data for *all* DMUs in order to evaluate the performance of *each* DMU in accordance with the following definition:

Efficiency – Extended Pareto-Koopmans Definition: Full (100%) efficiency is attained by any DMU if and only if none of its inputs or outputs can be improved without worsening some of its other inputs or outputs.

This definition has the advantage of avoiding the need for assigning *a priori* measures of relative importance to any input or output. In most management or social science applications the theoretically possible levels of efficiency will not be known. The preceding definition is therefore replaced by the following:

Relative Efficiency: A DMU is to be rated as fully (100%) efficient if and only if the performances of other DMUs does not show that some of its inputs or outputs can be improved without worsening some of its other inputs or outputs.

To implement this definition it is necessary only to designate any DMU_j as DMU_o with inputs x_{io} and outputs y_{ro} and then apply (1) to the input and output data recorded for the collection of DMU_j, $j = 1, \ldots, n$. Leaving this (1b) $DMU_j = DMU_o$ in the constraints ensures that solutions will always exist with an optimal $\theta_o = \theta_o^* \leq 1$. The above definition applied to (1) then gives

DEA Efficiency: The performance of DMU_o is fully (100%) efficient if and only if, at an optimum, both (i) $\theta_o^* = 1$, and (ii) all slacks $= 0$ in (1a) or, equivalently, $\Sigma_{r=1}^s \mu_r^* y_{ro} = 1$, in (1b), where * represents an optimal value for DMU_o.

A value of $\theta_o^* < 1$ shows (from the data) that a nonnegative combination of other DMUs could have achieved DMU_o's outputs at the same or higher levels while reducing all of its inputs. Non-zero slacks similarly show where input reductions or output augmentations can be made in DMU_o's performance without altering other inputs or outputs. These non-zero slacks show where changes in *mixes* could have improved performance in each of DMU_o's inputs or outputs while a $\theta^* < 1$ shows where *all* inputs could have been reduced in the same proportion. (This is a so-called "input-oriented" model. An output-oriented model can be similarly formulated by associating a variable φ_o to be maximized with all outputs. The measures are reciprocal, i.e., $\varphi_o^* \theta_o^* = 1$, so we do not develop this topic here.)

FARRELL MEASURE: The scalar θ_o^* is sometimes referred to as the "Farrell measure" after Farrell (1957). Notice, however, that a value of $\theta_o^* = 1$ does not satisfy the above definition of "Relative Efficiency" if any of the associated slacks, s_r^{-*} or s_r^{+*}, in (1) are positive, because any such non-zero slack provides an opportunity for improvement which may be used without affecting any other variable, as should be clear from the primal problem which is shown in (1a). Furthermore, there is a need to ensure that an optimum with $\theta_o^* = 1$ and all slacks zero is not interpreted to mean that full (100%) efficiency has been attained when an alternate solution with $\theta_o^* = 1$ and some slacks positive is also available.

To see how this is dealt with, we call attention to the fact that the slack variables s_i^- and s_r^+ in the objective of the primal (minimization)

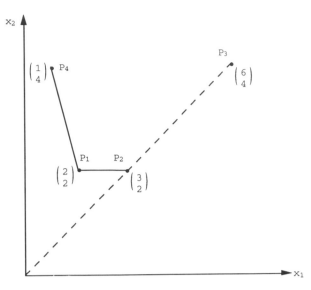

Fig. 1. DEA deficiencies.

problem, (1a), are each multiplied by $\varepsilon > 0$ which is a non-Archimedean infinitesimal – the reciprocal of the "big M" associated with the artificial variables in ordinary linear programming – so that choices of slack values *cannot* compensate for any increase they cause in θ_o. This accords preemptive status to the minimization of θ_o, and DEA computer codes generally handle optimizations in a two-stage manner which avoids the need for specifying ε explicitly. Formally, this amounts to minimizing the value of θ_o in stage 1. Then one proceeds in a second stage to maximize the sum of the slacks with the condition $\theta_o = \theta_o^*$ fixed for the primal in (1a). Since the sum of the slacks is maximized, one can be sure that a solution with all slacks at zero in the second stage means that DMU_o is fully efficient if the first stage yielded $\theta_o^* = 1$.

EXAMPLE: Figure 1 is a geometric portrayal off our DMUs interpreted as points $P_1, \ldots, P_4$ with coordinate values corresponding to the amounts of two inputs which each DMU used to produce the same amount of a single output. P_3 is evidently inefficient compared to P_2 because it used more of both inputs to achieve the same output. In fact, we can determine its Farrell measure of inefficiency relative to P_2 via the formula

$$\theta_0 = \frac{d(0, P_2)}{d(0, P_3)} = \frac{\sqrt{3^2 + 2^2}}{\sqrt{6^2 + 4^2}} = \frac{1}{2},$$

where $d(0, P)$ refers to the Euclidean, or ℓ_2 measure of distance.

We obtain this same value of θ by omitting the slacks and rewriting the primal problem in (1) in the following inequality form,

minimize θ_o (2)

subject to

$$6\theta_o \geq 2\lambda_1 + 3\lambda_2 + 6\lambda_3 + 1\lambda_4$$

$$4\theta_o \geq 2\lambda_1 + 2\lambda_2 + 4\lambda_3 + 4\lambda_4$$

$$1 \leq 1\lambda_1 + 1\lambda_2 + 1\lambda_3 + 1\lambda_4$$

$$0 \leq \lambda_1, \ldots, \lambda_4,$$

where the third constraint reflects the output $y = 1$ produced by each of these DMUs.

An optimum is achieved with $\theta_o^* = 1/2$, $\lambda_2^* = 1$ and this designates P_2 for the evaluation of P_3. However, we also need to take account of the slack possibilities. This is accomplished without specifying $\varepsilon > 0$ explicitly by proceeding to a second stage formed from (1) by using the thus obtained value of θ_o^* to form the following problem:

maximize $s_1^- + s_2^- + s^+$ (3)

subject to

$$0 = -6\theta_o + 2\lambda_1 + 3\lambda_2 + 6\lambda_3 + 1\lambda_4 + s_1^-$$

$$0 = -4\theta_o + 2\lambda_1 + 2\lambda_2 + 4\lambda_3 + 4\lambda_4 + s_2^-$$

$$-1 = -1\lambda_1 - 1\lambda_2 - 1\lambda_3 - 1\lambda_4 + s^+$$

$$0.5 = \theta_o$$

$$0 \leq \lambda_1, \ldots, \lambda_4, s_1^-, s_2^-, s^+$$

Following through in this second stage, we find that the optimum solution is $\theta_o^* = 1/2$, $\lambda_1^* = 1$ and $s_1^{-*} = 1$, with all other variables zero. This solution is interpreted to mean that the evidence from other DMUs (as exhibited by P_1's performance) shows that P_3 should have been able (a) to reduce *both* inputs to one-half their observed values as given by the value of θ_o and should also have been able (b) to reduce input one by the additional amount given by $s_1^{-*} = 1$.

This slack, $s_1^{-*} = 1$, represents the excess amount of the first input used by P_2 and it, too, must be accounted for if the above definition of relative efficiency is to be satisfied. In fact, using the primal in (1a) to evaluate P_2, it will be found that it also is inefficient with $\theta_1^* = 1$ and $\lambda_1^* = s_1^{-*} = 1$. The use of (1a) to determine whether the two conditions for relative efficiency are satisfied has a further consequence in that it ensures that only efficient DMUs enter into the solutions with positive coefficients in the basis sets that are used to effect efficiency evaluations. Computer codes that have been developed for DEA generally use this property to reduce the number of computations by identifying all such members of an optimal basis as efficient and, hence, not in need of further evaluation.

P_1 dominates P_2 and hence also dominates P_3, as can be seen from Figure 1. Only P_1 and P_4 are not dominated and hence can be regarded as efficient when DEA is restricted to dominance, as in Bowlin *et al.* (1993). However, if an assumption of continuity is added, then the entire line segment connecting P_1 and P_4 becomes available for use in effecting efficiency evaluations. This line segment is referred to as the "efficiency frontier." The term "efficient frontier" is appropriate because, unlike the frontier connecting P_1 and P_2, it is not possible to move from one point to another on the line connecting P_1 and P_4 without worsening one input to improve the other input.

Given the assumption of continuity, points not on the efficiency frontier are referred to it for evaluation. Even when not dominated by actually observed performances, the non-negative combinations of λ_j^* and slack values will locate points on the frontier which can be used for effecting efficiency evaluations of any DMU in the observation set.

The following formulas, called the "CCR projection formulas," may be used to move points up to the efficiency frontier:

$$\begin{cases} \hat{x}_{io} = \theta_0^* x_{io} - s_i^{-*} \leq x_{io}, & i = 1, \ldots, m \\ \hat{y}_{ro} = y_{ro} + s_r^{+*} \geq y_{ro}, & r = 1, \ldots, s \end{cases} \quad (4)$$

where $(\hat{x}_{io}, \hat{y}_{io})$ represents a point on the efficiency frontier obtained from (x_{io}, y_{ro}), DMU$_o$'s observed values. In fact, $(\hat{x}_{io}, \hat{y}_{io})$, the point on the efficiency frontier obtained from these CCR projections, is the point used to evaluate (x_{io}, y_{ro}), $i = 1, \ldots, m$; $r = 1, \ldots, s$, for any DMU$_o$.

RATIO FORM MODEL: The name "Data Envelopment Analysis" is derived from the primal (minimization) problem in (1) by virtue of the following considerations. The objective is to obtain as tight a fit as possible to the input-output vector for DMU$_o$ by enveloping its observed inputs from below and its observed outputs from above. As can be seen from the primal problem in (1a), an optimal envelopment will always involve a "touching" of the envelopment constraints to at least one of DMU$_o$'s inputs and one of its outputs.

The dual problem, (1b), is said to be in "multiplier" or "production function" form. The former refers to the values of μ and v as dual multipliers. The objective is to maximize y_o, which is called the "virtual output." This maximization is subject to the condition that the corresponding "virtual input" is unity, that is, $\Sigma_{i=1}^m v_i x_{io} = 1$, as given in the first constraint. The other constraints require that the virtual output cannot exceed virtual input for any of the DMU$_j$, $j = 1, \ldots, n$ – that is,

$$\sum_{r=1}^{s} \mu_r y_{rj} \leq \sum_{i=1}^{m} v_i x_{ij}, \quad j = 1, \ldots, n.$$

Finally, the conditions μ_r, $v_i \geq \varepsilon > 0$ mean that every input and every output is to be assigned "some" positive value in this "production function" form, where, as previously noted, the value of ε need not be specified explicitly.

We can multiply all of the variables in the right-hand (dual) problem of (1a) by $t > 0$ and then introduce new variables defined in the following manner:

$$u_r = t\mu_r \geq t\varepsilon, \quad v_i = tv_i \geq t\varepsilon, \quad t = \sum_{i=1}^{m} tv_i x_{io} \quad (5)$$

Multiplying and dividing the objective of the dual problem in (1b) by $t > 0$ then gives (6) and this accords a ratio form to the DEA evaluations. In accordance with the theory of fractional programming as given in Charnes and Cooper (1962), the optimal functional values in (1b) and (6) are equal.

$$\text{Max} \quad \frac{\displaystyle\sum_{r=1}^{s} u_r y_{r0}}{\displaystyle\sum_{i=1}^{m} v_i x_{io}}$$

subject to

$$\frac{\displaystyle\sum_{r=1}^{s} u_r y_{rj}}{\displaystyle\sum_{i=1}^{m} v_i x_{ij}} \leq 1, \quad j = 1, \ldots, n \quad (6)$$

$$\frac{u_r}{\displaystyle\sum_{i=1}^{m} v_i x_{io}} \geq \varepsilon, \quad r = 1, \ldots, s$$

$$\frac{v_i}{\displaystyle\sum_{i=1}^{m} v_i x_{io}} \geq \varepsilon, \quad i = 1, \ldots, m.$$

The formulation (6) has certain advantages. For instance, Charnes and Cooper (1985) used it to show that the optimal ratio value in (6) is invariant to the units of measure used in any input and any output and, hence, this property carries over to (1). Equations (6) also add interpretive power and provide a basis for unifying definitions of efficiency that stretch across various disciplines. For instance, as shown in Charnes, Cooper and Rhodes (1978), the usual single-output to single-input efficiency definitions used in science and engineering are derivable from (6). It follows that these definitions contain an implicit optimality criterion. The relation from (1) to (4), established via fractional programming, also relates these optimality conditions to the definitions of efficiency used in economics (see above discussion of Pareto-Koopmans efficiency). This accords a ratio form to the DEA evaluations. In accordance with the theory of fractional programming (as given in Charnes and Cooper, 1962), the optimal functional values in (1b) and (6) are equal.

As (6) makes clear, DEA also introduces a new principle for determining weights. In particular the weights are not assigned *a priori*, but are determined directly from the data. A "best" set of weights is determined for *each* of the $j = 1, \ldots, n$ DMUs to be evaluated. Given this set of best weights the test of inefficiency for any DMU$_o$ is whether any other DMU$_j$ achieved a higher ratio value than DMU$_o$ using the latter's best weights. (Care needs to be exercised in interpreting these weights, since (a) their values will in general be determined by reference to different collections of DMUs and (b) when determined via (1), allowance needs to be made for non-zero slacks. See the discussion in Charnes, Cooper, Divine, Ruefli and Thomas (1989) where dollar equivalents are used to obtain a complete ordering to guide the use of efficiency audits by the Texas Public Utility Commissions.)

DEA also introduces new principles for making inferences from empirical data. This flows from its use of n optimizations, to come as close as possible to *each* of n observations, in place of other approaches, as in statistics, for instance, which use a single optimization to come as close as possible to *all* of these points. It is not necessary to specify the functional forms explicitly. These forms may be nonlinear and they may be multiple (differing, perhaps, for each DMU) provided they satisfy the mathematical property of isotonicity (Charnes *et al.*, 1985).

The model (1) is only one of several DEA models that are now available. This makes clear that DEA now constitutes a body of concepts, models and methods which unite these models to each other. This comprehends extensions to identify "scale," and "allocative" and other efficiencies.

EXTENSIONS AND APPLICATIONS: Extensions beyond the extant literature have been made when the conditions encountered in applications of DEA required them. As described in Seiford and Thrall (1990), "assurance region" and "cone ratio" approaches were developed to replace the "allocatively most efficient point" concept by an "allocatively most preferred region" to obtain access to milder inequality conditions in place of the exact equality conditions on prices (or weights) which the concept of allocative efficiency requires. "Nondiscretionary variables" with exogenously fixed values were introduced by Banker and Morey (1986) to handle conditions which are beyond the control of DMU management, and these same authors introduced "categorical variables" to improve upon the ability of DEA to effect evaluations from efficient DMUs in the same category as the DMU_o being evaluated.

The bibliography compiled by L. Seiford (1993) now lists more than 600 references which have appeared since the publication by Charnes, Cooper and Rhodes (1978). This, however, is not the end of the trail. All of Seiford's references refer to uses of DEA in evaluating *past* performance. The paper by Arnold, Bardhan and Cooper (1993) has opened a new avenue which extends DEA for use in budgeting the resources needed to improve *future* performance by "DEA efficient" schools which, because of the conditions under which they are operating, had not been able to attain state-mandated levels of excellence.

This example (which formed part of a study supported by the state legislature to improve accountability for Texas public schools) shows DEA to be in an ongoing stage of development. It also allows us to elaborate on what was intended when our opening sentence referred to DEA as a "data oriented method." As part of this study, test scores for state-mandated excellence tests were statistically regressed against school and demographic characteristics. This assigned negative and highly significant coefficients to the proportions of Hispanic and Low English Proficiency students in "representative" schools. This can be regarded as an unsatisfactory answer to the preformed questions prescribed in the regression equations. Turning to a use of DEA, it was discovered that many of the individual schools with high proportions of such students were performing efficiently, while the "excellent schools" had uniformly failed to perform efficiently by virtue of excessive resource usages.

This raised a question as to the resources needed for attaining excellence by these efficiently performing schools. An extension of DEA to deal with this problem showed that in some cases very large resource augmentations would be required. This led to a need for developing parametric variation techniques for use in DEA to match improvements in performance against resource requirements. In turn, this raised a question as to whether uniform across-the-state (or across-the-U.S.) excellence standards might better be replaced by a more flexible and dynamic system involving (perhaps) categories of excellence to encourage progress in stages rather than in a single (perhaps impossible) jump.

One need not conclude from this example that statistical (regression) analyses should be shunned in favor of DEA. It is also possible to conclude that DEA and statistical regression approaches might be combined in various ways that have not heretofore been available. For instance, when the data from this study were restricted to schools which had performed efficiently, the regression coefficients associated with Hispanic and Low English Proficiency students were found to be positive or not statistically significant. This result is leading to further questions, some of which point to the need for research on strategies where DEA can be combined with other approaches in new and fruitful ways.

See **Dual linear programming problem; Fractional programming; Linear programming**.

References

[1] Ahn, T., A. Charnes and W.W. Cooper (1988), "A Note of the Efficiency Characterizations Obtained in Different DEA Models," *Socio-Economic Planning Sciences* 22, 253–257.

[2] Arnold, V., I. Bardhan and W.W. Cooper (1993), "DEA Models for Evaluating Efficiency and Excellence in Texas Secondary Schools," Working Paper, IC² Institute of the University of Texas, Austin.

[3] Arnold, V., I. Bardhan, W.W. Cooper and A. Gallegos (1994), "Primal and Dual Optimality in IDEAS (Integrated Data Envelopment Analysis Systems) and Related Computer Codes," *Proceedings of a Conference in Honor of G.L. Thompson*, Quorum Books.

[4] Banker, R.D. and R.C. Morey (1986), "Data Envelopment Analysis with Categorical Inputs and Outputs," *Management Science* 32, 1613–1627.

[5] Banker, R.D. and R.C. Morey (1986), "Efficiency Analysis for Exogenously Fixed Inputs and Outputs," *Operations Research* 34, 513–521.

[6] Banker, R.D. and R.M. Thrall (1992), "Estimation of Returns to Scale Using Data Envelopment Analysis," *European Jl. Operational Research* 62, 74–84.

[7] Bowlin, W.F., J. Brennan, W.W. Cooper and T. Sueyoshi (1993), "A DEA Model for Evaluating Efficiency Dominance," submitted for publication.

[8] Charnes, A. and W.W. Cooper (1962), "Programming with Linear Fractional Functionals," *Naval Research Logistics Quarterly* 9, 181–186.

[9] Charnes, A. and W.W. Cooper (1985), "Preface to Topics in Data Envelopment Analysis," in R. Thompson and R.M. Thrall, eds., *Annals Operations Research* 2, 59–94.

[10] Charnes, A., W.W. Cooper, D. Divine, T.W. Ruefli and D. Thomas (1989), "Comparisons of DEA and Existing Ratio and Regression Systems for Effecting Efficiency Evaluations of Regulated Electric Cooperations in Texas," *Research in Governmental and Nonprofit Accounting* 5, 187–210.

[11] Charnes, A., W.W. Cooper, B. Golany, L. Seiford and J. Stutz (1985), "Foundations of Data Envelopment Analysis and Pareto-Koopmans Efficient Empirical Production Functions," *Jl. Econometrics* 30, 91–107.

[12] Charnes, A., W.W. Cooper and E. Rhodes (1978), "Measuring Efficiency of Decision Making Units," *European Jl. Operational Research* 1, 429–444.

[13] Charnes, A., W.W. Cooper and R.M. Thrall (1991), "A Structure for Classifying and Characterizing Efficiency and Inefficiency in Data Envelopment Analysis," *Jl. Productivity Analysis* 2, 197–237.

[14] Farrell, M.J. (1957), "The Measurement of Productive Efficiency," *Jl. Royal Statistical Society*, Series A, 253–290.

[15] Kamakura, W.A. (1988), "A Note on the Use of Categorical Variables in Data Envelopment Analysis," *Management Science* 38, 1273–1276.

[16] Rousseau, J.J. and J.H. Semple (1993), "Notes: Categorical Outputs in Data Envelopment Analysis," *Management Science*, 39, 384–386.

[17] Seiford, L.M. and R.M. Thrall (1990), "Recent Developments in DEA," in *Frontier Analysis, Parametric and Nonparametric Approaches*, A.Y. Lewin and C.A. Knox Lovell, eds., *Jl. Econometrics* 46, 7–38.

[18] Seiford, L.M. (1993), "A Bibliography of Data Envelopment Analysis," Technical Report, Department of Industrial Engineering and Operations Research, University of Massachusetts, Amherst.

DEA

See **Data envelopment analysis**.

DECISION ANALYSIS

David A. Schum

George Mason University, Fairfax, Virginia

INTRODUCTION: The term *decision analysis* identifies a collection of technologies for assisting individuals and organizations in the performance of difficult inferences and decisions. Probabilistic inference is a natural element of any choice made in the face of uncertainty. No single discipline can lay claim to all advancements made in support of these technologies. Operations research, probability theory, statistics, economics, psychology, artificial intelligence, and other disciplines have contributed valuable ideas now being exploited in various ways by individuals in many governmental, industrial, and military organizations. As the term decision analysis suggests, complex inference and choice tasks are decomposed into smaller and presumably more manageable elements, some of which are probabilistic and others preferential or value-related. The basic strategy employed in decision analysis is "divide and conquer." The presumption is that individuals or groups find it more difficult to make *holistic* or global judgments required in undecomposed inferences and decisions than to make specific judgments about identified elements of these tasks. In many cases we may easily suppose that decision makers are quite unaware of all of the ingredients that can be identified in the choices they face. Indeed, one reason why a choice may be perceived as difficult is that the person or group charged with making this choice may be quite uncertain about the kind and number of judgments this choice entails. One major task in decision analysis is to identify what are believed to be the necessary ingredients of particular decision tasks.

The label *decision analysis* does not in fact provide a complete description of the activities of persons who employ various methods for assisting others in the performance of inference and choice tasks. This term suggests that the only thing accomplished is the decomposition of an inference or a choice into smaller elements requiring specific judgments or information. It is, of course, necessary to have some process by which these elements can be reassembled or aggregated so that a conclusion or a choice can be made. In other words, we require some method of *synthesis* of the decomposed elements of inference and choice. A more precise term for describing the emerging technologies for assistance in inference and choice would be the term *decision analysis and synthesis*. This fact has been noted in a recent account of progress in the field of decision analysis (Watson and Buede, 1987). As it happens, the same formal methods that suggest how to decompose an inference or choice into more specific elements can also suggest how to reassemble these elements in drawing a conclusion or selecting an action.

PROCESSES AND STAGES OF DECISION ANALYSIS: Human inference and choice are very rich intellectual activities that resist easy categorization. Human inferences made in natural settings (as opposed to contrived classroom examples) involve various mixtures of the three forms of reasoning that have been identified: *deduction* (showing that some conclusion is necessary), *induction* (showing that some conclusion is probable), and *abduction* (generating or discovering possible or plausible conclusions). There are many varieties of choice situations that can be discerned. Some involve the selection of an action or option such as where to locate a nuclear power plant or a toxic waste disposal site. Quite often one choice immediately entails the need for another and so we must consider entire sequences of decisions. It is frequently difficult to specify when a decision task actually terminates. Other decisions involve determining how limited resources may best be allocated among various demands for these resources. Some human choice situations involve episodes of bargaining or negotiation in which there are individuals or groups in some competitive or adversarial posture. Given the richness of inference and choice, analytic and synthetic methods differ from one situation to another as observed in several current surveys of the field of decision analysis (von Winterfeldt and Edwards, 1986; Watson and Buede, 1987; Clemen, 1991). Some general decision analytic processes can, however, be identified.

Most decision analyses begin with careful attempts to define and structure an inference and/or decision problem. This will typically involve consideration of the nature of the decision problem and the individual or group objectives to be served by the required decision(s). A thorough assessment of objectives is required since it is not possible to assist a person or group in making a wise choice in the absence of information about what objectives are to be served. It has been argued that the two central problems in decision analysis concern *uncertainty* and *multiple conflicting objectives* (von Winterfeldt and Edwards, 1986). A major complication arises when, as usually observed, a person or a group will assert objectives that are in conflict. Decisions in many situations involve multiple stakeholders and it is natural to expect that their stated objectives will often be in conflict. Conflicting objectives signal the need for various trade-offs that can be identified. Problem structuring also involves the generation of options, actions, or possible choices. Assuming that there is some element of uncertainty, it is also necessary to generate hypotheses representing relevant alternative states of the world that act to produce possibly different consequences of each option being considered. The result is that when an action is selected we are not certain about which consequence or outcome will occur.

Another important structuring task involves the identification of decision consequences and their attributes. The attributes of a consequence are measurable characteristics of a consequence that are related to a decision maker's asserted objectives. Identified attributes of a consequence allow us to express how well a consequence "measures up" to the objectives asserted in some decision task. Stated in other words, attributes form value dimensions in terms of which the relative preferability of consequences can be assessed. There are various procedures for generating attributes of consequences from stated objectives (Keeney and Raiffa, 1976). Particularly challenging are situations in which we have *multi-attribute* or *vector consequences*. Any conflict involving objectives is reflected in conflicts among attributes and signals the need for examining possible tradeoffs. Suppose, for some option O_i and hypothesis H_j, vector consequence C_{ij} has attributes $\{A_1, A_2, \ldots, A_r, \ldots, A_s, \ldots, A_t\}$. The decision maker may have to judge how much of A_r to give up in order to get more of A_s; various procedures facilitate such judgments. Additional structuring is necessary regarding the inferential element of choice under uncertainty. Given some exhaustive set of mutually exclusive hypotheses or action-relevant states of the world, we will ordinarily use any evidence we can discover that is relevant in determining how probable are each of these hypotheses at the time a choice is required. No evidence comes with already-established relevance, credibility, and inferential force credentials; these credentials have to be established by argument. The structuring of complex probabilistic arguments is a task now receiving considerable attention (Schum, 1987, 1990; Pearl, 1988; Neapolitan, 1990).

At the structural stage just discussed, the process of decomposing a decision is initiated. On some occasions such decomposition proceeds according to formal theories of probability and value taken to be normative. It may even happen that the decision of interest can be represented in terms of some existing mathematical programming or other formal technique common in operations research. In some cases the construction of a model for a decision problem proceeds in an iterative fashion until the decision-maker is satisfied that all ingredients necessary for a decision have been identified. When no new problem ingredients can be identified the model that results is said to be a *requisite model* (Phillips, 1982, 1984). During the process of decomposing the probability and value dimensions of a decision problem it may easily happen that the number of identified elements quickly outstrips a decision maker's time and inclination to provide judgments or other information regarding each of these elements. The question is: how far should we carry the process of divide and conquer? In situations in which there is not unlimited time to identify all conceivable elements of a decision problem, simpler or approximate decompositions at coarser levels of granularity have to be adopted (an example is the simplified multiattribute rating technique (SMART) discussed by von Winterfeldt and Edwards, 1986).

In most decision analyses there is a need for a variety of subjective judgments on the part of persons involved in the decision whose knowledge and experience entitles them to make such judgments. Some judgments concern probabilities and some concern the value of consequences in terms of identified attributes. Other judgments may involve assessment of the relative importance of consequence attributes. The study of methods for obtaining dependable quantitative judgments from people represents one of the most important contributions of psychology to decision analysis. A survey of these judgmental contributions was given in von Winterfeldt and Edwards (1986). After a decision has been structured and subjective ingredients elicited, the synthetic process in decision analysis is then exercised in order to identify the "best" conclusion and/or choice. In many cases such synthesis accomplished by an algorithmic process taken as appropriate to the situation at hand. Modern computer facilities allow decision makers to use these algorithms to test the consequences of various possible patterns of their subjective beliefs by means of *sensitivity analyses*. The means for defending the wisdom of conclusions or choices made by such algorithmic methods requires consideration of the formal tools used for decision analysis and synthesis.

THEORIES OF ANALYSIS AND SYNTHESIS:
Two major pillars upon which most of modern decision analysis rests are theories of probabilistic reasoning and theories of value or preference. It is safe to say that the conventional view of probability, in which Bayes' rule appears as a canon for coherent or rational probabilistic inference, dominates current decision analysis. For some body of evidence E, Bayes' rule is employed in determing a distribution of posterior probabilities $\Pr\{H_k|E\}$, for each hypothesis H_k in an exhaustive collection of mutually exclusive decision-relevant hypotheses. The ingredients Bayes' rule requires, prior probabilities (or prior odds) and likelihoods (or likelihood ratios), are in most cases assumed to be assessed subjectively by knowledgeable persons. In some situations, however, appropriate relative frequencies may be available. The subjectivist view of probability, stemming from the work of Ramsey and de Finetti, has had a very sympathetic hearing in decision analysis (Mellor, 1990; de Finetti, 1972).

Theories of coherent or rational expression of values or preferences stem from the work of Von Neumann and Morgenstern (1947). In this work appears the first attempt to put the task of stating preferences on an axiomatic footing. Adherence to the von Neumann and Morgenstern axioms places judgments of value on a cardinal or equal-interval scale and are often then called judgments of *utility*. These axioms also suggest methods for eliciting utility judgments and they imply that a coherent synthesis of utilities and probabilities in reaching a decision consists of applying the principle of *expected utility maximization*. This idea was extended in the later work of Savage (1954) who adopted the view that the requisite probabilities are subjective in nature. The canon for rational choice emerging from the work of Savage is that we should choose from among alternative actions by determining which one has the highest *subjective expected utility* (SEU). Required aggregation of probabilities is assumed to be performed according to Bayes' rule. In some works this view of action-selection is called *Bayesian decision theory* (Winkler, 1972; Smith, 1988).

Early works by Ward Edwards (1954, 1961) stimulated interest among psychologists in developing methods for probability and utility elicitation; these works also led to many behavioral assessments of the adequacy of SEU as a description of actual human choice mechanisms. In a later work, Edwards (1962) proposed the first system for providing computer assistance in the performance of complex probabilistic inference tasks. Interest in the very difficult problems associated with assessing the utility of multiattribute consequences stems from the work of Raiffa (1968). But credit for announcing the existence of the applied discipline we now call decision analysis belongs to Howard (1966, 1968).

DECISION ANALYTIC STRATEGIES:
There are now many individuals and organizations employed in the business of decision analysis. The

inference and decision problems they encounter are many and varied. A strategy successful in one context may not be so successful in another. In most decision-analytic encounters, an analyst plays the role of a *facilitator*, also termed "high priests" (von Winterfeldt and Edwards, 1986). The essential task for the facilitator is to draw out the experience and wisdom of decision-makers while guiding the analytic process toward some form of synthesis. In spite of the diversity of decision contexts and decision analysts, Watson and Buede (1987) were able to identify the following five general decision analytic strategies in current use. They make no claim that these strategies are mutually exclusive.

1) *Modeling*. In some instances decision analysts will focus upon efforts to construct a conceptual model of the process underlying the decision problem at hand. In such a strategy the decision-maker(s) being served not only provide the probability and value ingredients their decision requires but are also asked to participate in constructing a model of the context in which this decision is embedded. In the process of constructing these often-complex models, important value and uncertainty variables are identified.

2) *Introspection*. In some decision analytic encounters a role played by the facilitator is one of assisting decision-makers in careful introspective efforts to determine relevant preference and probability assessments necessary for a synthesis in terms of subjective expected utility maximization. Such a process places great emphasis upon the reasonableness and consistency of the often-large number of value and probability ingredients of action selection.

3) *Rating*. In some situations, especially those involving multiple stakeholders and multi-attribute consequences, any full-scale task decomposition would be decisionally paralytic or, in any case, would not provide the timely decisions so often required. In order to facilitate decision making under such circumstances, models involving simpler probability and value assessments are often introduced by the analyst. An example is the SMART technique mentioned above in which many of the difficult multi-attribute utility assessments are made simpler through the use of various rating techniques and by the assumption of independence of the attributes involved.

4) *Conferencing*. In a decision conference the role of the decision analyst as facilitator (or high priest) assumes special importance. In such encounters, often involving a group of persons participating to various degrees in a decision, the analyst promotes a structured dialogue and debate among participants in the generation of decision ingredients such as options, hypotheses and their probability, and consequences and their relative value. The analyst further assists in the process of synthesis of these ingredients in the choice of an action. The subject matter of a decision conference can involve action-selection, resource allocation, or negotiation.

5) *Developing*. In some instances, the role of the decision analyst is to assist in the development of strategies for recurrent choices or resource allocations. These strategies will usually involve computer-based *decision support systems* or some other computer-assisted facility whose development is justified by the recurrent nature of the choices. The study and development of decision support systems has itself achieved the status of a discipline (Sage, 1991). An active and exciting developmental effort concerns computer-implemented *influence diagrams* stemming from the work of Howard and Matheson (1981). Influence diagram systems can be used to structure and assist in the performance of inference and/or decision problems and have built-in algorithms necessary for the synthesis of probability and value ingredients (Shachter, 1986; Shachter and Heckerman, 1987). Such systems are equally suitable for recurrent and nonrecurrent inference and choice tasks.

CONTROVERSIES: As an applied discipline, decision analysis inherits any controversies associated with theories upon which it is based. There is now a substantial literature challenging the view that *the* canon for probabilistic inference is Bayes' rule (Cohen, 1977, 1989; Shafer, 1976). Regarding preference axioms, Shafer (1986) argued that no normative theories of preference have in fact been established and that existing theories rest upon an incomplete set of assumptions about basic human judgmental capabilities. Others have argued that the probabilistic and value-related ingredients required in Bayesian decision theory often reflect a degree of precision that cannot be taken seriously given the imprecise or fuzzy nature of the evidence and other information upon which such judgments are based (Watson, Weiss, and Donnell, 1979). Philosophers have recently been critical of contemporary decision analysis. Agreeing with Cohen and Shafer, Tocher (1977) argued against the presumed normative status of Bayes' rule. Rescher (1988) argued that decision analysis can easily show people how to decide in ways that are entirely consistent with objectives that turn out not to be in their best interests. Others like

Dreyfus (1984) have questioned whether or not decomposed inference and choice is always to be preferred over holistic inference and choice; this same concern is reflected in other contexts such as law (Twining, 1990). Thus, the probabilistic and value-related bases of modern decision analysis involve matters about which there will be continuing dialogue and, perhaps, no final resolution. This acknowledged, decision-makers in many contexts continue to employ the emerging technologies of decision analysis and find, in the process, that very complex inferences and choices can be made tractable and far less intimidating.

See **Choice theory; Decision support systems; Fuzzy sets; Group decision making; Multi-attribute utility theory; Utility theory.**

References

[1] Clemon, R.T. (1991), *Making Hard decisions: An Introduction to Decision Analysis*, PWS-Kent Publishing Co., Boston.

[2] Cohen, L.J. (1977), *The Probable and the Provable*, Clarendon Press, Oxford.

[3] Cohen, L.J. (1989), *An Introduction to the Philosophy of Induction and Probability*, Clarendon Press, Oxford.

[4] De Finetti, B. (1972), *Probability, Induction, and Statistics: The Art of Guessing*, John Wiley, New York.

[5] Dreyfus, S. (1984), "The Risks! and Benefits? of Risk-Benefit Analysis," *Omega*, 12, 335–340.

[6] Edwards, W. (1954), "The Theory of Decision Making," *Psychological Bulletin*, 41, 380–417.

[7] Edwards, W. (1961), "Behavioral Decision Theory," *Annual Review Psychology*, 12, 473–498.

[8] Edwards, W. (1962), "Dynamic Decision Theory and Probabilistic Information Processing," *Human Factors*, 4, 59–73.

[9] Howard, R. (1966), "Decision Analysis: Applied Decision Theory," in Hertz, D. B. and Melese, J., eds., *Proceedings of the Fourth International Conference on Operational Research*, Wiley-Interscience, New York.

[10] Howard, R. (1968), "The Foundations of Decision Analysis," *IEEE Transactions on Systems Science and Cybernetics*, SSC-4, 211–219.

[11] Howard, R. and Matheson, J. (1981), "Influence Diagrams," in Howard, R. and Matheson, J., eds., *The Principles and Applications of Decision Analysis, Vol. II*, Strategic Decisions Group, Menlo Park, California, 1984.

[12] Keeney, R. and Raiffa, H. (1976), *Decision With Multiple Objectives: Preferences and Value Trade-offs*, John Wiley, New York.

[13] Mellor, D.H. (1990), *F.P. Ramsey: Philosophical Papers*, Cambridge University Press, Cambridge.

[14] Neapolitan, R. (1990), *Probabilistic Reasoning in Expert Systems: Theory and Algorithms*, John Wiley, New York.

[15] Pearl, J. (1988), *Probabilistic Reasoning in Intelligent Systems: Networks of Plausible Reasoning*, Morgan Kaufmann Publishers, San Mateo, California.

[16] Phillips, L. (1982), "Requisite Decision Modelling: A Case Study," *Journal Operational Research Society*, 33, 303–311.

[17] Phillips, L. (1984), "A Theory of Requisite Decision Models," *Acta Psychologica*, 56, 29–48.

[18] Raiffa, H. (1968), *Decision Analysis: Introductory Lectures on Choices Under Uncertainty*, Addison-Wesley, Reading, Massachusetts.

[19] Rescher, N. (1988), *Rationality: A Philosphical Inquiry into the Nature and Rationale of Reason*, Clarendon Press, Oxford.

[20] Sage, A. (1991), *Decision Support Systems Engineering*, John Wiley, New York.

[21] Savage, L. J. (1954), *The Foundations of Statistics*, John Wiley, New York.

[22] Schum, D. (1987), *Evidence and Inference for the Intelligence Analyst* [two volumes], University Press of America, Lanham, Maryland.

[23] Schum, D. (1990), "Inference Networks and Their Many Subtle Properties," *Information and Decision Technologies*, 16, 69–98.

[24] Shachter, R. (1986), "Evaluating Influence Diagrams," *Operations Research*, 34, 871–882.

[25] Shachter, R. and Heckerman, D. (1987), "Thinking Backward for Knowledge Acquisition," *AI Magazine*, Fall, 55–61.

[26] Shafer, G. (1976), *A Mathematical Theory of Evidence*, Princeton University Press, New Jersey.

[27] Shafer, G. (1986), "Savage Revisited," *Statistical Science*, 1, 463–501 (with comments).

[28] Smith, J.Q. (1988), *Decision Analysis: A Bayesian Approach*, Chapman and Hall, London.

[29] Tocher, K. (1977), "Planning Systems," *Philosophical Transactions Royal Society London*, A287, 425–441.

[30] Twining, W. (1990), *Rethinking Evidence: Exploratory Essays*, Basil Blackwell, Oxford.

[31] von Neumann, J. and Morgenstern, O. (1947), *Theory of Games and Economic Behavior*, Princeton University Press, New Jersey.

[32] von Winterfeldt, D. and Edwards, W. (1986), *Decision Analysis and Behavioral Research*, Cambridge University Press, Cambridge.

[33] Watson, S.R. and Buede, D. (1987), *Decision Synthesis: The Principles and Practice of Decision Analysis*, Cambridge University Press, Cambridge.

[34] Watson, S.R., Weiss, J.J., and Donnell, M.L. (1979), "Fuzzy Decision Analysis," *IEEE Transactions on Systems, Man, and Cybernetics*, SMC-9(1), 1–9.

[35] Winkler, R.L. (1972), *Introduction to Bayesian Inference and Decision*, Holt, Rinehart, and Winston, New York.

DECISION MAKER (DM)

An individual (or group) who is dissatisfied with some existing situation or with the prospect of a future situation and who possesses the desire and authority to initiate actions designed to alter the situation. In the literature, the letters DM are often used to denote decision maker. See **Decision problem; Mathematical model.**

DECISION MAKING

Dennis M. Buede

George Mason University, Fairfax, Virginia

Decision making is a process undertaken by an individual or organization. The intent of this process is to improve the future position of the individual or organization in terms of one or more criteria. Most scholars of decision making define this process as one that culminates in an irrevocable allocation of resources to affect some chosen change or the continuance of the status quo. The most commonly allocated resource is money, but other scarce resources are goods and services, and the time and energy of talented people.

Three primary decision modes have been identified by Watson and Buede (1987): choosing one option from a list, allocating a scarce resource(s) amongst competing projects, and negotiating an agreement with one or more adversaries. Decision analysis is the common analytical approach for the first mode, optimization for the second, and a host of techniques have been applied to negotiation decisions (Jelassi and Foroughi, 1989).

The four major elements of a decision that make its resolution troublesome are the creative generation of options, the identification and quantification of multiple conflicting criteria, the causal linkage between options and criteria, and the assessment and analysis of uncertainty associated with the causal linkage. Many decision makers claim to be troubled by the feeling that there is an, as yet unidentified, option that must surely be better than those so far considered. The development of techniques for identifying such options has received considerable attention (Elam and Mead, 1990; Friend and Hickling, 1987; Keller and Ho, 1988; Keeney, 1992; McGoff *et al.*, 1990). Ample, additional research has been undertaken to identify the pitfalls in assessing probability distributions that represent the uncertainty of a decision maker

(von Winterfeldt and Edwards, 1986). Research has also focused on the identification of the most appropriate assessment techniques. Similar research by von Winterfeldt and Edwards (1986) focused on assessing value and utility functions. Keeney (1992) has advanced concepts for the development and structuring of a value hierarchy for key decisions.

The making of a good decision requires a sound decision making process because it is never possible to identify what would have been the best option. Tracing alternate worlds into the future for the purpose of following alternate courses of action to their conclusion is not possible. Numerous researchers have proposed multiphased processes for decision making (Dewey, 1933; Simon, 1965; Howard, 1968 and 1984; Witte, 1972; Mintzberg *et al.*, 1976; von Winterfeldt, 1980; Buede, 1992). The common phases include: intelligence or problem definition, design or analysis, choice, and implementation. A weakness in one phase in the decision making process often cannot be compensated for by strengths in the other phases.

See **Corporate strategy; Decision problem; Multi-attribute utility theory; Multi-objective programming; Multiple criterion decision making; Organization; Utility theory.**

References

[1] Buede, D. (1992), "Superior Design Features of Decision Analytic Software," *Computers and Operations Research*, 19, 43–57.

[2] Dewey, J. (1993), *How We Think*, Heath, Boston, Massachusetts.

[3] Elam, J. and Mead, M. (1990), "Can Software Influence Creativity?," *Information Systems Research*, 1, 1–22.

[4] Friend, J. and Hickling, A. (1987), *Planning Under Pressure: The Strategic Choice Process*, Pergamon Press, Oxford.

[5] Howard, R. (1968), "The Foundations of Decision Analysis," *IEEE Transactions on Systems, Science, and Cybernetics*, SSC-4, 211–219.

[6] Howard, R. (1989), "The Evolution of Decision Analysis," in Howard, R. and Matheson, J. eds., *The Principles and Applications of Decision Analysis*, Strategic Decisions Group, Menlo Park, California.

[7] Jelassi, M. and Foroughi, A. (1989), "Negotiation Support Systems: An Overview of Design Issues and Existing Software," *Decision Support Systems*, 5, 167–181.

[8] Keeney, R. (1992), *Value-Focused Thinking*, Harvard University Press, Boston.

[9] Keller, L. and Ho, J. (1988), "Decision Problem Structuring: Generating Options," *IEEE Trans-*

actions on *Systems, Man, and Cybernetics*, SMC-15, 715–728.

[10] McGoff, C., Vogel, D., and Nunamaker, J. (1990), "IBM Experiences with Group Systems," *DSS-90 Transactions*, 206–221.

[11] Mintzberg, H., Raisinghani, D. and Theoret, A. (1976), "The Structure of 'Unstructured' Decision Processes," *Administrative Sciences Quarterly*, 21, 246–275.

[12] Simon, H.A. (1965), *The Shape of Automation*, Harper & Row, New York.

[13] von Winterfeldt, D. (1980), "Structuring Decision Problems for Decision Analysis," *Acta Psychologica*, 45, 71–93.

[14] von Winterfeldt, D. and Edwards, W. (1986), *Decision Analysis and Behavioral Research*, Cambridge University Press, New York.

[15] Watson, S. and Buede, D. (1987), *Decision Synthesis: The Principles and Practice of Decision Analysis*, Cambridge University Press.

[16] Witte, E. (1972), "Field Research on Complex Decision-Making Processes – the Phase Theorem," *Int. Stud. Mgmt Organization*, 156–182.

DECISION PROBLEM

The basic decision problem is as follows: Given a set of r alternative actions $A = \{a_1 \ldots, a_r\}$, a set of q states of nature $S = \{s_1, \ldots, s_q\}$, a set of rq outcomes $O = \{o_1 \ldots, o_{rq}\}$, a corresponding set of rq payoffs $P = \{p_1, \ldots, p_{rq}\}$, and a decision criterion to be optimized, $f(a_j)$, where f is a real-valued function defined on A, choose an alternative action a_j that optimizes the decision criterion $f(a_j)$. See **Decision maker; Mathematical model; Multi-criteria decision making.**

DECISION SUPPORT SYSTEMS

Andrew Vazsonyi

University of San Francisco, California

REVIEW OF DECISION MAKING: Throughout history there has been a deeply embedded conviction that, under the proper conditions, some people are capable of helping others come to grips with problems in daily life. Such professional helpers are called counselors, psychiatrists, psychologists, social workers, and the like. In addition to these professional helpers, there are less formal helpers, such as ministers, lawyers, teachers, or even bartenders, hairdressers and cab drivers.

The proposition that science and quantitative methods, such as those used in OR/MS, may help people is relatively new, and is still received by many with deep skepticism. There are some disciplines overlapping and augmenting OR/MS. One called decision support systems, DSS, the subject matter of this article.

Before discussion of DSS, it is to be stressed that the expression is used in a different manner by different people, and there is no general agreement of what DSS really is. Moreover, the benefits claimed by DSS are in no way different from the benefits claimed by OR/MS. To appreciate DSS we must take a pluralistic view of the various disciplines offered to help managerial decision making.

FEATURES OF DECISION SUPPORT SYSTEMS: During the early 1970s, under the impact of new developments in computer systems, a new perspective about decision making appeared. Keen and Morton (1973) coined the expression *decision support systems*, DSS, to designate their approach to the solution of managerial problems. They postulated a number of distinctive characteristics of DSS, and we list five of them:

- A DSS is designed for specific decision makers and their decision tasks
- A DSS is developed by cycling between design and implementation
- A DSS is developed with a high degree of user involvement
- A DSS includes both data and models
- Design of the user-machine interface is a critical task in the development of a DSS

Fig. 1 shows the structure and major components of a DSS. The *database* holds all the relevant facts of the problem, whether they pertain to the firm or to the environment. The *database management system* (Fig. 2) takes care of the entry, retrieval, updating and deletion of data. It also responds to inquiries and generates reports.

The *modelbase* holds all the models required to work the problem. The *modelbase management system* (Fig. 3) assists in creating the mathematical model, in translating the human prepared mathematical model into computer understandable form. The critical process of the modelbase management system is finding the solution to the mathematical model. The system also generates reports and assists in the preparation of computer-human dialogs.

While OR/MS stresses the model, DSS stresses the computer-based data base. DSS emphasizes the importance of the user-machine interface, and the design of dialog generation and management software.

Advocates of DSS assert that by combining the power of the human mind and the computer, DSS

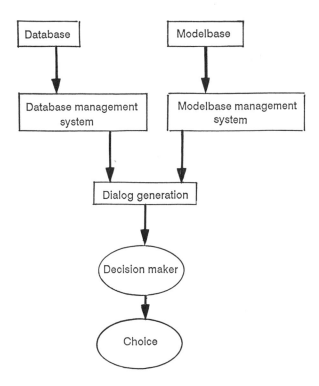

Fig. 1. Components of a DSS.

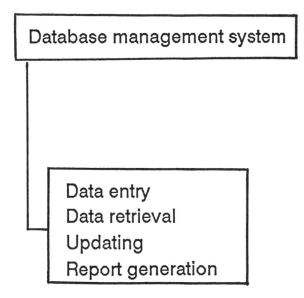

Fig. 2. Database management system.

is capable of enhancing decision making, and that DSS can grapple with problems not subject to the traditional approach of OR/MS.

Note that DSS stresses the role of humans in decision making, and explicitly factors human capabilities into decision making. A decision support system accepts the human as an essential subsystem. DSS does not usually try to optimize in a mathematical sense, and *bounded rationality* and *satisficing* provide guidance to the designers of DSS.

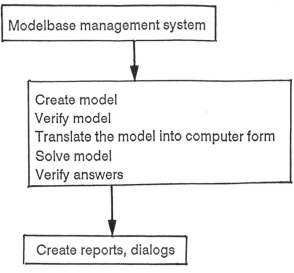

Fig. 3. Modelbase management system.

DESIGNING DECISION SUPPORT SYSTEMS: The design phases of DSS are quite similar to the phases of the design, implementation, and testing of other systems. It is customary to distinguish six phases, although not all six phases are required for every DSS.

1. *During the systems analysis and design phase*, existing systems are reviewed and analyzed with the objective of establishing requirements and needs of the new system. Then it is established whether meeting the specifications is feasible from the technical, economical, psychological and social point of view. Is it possible to overcome the difficulties, and are opportunities commensurate with costs? If the answers are affirmative, meetings with management are held to obtain support. This phase produces a conceptual design and master plan.

2. *During the design phase*, input, processing, and output requirements are developed and a logical (not physical) design of the system is prepared. After the logical design is completed and found to be acceptable, the design of the hardware and software is undertaken.

3. *During the construction and testing phase*, the software is completed and tested on the hardware system. Testing includes user participation to assure that the system will be acceptable both from the points of view of the user and management, if they are different.

4. *During the implementation phase*, the system is retested, debugged and put into use. To assure final user acceptance, no effort is spared in training and educating users. Management is kept up-to-date on the progress of the project.

5. *Operation and maintenance* is a continued effort during the life of the DSS. User satisfaction

is monitored, errors are uncovered and corrected, and the method of operating the system is fine-tuned.

6. *Evaluation and control* is a continued effort to assure the viability of the system and the maintenance of management support.

A FORECASTING SYSTEM: Connoisseur Foods is a diversified food company with several autonomous subdivisions and subsidiaries (adapted from Alter, 1980, and Turban, 1990). Several of the division managers are old-line managers relying on experience and judgment to make major decisions. Top management installed a DSS to provide quantitative help to establish and monitor levels of such marketing efforts as advertising, pricing and promotion. The DSS model was based on S-shaped response functions of marketing conditions to such decision functions as advertising. The curves were derived by using both historical data and marketing experts. The databases for the farm products division contained about 20 million data items on sales both in dollars and number of units for 400 items sold in 300 branches.

The DSS assisted management in developing better marketing strategies and more competitive positions. However, top management stated that the real benefit of the DSS was not so much the installation of isolated systems and models, but the assimilation of new approaches in corporate decision making.

A PORTFOLIO MANAGEMENT SYSTEM: The trust division of Great Eastern Bank employs 50 portfolio managers in several departments (adapted from Alter, 1980 and Turban, 1990). The portfolio managers control many small accounts, large pension funds, and provide advice to investors in large accounts. The on-line DSS portfolio management system provides information to the portfolio managers.

The DSS includes lists of stocks from which the portfolio managers can buy stocks, information and analysis on particular industries. It is basically a data retrieval system that can display portfolios as well as specific information on securities.

The heart of the system is the database that allows portfolio managers to generate reports with the following functions:
- directory by accounts
- table to scan accounts
- graphic display of breakdown by industry and security for an account
- tabular listing of all securities within an account

- scatter diagrams between data items
- summaries of accounts
- distribution of data on securities
- evaluation of hypothetical portfolios
- performance monitoring of portfolios
- warnings if deviations from guidelines occur
- tax implications.

The benefits of the systems were better investment performance, improved information, improved presentation formats, less clerical work, better communication, improved bank image, and enhanced marketing capability.

CONCLUSIONS: Advocates of DSS claim that DSS deals with *unstructured* or *semistructured* problems, while OR/MS is restricted to *structured* problems. Few workers in OR/MS would agree.

Bear in mind that at the onset, most of the time, a particular business situation is confusing, and to straighten it out a problem must be instituted and the problem must be structured. Thus, whether OR/MS or DSS or both are involved, attempts will be made to structure as much of the situation as possible.

The problem will be structured by OR/MS or DSS to the point that some part of the problem can be taken care of by quantitative methods and computers, and some others are left to human judgment, intuition and opinion. There may be a degree of difference between OR/MS and DSS: OR/MS may stress optimization, the model base; DSS the database. If there is a difference, it changes from person to person, case to case, and time to time.

Attempts to draw the line between DSS and OR/MS are counter productive. The point of view to accept is to look at the methodologies involved and results obtainable. Those who are dedicated to help management in solving hard problems need to be concerned with any and all theories, practices, principles, that can help. To *counsel* management in the most productive manner requires that no holds be barred, when a task is undertaken.

See **Decision analysis; Decision problem; Information systems/database design.**

References

[1] Alter, S.L. (1980). *Decision Support Systems: Current Practice and Continuing Challenges*, Addison-Wesley, Reading, Massachusetts.
[2] Bennett, J.L. (1983). *Building Decision Support Systems*, Addison-Wesley, Reading, Massachusetts.
[3] Keen, P.G.W. and Morton, S. (1973). *Decision Support Systems*, Addison-Wesley, Reading, Massachusetts.

[4] Simon, H.A. (1992), "Methods and Bounds of Economics," in *Praxiologies and the Philosophy of Economics*, Transaction Publishers, New Brunswick and London.

[5] Turban, E. (1990). *Decision Support and Expert Systems*, 2nd ed., Macmillan, New York.

DECISION TREES

Stuart Eriksen and L. Robin Keller

University of California, Irvine

A decision tree is a pictorial description of a well-defined decision problem. It is a graphical representation consisting of nodes (where decisions are made or chance events occur) and arcs (which connect nodes). Decision trees are useful because they provide a clear, documentable and discussable model of either how the decision will be made or how it was made.

The tree provides a framework for the calculation of the expected value of each available alternative. The alternative with the maximum expected value is the best choice path based on the information and mind-set of the decision makers at the time the decision is made. This best choice path indicates the best overall alternative, including the best subsidiary decisions at future decision steps, when uncertainties have been resolved.

The decision tree should be arranged, for convenience, from left to right in the temporal order in which the events and decisions will occur. Therefore, the steps on the left occur earlier in time than those on the right.

DECISION NODES: Steps in the decision process involving decisions between several choice alternatives are indicated by decision nodes, drawn as square boxes. Each available choice is shown as one arc (or "path") leading away from its decision node toward the right. When a planned decision has been made at such a node, the result of that decision is recorded by drawing an arrow in the box pointing toward the chosen option. As an example of the process, consider a pharmaceutical company president's choice of which drug dosage to market. The basic dosage choice decision tree is shown in Figure 1. Note that the values of the eventual outcomes (on the far right) will be expressed as some measure of value to the eventual user (specified by the patient or the physician).

CHANCE NODES: Steps in the process which involve uncertainties are indicated by circles

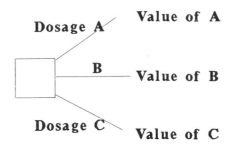

Fig. 1. The choice of drug dosage.

(called chance nodes), and the possible outcomes of these probabilistic events are again shown as arcs or paths leading away from the node toward the right. The results of these uncertain factors are out of the hands of the decision maker; chance or some other group or person will determine the outcome of this node. Each of the potential outcomes of a chance node is labeled with its probability of occurrence, and the sum of the potential outcome probabilities of a chance node must equal 1.0. Using the drug dose selection problem noted above, the best choice of dose depends on at least one probabilistic event: the level of performance of the drug in clinical trials, which is a proxy measure of the efficacy of the drug. A simplified decision tree for that part of the firm's decision is shown in Figure 2. Note that each dosage choice has a subsequent efficacy chance node similar to the one shown, so the expanded tree would have nine outcomes.

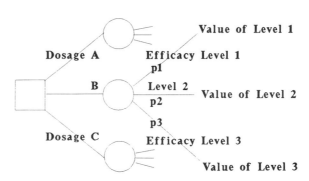

Fig. 2. The choice of drug dosage based on efficacy outcome.

There are often several nodes in a decision tree; in the case of the drug dosage decision, the decision will also depend on the toxicity as demonstrated by both animal study data and human toxicity study data as well as on the efficacy data. The basic structure of this more complex decision is shown in Figure 3. The completely expanded tree has 27 eventual outcomes and associated values.

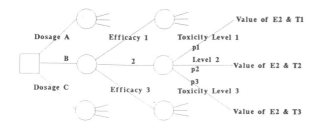

Fig. 3. The choice of dosage based on uncertain efficacy and toxicity.

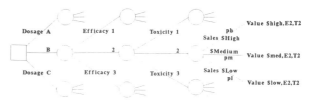

Fig. 4. The choice of dosage based on efficacy, toxicity and eventual sales.

One use of a decision tree is to clearly display the factors and assumptions involved in a decision. If the decision outcomes are quantified and the probabilities of chance events are specified, the tree can also be analyzed by calculating the expected value of each alternative.

PROBABILITIES: Estimates of the probabilities for each of the outcomes of the chance nodes must be made. In the simplified case of the drug dose decision above, the later chance node outcome probabilities are modeled correctly as being independent of the earlier chance nodes, since, with at least some drug classifications for example, the probability of high human toxicity is probably independent of the level of human efficacy. In the more general situation, however, for sequential steps, the later probabilities are often conditional probabilities, since their value depends on the earlier chance outcomes.

For example, consider the problem in Figure 4, where the outcome being used for the drug dose decision is based on the eventual sales of it. The values of the eventual outcomes now are expressed as profit to the firm.

High sales depends on the efficacy as well as on the toxicity, so the conditional probability of high sales is the probability of high sales given that the efficacy is high and toxicity is low,

which can be written as Pr{High|Efficacy high and Toxicity low}.

OUTCOME MEASURES: At the far right of the tree, the possible outcomes are listed at the end of each branch. To calculate the expected values for alternative choices, outcomes must be measured numerically and often monetary measures will be used. More generally, the "utility" of the outcomes can be calculated. Single or multiple attribute utility functions have been elicited in many decision situations to represent decision makers' preferences for different outcomes on a numerical scale.

THE TREE AS AN AID IN DECISION MAKING: The decision tree analysis method is called "foldback" and "prune." Beginning at a far right chance node of the tree, the expected value of the outcome measure is calculated and recorded for each chance node by summing, over all the outcomes, the product of the probability of the outcome times the measured value of the outcome. Figure 5 shows this calculation for the first step in the analysis of the drug dose decision tree.

This step is called "folding back the tree" since the branches emanating from the chance node are folded up or collapsed, so that the chance node is now represented by its expected

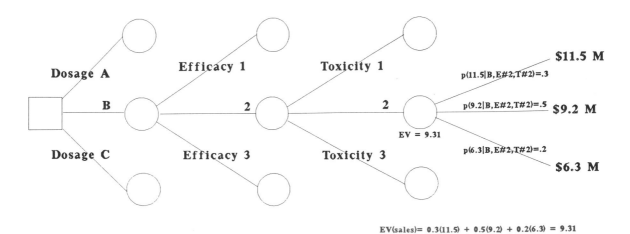

Fig. 5. The first step: calculating the expected value of the chance node for sales.

value. This is continued until all the chance nodes on the far right have been evaluated. These expected values then become the values for the outcomes of the decision nodes further to the left in the diagram. At a decision node, the best of the choices is the one with the maximum expected value which is then recorded by drawing an arrow towards that choice in the decision node box and writing down the expected value associated with the chosen option. This is referred to as "pruning the tree," as the less valuable choices are eliminated from each decision. The process continues from right to left, by calculating the expected value at each chance node and pruning at each decision node. Finally the best choice for the overall decision is found when the last decision node at the far left has been evaluated.

See **Decision analysis; Decision making; Decision problems; Group decision making.**

References

[1] Eriksen, Stuart P. and Keller, L. Robin (1993), "A Multi-Attribute Approach To Weighing The Risks and Benefitsof Pharmaceutical Agents," *J. Medical Decision Making*, 13, 118–125.

[2] Keeney, Ralph L. and Raiffa, Howard (1976), *Decisions with Multiple Objectives: Preferences and Value Tradeoffs*, John Wiley, New York.

[3] Raiffa, Howard (1968), *Decision Analysis*, Addison-Wesley, Reading, Massachusetts.

DECISION VARIABLES

The variables in a given model that are subject to manipulation by the specified decision rule. See **Controllable variables.**

DECOMPOSITION ALGORITHM

See **Bender's decomposition method; Dantzig-Wolfe decomposition algorithm.**

DEGENERACY

The situation in which a linear-programming problem has a basic feasible solution with at least one basic variable equal to zero. If the problem is degenerate, then an extreme point of the convex set of solutions may correspond to several feasible bases. As a result, the simplex method may move through a sequence of bases with no improvement in the value of the objective function. In rare cases, the algorithm may cycle repeatedly through the same sequence of bases and never converge to an optimal solution. Anticycling rules, and perturbation and lexicographic techniques prevent this risk, but usually at some computational expense. See **Bland's anticyclying rules; Linear programming; Simplex method.**

DEGENERATE SOLUTION

A basic (feasible) solution in which some basic variables are zero. See **Degeneracy.**

DEGREE OF A NODE

The number of edges incident with a given node in a graph.

DELAUNAY TRIANGULATION

See **Computational geometry; Voronoi constructs.**

DELAY

The time spent by a customer in queue waiting to start service. See **Waiting time.**

DELPHI METHOD

James A. Dewar and John A. Friel

The RAND Corporation, Santa Monica, California

INTRODUCTION: The Delphi method was developed at the RAND Corporation from studies on decision making that began in 1948. The seminal work, "An Experimental Application of the Delphi Method to the Use of Experts," was written by Dalkey and Helmer (1963).

The primary rationale for the technique is the age-old adage "two heads are better than one," particularly when the issue is one where exact knowledge is not available. It was developed as an alternative to the traditional method of obtaining group opinions – face-to-face discussions. Experimental studies had demonstrated several serious difficulties with such discussions. Among them were: (1) influence of the dominant individual (the group is highly influenced by the person who talks the most or has most authority); (2) noise (studies found that much "communication" in such groups had to do with individual and group interests rather than problem solving); and (3) group pressure for conformity (studies demonstrated the distortions of individual judgment that can occur from group pressure).

The Delphi method was specifically developed to avoid these difficulties. In its original formulation it had three basic features: (1) anonymous response – opinions of the members of the group are obtained by formal questionnaire; (2) iteration and controlled feedback – interaction is effected by a systematic exercise conducted in several iterations, with carefully controlled feedback between rounds; and (3) statistical group response – the group opinion is defined as an appropriate aggregate of individual opinions on the final round.

Procedurally, the Delphi method begins by having a group of experts answer questionnaires on a subject of interest. Their responses are tabulated and fed back to the entire group in a way that protects the anonymity of the irresponses. They are asked to revise their own answers and comment on the group's responses. This constitutes a second round of the Delphi. Its results are tabulated and fed back to the group in a similar manner and the process continues until convergence of opinion, or a point of diminishing returns, is reached. The results are then compiled into a final statistical group response to assure that the opinion of every member of the group is represented.

In its earliest experiments, Delphi was used for technological forecasts. Expert judgments were obtained numerically (e.g., the date that a technological advance would be made), and in that case it is easy to show that the mean or median of such judgments is at least as close to the true answer as half of the group's individual answers. From this, the early proponents were able to demonstrate that the Delphi method produced generally better estimates than those from face-to-face discussions.

One of the surprising results of experiments with the technique was how quickly in the successive Delphi rounds that convergence or diminishing returns is achieved. This helped make the Delphi technique a fast, relatively efficient, and inexpensive tool for capturing expert opinion. It was also easy to understand and quite versatile in its variations. By 1975, there were several hundred applications of the Delphi method reported on in the literature. Many of these were applications of Delphi in a wide variety of judgmental settings, but there was also a growing academic interest in Delphi and its effectiveness.

CRITIQUE: H. Sackman (1975), also of the RAND Corporation, published the first serious critique of the Delphi method. His book, *Delphi Critique*, was very critical of the technique – particularly its numerical aspects – and ultimately recommended (p. 74) "that . . . Delphi be dropped from institutional, corporate, and government use until its principles, methods, and fundamental applications can be experimentally established as scientifically tenable."

Sackman's critique spurred both the development of new techniques for obtaining group judgments and a variety of studies comparing Delphi with other such techniques. The primary alternatives can be categorized as statistical group methods (where the answers of the group are tabulated statistically without any interaction); unstructured, direct interaction (another name for traditional, face-to-face discussions); and structured, direct interaction (such as the Nominal Group Technique of Gustafson *et al.*, 1973). In his comprehensive review, Woudenberg (1991) found no clear evidence in studies done for the superiority of any of the four methods over the others. Even after discounting several of the studies for methodological difficulties, he concludes that the original formulation of the "quantitative" Delphi "is in no way superior to other (simpler, faster, and cheaper) judgment methods."

Another comprehensive evaluation of Delphi (Rowe *et al.*, 1991) comes to much the same conclusion that Sackman and Woudenberg did, but puts much of the blame on studies that stray from the original precepts. Most of the negative studies use non-experts with similar backgrounds (usually undergraduate or graduate students) in simple tests involving almanac-type questions or short-range forecasts. Rowe *et al.* (1991) point out that these are poor tests of the effects that occur when a variety of experts from different disciplines iterate and feed back their expertise to each other. They conclude that Delphi does have potential in its original intent as a judgment-aiding technique, but that improvements are needed and those improvements require a better understanding of the mechanics of judgment change within groups and of the factors that influence the validity of statistical and nominal groups.

APPLICATIONS: In the meantime, it is generally conceded that Delphi is extremely efficient in achieving consensus and it is in this direction that many of the more recent Delphi's have been used. Variations of the Delphi method, such as the policy Delphi and the decision Delphi, generally retain the anonymity of participants and iteration of responses. Many retain specific feedback as well, but these more qualitative variations generally drop the statistical group response. Delphi continues to be used in a wide variety of applications from its original purpose of technology forecasting (one report says that Delphi has been adopted in approximately 90% of the technological forecasts and studies of technological development strategy in China) to studying the future of medicine, examining possible shortages of strategic materials, regional planning of water and natural resources, analyzing national drug abuse policies, and identifying corporate business opportunities.

In addition, variations of Delphi continue to be developed to accommodate the growing understanding of its shortcomings. In a recent Delphi application, for example, a local area network (LAN) was constructed, composed of lap-top computers connected to a more capable workstation. Each participant had a dedicated spreadsheet available on a lap-top computer. The summary spreadsheet maintained by the workstation was displayed using a large-screen projector, and included the mean, media, standard deviation, and histogram of all the participants' scores. In real-time, the issues were discussed, the various participants presented their interpretation of the situation, presented their analytic arguments for the scores they believed to be appropriate, and changed their scoring as the discussion developed. Each participant knew their own scores, but not those of the other participants. When someone was convinced by the discussions to change a score they could do so anonymously. The score was transmitted to the workstation where a new mean, median, standard deviation, and histogram were computed and then displayed using a large screen projector. This technique retained all the dimensions of the traditional Delphi method and at the same time facilitated group discussion and real-time change substantially shortening the time typically required to complete a Delphi round.

See **Group decision computer technology; Group decision making.**

References

[1] Dalkey, N. and O. Helmer (1963), "An Experimental Application of the Delphi Method to the Use of Experts," *Management Science*, 9, 458–467.

[2] Gustafson, D.H., Shukla, R.K., Delbecq, A., and Walster, G.W. (1973), "A Comparison Study of Differences in Subjective Likelihood Estimates Made by Individuals, Interacting Groups, Delphi Groups, and Nominal Groups," *Organizational Behavior and Human Performance*, 9, 280–291.

[3] Rowe, G., Wright, G., and Bolger F. (1991), "Delphi: A Reevaluation of Research and Theory," *Technological Forecasting and Social Change*, 39, 235–251.

[4] Sackman, Harold (1975), *Delphi Critique*, Lexington Books, Lexington Massachusetts.

[5] Woudenberg, F. (1991), "An Evaluation of Delphi," *Technological Forecasting and Social Change*, 40, 131–150.

DENSITY

The proportion of the coefficients of a constraint matrix that are nonzero. For a given $(m \times n)$ matrix $A = (a_{ij})$, if k is the number of nonzero a_{ij}, then the density is given by $k/(m \times n)$. Most large-scale linear-programming problems have a low density of the order of 0.01. See **Sparse matrix; Super sparsity.**

DENSITY FUNCTION

When the derivative $f(x)$ of a cumulative probability distribution function $F(x)$ exists, it is called the density or probability density function (pdf). See **Probability density function**.

DEPARTURE PROCESS

Usually refers to the random sequence of customers leaving a queueing service center. More generally, it is the random point process or marked point process with marks representing aspects of the departure stream and/or the service center or node from which they are leaving. For example, the marked point process $(\mathbf{X}^d, \mathbf{T}^d)$ for departures from an M/G/1 queue takes $\mathbf{X}^d$ as the Markov process for the queue length process immdeiately after the departure time and $\mathbf{T}^d$ is the actual time of departure. See **Markov chains; Markov processes; Queueing theory.**

DESCRIPTIVE MODEL

A model that attempts to describe the actual relationships and behavior of a man/machine system: the "what is." For a decision problem, such a model seeks to describe how individuals make decisions. See **Decision problem; Expert systems; Mathematical model; Normative Model; Prescriptive model.**

DESIGN AND CONTROL

For a queueing system, design deals with the permanent, optimal setting of system parameters (such as service rate and/or number of servers), while control deals with adjusting system parameters as the system evolves to ensure certain performance levels are met. A typical example of a control rule is that a server is to be added when the queue size is greater than a certain number (say N_1) and when the queue size drops down to $N_2 < N_1$, the server goes to other duties. See **Dynamic programming**; **Markov decision processes**; **Queueing theory**.

DETAILED BALANCE EQUATIONS

A set of equations balancing the expected, steady-state flow rates or "probability flux" between each pair of states or entities of a stochastic process (most typically a Markov chain or queueing problem), for example written as:

$$\pi_j \, q(j, \, k) = \pi_k \, q(k,j)$$

where π_m is the probability that the state is m and $q(m,n)$ is the "flow rate" from states m to n. The states may be broadly interpreted to be multi-dimensional, as in a network of queues, and the entities might be individual service centers or nodes. Contrast this with global balance equations, where the average flow into a single state is equated with the flow out. See **Markov chains**; **Networks of queues**; **Queueing theory**.

DETERMINANT

See **Matrices and matrix algebra**.

DETERMINISTIC MODEL

A mathematical model in which it is assumed that all input data and parameters are known with certainty. See **Mathematical model; Model; Stochastic model**.

DEVELOPING COUNTRIES

Roberto Diéguez Galvão

COPPE/Federal University of Rio de Janeiro, Brazil

We discuss the main issues involving the use of OR in the developing countries, taking into account the social, political and technological environment in the developing world, including the presentation of different viewpoints about the role OR can play in these countries and of the organizations involved in promoting OR in them.

OR probably started to establish itself in the developing countries in the 1950s, approximately a decade after its post-war inception in Great Britain and the United States. Since then there has been a growing controversy on the role of OR in these countries. The central issue in this controversy is the following: Is there a separate OR for developing countries? If so, what makes it different from traditional OR? What steps could be taken to further OR in developing countries?

This issue has been discussed in different venues and several published papers have addressed it, for example Bornstein and Rosenhead (1990). Yet, as it seems natural, no conclusion has been reached, with advocates of two opposing viewpoints. At one end of the scale, there are those who think that there is nothing special about OR in developing countries, perhaps only less resources are available in these countries to conduct theoretical/applied work. The problem should resolve itself when each country reaches appropriate levels of development, and not much time should be dedicated to this issue. At the other end there are those who think that because of a different material basis and due to problems of infrastructure, OR does have a different role to play in these countries. In the latter case steps should be taken to ensure that OR plays a positive role in the development of their economies and societies.

THE SOCIAL, POLITICAL AND TECHNOLOGICAL ENVIRONMENT: To speak of developing countries in general may lead to erroneous conclusions, since the conditions vary enormously from one country to another. First of all, how to characterize a developing country? Which countries may be classified as "developing?" The United Nations has, for some years now, started to distinguish between more and less developed countries in the developing world. It has adopted the term "less developed countries" (*LDC's*) to address those developing countries that fall below some threshold levels measured by social and economic indicators. But these questions are clearly well beyond our scope here.

We adopt the view that developing countries are those in which large strata of the population live at or below the subsistence level, where social services are practically nonexistent for the majority of the population, where the educational and cultural levels are in general very low. The political consequence of this state of affairs is a high degree of instability for the institutions of these countries, at all levels.

The economy is generally very dependent on the industrialized nations. In many of the developing countries the State, in an attempt to promote development, is present in large sectors of the economy (though this is being reversed, given the emphasis on "market economics"). Bureaucracy and serious problems of infrastructure conspire against economic growth. In the technical sphere there is again a high level of dependency on the industrialized world, with very little technological innovation produced locally. It is against this difficult background that one must consider the role OR can play and how OR can be used as a tool for development.

THE USE OF OR: We consider here the existence of three different emphases in the development of OR, as outlined by ReVelle (private communication): (i) development of theory, which takes place mostly in the universities; (ii) development of methods for specific problems, which occurs both in the universities and in the practical world; (iii) applications, which occur mostly in the practical world. The problems of OR are therefore a continuum, and both developing and industrialized nations share in all these three aspects of the continuum. The more important aspect for the developing countries tends however to be applications, due to the nature of problems these nations have to face and their social, political and technological environment discussed above. According to Rosenhead (private communication), another important aspect is that existing theory and methods, grown in the developed world, are in many cases a poor "fit" for the problems facing the developing countries. Work on novel applications will be likely to throw up new methods and techniques of general interest.

The use of OR in the developing world is often seen as disconnected from the socio-economic needs of the respective countries, see Galvão (1988). Valuable theoretical contributions originate in these countries, but little is seen in terms of new theory and methods developed for the problems facing them.

A common situation in developing countries is a highly uncertain environment, which leads to the notion of "wicked" problems. These are for example problems for which there is often little or no data available, or where the accuracy of data is very poor. Complex decisions must nevertheless be made, against a background of competing interests and decision makers. There are not many tools available for solving these "wicked" problems, which are more common in developing countries.

One of the main characteristics of applied OR projects in developing countries is that a large majority of them is never implemented, see Löss (1981). This is due to a high degree of instability in institutions in these countries, to a lack of management education in OR, and to a tendency by OR analysts to attempt to use sophisticated OR techniques without paying due attention to the local environment and to the human factor in applied OR projects. These issues arise both in developed and developing countries, but our experience indicates that they are more often overlooked in the latter.

A number of papers address applications of OR in specific developing countries. An early account of OR in Peru is given by Sagasti (1972). Bandyopadhyay and Datta (1990) write about applications in India, while Papoulias and Darzentas (1990) analyze OR in Greece.

THE ORGANIZATIONAL BASIS OF OR: The main organizational basis of OR in the developing world are the national OR societies. These are in some cases well established, in other cases incipient. A number of them are members of the *International Federation of Operational Research Societies* (*IFORS*) and belong to regional groups within IFORS. In particular, *ALIO*, the *Latin-Iberian-American Association of Operational Research*, has the majority of its member societies belonging to developing countries. *APORS*, the *Asian Pacific* group, also represents OR societies from developing countries. In 1989 a *Developing Countries Committee* was established as part of the organizational structure of IFORS, with the objective of coordinating OR activities in the developing countries and promoting OR in these countries.

See **IFORS; Practice of OR/MS.**

References

[1] Bandyopadhyay, R. and Datta, S. (1990). "Applications of OR in Developing Economies: Some Indian Experiences," *European Jl. Operat. Res.* 49, 188–199.
[2] Bornstein, C.T. and Rosenhead, J. (1990). "The Role of Operational Research in Less Developed Countries: A Critical Approach," *European Jl. Operat. Res.* 49, 156–178.
[3] Galvão, R.D. (1988). "Operational Research in Latin America: Historical Background and Future Perspectives," in *Operational Research '87* (Proceedings of the 11th Triennial Conference on Op-

erations Research, edited by Rand, G.K.), North-Holland, 19–31.

[4] Löss, Z.E. (1981). "O Desenvolvimento da Pesquisa Operacional no Brasil (The Development of OR in Brazil), *Sc. Thesis*, COPPE/Federal University of Rio de Janeiro.

[5] Papoulias, D.B. and Darzentas, J. (1990). "OR in Greece: Myth and Reality." *European Jl. Operat. Res.* 49, 289–294.

[6] Sagasti, F.R. (1972). "Management Sciences in an Under-developed Country: The Case of Operations Research in Peru," *Mgmt. Sci.* 19, B1–B11.

DEVELOPMENT TOOL

Software used to facilitate the development of expert systems. The three types of tools are programming languages, shells, and integrated environments. See **Expert systems**.

DEVEX PRICING

A criterion for selecting the variable entering the basis in the simplex method. Devex pricing chooses the incoming variable with the largest gradient in the space of the initial nonbasic variables. This is contrasted with the usual simplex method entering variable criterion that chooses the incoming variable based on the largest gradient in the space of the current nonbasic variables. The Devex criterion tends to reduce greatly the total number of simplex iterations on large problems. See **Linear programming**; **Simplex method**.

DEVIATION VARIABLES

Variables used in goal programming models to represent deviation from desired goals or resource target levels. See **Goal programming**.

DFR

Decreasing failure rate. See **Probability distribution selection**; **Reliability of stochastic systems**.

DIAMETER

The maximum distance between any two nodes in a graph.

DIET PROBLEM

A linear program that determines a diet satisfying specified recommended daily allowance (RDAs) requirements at minimum cost. Stigler's diet problem was one of the first linear-programming problems solved by the simplex method. See **Stigler's diet problem**.

DIFFUSION APPROXIMATION

A heavy-traffic approximation for queueing systems in which the infinitesimal mean and variance of the underlying process are used to develop a Fokker-Planck diffusion-type differential equation which is then typically solved using Laplace transforms.

DIFFUSION PROCESS

A continuous-time Markov process on $\mathbb{R}$ or $\mathbb{R}^n$ which is analyzed similar to a continuous-time physical diffusion.

DIGRAPH

A graph all of whose edges have a designated one-way direction. See **Graph theory**.

DIJKSTRA'S ALGORITHM

A method for finding shortest paths (routes) in a network. The algorithm is a node labeling, greedy algorithm. It assumes that the distance c_{ij} between any pair of nodes i and j is nonnegative. The labels have two components $\{d(i),p\}$, where $d(i)$ is an upper bound on the shortest path length from the source (home) node s to node i, and p is the node preceding node i in the shortest path to node i. The algorithmic steps for finding the shortest paths from s to all other nodes in the network are as follows:

Step 1. Assign a number $d(i)$ to each node i to denote the tentative (upper bound) length of the shortest path from s to i that uses only labeled nodes as intermediate nodes. Initially, set $d(s) = 0$ and $d(i) = \infty$ for all $i \neq s$. Let y denote the last node labeled. Give node s the label $\{0, -\}$ and let $y = s$.

Step 2. For each unlabeled node i, redefine $d(i)$ as follows:
$d(i) = \min\{d(i),d(y) + c_{yi}\}$. If $d(i) = \infty$ for all unlabeled vertices i, then stop, as no path exists

from s to any unlabeled node. Otherwise, label the unlabeled node i with the smallest value of $d(i)$. Also, in the label, let p denote the node from which the arc that determined the minimum $d(i)$ came from. Let $y = i$.

Step 3. If all nodes have been labeled, stop, as the unique path of labels $\{d(i),p\}$ from s to i is a shortest path from s to i for all vertices i. Otherwise, return to *Step 2*.

See **Greedy algorithm; Minimum-cost network flow problem; Network optimization; Vehicle routing.**

DIRECTED GRAPH

See **Digraph.**

DIRECTION OF A SET

A vector d is a direction of a convex set if for every point x of the set, the ray $(x + \lambda d), \lambda \geq 0$, belongs to the set. If the set is bounded, it has no directions.

DIRECTIONAL DERIVATIVE

A rate of change at a given point in a given direction of the value function of an optimization problem as a function of problem parameters. See **Nonlinear programming.**

DISCRETE-PROGRAMMING PROBLEM

See **Combinatorial and integer optimization.**

DISCRETE-TIME MARKOV CHAIN (DTMC)

A discrete-time, countable-state Markov process. It is often just called a Markov chain. See **Markov chains; Markov processes.**

DISTRIBUTION SELECTION FOR STOCHASTIC MODELING

Donald Gross

The George Washington University, Washington, DC

INTRODUCTION: The choice of appropriate probability distributions is the most important step in any complete stochastic system analysis and hinges upon knowing as much as possible about the characteristics of the potential distribution and the "physics" of the situation to be modeled. Generally, we have first to decide which probability distributions are appropriate to use for the relevant random phenomena describing the model. For example, the exponential distribution has the Markovian (memoryless) property. Is this a reasonable condition for the particular physical situation under study? Let us say we are looking to describe the repair mechanism of a complex maintained system. If the service for all customers is fairly repetitive, we might feel that the longer a failed item is in service for repair, the greater the probability that its service will be completed in the next interval of time (non-memoryless). In this case, the exponential distribution would not be a reasonable candidate for consideration. On the other hand, if the service is mostly diagnostic in nature (we must find the trouble in order to fix it), or there is a wide variation of service required from customer to customer so that the probability of service completion in the next instant of time is independent of how long the customer has been in service, the exponential with its memoryless property might indeed suffice.

The actual shape of the density function also gives quite a bit of information, as do its moments. One particularly useful measure is the ratio of the standard deviation to the mean, called the coefficient of variation (CV). The exponential distribution has a $CV = 1$, while the Erlang or convolution of exponentials has a $CV < 1$, and the hyperexponential or mixture of exponentials has a $CV > 1$. Hence choosing the appropriate distribution is a combination of knowing as much as possible about distribution characteristics, the "physics" of the situation to be modeled, and statistical analyses when data are available.

HAZARD RATE: To help in characterizing probability distributions, we present an important concept that is most strongly associated with reliability modeling, namely, the hazard-rate (or as it is also called, the failure-rate) function. However, this concept can be useful in general when trying to decide upon the proper probability distribution to select. We shall relate the hazard-rate to the Markov property for the exponential distribution and point out its use as a way to gain insight about probability distributions.

Suppose we desire to choose a probability distribution to describe a continuous lifetime random variable T with a cumulative distribution function (CDF) of F(t). The density function, f(t) = $df(t)/dt$, can be interpreted as the approximate probability that the random time to failure will be in a neighborhood about a value t. The CDF is, of course, the probability that the time will be less than or equal to the value t. We then define the hazard rate $h(t)$ as the conditional probability that the lifetime will be in a neighborhood about the value t, given that the time is already at least t. That is, if we are dealing with failure times, $h(t)dt$ is the approximate probability that the device fails in the interval $(t, t + dt)$, given it is working at time t.

From the laws of conditional probability, we can show that

$$h(t) = \frac{f(t)}{1 - F(t)}.$$

This hazard or failure-rate function can be increasing in t (then called an increasing failure rate, or IFR), decreasing in t (then called a decreasing failure rate, or DFR), constant (considered to be both IFR and DFR), or a combination. The constant case implies the memoryless or ageless property, and we shall shortly show this to hold for the exponential distribution. If, however, we believe that our device ages and that the longer it has been operating, the more likely it is that the "device" will fail in the next dt, then we desire an $f(t)$ for which $h(t)$ is increasing in t; that is, we want an IFR distribution. This concept can be utilized for any stochastic modeling situation, for example, if instead of modeling lifetime of a device, we are concerned with describing the service time of a customer at a bank, then if service is fairly routine for each customer, we would probably desire an IFR distribution, but if customers required a variety of needs (say a queue where both business and personal transactions were allowed), then a DFR or perhaps the CFR exponential might be the best choice.

Note that we can reverse the algebraic calculations and uniquely obtain $F(t)$ from $h(t)$ by solving a simple linear, first-order differential equation and obtain

$$F(t) = 1 - \exp\left(-\int_0^t h(u)du\right).$$

Thus the hazard rate is another important information source (as is the shape of $f(t)$ itself) for obtaining knowledge concerning candidate probability distributions.

Now consider the exponential distribution

$$f(t) = \theta \exp(-\theta t).$$

From our earlier discussion, it is easily shown that $h(t) = \theta$. Thus the exponential distribution has a constant failure (hazard) rate and is memoryless. Suppose we feel in a particular situation that we need an IFR distribution for describing some random times. It turns out that the Erlang has this property. The density function is

$$f(t) = \theta^k t^{k-1} \exp(-\theta t)/(k-1)!$$

(a special form for the gamma), with its CDF determined in terms of the incomplete gamma function or equivalently as a Poisson sum. From these, it is not too difficult to calculate the Erlang's hazard rate, which also has a Poisson sum term, but is somewhat complicated to ascertain the direction of $h(t)$ with t without doing some numerical work. However, it does turn out that $h(t)$ increases with t and at a decelerating rate.

Now suppose that we were to desire the opposite IFR condition, that is, an accelerating rate of increase with t. There is a distribution called the Weibull for which we can obtain this condition. In fact, depending on how we pick the main parameter of the Weibull, we can obtain an IFR with decreasing acceleration, constant acceleration (linear with t), or increasing acceleration, as well as even obtaining a DFR or the constant failure rate exponential. The CDF of the Weibull is represented by

$$F(t) = 1 - \exp(-at^b)$$

and its hazard rate turns out to be the simple monomial $h(t) = abt^{b-1}$, with shape determined by the value of b (therefore, called the shape parameter).

As a further example in the process of choosing an appropriate candidate distribution for modeling, suppose that we are satisfied with an IFR that has a deceleration effect, such as the Erlang, but we believe that the CV might be greater than one. This latter condition eliminates the Erlang from consideration. But we know that a mixture of (k) exponentials (often denoted by H_k) does have a CV > 1. It is also known that any mixture of exponentials is DFR. In fact, it can be shown that all IFR distributions have CV < 1, while all DFR distributions have CV > 1 (Barlow and Proschan, 1975). Thus if we are convinced that we have an IFR situation, we must accept a CV < 1. Intuitively, this can be explained as follows. Situations that have CV > 1 often are cases where the random variables are mixtures (say, of exponentials). Thus, for example, if a customer has been in service a long time, chances are that it is of a type requiring "a lot of service," so the probability of completion in the next infinitesimal interval of width *dt* diminishes over time. Situations for which we have an IFR condition indicate a more consistent pattern among items, thus yielding a CV < 1.

RANGE OF THE RANDOM VARIABLE: Knowledge of the range of the random variable under study can also help narrow the possible choices in selecting an appropriate distribution. In many cases, there is a minimum value that the random variable can take on. For example, suppose we are trying to model the interarrival times between subway trains and we know that there is a minimum time for safety of γ. The distributions we have discussed thus far (and, indeed, many distributions) have zero as their minimum value. However, any such distribution can be made to have a minimum other than zero by adding a location parameter, say γ. This is done by subtracting from the random variable in the density function expression. Suppose we desire to use the exponential distribution, but we have a minimum value of γ. The density function would then become $f(t) = \theta \exp(-\theta[t - \gamma])$. It is not quite so easy to build in a maximum value if this should be the case. For this situation a distribution with a finite range would have to be chosen, such as the uniform, the triangular or the more general beta (Law and Kelton, 1991).

DATA: While much information can be gained from knowledge of the physical processes associated with the stochastic system understudy, it is very advantageous to obtain data if at all possible. For existing systems, data may already exist or can be obtained by observing the system. These data can then be used to gain further insight on the best distributions to choose for modeling the system. For example, the sample standard deviation and mean can be calculated, and it can be observed whether the sample CV is less than, greater than, or approximately equal to one. This would give an idea as to whether an IFR, DFR or the exponential distribution would be the more appropriate.

If enough data exist, just plotting a histogram can often provide a good idea of possible distributions from which to choose, since theoretical probability distributions have distinctive shapes (although some do closely resemble each other). The exponential shape of the exponential distribution is far different, for example, than the bell-shaped curve of the normal distribution.

There are rigorous statistical goodness of fit procedures to indicate if it is reasonable to assume that the data could come from a potential candidate distribution. These do, however, require a considerable amount of data and computation to yield satisfactory results. But, there are statistical packages, for example UNIFIT II (Law, 1993), which will analyze sets of data and recommend the theoretical distributions which are the most likely to yield the kind of data being studied.

We again make the point that choosing an appropriate probability model is a combination of knowing as much as possible about the characteristics of the probability distribution being considered and as much as possible about the physical situation being modeled.

See **Failure-rate function; Hazard rate; Markov chains; Markov processes; Reliability of stochastic systems; Simulation; Stochastic model**.

References

[1] Barlow, R.E. and Proschan, F. (1975). *Statistical Theory of Reliability and Life Testing*. Holt, Rinehart and Winston, New York.

[2] Law, A.M. (1993). *Unifit II*. Averill M. Law and Associates, Tucson, Arizona.

[3] Law, A.M. and Kelton, W.D. (1991). *Simulation Modeling and Analysis*, 2nd ed., McGraw-Hill, New York.

DMU

Decision making unit. See **Data envelopment analysis**.

DOCUMENTATION

Saul I. Gass

University of Maryland, College Park

As many operations research studies involve a mathematical decision model that is quite complex in its form, it is incumbent upon those who developed the model and conducted the analysis to furnish documentation that describes the essentials of the model, its use, and its results. Of especial concern are those computer-based models that are represented by a computer program and its input data files. The most serious weakness in the majority of OR model applications, both those that are successful and those that fail, is the lack of documents that satisfy the minimal requirements of good documentation practices (Gass *et al.*, 1981; Gass, 1984). The reasons for requiring documentation are manyfold and include, among others, "to enable system analysts and programmers, other than the originators, to use the model and program;" "to facilitate auditing and verification of the model and the program operations;" and "to enable potential users to determine whether the model and programs will serve their needs" (Gass, 1984).

The most acceptable view of model documentation is that which calls for documents that record and describe all aspects of the model development life-cycle. The life-cycle model documentation approach given in Gass (1979) calls for the production of 13 major documents. However, it is recognized that in terms of the basic needs of model users and analysts, these documents can be rewritten and combined into the following four manuals: *Analyst's Manual*, *User's Manual*, *Programmer's Manual*, and *Manager's Manual*. We briefly describe the contents of these manuals; detailed tables of contents for each are given in Gass (1984).

ANALYST'S MANUAL: The analyst's manual combines information from the other project documents and is a source document for analysts who have been and will be involved in the development, revisions, and maintenance of the model. It should include those technical aspects that are essential for practical understanding and application of the model, such as a functional description, data requirements, verification and validation tests, and algorithmic descriptions.

USER'S MANUAL: The purpose of the user's manual is to provide (nonprogramming) users with an understanding of the model's purposes, capabilities and limitations so they may use it accurately and effectively. This manual should enable a user to understand the overall structure and logic of the model, input requirements, output formats, and the interpretation and use of the results. This manual should also enable technicians to prepare the data and to set up and run the model.

PROGRAMMER'S MANUAL: The purpose of the programmer's manual is to provide the current and future programming staff with the information necessary to maintain and modify the model's program. This manual should provide all the details necessary for a programmer to understand the operation of the software, to trace through it for debugging and error correction, for making modifications, and for determining if and how the programs can be transferred to other computer systems or other user installations.

MANAGER'S MANUAL: The manager's manual is essential for computer-based models used in a decision environment. It is directed at executives of the organization who will have to interpret and use the results of the model, and support its continued use and maintenance. This manual should include a description of the problem setting and origins of the project; a general description of the model, including its purpose, objectives, capabilities, and limitations; the nature, interpretation, use, and restrictions of the results that are produced by the model; costs and benefits to be expected in using the model; the role of the computer-based model in the organization and decision structure; resources required; data needs; operational and transfer concerns; and basic explanatory material.

See **Implementation; Practice of Operations Research and Management Science.**

References

[1] Brewer, G.D. (1976). "Documentation: An Overview and Design Strategy," *Simulation & Games*, 7, 261–280.
[2] Gass, S.I. (1979). *Computer Model Documentation: A Review and An Approach*, National Bureau of Standards Special Publication 500-39, U.S. GPO Stock No. 033-003-02020-6, Washington, D.C.
[3] Gass, S.I., K.L. Hoffman, R.H.F. Jackson, L.S. Joel and P.B. Sanders (1981). "Documentation for a Model: A Hierarchical Approach," *ACM Communications*, 24, 728–733.
[4] Gass, S.I. (1984). "Documenting a Computer-Based Model," *Interfaces*, 14, 84–93.
[5] NBS (1976). *Guidelines for Documentation of Computer Programs and Automated Data Systems*, FIPS PUB 38, U.S. Government Printing Office, Washington, DC.
[6] NBS (1980). *Computer Model Documentation Guide*, NBS Special Publication 500-73, U.S. Government Printing Office, Washington, DC.

DOMAIN KNOWLEDGE

The knowledge that an expert has about a given subject area. See **Artificial intelligence.**

DSS

See **Decision support systems**.

DUAL LINEAR-PROGRAMMING PROBLEM

A companion problem defined by a linear-programming problem. Every linear-programming problem has an associated dual-programming program. When the linear-programming problem has the form

Minimize $c^T x$

subject to

$$Ax \geq b$$
$$x \geq 0$$

then its dual problem is also a linear-programming problem with the form

Maximize $b^T y$

subject to

$$A^T y \leq c$$

$$y \geq 0$$

The original problem is called the primal problem. The dual of the dual problem is the primal problem. If the primal minimization problem is given as equations in nonnegative variables, then its dual is a maximization problem with less than or equal to constraints whose variables are unrestricted (free). The optimal solutions to primal and dual problems are strongly interrelated. See **Complementary slackness theorem; Duality theorem; Symmetric dual problem; Unsymmetric dual problem**.

DUAL-SIMPLEX METHOD

An algorithm which solves a linear-programming problem by solving its dual problem. The algorithm starts with a dual feasible but primal infeasible solution, and iteratively attempts to improve the dual objective function while maintaining dual feasibility.

DUALITY THEOREM

A theorem concerning the relationship between the solutions of primal and dual linear-programming problems. One form of the theorem is as follows: If either the primal or the dual has a finite optimal solution, then the other problem has a finite optimal solution, and the optimal values of their objective functions are equal. From this it can be shown that for any pair of primal and dual linear programs, the objective value of any feasible solution to the minimization problem is greater than or equal to the objective value of any feasible solution to the dual maximization problem. This implies that if one of the problems is feasible and unbounded, then the other problem is infeasible. Examples exist for which the primal and its dual are both infeasible. Another form of the theorem states: If both problems have feasible solutions, then both have finite optimal solutions, with the optimal values of their objective functions equal. See **Strong duality theorem**.

DUALPLEX METHOD

A procedure for decomposing and solving a weakly-coupled linear-programming problem. See **Block-angular system**.

DUMMY ARROW

A dashed arrow used in a project network diagram to show relationships among project items, a logical dummy, or to give a unique designation to an activity, thus called a uniqueness dummy. A dummy or dummy arrow represents no time or resources. See **Network planning**.

DYNAMIC PROGRAMMING

Chelsea C. White, III

University of Michigan, Ann Arbor, Michigan

INTRODUCTION: Dynamic programming is both an approach to problem solving and a decomposition technique that can be effectively applied to mathematically describable problems having a sequence of interrelated decisions. Such decision making problems are pervasive. Determining a route from an origin (e.g., your house) to a destination (e.g., your school or place of employment) on a network of roads requires a sequence of turns. Managing a retail store (e.g., that sells, say, televisions) requires a sequence of wholesale purchasing decisions.

Such problems also share common characteristics. Each is invariably associated with a criterion. We may wish to choose the shortest or most scenic route from home to our place of employment; the retail store manager purchases televisions with the intent of selling them to maximize expected profit. Each is such that a currently determined decision has impact on the future decision making environment. In going from home to work, the turn currently selected will determine the geographical location of our next turn decision; in managing the retail store, the number of items ordered today will affect the level of inventory next week.

Roots and Key References: Bellman (1957) is usually credited with coining the phrase *dynamic programming* and seeing its broad potential for application. A brief history of dynamic programming that precedes Bellman's 1957 book can be found in Denardo (1982). In depth descriptions and applications of dynamic programming can be found in Bertsekas (1987), Denardo (1982), Heyman and Sobel (1984), Hillier and Lieberman (1990; see Chapter 11), and Ross (1983).

Central to the philosophy and methodology of dynamic programming is the *Principle of Optimality*, as related to the following multi-stage decision problem (Bellman, 1957). Let $\{q_1, q_2, \ldots, q_n\}$ be a sequence of allowable decisions called a *policy*; specifically, an n-stage policy. A policy that yields the maximum value of the related criterion function is called an *optimal policy*. Decisions are based on the state of the process, that is, the information available to make a decision. The basic property of optimal policies is expressed by the following:

> *Principle of Optimality:* An optimal policy has the property that, whatever the current state and decision, the remaining decisions must constitute an optimal policy with regard to the state resulting from the current decision.

The Principle of Optimality can be expressed as an optimization problem over the set of possible decisions by a recursive relationship, the application of which yields the optimal policy. We illustrate this below by two examples.

Examples:
1. *An itinerary selection problem.* We wish to find the shortest path from home to work. A map of the area describes the network of streets that includes home and work, intermediate intersections, connecting streets, and the distance from one intersection to any other intersection that is directly connected by a street. We model this problem as follows. Let N be the set composed of home, work, and all the intersections. We call an element of N a node. For simplicity, assume all of the streets are one-way. We describe a street as an ordered pair of nodes; that is, (n, n') is

the street going from node n to node n' (we say that n' is an immediate successor of node n). Let $m(n, n')$ be the distance from node n to node n'; that is, $m(n, n')$ represents the length of street (n, n').

We proceed to examine this problem recursively as follows. Let $f(n)$ equal the shortest distance from the start node *home* to the goal node *work*. Our objective is to find $f(\text{home})$, the minimum distance from home to work, and a path from home to work that has a distance equal to $f(\text{home})$, a minimum distance path. Note that $f(n) < m(n, n') + f(n')$ for any node n' that is an immediate successor of node n. It seems reasonable that if we can find an immediate successor n'' of n such that $f(n) = m(n, n'') + f(n'')$, then if we find ourselves at node n, we should traverse the street that takes us to node n''. Thus, determination of all of the values $f(n)$ determine both $f(\text{home})$ and a minimum distance path from home to work. Formally, determination of these values can proceed recursively from the equation $f(n) = \min\{m(n, n') + f(n')\}$, where the minimum is taken over all nodes n' that are immediate successors of node n and where $f(\text{work}) = 0$ is the initial condition.

2. *An inventory problem.* Let $x(t)$ be the number of items in stock at the end of week t, $d(t+1)$ the number of customers wishing to make a purchase during week $t+1$, and $u(t)$ the number of items ordered at the end of week t and delivered at the beginning of week $t+1$. Although it is unlikely that we know $d(t)$ precisely (without a crystal ball), we assume that we know the probability that $d(t) = n$, for all $n = 0, 1, \ldots$. Assuming that we keep backorders, we note that $x(t+1) = x(t) - d(t+1) + u(t)$. A reasonable objective is to minimize the expected cost accrued over the period from $t = 0$ till $t = T$ ($T > 0$) by choice of $u(0), \ldots, u(T-1)$, assuming that ordering decisions are made on the basis of the current inventory level, that is, assuming that the mechanism that determines $u(t)$ (e.g., the store manager) is aware of $x(t)$, for all $t = 0, \ldots, T-1$. Costs might include a shortage cost (a penalty if there is an insufficient amount of inventory in stock), a storage cost (a penalty if there is too much inventory in stock), an ordering cost (reflecting the cost necessary to purchase items wholesale), and a selling price (reflecting the income received when an item is sold; a negative cost). Let $c(x, u)$ represent the expected total cost to be accrued from the end of week t till the end of week $t+1$, given that $x(t) = x$ and $u(t) = u$. Then the criterion we wish to minimize is $E\{c[x(0), u(0)] + \ldots + c[x(T-1), u(T-1)]\}$, where E is the expectation operator associated with the random variables $d(1), \ldots, d(T)$.

We can also examine this problem recursively. Let $f(x, t)$ be the minimum expected cost to be accrued from time t till time T, assuming that $x(t) = x$. Clearly, $f(x, T) = 0$. Note also that $f[x(t), t] < c[x(t), u(t)] + E\{f[x(t) - d(t + 1) + u(t), t + 1]\}$ for any available $u(t)$. As was true for Example 1, the order number u'' which is such that $f[x(t), t] = c[x(t), u''] + E\{f[x(t) - d(t + 1) + u'', t + 1]\}$ is the order to place at time t when the current inventory is $x(t)$. Thus, the recursive equation determines both $f(x, 0)$ for all x but also the order number as a function of current inventory level.

COMMON CHARACTERISTICS: Two key aspects of dynamic programming are the notion of a state and recursive equations. The state of the dynamic programming problem is the information that is currently available on which to base the current decision. For example, in the itinerary selection problem, the state is the current node; in the inventory problem, the current number of items in stock represents the problem's state. We remark that in both examples, how the system arrived at its current state is inconsequential from the perspective of decision making. For the itinerary selection problem, all that is needed is the current node, and not the path that leads to that node, in order to determine the best next street to traverse. The determination of the number of items to order this week depends only on the current inventory level equations (other names include functional equations and optimality equations) that can be used to determine the minimum expected value of the criterion and an optimal sequence of decisions that depend on the current node or current inventory level. We observe that in both cases the recursive equations essentially decompose the problem into a series of subproblems, one for each node.

See **Dijkstra's algorithm; Networks; Markov decision processes.**

References

[1] Bellman, R.E. (1957), *Dynamic Programming*. Princeton University Press, Princeton, New Jersey.

[2] Bertsekas, D.P. (1987), *Dynamic Programming: Deterministic and Stochastic Models*. Prentice-Hall, Englewood Cliffs, New Jersey.

[3] Denardo, E.V. (1982), *Dynamic Programming: Models and Applications*. Prentice-Hall, Englewood Cliffs, New Jersey.

[4] Heyman, D.P., and Sobel, M.J. (1984), *Stochastic Models in Operations Research*, Vol. II. McGraw-Hill, New York.

[5] Hillier, F.S., and Lieberman, G.J. (1990), *Introduction to Operations Research* (fifth edition). McGraw-Hill, New York.

[6] Ross, S.M. (1983), *Introduction to Stochastic Dynamic Programming*, Academic Press, New York.

EARLIEST FINISH TIME

The earliest possible time an activity can be completed without reducing the duration of any of the preceding activities as described in a project network. It is simply the sum of the earliest start time for the activity and the duration of the activity. See **Critical path method**; **Program evaluation and review technique**; **Network planning**.

EARLIEST START TIME

The earliest possible time an activity can begin without reducing the duration of any of the preceding activities as described in a project network. It is calculated by summing the durations of all activities on the longest path leading to the event that identifies the beginning of the activity. See **Critical path method**; **Program evaluation and review technique**; **Network planning**.

ECONOMETRICS

Harry H. Kelejian and
Ingmar R. Prucha

University of Maryland, College Park

DEFINITION: Literally speaking, econometrics stands for measurement in economics. Broadly speaking, econometrics is concerned with the empirical analysis of economic relationships. While early empirical work goes back at least to Sir William Petty's political arithmetic in the seventeenth century, econometrics as a field was firmly established through the foundation of the Econometric Society in 1930. Publication of its journal, *Econometrica*, started in 1933. The scope of the society is defined as follows: "The Econometric Society is an international society for the advancement of economic theory in its relation to statistics and mathematics...". Samuelson, Koopmans and Stone (1954, p. 142) in a report on *Econometrica* define econometrics

"...as the quantitative analysis of actual economic phenomena based on the concurrent

development of theory and observation, related by appropriate methods of inference."

Similar definitions can be found in most econometric texts. For example, Goldberger (1964, p. 1) defines econometrics

"...as the social science in which the tools of economic theory, mathematics, and statistical inference are applied to the analysis of economic phenomena. Its main objective is to give empirical content to economic theory...".

SINGLE EQUATION REGRESSION MODELS: Much of the early work in econometrics is related to the classical linear regression model

$$y_t = x_t\beta + u_t, \quad t = 1, \ldots, n, \qquad (1)$$

where y_t is the tth observation on the dependent variable, x_t is the $1 \times k$ vector of observations on the explanatory variables, β is a $k \times 1$ vector of unknown parameters and u_t is the tth disturbance term. The assumptions of the classical model are: (i) $E(u_t) = 0$, (ii) $E(u_t^2) = \sigma^2$ and $E(u_t u_s) = 0$ for $t \neq s$, (iii) x_t is nonstochastic and (iv) $X = (x_1', \ldots, x_n')'$ has full column rank. Under these assumptions the Gauss-Markov theorem implies that the ordinary least squares estimator is best (in the sense of having the smallest variance-covariance matrix) within the class of linear unbiased estimators. If the disturbances are normally distributed, exact small sample inference is available. If normality is not maintained, then approximate inference is possible under additional assumptions on x_t and u_t.

The nature of economic data and models are such that the above assumptions are restrictive in certain applications and, hence, various extensions of the classical model have been considered. In particular, disturbances have been permitted to be autocorrelated and/or to have different variances, that is, to be heteroskedastic. Other extensions permit for the regressors to be stochastic. Stochastic regressors arise, for example, if the regressors are measured with error. They also arise in dynamic models in which one or several of the regressors depend on lagged values of the dependent variable. Models in which the parameters are permitted to vary deterministically or stochastically from observation to observation have also been considered. Still other extensions relate to sample selection issues. Text presentations of the issues discussed above are, for example, given in Amemiya (1985), Davidson and MacKinnon (1993), Judge *et al.* (1985), and Schmidt (1976).

SIMULTANEOUS EQUATION MODELS: The economy is a complex system of relationships. For this reason economic models often involve more than one equation and so more than one dependent variable. To see the issues involved consider the following system of m equations:

$$y_t = y_t B + z_t C + u_t, \quad t = 1, \ldots, n, \quad (2)$$

where y_t is a $1 \times m$ vector of the jointly dependent variables, $z_t = (y_{t-1}, \ldots, y_{t-h}, x_t)$ where x_t is a $1 \times k$ vector of nonstochastic variables, u_t is a $1 \times m$ vector of disturbances, and B and C are correspondingly defined matrices of parameters.

Basic assumptions for the model are: (i) u_t is i.i.d. with finite fourth moments and $E(u_t) = 0$, $E(u_t'u_t) = $ with nonsingular, (ii) $I - B$ is nonsingular and the diagonal elements of B are zero, (iii) $n^{-1}\Sigma x_t' x_{t-\tau} \to Q(\tau)$ where the matrices $Q(\tau)$ are finite, and nonsingular for $\tau = 0$, (iv) the system is dynamically stable. Since $I - B$ is invertible the system can be solved as

$$y_t = \Pi z_t + v_t, \quad \Pi = C(I-B)^{-1} \text{ and } v_t = u_t(I-B)^{-1}. \quad (3)$$

In the literature, equations (2) and (3) are called the structural and reduced form of the model, respectively. The parameters in B and C are generally not identified and hence not consistently estimable without additional parameter restrictions. These parameter restrictions often take the form of exclusion restrictions based on economic theory; that is, theory may suggest that every variable does not appear in every equation and so certain elements of B and C are specified to be zero.

As is obvious from (3), the elements of y_t depend in general on all of the elements of u_t. As a consequence, the structural equations in (2) cannot in general be estimated consistently by ordinary least squares. Fundamental work on estimation and identification of the model in (2) was done by the Cowles Foundation, which focused on the maximum likelihood technique based on the normal distribution; see Koopmans (1950) and Hood and Koopmans (1953). Estimation procedures developed later were typically based on instrumental variable techniques which do not require specific distributional assumptions; see Basmann (1957) and Theil (1953) for early fundamental contributions, and, for ex-

ample, Amemiya (1985), Davidson and MacKinnon (1993), Judge *et al.* (1985) and Schmidt (1976) for later text presentations.

In recent years the model in (2) has been generalized in ways that are similar to those mentioned above in reference to model (1). In addition, starting with the fundamental contributions of Jennrich (1969) and Malinvaud (1970), estimation theory has been developed for nonlinear counterparts to models (1) and (2); for recent presentations of estimation theory for dynamic nonlinear systems see, for example, Gallant and White (1988) and Pötscher and Prucha (1991a,b). Finally, Bayesian extensions of these models have been considered, see, for example, Zellner (1971) for early fundamental work and Judge *et al.* (1985) for a more recent text presentation.

OTHER MODELING TECHNIQUES: (a) *Time series models* – An important class of models used to describe economic data are autoregressive moving average (ARMA) models. These models have been popularized in economics by Box and Jenkins (1976); for a recent discussion of time series techniques see, for example, Brockwell and Davis (1991), and Harvey (1993).

A stationary stochastic process y_t (time series) that satisfies for every t

$$y_t = a_1 y_{t-1} + \ldots + a_p y_{t-p} + \varepsilon_t + b_1 \varepsilon_{t-1} + \ldots + b_q \varepsilon_{t-q} \quad (4)$$

where $E(\varepsilon_t) = 0$, $E(\varepsilon_t^2) = \sigma_\varepsilon^2$ and $E(\varepsilon_t \varepsilon_s) = 0$ for $t \neq s$ is called an ARMA(p,q) process. If y_t was obtained by differencing some process z_t, then z_t is called an autoregressive integrated moving average (ARIMA) process. If the specification in (4) also permits nonstochastic regressors, then the corresponding processes are called ARMAX and ARIMAX, respectively. Clearly, the reduced form in (3) can be viewed as an ARMAX model. Although ARMAX models do not describe the structure of the system, they have been found, for example, to be useful for prediction purposes.

An important recent development in the time series literature is the introduction of the concept of cointegration as an equilibrium relationship between integrated variables. This development has particular appeal to economists because many economic variables appear to have random walk representations but yet certain linear combinations of them appear to be stationary. The basic ideas were proposed by Granger (1981); recent extensions and developments are discussed in Davidson and MacKinnon (1993), and Engle and Granger (1991).

(b) *Qualitative and limited dependent variable models* – Economists often formulate models to explain events which are at least partially qualitative in nature. For example, one might be interested in the factors determining whether or not a bank fails, a firm undertakes an investment, etc. More generally, such models could relate to events that are described by more than one category. Models relating to occupational choice, firm structure, and travel mode fall in this class.

Another class of models are limited dependent variable models. In these models the range of the dependent variable is constrained in some way. As one example, suppose we have a model describing the selling price of a house. A limited dependent variable problem would arise if, for example, the only transactions that are recorded are those for which the selling price exceeds a certain dollar amount. The techniques involved for limited dependent variable models are similar to those in qualitative models. In recent years econometric models relating to qualitative and limited dependent variables have been generalized in ways that are similar to those described in the sections above. Excellent early reviews are given in McFadden (1974, 1976) and in Amemiya (1981). Later text presentations are given in Amemiya (1985), Maddala (1983), and Judge *et al.* (1985).

CONCLUDING NOTE: Of necessity, this review has been brief and so various issues have not been considered. Model specification tests, rational expectations models, and model simulation are just some of these omitted issues.

See **Economics; Regression; Time series.**

References

[1] Amemiya, T. (1985). *Advanced Econometrics.* Harvard University Press, Cambridge, Massachusetts.

[2] Amemiya, T. (1981). "Qualitative Response Models, A Survey." *Jl. Economic Literature,* 19, 1483–1536.

[3] Basmann, R.L. (1957). "A Generalized Classical Method of Linear Estimation of Coefficients in a Structural Equation." *Econometrica,* 25, 77–83.

[4] Box, G.E.P., and Jenkins, G.M. (1976). *Time Series Analysis, Forecasting and Control.* Holden Day, San Francisco.

[5] Brockwell, P.J., and Davis, R.A. (1991). *Time Series, Theory and Methods.* Springer Verlag, New York.

[6] Davidson, R., and MacKinnon, J.G. (1993). *Estimation and Inference in Econometrics.* Oxford University Press, New York.

[7] Engle, R.F., and Granger, C.W.J., eds. (1991). *Long-Run Economic Relationships, Reading in Cointegration.* Oxford University Press, Oxford.

[8] Gallant, A.R., and White, H. (1988). *A Unified Theory of Estimation and Inference for Nonlinear Dynamic Models.* Basil Blackwell, New York.

[9] Goldberger, A.S. (1964). *Econometric Theory.* Wiley, New York.

[10] Granger, C.W.J. (1981). "Some Properties of Time Series Data and their Use in Econometric Model Specification." *Jl. Econometrics,* 16, 121–130.

[11] Harvey, A.C. (1993). *Time Series Models.* MIT Press, Cambridge, Massachusetts.

[12] Hood, W.C., and Koopmans, T.C., eds. (1953). *Studies in Econometric Methods.* Cowles Commission Monograph 14. Wiley, New York.

[13] Jenrich, R.I. (1969). "Asymptotic Properties of Non-Linear Least Squares Estimators." *Annals Mathematical Statistics,* 40, 633–643.

[14] Judge, G.G., Griffiths, W.E., Hill, R.C., Lütkepohl, H., and Lee, T.C. (1985). *The Theory and Practice of Econometrics* (2nd ed.). Wiley, New York.

[15] Koopmans, T.C., ed. (1950). *Statistical Inference in Dynamic Economic Models.* Cowles Commission Monograph 10. Wiley, New York.

[16] Malinvaud, E. (1970). "The Consistency of Nonlinear Regressions." *Annals Mathematical Statistics,* 41, 956–969.

[17] McFadden, D. (1976). "Quantal Choice Analysis, A Survey." *Annals Economic and Social Measurement,* 5, 363–390.

[18] McFadden, D. (1974). "Conditional Logit Analysis of Qualitative Choice Behavior." In P. Zarembka (ed.), *Frontiers in Econometrics.* Academic Press, New York, 105–142.

[19] Pötscher, B.M., and Prucha, I.R. (1991a). "Basic Structure of the Asymptotic Theory in Dynamic Nonlinear Econometric Models. I, Consistency and Approximation Concepts." *Econometric Reviews,* 10, 125–216.

[20] Pötscher, B.M., and Prucha, I.R. (1991b). "Basic Structure of the Asymptotic Theory in Dynamic Nonlinear Econometric Models. II, Asymptotic Normality." *Econometric Reviews,* 10, 253–325.

[21] Samuelson, P.A., Koopmans, T.C., and Stone, J.R. (1954), "Report of the Evaluative Committee for Econometrica." *Econometrica,* 22, 141–146.

[22] Schmidt, P. (1976). *Econometrics.* Marcel Dekker, New York.

[23] Theil, H. (1953). "Estimation and Simultaneous Correlation in Complete Equation Systems." Central Planning Bureau, The Hague (mimeographed).

[24] Zellner, A. (1971). *An Introduction to Bayesian Inference in Econometrics.* Wiley, New York.

ECONOMIC ORDER QUANTITY

The policy for a simple, deterministic inventory model that tells how much to order so that the sum of ordering and holding costs is minimized.

See **Inventory modeling.**

ECONOMICS

Frederic H. Murphy

Temple University, *Philadelphia*, *Pennsylvania*

INTRODUCTION: To understand the relationship between operations research and economics, we need to understand some of the history of both fields. The founders of the field of operations research came from diverse backgrounds, including physics, mathematics, engineering and economics. Operations research as a field has tried to maintain its multidisciplinary character. Still if one looks at textbooks on the subject, one sees a common set of techniques: stochastic modeling, simulation, optimization, and game theory. The field of operations research also emphasizes certain application areas such as operations management and encompasses some, such as inventory management.

Operations research provides tools to analyze operations of and assist in decisionmaking in organizations. The use of algorithms in solving operations research models means the field is tightly linked to computer science, the use of data creates strong ties to information systems. Engineers cannot do their best work without operations research tools to optimize their designs. The military is the largest single consumer of operations research products and services. Given all these connections to fields that are unrelated to economics and are often directly connected to practical activities, it is hard to remember the historically close ties to a profession that mostly consists of theoreticians with relatively few practitioners outside of macroeconomists.

The subject areas of economics can be defined broadly as follows: macroeconomics, the study of economic aggregates; microeconomics, the study of economic agents, such as firms, and the market structures within which these agents operate, such as monopolies; and econometrics, the statistical techniques used for estimating the parameters of economic models. There is traditionally little overlap between macroeconomics and operations research. Indeed, there is little overlap between macroeconomics and microeconomics.

The rapid growth of mathematical economics and the field of operations research after World War II stemmed from the same root: the application of mathematics to build and understand models that only approximated the reality being studied. Although, mathematics has been a tool of science for centuries and the development of each fed off of the other, the systematic use of mathematics outside of science began only after World War II when it became clear that a model does not have to be "true" to be very useful.

THE COMMON HISTORY: Given this common starting point and the interest of operations researchers in finding the most economic solutions, there clearly has to be significant overlap in the fields. The connections were most prominent in the early days of operations research and they involved the areas of optimization and game theory.

Hitchcock (1941), a physicist, and Koopmans (1951), an economist, independently developed the first useful optimization model, the transportation problem. Kantorovitch (1939), a mathematician in the Russian central planning agency developed several linear programming models for production and distribution including the transshipment model. Stigler (1944), an economist, developed the diet/feed mix model. Dantzig (1951a; 1963), at the time, a mathematician in the US Air Force, invented the first generic linear programs and the simplex algorithm for solving them. The simplex algorithm has survived 50 years as the primary method for solving linear programs.

The collection of papers, Koopmans (1951), defines the beginning of the subject of optimization, game theory and the relationship between the two. The contributing authors were a mix of economists and mathematicians. It devotes a substantial amount of space to generalizing the input-output model of an economy. Dantzig (1963) points to the work of Leontief in input-output models of the US economy as an important starting point for his ideas. Another important early book, Dorfman, Samuelson and Solow (1958), was written by economists. Current texts on microeconomics continue to include chapters on optimization and game theory.

The early articles on optimization appeared in such journals as *Econometrica* (see the references in Dantzig, 1963). Charnes and Cooper, the developers of many of the first linear programming models, also published in the economics journals; for example, Charnes Cooper and Mellon (1952). Agricultural economists were quick to develop the feed mix model for farmers. Mathematical programming has become a mainstay for agricultural economists (Hazell, 1986).

In the early days of inventory theory, the links between economists and operations researchers were equally strong. This area involved using such optimization techniques as dynamic programming and traditional, calculus-based methods to find optimal inventory policies (Arrow, Karlin and Scarf, 1958; Whitin, 1957). However, the development of the field moved very quickly into the hands of operations researchers because the issues in inventory analysis evolved into implementation of inventory systems and situation-specific models, and away from the more broadly-based economic considerations.

Game theory was developed by von Neumann to study issues of conflict and cooperation at a theoretical level. The seminal work by von Neumann and Morgenstern (1944) was an application of game theory to economics. The RAND Corporation became an early center for game theory as applied to geopolitical and military strategy. For example, the famous prisoner's dilemma game was invented at RAND (Poundstone, 1992). The link between game theory and optimization was clear from very early on (Dantzig, 1951b; Gale, Kuhn and Tucker, 1951).

Operations researchers have contributed significantly to economics. Once Samuelson (1952) recognized the connection between mathematical programming and economic equilibrium models, mathematical programming became an important tool for economic analysis. In fact, the GAMS modeling language was developed by operations researchers at the World Bank for the purpose of solving computable general equilibrium models for evaluating national development plans (Brooke, Kendrick and Meeraus, 1993). The economist, Gustafson (1958), used dynamic programming, an operations research tool, to develop the first grain storage models to protect against famine. One of the most prominent microeconomic policy-analysis models of the 1970s, the Project Independence Evaluation System (PIES), was built by a team of operations researchers and economists led by Hogan (1975), an operations researcher, who went on to organize the International Association of Energy Economists.

THE DIFFERENT PERSPECTIVES OF ECONOMICS AND OPERATIONS RESEARCH:
Economics and operations research are distinct fields because the groups have different interests. Economists are primarily interested in qualitative analysis for policymaking, while operations researchers are more interested in assisting decisionmaking within the firm and have a strong computational orientation. For example, oil companies use the results of their mathematical programming models for operating their refineries. Even when economists are interested in numbers, they are looking to measure the impact of the sum of individual decisions, rather than determining the decisions. This distinction between the fields is not absolute. Econometricians are interested in computational issues. Scarf (1973) has developed algorithms for computing economic equilibria. The function of corporate planning and public policy studies produced by operations researchers is to provide insight rather than specific numbers. This is done through multiple scenarios, examining alternative policies and the sensitivity of the results to the underlying parameters.

The different interests can be seen in the study of inventories. For the past few decades operations researchers and computer scientists have been implementing inventory systems, while the economists have been focussing on the effect of inventories in the business cycle rather than inventory policies *per se*. The recent popularity of scientific inventory management in corporations and the desire to reduce inventories to free up capital and gain operational flexibility has led to a significant decline in the inventory-to-sales ratio. That is, inventories turn over more quickly and companies are able to adapt to fluctuations in demand more rapidly with less draconian changes in production levels. Economists measure this drop at the national level and factor this secular change into their macroeconomic models to explain the resultant dampening of business cycles. For example, the recession in the early 1990s was slow in coming and going, but it was also shallow relative to past recessions because of the cumulative impact of individual improvements of inventory systems and production management.

The economics perspective can be best understood by looking at the theory of the firm in economics. Here one postulates a production function $Q = F(K,L)$, where Q is the output of the firm, K is the capital input and L is the amount of labor used. A common production function is the Cobb-Douglas production function:

$$F(K,L) = aK^{\alpha}L^{1-\alpha}.$$

This view of the firm is very macro and cannot be used for decisionmaking within the firm. However, let us form the optimization problem for maximizing profits for the monopolist when the demand curve is $D(p)$ and the inverse of the demand curve is $p(q)$:

$$\max \quad p(q)q - p_K K - p_L L$$

subject to

$$q = F(K,L)$$

where p_K and p_L are the prices of capital and labor. We can draw the following conclusion from this optimization problem:

$$\frac{\partial F/\partial K}{\partial F/\partial L} = \frac{p_K}{p_L}.$$

That is, the ratio of the marginal rate of technical substitution of capital and labor must equal the ratio of the prices in the profit maximizing firm.

We illustrate this approach to understanding market behavior, using an early model from regulatory economics. Say we look at a firm that is a regulated monopoly. Here the model looks as follows:

$$\max\ p(q)q - p_K K - p_L L$$

subject to

$$p(q)q - sK - p_L L \le 0$$

$$q = F(K,L)$$

$$K, L \ge 0,$$

where s is the allowed rate of return. We now have two constraints. The added constraint says that profits cannot exceed the allowed return on capital (the allowed rate, s, times the amount of capital).

The Karush, Kuhn, Tucker (KKT) conditions imply that

$$\frac{\partial F/\partial K}{\partial F/\partial L} = \frac{p_K}{p_L} - \frac{\lambda^*(s - p_K)}{(1 - \lambda^*)p_L},$$

where λ is the Karush, Kuhn, Tucker multiplier for the rate of return constraint. One can show that $\lambda < 1$. From this we see that

$$\frac{\partial F/\partial K}{\partial F/\partial L} < \frac{p_K}{p_L}.$$

Consequently, we can deduce that capital is used beyond its marginal rate of technical substitution. That is, at the same level of production, regulated utilities use capital beyond the optimal amount for the unregulated firm when maximizing profits. This is known as the Averch-Johnson (1962) hypothesis. It is a hypothesis because it needs empirical testing to see if the model is valid. This succinct analysis shows the power of the KKT conditions for qualitative, marginal analysis of models. We see a potential significant distortion in the way the economy operates under traditional regulatory policies. However, this model illustrates the problem with this kind of analysis for policy formulation.

There were many studies trying to verify this result. When evaluating these studies using the electric utility industry, Murphy and Soyster (1983) found that the only one with strong statistical correlations had the dependent variable as an independent variable as well. The studies suffered because the firm is more complicated than a simple production function. The econometricians were looking for a bias in favor of capital expenditures versus of fuel purchases in utilities. The problem is that the cheaper fuels require more capital intensive plants: coal plants are significantly more expensive to build than oil plants and nuclear are the most expensive of all. So capital intensive utilities had low fuel costs and vice versa, whereas, the typical production function, such as Cobb-Douglas, says that when fuel costs are high, one substitutes plant for fuel and should see more expenditures on capital when fuel costs are high. That is, fuels are heterogeneous and there is no way to create a single aggregate fuel required by this simple production function.

Interestingly, the intuition that regulated firms are inefficient has proved correct. However, when one looks at what happened with deregulation in airlines, trucking and telecommunications, what one sees is a massive reduction in the wage bill through givebacks in wages by unionized workers (the regulated industries are heavily unionized) and employment reductions at both the worker and managerial levels. The implication is that the qualitative analysis was done with the wrong model rather than that the notion of qualitative analysis is wrong. This example illustrates the greater difficulty in validating these social science models versus models in the physical sciences.

At the same time economists were looking at the electric utility industry in aggregate, operations researchers were building models for utilities to use in their planning and operations. An example is the AGEAS model built at MIT for the Electric Power Research Institute and distributed to member utilities (Caramanis, Schweppe and Tabors, 1982). AGEAS is a suite of optimization and simulation models for capacity expansion analysis. Furthermore, the models of electricity flow and transmission line expansion use optimization. Models like AGEAS and the early model from which it is derived by Mass and Gibrat (1957), when operated at the planning level, are models of the utility's production function modeling the heterogeneous inputs at more detailed level than the standard production function in the economics literature.

WHERE ECONOMICS AND OPERATIONS RESEARCH HAVE COMMON INTERESTS: There are several areas where the two fields overlap.

We mention four: public policy analysis, finance, game theory and decision analysis. The convergence of the fields in policy analysis comes about because politicians want quantitative analyses of programs. Economic models have a lot to say about how economic agents behave and operations researchers have the computational skills and modeling expertise to implement the economic theories and solve for the economic impacts of policy alternatives.

Examples here include the previously mentioned activity at the World Bank. The close working relationships between economists and operations researchers have continued with the successor models to PIES, the Intermediate Future Forecasting System (Murphy, Conti, Sanders and Shaw, 1988) and the National Energy Modeling System (Energy Information Administration, 1994).

A key feature of these kinds of policy models is that in some sectors they model the decisions using optimization by representing the technology choices directly in the model. The main reason for using optimization is that the models need to have representations for policies and technologies that affect more than input and output prices and quantities, and there is no history to assess the resulting decisions for some sectors. Other reasons include the need to link more than one sector and a convoluted data history that muddy the econometric analysis for estimating such things as a production function for electric utilities. The optimization models are usually simplified versions of the planning models used by the industry with coefficients based on industry aggregates. They are treated as simulation models based on the result of Samuelson (1952) showing the connection between optimization and economic equilibrium models.

Policy models almost always include econometric components as well. For example, the above-mentioned energy models include econometrically estimated demand curves, along with optimization models for coal and electric utilities. In econometric models of production, one measures the inputs and outputs to statistically estimate the parameters of a production function. The model makes no statement about the actual decisions made. Instead, it models the outcomes of the decisions made by the actors in the economic sector. Econometric approaches dominate optimization when there is too much heterogeneity among participants to specify the parameters of their decision environment, as in demand modeling or the behavior of producers when the industry has a large number of independent, small firms.

The finance literature is dominated by economic studies of financial markets and their efficiency. An example is the book *A Random Walk Down Wall Street* by Malkiel (1973). This book showed that movements in stock prices are a random walk illustrating why stock pickers in general cannot beat the market. However, financial markets are not entirely efficient and the Black-Scholes (1973) model for pricing options created a whole new segment of the finance industry.

Optimization models have come to play an important role in determining the mix, the first one being the model by Markowitz (1952), which represents the beginning of computational finance. Also, Tobin's (1958) results on the relationship between risk and return were key to the development of decision models in finance. Markowitz's model led to the work of Sharpe (1964) in portfolio theory. His CAPM model made the computation of an optimal portfolio possible. Also, his analysis of optimal portfolios came up with the conclusion that the optimal portfolio should carry each stock in proportion to its value, relative to the total market. This qualitative analysis has led to the current index funds that do not try to beat the markets and has permanently changed financial management.

Because of the ability to solve far larger linear programs than in the past, stochastic programming models for building portfolios have made an important mark in the industry. Carino *et al.* (1994) describes the kind of operations research models used by the people known as "rocket scientists" in the financial press.

The interconnection between economics and operations research in game theory can be illustrated by returning to the Averch-Johnson hypothesis. This hypothesis proved false on the capital bias because it said nothing about the participants in the firm. Their model presumed that all of the members of the firm were focussed on maximizing profits for the benefit of stockholders. However, this is not necessarily the case. For example, the CEO, who makes the decisions, is not the owner of the firm and, therefore, has different incentives. The goal may be to maximize reliability rather than face hearings before the public utility commission over a power outage, leading to the same capital bias. The goal may be to minimize labor disruption and strikes by acceding to wage demands, since workers are voters with access to the public utility commissions, as well as employees. Figuring out the underlying incentives of the members of a firm and analyzing their behavior relative to the interests of stockholders is known as principal/agent theory, an important area of microeconomics. It essentially stops treating the firm as the atom when looking at the interaction of agents and looks further into the nature of the behaviors of the agents who make up the firm.

Studying the behavior of economic agents and other individuals has a long tradition in economics and is the essence of game theory. Since there are few data for numerically evaluating game models, almost everyone involved studies the qualitative properties of the resulting games. Economists have focussed mostly on markets (Shubik, 1959). Indeed, outside of von Neumann's early work on parlor games, the book by von Neumann and Morgenstern (1944) was the first major treatment of the subject and focussed on economics. Operations researchers have studied other types of games such as war games or some of the classic strategy games like the prisoner's dilemma game (Poundstone, 1992). The center for this work was RAND Corporation. An example of a strategic game was the stability of mutual assured destruction as a defense against nuclear war. Schelling (1980) presents an analysis of these strategic games. Shubik presents an interesting example of a researcher who does both strategic and economic games. As part of his examination of strategic issues, he used the dollar auction game to describe games of escalation such as war and lawsuits (Poundstone, 1992).

Political scientists and sociologists have become involved in game theory. The link between political science and games is direct through the games already mentioned and the use of game theory concepts in negotiation. Sociologists use games to understand social interactions. The prisoner's dilemma game has been used repeatedly to explain the behavior of individuals in social situations and social structures. Thus, the notions of game theory have moved beyond the disciplines in which they were developed and affected important areas of the social sciences. Part of the reason for the common interest of economists and operations researchers in game theory is its universality to understanding conflict and cooperation, as evidenced by its use in other fields.

Game theory is part of the subject of rational decisionmaking. Indeed, an invaluable work on the subject that treated both together is Luce and Raiffa (1957). What is a rational decision is subject to debate. To explore the subject, von Neumann and Morgenstern developed the concept of expected utility. This is a simple concept in many situations when the goal can be clearly stated as maximize profits. However, in real life we face many tradeoffs. Examples include our willingness to bear risk, how we value income versus leisure, what we value in the products we consume and how we value the future over the present. In Arrow's (1951) seminal work on social choice, he posited a set of axioms that define rationality and then shows how group interactions and voting processes lead to irrational decisions, even though the original actors have rational utility functions. In the decisionmaking literature, Keeney and Raiffa (1976) explore the issues associated with multi-attribute utility in decisionmaking. The economists' notion of utility is central to the study of negotiation (Raiffa, 1982).

As with mathematical programming, economists have not focussed on making actual decisions except in so far as there are general properties that can be understood from the decision-making process, as in Arrow (1951). Another example of this is the economics literature on rational expectations. In its most basic form, the question addressed in the context of macroeconomic models is: "How do the consequences of macroeconomic policy change when the participants in the economy have rational expectations about the effect of macroeconomic policies and adjust their decisions?" Redman (1992) and Sargent (1993) discuss this area of economics.

Operations research has come to dominate the subject so far as making actual decisions. Borrison (1994) is an example of a detailed decision analysis in a corporation. Some of the most important literature has come from psychologists trying to understand peoples' decision processes. The psychology literature is aimed squarely at the rational actor hypothesis of economics and finds it wanting. Bell, Raiffa, and Tversky (1988) examine the approaches of all three disciplines.

SUMMARY: Economics and operations research have common roots. The fields often use the same tools, such as the Karush, Kuhn, Tucker conditions. In economics, these conditions are used for marginal analysis, as with the search for institutional distortions of the marketplace in the Averch-Johnson hypothesis and for such uses as the derivation of cost functions from production functions. Operations researchers exploit these conditions to improve algorithms and use the actual duals and ranges for evaluating the stability of the model results, estimating the effects of uncertainty in the coefficients on the solution, and determining the costs of constraints with an eye towards adding or reducing resources.

Typically, the fields use these tools differently for different purposes. This reflects the varied professional goals of the individuals involved in these fields. Operations researchers focus on making decisions, while economists study the consequences of different market structures and policies through an assumption of rational decisionmaking. Both groups are interested in understanding rational decisionmaking and the consequences of rational decisions. This can be seen in the different views of the firm. The economic theory of the firm is really a theory of the interactions of firms. Operations research models provide a theory of decisionmaking within the firm and are an important component of a theory of the internals of the firm. The operations research models do not provide a complete theory of decisionmaking in the firm because operations researchers, although commenting on conflicts in the firm, tend to not focus on the incentives and structures that create these conflicts. This is where agency theory fits in and one of the places where game theory links both fields.

The fields are now distinct. Operations research takes an engineering perspective: the goal is to invent improved ways for making decisions, and in the process of doing this, inventing new models and algorithms as needed. Economics, instead, is a social science where the goal is to understand the existing world using the basic theme of exploring the consequences of rational self-interest. The two worlds come together when there is the need to change the rules of the marketplace or when the marketplace creates opportunities to engineer new products that provide a profit. Both fields have their distinct niches, yet will always be connected by their tools and history.

See **Banking; Corporate strategy; Decision analysis; Econometrics; Game theory; Input-output analysis; Portfolio theory; Public policy analysis; RAND Corporation; Utility theory.**

References

[1] Arrow, K. (1951), *Social Choice and Individual Values*, John Wiley, New York.

[2] Arrow, K., S. Karlin, and H. Scarf (1958), *Studies in the Mathematical Theory of Inventory and Production*, Stanford University Press, Stanford, California.

[3] Averch, H. and L. Johnson (1962), "Behavior of the Firm Under Regulatory Constraint," *American Economic Review*, 52, 369–372.

[4] Bell, D., H. Raiffa, and A. Tversky (1988), *Decisionmaking, Descriptive, Normative and Prescriptive Interactions*, Cambridge University Press, Cambridge.

[5] Black, F. and M. Scholes (1973), "The Pricing of Options and Corporate Liabilities," *Jl. Political Economy*, 81, 637–659.

[6] Borrison, A. (1995), "Oglethorpe Power Corporation Decides About Investing in a Major Transmission System," *Interfaces*, 25, to appear.

[7] Brooke, A., D. Kendrick and A. Meeraus (1993). *GAMS: A User's Guide*, Scientific Press, Redwood City, California.

[8] Caramanis, M., F. Schweppe, and R. Tabors (1982), *Electric Generation Expansion Analysis System*, Electric Power Research Institute EL-2561, Palo Alto California.

[9] Carino, D., T. Kent, D. Myers, C. Stacy, M. Sylvanus, A. Turner, K. Watanabe, and W. Ziemba (1994), "The Russell-Yasuda Kasai Model: An Asset Liability Model for a Japanese Insurance Company Using Multi-stage Stochastic Programming," *Interfaces*, 24(1), 29–49.

[10] Charnes, A., W.W. Cooper, and B. Mellon (1952), "Blending Aviation Gasolines – A Study in Programming Interdependent Activities in an Integrated Oil Company," *Econometrica*, 20, 2, 135–159.

[11] Dantzig, G. (1951a), "Maximization of a Linear Function of Variables Subject to Linear Inequalities," in T.C. Koopmans, ed., *Activity Analysis of Production and Allocation*, John Wiley, New York.

[12] Dantzig, G. (1951b), "A Proof of the Equivalence of the Programming Problem and the Game Problem," in T.C. Koopmans (ed.), *Activity Analysis of Production and Allocation*, John Wiley, New York.

[13] Dantzig, G. (1963), *Linear Programming and Extensions*, Princeton University Press, Princeton, New Jersey.

[14] Dorfman, R., P. Samuelson, and R. Solow (1958), *Linear Programming and Economic Analysis*, McGraw-Hill, New York.

[15] Energy Information Administration (1994), *The National Energy Modeling System: An Overview*, DoE/EIA-0581, May.

[16] Gale, D., H. Kuhn, and A. Tucker (1951), "Linear Programming and the Theory of Games," in T.C. Koopmans, ed., *Activity Analysis of Production and Allocation*, John Wiley, New York.

[17] Gustafson, R.L. (1958), *Carryover Levels for Grains*, US Dept. of Agriculture Technical Bulletin 1178.

[18] Hazell, P.B.R. (1986), *Mathematical Programming for Economic Analysis in Agriculture*, MacMillan, New York.

[19] Hitchcock, F. (1941), "The Distribution of a Product from Several Sources to Numerous Localities," *Jl. Mathematical Physics*, 20, 224–230.

[20] Hogan, W.W. (1975), "Energy Policy Models for Project Independence," *Computers and Operations Research*, 2, 251–271.

[21] Kantorovitch, L. (1939), "Mathematical Methods in the Organization and Planning of Production," Leningrad State University. Translated in *Management Science*, 6 (1960), 366–422.

[22] Keeney, R. and H. Raiffa (1976), *Decisions with Multiple Objectives*, John Wiley, New York. Reprinted in 1993 by Cambridge University Press, New York.

[23] Koopmans, T. (1951), *Activity Analysis of Production and Allocation*, John Wiley, New York.

[24] Luce, D. and H. Raiffa (1957), *Games and Decisions*, John Wiley, New York.

[25] Malkiel, B. (1973), *A Random Walk Down Wall Street*, W.W. Norton, New York.

[26] Markowitz, H. (1952), "Portfolio Selection," *Jl. Finance*, 7, 77–91.

[27] Mass, P. and R. Gibrat (1957), "Applications of Linear Programming to Investments in the Electric Power Industry," *Management Science*, 3(1), 149–166.

[28] Murphy, F.H., J. Conti, R. Sanders and S. Shaw (1988). "Modeling and Forecasting Energy Markets with the Intermediate Future Forecasting System," *Operations Research*, 36, 406–420.

[29] Murphy, F.H. and A.L. Soyster (1983), *Economic Behavior of Public Utilities*, Prentice-Hall, Englewood Cliffs, New Jersey.

[30] Poundstone, W. (1992), *Prisoner's Dilemma*, Doubleday, New York.

[31] Raiffa, H. (1982), *The Art and Science of Negotiation*, Harvard University Press, Cambridge, Massachusetts.

[32] Redman, D.A., (1992), *A Reader's Guide to Rational Expectations*, Edward Elgar, Hants, England.

[33] Samuelson, P.A. (1952), "Spatial Price Equilibrium and Linear Programming," *Amer. Economic Rev.*, 42, 283–303.

[34] Sargent, T.J. (1993), *Rational Expectations and Inflation*, Harper Collins, New York.

[35] Scarf, H. with T. Hansen (1973), *The Computation of Economic Equilibria*, Yale University Press, New Haven, Connecticut.

[36] Sharpe, W. (1964), "Capital Asset Prices: A Theory of Market Equilibrium Under Conditions of Risk," *Jl. Finance*, 19, 425–442.

[37] Schelling, T. (1980), *The Strategy of Conflict*, Harvard University Press, Cambridge, Massachusetts.

[38] Shubik, M. (1959), *Strategy and Market Structure*, John Wiley, New York.

[39] Stigler, G. (1945), "The Cost of Subsistence," *Jl. Farm Economics*, 27, 303–314.

[40] Tobin, J. (1958), "Liquidity Preference as Behavior Toward Risk," *Rev. Economic Studies*, 25, 65–86.

[41] von Neumann, J. and O. Morgenstern (1944), *Theory of Games and Economic Behavior*, John Wiley, New York.

[42] Whitin, T. (1957), *The Theory of Inventory Management*, 2nd edition, Princeton University Press, Princeton, New Jersey.

EDGE

(1) An edge is the line segment joining two extreme points of a polyhedron such that no point on the segment is the midpoint of two other points of the polyhedron not on the segment. See **Polyhedron**. (2) A line connecting two nodes in a graph (network). See **Graph theory**; **Network optimization**.

EFFICIENCY

(1) In statistics, an unbiased estimator's efficiency is the relative size of its variance compared to other unbiased estimators. (2) See **Efficient solution**. (3) See **Data envelopment analysis**.

EFFICIENCY FRONTIER

See **Data envelopment analysis**.

EFFICIENT ALGORITHM

See **Computational complexity**.

EFFICIENT POINT

See **Efficient solution**; **Multi-objective programming**.

EFFICIENT SOLUTION

For a maximizing multi-objective problem, a solution x^0 is efficient if x^0 is feasible and there exists no other feasible solution x such that $cx \geq cx^0$ and $cx \neq cx^0$. An alternative definition is that a feasible x^0 is efficient if and only if there exists no other feasible x such that $c_k x \geq c_k x^0$ for at least one k. An efficient solution is a feasible solution for which an increase in value of one objective can be achieved only at the expense of a decrease in value of at least one other objective. Efficient solutions are also called nondominated solutions or Pareto-optimal solutions. See **Multi-objective linear-programming problem**; **Multi-objective programming**; **Pareto-optimal solution**.

EIGENVALUE

See **Matrices and matrix algebra**.

EIGENVECTOR

See **Matrices and Matrix algebra**.

ELECTRIC POWER SYSTEMS

Oliver S. Yu

SRI International, Menlo Park, California

An electric power system is designed for reliable, economic, and socially acceptable production and delivery of electricity to individual customers. It involves many interrelated elements: generation stations, control centers, transmission lines, distribution substations, and distribution feeders. With an intricate system structure, complex economic and social objectives, and numerous reliability, safety, and resource constraints, power system planning and operation has long been an ideal field for the development and application of operations research and management science (OR/MS) techniques. These developments and applications continue to expand and evolve with the advancements in power technologies and changes in the utility industry. In the following sections, we shall use electric power generation system planning and operation as a prototype example to present some basic concepts, as well as a small glimpse of the enormous opportunity for OR/MS applications to power systems.

OVERVIEW: Electric power systems have been planned, constructed, and operated to supply electricity to the general public by regulated utilities. As regulated entities, these utilities are allowed to recover their capital investments and operating costs for supplying electricity with an allowance for reasonable returns by collecting revenues from customers in the form of electric rates. To assure the economic efficiency of the utilities, regulatory commissions in general have required utilities to minimize their total revenues required for electricity supply. Therefore, electric power system planning and operation has been a classical OR/MS problem of minimizing utility revenue requirements to meet projected electric demand growth over a future time period at a given level of reliability.

Electric power systems have three major parts: generation, transmission, and distribution. Because power flows in transmission and distribution are still difficult to be estimated accurately and economically, most applications of optimization techniques have been in generation system planning and operation. Specifically, the generation system planning and operation problem is to select a combination of power plants and unit dispatch schedules to minimize the present worth of the total capital, fuel, and operations and maintenance expenditures for meeting future electric demand while satisfying generally agreed-upon generation system reliability standards.

OPTIMAL GENERATION SYSTEM RELIABILITY: Setting generation system reliability standards is itself an optimization problem, because too low a standard would cause economic losses to the customers from frequent electric supply interruptions, while too high a standard would cause low capacity utilization of power plants and thus high electricity costs.

In the past, a commonly accepted empirical generation system reliability standard has been the one day in 10 year loss of load probability, that is, the daily electric peak load not to exceed available generating capacity of each day by more than one day in 10 years. However, with increasing technical capability, computationally efficient procedures have been developed to enable utilities to assess generation system reliability in detail (Yu, 1978). Furthermore, cost/benefit approaches have been used to derive the optimal reliability standard for a given socio-economic environment by determining the appropriate tradeoffs between the cost of power supply shortage and disruption and the cost of over-capacity to the customers (Kaufman, 1975; Telson, 1975; Keane and Woo, 1992).

OPTIMAL DISPATCH OF GENERATING UNITS: Because electric load varies by hour, day, and season in a year, the dispatching of generating plants to meet daily load requirements is also itself an optimization problem. This so-called production costing problem strives to determine the plant dispatch schedule that will minimize the fuel as well as operations and maintenance costs for meeting the load. In a broader context, the production costing problem also involves power purchases from neighboring utilities for either low cost or backup capacity. Therefore, for optimal generation system planning, we must first find a solution to the operational subproblem of production cost optimization.

A classical generation system operation optimization problem is the combined scheduling of hydroelectric and thermal power plants. Specifically, ineffective use of hydropower will increase the use of high operating cost thermal plants. The power system operator's problem is therefore to minimize the total cost of generation system operations for a given time period with uncertainties in load requirement and water availability.

GENERATION SYSTEM EXPANSION PLANNING:

In a simplified form, the generation system expansion planning problem for a time horizon $[0, T]$, may be expressed as follows (Anderson, 1972):

Minimize $\quad c_f(x_1, x_2, \ldots, x_n)$

$$+ c_v[y_1(t), y_2(t), \ldots, y_n(t)]$$

such that $\quad \sum_i y_i(t) \geq L(t) \quad$ for t in $[0, T]$

$$0 \leq y_i(t) \leq d_i x_i$$

where $\quad x_i$ is the capacity of plant i

$\quad\quad y_i(t)$ is the capacity of plant i used at time t

$\quad\quad L(t)$ is the load at time t

$\quad\quad d_i$ is the derating of plant i because of random forced outages

$\quad\quad c_f$ is the present worth of the fixed costs

$\quad\quad c_v$ is the present worth of the variable costs

In a more sophisticated formulation, the random nature of plant outage is taken into account by replacing the first constraint with:

$$\Pr\{\textstyle\sum_i y_i(t) < L(t)\} \leq p \quad \text{for } t \text{ in } [0, T].$$

Probabilistic simulation models have been used to determine the optimal dispatch schedule accurately (Stremel *et al.*, 1980; Sidenblad and Lee, 1981). With the solution of the production costing subproblem, a number of OR/MS techniques, including linear and nonlinear programming, can be used to solve the overall generation system planning problem.

Another level of sophistication is to require x_i to be integer-valued. In this case, either mixed integer programming (Benders, 1962) or dynamic programming (Jenkins and Joy, 1974) can be used for generation system planning solutions.

A comprehensive application of OR/MS and other engineering-economic analysis techniques to generation system expansion planning has been the Electric Generation Expansion Analysis System (EGEAS) developed by the Electric Power Research Institute (EPRI, 1983).

OPTIMAL MAINTENANCE SCHEDULING OF GENERATING UNITS:

Another optimization problem in generation system operations is unit maintenance scheduling. Each generating unit has a set period each year for preventive maintenance. The objective of optimal maintenance scheduling is to minimize the overall production cost while meeting the generation system reliability standards throughout the year. The problem is somewhat similar to the knapsack problem in OR/MS. Because utility business requirements vary, often a heuristic approach is required to find a solution for maintenance scheduling for a specific power system (Yu and Freddo, 1978).

FUEL INVENTORY PLANNING:

One other area in generation system planning amenable to OR/MS application is fuel inventory planning. Chao *et al.* (1989) have developed an optimization computer program that performs formal cost-benefit analysis of the following problems:

- Uncertain fuel deliveries and fuel burn
- Seasonality in fuel use and fuel supply
- Supply disruption of varying severity, warning times, and duration
- Nonlinear shortage costs

UTILITY RESOURCE PLANNING AND FUTURE CHALLENGES:

In the last 15 years, there have been major changes in the electric utility industry in the United States. As a result, utility resource planning objectives have also evolved (Yu and Chao, 1989).

First, in the 1970s, growing environmentalism imposed an additional tradeoff between environmental control cost and generation system reliability. A major application of OR/MS techniques was the development of the Over/Under Capacity Expansion Model funded by EPRI (EPRI, 1987).

Since the late 1980s, prevailing energy conservation ethics has given rise to the widespread adoption of the Least Cost Planning concept, also often referred to as Integrated Resource Planning. Under this planning concept, in addition to an economic comparison among themselves, electric supply alternatives are to be further compared with demand-side management options, which include energy conservation and load management. As the research management arm of the U.S. electric utility industry, EPRI has also funded the development of a number of major optimization tools in this area (EPRI, 1988), including the Multi-objective Integrated Decision Analysis and Simulation (MIDAS) model and the Utility Planning Model (UPM).

In recent years, there has been a strong trend towards generation deregulation. In an increasingly competitive market, generation resource investment planning will take on additional level of complexity and provide new challenges to OR/MS.

See **Combinatorial and integer programming; Linear programming; Nonlinear programming.**

References

[1] Anderson, D. (1972). "Models for Determining Least Cost Investments in Electricity Supply," *Bell Jl. of Economics and Management Sciences*, 3, 267–299.

[2] Benders, J.R. (1962). "Partitioning Procedures for Solving Mixed-Variable Programming Problems," *Numerische Mathematik*, 4, 238–252.

[3] Chao, H., Chapel, S.W., Morris, P.A., Sandling, M.J., Fancher, R.B., and Kohn, M.A. (1989). "EPRI Reduces Fuel Inventory Costs in the Electric Utility Industry," *Interfaces*, 19, 48–67.

[4] Electric Power Research Institute (1988). *EPRI Products*, *Volume 8, Planning*, Palo Alto, California.

[5] Electric Power Research Institute (1987). *OVER/UNDER Capacity Planning Model, Version 3*, P-5233-CCM, Palo Alto, California.

[6] Electric Power Research Institute (1983). *EGEAS, The Electric Generation Expansion Analysis System*, EL-2561, Palo Alto, California.

[7] Ikura, Y., Gross, G., and Hall, G.S. (1986). "PG&E's State-of-the-Art Scheduling Tool for Hydro Systems," *Interfaces*, 16, 65–82.

[8] Jenkins, R.T., and Joy, D.S. (1974). *WIEN Automatic System Planning Package (WASP) – An Electric Utility Optimal Generation Expansion Planning Computer Code*, Report ORNL-4945, Oak Ridge National Laboratory, Oak Ridge, Tennessee.

[9] Kaufman, A. (1975). *Reliability Criteria – A Cost Benefit Analysis*, Report 75-9, New York Department of Public Service, New York, New York.

[10] Keane, D.M., and Woo, C.K. (1992). "Using Customer Outage Cost to Plan Generation Reliability," *Energy*, 17, 823–827.

[11] Sidenblad, K.M., and Lee, S.T.Y. (1981). "A Probabilistic Production Costing Methodology for Systems With Storage," *IEEE Transactions on Power Apparatus and Systems*, 100, 3116–3124.

[12] Stremel, J.P., Jenkins, R.T., Babb, R.A., and Bayless, W.D. (1980). "Production Costing Using the Cumulant Method of Representing the Equivalent Load Curve," *IEEE Transactions on Power Apparatus and Systems*, 98, 1947–1956.

[13] Telson, M.L. (1975). "The Economics of Alternative Levels of Reliability for Electric Power Generation System," *Bell Jl. of Economics*, 6, 679–694.

[14] Yu, O.S., and Freddo, W. (1978). *An Efficient Electric Power Generation Maintenance Scheduling Procedure*, presentation at ORSA/TIMS National Meeting, San Francisco, California.

[15] Yu, O.S. (1977). *An Efficient Approximation Computational Procedure for Generation System Reliability*, Technical Report, Mid-American Interconnection Network, Chicago, Illinois.

[16] Yu, O.S., and Chao, H. (1989). "Electric Utility Planning is a Changing Business Environment: Past Trends and Future Challenges," *Proceedings of Stanford-NSF Workshop on Electric Utility Planning Under Uncertainty*, 253–272, Stanford, California.

ELEMENTARY ELIMINATION MATRIX

A square nonsingular matrix obtained by replacing a column of the identity matrix by some vector. Every pivot operation on a system of linear equations is equivalent to multiplication of the system from the left by an elementary elimination matrix. In the simplex method, such a pivot matrix in is called an eta-matrix. See **Matrices and matrix algebra**.

ELIMINATION METHOD

See **Gaussian elimination**.

ELLIPSOID ALGORITHM

The first polynomial-time algorithm for linear programming. The ellipsoid algorithm was originally developed by Shor, Yudin and Nemirovsky as a method for solving convex programming, but it was Khachian who showed that this method can be adapted to give a polynomial-bounded algorithm for linear programming. The basis of the ellipsoid algorithm is a method for finding a feasible solution to a set of linear inequalities. This method constructs a sequence of ellipsoids of shrinking volume, each of which contains a feasible point (if one exists). If the center of one of these ellipsoids is feasible to the system of inequalities, the algorithm terminates. If not, then after a known (polynomial) number of iterations the volume of the ellipsoid will be too small to contain a feasible point, and hence the system is infeasible. This method for solving inequalities can be used to solve linear-programming problems in polynomial time by writing the primal and dual feasibility constraints and the equality of the primal and dual objectives as a system of inequalities. Despite its tremendous theoretical importance, the ellipsoid algorithm appears to have little practical significance, since its computational performance has been very poor.

ELSP

Economic lot scheduling problem. See **Production management**.

EMBEDDING

(1) The drawing of a graph on a surface without edge crossings. (2) The use of a subsidiary stochastic process to solve a "larger" one in which the subsidiary is contained. See **Imbedded Markov chain**; **Queueing theory**.

EMERGENCY SERVICES

Kenneth Chelst

Wayne State University, Detroit, Michigan

INTRODUCTION: Police, fire and emergency medical services (EMS) all operate in a complex 24-hours-a-day, seven-days-a-week unpredictable environment. The planning and management of these services are complicated by uncertainty regarding
1. the time and location of each emergency;
2. the type of call in particular, the personnel and equipment needed to handle the emergency;
3. the amount of time spent at the emergency scene and in follow-up activities.

Typically, managers of these emergency services function within severely constrained budgets and face two common complex operational questions:
1. How many emergency service vehicles should be staffed each hour of the day?
2. Where should these vehicles and personnel be located?

In an ideal world, decision makers would focus on their ultimate goals when making these decisions. Police officials would evaluate strategies in terms of their relative effectiveness in reducing crime and the fear of crime. EMS managers would compare alternatives with regard to lives saved and disabilities avoided. Fire service administrators would allocate fire fighting equipment and personnel so as to reduce fire damage. Unfortunately, we do not fully understand the relationship between many of the decisions made and their ultimate impact. For example, we can neither predict the impact on crime levels of adding 10% more patrol cars to the streets nor estimate the number of lives saved as result of building one more fire station. In contrast in the past ten years, we have made significant progress in understanding the relationship between

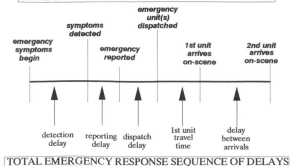

Fig. 1. Emergency response system.

the number and type of emergency medical units in a community and the likelihood that the ambulance service will save the life of a victim of a full cardiac arrest. However, we still do not understand the impact of the ambulance service in other types of medical emergencies such as automobile accidents.

As a result, operations research models designed to assist emergency service decision makers use the surrogate measure of response time when structuring resource allocation decisions. Figure 1 illustrates the total response pattern to an emergency. It begins with the onset of symptoms of an emergency (e.g., chest pains, smoke, suspicious persons). There is a delay until the symptoms are recognized as an emergency and a further delay until the emergency is reported. The dispatcher must process the call and find an emergency unit to dispatch. The final response time delay is the travel time to the scene of the call. For fire services and ambulance services, there could be an additional delay between the time the station is notified of the call and the time it takes for the vehicle to begin in motion on the streets. If multiple units are dispatched to the same call, there will be multiple arrival times. Operations research models focus on two components of response time: 1) the queueing delay – the time from call receipt until an emergency unit(s) is available for dispatch and 2) the travel time(s) to the scene of an emergency.

In the 1970s faculty at M.I.T. and researchers at the New York City RAND Corporation (a joint venture between New York City's government and the RAND Corporation) developed models that address the two basic deployment questions. The National Science Foundation and the Department of Housing and Urban Development funded this early work. One specific relationship that developed from this research was the square-root law, which estimates the average travel distance to a random call:

Average Travel Distance $= \{A/[N(1-b)]\}^{1/2}$

where A = area, N = the number of emergency service stations or vehicles, b = the proportion of time an emergency service unit is busy, and c = a constant of proportionality that is calibrated from actual data and has been found to be usually between 0.6 and 0.7. The parameters a and b are city specific, while N represents the key decision, namely, the number of units to deploy. This simple square-root formula is utilized to make aggregate decisions on deployment. It points out to decision makers that quadrupling the number of available emergency units reduces average travel distance by only 50%. The estimated travel distance is converted into travel time in a variety of ways. Either an average speed is assumed or regression analysis is used to define a nonlinear relationship between distance and time. Police deployment models use the average speed factor while fire models tend to use the regression relationship.

A major problem in planning emergency services is that the number of calls in an hour is random variable. This random number fits the classic assumptions of the Poisson process, which provides decision makers with a probabilistic model for forecasting the number of calls in one hour or an entire eight-hour shift. These forecasts are used as input into queueing models that estimate the average delay between the arrival of a call and the dispatch of an emergency unit. The average queueing delay is added to average travel time to determine total average response time.

FIRE SERVICES: Fire service deployment is the least complex of the three services to analyze. Fire call rates are typically low and firemen spend less than 5% of their time responding to calls. Response times are generally affected just by the number of fire stations that are staffed and are not sensitive to variations in the average call rate over the course of a day. Consequently, cities staff their fire equipment with the same manpower level around the clock to provide a constant level of fire protection.

This low workload simplifies the analysis of fire station location problems. Fire station planning models make the reasonable assumption that the nearest fire unit will be available to dispatch immediately when a need arises. Two classic deterministic optimal location models that have been used to locate fire stations are p-median models and coverage models. A coverage model locates fire stations so as to maximize the number of people or houses situated within, for example, two miles of the nearest fire station. The p-median model locates stations so as to minimize the average response time to a target population at risk.

In some complex environments, such as New York City, operations researchers have used a descriptive model that allows the decision maker to add or delete fire stations and assesses the impact. The descriptive model provided a wider range of performance statistics than the coverage and p-median models. This model could predict response times for the arrival of the second and third fire engines as well as differentiate between equipment with different roles (e.g., engine trucks and ladder trucks).

EMERGENCY MEDICAL SERVICES: Urban ambulance services operate at higher workloads than fire services. Utilization rates of 15% to 30% are not uncommon and in busy periods during the week workloads exceed 50%. Consequently, it is inappropriate for ambulance location models to make the simplifying assumption that the nearest stationed ambulance will not be busy when a call comes in. This has led to the development of more complex coverage models. These models incorporate concepts such as backup coverage and workload adjustment factors so as to implicitly approximate probabilistic concepts while maintaining the capability of using a deterministic optimal location model to place ambulances. An alternative to these optimal location models is the hypercube queueing model. This model has the advantage that it explicitly captures the potential unavailability of nearby ambulances. This model is primarily descriptive and can be used to evaluate ambulance placement plans. It does not, however, generate optimal or near optimal locations.

POLICE SERVICES: The allocation of police patrol resources is complicated by a number of factors. The number of police patrol units on the street is usually much larger than the number of fire stations or ambulances. These police units respond to a wide variety of calls with different levels of urgency (e.g., armed robbery in progress versus loud music next door). Lastly, police call rates vary significantly by time of day and day of week. Because of this added complexity, planning in a large city might proceed in three phases.

In the first phase, police officials would use aggregated data and a model such as PCAM (Patrol Car Allocation Model) to determine the number of patrol units to assign to different parts or precincts of the city for different time periods during the week. Next, police commanders would use a descriptive model such as the hypercube model or a simulation model to design patrol beats for individual patrol units. These models forecast response times for different priorities of calls as well as workloads of patrol units. Police managers could then apply a mathematical programming model to design a work schedule for personnel to insure that there was enough manpower on duty to staff the proposed street patrol plan.

In all emergency services decision making, there is a tension between efficiency and equity. A deployment plan that minimizes average response time will tend to concentrate resources in high risk areas leaving other areas with response times that are significantly above the overall average. This conflict has been addressed in a variety of ways. To achieve equity, PCAM allows the decision maker to establish minimum response time standards for each region of the city. Once enough patrol units have been assigned to each region to achieve these minimum standards, PCAM can allocate excess patrol resources to minimize citywide average response time. Alternatively, concepts of multiattribute utility and group decision making can be used to assist decision makers in exploring trade-offs between equitable deployment plans and efficient ones.

POLICY QUESTIONS: The models described above focus on developing plans to improve day-to-day operation of emergency services. OR models have also played an important role in exploring a number of policy questions:
1. Which is more cost effective: to spend public dollars on more fire stations to reduce response time or to subsidize the placement and maintenance of smoke detectors to reduce the delay until detection of the fire?
2. Which is more cost effective: more but less costly basic life support ambulances or fewer but more expensive advanced life support ambulances?
3. What are the relative benefits of one-officer and two-officer patrol units?

One collection of OR models was developed to study the merger of emergency services. A number of small-to-medium sized cities with populations as large as 100,000 have trained public safety officers to handle both police and fire emergencies. Although the concept has proven cost effective in many cities, the concept has failed in others. OR models enable decision makers to assess the potential impact of a merger on fire response times,

police response times and cost before implementing the politically sensitive concept.

OR models have played an important role in key operational and policy questions. Despite these documented successes, relatively few cities in the U.S. employ operations research models in their decision making on a regular basis. Even when studies are successfully completed, all too often political issues overwhelm cost effective recommendations. This problem arises in the private sector but is more pronounced in the public sector which does not have market forces to control wasteful practices. Whether or not OR plays a significant role in the struggle to maintain or improve emergency services in the face of our cities' shrinking revenue base will depend more on the political will of our cities' leaders and managers than on the technical sophistication of a new generation of emergency service models.

See **Crime and justice; Facility location; Hypercube queueing models; Location analysis; Queueing theory; RAND Corporation.**

References

[1] Chelst, K. R. (1981), "Deployment of One- vs. Two-officer Patrol Units: A Comparison of Travel Times," *Mgmt. Sci.* 27, 213–230.

[2] Cretin, S. and T.R. Willemain (1979), "A Model of Pre-Hospital Death from Ventricular Fibrillation following Myocardial Infarction," *Health Services Research* 14, 221–234.

[3] Halpern, J. (1979), "Fire Loss Reduction: Fire Detectors vs Fire Stations," *Mgmt. Sci.* 25, 1082–1092.

[4] Keeney, R. and H. Raiffa (1976), *Decisions with Multiple Objectives*, Wiley, New York.

[5] Larson, R.C. (1972), *Urban Police Patrol Analysis*, MIT Press, Cambridge, Massachusetts.

[6] Larson, R.C. (1974), "A Hypercube Queuing Model for Facility Location and Redistricting in Urban Emergency Services, *Computers and Opns. Res.* 1, 67–95.

[7] Matarese, L.A. and K.R. Chelst (1991), "Forecasting the Outcome of Police/Fire Consolidations," *MIS Report* 23, 4, 1–22. International City Management Association.

[8] ReVelle, C. (1991), "Siting Ambulances and Fire companies: New Tools for Planners," *Jl. of the American Planning Assn.* 57, 471–484.

[9] Taylor, P.E. and S.J. Huxley (1989), "A Break from Tradition for the San Francisco Police: Patrol Officer Scheduling Using an Optimization-Based Decision Support System," *Interfaces* 19, 4–24.

[10] Walker, W.E., J.M. Chaiken and E.J. Ignall, eds. (1979), *Fire Department Deployment Analysis*, North Holland.

EMS

Emergency medical services. See **Emergency services**.

ENTERING VARIABLE

The nonbasic variable chosen to become basic in an iteration of the simplex or similar linear-programming algorithm. See **Simplex method**.

ENVIRONMENTAL SYSTEMS ANALYSIS

Charles ReVelle

The Johns Hopkins University, Baltimore, Maryland

INTRODUCTION: Within a decade after the emergence of operations research at the end of World War II, civil and environmental engineers were already adapting the remarkable mathematical tools that had evolved in the defense sector during the war. They quickly applied these tools to the solution of important societal problems relating to environmental protection. Applications of operations research to urban and regional water management began in the late 1950s primarily under the leadership of Walter Lynn and Abraham Charnes. Solid wastes management using the tools of OR began in the mid-1960s led by Jon Liebman and engineers at Berkeley. The development of air pollution management models followed not long after. Parallel to these engineering-based investigations of environmental issues were applications in forestry/timber management/recreation as well as game and fisheries management. The three engineering-based areas of environmental OR application will be reviewed.

URBAN WATER MANAGEMENT: First, in historical sequence of environmental applications of operations research, was urban and regional water management. Urban and regional water management encompasses many activities, some of which have been approached with OR tools and others which have seen very little application activity. Water resources management, a parallel activity to urban and regional water management, focuses on the operation of reservoirs and systems of reservoirs for purposes of water supply, recreation, flood control, irrigation, hydropower, and navigation. It also deals

with aquifer management, conjunctive use of ground and surface waters, and interbasin transfers. The activities of urban and regional water management, in contrast, are principally concerned with the local delivery of water, with treatment of water, with disposal of wastewater, and with the quality of receiving waters, although quantity and quality intersect in a number of problem settings. Water resources management is discussed in this encyclopedia under its own heading.

We will follow water (1) out of the reservoir to the water treatment plant that produces drinking water, (2) through the distribution system to the consumer, (3) from the consumer through the sewer system (4) to the wastewater treatment plant (sewage treatment plant) and (5) into the receiving water body, where the pollution content of the treated wastewaters of many communities interacts with stream dissolved oxygen resources.

The water treatment plant is designed to produce drinking water that is free of disease-causing bacteria, viruses and protozoa. The water should be attractive (clear) and palatable with little in the way of objectionable tastes, odors, or color. The processes in a typical water treatment plant are designed to achieve these criteria. The design and arrangement of the component processes are probably susceptible to a cost optimization in which constraints are placed on the final concentrations of various contaminants. The design would constitute the first stage of applications of systems analysis to urban water management. However, very little in the way of OR/systems analysis has been done in water treatment plant design.

From the water treatment plant, the water enters a distribution system for delivery to customers. In this second stage of systems applications, linear programming and non-linear programming have been applied to the design of water distribution systems although the non-linearity in the equations has generally been approached by iterative application of linear programming to approximate an unknown multiplicative term. Decisions include which links of the system to build, the diameters of the pipes, the water flows in each link of the system, and the pressure heads at junctions of the system. A series of papers have appeared on this topic beginning in the late 1960s. A major difficulty in this design problem is the trade-off between cost and redundancy (needed for reliability) and the lack of a good operational measure of redundancy. A unique article that explicitly compares a number of the approaches to this problem was jointly prepared by many of the researchers in the area of pipe network optimization and appears under the title "Battle of the Network Models" (Walski *et al.*, 1987).

Residential consumers, as well as industry and commerce, receive the water from the distribution system, use it and abuse it for washing, bathing, lawn watering, irrigation and manufacturing processes. As a consequence of the use, the quality of the water is degraded, principally by the presence of organic contaminants, but also with micro-organisms and inorganic chemicals. Treatment at a sewage treatment plant is required to restore the water to a level of quality that will not impair the water in the receiving body.

To reach the sewage treatment plant, the wastes from residences, commerce and industry enter a wastewater collection system, the sewer system, which conveys them to the plant. The design of the sewer system represents a third stage of application of systems analysis to urban water management. A number of optimization models have been built to determine which links to build, diameters of individual sewer lines, the depth at which each line is placed and the slopes of each line. A representative work in this field is that of (Walters, 1985).

The sewers transport the wastes from their origins to the treatment plant which itself requires design. The design of the wastewater treatment plant represents a fourth stage of application of systems analysis to urban water management. The treatment plant typically consists of ordinary biological processes which remove organic wastes that are in solid form as well as organic wastes that are dissolved in the waste stream. Organic wastes are removed from wastewater because they will otherwise be degraded by microbes when they reach a lake or river and that biodegradation would remove dissolved oxygen from the water. Because fish and other aquatic organisms require adequate levels of oxygen to survive, it is imperative that sufficient amounts of organics be removed from wastewater to protect the dissolved oxygen resource that sustains the fish and other aquatic life.

Depending on the requirement of the receiving water, the treatment plant could also include physical-chemical processes to remove not only nitrates and phosphates, but also the small fraction of organics which is resistant to biological treatment. The design of treatment processes using optimization methodology was begun in the early 1970s. A recent research paper on the subject which usefully refers to past works is that of (Tang *et al.*, 1987).

From the treatment plant (where the removal of organics takes place), the restored wastewater may enter a river or lake where it mixes with receiving waters. The concentration of organic wastes ultimately discharged into the receiving waters will decrease as the degree of treatment/level of removal at the wastewater treatment plant increases. Without much treatment, the amount of dissolved oxygen in the lake or river consumed by oxidation of the organic wastes will be relatively large, making the water environment inhospitable to fish and other desirable aquatic organisms. Predictive models for the removal of the oxygen resource and biological decay of organic wastes were first developed in the 1910s and have become increasingly descriptive and encompassing since that time. A text reference that clearly describes these numerous models is that of (Thomann and Mueller, 1987). These differential and difference equation models describe the response of the receiving waters to inputs of organic and other wastes.

While the response of the receiving water to the input of a single stream of organic wastes had been largely modeled by the late 1950s, the response of a river or lake to a number of spatially separated waste streams had not been described analytically. If we consider that wastes from multiple treatment plants are entering a river, then an optimization problem arises in which one seeks the least cost set of treatment plant efficiencies (level of treatment or degree of removal) that can ensure the dissolved oxygen concentration everywhere in the river remains above a desired level or standard. The desired level or standard reflects the uses of the water body whether for fishing, swimming, boating, etc. Of course, it is possible and desirable to develop the trade-off between the system cost of wastewater treatment and the dissolved oxygen standard. The dissolved oxygen standard represents the value of the lowest level of dissolved oxygen that occurs anywhere along the length of the river. Linear and dynamic programming models have been developed for this optimization problem which is really a problem in linear optimal control – in the sense that the dissolved oxygen responses are governed by differential equations.

After manipulation, which can be extensive, of the governing equations that describe the system and the constraints on performance, these models can be converted to the optimization form

Water Pollution Abatement Model:

Minimize $\sum\limits_{i=1}^{n} c_i e_i$

s.t. $\sum\limits_{i \in I} a_{ij} e_i \geq S \quad \forall j \in J$

$0 \leq e_i \leq 1 \quad \forall i \in I$

where i, I = the sources and set of sources from which organic pollutants are discharged;

j, J = the points and set of points at which the dissolved oxygen standard must be met;

c_i = cost per unit of removal efficiency at source i;

e_i = the removal efficiency at source i;

a_{ij} = the amount of dissolved oxygen that is protected or is allowed to be present in the stream at point j per unit of removal efficiency at point i; and

S = the dissolved oxygen standard that must be met at all monitoring points in the river.

In an estuarine situation where tidal movements cause pollutants to mix upstream and downstream of their point of discharge, all a_{ij} coefficients are non-zero and positive. In contrast, in a non-tidal river, only those a_{ij} are non-zero and positive for which the point j is downriver from source i. That is, in a non-tidal river, pollution from a source i has negligible effect on a point of measurement upstream from that source.

It is useful to develop the multi-objective trade-off curve between total treatment cost and the dissolved oxygen standard because costs may increase rapidly in some portion of the curve, suggesting that further gains in quality can be obtained only at considerable expense. The river basin optimization model that chooses treatment efficiencies for each of many waste sources is a fifth stage in the application of systems analysis to urban water management (see ReVelle and Ellis, 1994).

While realistic and relatively complex, the treatment plant/river basin optimization models do not completely describe the options for designing a pollution abatement program for the waste sources on a river. Their lack centers around the assumption that each discharge of treated wastewater enters the receiving water body at a known and *prespecified* point somewhere along the river, usually at its point of origination. Thus, another fundamental problem,

and a sixth stage of application, is the siting of wastewater treatment plants along a river. That is, the previous model selected removal efficiencies but assumed that the flows from each of the wastewater treatment plants entered the river at the same geographical position as the community or industry that generated the flow. In contrast, the problem of siting wastewater treatment plants assumes a single prespecified and high removal efficiency for all the plants on the river, but seeks the *positions* for discharges which minimize the total treatment cost. The single treatment level is presumed to be sufficiently high that dissolved oxygen standards are not violated along the river. At one extreme, discharges may still occur at each community or industry along the river. At the other extreme, discharges may be consolidated into a single regional wastewater treatment plant. Most likely, however, is the partial consolidation of wastewater flows with some at-source discharges and some flows merging at regional plants for treatment and discharge.

The motivation of this problem setting is that economies of scale in treatment may be captured when wastewater flows are combined and treated together. Working against this cost advantage in consolidating flows are the additional costs of piping and pumping that are incurred when wastewater flows are merged at central points. Thus, the objective of the regional wastewater treatment plant problem is to minimize the sum of treatment costs and piping/pumping costs. The more dispersed that communities are along a river, the less likely will be consolidation in regional plants. Much work has been done on this problem since the early 1970s, most of it focussing on a fixed charge approximation of the concave costs of treatment. Zhu and ReVelle (1988) provide an efficient and exact solution to siting regional treatment plants along an essentially linear river via linear integer programming and refer to most previous published research on the problem.

A variation of and combination of the two previous problems has not been well studied; this is the problem which seeks the least cost set of treatment efficiencies *and* the sites for regional wastewater treatment plants given that a dissolved oxygen standard is honored along the length of the river. Multiplicative non-linearities and concave or fixed charge cost functions make the problem especially challenging.

A final and seventh stage of application of systems analysis to urban and regional water management is the problem of cost or burden sharing. The notion here is that a regional authority has been created whose goal is to stimulate cooperation in the solution of environmental problems. Cooperation takes the form of joint activities, e.g. regional wastewater plants *vs.* separate plants for each community – if the regional solution saves money. Since the authority is assumed to be unable to coerce the communities to cooperate, it must find the means to induce cooperation; the goal is to find an effective and attractive way to distribute the savings from joint undertakings. Such a distribution should be chosen in a way that makes every participating community better off than it would be if it were to treat its waste flow alone or to join some other non-optimal coalition. An article that refers to most prior work is Zhu and Re-Velle (1990).

SOLID WASTES MANAGEMENT: Management of the operation and of the design of urban and regional solid wastes systems constitutes an important environmental area for the application of optimization. Although the level of research activity in urban solid waste systems has decreased since the middle 1970s, challenging problems remain and would surely be addressed if research funding were to flow to this sector as it did twenty years ago. Regional solid waste management, especially with regard to hazardous wastes routing and siting, has been thriving.

The field of urban waste management may be roughly divided into two sectors: a collection/routing sector and a siting sector. In the first category, collection/routing, are two related classes of problem: routing within a district and the creation of districts. Routing within a pre-specified district means either visiting at least once every link within the district with the least total length route (minimal retracing of links) or visiting each link twice (corresponding to collection on both sides of the street) with the least total length route (no triple tracing required). The principle of a routing that includes every link at least once is minimal retracing, a property that is achieved by minimal length matching of odd nodes (nodes with an odd number of incident links). In the routing that covers each arc twice (two-sided collection) there are no odd nodes so that matching is not required and the route can be completed with a total length equal exactly to two times the total length of links in the district. Alternatively, the best routing might be the one with the least total cost, where time is a major factor in the determination of cost. As a consequence, a route that included many left turns against a high volume of oncoming traffic might be inferior to a longer route

with mostly right turns. Route design with consideration of cost, time and left turns remains an open and challenging problem.

The creation of collection districts from a large network of streets, however, requires a prior step in which links are assigned to each district in such a way that the total collection distance/time (and possibly volume or weight loading) in each district is within a preset bound. Once each district is created, the routing step would be undertaken next. However, the district creation step and the routing within a district step influence one another. That is, the length of the minimal length routing within the district i*l* only finally determined by routing so that in theory the assignment of links to a district cannot be completed without knowing the length of the minimal route within the district. Heuristics have been created by Liebman and his students to attack this problem (see Liebman, 1975).

In the second sector are siting problems. At least four types of siting problems can be identified: the central siting of collection vehicles, the siting of a central incinerator, the siting of sanitary landfills (of which there may be a number) and the siting of transfer stations (the stations where smaller trucks offload to larger vehicles for more distant hauling). All of these siting problems have been approached, but not all are well solved. A review of solid waste operations management is given in Liebman (1975) and a review of the siting of processing facilities and landfill sites is given in Gottinger (1988).

Regional solid waste management has received inspiration from the mounting problems of disposing of hazardous wastes. Issues of routing, scheduling and siting abound – with conflicting objectives. On the one hand, routes and sites should be chosen to minimize cost. On the other, routes and sites should be chosen to decrease risk and population exposure. A 1991 special issue of *Transportation Science*, edited by Turnquist and Zografos, takes up the issues in hazardous materials transportation in five articles.

CONTROL OF AIR POLLUTION: The application of systems analysis to the control of air pollution dates from the late 1960s. Modeling efforts drawn from air pollution meteorology were then applied for the first time to develop predictive equations that could be applied in constraint form for air pollutant concentrations downwind from emission sources. From these predictive equations could be derived what one refers to as transfer coefficients. Each transfer coefficient provides the unit increment of pollutant concentration downwind (in say milligrams/cubic meter) at a particular point of measurement for each unit of emission (say tons per day) at each pollutant source. Thus, each ton/day of sulfur dioxide emitted from a specified source has a quantifiable impact on the atmospheric sulfur dioxide concentration at each of a number of downwind sites. With such transfer coefficients as well as costs of abatement at the sources, and with an atmospheric concentration standard to be met, it is possible to structure an optimization model. This air pollution management model chooses the least cost set of removal efficiencies, one removal level for every source, that achieves atmospheric concentrations of the pollutant at all specified points of concern at or below the standard.

We next describe the basic model that has been suggested for the management of acid rain. The model follows the same form as the water pollution abatement optimization model described earlier. It should be noted that this model presumes no chemical reactions of pollutants but only gradual dissipation of the pollutant. Decision variables and parameters are as follows:

i, I = index and set of sources;

j, J = index and set of points at which concentrations are monitored;

R_i = fractional removal efficiency at source i;

E_i = tons per unit time of emissions at source i without any removal;

t_{ij} = transfer coefficients (mg/m^3/ton/day); the increment to the atmospheric concentration at j per unit of emissions at i;

S_j = atmospheric standard at receptor point j for the pollutant; and

c_i = cost per unit of pollutant removal.

Thus, the optimization problem is given as

Acid Rain Management Model:

$$\text{Minimize} \quad z = \sum_{i \in I} c_i R_i$$

$$\text{s.t.} \sum_{i \in I} t_{ij} E_i R_i \geq \sum_{i \in I} t_{ij} E_i - S_j \quad (j \in J)$$

$$0 \leq R_i \leq 1 \quad \forall i \in I.$$

This basic acid rain management problem can be manipulated in many ways. The transfer coefficients can be a single set of known numbers. They can also be random variables, and the constraints can then be either expected value constraints or chance constraints. Many possible sets of transfer coefficients can also be considered, leading to models which minimize maximum regret subject to investment in pollutant removal. The developments in air pollution management models in general and acid rain in particular are referred to in a recent review of water and air quality management by ReVelle and Ellis (1994).

CONCLUSION: The environmental models discussed here have much in common. Air and water pollution control models both have the equivalent of transfer coefficients which translate upstream and upwind discharges and emissions into downstream and downwind concentrations. Water pollution control facilities as well as landfills and incinerators require siting and regionalization to minimize costs. Where power plant air emissions are part of the control equations, siting of power plants also becomes an issue in air quality management. Finally, burden sharing and cost allocation issues are common to all these areas of environmental management. The problem set in environmental systems analysis is rich, diverse, challenging and important.

See **Integer programming; Linear programming; Location analysis; Natural resources; Vehicle routing; Water resources.**

References

[1] Walski, T., E. Brill, J. Gessler, I. Goulter, R. Jeppson, K. Lansey, H. Lee, J. Liebman, L. Mays, D. Morgan, and L. Ormsbee (1987), "Battle of the Network Models: Epilogue," *Journal for Water Resources Planning and Management*, Div. ASCE, 113 (2), 191.

[2] Walters, G. (1985), "The Design of the Optimal Layout for a Sewer Network," *Engineering Optimization*, 9, 37–50.

[3] Tang, C. Brill, E., and J. Pfeffer (1987), "Optimization Techniques for Secondary Wastewater Treatment Systems," *Journal of the Environmental Engineering Division*, ASCE, 113 (5), 935–951.

[4] Thomann, R., and J. Mueller (1987), *Principles of Surface Water Quality Modeling and Control*, Harper and Row, Inc.

[5] ReVelle, C., and J. Ellis (1994), "Models for Air and Water Quality Management," in *Operations Research and Public Systems*, edited by Pollock, S., Barnett, A. and M. Rothkopf, *Handbooks of Operations Research*, 7, Elsevier.

[6] Zhu, Z-P., and C. ReVelle (1988), "A Siting Model for Regional Wastewater Treatment Systems," *Water Resources Research*, 24 (1), 137–144.

[7] Zhu, Z-P., and C. ReVelle (1990), "A Cost Allocation Method for Facilities Siting with Fixed Charge Cost Functions," *Civil Engineering Systems*, 7 (1), 29–35.

[8] Liebman, J. (1975), "Models of Solid Waste Management," Chapter 5 in S. Gass and R. Sisson (editors), *A Guide to Models in Government Planning and Operations*, Sauger Books, Potomac, MD.

[9] Turnquist, M. and C. Zografos, eds. (1991), *Transportation Science*, Special Issue: Transportation of Hazardous Materials, 25 (1).

[10] Gottinger, H. (1988), "A Computational Model for Solid Waste Management with Application," *European Journal of Operation Research*, 35, 350–364.

EOQ

Economic order quantity. See **Economic order quantity; Inventory modeling.**

ERGODIC THEOREMS

Results giving the conditions for the time averages of a stochastic process to converge to its limiting or steady-state probabilities. See **Markov chains; Markov processes; Queueing theory.**

ERLANG

The unit of traffic load used in congestion analysis of telecommunication networks. The traffic load is the expected number of arrivals during an average service time. This quantity is dimensionless, but is referred to as the number of erlangs offered to the system. See **Offered load; Queueing theory.**

ERLANG B FORMULA

The probability that all servers are busy in the multiserver queueing system M/M/c/c with Poisson input, exponential service and no waiting space, and thus that an arriving customer will be unable to enter the system (i.e., is blocked). See **Queueing theory.**

ERLANG C FORMULA

The probability that all servers are busy in the multiserver queueing system M/M/c with Poisson input, exponential service, and infinite capacity. See **Queueing theory.**

ERLANG DELAY MODEL

The multi-server queueing system M/M/c with Poisson input and identical exponential service for each server. **Queueing theory.**

ERLANG DISTRIBUTION

A continuous random variable is said to have an Erlang distribution if its probability density may be written in the form $f(t) = a(at)^{k-1}e^{-at}/(k-1)!$ where k is a positive integer and a is a positive real number. The constant k is called the *shape parameter*, while a (or various equivalents) is called the *scale parameter*. The Erlang distribution is equivalent to a gamma density with integral shape parameter. See **Gamma distribution.**

ERLANG LOSS MODEL

The multiple-server queueing system M/M/c/c with Poisson arrivals, exponential service times, c servers, but no additional space for holding waiting customers. See **Erlang B formula; Queueing theory.**

ERROR ANALYSIS

See **Numerical analysis.**

ETA FILE

A sequential file storing the sequence of elementary elimination matrices used to obtain the *LU* decomposition of the basis matrix in the simplex method. Each elementary elimination matrix is represented by its eta vector. See **Revised simplex method.**

ETA MATRIX

See **Elementary elimination matrix.**

ETA VECTOR

The special column of a pivot (elementary elimination) matrix that is different from the corresponding column vector of the identity matrix. A pivot matrix is uniquely specified by its eta vector and its location in the matrix. See **Revised simplex method.**

ETHICS

Joseph H. Engel

Bethesda, Maryland

Ethics in the practice of operations research is the set of moral standards to which a practitioner of OR/MS should adhere in doing his or her work, so that the analysts can do relevant work responsibly and objectively, and be perceived as doing so.

The OR/MS worker must apply the basic principles of scientific methodology in such a way as to be "transparent" in the way the work is reported. A technically qualified but disinterested party should be able to verify that the work has been carried out in a valid manner, based on data that have been gathered and analyzed correctly.

Operations research, as distinguished from the physical sciences in general, deals with interactions between people and the systems they operate. With this in mind, we will discuss the OR/MS analyst's ethical requirements operationally, in terms of beginning, conducting and reporting a study (as covered in Caywood *et al.*, 1971, pp. 1129–1130).

IN BEGINNING A STUDY: The OR/MS analyst should discuss thoroughly with the client the nature of the problem to be solved, and should become familiar with the system, so that he or she and the client can reach agreement on the client's objectives in operating the system to be studied, measures of effectiveness in achieving the system objectives, and the boundaries of the system. Both parties need to "agree on what *will* and *will not* be done" (Caywood *et al.*, 1971).

The careful delineation of the objectives of a system in planning how to begin a study is important in all cases, and particularly where there are multiple objectives. In such cases, all component objectives and measures of performance must be carefully defined.

Of comparable ethical importance in the formulation of the problem, is the determination of the extent of the system to be studied. This means that analyst and client should agree on which portions of the system can be affected by the operation of the system, and what phenomena affected by the operation of the system are of concern to the client. Then, properly relevant analysis of possible operation of the system can lead to recommendations which should lead to the desired improvement.

It also is important for the OR/MS analyst to understand the general nature of all of the effects that the system can have on the total environment, regardless of whether or not the system operator is directly interested in some of these effects. Uncovering unexpected effects may make it possible for the system operator to acquire a better understanding of the overall relationship of the system to its surrounding. This, perhaps, may lead to more useful results than might have otherwise been possible.

IN CONDUCTING A STUDY: Having selected measures of performance and having defined the system boundaries, the OR/MS analyst must plan for data collection to ensure its maximum accuracy and relevance to the problem at hand, without interfering unduly with what the operating personnel are doing.

As in other people-oriented sciences, the OR/MS scientist often cannot conduct controlled experiments (because they inflict an undue burden of cost or damage on the operating personnel). He or she may have to be content with observing a series of "operational" trials under what are hoped to be relevant field conditions. But to the maximum extent possible, the number and nature of specific data collecting trials should be stipulated by the OR/MS analyst, with the concurrence of the client, so as to ensure that a statistically valid amount of data covering all relevant facets of the operation will be collected.

This depends to a great degree on the nature of the mathematical model being used to describe the system being studied, and how its performance is affected by factors under the control of the system operator as well as by uncontrollable environmental factors. The time, personnel and equipment available and costs of conducting trials must also be taken into account, in planning the quantity and extent of data collection.

The analyst should assure that qualified operators are trained in using appropriate data recording equipment, and that they do record the data from the proper number of trials under the desired range of conditions that are to be recorded, or that they tell the analyst how many trials were really conducted, and under what conditions.

Wherever possible, the OR/MS analyst should observe the data collecting trials directly, so that he or she can determine whether they are conducted under true field conditions, and to become aware of possible sources of error or inaccuracy in the data collection.

The OR/MS analyst should not operate equipment being evaluated during data collection trials, *because he or she is not part of the operating system*, and his or her participation can bias the results of the trials unexpectedly. This need not preclude the analyst from operating or observing the system on other convenient occasions, to help build a good understanding of how the system is supposed to operate. It is generally preferable for the client's operating personnel rather than the OR/MS analyst to collect the data. This is most desirable if the collection of such data is or ought to be part of the normal operating process (because such procedures are often valuable to the operators for training and self evaluation purposes).

Once the data are collected, the analyst must study it together with the mathematical model being used to describe system performance. The analysts must not arbitrarily omit data or add new or non-existent data in order to develop results that are more to the liking of the analyst (or the client), and should make proper statistical, mathematical and logical use of the data to derive valid conclusions. Rather, he or she should try to reach an understanding of the nature of the conclusions, and why they agree (or fail to agree) with any prior opinions that may exist concerning the system being studied. The analyst should conduct sensitivity analyses of the effects of variations in key parameters or assumptions, and should deal with possible limitations on the accuracy of observed data values, and the effect of these limitations on the conclusions.

In studying a single objective system, the OR/MS analyst uses a mathematical model appropriately to determine what combination of control variables will yield maximum performance effectiveness. Multi-objective systems are trickier.

In the comparatively simple case of a single type of cost versus a single measure of performance, the OR/MS analyst can treat the problem by first discovering how to maximize performance for a given cost, or, alternatively, by discovering the least expensive way to achieve a specified level of performance.

The decision maker (usually the client rather than the OR/MS worker) must decide on the maximum amount of money he or she wishes to spend, or the minimum level of performance desired. Then the OR/MS worker can solve the problem by recommending how to get the best performance at the desired expenditure level, or, alternatively, by recommending the smallest expenditure to achieve the desired level of performance.

The general solution in multi-objective systems is often difficult because there does not exist a mathematically rigorous (hence rational) method to find how to optimize the operation of any such system. In the special case when it can be shown that *each of the objectives* can be achieved more effectively when the system is operated in one certain way rather than in any other way, that way is a unique and dominating optimum solution to the problem. Similarly, multiple dominating solutions (each of them identically effective to all other dominating solutions with respect to each of the corresponding values of the component measures, and at least as effective in all of their components and more effective than some of the corresponding components of the non-dominated solutions) can be found. But dominating solutions do not always exist.

In such cases, it is necessary to consider possibly conflicting objectives in order to develop balanced procedures that deal with all important factors affecting performance. Such a problem might arise, for example, in optimizing the design of a military aircraft in terms of its range, cruising altitude, speed, payload weight, delivery accuracy, defensive ability, procurement and operating costs.

In general, a multi-objective index is used together with a sensibly designed mathematical model to be used to find out how to select values of control variables to maximize the value of the index. Such an index is usually structured to increase in value in a balanced way whenever any component measure indicates an improvement within an acceptable range (for example, the index might be a positively weighted sum of positive powers of each positive component measure). This is often hard to accomplish. If, for example, a system must be designed to optimize a time stream of expected short-run and long-term future costs and benefits, it is difficult to decide how to discount and balance long-term future costs and benefits against the short term. These long-range planning problems are not always dealt with properly, as witness frequent emphasis on ending each fiscal year in the black without paying enough attention to long-term profitability.

IN REPORTING A STUDY: Having performed the analysis and drawn conclusions, the OR/MS analyst must report the findings and their possible limitations to the client in as complete and understandable a manner as possible.

It is well worth reviewing the mechanics as well as the ethics of reporting a study. All aspects of the analysis, the data collection procedures, the fundamental assumptions and mathematical model used, including the values of any multi-objective index and each of its components, conclusions, recommendations and their limitations, should all be reported and explained to the client. When an analysis is conducted using a multi-objective index, the analyst should explain to the client how the values of the coefficients and exponents used in the index have been chosen, including a thorough discussion of the implications inherent in their selection. *At all costs, the analyst must avoid conducting and then reporting the analysis in such a way as to warp results intentionally so as to validate his or her own or anyone else's prior conclusions.* Further the analyst reports only to his client, and nobody else without prior permission from the client. "Leaks" are unethical (Caywood *et al.*, 1971).

The ethical problems connected with the reporting process revolve around the need to analyze and report relevantly, honestly, completely, clearly, and exclusively, so the client will understand what has been done as well as what has not been done. Failure to do so is an ethical failure, because the analyst will have failed to deliver what was contracted.

OTHER ETHICAL ISSUES: Beyond the ethical requirements of beginning, conducting and reporting an OR study, as discussed above, there are a number of other ethical issues that OR/MS analysts encounter in their research and applied activities. Many of these issues are similar to ones faced by all professionals, for example, data availability and computational reproducibility, peer review procedures, the handling of conflicts of interest. The book edited by Wallace (1994) covers the full range of such ethical issues. The Caywood *et al.* (1971) article offers a valuable discussion of the concept and ethical issues of the OR/MS analyst as an advocate.

See **Implementation; Validation, verification and testing of models; Multi-objective programming; Multi-attribute utility theory; Practice of operations research and management science.**

References

[1] Caywood, T.E., H.M. Berger, J.H. Engel, J.F. Magee, H.J. Miser, and R.M. Thrall (1971), "Guidelines for the Practice of Operations Research," *Operations Research* 19, 1123–1158.

[2] Wallace, W. A., ed. (1994), *Ethics in Operations Research*, Elsevier, New York.

EULER TOUR

In an undirected connected graph, an Euler tour is a cycle that starts at some node, visits each arc exactly once, and returns to the starting node. See **Chinese postman problem; Combinatorics; Combinatorial and integer optimization; Graph theory.**

EVALUATION

See **Model evaluation.**

EVENT-DRIVEN SIMULATION

A computer model in which each possible event is contained in a (logical) module of code. Each module is executed when and only when other code determines that event should occur. Generally, event-driven simulations have stochastic (random) decisions which determine whether and when (in model time) an event will occur. See **Discrete-event stochastic simulation.**

EVOP

Evolutionary operation. See **Statistical quality control.**

EX ANTE FORECASTS

Forecasts that are made without any knowledge of the period to be forecast.

EXCLUSIVE-OR NODE

In a network, an event (node) that will be realized if one and only one of the arcs leading to it is realized. See **Network planning.**

EXPECTED UTILITY THEORY

See **Decision analysis; Preference theory; Utility theory.**

EXPERT SYSTEMS

Clyde W. Holsapple

University of Kentucky, Lexington

and

Andrew B. Whinston

University of Texas at Austin

INTRODUCTION: Devising computer-based systems that can solve problems by reasoning about facts and assertions has been a central, ongoing quest in the artificial intelligence field. By the early 1970s, research into these reasoning systems had begun to focus on systems that could solve difficult problems in narrow problem domains such as diagnosing diseases, assessing chemical structures of unknown molecules, determining ore deposits in geological sites, and solving applied mathematical problems. These systems have come to be known as expert systems, because they solve problems that would otherwise require services of experts in their respective problem areas. Perhaps the best known of these early expert systems is MYCIN, whose approach to diagnosing blood infections is documented in Buchanan and Shortliffe (1984). Descriptions of other pioneering expert systems such as DENDRAL, MACSYMA, and PROSPECTOR can be found in Barr and Feigenbaum (1982).

By the early 1980s, the focus in expert system (ES) research had shifted from demonstrating the feasibility and efficacy of such systems to the identification of tools and methods that could facilitate their development. Each of the pioneering expert systems was custom-built, requiring considerable expense and years of development by specialists in artificial intelligence. If expert systems were to come into widespread use, it was clear that faster and less costly means for creating them had to be found. This search has been largely successful, spawning a host of commercially available, computer-based tools for ES development and leading to the creation of specific methods for guiding the process of ES development. These tools and methods have been instrumental in the growing number of expert systems used in such application areas as engineering, manufacturing, finance, and business administration (Blanning, 1984; Mockler, 1989).

Two prerequisites for assessing ES possibilities are an understanding of the nature of expert systems and an appreciation of how they can be developed. The general nature of expert systems is described first, including characterizations of ES functions, architecture, and operation. Then, ES development is examined in terms of methodological issues and classes of available tools.

GENERAL NATURE OF AN EXPERT SYSTEM: An expert system functions as a readily available substitute for some source of expertise that cannot always be consulted in a facile, timely, affordable manner. For instance, consider the case of a person who is an expert in some problem domain, such as financial planning. This human source of expertise about financial planning is able to accept requests for advice about specific problems in the domain, reason with the expertise to produce recommendations, communicate the resultant advice, and explain the rationale underlying that advice. A computer-based system that can perform these same functions, giving recommendations and explanations comparable to those of the expert, is called an expert system. Not all ESs substitute for an individual human expert. An ES can also be a surrogate for a group of experts, multiple individual experts, the expertise embodied in a set of historical data, or expertise revealed in the behavior of some non-human system.

Expert systems offer many potential advantages over relying on the original source of expertise for advice (Holsapple and Whinston 1986). Unlike a human expert, an ES does not sleep, become ill, take vacations, have a bad day, forget, require compensation, become tied up with more important matters, or retire. From an organization's viewpoint, the exercise of building an ES results in a formalization and preservation of expertise. It yields an advice giver that can be readily replicated for simultaneous use at geographically diverse sites, ensuring consistency of recommendations. Holsapple and Whinston (1990) have argued that ESs can be instrumental in implementing competitive strategies.

The functioning of an ES is based on three major components: a user interface, an inference engine, and a body of stored knowledge that forms the basis for reasoning about problems in some domain of interest. The user interface is that part of an ES that its user (e.g., a person seeking advice) directly experiences. It accepts a user's characterization of a specific problem. It asks for clarifications of that characterization, as needed. It presents the ES's advice to the user about treating the problem. The user interface also accepts user requests for justifying advice and presents those justifications to a user. Expert systems can vary widely in terms of the style and sophistication of their user interfaces, even when the other two architectural components are fixed.

A second component of ES architecture is the knowledge store it possesses. In ES parlance, this is often called a *knowledge base*. It typically holds two distinct kinds of knowledge: descriptive and reasoning. Descriptive knowledge is concerned with describing states of the world (e.g., that revenue was $10 million last year or is expected to be $15 million next year). Reasoning knowledge is concerned with specifying what conclusion is valid when a particular situation is known to exist (e.g., that an unemployment rate of over 8% warrants a certain reduction in revenue expectations). Additional types of knowledge can be found in the knowledge bases of some ESs.

There is more than one way to represent each type of knowledge stored in an ES. Descriptive knowledge may be simply represented as values of state variables, often called attribute-value pairs (e.g., the revenue attribute or variable has a value of $10 million). Or, such pieces of descriptive knowledge may be structured into data base records, frames, semantic nets, arrays, spreadsheet cells, and other computer-based organizations. Similarly, pieces of reasoning knowledge are subject to multiple representation modes. One commonly used approach involves the use of rules. Each rule has a premise, characterizing some situation, and a conclusion, indicating what actions can be taken (e.g., what changes can be made to state variable values) if the situation is determined to exist. A variety of rule representation languages exist. They differ in terms of style, flexibility, and power of representation. Some such differences are surveyed in Mockler (1989).

INFERENCE ENGINE: At the heart of general ES architecture is an *inference engine*. This is a software component that reasons with the stored knowledge of an ES to derive advice corresponding to a user's problem statement. It also tracks the flow of reasoning about the problem, as a basis for justifications presented via the user interface. Clearly, an ES's inference engine must be compatible with a) the particular representation language used to specify stored descriptive and reasoning knowledge, b) the user interface's interpretations of user requests, and c) the user interface's ability to package inference engine results. From one ES to another, inference engines vary not only to ensure compatibility with different user interface and knowledge representation conventions, but also with respect to how they reason.

Two prominent kinds of reasoning approaches are forward chaining and backward chaining. In either case, the inference engine uses rules to establish values for variables whose states are unknown at the outset of a consultation. These values constitute the raw form of the advice that is ultimately packaged for presentation to a user. The main difference between the two kinds of reasoning is the progression of processing whereby unknown variables become known. In either case, if an ES holds insufficient knowledge to solve a problem, the inference engine will fail in its attempt to establish values for unknown variables.

In the forward chaining case, an inference engine examines the premise of each rule. If the premise is true, then the actions specified in the rule's conclusion are performed (i.e., the rule is "fired"). Thus, firing a rule can result in changes to values of variables, including the assignment of values to previously unknown variables. After every rule has been examined in this way, the inference engine makes a second pass through the rules. If additional rules are fired in this pass, the process continues with a third pass, and so forth. Processing stops when no further rules are fired in a pass or when some other terminal condition is satisfied (e.g., a value has been established for some designated unknown variable).

In contrast, backward chaining is a more goal-directed approach to reasoning. It considers a rule's conclusion before trying to evaluate the premise. Establishing a value for a specific unknown variable is the inference engine's overall goal. In its effort to meet that goal, the inference engine identifies the subset of rules whose conclusions could affect the goal variable's value. These are called candidate rules. In considering a candidate rule, the inference engine attempts to evaluate the premise. If this evaluation is impossible because the premise involves unknown variables, then each of those variables successively becomes the new current goal. The inference engine performs backward chaining for the current goal variable, identifying its candidate rules and attempting to evaluate their premises. When a rule's premise is found to be true, the rule is fired. When a premise is found to be false, the inference engine proceeds to another candidate rule. This basic processing pattern continues recursively until a value is established for the consultation's overall goal variable or until that variable's candidate rules are exhausted (possibly without reaching a solution).

There are many variations to each of these two reasoning approaches, affecting both the speed and the results of inference engine operation. One kind of variation involves the degree of reasoning rigor. These are variations in how exhaustive inference engines are in making passes through a rule set or in considering candidate rules. Another variation concerns rule selection order. That is, in what sequence does an inference engine process rules within a pass or within a candidate rule subset? There is also considerable room for variation in the strategies inference engines use for evaluating a premise (e.g., the order for considering conditions in a compound premise). Inference engines can vary greatly in their treatments of uncertainties about variable values and rule efficacy. Some ignore the possibility of uncertainties, while others use specific algebras to combine certainty factors in an effort to qualify the resultant advice. Holsapple and Whinston (1986) gave an extensive discussion of such variations.

EXPERT SYSTEM DEVELOPMENT: Tools for building expert systems fall into three major categories: programming languages, shells, and integrated environments. In using the former, an ES developer, often called a *knowledge engineer*, designs and programs the inference engine and the user interface. Also, storage structures for holding reasoning and descriptive knowledge must be designed so their contents are accessible to the inference engine. Appropriate knowledge must then be stored in such structures. A shell removes much of this work from an ES developer, but also reduces the developer's flexibility. With a shell, the developer has a ready-made inference engine, user interface, and knowledge storage structure. Thus, the main development task consists of putting the appropriate knowledge into that structure. Shells also commonly give developers some facilities for customizing the user interface inputs and outputs. Some give developers a modicum of control over the inference engine's reasoning behavior (e.g., over the degree of rigor, selection order, treatment of uncertainty). Shell inference engines can often interface to other pieces of software (e.g., use spreadsheet data).

An integrated environment for ES development has all the facilities of a shell, plus other computing capabilities normally found in separate software tools. That is, the inference engine is enhanced to accomplish kinds of processing other than reasoning: data base management, spreadsheet processing, model management, forms handling, graph generation, and so forth. Such capabilities can be exercised in the course of rule processing. Conversely, consultation can occur in the context of one of these other kinds of processing. Such tools allow the creation of ESs that more closely approximate human experts, who are not limited to reasoning about a problem, but can also do data retrieval, extensive calculations, fancy presentations, and so on. A checklist for tool selection, plus an in-depth examination of an integrated environment, can be found in Holsapple and Whinston (1986).

Mockler (1989) provides a comparative feature survey of representative commercial expert systems.

Regardless of the tool used, a developer faces the task of managing the ES development project. The books of Buchanan and Shortliffe (1984) and Hayes-Roth, Lenat, and Waterman (1983) contain valuable insights into the methodology of ES development. The ES development cycle introduced by Holsapple and Whinston (1986) is typical of several that appear in the literature. One aspect of development that has received considerable attention is the phenomenon of knowledge acquisition (KA). This is concerned with the activity of eliciting reasoning knowledge from a source (e.g., human expert), perhaps structuring/analyzing it, and representing it in a form that can be directly stored in the knowledge base of an ES.

A representative overview of KA issues, methods, and tools is provided by Kidd (1987). The methods include such techniques as structured interviewing and protocol analysis. The KA tools are primarily induction mechanisms that attempt to acquire general domain knowledge from sets of specific examples of expert behavior. Dhaliwahl and Benbasat (1990) have introduced a variable-oriented framework for guiding empirical research that evaluates performance of such methods and tools. Holsapple, Raj, and Wagner (1992) provide a summary of recent theoretical and empirical KA developments.

FURTHER READING: Aside from the growing number of books dealing with expert systems, there are several scholarly journals devoted to ES topics. These include *Expert Systems; Expert Systems with Applications; Heuristics: The Journal of Knowledge Engineering; IEEE Expert; Intelligent Systems in Accounting, Finance, and Management;* and *Knowledge Acquisition.* ES topics are also covered in more general journals of artificial intelligence, as well as journals concerned with computer systems for business or engineering.

See **Artificial intelligence; Decision support systems.**

References

[1] A. Barr and E.A. Feigenbaum, eds. (1982). *The Handbook of Artificial Intelligence*, William Kaufmann, Los Altos, California.

[2] R.W. Blanning (1984). "Management Applications of Expert Systems," *Information and Management*, 6, 311–316.

[3] B.G. Buchanan and E.H. Shortliffe (1984). *Rule-Based Expert Systems: The MYCIN Experiments of the Stanford Heuristic Programming Project*, Addison-Wesley, Reading, Massachusetts.

[4] J.S. Dhaliwal and I. Benbasat (1990). "A Framework for the Comparative Evaluation of Knowledge Acquisition Tools and Techniques," *Knowledge Acquisition*, 2, 145–166.

[5] F. Hayes-Roth, D.B. Lenat and D.A. Waterman, eds. (1983). *Building Expert Systems*, Addison-Wesley, Reading, Massachusetts.

[6] C.W. Holsapple and A.B. Whinston (1986). *Manager's Guide to Expert Systems*, Dow Jones-Irwin, Homewood, Illinois.

[7] C.W. Holsapple and A.B. Whinston (1990). "Business Expert Systems – Gaining a Competitive Edge," *Proceedings of Hawaiian International Conference on Systems Sciences*, Kona, Hawaii, January.

[8] A. Kidd, ed. (1987). *Knowledge Elicitation for Expert Systems: A Practical Handbook*, Plenum Press, New York.

[9] R.J. Mockler (1989). *Knowledge-Based Systems for Management Decisions*, Prentice-Hall, Englewood Cliffs, New Jersey.

EXPLORATORY MODELING

Steve Bankes

RAND, Santa Monica, California

INTRODUCTION: Exploratory modeling is a research methodology that uses computational experiments to analyze complex and uncertain systems (Bankes, 1993). Exploratory modeling can be understood as search or sampling over an ensemble of models that are plausible given a priori knowledge or are otherwise of interest. This ensemble may often be large or infinite in size. Consequently, the central challenge of exploratory modeling is the design of search or sampling strategies that support valid conclusions or reliable insights based on a limited number of computational experiments.

Exploratory modeling can be contrasted with the use of models to predict system behavior where models are built by consolidating known facts into a single package. When experimentally validated, this single model can be used for analysis as a surrogate for the actual system. Examples of this approach include the engineering models that are used in computer aided design systems. Where successful, this "consolidative" methodology is a powerful technique for understanding the behavior of complex systems. Unfortunately, for many systems of interest, the construction of models that may be validly used

as surrogates is simply not a possibility. This may be due to a variety of factors including the unfeasibility of critical experiments, immaturity of theory, or the non-linearity of system behavior, but is fundamentally a matter of not knowing enough to make predictions. For such systems, a methodology based on consolidating all known information into a single model and using it to make best estimate predictions can be highly misleading.

When insufficient knowledge or unresolvable uncertainties preclude building a surrogate for the target system, modelers must make guesses at details and mechanisms. While the resulting model cannot be taken as a reliable image of the target system, it does provide a "computational experiment" that reveals how the world would behave if the various guesses were correct. Exploratory modeling is the use of series of such computational experiments to explore the implications of varying assumptions and hypotheses.

The focus of exploratory modeling cannot be a single model of interest, but rather must consider an ensemble of models, all of which are plausible or interesting in the context of the research or analysis being conducted. This ensemble is generated by the uncertainties associated with the problem of interest, and is constrained by available data and knowledge. Selecting a particular model out of the ensemble of plausible ones requires making suppositions about factors that are uncertain or unknown. One such computational experiment is typically not that informative (beyond suggesting the plausibility of its outcomes). Instead, exploratory modeling methodology must support reasoning about general conclusions through the examination of the results of numerous such experiments. Thus, exploratory modeling can be understood as search or sampling over the ensemble of models that are plausible given a priori knowledge.

CENTRAL PROBLEM: The problem of how to cleverly select the finite sample of models and cases to examine from the large or infinite set of possibilities is the central problem of exploratory modeling methodology. A wide range of research strategies are possible, including structured case generation by Monte Carlo or factorial experimental design methods, search for extremal points of cost functions, sampling methods that search for regions of "model space" with qualitatively different behavior, or combining human insight and reasoning with formal sampling mechanisms. Computational experiments can be used to examine ranges of possible outcomes, to suggest hypotheses to explain puzzling data, to discover significant phases, classes, or thresholds among the ensemble of plausible models, or to support reasoning based upon an analysis of risks, opportunities, or scenarios. Exploration can be over both real valued parameters and non-parametric uncertainty such as that between different graph structures, functions, or problem formulations.

APPLICATION TYPES: There are three general types of applications where exploratory modeling could be used, which can be labeled as data driven, question driven, and model driven. Data driven exploration starts with a dataset, and attempts to derive insight from it by searching over an ensemble of models to find those that are consistent with the data. Question driven exploration begins with a question we wish to answer (e.g. what policy should the government pursue regarding global warming) and addresses this question by searching over an ensemble of models and cases believed to be plausible in order to inform the answer. Model driven exploration involves neither a fixed data set, nor a particular question or policy choice, but rather is a theoretical investigation into the properties of a class of models, and is consequently a branch of experimental mathematics. Examples of data driven and model driven exploration can be found in all the sciences. Question driven exploration, on the other hand, has particular salience for policy analysis.

In making policy decisions about complex and uncertain problems, exploratory modeling can provide new knowledge even where validated models cannot be constructed. One example is the use of models as existence proofs or hypothesis generators. Demonstrating a single plausible model/case with counterintuitive properties can beneficially change the nature of a policy discussion. For another example, consider situations where risk aversion is prudent. Here an exploration that develops an assortment of plausible worst case failure modes can be very useful for designing hedging strategies. This is true even if models are not validated, and sensitivities are unknown. Other examples of useful research strategies include the search for special cases where small investments could (plausibly) produce large dividends, or extremal cases (either best or worst) where the uncertainties are all one sided, and *a fortiori* arguments can be used. All these examples depend on the fact that partial information can inform policy even when prediction and optimization are not possible. The space of models and associated computational experiments can be searched for examples with characteristics that are useful in choosing between alternative policies.

The search for information of use in answering policy questions can often be served by the discovery of thresholds, boundaries, or envelopes in a space of models that decompose the space into subensembles with different properties. For example, exploratory modeling could seek to discover which models have stable or chaotic dynamics. Or search could have the goal of discovering what combination of model properties favor either of two alternative policies.

Aggressive exploitation of exploratory modeling for complex models requires significant computational resources. Consequently, this approach has only recently had widespread use. As computer power continues to grow, this approach can be expected to become increasingly important.

Most existing technology to support modeling has evolved from the consolidative modeling tradition. The construction and use of single models is much better supported than is exploration through computational experiments involving multiple models. Facilitating the exploratory modeling process requires an exploratory modeling environment that allows users to efficiently navigate through the space of plausible models and model outcomes and to construct lines of reasoning and to learn about the implications of both knowledge and hypothesis.

See **Practice of Operations Research and Management Science; Public policy analysis; Validation.**

References

[1] Bankes, S. (1993), "Exploratory Modeling for Policy Analysis," *Operations Research*, 41, 435–449.

[2] Campbell, D., J. Crutchfield, D. Farmer and E. Jen (1985), "Experimental Mathematics: The Role of Computation in Nonlinear Science," *Communications of the ACM*, 28, 374–384.

[3] Hodges, J. (1991), "Six (or so) Things You Can Do With a Bad Model," *Operations Research*, 39, 355–365.

[4] E. Leamer (1978), *Specification Searches: Ad Hoc Inference with Nonexperimental Data*, John Wiley, New York.

[5] Miser, H.J. and E.S. Quade eds. (1985), *Handbook of Systems Analysis: Overview of Uses, Procedures, Applications, and Practice*, North-Holland, New York.

[6] Miser, H.J. and E.S. Quade, eds. (1988), *Handbook of Systems Analysis: Craft Issues and Procedural Choices*, North-Holland, New York.

[7] Quade, E.S. (1980), "Pitfalls in Formulation and Modeling," in Q. Majone and E.S. Quade (eds.), *Pitfalls of Analysis*, John Wiley, Chichester, England.

[8] Quade, E.S. (1985), "Predicting the Consequences: Models and Modeling," in H.J. Miser and E.S. Quade (eds.), *Handbook of Systems Analysis: Overview of Uses, Procedures, Applications, and Practice*, North-Holland, New York.

EXPONENTIAL ARRIVALS

When customer interarrival times to a queueing system are defined by a sequence of independent and identically distributed exponential random variables. If a system has exponential interarrival times with distribution function $A(t) = 1 - \exp(-\lambda t)$, then the number of arrivals to the system in any period of time t has the Poisson distribution with probability function $p_n(t) = \exp(-\lambda t)(\lambda t)^n/n!$. See **Poisson arrivals**; **Queueing theory**.

EXPONENTIAL-BOUNDED (-TIME) ALGORITHM

An algorithm for which it can be shown that the number of steps required to find a solution to a problem is an exponential function of the problem's data. The simplex algorithm is an exponential-bounded algorithm, although its use in practice belies that designation. See **Polynomially-bounded algorithm**; **Simplex method**.

EXPONENTIAL SMOOTHING

Robert G. Brown

Materials Management Systems, Vermont

Exponential smoothing is a technique for revising an estimate of the average of a time series to extrapolate as a forecast. It was first formalized by R.G. Brown (about 1944) with continuous variables in the analysis of a ball-disc integrator used in a naval fire control device. It was later applied, also by R.G. Brown (1959), with discrete observations in the early 1950s.

Exponential smoothing with discrete (usually monthly) observations of demand had considerable appeal for inventory control because it was possible to revise the forecasts for all products in an inventory in less than 30 days with unit record (punched card) equipment. The formula for revising the estimate of the average is

$$new\ forecast = old\ forecast + \alpha \times (error)$$

$$= old\ forecast$$

$$+ \alpha \times (latest\ observation$$

$$- old\ forecast)$$

$$= (1 - \alpha) \times old\ forecast$$

$$+ \alpha \times (latest\ observation).$$

The smoothing constant α was originally set to 0.1, not from any theoretical considerations, but because one can "multiply" on unit-record equipment merely by moving the wires on the plug board one place to the left and adding. But, clearly, better approaches for estimating a "good" value to use for would come over time.

Later, Weiner (1949) showed that an estimate of the average of a time series gave the minimum squared error when one uses some optimum set of weights. Box and Jenkins (1970) demonstrated a rigorous procedure for finding that optimum set of weights. The optimum weights often decline more or less geometrically with age. Indeed, for short series the wholly empirical technique of exponential smoothing was usually indistinguishable from the optimum weights.

In the late 1950s, Brown (1963) extended the model to include a secular trend. The first model used single smoothing and double smoothing. Brown (1967) showed that this was equivalent to smoothing the values of the coefficients in a polynomial of any degree. This led to the generalization to include complex polynomials that could be interpreted as Fourier series to approximate repeatable seasonal variation.

Winters (1962) developed an elaborate simulation to find the "best" value of three smoothing constants, for level, trend and seasonal profile respectively. Simulations with different models and different smoothing constants are usually misleading because the sampling error with short series is larger than the effect being sought. Since exponential smoothing "learns," several implementers let the coefficients in the model start from arbitrary values. There is a problem in that the rate of learning takes much longer than the normal span of patience of people who need a good forecast now.

The initial values of the coefficients should be estimated by regression on available history. (For new products there will be other products in the inventory that serve the same market which can be used as analogs.) Since the Fourier series is an orthogonal basis, one can fit all the terms up to the Nyquist frequency (at least two observations per cycle for the highest frequency) and reject harmonics that are not significant under a chi-square test with two degrees of freedom.

Decisions based on a forecast need information about the probability distribution of forecast errors. The form of the distribution may be Gaussian, but there are not infrequent cases where the errors are bounded below, with a long upper tail. Therefore it is advisable to check the distribution form that is appropriate to the actual data. Usually one parameter, the variance, is sufficient to develop the model of probabilities.

Brown (1959) proposed the use of the Mean Absolute Deviation (MAD) as a measure of dispersion. On unit-record equipment it is simple to measure the absolute deviation – leave out the wire that carries sign. If the form of the distribution is exactly normal the standard deviation is 1.25 times the MAD. However in actual data the ratio has been observed to be anywhere from 1 to 1.7. Thus it is prudent to measure the Mean Squared Error (MSE). The MSE can be revised with each new observation by applying exponential smoothing to the square of the error in the most recent forecast.

A theoretical model of a time series often is seriously different from actual data. Sales are distorted by promotions, federal regulations, competition, weather and errors in recording. During the process of revising the forecast it is advisable to produce exception reports. The demand filter reports data that are more than K standard deviations from the most recent forecast. The tracking signal reports significant bias in the forecast.

The head of forecasting should take these exception reports seriously. First find the assignable cause for the exception, and then take appropriate action. Don't wait until the exception is reported to start thinking about the assignable cause. Be aware of events in the operating environment that could cause exceptions and use the reports to confirm or deny hypotheses about the problems actually occurring. Look for patterns, where several series seem to show the same anomalies.

Several techniques have been proposed for the tracking signal. Brown (1959) originally used the cumulative sum of forecast errors. However the expected value of that sum in the future is the current sum. A large error can bias the signal so that a very small error later could cause an exception.

The next technique was the smoothed error tracking signal (SETS) which applied exponential smoothing to the error with sign. This technique is slow to react to a real change in the underlying process which generates the data.

Trigg and Leach (1966) proposed using the SETS to modify the smoothing constant(s) – make the forecasts more responsive when they are wrong and more stable when they are close to the data. They failed to show that the feedback system is critically damped in all regions where it may be applied.

Gardner (1985) has done a study of the comparative effectiveness of a variety of tracking signal techniques. Barnard (1959) proposed the V-mask based on Wald's sequential analysis (Wald, 1947) for quality control. Brown (1971) used a parabolic mask as the envelope of these V-masks with a range of likelihood ratios. The technique has been extended to monitor the MSE as well as the forecast, based on analogy with Shewhart's (1931) x-bar and R charts for statistical quality control.

During the course of fitting the initial model outliers are the analogs of demand filter exceptions. Brown (1990) used the term "significant event" to refer to history where there is evidence that not all the history came from a process that can be described by the same model. Significant serial correlation with a lag of one observation may be caused by such a significant event (there are other causes). Brown (1990) has evolved a method with cumulative sums for estimating the time when that event occurred, so that the model can be fitted only to observations since that time.

The whole idea of forecasting from a description of history is on the way out. By the end of this century it will be common to pass point-of-sale data quickly and accurately to each operation in the logistics chain, rather than to forecast what one enterprise will order from another enterprise.

See **Forecasting; Marketing; Regression analysis; Retailing; Time series analysis.**

References

[1] Barnard, G. (1959). "Control Charts and Stochastic Processes," *Jl. Royal Statist. Soc., Ser. B*, **21**, 239–271.

[2] Box, G. and Jenkins, G. (1970). *Time Series Analysis*, Holden-Day, San Francisco.

[3] Brown, R.G. (1959). *Statistical Forecasting for Inventory Control*, McGraw Hill, New York.

[4] Brown, R.G. (1963). *Smoothing Forecasting and Predicton*, Prentice Hall, Englewood Cliffs, New Jersey.

[5] Brown, R.G. (1967). *Decision Rules for Inventory Management*, Holt, Rinehart & Winston, New York.

[6] Brown, R.G. (1971). "Detection of Turning Points," *Decision Science*, **2**, 383–403.

[7] Brown, R.G. (1990). "Significant Events," *Proc. ISF*, New York.

[8] Gardner, E. (1985). "Exponential Smoothing: The State of the Art," *Jl. Forecasting*, **4**, 1–28.

[9] Shewhart, W.A. (1931). *Economic Control of Quality*, Van Nostrand, New York.

[10] Trigg, D.W. (1966). "Monitoring a Forecasting System," *Jl. Operational Research Society*, **15**, 211–274.

[11] Wald, A. (1947). *Sequential Analysis*, Wiley, New York.

[12] Weiner, N. (1949). *Extrapolation, Interpolation and Smoothing of Stationary Time Series*, Wiley, New York.

[13] Winters, P.R. (1962). "Constrained Rules for Production Smoothing," *Management Science*, **8**, 470–481.

EXTREMAL

Maximum or minimum.

EXTREMAL COLUMN

In the Dantzig-Wolfe decomposition algorithm, the extremal columns are the columns of the extremal (master) problem. See **Dantzig-Wolfe decomposition algorithm**.

EXTREMAL PROBLEM

In the Dantzig-Wolfe decomposition algorithm, the extremal problem is the original linear-programming problem expressed in terms of its extreme point solutions. See **Dantzig-Wolfe decomposition algorithm**.

EXTREME DIRECTION

A direction of a set that cannot be expressed as a nonnegative combination of two other directions of the set.

EXTREME POINT

A point in a convex set that cannot be expressed as a convex combination of two other distinct points in the set. Extreme points are also known as corner points or vertices. The extreme points of a rectangle are its four vertices, while the extreme points of a circular disc are the points on its circumference. For a linear-programming problem, the extreme points of its convex set of solutions correspond to basic feasible solutions, and it can be shown that, if the problem has a finite optimal solution, then one of the extreme points is optimal.

EXTREME POINT SOLUTION

A solution to a linear-programming problem that is an extreme-point of its convex set of solutions. Such solutions correspond to basic feasible solutions.

EXTREME RAY

A ray in a convex set whose direction is an extreme direction.

FACE VALIDITY

See **Verification, validation, and testing of models.**

FACILITIES LAYOUT

Bharat K. Kaku

University of Maryland, College Park

In both manufacturing and service operations, the relative location of facilities is a critical decision affecting costs and efficiency of operations. The facility layout problem (FLP) deals with the design of layouts wherein a given number of discrete entities are to be located in a given space. The definitions of entities and spaces can vary considerably, making solution techniques applicable in a wide variety of settings, as can be seen from the examples given below.

Entities	Space	Objective
Departments	Office building	Minimize cost of interactions
Departments	Factory floor	Minimize cost of material handling
Departments	Hospital	Minimize movement of patients and medical staff
Inter-dependent plants	Geographical market	Maximize profit
Indicators and controls	Control panel	Minimize eye/hand movement
Components	Electronic boards	Minimize cost of connections
Keys	Typewriter keyboard	Minimize typing time

We first discuss approaches used to model the FLP, followed by optimal algorithms and heuristic approaches to solving these problems, and end with some remarks concerning directions for future research.

THE QUADRATIC ASSIGNMENT FORMULATION: The FLP is most often treated in the OR/MS literature as the Quadratic Assignment Problem (QAP), which is a special case requiring identical area and shape requirements for the locations of all facilities. This allows pre-definition of the locations and calculation of the distances between them (typically center-to-center, either rectilinear or Euclidean). Suppose there are N facilities to be assigned to N locations.

Define four $N \times N$ matrices whose elements are, respectively:

c_{ij} = fixed cost of assigning facility i to location j
f_{ij} = level of interaction between facilities i and j
d_{ij} = cost of one unit of interaction (e.g., the distance) between locations i and j
x_{ij} = 1 if facility i is assigned to location j, and 0 otherwise.

Then the QAP is to:

$$\text{Min} \sum_{ij} c_{ij} x_{ij} + \sum_{i,p} \sum_{j,p} f_{ip} d_{jp} x_{ij} x_{pq} \quad (1)$$

subject to

$$\sum_{j} x_{ij} = 1 \quad \forall i \quad (2)$$

$$\sum_{i} x_{ij} = 1 \quad \forall j$$

$$x_{ij} \in \{0,1\} \quad \forall i,j \quad (3)$$

Alternatively, we can define $\rho(i)$ as the location to which facility i is assigned, leading to an equivalent but more compact statement of the problem. The QAP is then to find a mapping of the set of facilities into the set of locations so as to:

$$\text{Min} \sum_{i} c_{i,\rho(i)} + \sum_{i,p} d_{\rho(i),\rho(p)}. \quad (4)$$

The quadratic assignment problem was first formulated by Koopmans and Beckmann (1957) in the context of the location of interdependent plants. The c_{ij} elements represent the expected revenue of operating plant i at location j independent of other plant locations, the f_{ij} elements represent the required commodity flows from plant i to plant j, and the d_{ij} elements represent the transportation costs per unit between location i and location j. The objective function maximizes the net revenue, that is, the excess of expected revenue over the transportation costs.

It is the interdependence of facilities due to interactions between them that leads to the quadratic term in the objective function and makes the problem a difficult one. If the departments are independent of each other (i.e., all $f_{ij} = 0$), the QAP reduces to the familiar linear assignment problem, for which efficient solution techniques exist. Further, the traveling salesman problem is a special case of the QAP. To see this, consider the interaction matrix to be a cyclic permutation matrix with the following interpretation: A flow of one unit (the salesman) travels from the first city in the tour to the second city in the tour to the third city, and so on, finally returning to the first city in the tour. The distance matrix is simply the matrix of distances between cities, and the fixed costs are zero. A solution to this QAP can be interpreted as follows. If $x_{ij} = 1$, then city i is in the jth location in the tour. This shows that the QAP belongs to the class of NP-hard problems.

THE ADJACENCY REQUIREMENTS FORMULATION:

This approach is based on adjacency requirements and closeness ratings. The former stipulate the set of pairs of facilities that *must* be adjacent, or *must not* be adjacent, in any feasible solution, whereas the latter are measures of the *desirability* of locating a pair of facilities in adjacent locations, generally based on the interaction between them. The adjacency requirements must permit at least one feasible solution. If there is more than one feasible solution, then the closeness ratings are used to choose the optimal solution. In evaluating a solution, the closeness ratings are added only for facility pairs that are adjacent.

The QAP has received most of the attention in the literature for the following two reasons. First, it considers interaction costs for all pairs of facilities, whereas the adjacency requirements formulation maximizes the sum of closeness ratings for adjacent facilities only, while satisfying the adjacency requirements. Second, the adjacency requirements formulation does not consider fixed costs which can be important, especially when a layout is being re-designed, which is more common than design of a brand-new facility. In the rest of this paper we restrict discussion to the QAP. Further information on the adjacency requirements approach can be found in Foulds (1983) and a more extensive list of references for the QAP can be found in the review by Kusiak and Heragu (1987).

OPTIMAL ALGORITHMS FOR THE QAP:

Algorithms for obtaining exact solutions to the QAP can be classified under the categories of linearization and implicit enumeration.

Linearization – Several linearizations have been proposed for the QAP. The first one was by Lawler (1963) who linearized the problem by defining variables $y_{ijpq} = x_{ij} x_{pq}$. Compared to the original QAP, the resulting integer programming problem has N^4 additional binary variables and $N^4 + 1$ additional constraints. A linearization proposed by Kaufman and Broeckx (1978) is the most compact, adding only N^2 new continuous variables and N^2 new constraints. Bazaraa and Sherali (1980) suggested another linearization to which they applied Bender's Decomposition. None of these approaches has proved to be computationally effective. More details can be found in the survey paper by Kusiak and Heragu (1987).

Implicit Enumeration – Branch-and-bound algorithms have been the most successful in solving the QAP to optimality; problems with as many as 15 or 16 facilities can be solved in reasonable time. Some earlier implicit enumeration methods were pair-assignment algorithms where a node in the branch-and-bound tree corresponds to the assignment of a pair of facilities to a pair of locations (Land, 1963; Gavett and Plyter, 1966). These did not prove to be competitive with single-assignment algorithms which assign one facility to one location at each node.

Gilmore (1962) and Lawler (1963) independently developed a lower bound for use in a single-assignment branch-and-bound procedure. This lower bound forms the basis for the most successful implicit enumeration algorithms published (Bazaraa and Kirca, 1983; Burkard and Derigs, 1980). We now describe how the Gilmore-Lawler lower bound is calculated.

Suppose $\mathscr{F}$ is the set of facilities (possibly empty) that have already been assigned, and $\mathscr{L}$ is the set of locations to which these facilities have been assigned. Using the alternate formulation (4), a lower bound on completions of this partial assignment is given by

$$
\begin{aligned}
\text{Min} \sum_{i \in \mathscr{F}} c_{i,\rho(i)} &+ \sum_{i \in \mathscr{F}} \sum_{p \in \mathscr{F}} f_{ip} \, d_{\rho(i),\rho(p)} \\
&+ \sum_{i \in \mathscr{F}} \sum_{p \notin \mathscr{F}} [f_{ip} \, d_{\rho(i),\rho(p)} + f_{pi} \, d_{\rho(p),\rho(i)}] \\
&+ \sum_{i \notin \mathscr{F}} c_{i,\rho(i)} + \sum_{i \notin \mathscr{F}} \sum_{p \notin \mathscr{F}} f_{ip} \, d_{\rho(i),\rho(p)}.
\end{aligned} \tag{5}
$$

The first two terms in the expression (5) are the known fixed and interaction costs of assignments already made; the third term captures the interaction costs between assigned facilities and those yet to be assigned; and the last two terms represent the fixed and interaction costs of assignments not yet made. A minimum can be calculated for the last three terms as follows. Consider the assignment of any unassigned facility $i \notin \mathscr{F}$ to a free location $j \notin \mathscr{L}$. The incremental cost due to this assignment is

$$\sum_{p \in \mathscr{F}} [f_{ip} d_{j,\rho(p)} + f_{pi} d_{\rho(p),j}] + c_{i,j} + \sum_{p \notin \mathscr{F}} f_{ip} d_{j,\rho(p)}. \quad (6)$$

Now the first two terms of (6) are known, and we need to minimize the third term. Form a vector of flows consisting of the ith row of the flow matrix minus the diagonal element minus the elements corresponding to assigned facilities ($i \in \mathscr{F}$). Arrange the elements of this vector in decreasing order. Form a similar vector of distances consisting of the jth row of the distance matrix minus the diagonal element minus the elements corresponding to filled locations ($j \in \mathscr{L}$), and arrange it in increasing order. The scalar product of these vectors provides the necessary minimum cost. Essentially, the largest interaction with i incurs the lowest per unit cost, the second largest interaction incurs the second lowest cost, and so on. Repeat for all pairs (i, j) such that $i \notin \mathscr{F}$ and $j \notin \mathscr{L}$. A solution to the linear assignment problem (LAP) with these incremental costs as cost coefficients provides a lower bound on the three unknown terms of (5).

Let the value of this solution be z^*. The Gilmore-Lawler lower bound is then obtained as

$$LB = \sum_{i \in \mathscr{F}} c_{i,\rho(i)} + \sum_{i \in \mathscr{F}} \sum_{p \in \mathscr{F}} f_{ip} d_{\rho(i),\rho(p)} + z^*. \quad (7)$$

Any node at which the lower bound is greater-than-or-equal-to the upper bound can be fathomed in the usual way. However, an attractive feature of this lower bound is the fact that additional information is available to be used in the search process. Consider the solution to the LAP solved to obtain the lower bound, with facility $i \notin \mathscr{F}$ assigned to location $\rho(i)$. We can use the dual variables of the optimal solution to reduce the cost matrix so that every $(i, \rho(i))$ element is zero. Adding together the next smallest element in that row and in that column gives the *regret* or minimum additional cost if assignment $(i, \rho(i))$ is not made. The lower bound plus regret gives us the *alternate cost* of this assignment. We can use this cost as a branching rule, choosing next the assignment with the maximum alternate cost. Further, while backtracking, if the alternate cost at a node is greater than the upper bound, no more

nodes need to be evaluated at that level on the present branch.

HEURISTIC SOLUTION METHODS FOR THE QAP:

Given the limited size of problems that can be solved to optimality (smaller than most practical problems), there has been considerable interest in developing heuristic procedures for the QAP. Heuristics for the QAP can be classified as limited enumeration, construction methods, improvement methods, and hybrid methods.

Limited Enumeration – It has often been observed that an optimal solution is found fairly early in a branch-and-bound procedure, with the majority of the solution time then being spent in proving optimality. A heuristic based on limited enumeration takes advantage of this feature by setting a cut-off time to truncate the search process. The search can either be shortened or allowed to cover more of the search space in a fixed amount of time by fathoming a node at which the gap between the lower and upper bounds is sufficiently small. This gap can be set based on empirical evidence about the behavior of bounds, for example, in the QAP the lower bound rises rapidly at higher levels of the branch-and-bound tree and then more gradually. Thus a dynamic gap could be used, larger at higher levels and decreasing at lower levels of the tree.

Construction Methods – A constructive procedure starts with an empty assignment and adds assignments one at a time until a complete solution is obtained. The rule used to choose the next assignment can be a simple one such as assign the facility with the maximum interaction with a facility already assigned and place it as close as possible to that facility. Alternatively, we might employ a rule that takes into account assignments already made as well as future assignments to be made, which is likely to lead to better solutions. Examples of the latter type of rule would be the use of alternate costs obtained in the process of calculating lower bounds (see above) or the use of an evaluation function such as that devised by Graves and Whinston (1970). The Graves-Whinston method uses statistical properties to compute an expected value for the completion of any partial assignment using only basic arithmetic operations. The computation time is very reasonable, thus making it a good choice as a constructive heuristic. Suppose k assignments have already been made. The expected value of a complete assignment is given by expression (8) whose terms are analogous to those in (5):

$$EV = \sum_{i \in \mathscr{F}} c_{i,\rho(i)} + \sum_{i \in \mathscr{F}} \sum_{p \in \mathscr{F}} f_{ip} \, d_{\rho(i),\rho(p)}$$

$$+ \frac{\sum_{i \in \mathscr{F}} \sum_{p \notin \mathscr{F}} \sum_{j \notin \mathscr{L}} [f_{ip} \, d_{\rho(i),j} + f_{pi} \, d_{j,\rho(i)}]}{n-k}$$

$$+ \frac{\sum_{i \notin \mathscr{F}} \sum_{p \notin \mathscr{L}} c_{i,j}}{n-k} + \frac{\sum_{i,p \notin \mathscr{F}} f_{ip} \left(\sum_{j,q \notin \mathscr{L}} d_{jq} \right)}{(n-k)(n-k-1)} \qquad (8)$$

Improvement Methods – Improvement procedures start with some sub-optimal solution and attempt to improve it through partial changes in the assignments. The design of an improvement routine requires decisions concerning the following: type of exchange—pairwise, triple, or some higher order; number of exchanges to consider— should all possible exchanges be considered or a limited set; choice of exchange actually made— first improvement or best improvement; order of evaluation—random or predetermined. An effective strategy is to use pairwise exchanges in fixed order of decreasing interactions, considering all possible exchanges, and accepting the first improvement. Higher order exchanges are best used sparingly. More recently developed improvement techniques, such as simulated annealing (Connolly, 1990) and tabu search (Skorin-Kapov, 1990), that avoid the trap of local optima have been applied with success to the QAP.

Hybrid Methods – Some of the most successful heuristic solution methods for the QAP can be termed hybrid methods because they combine the power of improvement methods with some method for obtaining solutions to be improved, for example, construction methods (Liggett, 1981), limited enumeration (Bazaraa and Kirca, 1983), or cutting planes (Burkard and Bonniger, 1983). Kaku, Thompson, and Morton (1991) successfully combined constructed solutions with exchange improvement by systematically constructing solutions that were different from each other. This forces different areas of the search space to be examined.

Conclusion – The QAP formulation suffers from two drawbacks. First, it assumes identical area and shape requirements for all facilities. Unequal areas could be dealt with by dividing all facilities into equal-area modules which could be kept together in a solution by introducing very high artificial flows between them. However, this increases the size of the problem. Improvement methods can employ such a strategy since the number of facilities is less of a concern; however, exchanges are then limited to either equal-sized facilities or to adjacent facilities. Work by Bozer *et al.* (1994) incorporating the use of spacefilling curves in facility layout overcomes this handicap for improvement methods. Second, the QAP deals exclusively with interaction costs, generally material handling costs. This is not likely to be the only concern in facility layout. For this reason, the general practice is to allow a human designer to evaluate and fine tune a solution before implementation. For example, Fu and Kaku (1994) examined the effect of layout design on work-in-process (WIP) levels in a factory, an issue of great interest in these days of "lean manufacturing." They found that good QAP solutions generally reduce the levels of WIP, however, there are exceptions which the QAP approach cannot discern. We can now list some features that are desirable in a heuristic solution procedure for the facility layout problem: the ability to handle different area requirements; the ability to produce good solutions with reasonable computational requirements; and the ability to either consider multiple criteria or present the decision maker with good layout alternatives to choose from.

See **Branch and bound; Facility location; Location analysis; Quadratic assignment problem.**

References

[1] Bazaraa, M.S. and O. Kirca (1983). "A Branch-and-Bound-Based Heuristic for Solving the Quadratic Assignment Problem," *Naval Research Logistics Quarterly*, 30, 287–304.

[2] Bazaraa, M.S. and H.D. Sherali (1980). "Bender's Partitioning Scheme Applied to a New Formulation of the Quadratic Assignment Problem," *Naval Research Logistics Quarterly*, 27, 29–41.

[3] Bozer, Y.A., R.D. Meller, and S.J. Erlebacher (1994). "An Improvement-type Layout Algorithm for Single and Multiple-floor Facilities," *Management Science*, 40, 918–932.

[4] Burkard, R.E. and T. Bonniger (1983). "A Heuristic for Quadratic Boolean Programs with Applications to Quadratic Assignment Problems," *European Jl. Operational Research*, 13, 374–386.

[5] Burkard, R.E. and U. Derigs (1980). *Assignment and Matching Problems: Solution Methods with Fortran Programs*. Vol. 184 of *Lecture Notes in Economics and Mathematical Systems*, Springer-Verlag, Berlin.

[6] Connolly, D.T. (1990). "An Improved Annealing Scheme for the QAP," *European Jl. Operational Research*, 46, 93–100.

[7] Foulds, L.R. (1983). "Techniques for Facilities Layout: Deciding Which Pairs of Activities Should Be Adjacent," *Management Science*, 29, 1414–1426.

[8] Fu, M. and B.K. Kaku (1995). "Minimizing Work-in-process and Material Handling in the Facili-

ties Layout Problem," Technical Report TR 95-41, University of Maryland, College Park, Institute for Systems Research.

[9] Gavett, J.W. and N.V. Plyter (1966). "The Optimal Assignment of Facilities to Locations by Branch and Bound," *Operations Research*, 14, 210–232.

[10] Gilmore, P.C. (1962). "Optimal and Suboptimal Algorithms for the Quadratic Assignment Problem," *Jl. of SIAM*, 10, 305–313.

[11] Graves, G.W. and A.B. Whinston (1970). "An Algorithm for the Quadratic Assignment Problem," *Management Science*, 17, 453–471.

[12] Kaku, B.K., G.L. Thompson, and T.E. Morton (1991). "A Hybrid Heuristic for the Facilities Layout Problem," *Computers & Operations Research*, 18, 241–253.

[13] Koopmans, T.C. and M. Beckmann (1957). "Assignment Problems and the Location of Economic Activities," *Econometrica*, 25, 53–76.

[14] Kusiak, A. and S.S. Heragu (1987). "The Facility Layout Problem," *European Jl. Operational Research*, 29, 229–251.

[15] Land, A.H. (1963). "A Problem of Assignment with Inter-related Costs," *Operational Research Quarterly*, 14, 185–199.

[16] Lawler, E.L. (1963). "The Quadratic Assignment Problem," *Management Science*, 9, 586–599.

[17] Liggett, R.S. (1981). "The Quadratic Assignment Problem: An Experimental Evaluation of Solution Strategies," *Management Science*, 27, 442–458.

[18] Skorin-Kapov, J. (1990). "Tabu Search Applied to the Quadratic Assignment Problem," *ORSA Jl. Computing*, 2, 33–45.

FACILITY LOCATION

Dilip Chhajed

University of Illinois, Champaign

Richard L. Francis

University of Florida, Gainesville

Timothy J. Lowe

University of Iowa, Iowa City

INTRODUCTION: Location problems which can be quantified as optimization problems are natural candidates for operations research approaches, and many such problems have been studied using mathematical programming methodology. This article attempts to give an overview of some of this activity. Models of these location problems are classified as planar, network, and mixed integer programming models, and methodology for solving such types of models is outlined. Most of the contributions of operations research/management science (OR/MS) to location theory have occurred within the last 30 years. This period constitutes about 60% of the lifetime of the profession itself, which is little more than 50 years old. We think there is little doubt that the contributions consist principally of algorithms – well-defined computational procedures for solving quantifiable problems. These algorithms have built largely upon results in an area known as mathematical programming, which got its start soon after World War II with the now famous simplex method of George Dantzig for solving linear programming problems.

A location problem must be quantifiable in order for there to be any hope of solving it with an algorithm: there must be a well-defined objective to be optimized, for example, cost to be minimized, or profit to be maximized. Likewise there are usually well-defined constraints, for example, budget constraints, which limit the scope of the optimization. Location problems which are highly subjective or political in nature are thus usually not very good candidates for operations research approaches, although even for such problems there may be results which can help to reduce the scope of the problem under consideration, or identify basic trade-offs of interest.

Many of the best known OR algorithms for solving location problems involve choosing best locations from a finite collection of possible sites. Since a site either is chosen or is not, such problems are intrinsically discrete in nature and are candidates for being solved as integer programming problems. For example, a banking corporation might be uncertain as to how many branch banks there should be. The corporation would realize that the more banks it locates, the more convenient the branches would be to its customers in terms of travel time or travel cost. On the other hand, the more branches there are, the higher would be the operating expenses and fixed costs. Thus there is a trade-off between convenience and operating costs, which it would be important to analyze. Such trade-offs often occur in solving location problems.

When a location problem has substantial transport costs, and the fixed site costs are relatively independent of location, there are several other approaches to modeling it. Often it is assumed that transport costs are directly proportional to transport distances. When these distances are incurred on a transport network, such as a road network, the result is often a network model. Such models usually employ shortest path algorithms to compute travel distances. The focus of network model research has been principally upon two topics: 1) algorithms to solve the problems, and 2) localization results, such as vertex-optimality results, which reduce to a finite collection the set of locations which must be considered to obtain an optimal solution. Once such a finite set is obtained, the resulting remaining problem may well be modeled as an integer or mixed integer programming problem.

It is possible, for a network location problem, that it may be prohibitively expensive to obtain, or to work with, the necessary network data. In such cases network distances may well be approximated using planar distances, for example, Euclidean or rectilinear distances. The resulting problems are often easier to analyze, and can be helpful for providing insight. In this case we are dealing with what we call a planar model. Often results from nonlinear programming can be employed to help solve such a model.

In what follows we consider some planar models, network models, and mixed integer programming models. This classification of models is not exhaustive, but it does capture much of what has been dealt with in the literature. For further reading, see the texts by Handler and Mirchandani (1979); Love, Morris and Wesolowsky (1988); Mirchandani and Francis (1990); and Francis, McGinnis and White (1992).

SELECTED MODELS: In this section we present some of the basic but popular and useful models of location theory. After discussing the models, we discuss very briefly the solution approaches available for these models. In our discussion we will refer to the demand points as *existing facilities* and the facilities to be located as *new facilities*.

In developing the models we use the following notation:

p: number of new facilities. The value of p may be a decision variable or may be fixed;

m: number of existing facilities;

w_i: weight associated with existing facility i;

X: location of a single new facility;

$X = (X_1, \ldots, X_p)$: locations of p new facilities; and

$D_i(X)$: the distance between existing facility i and the *nearest* new facility.

MODEL 1, P-CENTER PROBLEM: The objective in this model is to locate p new facilities to minimize the maximum distance to an existing facility. Let $g(X) = \max_{i=1,\ldots,m} \{w_i D_i(X)\}$ represent the maximum (weighted) distance any person has to travel; then the problem can be posed as

Minimize$_X\, g(X)$.

This problem is known as the *p-Center problem* and, besides other applications, has been used to model locations of emergency medical facilities, the location of a helicopter to minimize the maximum time to respond to an emergency, and the location of a transmitter to maximize the lowest signal level received.

MODEL 2, COVERING PROBLEM: In this problem the number of facilities to be located is not fixed *a priori*. Each existing facility should be within a specified weighted distance from at least one new facility. The objective is to find the number of new facilities, p, and their locations, X, to minimize the cost of the new facilities.

The version of the covering problem with a finite set of candidate facility locations can be modeled as a *set-covering* problem. With S as the set of n candidate sites for facilities, denote the sites by $j = 1, \ldots, n$. Define variables y_j which take on a value 1 if a new facility is opened at site j and 0 otherwise. Let f_j represent the cost of locating (opening) a facility at j. Customers (existing facilities) are indexed by $i = 1, \ldots, m$. Let $a_{ij} = 1$ if a new facility located at site j can cover existing facility i, 0 otherwise, $i = 1, \ldots, m$ and $j = 1, \ldots, n$. Note that the a_{ij} values are constants and are determined prior to the formulation. An integer programming formulation of the set-covering problem can be written as:

$$\min \sum_{j \in S} f_j y_j$$

$$\text{subject to: } \sum_{j \in S} a_{ij} y_j \geq 1, \quad \forall i = 1, \ldots, m$$

$$y_j \in \{0, 1\} \quad \forall j = 1, \ldots, n.$$

The objective function sums up the fixed costs of locating the new facilities. When each $f_j = 1$, the objective function minimizes the number of new facilities to be located. The first constraint ensures that each existing facility is covered while the second constraint restricts the variables to be binary.

MODEL 3, SIMPLE PLANT LOCATION PROBLEM: In the simple plant location problem (SPLP), we want to open a number of new facilities (the actual number is a decision variable) to serve a given set of customers. There is a fixed cost of opening each facility. The objective is to minimize the sum of the fixed and variable costs of serving the demand points, and to determine the optimal allocation pattern for all customers.

The SPLP can be formulated as a mixed integer program as follows. In addition to S, f_j and y_j defined earlier, let $c_{ij} \geq 0$, $i = 1, \ldots, m$ and $j = 1, \ldots, n$, be the unit cost of servicing customer i from a new facility located at j. Letting x_{ij} denote the fraction of customer i's service provided by site j, a formulation of the SPLP is:

$$\min \sum_{j \in S} f_j y_j + \sum_{i=1}^{m} \sum_{j \in S} c_{ij} x_{ij} \qquad (1)$$

subject to: $\sum_{j \in S} x_{ij} = 1$, $i = 1, \ldots, m$ $\qquad (2)$

$$y_j \in \{0,1\}, j = 1, \ldots, n \qquad (3)$$

$$x_{ij} \geq 0, i = 1, \ldots, m, j = 1, \ldots, n \qquad (4)$$

and $x_{ij} \leq y_j$, $i = 1, \ldots, m, j = 1, \ldots, n$. $\qquad (5)$

Expression (1) totals site costs and service costs. The requirement that each customer be completely served is assured by (2). Expression (3) prevents a fractional opening of a site, and (4) assures nonnegative service. The condition that service cannot be provided from an unopened facility is guaranteed by (5).

If there are no facility related cost terms in the objective function (i.e., $f_j = 0$ for all j) and the number of facilities is restricted to exactly p, then the resulting problem is known as the *p-median problem*.

SOLVING THE MODELS: In order to describe selected contributions that operations research has made towards providing solution procedures for the above problems, we can divide the problem space into three classes: planar models, network models, and discrete models. Many of the above models can be posed on any one of the three spaces with some slight modifications, and almost all models can be put into one of the three classes.

The main difference in these three classes of models is the manner in which the distance between two points is defined. In planar models, the distance function $d(\cdot)$ is a "norm," often Euclidean, rectilinear, or some other norm, and the number of possible locations for new facilities is infinite. This renders the corresponding problems as continuous. If (a_i, b_i) are the coordinates of a point i, then the Euclidean distance between points i and j is given by $\sqrt{[(a_i - a_j)^2 + (b_i - b_j)^2]}$, while the rectilinear distance is given by $|a_i - a_j| + |b_i - b_j|$, where $|\cdot|$ is the absolute value function.

In network models we have a transport network on which travel occurs. The transport network, for example, may represent a system of major highways and/or roads. The distance between two points is usually defined as the shortest distance on the network. Distances are often more accurately represented in network models than in planar models, but the need for data is also higher in network models since the length of each segment is needed. For many models it becomes advantageous to work directly with the network, exploiting its properties in developing a solution procedure. The existing facilities are located on the nodes of the network, and the new facilities are to be located at points on the network. An additional advantage of network models is that they make problem visualization easier. Thus, even if the problem is not solved as a network problem, a solution presented in network form may assist the decision maker in understanding the problem and the issues involved.

In discrete models, the number of existing facilities and the number of potential sites for new facilities is finite. Distances may be derived from planar or network distances, or some more general type of transport cost which is proportional to distance. Discrete problems are often modeled as mixed integer programs and are often more difficult to solve. On the other hand, many realistic assumptions can be incorporated in discrete models which cannot be included in planar or network models.

As concerns algorithms/solution approaches to the three classes of problems, planar problems are usually solved using linear or nonlinear programming methods; network problems are usually solved using network and graph-theoretic methods; discrete problems are usually solved using integer programming methods. More information on these methods can be found in the references.

See **Combinatorial and integer optimization; Facility location; Location analysis; Networks; Shortest route problem; Stochastic programming.**

References

[1] Francis, R., L.F. McGinnis, and J.A. White (1992). *Facility Layout and Location: An Analytical Approach*, Prentice Hall, Englewood Cliffs, New Jersey.

[2] Handler, G.Y., and P.B. Mirchandani (1979). *Location on Networks: Theory and Algorithms*, MIT Press, Cambridge, Massachusetts.

[3] Love, R.F., J.G. Morris, and G.O. Wesolowsky (1988). *Facilities Location; Models and Methods*, North-Holland, Amsterdam.

[4] Mirchandani, P.B., and R.L. Francis, eds. (1990). *Discrete Location Theory*, John Wiley, New York.

FACTORABLE PROGRAMMING

Richard H. F. Jackson

National Institute of Standards and Technology, Gaithersburg, Maryland

Factorable programming problems are mathematical programming problems of the form

$$\text{minimize } f(x),$$
$$x \in R^n$$

$$\text{subject to } g_i(x) > 0,$$

for $i = 1, \ldots, m$, in which all the functions involved are factorable. Loosely, a factorable function is a multivariable function that can be written as the last of a finite sequence of functions, in which the first n functions in the sequence are just the coordinate variables, and each function beyond the nth is a sum, a product, or a single-variable transformation of previous functions in the sequence. More rigorously, let $[f_1(x), f_2(x), \ldots, f_L(x)]$ be a finite sequence of functions such that $f_i : R^n \to R$ where each $f_i(x)$ is defined according to one of the following rules:

Rule 1: For $i = 1, \ldots, n, f_i(x)$ is defined to be the ith Euclidean coordinate, or $f_i(x) = x_i$.

Rule 2: For $i = n + 1, \ldots, L, f_1(x)$ is formed using *one* of the following compositions:
  a) $f_i(x) = f_{j(i)}(x) + f_{k(i)}(x)$; or
  b) $f_i(x) = f_{j(i)}(x) \cdot f_{k(i)}(x)$; or
  c) $f_i(x) = T_i[f_{j(i)}(x)]$;

where $j(i) < i$, $k(i) < i$, and T_i is a function of a single variable. Then $f(x) = f_L(x)$ is a *factorable function* and $[f_1(x), f_2(x), \ldots, f_L(x)]$ is a *factored sequence*. Thus a function, $f(x)$, will be called factorable if it can be formed according to Rules 1 and 2, and the resulting sequence of functions will be called a factored sequence, or at times the function written in factored form.

Although it is not always immediately grasped, the concept of a factorable function is actually a very natural one. In fact it is just a formalization of the natural procedure one follows in evaluating a complicated function. Consider for example the function

$$f(x) = [a^T x] \sin[b^T x] \exp[c^T x],$$

where a, b, c, and x are (2×1) vectors. The natural approach to evaluating this function for specified values x_0^1 and x_2^0 is first to compute the quantities within the brackets, then to apply the sine and exponential functions, and finally to multiply the three resulting quantities. This might be done in stages as follows:

$f_1 = x_1^0$	$f_9 = c_1 f_1$
$f_2 = x_2^0$	$f_{10} = c_2 f_2$
$f_3 = a_1 f_1$	$f_{11} = f_9 + f_{10}$
$f_4 = a_2 f_2$	$f_{12} = \sin(f_8)$
$f_5 = f_3 + f_4$	$f_{13} = \exp(f_{11})$
$f_6 = b_1 f_1$	$f_{14} = f_5 \cdot f_{12}$
$f_7 = b_2 f_2$	$f_{15} = f_{13} \cdot f_{14}$
$f_8 = f_6 + f_7$	

This is one possible factored sequence for $f(x)$.

In order to understand what follows, the concept of an outer product matrix must be introduced. An $(m \times n)$ matrix A is called an *outer product matrix* if there exists a scalar α, an $(m \times 1)$ vector a, and an $(n \times 1)$ vector b such that

$$A = a\alpha b^T.$$

The expression $a\alpha b^T$ is called an *outer product* or a *dyad*. Note that a dyad is conformable since the dimensions of the product are $(m \times 1)(1 \times 1)$ $(1 \times n)$, which yields the $(m \times n)$ outer product matrix A as desired. A useful property of outer product matrices is that, if kept as dyads, matrix multiplication is simplified to inner products alone, saving the computations required to form the matrices involved. For example,

$$Ac = a\alpha[b^T c],$$

$$d^T A = [d^T a]\alpha b^T, \text{ and}$$

$$AF = a\alpha[b^T F],$$

where c is $(n \times 1)$, d is $(m \times 1)$ and F is $(n \times m)$.

It is well-known (McCormick, 1983) that factorable functions possess two very special properties that can be exploited to produce efficient (fast and accurate) algorithms: i) once written in factorable form, their gradients and Hessians may be computed exactly, automatically, and efficiently; and ii) their Hessians occur naturally as sums of dyads whose vector factors are gradients of terms in the factored sequence. The first of these properties eases the task of providing the derivatives of a nonlinear programming problem to a computer software solution routine, and has the potential eventually to trivialize it. The second, as noted above, changes the way we look at matrix multiplication, which in many cases results in less computational effort.

There are factorable problems whose structure is such that the factorable approach results in more work: small, dense problems, for example. For these problems, the factorable approach can still be used for easy input, but some of the matrix techniques would be replaced by classical approaches.

Software packages have been written that perform the factoring automatically from natural language input. See Jackson and McCormick (1987) for a history of such efforts, as well as Jackson, McCormick, and Sofer (1989). The latter paper describes a system that allows user input for nonlinear functions in a format similar to FORTRAN, without any requirement on the user to understand the details of factorable functions.

As mentioned above, one fundamental value of factorable functions lies in the simple and computationally efficient forms that result for their Hessians. In fact, factorable programming is based on the existence of, and the simplified operations that result from, these simple forms. The seminal result is that the Hessian of a factorable function can be written as the sum of dyads, or outer products, of gradients of functions in the factored sequence (Fiacco and McCormick, 1968, pp. 184–188). This basic result was generalized in Jackson and McCormick (1986). Before explaining the generalization, it is necessary to generalize the concepts of Hessian and dyad.

Let $A \in R^{(n_1 \times \ldots \times n_N)}$, and let $A_{i_1,\ldots,i_N}$ denote the $(i_1, \ldots, i_N)^{th}$ element of this array. For the purposes of this paper, A is called the Nth-order *tensor* of a multivariable function $f(x)$ if

$$A_{i_1,\ldots,i_N} = \partial^N f(x)/\partial x_{i_N} \cdots \partial x_{i_1}.$$

Note that gradients and Hessians are tensors of order 1 and 2 respectively.

An N-dimensional array A is called *a generalized outer product matrix* if there exists a scalar α, and an ordered set of vectors $a_1, \ldots a_N$ (where each a_k is $(n_k \times 1)$) such that each element of A is generated by the product of the scalar and certain specific elements of the vectors $a_1, \ldots, a_N$ as follows:

$$A_{i_1,\ldots,i_N} = \alpha * a_{1,i_1} * \cdots * a_{N,i_N}$$

for $i_1 = 1, \ldots, n_1; \ldots; i_N = 1, \ldots, n_N$, where a_{k,i_k} represents the $(i_k)^{th}$ element of the $(n_k \times 1)$ vector a_k.

The scalar and set of vectors which generate a generalized outer product matrix taken together are called a *polyad* and are written

$$(\alpha: a_1 \cdots a_N), \tag{1}$$

where order is important, i.e. the vector in position j is associated with the jth dimension. A polyad containing N vector factors is an N-*ad*. Also, an expression containing a sum of polyads is a *polyadic*, and an expression containing a sum of N-ads is an N-*adic*. (The actual addition here is performed as a sum of the associated generalized outer product matrices.) When vector factors in a polyad are repeated, exponential notation is used, as in the case of the symmetric N-ad, $(\alpha:[a]^N)$. Note that the representation of a generalized outer product matrix by a polyad is not unique. For example, $(\alpha|\gamma:[a_1\gamma] \cdots a_N)$ generates the same N-dimensional array of numbers as does (1) for any nonzero scalar γ. Finally, a 2-*ad* of the form $(\alpha:ab)$ is equivalent to the more familiar dyad of the form $a\alpha b^T$, and the two will be used interchangeably.

The generalization mentioned above is that *all* tensors (that exist) of factorable functions possess a natural polyadic structure. Furthermore, the vector factors that comprise the monads of the gradient are the same vector factors which comprise the dyads of the Hessian, the triads of the third order tensor, and so on. This has important computational implications in mathematical programming. It means that once the gradient of a factorable function is computed, a major portion of the work involved in computing higher-order derivatives is already calculated. Consequently, high-order minimization techniques, previously considered computationally intractable, are once again worthy of consideration (Jackson and McCormick, 1986).

It should be noted that, by their very nature, the tensors of factorable functions are ideally suited for computation on parallel processing and array processing computers. We know of few other such ideal applications in numerical optimization. Also, it has been shown (McCormick, 1985) that all factorable programming problems have an equivalent separable programming representation, and that efficient algorithms (Falk and Soland, 1969; Falk, 1973; Hoffman, 1975; McCormick, 1976; Leaver, 1984) exist for finding global solutions to these problems. Thus there exists the potential of finding global solutions to factorable programming problems fast and accurately.

The discovery and development of factorable functions and their uses in mathematical programming is credited to McCormick (1974). Since the discovery of these functions, the theory of factorable programming has been further developed and refined. Ghaemi and McCormick (1979) developed a computer code (FACSUMT), which processes the functions in a factorable program and provides the interface to the SUMT nonlinear programming code (see Mylander *et al.*, 1971). A preliminary version of this code is described in Pugh (1972).

Further extensions of factorable programming theory were provided by Shayan (1978), who developed an automatic method for computing the mth-order of a solution technique can be evaluated when the functions are factorable by counting basic operations and basic functions, a more accurate measure of efficiency than the popular technique of counting the number of "equivalent function evaluations" (Miele and Gonzalez, 1978).

The natural dyadic structure of the Hessian of a factorable function was exploited by Emami (1978) to develop a matrix factorization scheme for obtaining a generalized inverse of the Hessian of a factorable function. Ghotb (1980) also capitalized on this structure and provided formulae for computing a generalized inverse of a reduced Hessian when it is given in dyadic form. Sofer (1983) extended this last concept further by utilizing the dyadic structure to obtain computationally efficient techniques for constructing a generalized inverse of reduced Hessian and updating it from iteration to iteration.

Another direction was pursued by DeSilva and McCormick (1978), who developed the formulae and methodology to utilize the input to general nonlinear programs in factorable form to perform first-order sensitivity analysis on the solution vector. This was generalized in Jackson and McCormick (1988), where second order sensitivity analysis methods were developed, with formulae involving third order tensors used to compute second derivatives of components of a local solution with respect to problem parameters.

It is important to understand that the derivative calculations performed in factorable programming are not estimations, but mathematically exact calculations. Furthermore they are also compact, since factored sequences mimic hand calculations, and thus this technique is different from symbolic manipulation techniques for differentiation, which tend to produce large amounts of code. The techniques used in factorable programming are efficient exploitations of the special structure inherent in factorable functions and their partial derivative arrays. Moreover, while it is true that some symbolic differentiaters also can recognize functions which can be described similarly as a sequence of rules, each of which can be differentiated, the similarity ends there. Such symbolic differentiaters continue to differentiate the rules, without exploiting the polyadic structure of the result (Kedem, 1980; Rall, 1980; Wengert, 1964; Reiter and Gray, 1967; and

Warner, 1975). It is this latter effort which provides the real value of factorable functions and which therefore separates the two techniques.

See **Mathematical programming; Nonlinear programming.**

References

[1] A. DeSilva and G.P. McCormick (1978), "Sensitivity Analysis in Nonlinear Programming Using Factorable Symbolic Input," Technical Report T-365, The George Washington University, Institute for Management Science and Engineering, Washington, DC.

[2] G. Emami (1978), "Evaluating Strategies for Newton's Method Using a Numerically Stable Generalized Inverse Algorithm," Dissertation, Department of Operations Research, George Washington University, Washington, DC.

[3] J.E. Falk (1973), "Global Solutions of Signomial Problems," Technical report T-274, George Washington University, Department of Operations Research, Washington, DC.

[4] J.E. Falk, and R.M. Soland (1969), "An Algorithm for Separable Nonconvex Programming Problems," *Management Science*, 15, 550-569.

[5] A.V. Fiacco and G.P. McCormick (1968), *Nonlinear Programming: Sequential Unconstrained Minimization Techniques*, John Wiley, New York.

[6] A. Ghaemi and G.P. McCormick (1979), "Factorable Symbolic SUMT: What Is It? How Is It Used?," Technical Report No. T-402, Institute for Management Science and Engineering, George Washington University, Washington, DC.

[7] F. Ghotb (1980), "Evaluating Strategies for Newton's Method for Linearly Constrained Optimization Problems," Dissertation, Department of Operations Research, George Washington University, Washington, DC.

[8] K.L. Hoffman (1975), "NUGLOBAL-User Guide," Technical Report TM-64866, Department of Operations Research, George Washington University, Washington, DC.

[9] R.H.F. Jackson and G.P. McCormick (1986), "The Polyadic Structure of Factorable Function Tensors with Applications to High-order Minimization Techniques," *JOTA*, 51, 63-94.

[10] R.H.F. Jackson and G.P. McCormick (1988), "Second-order Sensitivity Analysis in Factorable Programming: Theory and Applications," *Mathematical Programming*, 41, 1-27.

[11] R.H.F. Jackson, G.P. McCormick and A. Sofer (1989), "FACTUNC, A User-friendly System for Optimization," Technical Report NISTIR 89-4159, National Institute of Standards and Technology, Gaithersburg, Maryland.

[12] G. Kedem (1980), "Automatic Differentiation of Computer Programs," *ACM Transactions on Mathematical Software*, 6, 150-165.

[13] S.G. Leaver (1984), "Computing Global Maximum Likelihood Parameter Estimates for Product Models for Frequency Tables Involving Indirect Observation," Dissertation, The George Washington University, Department of Operations Research, Washington, DC.

[14] G.P. McCormick (1974), "A Minimanual for Use of the SUMT Computer Program and the Factorable Programming Language," Technical Report SOL 74-15, Department of Operations Research, Stanford University, Stanford, California.

[15] G.P. McCormick (1976), "Computability of Global Solutions to Factorable Nonconvex Programs: Part I – Convex Underestimating Problems," *Mathematical Programming*, 10, 147–145.

[16] G.P. McCormick (1983), *Nonlinear Programming: Theory, Algorithms and Applications*, John Wiley, New York.

[17] G.P. McCormick (1985), "Global Solutions to Factorable Nonlinear Optimization Problems Using Separable Programming Techniques," Technical Report NBSIR 85-3206, National Bureau of Standards, Gaithersburg, Maryland.

[18] A. Miele and S. Gonzalez (1978), "On the Comparative Evaluation of Algorithms for Mathematical Programming Problems," *Nonlinear Programming*, 3, edited by O.L. Mangasarian *et al.*, Academic Press, New York, 337–359.

[19] W.C. Mylander, R. Holmes and G.P. McCormick (1971), "A Guide to SUMT-Version 4: The Computer Program Implementing the Sequential Unconstrained Minimization Technique for Nonlinear Programming," Technical Report RAC-P-63, Research Analysis Corporation, McLean, Virginia.

[20] R.E. Pugh (1972), "A Language for Nonlinear Programming Problems," *Mathematical Programming*, 2, 176–206.

[21] L.B. Rall (1980), "Applications of Software for Automatic Differentiation in Numerical Computations," *Computing*, Supplement, 2, 141–156.

[22] A. Reiter and J.H. Gray (1967), "Compiler for Differentiable Expressions (CODEX) for the CDC 3600," MRC Technical Report No. 791, University of Wisconsin, Madison, Wisconsin.

[23] M.E. Shayan (1978), "A Methodology for Comparing Algorithms and a Method for Computing m^{th} Order Directional Derivatives Based on Factorable Programming," Dissertation, Department of Operations Research, George Washington University, Washington, DC.

[24] A. Sofer (1983), "Computationally Efficient Techniques for Generalized Inversion," Dissertation, Department of Operations Research, The George Washington University, Washington, DC.

[25] D.D. Warner (1975), "A Partial Derivative Generator," Computing Science Technical Report No. 28, Bell Telephone Laboratories, Murray Hill, New Jersey.

[26] R.E. Wengert (1964), "A Simple Automatic Derivative Evaluation Program," *Communications of the ACM*, 7, 463–464.

FAILURE-RATE FUNCTION

The failure rate at time t of a "unit" with lifetime density $f(t)$ and lifetime CDF $F(t)$ is defined by the (approximate) probability $h(t)\Delta t$ that a random lifetime ends in a small interval of time Δt, given that it has survived to the beginning of the interval. For the continuous case, this is formally written as

$$h(t) = \lim_{\Delta t \to 0} \left[\frac{f(t)}{1 - F(t)} \right].$$

The function $h(t)$ is often also called the hazard rate or function, the force of mortality, or the intensity rate or function. See **Probability distribution selection; Reliability function; Reliability of stochastic systems.**

FARKAS' LEMMA

Given a matrix A and a column vector b, one and only one of the following two alternatives holds. Either: (1) there exists a column vector $x \geq 0$ with $Ax = b$, or (2) there exists an unrestricted row vector y for which $yA \geq 0$ and $yb < 0$. This lemma can be proved by defining appropriate primal and dual linear-programming problems and applying the duality theorem. See **Gordan's theorem; Strong duality theorem; Theorem of the alternatives.**

FARRELL MEASURE

See **Data envelopment analysis.**

FATHOM

To analyze a computational path in enough detail to logically conclude that the analysis of the path has provided as much information possible and/or required. See **Branch and bound.**

FCFS

The First-Come, First-Served queueing discipline in which customers are selected for service in the precise order in which they arrive to the queue. See **FIFO; Queueing theory.**

FEASIBLE BASIS

A basis to a linear-programming problem that yields a solution that satisfies all the constraints of the problem. See **Linear programming; Simplex method.**

FEASIBLE REGION

Set of points that satisfy prescribed restrictions (constraints) on a solution.

FEASIBLE SOLUTION

A solution to an optimization problem that satisfies its constraints. In linear programming, these are the conditions $Ax = b$ and $x \geq 0$. See **Infeasible solution**.

FEBA

Forward edge of a combat area. See **Battle models**.

FEEDBACK QUEUE

A system where customers may return upon completion of service. In many real problems, there is a nonzero probability that a customer just completing service returns to the end of the queue and is serviced again. See **Networks of queues**; **Queueing theory**.

FIELD ANALYSIS

Howard W. Kreiner

Center for Naval Analyses, Alexandria, Virginia

INTRODUCTION: *Field analysis* is the practice of operations research usually at the place where the operations occur. It uses observations and data from those operations as they are carried out by the people who normally conduct them. Its purpose may be the immediate modification of an unsatisfactory process, or at longer range, the elucidation of the critical steps in the process for further analysis of options and changes. Its vital importance to systems analysis lies in this latter alternative.

As the operations under study are real and current, they involve the use of equipment or machines already in place, and the practices of operators who have been trained in their use. Projections of possible future capabilities and alternative training methods are not a major part of the basic data on which the analysis must be based.

Problems that are visible are those that obviously interfere with the smooth functioning of the system. Their solution must make major differences to be considered useful. There is a need to seek a solution that can improve matters by hemibels (about a factor of 3). Anything less may be lost in the noise of the system. In investigating the causes of problems, many of the potential variables will not lie within the control of the operators or the analyst as attempts are made to identify them. He or she must think in heuristic terms, rather than those of full scientific rigor.

Case studies often are interesting for their problem-solving methodologies, but as the problems differ in detail, such case studies do not fall into clear groupings that can be characterized as the essence of field analysis. The mathematical content of many if not most field analyses is simple, usually not beyond the level of undergraduate mathematics. Therefore, it is methods of behavior, thought and exposition rather than mathematics that truly constitute the methodology of field analysis.

HISTORICAL ORIGINS: The term field analysis arose in the earliest activities of operations analysts in the U.S. Navy. The Navy set up the nation's first operations research organization in 1942. It did so because it was engaged with the Germans' submarine forces in a battle that was not going well. Forces, doctrine and tactics that had grown from experience were proving insufficient to defeat the enemy. The people it brought into the operations research organization, however, were civilian scientists with little or no direct experience of naval operations. There was only the hope that a fresh scientific view of the situation might produce new methods and the means of victory. The hope was based upon successes of operations research in the British effort against the German bombing campaign.

The scientists at first worked with the statistics of combat derived from action reports. They soon found, though, that full understanding of the action reports required that they have closer contact with the operating forces that engaged the enemy and wrote the reports.

To obtain this contact, they sent scientists to the naval commands deployed against the German submarines. Their purpose was to talk directly with the naval officers and men who performed the combat activities, and to the degree possible, to observe at first hand the circumstances of warfare. Initially, their efforts were intended to insure that the scientists at the home office made proper interpretations of the reports and the statistics they derived. As they became more experienced, and as the aims of the operations research progressed toward model development and predictions of effectiveness, the analysts' purposes and roles with the operating forces also broadened. This pattern of visits and later, longer assignments, became known as the "field program," and the analysis done by the analysts at the deployed commands, as the "field analyses."

At the end of the war, Morse and Kimball (1946) characterized the program's purposes as: (a) direct help to the service units, (b) securing difficult-to-obtain information for the headquarters organization, (c) providing to the individual analyst the practical education indispensable in avoiding the pitfalls to which the pure theorist may be subject. They also commented on the administrative factors that made for successful field work. They stressed the need for the command receiving the analyst to invite the assignment and to approve the individual. The analyst should be attached to the highest level of the field activity, take assignments from the commanding officer, and make reports at the same level. There should be regular rotation of analysts at the field command, to bring back to the central staff the experience gained in the field.

Morse and Kimball typified the nature of the scientific work of field analysts in six categories:

1. Analytical
2. Statistical
3. Liaison
4. Experimental
5. Educational
6. Publication

At any field assignment station, however, work would not be limited to just one of these categories; the analysts might do all these types of work in some proportion.

The system of field assignments worked well under the pressure and circumstances of the war. Problem areas were of vital urgency. Help from this promising source was generally welcome at the field commands (though there were instances where it took dramatic analytic successes to establish acceptance for the analyst), and in the senior levels of the service. One instance of initial reluctance overcome by analytic success was Steinhardt's development of barriers against South Atlantic blockade runners (Tidman, 1984).

At the end of World War II, the postwar successor to the U.S. Navy's Operations Research Groups, the Operations Evaluation Group continued the practice of field assignments and field analyses. Because of a reduction in the size of the group, and because the activities of the Navy's deployed forces also were greatly curtailed, the field assignments were limited to units of the Navy's test and evaluation forces. The start of the Korean War caused an increase in the size of the parent group, and a revival of the assignment of analysts to fleet staffs and combat operational units.

The activities of operations analysts in the field were not limited to the U.S. Navy. The U.S. Army, both ground and air forces (and later, the U.S. Air Force), also formed analytic groups that sent representatives to field forces. Not all followed precisely the same administrative procedures, but the principal purposes of the assignments were paralleled in each case. Postwar, also, these organizations continued the practice, and expanded their field activities as war and other circumstances required.

The first paper published by the *Journal of the Operations Research Society of America* to report on field analysis in a non-military subject was Thornthwaite (1953). Among operations research practitioners used to the military version of the profession, it was a great relief and cause for elation. It showed that there truly was a possibility that the kind of operations research, field analysis, with which they were most familiar in the military services also could be applied successfully in the nonmilitary world. More than forty years later, the methods it describes are in use in unchanged form at the site where they were developed. Kreiner (1994) revisits Thornthwaite's paper to fill and clarify gaps in its exposition and make explicit the qualities it exhibits as a fine example of field analysis.

Since then, many other examples of good field analyses have appeared in various formats and publications. As remarked above in the definition section of this article, they generally have been identified primarily by the subject matter of the problem, rather than as examples of field analysis viewed as a separately defined branch of operations research.

FIELD ANALYSIS IN AN ERA OF SYSTEMS ANALYSIS: During World War II, there was a very close tie between the work of the military field analyst and that of the headquarters staff. The initial motivation for creating a field program was exactly that close tie; headquarters staff interest was directed almost exclusively to the day-to-day problems and success of the deployed forces. The guiding principles of operations research formulated in the postwar summaries were identical for the field and the headquarters analyses. In the fifty years since the U.S. creation of formal operations research organizations, this has continued to be true when war dominates the activities of the military services from combat forces to the highest command levels. It also was true to a large extent in peacetime immediately after World War II. The field analysts' assignments were to operational test and evaluation commands concerned with individual combat systems whose procurement decisions depended upon those test results.

The tie has become less close in the military services with the trend to high-level systems analysis at headquarters operations research groups. The increasing complexity and interaction of systems, their cost, and the very long development time for newer combat systems made headquarters command levels more concerned with future systems. It elevated the procurement process to the strategic level, and concentrated the attention of the central military staff on long-term budgetary matters. Headquarters operations research groups necessarily altered their point of view as well. Field commands, however, retained their concern with training their forces to operate and integrate systems already in use. The problem of divergence of interest between the field analyst and the headquarters group did not disappear entirely even during the Vietnam War and the Gulf War. Those wars were limited in character, and the Cold War and the larger threat of nuclear war still tended to dominate budgetary and strategic interests.

For the field analyst, however, the main subjects for analysis continue to be the operations involving the use of equipment already designed, developed, produced and distributed. If the equipment is not quite at this stage, it is at least far enough along in the process to justify operational testing and tactical development. The field analyst's concerns in commercial, nonmilitary government or military activities, are with the practices for employing, and with the training of people to make the equipment and its use as effective and efficient as possible. To the extent that a central group, commercial or military, arranges to provide field analysts and uses the field assignments both for training analysts and to insure realism in describing current and possible future operations, it shares the same objectives. It can be difficult for a central group, however, to balance priorities for attention to field analysis with those of important future systems studies.

A CONTINUING ROLE FOR FIELD ANALYSIS:

Compared with the circumstances at the time operations research first was introduced, there now are greatly improved methods for data collection, and greatly enhanced ability of computers to model interactions of equipment and people. There have been theoretical developments in operations analysis and problem-solving techniques that are reported worldwide in the journals of numerous operations research societies. Yet, if operations research retains its focus on problem solving in operations, there continues to be an important mission for analysis to be done at the point, and in direct observation of the operations under study. This is particularly important when large scale modeling of operations is a major means of analysis in the headquarters groups. Both during the building of models and afterwards, it can be extremely difficult to review and test all the assumptions and possible omissions of critical factors.

Morse and Kimball made the point that it was a strength of the operations analyst to think in hemibels, to seek improvements in operations that multiply effectiveness by a factor of three or more. This differs qualitatively from the notion of improvements in small increments. The field analyst is in a unique position to see opportunities for hemibel improvements; he can observe at first hand the factors that control the operation. If there are differences between what has been assumed about those factors and what actually is occurring, he can document them, measure them, and propose changes to exploit the differences in favor of improved understanding, and ultimately, improved operations.

Kreiner (1992) noted an example in radar detection of small targets, in which the field analyst identified assumptions about the statistical character of radar returns. Current data at the field site proved the assumptions to be faulty. Ultimately, the original theory had to be abandoned, and alternate methods devised. The same book also reports on an analytical look at an operational plan that revealed unstated, unconsidered, and erroneous assumptions. The plan assumed static, fixed naval forces, assured of long warning times of possible attack, and manned by pilots expected to fly missions they considered suicidal, although safer, equally effective alternatives were available. When the analysis made these assumptions explicit, the entire plan had to be rewritten.

The field analyst also is in a better position to examine the choice of measures for evaluating operational effectiveness than his counterpart at headquarters. Larson (1988), though not assigned as a field analyst, nevertheless functioned as one as a customer of a queueing system when he attempted to buy a bicycle for his daughter. In his job as an analyst of such systems, he had accepted the standard measure of average customer waiting time, and the goal of minimizing this measure. As an actual customer, however, he discovered a major deficiency in the measure, a lack of perceived fairness to the individual customer. His article explores the ways to make the queues and the measures of their performance more responsive to the broader interpretation of effectiveness.

The field analyst has another important requirement, the need to develop results in terms that serve primarily the purposes of the customer. Scientific journals, including those of the operations research societies, require presentations that are concise, rigorous, and of enough generality to be of interest to a spectrum of fellow professionals. The field analyst has another audience entirely. He must initially establish a role as a participant in the operations analyzed, lest the activities lack credibility. As an outsider, he may not gain access to the intimate, seemingly tiny details that make up the operation. The field analyst similarly must make reports in operational terms. The audience will classify problems in their own terms, rather than by the methodologies used to solve them. The report must make the same close connection with the immediate problem. It may be important to the analyst to innovate in methodology. The client wants only assurance that the methodology addresses the correct aspects of the problem, that the analyst is competent to apply it, and that the results enable them to improve their operations.

See **Air Force operations analysis; Center for Naval Analyses; Implementation; Military operations research; Operations Research Office and Research Analysis Corporation; Practice of operations research and management science; RAND Corporation.**

References

[1] Kreiner, H.W. (1992), *Fields of Operations Research*, Operations Research Society of America, Baltimore, Maryland.

[2] Kreiner, H.W. (1994), "Operations Research in Agriculture: Thornthwaite's Classic Revisited," *Operations Research*, **42**, 987–997.

[3] Larson, R.C. (1988), "There's More to a Line Than Its Wait," *Technology Review*, **91–5**, 60–67.

[4] Morse, P.M. and Kimball, G.E. (1946), *Methods of Operations Research*, OEG Report 54, Office of the Chief of Naval Operations, U.S. Navy Department, Washington, D.C.

[5] Thornthwaite, C.W. (1953), "Operations Research in Agriculture," *Jl. Operations Research Society of America*, **1**, 33–38.

[6] Tidman, K.R. (1984), *The Operations Evaluation Group*, Naval Institute Press, Annapolis, Maryland.

FIFO

The First-In, First-Out queue discipline in which customers are taken out of the line for service in the exact order in which they arrived (meant to be equivalent to the first-come, first-served scheme). See **FCFS; Queueing theory.**

FINITE SOURCE

When the potential number of customers who could use a queueing system is finite, as in models of machine repair. See **Queueing theory.**

FIRE MODELS

Bernard Levin

Rockville, Maryland

INTRODUCTION: In the fields of fire science and fire protection engineering, the term fire models refers to mathematical/computer models that predict the environment in the vicinity of the burning material or item, and, therefore, the impact of the fire within the room of fire or within nearby rooms. There are a number of such models (Friedman, 1991). The growth of the fire itself is described in terms of the products of combustion such as heat produced and the generation of toxic and nontoxic gases: these characteristics are usually inputs to the fire model. The fire models predict the temperature of the room of fire origin over time at different heights in the room. Some models predict the temperatures in a limited number of other rooms. Similarly, the models will also predict the flow of toxic gases.

Data regarding the generation of heat and gases from the fire are usually based on data from controlled burns. These burns can range from large scale simulations of fires of interest to small scale test procedures. Developing data inputs to the fire models by extrapolating, modifying, combining and synthesizing the data from controlled burns is an art that requires considerable technical knowledge and experience, and requires understanding of the basic assumptions in the model. For some applications, a simple mathematical expression is satisfactory (e.g., heat generated $= at^n$, where $t =$ time; $a > 0$; and $n > 1$): attempts are underway to develop more sophisticated mathematical models.

Until the early 1980s, fire models were mainly of academic interest. Section A15-3.1.3 of the 1981 Edition of the Life Safety Code contained a procedure for determining the minimum acceptable exhaust fan capacity in an atrium that will maintain an acceptable level of smoke at and below a prescribed height. The specific application was for a multi-level prison cell block with the fire being in a cell heavily loaded with combustible materials. This was based on the ASET Model developed by Cooper (1982, 1984).

Many of the current applications of fire models are motivated by liability lawsuits. The models can be used to predict what is likely to have been the growth of the fire and spread of the products of combustion – and consequently the likelihood of damage, injuries and fatalities – if the defendant had used less combustible materials, included additional fire safety features in the building, or taken other actions that one might expect would have helped avoid the injuries or damage caused by the fire.

Fire models can be used in studying product fire risk by predicting the effect of changing fire related characteristics of combustible products (Bukowski, Stiefel, Clarke, and Hall, 1992). In these applications, a number of pairs of fire scenarios are modeled, where for each pair the scenarios are identical except for the two products being compared. This use of fire models has potential for use in developing fire safety regulations and for redesigning combustible products.

Most fire models that are being used in practical applications are Two Zone Models. These models use the simplifying assumption that a room can be considered as two zones separated by a horizontal neutral zone of a height that can be ignored. The temperature and gas concentrations are uniform within the hot upper layer, the space above the neutral zone or neutral plane. The same is true of the cooler and less smokey lower layer. This simplification makes it possible to handle the mathematical calculations for interesting and useful situations using personal computers. This assumption appears to be a reasonable approximation to fire conditions in a room based on visual observation of smoke (or artificial smoke) and of measured temperature at a number of heights during controlled burns.

The models are validated by comparing the flow of the products of combustion in controlled fires with the predictions of the models. Several types of fires are used. The most reproducible fire is produced by a gas burner: the heat produced by a gas burner can be rather consistent in repeated burns. Wood cribs also provide a rather reproducible fire. (Wood cribs are constructed by laying layers of separated wood strips, with the strips in a layer being perpendicular to the strips in adjacent layers.) Controlled burns using single items of furniture can also be used if repeated burns use items of furniture built to the same specifications.

When reproducible fires are used to study validity, the agreement between fire tests and model predictions is often sufficiently close to encourage use of the models when the fire is somewhat similar to that in the validation test, Jones and Peacock (1989), Nelson and Deal (1991), and Peacock et al. (1993). In applications, the validity of the results is limited by the data inputs to the fire model, that is, how well the fire of interest is described by the data inputs.

COMPUTER MODELS: HAZARD I is a user friendly set of computer models and supporting materials that can be used to model specific fires in buildings with a limited set of compartments or rooms (Bukowski *et al.*, 1989). It includes components for moving the occupants (the EXITT model described below) and for determining if the occupants are incapacitated or killed based on the cumulative exposure over time to heat and toxic gases. It is the first comprehensive application of fire modeling and is currently being used in litigation cases, and hazard and risk analysis studies.

HAZARD I supporting materials include data from controlled burns regarding heat and toxic gases produced. These data can be used as input to the fire model that distributes the products of combustion throughout the building.

There are two models of building evacuation of note. EVACNET determines the time required to evacuate a large building when the capacity of the arcs of the network (e.g. the stairs) are less than the demand, that is, buildings that have queues during total building evacuation (Francis and Saunders, 1979). The model determines the escape paths for the occupants that minimize the time for the last occupant to leave the building. The solutions require some occupants to take non-shortest-path routes to avoid queues. Since in real fire emergencies, occupants are unlikely to choose the most efficient route from the standpoint of total building evacuation, EVACNET provides a lower limit for evacuation time. The model can be used in designing capacity (i.e., width) and location of exit routes before the building is built and for providing occupants with recommended escape routes in occupied buildings prior to a fire emergency. EVACNET is based on "an advanced capacitated network flow transshipment algorithm, a specialized algorithm used in solving linear programming problems with network structure," (Kisko and Francis, 1983).

EXITT is a fairly straightforward "node" and "arc" model of a family sized residence (Levin, 1987, 1989). It simulates the decisions and actions, as well as the evacuation progress, of the residents during a fire. Occupants may be asleep, have limited travel speed, or need assistance to evacuate. Actions include waking up, investigation, rescue, and evacuation. The model contains a large number of behavioral (or decision) rules. These rules are based on the behavior of people in fires as determined through fire investigations. Data inputs include smoke characteristics throughout the building over time: some of the decision rules include response to smoke conditions. The model has face validity. Given the unavailability of detailed data on the evacuation times in real fires, determining validity empirically is a substantial methodological challange. There have been no rigorous validity studies. The model is deterministic but it can be expanded to include probabilistic characteristics in the behavioral rules. Developing the required probability distributions within the decision rules is much more of a challenge than expanding the computer model. A version of EXITT is part of HAZARD I.

Fire researchers have employed decision analysis and decision trees to estimate the impact of policy decisions. Each path through the tree represents: (1) the expected value of the fire loss if the path describes the fire scenario; and (2) the probability that a random fire scenario, in the defined population of fires, is described by the path. Each node in the path further describes the fire scenario and sets the probabilities of the alternatives. In a population of upholstered chair fires, successive nodes (points of branching) along a path might be: What is the ignition source? Is someone home? Is a responsible person awake? Is there a functional smoke detector? Is the fire discovered before spreading to a second item?

Probability assignments are based on data as available and on professional judgment (e.g.,

Delphi technique). Levinthal (1980) used this approach to study alternate fire safety standards for liquid insulated transformers. Helzer, Offensend and Buchbinder (1979) used this approach to estimate the effect of increasing the percentage of homes with operating smoke detectors and to estimate the effect of issuing a mandatory product standard that would require new furniture to have features to prevent misplaced cigarettes from causing flaming fires.

See **Decision analysis; Decision trees; Delphi method; Networks; Simulation.**

References

[1] Bukowski, R.W., R.D. Peacock, W.W. Jones, and C.L. Forney (1989), "Technical Reference Guide for the HAZARD I Fire Hazard Assessment Method," NIST Handbook 164, Volume II, National Institute of Standards and Technology, Gaithersburg, Maryland.

[2] Bukowski, R.W., S.W. Stiefel, F.B. Clarke, and J.R. Hall (1992), "Predicting Product Fire Risk: A Review of Four Cases," *Fire Hazard and Fire Risk Assessment*, ASTM STP 1150, M.M. Hirschler, American Society of Testing and Materials, Philadelphia, 136–160.

[3] Cooper, L.Y. (1982), "A Mathematical Model for Estimating Available Safe Egress Time in Fires," *Fire and Materials*, 6, 135–144.

[4] Cooper, L.Y. (1984), "Appendix B: An Interim Buoyant Smoke Control Approach (for Atrium-Like Arrangements)," in Nelson, H.E., A.J. Shibe, B.M. Levin, S.D. Thorne, and L.Y. Cooper, "Fire Safety Evaluation System for National Park Service Overnight Accommodations," NBSIR 84-2896, National Bureau of Standards, Gaithersburg, Maryland.

[5] Friedman, R. (1991), "Survey of Computer Models for Fire and Smoke, Second Edition," Factory Mutual Research Corp., Norwood, Massachusetts.

[6] Francis, R.L., P.B. Saunders (1979), "EVACNET: Prototype Network Optimization Models for Building Evacuation," NBSIR 79-1738, National Bureau of Standards, Washington, DC.

[7] Helzer, S.G., F.L. Offensend, and B. Buchbinder (1977), "Decision Analysis of Strategies for Reducing Upholstered Furniture Fire Loses," NBS Technical Note 1101, National Institute of Standards and Technology, Gaithersburg, Maryland.

[8] Jones, W.W., and R.D. Peacock (1989), "Refinement and Experimental Verification of a Model for Fire Growth and Smoke Transport," in Wakamatsu, T., Y. Hasemi, A. Sekizawa, P.G. Seeger, P.J. Pagni, and C.E. Grant, *Fire Safety Science: Proceedings of the Second Interna-*

tional Symposium, *International Association for Fire Science*, Hemisphere Publishing, New York.

[9] Kisko T.M., and R.L. Francis (1983), "EVAC-NET +: A Computer Program to Determine Optimal Building Evacuation Plans," preprint of paper presented at the SFPE Symposium, Computer Applications in Fire Protection: Analysis, Modeling, and Design, March 1984.

[10] Levin, B.M. (1987), "EXITT, a Simulation Model of Occupant Decisions and Actions in Residential Fires: Users Guide and Program Description," NBSIR 87-3591, National Bureau of Standards, Gaithersburg, Maryland.

[11] Levin, B.M. (1989), "EXITT, a Simulation Model of Occupant Decisions and Actions in Residential Fires," in Wakamatsu, T., Y. Hasemi, A. Sekizawa, P.G. Seeger, P.J. Pagni, and C.E. Grant, *Fire Safety Science: Proceedings of the Second International Symposium, International Association for Fire Science*, Hemisphere Publishing Corp., New York.

[12] Levinthal, D. (1980), "Application of Decision Analysis to Regulatory Problem: Fire Safety Standards for Liquid Insulated Transformers," NBS-GCR-80-198, National Bureau of Standards, Gaithersburg, Maryland.

[13] Nelson. H.E., and S. Deal (1991), "Comparing Compartment Fires with Compartment Fire Models," in Cox. G., and B. Langsford, *Fire Safety Science: Proceedings of the Third International Symposium, International Association for Fire Science*, Elsevier, New York.

[14] National Fire Protection Association (1981), *Life Safety Code*, National Fire Protection Association, Quincy, Massachusetts.

[15] Peacock, R.D., G.P. Forney, P. Reneke, R. Portier, W.W. Jones (1993), "CFAST, the Consolidated Model of Fire Growth and Smoke Transport," NIST Technical Note 1299, National Institute of Standards and Technology, Gaithersburg, Maryland.

FIRING A RULE

The activity of carrying out the actions in a rule's conclusion, once it has been established that the rule's premise is true. See **Artificial intelligence; Expert systems**.

FIRST FEASIBLE SOLUTION

The feasible (usually basic) solution used to initiate the Phase 2 procedure of the simplex method. The solution satisfies both $Ax = b$ and $x \geq 0$. The first feasible solution is often a product of the Phase I procedure of the simplex method, while in other instances, it is user-supplied or generated by previous solutions of the problem. See **Phase I procedure; Phase II procedure; Simplex method**.

FIRST-FIT DECREASING ALGORITHM

See **Bin packing**.

FIRST-ORDER CONDITIONS

Conditions involving first derivatives.

FIXED-CHARGE PROBLEM

A problem in which a one-time cost is incurred only if the associated variable is positive. The fixed cost is added to the linear variable cost. Problems with linear constraints and fixed charges are usually reformulated using subsidiary binary variables.

FLEXIBLE MANUFACTURING SYSTEMS

Kathryn E. Stecke

The University of Michigan, Ann Arbor, Michigan

INTRODUCTION: In the metal-cutting industry, a flexible manufacturing system (FMS) is an integrated system of machine tools linked by automated material handling. Because of the versatility of the machine tools and the quick (seconds) cutting tool interchange capability, these systems are quite flexible with respect to the number of part types that can be produced simultaneously and in low (sometimes unit) batch sizes. These systems can be almost as flexible as a job shop, while having the ability to attain the efficiency of a well-balanced assembly line.

An FMS consists of several computer numerically controlled machine tools, each capable of performing many operations. Each machine tool has a limited capacity tool magazine that holds all of the cutting tools required to perform each operation. Once the appropriate tools have been loaded in the tool magazines, the machines are under computer control. During system operation, the automatic tool interchange capability of each machine allows no idle set-up time in between consecutive operations or between the use of consecutive tools. When a new tool is required, the tool magazine rotates into position, and the changer automatically interchanges the new tool with the one that is in the spindle in seconds. Each part type that is machined is defined by several operations. Each operation requires several cutting tools (say, about 5–20). All tools for each operation need to occupy slots in one or more machine tool's tool magazine.

Each cutting tool takes 1, 3, or 5 slots in a machine's magazine. Magazines can have, say, 40–160 slots. Sixty is typical. Tools wear and break, so a computer needs to track the lives of all tools. A tool that breaks during the cut can severely damage the part and sometimes the machine or spindle. Tools can be delivered to the FMS either manually or automatically, for example, via automated guided vehicles. The delivered tools are loaded into the magazines either manually or automatically.

FMSs have an automated materials handling system that transports parts from machine to machine and into and out of the system. These may consist of wire-guided Automated Guided Vehicles (AGVs), a conveyor system, or tow-line carts, with a pallet interchange with the machines. The interfaces between the materials handling system and the part are pallets and fixtures. Pallets sit on the cart and fixtures hold and clamp the parts onto the pallets. Pallets are identical and fixtures are usually of different types. Fixtures are different in order to be able to hold securely different types of parts and in different orientations. The number of pallets in the FMS defines the amount of work-in-process inventory in the system.

After some machining, parts are often checked at the machine by interchanging automatically a probe into the spindle. The probe does some at-the-machine inspection of the cuts that were made. After several operations, a cart may bring the part to a washing station, to remove the chips before either further machining, refixturing, or inspection.

Detailed descriptions of several existing systems can be found in Stecke (1992). A decision to automate should be based on both economic comparisons and strategic considerations. Assuming that management has decided that flexible manufacturing is appropriate for a particular application, perhaps to increase capacity in a certain department producing changing products or for new families of part types, there are many design issues that have to be addressed. Additional details and descriptions of these problems can be found in Stecke (1985, 1992).

The amount of flexibility that is needed or desired has to be decided and this helps to determine the degree of automation and the type of FMS that is designed. Impacting this latter decision is the type of automated material handling system that will move the parts from machine to machine. See Browne *et al.* (1984) and Sethi and Sethi (1991) for information on a spectrum of flexibility options.

Efficient and accurate mathematical and other models are required to help narrow in on the appropriate FMS design. Following the development and subsequent implementation of the FMS design, models are also useful to help set up and schedule production through the system.

FMS PLANNING AND SCHEDULING: Because of the quick automated cutting tool capability, there is negligible set-up time associated with a machine tool in between consecutive operations, as long as all of the cutting tools required for that next operation have previously been loaded into the machine tool's limited capacity tool magazine. However, determining which cutting tools should be placed in which tool magazine and then loading the tools into the magazine requires some planning and system set-up time. Those set-up decisions that have to be made and implemented before the system can begin to manufacture parts are called *FMS planning problems* (Stecke, 1983). When the system has been set–up and can begin production, the remaining problems are those of *FMS scheduling*.

The first *FMS planning problem* is to decide which, of the part types that have production requirements (either forecasted demand or customer orders), should be those next manufactured during the same time over the immediate time period. This information can be used to help determine the amount of pooling among the identical machine tools that can occur. Pooling, or identically tooling all machines that are in the same machine group, has many system benefits. For example, alternative routes for parts are automatically allowed and also, machine breakdowns may not cause production to stop. This is because all machine tools in a group, being tooled identically, are able to perform the same operations.

Another *FMS planning problem* is to determine the relative ratios at which the selected part types should be on the system, to attain good utilization. The limited numbers of pallets and fixtures of each fixture type impact these production ratios. Finally, each operation and its associated cutting tools of the selected set of part types has to be assigned to one or more of the machine tools in an intelligent manner. Different loading objectives that can be followed are applicable in different situations. When all of these decisions have been made and the cutting tools loaded into the selected tool magazines, production can begin. Then the following *FMS scheduling problems* have to be addressed.

These problems are concerned with the operation of the system after it has been set up during the planning stage. One problem is to determine an appropriate policy to input the parts of the selected part types into the FMS, or efficient means to determine which parts to input next.

Then, applicable algorithms to schedule the operations of all parts through the system have to be determined. Real-time scheduling is usually more appropriate for these automated systems, as opposed to a fixed schedule. Tool breakage, down machine tools, etc., would totally disrupt a fixed schedule. However, a fixed schedule is useful as an initial guideline to follow. Potential scheduling methods range from simple dispatching rules to sophisticated algorithms having look-ahead capabilities. Machine breakdowns and the many other system disturbances should be considered when developing scheduling and control procedures. If the system is "set-up" during the planning phase with sufficient care and flexibility, the scheduling function will be much easier.

FMS control involves the continuous monitoring of the system to be sure that it's doing what was planned for it to do and is meeting the expectations set up for it. For example, during the *FMS design* phase, policies should be determined to handle breakdown situations of many types. In any case, it is desirable to reallocate operations and reload the cutting tools (if they have to be) so that the tool changing time is minimized. Monitoring procedures for both the processes and cutting tool lives has to be specified as well as methods to collect data of various types (monitoring and breakdown). Tool life estimates should be reviewed and updated. Reasons for process errors have to be found (i.e., machine or pallet misalignment, cutting tool wear and detection, chip problems) and the problems corrected.

Because the planning and scheduling problems are complex and require a lot of data consideration, many of these problems have been framed and subdivided within a hierarchy. The solution of each subproblem provides constraints on problems lower in the hierarchy. The partition of FMS problems into *planning* (before time zero) and *scheduling* (after production begins) is one example of a hierarchy. The *FMS planning problems* are another hierarchical decomposition of a system set-up problem. Stecke (1983, 1985) suggested hierarchical and iterative approaches to several of these problems.

FMS MODELS: Models are useful to identify key factors that will affect system performance and to provide insight into how a system behaves and how the system components interact. Models should be applied to help determine the appropriate procedures to design and set up a system or strategies to help run a system efficiently.

Depending on the amount of information that is built into a particular model, simulation has the potential to be the most detailed and flexible model, allowing as much detail as desired or necessary to mimic reality. Simulation can also potentially be the most expensive and time-consuming to develop, debug, and run. Many computer runs may be required to investigate the possibilities before a decision is made.

Both open and closed queueing networks have been used to model an FMS at an aggregate level of detail. These models can take into account the interactions and congestion of parts competing for the same machines and the uncertainty and dynamics of an FMS. Most simple queueing networks require as input, certain average values, such as the average processing time of an operation at a particular machine tool and the average frequency of visits to a machine. The outputs that are obtained which are useful for evaluating the performance of a suggested system configuration are also average values and include the steady state expected production rate, mean queue lengths, and machine utilizations.

Solberg (1977) was the first to suggest the use of a simple, single-class, multiserver, closed queueing network to model an FMS. His computer program, called CAN-Q, uses Buzen's efficient algorithm to analyze product-form queueing networks. A review of such analytical queueing network models can be found in Buzacott and Yao (1986).

Some of the problems have been formulated mathematically, either as nonlinear integer programs or using linear or integer programs (Stecke, 1983). Depending on the problems formulated, some formulations are detailed and tractable and, hence, immediately useful. Other formulations are detailed and untractable. However, heuristic or other algorithms can or have been developed from the exact formulations to solve the problems. Stecke (1992) described other FMS models. Each model is useful under different circumstances and for different types of problems. For some problems, a hierarchy of models are needed to solve them.

See **Job shops; Networks of queues; Production management; Simulation.**

References

[1] Browne, J., D. Dubois, K. Rathmill, S.P. Sethi and K.E. Stecke (1984), "Classification of Flexible Manufacturing Systems," *FMS Magazine*, 2, 114–117.

[2] Buzacott, J.A. and D.D.W. Yao (1986), "Flexible Manufacturing Systems: A Review of Analytical Models," *Management Science*, 32, 890–905.

[3] Sethi, A.K. and S.P. Sethi (1990), "Flexibility in Manufacturing: A Survey," *International Jl. Flexible Manufacturing Systems*, 2, 289–328.

[4] Solberg, J.J. (1977), "A Mathematical Model of Computerized Manufacturing Systems," *Proceedings of the 4th International Conference on Production Research*, Tokyo, Japan.

[5] Stecke, K.E. (1992), "Flexible Manufacturing Systems: Design and Operating Problems and Solutions," Chapter in Maynard's *Industrial Engineering Handbook*, 4th Edition, W.K. Hodson, ed., McGraw-Hill, New York.

[6] Stecke, K.E. (1985), "Design, Planning, Scheduling, and Control Problems of Flexible Manufacturing Systems," *Annals Operations Research*, 3, 3–12.

[7] Stecke, K.E. (1983), "Formulation and Solution of Nonlinear Integer Production Planning Problems for Flexible Manufacturing Systems," *Management Science*, 29, 273–288.

FLOAT

The amount of time a project job can be delayed without affecting the duration of the overall project. Total float is the difference between the time that is calculated to be available for a work item to be completed and the estimated duration of that item. See **Network planning**.

FLOW

The amount of goods or material that is sent from one node (source) in a network to another node (sink). See **Network optimization**.

FLOW SHOP

See **Sequencing and scheduling**.

FLOW TIME

See **Sequencing and scheduling**.

FMS

See **Flexible manufacturing systems**.

FORECASTING

J. Scott Armstrong

University of Pennsylvania, Philadelphia

INTRODUCTION: Formal forecasting procedures are needed only if there is uncertainty about the future. Thus, a forecast that the sun will rise tomorrow is of little value. Forecasts are also unnecessary when one can completely control the event. For example, predicting the temperature in one's home does not require the use of forecasting procedures because one has the ability to control it. Many decisions, however, involve uncertainty, and in these cases formal forecasting procedures (referred to simply as "forecasting") are useful. For example, reducing uncertainty about changes in the environment or about effects of policy changes can help managers to make better decisions.

In practice, forecasting is often confused with planning. Whereas planning is concerned with what the world *should* look like, forecasting is concerned with what it *will* look like. Figure 1 summarizes this relationship. Forecasting methods are used to predict outcomes for each plan. If forecasted outcomes are not satisfactory, plans should be revised. This process is repeated until forecasted outcomes are satisfactory. Revised plans are then implemented and

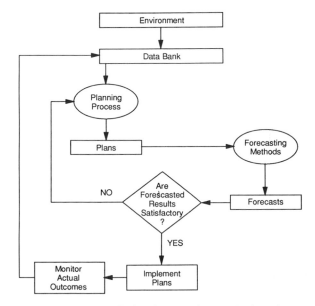

Fig. 1. Framework for forecasting and planning.

actual outcomes are monitored for use in the next planning period.

With some notable exceptions (primarily in psychology and economics), research on forecasting began in earnest around 1960 when Brown (1959) introduced exponential smoothing. Brown's method implemented the philosophy that the most recent data should have the greatest influence on forecasts.

Box and Jenkins (1970) advocated that forecasting models be based on the nature of the data, a philosophy that has had much influence on forecasting. Box and Jenkins had a strong influence on statisticians, and their procedures have been adopted by many companies. Nevertheless, these procedures have not led to measurable gains in accuracy in comparison with simpler quantitative methods (Armstrong, 1985).

The forecasting field grew rapidly since 1960, such that Fildes' (1981) bibliography contained 4,000 entries, although most of these were only indirectly related to forecasting. Growth since 1980 was aided by the founding of the International Institute of Forecasters, a multidisciplinary society of researchers on forecasting methodology. Two academic journals were founded in the 1980s, the *Journal of Forecasting* and the *International Journal of Forecasting*.

Basic forecasting methods are briefly described below. Next, guidelines are provided for selection of methods. Typical forecasting applications in OR/MS are then described along with suggestions on which methods are most appropriate for each situation. Major research findings are highlighted and citations are provided for key review papers. Finally, implementation is discussed.

FORECASTING METHODS: Forecasting methods differ in several ways. They may use subjective or objective processes to analyze information. Also, some methods use information other than the variable of interest; of particular importance here is the use of causal information. Finally, forecasting methods differ in that some use classifications whereas others use relationships. These factors allow us to describe methods using the forecasting tree of Figure 2. We describe each method briefly and summarize some of the more important findings about each. Makridakis and Wheelwright (1989) provide details on application of these methods.

Judgmental forecasts remain the most popular method used by managers for making important decisions. In an effort to improve judgmental procedures, researchers have obtained some important findings. Among these are to: 1) use more than one expert, preferably between five and 20;

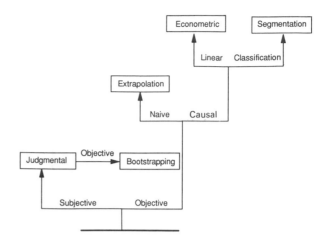

Fig. 2. Forecasting methodology tree.

2) use structured methods (e.g. decompose the problem in order to make best use of experts' information, and obtain anonymous predictions from each expert to avoid group pressures); and 3) frame the problem so as to avoid biases (e.g. state the forecasting problem in several different ways when asking experts to make forecasts). In some cases, one can forecast by using the intentions of key actors. Intentions surveys are used, with success that has been improving over the years, to forecast such things as political elections and adoption of new products. For example, people's intentions to purchase can be assessed by asking them to examine various alternative product offerings (a procedure known as "conjoint analysis"). Bunn and Wright (1991) provide a review of literature on judgmental forecasting and show how it relates to statistical forecasting methods.

Bootstrapping, a form of expert system, converts judgmental methods into objective procedures. One way to do this is to obtain protocols of experts, whereby the expert describes the process as he/she makes forecasts. Another approach is to create a series of situations and ask the expert to make forecasts for each. These judgmental forecasts are then regressed against data that describe the situations in order to develop a model. Once developed, bootstrapping models offer a low cost procedure for making forecasts. They almost always provide an improvement in accuracy, although these improvements are typically small. Bootstrapping is especially useful for forecasts when one does not have data, such as in forecasting new products, or when outcomes are difficult to observe, such as with personnel selection of managers.

Extrapolation methods are appropriate when one has historical time series. Such methods have been widely used for short-term forecasts of inventory and production. Some important findings are to: 1) seasonally adjust data, 2) use relatively simple methods, 3) extrapolate trend when the historical trend has been consistent and it is expected to continue, and 4) be more conservative in extrapolation of trends (e.g., forecast smaller changes from the most recent value) as uncertainty about the forecast increases. Gardner (1985) provides a review of research on exponential smoothing methods, and Fildes (1988) provides a broader review of extrapolation methods.

Econometric methods are appropriate when one needs to forecast what will happen under different assumptions about the environment, and what will happen given different strategies. Econometric methods are most useful when, 1) strong causal relationships are expected, 2) causal relationships can be estimated, 3) large changes are expected to occur in causal variables over the forecast horizon, and 4) changes in causal variables can be forecasted. When these conditions do not hold, such as is typically the case for short-range forecasts of the economy, econometric methods are not more accurate. Important findings about econometric methods are: 1) base the selection of causal variables upon theory and domain knowledge, rather than upon statistical fit to historical data, 2) use relatively simple models (e.g., break the problem into a series of smaller independent problems; do not use simultaneous equations; use only models that can be specified as linear in the parameters), and 3) use variables only if the estimated relationship is in the same direction as specified *a priori*. Fildes (1985) provides a review of econometric methods for forecasting.

Figure 2 represents idealized methods. In practice, analysts borrow from among these methods. An important finding from the research is to employ more than one method and combine forecasts. Clemen (1989) reviews research on combining forecasts.

SELECTION OF METHODS: Empirical literature provides guidance on choosing the method that is most appropriate for a given situation. Figure 3 summarizes this advice. It shows when to use subjective or objective methods, causal or naive methods, and linear or classification methods.

General advice from the forecasting literature (Figure 3) is not sufficient for the selection of a given model. Selection should be supplemented by 1) requirements of the forecasting task, 2) domain knowledge, 3) knowledge of forecasting methods, and 4) nature of the data. Until re-

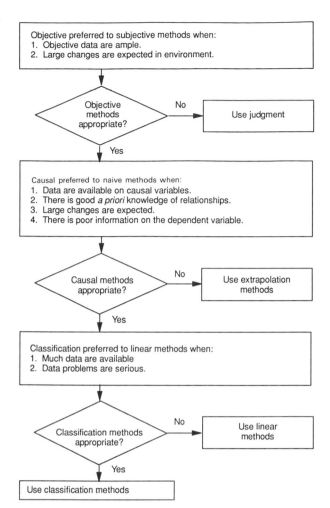

Fig. 3. Accuracy of methods by situation.

cently, it has been difficult to integrate these types of information. Managers have tended to rely upon their judgment and ignore statisticians, while statisticians have approached the problem through ever more complex approaches and have ignored managers.

Rule-based forecasting uses generalizations from the forecasting literature to integrate managers' domain knowledge with statisticians' quantitative methods. Rule-based forecasting relies upon a decision support approach to eliciting information from domain experts. It also incorporates the latest guidelines from research on forecasting. This information is used to apply differential weights to forecasts from different forecasting methods. Rule-based forecasting has produced substantial improvements in accuracy when applied to extrapolation methods (Collopy and Armstrong, 1992).

Selection of the appropriate forecasting method for a given situation depends to a large extent upon procedures used to compare alternative methods. Significant progress has been made in testing procedures. For example, statisticians have relied upon sophisticated procedures for analyzing how well models fit historical data; however, this has been of little value for the selection of forecasting methods. Instead, one should rely on *ex ante* forecasts from realistic simulations of the actual situation faced by the forecaster. Also, traditional error measures, such as mean square error, have not provided a reliable basis for comparison of methods. The median absolute percentage error is more appropriate because it is invariant to scale and is not overly influenced by outliers. For comparisons using a small set of series, it is desirable to also control for degree of difficulty in forecasting. One measure that does this is the median relative absolute error, which compares the error for a given model against errors for the naive 'no change' forecast (Armstrong and Collopy, 1992).

In addition to reducing uncertainty (that is, improving accuracy), forecasting is also concerned with *assessing* uncertainty. Although statisticians have given much attention to this problem, their efforts generally relied upon the fit to historical data as a way to infer forecast uncertainty. Empirical studies have shown that, often, over half of actual outcomes are well outside of the 95% confidence intervals. A better approach, then, is to simulate the actual forecasting procedure as closely as possible and use the distribution of the resulting *ex ante* forecasts to assess uncertainty (Chatfield, 1993).

APPLICATIONS OF FORECASTING IN MANAGEMENT: Forecasting methods can be applied to many areas of management. Figure 4 provides a description of some areas and outlines how these relate to one another. For example, econometric methods are often appropriate for long-range forecasting of the environment and industry. Extrapolation methods are useful for short-range forecasting of costs, sales, and market share. Forecasts of competitors' actions could be made judgmentally; for example, if a company had only one major competitor, their actions could be forecasted by role-playing. (People would act out the role of the competitor in a simulation of the actual situation.) Role-playing can also be useful to forecast actions of government regulators.

IMPLEMENTATION OF FORECASTS: Progress in forecasting has been achieved through application of structured and quantitative techniques. Despite this, managers continue to rely heavily on subjective forecasts for important decisions. Even when quantitative forecasts are made, managers use their judgment to revise them.

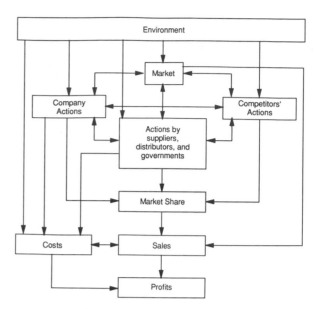

Fig. 4. Some needs for forecasts in firms.

Judgment sometimes aids the forecast, but often it harms it. As a general rule, experts are especially skilled at determining the current state of affairs, but are less competent in predicting change. Thus, judgment improves the forecast when it is used to reduce uncertainty about the current level, but it often harms the forecast of change. The challenge is to develop procedures that will most effectively blend quantitative and judgmental methods.

Implementation of forecasts depends not only on intrinsic merit of the forecast, but also upon its acceptability to the organization. This, in turn, may depend upon the capability of users and upon organizational norms. It is useful, then, to assess how well different procedures perform in realistic settings. Some encouraging research in this area has been conducted by Bretschneider *et al.* (1989). They concluded, for example, that the use of complex procedures by organizations harmed forecast accuracy.

Accurate forecasts are often ignored, particularly when they involve bad news. Scenarios (Armstrong, 1985), which are stories about alternative futures, can aid implementation in such cases by making the forecast seem plausible to the decision maker. The plausibility of the scenario can be increased by using vivid and concrete examples, showing a logical sequence of events, having the decision maker describe how he or she would act in the scenario, and using past tense when writing the scenario.

CONCLUSIONS: Research on forecasting since 1960 has relied heavily upon empirical testing of alternative approaches. It also has examined conditions under which various methods are most appropriate. These research strategies have provided useful findings. In many cases the findings conflict with expectations of statisticians and managers, and this has slowed adoption of new methods. As a result, many organizations have not yet adopted many of the developments.

See **Delphi method; Econometrics; Exponential smoothing; Regression analysis; Time series.**

References

[1] Armstrong, J. Scott (1985). *Long-Range Forecasting: From Crystal Ball to Computer* (2nd ed.), John Wiley, New York.

[2] Armstrong, J. Scott and F. Collopy (1992). "Error Measures for Generalizing About Forecasting Methods: Empirical Comparisons," *International Jl. Forecasting*, 8, 69–80.

[3] Box, George E. and G.M. Jenkins (1970). *Time Series Analysis for Forecasting and Control.* Holden Day, San Francisco.

[4] Bretschneider, Stuart I., W.L. Gorr, G. Grizzle, and E. Klay (1989). "Political and Organizational Influences on the Accuracy of Forecasting State Government Revenues," *International Jl. Forecasting*, 5, 307–319.

[5] Brown, Robert G. (1959). *Statistical Forecasting for Inventory Control.* McGraw Hill, New York.

[6] Bunn, Derek and G. Wright (1991). "Interaction of Judgmental and Statistical Methods: Issues and Analysis," *Management Science*, 37, 501–518.

[7] Clemen, Robert T. (1989). "Combining Forecasts: A Review and Annotated Bibliography," *International Jl. of Forecasting*, 5, 559–583.

[8] Chatfield, Chris (1993). "Calculating Interval Forecasts," *Jl. Business and Economic Statistics*, 11, 121–135.

[9] Collopy, Fred and J.S. Armstrong (1992). "Rule-based Forecasting: Development and Validation of an Expert Systems Approach to Combining Time Series Extrapolations," *Management Science*, 38, 1394–1414.

[10] Fildes, Robert (1988). "Recent Developments in Time Series Forecasting," *OR Spektrum*, 10, 195–212.

[11] Fildes, Robert (1981). *A Bibliography of Business and Economic Forecasting.* Gower Publishing, Farnborough, Hants, England. [An update was published by the Manchester Business School under the same title in 1984.]

[12] Fildes, Robert (1985). "Quantitative Forecasting – The State of the Art: Econometric Models," *Jl. Operational Research Society*, 36, 549–580.

[13] Gardner, Everette S. Jr. (1985). "Exponential Smoothing: The State of the Art," *Jl. Forecasting*, 4, 1–28.

[14] Makridakis, S. *et al.* (1982). "The Accuracy of Extrapolation (Time Series) Methods: Results of a Forecasting Competition," *Jl. Forecasting*, 1, 111–153.

[15] Makridakis, Spyros and S.C. Wheelwright (1989). *Forecasting Methods for Management* (3rd ed.). John Wiley, New York.

[16] Makridakis, Spyros and S.C. Wheelwright (1987). *The Handbook of Forecasting: A Manager's Guide.* John Wiley, New York.

[17] Special Issue (1988). "The Future of Forecasting," *International Jl. Forecasting*, 4, No. 3.

FORWARD CHAINING

An approach to reasoning in which an inference engine determines the effect of current known variable values on unknown variables by firing all rules whose premises can be established as being true. See **Artificial intelligence**; **Expert systems**.

FORWARD KOLMOGOROV EQUATIONS

In a continuous time Markov chain $\{X(t)\}$, define $p_{ij}(t)$ as the probability that $X(t+s)=j$, given that $X(s)=i$, for $s,t \geq 0$, and r_{ij} as the transition rate out of state i to state j. Then Kolmogorov's forward equations say that, for all states i,j and times $t \geq 0$, $dp_{ij}(t)/dt = \Sigma_{k \neq j} r_{kj} p_{ik}(t) - v_j p_{ij}(t)$ where v_k is the transition rate out of state k, $v_k = \Sigma_j r_{kj}$. See **Markov chains**; **Markov processes**.

FORWARD-RECURRENCE TIME

Suppose events occur at epochs $T_1, T_2, \ldots$ such that the interevent times $T_k - T_{k-1}$ are mutually independent, positive random variables with common cumulative distribution function. Then the forward recurrence time from an arbitrary time t is the time from t to the next occurrence. See **Point stochastic processes**; **Renewal processes**.

FOURIER-MOTZKIN ELIMINATION METHOD

A computational procedure for solving a system of linear inequalities.

FRACTIONAL PROGRAMMING

Siegfried Schaible

University of California, Riverside

INTRODUCTION: Certain decision problems in OR/MS, as well as other extremum problems, give rise to the optimization of ratios. Constrained ratio optimization problems are commonly called *fractional programs*. They may involve more than one ratio in the objective function.

One of the earliest fractional programs (though not called so) is an equilibrium model for an expanding economy in which the growth rate is determined as the maximum of the smallest of several output-input ratios (von Neumann 1937, 1945). Since then, but mostly after the classical paper by Charnes and Cooper (1962), some nine hundred publications have appeared in fractional programming; for comprehensive bibliographies, see Schaible (1982, 1993). Monographs solely devoted to fractional programming include Schaible (1978) and Craven (1988).

Almost from the beginning, fractional programming has been discussed in the broader context of generalized concave programming. Ratios, though not concave in general, are often still generalized concave in some sense. An introduction to fractional programming in this context is Avriel, Diewert, Schaible, and Zang (1988).

NOTATION AND DEFINITIONS: Suppose f, g and h_j ($j = 1, \ldots, m$) are real-valued functions which are defined on the subset X of the n-dimensional Euclidean space $\Re^n$ and let $\boldsymbol{h} = (h_1, \ldots, h_m)^{\mathrm{T}}$ where T denotes the transpose. We consider the ratio

$$q(\boldsymbol{x}) = f(\boldsymbol{x})/g(\boldsymbol{x}) \tag{1}$$

over the set

$$S = \{\boldsymbol{x} \in X | \boldsymbol{h}(\boldsymbol{x}) \le \boldsymbol{0}\}. \tag{2}$$

assuming $g(\boldsymbol{x}) > 0$ on X. The nonlinear program

$$sup\{q(\boldsymbol{x}) | \boldsymbol{x} \in S\} \tag{3}$$

is called a (single-ratio) fractional program.

In addition, the following three types of multi-ratio fractional programs are of interest:

$$sup\{\textstyle\sum_{i=1}^p q_i(\boldsymbol{x}) | \boldsymbol{x} \in S\}, \tag{4}$$

$$sup\{min_{1 \le i \le p} q_i(\boldsymbol{x}) | \boldsymbol{x} \in S\}, \tag{5}$$

and the multi-objective fractional program

$$sup_{\boldsymbol{x} \in S}\{q_1(\boldsymbol{x}), \ldots, q_p(\boldsymbol{x})\}. \tag{6}$$

Here $q_i(\boldsymbol{x}) = f_i(\boldsymbol{x})/g_i(\boldsymbol{x})$ ($i = 1, \ldots, p$) when f_i and g_i are real-valued functions on X with $g_i(\boldsymbol{x}) > 0$. Problem (5) is often referred to as a *generalized fractional program* (Schaible and Ibaraki, 1983).

The focus in fractional programming is the objective function and not the feasible region S. As in most publications, we assume that the h_j are convex functions on the convex domain X yielding a convex feasible region S.

Most of the theory for fractional programs (3)–(6) is developed under the assumption that ratios satisfy the following concavity/convexity condition: f is concave and g is convex on the convex set X (f is to be nonnegative if g is not affine-linear, as below). Such problems are called *concave fractional programs*. However, it is to be noted that the objective function in these problems is *not* concave in general; hence they are not concave programs. Problem (3) is called a *quadratic fractional program* if f and g are quadratic functions and S is a convex polyhedron.

A special case is the *linear fractional program* where f and g are affine-linear functions and S is a convex polyhedron

$$sup\{(\boldsymbol{c}^{\mathrm{T}}\boldsymbol{x} + \boldsymbol{\alpha})/(\boldsymbol{d}^{\mathrm{T}} + \beta) | A\boldsymbol{x} \le \boldsymbol{b}, \boldsymbol{x} \ge \boldsymbol{0}\}. \tag{7}$$

Here $\boldsymbol{c}, \boldsymbol{d} \in \Re^n$, $\boldsymbol{b} \in \Re^m$, $\boldsymbol{\alpha}, \boldsymbol{\beta} \in \Re$, A is an $m \times n$ matrix and the denominator is positive on the feasible region. We will see below that linear and concave fractional programs have still many properties in common with linear and concave programs.

SINGLE-RATIO FRACTIONAL PROGRAMS: We find the following types of single-ratio fractional programming applications in the literature: economic, non-economic and indirect applications.

Economic Applications – The efficiency of a system is sometimes characterized by a ratio of economical and/or technical terms. Then, maximizing the efficiency leads to a fractional program. Examples of such ratios are:

profit/capital, profit/revenue, cost/volume, productivity, relative usage of material, return/cost, return/risk, expected cost/beta-index, (expected) cost/time, profit/time, liquidity, earnings per share, dividend per share, weighted outputs/weighted inputs, income/(investment + consumption), mean/standard deviation.

Such ratios arise in resource allocation, transportation, production, maintenance, inventory, finance, data envelopment analysis, and macro-economics for example. No longer are these rates merely used to control past economic behavior. Instead, the optimization of rates is getting more attention in decision making for future projects. Depending on the form of the functions in the numerator and denominator, many of the ratio optimization problems above are linear, quadratic or concave fractional programs.

Non-economic Applications – In information theory, the capacity of a communication channel can be defined as the maximal transmission rate, thus giving rise to a (nonquadratic) concave fractional program. In numerical analysis, the eigenvalue problem can be reduced to constrained maximization of the Rayleigh quotient, and hence, leads to a (nonconcave) quadratic fractional program. In physics, maximization of the signal-to-noise ratio gives rise to a concave quadratic fractional program.

Indirect Applications – Fractional programs may also arise in the process of solving other optimization problems involving no ratios. Examples are:

> subproblems in large-scale mathematical programming, deterministic substitutes in stochastic mathematical programming, subproblems in nondifferentiable convex programming, problems in connection with interior-point methods for linear programming, dual location problems, approximations to numerically intractable portfolio selection problems, bounds on the trauma outcome function for emergency medical facilities.

Depending on the original optimization problem, often linear, quadratic or concave fractional programs are encountered.

PROPERTIES: Concave fractional programs have the following properties (Avriel *et al.*, 1988):

Proposition 1: A local maximum is a global maximum since the objective function $q(x) = f(x)/g(x)$ is semistrictly quasiconcave.

Proposition 2: A maximum is unique if the numerator $f(x)$ is strictly concave or the denominator $g(x)$ is strictly convex since in this case the objective function is strictly quasiconcave.

Proposition 3: In case of differentiable functions $f(x)$, $g(x)$, $h(x)$, a solution of the Karush-Kuhn-Tucker conditions is a maximum since the objective function $q(x)$ is pseudoconcave.

For *linear fractional programs*, the following additional property holds:

Proposition 4: A maximum is attained at a vertex in case of a (nonempty) bounded feasible region S, since the objective function is quasiconvex (in addition to quasiconcave).

Concave and linear fractional programs share not only the above properties with concave and linear programs, respectively, but they can also be related to these programs through transformations. The first transformation below changes the variables, whereas the second one maintains the same variables, but requires a parameter in the transformed problem.

Introducing the new variables

$$y = [1/g(x)]x, \quad t = 1/g(x), \tag{8}$$

we can show (Schaible, 1976):

Proposition 5: A concave fractional program (3) with an affine-linear denominator can be reduced to the concave program

$$sup\{t\, f(y/t)\,|\,t\, h(y/t) \le 0,\, t\, g(y/t) = 1,\, y/t \in X,\, t > 0\}. \tag{9}$$

If $g(x)$ is not affine-linear, an equivalent concave program is obtained for (9) by relaxing the equality in (9) to $t\, g(y/t) \le 1$.

In the special case of a linear fractional program (7), the equivalent concave program (9) becomes the linear program

$$sup\{c^{\mathrm{T}}y + \alpha t\,|\,Ay - bt \le 0,\, d^{\mathrm{T}}y + \beta t = 1,\, y \ge 0,\, t > 0\} \tag{10}$$

where $t > 0$ can be replaced by $t \ge 0$ if (7) has an optimal solution. The equivalence between (7) and (10) was first established by Charnes and Cooper (1962).

In the second transformation, variables and the feasible region are maintained and a parameter is introduced to separate numerator and denominator (Dinkelbach, 1967). Consider

$$sup\{f(x) - \lambda g(x)\,|\,x \in S\}, \quad \lambda \in \Re \text{ parameter.} \tag{11}$$

If (3) is a concave, linear, or quadratic fractional program, then (11) is a parametric concave, linear, or quadratic program, respectively.

Suppose $f(x)$, $g(x)$ are continuous and S is a (nonempty) compact set. Then we have:

Proposition 6: Problems (3) and (11) have the same optimal solutions where $\lambda = \bar{\lambda}$ is the unique zero of the strictly decreasing, continuous function

$$F(\lambda) = sup\{f(x) - \lambda g(x)\,|\,x \in S\}. \tag{12}$$

Turning now to duality for fractional programs, we note that standard concave programming duality relations are no longer true in case of concave, or even linear fractional programs. However, Proposition 5 can be used to introduce duality via the equivalent concave program (9) (Schaible, 1976). For a detailed presentation of various duality approaches as well as their use in sensitivity analysis, see Schaible (1978), Avriel *et al.* (1988), and Craven (1988).

The properties of concave and linear fractional programs in Proposition 1–6 allow for at least four different solution strategies (Martos, 1975; Schaible and Ibaraki, 1983; Craven, 1988):

a) direct solution of the quasiconcave (pseudoconcave) program (3);

b) solution of the equivalent concave (linear) program (9);

c) solution of the dual of (9);

d) solution of the parametric concave (linear) program (11).

In the case of d), rather than applying parametric programming techniques, the iterative method by Dinkelbach (1967) can be used. It turns out to be equivalent to Newton's classical method for finding the zero $\bar{\lambda}$ of $F(\lambda)$ in (12). Various modifications and computational results were discussed in Schaible and Ibaraki (1983).

MULTI-RATIO FRACTIONAL PROGRAMS:

Maximizing the Sum of Ratios – Problem (4) arises naturally in decision making when several of the rates above are to be optimized and a compromise is sought that optimizes the weighted sum of these rates. Other applications of this model were given by Schaible (1990).

Unfortunately, none of the above properties of concave fractional programs hold anymore if each ratio is a quotient of a concave and a convex function, even in the linear case. Only some preliminary theoretical and algorithmic results are known for this important, but difficult problem (Craven, 1988; Schaible, 1990).

Maximizing the Smallest of Several Ratios – Apart from the economic equilibrium model by von Neumann (1937, 1945), problems in financial planning and fund allocation under equity considerations give rise to a generalized fractional program (5) (Schaible, 1990). Furthermore, the same model is of interest in numerical mathematics in rational approximation involving the Chebychev norm. In all these examples, the ratios are quotients of concave and convex functions.

Starting with von Neumann (1937, 1945), several authors have proposed a duality theory for concave generalized fractional programs (Avriel *et al.*, 1988; Craven, 1988; Schaible, 1990). Though different approaches are employed, most duals coincide and are again a generalized fractional program. The objective function of the primal is semistrictly quasiconcave and the one of the dual is semi-strictly quasiconvex (Avriel *et al.*, 1988). Thus a local optimal solution is global in both the primal and the dual. A duality theory has been established for these nonconcave problems which is as rich as the one for concave and linear programs.

Concave generalized fractional programs can be solved in either the primal or the dual by an extension of Dinkelbach's algorithm. In case of more than one ratio, this is no longer identical with Newton's method. It gives rise to a sequence of concave (linear) programs as subproblems. The convergence properties are well analyzed. Computational results for various modifications of this algorithm have been encouraging (Schaible, 1990).

Multi-Objective Fractional Programs – Problem (6) arises when several rates above are to be maximized simultaneously and, in contrast to the problems in (4) and (5), a unifying objective function is not considered; instead, the decision maker is to be provided with the set of efficient alternatives, that is, all those feasible solutions for which none of the rates can be increased without decreasing another rate. Some theoretical results are known for (6) in case of ratios of concave and convex functions including duality relations (Craven, 1988; Schaible, 1990). Also, the connectedness of the set of efficient alternatives has been established under limiting assumptions. Additional theoretical and algorithmic results are known for the case of two ratios (Schaible, 1990).

CONCLUSION:

Concave single-ratio fractional programs, as well as concave generalized fractional programs, have been analyzed quite successfully, both theoretically and algorithmically. More research needs to be done for the nonconcave case, sum-of-ratios problem (4), and multi-objective fractional programming (6).

See **Data envelopment analysis; Linear fractional-programming problem; Linear programming; Multi-objective programming; Nonlinear programming; Quadratic programming.**

References

[1] Avriel, M., W.E. Diewert, S. Schaible and I. Zang (1988). *Generalized Concavity*, Plenum, New York.

[2] Charnes, A. and W.W. Cooper (1962). "Programming With Linear Fractional Functionals," *Naval Research Logistics Quarterly* 9, 181–186.

[3] Craven, B.D. (1988). *Fractional Programming*, Heldermann Verlag, Berlin.

[4] Dinkelbach, W. (1967). "On Nonlinear Fractional Programming," *Management Science* 13, 492–498.

[5] Martos, B. (1975). *Nonlinear Programming: Theory and Methods*, North-Holland, Amsterdam.

[6] Schaible, S. (1976). "Duality in Fractional Programming: A Unified Approach," *Operations Research* 24, 452–461.

[7] Schaible, S. (1978). *Analyse und Anwendungen von Quotientenprogrammen*, Mathematical Systems in Economics 42, Hain-Verlag, Meisenheim.

[8] Schaible, S. (1982). "Bibliography in Fractional Programming," *Zeitschrift für Operations Research* 26, 211–241.

[9] Schaible, S. (1990). "Multi-Ratio Fractional Programming – Analysis and Applications," *Proceedings of 13th Annual Conference of Associazione per la Matematica Applicata alle Scienze Economiche e Sociali*, Verona/Italy, September 1989, Mazzoleni, P., ed., Pitagora Editrice, Bologna, 47–86.

[10] Schaible, S. (1993). "Fractional Programming," *Handbook of Global Optimization*, Horst, R. and P. Pardalos, eds., Kluwer Academic Publishers, Dordrecht.

[11] Schaible, S. and T. Ibaraki (1983). "Fractional Programming," *European Jl. Operational Research* 12, 325–338.

[12] von Neumann, J. (1937). "Über ein ökonomisches Gleichungssystem und eine Verallgemeinerung des Brouwerschen Fixpunktsatzes," *Ergebnisse eines mathematischen Kolloquiums* 8, Menger, K., ed., Leipzig und Wien, 73–83.

[13] von Neumann, J. (1945). "A Model of General Economic Equilibrium," *Review Economic Studies* 13, 1–9.

FRAMING

Refers to how a problem is presented to decision makers, or how they formulate it in their minds. **Choice theory**.

FRANK-WOLFE METHOD

See **Quadratic programming**.

FREE FLOAT

The amount of time a designated activity can be delayed without affecting succeeding activities of a project. This will be the float for the final activity of a chain (or for a single activity which does not lie on a chain with equal float). See **Float**; **Network planning**.

FREE VARIABLE

A variable that can take on any value, as contrasted to a variable that must take on nonnegative values. In a linear-programming problem, a variable that is free can be expressed as the difference between two nonnegative variables. However, when using the simplex method to solve a linear-programming problem with free variables, it is more effective to eliminate those variables by means of constraints in which they appear. See **Unrestricted variable**.

FREIGHT ROUTING

The itinerary of a shipment through a logistics network.

FTRAN

The procedure for computing the updated version of the entering column in a simplex iteration, when the LU factors of the basis matrix are given in product form. The name FTRAN (forward transformation) derives from the fact that the eta file is scanned forward in the process. See **Eta file**.

FUZZY SETS

Saul I. Gass

University of Maryland, College Park

When analyzing a problem situation, we tend to impose a requirement for precision on the problem's description and its data. But, it soon becomes clear that precision in language and accuracy in data are unattainable goals. We must resort to interpretations and approximations. One way of doing this is to assume that the problem situation is fixed and the related data are deterministic. The analysis is then done by evaluating applicable scenarios, and the solutions described in terms of ranges of data, robustness of the underlying model, and associated sensitivity analyses. In contrast, the analysis may be accomplished by assuming standard statistical properties and related probability distributions for the data, that is, the problem is analyzed in terms of a known stochastic environment. Most mathematical modeling analyses are either deterministic or stochastic. However, there are some problems in which the imprecision of their description and data call for a different analysis approach, that is, the use of fuzzy set theory.

The concept of fuzzy sets was first introduced by Zadeh (1962), and further developed and reported on by Zadeh (1965, 1973), Bellman and Zadeh (1970), Gaines and Kohout (1977), and others. Zadeh's motivation for the development of fuzzy set theory is given in Zadeh (1962) where he stated a need for "... a mathematics of fuzzy or cloudy quantities which are not describable in terms of probability distributions." The concept of fuzziness is set forth by Zadeh and Bellman (1970) as follows:

> By fuzziness, we mean a type of imprecision which is associated with *fuzzy sets*, that is, classes in which there is no sharp transition from membership to nonmembership. For example, the class of *green objects* is a fuzzy set. So are classes of objects characterized by such commonly used adjectives as large, small, significant, important, serious, simple, accurate, approximate, etc. Actually, in sharp contrast to the notion of a class or set in mathematics, most of the classes in the real world do not have crisp boundaries which separate those objects which belong to a class from those that do not. In this connection, it is important to note that, in the discourse between humans, fuzzy statements such as "John is *several* inches taller than Jim," "x is *much larger* than y," "Corporation X has a bright future," "the stock market has suffered a *sharp decline*," convey information despite the imprecision of the italicized words. (pp. B141–B142)

Given a set X (finite or countable), let A be a subset of X. For every element $x \in X$, we can define the degree of inclusion (or grade) to which x is included in the subset A by a membership function, $\mu_A(x)$, with $\mu_A(x)$ taking on nonnegative values between 0 and 1. The function $\mu_A(x)$ is termed the grade of membership of $x \in A$. The subset A is then a fuzzy set and is defined by the set of ordered pairs $A = \{x, \mu_A(x) | x \in X\}$. In contrast, the classical mathematical notion of a set is defined by requiring that for a subset A of X and for an element $x \in X$, x either belongs to A (its grade is 1) or does not belong to A (its grade is 0). Such a set is referred to as a *crisp set*.

Fuzzy sets enable us to capture the meaning of vague and qualitative descriptions. For example, let x be the set of men and the fuzzy subset A be the set of *tall* men. We can define the set of tall men by the (nonunique) membership function as follows:

For $x \leq 5'\ 8''$, $\mu_A(x) = 0$.
For $5'8'' < x \leq 5'10''$, $\mu_A(x) = 0.2$
For $5'10'' < x \leq 6'$, $\mu_A(x) = 0.4$
For $6' < x \leq 6'2''$, $\mu_A(x) = 0.6$
For $6'2'' < x \leq 6'4''$, $\mu_A(x) = 0.8$
For $6'4'' < x$, $\mu_A(x) = 1.0$

Bellman and Zadeh (1970) give the following example of a fuzzy set A and corresponding membership function $\mu_A(x)$, where A is the set of numbers substantially larger than 10, with

$$\mu_A(x) = 0 \text{ if } x \leq 10$$

$$\mu_A(x) = [1 + (x - 10)^{-2}]^{-1} \text{ if } x > 10$$

For fuzzy sets, using the membership function, the concepts of union, intersection, complement and other logical relations and laws can be defined and established. Similarly, definitions and mathematical operations for fuzzy sets in terms of sum, product, convexity and mapping can be developed.

FUZZY LINEAR PROGRAMMING: The relationship of fuzzy sets to other operations research decision-aiding paradigms and models is discussed in Carlson (1984). In particular, the following describes the fuzzy set counterpart of the linear-programming model due to Zimmermann (1991). Let the standard linear-programming (LP) model be:

Maximize $z = \boldsymbol{cx}$

subject to

$$\boldsymbol{Ax} \leq \boldsymbol{b}$$
$$\boldsymbol{x} \geq \boldsymbol{0}$$

where $\boldsymbol{A}$ is an $(m \times n)$ matrix. The fuzzy set LP model is then given by:

Maximize $z \simeq \boldsymbol{cx}$

subject to

$$\boldsymbol{Ax} \simeq \boldsymbol{b}$$
$$\boldsymbol{x} \geq \boldsymbol{0}$$

where the symbol $\simeq$ indicates that the objective function and constraints are fuzzy and are characterized by membership functions of the following form:

$$\mu_i = \begin{cases} 1 & \text{if} \quad (A'X)_i \leq b'_i \\ 1 - \dfrac{(A'X)_i - b'_i}{\delta_i} & \text{if} \quad b'_i < (A'X)_i \leq b'_i + \delta_i \\ 0 & \text{if} \quad (A'X)_i > b'_i + \delta_i \end{cases}$$

where A' is matrix $\boldsymbol{A}$ augmented with the row vector $\boldsymbol{c}$, and $\boldsymbol{b}'$ the vector $\boldsymbol{b}$ augmented with the aspiration level $\boldsymbol{z}$, and index i denotes the *i*th row of the matrix. The δ_i are subjectively chosen tolerance intervals. The fuzzy LP can be restated as a normal LP as follows:

Maximize λ

subject to

$$\lambda \delta_i + (A'x)_i \leq b'_i + \delta_i \quad (i = 1, \ldots, m+1)$$

$$x \geq 0$$

As Carlson (1984) notes, the advantages of the fuzzy LP are: (i) it is more flexible than the regular LP model, (ii) conflicts between the constraints and aspiration levels can be analyzed and eliminated, and (iii) it allows for the incorporation of multiple objective functions.

Applications of fuzzy sets to OR/MS and decision making in general include multicriteria decision making in fuzzy environments (Bellman and Zadeh, 1970; Zimmermann, 1991), fuzzy linear programming (Zimmermann, 1979, 1991), fuzzy dynamic programming (Esogbue and Bellman, 1984), with applications in logistics, inventory control, expert systems, manufacturing, scheduling and location, media selection (Zimmermann, 1991). Fuzzy theory has been applied in the manufacture, design and operations of cameras (to determine when an object is in focus), washing machines (to determine what wash cycle to apply to a load of clothes), automobiles (to determine when to shift gears); and in robotics, medical diagnosis, process control, pattern recognition, information retrieval, and artificial intelligence (Sugeno, 1985; Terano *et al.*, 1992; Zimmermann, 1991). Also see Wiedey and Zimmermann (1978) for an application of fuzzy LP to media selection.

See **Choice theory; Decision analysis; Linear programming; Preference theory; Utility theory.**

References

[1] R.E. Bellman and L.A. Zadeh (1970), "Decision-making in a Fuzzy Environment," *Management Science*, 17(4), B141–B164.

[2] C. Carlson (1984), "On the Relevance of Fuzzy Sets in Management Science Methodology," pp. 11–28 in Zimmermann *et al.* (1984).

[3] A.O. Esogbue and R.E. Bellman (1984), "Fuzzy Dynamic Programming and its Extensions," pp. 147–167 in Zimmermann *et al.* (1984).

[4] A.O. Esogbue and R.C. Elder (1983), "Measurement and Valuation of a Fuzzy Mathematical Model for Medical Diagnosis," *FSS*, 10, 223–242.

[5] B.R. Gaines and L.J. Kohout (1977), "The Fuzzy Decade: A Bibliography of Fuzzy Systems and Closely Related Topics," in M.M. Gupta, G.N. Saridis and B.R. Gaines, eds., *Fuzzy Automata and Decision Processes*, North-Holland, New York.

[6] M.M. Gupta, R.K. Ragade and R.R. Yager, eds. (1979), *Advances in Fuzzy Set Theory and Applications*, North Holland, New York.

[7] A. Jones, A. Kaufmann and H.-J. Zimmermann, eds. (1985), *Fuzzy Sets Theory and Applications*, Reidel Publishing Company, Dordrecht, The Netherlands.

[8] McNeill, D. and P. Freiberger (1993). *Fuzzy Logic*, Touchstone Books, New York.

[9] M. Sugeno, ed. (1985), *Industrial Application of Fuzzy Control*, North-Holland, New York.

[10] T. Terano, K. Asai and M. Sugeno (1992), *Fuzzy Systems Theory and Its Applications*, Academic Press, Boston.

[11] G. Wiedey and H.-J. Zimmermann (1978), "Media Selection and Fuzzy Linear Programming," *Jl. Operational Research Society*, 29, 1071–1084.

[12] L.A. Zadeh (1962), "From Circuit Theory to Systems Theory," *Proceedings of Institute of Radio Engineers*, 50, 856–865.

[13] L.A. Zadeh (1965), "Fuzzy Sets," *Information and Control*, 8, 338–353.

[14] L.A. Zadeh (1973), "Outline of a New Approach to the Analysis of Complex Systems and Decision Processes," *IEEE Transactions Systems, Man and Cybernetics*, SMC-3, 28–44.

[15] H.-J. Zimmermann, L.A. Zadeh and B.R. Gaines, eds. (1984), *Fuzzy Sets and Decision Analysis*, North-Holland, New York.

[16] H.-J. Zimmermann (1991), *Fuzzy Set Theory*, 2nd. edition, Kluwer Academic Publishers, Boston.

GA

See **Genetic algorithms**.

GAME THEORY

William F. Lucas

Claremont Graduate School, California

INTRODUCTION: Game theory studies situations involving conflict and cooperation. The three main elements of a game are players, strategies, and payoffs. Games arise when two or more decisionmakers (*players*) select from various courses of action (called *strategies*) which in turn result in likely outcomes (expressed as *payoffs*). There must be at least two interacting participants with different goals in order to have a game. Game theory makes use of the vocabulary from common parlor games and sports. It is, nevertheless, a serious mathematical subject with a broad spectrum of applications in the social, behavioral, managerial, financial, system, and military sciences.

Game theory differs from classical optimization subjects in that it involves two or more players with different objectives. It also extends the traditional uses of probability and statistics beyond the study of one-person decisions in the realm of statistical uncertainty. This latter case is often referred to as "games of chance" or "games against nature" in contrast to the "games of skill" studied in game theory. Many aspects of social and physical science can be viewed as "zero-person" games since actions are frequently specified by various laws that are not under human control.

Game theory presumes that conflict is not an evil in itself and as such unworthy of study. Rather, this topic arises naturally when individuals have free will, different desires, and the freedom of choice. Furthermore, this subject often provides guidelines to aid in the resolution of conflict. Game theory also assumes that the players can quantify potential outcomes (as in measurement theory or utility theory), that they are rational in the sense that they seek to maximize their payoffs, and skillful enough to undertake the necessary calculations. The theory of games attempts to describe what is optimal strategic behavior, the nature of equilibrium outcomes, the formation and stability of coalitions, as well as fairness.

There are many different ways to classify games. A significant difference exists between the *two-person* games and the multiperson ones (also called the *n-person* games when $n \geq 3$). There is a major distinction depending upon whether games are played in a *cooperative* or *noncooperative* manner. The nature of the types or amount of information available to the players is very fundamental in the analysis of games, and this relates to whether the best way to play involves "pure" or "randomized" strategies.

INFORMATION AND STRATEGIES: Any possible way a player can play completely through a game is called a *pure* strategy for this player. It is an overall plan specifying the actions (moves) to be taken in all eventualities which can conceivably arise. In theory such pure strategies suffice to solve many popular recreational games such as checkers which have perfect information. A game has *perfect information* if throughout its play all the rules, possible choices, and past history of play by any player are known to all of the participants. In this case there are no unknown positions or hidden moves, and thus no need for secrecy, deception, or bluffing.

The first general theorem in game theory was published by the logician Ernst Zermelo in 1913. It states that there is an *optimal* pure strategy for playing any finite game with perfect information. An elementary game with perfect information such as tic-tac-toe soon becomes no real challenge. Each player soon discovers a strategy that prevents the other from winning. From then on this game always results in a draw. On the other hand, Zermelo's theorem is an example of an "existence theorem." It does not provide a practical way to determine an optimal pure strategy for many interesting but complex games with perfect information such as chess. Furthermore, one cannot even spell out one pure strategy for chess—one that lists a possible response to all legal moves by an opponent. The challenge of such games comes from the bewildering complexity and imagination involved.

Many other games like the card games known as poker, however, do not have perfect information. Secrecy, deception, randomness, and bluffing are in order. Another level of interest and a new notion of strategic choice enters. Pure strategies no longer suffice for optimal play. The main fundamental concept for such noncooperative games is that of a mixed strategy. A *mixed strategy* for a player is a probability distribution over his or her pure strategies. The idea is that a player will pick a particular pure strategy with some given probability. This greatly enlarges the realm of strategies from which each player can choose. The tradeoff, on the other hand, is that the players must now view their potential payoffs as "averages." They thus resort to maximizing their gains in terms of *expected values* in a statistical sense. These ideas are best illustrated by the theory of matrix games.

MATRIX GAMES: The best known class of games are the two-person, zero-sum games. Any game is called *zero-sum* when the particular payoffs to the players always sum to zero. In the case of two players this states that one's winnings equals the other's losses. There is clearly no room for cooperation in this case. These games are also referred to as *strictly competitive* or *antagonistic*. They arise in many sorts of duels, inspections, searches, business competitions and voting situations, as well as most parlor games and sports contests.

These games are characterized by an m by n table of numbers and are accordingly referred to as *matrix games*. The rows of the table correspond to the pure strategies for the first player, denoted by I. The columns are likewise identified with the pure strategies of the second player, II. The numbers within the table itself are the corresponding payoffs received by player I from player II. A negative number in the matrix means that I makes a (positive) payment to II. Each player seeks to select a strategy so as to maximize his or her payoff.

We can illustrate the theory of matrix games by the following two-by-two, zero-sum game of matching coins. Player I has two pure strategies: to show heads H or tails T. Player II can likewise select H or T. If the two players' coins "match" with either two heads or else two tails, then player I wins $3 or $1, respectively, from player II. If the coins do not match (one H and one T), then player II collects $2 from player I. This game can be represented by the following table.

The worse thing that can happen to player I is his *maximin value* of -2. (This is the largest of the smallest numbers from each row.) Similarly,

Table 1. A matrix game.

		Player II		
		H	T	Row Minima
Player I	H	3	-2	-2
	T	-2	1	-2
Column Maxima:		3	1	

the *minimax* value for player II is 1. This is the smallest loss player II can guarantee and it occurs when player II plays the second column T (while I plays the second row T). There is a "gap" of $3 between this maximin value of $1 and the minimax value of $-\$2$. Both players can "win" some of this gap of 3 units if they resort to mixed strategies and are willing to evaluate their payoffs in terms of expected values.

If either player uses the *optimal* mixed strategy of playing H with probability 3/8 and T with probability 5/8, he or she can ensure an average payoff of

$$3(3/8) - 2(5/8) = -1/8 = -2(3/8) + 1(5/8)$$

against any strategy by the opposing player. Using mixed strategies the players can "close the gap" between -2 and 1 to the game's (expected) *value* of $-1/8$. This game favors player II who should average a gain of 12.5¢ per play. This game is not *fair* in the sense that optimal play does not produce an expected outcome of 0. The optimal mixed strategies (3/8, 5/8) for players I and II along with the value $-1/8$ are called the *solution* of this matrix game. (In general, the two players will not have the same optimal mixed strategy as is the case for this game with a symmetric payoff matrix.)

The main theoretical result for matrix games is the famous *minimax theorem* proved by John von Neumann in 1928. It states that any matrix game has a solution in terms of mixed strategies. Each player has an optimal mixed strategy which guarantees that he or she will achieve the value of the game (in the statistical sense of expected values).

Von Neumann also observed in 1947 that the duality theorem in linear programming is equivalent to his minimax theorem. Furthermore, it is known that the subjects of matrix games and linear programming are entirely equivalent mathematically. Various algorithms are known for solving m by n matrix games. However, one typically expresses the solution for a matrix game in terms of a pair of dual linear programs and employs one of the popular algorithms used in the latter subject.

NONCOOPERATIVE GAMES: When games are not zero-sum or have more than two players, then it is essential to distinguish between whether they are played in a cooperative or noncooperative manner. To *cooperate* means the players are able to communicate (negotiate or bargain) and correlate their strategy choices before they play. Also, that any agreements made are binding (enforceable). In contrast, each player in a *noncooperative* game chooses a strategy unaware of the selection made by the other players.

For noncooperative games the primary ingredient to any notion of solution is that of an *equilibrium* point. A set of (pure or mixed) strategies, one for each player, is said to be in equilibrium if no one player can change his or her strategy unilaterally and thus obtain a higher payoff. Unfortunately, equilibrium outcomes do not always possess every property that one would desire for a satisfactory concept of solution. Nevertheless, this idea of equilibrium seems crucial to the very notion of what can be called a "solution" to a noncooperative game. It is the social science analogy to the idea of equilibrium or stability in mechanical systems.

The difficulties that might arise with equilibrium outcomes are illustrated by the following two 2 by 2, nonzero-sum, two-person games known as the *prisoner's* dilemma and *chicken*. These are the driving forces behind escalation (arms races and price wars) and confrontation, respectively. In these two games each player has two strategies: to compromise C or to defect D. The resulting payoffs are indicated in the following two tables where we assume that each player prefers the outcome of 4 over 3 over 2 over 1. The payoffs in these tables give a pair of numbers (a,b) where a is the payoff to player I (the row player) and b is for player II (the column player). For example, if players I and II select the strategies D and C, respectively, in Chicken (Table 3) they obtain the respective payoffs of 4 and 2.

In either of these games, the best overall outcome for the two players when taken together is the strategy pair (C,C) where each compromises and in turn receives the second best payoff of 3. This would be the likely outcome if these games were played cooperatively. This result, however, is *not* in equilibrium. Either player can achieve the higher payoff of 4 if he or she alone were to

Table 2. The prisoner's dilemma.

Player II

	C	D
C	(3,3)	(1,4)
D	(4,1)	(2,2)

Player I

Table 3. The game of chicken.

Player II

	C	D
C	(3,3)	(2,4)
D	(4,2)	(1,1)

Player I

switch from strategy C to D. In the prisoner's dilemma the "dominant" strategy for each player is D. One does better individually by selecting D, no matter what the other chooses. This leads to each receiving 2, their second worse payoff. In chicken the two (pure) strategy pairs (C,D) and (D,C) both lead to an equilibrium result. No one player can switch strategy and do better in either case. However, these two outcomes are not *interchangeable*. If both players select D in an attempt to reach the particular equilibrium that would pay them 4, then the resulting strategy pair (D,D) leads to their worst payoffs (1,1). Of the 78 possible 2 by 2 games, these two are the most troublesome.

Some noncooperative games have no equilibrium in *pure* strategies. In 1950 John F. Nash extended von Neumann's minimax theorem for two-person, zero-sum games to prove that every finite multiperson, general-sum game has at least one equilibrium outcome in *mixed* strategies. Algorithms to calculate equilibria involve nonlinear techniques and often use "path-following" approaches that may be approximate in nature. There are also many refinements and extensions of the idea of equilibrium described here, and these concepts are fundamental to quantitative approaches in modern economics and politics, as well as system analysis and operations research.

COOPERATIVE GAMES: If the players in a game are allowed to cooperate, they typically agree to undertake joint action for the purpose of mutual gain. In this case coalition formation is a common activity, and the additional worth that can accrue to any potential coalition is of primary interest. In practice, the players often solve some optimization problem or consider some noncooperative game in order to arrive at the amount of additional value available from cooperation. The problem that remains concerns how this newly obtained wealth will be, or should be, divided among the players. This latter aspect is again a competition as each participant seeks to maximize his or her own gain. This may involve negotiations, bargaining, threats, arbitration, coalitional realignments, attempts to arrive at "stable" allocations or coalition structures, as well as appeals to different ideas about fairness. It is thus not surprising that several different models and "solution concepts" have been proposed for multiperson cooperative games.

The first general model and idea of a solution for the multiperson cooperative game was presented in the monumental book by John von Neumann and Oskar Morgenstern in 1944 (3rd ed., 1953). Their approach is referred to as the *n*-person game in characteristic function form. One begins with a set $N = \{1, 2, \ldots, n\}$ of n players who are indicated by $1, 2, \ldots,$ and n. A *characteristic function* v assigns a value $v(S)$ to each subset S of N. This number $v(S)$ represents the worth achievable by the coalition S, independent of the remaining players in the complementary set $N - S$. In this context they proposed a notion of solution that they called a *solution* and which is often referred to now as a *stable set*. Stable sets proved to have some difficulties of both a theoretical and practical nature. They are also rather mathematically involved, and thus will not be presented here. They are, nevertheless, still a useful tool, especially for the class of games for which the "core" (see below) is nonexistent. Dozens of alternate solution concepts have since been proposed for these coalitions games, and five of these have received the most attention. Three of these solution concepts will be described in the context of the following three-person illustration.

Three neighboring towns A, B, and C plan to tap into an additional water source at O. The costs (in \$100,000) for installing the alternate segments of water pipe appear on the edges in Figure 1. The joint costs for the various subsets of the three-person coalition $\{A,B,C\}$ are obtained by finding the "minimal cost spanning

tree" for each such coalition. The total cost for the coalition $\{A,B,C\}$ is $c(ABC) = 18$ and it is realized by the link *OACB*. Similarly, the minimal costs for the six other coalitions are:

$$c(AB) = 15 \text{ via } OAB, \quad c(AC) = 11 \text{ via } OAC,$$

$$c(BC) = 16 \text{ via } OCB, \quad c(A) = 5 \text{ via } OA,$$

$$c(B) = 14 \text{ via } OB, \text{ and } c(C) = 9 \text{ via } OC.$$

(We shorten expressions like $c(\{A, B, C\})$ and $v(\{A, B, C\})$ to $c(ABC)$ and $v(ABC)$, respectively.)

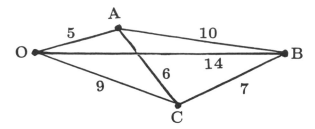

Fig. 1. A cost allocation game.

One can reformulate this problem in terms of the savings available by means of cooperation. Each coalition considers what it saves in a joint project over what it would have cost its members if they were each to make a separate connection to the source O. This "savings game" has the following characteristic function: $v(ABC) = 10$ ($= c(A) + c(B) + c(C) - c(ABC)$), $v(AB) = 4$, $v(AC) = 3$, $v(BC) = 7$, and $v(A) = v(B) = v(C) = 0$. The three towns can save $10 \times \$100,000 = \$1,000,000$ by acting together. The problem is: how should these savings be allocated to the individual towns? How does one select the three numbers (x_A, x_B, x_C) in the *imputation set* determined by the relations $x_A + x_B + x_C = 10 = v(ABC)$, $x_A \geq 0 = v(A)$, $x_B \geq 0 = v(B)$, and $x_C \geq 0 = v(C)$? These points are pictured by the large triangle in Figure 2.

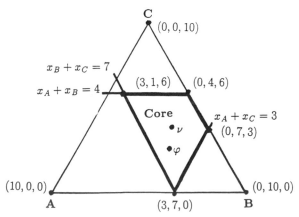

Fig. 2. Solutions to the cost game.

One solution concept for cooperative games is called the core. The *core* consists of those allocations in the imputation set for which every coalition S receives or exceeds its value $v(S)$. No coalition has the capability to improve its total allocation at a core point by going off on its own. For our savings game the core consists of all imputations (x_A, x_B, x_C) that satisfy the inequalities $x_A + x_B \geq 4 = v(AB)$, $x_A + x_C \geq 3 = v(AC)$, and $x_B + x_C \geq 7 = v(BC)$. This is the four-sided region in Figure 2. Note that the core is not a unique allocation, and for some games it can be the empty set. However, the core is always nonempty for such cost allocation games.

Another popular solution concept is called the nucleolus. The *nucleolus* is the one allocation in the "center" of the core. For our savings game the nucleolus is the imputation = (6/4, 19/4, 15/4). (The nucleolus is also defined for those games with empty cores as the unique imputation where the core would "first appear" when each proper coalition S has its value $v(S)$ decreased uniformly.) If we translate the nucleolus for our savings game back to a cost allocation for our original problem we obtain the allocation

$$\$100{,}000[(5{,}14{,}9) - v] = (\$350{,}000;\ \$925{,}000;\ \$525{,}000).$$

In 1951 Lloyd S. Shapley introduced a solution concept that also provides a "fair" and unique outcome for our savings game. The *Shapley value* in general gives the average of each player's marginal contribution taken over all possible orderings of a set N of n players. Each one of the $n!$ orderings (permutations) is a way the full coalition N could build up, one player at a time. There are six orderings of the three towns A, B, and C in our example: (CBA), (BCA), (CAB), (BAC), (ACB), and (ABC). The Shapley value φ_A for town A is accordingly $6\varphi_A = 2[v(ABC) - v(BC)] + [v(AB) - v(B)] + [v(AC) - v(C)] + 2[v(A) - 0] = (2 \cdot 3) + 3 + 4 + (2 \cdot 0) = 13$. A similar calculation gives $6\varphi_B = (2 \cdot 7) + 4 + 7 + (2 \cdot 0) = 25$ and $6\varphi_C = (2 \cdot 6) + 3 + 7 + (2 \cdot 0) = 22$. The Shapley value φ for our savings game is $\varphi = (\varphi_A, \varphi_B, \varphi_C) = (13/6, 25/6, 22/6)$. Note that this point is in the core of this game, although this is not always the case for cost allocation problems. In our original cost problem, corresponds to the result ($283,333; $983,333; $533,333).

The various solution concepts for multiperson cooperative games have been applied throughout economics, political science, and operations research. The core is important in the study of economic markets. The nucleolus is viewed as a fair outcome for many bargaining situations. The Shapley value has also been employed as a measure of power for voting systems, where the core is typically the empty set.

See **Decision analysis; Duality theorem; Graph theory; Linear programming; Minimum spanning tree problem; Prisoner's dilemma game; Utility theory.**

References

[1] Aumann, R.J. and S. Hart, editors (1992, 1994, 1995), *Handbook of Game Theory: With Application to Economics*, Volumes 1, 2 and 3, North-Holland, Amsterdam.

[2] Lucas, W.F. (1971), "Some Recent Developments in n-person Game Theory," *SIAM Review*, 13, pp. 491–523.

[3] Luce, R.D. and H. Raiffa (1957), *Games and Decision*, Wiley, NY (reprinted by Dover, 1989).

[4] McDonald, J. (1975), *The Game of Business*, Doubleday, Garden City, NY (reprinted by Anchor, 1977).

[5] Ordeshook, P.J., editor (1978), *Game Theory and Political Science*, New York University Press, New York.

[6] von Neumann, J. and O. Morgenstern (1953), *Theory of Games and Economic Behavior*, 3rd ed., Princeton University Press, New Jersey.

[7] Williams, J. D. (1954), *The Compleat Strategyst* (sic), McGraw-Hill, New York (revised edition, 1966; Dover, 1986).

GAMING

William Schwabe

The RAND Corporation, Santa Monica, California

INTRODUCTION: Abt (1970) broadly defines a game as "an activity among two or more independent decision-makers seeking to achieve their objectives in some limiting context." Gaming involves the activity itself, whereas game theory uses mathematics to seek the best strategies, the sets of decisions that decisionmaking players might make.

Games are played for entertainment, sport, teaching, training, and research. As a research method, gaming is used by psychologists, educators, and sociologists, interested in how people learn and play games, and by operations researchers, other analysts, and decision makers, interested in developing, exploring, and testing policies, strategies, hypotheses, and other ideas.

As an OR/MS method, gaming is controversial, often practiced more as an art than a science. Few methods have been so inadequately named, prompting ridicule from skeptics and attempts by adherents to call it something more serious sounding or descriptive, such as "operational gaming," "simulation gaming," "free-form gaming," and, in defense analysis, "wargaming" and "political-military gaming." Although gaming (along with several other techniques of OR/MS) has not been made as scientifically rigorous nor as universally accepted as adherents hoped for decades ago, it has helped importantly in developing strategy, in pretesting policies before actual implementation, and in communicating understanding of operational complexities.

Research games are often played as part of the planning process in developing important policies and strategies for organizations in competitive situations. Accordingly, the results – and sometimes the existence of gaming – is not publicized. For example, several Iraq-Kuwait scenarios were gamed in 1990 before Iraq actually attacked, but they are not fully documented in the open (unclassified) literature. Sources of information on research gaming include reports and bibliographies published by organizations with a tradition of gaming (such as the Naval War College, RAND, and others), the journal *Simulation & Games*, and various books and articles. Shubik (1975) is one of the most comprehensive discussions of gaming, including a game theory background for gaming, analytical and behavioral models, and examples of games used for a variety of purposes. Brewer and Shubik (1979) provided an historical and contemporary review of the use of military war games.

LEARNING FROM GAMING: People can learn from gaming by designing the game, by playing it, or by analyzing the play or results (Perla and Barrett, 1970). Greenblat (1988) discusses game design as a five-stage process: (1) setting objectives of and constraints on the game, (2) conceptual model development, (3) decisions about representation, (4) construction and refinement, and (5) documentation. Because a game is meant to model one or more important aspects of something that is operationally complex, game design is usually an intense intellectual exercise in analysis. Analysts commonly learn a great deal from the process of designing a game, as is true with other types of model design.

Most games have two or more teams, each representing a decisionmaking entity, such as a country, a military command, or a business firm, with from one to hundreds of players on a team.

Players may be assigned specific roles – a leader of a country, a CEO in a regulated industry, a local warlord – in which they can criticize or embrace the policies of their own or competing governments or organizations. Formal games have rigid rules for play, while seminar or free-form games, have few rules. Play of a game is usually divided into moves, each being a period of real time during which game time (often posited to be in the future) is assumed to be frozen. Moves usually begin with teams being presented with information players are asked to accept as true for the purposes of the game and use as a basis for their deliberations and decisions; this information is often in the form of a scenario. The set of decisions made by a player or team during a move period is sometimes called its move. Game administrators, who usually include researchers who designed the game and will analyze its results, are commonly called controllers or referees. Games are most commonly played with all participants at one site; however, distributed games can be played with remotely located players communicating via electronic mail or other means.

Despite the "make believe" aspects of gaming, players often become intellectually (and sometimes emotionally) caught up in the game, engaging in intense, goal-focused thought and discussion. In the process they learn about the issues, about their teammates, and about themselves. Controllers often learn what can go wrong operationally and how "signals" and other forms of communications between teams can be misunderstood. Analysis usually begins with a critique at the end of the last game move, attended by players, controllers, and observers. The game director may ask each team leader to present his or her analysis of what the team saw as the major issues, how they analyzed their options, what they decided, and what results they expected. Whether analysis is more formal than this depends in part on whether the game was designed as an experiment, to yield data or other observations suitable for analysis. If a series of games is played, then there is opportunity for comparative analysis.

WHY GAME? Unlike many other methods of OR/MS, gaming is not a solution method. The output of a game is not a forecast or prediction, solution, or rigorous validation. The output of a good game is increased understanding.

Gaming can do several things: reveal errors or omissions in concept; explore assumptions and uncover the implicit ones; draw out divided opinion; examine the feasibility of an operational concept; identify areas that are particularly sensitive or in which information is lacking (Quade, 1975); pool the knowledge of several experts; suggest questions or hypotheses for further study; identify the values or measures of effectiveness (MOE) that people care about; breadboard approval-winning or implementation of policies; or test strategies for long-term consequences. Gaming can do something an individual person cannot do, no matter how vigorous his or her analysis, to list the things that would never occur to him. It can help identify all the ways that a carefully composed statement can be misinterpreted. It can "generate the phenomena of understanding and misunderstanding, perception and misperception, bargaining, demonstrations, dares and challenges accommodation, coercion and intimidation, conveyance of intent, and uncertainty about what each other has already done or decided on. There are some things that just cannot be done by a single person or by a team that works together." (Levine, Schelling, and Jones, 1991).

PROSPECTS FOR GAMING: The popularity of gaming is cyclical; its use, however, appears to be on an upward path. Video-conferencing and electronic mail networks open possibilities for less expensive games with broader participation, including international play. Advances in computers and software make it easier to develop models to support games, to use them on the fly during games to update scenarios, to query data files in response to player questions during games, and to prepare presentation graphics during the games and for post-game critiques. Video-taping has been used to present scenario updates to players in "newscast" format and to present pre-taped briefings by experts to players. Expert systems are used to support some games, but the use of artificial intelligence, rule-based "agents" in gaming, is not as active as it was in the 1980s.

Gaming has often not been as well integrated into studies using other methodologies as might be warranted. Gaming is but one form of analysis to inform policy, managerial, or operational decisions. Figure 1, adapted from Paxson (1963), summarizes some of the relationships between gaming and other analysis.

Regardless of whether gaming ever achieves the rigor early proponents sought, it appears to have continuing value as a tool of OR/MS. Gaming can often respond to changing operational or strategic contexts more rapidly than other methods. Challenges remain in making

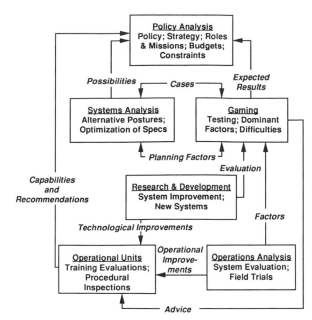

Fig. 1. Gaming and analysis relationships.

games less demanding of player time (especially important in enlisting senior officials as players), in reducing costs of games (including travel costs), and in using game results responsibly and effectively in analyses to inform decisions.

See **Game theory; Military operations research; RAND Corporation.**

References

[1] Abt, C.C. (1970). *Serious Games.* Viking Press, New York.

[2] Brewer, G. and M. Shubik (1979). *The War Game: A Critique of Military Problem Solving.* Harvard University Press, Cambridge, Massachusetts.

[3] Greenblat, C.S. (1988). *Designing Games and Simulations: An Illustrated Handbook.* Sage Publications, Newbury Park, California.

[4] Levine, R., T. Schelling, and W. Jones (1991). *Crisis Games 27 Years Later.* Report P-7719, The RAND Corporation, Santa Monica, California.

[5] Paxson, E.W. (1963). *War Gaming.* Report RM-3489-PR, The RAND Corporation, Santa Monica, California.

[6] Perla, P. and P.R.T. Barrett (1985). *An Introduction to Wargaming Its and Uses.* Report CRM 85-91, Center for Naval Analyses, Alexandria, Virginia.

[7] Quade, E.S. (1975). *Analysis for Public Decisions.* Elsevier, New York.

[8] Shubik, M. (1975). *Games for Society, Business and War: Towards a Theory of Gaming.* Elsevier, New York.

GAMMA DISTRIBUTION

A continuous random variable is said to have a gamma distribution if its probability density can be written in the form $f(t) = a(at)^{b-1} e^{-at}/\Gamma(b)$ where a and b are any positive real numbers and $\Gamma(b)$ is the gamma function evaluated at b. The constant b is called the *shape parameter*, while a (or various equivalents) is called the *scale parameter*. If b happens to be a positive integer, then $\Gamma(b) = (b-1)!$ and this gamma distribution is also called an Erlang distribution. Furthermore, if b is either an integer or half-integer (1/2, 3/2, etc.) and $a = 1/2$, the resultant gamma distribution is equivalent to the classical χ^2 distribution of statistics. See **Erlang distribution**.

GANTT CHARTS

Steven Nahmias

Santa Clara University, California

DEFINITION: There are three well-known types of Gantt charts: the Gantt load chart, the Gantt layout chart, and the Gantt project chart. A Gantt chart is essentially a bar chart laid on its side. The horizontal axis corresponds to time and the vertical axis to a collection of related activities, machines, employees, or other resource. Bars are used to represent load durations or activity starting and ending times. The Gantt chart is appealing in that it is easy to interpret and can provide a visual summary of a complex schedule.

In principal, the three types of Gantt charts are similar, but each has a somewhat different application. The load chart is used to show the amount of work assigned to resources (typically equipment) over a given amount of time. Sequencing issues are ignored here. Load charts are useful for showing work assigned to a project, but do not show the progress of an on-going project. The Gantt layout chart is used to block out reserved times on facilities and is one means of keeping track of the progress of an ongoing project.

The most popular type of Gantt chart is the Gantt project chart. A Gantt project chart is used to show the starting and ending times of all the activities comprising a project. It can be used to monitor the progress of a project and determine where stumbling blocks may be. Below, we provide an example of a Gantt project chart.

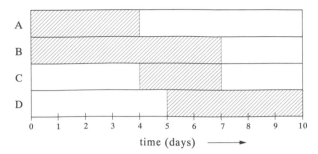

Fig. 1. Four-activity Gantt chart.

EXAMPLE: Suppose that a project consists of four activities: A, B, C, and D requiring respectively 4, 7, 3, and 5 days. Figure 1 is a Gantt chart representing the starting and ending times for these activities.

According to this chart, A and B are started at day 0 and are scheduled to be completed respectively at the start of days 4 and 7. C is begun when A ends on day 4 and is completed at the start of day 7. D is started on day five and completed on day 10. While the chart shows the start and finish times of each activity, it has the shortcoming of not showing *precedence relationships*. Specifically, did we require C to wait for the completion of A or could C have been scheduled earlier? Does D require A to be completed before it could start? Should D have been started on day 4 instead of day 5? Because of this significant limitation, years later professionals recognized that networks were much more powerful ways of representing projects, since precedence constraints could be incorporated directly into a network structure. Both Critical Path Methods (CPM) and PERT (Project Evaluation and Review Technique) overcome this shortcoming of Gantt charts. Even though it has limitations as a planning tool, the Gantt chart is still one of the most convenient ways to represent a schedule once it is determined.

While Gantt charts vary considerably in format and structure, this example contains the basic elements of all Gantt project charts. Invariably, the horizontal axis corresponds to time, and the vertical axis to a set of activities (or machines or resources for the other types of Gantt charts). Interpret "activities" very broadly here. They may be parts of a project, they may be part numbers, or machines or personnel. The bars generally correspond to beginning and ending times for activities, but might have different interpretations in other contexts. For example, they may correspond to work shifts for personnel or delivery and shipment times for parts.

IMPLEMENTATION ISSUES: There are some issues one must be concerned with when trying to implement a Gantt chart. One is the way time is measured and scaled. In the example above, time is shown in numbers of elapsed days from an arbitrary day labelled day 0. In practice, it is more common, however, for time to be measured in calendar days. The horizontal axis would correspond to specific calendar dates. While calendar dating makes starting and ending times more explicit, there are other issues to consider as well. How long is a day? In most work environments, a work day is 8 hours. In other contexts, a day may be 24 hours. Another issue is whether operations continue during weekends. There are several ways to handle this problem. The easiest is just to exclude weekend dates from the chart. The interested reader should refer to Battersby (1967) and Clark (1952) where these and related issues are discussed in detail.

EXTENSIONS: An extension of the Gantt project chart which was the precursor of modern networks is the milestone chart. Networks are collections of nodes and directed arcs. In the context of project planning, nodes represent completion of a collection of activities, and directed arcs to the durations of specific activities. Developed in the 1940s by the U.S. Navy, the milestone chart is a Gantt chart with circles representing key time periods which occur during the completion of an activity. The milestones could then be linked, in much the same way that nodes are linked on a project network. An example of a milestone Gantt chart is shown in Figure 2.

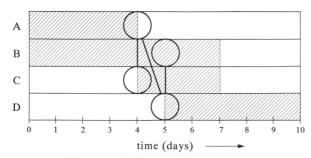

Figure 2. Milestone Gantt chart.

The vertical line linking activities A and C means that C cannot be started before A is complete. In the case of activities B and D, the vertical line there implies that D cannot be started until five days after B has started. (This suggests that B should be represented as two activities.) The diagonal line connecting A and D means that D cannot be started until A is completed.

USES: The Gantt chart was the precursor to several commercial graphical control systems, many of which are found today adorning the walls of manufacturing facilities throughout the United States. In his classic work, Moore (1967) noted several commercial variations of the Gantt chart available in the 1960s including Productrol boards, Schedugraphs, and Boardmasters. All use a time scale across the top and horizontal lines to picture machines, schedules or orders or whatever is being graphed. Although very popular in the 1950s, these manual techniques have lost favor because computers can quickly update and print progress charts.

HISTORY: The concept was originally developed by Henry L. Gantt, a contemporary of Frederick Taylor's, a major force in the development of scientific methods for operations and production control. Gantt developed the idea of a bar chart to monitor project status while he was affiliated with the Army Bureau of Ordnance during World War I. His original intent was to display graphically the status of munitions programs for that day. Gantt recognized that time was a key variable against which the progress of a program could be assessed. Gantt's development was certainly a first key step in the development of scientific methods for project management, a powerful tool for project planning. Both CPM and PERT, mentioned earlier, were consequences of the kind of planning recommended by Gantt. Dozens of texts have been written on the subject and the methods have been applied to a large variety of industries. Personal computer software products are widely available which make extensive use of Gantt charts to display schedules. An overview of project planning is given in Nahmias (1993), and more details on project planning techniques can be found in Moder, Phillips, and Davis (1983).

See **Critical path method (CPM); Network planning; Program evaluation and review technique (PERT).**

References

[1] Battersby, A. (1967), *Network Analysis for Planning and Scheduling*, Second Edition, Macmillan, London.
[2] Clark, W. (1952), *The Gantt Chart: A Working Tool for Management*, Pitman Publishing, New York.
[3] Moder, J.J., C.R. Phillips, and E.W. Davis (1983), *Project Management with CPM, PERT, and Precedence Diagramming*, Third Edition, Van Nostrand Reinhold, New York.

[4] Moore, F.G. (1967), *Manufacturing Management*, Fourth Edition, Richard D. Irwin, Homewood, Illinois.

[5] Nahmias, S. (1993), *Production and Operations Analysis*, 2nd ed., Richard D. Irwin, Homewood, Illinois.

GAUSSIAN ELIMINATION

A computational procedure for reducing a set of $(m \times m)$ linear equations $Ax = b$ to the form $MAx = Mb$, where $MA = U$ is an upper triangular matrix. The variables of the solution vector are found by solving the resulting triangular system for one variable in the last equation, and back-substituting in the next to last equation, and so on. Some form of elimination is central to the simplex method for solving linear-programming problems. See **Matrices and matrix algebra**; **Simplex method**.

GAUSS-JORDAN ELIMINATION METHOD

A computational procedure for reducing a set of $(m \times m)$ linear equations $Ax = b$ to the explicit solution form of $x = A^{-1}b$. See **Gaussian elimination**.

GENE

In genetic algorithms, the unit of inheritance, carried by chromosomes (i.e., solutions); a piece of the genetic material that determines the inheritance of a particular characteristic.

GENERALIZED ERLANGIAN DISTRIBUTION

The probability distribution of a finite sum of independent, exponentially distributed random variables whose parameters may not be the same. Sometimes, the term is also used for a convex sum of Erlang distributions, which, however, is more often called a mixture.

GENERALIZED UPPER-BOUNDED (GUB) PROBLEM

A linear-programming problem with a set of constraints of the form $\Sigma x_J = 1$, where J is a subset of the indices $j = 1, 2, \ldots, n$ and each j can appear at most once in some J. This problem is called a GUB problem and a special adaptation of the simplex method is available that reduces the computational burden of having a large number of GUB constraints.

GENERATOR (OF A MARKOV PROCESS)

The matrix of state-transition rates (intensities).

GENETIC ALGORITHMS

Zbigniew Michalewicz

University of North Carolina at Charlotte

INTRODUCTION: During the last thirty years there has been a growing interest in problem solving systems based on principles of evolution and heredity: such systems maintain a population of potential solutions, they have some selection process based on fitness of individuals, and some recombination operators. One type of such systems is a class of Evolution Strategies i.e., algorithms which imitate the principles of natural evolution for parameter optimization problems (Schwefel, 1981). Fogel's Evolutionary Programming is a technique for searching through a space of small finite-state machines (Fogel *et al.*, 1966). Glover's Scatter Search techniques maintain a population of reference points and generate offspring by weighted linear combinations (Glover, 1977). Another type of evolution-based system is Holland's Genetic Algorithms (GAs) (Holland, 1975). In 1990, Koza proposed an evolution-based system to search for the most fit computer program to solve a particular problem (Koza, 1990). Michalewicz's Evolution Programs generalize the idea of genetic algorithms allowing complex data structures (for chromosomal representation) and complex operators (Michalewicz, 1992).

However, because of their simplicity and broad applicability, genetic algorithms are the best known techniques based on analogy between computing and biology. The beginnings of genetic algorithms can be traced back to the early 1950s when several biologists used computers for simulations of biological systems. However, the work done in late 1960s and early 1970s at the University of Michigan under the direction of Holland led to genetic algorithms as they are known today.

Genetic algorithms aim at complex problems. They belong to the class of probabilistic algorithms, yet they are very different from random algorithms as they combine elements of directed and stochastic search. Because of this, GAs are also more robust than existing directed search methods. In GAs each chromosome represents a potential solution to a problem; an evolution process run on a population of chromosomes corresponds to a search through a space of potential solutions. Such a search requires balancing two (apparently conflicting) objectives: exploiting the best solutions and exploring the search space. Hill climbing is an example of a strategy which exploits the best solution for possible improvement; on the other hand, it neglects exploration of the search space. Random search is a typical example of a strategy which explores the search space ignoring the exploitations of the promising regions of the space. Genetic algorithms are a class of general purpose (domain independent) search methods which strike a remarkable balance between exploration and exploitation of the search space.

THE IDEA: As stated in Davis (1987), ". . . the metaphor underlying genetic algorithms is that of natural evolution. In evolution, the problem each species faces is one of searching for beneficial adaptations to a complicated and changing environment. The 'knowledge' that each species has gained is embodied in the makeup of the chromosomes of its members." So the idea behind genetic algorithms is to do what nature does. Let us take rabbits as an example: at any given time there is a population of rabbits. Some of them are faster and smarter than other rabbits. These faster, smarter rabbits are less likely to be eaten by foxes, and, therefore, more of them survive to do what rabbits do best: make more rabbits. Of course, some of the slower, dumber rabbits will survive just because they are lucky. This surviving population of rabbits starts breeding. The breeding results in a good mixture of rabbit genetic material: some slow rabbits breed with fast rabbits, some fast with fast, some smart rabbits with dumb rabbits, and so on. And on the top of that, nature throws in a 'wild hare' every once in a while by mutating some of the rabbit genetic material. The resulting baby rabbits will (on average) be faster and smarter than these in the original population because more faster, smarter parents survived the foxes. (It is a good thing that the foxes are undergoing a similar process – otherwise the rabbits might become too fast and smart for the foxes to catch any of them.)

```
procedure genetic algorithm
begin
      t ← 0
      initialize P(t)
      evaluate P(t)
      while (not termination-condition) do
      begin
            t ← t + 1
            select P(t) from P(t − 1)
            recombine P(t)
            evaluate P(t)
      end
end
```

Fig. 1. The structure of a genetic algorithm.

THE ALGORITHM: A genetic algorithm follows a step-by-step procedure that closely matches the story of the rabbits. The structure of a genetic algorithm is shown in Figure 1.

During iteration t, the genetic algorithm maintains a population of potential solutions (called chromosomes following the natural terminology) $P(t) = \{x_1{}^t, \ldots, x_n{}^t\}$. Each solution $x_i{}^t$ is evaluated to give some measure of its "fitness." Then, a new population (iteration $t + 1$) is formed by selecting the more fitted individuals. Finally, random members of this new population undergo reproduction by means of crossover and mutation – to form new solutions.

Crossover combines the features of two parent chromosomes to form two similar offspring by swapping corresponding segments of the parents. For example, if the two parents are $\langle v_1, \ldots, v_m \rangle$ and $\langle w_1, \ldots, w_m \rangle$ (with each element called a *gene*), then crossing the chromosomes after the kth gene ($1 \leq k \leq m$) would produce the offspring $\langle v_1, \ldots, v_k, w_{k+1}, \ldots, w_m \rangle$ and $\langle w_1, \ldots, w_k, v_{k+1}, \ldots, v_m \rangle$. The intuition behind the applicability of the crossover operator is information exchange between different potential solutions.

Mutation arbitrarily alters one or more genes of a selected chromosome by a random change with a probability equal to some mutation rate. The intuition behind the mutation operator is the introduction of some extra variability into the population.

THEORETICAL FOUNDATIONS: The theoretical foundations of genetic algorithms rely on a binary string representation of solutions and on the notion of a schema – a template allowing exploration of similarities among chromosomes (Holland, 1975). In a population of size n of chromosomes of length m, between 2^m and $n2^m$ different schemata may be represented; at least n^3 of them are processed usefully. Holland has called this property an *implicit parallelism*, as it is obtained without any extra memory/processing requirements.

A growth equation for schemata shows that selection increases sampling rates of the above-average schemata, and that this change is exponential. However, no new schemata (not represented in the initial $t = 0$ sampling) can be formed, which prohibits the application of selection alone. This is exactly why the crossover operator is introduced – to enable structured, yet random information exchange. Additionally, the mutation operator introduces greater variability into the population. It has been shown that the negative effect of these two operators on the growth of a schema is minimal for a specific kind of schemata – called the *building blocks* (Goldberg, 1989). Therefore, the coding of a problem should encourage the formation of such blocks.

The binary alphabet offers the maximum number of schemata per bit of information of any coding and consequently the bit string representation of solutions has dominated genetic algorithm research (Goldberg, 1989). This coding also facilitates theoretical analysis and allows elegant genetic operators.

APPLICATIONS: A genetic algorithm for a particular problem must have the following five components:
- a genetic representation for potential solutions to the problem,
- a way to create an initial population of potential solutions,
- an evaluation function that plays the role of the environment, rating solutions in terms of their "fitness,"
- genetic operators that alter the composition of children during reproduction,
- values for various parameters that the genetic algorithm uses (population size, probabilities of applying genetic operators, etc.).

GAs have been quite successfully applied to optimization problems like wire routing, scheduling, adaptive control, game playing, cognitive modeling, transportation problems, traveling salesman problems, optimal control problems, database query optimization, constrained nonlinear optimization, machine learning, etc. (Proceedings, 1985, 1987, 1989, 1991).

FURTHER READING: There are several texts on genetic algorithms. An introductory material on genetic algorithms and simulated annealing is given in Davis (1987); newer GA methods are covered in Davis (1991). The book by Goldberg (1989) provides a complete reference to all aspects of genetic algorithms. Additional reference on genetic algorithms and their generalizations is given in Michalewicz (1992). From the historical perspective, the most important work was done by Holland (1975) and DeJong (1975); however, it is worthwhile to compare these with the work of Fogel (1966), Glover (1977), and Schwefel (1981). An interesting approach to evolution programming methodology was developed recently by Koza (1992), who suggests that the desired programs to solve the problem (as opposed to the solutions themselves) should evolve during the evolution process. A rich material on recent GA advances in theory and applications is given in four volumes of proceedings from the international conferences on genetic algorithms (Proceedings, 1985, 1987, 1989, 1991).

See **Artificial intelligence; Combinatorial and integer optimization; Simulated annealing.**

References

[1] L. Davis, ed. (1987). *Genetic Algorithms and Simulated Annealing*, Morgan Kaufmann Publishers, Inc., Los Altos, California.

[2] L. Davis, ed. (1991). *Handbook of Genetic Algorithms*, Van Nostrand Reinhold, New York.

[3] K.A. DeJong (1975). *An Analysis of the Behavior of a Class of Genetic Adaptive Systems*, Doctoral Dissertation, University of Michigan.

[4] L.J. Fogel, A.J. Owens, and M.J. Walsh (1966). *Artificial Intelligence Through Simulated Evolution*, John Wiley, New York.

[5] F. Glover (1977). "Heuristics for Integer Programming Using Surrogate Constraints," *Decision Sciences*, 8, 156–166.

[6] D.E. Goldberg (1989). *Genetic Algorithms in Search, Optimization and Machine Learning*, Addison-Wesley, Reading, Massachusetts.

[7] J. Holland (1975). *Adaptation in Natural and Artificial Systems*, University of Michigan Press, Ann Arbor, Michigan.

[8] J.R. Koza (1992). *Genetic Programming*, MIT Press, Cambridge, Massachusetts.

[9] Z. Michalewicz (1992). *Genetic Algorithms + Data Structures = Evolution Programs*, Springer-Verlag, New York.

[10] *Proceedings of the First* (1985), *Second* (1987), *Third* (1989), *and Fourth* (1991) *International Conferences on Genetic Algorithms*. Lawrence Erlbaum Associates, Hillsdale, New Jersey (First and Second), and Morgan Kaufmann Publishers, Los Altos, California (Third and Fourth).

[11] H.-P. Schwefel (1981). *Numerical Optimization for Computer Models*, Wiley, Chichester, United Kindom.

GEOGRAPHIC INFORMATION SYSTEMS (GIS)

Paul Gray and Susan Suchocki

The Claremont Graduate School, California

A Geographical Information System (GIS) is a software tool linking electronic maps and databases. Geographic information technology is emerging as an operations research tool for exploiting massive amounts of geographic data in a form easily comprehended by analysts and decision makers (Hanigan, 1988). Among the problems to which GIS has been applied are:

- interactive site selection;
- election administration and redistricting;
- infrastructure management;
- map and database publishing (particularly among federal mapping agencies, such as The Defense Mapping Agency, The Department of the Interior, and The United States Bureau of the Census);
- mineral exploration;
- public health and safety (e.g., tracking communicable diseases, dispatching emergency vehicles);
- real estate management;
- renewable resources management (e.g., forestry);
- surveying and mapping;
- transportation and other logistics; and
- urban and regional planning (Hanigan, 1988).

A particularly important data source in the United States is TIGER (Topological Integrated Geographic Encoding and Referencing System, 1990), a set of comprehensive digitized street maps available on a scale of 1:100,000.

GIS BACKGROUND AND CAPABILITIES:
The term geographic information system was coined in 1965 by Dacey and Marble (1965). It now applies to generic capabilities for studying and analyzing spatial phenomena (Attenucci *et al.*, 1991). Hanigan (1986) defined its capabilities as including:

- collecting, storing, and retrieving spatial location data;
- identifying locations which meet specified criteria;
- exploring relationships among data sets;
- analyzing related spatial data to aid in making decisions;
- facilitating assessment of alternatives and their impacts; and
- displaying selected environments both visually and numerically.

Specialized software links attributes (discrete object values) and electronic maps through common identifiers. This analysis function permits displaying multiple spatial relationships:

1) overlaying combinations of features and recording resulting conditions,
2) analyzing networks, and
3) defining areas in terms of specified criteria.

In practice, a geographic information system consists of a series of layers, each presenting a particular feature, which can be superimposed accurately on top of one another. Each feature constitutes a distinct layer which can be displayed or not, as desired by the analyst. Basic features take the form of points, lines, or polygons, representing the full spectrum of spatial phenomena. For example, lines may represent transportation options, such as railroads or highways. Because the typical map contains vector-based longitude and latitude data, distances traversed along a route are easily calculated and available for inclusion in analysis.

A feature supporting analysis of spatial distributions is electronic street addresses, either defined for the ends of blocks or positioned at regular intervals along major thoroughfares. The software contains interpolation algorithms enabling the analyst to pinpoint specific phenomena by address. In marketing applications, for example, this capability is used to locate customers or potential customers. In school districting, the location of students can be represented.

The display of variable densities or thematic mapping displays (e.g., expenditures, activity levels, or incidents by selected administrative districts) is also possible. For example, the incidence of people aged 60 and over, based on census data, can be displayed by location using different gray scales or colors.

NCGIA INITIATIVE 6 – THE CALL FOR SPATIAL DECISION SUPPORT SYSTEMS:
Research Initiative 6 sponsored in 1990 by the National Center for Geographic Information and Analysis (NCGIA) sought to identify the future needs of Geographic Information Systems. The initiative involved a workshop attended by university and government people. A major objective was to define the roles of Spatial Decision Support Systems (SDSS) for research purposes (Fotheringham, 1990). The participants adopted Geoffrion's (1983) definition of Decision Support Systems (DSS) as semi-structured problems investigated by software which contained one or more well-structured models used as decision aids. They discuss the application of these aids to marketing, retailing, and location theory.

A key objective was to describe an ideal database management system tied to electronic mapping which would support cartographic display, spatial query and analytical modeling. This system would integrate locational, topological and thematic data and provide the ability to construct and exploit complex spatial relations among data types at a variety of scales and levels of aggregation. Microcomputer-based GIS systems having these capabilities are currently available.

SDSS's can be divided into those systems that rearrange existing information and those that generate new information (Fotheringham, 1990). To rearrange is to observe data presented in different ways, as in looking at a map of income distribution by census tract to locate retail outlets or target an advertising campaign. New information can be the result of overlaying regions and combining their features in new ways as well as by noting spatial relationships not evident in spreadsheet or tabular data.

USE OF AN SDSS FOR DECISION MAKING:
Critical conference questions were
- How would SDSS benefit decision making?
- What impact would SDSS have on the decision making process?

The participants identified four answers:
- Decision making should be improved because the SDSS has access to more information to develop a better solution;
- The flexibility and efficiency of the decision maker(s) would also improve from increased access to analytical tools;
- The political (or ethical/moral) underpinnings become more explicit, helping decision makers to understand their impacts; and
- The solutions or selections would be better substantiated.

USE OF GIS IN URBAN PLANNING AND OTHER POLICY MAKING ACTIVITIES:
"... No wide consensus concerning the use of computer models (exists) in policy making, nor is there any agreement on the way spatial decision problems should be defined and approached" (Batty, 1989).

Typical planning models tend to be displays of existing facilities, such as retailing and transportation. The applications tend to involve structured issues, such as allocation, rather than the indefinite set of options policy makers face.

GIS used today in urban planning depend upon the paradigm of general systems theory (Harris and Batty, 1992). Cities and regions are taken to be complex systems with structures composed of hierarchical subsystems, primarily spatial and nominally static. Planning for these systems consists of optimizing general systems properties, such as idealized population distributions by type. Difficulties arise from the interactions of systems at the periphery of any region under study.

The early interest in applying computerized capabilities to urban planning issues faced the obstacles of difficulties in collecting data, task size in terms of data representation, and difficulty in developing appropriate system models (Brewer, 1973; Lee, 1973). These operational problems coincided with a changing planning philosophy shifting practitioners' interest toward more pragmatic approaches. Today there is less emphasis on optimization and more concern with broader-based issues of equity. This transformation in urban planning thrust is evidenced in current demand for data systems for facility location, emergency services planning, resource management and conservation, and property and tax register recordings. Forrest (1990) listed over 60 distinct systems and problem areas to which a GIS might be applied, ranging across such apparently disparate issues as navigation, political redistricting, hazardous waste management, and wildlife protection. Presently available GIS systems are toolkits whose designs include enough flexibility to accommodate these multiple dimensions.

Although quite sophisticated software is available in the marketplace, the reality for many city and regional planning departments is quite primitive technological support. The urban planning profession is not very far along the GIS learning curve. Survey results show that most planners in local agencies have extremely limited access to GIS technology (Wiggins, 1989). The predominant platform is a stand alone microcomputer. Planners report that their most severe problems involve lack of training, funding, and planning-related software (French and Wiggins, 1989a, 1989b).

AN EXAMPLE: MODELING COMMUNICATIONS NETWORKS:
An example of the use of GIS and visual interactive modeling is given by Anghern and Lüthi (1990). They built a GIS system, named Tolomeo, in a Macintosh environment, which incorporates elements of modeling-by-example, an expert systems technique. They applied it to the problem of selecting the number and location of switching centers and routes in a communications network. The model displays the geographical area under consideration, the location of existing transmitting and receiving stations, proposed switching centers (nodes), and

the transmission channels (arcs). By making the model object oriented, it is possible to attach data defining such quantities as traffic, cost, and transmission times, to each node and arc. Users can redefine the network model interactively on the screen. Furthermore, they can create multiple views of the situation, using different visual metaphors. As the user modifies the model, the underlying data are recalculated to show the implications of proposed changes. Constraints and goals can be introduced so that the calculations take into account constraints and show where the current solution fails to meet goals. The modeling-by-example capability provides suggestions to the user on directions for improvement. These suggestions are based on applying optimization.

CONCLUSION: Geographic Information Systems provide the base for powerful operations research analyses that integrate the visual capabilities of the microcomputer with available spatial information and optimization techniques. The GIS capabilities offer the opportunity to change, fundamentally, the way two-dimensional problems, such as location-allocation, are approached.

See **Database design; Decision support systems; Locational analysis; Logistics; Information systems; Vehicle routing.**

References

[1] Anghern, A.A. and Lüthi, H.-J. (1990). "Intelligent Decision Support Systems: A Visual Interactive Approach." *Interfaces* 20, 17–28.

[2] Anselin, L. (1992). "Spatial Analysis with GIS: An Introduction to Application in the Social Sciences." National Center for Geographic Information and Analysis Technical Paper 92-10.

[3] Antenucci, J.C., Brown, K., Croswell, P.L., Kevany, M.J., and Archer, H. (1991). *Geographic Information Systems: A Guide to the Technology.* Van Nostrand Reinhold, New York.

[4] Batty, M. (1989). "Urban Modelling and Planning," in *Remodelling Geography*, B. Macmillan (ed.), 147–169, Oxford, United Kingdom.

[5] Brewer, G.D. (1973). *Politicians, Bureaucrats and the Consultant: A Critique of Urban Problem Solving.* Basic Books, New York.

[6] Dacey, M. and Marble, D. (1969). "Some Comments On Certain Aspects of Geographic Information Systems." Technical Report No. 2, Department of Geography, Northwestern University, Evanston, Illinois.

[7] Fotheringham, A.S. (1990). "Some Random(ish) Thoughts On Spatial Decision Support Systems." National Center for Geographic Information and Analysis Technical Paper 90-5.

[8] Forrest, E. (ed.) (1990). "Intelligent Infrastructure Workbook: A Management-Level Primer of GIS." *A-E-C Automation Level Newsletter*, Fountain Hills, Arizona.

[9] Geoffrion, A.M. (1983). "Can OR/MS Evolve Fast Enough?" *Interfaces* 13, 10–25.

[10] Hanigan, F.L. (1989). "GIS Recognized As Valuable Tool For Decision Makers," *The GIS Forum*, 1, 4.

[11] Harris, B. and Batty, M. (1992). "Locational Models, Geographic Information and Planning Support Systems." National Center for Geographic Information and Analysis Technical Paper 92-1.

[12] Lee, D.B. (1973). "Requiem for Large-Scale Models," *Jl. American Institute of Planners* 39, 163–178.

GEOMETRIC PROGRAMMING

Joseph G. Ecker

Rensselaer Polytechnic Institute, Troy, New York

INTRODUCTION: Early work in geometric programming was stimulated by Zener (1961, 1962) in his investigation of cost minimization techniques for engineering design problems. Subsequent work by Duffin (1962), Duffin and Peterson (1966), and Duffin, Peterson, and Zener (1967) provided the fundamental groundwork of the subject. Geometric programming refers to a class of optimization problems that have the form

(**P**)   minimize $g_o(t)$ subject to $g_k(t) \leq 1$ and $t > 0$

where $t = (t_1, t_2, \ldots, t_m)$ is a vector of variables and, for $k = 0, 1, \ldots, p$, the functions $g_k(t)$ are sums of terms having the form

$$u_i(t) = c_i t_1^{a_{i1}} t_2^{a_{i2}} \ldots t_m^{a_{im}}$$

where the coefficients $\{c_i\}$ and the exponents $\{a_{ij}\}$ are arbitrary real numbers. The following is an example of a possible geometric program with three variables:

$$\text{minimize} \quad g_o(t) = \frac{40}{t_1 t_2 t_3} + 40 t_2 t_3$$

$$\text{subject to} \quad g_1(t) = \frac{1}{2} t_1 t_3 + \frac{1}{4} t_1 t_2 \leq 1$$

and $t_1 > 0$, $i = 1, 2, 3$.

The term geometric programming was adopted because of the role that the geometric-arithmetic mean inequality played in the initial development of a duality theory for problems having the above form. Initially, the class of problems was restricted by requiring that the coefficients be positive and the corresponding terms $u_i(t)$ were called "posynomials." Thus, the term posynomial programming might well have been chosen instead of geometric programming. Many engineering design problems do have the form of a geometric program where the coefficients are positive. Several examples of such problems are given in Duffin, Peterson, and Zener (1967), in the paper on methods, computations, and applications of geometric programming by Ecker (1980), and in the references of the latter paper.

Geometric programs where some of the $\{c_i\}$ coefficients can be negative are called signomial programs and this class of optimization problems was first studied by Passy and Wilde (1967) and Blau and Wilde (1969). The initial theory of geometric programming has been generalized to a much broader class of optimization problems. The review article by Peterson (1976) shows how the approach to developing a duality theory through the use of inequalities can be generalized to a very broad class of problems.

EQUIVALENCE OF POSYNOMIAL AND CONVEX PROGRAMS:

Posynomial programs can be reformulated so that the objective function g_0 and the constraint functions g_i are convex. The simple transformation

$$t_j = e^{z_j} \text{ for } j = 1, 2, \ldots, m$$

allows each posynomial term to be rewritten in the form

$$u_i(t) = c_i\, e^{a_{i1} z_1 + a_{iZ} + \ldots + a_{im} z_m}.$$

If we let A be the matrix whose ith row A_i gives the exponents of the ith posynomial term, then $u_i(t)$ can be written as

$$u_i(t) = c_i\, e^{A_i z}$$

where z is the column vector with entries z_i. The matrix A is usually called the exponent matrix. Notice that A is $n \times m$ where m is the number of variables and n is the number of posynomial terms. For our example three-variable problem, the exponent matrix A is given by

$$A = \begin{bmatrix} -1 & -1 & -1 \\ 0 & 1 & 1 \\ 1 & 0 & 1 \\ 1 & 1 & 0 \end{bmatrix}$$

If we now define $x = Az$, then we see that each geometric program with positive coefficients can be written so that the objective function and all of the constraints are convex functions of the variables x because then each posynomial term can be written as

$$u_i(t) = c_i e^{x_i}$$

where we add the linear constraints $x = Az$.

THE DUAL OF A POSYNOMIAL PROGRAM:

Through the use of the geometric-arithmetic mean inequality a maximization problem can be generated from the posynomial program (P) above. The maximization problem has a dual variable d_i for each posynomial term u_i so that dual vector is given by

$$d = (d_1, d_2, \ldots, d_n)^T.$$

We will let

$$L_k = \text{the sum of the variables } d_i$$

corresponding to the kth function $g_k(t)$.

The dual program has the form

$$\max v(d) = \frac{c_1}{d_1}\frac{c_2}{d_2} \cdots \frac{c_n}{d_n} L_1^{L_1} L_2^{L_2} \ldots L_p^{L_p}$$

subject to

$$L_0 = 1$$

$$A^T d = 0 \text{ and } d \geq 0.$$

The three-variable example above has the following dual program:

$$\max v(d) = \left(\frac{40}{d_1}\right)\left(\frac{40}{d_2}\right)\left(\frac{1}{2d_3}\right)\left(\frac{1}{4d_4}\right)(d_3 + d_4)^{d_3 + d_4}$$

subject to

$$A^T d = 0$$

$$d_1 + d_2 = 1$$

$$d \geq 0$$

The duality theory showing how to use a solution to the dual program to obtain a solution to the original primal program (P) is developed in Duffin, Peterson, and Zener (1967). The problem (P) is called canonical if there is a dual vector d satisfying

$$d > 0 \text{ with } A^T d = 0.$$

Canonical problems always have a minimizing point t^* and, if the set of all points satisfying the constraints in (P) has a non-empty interior, then the following duality results hold:

(i)   the dual problem has a maximizing vector d^*;
(ii) the maximum value of the dual is equal to the minimum value for the primal program (P);
(iii) each minimizing point t for (P) satisfies $u_i(t) = d_i^* v(d^*)$ for each i corresponding to the terms $u_i(t)$ in the objective function, and
      $u_i(t) = d_i^*/L_k(d^*)$ for all i when $L_k(d^*) > 0$.

The right-hand side of each equation in (iii) is a positive constant and, given a solution d^*, we can take common logarithms of both sides of the equations to obtain a linear system in the variables $\log(t_i)$. Typically, this linear system has more equations than variables so it uniquely determines a minimizing vector t^*.

COMPUTATIONAL METHODS: The first published algorithm for solving posynomial programs was a method by Frank (1966) that solves the dual problem and then uses the above duality relations to obtain a minimizing point for (**P**). Blau and Wilde (1971) and Rijckaert and Martens (1976) developed similar methods that solve the Karush-Kuhn-Tucker optimality conditions for the dual problem. Other dual methods have been investigated, as for example in Dinkel, Kochenberger, and McCarl (1974) and in Beck and Ecker (1975).

A class of computational methods that solve (**P**) directly are based on the idea of linearizing geometric that was initially proposed by Duffin (1970). Avriel and Williams (1970) and Avriel, Dembo, and Passy (1975) use the idea of condensing each function into a single posynomial term to formulate a linear program that can be used to obtain an approximate solution to (**P**) even if some of the coefficients are negative. For more details on these types of approaches were given in Dembo (1978).

See **Nonlinear programming; Optimization.**

References

[1] Avriel, M., R. Dembo and U. Passy (1975). "Solution of Generalized Geometric Programs," *Internat. Jl. Numer. Methods Engrg.*, **9**, 149–169.

[2] Avriel, M. and A.C. Williams (1970). "Complementary Geometric Programming," *SIAM Jl. Appl. Math.*, **19**, 125–141.

[3] Beck, P.A. and J.G. Ecker (1975). "A Modified Concave Simplex Algorithm for Geometric Programming," *Jl. Optimization Theory Appl.*, **15**, 189–202.

[4] Blau, G.E. and D.J. Wilde (1971). "A Lagrangian Algorithm for Equality Constrained Generalized Polynomial Optimization," *AI Ch. E. Jl.*, **17**, 235–240.

[5] Blau, G.E. and D.J. Wilde (1969). "Generalized Polynomial Programming," *Canad. Jl. Chemical Engineering*, **47**, 317–326.

[6] Dembo, R.S. (1978). "Current State of the Art of Algorithms and Computer Software for Geometric Programming," *Jl. Optimization Theory Appl.*, **26**, 149–184.

[7] Dinkel, J., J. Kochenberger and B. McCarl (1974). "An Approach to the Numerical Solution of Geometric Programming," *Mathematical Programming*, **7**, 181–190.

[8] Duffin, R.J. (1962). "Cost Minimization Problems Treated by Geometric Means," *Operations Res.*, **10**, 668–675.

[9] Duffin, R.J. (1970). "Linearizing Geometric Programs," *SIAM Review*, **12** 211–227.

[10] Duffin, R.J. and E.L. Peterson (1966). "Duality Theory for Geometric Programming," *SIAM Jl. Appl. Math.*, **14**, 1307–1349.

[11] Duffin, R.J., E.L. Peterson and C.M. Zener (1967). *Geometric Programming*. John Wiley, New York.

[12] Ecker, J. G. (1980). "Geometric Programming: Methods, Computations, and Applications," *SIAM Review*, **22**, 338–362.

[13] Frank, C.J. (1966). "An Algorithm for Geometric Programming," in *Recent Advances in Optimization Techniques*, A. Lavi and T. Vogl, eds., John Wiley, New York, 145–162.

[14] Passy, U. and D.J. Wilde (1967). "Generalized Polynomial Optimizations," *SIAM Jl. Appl. Math.*, **15**, 1344–1356.

[15] Peterson, E.L. (1976). "Geometric Programming – A Survey," *SIAM Review*, **18**, 1–51.

[16] Ricjkaert, M.J. and X.M. Martens (1976). "A Condensation Method for Generalized Geometric Programming," *Math. Programming*, **11**, 89–93.

[17] Zener, C. (1961). "A Mathematical Aid in Optimizing Engineering Design," *Proc. Nat. Acad. Sci. U.S.A.*, **47**, 537–539.

[18] Zener, C. (1962). "A Further Mathematical Aid in Optimizing Engineering Design," *Ibid.*, **48**, 518–522.

GERT

Graphical Evaluation and Review Technique – model of a network where all nodes are of the "exclusive-or" type on their receiving side. See **Network planning; Project management; Research and development**.

GIS

See **Geographic information systems**.

GLOBAL BALANCE EQUATIONS

A system of steady-state equations for a Markov chain (typically a queueing problem) obtained by balancing the mean flow rates or probability flux in and out of each individual state, symbolically written as $\pi Q = 0$. See **Networks of queues; Queueing theory**.

GLOBAL MAXIMUM (MINIMUM)

For an optimization problem, the largest (smallest) value that the objective function can achieve over the feasible region. See **Local maximum (minimum)**; **Maximum (minimum)**. **Nonlinear programming**; **Quadratic programming**.

GLOBAL MODELS

Saul I. Gass

University of Maryland, College Park

Global or world models are concerned with the application of systems analysis to policy problems of intra- and international interest. Typical problems of concern include population growth, ecological issues (forestry, fisheries, pesticides, insect infestation), energy and water resource availability and uses, the spread of diseases, and environmental models (acid rain, air pollution), Clark and Cole (1975), Holcomb (1976). Global models are usually highly aggregated in their structure and in their data requirements. However, such models can be developed by integrating lower-level and more detailed national or regional models. Of related interest are global and regional predictive models that deal with long-range weather or macro-economic activity.

Although the trail of global models leads back to Malthus and his publication of *An Essay on the Principle of Population* (1798), the modern development of global models begins with the use of systems analysis in the study of global problems, and the availability of specific tools for analysis such as Forrester's *System Dynamics*, Leontief's *Input-Output Interindustry Structure*, and Dantzig's *Linear Programming Model*. In particular, Forrester and his associates brought the use of global models to the attention of governmental officials and to the scientific community by their application of the *World* 2 and *World* 3 system dynamics models that are described in *World Dynamics* (Forrester, 1971) and *Limits to Growth* (Meadows *et al.*, 1972), respectively.

The *World* 3 model considers the world as a whole and evaluates five global indicators and their interactions: population, consumption of nonrenewable resources, pollution, food production, and industrialization. The model's calculations lead to the conclusion that sometime in the twenty-first century, the world will witness a steep decline in food per capita and in population. The general pessimistic conclusion reached by the *World* 3 model (under varying assumptions such as availability of resources) was that the world will soon be hitting resource, economic and population limits to growth, and measures must be initiated by the world community to avoid calamity. The model indicated a stable future only if such stringent measures as maintaining a stable (zero growth) world population and capital base, and such measures are applied soon (Meadows *et al.*, 1972; Clark and Cole, 1975). Criticisms of this conclusion abound and they address the issues of the model's structure, data, aggregation, and methodological approach. (See, for example, Cole and Curnow, 1973, Schwartz and Foin, 1972, and a rebuttal by Forrester, 1976.)

Other global models have been developed in an attempt to overcome some of the limitations and criticisms of the *World* 3 model. In particular, we note the one by Mesarovic and Pestel (1974). This model divided the world into ten regions and enabled some policy options (e.g., energy resource utilization) to be evaluated. Research in global models continues, with one center for such investigations being the International Institute for Applied Systems Analysis (IIASA). IIASA has initiated a database collection for environmental analyses, developed an acid rain model for Europe, forest resource and pest management models, plus econometric and linear-programming based approaches to global policy modeling (Bruckmann, 1980).

Our ability to encompass the complex interactions of the global system into a computer-based model will always be open to criticisms. As any model is an approximation of the real-world, surely a model that attempts to encompass the whole world or even major subelements can not do so with much exactitude. We should not expect it to be so. As noted by Mason (1976, p. 4): "We have seen that ultimately there is no objective way to assess world models." But, there is no reason why investigators, building on such past efforts as those described above and others, cannot develop global models that would be of value to the world's policy-makers.

See **Environmental systems analysis; Input-output analysis; Systems dynamics; Validation**.

References

[1] G. Bruckmann, ed. (1980), *Input-Output Approaches in Global Modeling*, Pergamon Press, Oxford.

[2] C.W. Churchman and R.O. Mason, eds. (1976), *World Modeling: A Dialogue*, North-Holland, New York.

[3] J. Clark and S. Cole (1975), *Global Simulation Models: A Comparative Study*, John Wiley, New York.

[4] H.S.D. Cole, C. Freeman, M. Tahoda, and K.L.R. Pavitt (1973), *Models of Doom*, Universe Books, New York.

[5] J.W. Forrester (1961), *Industrial Dynamics*, MIT Press, Cambridge, Massachusetts.

[6] J.W. Forrester (1971), *World Dynamics*, MIT Press, Cambridge, Massachusetts.

[7] J.W. Forrester (1976), "Educational Implications of Responses to System Dynamics Models," pp. 27–35 in *World Modeling: A Dialogue*, C.W. Churchman and R.O. Mason, eds., North-Holland, New York.

[8] Holcomb Research Institute (1976), *Environmental Modeling and Decision Making*, Praeger, New York.

[9] R.O. Mason (1976), "The Search for a World Model," pp. 1–9 in *World Modeling: A Dialogue*, C.W. Churchman and R.O. Mason, eds., North-Holland, New York.

[10] D.H. Meadows, D.L. Meadows, J. Randers, and W.W. Behrens III (1972), *The Limits to Growth*, Signet Books, Washington, DC.

[11] M. Mesarovic and E. Pestel (1974), *Mankind at the Turning Point*, E.P. Dutton, Reader's Digest Press, New York.

[12] S.I. Schwartz and T.C. Foin (1972), "A Critical Review of the Social Systems Models of Jay Forrester," *Human Ecology*, 1, 2, 161–173.

GLOBAL SOLUTION

An optimal solution over the entire feasible region.

GOAL CONSTRAINTS

Mathematical expressions consisting of resource utilization rates, decision variables, deviation variables, and targeted minimum and maximum resources levels. They are used to model individual resource goals in a goal programming model. See **Goal programming**.

GOAL PROGRAMMING

Marc J. Schniederjans

University of Nebraska-Lincoln

Goal Programming (GP), also called *linear goal programming* (LGP), can be categorized as a special case of linear programming (LP). The origin of GP as a means of resolving infeasible LP problems attests to its characterization as a LP methodology (Charnes and Cooper, 1961). GP is now considered a *multi-criteria decision making* (MCDM) method (Steuer, 1986); it is used to solve multi-variable, constrained resource and similar problems that have multiple goals.

GP MODELING: Similar to LP, the GP model has an *objective function*, constraints (called *goal constraints*), and *nonnegativity requirements*. The GP *objective function* is commonly expressed as a minimization function (Schniederjans, 1984):

$$\text{Minimize } Z = \sum w_{kl}P_k(d_i^- + d_i^+) \text{ for all } k, l, i.$$

In this objective function, Z is the summation of all deviations, the w_{kl} are optional mathematical weights used to differentiate *deviation variables* within a kth priority level, the P_k are optional rankings of deviation variables within goal constraints, the d^- values are the negative deviational variables, and the d^+ values are the positive deviational variables. The P_k rankings are called *preemptive priorities* because they establish an ordinal priority ranking (where $P_1 > P_2 > P_3 > \ldots$ etc.) that orders the systematic optimization of the deviation variables.

The fact that the optimization of the variables in the objective function is ordered by the preemptive priority ranking has given rise to the use of the term *satisficing*. This term results from the fact that a solution in a GP model satisfies the ranking structure while minimizing deviation from goals. One of the best features of GP is that the P_k permit the decision makers to rank goals in accordance with their personal preferences, and even weight the importance of those preferences within goals using w_{kl}. The greater the w_{kl} mathematical weighting, the greater the importance attached to its related deviation variable.

The *goal constraints* in a GP model can be expressed as:

$$a_{ij}x_j + d_i^- - d_i^+ = b_i \text{ for all } i, j.$$

Here, the a_{ij} values are the resource utilization rates representing the per unit usage of the related resource b_i, and the x_j are *decision variables* we seek to determine. The goal constraints in a GP model are used to express goals that we seek to achieve by minimizing deviation (in the form of deviation variables) from their right-hand-side b_i goal targets. In essence, this use of deviation variables minimizes the absolute difference between the right- and left-hand sides of each constraint. The d_i^- are termed *underachievement* variables and the d_i^+ are *over-*

achievement variables. The *nonnegativity require-ments* in a GP model are usually expressed as:

$$x_j, d_i^-, d_i^+ \geq 0 \text{ for all } i, j.$$

GP SOLUTION METHODS: Different solution methodologies exist to solve a variety of types of GP models. The type of GP model depends on special requirements placed on the decision variables in the model. Borrowing from LP, most of the solution methodologies for GP models are based on the *revised simplex method*. Simplex based-solution methods for GP problems originate from the sequential goal (preemptive priority) procedure of Lee (1972). There are additional methodologies for solving *integer* GP problems, *zero-one* GP problems and *nonlinear* GP problems. Like LP, these special types of GP solution methods are based on revised simplex methods, enumeration methods, and the calculus.

One can also get *duality* and *sensitivity analysis* information from the simplex based GP solution methods (Ignizio, 1982). Duality in GP models is focused on examining trade-offs in deviation between priorities. The software system by Lee and Shim (1993) computes the marginal trade-offs of revising right-hand-side b_i goal targets to reduce deviation from lower priority goals. There are a variety of LP-based sensitivity analysis procedures for GP. Unique to GP is P_k-sensitivity analysis. In P_k-sensitivity analysis, alterations in sets of k priority level goals are implemented to examine their ordering effect upon the model's solution.

GP RESEARCH AND APPLICATIONS: While both GP modeling and GP solution methods share LP origins, there are two characteristics of GP that differentiate the application of GP from LP problems: multiple goals and an ordinal ranking of the goals to deal with conflict. Since many business and governmental problems contain the same two characteristics, GP quickly became a very popular methodology.

During the 1960s and early 1970s, most research on GP focused on revisions of prior LP-type models, but with a ranking of conflicting goals. The series of case applications presented in Lee (1972) typify the work during this period. The most common applications followed functional areas in business, such as budgeting in accounting, portfolio analysis in finance, production planning in management, and advertising resource allocation in marketing. Having exhausted these areas of application, interest in GP started decreasing. In the late 1970s and early 1980s, integer (particularly zero-one) GP

methodologies appeared and caused a renewed interest in the use of GP. Zero-one GP solution methods permitted the model to be applied in binary outcome situations and broadened the potential application base of GP. Zero-one GP applications include project selection, personnel selection, and logistics. This period also saw the combining of other operations research and management science methodologies within a GP model: *the transportation simplex method, assignment method, network models, dynamic programming, simulation, game theory, fuzzy programming* and *heuristic procedures* (Steuer, 1986).

In the late 1980s and early 1990s, GP microcomputer software appeared, placing fairly powerful solution capabilities in the hands of practitioners, causing another burst of interest in GP applications: planning in small businesses, improving productivity in service operations, and planning product development. GP engineering applications include: planning flexible manufacturing systems, robot selection, strategic planning, metal cutting and inventory lot-sizing. Improvements in GP weighting strategies have been developed using the *analytic hierarchy process* (AHP) and *regression analysis*.

A GP model's ability to use personal preference information has made it a very useful tool in dealing with socially sensitive issues. Throughout GP's history, applications and models have illustrated how GP is a powerful tool for analyzing public policy issues. Applications include: scheduling school busing to achieve racial balance, waste-management planning, health-care planning, and reemployment allocation.

See **Analytic hierarchy process; Assignment problem; Dynamic programming; Fuzzy systems; Game theory; Linear programming; Multi-objective programming; Multiple criteria decision making; Regression analysis; Simplex method; Transportation problem.**

References

[1] A. Charnes and W.W. Cooper (1961). *Management Models and Industrial Applications of Linear Programming*, John Wiley, New York.

[2] J. Ignizio (1982). *Linear Programming in Single- and Multiple-Objective Systems*, Prentice-Hall, Englewood Cliffs, New Jersey.

[3] S.M. Lee and J.P. Shim (1993). *Micro Management Science*, 3rd ed., Allyn and Bacon, Boston.

[4] S. M. Lee (1972). *Goal Programming for Decision Analysis*, Auerbach Publishers, Philadelphia.

[5] H. Min and J. Storbeck (1991). "On the Origin and Persistence of Misconceptions in Goal Programming," *Jl. Operational Research Society*, 42, 301–312.

[6] M.J. Schniederjans (1984). *Linear Goal Programming*, Petrocelli Books, Princeton, New Jersey.

[7] M.J. Schniederjans (1995). *Goal Programming: Methodology and Applications*, Kluwer Academic Publishers, Norwell, Massachusetts.

[8] R.E. Steuer (1986). *Multiple Criteria Optimization: Theory, Computation, and Application*, John Wiley, New York.

[9] S.H. Zanakis and S. Gupta (1985). "A Categorized Bibliographic Survey of Goal Programming," *OMEGA*, 13, 211–222.

GOMORY CUT

A linear constraint that is added to a linear-programming problem to reduce the solution space without cutting off any integer-valued points. Such cutting planes are the basis of many solution procedures that find integer solutions to a linear constrained optimization problem. The idea is to eventually reduce the solution space so that its optimal integer solution corresponds to an extreme point of the reduced solution space.

GORDAN'S THEOREM

Let A be an $m \times n$ matrix, then exactly one of the following systems has a solution: (i) $Ax < 0$ or (ii) $A^T y = 0$, $y \geq 0$, $y \neq 0$.

GP

See **Goal programming**.

GRADIENT VECTOR

For the function $f(\mathbf{x})$ of the vector $\mathbf{x}$, the gradient is the vector of first partial derivatives (if they exist) evaluated at a specific point x^0 and is written as

$$\nabla f(x^0) = \left[\frac{\partial f(x^0)}{\partial x_1}, \frac{\partial f(x^0)}{\partial x_2}, \dots, \frac{\partial f(x^0)}{\partial x_n} \right].$$

It is normal or perpendicular to the tangent of the contour of $f(\mathbf{x})$ that passes through x^0. Its direction is the direction of maximum increase of $f(\mathbf{x})$ and its length is the magnitude of that maximum rate of increase.

GRAECO-LATIN SQUARE

See **Combinatorics**.

GRAPH

A graph $G = (V, E)$ consists of a finite set V of vertices (nodes, points) and a set E of edges (arcs, lines) joining different pairs of distinct vertices.

GRAPHICAL EVALUATION AND REVIEW TECHNIQUE

See **GERT**.

GRAPHICS

See **Visualization**.

GRAPH THEORY

Douglas R. Shier

Clemson University, South Carolina

INTRODUCTION: Graph theory is the general study of the interconnection of various elements. While the origins of graph theory can be traced back to the eighteenth century, this area of discrete mathematics has experienced most of its tremendous growth during the past few decades. This rapid growth, both in the development of new theory and applications, reflects the fact that graphs can model a wide variety of natural and technological systems.

A number of physical systems can be viewed as *graphs*, composed of *nodes* (or vertices) connected together by *edges* (or arcs). For example, a local area computer network defines a graph whose nodes represent individual computers (or peripheral devices) and whose edges represent the physical cables connecting such computers. A telecommunication network consists of telephone locations (and central switching stations) joined by sections of copper wire (and optical fibers); an airline system has airports as its nodes and direct flights as its edges; a street network involves road segments (edges) whose intersections define its nodes; and an electronic switching circuit contains logic gates whose input and output leads form a graph.

In addition, graphs can with equal ease represent logical relationships between elements. For example, the subroutines of a computer program might form the nodes of a graph, with edges indicating the flow of control or data between subroutines. A project involving a large number of tasks can be modeled by a graph, with the tasks being nodes and logical precedence relations defining the edges. In an ecological system the edges could indicate which species (nodes) feed upon other species. Examination scheduling at a university can be studied using a graph whose nodes are courses and whose edges indicate whether two courses contain students in common; examinations for such *adjacent* courses should not be scheduled at the same time.

As suggested by the applications above, the direct connections between nodes can be bidirectional (as in making a telephone call, or in traversing a major highway) or there can be a specific orientation implied by the relationship (as the precedence relation in a project graph, or the predator-prey relation in an ecological graph). Consequently, graph theory treats both undirected graphs (in which the underlying relationship between nodes is symmetric) and directed graphs, or *digraphs* (in which the relationship need not be symmetric). These two graph models are pictured in Figures 1 and 2, respectively. In this exposition, we focus on undirected graphs, since the analogous concepts for digraphs are usually apparent. Throughout $G = (N, E)$ will indicate an undirected graph with node set N and edge set E.

One of the earliest applications of graph theory was to the structure of molecular compounds, in which atoms (nodes) are joined by chemical bonds (edges). The task of identifying which chemical compounds are structurally the same (isomers) is reflected in the graph-theoretic concept of *isomorphism*, meaning that two given graphs are the same up to relabeling of their nodes. In addition, each atom has a "valency" which indicates the number of other atoms to which it is connected. In graph-theoretic terms this is called the *degree* of the node, the number of edges with which it is *incident*. This concept provides a quantifiable measure of local connectivity. For instance, the degree of a node in a communication network indicates the relative burden on that node in transporting information, so a robust communication system would be designed to avoid nodes with large degrees. Since such systems support point-to-point com-

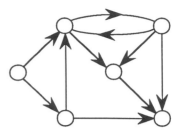

Fig. 2. A directed graph.

munication, a more global measure of connectivity is also needed. Thus, a fundamental concept is that of a *path* between nodes i and j: an alternating sequence of nodes and incident edges leading from node i to node j. A *cycle* is a closed path. The graph G is *connected* if every pair of distinct nodes is joined by a path in G. The *distance* between two nodes of G is defined as the smallest number of edges in a path joining the nodes. Then an overall measure of compactness of the graph is its *diameter*: the maximum distance between any two of its nodes.

EULERIAN AND HAMILTONIAN CYCLES: In certain applications specific types of paths or cycles are sought in the graph. For example, an *Eulerian cycle* is a cycle in the graph G which traverses each edge of G exactly once. This concept models the task of efficiently routing trucks for collection of trash throughout a city, since multiple passes along a street are not desirable. A *Hamiltonian cycle* in G is a cycle in the graph that visits each node exactly once. This concept has been applied to the sequencing of artifacts found at archaeological sites as well as to the manufacture of electronic circuit boards. The concepts of *Eulerian paths* and *Hamiltonian paths* are defined analogously.

In designing a logistics system it seems prudent to require several paths joining nodes i and j, thus providing redundant routes for sending messages in case of node or edge failures. For example, an adversary might select various edges (bridges, roads) for destruction in order to disrupt the flow of material from node i to node j. An i–j *cutset* is a minimal subset of edges whose removal disconnects i from j in G. To disrupt communication between i and j in an efficient manner, the adversary might then attack an i–j cutset having the minimum size $\lambda_{ij}(G)$. A theorem of Menger establishes a min-max relationship between these "dual" viewpoints. Namely, the maximum number of edge-disjoint paths joining i and j equals the minimum number of edges in an i–j cutset. (The same conclusion holds if the paths are node-disjoint and the cutsets are defined in terms of nodes.)

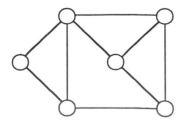

Fig. 1. An undirected graph.

TREES: A related concept addresses the connectivity of all nodes, rather than just a specified pair. A *tree* is a connected graph containing no cycles, and a *spanning tree* of the graph $G = (N, E)$ is a tree with node set N and whose edge set is a subset of E. Any spanning tree thus supports communication among all nodes of the graph. On the other hand, a *cutset* of G is a minimal set S of edges whose removal disconnects some pair of nodes in the graph. The edges of S must intersect the edge set of every spanning tree of G. An overall connectivity measure for the graph G is the minimum size $\lambda(G)$ of a cutset in G.

Trees find many other applications in the theory of graphs. Trees can model the organizational hierarchy of a corporation, the table of contents of a book, or the syntactic structure of languages. Trees also serve as useful data structures for organizing elements of a data base for subsequent retrieval and updating. Moreover they are used by compilers of computer languages to provide a concise representation of arithmetic expressions. Of particular relevance to operations research is the fact that spanning trees correspond exactly to basic solutions for linear programming problems formulated on graphs.

EMBEDDING AND COLORING: Special types of graphs find application in the layout of circuit boards, in which it is desired to place the components and their connections so that no two wires meet except at a component. This corresponds to an *embedding* of the graph in the plane so that edges only intersect at nodes. Kuratowski's theorem provides an elegant characterization of which graphs are in fact planar. More generally, any graph G can be decomposed into a number of edge-disjoint planar subgraphs, and the minimum number of such subgraphs is termed the *thickness* $\theta(G)$ of the graph. Planar graphs G are also of interest because a dual G^* of such graphs can be defined. In particular, the cycles of G are in one-to-one correspondence with the cutsets of G^*.

The coloring of the nodes of a graph also arises in several applications. A *proper coloring* of G with k colors is an assignment of these colors to the nodes of G such that adjacent nodes are colored differently. For example, if the nodes of G represent courses and edges represent conflicts (courses that cannot have examinations scheduled at the same time), then a proper coloring of G with k colors defines a conflict-free schedule using k time periods. In another application, suppose the graph G indicates a compatibility relationship between tasks. A k-coloring of the *complement* of G (whose edges are those node pairs not appearing as edges of G) then yields a partitioning of the nodes of G into k groups of mutually compatible tasks. The minimum number of colors $\chi(G)$ needed to color G properly is termed the *chromatic number* of G. The famous four-color theorem, finally proved in 1976, states that $\chi(G) \leq 4$ holds for any planar graph G.

OPTIMIZATION: One important aspect of graphs is that they distill the essential adjacency relationships between objects. Another is that they suggest certain optimization problems. It is apparent that determining the connectivity $\lambda(G)$, thickness $\theta(G)$, and chromatic number $\chi(G)$ are graph optimization problems. Other graph optimization problems arise directly from applications, in which it might be required to optimally schedule courses, allocate facilities, route goods, or design computer systems, relative to some objective function and subject to appropriate constraints. As a specific example, we might wish to design a minimum diameter communication graph with $\lambda_{ij}(G) \geq k$ for all distinct node pairs i and j, using a fixed number of edges.

More generally, quantitative information may be associated with the nodes and/or edges of a graph, reflecting the cost, time, distance, capacity, or desirability of these components. A variety of graph optimization problems are then apparent: (1) find a spanning tree of G having minimum cost; (2) find a minimum distance path joining two nodes of G; (3) find the maximum amount of material that can feasibly flow from an origin node to a destination node; (4) find a minimum cost Hamiltonian cycle in G; (5) optimally locate facilities on the edges of G to serve demands arising at the nodes; (6) find a maximum weight set of nonadjacent edges in G.

There are additional application areas in which graphs are used to model a variety of physical and logical systems. While our description has concentrated on undirected graphs, directed graphs are pertinent in other areas, such as in representing the state diagram of a Markov chain. Connectivity in this digraph G can be used to classify the states of the chain, and the (directed) cycle lengths in G define the periodicity of the chain. The study of optimization problems on graphs and digraphs has in turn stimulated research into the design of effective algorithms for solving such problems, as well as determining when these problems belong to an inherently difficult class of problems (NP-hard problems). In the latter case, it is important to identify special types of graphs (e.g., planar graphs) for which the computation can be carried out efficiently, even though there may not exist an efficient solution method applicable to all graphs.

FURTHER READING: The book by Biggs *et al.* (1976) provides an excellent reference for the history of graph theory. Fulkerson (1975), Roberts (1976), and Michaels and Rosen (1991) discuss a variety of applications of graphs. Wilson and Watkins (1990) give a nice introduction to the theory of graphs, with more advanced treatment provided in Harary (1971) and Berge (1973). Algorithmic aspects of graph theory are discussed in Evans and Minieka (1992).

See **Combinatorial and integer optimization; Combinatorics; Computational complexity; Linear programming; Markov chains; Matching; Networks; Project management; Traveling salesman problem.**

References

[1] Berge, C. (1973). *Graphs and Hypergraphs*. North-Holland, Amsterdam.
[2] Biggs, N.L., E.K. Lloyd, and R.J. Wilson (1976). *Graph Theory* 1736–1936. Clarendon Press, Oxford.
[3] Evans, J.R. and E. Minieka (1992). *Optimization Algorithms for Networks and Graphs*. Marcel Dekker, New York.
[4] Fulkerson, D.R. (1975). *Studies in Graph Theory, Parts I-II*. Volumes 11–12, MAA Studies in Mathematics. Mathematical Association of America, Washington, DC.
[5] Harary, F. (1971). *Graph Theory*. Addison-Wesley, Reading, Massachusetts.
[6] Michaels, J.G. and K.H. Rosen (1991). *Applications of Discrete Mathematics*. McGraw-Hill, New York.
[7] Roberts, F. (1976). *Discrete Mathematical Models, with Applications to Social, Biological, and Environmental Problems*. Prentice-Hall, Englewood Cliffs, New Jersey.
[8] Wilson, R.J. and J.J. Watkins (1990). *Graphs: An Introductory Approach*. John Wiley, New York.

GREEDY ALGORITHM

A heuristic algorithm that at every step selects the best choice available at that step without regard to future consequences. A greedy method never rescinds its choices or decisions made earlier. A greedy method is usually applied to an optimization problem for which the method attempts to determine an optimal solution (least cost, maximum value), with no guarantee that the optimal solution will be found. Kruskal's and Prim's minimum spanning tree algorithms are greedy methods that do produce an optimal solution. See **Algorithm; Heuristic procedure; Kruskal's algorithm; Prim's algorithm.**

GRG METHOD

Generalized reduced gradient method. See **Quadratic programming**.

GROUP DECISION COMPUTER TECHNOLOGY

Dennis M. Buede

George Mason University, Fairfax, Virginia

With the rise of computer technology and the success of quantitative decision support techniques there has been a great deal of interest in moving these technologies into the board room, so to speak. At this point there is no commonly accepted approach for supporting group decision making (see DeSanctis and Gallupe, 1987; Huber, 1984). The oldest approach is decision conferencing (Watson and Buede, 1987), which started in 1979 and has spread but not exploded. Decision conferencing is a group process that is led by a decision analytic facilitator. The facilitator employs simple decision analysis models to focus the group's discussion on the options available to them in light of their objectives and uncertainties. The facilitator also mixes analytic activities with creative problem structuring and option generation activities. The decision conference can be as short as two days or many involve several two to three day sessions. References for decision conferences are: McCartt and Rohrbaugh (1989), Phillips (1984; 1990), Reagan-Cirincione (1992), and Rohrbaugh (1989).

There are a range of approaches for group decision support that place a computer in the hands of each participant. This approach still employs a group process facilitator although there is some disagreement about the importance of the facilitator amongst the researchers in this area. The computer technology is designed to enhance the productivity of the individual and the communication of information between individuals. The effectiveness of computer technology as a communication medium when the group has a single decision focus is questioned by some. However, computer technology opens the group's options in terms of whether they meet at the same place or even at the same time. The major options of the group have been named: same time, same place; same time, different place; different time, same place; and different time, different place. Much of the work in this area is characterized as research or demonstrations. A recent reference of this work is Nunamaker *et al.* (1991).

Group process support is a rapidly expanding area. In order for the group process to be considered successful researchers must show that the group acting with decision support can be more effective than the second most effective member of the group (Reagan-Cirincione, 1992). The group must be provided with both cognitive support and social support in their activities; there is no lack of options for providing this support.

See **Group decision making**.

References

[1] DeSanctis, G. and Gallupe, R. (1987), "A Foundation for the Study of Group Decision Support Systems," *Management Science*, 33, 589–609.

[2] Huber, G. (1984), "Issues in the Design of Group Decision Support Systems," *MIS Quarterly*, 195–204.

[3] McCartt, A. and Rohrbaugh, J. (1989), "Evaluating Group Decision Support System Effectiveness: A Performance Study of Decision Conferencing," *Decision Support Systems*, 5, 243–253.

[4] Nunamaker, J., Dennis, A., Valaich, J., Vogel, D., and George, J. (1991), "Electronic Meeting Systems to Support Group Work," *Communications of the ACM*, 34, 40–61.

[5] Phillips, L. (1984), "A Theory of Requisite Decision Modeling," *Acta Psychologica*, 56, 29–48.

[6] Reagan-Cirincione, P. (1992), "Combining Group Facilitation, Decision Modeling, and Information Technology to Improve the Accuracy of Group Judgment," in Nunamaker, J. and Sprague, R., eds., *Proceedings of the Hawaii International Conference on System Sciences*, Vol. IV, IEEE Computer Society Press, Los Alamitos, California.

[7] Rohrbaugh, J. (1989), "Demonstration Experiments in Field Settings: Assessing the Process, not the Outcome, of Group Decision Support," in Bengasat, I., ed., *The Information Systems Research Challenge: Experimental Research Methods*, Vol. 2, Harvard Business School Publishing, Cambridge, Massachusetts.

GROUP DECISION MAKING

Fatemeh (Mariam) Zahedi

University of Wisconsin, Milwaukee

Group decision making focuses on problems in which there is more than one decision maker and more than one choice. The choices or alternatives have multiple attributes. In other words, the decision makers must consider more than one objective or criterion in their decision. Hence, group decisions involve multiple criteria and multiple decision makers. Since preferences and objectives of individual decision makers vary and may be in conflict, arriving at a decision is far more complex in a group setting than in individual cases.

Group decision covers a wide range of collective decision processes and encompasses numerous methods designed under various assumptions and for different circumstances. One can divide group decision approaches into the following categories: group utility analysis, group consensus, group analytic hierarchy process, social choice theory, and game theory.

GROUP UTILITY ANALYSIS: Group utility analysis is based on the von Neumann-Morgenstern utility function. This method assumes that there is a multicriteria utility function $U_i(x_1, x_2, \ldots, x_m)$, where i represents member number, x_m represents the mth attribute, and m is the number of attributes. Based on the assumption that the utilities of the members are functionally independent, the group utility function is computed as the aggregation of the member utility functions by one of the following two function types.

The additive form of the function is:

$$U = \sum_{i=1}^{n} W_i U_i$$

where n is the number of attributes, and the multiplicative form is:

$$wU + 1 = \sum_{i=1}^{n} (ww_i U_i + 1),$$

where w and w_i are scaling constants satisfying $0 < w_i < 1$, $w > -1$, and $w \neq 0$. The estimation of member utility functions follows the assumptions and procedures used in estimating the individual multicriteria utility functions. The important question in the group utility estimation is the determination of the scaling constants. Keeney and Kirkwood (1975) suggest that these constants could be determined either by a "benevolent dictator" or internally by group members.

To resolve the problem of assigning weights to the utility functions of group members, two methods have been proposed. First, Bodily (1979) suggested the *delegation process*. This method is an iterative process for combining the utility functions of group members. The idea is that each member should assign weights or relative importance to other members. This process assumes that each member is adequately familiar with the views and utilities of other members. Each member replaces his or her utility by linearly combining other members' utilities. Members will not know the weights assigned to them by others. The method consists of the following steps.

Step 1. A delegation subcommittee for member i is formed consisting of the remaining $n-1$ members. Member i assigns a weight w_{ij} (a value between 0 and 1) to member j, and repeats the weight assignment for all $n-1$ members. The $n-1$ assigned weights should sum to 1. The weight for member i is zero; that is, $w_{ii} = 0$.

Step 2. The combined utilities of the delegation subcommittee is computed as:

$$u_i^1 = \sum_{j=1}^{n} w_{ij} u_j,$$

and replaces member i's utility. This process is repeated for all members.

Step 3. Step 2 is iterated for a second time as:

$$u_i^2 = \sum_{j=1}^{n} w_{ij} u_j^1$$

and for the rth time as:

$$u_i^r = \sum_{j=1}^{n} w_{ij} u_j^{r-1}.$$

In matrix form, the iteration can be represented as:

$$U^r = PU^{r-1},$$

where

$$U^r = [u_1^r, u_2^r, \ldots, u_n^r].$$

If the process is repeated adequately, from a theorem in Markov processes, one can show that under certain conditions, U^r converges and represents the group utility function.

BROCK METHOD: Brock (1980) developed an alternative method for estimating the weights for aggregating the utilities. The Brock Method is based on the assumptions that the solution for the group decision should be Pareto optimal, obtained from the additive combination of member utilities, and that utility gains should be distributed based on the needs of the affected parties. The needs are defined as the intensity of desire, computed as:

$$\frac{u_i - d_i}{u_j - d_j} = -\frac{du_i}{du_j} \quad \forall i,j.$$

Brock shows that the relative weights of members' utility functions are the reciprocals of the above coefficients.

GROUP CONSENSUS: Group consensus methods combine the observed preferences of members to create consensus points. These points are used to estimate the consensus function for the group. Group consensus methods do not require explicit estimation of member utility functions and may not necessarily lead to the estimation of a function for the group. This approach is in contrast with the utility approach in which the utility functions of members are estimated, then combined to arrive at a group utility function.

Krzysztofowicz Method. This method (1979) is based on the following assumptions:

1. The group utility function can be decomposed into functions (W_i) of its attributes (x_i) and these functions could be combined via another function (H) such that:

$$W(x_1, x_2, \ldots, x_n) = H(W_1(x_1), W_2(x_2), \ldots, W_n(x_n))$$

where $W_i(x_i)$ is the group marginal utility of attribute x_i and n is the number of attributes relevant to the group decision.

2. The group's observed preference is the result of combining members' observed preferences by the decision rule d.

3. The group members are divided into subgroups of experts. Each subgroup has expertise in a subset of attributes. Each subgroup is responsible for estimating $W_i(x_i)$.

4. Each member and subgroup behaves according to the axioms of utility theory.

In this method, the group is divided into subgroups. Each subgroup estimates $W_i(x_i)$ based on its expertise. The group's marginal utility functions of attributes are then combined by H, which is either an additive or multiplicative function, similar to those in the group utility theory.

In the subgroup estimation of $W_i(x_i)$, the expressed preferences of members are combined by the decision rule d. This leads to a series of consensus points from which the function $W_i(x_i)$ is estimated.

Zahedi Group Consensus Method. This method (1986a) is based on the following assumptions:

1. Preferences of individual members are uncertain.

2. The relative weight (or importance) of a member is inversely proportional to his or her degree of uncertainty in his or her response.

3. Standard deviation is the measure of uncertainty.

4. A member's preference response has a normal probability distribution.
5. Correlations among members remain constant over various alternatives.
6. A consensus point is generated by combining the members' expressed preference responses such that the combined point has the minimum variance or uncertainty.
7. The consensus function is estimated based on the generated consensus points.

Based on the above assumptions, the following steps lead to the estimation of the group consensus function:

Step 1. For each multicriteria alternative a, member i assigns an interval score $[x_{ai}, y_{ai}]$.

Step 2. Estimate the mean and standard deviation of the interval by:

$$\hat{U}_{ai} = \frac{y_{ai} + x_{ai}}{2}$$

$$\hat{\sigma}_{ai} = \frac{y_{ai} - x_{ai}}{6}$$

Step 3. Compute group member correlations by

$$\hat{\rho}_{ik} = \frac{Cov(\hat{U}_i, \hat{U}_k)}{\sqrt{Var(\hat{U}_i) \cdot Var(\hat{U}_k)}},$$

where i and k are members. Form a covariance matrix among group members for alternative a by using the standard deviations obtained in Step 2 and pairwise covariance obtained in Step 3. This matrix is symmetric of size n, where n is the number of group members. The main diagonal elements are the variances of n members. The off-diagonal element of row i and column j is $\hat{\sigma}_{ai}\hat{\sigma}_{aj}\hat{\rho}_{ik}$.

Step 4. Compute member i's weight for alternative a (w_{ia}s) by:

$$w_{ia} = \frac{\sum_{k=1}^{n} \alpha_{ika}}{\sum_{h=1}^{n} \sum_{k=1}^{n} \alpha_{hka}}$$

where α_{hka} is the element of the inverse of the covariance matrix computed at Step 3.

Step 5. Compute the consensus point for alternative a by using the results of Steps 2 and 4 in:

$$\hat{U}_a = \sum_{i=1}^{n} w_{ia} \hat{U}_{ia}.$$

Step 6. The consensus point could be used directly for selecting the alternative with the highest consensus value. Furthermore, one can estimate the group consensus function by using the consensus points as dependent variable in a regression analysis in which the independent variables are the attribute values.

In the Zahedi method, consensus values and the consensus function are obtained directly from the preference responses of members. It does not assume the existence of utility axioms and does not require members' utility estimation.

Nominal Group Technique. The nominal group technique was first proposed by Delbecq and Van de Ven (1971). The idea of nominal group technique became one of the methods of consensus generation in total quality management (TQM). In this technique, the ideas are generated in silence and recorded, then they are discussed in group, their importance is voted upon, and the final vote is taken. It has the following steps:

Step 1. The team leader presents the group with the description of the problem and each member records his or her idea or solution individually in silence.

Step 2. The leader asks members to express their ideas one at a time and record them on a chart.

Step 3. Members discuss the recorded ideas so that all members understand each idea.

Step 4. Each idea is voted upon and ranked by members and the average ranking for each idea is computed.

Step 5. Another round of discussions clarifies the position of various members.

Step 6. The final vote is taken by a procedure similar to that of Step 4.

Delphi Method. The Delphi method, developed by Dalkey (1967), is used for generating consensus among members who are not in the same location. It involves written questionnaires and written answers. In this method, the group leader identifies the problem or the question, identifies group members, and contacts them. A sample of group members are selected by the group leader. The method goes through iterations of the following steps:

Step 1. Design the questionnaire to be answered by the selected group members.

Step 2. Have members complete the questionnaire.

Step 3. Analyze responses, make changes in the questionnaire, and include the aggregated responses in the previous round. Ask the members to react to the results of the previous round.

After a number of iterations, the final results are computed, and alternatives are ranked accordingly.

Iterative Open Planning Process. Ortolano (1974) suggested the open planning process. In this method, the activities are divided into four stages: problem identification, plan formulation, impact assessment, and evaluation. There are two sets of decision makers: planners and the affected public. Planners and the public interact at each stage.

- At the problem identification stage, planners determine and evaluate factors from many perspectives and the public articulate problems and concerns.
- At the plan formulation stage, planners delineate alternatives and the affected public suggests alternatives.
- At the impact assessment stage, planners forecast and describe the impacts while the affected public assists them in describing the impacts.
- At the evaluation stage, planners organize and display information on alternatives and impacts and the affected public evaluates impacts, makes tradeoffs, and expresses preferences.

These stages take place concurrently and the planners and the affected public repeat the process a number of times.

GROUP ANALYTIC HIERARCHY PROCESS: The analytic hierarchy process (AHP) method was developed by Saaty (1977) and extended to group decision making by Aczel and Saaty (1983). In this method, the alternatives receive a score computed via AHP. This method does not require estimation of utility functions and does not assume axioms of utility analysis. The group AHP method consists of the following steps (Zahedi, 1986b).

Step 1. A decision hierarchy is created based on the nature of the decision problem. This hierarchy has multiple levels. At the upper-most level, the decision goal is specified as selecting the best alternative. The next level consists of categories of the attributes that are of importance in the group decision. The next level details each category of attributes into finer and more tangible features. The lowest level of the hierarchy contains the decision alternatives.

For example, for selecting the best car, the highest level of the hierarchy has the selection of the best car as its only element. The second level of the hierarchy includes cost, safety, and design attributes. At the third level of the hierarchy, these attributes are divided into more specific attributes. For example, the cost attribute may be divided into purchase price, preventive maintenance costs, and repair costs at the third level. The safety attribute may be divided into accident outcomes and frequency of breakdowns. The design attribute may be divided into esthetics, driver comfort, and space. The fourth level of the hierarchy includes the decision alternatives – cars to be selected – say, Toyota, Ford, and GM.

Step 2. At each level of the hierarchy, elements are compared pairwise for their role or importance in each element on the level immediately above. The input matrix for the pairwise comparisons has the following form:

$$A = \begin{pmatrix} a_{11} & a_{12} & a_{13} & \ldots & a_{1n} \\ a_{21} & a_{22} & a_{23} & \ldots & a_{2n} \\ a_{31} & a_{32} & a_{33} & \ldots & a_{3n} \\ . & . & . & \ldots & . \\ . & . & . & \ldots & . \\ a_{n1} & a_{n2} & a_{n3} & \ldots & a_{nn} \end{pmatrix}$$

where $a_{ij} = 1/a_{ji}$ for all $i,j = 1, 2, \ldots, n$, $a_{ii} = 1$, and n are the elements of one level being compared pairwise for their role in accomplishing one of the elements on the upper level.

For example, Toyota, Ford, and GM cars could be compared pairwise in their purchase price to get a pairwise comparison matrix A. The size of this matrix will be 3. One such matrix is needed for each of the elements of the prior level – purchase price, preventive maintenance costs, repair costs, accident outcomes, frequency of breakdowns, esthetics, driver comfort, and space.

Step 3. At this step, a computational method is used to reduce the matrix of pairwise comparisons into a vector of local relative weights. The best-known and most widely used computational method is the eigenvalue method, in which the local relative weights are computed as $AW = \lambda W$, were W is the vector of local relative weights, which is the largest eigenvector of A, and λ is the largest eigenvalue of matrix A.

Step 4. At this step, the local relative weights are combined to arrive at one vector of global relative weights for alternatives at the lowest level of the hierarchy in accomplishing the goal specified at the highest level of the hierarchy. The alternative with the highest global relative weight is the best choice, according to the AHP.

Applied to the group decision setting, the group must reach a consensus as to the structure of the hierarchy at Step 1. At Step 2, there will be a matrix of pairwise comparison elicited for each member of the group. The group pairwise matrix is computed from combining the member matrices. Each element of this matrix is the geometric average of the corresponding elements of the member matrices.

For example, assume that there are four decision makers involved in the decision to purchase a car. When three cars are compared pairwise for purchase price, one matrix of pairwise comparison is created when there is only one decision maker. When there are four decision makers, there will be four such matrices. To compute the group matrix for comparing cars, g_{ij}, the ijth elements of the four matrices are multiplied and then raised to one-fourth power, so that $g_{ij} = (a_{ij}^1 a_{ij}^2 a_{ij}^3 a_{ij}^4)^{1/4}$, where the superscript on a_{ij} represents the decision maker and g_{ij} is the geometric mean of the four decision makers' pairwise values.

Steps 3 and 4 of the group AHP are the same as those of the single-decision maker AHP.

SOCIAL CHOICE THEORY: The group decision problem was of interest as early as the eighteenth century, when Borda studied voting problems in the 1770s and Marquis de Condorcet had noticed the paradoxes and problems of majority rule in 1780s. One example of such problems is that of three alternatives a, b, and c, where a is preferred to b, b is preferred to c, and c is preferred to a. The attention to the methods for making a social choice continued in nineteenth century and greatly intensified in this century.

One way to arrive at a group decision is through voting, which also falls under the heading of social choice theory. The social choice theory investigates the process of arriving at a group decision in democratic societies through the expression of the majority's will. Voting involves selecting an alternative or candidate based on multiple criteria. It involves two processes: voting and the aggregation method for determining the winner, i.e.; voting and counting the vote. There are a number of methods for voting, such as bivalue (yes, no), rating, and ranking the alternatives. Counting could be a simple counting of yes or no votes, averaging the rates, or a more complex aggregation method using ranks.

Social Welfare Function. In the social welfare function, the voting and counting processes are given a formal mathematical structure. Each member has a utility function based on which he or she determined the ordinal ranking of all alternatives. A member's ordering of alternatives is called the preference profile. The social welfare function is a rule for structuring group preference orderings of alternatives from the members' preference profiles. Obviously, there are numerous ways to arrive at group ordering alternatives. Arrow sought to limit the number of possible group orderings of alternatives, which led to his famous impossibility theorem.

Arrow's Impossibility Theorem. Arrow (1951) observed that by imposing rational conditions, one can reduce the number of solutions in the social welfare function. He postulated that:

1. All possible choices are already included in the problem set.
2. If we drop one alternative such that the preference relations remain unchanged, the group preference will not change.
3. For any given two alternatives, the group can express its preference of one over the other.
4. There is no individual in the group whose preference represents the group preference.
5. Assume the group prefers alternative 1 to 2. If one member's preference for alternative 1 increases without affecting the pairwise preference ordering of other alternatives, the group will continue to prefer alternative 1 to 2.

Arrow's famous impossibility theorem states that there is no social welfare function that satisfies all of the above five properties.

Note that in voting and the social welfare function, the strength of members' preferences is not taken into account. Whereas, in utility theory, consensus generation methods, and AHP, the strength of members' preferences is incorporated into the method. In utility theory and consensus generation methods, members' preferences are assigned relative weights, relative to their importance in the group decision.

GAME THEORY: Game theory has been developed in the context of decision makers (or players) who are in conflict. However, game theory has been extended to include a cooperative n-person game, in which the players cooperate with one another in order to maximize their own gain or payoff. The Nash-Harsanyi and Shapley methods are among the game-theoretic methods of group decision.

Nash-Harsanyi bargaining method. Nash (1950, 1953) developed the two-person cooperative game, which was generalized to the n-person cooperative game by Harsanyi (1963). In this model, one can find a unique solution to the n-person cooperative game problem by solving the following:

$$\max_{x_i} \prod_{i=1}^{n} (x_i - d_i)$$

$$s.t.$$

$$x_i \geq d_i \quad x \in P, D = (d_1, d_2, \ldots, d_n),$$

where P is the set of payoff vectors, and D is the payoff when disagreement exists.

The above formulation is based on the following assumptions:

(1) No payoff is better than the solution of the above formulation.
(2) The players' payoffs are the same.
(3) The linear transformation of all payoffs does not change the solution to the above problem.

(4) Assume there are two games 1 and 2 with the same payoffs for disagreement, and the payoff vector of game 1 is a subset of the payoff vector of game 2. If the solution of game 2 is in the payoff vector of game 1, then it is also the solution to game 1. This assumption ensures that adding nonoptimal payoffs does not change the optimal solution.

Harsanyi (1977) has shown that the above formulation can also be derived from the Zeuthen's principle that the player who has the highest risk-aversion towards conflict always makes the next concession.

The Shapley Value. If the utility of the players is transferable – that is, one player can transfer money, goods, or services to another player such that the sum of the two players' utilities remains the same – then the Nash-Harsanyi solution does not hold. This is due to the fact that there would not be a unique payoff vector for disagreement. Shapley's value (1953) provides a solution for an n-person cooperative game with a transferable utility function.

A *coalition* is defined as a subgroup of members. The grand coalition consists of all the group members. If a member i joins a coalition, his or her marginal contribution to the coalition C is specified as $V(C) - V(C - i)$.

The payoff to member i should be the average marginal contribution of the player to the grand coalition. Assume that the grand coalition is formed by members gradually joining the coalition and the order of members joining the coalition is equally likely, then the payoff of each player ($P_i, i = 1, 2, \ldots, n$), or the Shapley value, is

$$P_i = \sum_{C \subseteq N} \frac{(c-1)! \, (n-c)!}{n!} \, V(C) - V(C - i),$$

where c is the number of players in C, n is the number of players, and N is the set of players.

The above solution is based on the following assumptions:
1. The value of the entire game is the sum of the payoffs to members.
2. All members receive an equal payoff.
3. If a game consists of two subgames, the payoff of the game is the sum of the payoffs of the two subgames.

COMPUTER-BASED GROUP DECISION PROCESS: A number of computer-based methods have been developed to facilitate the group decision process in various circumstances. These systems can divided into two groups: intelligent systems and group decision support systems.

Intelligent Systems for Group Decisions. In a sequential group process, there is more than one party involved in a negotiation process that takes place sequentially through time. One can use the artificial intelligence and expert system techniques to facilitate the process. A number of such systems have been developed.

Sycara developed PERSUADER, which simulates the labor-management negotiation process (Sycara, 1991). This system uses frame-based knowledge representation and case-based reasoning of artificial intelligence with graph search and multi-attribute utilities to propose problem restructuring for simulated negotiations. The system restructures the problem by (1) introducing new goals, (2) substituting goals, and (3) abandoning goals.

The logical representation of the negotiation process using the framework of mathematical logic is another way to model the group negotiation process. Kersten, Michalowski, Szpakowicz, and Koperczak (1991) show how one can model negotiation and restructure it for arriving at a negotiated solution.

Group Decision Support Systems. Group decision support systems refer to computer-based systems and methods developed to facilitate group decision making. One category of such systems is the electronic meeting system (EMS), which consists of a collection of hardware, software, audio and video equipments, and group procedures to create a supportive environment for the group decision process (Dennis, George, Jessup, Nunamaker, and Vogel, 1988).

These systems are designed for various purposes:
- Generating group options and brainstorming.
- Supporting and improving communication among the group members.
- Increasing participation.
- Providing computational and procedural support for the group process.

The existence and extent of positive contributions of such systems are under study at present.

See **Analytic hierarchy process; Decision problem; Delphi method, Game theory; Group decision computer technology; Markov processes; Multi-criteria decision making; Total quality management; Utility theory.**

References

[1] Arrow, K.J. (1951). "Social Choice and Individual Values," *Cowles Commission Monograph* 12, Wiley, New York.

[2] Aczel, J. and T.L. Saaty (1983). "Procedures for Synthesizing Rational Judgements," *Jl. Mathematical Psychology*, 27, 93–102.

[3] Bodily, S.E. (1979). "A Delegation Process for Combining Individual Utility Function," *Management Science*, 25, 1035–1041.

[4] Brock, H.W. (1980). "The Problem of Utility Weights in Group Preference Aggregation," *Operations Research*, 28, 176–187.

[5] Dalkey, N.C. (1967). *Delphi*, Rand Corporation.

[6] Delbecq A.L. and A.H. Van de Ven (1971). "A Group Process Model for Problem Identification and Program Planning," *Jl. Applied Behavior Sciences*, 7, 466–492.

[7] Dennis, A.R., George, J.F., Jessup, L.M., Nunamaker, Jr, J.F., and Vogel, D.R. (1988). "Information Technology to Support Electronic Meetings," *MIS Quarterly*, 12, 591–624.

[8] Harsanyi, J.C. (1963). "A Simplified Bargaining Model for the n-person Cooperative Game," *International Economic Review*, 4, 194–220.

[9] Harsanyi, J.C. (1977). *Rational Behavior and Bargaining Equilibrium in Games and Social Situations*, Cambridge University Press, Cambridge, England.

[10] Keeney, R.L. and C.W. Kirkwood (1975). "Group Decision Making Using Cardinal Social Welfare Functions," *Management Science*, 22, 430–437.

[11] Kersten, G., Michalowski, W., Szpakowicz, S., and Koperczak. Z. (1991). "Restructurable Representations of Negotiation," *Management Science*, 37, 1269–1290.

[12] Krzysztofowicz, R. (1979). "Group Utility Assessment Through a Nominal-Interacting Process," Unpublished working paper, Department of Civil Engineering, MIT, Cambridge, Massachusetts.

[13] Mirkin, B.G. (1979). *Group Choice*, V.H. Winston & Sons, Washington, D.C.

[14] Nash, J. (1950). "The Bargaining Problem," *Econometrica*, 18, 155–162.

[15] Nash, J. (1953). "Two-Person Cooperative Games," *Econometrica*, 21, 128.

[16] Ortolano, L. (1974). "A Process for Federal Water Planning at the Field Level," *Water Resources Bulletin*, 10(4), 776–778.

[17] Saaty, T.L. (1977). "A Scaling Method for Priorities in Hierarchical Process," *Jl. Mathematical Psychology*, 15, 234–281.

[18] Shapley, L.S. (1953). "A Value for n-person Games," in *Contributions to the Theory of Games*, H. W. Kuhn and A. W. Tucker (eds.), Princeton University Press, 307–317.

[19] Sycara, Katia P. (1991). "Problem Restructuring in Negotiation," *Management Science*, 37, 1248–1268.

[20] Zahedi, F. (1986a). "Group Consensus Function Estimation When Preferences Are Uncertain," *Operations Research*, 34, 883–894.

[21] Zahedi, F. (1986b). "The Analytic Hierarchy Process – A Survey of the Method and its Applications." *Interfaces*, 16, 96–108.

GUB

See **Generalized upper-bounded problem**.

HALF SPACE

See **Linear inequality**.

HAMILTONIAN TOUR

In an undirected connected graph, a Hamiltonian tour is a sequence of edges that passes through each node of the graph exactly once. See **Graph theory**; **Traveling salesman problem**.

HAZARD RATE

See **Failure-rate function**; **Probability distribution selection**; **Reliability of stochastic systems**.

HEALTH CARE SYSTEMS

Viviane Gascon

Collège militaire royal de St-Jean, Québec

William P. Pierskalla

University of California at Los Angeles

INTRODUCTION: The field of operations research applied to health care experienced its major growth in the 1960s and the 1970s. The 1980s saw refinements and/or reapplications of this earlier work except for the area of medical diagnostic and therapeutic decision-making which caught the attention of many research physicians, and which is continuing to grow substantially. With the current trend to nationally structured quasi public health care systems in the world, interest in operations research in health care delivery is again on the rise.

The different studies in operations research applied to health care cover a broad variety of subjects most of which fall into five areas: scheduling, allocation, forecasting demand, supplies/materials planning, medical-decision making. We will examine each area by describing some of the different topics they include, and the operations research techniques used to approach the problems.

For extensive reviews of operations research in health care, see Fries (1976, 1979) and Pierskalla *et al.* (1993).

SCHEDULING: Scheduling problems in the health care sector apply to situations involving patients, nurses, physicians, facilities, vehicles, and other health specific providers and settings. Many models for nurse scheduling have been constructed. Maier-Rothe and Wolfe (1973) proposed a cyclical model where the nurses would have the same schedule on a regular basis. Mathematical programming approaches were also used to solve the nurse scheduling problem (Warner, 1976; Miller *et al.*, 1976). A comprehensive nurse scheduling model should include most of the following features: the preferences of the nurses, the minimum staff requirements, frequence of weekends off, maximum number of consecutive working days, varying shift lengths, shift rotation, but few attempts were made to include all of them because of the complexity of the problem. Goal programming is also used to meet more than one objective such as desired patient care staffing requirements, nurses' preferences, nurses' special requests, and minimum costs. Heuristic algorithms taking into account the combinatorial structure of the problem were also considered. Nurse scheduling problems still continue to arouse interest because there is not yet a model that can gather all the elements of the whole problem.

Most of the work on vehicle location and scheduling in health care delivery involves dispatching EMS (emergency medical service) vehicles, blood donor mobiles, and blood delivery vehicles. The EMS scheduling solution primarily assigns the closest vehicle to best satisfy the demand (Liu and Lee, 1988; Trudeau *et al.*, 1989). The blood delivery solution is more closely related to the multiple traveling salesman problem. In both areas, however, the problem and solution are very closely connected to the vehicle locations, and also to the demands which must be forecasted. Simulation was the technique of choice to help determine solutions to these complex comprehensive problems.

The scheduling problem of patients for outpatient services consists in constructing an appointment system that will reduce the staff idle time, and the patients' waiting time. A good coordinated approach in outpatient scheduling leads to a solution that maximizes a function which trades off the satisfaction and utilization of both patients and providers (Fries and Marathe, 1981). Inpatient scheduling systems are more complicated since they have more constraints and more objectives to achieve in addition to those faced by outpatient systems. Many approaches have been used for these problems such as simulation (Dumas, 1985), mathematical programming, and heuristic methods (Hancock and Isken, 1992).

ALLOCATION: Health care resources are increasingly limited. It is thus necessary for the providers of health services to determine how to allocate the different services, rearrange the capacities of different services and/or choose the site of the services. Most service capacity planning studies are concerned with allocating beds to services within a hospital to meet demand. The factors considered are categories of patients based on disease and sex, categories of beds, type of physicians, etc. The primary objective is to maximize the utilization of available beds while being able to admit immediately the emergency patients and to reduce the length of the waiting list for elective admissions (Dumas, 1985). Simulation is the approach most often used. Service capacity planning which is closely related to allocation also deals with patient flow through a health care service. Queueing theory and simulation techniques are used in many studies to determine the patient population movement from one service care level to another in order to minimize the waiting time.

The site/location problem in health care was formerly related to constructing new hospitals. Now this construction is a much less frequent event. So location problems currently involve locating expensive facilities such as scanners, outpatient facilities and other high cost technologies to minimize the costs or involve the location of ambulances to minimize the response time and to serve a maximum number of emergency calls (Pirkul and Schilling, 1988; Trudeau *et al.*, 1989). Mathematical programming approaches such as goal programming and integer programming as well as heuristic algorithms and simulation are used for these problems.

FORECASTING DEMAND: Forecasting demand in health care is necessary to estimate the future occurrence of a disease and demands for diagnosic services and/or treatment. It has been used to make better decisions for allocating resources and also to study the progression of diseases. The forecasting models used in health care are numerous. Among them are multiple linear regression, exponential smoothing, ARIMA, Box Jenkins models and Markov processes (Kao and Pokladnik, 1978; Kao and Tung, 1980). A given model will be more appropriate for a certain type of database accuracy, time constraints and other factors. There have been many applications of forecasting in health care to forecast admissions, census and discharge from hospitals, the need and demand for emergency transportation, for services in nursing home and specialized clinics (Lane *et al.*, 1985), or for specific services in hospitals or clinics. The large volume of research studies involving forecasting models indicates how essential they are in health care. Morever, the need to reduce the costs in health care will encourage operations researchers to provide even better forecasting solutions.

SUPPLIES/MATERIALS PLANNING: The supplies in health delivery situations consist of perishable and non-perishable items many of which are very time and location dependent. In these cases, it is very important to prevent shortages. Hence, most applications of inventory models in the health care sector concentrate on preventing shortages rather than on minimizing the costs of inventories. However, some studies considered costs in determining the inventory levels for a multi-item pharmacy warehouse or determining the amount of inventory needed for sterile supplies (Ebrahimzadeh *et al.*, 1985). More work is currently underway on stockless just-in-time inventories. Because blood is a perishable and critical product, most of the articles on inventory models in health care are on managing blood bank inventories (Prastacos, 1984).

MEDICAL-DECISION MAKING: Medical-decision making in detecting and treating disease has flourished since the early introduction of artificial intelligence and expert systems for cardio-pulmonary diseases in the 1970s. Paralleling this work was work on Bayesian decision trees to detect gastrointestinal diseases. Both approaches are now used extensively for the detection, prevention and treatment of hundreds of diseases. Many medical schools have introduced decision analysis in their regular curriculum (Pauker and Kassirer, 1987).

Data envelopment analysis (DEA), first proposed by Charnes and Cooper (1978), has seen a growing interest in applications in health care. Chiligerian and Sherman (1990) used DEA to evaluate physician decision-making efficiency in the provision of hospital services. Other researchers applied DEA to compare quality of care in different hospitals, and the efficiency of similar services in teaching hospitals.

See **Emergency services; Facility location; Forecasting; Goal programming; Hospitals; Medicine and medical practice; Practice of OR/MS; Queueing theory; Scheduling and sequencing; Simulation of discrete-event stochastic systems.**

References

[1] A. Charnes, W.W. Cooper, and E. Rhodes (1978), "Measuring Efficiency of Decision Making Units," *Eur. J. Opl. Res.* 2, 429–444.

[2] J.A. Chiligerian and H.D. Sherman (1990), "Managing Physician Efficiency and Effectiveness in Providing Hospital Services," *Health Services Management Research* 3, 3–15.

[3] M.B. Dumas (1985), "Hospital Bed Utilization: An Implemented Simulation Approach to Adjusting and Maintaining Appropriate Levels," *Health Services Research* 20, 43–61.

[4] M. Ebrahimzadeh, S. Barnoon, and Z. Sinuani-Stern (1985), "A Simulation of a Multi-Item Drug Inventory System," *Simulation* 45, 115–121.

[5] B.E. Fries and V.P. Marathe (1981), "Determination of Optimal Variable-Sized Multiple-Block Appointment Systems," *Operations Research* 29, 324–345.

[6] B.E. Fries (1979), "Bibliography of Operations Research in Health-Care Systems: An Update," *Operations Research* 27, 409–419.

[7] B.E. Fries (1976), "Bibliography of Operations Research in Health-Care Systems," *Operations Research* 24, 801–814.

[8] A.M. Hancock and M.A. Isken (1992), "Patient Scheduling Methodologies," *Jl. Society for Health Systems* 3(4), 83–94.

[9] E.P.C. Kao and F.M. Pokladnik (1978), "Incorporating Exogenous Factors in Adapting Forecasting of Hospital Census," *Management Science* 24, 1677–1686.

[10] E.P.C. Kao and G.C Tung (1980), "Forecasting Demands for Inpatient Services in a Large Public Health Care Delivery System," *Socio-Econ. Plan. Sci.* 14, 97–106.

[11] D. Lane, D. Uyeno, A. Stark, E. Kheiver, and G. Gutman (1985), "Forecasting Demand for Long Term Care Services," *Health Services Research* 20, 435–460.

[12] M.Liu and J. Lee (1988), "A Simulation of a Hospital Emergency Call System Using SLAM II," *Simulation* 51, 216–221.

[13] C. Maier-Rothe and H.B. Wolfe (1973), "Cyclical Scheduling and Allocation of Nursing Staff," *Socio-Econ. Plan. Sci.* 7, 481–487.

[14] H.E. Miller, W.P. Pierskalla, and G.J. Rath (1976), "Nurse Scheduling Using Mathematical Programming," *Operations Research* 24, 857–870.

[15] S.G. Pauker and J.P. Kassirer (1987), "Decision Analysis," *New England Jl. Medicine* 316(5), 250–258.

[16] W.P. Pierskalla, D. Wilson, and V. Gascon (1993), "Review of Operations Research Improvements in Patient Care Delivery Systems," Working Paper, University of Pennsylvania, Philadelphia.

[17] H. Pirkul and D.A. Schilling (1988), "The Siting of Emergency Service Facilities with Workload Capacities and Backup Service," *Management Science* 34, 896–908.

[18] G.P. Prastacos (1984), "Blood Inventory Management: An Overview of Theory and Practice," *Management Science* 30, 777–800.

[19] P. Trudeau, J.-M. Rousseau, J.A. Ferland, and J. Choquette (1989), "An Operations Research Approach for the Planning and Operation of an Ambulance Service," *INFOR* 27, 95–113.

[20] D.M. Warner (1976), "Scheduling Nursing Personnel According to Nursing Preference: A Mathematical Programming Approach," *Operations Research* 24, 842–856.

HEAVY-TRAFFIC APPROXIMATION

As the traffic intensity of a queueing problem approaches 1 (from below), the measures of effectiveness for the system often take on patterns which become essentially insensitive to the exact form of the input and service processes defining the system and, for example, may depend only on expectations and variances. As an illustration, the distribution for line delay of the general G/G/1 queue with utilization rate $\rho = 1 - \varepsilon$ can be well approximated by $W_q(t) = 1 - \exp(-at)$, where $a = (1/2)$(interarrival time variance + service-time variance)/(mean interarrival time − mean service time). See **Queueing theory**.

HESSENBERG MATRIX

A matrix that would be upper triangular except for having nonzero elements immediately below the main diagonal. Such matrices arise when trying to preserve sparsity in computing a matrix inverse. See **Matrices and matrix algebra**.

HESSIAN MATRIX

For a function $f(x)$ of the vector variable x, the Hessian matrix is a square matrix formed by the set of second-order partial derivatives evaluated at a specific point x^0 (if they exist) and is denoted by $\nabla^2 f(x)$. It is an $n \times n$ matrix whose i,j element is

$$\nabla^2 f(x)_{i,j} = \frac{\partial^2 f(x^0)}{\partial x_i \, \partial y_j}.$$

If the second partial derivatives are continuous at x^0, then the Hessian is a symmetric matrix. See **Nonlinear programming**; **Quadratic programming**.

HETEROGENEOUS LANCHESTER EQUATIONS

Differential (or difference) equations equating force size changes for each of several weapons systems (components) on each side to sums of the products of coefficients and component force sizes. The concept is that each component is attrited to some degree by each component of the opposing side; however, the killing mechanism and rate depend on the pairing. Hence, each term defines the mechanism (such as square law or linear law) and rate (the coefficient) and the sum of the terms defines the total attrition for the system. Therefore, rather than the two equations of a homogeneous Lanchester law, there is one equation for each component of each side. See **Battle models**.

HEURISTIC PROCEDURE

For a given problem, a collection of rules or steps that guide one to a solution that may or may not be optimal. The rules are usually based on the problem's characteristics, intuition, hunches, good ideas, or reasonable processes for searching. See **Greedy algorithms**; **Tabu search**; **Simulated annealing**.

HIERARCHICAL PRODUCTION PLANNING

Arnoldo C. Hax

Massachusetts Institute of Technology, Cambridge

INTRODUCTION: Production management encompasses a large number of decisions that affect several organizational echelons. These decisions can be grouped into three broad categories:
1. Strategic decisions, involving policy formulation, capital investment decisions, and design of physical facilities.
2. Tactical decisions, dealing primarily with aggregate production planning.
3. Operational decisions, concerning detailed production scheduling issues.

These three categories of decisions differ markedly in terms of level of management responsibility and interaction, scope of the decision, level of detail of the required information, length of the planning horizon needed to assess the consequences of each decision, and degree of uncertainties and risks inherent in each decision. These considerations have led us to favor a hierarchical planning system to support production management decisions, which guarantees an appropriate coordination of the overall decision-making process but, at the same time, recognizes the intrinsic characteristics of each decision level.

HIERARCHICAL PRODUCTION PLANNING: The basic design of a hierarchical planning system includes the partitioning of the overall planning problem, and the linkage of the resulting subproblems. An important input is the number of levels recognized in the product structure. We identify three different levels:
1. *Items* are the final products to be delivered to the customers. They represent the highest degree of specificity regarding the manufactured products. A given product may generate a large number of items differing in characteristics such as color, packaging, labels, accessories, size, and so on.
2. *Families* are groups of items which share a common manufacturing setup cost. Economies of scale are accomplished by jointly replenishing items belonging to the same family.
3. *Types* are groups of families whose production quantities are to be determined by an aggregate production plan. Families belonging to a type normally have similar costs per unit of production time, and similar seasonal demand patterns.

These three levels are required to characterize the product structure in many batch-processing manufacturing environments. In this section, we propose hierarchical planning systems based on these three levels of item aggregation.

The first step in our hierarchical planning approach is to allocate production capacity among product types by means of an aggregate planning model. The planning horizon of this model normally covers a full year in order to take into proper consideration the fluctuation demand requirements for the products. We advocate the use of a linear programming model at this level.

The second step in the planning process is to allocate the production quantities for each product type among the families belonging to that type by disaggregating the results of the aggregate planning model only for the first period of the planning horizon. Thus, the required amount of data collection and data processing is reduced substantially. The disaggregation assures consistency and feasibility among the type and family production decisions and, at the same time, attempts to minimize the total setup costs incurred in the production of families. It is only at this stage that setup costs are explicitly considered.

Finally, the family production allocation is divided among the items belonging to each family. The objective of this decision is to maintain all items with inventory levels that maximize the time between family setups. Again, consistency and feasibility are the driving constraints of the disaggregation process. Figure 1 shows the overall conceptualization of the hierarchical planning effort.

AGGREGATE PRODUCTION PLANNING FOR PRODUCT TYPES:

Aggregate production planning is the highest level of planning in the production system, addressed at the product-type level. Any aggregate production planning model can be used as long as it adequately represents the practical problem under consideration. We consider the following simplified linear program at this level:

Problem P

$Minimize \quad \sum_{i=1}^{I} \sum_{t=1}^{T} (c_{it} X_{it} + h_{i,t+L} I_{i,t+L})$
$\qquad + \sum_{t=1}^{T} (r_t R_t + o_t O_t)$

subject to:

$$X_{it} - I_{i,t+L} + I_{i,t+L-1} = d_{i,t+L} \quad i=1,\ldots,I; t=1,\ldots,T$$

$$\sum_{i=1}^{I} m_t X_{it} = O_t + R_t \quad t=1,\ldots,T$$

$$R_t \le (rm)_t \quad t=1,\ldots,T$$

$$O_t \le (om)_t \quad t=1,\ldots,T$$

$$X_{it}, I_{i,t+L} \ge 0 \quad i=1,\ldots,I; t=1,\ldots,T$$

$$R_t, O_t \ge 0 \quad t=1,\ldots,T.$$

The decision variables of the model are: X_{it}, the number of units to be produced of type i during t; $I_{i,t+L}$, the number of units of inventory of type i left over at the end of period $t+L$: and

R_t and O_t, the regular hours and the overtime hours used during period t, respectively.

The parameters of the model are: I, the total number of product types; T, the length of the planning horizon; L, the length of the production lead time; c_{it}, the unit production cost (excluding labor); h_{it}, the inventory carrying cost per unit per period; r_t and o_t, the cost per manhour of regular labor and of overtime labor; $(rm)_t$ and $(om)_t$, the total availability of regular hours and of overtime hours in period t, respectively; and m_t, the inverse of the productivity rate for type i in hours/unit, $d_{i,t+L}$ is the effective demand for type i during period $t+L$.

Because of the uncertainties present in the planning process, only the first time period results of the aggregate model are implemented. At the end of every time period, new information becomes available that is used to update the model with a rolling planning horizon of length T. Therefore, the data transmitted from the type

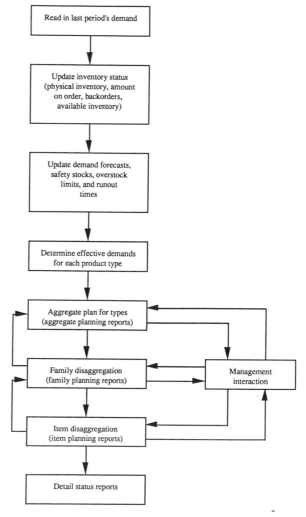

Fig. 1. Conceptual overview of hierarchical planning system.

level to the family level are the resulting production and inventory quantities for the first period of the aggregate model. These quantities will be disaggregated among the families belonging to each corresponding type.

THE FAMILY DISAGGREGATION MODEL: The central condition to be satisfied at this level for a coherent disaggregation is the equality between the sum of the productions of the families in a product type and the amount dictated by the higher level for this type. This equality will assure consistency between the aggregate production plan and the family disaggregation process. This consistency is achieved by determining run quantities for each family that minimize the total setup cost among families.

Bitran and Hax (1977, 1981) propose the following model for family disaggregation which has to be solved for every product type i and gives rise to a continuous knapsack problem:

Problem P_i

$\quad$ *Minimize* $\quad \sum_{j \in J^0} (s_j d_j / Y_j)$

subject to:

$\quad \sum_{j \in J^0} Y_j = X_i^*$

$\quad lb_j \leq Y_j \leq ub_j \qquad (j \in J^0) \qquad\qquad (1)$

where Y_j is the number of units of family j to be produced; s_j is the setup cost for family j; d_j is the forecast demand (usually annual) for family j; lb_j and ub_j are lower and upper bounds for the quantity Y_j; and X_i^* is the total amount to be allocated among all the families that belong to type i. The quantity X_i^* has been determined by the aggregate planning model and corresponds to the optimum value of the variable X_{i1} since only the first-period result of the aggregate model is to be implemented.

The lower bound lb_j, which defines the minimum production quantity for family j, is given by:

$\quad lb_j = \max[0, (d_{j,1} + d_{j,2} + \ldots + d_{j,L+1}) - AI_j + SS_j],$

where $d_{j,1} + d_{j,2} + \ldots + d_{j,L+1}$ is the total forecast demand for family j during the production lead time plus the review period (assumed equal to one); AI_j is the current available inventory for family j (equal to the sum of the physical inventory and the amount on order minus the backorders); and SS_j is the required safety stock. The lower bound lb_j guarantees that any backorder will be caused by forecast errors beyond those absorbed by the safety stock SS_j.

The upper bound ub_j is given by:

$\quad ub_j = OS_j - AI_j,$

where OS_j is the overstock limit of family j.

The objective function of problem P_i assumes that the family run quantities are proportional to the setup cost and the annual demand for a given family. This assumption, which is the basis of the economic order quantity formulation, tends to minimize the average annual setup cost. Notice that the total inventory carrying cost has already been established in the aggregate planning model; therefore, it does not enter into the current formulation.

The first constraint of problem P_i,

$\quad \sum_{j \in J^0} Y_j = X_i^*$

assures the equality between the aggregate model input X_i^* and the sum of the family run quantities.

Initially, J^0 contains only those families which trigger during the current planning period. A family is said to trigger whenever its current available inventory cannot absorb the expected demand for the family during the production lead time plus the review period; that is, those families whose current available inventory is such that

$\quad AI_j < (d_{j,1} + d_{j,2} + \ldots + d_{j,L+1}) + SS_j.$

Equivalently, one can define J^0 as containing all those families whose runout times are less than one time period, that is,

$$ROT_j = \frac{AI_j - SS_j}{\sum_{t=1}^{L+1} d_{j,t}} < 1.$$

It is necessary to start production for these families in order to avoid future backorders. All other families are put on a secondary list and will be scheduled only if extra capacity is available. Bitran and Hax (1977) propose an efficient algorithm to solve problem P_i through a relaxation procedure.

THE ITEM DISAGGREGATION MODEL: For the period under consideration, all the costs have already been determined in the former two levels, and any feasible disaggregation of a family run quantity has the same total cost. However, the feasible solution chosen will establish initial conditions for the next period and will affect future costs. In order to save setups in future periods, one could distribute the family run quantity among its items in such a way that the runout times of the items coincide with the runout time of the family. A direct consequence is that all items of a family will trigger simultaneously. To attain this objective, we propose the following strictly convex knapsack problem for each family j.

Problem P_j

Minimize

$$\frac{1}{2} \sum_{k \in K^0} \left[\frac{Y_j^* + \sum\limits_{k \in K^0} (AI_k - SS_k)}{\sum\limits_{k \in K^0} \sum\limits_{t=1}^{L+1} d_{k,t}} - \frac{Z_k + AI_k - SS_k}{\sum\limits_{t=1}^{L+1} d_{k,t}} \right]^2$$

subject to:

$$\sum_{k \in K^0} Z_k = Y_j^*$$

$$Z_k \leq OS_k - AI_k$$

$$Z_k \geq \max \left[0, \sum_{t=1}^{L+1} d_{k,t} - AI_k + SS_k \right]$$

where Z_k is the number of units to be produced of item k; AI_k, SS_k, and OS_k are, respectively, the available inventory, the safety stock, and the overstock limit of item k; $d_{k,t}$ is the forecast demand for item k in period t; $K^0 = \{1, 2, \ldots, j\}$; and, Y_j^* is the total amount to be allocated for all items belonging to family j. The quantity Y^* was determined by the family disaggregation model.

The first constraint of problem P_j requires consistency in the disaggregation from family to items. The last two constraints are the upper and lower bounds for the item run quantities. These bounds are similar to those defined for the family disaggregation model in the previous section.

The two terms inside the square bracket of the objective function represent, respectively, the runout time for family j and the runout time for an item k belonging to family j (assuming perfect forecast). The minimization of the square of the differences of the runout times will make those quantities as close as possible. (The term $1/2$ in front of the objective function is just a computational convenience.)

For a description of the algorithm recommended to solve this problem, as well as a discussion on performance of the hierarchical production planning model, the reader is referred to Hax and Candea (1984).

See **Operations management; Production planning.**

References

[1] Bitran, G.R., E.A. Haas, and A.C. Hax (1982), "Hierarchical Production Planning: A Single Stage System," *Operations Research*, 29, 717–743.

[2] Bitran, G.R., E.A. Haas, and A.C. Hax (1982), "Hierarchical Production Planning: A Two Stage System," *Operations Research*, 30, 232–251.

[3] Bitran, G.R. and A.C. Hax (1977), "On the Design of Hierarchical Production Planning Systems," *Decision Sciences*, 8, 28–54.

[4] Bitran, G.R. and A.C. Hax (1981), "Disaggregation and Resource Allocation Using Convex Knapsack Problems with Bounded Variables," *Management Science*, 27, 431–441.

[5] Hax, A.C. and D. Candea (1984), *Production and Inventory Management*, Prentice Hall, Englewood Cliffs, New Jersey.

[6] Hax, A.C. and J.J. Golovin (1978), "Hierarchical Production Planning Systems," A.C. Hax (editor), in *Studies in Operations Management*, North Holland, Amsterdam.

[7] Hax, A.C. and J.J. Golovin (1978), "Computer Based Operations Management System (COMS)," A. C. Hax (editor), in *Studies in Operations Management*, North Holland, Amsterdam.

[8] Hax, A.C. and H.C. Meal (1975), "Hierarchical Integration of Production Planning and Scheduling," in M. Geisler (editor), *TIMS Studies in Management Science*, 1: Logistics, North Holland/American Elsevier, New York.

[9] Holt, C.C., F. Modigliani, J.F. Muth, and H.A. Simon (1960), *Planning Production Inventories and Work Force*, Prentice Hall, Englewood Cliffs, New Jersey.

[10] Lasdon, L.S. and R.C. Terjung (1971), "An Efficient Algorithm for Multi-Item Scheduling," *Operations Research*, 19, 946–969.

[11] Winters, P.R. (1962), "Constrained Inventory Rules for Production Smoothing," *Management Science*, 8, 470–481.

HIGHER EDUCATION

James C. Hearn and John R. Lough

University of Georgia, Athens, Georgia

Until the mid-1960s, rigorous planning techniques and serious attention to campus resource allocations were relatively uncommon in higher education. Expansionist institutions were flush with both students and public support. Nevertheless, in the ensuing years, academic administrators' interest in better management techniques was heightened by the increasing size and complexity of academic organizations and by persuasive warnings of shrunken public financing and reduced numbers of college-going young people. By the early 1980s, college and university operations were a well-established locus of interest among OR/MS professionals. Since that time, however, the trend appears to have reversed. Although no hard evidence is available, a scan of literature and professional activity in the field suggests that the interest in using management-science approaches on campus has waned somewhat.

HISTORICAL BACKGROUND: The earliest appearance of modern OR/MS applications in higher education came in the development of formal planning models for institutions and systems. The first of these was CAMPUS (Comprehensive Analytical Methods for Planning University Systems). Begun in 1964 at the University of Toronto, CAMPUS was part of an attempt to build a computer-based econometric model simulating cost patterns in Canadian universities. Early versions of CAMPUS required extensive data input and placed great demand on computer capacity, making widespread use by a number of colleges and universities virtually impossible. CAMPUS clearly demonstrated, however, the feasibility of developing useful planning and decision-making tools for postsecondary institutions.

Around 1969, through funding from the United States Office of Education, a similar effort was begun by the National Center for Higher Education Management Systems in Colorado. The ultimate product was another computer-simulation model known as RRPM (Resource Requirements Prediction Model). The major purpose of the various RRPM versions was to provide institutions detailed information on costs and resource requirements for establishing and maintaining academic programs.

At about the same time, two significant European models were developed. The first of these, HIS (Hochschule Information System), was produced by a West German firm and supported by the Volkswagen Foundation. The other was TUSS (Total University Simulation System), developed and installed at the University of Utrecht in Holland. These two planning tools were limited to academic operations and focused mainly on the efficient use of instructional space. As such, they were less generalizable than the North American models of the time, and much less powerful than the models soon to come.

In 1977, to assist administrators in forecasting income and expenses, Stanford University developed a computer-based financial planning model called TRADES. Because it was interactive, fast, and extraordinarily comprehensive in scope, the TRADES model represented a distinct advance for the field. True to its name, the TRADES model focused on the trade-offs facing campus leaders. Specifically, it allowed the user to manipulate certain "Primary Planning Variables" (such as the number of faculty, the number of admitted students, or the level of utility rates) plus approximately 200 other variables to create "what-if" scenarios for any variety of campus and environmental conditions. Although

TRADES was created solely for use at Stanford, a generalized version, GENTRA (GENeralized TRAdes), was soon made available through the institution's Academic Planning Office. The TRADES and GENTRA models were the first to handle the large volume of tedious calculations involved in systematic, holistic financial forecasting. As a consequence, the models gave their users new freedom to focus directly on core issues in institutional planning.

In the years since their initial development, Stanford's models have been both adapted and imitated. Perhaps the best known of the "descendant" models is EFPM, the EDUCOM Financial Planning Model. EFPM was developed in the late 1970s by EDUCOM, a nonprofit consortium of more than 350 colleges and universities in the U.S. Largely a trend-projection model providing institutions the ability to estimate future financial position by either extending past budgetary conditions or by hypothetically imposing new policies, EFPM is probably the most commonly used decision-support system [DSS] in higher education today.

THE LITERATURE: The literature in this field may be organized into three general categories. First, there are *works providing general reviews of OR/MS applications in higher education*. An excellent early example is Schroeder (1973), which critically surveyed work in four areas: Planning, Programming and Budgeting Systems [PPBS]; management information systems [MIS]; resource-allocation models; and mathematical models for enrollment planning, faculty staffing, and optimization of resource use. A less technical piece in this same vein is Wilson (1981), a collection of eight short but well-documented articles describing actual and proposed applications. Significantly, this book contains contributions by several central figures in the field, including Cyert, Balderston, and Updegrove. A final work, by White (1985) serves well as a summary of research and modeling efforts through the mid-1980s. White classifies his bibliographic entries along six dimensions: (1) administrative level, (2) primary purpose of the model, (3) program type, (4) techniques used, (5) resources being allocated, and (6) implementation. Although White's effort is becoming somewhat dated, it still may be the most contemporary of the comprehensive reviews of OR/MS applications in higher-education administration.

A second general category of relevant literature is *surveys of decision-making and planning models*. Two researchers (Hussain, 1976, and Bleau, 1981b) produced thorough reviews of the "packaged" planning models developed for higher education in the 1960s and 1970s. The Hussain book focuses on the CAMPUS, RRPM, HIS, and TUSS models, while the Bleau article focuses primarily on models developed in North America: CAMPUS, TRADES, RRPM, and EFPM, as well as SEARCH (by Peat, Marwick, Mitchell & Company) and HELP/PLANTRAN (developed at the Midwest Research Institute of Kansas City). In the years since the Hussain and Bleau reviews, the improvement in decision-support systems has made the use of complex planning models more palatable to administrators not trained in technical applications. Rohrbaugh and McCartt (1986) present a solid overview of the emerging uses of DSS in higher education. Among the topics considered in their volume are Markov-based decision-support applications, formal decision models, tactical and strategic decision making, system-dynamics simulation models, and approaches to evaluating alternative decision processes.

For those seeking a more comprehensive examination of decision-making and planning models, two classic works stand out: Halstead (1974) and Hopkins and Massy (1981). The Halstead book focuses on the planning efforts of state-level postsecondary officials, and pays special attention to forecasting revenues and costs and examining alternative possible uses of scarce state resources. The Hopkins and Massy book, in contrast, addresses central financial issues facing individual institutions. The authors define financial planning models, outline what these models can reasonably be expected to accomplish, specify how to build the necessary models, and offer helpful historical background on Stanford's experience in developing and applying planning models. Contained in this volume are 12 appendices, an exhaustive list of references, and a useful glossary and index.

An impressive entry in the literature focusing on decision-making and planning is the book edited by Hoenack and Collins (1990). Employing concepts from both economics and OR/MS, the authors review recent thinking on resource allocation, decision processes and priorities, incentive structures, fiscal environments, and cost functions. An especially valuable chapter, by Becker, reviews the extensive econometric research on students' sensitivity to institutional prices and considers the implications of that research for planning efforts.

The third and final category of relevant literature in OR/MS in higher education is composed of *specialized essays and research reports*. Prominent here is work on effective resource allocation, and particularly notable within that domain is Lee and Van Horn (1983). The heart of this book is a proposal for improving institutional management through joining administration-by-objectives with goal programming, but the volume also includes a useful technical review of several resource-allocation modeling approaches. Quality assurance and improvement is a second prominent topic in this literature domain. In an era in which institutions are being held to increasingly difficult standards of accountability, academic-program evaluation and assessment have become important aspects of administration on virtually every campus. A number of writers have addressed these questions from an OR/MS perspective. Dressel (1976) argued that techniques such as program budgeting, MBO, cost-benefit analysis, MIS, PERT, and PPBS can provide logical, systematic, comprehensive, and, above all, rational support for evaluation and assessment processes. Lewis (1988) reviewed not only the content of assessment efforts but also the analysis of their attractiveness in cost-effectiveness terms.

One of the first specialized uses of OR/MS on campus was in facilities management, i.e., the efficient use of buildings and grounds, their maintenance, and the identification of new facility requirements. Primary related issues include space utilization, inventory, demand projections, and energy consumption. Among the most intriguing recent developments in this area have been the use of integrated multi-objective decision models and mixed-integer goal programming models to help facilities planners act in the context of the sometimes conflicting goals of teaching, research, and public service activities (e.g., see Ritzman, Bradford, and Jacobs, 1979). A related specialized use for OR/MS applications, especially at large universities, is the scheduling, loading, and controlling of students' course enrollments. Typical management-science approaches to this dilemma extrapolated historical data into the near and long-term future. Some institutions have experienced success with material-requirements planning [MRP] in addressing this problem (Cox and Jesse, 1981). In addition, specialized OR/MS applications have been applied frequently to staffing issues. Planning for short and long-term faculty needs, for example, is a continuous challenge because of ongoing changes in tenure conditions, retirement legislation, and other factors associated with academic employment. Markov-chain methods are one approach frequently used to deal with this complexity (Bleau, 1981a).

Locally generated, limited-use techniques have found their way into the administrative portfolios of many institutions. Focusing solely on their own institutions, offices of institutional research [IR] on campuses often develop sophisticated OR/MS applications, including 1) "enrollment management" techniques for monitoring and shaping the characteristics of student bodies; 2) models for assessing statistically the relative importance of various student characteristics in predicting whether newly admitted freshmen will register; 3) Markov projections and related predictive models for forecasting enrollments; and 4) student-flow models for analyzing the movement of students into or out of specific programs. Unfortunately, these efforts usually remain unknown to those off campus, unless special dissemination efforts are undertaken. One such effort worth noting here is Yanceys (1988) comprehensive overview of some of the more useful statistical methods employed by IR offices on campus. The annual meetings of professional organizations for higher-education administrators, such as the Society for College and University Planning [SCUP] and the Association for Institutional Research [AIR], also provide outlets for disseminating work of this kind.

OR/MS ON CAMPUS: Definitive conclusions on the extent and nature of the use of OR/MS approaches in contemporary postsecondary education are probably impossible. Much applied work is done on a scale smaller than that of the well-known comprehensive planning models like EFPM, and much of that work remains unknown to those not directly involved. For example, decisions such as choosing among several options for meeting campus heating needs or designing an approach for using residence halls more efficiently are important and are clearly soluble in the OR/MS tradition, but are also quite constrained in scope. Work of this kind is usually not much shared even within institutions, much less across institutions. Managerial and political demands on campuses, the need to keep certain kinds of analyses "in-house," and the limited professional rewards and outlets for publicly disseminating educationally oriented OR/MS work each contribute to an element of invisibility in the enterprise, and that invisibility unquestionably limits our knowledgability regarding the use of OR/MS approaches.

On the other hand, to all outward appearances, we at least *seem* to have come to the end of the era of surging interest in applications of OR/MS to higher-education administration. The movement begun with great hopes in the 1960s and the 1970s apparently started to lose steam in the early 1980s. Since that time, there has been a clear decline of publicly disseminated essays, reviews, and research in the area. Illustrative of this is the fact that the 1992 and 1993 ORSA/TIMS joint national meetings did not include a single session on this category of application, but featured several such sessions a decade earlier. Even more striking is the decline in published articles on the topic. From 1973 to 1982, *Decision Sciences* published 38 articles on the topic, then published only 6 from 1983 to 1992. Similar figures for *Management Science* (going from 22 to 3 articles in the same period) reinforce the perception that the robustness of the field has declined. Although other outlets for publishing on the topic do exist (e.g., EDUCOM publications, specialized education journals), one cannot deny the decline in disseminated evidence of attention among mainstream OR/MS professionals.

Of course, the absence of attention to OR and MS in professional forums is not a perfect indicator of the state of the field. Perhaps these ideas and techniques have become so institutionalized that they are hardly noticed, or trumpeted. Perhaps their legitimacy is now firmly established, even taken for granted. That view is not one frequently expressed among authorities in the area, however. A more likely interpretation of the dropoff in professional interest is the possibility that OR/MS approaches have proved too complex, too user-hostile, or too "foreign" to elements of the academy's organizational culture for truly widespread implementation. Openness to OR/MS approaches may vary widely across institutions, in keeping with long acknowledged institutional differences in organizational forms and processes (Baldridge *et al.*, 1978; Baldridge and Tierney, 1979). For example, compared to other institutions, research universities are relatively decentralized, loosely coupled organizations in which significant decision making takes place at relatively low levels in the hierarchy. On *academic* matters, at least, institution-level, and even college-level, authority is resisted on many such campuses. What is more, many of those doing the deciding are reluctant to define themselves as "managers," and tend to distrust managerial techniques used successfully in other settings. Some aspects of OR/MS may not sit well in such settings. At community colleges and other primarily undergraduate institutions, however, central authority tends to be stronger, and the potential for more aggressive implementation of OR/MS in academic matters may be greater.

In any event, whatever the contemporary level of acceptance of OR/MS in higher education might be, there can be no denying that postsecondary institutions are complex organizations now experiencing severe economic, social, political, and cultural threats. As a consequence, they are increasingly dependent on good information to frame and support their decision making. The emergence of expanded telecommunications, lower-cost computing technology, and improved decision-support systems brings promise for help. With all due caution, therefore, one might reasonably predict a new wave of interest in OR/MS applications on campus.

See **Cost analysis; Decision support systems; Forecasting; Goal programming; Markov chains; Multiobjective programming; PERT; Statistical quality control.**

References

[1] Baldridge, J.V. and M.L. Tierney (1979). *New Approaches to Management: Creating Practical Systems of Management Information and Management by Objectives.* Jossey-Bass, San Francisco.

[2] Baldridge, J.V., D.V. Curtis, G. Ecker, and G.L. Riley (1978). *Policy Making and Effective Leadership: A National Study of Academic Management.* Jossey-Bass, San Francisco.

[3] Bleau, Barbara L. (1981a). "The Academic Flow Model: A Markov-Chain Model for Faculty Planning," *Decision Sciences*, 12, 294–309.

[4] Bleau, Barbara L. (1981b). "Planning Models in Higher Education: Historical Review and Survey of Currently Available Models," *Higher Education*, 10, 153–168.

[5] Cox, J.F. and R.R. Jesse Jr. (1981). "An Application of Material Requirements Planning to Higher Education," *Decision Sciences*, 12, 240–260.

[6] Halstead, D. Kent (1974). *Statewide Planning in Higher Education.* U.S. Government Printing Office, Washington, DC.

[7] Hoenack, Stephen A. and Eileen L. Collins., eds. (1990). *The Economics of American Universities: Management, Operations, and Fiscal Environment.* State University of New York Press, Albany, New York.

[8] Hopkins, David S.P. and William F. Massy (1981). *Planning Models for Colleges and Universities.* Stanford University Press, Stanford, California.

[9] Hussain, K.M. (1976). *Institutional Resource Allocation Models in Higher Education.* Paris: The Organization for Economic Co-operation and Development.

[10] Lee, Sang M. and James C. Van Horn (1983). *Academic Administration: Planning, Budgeting, and Decision Making with Multiple Objectives.* University of Nebraska Press, Lincoln, Nebraska.

[11] Lewis, Darrell R. (1988). Costs and Benefits of Assessment: A Paradigm, T.W. Banta, ed., *Implementing Outcomes Assessment: Promise and Perils*, 69–80. New Directions for Institutional Research, 59, Fall. Jossey-Bass, San Francisco.

[12] Ritzman, L., J. Bradford, and R. Jacobs (1979). "A Multiple-Objective Approach to Space Planning for Academic Facilities," *Management Science*, 25, 895–906.

[13] Rohrbaugh, John and Anne T. McCartt, eds. (1986). *Applying Decision Support Systems in Higher Education.* New Directions for Institutional Research, 49, March. Jossey-Bass, San Francisco.

[14] Schroeder, Robert G. (1973). "A Survey of Management Science in University Operations," *Management Science*, 19, 895–906.

[15] White, Gregory P. (1985). *Management Science Applications to Academic Administration: An Annotated and Indexed Bibliography.* CPL Bibliography, 157, CPL Bibliographies, Chicago.

[16] Wilson, James A., ed. (1981). *Management Science Applications to Academic Administration.* New Directions for Higher Education, 35, September. Jossey-Bass, San Francisco.

[17] Yancey, Bernard D., ed. (1988). *Applying Statistics in Institutional Research.* New Directions for Institutional Research, 58, Summer. Jossey-Bass, San Francisco.

HIRSCH CONJECTURE

This conjecture is a long-standing one in linear programming and concerns how many simplex iterations (basis changes) are necessary in going from one extreme point to another. Specifically, for any linear-programming problem, does there exist a sequence of m or fewer simplex iterations, each generating a new basic feasible solution, which starts with a given basic feasible solution and ends with some other given basic feasible solution, where m is the number of equations (constraints) in the linear-programming problem? If the conjecture is true, then the optimal solution to the problem could be found in m or less simplex iterations.

HOMOGENEOUS LANCHESTER EQUATIONS

Simple Lanchester equations with one equation for each side. These equations are used when the weapons for each side are homogeneous in nature (all small-arms) or as a simplified approximation of a heterogenous situation. See **Lanchester's equations.**

HOMOGENEOUS LINEAR EQUATIONS

A set of linear equations of the form $Ax = 0$.

HOMOGENEOUS SOLUTION

A solution to the set of equations $Ax = 0$. The solution $x = 0$ is called a trivial solution, while a solution $x \neq 0$ is called a nontrivial solution.

HORN CLAUSE

A logical expression of the form $A \to C$, where A (the antecedent) is a simple conjunction of basic (atomic) propositions and C (the consequent) is either null or is a single atomic proposition. See **Artifical intelligence**.

HOSPITALS

Richard C. Morey

Cornell University, Ithaca, New York

The modern hospital, in the words of Dowling (1984), is "the key resource and organizational hub of the American health care system, central to the delivery of patient care, to the training of health personnel, and to the conduct and dissemination of health related research." It represents a growing 320 billion dollar industry (in 1993), employing about three-fourths of all health care personnel, and is responsible for: about 40% of the nation's health expenditures; 58% of all Federal expenditures; and 40% of all state expenditures.

To begin with, there are many different types of hospitals: acute care (defined to be those hospitals with an average length of stay of less than 30 days); psychiatric; chronic (long-term) rehabilitation; nursing homes; and Federal (e.g., Veterans Administration Hospitals). The emphasis in this note is on the approximately 5,600 acute care hospitals operating in the 1990s.

The key aspects surrounding hospitals can be summarized as: i) access (or availability), ii) costs, iii) and quality of delivered care. We consider each in turn.

ACCESS: There are now close to one million acute care beds (1990), or about 4.16 beds per 1,000 of population. This is up from 1946 when there were about 3.2 hospital beds available for every 1,000 Americans. Great advances in medicine (discovery of the polio vaccine and widespread use of penicillin) led to the notion that health care is now a necessity and no longer just a privilege but has become a "right." As part of President Lyndon Johnson's Great Society, Medicare (federal financing for the health care cost for the aged), and Medicaid (grants to states to assist with medical care for the poor) were created in 1965. These programs brought health care to 12 million Americans previously unable to afford health care. There are still some 37 million Americans without health insurance.

A second key situation surrounding hospital access is the dramatic increase in outpatient (ambulatory) services where e.g. from 1980 to 1988, outpatient visits per year increased 40% (from 202 million to 285 million). Because of this increase and the tendency for less acute care to be rendered on an outpatient basis, those admitted to hospitals tend to be sicker than in the past.

Finally, a third development surrounding access is the possible need for "rationing" of care whereby the need for services would be prioritized, with not all possible services (e.g. transplants, etc.) being rendered. The State of Oregon is testing the feasibility and equity of this concept.

COST: Hospital inpatient costs per patient day have for several years been increasing at an annual rate of over 10%. This increase is due to technological advances, the aging of the American population, and the fact that insurance companies, rather than the patient, pay most of the hospital's bill. Prior to 1983, government programs reimbursed the hospital's actual cost, plus an allowed return on equity. However, in 1983, Medicare switched to the Prospective Payment System (PPS) under which Medicare paid fixed fees per patient for each diagnosis-related group (DRG's), regardless of the hospital's actual cost. The hope was that hospitals would have an incentive to reduce actual lengths of stay, ancillary services, etc, to maximize their profits. By means of a "ratchet" effect, it was hoped the spiraling hospital inflation would be somewhat checked. This same approach was soon adopted by Medicaid and many insurance programs. A similar approach involving a fee schedule, known as the Resource Based Relative Value System (RBRVS), is expected to be implemented for doctor's services. Unfortunately, PPS has not had the kind of cost containment successes hoped for.

Another key element in the cost area is the differential paid to teaching hospitals, i.e. those 17% of the hospitals training interns and residents. They currently receive some $3 billion extra from Medicare in recognition of their teaching mission. This amount consists of a direct medical education (DME) component (in recognition of the hospital's outlays for resident salaries, teaching physician salaries, classroom, lab equipment, etc.) and a so-called indirect medical education (IME) component (in recognition of the fact that a teaching hospital's cost per patient day is substantially higher than that for a non-teaching hospital, even after adjustment for cost of living, case mix severity, direct medical education, etc.). The size of this differential, as well as the huge variations over medical schools in their cost per resident, are fruitful areas for research.

QUALITY OF CARE: Partially due to the introduction of PPS and fear that hospitals might compromise quality of care to enhance their profitability, there was a surge in efforts to measure, monitor, and improve a hospital's delivered level of quality of care. This was exacerbated by the spiraling cost of insurance so that employers sought relationships with hospitals providing low cost but high quality care. The key quality outcomes of interest include a hospital's mortality rate, infection rate, complications rate, and readmission rate. As it's recognized these measures depend strongly on the diagnosis mix of the patients, as well as on patient characteristics (age, sex, transfer status, education, socio-economic level, etc.), much analytical effort has been expended of late to derive and validate quality of care indices which reflect these considerations. The U.S. Health Care Financing Administration has taken the lead to rank all acute care hospitals on the level of quality care delivered on the inpatient side. Their rankings, while controversial, have helped stimulate research in this important area and to build indices of quality on the outpatient side as well. Many countries with socialized health care programs (e.g., Brazil and India), are also keenly interested in such measurement systems in order to hold hospitals more accountable for the dollars they receive.

NEED FOR OR: At this point it must be emphasized (as did Pierskalla and Wilson, 1989) that "with few exceptions, the great progress in the past decade made in OR theory and methods and in the use of OR to solve major problems in industry, government, and the military, has *not* carried over to significant new progress in the use of OR in health care delivery... The major exception... is in the area of medical diagnostic decision making, particularly in screening and medical diagnostic decisions."

Pierskalla and Wilson (1989) also pointed out that "the hospital industry capturing 5–6% of the U.S. GNP and attracting some of the nation's brightest and best young people to the medical (and administrative) professions with handsome financial and social rewards, has little or no interest in structuring the incentives to attract similarly bright and capable people to many of the other jobs in hospitals."

The areas in which operations research and management science techniques have had some success are: outpatient/inpatient scheduling; service capacity planning; service demand forecasting; service system design; site location selection; supplies/material planning; vehicle scheduling; staffing and scheduling; strategic planning; and service delivery (Pierskalla and Wilson, 1989). We suggest some other promising possibilities for research and application:

i) Efforts to measure quality of care on the outpatient side;

ii) Efforts to measure effectiveness in hospitals, including economies of scale and economies of scope. For example what does an efficient hospital look like, how does it apportion its resources, what procedures, process, etc, does it employ to set itself apart from its poorly performing counterparts?

iii) Reimbursement issues related to the large Federal differentials paid to hospitals for disproportionate share adjustment (i.e., hospitals who treat a larger than normal share of poor patients), for large city adjustment, for teaching, etc.

iv) How does the volume of services in a hospital affect cost and quantity? Brailer and Pierskalla (1992) showed that adjusted mortality rates went up as congestion in the hospital increased over the design capacity level.

v) Are adjusted death rates, etc. reasonable proxies for quality of care assessment?

vi) What is the cost of quality in a hospital? Is it "free" as in manufacturing?

vii) What sort of substitutions are possible for different types of labor, capital, etc? Pierskalla and Wilson (1989) found that two-thirds of an RN's time was spent on non-nursing activities in which RN qualifications were neither needed nor used.

Only when the answers (or partial insights) to some of the above issues are available will hospital decision makers have the information and tools necessary to make informed trade-offs and to improve operations, both tactically and strategically, in hospitals. Probably no other area of the economy can benefit more from application of OR/MS in the future than the health care area in general, and the hospital sector in particular.

See **Health care systems; Medicine and medical practice.**

References

[1] Brailer, D.J. and Pierskalla, W.P. (1992), "The Impact of Hospital Congestion on Mortality," *Proceedings of TIMS International Conference*, Helsinki, Finland, July.

[2] Dowling, William L. (1984), "The Hospital," chapter in *Introduction to Health Services*, S.J. Williams and P.R. Torrens, eds., John Wiley, New York.

[3] American Hospital Association (1990), *Hospital Statistics*.

[4] Pierskalla, W.P. and Wilson, D. (1989), "Review of Operations Research Improvement in Patient Care Delivery Systems," University of Pennsylvania study for R.W. Johnson Foundation.

HUNDRED PERCENT RULE

Given an optimal basic feasible solution to a linear-programming problem, this rule allows for simultaneous changes in objective function coefficients or right-hand-side values of a linear-programming problem that maintains the optimality of the current basis. The name comes from the fact that the sum of the ratios of the proposed changes over their respective possible ranges must sum to one or less. See **Sensitivity analysis; Tolerance analysis.**

HUNGARIAN METHOD

An algorithm for solving the assignment problem that is based on the following version of a theorem that was first stated by the Hungarian mathematician König and later generalized by the Hungarian mathematician Egerváry: If A is a matrix and m is the maximum number of independent zero elements of A, then m lines can be drawn in the rows and columns of the matrix that contain all the zero elements of A. (A set of elements of a matrix is said to be independent if no two elements lie in the same row or column.)

HYPERCUBE QUEUEING MODEL

Richard C. Larson

Massachusetts Institute of Technology, Cambridge

The hypercube queueing model was developed in the late 1960s and early 1970s, a period driven by a national commitment to devote scientific energies to our country's urban ills. The initial application focus for the model was the deployment of urban police patrol cars. Issues that could be examined with the model involved determining appropriate numbers of cars to allocate in each part of the city, spatially deploying the cars to police "beats" or other territories and evaluating the impact of alternative dispatch policies. Over the years, the model has been applied to a large number of police departments and to other services as well, both public and private.

In this article, we review the history of the model's development, the key ideas of the model and its implementation. Since the technology is more than 20 years old, there are numerous references in the literature providing technical details of various aspects of the model. Our purposes here are to describe the historical evolution and framing of the model, to provide a roadmap through the references and to outline implementation impact.

EARLY WORK: The hypercube model's roots started with our work with the Crime Commission and with MIT-affiliated work with the Boston Police Department. From numerous hours riding around in the rear seats of police patrol cars and standing behind police radio dispatchers, we learned that the fleet of police cars in an area of the city can be viewed as "spatially distributed servers" in a queueing system. "Customer inputs" to this queueing system are generated by citizens calling "911" and asking for emergency on-scene service. Unlike most multiserver queues, the police queueing system has a heterogeneous pool of servers. Each server faces his or her own workload situation, dependent of local geography, patterns of "customer demand" and workloads of near-by servers. While writing a PhD thesis in 1969, we recognized the need for a multi-server queueing model whose state space retained knowledge of which servers were available and which were busy: "If the state of a server is either 'busy' or 'idle,' then there are 2^N possible states of the system, corresponding to all possible combinations of servers busy and idle. It is convenient to represent a particular state i by a binary number, the "ones" corresponding to the busy servers and the "zeros" to idle servers. The state of server N is given by the Nth binary digit counted from the right. For instance, state $i = 01000\ldots$ corresponds to server $N-1$ busy and all others idle. State $i = 2^N - 1$ implies that all servers are busy and that a queue may exist, (Larson, 1969, p. 124). The thesis discussion on these "hypercube" issues continues with numerical examples worked out for "small *N*." But the algorithmic implementation for arbitrary *N* required additional work.

Simultaneously with the thesis, we sought to confirm the nature of the spatial queueing by conducting a two week data gathering study in the New York Police Department, "NYPD" (Larson, 1971). Data were collected by the "passenger officer" in 54 precinct tours, where a precinct tour is defined to be a full set of operational data from one eight-hour tour or shift gathered over all police vehicles (typically 12 or less) fielded in a precinct or local area police command. While queueing in the usual sense was rare, queueing in the sense of *probabilistic congestion* was common.

To understand probabilistic congestion, suppose that you live in police beat "A" and that you call 911 requesting rapid on-scene police response. Suppose the police car assigned to beat A is busy with customers a fraction of time ρ_A, representing the utilization factor of the car ostensibly assigned to beat A, the so-called "A car." We assume that the time you need police service is independent of the real-time status of car A, busy or free. Thus, when you call 911, there is a probability ρ_A that "your car" is currently unavailable for immediate dispatch to your address; in that event, the dispatcher will select a near-by car that is available and dispatch that one. Such interbeat dispatches are sometimes called "workload sharing" dispatches, because car B, say, will respond when available into beat A and, conversely, car A will occasionally respond into beat B, when needed. In that way, cars A and B share each other's workload. In general, a large number of cars share each other's workload in complex ways. Now suppose for the sake of argument that the utilization factors of all cars A, B, C, etc. are all equal, that is, $\rho_A = \rho_B = \rho_C = \ldots = \rho$. In that case, whenever anyone in the service region calls 911, the chance that the responding car will be "their own" beat car will be $1 - \rho$. Consequently, the fraction of dispatches that are interbeat or workload sharing dispatches is equal to ρ. In urban America a typical value for in the 1960s was 0.5; in the 1990s, a typical value is 0.8. We checked the prediction of this simple aggregate "queueing model" for interbeat dispatching for NYPD Division 16, Tour 3, Friday, February 28, 1969. The results were as follows:

Precinct	Percentage of time unavailable	Percentage of dispatches that are Interbeat
103	48	55
105	59	57
107	38	48
109	38	37
111	36	48

As we can see, the extent of interbeat dispatching is never significantly less than the percentage of time unavailable, and it may be significantly more. There are sound theoretical arguments for suggesting that the simple "Poisson model" above represents a lower bound on the amount of interbeat dispatches (Larson, 1969).

The results were important for two reasons. First, the percentage of dispatches that are interbeat dispatches is a useful performance measure of the fielded police force. The officers in each police car are in theory supposed to build an identity with the beat that they are assigned to, "their patrol beat." This beat identity is supposed to cause the officer to feel personally responsible for public order in that beat. However, as we have seen empirically and argued theoretically, a patrol car is quite frequently dispatched to incidents in beats other than its designated beat, a phenomenon known in police circles as "flying." The more flying there is, the less the officer builds a strong beat identity. Prior to our work, police commanders in general had no idea that flying was as rampant as it in fact was. In the 1990s, it is far worse. Second, the results, theoretical and empirical, demonstrated that the fielded police force is a complex spatially distributed queuing system, with the statuses and workloads of the various "servers" heavily dependent on one another.

Armed with these results and the preliminary hypercube described in Larson (1969), we then proceeded to develop a more general model. Prior to this work there was very little in terms of analytical guidance for the police planner who wanted to design police beats. A common practice had been to design each beat to have equal internally generated workload. It was thought that equal internal workload would result in equal workloads experienced by the officers in the police cars. Our empirical work and our subsequent hypercube work showed that this rule of thumb can be very wrong (Larson, 1974b).

CENTRAL IDEAS ON STATE, TRANSITION AND PROBABILITIES: *State* – As discussed above, the hypercube model can be visualized as the corners and edges of a regular cube. For an $N = 3$ police car system, for example, one state of the system could be specified in words: Unit 1 is free or available; Unit 2 is busy; Unit 3 is busy. This state would be depicted by the binary set $\{0, 1, 1\}$, which is a corner of the three-dimensional cube. The state $\{0, 0, 0\}$ represents the situation in which all three units are simultaneously free. The state $\{1, 1, 1\}$ represents a situation in which all units are simultaneously busy and in which a queue of waiting 911 callers may exist. If there is a queue, the augmentation to the cubic state space may be thought of as an infinitely long tail emanating from state $\{1, 1, 1\}$, a situation resembling a "Chinese kite."

In generalizing the three-dimensional cube, the analogous figure for an $N = 2$ unit system is a square. For N greater than three, we must extend our visualization into hyperspace and imagine a unit volume cube residing in the positive orthant of an N dimensional hyperspace. This is the motivation for calling the model the hypercube model, a model having 2^N states.

Transitions – A state transition occurs whenever a server changes status from free to busy or from busy to free. Each such transition occurs only along a given edge of the hypercube. This requirement imposes the assumption that only one server (e.g., police car) is assigned to each customer, that is, there are no "bulk services" of customers.

Transitions occur probabilistically. Downward transitions corresponding to completions of service on customers occur for server j with rate μ_j. It is assumed that the service time distribution for server j is negative exponential. The rate of upward transitions from a given state to another adjacent state is determined by a complex set of "dispatching rules" or "server assignment policies." Computation of the upward transition rates is itself a daunting task for a human user of a large system, and has to be automated. It is assumed that from each area within the service territory, customers arrive as in a Poisson process, each process in nonoverlapping neighborhoods operating independently. Thus, once the set of upward transition rates is known, the process governing upward transitions from any given state is Poisson. Hence, the entire model is a continuous-time Markov model.

State Probabilities – The system performance measures of the hypercube model can be obtained once the limiting probabilistic behavior is determined. To do this, we must compute the limiting or steady state probability that the system is operating in some state, $i = 0, 1, \ldots, 2^N - 1$, where we have indexed the hypercube vertices in some convenient way. This is simply done by employing a balance of flow argument: In the steady state, the probability that the system will enter state i in any small interval of time Δt must equal the probability that the system will exit that state in a time interval of length Δt. That is, inward and outward probability fluxes must be equal. For if they were not, then there would be a net buildup or builddown of probability in state i, a contradiction to the steady state hypothesis. We generate the balance of flow equations by constructing an N-dimensional sphere around each hypercube vertex, and then equating outward flow to inward flow. In general, there are 2^N equations to solve, with one being redundant and replaced by the condition that the sum of all probabilities must equal unity.

Campbell was the first to create a general computer code for the N-server hypercube model (Campbell, 1972). Larson generalized that code, a program written in PL/I, and released it into the public domain in 1975 (Larson, 1975b). That version contained several algorithms that sped the execution time of the model, including an enhanced Gauss-Seidel procedure, a general method for performing a complete unit step tour of the hypercube and more. The code implemented all of the ideas discussed in the first journal paper describing the model (Larson, 1974a).

THE PHYSICAL ASSUMPTIONS OF THE ORIGINAL MODEL: The original model, sometimes now called the "basic hypercube model," requires the following assumptions (Larson, 1978; Larson and Odoni, 1981):

1. *Geographical atoms.* The area in which the system provides service can be broken down into a number N_A of "statistical reporting areas" or "geographical atoms." These might correspond, for instance, to census blocks, small collections of city blocks, or police reporting areas. In the model, each atom is modeled as a single point located in the center of the atom. Each can also be viewed as a node or vertex of a transportation network over which the servers operate.

2. *Independent Poisson Arrivals.* Each atom is viewed as an independent Poisson generator of customers requiring on-scene service, with the rate λ_j being the Poisson arrival rate from atom j.

3. *Travel times.* Data are available to estimate the mean travel time t_{ij} from each atom i to each

atom j. In the absence of such data, plausible approximations for travel times can be made using analytical models and/or transportation network algorithms.

4. *Servers.* There are N spatially distributed servers or response units, each of which can travel to any geographical atom in the service region.

5. *Server locations.* The server location methodology includes both the probabilistic locations of patrolling police cars and the deterministic locations of ambulances. Define a probability l_{nj} = probability that server n is located in atom j at a random time during which server n is known to be free or idle. For an ambulance n with a known fixed location (when idle), there is one l_{nj} having value unity and all others $\{l_{nj}\}$ (for fixed n) equal to zero. For a police car, which may have to patrol several atoms, we would have several $\{l_{nj}\}$ nonzero, corresponding to the atoms in which the car patrols.

6. *Server assignment.* In response to each customer call, exactly one server is dispatched to the customer, assuming that at least one server is currently available in the service region. If no unit is currently available, there are options of queueing or forwarding the customer to some "backup service," for example, a private ambulance service.

7. *Fixed preference dispatching.* Server assignment takes place according to a fixed preference procedure. By this we mean for each atom there is an ordered list of preferred servers to dispatch to that atom. The dispatcher will search that list in order and always dispatch the first idle server. Usually the list is generated by concerns of geography, such as travel time minimization, but on occasion other concerns such as assigning bilingual personnel could be important.

8. *Service times.* The service time associated with servicing a customer, including travel time, on-scene time, and possible related follow-up time, has a known average value. In general each server may have its own average value. Service-time distribution, as discussed above, is assumed to be negative exponential, an obvious crude approximation in some cases.

9. *Service-time dependence on travel time.* Variations in service times that are due solely to variations in travel time are assumed to be second order compared to variations of on-scene time and related off-scene time.

Given the assumptions above, the model is used to generate a variety of useful performance measures related to server workloads, travel times throughout the service region and dispari-

ties among neighborhoods in quality of service received (Larson, 1974a; Larson, 1974b).

In practice, no actual system will ever conform exactly to all the model's assumptions. There is always a balance to be struck between modeling simplicity and operational reality, with the determining factor being the quality of decisions that can be derived from the model with limited expenditure of effort.

APPROXIMATIONS: Early on in our hypercube work (1973), we received a phone call from the New Haven, Connecticut police department. The planners there wanted to use the model. We were excited as this was likely to be our first real testbed application. We had programmed the hypercube model in PL/I to accommodate up to 15 servers, a limit imposed at the time by the number of bits in a computer word. If New Haven were like New York City or Boston, it would be divided into a number of independently operating precincts or commands, with typically 8 to 12 servers (police cars) in each. In New York City and in many other large U.S. cities, police cars do not routinely cross over precinct boundary lines, so each precinct can be modeled independently with the hypercube model. So we asked the New Haven planner about precincts and were informed that any police car in New Haven can be assigned to virtually any address in the city. In fact all of New Haven was one big precinct, with $N = 48$ police cars for the hypercube model. We do not think that even today's computers could solve in finite time an $N = 48$ hypercube model, requiring solution to 2^{48} simultaneous linear equations.

So we were motivated to solve the "curse of dimensionality" imposed by the state-space structure of the hypercube model, a structure that doubled the size of the state space with each additional server. Our idea was simple: the equations used to compute performance measures suggested that it was not necessary to compute the fine grain 2^N state probabilities in order to evaluate the system performance measures. All we really needed were the workloads (utilization factors) of the respective units and the dispatch frequencies in the form, "the fraction of dispatches that send unit n to atom j." However, the logic behind this sort of argument was wrong because there is an implicit assumption that the units operate *independently*.

In 1975, a probabilistically valid way of dealing with this lack-of-independence problem was derived (Larson, 1975a). Using an M/M/N queueing model to represent the aggregate probabilistic behavior of the system, we found a set of "correction factors" to make our prior approximation precisely correct for an M/M/N system having a homogeneous pool of servers with a random dispatch policy and approximately correct for the heterogeneous server system we confront with the hypercube model.

Armed with the correction factors, one can write a set of N simultaneous nonlinear equations whose solution provides the (approximate) utilization factors of all the N units. The nonlinear equations have a nice geometrically decreasing quality that results in solutions usually within 3 or 4 Gauss-Seidel type iterations. From this result, using the correction factors again, we can compute the fraction of dispatches that send server n to atom j. From that, the problem is solved and all required performance measures can be found. For a period of two years we ran the "exact hypercube model" and the "approximate model" concurrently with numerous different data sets. In almost all of our runs, the approximate model was within about 2% of the exact model's results. We judged this accuracy to be within the modeling accuracy of the exact model and essentially decided to proceed from that point on only with the approximate model in implementations. The results of the use of the approximate model in New Haven were documented in Chelst (1975).

IMPLEMENTATIONS: The hypercube decision technology has been considerably tested (Chelst and Barlach, 1981) and generalized over the years, with almost all such upgrades due to implementation experience and suggestions. Space constraints here preclude a detailed description of these improvements, but the reader is referred to Larson (1979) for a detailed technical description.

The hypercube model has been implemented by police departments in many cities, including Hartford, Connecticut; Orlando, Florida (Sacks and Grief, 1994); Rotterdam, the Netherlands (Larson and McEwen, 1974); Chapel Hill, North Carolina; Dallas, Texas; New York City (Larson and Rich, 1987; Larson, 1979); and Cambridge, Massachusetts. It has also been implemented by ambulance services in Boston (Brandeau and Larson, 1986; Hill *et al.*, 1981; Larson, 1982) and New York City. In Hartford, for instance, the focus was to redesign the spatial deployment of the police cars so that a number of them could be freed from the usual 911 responding force and reassigned to special drug fighting units; this was successfully done in 1991. The Orlando Police Department in 1992 essentially redesigned

the deployment of its entire force using the model, within a project that implemented a new "downtown" police precinct. The Cambridge Police Department used the model to demonstrate to city management the deleterious consequences of reducing the size of the force in response to the tax cutting required from Massachusetts' "Proposition 2 1/2."

See **Markov chains; Markov processes; Queueing theory; RAND Corporation.**

References

[1] Bodily, S.E. (1978), "Police Sector Design Incorporating Preferences of Interest Groups for Equality and Efficiency," *Management Science*, 24, 1301–1313.

[2] Brandeau, M., and R.C. Larson (1986), "Extending and Applying the Hypercube Queueing Model to Deploy Ambulances in Boston," in *Delivery of Urban Services*, A. Swersey and E. Ignall, eds., North Holland, New York.

[3] Campbell, G.L. (1972), "A Spatially Distributed Queueing Model for Police Patrol Sector Design," S.M. thesis, M.I.T., Cambridge, Massachusetts.

[4] Chelst, K. (1975), "Implementing the Hypercube Model in the New Haven Department of Police Services," The New York City Rand Institute, R-1566/7.

[5] Chelst, K. (1978), "An Interactive Approach to Police Sector Design," in R.C. Larson, ed., *Police Deployment, New Tools for Planners*, D. C. Heath, Lexington, Massachusetts.

[6] Chelst, K. and Z. Barlach (1981), "Multiple Unit Dispatches in Emergency Services: Models to Estimate System Performance," *Management Science*, 27, 1390–1409.

[7] Heller, N. (1977), "Field Evaluation of the Hypercube System for the Analysis of Police Patrol Operations: Final Report," The Institute for Public Program Analysis, St. Louis, Missouri.

[8] Hill. E.D. *et al.* (1981), "Planning for Emergency Ambulance Service Systems, City of Boston," Department of Health and Hospitals, Massachusetts.

[9] Jarvis, J.P. (1975), "Optimization in Stochastic Service Systems with Distinguishable Servers," M.I.T. Ph.D. thesis, Cambridge, Massachusetts.

[10] Larson, R.C. (1969), "Models for the Allocation of Urban Police Patrol Forces," Tech. Rep. #44, Operations Research Center, M.I.T., Cambridge, Massachusetts.

[11] Larson, R.C. (1971), "Measuring the Response Patterns of New York City Police Patrol Cars," New York City Rand Institute R-673-NYC/HUD.

[12] Larson, R.C. (1974), "A Hypercube Queueing Modeling for Facility Location and Redistricting in Urban Emergency Services," *Jl. Computers and Operations Research*, 1, 67–95.

[13] Larson, R.C. (1974), "Illustrative Police Sector Redesign in District 4 in Boston," *Urban Analysis*, 2(1), 51–91.

[14] Larson, R.C. (1975), "Approximating the Performance of Urban Emergency Service Systems," *Operations Research*, 23, 845–868.

[15] Larson, R.C. (1975), "Computer Program for Calculating the Performance of Urban Emergency Service Systems: User's Manual (Batch Processing)," Innovative Resource Planning in Urban Public Safety Systems, Report TR-14-75, M.I.T., Cambridge, Massachusetts.

[16] Larson, R.C. (1979), "Structural System Models for Locational Decisions: An Example Using the Hypercube Queueing Model," in *Operational Research '78*, Proceedings of the Eighth IFORS International Conference on Operations Research, K. B. Haley, ed., North-Holland, Amsterdam.

[17] Larson, R.C., ed. (1978), *Police Deployment: New Tools for Planners*. Lexington Books, Massachusetts.

[18] Larson, R.C. (1982), "Ambulance Deployment with the Hypercube Queueing Model," *Medical Instrumentation*, 16(4), 199–201.

[19] Larson, R.C. and E. Franck (1978), "Evaluating Dispatching Consequences of Automatic Vehicle Location in Emergency Services," *Jl. Computers and Operations Research*, 5, 11–30.

[20] Larson, R.C. and V.O.K. Li (1981), "Finding Minimum Rectilinear Distance Paths in the Presence of Barriers," *Networks*, 11, 285–304.

[21] Larson, R.C. and T. McEwen (1974), "Patrol Planning in the Rotterdam Police Department," *Jl. Criminal Justice*, 2, 235–238.

[22] Larson, R.C. and M.A. McKnew (1982), "Police Patrol-Initiated Activities within a System Queueing Model," *Management Science*, 28, 759–774.

[23] Larson, R.C. and A.R. Odoni (1981), *Urban Operations Research*. Prentice-Hall, Englewood Cliffs, New Jersey.

[24] Larson, R.C. and T. Rich (1987), "Travel Time Analysis of New York City Police Patrol Cars," *Interfaces*, 17(2), 15–20.

[25] Li, Victor on-Kwok (1977), "Testing the Hypercube Model in the New York City Police Department," S.B. thesis, EE, M.I.T., Cambridge, Massachusetts.

[26] McKnew, Mark (1978), "The Performance of Initiated Activities and Their Impact on Resource Allocation," M.I.T. Ph.D. thesis, Cambridge, Massachusetts.

[27] Sacks, Stephen R. and Shirley Grief (1994), "Orlando Magic," *OR/MS Today*, 21(1), 30–32.

HYPEREXPONENTIAL DISTRIBUTION

A continuous random variable is said to be hyper-exponential (or mixed exponential) when its probability density function is the convex sum of exponential density functions. Often, the term hyperexponential is used only when the mixture has two terms, written in such a way that there are just two parameters instead of the usual $2n - 1$ for a mixed exponential with n terms.

HYPERGAME ANALYSIS

A problem structuring method which addresses situations of conflict and cooperation between independent actors. A key feature is its ability to represent differing perceptions of the situation which may be held by different actors. See **Problem structuring methods**.

HYPERPLANE

A hyperplane in n-dimensional space is defined by the set of vectors $X = (x_1, \ldots, x_n)$ that satisfy a linear function of the form $a_1 x_1 + a_2 x_2 + \ldots + a_n x_n = b$ for given numbers a_j and b. This can be written as $ax = b$, $a = (a_1, \ldots, a_n)$. For $n = 2$, the function defines a line, and for $n = 3$, the function defines a plane.

I

IDENTITY MATRIX

A square matrix $A = (a_{ij})$ with $a_{ii} = 1$ and all $a_{ij} = 0$ for $i \neq j$. See **Inverse matrix**; **Matrices and matrix algebra.**

IFORS

See **International Federation of Operational Research Societies.**

IFR

Increasing failure rate. See **Failure-rate function**; **Distribution selection for stochastic modeling**; **Reliability of stochastic systems.**

IIASA

See **International Institute for Applied Systems Analysis.**

IID

Independent and identically distributed (random variables).

IMBEDDED MARKOV CHAIN

An analysis technique used to analyze a queueing system that is not a continuous-time Markov chain. It appraises the system at selected time points which allow the system to be analyzed via a discrete-parameter Markov chain. The queue length process in the M/G/1 queueing system is not Markovian, but can be analyzed via a Markov chain at service completion time points. See **Markov chains**; **Markov processes**; **Queueing theory.**

IMPLEMENTATION

R.E.D. Woolsey

Colorado School of Mines, Golden

This writer would assert that there exist two distinct schools of thought concerning implementation of OR/MS. The first is the "watch it and model it" approach of eminent universities populating the Northeast and Western coasts of the United States, while the second approach is the "get down and *do* it approach" of other, sometimes less distinguished schools around the country.

THE "WATCH IT AND MODEL IT" APPROACH: The primary assumption of this approach to OR Implementation is that if the OR/MS person is sufficiently educated in the theoretical constructs and methodology that only minimum exposure to the actual situation is required. This assumption often works startlingly well in practice because the graduates of such programs are customarily drawn from the already rich and/or extremely bright quartile of our population. In short, entrance to these schools requires either massive amounts of money or outstanding academic performance which generates a scholarship. The argument is as follows. The product of this approach, when confronted with a real-world problem, could do a memory search from their conceptual education and unerringly choose the proper model for the solution of the problem. It is often implied that the rest is dog work which can be safely handed to others.

The good news about this approach is that if the OR/MS person is quite bright, quick and politically aware, excellent results *usually* obtain in spite of lack of knowledge of the system. Any tailoring of the process again is accomplished quickly due to the acuteness of the intellect of the person. It must also be pointed out that the customer is often sufficiently in awe of the educational, cultural, and economic background of the consultant that Gestalt psychology plays no small part in acceptance of models. This approach has been found to be particularly effective in strategic and high-level corporate planning with correspondingly high acceptance by top management. Another way to characterize situations where this approach does well is to say that the less measurable the results, the better the acceptance.

The bad news about this approach is that it almost uniformly fails in the tactical world. Manufacturing managers are justly famous for having little time for academic "experts" with no shop floor experience. This often supports the argument about how little OR/MS has actually been used in the manufacturing workplace as opposed to the corporate levels mentioned above. The time it takes to accomplish a Ph.D. militates strongly against a person having also the shopfloor experience in a manufacturing situation. A story going the rounds of the profession is of interest here. It is alleged that a Lanchester Prize winner for nonlinear optimization was suddenly thrown out of work by the Army pulling the monetary plug on his particular "Beltway Bandit" employer. He well knew that refineries had a multitude of nonlinear problems in the production of hydrocarbons. He therefore hied himself off to the nearest refinery and offered his services to the refinery manager to solve his nonlinear refinery optimization problems. He named a price for his services, and the refinery manager then asked him how much he knew about chemical engineering. When he confessed his total ignorance in this area, the refinery manager politely asked him how much he was prepared to pay the oil company for him to learn enough about chemical engineering to help *them*! With this cautionary tale we plumb the difference between conceptual excellence and practical reality.

The principal reason for failure at the tactical level is that the customer must perceive that you know enough about their area so that you

(a) understand that politics wins over optimality *all* the time, and

(b) that such knowledge will create respect for what they have to put up with.

Further, people at the tactical level of companies are *not* impressed by anyone unless they have gone through the same bootcamp learning process that *they* have. In short, I believe that an education from an eminent institution may be actually more of something to be overcome than an asset in the milieu of tactics.

THE "GET DOWN AND *DO* IT" APPROACH: It is my custom to encourage implementation in OR by attending conventions and asking the presenters the following questions.

(1) **Did you know what was happening on the project *before* you modeled it?**

If your answer is yes, I will then politely ask you:

(2) **How do you *know*?**

The only acceptable answer is that you found out by *doing* the work being modeled under the conditions of the people who are presently *doing* it until they had enough confidence in you to take a day off and let you do it alone. Anyone that believes that you can learn enough by watching should be treated with the amusement they deserve.

The next question is:

(3) **Did they accept and *use* your model?**

If your answer is yes, I proceed to the last question which is:

(4) **Do you have *measurable* results in, for example, dollars at present worth, after tax, adjusted for inflation?**

In my opinion, people who answer "no" to any of the above questions have failed the test of operations research implementation.

See **Field analysis; Practice of OR/MS.**

IMPLICIT ENUMERATION

A process for solving integer-programming problems in which all possible integer solutions need not be investigated (enumerated) due to information obtained in the process that relates to problem feasibility and value of the objective function. That is, certain solutions need not be pursued as it can be shown that they would lead to infeasible solutions or values of the objective function that are worse than those that are known to be possible. See **Branch and bound.**

IMPLICIT PRICE

See **Marginal value.**

IMPORTANCE SAMPLING

See **Monte Carlo sampling and variance reduction techniques.**

IMPOSSIBILITY THEOREM

See **Group decision making.**

INACTIVE CONSTRAINT

An inequality constraint of an optimization problem that is satisfied as a strict inequality. See **Active constraint; Slack variable; Surplus variable.**

INCIDENCE MATRIX

See **Node-arc incidence matrix**.

INCIDENT

An edge of a graph is said to be incident with the two nodes it connects, and conversely. See **Adjacent**; **Node-arc incidence matrix**.

INDEPENDENT FLOAT

The amount of time that an activity can be delayed without affecting the earliest start of the preceding activity and the latest finish of the succeeding activity in a project network. See **Network planning**.

INDEPENDENT PRIVATE VALUES BIDDING MODEL

A bidding model in which bidder's estimate of value for what is being auctioned is statistically independent of the value estimate for any other bidder. In such a model, no bidder has any reason to adjust an estimate of value upon learning the information of any other bidder. See **Bidding**.

INDIRECT COSTS

In the simplex method, the indirect costs are found by taking the inner products of the multiplier (pricing) vector with each column of the problem's defining A matrix. This product, for column j, is usually denoted by z_j. For c_j, the original objective function coefficient for column j, the term $(z_j - c_j)$ or $(c_j - z_j)$ is used to determine if the associated variable is a candidate to enter the basic feasible solution. For any basic variable x_k, $(z_k - c_k) = 0$. The $(z_j - c_j)$ terms are called relative costs (relative to the basis) or reduced costs. See **Prices**.

INDUSTRIAL APPLICATIONS

Leonard Fortuin

Eindhoven University of Technology, The Netherlands

Paul van Beek

Agricultural University, Wageningen, The Netherlands

Luk Van Wassenhove

INSEAD, Fontainebleau, France

SOME HISTORY: Although this article is based largely on the authors' experiences and views of European OR/MS, it is of direct importance to the worldwide OR/MS community. The field is better known in Europe as OR for Operational Research, and occupies itself with quantitative methods for the analysis and solution of management problems. Its origins lie in military organizations during World War II: first the Royal Air Force (UK) preparing for the "Battle of Britain" and later on the US Navy fighting German *U-Boote* (submarines). After the war there was a general feeling that OR/MS could be helpful to managers in industry, government, public services and financial institutions also. The logic was obvious: industrial activities such as production planning, inventory control and physical distribution were very suitable for model building and other forms of abstraction that lead to challenging mathematical problems, whereas trained OR workers were available. But soon it became evident that solutions capable of being applied in practice were not so numerous as expected.

For this phenomenon we see two causes. On the one hand models running on computers then available, were so strongly a simplification of reality that the managers did not recognize their problem any longer. On the other hand, OR workers in academia moved their attention to the basics of their discipline. Their theoretical results were very impressive, especially in the field of mathematical programming, combinatoric analysis and queueing theory. But for managerial problems of daily life these OR workers had little interest. Consequently, the decision makers felt disappointed and lost their confidence in "mathematical decision theory," another name for OR/MS, and returned to simple, often too simple, "rules of thumb." In this way a "practicality gap" came into existence, a gap between (1) managers with real, urgent decision problems demanding simple solutions, and (2) OR scientists in their "ivory towers" obsessed by finding elegant solutions to abstract problems of own invention. For a discipline aimed at application, as OR claims to be, the gap was a highly unsatisfactory situation. Hence, after a while, papers in professional journals were trying to find a remedy and outstanding OR workers attempted to regain managers' interest for "management science." But all efforts seemed in vain. One of the *gurus* of OR/MS even concluded that "The future of OR is past" (Ackoff, 1979).

This is how OR/MS lost the good reputation earned during the war. Even OR workers in staff departments of industrial companies had to fight for their existence and often lost their jobs. Many departments were dissolved or put at work on other tasks, for example, on automation projects. Mainly, the so-called "loners," working in decentralized positions, continued to do OR work (see Fortuin and Lootsma, 1985).

But OR workers never lost faith in their discipline. Gradually they improved their position by rediscovering real-life problems. In the eighties two developments fostered this process: the availability of cheap and versatile computer power (PCs) and the establishment of special university chairs for OR/MS and other quantitative methods. Ten years after Ackoff, a completely different sound could be heard: "The future of OR is bright!" (Rinnooy Kan, 1989).

OR/MS TODAY: By definition OR has two faces: on the one hand it concentrates on *operations* and as such it tries to be practical and to provide solutions to real-life problems; on the other hand, OR means *research*, involving theoretical studies of problems that at best may be regarded as abstract versions of problems that actually exist in real life. These two faces of OR/MS have brought into existence two types of OR workers: the *practitioners* and the *theoreticians*. The practitioners are to be found primarily in consultancy bureaus, but also at universities, for example in departments such as "industrial engineering" and "industrial mathematics." In large companies, "loners" can still be found. As for the theoreticians, they work at universities and related institutions only.

The two types of OR workers are carrying out their tasks independently, but contact between them is improving, with exchange of ideas at conferences and seminars, and bilaterally. This situation originated in a natural way:

- Most consultants graduated from university. They maintain their "university network" to learn about theoretical breakthroughs. In return, they inform their fellow OR workers in academia about the problems their clients in industry are grappling with.
- Many consultants are working as *part-time professors* at a university. They use their experiences and practical knowledge as a consultants to keep their teaching up to date and use the results of their academic studies to support their consultancy work.

In this way, opportunities for OR/MS have improved considerably. Other factors have enhanced this process:

- Modern managers in industry have an academic background. During their studies they have become acquainted with the basics of OR/MS and so they are easier to convince that OR/MS can do something for them. As they no longer have OR/MS staff departments in their organization, they become clients of OR/MS consultants.
- Universities have discovered the importance of good relations with business companies:
  (1) It makes academic OR/MS workers more practical, and taught them to cooperate with managers that are aware of the importance of giving future industrial engineers a proper training.
  (2) It enhances their cash flow by doing contract research in OR/MS.
  (3) It offers students an opportunity for working temporarily in an industry as part of their program, to the benefit of the quality of their education.
  (4) It helps universities to assign priorities to the items on their research program.
- The pure theoreticians no longer can select in isolation the subjects of their investigations. Instead, they have to pay attention to the signals that reach them from their colleagues operating with their students in industry.
- Information technology has produced powerful computer software and hardware. Consequently, model building has become very realistic, all relevant details are taken into account, and animated graphics more easily than words convince the manager that indeed *his* problem is being analyzed.

This improvement process is reflected in professional journals on OR/MS. Many successful applications of OR/MS in practice have been reported in the literature. These are the so called *case studies*, in which usually the following sequence of subjects can be found: (1) the problem and its environment; (2) the OR/MS approach towards a solution; (3) results of the OR/MS analysis; (4) selection by management of a solution from a set of alternatives; (5) implementation of that solution; (6) results, in terms of improvements with respect to the situation before the OR/MS intervention, often endorsed by the Board of Management. Examples can be found in Bell (1989), Lootsma (1991), and Fortuin and Korsten (1988).

But there are still "problem owners," that is, managers that have to make decisions in complex and complicated situations, who tend to continue the tradition of their predecessors, some of which that were so disappointed in OR/MS. Often they are facing problems with more than one solution, each of them with far-reaching consequences. Then they are obliged, mostly under time pressure, to select the best. Here they could be supported nicely with OR/MS in its modern version, if only they were aware of (1) the powerful methods and tools that OR practitioners have nowadays at hand, and (2) the changed attitude of OR/MS towards real-life problems. In general it can be stated that in the 1990s, OR/MS has become very useful to managers in industry, once it got a chance to show its true character. The main problem lies in gaining the confidence of managers so that they are prepared to give OR/MS that chance. Obviously, the "practicality gap" still exists. In order to bridge it, OR consultants and academic OR workers have to act as missionaries, to the benefit of their profession, their career perspectives, and, last but not least, to managers trying desperately to improve their business in the face of an ever increasing global competition.

OR/MS in Industry: Where and What?

Fortuin and Zijlstra (1989) reported on the experiences of an OR group within Philips Electronics, a multi-national company producing consumer products (e.g., domestic appliances, lighting, television sets, high-fidelity consumer electronics, razors) and professional products (e.g., medical systems, telephone exchanges, lighting systems). They analyzed over 200 projects in OR/MS and showed which areas are most important for

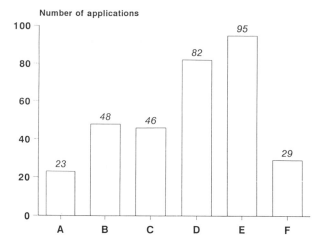

Fig. 2. OR/MS tools applied in industry. Legend: A = Mathematical programming, with emphasis on LP; B = Combinatoric analysis; C = Inventory models; D = Waiting theory models; E = Discrete event simulation; F = Miscellaneous, such as the structuring of facts and figures (many projects start with this type of OR; sometimes it is all the client desires).

OR/MS application in industry and which OR/MS tools are most frequently used. A recent update of these investigations, in 1992, confirms these results. See Figures 1 and 2. Apparently, the most frequently occurring project is one on the design of a production system, whereby discrete computer simulation is the OR tool employed to take all the complicated interactions duly into account. This conclusion holds for a large multi-national company in Europe. In the USA the picture seems to be slightly different. A longitudinal survey in *Interfaces*, for instance, mentions statistics, linear programming and discrete simulation as the top three OR tools, in that order (Harpell, Lane and Mansour, 1989).

In most OR/MS projects in industry, an important part of the work is model building. In short it means the description of a piece of reality that has to be analyzed in the course of the project, leaving out all irrelevant details while maintaining essential characteristics. This gives model building the character of an art rather than of a science. (See Figure 3.)

As for the position of OR/MS in industry, times have changed. Some ten years ago, it primarily was in the hands of staff departments. Nowadays, companies are withdrawing on their core business, a process that started in the mid 1980s. Consequently staff departments are reduced to a bare minimum, if not completely eliminated. It also happened that they were disconnected from their original company. The OR/MS department that figured in Fortuin and Zijlstra (1989), for

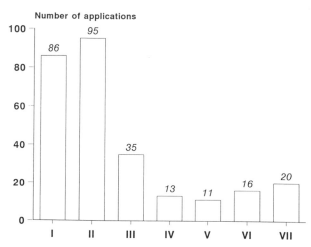

Fig. 1. Application areas for OR/MS in industry. Legend: I = Production; II = Design of production systems; III = Transport and storage; IV = Design of systems for transport and storage; V = Performance of systems; VI = Training and courses; VII = Miscellaneous, such as: portfolio analysis, measuring the quality of information systems, and performance indicators.

Model building plays an important part in modern OR projects, often in combination with discrete simulation and optimization. Models offer insight and the possibility to compare decision scenarios with each other, both in the qualitative and the quantitative sense. Almost always the computer is an indispensable tool in this. The driving force exerted by developments in informatics cannot easily be overestimated. A large part of the arsenal of OR techniques can be used on the PC, thanks to software that is becoming more and more user-friendly and also cheaper. In recent years, the ease with which a problem area can be represented by a model that the problem owners consider sufficiently realistic, has grown enormously. Large quantities of data can easily be stored in databases that are simple to access. The opportunities for OR to really contribute to the reduction of uncertainty in complex industrial situations and to increasing control of business processes are large. But there still are managers who are unaware of the help they can get from OR, like quickly calculating the consequences of decision variants, the preparation of decisions, and decision support. These managers are ignorant of the fact that with this help they can save considerable amounts of money.

Fig. 3. Model building.

instance, has become an independent consultancy bureau with, not surprisingly, Philips Electronics as its main client. The same tendency can be seen at other big industrial companies, at least on the European side of the Atlantic Ocean.

In Europe, OR/MS support is offered to industry from two sources. First of all there are the consultancy bureaus already mentioned. They usually follow the project approach, according to which the work is done in phases. To a certain extent they have to compete with the second source, that is, OR practitioners at universities who try to find for their students a company willing to spend some money on an OR project. Both parties profit from this alliance: the student learns to practice the "profession," the company obtains an inexpensive solution to a business problem, or the beginning of it, by using the workforce of both professor and student. As these academic OR workers show managers how advantageous the support of external consultants can be, eventually, the commercial bureaus can benefit from their efforts.

Figure 4 gives an overview of the project approach. Client and OR consultant make a contract, stating what problem will be studied and possibly solved, at what costs, and during which time interval. It also indicates what "deliverables" the client may expect and which efforts he and his staff have to contribute. More details can be found in Fortuin, van Beek and van Wassenhove (1992). This approach has proven to be very successful, for a number of reasons:

- The preliminary phase is usually short. Its aim is reconnaissance of the problem area and a problem description that the problem owner can agree with. Costs are relatively modest, so that financial risk is low for the manager. This facilitates the process of making the manager confident that the external consultant is indeed able to help him improve his business.
- It may happen that during the preliminary phase the problem becomes so transparent that a solution can be seen immediately. Then a follow-up phase is not necessary at all.

	ACTIVITIES	DETAILS
Phase 1	General survey	Discussions with client and staff. Interviews, study of documents. Global problem description. Generation of ideas for a possible approach.
	Reporting	Outline of results to be expected. Proposal for Phase 2.
Phase 2	Model building	Systematic description of the problem area. In order not to make the model too complicated, only the most relevant factors are taken into account.
	Verification	Discussions with client and staff: Is the model correctly presenting the problem area, the organization, the methods, processes and procedures?
	Experiments	Translation of the model into a computer program. Calculations under various circumstances.
	Analysis	Investigation of results.
	Reporting	Presentation of the most important results, conclusions and recommendations. A proposal for Phase 3.
Phase 3	Implementation	Working out and implementation of recommendations. Teaching client and/or staff to work with the new method.

Fig. 4. Summary of the steps in an OR/MS project. In Phases 1 and 2 the consultant is heavily involved in the project. Usually, Phase 3 is carried out by the problem owner and/or his staff.

CONCLUSION: It can be stated that the prospects for the application of OR/MS in industry continue to look good:

1. Managing an industrial company is becoming ever more complicated. Global competition, surplus capacity, demanding customers, decreasing profit margins, new markets, and fluctuating exchange rates are just a few causes. The time for simple solutions is over, only fundamental and theoretically sound analyses can justify management decisions. Managers lack the time and the expertise for such analyses.

2. Practitioners now working in independent consultancy bureaus can make their own business plans and follow their own strategy when promoting OR/MS, rather than being ruled, or overruled, by a general company policy.

3. Computer power is widespread in industry, which facilitates the implementation of OR solutions, even if these are complex and require large data bases.

See **Decision making; Mathematical model; Practice of OR/MS.**

References

[1] R.L. Ackoff (1979), "The future of OR is past," *Jl. Operational Research Society*, **30**, 93–104.

[2] P. Bell, "Successful Operational Research in Canada," *CORS-SCRO Brochure* (year not mentioned, but later than 1989).

[3] L. Fortuin and F.A. Lootsma (1985), "Future directions in Operations Research," in A.H.G. Rinnooy Kan (ed.), *New Challenges for Management Research*, North-Holland, Amsterdam/New York/Oxford.

[4] L. Fortuin and A.T.M. Korsten (1988), "Quantitative methods in the field: two case studies," *European Jl. Operational Research*, **37**, 187–193.

[5] L. Fortuin and M. Zijlstra (1989), "Operational Research in practice: experiences of an OR group in industry," *European Jl. Operational Research*, **41**, 108–121.

[6] L. Fortuin, P. van Beek and L. Van Wassenhove (1992), "Operational research can do more for managers than they think!," *OR Insight*, **5**, 1, 3–8.

[7] J.L. Harpell, M.S. Lane and A.H. Mansour (1989), "Operations Research in Practice: A Longitudinal Study," *Interfaces*, **19**, 65–74.

[8] F.A. Lootsma (1991), "Perspectives on Operations Research in long-term planning," *European Jl. Operational Research*, **50**, 76–84.

[9] A.H.G. Rinnooy Kan (1989), "The future of OR is bright," *European Jl. Operational Research*, **38**, 282–285.

INDUSTRIAL DYNAMICS

See **System dynamics.**

INDUSTRIAL ENGINEERING AND OPERATIONS RESEARCH

John J. Jarvis

Georgia Institute of Technology, Atlanta

THE WORLD OF INDUSTRIAL ENGINEERING: Industrial Engineering (IE) and Operations Research (OR) have had a very close relationship throughout their development. The basis for industrial engineering is "improvement." Another term used early in the development of IE is "efficiency;" although, in the last thirty years, this word and the industrial engineer as "efficiency expert" have taken on a bad connotation. The Institute of Industrial Engineers (IIE), the professional society for industrial engineering, gives the following definition of IE. "Industrial engineering is concerned with the design, improvement, and installation of integrated systems of people, material, information, equipment, and energy. It draws upon specialized knowledge and skills in the mathematical, physical, and social sciences together with the principles and methods of engineering analysis and design to specify and evaluate the results to be obtained from such systems."

People – Frederick W. Taylor, the father of "scientific management," is credited with creating industrial engineering to increase the productivity of factories in the late eighteen hundreds. In the beginnings of industrial engineering, the focus was on improvement of "methods" and "standards" for people. IEs sought to make workers and workplaces more efficient. Methods dealt with the way people did their jobs, while standards dealt with the speed at which jobs were done. OR had not yet developed and the tools IEs used were "common sense," "ergonomics," and "statistics."

Processes – OR developed out of military necessity during World War II. As OR came on the academic scene in the 1950s, it was rapidly embraced by IE. Many OR programs developed within IE departments. Using the tools and techniques of OR – especially linear programming, networks, and simulation – industrial engineers were able to address improvement of "processes" from a quantitative modeling point of view. These new methodologies were quickly applied to inventory, production, and scheduling, to name a few. In many cases, these "pieces of the system" lent themselves nicely to OR modeling. The computing of the time severely limited the size of models that could be manipulated and, therefore, the focus was limited.

Systems – Computing capability increased rapidly during the 1970s and 1980s. The combination of new flexible simulation languages and available large-scale computing resulted in comprehensive analyses of manufacturing systems.

IE FROM MANUFACTURING TO SERVICE:
Early industrial engineering dealt almost entirely with the factory. It had not been that many years since Henry Ford created his first assembly line. This change created a new set of problems and a new breed of engineer was needed to address them.

It was natural for industrial engineering to grow out of mechanical engineering (ME). MEs designed and built the machines operating in the factories. They understood the capabilities of machines. They would be the best engineers to deal with the complexities of workers interfacing with machines. IE courses sprang up within ME departments. Then in the 1940s, IE began to split off into a separate department.

In the late 1950s, IE began to branch out into the service sector. A number of IEs moved into the health arena. Industrial engineering was embraced by hospital administrators as a way to analyze people, processes and systems associated with health care delivery. Addressing problems

tends to put IE into a well-defined "bucket" within the organization. It is common to hear "our IEs work on factory problems" or "we have *other types* (substitute your favorite words, e.g., *systems analysts*) at the corporate level." And, in fact, this is the source of an important distinction between IE and OR. While nothing is universal, IEs, and many tasks that IEs do that do not involve OR, tend to focus on the plant or operational activity floor. Although much OR activity can also be found at the factory floor level, more industrial OR than IE would be found at the corporate management level. The activities that corporate personnel undertake are often associated with acquisitions and strategies requiring very large, complex OR models.

INDUSTRIAL ENGINEERING CURRICULA: IE
and OR are not the same! There are not many OR programs at the undergraduate level. However, it is instructive to examine an undergraduate IE program to understand the distinction between OR and IE. (The first continuing IE curriculum was established at Pennsylvania State College in 1908.) The undergraduate IE program at Georgia Tech (GT) probably has quite a large overlap between OR and IE. But even this undergraduate IE program has major distinctions. In the mid-1990s, the GT IE programmatic requirement, in quarter hours, was:

Mathematics – 28 quarter hours
Physical Science – 25
English – 12
Humanities & Social Science – 21
Engineering Science – 13
Computing – 12
General Electives – 6
Misc. – 4

Operations Research – 9 quarter hours
Probability & Statistics – 9
Quality – 6
Human Sciences – 9
Economics – 9
Management – 6
IE Applications – 8
IE Electives – 12
Design – 6

---

of hospitals was less complex than other service areas, for example, insurance. In many ways a hospital resembles a factory, with patients replacing products in the system. Eventually, IE moved into other areas, including transportation/logistics, and government.

A visit to a manufacturing plant will reveal the fact that it probably has an IE department by name, but generally no other engineering discipline will have such a distinction. Like manufacturing, many hospitals have IE departments. This is a two-edged sword. It provides name recognition for industrial engineering in the workplace. But it also focuses blame. Worst, it

To highlight the differences between an IE and an OR curriculum, consider the table above. Let us assume that the course work in the first column, together with the first two courses and last course in the second column, are all common to an OR curriculum within an engineering school. (There might exist more OR hours and less computing, probability and statistics hours; but it would probably even out.) The essential differences appear in items three through eight in the second column. Industrial engineers are introduced to quality and quality control, economics and engineering economy, management and accounting, operations planning and scheduling, and operations and facilities design. In addition, IEs are required to select from a variety of restricted electives, including project management, storage and distribution systems, material handling, ergonomics, measurement, robotics, systems dynamics, technological forecasting, technology assessment and ethics. As can be seen from these requirements, industrial engineers take a significant amount of course work in process/systems understanding and improvement. While many of these courses involve the application of OR, most of them are not likely to be singled out for offering to OR students. Interestingly, as the curriculum moves from BS to MS to PhD, the distinction between IE and OR becomes less and less.

THE FUTURE OF IE AND OR: While industrial engineering and operations research will continue to enjoy a symbiotic relationship, it is unlikely that they will become one and the same. Taking Daellenback and George's (1978) definition of operations research, "the systematic application of quantitative models, techniques, and tools to the analysis of problems involving the operation of systems," there are elements of IE that will largely not be addressed by OR. The "people" part of IE is particularly important (in ergonomics, man-machine interfaces, quality management, etc.), while such "people problems" have not traditionally played as large a role in OR. But "never say never!"

See **Industrial applications; Practice of Operations Research and Management Science.**

References

[1] Churchman, C.W., R.L. Ackoff and E.L. Arnoff (1957), *Introduction to Operations Research*, John Wiley, New York.
[2] Copley, F.B. (1923), *Frederick W. Taylor, Father of Scientific Management*, vol. I, Harper and Brothers, New York.
[3] Daellenback, H.G. and J.A. George (1978), *Introduction to Operations Research Techniques*, Allyn and Bacon, Boston.
[4] Emerson, H.P. and D.C.E. Naehring (1988), *Origins of Industrial Engineering: The Early Years of a Profession*, Industrial Engineering and Management Press, Institute of Industrial Engineers, Atlanta, Georgia.
[5] Lehrer, R.N., "Organization of Industrial Engineering Curricula," Final Report on MSA Project TA-31-91, School of Industrial Engineering, Georgia Institute of Technology, Atlanta, 1952.
[6] Maynard, H.B., ed. (1956), *Industrial Engineering Handbook*, McGraw-Hill, New York.
[7] Moder, J.J. and S.E. Elmaghraby, eds. (1978), *Handbook of Operations Research: Models and Applications*, Van Nostrand Reinhold, New York.
[8] Salvandy, G., ed. (1982), *Handbook of Industrial Engineering*, John Wiley, New York.

INFEASIBLE SOLUTION

In general, a proposed solution to an optimization problem that does not satisfy all the constraints. For the linear-programming problem $Ax = b$, $x \geq 0$, a vector x^0 is an infeasible solution if it does not fully satisfy the equations or the nonnegativity conditions. See **Feasible solution**.

INFERENCE ENGINE

A piece of software or a computational strategy that is based on a problem statement from the user, uses reasoning knowledge about the problem area in attempting to derive a solution, gathers needed problem-specific information (e.g., from the user) in the course of reasoning, explains why it needs this added information, presents the solution to the user, and explains the line of reasoning used in reaching the solution. See **Artificial intelligence; Expert systems**.

INFLUENCE DIAGRAMS

A procedure for constructing a diagram that shows the interrelationships of a decision maker's alternatives, uncertainties and values. See **Decision analysis; Decision trees**.

INFORMATION SYSTEMS AND DATABASE DESIGN IN OR/MS

Heiner Müller-Merbach

Universität Kaiserslautern, Germany

There are many close relations between *information systems*, *database structures*, and *operations research* (OR). The models and algorithmic procedures of OR will more and more become integrated parts of information systems, and the task of OR may continually shift towards the comprehensive design of information systems, database structures included.

ARCHITECTURE OF COMPREHENSIVE INFORMATION SYSTEMS: *Traditional* data processing was based on collections of *individual* programs, separated from one another, each with its own individual data organization. Similarly, the characteristic OR packages were stand-alone solutions for singular types of problems, be it mathematical programming, network analysis, and simulation or even more specialized packages for the "knapsack problem," "traveling salesman problem," "set covering problem," etc.

Future information systems, in contrast, will have a *comprehensive* architecture. The vast majority of data will be stored and maintained centrally, that is, on a data management computer or on a network of such computers. Most of the programs will mainly process such centralized data, and the programs themselves will be available from the comprehensive information system.

A particular feature of the comprehensive information systems is the *client-server* structure, that is, a network with a huge number of clients (client computers) being provided with data and programs from a server (server computer) or networks of servers.

RELATIONAL DATABASES: The design of such comprehensive information systems and their databases requires *standards*, in particular those for data structures. A quite common standard today is that of *relational databases*, such as designed by Codd (1970). The main principles of relational databases are (in non-technical terms):

- All the information is organized in terms of *attributes* to *entity sets*. Entity sets are collections of entities with identical attributes – but individual attribute *values*.
- There is no hierarchy between the entity sets. All the entity sets are at the same level and allow for immediate access. However, it is sometimes advantageous to distinguish between *elementary* and *connecting* entity sets. The first ones are self-contained, while the latter ones connect other entity sets and, therefore, depend partly on them.
- Any information is only stored *once*, and *no redundancy* is allowed. Any attribute, therefore, has to be attached to its corresponding (elementary or connecting) entity set. This is the essence of the "normalization" concept of relational database structures.

MODELS AND DATABASES: There exists a narrow correspondence between mathematical models and relational database structures. *Indices* of a mathematical model indicate the individual entities of an entity set, *single* indices those of *elementary*, *multiple* indices those of *connecting* entity sets. The *constants* and *variables* of a mathematical model correspond with the *attributes* of the entity sets (Müller-Merbach, 1983, 1989; Geoffrion, 1989).

This correspondence can easily be shown by a production function, connecting the quantities of production factors with the quantities of products:

$$r_j = \sum_k a_{jk} x_k$$

with j and k indicating the entities of the entity sets **FACTOR**(j) and **PRODUCT**(k), respectively, and

r_j = quantity of production factor j required,
x_k = quantity of product k to be produced, and
a_{jk} = production coefficient, representing the quantity of factor j required per unit of product k.

The relational database structure corresponding with the production function is given in Fig. 1. There are the elementary entity sets **FACTOR**(j) and **PRODUCT**(k), as well as the connecting entity set $F \times P(j,k)$. This database structure is to be considered as a subset of the comprehensive database of the corresponding enterprise with many more entity sets and many more attributes to the entity sets.

Any mathematical model (be it from OR, statistics, etc.) should have such an immediate correspondence with the database. The entity sets correspond with the mathematical indices, the attributes to the entity sets correspond with the constants and variables of the model.

Therefore, model design and database design follow the same logical structure. Either one can proceed the other. However, normally database design is prior to model design, and the attributes required for a model can be derived from the database. Should, however, the attributes required for a model not be available in the database, an appropriate extension of the database may become necessary.

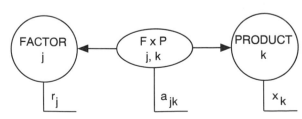

Fig. 1. Relational database structure for a production function. (The attributes are attached to the entity sets **FACTOR**, **PRODUCT**, and $F \times P$.)

MNEMONIC NOTATION: Large-scale mathematical models and – even more so – databases in general tend to cover huge numbers of entity sets and attributes. In order to cope with them, a mnemonic notation of the attributes is useful. The notation should (i) refer to the entity set, (ii) specify the content of the attribute, and (iii) indicate the formal property of the attribute.

Considered be a two-stage production and cost function, respectively, connecting the three elementary entity sets **LABOR, MACHINE,** and **PRODUCT** (Fig. 2). The indices indicate the qualification class of labor (i), the machines (j) and the products (k). All the attributes of the entity set **LABOR** start with an L, the others with an M or a P, respectively, referring to the entity sets. The content is represented by a Q (quantity), a T (time required), or a C (cost) in the second position. The third letter indicates constants (C), variables (V) and other formal properties such as discrete variables (D), Boolean variables (B), etc.

Thus, the constants vector PQC_k represents the known (therefore, C) quantities (Q) of the single products (P). The variables vector MTV_j represents the times (T) of the machines (M) required for producing the given quantities of the products. The variables vector LQV_i stands for the quantity (Q) of labor (L) necessary for running the machines.

In addition, the production coefficients have to be introduced. They are attributes of the dependent entity sets, connecting the elementary entity sets **MACHINE** and **PRODUCT** as well as **LABOR** and **MACHINE**. It is convenient that the attributes of the dependent entity sets refer immediately to the attributes of the elementary entity sets. Thus, the constants matrix $MTPQC_{jk}$ represents the machine time (MT) per unit of the product quantities (PQ). This leads immediately to the production function for the machine times required for the given product quantities:

$$MTV_j = \sum_k MTPQC_{jk}PQC_k.$$

In a similar way, the quantity of labor hours (LQ) per unit of the machine times (MT) is represented by the constants matrix $LQMTC_{ij}$, the basis for the production function for the labor quantities required for the computed machine times:

$$LQV_i = \sum_j LQMTC_{ij}MTV_j.$$

The cost functions (here only labor costs) are dual to the production functions. They use the same production coefficients matrices as the production functions, but the attributes of the elementary entity sets are different:

$LCC_i =$ cost of quantity unit of labor (qualification class i),

$MCV_j =$ labor cost per time unit of machine j, and

$PCV_k =$ labor cost per quantity unit of product k.

By the first cost function, the labor costs are assigned to the machines:

$$MCV_j = \sum_i LCC_i LQMTC_{ij}.$$

By the second cost function, the resulting labor costs per machine time unit are assigned to products:

$$PCV_k = \sum_j MCV_j MTPQC_{jk}.$$

OBJECT-ORIENTED MODELING: There is a tendency from *relational* databases and modeling towards *object-oriented* databases and modeling. One of the object-oriented features is the integration of functions and data. Even if there is no unique standard as yet for object-oriented databases, the idea of object-oriented mathematical models can be presented (Fig. 3), with all the production functions integrated into the database structure.

ADVANTAGE OF INTEGRATION: The integration of models and algorithmic procedures of OR into comprehensive information systems has

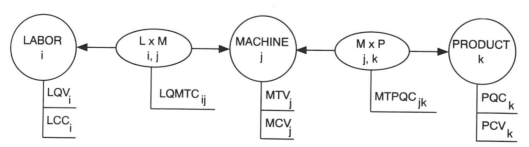

Fig. 2. Relational database structure for a two-stage production function and cost function. (The mnemonic attributes are attached to the elementary entity sets **LABOR, MACHINE,** and **PRODUCT,** as well as to the dependent entity sets $L \times M$ and $M \times P$.)

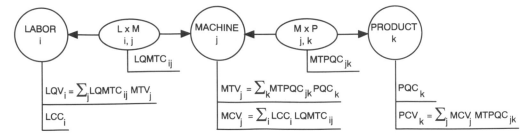

Fig. 3 Object-oriented database structure for a two-stage production function and cost-function. (The mnemonic attributes and the functions for the variables are attached to the elementary entity sets **LABOR**, **MACHINE** and **PRODUCT**, as well as to the dependent entity sets $L \times M$ and $M \times P$ – same example as of Fig. 2).

many convenient properties. The main advantage is: the data required for a model can immediately be taken from the centralized database, and the results derived from the model can immediately be transferred back to the database and is then available to other users.

See **Model management; Structured modeling; Systems analysis.**

References

[1] Codd, E.F. (1970). "A Relational Model of Data for Large Shared Data Banks." *Comm. ACM* **13**, 377–387.
[2] Geoffrion, A.M. (1989). "Computer-Based Modeling Environments." *Euro. Jl. Operational Res.* **41**, 33–43.
[3] Müller-Merbach, H. (1983). "Model Design Based on the Systems Approach." *Jl. Operational Res. Soc.* **34**, 739–751.
[4] Müller-Merbach, H. (1989). "Database-Oriented Design of Planning Models." *IMA J. Math. Applied in Business and Industry* **2**, 141–155.

INFORMS

See **Institute for Operations Research and the Management Sciences.**

INITIAL FEASIBLE SOLUTION

See **First feasible solution.**

INPUT PROCESS

The stochastic point process representing some aspect of customers actually entering a queueing system (or nodal part of one) or some aspect of the state of the node at the instant of input, with points representing the instants of entrance. For

example, in finite capacity queues, an $(\mathbf{X}^a, \mathbf{T}^a)$ process has the $\mathbf{X}^a$ process as a sequence of 1s and 0s representing whether the queue is full or not at an arrival and the $\mathbf{T}^a$ process represents the times of arrivals. The subset of the $\mathbf{T}^a$ process for which $\mathbf{X}^a = 0$ represents the set of arrival epochs at which a customer actually enters the node, while the $\mathbf{T}^a$ for which $\mathbf{X}^a = 1$ represents the set of arrival times at which customers do not gain access to the node but overflow. See **Arrival process; Networks of queues; Queueing theory.**

INPUT-OUTPUT ANALYSIS

The economic theory developed by the economist W.W. Leontief to study a national economy. The approach requires the development of an input-output table (matrix) in which the coefficients in a row indicate how much of the industry designated by that row is required to produce a unit of output for itself and all other industries, and the coefficients in a column represent the amounts of each industry required to produce one unit of output for the industry designated by that column. Under the assumption that the input-output coefficients are stable over the near future and reflect a constant return to scale (linear) relationship, a square set square of equations can be established to determine production levels for the industries that meet projected demand. See **Input-output coefficients.**

INPUT-OUTPUT COEFFICIENTS

For some linear-programming and other production problems, we can interpret the $A = (a_{ij})$ coefficients of the constraints $Ax = b$ as the amount of resource i required (input) to produce one unit of product j (output). More generally, an input-output matrix of American industries formed the basis of the economist Leontief's contribution to economic theory. See **Activity-analysis problem; Input-output analysis**.

INSENSITIVITY

A property of a queueing system wherein some measure of effectiveness does not depend on a particular distribution assumption except through its mean value. The classical example is the Erlang loss call formula in the multi-server M/G/c/c queue which depends on the service-time process only through its mean value. See **Erlang B formula; Queueing theory**.

INSTITUTE FOR OPERATIONS RESEARCH AND THE MANAGEMENT SCIENCES (INFORMS)

The main organization for operations research and the management sciences in the United States begun officially on January 1, 1995, upon the merger of the Operations Research Society of America (ORSA) and The Institute of Management Sciences (TIMS).

INTEGER GOAL PROGRAMMING

A goal programming methodology that generates an integer solution for decision variables.

INTEGER-PROGRAMMING PROBLEM

A mathematical-programming problem in which some or all of its variables are restricted to integer values. See **Binary variables; Combinatorial and integer optimization; Mathematical-programming problem**.

INTENSITY FUNCTION

See **Failure-rate function; Point stochastic processes; Renewal processes**.

INTERACTIVE OPTIMIZATION

A solution approach involving human-machine interaction.

INTERCHANGE HEURISTIC

A type of local improvement heuristic.

INTERFERING FLOAT

Float which is shared among the activities on a chain or path in a project network, that is, all the activities on the chain have the same float. See **Network planning**.

INTERIOR POINT

In a constrained optimization problem, an interior point is a solution point that is not on the boundary of the solution space S. If S is defined by the set of constraints $\{g_i(x) \leq 0\}$, then x^0 in S is an interior point if $g_i(x^0) < 0$ for all i.

INTERIOR-POINT METHODS

Paul T. Boggs

National Institute of Standards & Technology, Gaithersburg, Maryland

INTRODUCTION: Notwithstanding the success of the simplex method for linear programming, there was, even from the earliest days of operations research, a desire to create an algorithm for solving linear programming problems that proceeded on a path through the polytope rather than around its perimeter. Indeed, methods were considered, but until recently, none was as effective as the simplex method. In this article the motivation for desiring an "interior" path, the concept of the complexity of solving a linear programming problem, a brief history of the developments in the area, and the status of the subject as of this writing are discussed. More complete surveys are given in Wright (1992), Gonzaga (1991) and Goldfarb and Todd (1989).

The linear programming problem in standard form is

$$\begin{aligned}
\text{minimize}_x \;\; & c^t x \\
\text{s.t.} \quad & Ax = b \\
& x \geq 0
\end{aligned} \qquad \text{(LP)}$$

where $c, x \in R^n$, $b \in R^m$, and $A \in R^{m \times n}$. The feasible region for the problem (LP) in general has a strictly feasible point, that is, a point x^0 such that $Ax^0 = b$ and $x^0 > 0$ (i.e., each component of

x^0 is strictly positive). The simplex method proceeds on a path from vertex to vertex on the boundary of this region, a process that could require many steps to go around a multifaceted feasible region. Intuitively, a more direct path through the interior of the region is appealing since there exists the possibility of moving through the polytope in very few steps.

Despite the contention that the simplex method could require a large number of steps, in actual practice it was observed to be quite efficient. A formal analysis of its complexity, however, was elusive. A major breakthrough in the understanding of the complexity of the simplex method was the famous result of Klee and Minty (1972) who showed with a simple example that the worst case complexity is exponential. Their example is a slightly out of kilter cube (in n dimensions) in which all 2^n vertices can be visited by a simplex method, that is, starting at the origin, there is a path through all of the vertices such that the objective function is decreased at each step. It was immediately recognized, however, that no practical simplex method would use this path; thus there was a desire to explain the efficiency of practical simplex methods. Later analyses have shown that a simplex method could expect linear performance, thus partially explaining its behavior (Goldfarb and Todd, 1989).

The first algorithm for (LP) that was proven to have a worst-case polynomial complexity is the ellipsoid algorithm of Khachiyan (1979). Briefly, in Khachiyan's algorithm one first constructs an ellipsoid that is large enough to contain the feasible region. At subsequent iterations that ellipsoid is shrunk to a point that is the solution to (LP). For his method Khachiyan proved that the complexity is $O(n^4 L)$ where L is the number of bits necessary to specify the problem. Unfortunately, the algorithm also seemed to have an expected performance of similar complexity, and was quickly shown to be noncompetitive in practice. Note that Khachiyan's algorithm is not an interior-point method.

BRIEF HISTORY: Interior-point methods seek to approach the optimal solution through a sequence of points that is always strictly feasible. Such methods have been known for a long time, but, for reasons explained below, they were not considered to be effective.

One of the earliest interior-point methods is the barrier method originally proposed in the 1950s. When applied to problems with only inequality constraints, it is typically used in conjunction with the sequential unconstrained minimization technique that is described more generally in Fiacco and McCormick (1968). In the barrier method for (LP), one creates the barrier function with equality constraints

$$B(\boldsymbol{x}, \mu) = \boldsymbol{c}^t \boldsymbol{x} - \mu \sum_{i=1}^{m} \log x_i$$

subject to: $\boldsymbol{Ax} = \boldsymbol{b}$

where μ is a positive parameter. One then selects a value of μ and a strictly feasible point $\boldsymbol{x}^0$ and solves the equality constrained problem

minimize$_x$ $B(\boldsymbol{x},\mu)$

subject to: $\boldsymbol{Ax} = \boldsymbol{b}$ (1)

calling the result $\boldsymbol{x}^1$. (Note that solving a problem with linear equality constraints is not difficult.) Clearly, $\boldsymbol{x}^1$ will remain strictly feasible since the log-barrier function becomes infinite at the boundary of the feasible region. The parameter μ is reduced and (1) is solved again using $\boldsymbol{x}^1$ as the initial start. It can be shown that if μ is reduced to zero, for example, by setting $\mu = \mu/2$ at the beginning of each iteration, the resulting sequence $\{\boldsymbol{x}^i\}$ will converge to $\boldsymbol{x}^*$, an optimal solution to (LP).

Another interior-point approach, called the method of centers, was suggested by Huard (1967). In this method, one starts with a strictly feasible point $\boldsymbol{x}^0$ and finds the center of the polytope formed by the intersection of the original polytope and the half space of points corresponding to an objective function value less than $\boldsymbol{c}^t \boldsymbol{x}^0$. The center of this bounded polytope is defined to be the maximum of the centering function

$$C(\boldsymbol{x}) = \prod_{i=0}^{m} r_i(\boldsymbol{x})$$

where $r_i(\boldsymbol{x}) = (\boldsymbol{b} - \boldsymbol{Ax})_i$, $i = 1, \ldots, m$ and $r_0(\boldsymbol{x}) = \boldsymbol{c}^t \boldsymbol{x}^0 - \boldsymbol{c}^t \boldsymbol{x}$. The function $C(\boldsymbol{x})$ is clearly zero on the boundary and positive in that part of the interior of the polytope that corresponds to lower values of $\boldsymbol{c}^t \boldsymbol{x}$, and thus has a maximum, say $\boldsymbol{x}^1$. The function $r_0(\boldsymbol{x})$ is then redefined using $\boldsymbol{x}^1$ in place of $\boldsymbol{x}^0$, and the process repeated. Again, it can be shown that the sequence $\{\boldsymbol{x}^i\}$ converges to $\boldsymbol{x}^*$.

Early interior point algorithms also include that of Dikin (1967). In his method, one begins each iteration by scaling the variables so that the current, strictly feasible point is transformed to the vector of all ones, a point well away from the boundary in the transformed space. Steepest descent is then applied to this scaled problem, and the resulting point is transformed back to the original space to obtain the next iterate. The advantage of this idea is that the steepest descent step in the original space can be extremely short if the current iterate is close to the boundary, whereas "long" steps are always possible in the transformed space. This method is known as the affine scaling algorithm, and is related to Karmarkar's method discussed below.

While these methods were tried and compared with the simplex method, none was seen to be competitive for two principal reasons. First, almost all interior-point methods require at each step the solution to a linear system of equations of the

$$(A^tA)d = v, \qquad (2)$$

where d and v are n-vectors. It was not until the 1970s that there were sufficiently powerful linear algebra routines that exploited the sparsity structure of (A^tA) to solve such systems efficiently. Second, interior-point methods tend to be competitive with the simplex method only on larger problems that were well beyond the capabilities of the computers of the 1960s.

While almost nothing was done in interior-point methods in the 1970s, significant advances in numerical linear algebra were made and, of course, in computational capacity and speeds. Interest in interior-point algorithms was then revitalized by the announcement of Karmarkar (1984) that he had developed an interior-point method that had provable polynomial complexity and was competitive with the simplex method. Karmarkar's procedure begins with a strictly feasible point and then embeds (LP) in a space of one higher dimension in which the feasible point is in the center of the higher dimensional polytope. As in the affine scaling algorithm, a "good" step can then be taken in this space and the new point projected back to the original space to obtain the next iterate. Karmarkar's method and its relatives, including the affine scaling algorithm and barrier methods, were studied intensively. Since then, hundreds of papers have been written on both the theoretical and computational aspects of interior-point methods for (LP) and on the extension of these ideas to quadratic and more general nonlinear programming problems.

STATUS AND EXTENSIONS: In the theoretical arena, there has been considerable interest in improving the bound on the number of iterations required to solve (LP). Early on it was observed that interior-point algorithms could be cast in such a way as to generate a continuous path, or trajectory, from any initial feasible point to an optimal solution. In particular, a path, called the central trajectory, was defined that has certain desirable properties. Much analysis of these trajectories and of algorithms based on following the central trajectory has been performed. These are surveyed in Gonzaga (1991a, 1991b). Simplified versions of the algorithms described below have also been extensively analyzed. To date, the best theoretical results for these methods demonstrate a complexity that is $O(\sqrt{n}L)$ steps with a quadratic asymptotic rate of convergence.

The most computationally successful interior-point methods for solving (LP) are based on using a primal-dual formulation and applying Newton's method to the system of equations arising from the barrier method. Specifically, the dual problem to (LP) is

maximize$_y$ b^ty

subject to: $A^ty + z = c$ \qquad (DP)
$\qquad z \geq 0$

where $y \in R^m$ and $z \in R^n$. Thus the primal-dual problem can be formulated as

minimize$_{x,y,z}$ $c^tx - b^ty$

subject to: $Ax = b$
$\qquad A^ty + z = c$ \qquad (PD)
$\qquad x \geq 0$
$\qquad z \geq 0.$

The barrier function for (PD) can be easily specified and, by making certain identifications in the first-order conditions for optimality, one arrives at a nonlinear system of equations in (x,y,z). The system also contains the penalty parameter which is reduced in the course of applying Newton's method. Because of the adaptive manner in which certain parameters, including μ, are adjusted, it has not been possible to provide a complexity analysis for practical implementations of these methods.

The affine scaling methods, in particular, the dual affine method, have also enjoyed some success, but, in general, they have not been as efficient as the primal-dual, barrier methods. In addition, since it was observed that Karmarkar's method is a certain linear combination of the affine scaling direction and a so-called "recentering" direction, studies were directed to other combinations of these directions. The most successful of these methods, called optimizing on subspaces, computes these two (and possibly other) directions and solves (LP) restricted to these directions, thus making "optimal" use of them. New directions are then computed at the new point and the process continued. Algorithms based on this strategy are nearly competitive with the primal-dual methods on linear-programming problems and have shown great promise on quadratic programming problems.

Interior-point methods have been extremely successful in solving some very large linear programs, but they do not completely replace the simplex method. The "best" algorithm is, of course, problem-dependent, but generally speaking the interior-point methods perform better on larger problems and on problems that allow efficient exploitation of the numerical linear algebra. Specifically, as noted above, if the structure of $(A^t A)$ can be exploited to solve systems of the form (2) quickly, then the interior-point methods have an advantage. An example of such an A matrix arises in multi-period resource planning problems where A has a "staircase" structure. The matrix $(A^t A)$ is then block diagonal and can usually be factored efficiently. Interior-point methods also perform better on highly degenerate problems that often arise in large-scale applications. Simplex methods typically have severe difficulties in this case. One of the shortcomings of the interior-point methods is that, without additional work, they do not directly provide a basis. Thus if an optimal basis is needed, say for sensitivity analysis, the simplex method has an advantage. There have been successful procedures that combine simplex and interior-point methods to achieve the best of both approaches.

Because of the success of interior-point methods on (LP), research is ongoing to extend them to quadratic programming problems and to general nonlinear programming problems.

See **Computational complexity; Large-scale systems; Linear programming; Nonlinear programming; Optimization; Quadratic programming; Simplex method.**

References

[1] I.I. Dikin (1967). "Iterative solution of problems of linear and quadratic programming." *Soviet Mathematics Doklady*, 8, 674–675.

[2] A.V. Fiacco and G.P. McCormick (1968). *Nonlinear Programming: Sequential Unconstrained Minimization Techniques.* John Wiley, New York.

[3] C.C. Gonzaga (1991a). "Large-steps path-following methods for linear programming, part i: Barrier function method." *SIAM Jl. Optimization*, 1, 268–279.

[4] C.C. Gonzaga (1991b). "Large-steps path-following methods for linear programming, part ii: Potential reduction method." *SIAM Jl. Optimization*, 1, 280–292.

[5] C.C. Gonzaga (1992). "Path following methods for linear programming." *SIAM Review*, 34, 167–224.

[6] D. Goldfarb and M.J. Todd (1989). "Linear programming," In G.L. Nemhauser, A.H.G. Rinnooy Kan, and M.J. Todd, eds., *Optimization*, 73–170, North Holland, Amsterdam and New York.

[7] P. Huard (1967). "Resolution of mathematical programming with nonlinear constraints by the method of centres." In J. Abadie, ed., *Nonlinear Programming*, pages 209–219. North Holland, Amsterdam.

[8] N.K. Karmarkar (1984). "A new polynomial-time algorithm for linear programming." *Combinatorica*, 4, 373–395.

[9] L.G. Khachiyan (1979). "A polynomial algorithm in linear programming." Translated in *Soviet Mathematics Doklady*, 20, 191–194.

[10] V. Klee and G. Minty (1972). "How good is the simplex algorithm?" In O. Sisha, ed., *Inequalities III*. Academic Press, New York.

[11] M.H. Wright (1992). "Interior methods for constrained optimization." In A. Iserles, ed. *Acta Numerica*, 341–407, Cambridge University Press, New York.

INTERNATIONAL FEDERATION OF OPERATIONAL RESEARCH SOCIETIES (IFORS)

The international society whose members are national operational research societies. IFORS, founded in 1959, is dedicated to the development of operational research as a unified science and its advancement in all nations of the world. It sponsors a triennial international conference and other meetings, provides a means of exchanging information, encourages the establishment of national operational research societies, and encourages the development and teaching of operational research.

INTERNATIONAL INSTITUTE FOR APPLIED SYSTEMS ANALYSIS (IIASA)

The International Institute of Applied Systems Analysis (IIASA) is a nongovernmental research institution located in Laxenburg, Austria. IIASA was founded in 1972 on the initiative of the academies of science or equivalent institutions of 12 nations. As of January 1994, the following countries were national member organizations: Austria, Bulgaria, Canada, Czech and Slovak Republics, Finland, Germany, Hungary, Italy, Japan, Netherlands, Poland, Russia, Sweden,

Ukraine and United States of America. The original motivation for the establishment of IIASA was to enable scientists from East and West to work together on problems of common concern. Although this is still an objective of the Institute, it has been broadened to encompass joint work by scientists from most countries. The current goal of IIASA is "To conduct international and interdisciplinary scientific studies to provide timely and relevant information and options, addressing critical issues of global environmental, economic, and social change, for the benefit of the public, the scientific community, and national and international institutions" (IIASA *Agenda for the Third Decade*). Resident scientists at IIASA coordinate research projects, working in collaboration with worldwide networks of researchers, policymakers, and research organizations. IIASA has been instrumental in the development of global (world) models that are concerned with environmental, energy and other resource, economic and population issues. See **Environmental systems analysis; Global models**.

INTERVENTION MODEL

See **Time series**.

INVENTORY MODELING

Edward A. Silver

The University of Calgary, Alberta, Canada

INTRODUCTION: The acquisition, production and/or distribution of inventories are issues of concern to all organizations. From a national or international perspective there are huge amounts (billions of dollars) of capital tied up in stocks. Moreover, there are very large costs incurred as a result of replenishment actions, shortages (caused by inadequate stock levels), and utilization of managerial and clerical time in making and routinely implementing inventory management decisions. Thus, properly designed decision rules, based on mathematical modeling, can lead to substantial benefits. Here are two illustrative practical applications:

i) Pfizer Pharmaceuticals developed and implemented an integrated system to manage inventories in its US pharmaceutical business. A series of management science models contributed to a reduction of $23.9 million in inventories and a concurrent 95% drop in

backorders over a three-year period (Kleutghen and McGee, 1985).

ii) Annual cost savings of $2 million were realized by the US Navy at its supply centers through the use of trade-off curves based on inventory modeling. Gardner (1987) describes the study and associated model.

Inventory decisions can often interact with decisions in other areas of the organization. Examples include: i) preventive maintenance (determination of inventory levels of spare parts); ii) marketing (effects of pricing and promotion on demand, hence stock requirements), iii) quality assurance (higher quality levels reduce the need for buffer or safety stocks), and iv) production scheduling (provision of supporting raw materials and supplies). Here we do not explicitly present models that deal with these complexities, rather the intention is to provide an introduction to inventory models per se, thus whetting the appetite of the reader for the general subject area.

In the next section a listing is provided of the generic reasons why organizations carry inventories. This is followed by a discussion of the types of costs that are of relevance in the development and use of inventory models. Then, the subsequent four sections deal with illustrative models. These are followed by a general classification scheme of inventory models. The coverage of the topic area concludes with a discussion of the increasingly important possibility of changing some of the parameters (givens) in inventory models.

REASONS FOR CARRYING INVENTORIES: There are basically five generic reasons for organizations to carry some inventories. Most situations involve a mix of these reasons, but we discuss each separately, under rather extreme conditions, to emphasize the associated rationale:

i) *Cycle stock* – when the demand pattern is level and known and there is no uncertainty in supply it still may make sense to not have the replenishment inflow precisely match the steady outflow. There can be physical limits on replenishment sizes (e.g., batch container sizes in chemical processes), major fixed costs associated with each replenishment action, or quantity discounts on purchase price and/or transportation costs. Each of these reasons leads to the repeated (or cyclic) use of a significant replenishment size.

ii) *Congestion stock* – even when the usual reasons (discussed above) for holding cycle stock are not present and there is still no uncertainty in supply or demand, it may be necessary to have inventories of items when they are produced on the same piece of equipment and it takes an appreciable amount of time to change over from production of one item to another. One has to produce more than the immediate needs of an item in that the congestion on the equipment prevents returning to that item for an appreciable amount of time.

iii) *Buffer or safety stock* – when there is uncertainty in demand and/or supply and the customer response time is lower than the time required to acquire/produce the demanded goods, it is necessary to have extra stock on hand to ensure an adequate level of customer service. Note that the "customer" can be internal to the organization, for example, spare parts needed to repair equipment that has broken down.

iv) *Pipeline stock* – if an item has to be moved an appreciable distance before being delivered to the customer, then there has to be stock in the pipeline. More generally, if units must go through a process (transportation is a special case) and this requires a positive time, then there will be associated pipeline stock equal to the throughput rate multiplied by the unit process time.

v) *Anticipation stock* – where factors such as demand levels, raw material availability or raw material prices are expected to change appreciably with time it may make sense to build up (and deplete) inventory levels in anticipation of these changes.

THE CATEGORIES OF INVENTORY-RELATED COSTS:

The costs are not easy to estimate in practice. Moreover, only so-called relevant costs, that is, those that can be influenced by inventory management decisions, should be considered. In particular, care must be taken with respect to overhead costs that are often not affected by inventory decisions. Considerably more detail is provided in Silver and Peterson (1985).

We consider five different categories of costs.

a) *Costs of the Material Itself:* These costs are relevant only insofar as they are affected by the size of the replenishments used. If there are no quantity discounts in acquisition cost (including the transportation component), then, over a given time period (such as a year) the costs of the material will be a constant, independent of the replenishment sizes used. Specifically, if the so-called unit variable cost (raw material plus value added to the inventorying stage under consideration), denoted by v and in dollars/unit, is independent of the replenishment sizes used,

then the total cost of the material is Dv per year where D is the demand rate in units/year.

b) *Fixed Cost of Each Replenishment Action:* We denote by A the fixed cost of a replenishment action, that is, the cost component that is independent of the size of the replenishment. In a production context, A is often referred to as the setup or changeover cost.

c) *The Costs of Having Material in Inventory (Inventory Carrying Costs):* The common way of modeling the costs of having material in inventory is as follows:

$$\text{Cost/year} = \bar{I}vr$$

where $\bar{I}$ is the *average* inventory, in convenient units of the item under consideration, v is as defined above, in \$/unit, and r is the carrying charge in \$/unit/year. The cost element r encompasses out-of-pocket expenses (e.g., insurance, taxes, operating the warehouse, etc.) and the lost opportunity of having capital tied up in the stock (it could be invested elsewhere or used to pay off debt). Some models use the symbol $h \equiv vr$ to represent the cost *per unit* in inventory per year.

d) *The Costs of Insufficient Stock in the Short Run:* If the stock is inadequate to meet pending demand, then costs are incurred. There are basically two types of costs, those associated with stockouts (lost sales, backorders, loss of goodwill) and those of emergency actions to avoid stockouts (e.g., expediting, use of an emergency high cost local supplier, etc.). There is no universally appropriate way of modeling such costs as a function of the occurrence and magnitude of the shortage. Possibilities include a fixed cost per stockout occasion, a cost proportional to the number of units short, and so on.

In lieu of prescribing a cost of insufficient stock many organizations impose a service constraint on the inventory policy. Again, there is a wide variety of possible service measures. Two of the more common ones are a specified high probability of no stockout prior to the receipt of each replenishment, and a specified high fraction of the demand to be routinely met from stock.

e) *The Costs of the Inventory Control System:* Many models are concerned with minimizing the total of two or more of the preceding four cost categories. However, there is a fifth category that should be considered in selecting among different inventory control systems, namely the costs of the system itself. These include the costs of acquiring and updating the data required for the operation of the system and its associated decision rules (e.g., demand rate, measure of demand variability, cost parameters, etc.); numerical calculations; and training and other aspects of implementation.

THE ECONOMIC ORDER QUANTITY (WILSON LOT-SIZE):

This is one of the earliest results developed in inventory modeling (Harris, 1913). It addresses the issue of how much to replenish under very stable conditions where there is a significant fixed cost (A) per replenishment, that is, the economic order quantity (EOQ) is concerned with cycle stock. Strictly speaking, it is based on a number of rather severe assumptions but it is still an important result for two reasons: (1) the costs tend to be insensitive to some of the assumptions; and (2) many of the assumptions can be relaxed leading to somewhat more complicated results, but the EOQ or an obvious variation thereof often still plays a central role.

a) *Assumptions* – There are eight underlying assumptions:

i) the demand rate is constant and known,

ii) there are no restrictions on the size of the replenishment quantity (including that it need not be an integer number of units),

iii) there are no quantity discounts,

iv) the cost factors do not change appreciably with time,

v) each item is treated independently of others (i.e., we choose to ignore any possible benefits of coordination),

vi) the replenishment lead time (the time interval from when we decide to place a replenishment order until the moment that the associated material is on the shelf ready to satisfy demand) has a known value,

vii) the entire replenishment arrives at the same time (unlike in a production context where there may be a gradual buildup of stock),

viii) no shortages are permitted.

b) *Derivation of the EOQ* – Under the above set of assumptions there is no uncertainty and nothing is changing appreciably with time. Therefore, it is appropriate to restrict attention to a policy of ordering the *same* quantity Q (in units) over and over again with each replenishment arriving just as the onhand inventory goes to zero (each order is placed exactly a lead time before such a moment). The resulting pattern of inventory versus time is as shown in Figure 1, where the slope is the demand rate, D, in units/year.

From the set of assumptions it follows that there are only two categories of relevant costs (relevant in the sense that they will be affected by the choice of Q), namely the fixed costs of replenishments and the inventory carrying costs.

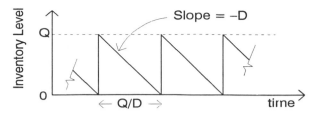

Fig. 1. Inventory level versus time.

The total relevant costs per year are given by

$$TRC(Q) = \frac{AD}{Q} + \frac{Qvr}{2}. \qquad (2)$$

The first term is the product of the fixed cost per replenishment and the number of replenishments per year while the second term comes from Equation (1) and the fact that $\bar{I}$ for the triangles of Figure 1 is $Q/2$.

Setting $\mathrm{d}TRC(Q)/\mathrm{d}Q = 0$ leads to the optimum Q-value as

$$\mathrm{EOQ} = \sqrt{\frac{2AD}{vr}}. \qquad (3)$$

Moreover,

$$\frac{\mathrm{d}^2 TRC(Q)}{\mathrm{d}Q^2} = \frac{2AD}{Q^3} > 0$$

so that we have indeed found the minimizing value of Q.

c) *Some Remarks:*

i) At the EOQ value, one can show that the two components of $TRC(Q)$ in Equation (2) are equal.

ii) The EOQ expressed as a time supply is

$$\frac{\mathrm{EOQ}}{D} = \sqrt{\frac{2A}{Dvr}}. \qquad (4)$$

Many organizations have tended to use a very simple decision rule, namely the same time supply replenishment quantity for a broad range of items. Equation (4) shows that this is inappropriate in that any of A, D and v are likely to vary between items.

iii) An important relaxation of the EOQ model is to permit quantity discounts, specifically the socalled all units discount situation where the unit variable cost

$$v = \begin{cases} v_0 & if \quad Q < Q_b \\ v_1 & if \quad Q \geq Q_b \end{cases}$$

where Q_b is the breakpoint order quantity and $v_1 < v_0$. Under such circumstances, it can be shown that the best order must be at one of three positions: the EOQ using v_0, Q_b, or the EOQ using v_1 (Silver and Peterson, 1985).

iv) Another important extension is where the demand pattern is still known but now varies with time for at least two situations, namely lumpy demand (as in Material Requirements Planning) and trended demand. It no longer follows that repetitive use of the same Q value is appropriate; hence it is inadequate to look at average costs in a typical year and an exact analysis becomes much more complicated. There is a rather extensive literature on this so-called lotsizing problem (Ritchie and Tsado, 1986).

AN ILLUSTRATIVE MODEL FOR THE CASE OF CONGESTION STOCK:

Here we consider a group of n items (numbered $i = 1, 2, \ldots, n$) satisfying all but two of the EOQ assumptions, specifically now the items are produced on the same piece of equipment (i.e., coordination is necessary) and there is a gradual buildup of the stock of the item being replenished (m_i units/year for item i). Furthermore, we assume that a so-called cyclic production schedule is used, that is, we produce item 1, then item 2, $\ldots$, then item n and return to item 1 to begin a new cycle. There is idle time in each cycle, as appropriate. Item i has parameters D_i, A_i, v_i and m_i and let Q_i be its replenishment quantity. Moreover, assume that there is a setup time of τ_i at the beginning of the replenishment of item i. The single decision variable is the duration of each cycle, T, in years. The associated replenishment quantities are given by

$$Q_i = D_i T \qquad i = 1, 2, \ldots, n. \tag{5}$$

Production of item i begins just as its inventory level is depleted, that is, the setup must be commenced τ_i before that moment. Production of i continues for $D_i T / m_i$ units of time and the inventory reaches a maximum level of $Q_i(1 - D_i/m_i)$, not Q_i, because usage at rate D_i continues during the production. Thus the average inventory level of item i is

$$\bar{I}_i = \frac{D_i T}{2}(1 - D_i/m_i) \tag{6}$$

The total relevant costs per year are

$$TRC(T) = \sum_{i=1}^{n} \frac{A_i}{T} + \sum_{i=1}^{n} \frac{D_i T}{2}(1 - D_i/m_i)v_i r. \tag{7}$$

One wishes to minimize this expression but subject to having adequate capacity, viz.

$$\sum_{i=1}^{n} \left(\tau_i + \frac{D_i T}{m_i} \right) \leq T$$

$$\text{or} \quad T \geq \frac{\sum_i \tau_i}{1 - \sum_i D_i/m_i}. \tag{8}$$

Again, one can show that $d^2 TRC(T)/dT^2 > 0$, so

that setting $dTRC(T)/dT = 0$ will give a minimum of $TRC(T)$. This turns out to be where

$$T_{\text{opt}} = \sqrt{\frac{2 \sum_i A_i}{r \sum_i D_i v_i (1 - D_i/m_i)}}. \tag{9}$$

Because of the convexity of $TRC(T)$ we use T_{opt} if it satisfies condition (8), otherwise we set T equal to the righthand side of (8).

A more complicated problem to analyze is where not every item is produced on each cycle. Rather we now let $Q_i = k_i D_i T$ where $k_i = 1, 2, 3, \ldots$.

THE NEWSVENDOR (OR SINGLE PERIOD) PROBLEM:

Typically, some type of forecasting model is used to forecast demand in the period of interest, and there is an associated probability distribution of forecast errors or equivalently of actual demand that will result. Let the continuous probability distribution of demand x in the period of interest be denoted by $f(x)$ with a cumulative distribution

$$F(x) = \int_0^x f(y)\, dy. \tag{10}$$

The decision to be made is how large a quantity, Q, of the item to have available to meet demand in the period. Suppose that there is an underage cost of c_u for each unit of demand not satisfied (case of $Q < x$) and an overage cost of c_o for each unit of stock that is not demanded, that is, remaining at the end of the period (case of $Q > x$).

We use a marginal, as opposed to a total, cost argument. Specifically consider the Qth unit made available. It will save an underage cost anytime $x \geq Q$. The probability of this event is $1 - F(Q)$. Hence, the expected marginal cost savings of the Qth unit are

$$EMS(Q) = c_u[1 - F(Q)]. \tag{11}$$

The Qth unit will incur an avoidable cost of c_o if demand turns out to be less than Q. Therefore, the expected marginal cost increase of the Qth unit is

$$EMI(Q) = EMI(Q) \tag{12}$$

One can argue that for optimality we want to stop with the Q value where

$$EMS(Q) = EMI(Q)$$

or, using equations (11) and (12), the best Q, denoted by Q^*, must satisfy

$$F(Q^*) = \frac{c_u}{c_u + c_o}. \tag{13}$$

Note that (13) is a general result for *any* continuous distribution of demand.

A multi-item extension of this problem, including a budget constraint on the total amount that can be spent on the set of items, has been modeled and solved (Silver and Peterson, 1985, pp. 406–410).

AN ILLUSTRATION OF DEALING WITH UNCERTAIN DEMAND IN AN ON-GOING SITUATION:

In contrast with the previous section, consider the case where demand continues on indefinitely so that unused material can be kept in stock until it is used up by future demand. There is typically a nonzero replenishment lead time. The combination of random demand and a nonzero lead time forces us to more carefully define what is meant by inventory level. In fact, there are at least four different definitions:

i). On-hand stock — material physically present.
ii). Backorders — unsatisfied demand that will be met when stock becomes available.
iii). Net stock = (On-hand) − (Backorders)
iv). Inventory position = (On-hand)
$$+ \left(\begin{array}{c} \text{On-order} \\ \text{from supplier} \end{array} \right) - \left(\begin{array}{c} \text{Backordered} \\ \text{customer demands} \end{array} \right)$$

Reordering decisions are based on this last quantity.

a) *Common Individual Item Control Systems:* When demand is uncertain, there are really three decision variables regarding the inventory management of a particular item at a specific location, namely,

i) how often to review the status of the item (continuous review, sometimes called transactions reporting, versus periodic review and, if the latter, what review interval (R) to use),
ii) when to initiate a replenishment, and
iii) how much to replenish.

The three most common individual item control policies are
i) (s, Q) — continuous review ($R = 0$) with an order for a fixed quantity Q being placed when the inventory position drops to the reorder point s or lower,
ii) (R, S) — every R units of time enough is ordered to raise the inventory position to the order-up-to-level S, and
iii) (R, s, S) — every R units of time a review is made. If the inventory position is at s or lower, enough is ordered to raise it to S.

It should be emphasized that here we are dealing with a so-called independent demand situation where the demand for the item under consideration is not a function of replenishment decisions for other items. In particular, where an item is a component of other items, its demand is dependent and the control procedures of Material Requirements Planning are more appropriate than any of the above control policies.

b) *Selecting s in an (s, Q) System:* We illustrate for the case of normally distributed demand during a constant lead time (of duration L) and for a particular service constraint, namely, where there is a specified probability P of no stockout during each lead time. [A variety of other combinations of control policy, demand distribution and service measure/shortage costing method can be treated (Brown, 1982; Hax and Candea, 1984; Nahmias, 1989; Silver and Peterson, 1985).

It is assumed that each replenishment is triggered when the inventory position is *exactly* at the level of s (undershoots, caused by large transactions, substantially complicate the mathematics). If we let $f(x)$ be the general probability distribution of the lead time demand x, then the probability of no stockout must satisfy

$$P = \int_{-\infty}^{s} f(x) \, dx. \tag{14}$$

For the special case of $f(x)$ being normal (with mean μ_L and standard deviation σ_L), if we let

$$s = \mu_L + k\sigma_L, \tag{15}$$

then a substitution of $u = (x - \mu_L)/\sigma_L$ in Equation (14) leads to

$$P = \Phi(k) \tag{16}$$

where $\Phi(k) = \int_{-\infty}^{k} \phi(u) du$ is the widely tabulated, unit normal distribution function and $\phi(u)$ is the probability density function of that distribution.

In summary, the procedure is as follows: The specified value of P gives $\Phi(k)$ from (16). A table lookup (or approximation for computer use) provides the associated k value. Then s is determined from condition (15).

A few remarks are in order:
i) The above choice of s is independent of Q. Other service measures/shortage costing methods lead to a decision rule for s that depends upon Q. Such a dependence was ignored in the derivation of the EOQ. There are optimization procedures for *simultaneously* choosing values of Q and s (Hadley and Whitin, 1963; Naddor, 1966; Nahmias, 1989; Silver and Peterson, 1985).
ii) The above analysis was based on a constant lead time L. In fact, at least approximately, the same type of analysis can be used for a

variable lead time where x is now the total demand during the lead time and its distribution must reflect variability of demand per unit time *plus* variability of the lead time.

THE WIDE VARIETY OF POSSIBLE INVENTORY MODELS: There are a large number of structural parameters that can take on two or more values in actual inventory systems. In principle, each combination of these parameters leads to a different inventory model. In this section, we list most of the important parameters (many of the possible combinations have been modeled in the literature but often tailormade adaptations or approximations are needed to end up with a usable model in the context of a specific organization, Silver, 1981; Zanakis *et al.*, 1980):

Nature of Demand
- deterministic vs probabilistic (in the latter case, known versus uncertain probability distribution)
- stationary vs varying with time
- influenced by on-hand inventory?
- consumables vs returnables/repairables
- independent of vs dependent on replenishment decisions of other items

Time Horizon
- single period vs multiperiod
- discrete vs continuous time
- use of discounting or not?

Supply Issues
- quantity discounts (economies of scale)
- minimum order size or fixed batch size
- supply available or not in certain periods
- fixed or random lead time
- can orders cross in time?
- possible random yield (acceptable quantity received is not the same as that ordered)
- capacity restrictions

Time-Dependent Parameters (*other than demand*)
- inflation
- one-time special prices
- lead time varies with time
- capacity varies with time

What Happens Under a Stockout Situation
- lost sales vs backorders vs mix of these

Shelf-Life Considerations
- obsolescence
- perishability (deteriorating inventory)

Single vs. Multiple Items
- group budget or space constraint
- coordinated control (or joint replenishment) because of common supplier, mode of transport or production equipment
- substitutable or complementary items

Single vs Multiechelon (Schwarz, 1981)
- in multiechelon (multistage) – serial vs convergent (e.g., assembly) vs divergent (e.g., distribution)

Knowledge of Status of Stock (*and other parameter values*)
- known exactly or not?
- continuously vs at discrete points in time

CHANGING THE GIVENS IN INVENTORY MODELS: Traditionally the values of the parameters, discussed above, have been accepted as givens in inventory modeling. The philosophy of continuous improvement challenges this assumption and argues that items like the setup cost, the replenishment lead time and so on can be changed, often with much more substantial benefits than simply optimizing subject to the given parameter values. Another way of saying this is that it may be better to at least partially eliminate the causes of inventories rather than just choosing the best inventory level (Silver, 1992).

See **Hierarchical production planning; Logistics; Production management.**

References

[1] Brown, R.G. (1982). *Advanced Service Parts Inventory Control*, 2nd ed., Materials Management Systems Inc., Norwich, Vermont.

[2] Gardner, E.S. (1987). "A Top-Down Approach to Modeling US Navy Inventories," *Interfaces*, 17(4), 1–7.

[3] Hadley, G. and T. Whitin (1963). *Analysis of Inventory Systems*. Prentice-Hall, Englewood Cliffs, New Jersey.

[4] Harris, F.W. (1913). "How Many Parts to Make at Once," *Factory, the Magazine of Management*, 10, 2, 13–56 and 152 (reprinted in *Operations Research*, 38(6), 947–950).

[5] Hax, A.C. and D. Candea (1984). *Production and Inventory Management*. Prentice-Hall, Englewood Cliffs, New Jersey.

[6] Kleutghen, P.P. and J.C. McGee (1985). "Development and Implementation of an Integrated Inventory Management Program at Pfizer Pharmaceuticals," *Interfaces*, 15(1), 69–87.

[7] Naddor, E. (1966). *Inventory Systems*. John Wiley, New York.

[8] Nahmias, S. (1989). *Production and Operations Analysis*. Irwin, Homewood, Illinois.

[9] Ritchie, E. and A. Tzado (1986). "A Review of Lot-Sizing Techniques for Deterministic Time-Varying Demand," *Production and Inventory Management*, 27(3), 65–79.

[10] Schwarz, L.B., ed. (1981). *Multi-Level Production/ Inventory Control Systems: Theory and Practice*. Vol. 16, *Studies in the Management Sciences*. North-Holland, Amsterdam.

[11] Silver, E.A. (1992). "Changing the Givens in Modelling Inventory Problems: The Example of Just-in-Time Systems," *International Jl. Production Economics*, 26, 347–351.

[12] Silver, E.A. (1981). "Operations Research in Inventory Management: A Review and Critique," *Operations Research*, 29, 628–645.

[13] Silver, E.A. and R. Peterson (1985). *Decision Systems for Inventory Management and Production Planning*, 2nd ed., John Wiley, New York.

[14] Zanakis, S.H., L.M. Austin, D.C. Nowading and E.A. Silver (1980). "From Teaching to Implementing Inventory Management: Problems of Translation," *Interfaces*, 10(6), 103–110.

INVERSE MATRIX

For a square $m \times m$ matrix A, the inverse matrix A^{-1} is also an $m \times m$ matrix such that $A^{-1}A = I = AA^{-1}$, where I is the identity matrix. If a matrix has an inverse, then its inverse is unique and the matrix is said to be nonsingular. If an inverse does not exist, the matrix is said to be singular. A nonsingular matrix has a nonzero value for its determinant; a singular matrix has a determinant value equal to zero. See **Matrices and matrix algebra**.

IP

Integer programming. See **Combinatorial and integer optimization**.

IS

Information systems. See **Information systems and database design**.

ISOMORPHIC GRAPH

Graphs that have identical structure.

ISOP 9000 STANDARD

See **Quality control**.

ISOQUANT

For a function $f(x)$, the graph or contour $f(x) = C$, where C is a constant, is called an isoquant. If $f(x)$ is a profit (cost) function, then the isoquant is termed an isoprofit (isocost) line.

ITERATION

The cycle of steps of an algorithm is called an iteration. For example, in the simplex algorithm for solving linear-programming problems, one iteration is given concisely by the steps: (1) select a nonbasic variable to replace a basic variable, (2) determine the inverse of the new feasible basis, and (3) determine if the new basic feasible solution is optimal.

IVHS

Intelligent vehicle-highway system. See **Traffic analysis**.

J

JACKSON NETWORK

A collection of multi-server queueing systems or nodes with exponential service and Markovian or memoryless probabilistic routing of departures from one node to the others. If there are customers arriving from outside the network to individual nodes in Poisson streams, the network is said to be *open*; otherwise, it is *closed*. All customers who arrive from outside to an open network must eventually leave after receiving service at one or more systems within the network. See **Networks of queues**; **Queueing theory**.

JIT

Just-in-time. See **Just-in-time manufacturing**.

JOB SHOP SCHEDULING

Albert Jones, Luis Rabelo, and Yuehwern Yih

National Institute of Standards & Technology, Gaithersburg, MD

INTRODUCTION: In the United States today, there are approximately 40,000 factories producing metal fabricated parts. These parts end up in a wide variety of products sold here and abroad. These factories employ roughly 2 million people and ship close to $3 billion worth of products every year. The vast majority of these factories are what we call "job shops," meaning that the flow of raw and unfinished goods through them is completely random. Over the years, the behavior and performance of these job shops have been the focus of considerable attention in the operations research and management science (OR/MS) literature. Research papers on topics such as factory layout, inventory control, process control, production scheduling, and resource utilization can be found in almost every issue of every OR/MS journal on the market. The most popular of these topics is production (often referred to as job shop) scheduling. Job shop scheduling can be thought of as the allocation of resources over a specified time to perform a predetermined collection of tasks. Job shop scheduling has received this large amount of attention because it has the potential to dramatically decrease costs and increase throughput, thereby, profits.

A large number of approaches to the modeling and solution of these job shop scheduling problems have been reported in the OR literature, with varying degrees of success. These approaches revolve around a series of technological advances that have occurred over the last 30 years. These include mathematical programming, dispatching rules, expert systems, neural networks, genetic algorithms, and inductive learning. In this article, we take an evolutionary view in describing how these technologies have been applied to job shop scheduling problems. To do this, a few of the most important contributions in each of these technology areas are discussed. We close by looking at the most recent trend which combines several of these technologies into a single hybrid system.

MATHEMATICAL PROGRAMMING: Mathematical programming has been applied extensively to job shop scheduling problems. These problems, which belong to the class of NP-complete problems, have been formulated and solved using integer programming, mixed integer programming, and dynamic programming (Panwalker and Iskander, 1977). Because of the difficulties in formulating material flow constraints as mathematical relationships and developing generalized solution techniques, these approaches are seldom used outside the classroom. To overcome these deficiencies, researchers began to decompose the job shop scheduling problem into a number of subproblems, proposing a number of techniques to solve them.

Davis and Jones (1988) described a methodology based on the decomposition of mathematical programming problems which used both Benders-type and Dantzig/Wolfe-type decompositions (Benders, 1960; Dantzig and Wolfe, 1960). The methodology was part of a closed-loop, real-time, two-level hierarchical shop floor control system. The top level scheduler (i.e., the supremal) specified the earliest start time and the latest finish time for each job. The lower level scheduling modules (i.e., the infimals) would refine these limit times for each job by detailed sequencing of all operations. A multi-criteria objective function was specified that included tardiness, throughput, and process utilization costs. The decomposition was achieved by first reordering the constraints of the original problem to generate a block angular form, then transforming that block angular form into a hierarchical tree structure. In general, N subproblems would result plus a constraint set which contained partial members of each of the subproblems. The latter constraint set was termed the "coupling" constraints which included precedence relations and material handling. The supremal unit explicitly considered the coupling constraints while the infimal units considered their individual decoupled constraint sets. The authors pointed out that the inherent stochastic nature of job shops and the presence of multiple but often conflicting objectives made it difficult to express the coupling constraints using exact mathematical relationships. This made it almost impossible to develop a general solution methodology. To overcome this, a new real-time simulation methodology was proposed by Davis and Jones (1988) to solve the supremal and infimal problems.

Gershwin (1989) used the notion of temporal decomposition in a mathematical programming framework for analysis of production planning and scheduling. A multi-layer hierarchical model was proposed in which 1) formulations needed to control events at higher layers ignore the details of the variations of events occurring at lower layers, and 2) formulations at the lower layers view the events at the higher layers as static, discrete events. Scheduling is carried out in the bottom layers so that the production requirements imposed by the planning layers can be met. First, a hedging point is found by solving a complex dynamic programming problem. This hedging point is the number of excess goods that should be produced to compensate for future equipment failures. This hedging point is used to formulate a linear programming problem to determine instantaneous production rates. These rates are then used to determine the actual schedule (what parts to make and when). A variety of approaches are under investigation to generate that schedule.

DISPATCHING RULES: Dispatching rules have been applied consistently to job shop scheduling problems. They are procedures designed to provide good solutions to complex problems in real-time. The term dispatching rule, scheduling rule, sequencing rule, or heuristic are often used synonymously (Panwalker and Iskander, 1977). Dispatching rules have been classified mainly according to the performance criteria for which

they have been developed. Wu (1987) categorized dispatching rules into several classes. Class one contains simple priority rules, which are based on information related to the jobs. Sub-classes are made based on which particular piece of information is used. Example classes include those based on processing times (such as Shortest Processing Time – SPT), due dates (such as Expected Due Date – EDD), slack (such as Minimum Slack – MINSLACK), and arrival times (such as First-In, First Out – FIFO).

Class two consists of combinations of rules from class one. The particular rule that is implemented can now depend on the situation that exists on the shop floor. A typical example of a rule in this class is: "Use SPT until the queue length exceeds 5, then switch to FIFO." This prohibits jobs with large processing times from staying in the queue for unusually long periods of time.

Class three contains rules which are commonly referred to as Weight Priority Indexes. The idea is to use more than one piece of information about the jobs to determine the schedule. These pieces of information can be assigned weights to reflect their relative importance. Usually, we define a function such as

$$f(x) = \text{weight}_1 * \text{Processing Time Job}(x)$$
$$+ \text{weight}_2 * (\text{Current Time}$$
$$- \text{Due Date Job}(x)).$$

Then any time a sequencing decision is to be made, the function $f(x)$ is calculated for each job x in the queue. The job x with the lowest value of $f(x)$ is processed next.

During the last 30 years, the performance of a large number of these rules has been studied extensively using simulation techniques. These studies have been aimed at answering the question: "If you want to optimize a particular performance criteria, which rule should you choose?" Most of the early work concentrated on the shortest processing time rule (SPT). Conway and Maxwell (1967) were the first to study the SPT rule and its variations. They found that, although some individual jobs could experience prohibitively long flow times, the SPT rule minimized the mean flow time for all jobs. They also showed that SPT was the best choice for optimizing the mean value of other basic measures such as waiting time and system status utilization. Many similar investigations have been carried out to determine the dispatching rule which optimizes a wide range of job-related (such as due date and tardiness) and shop-related (such as throughput and utilization) performance measures. This problem of selecting the best dispatching rule for a given performance measure continues to be a very active area of research.

ARTIFICIAL INTELLIGENCE (AI) TECHNIQUES: Starting in the early 1980s, a series of new technologies were applied to the job shop scheduling problem. They all fall under the general title of Artificial Intelligence and include expert systems, knowledge-based systems, neural networks, genetic algorithms and inductive learning. There are three main advantages of these techniques. First, and perhaps most important, is that they can use both quantitative and qualitative knowledge in the decision-making process and capture complex relationships in elegant new data structures. Second, they are capable of generating heuristic rules which are significantly more complex than the simple dispatching rules described above. Third, the selection of a heuristic rule can be based on a range of information about the entire job shop including the current jobs, expected new jobs, status of machines and material transporters, and status of inventory and personnel.

Expert/knowledge-based Systems – Expert or knowledge-based systems consist of two parts: knowledge base and inference engine to operate on that knowledge. The knowledge-base attempts to formalize the knowledge that human experts use into rules, procedures, heuristics, and other types of abstractions. Three types of knowledge are usually included: procedural, declarative, and meta. Procedural knowledge is domain-specific problem solving knowledge. Declarative knowledge provides the input data defining the problem domain. Meta knowledge is knowledge about how to use the procedural and declarative knowledge to actually solve the problem. Several data structures have been utilized to represent the knowledge in the knowledge base including semantic nets, frames, scripts, predicate calculus, and production rules. The inference engine selects the strategy of how to use the knowledge bases to solve the problem at hand. It can be forward chaining (data driven) or backward chaining (goal driven).

ISIS was the first major expert system aimed specifically at the job shop scheduling problem (Fox, 1983). ISIS used a constraint-directed reasoning approach with three constraint categories: organizational goals, physical constraints and casual restrictions. Organizational goals considered objective functions based on due-date and work-in-progress. Physical constraints referred to situations where a machine had limited processing capability. Procedural constraints

and resource requirements were typical examples of the third category. Several issues with respect to constraints were considered such as constraints in conflict, importance of a constraint, interactions of constraints, constraint generation and constraint obligation. ISIS uses a three level, hierarchical, constraint-directed search. Orders were selected at level 1. A capacity analysis was performed at level 2 to determine the availability of the resources required by the order. Detailed scheduling was performed at level 3. ISIS also provided for the capability to interactively construct and alter schedules. In this capacity, ISIS utilized its constraint knowledge to maintain the consistency of the schedule under development and identified scheduling decisions that resulted in poorly satisfied constraints.

Wysk *et al.* (1986) developed an integrated expert system/simulation scheduler called MPECS. The expert system used both forward and backward chaining to select a small set of potentially good rules from a predefined set of dispatching rules and other heuristics in the knowledge base. These rules were selected to optimize a single performance measure, although that measure could change from one scheduling period to the next. The selected rules were then evaluated one at a time using a deterministic simulation of a laboratory manufacturing system. After all of the rules were evaluated, the best rule was implemented on the laboratory system. Information could be gathered about how the rule actually performed and used to update the knowledge base in an off-line mode. They were able to show that periodic rescheduling makes the system more responsive and adaptive to a changing environment. MPECS was important for several reasons. It was the first hybrid system to make decisions based on the actual feedback from the shop floor. It incorporated some learning into its knowledge base to improve future decisions. The same systems could be used to optimize several different performance measures. And, finally, it utilized a new multi-step approach to shop floor scheduling.

Artificial Neural Networks – Neural networks, also called connectionist or distributed parallel processing models, have been studied for many years in an attempt to mirror the learning and prediction abilities of human beings. Neural network models are distinguished by network topology, node characteristics, and training or learning rules. An example of a three-layer feed-forward neural network is shown in Figure 1.

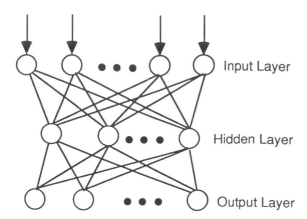

Fig. 1. An example of a three-layer feed-forward neural network.

Through exposure to "training patterns," neural networks attempt to capture the desired relationships between inputs and the outputs. Back-propagation applies the gradient-descent technique in the feed-forward network to change a collection of weights so that some cost function can be minimized (Rumelhart *et al.*, 1986). The cost function, which is only dependent on weights and training patterns, is defined by:

$$C(W) = \frac{1}{2} \sum (T_{ip} - O_{ip}) \tag{1}$$

where the T is the target value, O is the output of network, i is an output node, and p is the training pattern.

After the network propagates the input values to the output layer, the error between the desired output and actual output will be "back-propagated" to the previous layer. In the hidden layers, the error for each node is computed by the weighted sum of errors in the next layer's nodes. In a three-layered network, the next layer means the output layer. The activation function is usually a sigmoid function with the weights modified according to

$$\Delta W_{ij} = \eta X_j (1 - X_j)(T_j - X_j) X_i \tag{2}$$

or

$$\Delta W_{ij} = \eta X_j (1 - X_j)(\sum \delta_k W_{jk}) X_i \tag{3}$$

where W_{ij} is weight from node i to node j, η is the learning rate, X_j is the output of node j, T_j is the target value of node j, and δ_k is the error function of node k.

If j is in the output layer, relation (2) is used; if j is in the hidden layers, relation (3) is used. The weights are updated to reduce the cost function at each step. The process continues until the error between the predicted and the actual outputs is smaller than some predetermined tolerance.

Rabelo (1990) was the first to use back propagation neural nets in a job shop scheduling system with several job types, exhibiting different arrival patterns, process plans, precedence sequences and batch sizes. Training examples were generated to train the neural network to provide the correct characterization of the manufacturing environments suitable for various scheduling policies and the chosen performance criteria. In order to generate training samples, the performance simulation of the dispatching rules available for the manufacturing system was carried out. The neural networks were trained for problems involving 3, 4, and 5 machines. To carry out this training, a special input feature space was developed. This input feature space contained both job characteristics (such as types, number of jobs in each type, routings, due dates, and processing times) and shop characteristics (such as number of machines and their capacities). The output of the neural network represented the relative ranking of the available dispatching rules for that specific scheduling problem and the selected performance criteria. The neural networks were tested in numerous problems and their performance (in terms of minimizing Mean Tardiness) was always better than each single dispatching rule (25% to 50%).

Hopfield neural nets (Hopfield and Tank, 1985) have also been used to solve some classic, textbook job shop scheduling problems. These implementations have been based on relaxation models (i.e., pre-assembled systems which relaxes from input to output along a predefined energy contour). The neural networks were used in Foo and Takefuji (1988) and Zhou *et al.* (1990) to solve 4-job 3-machine and 10-job 10-machine job shop scheduling problems, respectively. The two-dimensional Hopfield network was extended by Lo and Bavarian (1990) to 3 dimensions to represent jobs, machines, and time. Their objective, minimize makespan, was defined as

$$E_t = \frac{1}{2} \sum_{j=1} \sum_{i=1} \sum_{l=1} (v_{ijl}/C_k)(1 + T_{ij} - 1)$$

where C_k is a scaling factor, v_{ijl} is the output of neuron ijl, and T_{ij} is the time required by jth machine to complete the ith job.

Due to a large number of variables involved in generating a feasible schedule, these approaches tend to be computationally inefficient and frequently generate infeasible solutions. Consequently, they have not been used to solve realistic job shop scheduling problems.

Genetic Algorithms – Genetic algorithms are an optimization methodology based on a direct analogy to Darwinian natural selection and mutations in biological reproduction. In principle, genetic algorithms encode a parallel search through concept space, with each process attempting coarse-grain hill-climbing (Goldberg, 1988). Instances of a concept correspond to individuals of a species. Induced changes and recombinations of these concepts are tested against an evaluation function to see which ones will survive to the next generation. The use of genetic algorithms requires five components:

1. A way of encoding solutions to the problem – fixed length string of symbols;
2. An evaluation function that returns a rating for each solution;
3. A way of initializing the population of solutions;
4. Operators that may be applied to parents when they reproduce to alter their genetic composition such as crossover (i.e., exchanging a randomly selected segment between parents), mutation (i.e., gene modification), and other domain specific operators; and
5. Parameter setting for the algorithm, the operators, and so forth.

A number of approaches have been utilized in the application of genetic algorithms (GAs) to job shop scheduling problems including the use of blind recombination operators (Davis, 1985; Goldberg and Lingle, 1985), and the mapping of constraints to a Boolean satisfiability problem (De Jong and Spears, 1989).

Starkweather *et al.* (1993) were the first to use genetic algorithms to solve a dual criteria job shop scheduling problem in a real production facility. Those criteria were the minimization of average inventory in the plant and the minimization of the average waiting time for an order to be selected. These criteria are negatively correlated (i.e., the larger the inventory, the shorter the wait; the smaller the inventory, the longer the wait). To represent the production/shipping optimization problem, a symbolic coding was used for each member (chromosome) of the population. In this scheme, customer orders are represented by discrete integers. Therefore, each member of the population is a permutation of customer orders. The GA used to solve this problem was based on a modification to the blind recombinant operator. This recombination operator emphasizes information about the relative order of the elements in the permutation, because this impacts both inventory and waiting time. A single evaluation function (a weighted sum of the two criteria) was utilized to rank each member of the population. That

ranking was based on an on-line simulation of the plant operations. This approach generated schedules which produced inventory levels and waiting times which were acceptable to the plant manager. In addition, the integration of the genetic algorithm with the on-line simulation made it possible to react to the changing system dynamics.

Learning – The first step in developing a knowledge base is knowledge acquisition. This itself is a two step process: get the knowledge from knowledge sources and store that knowledge in a digital form. Much work has been done in the area of knowledge acquisition, such as protocol analysis, interactive editing, and so on. Knowledge sources may be human experts, simulation data, experimental data, databases, text, etc. In job shop scheduling problems, the knowledge sources are likely to be human experts or simulation data. To extract knowledge from these two sources, the machine learning technique that learns from examples (data) becomes a promising tool. Inductive learning is a state classification process. Viewing the state space as a hyperplane, the training data (consisting of conditions and decisions) can be represented as points on the hyperplane. The inductive learning algorithm is to draw lines on the hyperplane based on the training data to divide the plane into several areas within which the same decision (conclusion) will be made.

One algorithm that has been implemented in inductive aids and expert system shells is that developed by Quinlan (1986), called Iterative Dichotomister 3 or ID3. ID3 uses examples to induce production rules (e.g., IF...THEN...), which form a simple decision tree. Decision trees are one way to represent knowledge for the purpose of classification. The nodes in a decision tree correspond to attributes of the objects to be classified, and the arcs are alternative values for these attributes. The end nodes of the tree (leaves) indicate classes which groups of objects belong to. Each example is described by attributes and a resulting decision. To determine a good attribute to partition the objects into classes, entropy is employed to measure the information content of each attribute, and then rules are derived through a repetitive decomposition process that minimizes the overall entropy. The entropy value of attribute A_k can be defined as

$$H(A_k) = \sum_{j=1}^{M_k} P(a_{kj}) \left\{ - \sum_{i=1}^{N} P(c_i|a_{kj}) \log_2 P(c_i|a_{kj}) \right\}$$

where

$H(A_k)$ = entropy value of attribute A_k;

$P(a_{kj})$ = probability of attribute k being at its jth value;

$P(c_i | a_{kj})$ = probability that the class value is c_i when attribute k is at its jth value;

M_k = total number of values for attribute A_k; and

N = total number of different classes (outcomes).

The attribute with the minimum entropy value will be selected as a node in the decision tree to partition the objects. The arcs out of this node represent different values of this attribute. If all the objects in an arc belong to one class, the partition process stops. Otherwise, another attribute will be identified using entropy values to further partition the objects that belong to this arc. This partition process continues until all the objects in an arc are in the same class. Before applying this algorithm, all attributes that have continuous values need to be transformed to discrete values.

In the context of job shop scheduling, the attributes represent system status and the classes represent the dispatching rules. Very often the attribute values are continuous. In 1990, Yih proposed a trace-driven knowledge acquisition (TDKA) methodology to deal with continuous data and to avoid the problems occurring in verbally interviewing human experts. TDKA learns scheduling knowledge from expert schedulers without a dialogue with them. There are three steps in this approach. In step one, an interactive simulator is developed to mimic the system of interest. The expert will interact with this simulator and make decisions. The entire decision making process will be recorded in the simulator and can be repeated for later analysis. The series of system information and the corresponding decision collected is called a "trace." Step two analyzes the "trace" and forms classification rules to partition the trace into groups. The partition process stops when most of the cases in each group use the same dispatching rule (error rate is below the threshold defined by the knowledge engineer). Then, the decision rules are formed. The last step is to verify the generated rules. The resulting rule base is used to schedule jobs in the simulator. If it performs as well as or better than the expert, the process stops. Otherwise, the threshold value is increased, and the process returns to step two.

NEW TRENDS: During the last 10 years, many manufacturing companies have invested heavily in modern computer-based technologies. These technologies have made it possible to gather data about the events on the shop floor literally as they are happening. The availability of this data has spawned a new challenge to the OR community. That challenge is to develop job shop scheduling systems that can reschedule jobs *in real-time* whenever that data indicates that problems exist on the shop floor. In addition, many manufacturers are demanding that these systems have the ability to 1) handle multiple performance objectives and priority jobs, 2) freeze a specific set of jobs during rescheduling, and 3) be integrated with the rest of their hierarchical and distributed shop floor control architectures.

To address these requirements, many researchers have begun building hybrid systems (a few of them are discussed in the preceding sections) which combine two or more of the technologies listed above. One of the most ambitious of these hybrid systems is described in Jones *et al.* (1994). This system is based on the ideas described in Davis and Jones (1988), Gershwin (1989), Rabelo (1990), and Yih (1990). It is designed to implement a two level hierarchical scheduler by integrating artificial neural networks, real-time simulation, genetic algorithms, and an inductive learning technique. It is a multi-step, multi-criteria approach which has been designed specifically to address all of these new requirements. The first step, as in MPECS, is to quickly generate a small set of candidate scheduling rules from some larger set of heuristics. This is done using single-performance, artificial neural nets (Rabelo, 1990). In step 2, a more detailed evaluation of these candidates is carried out using the real-time simulation approach from Davis and Jones (1988). This evaluation is necessary to generate a ranking that specifies how each rule performs against all of the performance criteria. Even though a further reduction in the number of candidates can be achieved from this ranking, it has been shown in Davis *et al.* (1991) that, in general, it is not possible to simultaneously optimize several performance criteria using a single heuristic rule.

To overcome this, one would like to create a new "rule" which 1) combines the best features of the most attractive candidates, 2) eliminates the worst features of those candidates, and 3) achieves satisfactory levels of performance for all objectives. As proposed in Jones *et al.* (1991), this is a two step process. First, the actual schedules that result from applying the best candidate rules are used as input into a genetic algorithm. The output from the genetic algorithm is the "best" schedule. The last step is to capture the knowledge contained in that schedule in a new rule which can be added to the original set of heuristics for future use. The trace-driven knowledge acquisition technique is used to generate that rule (Yih, 1990). In the experiments completed to date, which use the sum of mean flow time and maximum tardiness as its objective function, this process takes less than 1 second on a 486 PC-class machine running at 33MHz and obviously even less on a Pentium-class machine.

SUMMARY: Through an evolutionary approach and a number of key examples, we have summarized the major strategies used in solving job shop scheduling problems. These strategies have been based on a number of technologies ranging from mathematical programming to genetic algorithms. In addition, a new trend was described which attempts to integrate two or more of those technologies.

Since job shop scheduling problems fall into the class of NP-complete problems, they are among the most difficult to formulate and solve. Operations research analysts have been pursuing solutions to these problems for more than 30 years, with varying degrees of success. On the other hand, they are among the most important problems because of their impact on the ability of manufacturers to meet customer demands and make a profit. For this reason, operations research analysts will continue this pursuit well into the next century.

See **Artificial intelligence; Computational complexity; Decision trees; Flexible manufacturing systems; Gantt charts; Genetic algorithms; Hierarchical production planning; Linear programming; Neural networks; Operations management; Project management; Scheduling and sequencing.**

References

[1] Adams, J., E. Balas and D. Zawack (1988), "The Shifting Bottleneck Procedure for Job Shop Scheduling," *Management Science*, 34, 391–401.

[2] Baker, K. (1974), *Introduction to Sequencing and Scheduling*, John Wiley, New York.

[3] Bean, J. and J. Birge (1986), "Match-up Real-time Scheduling," NBS Special Publication 724, 197–212.

[4] Benders, J. (1960), "Partitioning Procedures for Solving Mixed-Variables Mathematical Programming Problems," *Numersche Mathematik*, 4, 238–252.

[5] Chiu, C. (1994), "A Learning-Based Methodology for Dynamic Scheduling in Distributed Manufacturing Systems," PhD Dissertation, Purdue University, West Lafayette, Indiana.

[6] Conway, R. and W. Maxwell (1967), *Theory of Scheduling*, Addison-Wesley, Reading, Massachusetts.

[7] Dantzig, G. and P. Wolfe (1960), "Decomposition Principles for Linear Programs," *Operations Research*, 8, 101–111.

[8] Davis, L. (1985), "Job Shop Scheduling with Genetic Algorithms," *Proceedings of an International Conference on Genetic Algorithms and Their Applications*, Carnegie Mellon University, 136–140.

[9] Davis, W. and A. Jones (1988), "A Real-Time Production Scheduler for a Stochastic Manufacturing Environment," *International Journal of Computer Integrated Manufacturing*, 1, 101–112.

[10] Davis, W., H. Wang and C. Hsieh (1991), "Experimental Studies in Real-Time Monte Carlo Simulation," *IEEE Transactions on Systems, Man and Cybernetics*, 21, 802–814.

[11] De Jong, K. and W. Spears (1989), "Using Genetic Algorithms to solve NP-Complete Problems," *Proceedings of the Third International Conference on Genetic Algorithms*, Carnegie Mellon University, 124–132.

[12] Foo, Y. and Y. Takefuji (1988), "Stochastic Neural Networks for Solving Job-Shop Scheduling: Part 2. Architecture and Simulations," *Proceedings of the IEEE International Conference on Neural Networks*, published by IEEE TAB, II283–II290.

[13] Fox, M. (1983), "Constraint-Directed Search: A Case Study of Job Shop Scheduling," PhD Dissertation, Carnegie-Mellon University.

[14] Gershwin, S. (1989), "Hierarchical Flow Control: a Framework for Scheduling and Planning Discrete Events in Manufacturing Systems," *Proceedings of IEEE Special Issue on Discrete Event Systems*, 77, 195–209.

[15] Goldberg, D. (1988), *Genetic Algorithms in Search Optimization and Machine Learning*, Addison-Wesley, Menlo Park, California.

[16] Goldberg, D. and R. Lingle (1985), "Alleles, Loci, and the Traveling Salesman Problem," *Proceedings of the International Conference on Genetic Algorithms and Their Applications*, Carnegie Mellon University, 162–164.

[17] Hopfield, J. and D. Tank (1985), "Neural Computation of Decisions in Optimization Problems," *Biological Cybernetics*, 52, 141–152.

[18] Jones, A., L. Rabelo and Y. Yih (1994), "A Hybrid Approach for Real-Time Sequencing and Scheduling," *International Journal of Computer Integrated Manufacturing*, to appear.

[19] Lo, Z. and B. Bavarian (1991), "Scheduling with Neural Networks for Flexible Manufacturing Systems," *Proceedings of the IEEE International Conference on Robotics and Automation*, Sacramento, California, 818–823.

[20] McKenzie, L., (1976), "Turnpike Theory," *Econometrics*, 44, 841–864.

[21] Panwalker, S. and W. Iskander (1977), "A Survey of Scheduling Rules," *Operations Research*, 25, 45–61.

[22] Quinlan, J. (1986), "Induction of Decision Trees," *Machine Learning*, 1, 81–106.

[23] Rabelo, L. (1990), "A Hybrid Artificial Neural Networks and Knowledge-Based Expert Systems Approach to Flexible Manufacturing System Scheduling," PhD Dissertation, University of Missouri-Rolla.

[24] Rumelhart, D., J. McClelland and the PDP Research Group (1986), *Parallel Distributed Processing: Explorations in the Microstructure of Cognition Vol. 1: Foundations*, MIT Press, Cambridge, Massachusetts.

[25] Saleh, A. (1988), "Real-Time Control of a Flexible Manufacturing Cell," PhD Dissertation, Lehigh University.

[26] Starkweather, T., D. Whitely and B. Cookson (1993), "A Genetic Algorithm for Scheduling with Resource Consumption," *Proceedings of the Joint German/US Conference on Operations Research in Production Planning and Control*, Springer-Verlag, 567–583.

[27] Wysk, R., D. Wu and R. Yang (1986), "A Multi-Pass Expert Control System (MPECS) for Flexible Manufacturing Systems," NBS Special Publication 724, 251–278.

[28] Wu, D. (1987), "An Expert Systems Approach for the Control and Scheduling of Flexible Manufacturing Systems," PhD Dissertation, Pennsylvania State University.

[29] Yih, Y. (1990), "Trace-Driven Knowledge Acquisition (TDKA) for Rule-Based Real-Time Scheduling Systems," *Journal of Intelligent Manufacturing*, 1, 217–230.

[30] Zhou, D., V. Cherkassky, T. Baldwin and D. Hong (1990), "Scaling Neural Network for Job Shop Scheduling," *Proceedings of the International Conference on Neural Networks*, 3, 889–894.

JOHNSON'S THEOREM

See **Sequencing and scheduling**.

JUST-IN-TIME (JIT) MANUFACTURING

A manufacturing philosophy focusing on the elimination of waste (non-value added activities) in the manufacturing process by the more timely sequencing of operations. See **Flexible manufacturing**.

K

KARMARKAR'S ALGORITHM

An algorithm devised by N. Karmarkar for solving a linear-programming problem by generating a sequence of points that lies in the strict interior of the problem's solution space and that converges to an optimal solution. Karmarkar's algorithm, and its many variations, have been shown to be polynomial-time algorithms that solve large-scale linear-programming problems in a computationally efficient manner. See **Interior point methods**; **Polynomial-time algorithm**.

KARUSH-KUHN-TUCKER (KKT) CONDITIONS

The Karush-Kuhn-Tucker (KKT) conditions are necessary conditions that a solution to a general nonlinear programming problem must satisfy, provided that the problem constraints satisfy a regularity condition called a constraint qualification. If the problem is one in which the constraint set (i.e., solution space) is convex and the maximizing (minimizing) objective function is concave (convex), the KKT conditions are sufficient. Applied to a linear-programming problem, the KKT conditions yield the complementary slackness conditions of the primal and dual problems. See **Nonlinear programming**.

KENDALL'S NOTATION

A shorthand notation of the form A/S/c/K/Q used to describe queueing systems. The A refers to an acronym for the interarrival-time distribution, S the service-time distribution, c the number of parallel servers, K the maximum allowable system size, and Q the queue discipline. Common designations for A and S include: M for Markovian or exponential; E_k for Erlang (k); D for deterministic or constant; G for general; etc. See **Queueing theory**.

KILTER CONDITIONS

For the minimum cost flow network problem, the complementary slackness optimality conditions are called kilter conditions. See **Out-of-kilter algorithm**.

KKT CONDITIONS

See **Karush-Kuhn-Tucker conditions**; **Nonlinear programming**; **Quadratic programming**.

KLEE-MINTY PROBLEM

The Klee-Minty problem is a linear-programming problem designed to demonstrate that a problem exists that would require the simplex algorithm to generate all extreme point solutions before finding the optimal. This problem demonstrated that, although the simplex algorithm (under a nondegeneracy assumption) would find an optimal solution in a finite number of iterations, the number of iterations can increase exponentially. Thus, the simplex method is not a polynomially bounded algorithm. One form of the Klee-Minty problem, which defines a slightly perturbed hypercube, is the following:

Minimize $-x_d$

subject to

$x_1 \geq 0$

$x_1 \leq 1$

$-\varepsilon x_1 + x_2 \geq 0$

$\varepsilon x_1 + x_2 \geq 1$

....

$-\varepsilon x_{d-1} + x_d \geq 0$

$\varepsilon x_{d-1} + x_d \geq 1$

$x_j \geq 0$

with $0 < \varepsilon < 1/2$.

KNAPSACK PROBLEM

The following optimization problem is called the knapsack problem:

Maximize $c_1 x_1 + c_2 x_2 + \ldots + c_n x_n$

subject to

$a_1 x_1 + a_2 x_2 + \ldots + a_n x_n \leq b$

with each x_j equal to 0 or 1, with all (a_j, c_j, b) usually taken to be positive integers. The name is due to interpreting the problem as one in which a camper has a knapsack which can carry up to b pounds. The camper has a choice of packing up to n items, with $x_j = 1$ if the item is packed and $x_j = 0$ if the item is not packed. Item j weighs a_j pounds. Each item has a "value" c_j to the camper if it is packed. The camper wishes to choose that collection of items having the greatest total value subject to the weight condition. The knapsack problem arises in many applications such as selecting a set of projects and as a subproblem of other problems. It can be solved by dynamic programming or by integer-programming methods. If the x_j are ordered such that $c_1/a_1 \geq c_2/a_2 \geq \ldots \geq c_n/a_n$ and the integer restrictions on the variables are replaced by $0 \leq x_j \leq 1$, then an optimal solution to the relaxed problem is to just pack all the items starting with the first until the weight restriction is violated. The item that caused the violation is then chosen at a fractional value so that the total weight of the selected set is equal to b.

KNOWLEDGE ACQUISITION

The activity of eliciting, structuring, analyzing knowledge from some source of expertise and representing it in a form that can be used by an inference engine. See **Artificial intelligence**; **Expert systems**; **Inference engine**.

KNOWLEDGE BASE

That part of an expert system containing application-specific reasoning knowledge that the inference engine uses in the course of reasoning about a problem. In expert systems whose reasoning knowledge is represented as rules, the knowledge base is a rule set or rule base. A knowledge base can also contain other kinds of knowledge. See **Artificial intelligence**; **Expert systems**.

KNOWLEDGE ENGINEER

One who develops an expert system, or one who elicits reasoning knowledge from a human expert for use in an expert system. See **Expert system**.

KÖNIGSBERG BRIDGE PROBLEM

See **Chinese postman problem**; **Combinatorics**.

KÖNIG'S THEOREM

See **Hungarian method**.

KRUSKAL'S ALGORITHM

A procedure for finding a minimum spanning tree in a network. The method selects the lowest cost arcs in sequence, while ensuring that no cycles are allowed. Ties are broken arbitrarily. For a network with n nodes, the process stops when $n - 1$ arcs are selected. See **Greedy algorithm**; **Minimal spanning tree**; **Prim's algorithm**.

KUHN-TUCKER (KT) CONDITIONS

See **Karush-Kuhn-Tucker conditions**.

LACK OF MEMORY

See **Exponential arrivals**; **Markov processes**; **Memoryless property**.

LAGRANGE MULTIPLIERS

The multiplicative, linear-combination constants that appear in the Lagrangian of a mathematical programming problem. They are generally dual variables if the dual exists, so-called shadow prices in linear programming, giving the rate of change of the optimal value with constraint changes, under appropriate conditions. See **Lagrangian function**; **Nonlinear programming**.

LAGRANGIAN DECOMPOSITION

See **Combinatorial and integer optimization**.

LAGRANGIAN FUNCTION

The general mathematical-programming problem of minimizing $f(X)$ subject to a set of constraints $\{g_i(x) \leq b_i\}$ has associated with it a Lagrangian function defined as $L(x,\lambda) = f(x) + \Sigma_i \lambda_i [g_i(x) - b_i)]$, where the components λ_i of the nonnegative vector λ are called Lagrange multipliers. For a primal linear-programming problem, the Lagrange multipliers can be interpreted as the variables of the corresponding dual problem. See **Nonlinear programming**.

LAGRANGIAN RELAXATION

An integer programming decomposition method. See **Combinatorial and integer optimization**.

LANCHESTER ATTRITION

The concept of an explicit mathematical relationship between opposing military forces and casualty rates. The two classical "laws" are the linear law, which gives the casualty rate (derivative of force size with respect to time) of one side as a negative constant multiplied by the product of the two sides' force sizes, and the square law, which gives the casualty rate of one side as a negative constant multiplied by the opposing side's force size. See **Battle modeling**; **Homogeneous Lanchester equations**; **Lanchester's equations**.

LANCHESTER'S EQUATIONS

Joseph H. Engel

Bethesda, Maryland

HISTORICAL BACKGROUND: Lanchester's equations are named for the Englishman, F.W. Lanchester, who formulated and presented them in 1914 in a series of articles contributed to the British journal, *Engineering*, which then were printed in toto in Lanchester (1916). More recent presentation of these results appeared in the 1946 Operations Evaluation Group Report No. 54, *Methods of Operations Research* by Philip M. Morse and George E. Kimball, which was published commercially by John Wiley and Sons (Morse and Kimball, 1951). In addition, a reprint of the original 1916 Lanchester work, "Mathematics in Warfare," appeared in *The World of Mathematics*, *Volume 4*, prepared by James R. Newman and published by Simon and Schuster in 1956.

The significance of these equations is that they represented possibly the first mathematical analysis of forces in combat, and served as the guiding light (for the U.S.A. and its allies) behind the development, during and after World War II, of all two sided combat models, simulations and other methods of calculating combat losses during a battle.

It appears that M. Osipov developed and published comparable equations in a Tsarist Russian military journal in 1915, perhaps independent of Lanchester's results. A translation of his work into English, prepared by Robert L. Helmbold and Allen S. Rehm, was printed in September 1991 by the U.S. Army Concepts Analysis Agency.

Lanchester's equations present a mathematical discussion of concepts such as the relative strengths of opposing forces in battle, the nature of the weapons, the importance of concentration, and their effects on casualties and the outcome of the battle. His arguments are paraphrased here, preserving much of his original symbolism. The equations deal with "ancient warfare" and "modern warfare."

ANCIENT WARFARE: Lanchester explained that, because of the limited range of weapons in ancient warfare (like swords), the number of troops on one side of a battle (which we call the Blue force) that are actively engaged in hand-to-hand combat on the combat front at any time during the battle must equal approximately the number of troops responding to them on the other side (the Red force). For this reason, one may assume that the rate at which casualties are produced is constant, because the number of troops actively engaged on each side is constant (until very near the end of the battle), and the rate c (> 0) at which Blue combatants become casualties is a product of the fixed number of Red troops engaged and their average individual casualty producing effectiveness (dependent on the average strength of Red's weapons and the effectiveness of the Blue defenses). Similar results apply to k (> 0), the Red casualty rate. The two casualty rates need not be the same, as the weapons and defenses of the two sides may differ.

If $b(t)$ is the number of effective Blue troops at time t after the battle has started and $r(t)$ is the number of effective Red troops, the following equations may be written

$$db/dt = c, \qquad dr/dt = -k \qquad (1)$$

The relationship between the sizes of the two forces may easily be ascertained by observing from (1) that

$$db/dr = c/k, \qquad (2)$$

from which we deduce that

$$k[b(0) - b(t)] = c[r(0) - r(t)] \qquad (3)$$

In the above equations, $b(0)$ and $r(0)$ are assumed to be the initial (positive) sizes of the forces at time 0, the beginning of the battle, and the equations are valid only as long as $b(t)$ and $r(t)$ remain greater than zero. Assuming the combatants battle until all the troops on one side or the other are useless for combat, having become casualties, the battle ends at the earliest time when $b(t)$ or $r(t)$ becomes equal to zero. Thus, solving for r in (3) when b becomes zero (or vice versa), we get

when $b(t) = 0$, $r(t) = [c*r(0) - k*b(0)]/c$,

when $r(t) = 0$, $b(t) = [k*b(0) - c*r(0)]/k$. $\qquad (4)$

Thus we see that if $c*r(0) > k*b(0)$, the Red force wins the battle, while if $k*b(0) > c*b(0)$ the Blue force wins the battle. We can summarize these observations by designating the initial effectiveness of the Blue force to be $k*b(0)$, and that of the Red force $c*r(0)$, and we see that the force with the larger initial effectiveness wins, while equal initial effectiveness ensures a draw.

It is also simple to return to the original differential equations (1) and to solve them to determine the number of effective troops of either force as a linear function of time. This essentially completes Lanchester's modeling of "ancient warfare."

MODERN WARFARE: Lanchester postulated that the major difference between modern and ancient warfare is the ability of modern weapons (such as rifles and, to a lesser degree bows and arrows, cross bows, etc.) to produce casualties at long range. As a result, the troops on one side of an engagement can, in principle, be fired upon by the entire opposing force. Consequently, assuming that all of each of the troops on a side have the same (average) ability to produce casualties at a fixed rate, the combined casualty rate against a given side is proportional to the number of effective troops on the other side.

This leads directly to the following differential equations constituting Lanchester's model of modern warfare:

$$db/dt = -c*r , \qquad dr/dt = -k*b. \qquad (5)$$

As in the ancient warfare case, the individual casualty producing rates, c and k, are assumed to be known constants for the duration of the battle.

We can now combine these two equations (as was done in the ancient warfare case) and obtain

$$db/dr = (c*r)/(k*b). \qquad (6)$$

We solve (6) to obtain the relationship between the numbers of effective forces on the two sides as the battle progresses. This leads to

$$k[b^2(0) - b^2] = c[r^2(0) - r^2]. \qquad (7)$$

Since these equations are valid only when b and r equal or exceed zero, we may observe, as in the ancient warfare case, that (with the battle ending when the losing side has been reduced through casualties to no effective troops) and the victor has a positive number of effective troops, the force with the larger initial effectiveness ($k*b^2(0)$ for Blue and $c*r^2(0)$ for Red) will win the battle while (as in the ancient warfare case) equal initial effectiveness produces a draw. Equation (7) and this paragraph constitute

Lanchester's "Square Law" for his model of modern warfare.

Again, as in the ancient warfare case, it is possible to solve the initial differential equations in (5) to obtain the specific functions that describe the behavior of the side of either force as a function of time. These results also appear in Morse and Kimball (1951), and this essentially completes Lanchester's modeling of "modern warfare."

EXTENSIONS: In presenting his results, Lanchester used many techniques that we tend to take for granted in contemporary OR practice. He formulated clear assumptions about the operation of the system he was studying, derived the mathematical consequences of his assumptions, and discussed how variation of assumptions affected results. Consequently he was able to provide specific numerical insights into characteristics of the system that could be translated into useful ways of improving a system that operated in accordance with the specified assumptions.

It was possible for Lanchester to accomplish his mathematical modeling by using what is often referred to as the First Theorem of Operations Research:

A function of the average equals the average of the function.

The above result applies only in very special circumstances; nevertheless, there are many cases in which use of this theorem allows deterministic results to be derived easily. Such results will usually provide a good approximation of average results occurring in reality. It is through this technique that various chemical formulas or formulas in the physical sciences pertaining to concepts such as temperature, thermodynamics, etc. were derived.

In those formulations, it is assumed that a group of many small objects moving at various speeds with a known average speed will function in the same manner as if all the objects moved at the same (average) speed. Similarly, in his warfare modeling, Lanchester assumed that the casualty producing rate of every one of the troops on one side of a battle was constant and equal to the average (per troop) casualty producing rate of the entire force, and the same is true of the troops on the other side.

The usefulness of Lanchester's work is primarily in its demonstration of the fact that it is possible to draw mathematical and numerical conclusions concerning the occurrence of casualties in certain battles that can be described, *a*

priori, as conforming to certain specified assumptions concerning how the battle is conducted. From such an observation, it is possible to generalize and derive other models that conform to other sets of assumptions, so that a wider range of combat situations can be dealt with. This has led to all sorts of models that can be handled through generalizations of Lanchester's techniques.

The analyst can take into account other factors not specifically covered by Lanchester, such as addition or withdrawal of troops in the course of an engagement. Movement of forces can be considered. Different weapons and defensive techniques can be studied.

Dispersing and hiding the troops on one side of a battle (as in guerrilla warfare) affects the rate at which they can be hit by the other side, which led Lanchester to present another differential equation for such a force. This leads to analyses in which one or the other or both forces engage in ancient, modern or guerrilla warfare. There are nine kinds of battles that an analyst can deal with just by adding the consideration of the possibility of guerrilla warfare to his bag of tricks (Deutschman, 1962).

Clearly there is a great deal of flexibility in deriving models involving the use of deterministic differential equations that predict specific "average" results. The *probabilistic* events that take place during the course of a battle can also be dealt with in comparatively simple cases as demonstrated by B.O. Koopman as described in Morse and Kimball (1951). Regrettably, the mathematics of probabilistic systems is frequently much more difficult than that of deterministic systems, and the need to recognize the existence of all sorts of complications in a battle, frequently leads to rather complicated and abstruse mathematics which can best be handled through the use of computers for the required numerical calculations.

The field of "combat simulation" is recognized as a direct descendant of the Lanchester approach. Of historic interest in this connection is the fact that Lt. Fiske of the U.S. Navy presented, in 1911, a model of warfare consisting of a salvo by salvo table that computed casualties on two sides of a battle. This material was brought to the attention of contemporary analysts by H.K. Weiss (1962).

Engel (1963) showed that the equations of the Fiske model were difference equations which became, in the limit as the time increment between successive salvos approached zero, identical to the Lanchester differential equations of modern warfare. In a sense, this validated the use of discrete time models that approximated combat models for computer calculations, allowing greater confidence on the part of the analyst that no great surprises would result from a use of such discrete time approximations of combat models.

A cautionary note must be sounded at this point. Before using whatever mathematical model the analyst may have derived in discussing any past or future battles, the analyst must be certain that the assumptions of the model on how the battle will be conducted and terminated pertain to the battle being analyzed. The analyst should be able to derive the appropriate values of any parameters (such as $b(0)$, $r(0)$, c and k) to be used in the Lanchester or other models believed to apply in the case under study. Thought experiments do not suffice. The analyst must examine data to determine whether the assumptions provide a valid description of the way the battle proceeds, and to ascertain from relevant combat and experimental data that the model's numerical values for the parameters are appropriate.

VALIDATION OF EQUATIONS: Lanchester did not provide any demonstration of the relevance of his models to any specific historic battles, although he did discuss "examples from history" in which he suggested that the results of certain tactical actions were consistent with results that could be derived from his models. A validation of Lanchester's modern warfare equations was first given by Engel (1954), based on an analysis of the Battle of Iwo Jima during World War II. The analysis showed that the daily casualties inflicted on the U.S. forces over the approximately forty days of the battle were consistent with Lanchester's model for modern warfare. Since that time, additional analyses of combat results and experiments have demonstrated that the values of various parameters can be estimated for use in specified combat situations, and that appropriate combat models can be used in conjunction with those parameter values to obtain results of interest to military planners and decision makers.

The modeling methodology pioneered by Lanchester in the field of combat casualty analysis has served as a most important guide for analysts of military problems. He showed how application of these techniques can be used in developing mathematical models of combat that can be applied in forecasting the results of hypothetical battles. This enables operations research analysts to predict outcomes of these battles, plan tactics and strategy, develop weapons requirements, determine force requirements, and otherwise assist planners and decision makers concerned with the effective use of military forces.

See **Battle modeling; Military operations research; Model validation, verification, and testing; Simulation.**

References

[1] Deitchman, S.J. (1962), "A Lanchester Model of Guerrilla Warfare," *Operations Research*, 10, 818–827.
[2] Engel, J.H. (1963), "Comments on a Paper by H.K. Weiss," *Operations Research*, 11, 147–150.
[3] Engel, J.H. (1954), "A Verification of Lanchester's Law," *Jl. Operations Research Soc. Amer.*, 2, 163–171.
[4] Lanchester, F.W. (1916), *Aircraft in Warfare: The Dawn of the Fourth Arm*, Constable and Company, London.
[5] Morse, P.M. and G.E. Kimball (1951), *Methods of Operations Research*, John Wiley, New York.
[6] Weiss, H.K. (1962), "The Fiske Model of Warfare," *Operations Research*, 10, 569–571.

LAPLACE-STIELTJES TRANSFORM

For any function $G(t)$ defined $t \geq 0$ (like a cumulative probability distribution function), its Laplace-Stieltjes transform (LST) is defined as $\int_0^\infty e^{-st} dG(t)$, $\mathrm{Re}(s) > 0$. When the function $G(t)$ is differentiable, it follows that the LST is equivalent to the regular Laplace transform of the derivative, say $g(t) = dG(t)/dt$.

LAPLACE TRANSFORM

For any continuous function $g(t)$ defined on $t \geq 0$ (like a probability density), its Laplace transform is defined as $\int_0^\infty e^{-st} g(t) dt$, $\mathrm{Re}(s) > 0$.

LARGE-SCALE SYSTEMS

James K. Ho

The University of Illinois at Chicago

In OR/MS, large-scale systems refer to the methodology for the modeling and optimization of problems that, due to their size and information content, challenge the capability of existing solution technology. There is no absolute measure to classify such problems. In any given computing environment, the cost-effectiveness of problem solving generally depends on the dimensions and the volume of data involved. As problems get larger, the cost tends to go up, lowering effectiveness. Even before the physical limits of the hardware or the numerical resolution of the software are exceeded, the effectiveness of the solution environment may have become unacceptable. Efforts to improve on any of the relative performance measures such as solution time, numerical accuracy, memory and other resource requirements, are subjects in the topic of large-scale systems. Since solving larger problems more effectively is also an obvious goal in all specializations of operations research, there are natural linkages and necessary overlaps with most other area in the field.

All known methodology for large-scale systems can be viewed as the design of computational techniques to take advantage of various structural properties exhibited by both the problems *and* known solution algorithms. Broadly speaking, such special properties can be regarded as either *micro-structures* or *macro-structures*. Micro-structures are properties that are independent of permutations in the ordering of the variables and constraints in the problem. An example is sparsity in the constraint coefficients. Macro-structures are those that depend on such orderings. An example is the block structure of loosely coupled or dynamic systems.

USING MICRO-STRUCTURES OF PROBLEMS:

In the modeling of real systems, the larger the problem, the less likely it is for a variable to interact with all the others. If each variable is coupled only to a small subset of the total, the resulting constraints will be *sparse*. Techniques that eliminate the representation of the nonexistent interactions can reduce storage requirement significantly. For example, a linear program with 10,000 variables and 10,000 constraints has potentially 10^8 coefficients. If on the average, each variable appears in 10 constraints, there will be only 10^5 nonzero coefficients, implying a density of 0.1%. Sparse matrix methods from numerical analysis have been used with great success here. Furthermore, the nonzero coefficients may come from an even smaller pool of unique values. This feature is known as *supersparsity* and allows additional economy in data storage. Large, complex models are usually generated systematically by applying the logic of the problem iteratively over myriad parameter sets. This may lead to formulations with redundant variables and constraints. Examples include flow balance equations that produce a redundant constraint when

total input equals total output; lower and upper bounds that are equal imply the variable can be fixed. Methods to simplify the problem by identifying and removing such redundancies are incorporated into the procedure of *preprocessing*. It is not unusual to observe reductions of problem dimensions by 10 to 50 percent with this approach.

USING MICRO-STRUCTURES OF ALGORITHMS:

Algorithms may have steps that are adaptable to advanced computing architecture at the microprocessing level. An example is the vectorization of inner-product calculations in the simplex method. A completely different exploit is the relatively low number of iterations required by interior-point methods. As the number of iterations seems to grow rather slowly with problem size, it is a micro-structure of such algorithms that automatically sheds light on the optimization of large-scale systems. Yet another promising approach that falls under this heading is the use of sampling techniques in stochastic optimization.

USING MACRO-STRUCTURE OF PROBLEMS:

Most large-scale systems are comprised of interacting subsystems. Examples are multidivisional firms with a headquarter coordinating the activities of the semi-autonomous divisions; time-phased models of dynamic systems with linkages only among adjacent time periods; capital investment or financial planning models with each period linked to all subsequent periods. Linear programming modeling of the above examples gives rise to problems with the *block-angular*, *staircase* and *block-triangular* structures, respectively (see Figures 1, 2, and 3). Other variations and combinations are also possible. Two major approaches to take advantage of such structures are *decomposition* and *factorization*. Decomposition relies on algorithms that transform the problem into a sequence of smaller subproblems that can be solved independently. Various schemes are devised to coordinate the subproblems and steer

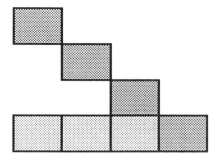

Fig. 1. Block-angular structure.

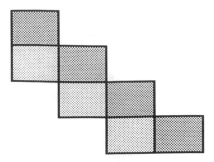

Fig. 2. Staircase structure.

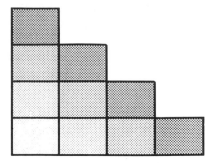

Fig. 3. Block-triangular structure.

them towards the overall solution. Many algorithms are derived from the Dantzig-Wolfe decomposition principle which provides a rigorous framework for this approach. Factorization is the adaptation of existing algorithms to take advantage of the problem structure. In the case of the simplex method, the representation of the basis matrix required at each step can be partitioned into blocks and updated separately. It has been shown that all of the simplex-based techniques proposed over the years under somewhat confusing guises of partitioning and decomposition are indeed special cases of the factorization approach.

USING MACRO-STRUCTURE OF ALGORITHMS:
Both decomposition and factorization algorithms are natural candidates for parallel and distributed computation since they involve the solution of independent subproblems. The latter can be solved concurrently on multiprocessor computers of various architectures. Particularly suitable is the class of Multiple-Instruction-Multiple-Data (MIND) machines that are essentially networks of processors that can execute independent instructions. They represent a cost-effective way to harness tremendous computing power from relatively modest and economical components. One processor can be programmed as the coordinator of the algorithmic procedures. Each of the other processors can be assigned a subproblem and programmed to communicate with the coordinating process. As the gain in overall

efficiency is bounded by the number of processors used, the intent of this approach is to realize the full potential of certain algorithms rather than fundamentally enhancing their performance. However, it is becoming an essential aspect of large-scale systems as multiprocessor computers are expected to be prevalent in the future.

STATE-OF-THE-ART: Linear and mixed integer programming remain the primary focus in the optimization of large-scale systems. New computer architectures with ever-increasing processing power and memory capacities have facilitated the empirical approach to algorithmic development. Experimentation with large-scale problems becomes a viable strategy to identify, test, and fine tune ideas for improvement. This has been especially successful in recent commercial implementations of both the simplex and interior-point methods exploiting mainly the micro-structures of problems and algorithms. Problems with hundreds of thousands of constraints and millions of variables are solvable on workstation–grade computers. Previous experiences with macro techniques in decomposition and factorization did not have the benefits of recent technological advances. The results are either inconclusive or less than promising. Future work, especially in hybrid schemes using advanced hardware, may lead to significant contributions to large-scale nonlinear, integer and stochastic optimization.

See **Combinatorial and integer programming; Dantzig-Wolfe decomposition algorithm; Density; Linear programming; Nonlinear programming; Parallel computing; Sparsity; Supersparsity.**

References

[1] Lasdon, L.S. (1970), *Optimization Theory for Large Systems*, Macmillan, New York.
[2] Dantzig, G.B., M.A.H. Dempster and M.J. Kallio, eds. (1981), "Large-Scale Linear Programming," IIASA CP-81-S1, Laxenburg, Austria.
[3] Ho, J.K. (1987), "Recent advances in the decomposition approach to linear programming," *Mathematical Programming Study* 31, 119–128.
[4] Eckstein, J. (1993), "Large-Scale Parallel Computing, Optimization and Operations Research: A Survey," *ORSA Computer Science Technical Section Newsletter*, 14, Fall.
[5] Nemhauser, G.L. (1994), "The Age of Optimization: Solving Large-Scale Real-World Problems," *Operations Research* 42, 5–13.

LATEST FINISH TIME

The latest time an activity must be completed without delaying the end of a project. It is simply the sum of the latest start time of the activity and its duration. See **Network planning**.

LATEST START TIME

The latest time an activity can start without delaying the end of a project. A delay of an activity beyond the latest start time will delay the entire project completion by a corresponding amount. These times are calculated on the basis of a reverse pass through the network. See **Network planning**.

LATIN SQUARE

See **Combinatorics**.

LCFS

A queueing discipline wherein customers are selected for service in reverse order of their order of their arrival, that is, on a Last-come, First-served basis. See **LIFO**; **Queueing theory**.

LCP

Linear complementarity problem. See **Complementarity problem**; **Quadratic programming**.

LDU MATRIX DECOMPOSITION

For a nonsingular square matrix A, the transformation by Gaussian elimination of A into the form LDU, where L is a lower triangular matrix, D is a diagonal matrix, and U is an upper triangular matrix. It can be written so that the diagonal elements of L and U are 1s and D is the diagonal matrix of pivots. See **LU matrix decomposition**; **Matrices and matrix algebra**.

LEARNING

James R. Buck

The University of Iowa, Iowa City

INTRODUCTION: Learning is a human phenomenon where *performance improves with experience*. There are a number of reasons for task improvement. As tasks are repeated, elements of the task are: better remembered, cues are more clearly detected, skills are sharpened, eye-hand coordinations are more tightly coupled, transitions between successive tasks are smoothed, and relationships between task elements are discovered. Barnes and Amrine (1942), Knowles and Bell (1950), Hancock and Foulke (1966), Snoddy (1926), and Wickens (1992) describe these and other sources of human performance change. All these causes of individual person improvement manifest themselves in faster performance times, fewer errors, less effort, and there is often a better disposition of the person as a result.

Learning is implied by performance changes due primarily to experience. Changes in the methods of performing a task, replacing human activities with machines, imparting information about the job, training, acquiring performance changes with incentive systems, and many other things can cause performance changes other than learning. Thus, detection involves the identification of an improvement trend as a function of more experience. It also involves the elimination of other explanations for this improvement. Analogous to a theory, learning can never be proved; it can only be disproved.

After detecting learning, measurement and prediction follows. These activities involve fitting mathematical models, called *learning curves*, to performance data. First, there is the selection of an appropriate model. Following the selection of a model, there is the matter of fitting the selected model to performance data. In some cases alternative models are fit to available data and the quality of fit is a basis in the choice of a model.

Some of those sources which contribute to an individual person's improvement in performance with experience are similar to the causes of improvement by crews, teams, departments, companies, or even industries with experience. As a result, similar terms and descriptions of performance change are often fit to organizational performance changes. However, the term *progress curves* (Konz, 1990) is more often applied to cases involving: assembly lines, crews, teams, departments, and other smaller groups of people whereas the term *experience curves* is sometimes applied to larger organizational groups such as companies and industries (Hax and Majluf, 1982). A principal distinction between these different types of improvement curves is that between-person activities (e.g., coordination) occur as well as within-person learning. In the case of progress curves, there are improvement effects due to numerous engineering changes. Experience curves also embody scientific and technological improvements, as well as progressive engineering changes and individual-person learning. Regardless of the person, persons, or thing which improves or the causes of improvement, the same learning curve models are frequently applied. Progress and experience curves are really forms of personification.

Learning occurs in a number of important applications. One of these applications is the prediction of direct labor changes in production. Not only is this application important to cost estimation, it is also important in production planning and manning decisions. Another application is the selection of an operational method. If there are alternative methods of performing particular operations which are needed, then one significant criterion in the selection of an appropriate method is learning because the average cost can favor one method over another that has lower initial performance costs. In other cases, one operation can cause bottlenecks in others unless the improvements with experience are sufficient over time. Also, production errors can be shown to decrease with experience as another form of learning and so learning is important in quality engineering and control.

PERFORMANCE CRITERIA AND EXPERIENCE UNITS:

Performance *time* is the most common criterion used for learning curves in industry today. Production cycles are also the most commonly used variable for denoting experience. If t_i is the performance time on the ith cycle, then a learning curve should predict t_i as a function of n cycles. Since learning implies improvement with experience, then one would expect $t_i \leq t_{i+1}$ and $t_i - t_{i+1} \leq 0$ for the average case of $i = 1, 2, \ldots, n$ cycles.

An associated time criterion on the ith cycle is the cumulative average performance time on the ith cycle or A_i. Cumulative average times consist of the sum of all performance times up to and including the nth cycle divided by n. In the first cycle, $A_1 = t_1$. With learning, t_i tends to decrease with i and so does A_i. However, A_i decreases at a slower rate than t_i. This effect can be shown by the first-forward difference of A_i which is:

$$\Delta A_n = A_{n+1} - A_n = \frac{\sum_{i=1}^{n+1} t_i}{n+1} - \frac{\sum_{i=1}^{n} t_i}{n} = \frac{t_{n+1} - A_n}{n+1}. \quad (1)$$

So long as t_{n+1} is less than A_n, then ΔA_n is negative and the cumulative average time continues to decrease. It is also noted in equation

(1) that with sequential values of A_i for $i = 1, 2, \ldots, n$, the values of t_i can be found. On the other hand, A_i can be predicted directly rather than t_i.

Another criterion of interest is *accuracy*. However, it is usually easier to measure errors in production as the complement of accuracy. Thus, the sequence of production errors are $e_1, e_2, \ldots, e_i, \ldots, e_n$ over n serial cycles where e_i is the number of errors found in a product unit as in typing errors per page (Hutchings and Towill, 1975). If the person is doing a single operation on a product unit, then either an error is observed with a unit of production or it is not and observation over a production sequence is a series of zeros and ones. A more understandable practice is to define e_i as the fraction of the possible errors where the observed number of errors is divided by the m possible errors at an operation (Fitts, 1966; Pew, 1969). In this way e_i is 0, some proper fraction, or 1. It also follows that a learning curve could be fit to the series of e_i values over the n observations sequential units of production or to the cumulative average errors. If learning is present, then one would expect to see a general decrease in e_i with increases in $i = 1, 2, \ldots, n$ and also the cumulative average errors would similarly decrease but with a rate lag compared to the serial errors.

Pew (1969) invented the speed-accuracy-operating-characteristic graph which provides simultaneous analyses of correlated criteria. This operating characteristic consists of a bivariate graph where one axis denotes performance time per unit (complement is the speed) and the other axis denotes the number of errors per unit (complement is the accuracy). Simultaneous plots of speeds and accuracies with experience would be expected to show increases in both criteria with more experience. The slope of these plots with increases of experience describes "bias" between these criteria. It should be noted that when the power learning curve model is used for a prediction of learning performance, then logarithmic measurements will linearize the plots.

OTHER LEARNING METRICS:

Most applications of learning description, usually known as learning curves, use the production units as experience units, either as single units or lots. The time required to produce that product unit is the corresponding performance units. An alternative approach to predicting learning effects is to describe cumulative time as the experience unit (i.e., hours or days) and the number of production units produced during that experience unit. Thus, for cumulative production time

$t = 1, 2, 3, \ldots, k, \ldots, m$ and corresponding production of $n_1, n_2, n_3, \ldots, n_k, \ldots, n_m$. Most learning curve models merely relate n_k to k. An alternative model of learning which is not often shown is the discrete exponential model which relates pairs of n_k values as:

$$n_k = an_1 + b \qquad (2)$$

where a and b are parameters. This model was originally proposed by Pegels (1969) for startup cost prediction. Later, Buck, Tanchoco, and Sweet (1976) showed that this model was really a first-order forward-difference equation as described by Goldberg (1961). It follows in this model that:

$$n_k = a^k[n_1 - n^*] + n^* \qquad (3)$$

where $n^* = b/(1-a) > n_1$ and $0 < a < 1$. Since the parameter a is a fraction, the first term of (3) approaches zero with increasing k and so n^* is the asymptote. Accordingly, n_k approaches n^* exponentially with each discrete unit of time. Bevis, Finnicat, and Towill (1970) provided a similar model as:

$$n_k = n^* + [n_1 - n^*]e^{-ck} \qquad (4)$$

where k is a continuous measure to time and c is a parameter. Buck and Chen (1993) used the discrete form in traditional format but they showed that this model can be more difficult to fit to data than the more-common power model but that it can give a more accurate description of human learning.

CONCLUSIONS: Learning provides performance changes with experience which can be described in quantitative terms as a learning curve. Discrete learning curve models for performance time prediction associated with each production unit are shown in **Learning Curves**. There, a more general viewpoint of learning is given, where other criteria can be similarly described and these criteria can be described as a function of time, using either discrete time values or time as a continuous value.

See **Cost analysis; Cost effectiveness analysis; Learning curves.**

References

[1] Barnes, R. and Amrine, H. (1942), "The Effect of Practice on Various Elements Used in Screw-Driver Work," *Jl. Applied Psychology*, 197–209.

[2] Bevis, F.W., Finnicat, C. and Towill, D.R. (1970), "Prediction of Operator Performance During Learning of Repetitive Tasks," *International Jl. Production Research*, 8, 293–305.

[3] Buck, J.R., Tanchoco, J.M.A., and Sweet, A.L. (1976), "Parameter Estimation Methods for Discrete Exponential Learning Curves," *AIIE Transactions*, 8, 184–194.

[4] Buck, J.R. and Cheng, S.W.J. (1993), "Instructions and Feedback Effects on Speed and Accuracy with Different Learning Curve Functions," *IIE Transactions*, 25, 6, 34–47.

[5] Fitts, P.M. (1966), "Cognitive Aspects of Information Processing III: Set for Speed Versus Accuracy," *Jl. Experimental Psychology*, 71, 849–857.

[6] Goldberg, S. (1961), *Introduction to Difference Equations*, John Wiley, New York.

[7] Hancock, W.M. and Foulke, J.A. (1966), "Computation of Learning Curves," *MTM Journal*, XL, 3, 5–7.

[8] Hax, A.C. and Majluf, N.S. (1982), "Competitive Cost Dynamics: The Experience Curve," *Interfaces*, 12, 5, 50–61.

[9] Hutchings, B. and Towill, D.R. (1975), "An Error Analysis of the Time Constraint Learning Curve Model," *International Jl. Production Research*, 13, 105–135.

[10] Knowles, A. and Bell, L. (1950), "Learning Curves Will Tell You Who's Worth Training and Who Isn't," *Factory Management*, June, 114–115.

[11] Konz, S. (1990), *Work Design and Industrial Ergonomics*, 3rd ed., John Wiley, New York.

[12] Pegels, C.C. (1969), "On Startup of Learning Curves: An Expanded View," *AIIE Transactions*, 1, 216–222.

[13] Pew, R.W. (1969), "The Speed-Accuracy Operating Characteristic," *Acta Psychologia*, 30, 16–26.

[14] Snoddy, G.S. (1926), "Learning and Stability," *Jl. Applied Psychology*, 10, 1–36.

[15] Wickens, C.D. (1992), *Engineering Psychology and Human Performance*, 2nd ed., Harper Collins, New York.

LEARNING CURVES

Andrew G. Loerch

U.S. Army Concepts Analysis Agency, Bethesda, Maryland

With experience and training, individuals and organizations learn to perform tasks more efficiently, reducing the time required to produce a unit of output. This simple and intuitive concept is expressed mathematically through the use of the learning curve.

The learning curve was introduced in the literature by Wright (1936) who observed the learning phenomenon through his study of the construction of aircraft prior to World War II. Since then, these models have been used in the areas of work measurement, job design, capacity planning, and cost estimation in many industries. Yelle (1979) summarized 90 articles dealing with learning curves. Dutton, Thomas, and Butler (1984) traced the history of progress functions by examining 300 articles. They note that the terms "learning curve," "progress function," and "experience curve" are often used interchangeably. However, many authors differentiate between them in the following way. Learning curves are used to describe only direct-labor learning, while progress functions also incorporate learning by managerial and technical personnel, as well as improvements due to technological change. The term experience curve is used to describe learning or progress at the industry level. Experience curves often use price as a surrogate measure for progress or learning. In the discussion below, no distinctions are made between these terms.

Dutton *et al.* (1984) also note that learning curves are frequently confused with economies of scale. Although they are observed together in many cases, the two are separate effects with different causes. Progress and learning can occur in the absence of changes in size or scale of operations.

Basic learning curve theory is described below, with emphasis given to the so-called "power model." Other models are then introduced. Finally, issues regarding the estimation of learning curve parameters are presented.

THE POWER MODEL: Also known as the log-linear model, the power model is the most frequently encountered implementation of the various learning curve models. Wright observed that as the quantity of units manufactured doubles, the number of direct labor hours it takes to produce an individual unit decreases at a uniform rate. So, after one doubling of the cumulative production, direct-labor hours may have declined to, say 80% of its previous value. After an additional doubling there is another decline to 80% of that value, or 64% of the original. The learning rate, which is the actual decline per doubling, 80% in the above example, is assumed to be a characteristic of each particular type of manufacturing process.

In this model, learning curves have the following mathematical form:

$$L(y) = Ay^b,$$

where $L(y) =$ the number of hours needed to produce the yth unit, $A =$ the number of hours needed to produce the first unit, $y =$ the cumulative unit number, and $b =$ the learning index, the learning curve parameter, or the learning curve slope parameter. To account for the effect of

doubling, the learning curve index is computed as follows:

$$b = (\log r)/(\log 2),$$

where r is the learning rate. Figure 1 shows graphs of three such curves with different learning rates.

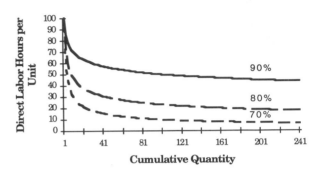

Fig. 1. Learning curves with different rates.

Note that this model is also applicable to cost in addition to direct-labor hours. In a cost application, the parameter A would represent the cost of the first unit produced. The use of learning curve costing is complicated by the problem of accounting for inflation and the change in hourly wages over time. In any event, labor hours can be easily converted into cost.

In the above model, the number of direct-labor hours required to produce the yth unit, or the cost of producing the yth unit is computed. Thus, the model is referred to as the "Unit Formulation," and it is attributed to James Crawford who introduced its use to the Lockheed Corporation in 1944 (Smith, 1989). A related model based on the original work of Wright is the so-called "Cumulative Formulation," where, in the above notation, $L(y)$ would represent the average labor hours or cost of all the units produced through the yth unit. Note that the cumulative formulation tends to smooth the effects of unusually high or low labor hours or costs for individual or groups of units, and it has been found to be more useful for application to batch-type production processes. Although much of the work on learning curves has been directed at specifying the functional relation between unit costs or direct-labor hours and cumulative output, the range of output measures has been expanded to include, for example, industrial accidents per unit output, defects and complaints to quality control per unit output, and service requirements during warranty periods.

VARIATIONS OF THE POWER MODEL: While the log-linear model has been, and still is the most widely used model, several other geometries have

been found to provide better fits in particular sets of circumstances. Some of the more well known models are:
1. Plateau model,
2. Stanford-B model, and
3. S-model.
Figure 2 depicts these models on a logarithmic scale.

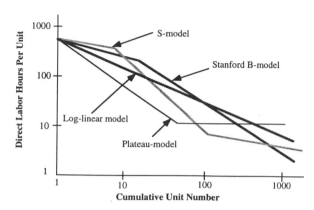

Fig. 2. Log-log plot of learning curve geometries.

The plateau model was first described by Conway and Schultz (1959), and it is used to represent the phenomenon that the learning phase of a process is finite, and is followed by a steady state phase. This model is often associated with machine-intensive manufacturing.

The Stanford-B model, expressed symbolically as

$$L(y) = A(B + y)^b,$$

represents a process that experiences accelerated learning after B units are produced (other notation as previously defined). This model was developed at the Stanford Research Institute and is useful for processes with design changes (Garg and Milliman, 1961).

The S-model, described by Cochran (1960), combines reduced learning at the outset of production, with another slackening of learning later in the production process. This model is usually approximated as a three-segment straight line on a log-log graph, and is sometimes used for heavy labor-intensive industries.

The choice of the appropriate model is usually based on empirical studies of the process in question and historical experience with similar processes. The utilization of these more complex representations involves increased difficulty in parameter estimation, coupled with limited improvement in accuracy. As such, the basic log-linear model continues to find favor among practitioners.

OTHER FACTORS AFFECTING LEARNING: Frequently, other factors affect production that, if ignored, could bias the estimation of the rate of learning. As mentioned, the presence of economies of scale would result in the situation where a more than proportional increase in output would be obtained due to an increase in inputs. If the effects of this variable are not controlled for in the estimation of learning rates, and the scale of the operation is gradually increased over time, the amount of learning would be overestimated. Other such factors that are independent of direct labor learning include increased capital investment, multiple shifts, time lapses between performance of operations, and production rate. Argote and Epple (1990) provides a review of the literature regarding the incorporation of factors that affect learning.

ESTIMATION OF LEARNING CURVE PARAMETERS: Most estimation schemes rely on the logarithmic representation of the learning curve, written as the following:

$$\log L = \log A + b \log y.$$

The learning curve parameters, A and b, are estimated either by plotting historical values on a log-log graph and visually fitting a line, or by computing the least squares regression line through the log-log data. Several computer programs are commercially available to estimate the learning curve parameters.

Frequently, organizations collect historical data for batches or lots, as opposed to discrete units. In order to estimate the parameters in this case, the batch's average labor or cost and the unit whose labor or cost corresponds to that average, the lot midpoint, must be known. The logarithm of this value is then used as the independent variable in the regression with the log of the average unit cost of the lot as the dependent variable. Note that the unit expressed by the batch size divided by two is not the lot midpoint since the learning curve is nonlinear. The actual lot midpoint, Q, is represented as the following:

$$Q = \left[\frac{(y_l - y_f + 1)(1 + b)}{(y_l + .5)^{1+b} - (y_f - .5)^{1+b}} \right]^{-\frac{1}{b}},$$

where y_f = the first unit of the batch, and y_l = the last unit of the batch. Observe that this value cannot be computed without first knowing the learning curve index, b. As such, the approximate algebraic lot midpoint is used. This value is computed as the following:

$$Q = \frac{y_f + y_l + 2\sqrt{y_f y_l}}{4}.$$

The learning curve parameters are estimated first using the approximate value of Q for each lot. The value of b is then used to calculate the actual lot midpoint, and the parameters are estimated again. The steps are repeated until the desired accuracy is obtained.

CONCLUSIONS: Research in the area of learning curves has been extensive and many models have been hypothesized to describe the learning process. Learning curve models have proven to be useful tools in many business and government applications. These include cost estimation, bid preparation and evaluation, labor requirement estimation, establishment of work standards, and financial planning.

See **Cost analysis; Cost effectiveness analysis; Learning.**

References

[1] Argote, L. and D. Epple (1990), "Learning Curves in Manufacturing," *Science*, 247, 920–924.
[2] Cochran, E.B. (1960), "New Concepts of the Learning Curve," *Jl. Industrial Engineering*, July-August.
[3] Conway, R.W. and A. Schultz (1959), "The Manufacturing Progress Function," *Jl. Industrial Engineering*, 10, January-February.
[4] Dutton, J.M., A. Thomas and J.E. Butler (1984), "The History of Progress Functions as a Management Technology," *Business History Review*, 58, 1984.
[5] Garg, A. and P. Milliman (1961), "The Aircraft Progress Curve Modified for Design Changes," *Jl. Industrial Engineering*, 12, 23.
[6] Smith, J. (1989), *Learning Curve for Cost Control*, Industrial Engineering and Management Press, Institute of Industrial Engineers, Norcross, Georgia.
[7] Yelle, L.E. (1979), "The Learning Curve: Historical Review and Comprehensive Survey," *Decision Sciences*, 10, 302–328.
[8] Wright, T.P. (1936), "Factors Affecting the Cost of Airplanes," *Jl. Aeronautical Sciences*, 3, 4, 122–128.

LEAST-SQUARES ANALYSIS

See **Quadratic programming; Regression analysis**.

LEONTIEF MATRIX

See **Input-output analysis**.

LEVEL CROSSING METHODS

Percy H. Brill

University of Windsor, Ontario, Canada

INTRODUCTION: Level crossing methods for obtaining probability distributions in stochastic models such as queues and inventories were originated by Brill (1975, 1976, 1979) and elucidated further in Brill and Posner (1974, 1975, 1977, 1981), and Cohen (1976, 1977). These methods began as an essential part of system point theory and are also known as system point analysis, sample path analysis, level crossing technique, level crossing approach, theory, or analysis in the literature (Brill, 1975). Level crossing methods are very useful rate conservation techniques for stochastic models (Miyazawa, 1994).

MODEL AND STATIONARY DISTRIBUTION: Consider a stochastic process $\{W(t), t \geq 0\}$ where both the parameter set and state space are continuous. The random variable $W(t)$ at time point t may denote the content of a dam with general efflux, the stock on hand in an $\langle s, S \rangle$ or $\langle r, nQ \rangle$ inventory system with stock decay, or the virtual wait or workload in a queue. Assume that upward jumps of $\{W(t)\}$ occur at Poisson rate λ_u and downward jumps at Poisson rate λ_d. Let upward and downward jump magnitudes have cumulative distribution function (CDF) B_u and B_d, respectively. We assume that the model parameters are such that the stationary distribution of $W(t)$ exists as $t \to \infty$. Let G and g denote the stationary CDF and probability density function (PDF), respectively, and our aim is to obtain expressions for g and G in terms of the model parameters by using a level crossing approach.

SAMPLE PATHS: A sample path of the $\{W(t)\}$ process is a right-continuous, real-valued function on the nonnegative reals whose value at time-point t is the realized value of random variable $W(t)$. We denote an arbitrary sample path by the function $X(t)$, $t \geq 0$. The function X has either jump or removable discontinuities on a sequence of strictly increasing time points $\{\tau_n | n = 0, 1, \ldots\}$, where $\tau_0 = 0$ without loss of generality. Typically, the time points $\{\tau_n\}$ may represent input or output epochs in dams, arrival epochs in queues, or demand or replenishment epochs in inventories. We assume that when a sample path is positive valued, it decreases continuously on time segments between jump points, described by $dX(t)/dt = -rX(t)$, $X(t) > 0$, $\tau_n \leq t < \tau_{n+1}$, $n = 0, 1, 2, , \ldots$ wherever the derivative exists, and where $r(x) > 0$ for $x > 0$. Note that for the virtual wait process in queues, $r(x) = 1$ $(x > 0)$ and $r(0) = 0$. In an $\langle s, S \rangle$ continuous review inventory system where the stock on hand decays at constant rate k, we have $r(x) = k$ for all x between the reorder level s and order-up-to-level S.

LEVEL CROSSINGS BY SAMPLE PATHS: Let x denote a fixed state space level and t_0 an arbitrary positive time point. Let t_0 be one of the jump time points $\{\tau_n\}$, $n = 1, 2, \ldots$ and let d_0 and u_0 denote the corresponding downward and upward jump magnitudes, respectively, where at least one of u_0, d_0, is strictly positive. The sample path may downcross level x at $t_0 > 0$ if t_0 is any positive epoch, but it can upcross level x at t_0 only if t_0 is one of the $\{\tau_n\}$.

If a sample path downcrosses level x at t_0 which is not one of the $\{\tau_n\}$, then the downcrossing is a continuous downcrossing, since the sample path is continuous at t_0. If a sample path downcrosses level x at a t_0 which is one of the $\{\tau_n\}$, then the downward jump of magnitude d_0 brings it from above x to level below x. If a sample path upcrosses level x at t_0, then necessarily t_0 is one of the epochs $\{\tau_n\}$, and the upward jump of magnitude u_0 brings it from below x to a level above x.

If both u_0 and d_0 are strictly positive at t_0 which is one of the $\{\tau_n\}$, the model mechanism would determine whether the downward or upward jump is considered to "precede" the other. In inventories without lead time, for example, stock depletions due to demands (downward jumps) "precede" stock replenishments (upward jumps). The jumps are not part of the sample path *per se*, but serve only to construct the path. Note that we may also define level crossings at some time point t_0 by considering the net jump which has magnitude $|u_0 - d_0|$ and upward (downward) direction if $u_0 > d_0$ ($u_0 < d_0$).

LEVEL CROSSINGS AND THE STATIONARY DISTRIBUTION:

Downcrossings. Let $D_t^c(x)$ denote the number of continuous downcrossings of level x and $D_t^j(x)$, the number of jump downcrossings of level x during $(0, t)$, $t > 0$. Then, for $r(x) = 1$, $x > 0$ and $r(0) = 0$, it follows with probability 1 that

$$\lim_{t \to \infty} \frac{D_t^c(x)}{t} = r(x)g(x) \text{ (for all } x\text{),} \quad (1)$$

(Brill, 1974). The following also holds with probability 1:

$$\lim_{t \to \infty} \frac{D_t^j(x)}{t} = \lambda_d \int_{y=x}^{\infty} \bar{B}_d(y - x)\, g_d(y)\, dy \text{ (for all } x\text{),} \quad (2)$$

where g_d is the limiting PDF at embedded downward jump points as $t \to \infty$ and $\bar{B} \equiv 1 - B$.

Both (1) and (2) also hold upon replacing $D_t^c(x)$ and $D_t^j(x)$ by their expectations, denoted by $E[D_t^c(x)]$ and $E[D_t^j(x)]$, respectively, and deleting "with probability 1." For exponentially distributed interarrivals between downward jumps (Poisson downward jumps), we have $g_d \equiv g$, which is the *PASTA* principle.

Upcrossings. Let $U_t^j(x)$ denote the number of jump upcrossings of level x during $(0, t)$. Then, with probability 1,

$$\lim_{t \to \infty} \frac{U_t^j(x)}{t} = \lambda_u \int_{-\infty}^{x} \bar{B}_u(x - y)\, g_u(y)\, dy \text{ (for all } x\text{),} \quad (3)$$

where g_u is the limiting PDF at embedded upward-jump time points as $t \to \infty$ (Brill, 1974). Formula (3) gives an expression for the long-run upcrossing rate of level x by any typical sample path at upward jump points, in terms of an integral of the density g_u. For Poisson upward jumps, we have $g_u \equiv g$, by the PASTA principle.

A CONSERVATION LAW FOR LEVEL CROSSINGS: For each state space level, the following conservation law holds:

long run total downcrossing rate
  = long run total upcrossing rate.

This conservation law, together with (1), (2) and (3), enables us to write an integral equation for the PDF g in which every term has a precise interpretation as a sample-path down or upcrossing rate, namely,

$$r(x)g(x) + \lambda_d \int_{y=x}^{\infty} \bar{B}_d(y - x)g(y)dy$$

$$= \lambda_u \int_{y=-\infty}^{x} \bar{B}_u(x - y)g(y)dy \text{ (for all } x\text{).} \quad (4)$$

In (4), the left-hand side depicts the total sample path long-run downcrossing rate of level x, while the right-hand side depicts the long-run upcrossing rate of the level x. Equation (4) is then solved for g by using standard applied mathematics techniques.

APPLICABILITY: The level crossing technique is applicable to dams with limited capacity, blocked-input rules, various control level policies, etc; to complex variants of M/G/1, M/M/c, G/M/1 queues with reneging, bounded virtual wait, server vacations, various state dependencies, cyclic-service queues; and to a wide class of inventory, production/inventory, counter, risk reserve, and related models.

The same level crossing ideas as in (1), (2) and (3) have been applied to cycles in regenerative processes by Cohen (1976, 1977). Upon combining the regenerative-processes level crossing approach and the embedded level crossing technique of Brill (1976, 1979) with the previously widely known "bubble diagram" method (rate into a state = rate out of that state) for discrete state continuous time Markov chains, we can apply level crossing methods to obtain probability distributions and other characteristics in a broad class of stochastic models.

LEVEL CROSSING ESTIMATION: The principle established in formula (1) motivates the idea of using $D_t^c(x)/tr(x)$ as an estimate for $g(x)$ when t is large. Level crossing estimation (also known as system point estimation) consists of three main steps: (i) simulating a single sample path over a large simulated time t; (ii) enumerating the continuous downcrossings of all state space levels over $(0,t)$; and (iii) computing both point and interval estimates of $\dot{g}$, G and the moments (Brill, 1991).

See **Inventory modeling; Markov chains; PASTA; Queueing theory.**

References

[1] Azoury, K. and Brill, P.H. (1986) "An Application of the System-Point Method to Inventory Models under Continuous Review." *Jl. Applied Probability*, 23, 778–789.

[2] Brill, P.H. (1975) "System Point Theory in Exponential Queues," Ph.D. Dissertation, University of Toronto.

[3] Brill, P.H. (1976) "Embedded Level Crossing Processes in Dams and Queues." WP #76-022, Department of Industrial Engineering, University of Toronto.

[4] Brill, P.H. (1979) "An Embedded Level Crossing Technique for Dams and Queues." *Jl. Applied Probability*, 16, 174–186.

[5] Brill, P.H. (1991) "Estimation of Stationary Distributions in Storage Processes Using Level Crossing Theory." *Proc. Statist. Computing Section*, Amer. Statist. Assn., 172–177.

[6] Brill, P.H. and Posner, M.J.M. (1974) "On the Equilibrium Waiting Time Distribution for a Class of Exponential Queues." WP #74-012, Department of Industrial Engineering, University of Toronto.

[7] Brill, P.H. and Posner, M.J.M. (1975) "Level Crossings in Point Processes Applied to Queues." WP #75-009, Department of Industrial Engineering, University of Toronto.

[8] Brill, P.H. and Posner, M.J.M. (1977) "Level Crossings in Point Processes Applied to Queues: Single Server Case." *Operations Research*, 25, 662–673.

[9] Brill, P.H. and Posner M.J.M. (1981) "The System Point Method in Exponential Queues: A Level Crossing Approach." *Mathematics Operations Research*, 6, 31–49.

[10] Cohen, J.W. (1976) *On Regenerative Processes in Queueing Theory*. Lecture Notes in Economics and Mathematical Systems 121, Springer-Verlag. New York.

[11] Cohen, J.W. (1977) "On Up and Down Crossings." *Jl. Applied Probability*, 14, 405–410.

[12] Miyazawa, M. (1994) "Rate Conservation Laws: A Survey," *QUESTA*, 18, 1–58.

[13] Ross, S. (1985) *Introduction to Probability Models*. 4th edition. Academic Press. Inc.

LEVEL CURVE

Also called isovalue contour, a curve along which the values of a given associated function remain constant. See **Isoquant**.

LEXICOGRAPHIC ORDERING

An ordering of a set of vectors based on the lexico-positive (negative) properties of the vectors. For example, the sequence of vectors $\{\boldsymbol{x}_1, \ldots, \boldsymbol{x}_q\}$ is ordered in a lexicographic sense if $\boldsymbol{x}_i - \boldsymbol{x}_j$ is lexico-positive for $i > j$. Such orderings are similar to dictionary ordering of words and are used to prove finiteness of the simplex algorithm. See **Cycling; Lexico-positive vector**.

LEXICO-POSITIVE (NEGATIVE) VECTOR

A vector $\boldsymbol{x} = (x_1, \ldots, x_n)$ is called lexico-positive (negative) if $\boldsymbol{x} \neq 0$ and the first nonzero term is positive (negative). The vector $\boldsymbol{x}$ is lexico-negative if $-\boldsymbol{x}$ is lexico-positive. A vector $\boldsymbol{x}$ is greater than a vector $\boldsymbol{y}$ in a lexico-positive sense if $\boldsymbol{x} - \boldsymbol{y}$ is lexico-positive. See **Lexicographic ordering**.

LGP

Linear goal programming. See **Goal programming**.

LIBRARIES

Arnold Reisman and Xiaomei Xu

Case Western Reserve University, Cleveland, Ohio

INTRODUCTION: According to *The American Heritage Dictionary of the English Language* (p. 753), a library is "a repository for literary and artistic materials such as books, periodicals, newspapers, pamphlets, and prints kept for reading or reference." This rather classical notion of a library does not recognize the fact that libraries are now a subset of the broader field known as *Information Systems* (IS). Nevertheless, we shall delimit the scope of this article to institutions which can be defined as above, albeit with some leeway.

The history of the application of operations research/management science to libraries is not very distinguished. Contributions in the library field were constrained up to and through the decade of the 1970s by the fact that few operations researchers chose libraries as a field of interest. Moreover, librarians have not sought out operations researchers to help in their problem solving, nor did they offer a particularly fertile environment for doing OR studies (Chen, 1974). On the other hand, since the 1970s, computer science has made significant inroads into the library field by merging with library science to create today's local and extended area computer networks linking users with comprehensive databases.

Early work in OR library applications is given in Bacon and Machol (1958) and Bacon *et al.* (1958). The 1960s recorded a more widespread interest (Leimkuhler and Cox, 1964; Cox, 1964; Morse, 1968; Cook, 1968). A comprehensive review on library operations research was done by Kantor (1979). In that review, Kantor summarized all of the previous review articles. Most noteworthy of these from the OR point of view are the bibliographies by Slamecka (1972) and Kraft and McDonald (1977), and surveys and/or assessments by Bommer (1975), Kraft and McDonald (1976), Leimkuhler (1970, 1972, 1977a, 1977b), Churchman (1972) and Morse (1972).

Literature on utilization of OR in libraries has classified the field in several different ways. Kantor (1979) classified papers and projects into the following groups according to the purpose of the research: *system description*; *modeling the system*; *parameter identification*; *optimization or multi-valuation*; *and application*. Rowley and Rowley (1981) classified the work by the *nature* of the *research* (recurrent problems, on/off decisions, etc.). Based on the type of problems being analyzed, the application areas are *operational or recurrent problems*, such as book storage problems; *strategies or on/off decisions*, such as library location problems; and *control/design problems*, such as loan policy problems (Rowley and Rowley, 1981).

The applications of OR to library management can also be classified according to the type of OR technique used:

(1) *Queueing models*

Given the average book circulation time $(1/\mu)$ and the mean number of persons who borrow the book (λ), the expected circulation rate of that particular book is derived using queueing theory (Morse, 1968).

(2) *Simulation*

With the number of staff, the volumes of various jobs (users' requests, new issues, overdue fees, etc.) and the job processing times specified, simulation is used to estimate the delays, processing times and utilization of each member of staff and the whole facility (Thomas and Robertson, 1975).

(3) *Facility location algorithms*

The library facilities and relocation problems are discussed by Min (1988).

(4) *Mathematical programming*

If there are two types of information services, both of which share the same set of resources (staff time in scanning, indexing, abstracting, etc.), and each of them has a different "unit profit," a linear programming problem is used to find out how many services of each type to produce in order to maximize the total profit (Rowley and Rowley, 1981, 58–64).

(5) *Network flow models*

Given the heights and thicknesses of a given collection of books and the cost of different shelf heights, a network model is developed to determine the optimal number of shelf heights for minimizing shelving costs through finding the shortest path in a directed network (Gupta and Ravindram, 1974).

(6) *Decision theory*

A decision regarding whether or not to install a library security system is addressed given the installation cost and the probabilities of success and failure (Rowley and Rowley, 1981).

(7) *Search theory*

Patterns of browsing in libraries are addressed in Morse (1970).

(8) *Transportation models*

A routing problem is explored for a vehicle delivering materials to branches (Heinritz and Hsiao, 1969; McClure, 1977).

(9) *Inventory control theory*

An EOQ model is used to determine the optimal order quantity for the stock of a certain library supply (Rowley and Rowley, 1981, 111–116).

(10) *Probability and statistics*

Library book circulation and individual book popularities are considered as probabilistic processes by Gelman and Sichel (1987) who demonstrated the superiority of beta over the negative binomial distribution.

(11) *Benefit cost analysis*

Library planning is addressed by Leimkuhler and Cooper (1971).

Each of these categories could be, in turn, further characterized by whether or not the research work was *grounded*, for example, based on "real world" library systems involving real data and/or bonafide librarians in the study, as opposed to models which were basically what might be called *logico/deductive*. A more thorough discussion of the problem area can be found in Reisman and Xu (1994), where Table I, page 37, provides a taxonomic review of the vast bulk of the literature in the field.

As can be seen from the above delineation and the referenced table, the utilization of OR in libraries is far from achieving its full potential. Except for simulation, probability and statistics based applications, the bulk of the literature is not well grounded in real life settings. The literature reflects the gap between the complex mathematical models in OR and the not very quantitatively educated library workers (Stueart and Moran, 1987). To enhance the application of OR in libraries, Bommer (1975) suggested a closer working relationship between operations researchers and library managers.

References

[1] Bacon, F.R. Jr. and Machol, R.E. (1958), "Feasibility Analysis and Use of Remote Access to Library Card Catalogs," paper presented at the Fall meeting of ORSA (unpublished).

[2] Bacon, F.R. Jr., Churchill, N.C., Lucas, C.J., Maxfield, D.K., Orwant, C.J. and Wilson, R.C. (1958), "Applications of a Teller Reference System to Divisional Library Card Catalogues: A Feasibility Analysis," Engineering Research Institute, University of Michigan, Ann Arbor, Michigan.

[3] Bommer, M. (1975), "Operations Research in Libraries: A Critical Assessment," *Jl. American Society for Information Science*, 26, 137–139.

[4] Chen, Ching-chih (1974), "Applications of Operations Research Models to Libraries: A Case Study of the Use of Monographs in the Francis A. Countway Library of Medicine, Harvard University," unpublished Ph.D. dissertation, Case Western Reserve University, School of Library Science, Cleveland, Ohio.

[5] Churchman, C.W. (1972), "Operations Research Prospects for Libraries: The Realities and Ideals," *Library Quarterly*, 42, 6–14.

[6] Cook, J.J. (1968), "Increased Seating in the Undergraduate Library: A Study in Effective Space Utilization," in *Case Studies in Systems Analysis in a University Library*, B.R. Burkhalter (ed.), Scarecrow Press, 142–170.

[7] Cox, J.G. (1964), "Optimal Storage of Library Material," unpublished Ph.D. dissertation, Purdue University Libraries, Lafayette, Indiana.

[8] Gelman, E. and Sichel, H.S. (1987), "Library Book Circulation and the Beta-Binomial Distribution," *Jl. American Society for Information Sciences*, 38, 4–12.

[9] Gupta, S.M. and Ravindram, A. (1974), "Optimal Storage of Books by Size: An Operations Research Approach," *Jl. American Society for Information Science*, 25, 354–357.

[10] Heinritz, F.J. and Hsiao, J.C. (1969), "Optimum Distribution of Centrally Processed Material," *Library Resources and Technical Services*, 13, 206–208.

[11] Kantor, P. (1979), "Review of Library Operations Research," *Library Research*, 1, 295–345.

[12] Kraft, D.H. and McDonald, D.D. (1976), "Library Operations Research: Its Past and Our Future," in *The Information Age*, D.P. Hammer (ed.), Scarecrow Press, Metuchen, New Jersey, 122–144.

[13] Kraft, D.H. and McDonald, D.D. (1977), "Library Operations Research: A Bibliography and Commentary of the Literature," *Information, Reports and Bibliographies*, 6, 2–10.

[14] Leimkuhler, F.F. (1970), "Library Operations Research: An Engineering Approach to Information Problems," *Engineering Education*, 60, 363–365.

[15] Leimkuhler, F.F. (1972), "Library Operations Research: A Process of Discovery and Justification," *Library Quarterly*, 42, 84–96.

[16] Leimkuhler, F.F. (1977a), "Operational Analysis of Library Systems," *Information Processing and Management*, 13, 79–93.

[17] Leimkuhler, F.F. (1977b), "Operations Research and Systems Analysis," in *Evaluation and Scientific Management of Libraries and Information Centres*, F.W. Lancaster and C.W. Cleverdon (eds.), Nordhoff, Leyden, The Netherlands, 131–163.

[18] Leimkuhler, F.F. and Cooper, M.D. (1971), "Analytical Models for Library Planning," *Jl. American Society for Information Science*, 22, 390–398.

[19] McClure, C.R. (1977), "Linear Programming and Library Delivery Systems," *Library Resources and Technical Services*, 21, 333–344.

[20] Min, H. (1988), "The Dynamic Expansion and Relocation of Capacitated Public Facilities: A Multi-Objective Approach," *Computers and Operations Research (UK)*, 15, 243–252.

[21] Morse, P.M. (1972), "Measures of Library Effectiveness," *Library Quarterly*, 42, 15–30.

[22] Morse, P.M. (1970), "Search Theory and Browsing," *Library Quarterly*, 40, 391–408.

[23] Morse, Philip M. (1968), *Library Effectiveness: A Systems Approach*, MIT Press, Cambridge, Massachusetts.

[24] Rowley, J.E. and Rowley, P.J. (1981), *Operations Research: A Tool for Library Management*, American Library Association, Chicago, 3–4.

[25] Slamecka, V. (1972), "A Selective Bibliography on Library Operations Research," *Library Quarterly*, 42, 152–158.

[26] Stueart, R.D. and Moran, B.B. (1987), *Library Management*, 3rd ed., Libraries Unlimited, Inc., Littleton, Colorado, 200–202.

[27] Thomas, P.A. and Robertson, S.E. (1975), "A Computer Simulation Model of Library Operations," *Journal of Documentation*, 31, 1–16.

LIFO

A queueing discipline in which customers are selected for service in reverse order of their order of their arrival, that is, on a Last-in, First-out basis. See **LCFS**; **Queueing theory**.

LIKELIHOOD RATIO

See **Score function method**.

LIMITING DISTRIBUTION

Let $p_{ij}(t)$ be the probability that a stochastic process takes on value j at "time" t (discrete or continuous), given that it began at time 0 from state i. If $p_{ij}(t)$ approaches a limit p_j independent of i at $t \to \infty$ for all j, the set $\{p_j\}$ is called the limiting or steady-state distribution of the process.

LINDLEY'S EQUATION

An integral equation for the steady-state waiting-time distribution in the general single-server G/G/1 queueing system. If $W_q(x)$, $x \geq 0$, is the steady-state distribution function of the delay or waiting time in the queue, then, for $x \geq 0$,

$$W_q(x) = \int\limits_{-\infty}^{x} W_q(x - y) dU(y)$$

with $W_q(x) = 0$ for $x < 0$, where the function $U(y)$ is the distribution function of the random variable defined as the service time minus the inter-arrival time. See **Queueing theory**.

LINE

A line is the set of points $\{x | x = (1 - \lambda)x_1 + \lambda x_2\}$, where x_1 and x_2 are points in n-dimensional space and λ is a real number. The line passes through the points x_1 and x_2, $x_1 \neq x_2$.

LINEAR COMBINATION

For a set of vectors $(x_1, \ldots, x_q)$, a linear combination is another vector $y = \Sigma_j \alpha_j x_j$, where the scalar coefficients α_j can take on any values.

LINEAR EQUATION

The mathematical form $a_1x_1 + a_2x_2 + .. + a_nx_n = b$ is a linear equation, where the a_j and b can take on any values. See **Hyperplane**.

LINEAR FRACTIONAL-PROGRAMMING PROBLEM

The linear-fractional programming problem is one in which the objective to be maximized is of the form $f(x) = (cx + \alpha)/(dx + \beta)$ subject to $Ax \leq b$, $x \geq 0$, where α and β are scalars, c and d are row vectors of given numbers, and b is the right-hand-side vector. The problem can be converted to an equivalent linear programming problem by the translation $y = x/(dx + \beta)$, provided that $dx + \beta$ does not change sign in the feasible region. See **Fractional programming**.

LINEAR FUNCTIONAL

A linear functional $f(x)$ is a real-valued function defined on an n-dimensional vector space such that, for every vector $x = \alpha u + \beta v$, $f(x) = f(\alpha u + \beta v) = \alpha f(u) + \beta f(v)$ for all n-dimensional vectors u and v and all scalars α and β.

LINEAR INEQUALITY

The mathematical form $a_1x_1 + a_2x_2 + .. + a_nx_n \leq b$ or $a_1x_1 + a_2x_2 + .. + a_nx_n \geq b$ is a linear inequality, where the numbers a_j and b can take on any values. The set of vectors $x = (x_1, \ldots, x_n)$ that satisfy the inequality form a solution half space. See **Hyperplane**.

LINEAR PROGRAMMING

Frederick S. Hillier

Stanford University, Stanford, California

Linear programming is one of the most widely used techniques of operations research and management science. Its name means that planning (*programming*) is being done with a mathematical model (called a *linear programming model*) where all the functions in the model are *linear* functions.

LINEAR PROGRAMMING MODELS: Linear programming models come in a variety of forms. To illustrate one common form, consider the problem of determining the most profitable mix of products for a manufacturer. Let n be the number of possible products. For each product j ($j = 1, 2, \ldots, n$), a *decision variable* x_j is introduced to represent the decision on its production rate (≥ 0). Let c_j be the profit per unit of product j produced, and let Z be the total rate of profit resulting from the choice of product mix. This choice is constrained by the limited capacities of the production facilities available for these products. Let m be the number of different types of facilities needed. For each type i ($i = 1, 2, \ldots, m$), let b_i be the amount of capacity available per unit time and let a_{ij} be the amount of capacity used by each unit produced of product j ($j = 1, 2, \ldots, n$). The resulting linear programming model then is to choose x_1, $x_2, \ldots, x_n$ so as to

Maximize $Z = c_1 x_1 + c_2 x_2 + \cdots + c_n x_n,$

subject to

$$a_{11}x_1 + a_{12}x_2 + \cdots + a_{1n}x_n \leq b_1$$
$$a_{21}x_1 + a_{22}x_2 + \cdots + a_{2n}x_n \leq b_2$$
$$\vdots$$
$$a_{m1}x_1 + a_{m2}x_2 + \cdots + a_{mn}x_n \leq b_m$$

and

$$x_1 \geq 0, \; x_2 \geq 0, \cdots, x_n \geq 0.$$

The linear function being maximized in this model is called the *objective function*. The m inequalities with a linear function on the left-hand side are referred to as *functional constraints* (or *structural constraints*), and the inequalities in the bottom row are *nonnegativity constraints*. The constants (the c_j, b_i, and a_{ij}) are the *parameters* of the model. Any choice of values of $(x_1, x_2, \ldots, x_n)$ is called a *solution*, whereas a solution satisfying all the constraints is a *feasible solution*, and a feasible solution that maximizes the objective function is an *optimal solution*.

Many other applications of linear programming having nothing to do with product mix also fit this same form for the model. In these cases, *activities* of some other kind replace the production of products and *resources* of some other kind replace production facilities. For each activity j ($j = 1, 2, \ldots, m$), the decision variable x_j represents the decision on the level of that activity. The problem then is to allocate these limited resources to these interrelated activities so as to obtain the best mix of activities (i.e., an optimal solution) according to the *overall measure of performance* adopted for the objective function.

Another common form for a linear programming model is to *minimize* the objective function, subject to functional constraints with $\geq$ signs and nonnegativity constraints. A typical interpretation then is that the objective function represents the total *cost* for the chosen mix of activities and the functional constraints involve different kinds of *benefits*. In particular, the function on the left-hand side of each functional constraint gives the level of a particular kind of benefit that is obtained from the mix of activities, and the constant on the right-hand represents the minimum acceptable level for that benefit. The problem then is to determine the mix of activities that gives the best tradeoff between cost and benefits according to the model.

Still other linear programming models have an equality instead of inequality sign in some or all of the functional constraints. Such constraints represent fixed requirements for the value of the function on the left-hand side.

It is fairly common for large linear programming models to include a mixture of functional constraints – some with $\leq$ signs, some with $\geq$ signs, and some with $=$ signs. Nonnegativity constraints always have a $\geq$ sign, but it occasionally is appropriate to delete this kind of constraint for some or all of the decision variables.

Successful applications of linear programming sometimes use *very* large models. A model with a few hundred functional constraints and a few hundred decision variables is considered to be of moderate size. Having a few thousand functional constraints and even more decision variables is not considered unusually large.

With large models, it is inevitable that mistakes and faulty decisions will be made initially in formulating the model and inputting it into the computer. Therefore, a thorough process of testing and refining the model (*model validation*) is needed. The usual end-product is not a single static model, but rather a long series of variations on a basic model to examine different scenarios as part of post-optimality analysis. A sophisticated software package or algebraic modeling language may be needed to assist in *model management* (inputting or modifying blocks of constraints, solution analysis, report writing, etc.).

SOME APPLICATIONS OF LINEAR PROGRAMMING:

The applications of linear programming have been remarkably diverse. They all involve determining the best mix of activities, where the decision variables represent the levels of the respective activities, but these activities arise in a wide variety of contexts. In the context of *financial planning*, the activities might be investing in individual stocks and bonds (portfolio selection), or undertaking capital projects (capital budgeting), or drawing on sources for generating working capital (financial-mix strategy). In the context of *marketing analysis*, the activities might be using individual types of advertising media, or performing marketing research in segments of the market. In the context of *production planning*, applications range widely from the product-mix problem to the blending problem (determining the best mix of ingredients for various individual final products), and from production scheduling to personnel scheduling.

In addition to manufacturing, these kinds of production planning applications also arise in agricultural planning, health-care management, the planning of military operations, policy development for the use of natural resources, etc.

One particularly important special type of linear programming problem is the *transportation problem*. A typical application of the transportation problem is to determine how a corporation should distribute a product from its various factories to various distributors. In particular, given the amount of the product produced at each factory and the amount needed by each distributor, one can determine how much to ship from each factory to each distributor in order to minimize total shipping cost. Other applications extend to areas such as production scheduling.

The transportation problem is a special case of another key type of linear programming problem, called the *minimum cost network flow problem*, which involves determining how to distribute goods through a distribution network at a minimum total cost. One example of an application of this type is the Supply, Distribution, and Marketing Modeling System developed by the Citgo Petroleum Corporation in the mid-1980s (Klingman *et al.*, 1986). This modeling system deals with the distribution of petroleum products through a distribution network consisting of pipelines, tankers, barges, and hundreds of terminals. This application of linear programming is credited with saving the company well over $15 million annually. Another application involving its refinery operations was implemented at about the same time, and achieved additional savings of about $50 million per year. In a similar manner, linear programming has made a major contribution to improving the efficiency of numerous companies and organizations around the world.

Another important kind of application of linear programming arises from its close relationship to several other important areas of operations research and management science, including *integer programming*, *nonlinear programming*, and *game theory*. Linear programming often is useful to help solve problems in these other areas as well.

SOLVING LINEAR PROGRAMMING MODELS:

Two crucial events have been primarily responsible for the great impact of linear programming since its emergence in the middle of the 20th century. One was the invention in 1947 by George Dantzig of a remarkably efficient algorithm, called the *simplex method*, for finding an optimal solution for a linear programming model. The second crucial event was the *computer revolution* that makes it possible for the simplex method to solve huge problems.

The simplex method exploits some basic properties of optimal solutions for linear programming models. An optimal solution is a feasible solution that has the most favorable value of the objective function. Because all the functions in the model are linear functions, the set of feasible solutions (called the *feasible region*) is a convex polyhedral set. The *vertices* (*extreme points*) of the feasible region play a special role in finding an optimal solution. A model will have an optimal solution if it has any feasible solutions (all the constraints can be satisfied simultaneously) and the constraints prevent improving the value of the objective function indefinitely. Any such model must have either exactly one optimal solution or an infinite number of them. In the former case, the one optimal solution must be a vertex of the feasible region. In the latter case, at least two vertices must be optimal solutions, and then all convex linear combinations of these vertices also are optimal. Therefore, it is sufficient to find the one or more vertices with the most favorable value of the objective function in order to identify all optimal solutions.

Based on these facts, the simplex method is an iterative algorithm that only examines vertices of the feasible region. At each iteration, it uses algebraic procedures to move along an outside edge of the feasible region from the current vertex to an "adjacent" vertex that is better. The algorithm terminates (except perhaps for checking ties) when a vertex is reached that has no better adjacent vertices, because the convexity of the feasible region then implies that this vertex is optimal.

The simplex method is an *exponential-time algorithm*. However, it consistently has proven to be very efficient in practice. Running time tends to grow approximately with the cube of the number of functional constraints, and less than linearly with the number of variables. Problems with even a few thousand functional constraints and a larger number of decision variables are routinely solved. One key to its efficiency on such large problems is that the path followed generally passes through only a tiny fraction of all vertices before reaching an optimal solution. The number of iterations (vertices traversed) generally is of the same order of magnitude as the number of functional constraints.

In 1984, Karmarkar created great excitement in the operations research/management science community by announcing a new *polynomial-time algorithm* for linear programming, along with claims of being many times faster than the simplex method. (Actually, the first polynomial-time algorithm for linear programming had been announced by Khachiyan in 1979, but this "ellipsoid method" proved to be not nearly competitive with the simplex method in practice.) Karmarkar's algorithm moves through the *interior* of the feasible region until it converges to an optimal solution, and so is referred to as an *interior-point method*. The announcement did not include details needed for computer implementation (Karmarkar, 1984).

Since 1984, there has been an ongoing flurry of research activity to fully develop and refine similar *interior-point methods*, along with sophisticated computer implementations. A key feature of this approach is that both the number of iterations (trial solutions) and total running time tend to grow very slowly (even more slowly than for the simplex method) as the problem size is increased. The crossover point above which the best implementations of this interior-point approach tend to become competitive to and faster than the simplex method is where the number of functional constraints plus the number of decision variables is of the order of ten thousand (with considerable variability depending on problem type). Interior-point methods tend to be more efficient for very large problems whose matrices are very sparse, that is, have few nonzero elements (Lustig *et al.*, 1994; Bixby, 1994).

Software packages for the simplex method and its extensions now are widely available for both mainframe and personal computers. Packages for interior-point methods gradually are becoming available as well.

DUALITY THEORY AND POST-OPTIMALITY ANALYSIS: Associated with any linear programming problem is another linear programming problem called the *dual*. Furthermore, the relationship between the original problem (called the *primal*) and its dual is a symmetric one, so that the dual of the dual is the primal. For example, consider the two related linear programming models shown below in matrix notation (where A is a matrix, c and y are row vectors, b, x and the null vector 0 are column vectors, all with compatible dimensions, and x and y are the decision vectors).

Maximize cx	Minimize yb
subject to $Ax \leq b$	subject to $yA \geq c$
and $x \geq 0$.	and $y \geq 0$.

For each of these problems, its dual is the other problem.

There are many useful relationships between the primal and dual problems, so the dual provides considerable information for analyzing the primal. This is especially helpful when conducting *post-optimality analysis*, that is, analysis done after finding an optimal solution for the initial validated version of the model. A key part of most linear programming studies, this analysis addresses a variety of "what-if questions" of interest to the decision makers. The purpose is to explore various scenarios about future conditions that may deviate from the initial model.

Although the parameters of the model are treated as constants, they frequently represent just "best estimates" of a quantity whose true value may turn out to be quite different. A key part of post-optimality analysis is *sensitivity analysis*, which involves investigating the parameters, determining which ones are *sensitive parameters* (those that change the optimal solution if a small change is made in the value of the parameter), and exploring the implications. For certain parameters, the decision makers may have some control over its value (e.g., the amount of a resource to be made available), in which case sensitivity analysis guides the decision on which value to choose. An extension of sensitivity analysis called *parametric programming* enables systematic investigation of simultaneous changes in various parameters over ranges of values.

Extensions of the simplex method are well suited for performing these kinds of post-optimality analysis. However, efforts to extend interior-point methods to perform post-optimality analysis successfully have been limited. Therefore, even when an interior-point method is used to find an optimal solution, a switch to the simplex method is necessary for determining extreme point solutions and for subsequent analysis.

When there is substantial uncertainty about what the true values of the parameters will turn out to be, it may be necessary to use a different approach, called *linear programming under uncertainty* or *stochastic programming*, that explicitly treats some or all of the parameters as random variables. This is especially pertinent when planning must be done for multiple time periods into an uncertain future. Linear programming under uncertainty is an active area of research (Infanger, 1993).

FURTHER READING: Dantzig (1982) described some of the early history of linear programming. Gass (1985) gave an entertaining introduction to the field. Hillier and Lieberman (1994) expanded on all the topics mentioned here. Dantzig (1963) and Schriver (1986) provided advanced books on the theory of linear programming. Marsten *et al.* (1990) discussed the basic concepts underlying interior-point methods. The case studies at Citgo Petroleum Corporation mentioned above were presented by Klingman *et al.* (1986, 1987). Salkin and Saha (1975) provided a collection of papers describing a variety of applications of linear programming.

See **Algebraic modeling languages for optimization; Computational complexity; Density; Duality theorem; Game theory; Hierarchical production planning; Integer programming; Interior-point methods; Mathematical model; Network analysis; Nonlinear programming; Parametric programming; Sensitivity analysis; Transportation problem.**

References

[1] Bixby, R.E. (1994). "Progress in Linear Programming," *ORSA Jl. Computing* **6**, 15–22.

[2] Dantzig, G.B. (1982). "Reminiscences About the Origins of Linear Programming," *Opns. Res. Letters* **1**, 43–48.

[3] Dantzig, G.B. (1963). *Linear Programming and Extensions*, Princeton University Press, Princeton, New Jersey.

[4] Gass, S.I. (1985). *Decision Making, Models and Algorithms*, John Wiley, New York.

[5] Hillier, F.S., and G.J. Lieberman (1994). *Introduction to Mathematical Programming*, 2nd ed., McGraw-Hill, New York.

[6] Infanger, G. (1993). *Planning Under Uncertainty: Solving Large-Scale Linear Programs*, Boyd and Fraser publishing, Danvers, Massachusetts.

[7] Karmarkar, N. (1984). "A New Polynomial-Time Algorithm for Linear Programming," *Combinatorica*, **4**, 373–395.

[8] Klingman, D., N. Phillips, D. Steiger, R. Wirth, and W. Young (1986). "The Challenges and Success Factors in Implementing an Integrated Products Planning System for Citgo," *Interfaces* **16**(3), 1–19.

[9] Klingman, D., N. Phillips, D. Steiger, and W. Young (1987). "The Successful Deployment of Management Science Throughout Citgo Petroleum Corporation," *Interfaces* **17**(1), 4–25.

[10] Lustig, I., R.E. Marsten, and D. Shanno (1994). "Interior Point Methods for Linear Programming: Computational State of the Art," *ORSA Jl. Computing* **6**, 1–14.

[11] Marsten, R., R. Subramanian, M. Saltzman, I. Lustig, and D. Shanno (1990). "Interior Point Methods for Linear Programming: Just Call Newton, Lagrange, and Fiacco and McCormick!" *Interfaces* **20**(4), 105–116.

[12] Salkin, H.M., and J. Saha, eds. (1975). *Studies in Linear Programming*, North-Holland, Amsterdam.

[13] Schrijver, A. (1986). *Theory of Linear and Integer Programming*, John Wiley, New York.

LINE SEGMENT

The straight line joining any two points in n-dimensional real space is a line segment. More specifically, if x_1 and x_2 are the two points, then the set of points $\{x \mid x = (1 - \lambda)x_1 + \lambda x_2, 0 \leq \lambda \leq 1\}$ is the line segment joining x_1 and x_2. See **Line**.

LIPSCHITZ

A function $f(x)$ is said to be Lipschitz, or Lipschitz continuous, if for every pair of points x_1, x_2 and for some $0 < K < 1$, we have $\|f(x_1) - f(x_2)\| \leq K\|x_1 - x_2\|$.

LITTLE'S LAW

Susan Albin

Rutgers University, Piscataway, New Jersey

Little's Law, among the most fundamental and useful formulas in queueing theory, relates the number of customers in a queueing system to the waiting time of customers for a system in steady state:

$$L = \lambda W$$

where

> L = the average number of customers in the system including customers in service;
> λ = the average arrival rate of customers to the system; and
> W = the average time a customer spends in the system including the time in service.

An alternate form of Little's Law addresses only the customers in the waiting line, or queue; that is,

$$L_q = \lambda W_q$$

where

> L_q = the average number of customers in the queue (excluding customers in service);
> λ = the average arrival rate of customers to the queueing system; and
> W_q = the average time that a customer spends in the queue (excluding the time in service).

Little's Law, formally proven in Little (1961) and simplified in Stidham (1974), is remarkably general requiring only that the queue is ergodic. The result holds for any arrival process, service time distribution, and number of servers. It holds for all queue disciplines; the customers need not be served in the order of arrival. The result holds for a specific class of customers that are distinguished from others by priority or some other characteristic.

Little's Law holds for every infinite sample path realization of the queueing system and it holds approximately in finite intervals with the accuracy increasing as the interval increases.

In the study of queues, whether by mathematical analysis, simulation or direct data collection, it is often simpler to find either the average number in system or the average waiting time. Once the simpler one has been found, Little's Law gives the other. For example, in an operating manufacturing system, if average time in the system (lead time) is simpler to estimate from data, Little's Law can be used to estimate the average number of parts in the system (in-process inventory).

An outline of a proof of Little's Law is based on the following figure depicting a sample path of the number in the system over an interval of time T for a steady-state queueing system with arrival rate λ. The number of customer-minutes spent in the system equals A, the area under the

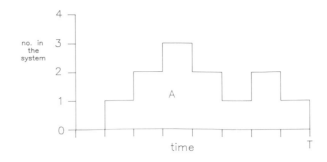

Fig. 1. Sample path realization of the number in the system over time.

curve. The average number of customers that arrive in the interval is λT (approximately); thus the average number of minutes in the system per customer is $W = A/\lambda T$. The average number of customers in the system $L = A/T$. Manipulating the two equations, taking limits, and accounting for end effects yields Little's Law.

See **Queueing theory.**

References

[1] Little, J.D.C. (1961), "A Proof for the Queueing Formula: $L = \lambda W$," *Operations Research*, 9, 383–387.

[2] Stidham, S., Jr. (1974), "A Last Word on $L = \lambda W$," *Operations Research*, 22, 417–421.

LOCAL BALANCE EQUATIONS

See **Detailed balance equations; Queueing theory.**

LOCAL IMPROVEMENT HEURISTIC

A heuristic rule which examines all the solutions that are closely related to a given initial solution and is guaranteed to reach at least a local optimum. See **Heuristic procedure; Local optimum.**

LOCAL MAXIMUM

A function $f(x)$ defined over a set of points S is said to have a local maximum at a point x_0 in S if $f(x_0) \geq f(x)$ for all x in a neighborhood of x_0 in S. The point x_0 is referred to as a local optimum (maximum). See **Global maximum; Nonlinear programming; Quadratic programming.**

LOCAL MINIMUM

A function $f(x)$ defined over a set of points S is said to have a local minimum at a point x_0 in S if $f(x_0) \leq f(x)$ for all x in a neighborhood of x_0 in S. The point x_0 is referred to as a local optimum (minimum). See **Global maximum**; **Nonlinear programming**; **Quadratic programming**.

LOCAL OPTIMUM

See **Local maximum**; **Local minimum**.

LOCAL SOLUTION

A best solution in a feasible neighborhood.

LOCATION ANALYSIS

Charles ReVelle

The Johns Hopkins University, Baltimore, Maryland

The term *location analysis* refers to the development of formulations and algorithms/methodologies to site facilities of diverse kinds in a spatial or geographic environment. The facilities may be sited with relation to demand points, supply points, or with respect to one another. Although facility layout falls within this definition, this topic is not generally considered under the rubric of location analysis. Common descriptive terms for location analysis are *deployment*, *positioning*, and *siting*, although these terms are actually the outcome that follows the execution of a formulation or algorithm.

Location settings may be classified into two broad categories: planar problems and network problems. Planar problems typically assume that the distances between facilities and demand points, supply points or other facilities are given by a metric, a formula that calculates distance between points based on their coordinates in space. Network problems, in contrast, assume that travel can only occur on an underlying network and that distances are the shortest distances between the particular points on the network. Shortest path algorithms must be applied to calculate those distances. A further distinction between these categories is provided by the assumption in most planar problems of an infinite solution space, that is, that facilities can be sited anywhere on the plane, perhaps subject to exclusion areas or regions. These problems are

most often nonlinear optimization problems and more abstract in their application than network-based problems. In contrast to the infinite solution space assumed by most planar problems, all but a few network problems restrict facilities to sites that have been pre-specified as eligible to house those facilities. The network problems tend to be linear zero-one optimization problems and so pose challenges in their resolution to integers. First, we will discuss planar problems and approaches to them; then we will discuss network location formulations and their solution.

PLANAR LOCATION PROBLEMS: The most famous of the planar problems and the first location problem to be posed historically is the Weber problem, named after the earliest modern investigator to pose the problem and propose a solution method for it. The Weber problem considers points dispersed on the plane which send items to or receive finished product from some central factory or facility. The problem seeks the central point which minimizes the sum of weights times distances over all dispersed points. The problem assumes that the Euclidean distances separate the dispersed points and the central point, that the central point can be anywhere on the plane, and that a weight or loading is associated with each of the dispersed points. An iterative solution method that can be shown to converge to an optimal solution was offered in the 1930s, lost to view, and rediscovered in the early 1960s by several independent investigators. In the multi-Weber problem, a number of central facilities are to be sited, each one associated with a cluster or partition of the dispersed points. Only in the early 1990s has this problem yielded to exact methods. Rosing (1992) described both an exact methodology and the history of these problems.

While the Weber problem in its single and multi-facility forms utilizes the Euclidean metric for distances, the minisum rectilinear problem utilizes the Manhattan or rectilinear metric for distances and minimizes the sum of weights times these distances to the central point. The rectilinear distance between two points is the sum of horizontal and vertical separation of the points. Because the problem can be reduced to the choice among a set of eligible points, the multi-facility rectilinear minisum problem yields either to heuristics or to the linear integer programming formulation used for the p-median problem, a problem which we will discuss under network location models. When the classic metrics are set aside, solution of the minisum problem generally becomes more difficult, except in the case of minimizing the weighted sum of squared distances, in which case the single facility minisum solution is simply the centroid.

A second important objective setting in planar problems is the siting of a single facility under the objective of minimizing the maximum distance that separates any demand/supply point from the central facility. The problem may utilize either of the two classic metrics, Euclidean or rectilinear. No matter the number of dispersed points, the minimax single facility location problem with rectilinear distances yields to either a geometric solution or to a four-constraint linear program. The minimax single facility location problem with Euclidean distances is a nonlinear programming problem but can also be solved by a geometric argument. Multi-facility versions of the planar minimax location problems may yield to heuristics resembling those applied to the p-median problem. The most complete and understandable general reference dealing with planar location problems is the text of Love, Morris, and Wesolowsky (1988).

NETWORK LOCATION PROBLEMS: In contrast to the use of formula-based metrics for the siting of facilities on a plane, network location problems always measure distance as the shortest distance across links of the network. Interestingly, the assumption of an infinite solution space can be made in network-based location problems as well. That is, the infinite solution space would consist of all the points on every arc of the network. For some problems, notably the p-median, however, the solution space can be reduced without loss of generality from all the points on all the arcs to a limited number of eligible points. Many network problems simply assume a pre-specified set of eligible facility sites based on needed characteristics of such points, such as transportation infrastructure, etc.

Within network location research are two distinct foci. The first is cost minimizing/profit maximizing siting that is goods-oriented, an activity especially of the manufacturing and distribution industries. The second is people or public service-oriented siting, an activity of government at a number of levels from local to national. The divisions are not perfect, as we shall see, but are, at the least, useful for discussion purposes. We will take up these two settings in that order.

GOODS-ORIENTED SITING: By far, the problem setting considered most extensively in the goods oriented location category is the *simple plant location problem* (SPLP). The problem assumes that an unknown number of plants are to be sited to manufacture product for distribution to a number of spatially dispersed demand points. The plants have no limit as to the amount manufactured, and each point must be fully supplied with its demand. The objective is the minimization of the total of manufacturing cost and distribution cost. Manufacturing includes a fixed opening cost and possibly a linear expansion cost. The problem may be stated mathematically as:

$$\text{minimize} \quad z = \sum_{i=1}^{m} \sum_{j=1}^{n} c_{ij} x_{ij} + \sum_{i=1}^{m} f_i y_i$$

subject to:

$$\sum_{i=1}^{m} x_{ij} = 1, \qquad j = 1, \ldots, n,$$

$$y_i - x_{ij} \geq 0, \quad i = 1, \ldots, m; j = 1, \ldots, n,$$

$$x_{ij}, y_i \in \{0,1\}, \quad i = 1, \ldots, m; j = 1, \ldots, n.$$

where i = the index of eligible plant sites of which there are m;

   j = index of demand points of which there are n;

   f_i = opening cost for a plant at i;

   c_{ij} = cost to deliver j's full demand from i, including the production cost at i;

   y_i = 0,1; it is 1 if a plant opens at i and 0 otherwise; and

   x_{ij} = 0,1; it is 1 if i delivers j's full demand, and 0 otherwise.

The problem formulation given above is due to Balinski (1965), and is one of several formulations possible for the SPLP. It is presented here because it is the basis for a number of solution methods.

The SPLP has attracted attention since the 1950s when heuristics were first suggested. In the 1960s, Balinski offered his formulation of the problem but dismissed it as unreliable. In addition, several branch and bound algorithms were created to solve the SPLP, but these algorithms proved impractical for large problems. In the mid-1970s, Bilde and Krarup (1977) and Erlenkotter (1978) both proposed dual ascent algorithms for the SPLP; the basic algorithm proposed by these two sets of investigators has proved to be capable of handling relatively large problems. Morris (1978) investigated 500 randomly generated plant location problems and found that if the formulation above were solved as a linear program (without integer requirements on any of the variables) that 96% of the problems so solved presented with all zero-one variables. Morris' experience thus suggested that linear programming alone was a powerful technique for the SPLP formulation that Balinski had abandoned. The problem has since been successfully pursued by Lagrangean relaxation by Galvao (1989) and Korkel (1989), who modified the dual ascent algorithm referred to above to solve remarkably large problems.

While the SPLP has attracted considerable attention, a related form, the *capacitated plant location problem* (CPLP), languished until the late 1980s. The CPLP set limits on the amount that could be manufactured at any site, but in all other respects is the same as the SPLP. First attacked by Davis and Ray (1969), the problem has lately received attention by Pirkul (1987), who provided both references to prior work and a solution algorithm based on Lagrangean relaxation. The CPLP also describes a problem in solid wastes management in which wastes are generated at population nodes and must be disposed of at sanitary landfills with limited capacities. Landfills are to be sited in this problem statement.

Many other plant location style problems can be stated. A maximum profit version of the SPLP is one such statement. The time dimension has been incorporated in a number of models as well. Another problem is siting plants between sources of raw material and demand points; the original SPLP assumed away the supply component of the problem. Another is the siting of warehouses between plants and demand points, a statement that resembles the previous problem but with capacities on the supply points. In addition, the problem could be formulated to site both plants and warehouses between sources of supply and points of demand. Multiple products can be treated as well. Finally, demands, prices, and costs can be viewed as random, leading to stochastic versions of the plant location problem. Many of these additional plant/warehouse location problems are treated in a fairly comprehensive review by Aiken (1985).

PUBLIC SERVICE ORIENTED SITING: Nearly all of the plant location problems emphasize the flow/movement of goods. In contrast, the public service oriented siting problems focus on the accessibility of people to services or services to people. Flow/movement is part of the equation in some of the models, but simple geographic coverage can suffice in others.

The same two objectives treated under planar problems, minisum and minimax, have also been considered for network location problems of public service siting. The minisum network location problem is known as the *p-median problem*; the minimax network location problem is known as the *p-center problem*.

The earliest of this lineage of models is the *p*-median problem, first solved exactly by ReVelle and Swain (1970). The *p-median problem*, which seeks the minimum cost assignment of each population node to one of *p* facilities, resembles the SPLP in all but one modelling aspect. Indeed, so strong is the resemblance of *p*-median to the simple plant location model that the same algorithms may be used for solution of both with minor adaptation, as Galvao (1989) shows. The single difference between the two models is easy to explain once a mathematical programming formulation of the *p*-median is offered. The *p*-median problem seeks to site *p* facilities in such a way that the least total of people times distance travelled to the assigned facility is achieved. Division of this objective by the total of population reveals that minimization of the total population-miles objective also minimizes the average distance that people travel to service. Travel/assignment is always assumed to be the closest among the *p* facilities.

In mathematics, the *p*-median problem may be formulated as:

$$\text{minimize} \quad Z = \sum_{i=1}^{n} \sum_{j=1}^{n} a_i d_{ij} x_{ij}$$

subject to

$$\sum_{j=1}^{n} x_{ij} = 1 \qquad i = 1, 2, \ldots, n$$

$$x_{ij} - x_{ij} \geq 0 \qquad i, j = 1, 2, \ldots, n; i \neq j$$

$$\sum_{j=1}^{n} x_{jj} = p$$

$$x_{ij} = (0,1) \qquad i, j = 1, 2, \ldots, n$$

where a_i = relevant population at demand node i;

d_{ij} = shortest distance from node i to node j;

n = number of nodes;

p = number of facilities; and

$x_{ij} = (0, 1)$; it is 1 if node i assigns to a facility at j, 0 otherwise.

It can be seen from a comparison of the *p*-median formulation and that of the SPLP that the objectives differ only in the presence or absence of fixed opening costs and their opening variables, and that the constraints differ only in the presence or absence of a constraint on the number of facilities. In all other respects, the formulations look virtually identical. If the constraint on the number of facilities in the p-median formulation is brought to the objective with a multiplier, the objective becomes

$$\sum_{i=1}^{n} \sum_{j=1}^{n} a_i d_{ij} x_{ij} + \sum_{j=1}^{n} \lambda x_{jj}$$

The subscripts, of course, reflect flow to central facilities rather than from plant to demand points, but beyond this difference, the p-median is now fully equivalent to an SPLP with equal opening costs, thus making all the techniques for solution of the SPLP available for solution of the p-median. Ranging the multiplier λ in the p-median is equivalent to trading off people-miles against the number of facilities by use of the weighting method of multi-objective programming. Among the methods available for the SPLP that can be used for the p-median are relaxed linear programming (ReVelle and Swain, 1970; Morris, 1978), the dual ascent methodology (Bilde and Krarup, 1977; Erlenkotter, 1978) and Lagrangean relaxation (Galvao, 1989). A number of other methods have been suggested for the p-median problem; a listing of many of the early methods for the p-median problem appeared in ReVelle *et al.* (1977).

While the p-median problem attracted considerable attention, researchers found its focus on the average condition of population accessibility to be limiting. Concern for those worst off relative to their distance to the nearest facility, that is, for the maximum distance or time separating population centers from service, gave rise to another concept, that of coverage. A population node is considered to be covered if it has a facility sited within some maximum distance or time, that is, sited within a time standard. Coverage can either be required within the standard or sought, giving rise to a host of new problems, the earliest of which is the *location set covering problem* (LSCP).

The LSCP seeks to position the least number of facilities so that every point of demand has one or more facilities sited within the time or distance standard. The problem can be couched as a linear zero-one programming problem as follows:

minimize $z = \sum_{j \in J} x_j$

subject to: $\sum_{j \in N_i} x_j \geq 1 \quad \forall i \in I$

$x_j = 0, 1 \quad \forall j$

where i, I = index and set of demands;

j, J = index and set of eligible sites for facilities;

$x_j = 1, 0$; 1 if a facility placed at j; 0 otherwise;

d_{ji} = the shortest distance (or time) from site j to demand point i;

S = the maximum distance (or time) that a demand point can be from its nearest facility; and

$N_i = \{j \,|\, d_{ji} \leq S\}$ = the set of facility sites eligible to serve demand point i, by virtue of being within S of i.

While general set covering problems may require integer programming algorithms to solve them, the LSCP appears to possess special properties. In particular, solution of the linear programming formulation on data from a geographic problem without any zero-one requirements, produces all zero-one answers with remarkable regularity (over 95% of the time). If a set of eligible facility sites is specified in advance, the LSCP can be used to derive solutions to the p-center problem as well. The p-center problem seeks to position p facilities in such a way that the maximum distance that separates any population node from its nearest facility is as small as possible. Solutions to this problem are determined by observing the location patterns that obtain at the left most corner points of the tradeoff curve that displays facilities required against maximum distance – a trade-off curve derived by successive solutions of the LSCP (Minieka, 1970). If, however, any point on any link of the network is eligible to house a facility (the infinite solution space case), the solution of the p-center problem remains open and challenging.

The LSCP, however, has several shortcomings as a meaningful problem statement. First, population is absent from the problem statement; proximity and population are not linked even though they should be. Second, all population nodes require coverage within the standard, a requirement that could and often proves very costly in terms of the number of facilities/servers required.

Recognizing these shortcomings of the LSCP, several researchers have created new models for siting that utilized the coverage concept not as a requirement but as a goal. The most widely known of these models is referred to as the maximal covering location problem (MCLP) or the partial covering problem depending on the specific formulation. The MCLP seeks the positions for p facilities among a prespecified set of eligible points that maximize the population that has a facility sited within a distance or time standard S, that is, that maximizes the population covered. The MCLP can be stated as:

$$\text{maximize } z = \sum_{i \in I} a_i y_i$$

$$\text{subject to: } y_i \leq \sum_{j \in N_i} x_j \qquad \forall i \in I$$

$$\sum_{j \in J} x_j = p$$

$$x_j, y_i = 0, 1 \quad \forall i, j$$

where additional notation is

a_i = the population at demand node i;

$y_i = 1, 0$; it is 1 if demand i is covered by a facility within N_i; 0 otherwise; and

p = the number of facilities that can be sited.

Basically, while the LSCP is attempting to find the least resources to cover all demand nodes within the distance goal, the MCLP is attempting to distribute lesser and limited resources to achieve as much population coverage as possible (Church and ReVelle, 1974).

The notion of partial coverage has given rise to a host of other formulations. Among these are a goal programming coverage formulation, backup and redundant covering problems, and a maximal expected covering formulation (Daskin, et al., 1987). Additionally, covering problems with multiple vehicle types and multiple objectives have been investigated.

A related area is probabilistic covering models, models in which the presence or availability of a vehicle or server within a time standard is not guaranteed. The probabilistic models suggest a chance constraint on vehicle availability, that is, a requirement that a vehicle be available within the time standard with a specified level of reliability. The chance constraint may be a strict requirement or may be treated as a goal for each population demand node. Many of the probabilistic as well as redundant/backup coverage models, and multiple vehicle type models were reviewed in ReVelle (1989) and ReVelle (1991). A review of location covering problems in general was provided by Schilling et al. (1993).

A number of other lines of research within the network location setting have been pursued. Among these are hierarchical location models, models in which a hierarchy of interacting/interrelated facility types are sited. One example is the health care hierarchy in developing nations, which consists of hospitals, clinics and remote doctors. Another is a banking system consisting of central banks, branch banks and teller machines. A review of morphological relations in hierarchical systems was provided by Narula (1987). Church and Eaton (1987) provided an interesting set of hierarchical models with referral between levels. Another significant line of siting research is embodied in the competitive location models, in which facilities are sited in a competitive market environment with goals of capturing market share from other retailers/manufacturers (Friesz, Miller and Tobin, 1988).

Another line of location research involves the siting of noxious facilities. Such facilities may be undesirable in and of themselves and should be distant from population centers or may be required to be distant from one another. A review of obnoxious facility siting is found in Erkut and Newman (1989).

CONCLUSION: Location modeling has proven an exciting field because of the presence of a wide variety of important applications and the challenging nature of the mathematics needed to approach the problem. Journals with significant numbers of location articles include *Transportation Science* and *European Journal of Operational Research*, plus *Location Science* whose inaugural issue occurred in 1993. The latter journal is devoted to the location question in its many forms. In addition, the proceedings of the triennial International Symposium on Locational Decisions (ISOLDe) appear in separate volumes of *Annals of Operations Research*, beginning with 1984 Boston/Martha's Vineyard conference.

See **Facility location; Networks; Shortest-route problem; Stochastic programming.**

References

[1] Aikens, C.H. (1985), "Facility location models for distribution planning," *European Jl. of Operational Research*, 22, 1985, 263–279.

[2] Balinski, M. (1965), "Integer Programming: Methods, Uses and Computations," *Management Science*, 12, 253–313.

[3] Bilde, O. and J. Krarup (1977), "Sharp lower bounds and efficient algorithms for the simple plant location problem," *Annals Discrete Mathematics*, 1, 79–97.

[4] Church, R. and D. Eaton (1987), "Hierarchical location analysis using covering objectives," in *Spatial Analysis and Location Models*, Van Nostrand-Rheinhold, New York, 163–185.

[5] Church, R. and C. ReVelle (1974), "The maximal covering location problem," *Papers Regional Science Association*, 32, 101–118.

[6] Daskin, M., K. Hogan and C. ReVelle (1987), "Integration of multiple, excess, backup and expected covering models," *Environment and Planning*, 15, 15–35.

[7] Davis, P. and T. Ray (1969), "A Branch-and-Bound Algorithm for the Capacitated Facilities Location Problem," *Naval Research Logistics Quarterly*, 16, 331–344.

[8] Erkut, E. and S. Newman (1989), "Analytical models for locating undesirable facilities," *European Jl. Operational Research*, 40, 275–291.

[9] Erlenkotter, D. (1978), "A Dual-Based Procedure for Uncapacitated Facility Location," *Operations Research*, 26, 992–1009.

[10] Friesz, T., T. Miller and R. Tobin (1988), "Competitive network facility location models: a survey," *Papers Regional Science Association*, 65, 45–57.

[11] Galvão, R. (1989), "A method for solving optimality uncapacitated location problems," *Annals Operations Research*, 18, 225–244.

[12] Korkel, M. (1989), "On the exact solution of large-scale simple plant location problems," *European Jl. Operational Research*, 39, 157–173.

[13] Love, R., Morris, J. and G. Wesolowsky (1988), *Facilities Location: Models and Methods*, North Holland, New York.

[14] Minieka, E. (1970), "The M-Centre Problem," *SIAM Review*, 12, 138–141.

[15] Morris, J., "On the extent to which certain fixed charge depot location problems can be solved by LP," *Jl. Operational Research Society*, 29(1), 71–76.

[16] Narula, S. (1986), "Minisum hierarchical location-allocation problems on a network: A survey," *Annals Operations Research*, 6, 257–272.

[17] Pirkul, H. (1987), "Efficient algorithms for the capacitated concentrates location problem," *Computers and Operations Research*, 14, 197–208.

[18] ReVelle, C. (1989), "Review, extension and prediction in emergency service siting models," *European Jl. Operational Research*, 40, 58–69.

[19] ReVelle, C. (1991), "Siting ambulances and fire companies," *Jl. American Planning Association*, Autumn, 471–484.

[20] ReVelle, C., Bigman, D., Schilling, D., Cohon, C. and R. Church (1977), "Facility location: A review of context-free and EMS models," *Health Services Research*, summer, 129–146.

[21] ReVelle, C. and R. Swain (1970), "Central facilities location," *Geographical Analysis*, 2, 30–42.

[22] Rosing, F. (1992), "An optimal method for solving the (generalized) multi-Weber problem," *European Jl. Operational Research*, 58, 414–426.

[23] Schilling, D., Jayaraman, V. and R. Barkhi (1993), "A review of covering problems in facility location," *Location Science*, 1(1).

LOGICAL VARIABLES

In a linear-programming problem, the set of variables that transform a set of inequalities to a set of equations are called logical variables. See **Linear inequality**; **Slack variable**; **Structural variables**; **Surplus variable**.

LOGIC PROGRAMMING

Use of symbolic logic for problem representation and inferential reasoning. A common logic programming language is PROLOG, developed by French scientists in the early 1970s. It has been adopted by Japan as a standard. See **Artificial intelligence**.

LOGISTICS

Marius M. Solomon

Northeastern University, Boston, Massachusetts

INTRODUCTION: For quite some time, logistics has accounted for a significant percentage of GNP (11% in 1990). In addition, a number of economic developments have highlighted logistics as an area amenable to large productivity improvements. To achieve them, given the intrinsic complexity of most logistics problems, has naturally led to the use of operations research/management science (OR/MS) methodology.

OR/MS models and algorithms developed for strategic, tactical and operational logistics decisions date back to the 1950s (Dantzig and Fulkerson, 1954). The work of Geoffrion and Graves (1974) on strategic logistics system planning is representative of the forefront of the research and implementation efforts undertaken in the early 1970s. It involved the development of a model and an algorithm that determine the least cost design of a broad class of two-stage multicommodity distribution systems. This consisted of determining the facility locations, transportation flows and plant loadings that minimize the total system cost subject to a variety of constraints. A Benders decomposition based algorithm was successfully implemented for a major food company.

Over the last twenty years, fueled by important developments in modeling and algorithmic methodology and constant breakthroughs in computer technology, operations researchers have found logistics a very fertile application area. The body of applications of OR/MS techniques to logistics has been increasing at a swift pace. In what follows, we discuss some of the more recent OR/MS applications in large-scale logistics systems.

NETWORKING AND ROUTING: Network design and freight routing have been addressed by Braklow *et al.* (1992) in the context of less-than-truckload (LTL) transportation. The authors formulate the problem as a nonlinear, multicommodity network design problem. Its solution is based on a hierarchical decomposition of the overall problems into a series of optimization subproblems. The network design problem is solved using interactive optimization, where the user guides the search performed by a local improvement heuristic which adds (drops) links to (from) the load planning network. The subproblems involve the routing of the LTL shipments, of truckload shipments and of empty trailers. The former two problems are solved using shortest path algorithms, while the latter problem involves the solution of a classical linear transshipment problem. They must be reoptimized every time a change is made in the load planing network. This is to be performed sufficiently fast to make interactive optimization possible. The model has been used as a tactical decision tool for load planning by one of the largest LTL motor carriers. It has also been used at the strategic level to determine the location and size of new terminals.

While logistics encompasses a broad set of activities, we are going to focus in the rest of the paper on two of its key elements: transportation and storage. Transportation is in fact the most costly component of most logistics systems. A very important segment of transportation management is the routing and scheduling of vehicles. This area has been reviewed in several insightful surveys, including that written by Bodin *et al.* (1983).

A very successful application involving the scheduling of a fleet of vehicles that make bulk deliveries over a multi-day horizon was undertaken by Bell *et al.* (1983). The authors present a mixed integer programming formulation and an algorithm based on Lagrangian relaxation coupled with a multiplier adjustment method to set dual variables (Fisher, 1985). The problem involves both routing and inventory aspects as the frequency of delivery and the amount to be delivered to a customer are decision variables. This route generation/route selection approach was implemented for a major manufacturer of liquid oxygen and nitrogen.

The application of OR/MS methods in this area has led to significant practical success. Golden and Wong (1992) have captured the breadth of the state of implementation by bringing together a number of routing applications in the private and the public sectors. These involved vehicle types as varied as planes, helicopters, ships, cars and trucks.

A variety of routing settings also involve the temporal aspect in the form of customer imposed time windows. A unified framework for all time constrained vehicle routing and crew scheduling problems is presented in the extensive survey by Desrosiers *et al.* (1993). The common structure of these problems is a multicommodity network flow model with additional resource constraints. Time is one example of a resource. Dantzig-Wolfe decomposition is applied to the model to provide the classical set partitioning formulation which is in turn solved by column generation. Resource variables help manage complex nonlinear cost functions and difficult local constraints (e.g., time windows, vehicle capacity, and union rules). Forward dynamic programming algorithms handle these resources in shortest path computations. The decomposition process is embedded in a branch and bound tree. Branching decisions and cuts appear either in the master problem or in the subproblem structures.

CREW SCHEDULING: Two notable application areas of the above framework are the urban transit crew scheduling problem and the airline crew scheduling problem. Blais *et al.* (1990) describe a software package to handle the former problem. It consists of several modules. The first uses standard network flow methodology to solve the bus scheduling problem. Next, crew scheduling is handled in two steps. In the first, several approximations are used to permit the fast derivation of a linear programming solution. Using this solution, specific driver assignments are then obtained in step two by means of solving a quadratic integer program heuristically and using an optimal matching algorithm. Finally, a shortest path algorithm utilizing the marginal costs from the matching problem is used to improve the solution. The software has been successfully implemented in a number of cities. An algorithm based on the above Dantzig-Wolfe/column generation framework is also being utilized for crew scheduling (Desrosiers *et al.*, 1993).

Anbil *et al.* (1991) provide a historical account of the techniques applied for crew scheduling at a major airline over the last twenty years. The evolution from the manual methods of the early 1970s to the powerful OR/MS based software in use today mirrors the developments that have occurred in other logistics areas. In addition, research in crew scheduling is part of the stream of research spearheading the development of optimization methods capable of handling practical size problems. This new generation of optimal algorithms blends the effectiveness of advanced optimization methods, designed to take advantage of special problem structures, with the efficiency of sophisticated computer science techniques, and the computing power of workstations. The computational breakthroughs in this area are discussed in Desrosiers *et al.* (1993).

DISPATCHING: While the size of problems solved by optimization algorithms increases constantly, heuristics remain a viable tool for very large-scale and/or very complex problems. Dispatching, an intricate activity given the need for a solution in real-time to large-scale problems, lends itself naturally to heuristic solutions. Brown *et al.* (1987) describe an interesting application involving the delivery by truck of light petroleum products to customers located within a wide area. The system typifies the use of route construction/route improvement heuristics to deal with the practical complexities of the problem. Fast interchange heuristics permit the system to provide decision support to human dispatchers in real-time. The system, implemented for a major oil company, manages all aspects of distribution and marketing. It is illustrative of the move toward computer integrated logistics (CIL) and toward horizontal management across functional areas through computer integration.

The highly dynamic character of dispatching is also apparent in truckload transportation. In this environment characterized by high demand uncertainty, a motor carrier must continuously manage the assignment of drivers to loads across the country. Powell *et al.* (1988) have developed a stochastic network optimization model to analyze this dynamic vehicle allocation problem, while Powell *et al.* (1993) provided an extensive survey of the problem area. It explicitly differentiates known from forecasted information. The model is solved with an efficiently modified network simplex code. The model has been implemented for one of the largest truckload motor carriers. Other application areas include rail car distribution and rental vehicle management.

When shipments could not be forecasted with accuracy, Moore *et al.* (1991) report having built mixed integer programming (MIP) and simulation models. The use of these techniques for operational purposes has stemmed from the successful solution of a strategic decision through similar methods. This decision involved the significant reduction in the number of carriers used and the creation of "partnerships" with them. To solve the carrier selection problem for a global, integrated aluminum company, the authors developed a MIP and further analyzed its results using simulation. This problem represented an important part of a redesign effort aimed at centralizing previously decentralized transportation and purchasing decisions. In turn, this was the reflection in logistics of the just-in-time manufacturing philosophy. Furthermore, in the direction of CIL, partnerships with carriers are frequently beginning to involve electronic data interchange (EDI).

LOGISTICS AND INVENTORY: The trade-offs between transportation and inventory costs are a central issue in materials management. Blumenfeld *et al.* (1987) present an ingenious analysis of the production network of a manufacturer of vehicle components. Their bottom-up approach begins with the analysis of the trade-offs on a single link. These are obtained using a standard economic order quantity (EOQ) model. Using several realistic approximations, the authors are then able to extend their analysis to much more complex networks. In particular, one approximation allows the decomposition of a large network into a number of small independent subnetworks, where shipment sizes can be computed using the single link model. This work involving simple, easy to understand models, supplemented by insightful graphical information, is representative of a line of research complementary to combinatorial optimization.

In light of intense global competitive pressures, many companies have tried to decrease their inventory investment while maintaining or improving customer service in their vital business processes. Yet, the implementation of just-in-time manufacturing has led to significant increases in product variety. In turn, this has augmented the complexity of the after-sales service logistics networks. Cohen *et al.* (1990) describe the design of a spare parts inventory control system capable of supporting multiple service levels. The building block of their approach is a periodic review, stochastic model for the one-part, one-location case. This model is then extended to a multi-product, one-location case, called the service allocation problem. This is solved using a greedy heuristic. A decomposition approach is utilized for the overall multi-product, multi-echelon problem. It involves a bottom-up procedure which begins by solving the service allocation problems at the lowest echelon. The solutions are then used to deal with the next higher echelon. The algorithm proceeds in this fashion, level-by-level up to the highest echelon. The model has been implemented by a global computer manufacturer. It has found applicability both as a strategic network redesign tool and as a weekly operational device.

Integrated Logistics Systems: The implementation of a comprehensive set of OR/MS tools in a variety of business areas of a large oil company are presented by Klingman *et al.* (1987). It is not surprising to see that this industry is at the leading edge of computer integrated horizontal management across functional areas, in general, and CIL, in particular. OR/MS techniques such as linear programming have been utilized in the oil industry as early as the 1950s. The work of Klingman *et al.* (1987) included such tools as mathematical programming, statistics, forecasting, expert systems, artificial intelligence, organizational theory, cognitive psychology and information systems. A core element is the optimization-based integrated system for supply, distribution and marketing. This strategic tool is used to make a number of decisions including how much product to buy or trade, how much to hold in inventory and how much product to ship by each mode of transportation. The system is based on the minimum cost flow network model. Ahuja *et al.* (1989) provide a comprehensive survey of network flow methodology.

See **Facility location; Minimum cost network flow problem; Multicommodity network flows; Networks; Transportation problem; Vehicle routing.**

References

[1] Ahuja R., T. Magnanti, and J. Orlin (1989). "Network Flows." In *Handbooks in Operations Research and Management Science, Vol.* 1, G. Nemhauser, A. Rinnooy Kan and M. Todd (eds.). Elsevier Science Publishers B.V., Amsterdam, The Netherlands, 211–369.

[2] Anbil R., E. Gelman, B. Patty, and R. Tanga (1991). "Recent Advances in Crew-Pairing Optimization at American Airlines." *Interfaces* 21, 62–74.

[3] Assad A., E. Wasil, and G. Lilien (1992). *Excellence in Management Science Practice.* Prentice Hall, Englewood Cliffs, NJ.

[4] Bell W., L. Dalberto, M. Fisher, A. Greenfield, R. Jaikumar, P. Kedia, R. Mack, and P. Prutzman (1983). "Improving the Distribution of Industrial Gases with an Online Computerized Routing and Scheduling Optimizer." *Interfaces* 13, 4–23.

[5] Blais J.-Y., J. Lamont, and J.-M. Rousseau (1990). "The HASTUS Vehicle and Manpower Scheduling System at the Societé de Transport de la Communauté Urbaine de Montréal." *Interfaces* 20, 26–42.

[6] Blumenfeld D., L. Burns, C. Daganzo, M. Frick, and R. Hall (1987). "Reducing Logistics Costs at General Motors." *Interfaces* 17, 26–47.

[7] Bodin L., B. Golden, A. Assad, and M. Ball (1983).

"Routing and Scheduling of Vehicles and Crews: The State of the Art." *Computers and Operations Research* 10, 62–212.

[8] Braklow J., W. Graham, S. Hassler, K. Peck and W. Powell (1992). "Interactive Optimization Improves Service and Performance for Yellow Freight System." *Interfaces* 22, 147–172.

[9] Brown G., C. Ellis, G. Graves, and D. Ronen (1987). "Real-Time, Wide Area Dispatch of Mobil Tank Trucks." *Interfaces* 17, 107–120.

[10] Cohen M., P. Kamesam, P. Kleindorfer, H. Lee, and A. Tekerian (1990). "Optimizer: IBM's Multi-Echelon Inventory System for Managing Service Logistics." *Interfaces* 20, 65–82.

[11] Dantzig G. and D. Fulkerson (1954). "Minimizing the Number of Tankers to Meet a Fixed Schedule." *Naval Research Logistics Quarterly* 1, 217–222.

[12] Desrosiers J., Y. Dumas, M. Solomon and F. Soumis (1993). "Time Constrained Routing and Scheduling," *Handbooks in Operations Research, Volume on Networks*, 35–139. Elsevier Science Publishers B.V., Amsterdam, The Netherlands.

[13] Fisher M. (1985). "An Applications-Oriented Guide to Lagrangian Relaxation." *Interfaces* 15, 10–21.

[14] Geoffrion A. and G. Graves (1974). "Multicommodity Distribution System Design by Benders Decomposition." *Management Science* 20, 822–844.

[15] Golden B. and R. Wong, eds. (1992). "Vehicle Routing by Land, Sea and Air." *Interfaces* 22.

[16] Klingman D., N. Phillips, D. Steiger, and W. Young (1987). "The Successful Deployment of Management Science Throughout Citgo Petroleum Corporation." *Interfaces* 17, 4–25.

[17] Moore E., J. Warmke, and L. Gorban (1991). "The Indispensable Role of Management Science in Centralizing Freight Operations at Reynolds Metals Company." *Interfaces* 21, 107–129.

[18] Powell W., Y. Sheffi, K. Nickerson, K. Butterbauch, and S. Atherton (1988). "Maximizing Profits for North American Van Lines' Truckload Division: A New Framework for Pricing and Operations." *Interfaces* 18, 21–41.

[19] Powell W., P. Jaillet, and A. Odoni (1993). "Stochastic and Dynamic Networks and Routing." *Handbooks in Operations Research, Volume on Networks*, 141–295. Elsevier Science Publishers, B.V., Amsterdam, The Netherlands.

LOG-LINEAR MODEL

See **Learning curves; Regression analysis.**

LONGEST-ROUTE PROBLEM

In a directed network, the finding of the longest route between two nodes is the longest-route problem. In an acyclic network, one that represents the precedence relationships between activities in a project, the longest route in the network represents the critical path, with the value of the longest route equal to the value of the earliest completion time of the project. See **Critical path method (CPM)**; **Program evaluation and review technique (PERT)**.

LOSS FUNCTION

See **Decision analysis**; **Total quality management**.

LOTTERY

In utility theory and decision analysis, a lottery consists of a finite number of alternatives or prizes $A_1 \ldots A_n$ and a chance mechanism such that prize A_i will be an outcome of the random experiment with probability $p_i \geq 0$, $\Sigma_i p_i = 1$. See Decision analysis; **Utility theory**.

LOWER-BOUNDED VARIABLES

The condition $\ell_j \leq x_j$, $\ell_j \neq 0$, defines x_j as a lower-bounded variable. Such conditions are often part of the constraint set of an optimization problem. For linear-programming, these conditions can be removed explicitly by appropriate transformations, given that the problem is feasible when $x_j = \ell_j$ for each j.

LOWEST INDEX ANTICYCLING RULES

See **Bland's anticycling rules**.

LP

See **Linear programming**.

LU MATRIX DECOMPOSITION

The decomposition of a matrix into the product of a lower- and an upper-triangular matrix. This is similar to an LDU decomposition in which the D and U matrices have been combined. See **LDU matrix decomposition**.

MAD

Mean absolute deviation.

MAINTENANCE

Igor Ushakov

SOTAS, Rockville, Maryland

Maintenance is the support of successful system operation during long periods of usage by means of: (1) regular or sample check-ups; (2) planned or preventive replacement of the system's units; (3) failure diagnosis; and/or (4) spare units supply. Operations research models for a system maintenance analysis are represented mainly by optimization models for the improvement of system and equipment reliability.

For (1) and (2), one usually uses methods of controlled stochastic processes. For (3), one uses special methods based on mathematical logic, while (4) is considered in the scope of *optimal redundancy* and *inventory control*.

References

[1] Ushakov, I.A., ed. (1994). *Handbook of Reliability Engineering*, Wiley, New York.

MAKESPAN

See **Sequencing and scheduling**.

MALCOLM BALDRIGE AWARD

See **Total quality management**.

MANHATTAN METRIC

See **Location analysis**.

MANPOWER PLANNING

David J. Bartholomew

The London School of Economics and Political Science

INTRODUCTION: Manpower (or, human resource) planning is concerned with the quantitative aspects of the supply of and demand for people in employment. At one extreme this might include the whole working population of a country but it has been most successful when applied to smaller, more homogeneous systems like individual firms or professions. The term *manpower planning* appears to date from the 1960s though many of the ideas can be traced back much further. A history of the subject from a UK perspective will be found in Smith and Bartholomew (1988). The literature of the subject is very scattered reflecting the diverse disciplinary origins of the practitioners but most of the technical material is to be found in the journals of operations research, probability and statistics. There was an initial surge of publication in the late 1960s and early 1970s and since then book length treatments include Grinold and Marshall (1977), Vajda (1978) and Bennison and Casson (1984). Bartholomew, Forbes and Mc-Clean (1991) gives a thorough coverage of the technical material and contains an extensive bibliography.

The essence of manpower planning is summed up in the aphorism that its aim is to have the right numbers of people of the right kinds in the right places at the right time. The basic approach is first to classify the members of a system in relevant ways. These will often be on the basis of such things as grade, salary level, sex, qualifications and job title. The state of the system at any point in time can then be described by the numbers in these categories, often referred to as the *stocks*. Over time, changes occur as individuals join, leave the system or move within it. The numbers making these transitions are called the *flows*. The factors giving rise to change may be predictable or unpredictable but will include such things as individual decisions to leave, changes in demand for goods, management decisions on promotion or organizational structure and so on. The operations researcher's role is to describe and model the system as a basis for optimizing its performance.

STOCHASTIC MODELS: The presence of uncertainty in so many aspects of the functioning of a manpower system means that any adequate model has to be stochastic. Two probability processes, in particular, have proved to be both flexible and realistic. These are the absorbing Markov chain and the renewal process. The former is appropriate in systems where the stocks are free to vary over time under the impact of constant flow rates, or probabilities. The art of successful application is so to define the classification of individuals that all those within a category have approximately the same probability of moving to any other category. Loss from the system corresponds to 'absorption' and the theory of Markov chains can then be used to predict future stock numbers for various sets of transition probabilities. Recent work has extended these methods by allowing the intervals between transitions to be random variables in which case we have a semi-Markov process or a Markov renewal process.

When the numbers in the categories are fixed, as they often are when the categories are grades or based on job function, a different approach must be used. Transitions cannot then be regarded as generated by fixed probabilities but arise in response to the occurrence of vacancies. We then have a replacement, or renewal process, where movement is driven by wastage (or the creation of new places).

If a system is relatively small or if the rules governing its operation are complex, the only realistic way to model it may be to use a computer-based simulation model. The term simulation is commonly used in two distinct senses in this context. Primarily it means that each individual movement is generated in the model by a random mechanism. Secondly, it is sometimes used of any algorithm for computing the aggregate properties of a system treated deterministically.

FORECASTING AND CONTROL: Broadly speaking all models may be used in two modes: for *forecasting* or *control*. In the early stages of a study one usually wishes to forecast the future state of the system if current trends continue. Next, it will usually be desirable to carry out a sensitivity analysis to explore the consequences of variations from present conditions. This leads on to questions of control where we ask, how those parameters under management control should be chosen to achieve some desired goal. The distinction between forecasting and control can be illustrated using a simple form of the Markov model. According to this model successive vectors of expected stocks are related by an equation of the form

$$n(T+1) = n(T)P + R$$

where T represents time, P is a matrix of transition probabilities and R is a vector of recruitment numbers. In forecasting mode we would use estimated or guessed values of P and R to predict future values of $n(T)$. In principle, P and R could both depend on T. In control mode we would be asking how some or all of the elements of P and R should be chosen to attain a given n within a specified time. This gives rise to questions of attainability (whether the problem is solvable) and *maintainability* (whether an n can be maintained once it is reached). These matters have led to an interesting set of theoretical questions about the solvability of such problems in deterministic or stochastic environments. At a more practical level it has led to the formulation of optimization problems expressed in goal programming and/or network analysis terms (Gass, 1991; Klingman and Phillips, 1984).

The wastage flow (also known as attrition or turnover) is an important element in a manpower system both because it is highly variable and, largely, beyond the control of management. It has been intensively studied mainly through the survivor function or, equivalently, the frequency distribution of completed length of service. In practice the analysis is complicated by the fact that the data are usually censored and sometimes truncated also. This work has three main objectives: measurement, prediction and gaining insight into the factors determining wastage.

The demand side of the manpower equation has proved to be less tractable. Demand for people is equivalent to the supply of jobs and this depends on technological, political, social and economic factors many of which may be specific to particular organizations or industries. To take only one example, the demand for qualified medical manpower will depend on such varied things as demographic changes, the willingness of government or users of the service to pay and the appearance and spread of new diseases like AIDS. The methods used have been, and have to be, as diverse as the fields of application. Because of the considerable uncertainties involved it is important to monitor constantly the changing environment and to adjust plans accordingly. A once-and-for-all plan has no place in manpower planning.

See **Goal programming; Markov chains; Markov processes; Networks.**

References

[1] Bartholomew, David J., A.F. Forbes and S.I. McClean (1991). *Statistical Techniques for Manpower Planning*, 2nd ed., John Wiley, Chichester.

[2] Bennison, Malcom and J. Casson (1984). *The Manpower Planning Handbook*. McGraw-Hill, London.

[3] Gass, S.I. (1991). "Military Manpower Planning Models," *Computers and OR*, 18(1), 65–73.

[4] Grinold, R.C. and K.T. Marshall, 1977. *Manpower Planning Models*. New York and Amsterdam: North-Holland.

[5] Klingman, D. and N. Phillips (1984). "Topological and Computational Aspects of Preemptive Multi-criteria Military Personnel Assignment Problems," *Management Science*, 30, 1362–1375.

[6] Smith, A.R. and D.J. Bartholomew (1988). "Manpower Planning in the United Kingdom: An Historical Review," *Jl. Operational Research Society*, 9, 235–248.

[7] Vajda, Steven (1978). *Mathematics of Manpower Planning*, John Wiley, Chichester.

MAP

Markov arrival process. See **Matrix-analytic stochastic models**.

MARGINAL VALUE (COST)

The marginal value is the extra cost of producing one extra unit of output. Similarly, marginal revenue is the extra revenue resulting from selling an extra unit of goods. From the economics of a firm, when marginal revenue equals marginal costs, the firm is in an equilibrium optimal condition in terms of maximizing profits. Depending on the application, the dual variables of a linear-programming problem can be interpreted as marginal values. The economic interpretation of the dual variables is complicated by alternate optimum solutions (corresponding to different bases) that may yield different values of the dual variables. Thus, there may be two or more marginal values for the same constraint. Such multiple values must be interpreted with care. See **Dual problem**; **Duality theory**.

MARKETING

Jehoshua Eliashberg

University of Pennsylvania, Philadelphia

Gary L. Lilien

Pennsylvania State University, University Park

Yoram (Jerry) Wind

University of Pennsylvania, Philadelphia

INTRODUCTION: As a *management function*, marketing includes such activities as advertising, sales, and marketing research. It is also a critical participant in cross-functional processes aimed at generating and launching new products and services that create customer value. As a *philosophy*, marketing views the need to understand, anticipate and meet customer needs as the key to organizational success. Here, the customer is the final arbitrator of the value of any product or service offering. Marketing philosophy also extends the concept of customer orientation to internal customers and other stakeholders.

Thus, marketing is concerned with anticipating and understanding human needs and wants and translating those needs and wants into the demand (as economists use the term) for products and services. Those needs and wants are satisfied with products and services. Those products and services in turn have particular physical as well as imagery characteristics. They are made available to the customer through a variety of channels ranging from direct sales to retail stores to mail order to interactive television programs. In order to effect an exchange, individuals have to be aware of and understand the product (through advertising or other communication media), find the product "worth the money" (by comparing the product's total cost—its purchase price adjusted by any promotional offerings plus the cost of maintaining, using and disposing of the product—with the benefits promised in terms of performance and image), and participate in the exchange process.

Exchanges take place in a market, which consists of the potential customers sharing a particular need and who might be willing to engage in exchange to satisfy that need. Finally, to summarize the above, the American Marketing Association defines marketing as: "...the process of planning and executing the conception, pricing, promotion, and distribution of ideas, goods, and services to create exchanges that satisfy individual and organizational objectives."

If we look at the above definition, we see a clear role for OR/MS in addressing marketing problems. Since customers are at the heart of the marketing system, OR/MS modeling approaches can help characterize, understand and predict behavior. For consumers and organizational buyers, that behavior can involve the search for solutions to a want or desire, the screening or evaluation of alternatives, the selection of a best alternative, the act of purchase, and the post-purchase feedback and learning that affects future purchasing behavior.

Firms and other organizations (such as museums, governmental agencies) can capitalize on that knowledge or model of individual behavior by focusing on such decisions as product design, pricing, distribution, promotion, advertising, personal selling, and the likely response to them. In addition, at a higher level, these decisions must be integrated and coordinated with the activities of other management functions (finance, manufacturing, R&D, etc.) and linked to other product and market decisions of the firm and organization, including the critical resource allocation decisions among products, markets, distribution options and businesses.

OR/MS Marketing Model Types: There are essentially three purposes for OR/MS in marketing: measurement, decision making, and theory-building. We call the corresponding models: measurement models, decision-making models, and stylized theoretical models, respectively (although it may be equally helpful to interpret these "categories" as classification dimensions for interpreting the multiple purposes of models).

Measurement Models. The purpose of measurement models is to measure the existing or anticipated "demand" for a product as a function of various independent variables. The word "demand" here should be interpreted broadly. It is not necessarily units demanded but could be some other related variables. For example, in Guadagni and Little's (1983) model, the dependent variable is the probability that the individual will purchase a given brand on a given purchase occasion. Other choice models have several independent variables including whether the brand was on deal at a given purchase occasion, regular price of the brand, deal price (if any), brand loyalty of the individual, etc. In addition, sometimes the focus of such models may be on certain variables preceding the steady-state demand (e.g., awareness, first-trial, repeat purchase). These examples suggest that measurement models can deal with individual (disaggregate) demand or aggregate (segment or market-level) demand as well as transitory or steady-state demand. Note that advances in measurement models can be due to better data (scanner data, for example) or better calibration methods and procedures (maximum-likelihood methods for generalized logit models, for example). With the increased recent interest in customer satisfaction and customer defined quality, OR/MS measurement models can greatly enhance the relatively simplistic survey-based approaches to the measurement of these constructs. For example, Green and Krieger (1993)

developed a conjoint analysis based approach to the measurement of customer satisfaction.

Decision-Making Models. These models are designed to help marketing managers make better decisions. They incorporate measurement models as building blocks, but go beyond measurement models in recommending optimal marketing-mix decisions for the manager. The techniques used to derive the optimal policies vary across applications, and include calculus, dynamic programming, optimal control and calculus of variations techniques, as well as linear and integer programming. These models have been developed for each marketing variable and for the entire marketing mix program (i.e., a product and service offering including pricing, distribution). Little's (1975) BRANDAID is an example of such a model.

Stylized Theoretical Models. The purpose of stylized theoretical models is to explain and provide insights into marketing phenomena: a stylized theoretical model typically begins with a set of assumptions that describes a particular marketing environment. Some of these assumptions may be purely mathematical but intuitively logical, designed to make the analysis tractable. Others are substantive assumptions with real empirical grounding. Two well-known theoretical modeling efforts are Bell, Keeney and Little (1975), who show what functional forms of market share models are consistent with a certain set of "reasonable" criteria, and Basu, Lal, Srinivasan and Staelin (1985), who show what form of salesforce compensation plan is optimal under a set of assumptions about firm and salesperson objectives and behavior.

The Emergence of Marketing Science: OR/MS in marketing began its growth in the 1960s and 1970s by most accounts. The literature used a variety of OR/MS methods to address marketing problems: those problems included product design/development decisions, distribution system decisions, salesforce management decisions, advertising and mass communication decisions, and promotion decisions (Kotler, 1971). The OR/MS tools that were most prevalent in the 1960s and earlier included math programming, simulation, stochastic processes applied to models of consumer choice behavior, response function analysis, and various forms of dynamic modeling (difference and differential equations, usually of the first order). Some uses of game theory were reported, but most models that included competition used decision analysis, risk analysis, or market simulation games.

Nearly three times the number of marketing articles appeared in the OR/MS literature in the 1970s as appeared in the period from 1952 through 1969 (Eliashberg and Lilien, 1993). In addition to the increase in the number of articles, reviews by Schultz and Zoltners (1981) and Lilien and Kotler (1983) show that a number of new areas had begun to emerge. These included descriptive models of marketing decisions, the impact of and interaction of marketing on organizational design, subjective decision models, strategic planning models, models for public and non-profit organizations, organizational buying models, and the emergence of the concept of the Marketing Decision Support System (MDSS). In addition, while the number of published articles rose dramatically, the impact on organizational performance did not appear to be equally significant, raising questions about effective implementation. Much of the literature of the 1970s pointed to the need to expand the domain of application. The "limitations" sections of some of the papers in the 1970s pointed out that many important phenomena that were being overlooked (such as competition, dynamics, and interactions amongst marketing decision variables) were both important and inherently more complex to model. Hence, the level of model-complexity and the insightfulness of the analyses in marketing seemed destined to escalate in the 1980s and beyond.

The 1980s saw another more-than doubling of the number of published OR/MS articles in marketing compared to the earlier decade. Two of the areas that produced much of this growth were stylized theoretical models (above) and process-oriented models. The shortening of product life cycles and the impact of competitive reactions in the market place preclude most markets from approaching steady-state or equilibrium. Hence, new areas of emerging research have included new domains of consumer behavior modeling (where the temporal nature of the stimuli that affect consumer reactions has been the focus of recent research), the new product area (where the moves and countermoves of competitors keep the marketplace in a constant state of flux), and negotiations (where the actions of one party provide information to the other party, directing the flow of the negotiation process) have seen much recent OR/MS modeling.

TRENDS IN OR/MS IN MARKETING:
The OR/MS literature in marketing is vast. Comprehensive reviews are given in Lilien, Kotler and Moorthy (1992), and Eliashberg and Lilien (1993). Models have been used to explore most facets of marketing and the marketplace, and increasingly marketing research is integrated with appropriate modeling. Developments in the past have been extensive and the future is challenging. Some of the key trends that we see are:

1. *OR/MS in marketing is having important impact both on academic development in marketing and in marketing practice.* During the 1980s two new and important journals were started that emphasize the OR/MS approach: *Marketing Science* and the *International Journal of Research in Marketing* (*IJRM*). Both are healthy, popular, and extremely influential, especially among academics. And both reflect the developments of marketing models. Two excellent discussions on the application and impact of market models on practice are given in Little *et al.* (1994) and Parsons *et al.* (1994). Yet, despite the documented impact of many of the marketing science models, their adoption by potential beneficiaries, especially top management, has still been limited (Wind, 1993 and Wind and Lilien, 1993). Better OR/MS model adoption represents one of the great opportunities facing marketing scientists and OR/MS professionals.

2. *New data sources are having a major impact on marketing modeling.* One of the most influential developments of the 1980s has been the impact of scanner data on the marketing models field. There are typically two or more special sessions at national meetings on the use of scanner data, a special interest conference on the topic was held recently, and a special issue of *IJRM* was devoted to the topic. Scanner data and the closely related single source data (where communication consumption data are tied into diary panel data collected by means of scanners) have enabled marketing scientists to develop and test models with much more precision than ever before. Indeed, the very volume of new data has helped spawn tools to help manage the flow of new information inherent in such data (Schmitz, Armstrong, and Little, 1990).

3. *Stylized theoretical modeling has become a mainstream research tradition in marketing.* While the field of microeconomics has always had a major influence on quantitative model developments in marketing, that influence became most profound in the 1980s. The July 1980 issue of the *Journal of Business* reported on the proceedings of a conference on the interface between marketing and economics. In January 1987, the European Institute for Advanced Studies in Management held a conference on the same topic and reported that "the links between

the two disciplines were indeed strengthening" (Bultez, 1988). Major theoretical modeling developments, primarily in areas of pricing, consumer behavior, product policy, promotions, and channel decisions are covered in detail in Lilien *et al.* (1992); the impact on marketing academics has been dramatic; the opportunities for testing those theories in practice is equally significant.

4. *New tools and methods are changing the content of marketing models.* The November 1982 issue of the *Journal of Marketing Research* was devoted to causal modeling. A relatively new methodology at the time, causal modeling has become a mainstream approach for developing explanatory models of behavioral phenomena in marketing. New developments have also occurred in psychometric modeling. As the August 1985 special issue of *Journal of Marketing Research* on competition in marketing pointed out, techniques like game theory, optimal control theory, and market share/response models are essential elements of the marketing modeler's tool kit. And the explosion of interest in and the potential of artificial intelligence and expert systems approaches to complement traditional marketing modeling approaches have the potential to change the norms and paradigms in the field; expert systems in marketing are discussed in the April 1991 issue of *IJRM* and in Rangaswamy (1991).

5. *Competition and interaction are major thrusts of marketing models today.* The saturation of markets and the economic fights for survival have changed the focus of interest in marketing models, probably forever. A key-word search of the 1989 and 1990 volumes of *Marketing Science, Journal of Marketing Research*, and *Management Science* (marketing articles only) reveals multiple entries for "competition," "competitive strategy," "non-cooperative games," "competitive entry," "late entry," and "market structure." These terms are largely missing in a comparable analysis of the 1969 and 1970 issues of *Journal of Marketing Research, Management Science*, and *Operations Research* (which dropped its marketing section when *Marketing Science* was introduced, but was a key vehicle for marketing papers at that time).

In sum, the OR/MS approach has had a major impact on marketing theory and marketing practice. Marketing science is a respected discipline that is capable of addressing many of the critical issues facing management today. The future is ripe for exciting new developments in constructing, testing and applying new marketing science models to the benefit of management and society.

See **Decision analysis; Game theory; Linear programming; Retailing.**

References

[1] Basu, A., R. Lal, V. Srinivasan and R. Staelin (1985), "Salesforce Compensation Plans: An Agency Theoretic Perspective," *Marketing Science*, 4, 267–291.

[2] Bell, D.E., R.L. Keeney and J.D.C. Little (1975), "A Market Share Theorem," *Jl. Marketing Research*, 12, 136–141.

[3] Bultez, A. (1988), "Editorial for special issue on marketing and microeconomics," *International Jl. Research in Marketing*, 5, 221–224.

[4] Eliashberg, J. and G.L. Lilien, eds. (1993), *Handbooks In Operations Research and Management Science: Marketing*, Elsevier, Amsterdam.

[5] Green, P.E. and A. Krieger (1993), "Voice: An Analytical and Predictive Model for Measuring Customer Satisfaction," Wharton School working paper, University of Pennsylvania, Philadelphia.

[6] Guadagni P. and J.D.C. Little (1983), "A Logit Model of Brand Choice Calibrated on Scanner Data," *Marketing Science*, 2, 203–238.

[7] Kotler, P. (1971), *Marketing Decision Making: A Model Building Approach*, Holt, Rinehart and Winston, New York.

[8] Lilien, G.L. and P. Kolter (1983), *Marketing Decision Making: A Model Building Approach*, Harper & Row, New York.

[9] Lilien, G.L., P. Kotler, and K.S. Moorthy (1992), *Marketing Models*, Prentice-Hall, Englewood Cliffs, New Jersey.

[10] Little, J.D.C. (1975), "BRANDAID: A Marketing Mix Model. Part I: Structure; Part II: Implementation," *Operations Research*, 23, 628–673.

[11] Little, J.D.C. *et al.* (1994), Commentary on "Marketing Science's Pilgrimage to the Ivory Tower," by Hermann Simon. In G. Laurent, G.L. Lilien, and B. Pras, eds., *Research Traditions in Marketing*, 44–51, Kluwer Academic Publishers, Norwell, Massachusetts.

[12] Parsons *et al.* (1994), "Marketing Science, Econometrics, and Managerial Contributions," Commentary on "Marketing Science's Pilgrimage to the Ivory Tower," by Hermann Simon. In G. Laurent, G.L. Lilien, and B. Pras, eds., *Research Traditions in Marketing*, 52–78, Kluwer Academic Publishers, Norwell, Massachusetts.

[13] Rangaswamy, A. (1993), "Marketing Decision Models: From Linear Programs to Knowledge-based Systems." In J. Eliashberg and G.L. Lilien, eds., *Handbooks in Operations Research and Management Science: Marketing*, 16, 733–771, Elsevier, Amsterdam.

[14] Schmitz, J.D., G.D. Armstrong, and J.D.C. Little (1990), "CoverStory: Automated News Finding in Marketing," *Interfaces*, 20(6), 29–38.

[15] Schultz, R.L. and A.A. Zoltners (1981), *Marketing Decision Models*, North Holland, New York.

[16] Wind, Y. (1993), "Marketing Science at a Crossroad," inaugural presentation of the Unilever Erasmus Visiting Professorship in Marketing at the Erasmus University, Rotterdam.

[17] Wind, Y. and G.L. Lilien (1993), "Marketing Strategy Models," In J. Eliashberg and G.L. Lilien, eds., *Handbooks in Operations Research and Management Science: Marketing*, 17, 773–826, Elsevier, Amsterdam.

MARKOV CHAINS

Carl M. Harris

George Mason University, Fairfax, Virginia

INTRODUCTION: A Markov chain is a Markov process $\{X(t), t \in T\}$ whose state space S is discrete, while its time domain T may be either continuous or discrete. In the following discussion, we shall focus only on the countable state-space problem; continuous-time chains are described in **Markov processes**. There is a vast literature on the subject and we recommend Breiman (1986), Cinlar (1975), Chung (1967), Feller (1968), Heyman and Sobel (1982), Isaacson and Madsen (1976), Iosifescu (1980), Karlin and Taylor (1975), Kemeny and Snell (1976), Kemeny, Snell and Knapp (1966), and Parzen (1962).

As a stochastic process of the Markov type, chains possess the Markov or "lack-of-memory" property. The Markov property means that the probabilities of future events are completely determined by the present state of the process and the probabilities of its behavior from the present point on. That is to say, the past behavior of the process provides no additional information in determining the probabilities of future events if the current state of the process is known. Thus we can write that the discrete process $\{X(t), t \in T\}$ is a Markov chain if, for any $n > 0$, any $t_1 < t_2 < \ldots < t_n < t_{n+1}$ in the time domain T, any states $i_1, i_2, \ldots, i_n$ and any state j in the state space S,

$$\Pr\{X(t_{n+1}) = j \mid X(t_1) = i_1, \ldots, X(t_n) = i_n\}$$
$$= \Pr\{X(t_{n+1}) = j \mid X(t_n) = i_n\}.$$

The conditional, transition probabilities on the right-hand side of this equation can be greatly simplified by mapping the n time points directly into the nonnegative integers and renaming state i_n as i. Then, the probabilities are only a function of the pair (i, j) and the transition number n. We (generally) simplify further by assuming that the transition probabilities are stationary, that is, that they are time invariant and thus the same for all transitions. What is left, therefore, is a set of $i \cdot j$ numbers, $[p_{ij}]$, which can be arrayed as a (potentially infinite) square matrix (viz., the *single-step transition matrix*) which gives us all possible conditional probabilities (even infinitely many) of moving to state j in a transition, given that the chain was in state i immediately prior. We typically call this matrix **P**. (Note that any matrix with the property that its rows are nonnegative numbers summing to one is called a *stochastic matrix*, whether or not it is associated with a particular Markov chain.)

EXAMPLES OF MARKOV CHAINS:

(1) *Random Walk*. In simplest form, we have an object that moves to the left one space at each transition time with probability p or to the right with probability $1 - p$. The problem can be kept finite by requiring *reflecting barriers* at fixed left- and right-hand points, say M and N, such that the transition probabilities send the chain back to states $M + 1$ and $N - 1$, respectively, whenever it reaches M or N. One important variation on this problem allows the object to stay put with non-zero probability.

(2) *Gambler's Ruin*. A gambler makes repeated independent bets, and wins \$1 on each bet with probability p or loses \$1 with probability $1 - p$. The gambler starts with an initial stake and will play repeatedly until all money is lost or until the fortune increases to \$$M$. Let X_n equal the gambler's wealth after n plays. The stochastic process $\{X_n, n = 0,1,2,\ldots\}$ is a Markov chain with state space $\{0,1,2,\ldots,M\}$. The Markov property follows from the assumption that outcomes of successive bets are independent events. The Markov model can be used to derive performance measures of interest for this situation, such as the probability of losing all the money, the probability of reaching the goal of \$$M$, and the expected number of bets before the game terminates. All these performance measures are functions of the gambler's initial state x_0, probability p and goal \$$M$. (The gambler's fortune is thus a random walk with *absorbing* boundaries 0 and M.) The gambler's ruin problem is a simplification of more complex systems that experience random rewards, risk, and possible ruin, such as insurance companies.

(3) *Coin Toss Sequence*. Consider a series of independent tosses of a fair coin. Then we have a Markov chain if we say that we are in state 1, 2, 3 or 4 at time n depending on whether the outcomes of tosses $n - 1$ and n are (H,H), (H,T), (T,H) or (T,T), respectively.

We define the n-step transition probability $p_{ij}^{(n)}$ as the probability that the chain moves from state i to state j in n steps, and write

$$p_{ij}^{(n)} = \Pr\{X_{m+n} = j \mid X_m = i\} \quad \text{for all } m \geq 0, n > 0.$$

Then it follows that the n-step transition probabilities can be computed using the *Chapman-Kolmogorov equations*

$$p_{ij}^{(n+m)} = \sum_{k=0}^{\infty} p_{ik}^{(n)} p_{kj}^{(m)} \quad \text{for all } n, m, i, j \geq 0.$$

In particular, we see for $m = 0$ that

$$p_{ij}^{(n)} = \sum_{k=0}^{\infty} p_{ik}^{(n-1)} p_{kj} = \sum_{k=0}^{\infty} p_{kj}^{(n-1)},$$

$$n = 2, 3, \ldots, i, j \geq 0.$$

If we denote the matrix of n-step probabilities by $\mathbf{P}^{(n)}$, then it follows that $\mathbf{P}^{(n)} = \mathbf{P}^{(n-k)} \mathbf{P}^{(k)} = \mathbf{P}^{(n-1)} \mathbf{P}$ and that $\mathbf{P}^{(n)}$ can be calculated as the nth power of the original single-step transition matrix $\mathbf{P}$.

To calculate the unconditional distribution of the state at time n, we must specify the initial probability distribution of the state, namely, $\Pr\{X_0 = i\} = p_i$, $i \geq 0$. Then we calculate the unconditional distribution of X_n as

$$\Pr\{X_n = j\} = \sum_{i=0}^{\infty} \Pr\{X_0 = j \mid X_0 = i\} \cdot \Pr\{X_0 = i\}$$

$$= \sum_{i=0}^{\infty} p_i p_{ij}^{(n)}$$

which is equivalent to multiplying the row vector p by the jth column of $\mathbf{P}$.

PROPERTIES OF A CHAIN: The ultimate long-run behavior of a chain is fully determined by the location and relative size of the entries in the single-step transition matrix. We use these numbers to determine what states can be reached from which other ones and how long it takes on average to make those transitions. More formally, we say that state j is reachable from state i ($i \to j$) if it is possible for the chain to proceed from i to j in a finite number of transitions, that is, if $p_{ij}^{(n)} > 0$ for some $n \geq 0$. If, in addition, $j \to i$, then we say that the two states *communicate* with each other and write $i \leftrightarrow j$. If, now, every state is reachable from every other state in the chain, we say that the chain is *irreducible* (i.e., it is not reducible into sub-classes of states that do not communicate with each other).

Furthermore, we define the *period* of state i as the greatest common divisor, $d(i)$, of the set of positive integers n such that $p_{ii}^{(n)} > 0$ (with $d(i) \equiv 0$ when $p_{ii}^{(n)} = 0$ for all $n \geq 1$). If $d(i) = 1$, then i is said to be *aperiodic*; otherwise, it is periodic. Clearly, any state with $p_{ii} > 0$ is an aperiodic state. All states in a single communicating class must have the same period, and the full Markov chain is said to be aperiodic if all of its states have period 1.

For each pair of states (i,j) of a Markov chain, we define $f_{ij}^{(n)}$ as the probability that a first return from i to j occurs in n transitions and f_{ij} as the probability of ever returning to j from i. If $f_{ij} = 1$, the expectation m_{ij} of this distribution is called the *mean first passage time from i to j*. When $j = i$, we write the respective probabilities as $f_i^{(n)}$ and f_i, and the expectation as m_i, which we now call the *mean recurrence time of i*. If $f_i = 1$ and $m_i < \infty$, the state i is said to be *positive recurrent* or *nonnull recurrent*; if $f_i = 1$ and $m_i = \infty$, the state i is said to be *null recurrent*. If, however, $f_i < 1$, then i is said to be a *transient state*.

A major result that follows from the above is that if $i \leftrightarrow j$ and i is recurrent, then so is j. Furthermore, if the chain happens to be finite, then all states cannot be transient and at least one must be recurrent; if all the states in the finite chain are recurrent, then they are all positive recurrent. More generally, all the states of an irreducible chain are either positive recurrent, null recurrent, or transient.

REFLECTING RANDOM WALK EXAMPLE: Consider such a chain with movement between its four states governed by the single-step transition matrix

$$\begin{bmatrix} 0 & 1 & 0 & 0 \\ 1/3 & 1/3 & 1/3 & 0 \\ 0 & 2/3 & 0 & 1/3 \\ 0 & 0 & 1 & 0 \end{bmatrix}.$$

All the states communicate since we can construct a path with non-zero probability to carry us from state 1 back to state 1 hitting all the other states in the interim. All the states are recurrent and aperiodic as well.

If the random walk were infinite instead and without reflecting barriers, then the chain would be recurrent if and only if it is equally probable to go from right to left from each state, for otherwise the system would drift to $+\infty$ or $-\infty$ without returning to any finite starting point.

LIMITING BEHAVIOR: The major characterizations of the stochastic behavior of a chain are typically stated in terms of its long-run or limiting behavior. To do this, we define the probability that the chain is in state j at the nth transition as $\pi_j^{(n)}$, with the initial distribution written as $\pi_j^{(0)}$. A discrete Markov chain is said to have a *stationary distribution* $\pi = (\pi_0, \pi_1, \ldots)$ if these (legitimate) probabilities satisfy the vector-matrix equation $\pi = \pi \mathbf{P}$. When written out in simultaneous equation form, the problem is equivalent to solving

$\pi_j = \sum_i \pi_i p_{ij} \quad j = 0, 1, 2, \ldots,$ with

$\sum_i \pi_i = 1.$

The chain is said to have a *long-run, limiting, equilibrium,* or *steady-state probability distribution* $\boldsymbol{\pi} = (\pi_0, \pi_1, \ldots)$ if

$\lim_{n \to \infty} \pi_j^{(n)} = \lim_{n \to \infty} \Pr\{X_n = j\} = \pi_j, \quad j = 0, 1, 2, \ldots$

A Markov chain which is irreducible, aperiodic and positive recurrent is said to be *ergodic*, and the following theorem relates these properties to the existence of stationary and/or limiting distributions.

THEOREM: If $\{X_n\}$ is an irreducible, aperiodic, time homogeneous Markov chain, then limiting probabilities

$\pi_j = \lim_{n \to \infty} \Pr\{X_n = j\}, \quad j = 0, 1, 2, \ldots$

always exist and are independent of the initial state probability distribution. If all the states are either null recurrent or transient, then $\pi_j = 0$ for all j and no stationary distribution exists; if all the states are instead positive recurrent (thus the chain is ergodic), then $\pi_j > 0$ for all j the set $\{\pi_j\}$ also forms a stationary distribution, with $\pi_j = 1/m_j$.

It is important to observe that the existence of a stationary distribution does not imply that a limiting distribution exists. An example is the simple Markov chain

$\boldsymbol{P} = \begin{bmatrix} 0 & 1 \\ 1 & 0 \end{bmatrix}.$

For this chain, it is easy to show that the vector $\boldsymbol{\pi} = (1/2, 1/2)$ solves the stationary equation. However, since the chain is oscillating between states 1 and 2, there will be no limiting distribution. It is the fact that the chain has period 2 which violates the sufficient conditions for the above ergodic theorem. Combined with our earlier discussion, this tells us that an irreducible finite-state chain needs to be aperiodic to be ergodic. Note that probabilities $(1/2, 1/2)$ still have meaning because they tell us what the chances are of finding the chain in either state in the limit, even though there is periodic oscillation.

MORE ON THE REFLECTING RANDOM WALK: Let us consider the chain given earlier and derive its steady-state probabilities. We have already shown that the chain is ergodic, so we need now to solve $\boldsymbol{\pi} = \boldsymbol{\pi}\boldsymbol{P}$, which we can write out as the simultaneous system

$\begin{cases} \pi_1 = \dfrac{1}{3}\pi_2 \\ \pi_2 = \pi_1 + \dfrac{1}{3}\pi_2 + \dfrac{1}{3}\pi_3 \\ \pi_3 = \dfrac{1}{3}\pi_2 + \pi_4 \\ \pi_4 = \dfrac{2}{3}\pi_3 \end{cases}$

When these equations are solved in terms of (say) π_1 and then the summability to 1 requirement used, we find that $\boldsymbol{\pi} = (1/9, 3/9, 3/9, 2/9)$. Note that this also means that the limiting n-step matrix here, $\lim_{n \to \infty} \boldsymbol{P}^n$, would have identical rows all equal to the vector $\boldsymbol{\pi}$.

MORE ON THE GAMBLER'S RUIN PROBLEM: Here, we recognize that we have three classes of states, $\{0\}$, $\{1, 2, \ldots, M-1\}$, and $\{M\}$. After a finite time, the gambler will either reach the goal of M units or lose all the money. The key question then is what is the probability that the gambler's fortune will grow to M before all the resources are lost, which we shall call $p_i, i = 0, 1, \ldots, M$. It is not too difficult to show that

$p_i = \begin{cases} \dfrac{1 - [(1-p)/p]^i}{1 - [(1-p)/p]^M} & \text{if } p \neq \dfrac{1}{3} \\ \dfrac{i}{M} & \text{if } p = \dfrac{1}{2} \end{cases}$

MORE ON THE COIN TOSS SEQUENCE PROBLEM: Here, we can show that the single-step transition matrix is given by

$\begin{bmatrix} 1/2 & 1/2 & 0 & 0 \\ 0 & 0 & 1/2 & 1/2 \\ 1/2 & 1/2 & 0 & 0 \\ 0 & 0 & 1/2 & 1/2 \end{bmatrix}$

This particular matrix is very special since its columns also add up to 1; such a matrix is said to be *doubly stochastic*. It can be shown that any doubly stochastic transition matrix coming from a recurrent and aperiodic finite chain has the discrete uniform steady-state probabilities $\pi_j = 1/M$.

CONCLUDING REMARKS: We emphasize that the more complete context of the discrete-time Markov chain is as a special case of a Markov process. Markov chains in continuous time include birth-death processes (and their application in such areas as queueing) and the Poisson process.

See **Markov processes; Queueing theory; Stochastic processes.**

References

[1] Breiman, L. (1986). *Probability and Stochastic Processes, With a View Toward Applications*, Second Edition. The Scientific Press, Palo Alto, California.

[2] Çinlar, E. (1975). *Introduction to Stochastic Processes*. Prentice-Hall, Englewood Cliffs, New Jersey.

[3] Chung, K.L. (1967). *Markov Chains with Stationary Transition Probabilities*. Springer-Verlag, New York.

[4] Feller, W. (1968). *An Introduction to Probability Theory and Its Applications, Volume I*, Third Edition. Wiley, New York.

[5] Heyman, D.P. and M.J. Sobel (1982). *Stochastic Models in Operations Research, Volume I: Stochastic Processes and Operating Characteristics*. McGraw-Hill, New York.

[6] Iosifescu, M. (1980). *Finite Markov Processes and Their Application*. Wiley, New York.

[7] Isaacson, D.L. and R.W. Madsen (1976). *Markov Chains: Theory and Applications*. Wiley, New York.

[8] Karlin, S. and H.M. Taylor (1975). *A First Course in Stochastic Processes*, Second Edition. Academic Press, New York.

[9] Kemeny, J.G. and J.L. Snell (1976). *Finite Markov Chains*. Springer-Verlag, New York.

[10] Kemeny, J.G., J.L. Snell, and A.W. Knapp (1966). *Denumerable Markov Chains*. Van Nostrand, Princeton.

[11] Parzen, E. (1962). *Stochastic Processes*. Holden-Day, San Francisco.

MARKOV DECISION PROCESSES

C. C. White, III

University of Michigan, Ann Arbor

The finite-state, finite-action Markov decision process is a particularly simple and relatively tractable model of sequential decision making under uncertainty. It has been applied in such diverse fields as health care, highway maintenance, inventory, machine maintenance, cashflow management, and regulation of water reservoir capacity. Here we present a definition of a Markov decision process and illustrate it with an example, followed by a discussion of the various solution procedures for several different types of Markov decision processes, all of which are based on dynamic programming.

PROBLEM FORMULATION: Let $k \in \{0, 1, \ldots, K-1\}$ represent the kth stage or decision epoch, that is, when the kth decision must be selected;

$K < \infty$ represents the planning horizon of the Markov decision process. Let s_k be the state of the system to be controlled at stage k. This state must be a member of a finite set S, called the state space, where $s_k \in S$, $k = 0, 1, \ldots, K$. The state process $\{s_k, k = 0, 1, \ldots, K\}$ makes transitions according to the conditional probabilities

$$p_{ij}(a) = \text{Prob}(s_{k+1} = j \mid s_k = i, a_k = a),$$

where a_k is the action selected at stage k. The action selected must be a member of the finite action space A, which is allowed to depend on the current state value, that is, $a_k \in A(i)$, when $s_k = i$. We allow a_k to be selected on the basis of the current state s_k for all k. Let δ_k be a mapping from the state space into the action space satisfying $\delta_k(s_k) \in A(s_k)$. Then δ_k is called a policy and a sequence of policies $\pi = \{\delta_0, \ldots, \delta_{K-1}\}$ is known as a strategy.

Let $r(i,a)$ be the one-stage reward accrued at stage $k = 0, 1, \ldots, K-1$ if $s_k = i$ and $a_k = a$. Assume $r(i)$ is the terminal reward accrued at stage K (assuming $K < \infty$) if $s_k = i$. The total discounted reward over the planning horizon accrued by strategy $\pi = \{\delta_0, \ldots, \delta_{K-1}\}$ is then given by

$$\sum_{k=0}^{K-1} \beta^k r(s_k, a_k) + \beta^K \bar{r}(s_K)$$

if $a_k = \delta_k(s_k)$, $k = 0, 1, \ldots, K-1$, where β is the nonnegative real-valued discount factor. The problem objective is to select a strategy, called the optimal strategy, that maximizes the expected value of the total discounted reward, with respect to the set of all strategies.

An Example – Assume that an inspector must decide at each stage, on the basis of a machine's current state of deterioration, whether to replace the machine, repair it, or do nothing. We assume that the machine can be in one of M states, and we may write $S = \{1, \ldots, M\}$, where 1 represents the "perfect" machine state, M represents the "failed" machine state, and $1 < m < M$ represents an imperfect but functioning state of the machine. Assume that each week the machine inspector can choose to let the machine produce (the "do nothing" decision $a = 1$), completely replace the machine (the "replace" decision $a = R$), or perform some sort of maintenance on the machine, $1 < a < R$. Thus, $A = \{1, \ldots, R\}$. Let $c(i,a)$ be the cost accrued over the following week if at the beginning of the week the machine is in state i and the machine inspector selects action a. Let β be the current value of a dollar to be received next week. Assume the transition probabilities $p_{ij}(a)$

are known for all $i, j \in S$ and $a \in A$. We would expect $p_{ij}(1) = 0$ if $j > i$.

DYNAMIC PROGRAMMING FORMULATION (FINITE STAGE CASE):

We next formulate the Markov decision process as a dynamic program for the finite planning horizon case. Such an assumption models situations where an optimal strategy is sought for only a finite period of time, such as the length of time until retirement for an individual. Let $f_k(i)$ be the optimal expected total discounted reward to be accrued from stage k through the terminal stage K, assuming $s_k = i$. We note that $f_k(i)$ should differ from $f_{k+1}(s_{k+1})$ only by the reward accrued at stage k. In fact, it is easily shown that f_k and f_{k+1} are related by the dynamic programming equation

$$f_k(i) = \max_{a \in A(i)} \left\{ r(i, a) + \beta \sum_{j \in S} p_{ij}(a) f_{k+1}(j) \right\},$$

which has boundary condition $f_K(i) = r(i)$. Note also that an optimal strategy $\pi^* = \{\delta_0^*, \ldots \delta_{K-1}^*\}$ necessarily and sufficiently satisfies

$$f_k(i) = r[i, \delta_k^*(i)] + \beta \sum_j p_{ij}[\delta_k^*(i)] f_{k+1}(j)$$

for all $k = 0, 1, \ldots, K - 1$. Thus, the action that should be taken at stage k, given $s_k = i$, is any action that achieves the maximum in

$$\max_{a \in A(i)} \left\{ r(i, a) + \beta \sum_j p_{ij}(a) f_{k+1}(j) \right\}.$$

THE INFINITE HORIZON DISCOUNTED REWARD CASE:

Assume $K = \infty$, which is an appropriate assumption for an institutional investor, for example, rather than an individual investor. Clearly, there may exist strategies that could be expected to generate an infinite reward. However, if the discount factor β is strictly less than 1, no such strategy exists. This fact can be verified by noting that

$$\sum_{k=0}^{\infty} \beta^k r(s_k, a_k) \le \sum_{k=0}^{\infty} \beta^k \max_{i,a} |r(i, a)| = \frac{\max_{i,a} |r(i, a)|}{1 - \beta}.$$

It seems reasonable that the dynamic program for the infinite horizon case could be related to the dynamic program for the finite horizon case. Let us define m as the number of stages to go until the terminal stage of the finite horizon case. The dynamic program for the finite horizon problem can then be rewritten as

$$g_{m+1}(i) = \max_{a \in A(i)} \left\{ r(i, a) + \beta \sum_j p_{ij}(a) g_m(j) \right\}$$

where $f_k(i) = g_{K-k}(i)$. Now the optimal expected total discounted reward should be $g(i) = \lim_{m \to \infty} g(i)$ for initial state i, which should satisfy

$$g(i) = \max_{a \in A(i)} \left\{ r(i, a) + \beta \sum_j p_{ij}(a) g(j) \right\} \quad (1)$$

if the limit and maximization operators can be interchanged. It so happens that this interchange is possible under the conditions considered here, and hence the optimal expected total discounted reward uniquely satisfies (1). It can also be shown that an optimal strategy exists that is stage invariant and that this strategy, or equivalently, policy, satisfies

$$g(i) = r[i, \delta^*(i)] + \beta \sum_j p_{ij}[\delta^*(i)] g(j) \quad (1a)$$

for all $i \in S$.

SOLUTION PROCEDURES:

We now present three different computational approaches for determining g and δ^* in (1).

Linear Programming – The following linear program can solve the infinite discounted Markov decision process:

minimize $\sum_{i \in S} g(i)$

subject to

$$g(i) - \beta \sum_j p_{ij}(a) g(j) \ge r(i, a)$$

where the constraint inequality must be satisfied for all $i \in S$ and $a \in A(i)$, $i \in S$.

Successive Approximations – This procedure, in its simplest form, involves determining $g_m(i)$ for large m, using the iteration equation

$$g_m(i) = \max_{a \in A(i)} \left\{ r(i, a) + \beta \sum_j p_{ij}(a) g_{m-1}(j) \right\},$$

where $g_0(i)$ can be arbitrarily selected. (Of course, we should attempt to select g_0 as close to g as possible if we have some way of estimating g a priori.)

Policy Iteration – This computational procedure involves the following iterative approach.

Step 0: Select δ.
Step 1: Determine g_δ, where g_δ satisfies

$$g_\delta(i) = r[i, \delta(i)] + \beta \sum_j p_{ij}[\delta(i)] g_\delta(j).$$

Note that

$$g_\delta = (I - \beta P_\delta)^{-1} r_\delta$$

where $P_\delta = \{p_{ij}[\delta(i)]\}$, $g_\delta = \{g_\delta(i)\}$, $r_\delta = \{r[i, \delta(i)]\}$,

I is the identity matrix, and the inverse is guaranteed to exist since $\beta < 1$.

Step 2: Determine δ' that satisfies

$$r[i, \delta'(i)] + \beta \sum_j p_{ij}[\delta'(i)]g_\delta(j)$$

$$= \max_{a \in A(i)} \left\{ r(i, a) + \beta \sum_j p_{ij}(a)g_\delta(j) \right\}.$$

Step 3: Set $\delta = \delta'$ and return to Step 1 until g_δ and $g_{\delta'}$ are sufficiently close.

Note that each of the above solution procedures is far more efficient than exhaustive enumeration. Much current research is underway to develop procedures for combining policy iteration and successive approximations into particularly efficient computational procedures for large-scale discounted Markov decision processes with infinite horizon.

MARKOV DECISION PROCESSES WITHOUT DISCOUNTING (THE AVERAGE REWARD CASE):

Let us assume that the criterion is

$$\lim_{K \to \infty} \left(\frac{1}{K+1} \right) E \left\{ \sum_{k=0}^K r(s_k, a_k) \right\}$$

which is the expected average reward criterion. When the system operates under stationary policy δ, it can be shown that there exist values $v_\delta(i)$, $i \in S$, and a state independent gain γ_δ, which satisfy

$$\gamma_\delta + v_\delta(i) = r[i, \delta(i)] + \sum_j p_{ij}[\delta(i)]v_\delta(j) \qquad (2)$$

if P_δ is ergodic. Let γ^*, δ^* and v be such that

$$\gamma^* + v(i) = \max_{a \in A(i)} \left\{ r(i, a) + \sum_j p_{ij}v(j) \right\}$$

$$= r[i, \delta^*(i)] + \sum_j p_{ij}[\delta^*(i)]v(j)$$

where we assume P_δ is ergodic for all δ. Then, γ^* is the value of the criterion generated by an optimal strategy and δ^* is an optimal strategy. We now present a policy iteration procedure for determining γ^*, δ^* and v, where it is necessary only to know v up to a positive constant due to the sum-to-one characteristic of the probabilities.

Step 0: Choose δ.
Step 1: Solve equation (2) for v_δ and γ_δ, where for some i, $v_\delta(i) = 0$.
Step 2: Determine a policy δ' that achieves the maximum in

$$\max_{a \in A(i)} \left\{ r(i, a) + \sum_j p_{ij}v_\delta(i) \right\}.$$

Step 3: Set $\delta = \delta'$ and go to Step I until γ_δ and $\gamma_{\delta'}$ are sufficiently close.

SUMMARY: We have briefly examined the Markov decision process when the state and action spaces are finite; the reward is temporarily separable; all rewards, the discount factor, and all transition probabilities are known precisely and the current state can be accurately made available to the decision maker before selection of the current alternative. All of these restrictions have been weakened in the current literature. Much research effort is devoted to improving the computational tractability of large-scale Markov decision processes so as to improve both the validity and tractability of this modeling tool.

See **Dynamic programming; Markov processes.**

References

[1] Bertsekas, D.P. (1976). *Dynamic Programming and Stochastic Control*. Academic Press, New York.
[2] Derman, C. (1970). *Finite State Markovian Decision Processes*. Academic Press, New York.
[3] Howard, R. (1971). *Dynamic Programming and Markov Processes*. MIT Press, Cambridge, Massachusetts.
[4] Ross, S.M. (1970). *Applied Probability Models with Optimization Applications*. Holden-Day, San Francisco.
[5] White, D.J. (1969). *Dynamic Programming*. Holden-Day, San Francisco.

MARKOVIAN ARRIVAL PROCESS (MAP)

See **Matrix-analytic stochastic models.**

MARKOV PROCESSES

Douglas R. Miller

George Mason University, Fairfax, Virginia

INTRODUCTION: A *Markov process* is a stochastic process $\{X(t), t \in T\}$ with state space S and time domain T that satisfies the *Markov property*. The Markov property is also known as *lack of memory*. For a stochastic process, probabilities of behavior of the process at future times usually depend on the behavior of the process at times in the past. The Markov property means that probabilities of future events are completely determined by the present state of the process: if the current state of the process is known, the past behavior of the process provides no additional information in determining the probabilities of future events. Mathematically, the process $\{X(t),\ t \in T\}$ is Markov if, for any $n > 0$, any $t_1 < t_2 < \ldots < t_n < t_{n+1}$ in the time domain T, and any states $x_1, x_2, \ldots, x_n$ and any set A in the state space S,

$$\Pr\{X(t_{n+1}) \in A \mid X(t_1) = x_1, \ldots, X(t_n) = x_n\}$$
$$= \Pr\{X(t_{n+1}) \in A \mid X(t_n) = x_n\}.$$

The conditional probabilities on the right hand side of this equation are the *transition probabilities* of the Markov process; they play a key role in the study of Markov processes. The transition probabilities of the process are presented as a *transition function* $p(s,x;t,A) = \Pr\{X(t) \in A \mid X(s) = x\}$, $s < t$, for $s,t \in T$, $x \in S$, and $A \subset S$. The *initial distribution* of the process is $q(A) = \Pr\{X(0) \in A\}$, for $A \subset S$. The distribution of a Markov process is uniquely determined by an initial distribution $q(\cdot)$ and a transition function $p(.,\ldots,.)$: for $0 = t_0 < t_1 < \ldots < t_n$ in the time domain, and subsets $A_1, A_2, \ldots, A_n$ of the state space S,

$$\Pr\{X(t_1) \in A_1, \ldots, X(t_n) \in A_n\}$$
$$= \int\limits_{x_o \in S} q(dx_0) \int\limits_{x_1 \in A_1} p(t_0, x_0; t_1, dx_1) \ldots$$
$$\int\limits_{x_{n-1} \in A_{n-1}} p(t_{n-2}, x_{n-2}; t_{n-1}, dx_{n-1})$$
$$p(t_{n-1}, x_{n-1}; t_n, A_n)$$

An equivalent interpretation of the Markov property is that the past behavior and the future behavior of the process are conditionally independent given the present state of the process: for any $m > 0$, any $n > 0$, any $t_{-m} < \ldots < t_{-1} < t_0 < t_1 < t_n$ in the time domain, and any state x_0 and any sets $A_1, A_2, \ldots, A_m$ and $B_1, B_2, \ldots, B_n$ in the state space S,

$$\Pr\{X(t_{-m}) \in A_m, \ldots, X(t_{-1}) \in A_1, X(t_1) \in B_1, \ldots, X(t_n)$$
$$\in B_n \mid X(t_0) = x_0\} = \Pr\{X(t_{-m}) \in A_m, \ldots, X(t_{-1})$$
$$\in A_1 \mid X(t_0) = x_0\} \cdot \Pr\{X(t_1) \in B_1, \ldots, X(t_n)$$
$$\in B_n \mid X(t_0) = x_0\}.$$

A Markov process has *stationary transition probabilities* if the transition probabilities are time-invariant, i.e., for s, $t > 0$, $\Pr\{X(s + t) \in A \mid X(s) = x\} = \Pr\{X(t) \in A \mid X(0) = x\}$. In this case the transition function takes the simplified form $p_t(x, A) = \Pr\{X(t) \in A \mid X(0) = x\}$. Most Markov process models assume stationary transition probabilities.

CLASSIFICATION OF MARKOV PROCESSES:

There is a natural classification of Markov processes according to whether the time domain T and the state space S are denumerable or non-denumerable. This yields four general classes. Denumerable time domains are usually modeled as the integers or non-negative integers. Non-denumerable time domains are usually modeled as the continuum ($\mathbb{R}$ or $[0,\infty]$). Denumerable state spaces can be modeled as the integers, but it is often useful to retain other descriptions of the states rather than simply enumerating them. Non-denumerable state spaces are usually modeled as a one or higher dimensional continuum. Roughly speaking, "discrete" is equivalent to denumerable and "continuous" is equivalent to non-denumerable. In 1907 Markov considered a discrete time domain and a finite state space; he used the word "chain" to denote the dependence over time, hence the term *Markov chain* for Markov processes with discrete time and denumerable states. See Maistrov (1974) for some historical discussion and see Appendix B of Howard (1971) for a reprint of one of Markov's 1907 papers. There is no universal convention for the scope of definition of "Markov chain." Chung (1967) defines Markov processes with denumerable state spaces to be Markov chains. Iosifescu (1980) and the Rumanian school use the convention that "Markov chain" applies to discrete time and any state space while "Markov process" applies to continuous time and any state space. The terminology varies in popular texts: Karlin and Taylor (1975, 1981) agree with Chung; Breiman (1968) agrees with the Rumanians. The terms *discrete time Markov chain* (*DTMC*) and *continuous time Markov chain* (*CTMC*) are sometimes used to clarify the situation.

Here are four examples of Markov processes representing the four classes with respect to discrete or continuous time and denumerable or continuous state space.

a. *Gambler's Ruin* (*discrete time/denumerable states*). A gambler makes repeated bets. On each bet he wins \$1 with probability p or loses \$1 with probability $1 - p$. Outcomes of successive bets are independent events. He starts with a certain initial stake and will play repeatedly until he loses all his money or until he increases his fortune to \$M. Let X_n equal the gambler's wealth after n plays. The stochastic process $\{X_n, n = 0,1,2, \ldots\}$ is a discrete time Markov chain (DTMC) with state space $\{0,1,2,\ldots, M\}$. The Markov property follows from the assumption that outcomes of successive bets are independent events. The Markov model can be used to derive performance measures of interest for this situation: *e.g.*, the probability he loses all his money, the probability he reaches his goal of $M, and the expected number of times he makes a bet. All these performance measures are functions of his initial stake x_0, probability p and goal M. (The gambler's fortune is a random walk with absorbing boundaries 0 and M.) The gambler's ruin is a simplification of more complex systems that experience random rewards, risk, and possible ruin; for example, insurance companies.

b. *A Maintenance System* (*continuous time/denumerable states*). A system consists of two machines and one repairman. Each machine operates until it breaks down. The machine is then repaired and put back into operation. If the repairman is busy with the other machine, the just broken machine waits its turn for repair. So, each machine cycles through the states: operating (O), waiting (W), and repairing (R). Labelling the machines as "1" and "2" and using the corresponding subscripts, the states of the system are (O_1,O_2), (O_1,R_2), (R_1,O_2), (W_1,R_2) and (R_1,W_2). We assume that all breakdown instances and repairs are independent of each other and that the operating times until breakdown and the repair times are random with exponential distributions. The mean operating times for the machines are $1/\alpha_1$ and $1/\alpha_2$, respectively (so the machines break down at rates α_1 and α_2). The mean repair times for the machines are $1/\beta_1$ and $1/\beta_2$, respectively (so the machines are repaired at rates β_1 and β_2). Letting $X_i(t)$ equal the state of machine "i" at time t, the stochastic process $\{(X_1(t),X_2(t)), 0 \leq t\}$ is a continuous time Markov chain (CTMC) on a state space consisting of five states. The Markov property follows from the assumption about independent exponential operating times and repair times. (The exponential distribution is the only continuous distribution with lack-of-memory.) For this type of system there are several performance measures of interest: for example, the long-run proportion of time both machines are broken or the long-run average number of working machines. This maintained system is a simplified example of more complex maintained systems.

c. *Quality Control System* (*discrete time/continuous states*). A manufacturing system produces a physical part that has a particularly critical length along one dimension. The specified value for the length is α. However, the manufacturing equipment is imprecise. Successive parts produced by this equipment vary randomly from the desired value, α. Let X_n equal the size of the nth part produced. The noise added to the system at each step is modeled as $D_n \sim \text{Normal}(0,\delta^2)$. The system can be controlled by attempting to correct the size of the $(n + 1)^{\text{st}}$ part by adding $c_n = -\beta(x_n - \alpha)$ to the current manufacturing setting after observing the size x_n of the nth part; however, there is also noise in the control so that, in fact, $C_n \sim \text{Normal}(c_n,(\gamma c_n)^2)$ is added to the current setting. This gives $X_{n+1} = X_n + C_n + D_n$. The process $\{X_n, n = 0,1,2, \ldots\}$ is a discrete-time Markov process on a continuous state space. The Markov property will follow if all the noise random variables (D_n's) are independent and the control random variables (C_n's) depend only on the current setting (X_n) of the system. Performance measures of interest for this system include the long-run distribution of lengths produced (if the system is stable over the long-run). There is also a question of determining the values of β for which the system is stable and then finding the optimal value of.

d. *Brownian Motion* (*continuous time/continuous states*). In 1828, English botanist Robert Brown observed random movement of pollen grains on the surface of water. The motion is caused by collisions with water molecules. The displacement of a pollen grain as a function of time is a two-dimensional Brownian motion. A one-dimensional Brownian motion can be obtained by scaling a random walk: Consider a sequence of independent, identically-distributed random variables, Z_i, with $P\{Z_i = +1\} = P\{Z_i = 1\} = 1/2$, $i = 1,2,\ldots$. Let $S_n = \Sigma_{i=1}^n Z_i$, $n = 0,1,2,\ldots$. Then, let $X_n(t) = n^{1/2}S_{[nt]}$, $0 \leq t \leq 1$, $n = 1,2,\ldots$, where $[nt]$ is the greatest integer $n \leq t$. As $n \to \infty$, the sequence of processes $\{X_n(t), 0 \leq t \leq 1\}$ converges to $\{W(t), 0 \leq t \leq 1\}$, *standard Brownian motion* or the *Wiener process*; see Billingsley (1968). The Wiener process is a continuous-time, continuous-state Markov process. The sample paths of the Wiener process are continuous. *Diffusions* are the general class of continuous-time, continuous-state Markov processes with continuous sample paths. Diffusion models are useful approximations to discrete processes analogous to how the Wiener process is an approximation to the above random walk process $\{S_n, n = 0,1,2, \ldots\}$; see Glynn (1990). *Geometric Brownian motion* $\{Y(t), 0 \leq t\}$ is defined as $Y(t) = \exp(\sigma W(t))$, $0 \leq t$; it is a diffusion. Geometric Brownian motion has been suggested as a model for stock price fluctuations; see Karlin and Taylor (1975). A performance measure of interest is the distribution of the maximum value of the process over a finite time interval.

There are various performance measures that can be derived for Markov process models. Some specific performance measures were mentioned for the above examples. Some general behavioral properties and performance measures are now described. The descriptions are for a discrete-time Markov chain $\{X_n, n = 0, 1, 2, \ldots\}$ but similar concepts apply to other classes of Markov processes. A Markov chain is *strongly ergodic* if X_n converges in distribution as $n \to \infty$, independent of the initial state x_0. A Markov chain is *weakly ergodic* if $n^{-1} \Sigma_{i=1}^n X_i$ converges to a constant as $n \to \infty$, independent of the initial state x_0. Also as $n \to \infty$, under certain conditions and for real-valued functions $f: S \to \mathbb{R}$, $f(X_n)$ converges in distribution, $n^{-1} \Sigma_{i=1}^n f(X_i)$ converges to a constant, and $n^{-1/2} \Sigma_{i=1}^n [f(X_i) - Ef(X_i)]$ is asymptotically normal. Markov process theory identifies conditions for ergodicity, conditions for the existence of limits, and provides methods for evaluation of limits when they exist. For example, in the above maintained system example, $f(\cdot)$ might be a cost function and the performance measure of interest is long-run average cost. The above performance is long-run (or *infinite-horizon*, or *steady-state*, or *asymptotic*) behavior. Short-run (or *finite-horizon*, or *transient*) behavior and performance is also of interest. For a subset A of the state space S, the *passage time* T_A is the time of the first visit of the process to A: $T_A = \min\{n : X_n \in A\}$. The *hitting probability* $\Pr\{T_A < \infty\}$, the distribution of T_A, and $E(T_A)$ are of interest. In the gambler's ruin example, the gambler wants to know the hitting probabilities for sets $\{0\}$ and $\{M\}$. Transient analysis of Markov processes investigates these and other transient performance measures. The analysis of performance measures takes on different forms for the four different classes of Markov processes.

Evaluation of performance measures for Markov process models of complex systems may be difficult. Standard numerical analysis algorithms are sometimes useful, and specialized algorithms have been developed for Markov models; for example, see Grassmann (1990). Workers in the *field of computational probability* have developed and evaluated numerical solution techniques for Markov models by exploiting special structure and probabilistic behavior of the system or by using insights gained from theoretical probability analysis. In this spirit, Neuts (1981) has developed algorithms for a general class of Markov chains. A structural property of Markov chains called "reversibility" leads to efficient numerical methods of performance evaluation; see Keilson (1979), Kelly (1979), and Whittle (1986). There is a relationship between discrete-time and continuous-time Markov chains called "uniformization" or "ran-domization" that can be used to calculated performance measures of continuous-time Markov chains; see Keilson (1979) and Gross and Miller (1984). For Markov chains with huge state spaces, Monte Carlo simulation can be used as an efficient numerical method for performance evaluation; see, for example, Hordijk, Iglehart and Schassberger (1976) and Fox (1990).

There are classes of stochastic processes related to Markov processes. There are stochastic processes that exhibit some lack of memory but are not Markovian. Regenerative processes have lack of memory at special points (regeneration points) but at other times the process has a memory; see Çinlar (1975). A semi-Markov process is a discrete-state continuous-time process that makes transitions according to a DTMC but may have general distributions of holding times between transitions; see Çinlar (1975). It is sometimes possible to convert a non-Markovian stochastic process into a Markov process by expanding the state description with supplementary variables; i.e., $\{X(t), 0 \le t\}$ may be non-Markovian but $\{(X(t), Y(t)), 0 \le t\}$ is Markovian. Supplementary variables are often elapsed times for phenomena with memory; in this way very general discrete state stochastic systems can be modeled as Markov processes with huge state spaces. The general model for discrete-event dynamic systems is the generalized semi-Markov process (GSMP); see Whitt (1980) and Cassandras (1993).

The index set T of a stochastic process $\{X(t), t \in T\}$ may represent "time" or "space" or both. We have temporal processes, spatial processes, or spatial-temporal processes when the index set is time, space, or space-time, respectively. Stochastic processes with multi-dimensional index sets are called *random fields*. The Markov property can be generalized to the context of multi-dimensional index sets resulting in *Markov random fields*; see Kelly (1978), Kindermann and Snell (1976) and Whittle (1986). Markov random fields have many applications. They are models for statistical mechanical systems (interacting particle systems). They are useful in texture analysis and image analysis; see Chellappa and Jain (1993).

See **Markov chains; Markov decision processes.**

References

[1] Billingsley, P. (1968). *Convergence of Probability Measures*. Wiley, New York.

[2] Breiman, L. (1968). *Probability*. Addison-Wesley, Reading, Massachusetts.

[3] Breiman, L. (1986). *Probability and Stochastic Processes, With a View Toward Applications*, Second Edition. The Scientific Press, Palo Alto, California.

[4] Cassandras, C.G. (1993). *Discrete Event Systems: Modeling and Performance Analysis*. Irwin, Boston.

[5] Chellappa, R. and A. Jain, eds. (1993). *Markov Random Fields: Theory and Application*. Academic Press, San Diego.

[6] Çinlar, E. (1975). *Introduction to Stochastic Processes*. Prentice-Hall, Englewood Cliffs, New Jersey.

[7] Chung, K.L. (1967). *Markov Chains with Stationary Transition Probabilities*. Springer-Verlag, New York.

[8] Feller, W. (1968). *An Introduction to Probability Theory and Its Applications, Volume I*, Third Edition. Wiley, New York.

[9] Feller, W. (1971). *An Introduction to Probability Theory and Its Applications, Volume II*, Second Edition. Wiley, New York.

[10] Fox, B.L. (1990). "Generating Markov-Chain Transitions Quickly." *ORSA J. Comput.* **2**, 126–135.

[11] Glynn, P.W. (1989). "A GSMP Formalism for Discrete Event Systems." *Proc. IEEE* **77**, 14–23.

[12] Glynn, P.W. (1990). "Diffusion Approximations." In *Handbooks in OR and MS, Volume 2*, D.P. Heyman and M.J. Sobel (eds.). Elsevier Science Publishers, Amsterdam, 145–198.

[13] Grassman, W.K. (1990). "Computational Methods in Probability." In *Handbooks in OR and MS, Volume 2*, D.P. Heyman and M.J. Sobel (eds.). Elsevier Science Publishers, Amsterdam, 199–254.

[14] Gross, D. and D.R. Miller (1984). "The Randomization Technique as a Modelling Tool and Solution Procedure for Transient Markov Processes." *Oper. Res.* **32**, 343–361.

[15] Heyman, D.P. and M.J. Sobel (1982). *Stochastic Models in Operations Research, Volume I: Stochastic Processes and Operating Characteristics*. McGraw-Hill, New York.

[16] Hordijk, A., D.L. Iglehart, and R. Schassberger (1976). "Discrete-time methods for simulating continuous-time Markov chains." *Adv. Appl. Probab.* **8**, 772-778.

[17] Howard, R.A. (1971). *Dynamic Probabilistic Systems, Volume I: Markov Models*. Wiley, New York.

[18] Iosifescu, M. (1980). *Finite Markov Processes and their Application*. Wiley, New York.

[19] Isaacson, D.L. and R.W. Madsen (1976). *Markov Chains: Theory and Applications*. Wiley, New York.

[20] Karlin, S. and H.M. Taylor (1975). *A First Course in Stochastic Processes*, Second Edition. Academic Press, New York.

[21] Karlin, S. and H.M. Taylor (1981). *A Second Course in Stochastic Processes*. Academic Press, New York.

[22] Keilson, J. (1979). *Markov Chain Models – Rarity and Exponentiality*. Springer-Verlag, New York.

[23] Kelly, F.P. (1979). *Reversibility and Stochastic Networks*. Wiley, New York.

[24] Kemeny, J.G. and J.L. Snell (1976). *Finite Markov Chains*. Springer-Verlag, New York.

[25] Kemeny, J.G., J.L. Snell, and A.W. Knapp (1966). *Denumerable Markov Chains*. Van Nostrand, Princeton.

[26] Kindermann, R. and J.L. Snell (1980). *Markov Random Fields and their Applications*. American Mathematical Society, Providence, Rhode Island.

[27] Maistrov, L.E. (1974). *Probability Theory: A Historical Sketch*. Academic Press, New York.

[28] Neuts, M.F. (1981). *Matrix-Geometric Solutions in Stochastic Models*. The Johns Hopkins University Press, Baltimore.

[29] Parzen, E. (1962). *Stochastic Processes*. Holden-Day, San Francisco.

[30] Snell, J.L. (1988). *Introduction to Probability*. Random House, New York.

[31] Whitt, W. (1980). "Continuity of Generalized Semi-Markov Processes." *Math. Opns. Res.* **5**, 494–501.

[32] Whittle, P. (1986). *Systems in Stochastic Equilibrium*. Wiley, New York.

MARKOV PROPERTY

When the behavior of a stochastic process $\{X(t), t \in T\}$ at times in the future depends only on the present state of the process (past behavior of the process affects the future behavior only through the present state of the process); that is, for any $n > 0$, any set of time points $t_1 < t_2 < \ldots < t_n < t_{n+1}$ in the time domain T, and any states $x_1, x_2, \ldots, x_n$ and any set A in the state space, $\Pr\{X(t_{n+1}) \in A \,|\, X(t_1) = x_1, \ldots, X(t_n) = x_n\} = \Pr\{X(t_{n+1}) \in A \,|\, X(t_n) = x_n\}$. See **Markov chains**; **Markov processes**.

MARKOV RANDOM FIELD

A random field that satisfies a generalization of the Markov property.

MARKOV RENEWAL PROCESS

When the times between successive transitions of a Markov chain are independent random variables indexed on the to and from states of the chain. See **Markov chains**; **Markov processes**; **Networks of queues**; **Renewal processes**.

MARKOV ROUTING

The process of assigning customers to nodes in a queueing network according to a Markov chain over the set of nodes, where $p(j,k)$ is the probability that a customer exiting node j proceeds next to node k, with $1 - \Sigma p(j,k)$ being the probability a customer leaves the network from node j (the sum is over all nodes of the network, including leaving the network altogether). See **Networks of queues**.

MARRIAGE PROBLEM

Given a group of m men and m women, the marriage problem is to couple the men and women such that the "total happiness" of the group is maximized when the assigned couples marry. The women and men determine an $m \times m$ table of happiness coefficients, where the coefficient a_{ij} represents the happiness rating for the couple formed by woman i and man j if they marry. The larger the a_{ij}, the higher the happiness. The problem can be formulated as an assignment problem whose solution matches each woman to one man. This result, which is due to the fact that the assignment problem has a solution in which the variables can take on only the values of 0 or 1, is sometimes used to "prove" that monogamy is the best form of marriage. See **Assignment problem**.

MASTER PROBLEM

The transformed extreme-point problem that results when applying the Dantzig-Wolfe decomposition algorithm. See **Dantzig-Wolfe decomposition algorithm**.

MATCHING

Richard W. Eglese

The Management School,
Lancaster University, UK

Matching problems form an important branch of *graph theory*. They are of particular interest because of their application to problems found in operations research. Matching problems also form a class of integer linear programming problems which can be solved in polynomial time. A good description of the historical development of matching problems and their solutions is contained in the preface of Lovasz and Plummer (1986).

Given a simple non-directed graph $G = [V,E]$ (where E is a set of edges), then a *matching* is defined as a subset of edges M such that no two edges of M are adjacent. A matching is said to *span* a set of vertices X in G if every vertex in X is incident with an edge of the matching. A *perfect matching* is a matching which spans V. A *maximum matching* is a matching of maximum cardinality, i.e. a matching with the maximum number of members in the set.

A graph is called a *bipartite graph* if the set of vertices V is the disjoint union of sets V_1 and V_2 and every edge in E has the form (v_1, v_2) where v_1 is a member of V_1 and v_2 is a member of V_2.

MATCHING ON BIPARTITE GRAPHS: The first type of matching problems are those which can be formulated as matching problems on a bipartite graph. For example, suppose V_1 represents a set of workers and V_2 represents a set of tasks to be performed. If each worker is able to perform a subset of the tasks and each task may be performed by some subset of the workers, the situation may be modeled by constructing a bipartite graph G, where there is an edge between v_1 in V_1 and v_2 in V_2 if and only if worker v_1 can perform task v_2. If it is assumed that each worker may only be assigned one task and each task may only be assigned to be carried out by one worker we have a form of assignment problem. To find the maximum number of tasks which can be performed, the maximum matching on G must be found. If a measure of effectiveness can be associated with assigning a worker to a task, then the question may be asked as to how the workers should be assigned to tasks to maximize the total effectiveness. This is a maximum weighted matching problem. If costs are given in place of measures of effectiveness, the minimum cost assignment problem can be solved as a maximum weighted matching problem after replacing each cost by the difference between it and the maximum individual cost.

Both forms of assignment problem can be solved by a variety of algorithms. For example, a maximum matching on a bipartite graph can be found by modeling the problem as a network flow problem and finding a maximum flow on the model network. A well known algorithm for solving the maximum weighted matching problem (for which the maximum matching problem can be considered a special case) on a bipartite graph is often referred to as the Hungarian method and was introduced by Kuhn (1955, 1956). He casts the procedure in terms of a primal-dual linear program. The algorithm can be implemented so as to produce an optimal matching in $O(m^2n)$ steps, where n is the number of vertices and m is the number of edges in the graph. The details are given in Lawler (1976). Although this is an efficient algorithm, it may be necessary to find faster implementations for problems of large size or when the algorithm is used repeatedly as part of a more complex procedure. Various methods have been proposed including those due to Jonker and Volgenant (1986) and Wright (1990).

JOB SCHEDULING: Another example of a problem which can be modeled as a matching problem arises from job scheduling (Coffman and Graham, 1972). Suppose n jobs are to be processed and there are two machines available. All jobs require an equal amount of time to complete and can be processed on either machine. However there are precedence constraints which mean that some jobs must be completed before others are started. What is the shortest time required to process all n jobs?

This example can be modeled by constructing a graph G with n vertices representing the n jobs and where an edge joins two vertices if and only if they can be run simultaneously. An optimum schedule corresponds to one where the two machines are used simultaneously as often as possible. Therefore the problem becomes one of finding the maximum matching on G, from which the shortest time can be derived. In this case though, the graph G is no longer bipartite and so an algorithm for solving the maximum matching problem on a general graph is required.

The first efficient algorithm to find a maximum matching in a graph was presented by Edmonds (1965a). Most successful algorithms to find a maximum matching have been based on Edmonds' ideas. Gabow (1976) and Lawler (1976) show how to implement the algorithm in a time of $O(n^3)$. It is possible to modify the algorithm for more efficient performance on large problems. For example, Even and Kariv (1975) present an algorithm running in a time of $O(n^{5/2})$ and Micali and Vazirani (1980) describe an algorithm with running time of $O(mn^{1/2})$.

ARC ROUTING: There is a close connection between arc routing problems and matching. Suppose a postal delivery person must deliver mail along all streets of a town. What route will traverse each street and return to the starting point in minimum total distance? This problem is known as the *Chinese Postman Problem* as it was first raised by the Chinese mathematician Mei-

Ko Kwan (1962). It may be formulated as finding the minimum length tour on a non-directed graph G whose edges represent the streets in the town and whose vertices represent the junctions, where each edge must be included at least once. Edmonds and Johnson (1973) showed that this problem is equivalent to finding a minimum weighted matching on a graph whose vertices represent the set of odd nodes in G and whose edges represent the shortest distances in G between the odd nodes. Odd nodes are vertices where an odd number of edges meet. This minimum weighted matching problem can be solved efficiently by the algorithm introduced by Edmonds (1965b) for maximum weighted matching problems where the weights on each edge are the distances multiplied by minus one. Gabow (1976) and Lawler (1976) show how the algorithm can be implemented in $O(n^3)$ steps. The Chinese Postman Problem is therefore easier to solve than the Traveling Salesman Problem where a polynomially bounded algorithm has not so far been established.

For large problems, faster versions of the weighted matching algorithm have been developed by Galil, Micali and Gabow (1982) and Ball and Derigs (1983) which require $O(mn\log n)$ steps. A starting procedure which significantly reduces the computing time for the maximum matching problem is described by Derigs and Metz (1986) and involves solving the assignment problem in a related bipartite graph.

See **Assignment problem; Chinese postman problem; Combinatorial and integer optimization; Dual-programming problem; Graph theory; Hungarian method; Maximum-flow network problem; Networks; Transportation problem; Traveling salesman problem.**

References

[1] Ball, M.O. and U. Derigs (1983). "An Analysis of Alternate Strategies for Implementing Matching Algorithms," *Networks* 13, 517–549.

[2] Coffman, E.G., Jr. and R.L. Graham (1972). "Optimal Scheduling for Two Processor Systems," *Acta Inform.* 1, 200–213.

[3] Derigs, U. and A. Metz (1986). "On the Use of Optimal Fractional Matchings for Solving the (Integer) Matching Problem," *Computing*, 36, 263–270.

[4] Edmonds, Jl. (1965a). "Paths, Trees, and Flowers," *Canad. Jl. Math.*, 17, 449–467.

[5] Edmonds, Jl. (1965b). "Maximum Matching and a Polyhedron with (0,1) Vertices," *Jl. Res. Nat. Bur. Standards Sect. B*, 69B, 125–130.

[6] Edmonds, Jl. and E.L. Johnson (1973). "Matching, Euler Tours and the Chinese Postman," *Math. Programming*, 5, 88–124.

[7] Even, S. and O. Kariv (1975). "An $O(n^{5/2})$ Algorithm for Maximum Matching in General Graphs," 16*th Annual Symposium on Foundations of Computer Science*, IEEE Computer Society Press, New York, 100–112.

[8] Galil, Z., S. Micali and H. Gabow (1982). "Priority Queues with Variable Priority and an $O(EV \log V)$ Algorithm for finding a Maximal Weighted Matching in General Graphs," 23*rd Annual Symposium on Foundations of Computer Science*, IEEE Computer Society Press, New York, 255–261.

[9] Gabow, H.N. (1976). "An Efficient Implementation of Edmond's Algorithm for Maximum Matching on Graphs," *Jl. Assoc. Comput. Mach.*, 23, 221–234.

[10] Gondran, Michel and Michel Minoux (1984). *Graphs and Algorithms*. John Wiley, Chichester.

[11] Jonker, R. and A. Volgenant (1986). "Improving the Hungarian Assignment Algorithm," *Opl Res. Lett.*, 5, 171–175.

[12] Kuhn, H.W. (1955). "The Hungarian Method for the Assignment Problem," *Naval Res. Logist. Quart.*, 2, 83–97.

[13] Kuhn, H.W. (1956). "Variants of the Hungarian Method for Assignment Problems," *Naval Res. Logist. Quart.*, 3, 253–258.

[14] Kwan, Mei-Ko (1962). "Graphic Programming Using Odd and Even Points," *Chinese Math.*, 1, 273–277.

[15] Lawler, E.L. (1976). *Combinatorial Optimization, Networks and Matroids*. Holt, Rinehart and Winston, New York.

[16] Lovasz L. and M.D. Plummer (1986). *Matching Theory*. Annals of Discrete Mathematics, vol. 29, North-Holland, Amsterdam.

[17] McHugh, James A. (1990). *Algorithmic Graph Theory*. Prentice-Hall, London.

[18] Micali, S. and V.V. Vazirani (1980). "An $O(V^{1/2}E)$ Algorithm for Finding Maximum Matching in General Graphs," 21*st Annual Symposium on Foundations of Computer Science*, IEEE Computer Society Press, New York, 17–27.

[19] Wright, M.B. (1990). "Speeding Up the Hungarian Algorithm," *Computers Opns Res.*, 17, 95–96.

MATERIAL HANDLING

Meir J. Rosenblatt

Washington University, St. Louis, Missouri; Technion, Haifa, Israel

INTRODUCTION: Material handling is concerned with moving raw materials, work-in-process, and finished goods into the plant, within the plant, and out of the plant to warehouses, distribution networks, or directly to the customers. The basic objective is to move the right combination of tools and materials (raw materials, parts and finished products) at the right time, to the right place, in the right form, and in the right orientation. And to do it with the minimum total cost.

It is estimated that 20% to 50% of the total operating expenses within manufacturing are attributed to material handling (Tompkins and White, 1984). Material handling activities may account for 80% to 95% of total overall time spent between receiving a customer order and shipping the requested items (Rosaler and Rice, 1983). This indicates that improved efficiencies in material handling activities can lead to substantial reductions in product cost and production leadtime; better space and equipment utilization, improved working conditions and safety, improvements in customer service; and, eventually to higher profits and larger market share. Material handling adds to the product cost but contributes nothing to the value added of the products.

With the growing popularity of Just-in-Time (JIT), the design of material handling systems has become even more important. Under JIT, production is done in small lots so that production leadtimes are reduced and inventory holding costs are minimized, requiring the frequent conveyance of material. Thus, successful implementation of JIT needs a fast and reliable material handling system as a prerequisite.

Production lot-sizing decisions have a direct impact on the assignment of storage space to different items (products) and consequently on the material handling costs. Therefore, lot sizing decisions must take into account not only setup and inventory carrying costs but also warehouse and material handling costs. In other words, production lot sizing, warehouse storage assignment, and material handling equipment decisions must be made simultaneously.

Also, in a flexible manufacturing environment, where batches of products may have several possible alternative routes, the choice of routing-mix can have a significant effect on shop throughput and work-in-process inventory. However, for such a system to be efficient, an appropriate material handling system needs to be designed. This design issue is especially important when expensive machines are being used, and major waste can be caused by a material handling system that is inappropriate and becomes a bottleneck.

Finally, it should be recognized that facility layout determines the overall pattern of material flow within the plant and therefore has a significant impact on the material handling activities and costs. It is estimated that effective facilities planning and layout can reduce material handling costs by at least 10% to 30% (Tompkins and White, 1984). However, an effective layout requires an effective material handling system. Therefore, it is critical that these decisions are made simultaneously.

MATERIAL HANDLING EQUIPMENT: There are several ways of classifying material handling equipment: 1. by the type of control (operator controlled vs. automated); 2. by where the equipment works (on the floor vs. suspended overhead); 3. by the travel path (fixed vs. flexible). We will use the fixed vs. flexible travel path classification as in Barger (1987). Flexible path equipment can be moved along any route, and in general is operator-controlled. Trucks are the most common mode of operations. There are several types of trucks depending on the type of handling which is needed. The most common ones are:

Counterbalanced fork trucks – used both for storage at heights of 20 feet or more, as well as for fast transportation;

Narrow-aisle trucks – mainly used for storage applications;

Walkie Pallet trucks – mainly used for transportation over short hauls; and

Manual trucks – mainly used for short hauls and auxiliary services.

There are three important types of fixed-path equipment:

Conveyors – Conveyors are one of the largest families of material handling equipment. They can be classified based on the load-carrying surface involved: roller, belt, wheel, slat, carrier chain; or on the position of the conveyor: on-floor or overhead;

Automatic Guided Vehicles (AGVs) – these are electric vehicles with on-board sensors that enable them to automatically track along a guidepath which can be an electrified guide wire or a strip of (reflective) paint or tape on the floor. The AGVs follow their designated path using their sensors to detect the electromagnetic field generated by the electric wire or to optically detect the path marked on the floor. AGVs can transport materials between any two points connected by a guidepath – without human intervention. Most of today's AGVs are capable of loading and unloading materials automatically. Most applications of AGVs are for load transportation, however, they could also be used in flexible assembly operations to carry the product being assembled through the various stages of assembly. While AGVs have traditionally been fixed path vehicles, recent advances in technology permit them to make short deviations from their guidepath. Such flexibility may considerably increase their usefulness; and

Hoists, Monorails, and Cranes – Hoists are a basic type of overhead lifting equipment and can be suspended from a rail, track, crane bridge or beam. A hoist consists of a hook, a rope or chain used for lifting, and a container for the rope/chain. Monorails consist of individual wheeled trolleys that can move along an overhead track. The trolleys may be either powered or non-powered. Cranes have traditionally found wide application in overhead handling of materials, especially where the loads are heavy. Besides the overhead type, there are types of cranes that are wall or floor mounted, portable ones and so on. Types such as stacker cranes are useful in warehouse operations.

INTERACTION WITH AUTOMATED STORAGE AND RETRIEVAL SYSTEMS (AS/RS): AS/RS consist of high density storage spaces, computer controlled handling and storage equipment (operated with minimal human assistance) and may be connected to the rest of the material handling system via some conveying devices such as conveyors and AGVs. Several types of AS/RS are available including: Unit Load, Miniload, Man-On-Board, Deep Lane and Carousels. The AS/RS systems help achieve very efficient placement and retrieval of materials, better inventory control, improved floor space utilization, and production scheduling efficiency. They also provide greater inventory accountability and reduce supervision requirements. Normally, stacker cranes, which can move both horizontally and vertically at the same time, are used for material handling. Typically, a crane operates in a single aisle, but can be moved between aisles (Rosenblatt *et al.*, 1993). Items to be stored or retrieved are brought to/picked from the AS/RS by a conveyor or an AGV. Such integration can be used to automate material handling throughout the plant and warehouse. A great deal of research has been done on scheduling jobs and assigning storage space in the AS/RS (Hausman *et al.*, 1976).

ISSUES IN MATERIAL HANDLING SYSTEM DESIGN:

Unit load concept – Traditional wisdom is that materials should be handled in the most efficient, maximum size using mechanical means to reduce the number of moves needed for a given amount of material. While reducing the number of trips required is a good objective, the drawback of this approach is that it tends to encourage the acceptance of large production lots, large material handling equipment, and large space requirements. Small unit loads allow for more responsive, less expensive, and less consuming material handling systems. Also, the current trend toward continuous manufacturing flow processes and the strong drive for automation necessitate the use of smaller unit loads (Apple and Rickles, 1987).

Container size and standardization – This is an issue related to the unit load concept. Container size has an obvious correlation with the size of unit load. Hence, not surprisingly, the current trend is to employ smaller containers. The benefits of smaller containers include compact and more efficient workstations, improved scheduling flexibility due to smaller transfer batch size, smaller staging areas, and lighter duty handling systems. Another consideration that strongly influences the optimal container size is the range of items served by one container. In warehouse operations, unless items vary widely in their physical characteristics, the cost of employing two or more container sizes is almost always higher than in the one-size case (Roll *et al.*, 1989). Use of standard containers eliminates the need for container exchanges between operation sites.

Capacity of the system or number of pieces of equipment – The margins in the design of material handling system require a careful examination of the relative costs of acquiring and maintaining of workcenters and handling equipment. In the design of the material handling system for an expensive job shop, enough excess capacity should be provided so that the handling system never becomes the bottleneck.

OR MODELS IN MATERIAL HANDLING:

OR tools have been applied to model and study a variety of problems in the area of material handling. One example, dealing with the initial design phase of material handling, used a graph-theoretic modeling framework (Kouvelis and Lee, 1990). Other examples include conveyor systems problems using queueing theory, and transfer lines where dynamic programming techniques were applied. Most of the theoretical work has focused on AGVs and AS/RS. The design and control of AGVs are extremely complex tasks. The design decisions include determining the optimal number of AGVs (Maxwell and Muckstadt, 1982), as well as determining the optimal flow paths (Kim and Tanchoco, 1993). Factors to be considered in the design decisions include hardware considerations, impacts on facilities layout, material procurement policy, and production policy. Resulting problems tend to be intractable for any realistic scenario and hence heuristics and simulation are the most used techniques in addressing design issues. Control problems including dispatching and routing tasks require real time decisions, making it difficult to obtain optimal solutions. Researchers have attempted to solve simplified problems, for example, by examining static versions instead of dynamic systems (Han and McGinnis, 1989), and using simple single-loop layouts (Egbelu, 1993).

In the study of warehousing in general, and AS/RS in particular, many different measures of effectiveness of warehouse designs have been considered. The most common ones are throughput as measured by the number of orders handled per day, average travel time of a crane per single/dual command and average waiting time per customer/order (Hausman *et al.*, 1976). Researchers have considered either simulation or optimization models, usually of the non-linear integer form, to solve these problems. Yet others have combined optimization and simulation techniques to obtain solutions that are both cost effective and operationally feasible (reasonable service time) (Rosenblatt *et al.*, 1993).

Since factories in the future will be increasingly automated, numerical control of machine tools and flexible manufacturing systems will become more common. Material handling systems will increasingly involve the use of robots. In the absence of an effective material handling system, an automated factory would be reduced to a set of "islands of automation" (White, 1992). In the integrated and fiercely competitive global economy of the future, material handling systems will play a crucial role in the battle to cut costs and improve productivity and service levels.

See **Flexible manufacturing; Inventory modeling; Job shop scheduling; Layout problems.**

References

[1] Apple, J.M. and H.M. Rickles (1987). "Material Handling and Storage," *Production Handbook*, John A. White, ed., Wiley, New York.

[2] Barger, B.F. (1987). "Materials Handling Equipment," *Production Handbook*, John A. White, ed., Wiley, New York.

[3] Egbelu, P.J. (1993). "Positioning of Automated Guided Vehicles in a Loop Layout to Improve Response Time," *European Journal of Operational Research*, 71, 32–44.

[4] Han, M.-H. and L.F. McGinnis (1989). "Control of Material Handling Transporter in Automated Manufacturing," *IIE Transactions*, 21, 184–190.

[5] Hausman, W.H., L.B. Schwarz, and S.C. Graves (1976). "Optimal Assignment in Automatic Warehousing Systems," *Management Science*, 22, 629–638.

[6] Kim, K.H. and J.M.A. Tanchoco (1993), "Economical Design of Material Flow Paths," *International Journal of Production Research*, 31, 1387–1407.

[7] Kouvelis, P. and H.L. Lee (1990). "The Material Handling Systems Design of Integrated Manufacturing System," *Annals Operations Research*, 26, 379–396.

[8] Maxwell, W.L., and J.A. Muckstadt (1982). "Design of Automated Guided Vehicle Systems," *IIE Transactions*, 14, 114–124.

[9] Roll, Y., M.J. Rosenblatt, and D. Kadosh (1989). "Determining the Size of a Warehouse Container," *International Journal of Production Research*, 27, 1693–1704.

[10] Rosaler, R.C. and J.O. Rice, eds. (1983). *Standard Handbook of Plant Engineering*," McGraw Hill, New York.

[11] Rosenblatt, M.J., Y. Roll and V. Zyser (1993). "A Combined Optimization and Simulation Approach to Designing Automated Storage/Retrieval Systems," *IIE Transactions*, 25, 40–50.

[12] Tompkins, J.A. and J.A. White (1984). *Facilities Planning*, John Wiley, New York.

[13] White, J.A (1982). "Factory of Future Will Need Bridges Between Its Islands of Automation," *Industrial Engineering*, 14, 4, 60–68.

MATERIAL REQUIREMENTS PLANNING

A material requirements planning (MRP) system is a collection of logical procedures for managing, at the most detailed level, inventories of component assemblies, subassemblies, parts and raw materials in a manufacturing environment. It is an information system and simulation tool that generates proposals for production schedules that managers can evaluate in terms of their feasibility and cost effectiveness. See **Hierarchical production planning; Production management**.

MATHEMATICAL MODEL

The mathematical description of (usually) a real-world problem. In operations research/management science, mathematical models take on varied forms (linear programming, queueing, Markov systems, etc.), many of which can be applied across application areas. The basic OR/MS mathematical model can be described as the decision problem of finding the maximum (minimum) of a measure of effectiveness (objective function) $E = F(X, Y)$, where X represents the set of possible solutions (alternative decisions) and Y the given conditions of the problem. Although a rather simple model in its concept, especially since it optimizes a single objective, this mathematical decision model underlies most of the problems that have been successfully formulated and solved by OR/MS methodologies. See **Decision problem; Deterministic model; Stochastic model**.

MATHEMATICAL PROGRAMMING

Mathematical programming is a major discipline in operations research/management science and, in general, is the study of how one optimizes the use and allocation of limited resources. Here programming refers to the development of a plan or procedure for dealing with the problem. It is considered a branch of applied mathematics as it deals with the theoretical and computational aspects of finding the maximum (minimum) of a function $f(x)$ subject to a set of constraints of the form $g_i(x) \le b_i$. The linear-programming model is the prime example of such a problem.

MATHEMATICAL-PROGRAMMING PROBLEM

A constrained optimization problem usually stated as Minimize (Maximize) $f(x)$ subject to $g_i(x) \le 0$, $i = 1, \ldots, m$. Depending on the form of the objective function $f(x)$ and the constraints $g_i(x)$ the problem will have special properties and associated algorithms. See **Combinatorial and integer optimization; Convex-programming problem; Fractional programming; Geometric programming; Integer-programming problem; Linear programming; Nonlinear programming; Quadratic programming; Separable-programming problem**.

MATHEMATICAL PROGRAMMING SOCIETY

This society is an international organization dedicated to the support and development of the application, computational methods, and theory of mathematical programming. The society sponsors the triennial International Symposium on Mathematical Programming and other meetings throughout the world.

MATHEMATICAL-PROGRAMMING SYSTEM (MPS)

An integrated set of computer programs that are designed to solve a range of mathematical-programming problems is often referred to as a mathematical-programming system (MPS). Such systems solve linear programs, usually by some form of the simplex method, and often have the capability to handle integer-variable problems and other nonlinear problems such as quadratic-programming problems. To be effective, an MPS must have procedures for input data handling, matrix generation of the constraints, reliable optimization, user and automated control of the computation, sensitivity analysis of the solution, solution restart, and output reports.

MATRICES AND MATRIX ALGEBRA

Alan Tucker

The State University of New York at Stony Brook

A matrix is an $m \times n$ array of numbers, typically displayed as

$$A = \begin{bmatrix} 4 & 3 & 8 \\ 1 & 2 & 3 \\ 4 & 5 & 6 \end{bmatrix},$$

where the entry in row i and column j is denoted as a_{ij}. Symbolically, we write $A = (a_{ij})$, for $i = 1, \ldots, m$ and $j = 1, \ldots, n$. A vector is a one-dimensional array, either a row or a column. A column vector is an $m \times 1$ matrix, while a row vector is a $1 \times n$ matrix. For a matrix A, its ith row vector is usually denoted by a_i' and its jth column by a_j. Thus an $m \times n$ matrix can be decomposed into a set of m row n-vectors or a set of n column m-vectors. Matrices are a natural generalization of single numbers, or scalars.

They arise directly or indirectly in most problems in operations research and management science.

BASIC OPERATIONS AND LAWS OF MATRIX ALGEBRA: The language for manipulating matrices is matrix algebra. Matrix algebra is a multivariable extension of the single-variable algebra learned in high school. The basic building block for matrix algebra is the scalar product. The scalar product $a \cdot b$ of a and b is a single number (a scalar) equal to the sum of the products $a_i b_i$, that is, $a \cdot b = \sum_{i=1}^{n} a_i b_i$, where both vectors have the same dimension n. Observe that the scalar product is a linear combination of the entries in vector a and also a linear combination of the entries of vector b.

The product of an $m \times n$ matrix A and a column n-vector b is a column vector of scalar products $a_i' \cdot b$, of the rows a of A with b. For example, if

$$A = \begin{bmatrix} a_{11} & a_{12} & a_{13} \\ a_{21} & a_{22} & a_{23} \end{bmatrix}$$

is a 2×3 matrix and

$$b = \begin{bmatrix} b_1 \\ b_2 \\ b_3 \end{bmatrix}$$

is a column 3-vector, then

$$Ab = \begin{bmatrix} a_1' \cdot b \\ a_2' \cdot b \end{bmatrix} = \begin{bmatrix} a_{11}b_1 + a_{12}b_2 + a_{13}b_3 \\ a_{21}b_1 + a_{22}b_2 + a_{23}b_3 \end{bmatrix},$$

so that Ab is a linear combination of the columns of A. Moreover, for any scalar numbers r, q, any $m \times n$ matrix A, and any column n-vectors b, c:

$$A(rb + qc) = rAb + qAc.$$

The product of a row m-vector c and an $m \times n$ matrix A is a row vector of scalar products $c \cdot a_j$, of c with the columns a_j of A. For example, if

$$A = \begin{bmatrix} a_{11} & a_{12} & a_{13} \\ a_{21} & a_{22} & a_{23} \end{bmatrix}$$

is a 2×3 matrix and $c = [c_1, c_2]$ is a row 2-vector, then

$$cA = [ca_1, ca_2]$$
$$= [a_{11}c_1 + a_{21}c_2, a_{12}c_1 + a_{22}c_2, a_{13}c_1 + a_{23}c_2].$$

If A is an $m \times r$ matrix and B is an $r \times n$ matrix, then the matrix product AB is an $m \times n$ matrix obtained by forming the scalar product of each row a_i' in A with each column b_j in B. That is, the (i,j)th entry in AB is $a_i' \cdot b_j$. Column j of AB is the matrix-vector product Ab_j and each column of AB is a linear combination of the columns of A. Row i of AB is vector-matrix product $a'_i B$ and each row of AB is a linear combination of the rows of B. The matrix-vector product Ab is a special case of the matrix-matrix product in which the second matrix has just one column; the analogous statement holds for the vector-matrix product bA.

Matrix multiplication is not normally commutative. Otherwise it obeys all the standard laws of scalar multiplication.

Associative Law. *Matrix addition and multiplication are associative*: $(A + B) + C = A + (B + C)$ *and* $(AB)C = A(BC)$.

Commutative Law. *Matrix addition is commutative*: $A + B = B + A$. *Matrix multiplication is not commutative* (except in special cases): $AB \neq BA$.

Distributive Law. $A(B + C) = AB + AC$ *and* $(B + C)A = BA + CA$.

Law of Scalar Factoring. $r(AB) = (rA)B = A(rB)$.

For $n \times n$ matrices A, there is an identity matrix I with ones on the main diagonal and zeros elsewhere, with the property that $AI = IA = A$. The transpose of an $m \times n$ matrix A, denoted by A^T, is an $n \times m$ matrix such that the rows of A are the columns of A^T.

If matrices are partitioned into submatrices in a regular fashion, say, a 4×4 matrix A is partitioned into four 2×2 submatrices,

$$A = \begin{bmatrix} A_{11} & A_{12} \\ A_{21} & A_{22} \end{bmatrix},$$

and a 4×4 matrix B is similarly partitioned, then the matrix product AB can be computed in terms of the partitioned submatrices:

$$AB = \begin{bmatrix} A_{11}B_{11} + A_{12}B_{21} & A_{11}B_{12} + A_{12}B_{22} \\ A_{21}B_{11} + A_{22}B_{21} & A_{21}B_{12} + A_{22}B_{22} \end{bmatrix}.$$

Solving Systems of Linear Equations:

Matrices are intimately tied to linear systems of equations. For example, the system of linear equations

$$4x_1 + 2x_2 + 2x_3 = 100$$

$$2x_1 + 5x_2 + 2x_3 = 200 \qquad (1)$$

$$1x_1 + 3x_2 + 5x_3 = 300$$

can be written as

$$Ax = b, \quad \text{where } A = \begin{bmatrix} 4 & 2 & 2 \\ 2 & 5 & 2 \\ 1 & 3 & 5 \end{bmatrix}, \quad x = \begin{bmatrix} x_1 \\ x_2 \\ x_3 \end{bmatrix},$$

$$b = \begin{bmatrix} 100 \\ 200 \\ 300 \end{bmatrix}. \qquad (2)$$

Essentially, the only way to solve an algebraic system with more than one variable is by solving a system of linear equations. For example, nonlinear systems must be recast as linear systems to be numerically solved. Since operations research and management science is concerned with complex problems involving large numbers of variables, matrix systems are pervasive in OR/MS.

Observe that the system (1) can be approached from the row point of view as a set of simultaneous linear equations and solved by row operations using Gaussian elimination or Gauss-Jordan elimination. The result of elimination will be either no solution, a unique solution or an infinite number of solutions. In linear programming, one typically wants to find a vector x maximizing or minimizing a linear objective function $c \cdot x$ subject to a system $Ax = b$ of linear constraints. The simplex method finds an optimal solution by a sequence of pivots on the *augmented* matrix $[A \ b]$. A pivot on non-zero entry (i, j) consists of a collection of row operations (multiplying a row by a scalar or subtracting a multiple of one row from another row) producing a transformed augmented matrix $[A' \ b']$ in which entry (i,j) equals 1 and all other entries in the jth column are 0. The pivot step can be accomplished by premultiplying A by a pivot matrix P, which is an identity matrix with a modified ith column.

System (1) can also be approached from the column point of view as the following vector equation:

$$x_1 \begin{bmatrix} 4 \\ 2 \\ 1 \end{bmatrix} + x_2 \begin{bmatrix} 2 \\ 5 \\ 3 \end{bmatrix} + x_3 \begin{bmatrix} 2 \\ 2 \\ 5 \end{bmatrix} = \begin{bmatrix} 100 \\ 200 \\ 300 \end{bmatrix}. \qquad (3)$$

Writing the system as (3) raises questions like, which right-hand side vectors b are expressible as linear combinations of the columns of A? The set of such b vectors is called the range of the matrix A. For a square matrix, the system $Ax = b$ will have a unique solution if and only if no column vector of A can be written as a linear combination of other columns of A, or equivalently, if and only if $x = 0$ is the only solution to $Ax = 0$, where 0 denotes a vector of all zeroes. When this condition holds, the columns are said to be linearly independent. When $Ax = 0$ has non-zero solutions (whether A is square or not), the set of such nonzero solutions is called the kernel of A. Kernels, ranges and linear independence are the building blocks of the theory of linear algebra. This theory plays an important role in the uses of matrices in OR/MS. For example, if x^* is a solution to $Ax = b$ and x° is in the kernel of A (that is, $Ax^\circ = 0$), then $x^* + x^\circ$ is also a solution of $Ax = b$, since $A(x^* + x^\circ) = Ax^* + Ax^\circ = b + 0 = b$, and one can show that all solutions to $Ax = b$ can be written in the form of a particular solution x^* plus some kernel vector x°. In a linear program to maximize or minimize $c \cdot x$ subject to $Ax = b$, once one finds one solution x^* to $Ax = b$, improved solutions will be obtained by adding appropriate kernel vectors to x^*.

MATRIX INVERSE: The inverse A^{-1} of a square matrix A has the property that $A^{-1}A = AA^{-1} = I$. The inverse can be used to solve $Ax = b$ as follows: $Ax = b \Rightarrow A^{-1}(Ax) = A^{-1}b$, but $A^{-1}(Ax) = (A^{-1}A)x = (I)x = x$. Thus $x = A^{-1}b$.

The square matrix A has an inverse if any of the following equivalent statements hold:

  (i) For all b, $Ax = b$ has a unique solution;
 (ii) The columns of A are linearly independent;
(iii) The rows of A are linearly independent.

The matrix A^{-1} is found by solving a system of equations as follows. The product $AA^{-1} = I$ implies that if x_j is the jth column of A^{-1} and i_j is the jth column of I (i_j has 1 in the jth entry and zeroes elsewhere), then x_j is the solution to the matrix system $Ax_j = i_j$. An impressive aspect of matrix algebra is that even when a matrix system $Ax = b$ has no solution, that is in (3), no linear combination of the columns of A equals b, there is still a "solution" y in the sense of a linear combination Ay of the columns of A that is as close as possible to b. By as close as possible, we mean that the Euclidean distance in n-dimensional space between the vectors Ay and b is minimized. There is even an "inverse"-like matrix A^*, called the pseudoinverse or generalized inverse, such that $y = A^*b$. The matrix A^* is given by the matrix formula $A^* = (A^TA)^{-1}A^T$, where A^T is the transpose of A, obtained by interchanging rows and columns.

EIGENVALUES AND EIGENVECTORS: A standard form of a dynamic linear model is $p' = Ap$, where A is an $n \times n$ matrix and p is a n-column vector of populations or probabilities (in the case of probabilities, it is the convention to use row vectors: $p' = pA$). For some special vectors e, called eigenvectors, we have $Ae = \lambda e$, where λ is a scalar called an *eigenvalue*. That is, premultiplying e by A has the effect of multiplying e by a scalar. It follows that $A^n e = \lambda^n e$. This special situation is very valuable because it is obviously much easier to compute $\lambda^n e$ than $A^n e$.

Most $n \times n$ matrices have n different (linearly independent) eigenvectors. If we express the vector p as a linear combination $p = ae_1 + be_2$ of, say, two eigenvectors e_1 and e_2, with associated eigenvalues λ_1, λ_2, then by the linearity of matrix-vector products, Ap and A^2p can be calculated as

$$Ap = A(ae_1 + be_2) = aAe_1 + bAe_2 = a\lambda_1 e_1 + b\lambda_2 e_2$$

and

$$A^2 p = A^2(ae_1 + be_2) = aA^2 e_1 + bA^2 e_2 = a\lambda_1^2 e_1 + b\lambda_2^2 e_2.$$

More generally,

$$A^k p = A^k(ae_1 + be_2) = aA^k e_1 + bA^k e_2 = a\lambda_1^k e_1 + b\lambda_2^k e_2.$$

If $|\lambda_1| > |\lambda_i|$, for $i \geq 2$, then for large k, λ_1^k will become much larger in absolute value than the other λ_i^k, and so $A^k p$ approaches a multiple of the eigenvector associated with the eigenvalue of largest absolute value. For ergodic Markov chains, this largest eigenvalue is 1 and the Markov chain converges to a steady-state probability p^* such that $p^* = p^*A$.

MATRIX NORMS: The norm $|v|$ of a vector v is a scalar value that is nonnegative, satisfies scalar factoring, that is, $|rv| = r|v|$, and the triangle inequality, that is, $|u + v| \leq |u| + |v|$. There are three common norms used for vectors:

 1. The Euclidean, or l_2, norm of $v = [v_1, v_2, \ldots, v_n]$ is defined as
    $$|v|_e = \sqrt{v_1^2 + v_2^2 + \cdots + v_n^2}.$$

 2. The sum, or l_1, norm of $v = [v_1, v_2, \ldots, v_n]$ is defined
    $$|v|_s = |v_1| + |v_2| + \cdots + |v_n|.$$

 3. The max, or l_∞, norm of $v = [v_1, v_2, \ldots, v_n]$ is
    $$|v|_m = \max\{|v_1|, |v_2|, \ldots, |v_n|\}.$$

The matrix norm $\|A\|$ is the (smallest) bound such that $|Ax| \leq \|A\| \, |x|$, for all x. Thus

$$\|A\| = \max_{x \neq 0} \frac{|Ax|}{|x|}. \tag{4}$$

It follows that $|A^k x| \leq \|A\|^k |x|$.

The Euclidean, sum, and max norms of the matrix are defined by using the Euclidean, sum, and max vector norms, respectively, in (4). When A is a square, symmetric matrix ($a_{ij} = a_{ji}$), the Euclidean norm $\|A\|_e$ equals the absolute value of the largest eigenvalue of A. When A is not symmetric, $\|A\|_e$ equals the positive square root of the largest eigenvalue of $A^T A$. The sum and max norms of A are very simple to find and for this reason are often preferred over the Euclidean norm: $\|A\|_s = \max_j\{|A_j|_s\}$ and $\|A\|_m = \max_i\{|A_i'|_s\}$, where A_j denotes the jth column of A and A_i' denotes the ith row of A. In words, the sum norm of A is the largest column sum (summing absolute values) and the max norm of A is the largest row sum.

Norms have many uses. For example, in a linear growth model $p' = Ap$, the kth iterate $p^{(k)} = A^k p$ is bounded in norm by $|p^{(k)}| \le \|A\|^k |p|$. One can show that if the system of linear equations $Ax = b$ is perturbed by adding a matrix E of errors to A, and if x^* is the solution to the original system $Ax = b$ while $x^* + e$ is the solution to $(A + E)x = b$, then the relative error $|e|/|x^* + e|$ is bounded by a constant $c(A)$ times the relative error $\|E\|/\|A\|$, that is, $|e|/|x^* + e| \le c(A)\|E\|/\|A\|$. The constant $c(A) = \|A\| \|A^{-1}\|$ and is called the condition number of A.

A famous linear input-output model due to Leontief has the form $x = Ax + b$. Here x is a vector of production of various industrial activities, b is a vector of consumer demands for these activities, and A is an inter-industry demand matrix in which entry a_{ij} tells how much of activity i is needed to produce one unit of activity j. Here, Ax is a vector of the input for the different activities needed to produce the output vector x. The model $x = Ax + b$ can be shown to have a solution if $\|A\|_s < 1$, that is, if the columns sums are all less than one. This condition has the natural economic interpretation that all activities must be profitable, that is, the value of the inputs to produce a dollar's worth of any activity must be less than one dollar.

Algebraically, $x = Ax + b$ is solved as follows:

$$x = Ax + b \Rightarrow x - Ax = b \Rightarrow (I - A)x = b$$

$$\Rightarrow x = (I - A)^{-1}b.$$

When $\|A\| < 1$, the geometric series $I + A + A^2 + A^3 + \cdots$ converges to $(I - A)^{-1}$, guaranteeing not only the existence of a solution to $x = Ax + b$ but also a solution with nonnegative entries, since when A has nonnegative entries, then all the powers of A will have nonnegative entries implying that $(I - A)^{-1}$ has nonnegative entries and hence so does $x = (I - A)^{-1}b$.

HISTORICAL SKETCH: The word "matrix" in Latin means "womb." The term was introduced by J.J. Sylvester in 1848 to describe an array of numbers that could be used to generate ("give birth to") a variety of determinants. A few years later, Cayley introduced matrix multiplication and the basic theory of matrix algebra quickly followed. A more general theory of linear algebra and linear transformations pushed matrices into the background until the 1940s and the advent of digital computers. During the 1940s, Alan Turing, father of computer science, introduced the LU decomposition and John von Neumann, father of the digital computer, working with Herman Goldstine, started the development of numerical matrix algebra and introduced the condition number of a matrix. Curiously, at the same time Cayley and Sylvester were developing matrix algebra, another Englishman, Charles Babbage, was building his analytical engine, the forerunner of digital computers which are critical to the use of modern matrix models.

See **Analytic hierarchy process; Gauss-Jordan elimination method; Gaussian elimination; Linear programming; Markov chains; Simplex method; Trivial solution.**

References

[1] Lay, D.C. (1993). *Linear Algebra and its Applications*, Addison-Wesley, Reading, Massachusetts.

[2] Strang, G. (1988). *Linear Algebra and its Applications*, 3rd. ed., Harcourt Brace Jovanovich, Orlando, Florida.

MATRIX-ANALYTIC STOCHASTIC MODELS

Marcel F. Neuts

The University of Arizona, Tucson

A rich class of models for queues, dams, inventories, and other stochastic processes has arisen out of matrix/vector generalizations of classical approaches. We present three specific examples in the following, namely, matrix-analytic solutions for M/G/1-type queueing problems, matrix-geometric solutions to GI/M/1-type queueing problems, and finally, the Markov arrival process (MAP) generalization of the renewal point process.

MATRIX-ANALYTIC M/G/1-TYPE QUEUES: The unifying structure that underlies these models is an imbedded Markov renewal process whose transition probability matrix is of the form:

$$\tilde{Q}(x) = \begin{vmatrix} B_0(x) & B_1(x) & B_2(x) & B_3(x) & B_4(x) & \cdots \\ C_0(x) & A_1(x) & A_2(x) & A_3(x) & A_4(x) & \cdots \\ 0 & A_0(x) & A_1(x) & A_2(x) & A_3(x) & \cdots \\ 0 & 0 & A_0(x) & A_1(x) & A_2(x) & \cdots \\ \cdot & \cdot & \cdot & \cdot & \cdot & \cdots \end{vmatrix},$$

where the elements are themselves matrices of probability mass functions. If the matrix

$$A = \sum_{k=0}^{\infty} A_k(\infty)$$

is irreducible and has the invariant probability vector π, then the Markov renewal process is positive recurrent if and only if some natural moment conditions hold for the coefficient matrices and if

$$\rho = \pi \sum_{k=1}^{\infty} k A_k e < 1 \quad \text{for } e = (1, \ldots, 1)^T.$$

The quantity ρ is the generalized form of the *traffic intensity* for the elementary queueing models.

The state space is partitioned in levels i, which are the sets of m states (i, j), $1 \leq j \leq m$. The crucial object in studying the behavior of the Markov renewal process away from the boundary states in the level 0 is the *fundamental period*, the first passage time from a state in $i + 1$ to a state in i. The joint transform matrix $\tilde{G}(z; s)$ of that first passage time, measured in the number of transitions to lower levels (completed services in queueing applications) and in real time, satisfies a nonlinear matrix equation of the form

$$\tilde{G}(z; s) = z \sum_{k=0}^{\infty} \tilde{A}(s)[\tilde{G}(z; s)]^k.$$

That equation can be analyzed by methods of functional analysis. It leads to many explicit matrix formulas for moments. In terms of the matrix $\tilde{G}(z; s)$, the boundary behavior of the Markov renewal process can be studied in an elementary manner. In queueing applications, that leads to equations for the busy period and the busy cycle. Waiting time distributions under the first-come, first-served discipline are obtained as first passage time distributions. Extensive generalizations of the Pollaczek-Khinchin integral equation for the classical M/G/1 queue have been obtained (see Neuts, 1986).

Applications of Markov renewal theory lead to a matrix formula for the steady-state probability vector x_0 for the states in level 0 in the imbedded Markov chain. Next, a stable numerical recurrence due to Ramaswami (1988) permits computation of the steady-state probability vectors x_i of the other levels i, $i \geq 1$.

There is an interesting duality between the random walks on the infinite strip of states (i, j), $-\infty < i < \infty$, $1 \leq j \leq m$, that underlie the Markov renewal processes of M/G/1 type and those of GI/M/1 type (which lead to matrix-geometric solutions). That duality is investigated in

Asmussen and Ramaswami (1990) and Ramaswami (1990).

The class of models with an imbedded Markov renewal process of M/G/1-type is very rich. It is useful in the analysis of many queueing models in continuous or discrete time that arise in communications engineering and other applications. In queueing theory, results for a variety of classical models have been extended to versatile input processes and to semi-Markovian services. These generalizations often lead to natural matrix generalizations of familiar formulas. For a transparent comparison of the *MAP/G/1* queue to the M/G/1 model, see Lucantoni (1993). For an analysis of the MAP/SM/1 queue with Markovian arrival imput and semi-Markovian services, see Lucantoni and Neuts (1994). A treatment of cycle maxima for the MAP/G/1 queue is found in Asmussen and Perry (1992). A mathematically rigorous discussion of the complex analysis aspects of the models of M/G/1-type is found in Gail, Hantler, and Taylor (1994) and in another, as yet unpublished manuscript by the same authors, cited therein. Asymptotic results on the tail probabilities of queue length and waiting time distributions are discussed in Abate, Choudhury and Whitt (1994), and Falkenberg (1994).

The implementation of the matrix-analytic results presents a variety of challenging numerical problems and has resulted in algorithms of considerable practical utility.

MATRIX-GEOMETRIC SOLUTIONS: Under ergodicity conditions, discrete-time Markov chains with transition probability matrix P of the form

$$P = \begin{vmatrix} B_0 & A_0 & 0 & 0 & 0 & \cdots \\ B_1 & A_1 & A_0 & 0 & 0 & \cdots \\ B_2 & A_2 & A_1 & A_0 & 0 & \cdots \\ B_3 & A_3 & A_2 & A_1 & A_0 & \cdots \\ \cdot & \cdot & \cdot & \cdot & \cdot & \cdots \end{vmatrix},$$

where the A_k are $m \times m$ nonnegative matrices summing to a stochastic matrix A, and the B_k are nonnegative matrices such that the row sums of P are one, have an invariant probability vector x of a *matrix-geometric* form. That is, the unique probability vector x which satisfies $xP = x$, can be partitioned into row vectors x_i, $i \geq 0$, which satisfy $x_i = x_0 R^i$. The matrix R is the unique minimal solution to the equation

$$R = \sum_{k=0}^{\infty} R^k A_k,$$

in the set of nonnegative matrices. All eigenvalues of R lie inside the unit disk. The matrix,

$$B[R] = \sum_{k=0}^{\infty} R^k B_k,$$

is an irreducible stochastic matrix. The vector x_0 is determined as the unique solution to the equations

$$\begin{cases} x_0 = x_0 B[R] \\ 1 = x_0 (I - R)^{-1} e \end{cases}$$

where e is the column m-vector with all components equal to one. If the matrix A is irreducible and has the invariant probability vector π, the Markov chain is positive recurrent if and only if

$$\pi \sum_{k=1}^{\infty} k A_k e > 1.$$

Analogous forms of the matrix-geometric theorem hold for Markov chains with a more complicated behavior at the boundary states and for continuous Markov chains with a generator Q of the same structural form. A comprehensive treatment of the basic properties of such Markov chains and a variety of applications are given in Neuts (1981).

The stated theorem has found many applications in queueing theory. The subclass where the matrix P or the generator Q are *block-tridiagonal* are called *quasi-birth and death* (QBD) processes. These arise naturally as models for many problems in communications engineering and computer performance. The matrix-geometric form of the steady-state probability vector of a suitable imbedded Markov chain leads to explicit matrix formulas for other descriptors of queues, such as the steady-state distributions of waiting times, the distribution of the busy period and others.

In addition to its immediate applications, the theorem has also generated much theoretical interest. Its generalization to the operator case was established in Tweedie (1982).

The largest eigenvalue η of the matrix R is important in various asymptotic results. Graphs of η as a function of a parameter of the queue are *caudal characteristic curves*. Some interesting behavioral features of the queues can be inferred from them (Neuts and Takahashi, 1981; Neuts, 1986; Asmussen and Perry, 1992). A matrix-exponential form for waiting-time distributions in queueing models was obtained in Sengupta (1989). Its relation to the matrix-geometric theorem was discussed in Ramaswami (1990). A matrix-analytic treatment, covering all cases of reducibility, of the equation for R, is given in Gail, Hantler and Taylor (1994).

The matrix R, which is crucial to all applications of the theorem, must be computed by an iterative numerical solution of the nonlinear matrix equation

$$R = \sum_{k=0}^{\infty} R^k A_k.$$

A major survey and comparisons of various computational methods are found in Latouche (1993). For the block tri-diagonal case (QBD-processes), a particularly efficient algorithm was developed by Latouche and Ramaswami (1993).

MARKOVIAN ARRIVAL PROCESSES: The analytic tractability of models with Poisson or Bernoulli input is due to the lack-of-memory property, an extreme case of Markovian simplification. At the expense of performing matrix calculations, more versatile arrival processes can be used in a variety of models. The *Markovian arrival process* (MAP) is a point process model in which only one of a finite number of phases must be remembered to preserve many of the simplifying Markovian properties. It can be incorporated in many models which remain highly tractable by matrix-analytic methods. The MAP has found many applications in queueing and teletraffic models to represent bursty arrival streams. Many queueing models for which traditionally Poisson arrivals were assumed are also amenable to analysis with MAP input.

It was first introduced in Neuts (1979), but a more appropriate notation was proposed by David Lucantoni in conjunction with the queueing model discussed in Lucantoni, Meier-Hellstern, and Neuts (1990). The MAP has found many applications in queueing and teletraffic models to represent bursty arrival streams. Many queueing models for which traditionally Poisson arrivals were assumed are also amenable to analysis with MAP input. Although discrete-time versions of the MAP, as well as processes with group arrivals have been defined, their discussion requires only more elaborate notation than the single-arrival MAP in continuous time described here. Expositions of the basic properties and many examples of the MAP are found in Neuts (1989, 1992) and Lucantoni (1991).

We begin with an irreducible infinitesimal generator D of dimension m with stationary probability vector θ. We write D as the sum of matrices D_0 and D_1. D_1 is nonnegative and D_0 has nonnegative off-diagonal elements. The diagonal elements of D_0 are strictly negative and D_0 is nonsingular. We consider an m-state Markov renewal process $\{(J_n, X_n), n \geq 0\}$ in which each transition epoch has an associated arrival. Its transition probability matrix $F(\cdot)$ is given by

$$F(x) = \int_0^x \exp(\boldsymbol{D}_0 u)\, du\, \boldsymbol{D}_1, \quad \text{for } x \geq 0.$$

The most familiar *MAP*s are the *PH*-renewal process and the *Markov-modulated Poisson Process* (*MMPP*). These respectively have the pairs of parameter matrices $\boldsymbol{D}_0 = \boldsymbol{T}$, $\boldsymbol{D}_1 = \boldsymbol{T}^\circ \alpha$, where $(\alpha, \boldsymbol{T})$ is the (irreducible) representation of a phase-type distribution and the column vector $\boldsymbol{T}^\circ = -\boldsymbol{T}\boldsymbol{e}$, and $\boldsymbol{D}_0 = \boldsymbol{D} - \Lambda$, $\boldsymbol{D}_1 = \Lambda$, where Λ is a diagonal matrix and $\boldsymbol{e}$ is the column m-vector with all components equal to one.

The matrix-analytic tractability of the *MAP* is a consequence of the matrix-exponential form of the transition probability matrix $F(\cdot)$. It, in turn, follows from the Markov property of the underlying chain with generator $\boldsymbol{D}$, in which certain transitions are labeled as arrivals. A detailed description of that construction is found in Lucantoni (1991).

The initial conditions of the *MAP* are specified by the initial probability vector of the underlying Markov chain with generator $\boldsymbol{D}$. For $\gamma = \theta$, the stationary probability vector of $\boldsymbol{D}$, we obtain the *stationary version* of the *MAP*. The rate λ^* of the stationary process is given by $\lambda^* = \theta \boldsymbol{D}_1 \boldsymbol{e}$. By choosing $\gamma = (\lambda^*)^{-1} \theta \boldsymbol{D}_1 = \theta_{\text{arr}}$, the time origin is an arbitrary arrival epoch.

Computationally tractable matrix expressions are available for various moments of the *MAP*. These require little more than the computation of the matrix $\exp(\boldsymbol{D}t)$. A comprehensive discussion of these formulas is found in Neuts and Narayana (1992). For example, the Palm measure, $H(t) = E[N(t)|\text{arrival at } t = 0]$, the expected number of arrivals in an interval $(0,t]$ starting from an arbitrary arrival epoch, is given by

$$H(t) = \lambda^* t + \theta_{\text{arr}}(\boldsymbol{I} - \exp(\boldsymbol{D}t))(\boldsymbol{e}\theta - \boldsymbol{D})^{-1}\boldsymbol{D}_1\boldsymbol{e}.$$

Other *MAP*s are constructed by considering selected transitions in Markov chains, by certain random time transformations or random thinning of a given *MAP*, and by superposition of independent *MAP*s. Statements and examples of these constructions are found in Neuts (1989, 1992). Specifically, the *superposition* of two (or more) independent *MAP*s is again an *MAP*. If two continuous-time *MAP*s have the parameter matrices $\{\boldsymbol{D}_k(i)\}$ for $i = 1, 2$, the parameter matrices for their superposition are given by $\boldsymbol{D}_k = \boldsymbol{D}_k(1) \otimes \boldsymbol{I} + \boldsymbol{I} \otimes \boldsymbol{D}_k(2) = \boldsymbol{D}_k(1) \otimes \boldsymbol{D}_k(2)$, for $k \geq 1, 2$, where $\otimes$ is the Kronecker pairwise matrix product.

See **Markov chains; Markov processes; Matrices and matrix algebra; Phase-type distribution; Queueing theory.**

References

[1] Abate, J., Choudhury, G.L. and Whitt, W. (1994), "Asymptotics for steady-state tail probabilities in structured Markov queueing models," *Stochastic Models*, 10, 99–143.

[2] Asmussen, S. and Perry, D. (1992), "On cycle maxima, first passage problems and extreme value theory for queues," *Stochastic Models*, 8, 421–458.

[3] Asmussen, S. and Ramaswami, V. (1990), "Probabilistic interpretation of some duality results for the matrix paradigms in queueing theory," *Stochastic Models*, 6, 715–733.

[4] Falkenberg, E. (1994), "On the asymptotic behavior of the stationary distribution of Markov chains of M/G/1-type," *Stochastic Models*, 10, 75–97.

[5] Gail, H.R., Hantler, S.L. and Taylor, B.A. (1994), "Solutions of the basic matrix equations for the M/G/1 and G/M/1 Markov chains," *Stochastic Models*, 10, 1–43.

[6] Latouche, G. (1985), "An exponential semi-Markov Process, with applications to queueing theory," *Stochastic Models*, 1, 137–169.

[7] Latouche, G. (1993), "Algorithms for infinite Markov chains with repeating columns," in *Linear Algebra, Markov Chains and Queueing Models*, Meyer, C.D. and Plemmons, R.J., eds., Springer Verlag, New York, 231–265.

[8] Latouche, G. and Ramaswami, V. (1993), "A logarithmic reduction algorithm for quasi-birth-and-death processes," *Jl. Appl. Prob.*, 30, 650–674.

[9] Lucantoni, D.M. (1991), "New results on the single server queue with a batch Markovian arrival process," *Stochastic Models*, 7, 1–46.

[10] Lucantoni, D.M. (1993), "The BMAP/G/1 queue: a tutorial," in *Models and Techniques for Performance Evaluation of Computer and Communications Systems*, L. Donatiello and R. Nelson, eds., Springer-Verlag, New York.

[11] Lucantoni, D.M., Meier-Hellstern, K.S., and Neuts, M.F. (1990), "A single server queue with server vacations and a class of non-renewal arrival processes," *Adv. Appl. Prob.*, 22, 676–705.

[12] Neuts, M.F. (1979), "A versatile Markovian point process," *Jl. Appl. Prob.*, 16, 764–779.

[13] Neuts, M.F. (1981), *Matrix-Geometric Solutions in Stochastic Models: An Algorithmic Approach*, The Johns Hopkins University Press, Baltimore. Reprinted by Dover Publications, 1994.

[14] Neuts, M.F. (1986), "The caudal characteristic curve of queues," *Adv. Appl. Prob.*, 18, 221–254.

[15] Neuts, M.F. (1986), "Generalizations of the Pollaczek-Khinchin integral equation in the theory of queues," *Adv. Appl. Prob.*, 18, 952–990.

[16] Neuts, M.F. (1989), *Structured Stochastic Matrices of M/G/1 Type and Their Applications*, Marcel Dekker, New York.

[17] Neuts, M.F. (1992), "Models Based on the Markovian Arrival Process," *IEEE Trans. Communications*, Special Issue on Teletraffic, E75-B, 1255–1265.

[18] Neuts, M.F. and Narayana, S. (1992), "The first two moment matrices of the counts for the Markovian arrival process," *Stochastic Models*, 8, 459–477.

[19] Neuts, M.F. and Takahashi, Y. (1981), "Asymptotic behavior of the stationary distributions in the GI/PH/c queue with heterogeneous servers," *Z. f. Wahrscheinlichkeitstheorie*, 57, 441–452.

[20] Ramaswami, V. (1988), "A stable recursion for the steady state vector in Markov chains of M/G/1 type," *Stochastic Models*, 4, 183–188.

[21] Ramaswami, V. (1990), "A duality theorem for the matrix paradigms in queueing theory," *Stochastic Models*, 6, 151–161.

[22] Ramaswami, V. (1990), "From the matrix-geometric to the matrix-exponential," *Queueing Systems*, 6, 229–260.

[23] Sengupta, B. (1989), "Markov processes whose steady state distribution is matrix-exponential with an application to the GI/PH/1 queue," *Adv. Appl. Prob.*, 21, 159–180.

[24] Schellhaas, H. (1990), "On Ramaswami's algorithm for the computation of the steady state vector in Markov chains of M/G/1-type," *Stochastic Models*, 6, 541–550.

[25] Tweedie, R.L. (1982). "Operator-geometric stationary distributions for Markov chains with application to queueing models," *Adv. Appl. Prob.*, 14, 368–391.

MATRIX GEOMETRIC

When the solution to a stochastic model is (vector) proportional to a geometric distribution whose parameter is a matrix instead of the usual scalar. See **Matrix-analytic stochastic models**.

MATRIX GAME

See **Game theory**.

MAUT

See **Multi-attribute utility theory**.

MAX-FLOW MIN-CUT THEOREM

For a maximum-flow network problem, it can be shown that the maximum flow through the network is equal to the minimum capacity of all the cuts that separate the source (origin) and the sink (destination) nodes, where the capacity of a cut is the sum of the capacities of the arcs in the cut. See **Maximum-flow network problem**.

MAXIMUM

A function $f(x)$ is said to have a maximum on a set S when the least upper bound of $f(x)$ on S is assumed by $f(x)$ for some x^0 in S. Thus, $f(x^0) \geq f(x)$ for all x in S. See **Global maximum (minimum)**.

MAXIMUM FEASIBLE SOLUTION

See **Minimum feasible solution**.

MAXIMUM-FLOW NETWORK PROBLEM

For a directed, capacitated network with source and sink nodes, the problem is to find the maximum amount of goods (flow) that can be sent from the source to the sink.

MAXIMUM MATCHING PROBLEM

Involves finding in a graph a maximal set of links which meet each node at most once.

MCDM

See **Multi-criteria decision making; Multi-objective programming**.

MEASURE OF EFFECTIVENESS (MOE)

In a decision problem, the single objective that is to be optimized is called the measure of effectiveness (MOE). In a linear-programming problem, the MOE is the objective function. See **Mathematical model**.

MEDICINE AND MEDICAL PRACTICE

Charles D. Flagle

The Johns Hopkins University, Baltimore, Maryland

INTRODUCTION: The techniques of operations research have found their way into medical practice, not just in the logistical and managerial support of clinical services, but in the central decision processes of disease screening, diagnosis and therapy, and in medical education. Hundreds of citations to operations research and its associated analytical techniques are to be found in the medical literature and are accessible in the U.S. National Library of Medicine's on-line MEDLARS system. The early applications spawned new professional organizations and journals, now thriving in the medical arena. Many operations research applications are indexed to the near synonymous term, "Medical Informatics," the application of computers and communications technology to the broad field of health care. To understand the position of operations research in the medical literature, a simple rule helps: medical informatics relates to medicine and health care as operations research relates to the work of business and industry. Both place a heavy emphasis on exploitation of the potentials of computer and communication technologies.

Some major traditional operations research methods predominate in medical care applications: stochastic models, computer simulation, mathematical programming, and decision analysis. Some examples follow.

COMPUTER SIMULATION, MONTE CARLO METHODS, AND STOCHASTIC MODELS:

Computer simulation plays an important role in teaching, research, and development of medical practice, expanding beyond its early role as a mimic of complex stochastic processes. A review of over a hundred documents indexed both to simulation and medicine reveals a range of applications from the traditional Monte Carlo representation of physiological processes to three dimensional imaging. An example of use of computer simulation to develop a protocol for burn care is given by Roa and Gomez-Cia (1994). Many applications are devoted to medical education by simulation of clinical problems; either through random sequences of events in diagnosis and therapy or the responsive behavior of images or a mannikin, a move toward virtual reality. Examples of simulation as an instructional aid in a clinical setting are: handling emergencies in cardiac care (Tanner and Gitlow, 1991; Sajid *et al.*, 1990; Bergeron and Greenes, 1989), ophthalmology care (Lonwe and Heiji, 1993), and training radiologists (Martin, Bruidley, and Awad, 1993). The use of simulation in anesthesiology is extensive and is reviewed by several authors (Swank and Vahr, 1992; Garfield *et al.*, 1992; Sciwa, 1992). Software involving simulation is available for continuing medical education (Chiao, 1992).

The field of epidemiology, lying within the domain of medicine, has been attractive to stochastic model building and simulation. Understanding the origin, spread, and decline of epidemics is essential to recognition of causal agents and vectors of transmittal and the development and evaluation of prevention measures. The complexity of epidemics and interventions to control them often defies a purely mathematical analysis. In early studies of mathematical epidemiology, Bailey (1967) used computer simulation to predict both temporal and spatial progress of epidemics, expressed in stochastic models. Current examples of stochastic models and simulation are found in epidemic models of HIV/AIDS (Bailey, 1991; Kaplan , 1991; Harris and Rattner, 1992; Kaplan and Brandeau, 1994). The problem of mixed susceptibility to disease in a population is addressed by Boylan (1991).

MATHEMATICAL PROGRAMMING:

Opportunities for optimization of treatment strategies have emerged most frequently in situations where there is inherent trade-off between the intended benefit of a therapeutic intervention and its potentially damaging side effects. Examples occur in therapeutic radiology (Rosen *et al.*, 1991), where the objective is to maximize target volume dosage while imposing constraints on dose-volume delivered to surrounding normal tissue. In a refinement, the pattern of constraints on exposure to normal tissue has been expressed as a function of distance of normal tissue from the target area (Morrill *et al.*, 1991).

Nuclear medicine presents challenges to unconstrained optimization, where concentration in a target organ is simultaneously a function of the rate of diminishing strength of a decaying isotope and increasing strength of target organ concentration of the isotope with the passage of time. Both dose volume and time lapse to scanning can be optimized (Emmons, 1968). A dynamic programming model has been applied to determination and automation of optimal procedures in anesthesia to extend the productivity of anaesthesia professionals (Esogbue *et al.*, 1976).

DECISION ANALYSIS IN MEDICAL DECISION MAKING:

The notion of trade-off among benefits or losses carries through to decisions made under uncertainty in screening and clinical diagnosis, where the costs of missing a case – a false negative – are balanced against the costs of interpreting as present a condition not there – a false positive. The expression in the normal form of search for a Bayesian solution, that is, as a minimization of expected loss, bears strong resemblance to linear programming, but the evolution of medical decision theories has yielded a number of new approaches. Nearly parallel in time to the early applications of operations research to the logistical and organizational problems of health services, examples of relevance of the techniques appeared directly in the procedures of diagnosis and therapy. The availability of large data bases linking patient signs, symptoms, and other descriptors to disease states has led to applications of statistical analysis and value theory to diagnostic processes and choice of therapy. A review (Barnoon and Wolfe, 1972) cites early work on the logical foundations of diagnosis (Ledley and Lusted, 1959) and on disease screening (Flagle, 1967) demonstrating that the optimal screening level of a test is a specific function of prevalence of undetected disease and the relevant costs, or regrets, of false negative and false positive determinations. The applications of linear programming, pattern recognition and decision support systems for breast cancer diagnosis are described in Mangasarian *et al.* (1990, 1994) and Wolberg and Mangasarian (1993).

Improvement in screening through increases in sensitivity and specificity of test remains an objective, and is aided by multivariate analysis made possible by large data bases and clinical trials. Emergence of the term, "Computer-Aided Diagnosis," has accompanied many efforts to sharpen the statistical relationship between symptoms and disease (e.g., Gorry, 1968; Gorry and Barnett, 1968). Turning attention to treatment, knowledge gained from statistical analysis and systematic compilation of outcomes has led beyond estimation of diagnostic probabilities to a large enterprise in expert systems, in which the processes of successful therapy are expressed in algorithmic form (Warner, 1964) built around a patient data base interfaced to a knowledge base. Beyond expert systems to diffuse and implement protocols, efforts have been made to understand and emulate the decision process itself – an approach to artificial intelligence (AI). Early development in AI related to specific diseases are reviewed by Szolovits and Pauker (1978). Generalization of AI processes in medical consultation, such as the One MYCIN program (Shortliffe, 1976), knowledge coupling (Weed, 1986) and Expert (Weiss and Kulikowski, 1979) characterize the development of computer-aided decision processes.

DIRECTIONS OF DEVELOPMENT: Two major patterns are discernible in the evolution of medical practice aided by growing data bases, improving analytic techniques, and new communications technologies. First is the formalization of decision processes in multi-disciplinary protocols or practice guidelines based on outcomes research and technology assessment. The dissemination of such guidelines, which often bear the imprimatur or approval of relevant specialty groups (American College of Physicians, 1994), is enhanced by electronic publication, storage and retrieval systems.

The format of practice guidelines, which often contain a prescriptive algorithm familiar to operations researchers, often contains also a version for patients. This marks the second emerging direction – the increasingly enlightened involvement of patients in decisions about choice of therapeutic strategies (Reiser, 1993). Practitioners of operations research have been among the pioneers in development of interactive query and support programs for patients (Gustafson, 1993), and continue to apply such techniques as the analytic hierarchy process (Saaty, 1980; Dolan and Bordley, 1992) to the development of medical decision processes. Recent developments, very much like the earliest ventures of OR/MS into the medical field, have begun with a seemingly spontaneous collaboration of a physician and OR analyst; and in time the concepts become internalized in the medical decision process.

See **Artificial intelligence; Dynamic programming; Expert systems; Health care systems; Hospitals; Linear programming.**

References

[1] American College of Physicians (1994). "Guidelines for Medical Treatment for Stroke Prevention," *Annals Internal Medicine*, 121, 54–55.

[2] Bailey, N.T.J. (1967). *The Mathematical Approach to Biology and Medicine*, John Wiley, London.

[3] Bailey, N.T. (1991). "The Use of Operational Modeling of HIV/AIDS in a Systems Approach to Public Health Decision Making," *Mathematical Biosciences*, 107, 413–430.

[4] Barnoon, S. and Wolfe, H. (1972). *Measuring the Effectiveness of Medical Decisions: An Operations Research Approach*, Clarke C. Thomas, Springfield, Illinois.

[5] Bergeron, B.P. and Greenes, R.A. (1989). "Clinical Skill-building Simulations in Cardiology: Heartlab and Eklab," *Computer Methods and Programs in Biomedicine*, 30(2–3), 111–126.

[6] Boylan, R.D. (1991). "A Note on Epidemics in Heterogeneous Populations," *Mathematical Biosciences*, 105, 133–137.

[7] Dolan, J.G. and Bordley, D.R. (1991). "Should Concern Over Gastric Cancer Influence the Choice of Diagnostic Tests in Patients with Acute Upper Gastrointestinal Bleeding?" *Proceedings of the Second International Symposium on the Analytic Hierarchy Process*, Pittsburgh, Pennsylvania, 391–404.

[8] Emmons, H. (1968). "The Optimal Use of Radioactive Pharmaceuticals in Medical Diagnosis," Doctoral Dissertation, The Johns Hopkins University, Baltimore.

[9] Esogbue, A.C., Aggarwal, V. and Kaujalgi, V. (1976). "Computer-Aided Anesthesia Administration," *Int. Jl. Biomedical Computing*, 7, 271–288.

[10] Flagle, C.D. (1967). "A Decision Theoretical Comparison of Three Procedures of Screening for a Single Disease," *Proceedings of the Fifth Berkeley Symposium on Mathematical Statistics and Probability*, University of California Press, Berkeley.

[11] Garfield, D.A., Rapp, C., and Evens, M. (1992). "Natural Language Processing in Psychiatry: Artificial Intelligence Technology and Psychopathology," *Jl. Nervous and Mental Disease*, 180, 227–237.

[12] Gorry, G.A. (1968). "Strategies for Computer Aided Diagnosis," *Mathematical Biosciences*, 2, 293–318.

[13] Gorry, G.A. and Barnett, G.O. (1968). "Experience with a Model of Sequential Diagnosis," *Computers in Biomedical Research*, 1, 490–507.

[14] Gustafson, D.H., Taylor, J.O., Thompson, S. and Chesney, P. (1993). "Assessing the Needs of Breast Cancer Patients and Their Families," *Quality Management in Health Care*, 2(1), 6–17.

[15] Harris, C.M. and Rattner, E. (1992). "Forecasting the Extent of the HIV/AIDS Epidemic," *Socio-Econ. Plann. Sci*, 26(3), 149–168.

[16] Kaplan, E.H. (1991). "Mean-max Bounds for Worst Case Endemic Mixing Models," *Matm. Bioscience*, 105, 97–109.

[17] Kaplan, E.H. and Brandeau, M.L., eds. (1994). *Modeling the AIDS Epidemic: Planning, Policy, and Prediction*, Raven, New York.

[18] Ledley, R.S. and Lusted, L.B. (1959). "Reasoning Foundation of Medical Diagnosis," *Science*, 130, 9–29.

[19] Lonwe, B. and Heiji, A. (1993). "Computer-assisted Instruction in Emergency Ophthalmological Care," *Acta Ophthalmologica*, 71, 289–295.

[20] Mangasarian, O.L., Setiono, R., and Wolberg, W.H. (1990). "Pattern Recognition via Linear Programming: Theory and Application to Medical Diagnosis," in *Large-Scale Numerical Optimization*, Thomas F. Coleman and Yuying Li, eds., SIAM, Philadelphia, 22–30.

[21] Mangasarian, O.L., Street, W.N., and Wolberg, W.H. (1994). "Breast Cancer Diagnosis and Prognosis via Linear Programming," University of Wisconsin Computer Sciences Mathematical Programming Technical Report 94-10, Madison.

[22] Morrill, S.M., Lane, R.G., Wong, J.A. and Rosen, I.I. (1991). "Dose-volume Considerations with Linear Programming Optimization," *Medical Physics*, 18, 1201–1210.

[23] Reiser, S.J. (1993). "The Era of the Patients," *Jl. Amer. Med. Assn.*, 269, 1012–1017.

[24] Roa, L. and Gomez-Cia, T. (1994). "A Burn Patient Resuscitation Therapy Designed by Computer Similation," *Yearbook of Medical Informatics*, Schattauer Verlagsgesellschaft, Stuttgart.

[25] Rosen, I.I., Lane, R.G., Morrill, S.M., and Belli, J.A. (1991). "Treatment Plan Optimization Using Linear Programming," *Medical Physics*, 18, 141–152.

[26] Saaty, T.L., (1981). "The Analytic Hierarchy Process and Health Care Problems," *Proceedings of International Conference on Systems Science in Health Care*, Montreal, 1980.

[27] Shortliffe, Edward H. (1976). *Computer-Based Medical Consultations MYCIN*, American Elsevier, New York.

[28] Szolovits, P. and Pauker, S.C. (1978). "Categorical and Probabilistic Reasoning in Medical Diagnosis," *Artificial Intelligence*, 11, 115–144.

[29] Tanner, T.B. and Gitlow, S. (1991). "A Computer Simulation of Cardiac Emergencies," *Proceedings Annual Symposium on Computer Applications in Medical Care*.

[30] Warner, H.R. (1979). *Computer Assisted Medical Decision Making*, Academic Press, New York.

[31] Weed, L.L. (1986). "Knowledge Coupling Medical Education and Patient Care," *Crit. Ref. Med. Infomatics*, 1, 55–79.

[32] Wolberg, W. H. and O. L. Mangasarian (1993). "Computer-designed Expert Systems for Breast Cytology Diagnosis," *Analytical and Quantitative Cytology and Histology*, 15, 67–74.

MEMORYLESS PROPERTY

For stochastic processes, lack-of-memory is synonymous with the Markov property. For a positive random variable T that models the duration of some phenomenon, lack-of-memory means that the time remaining is independent of the time already passed, that is, $\Pr\{T > t + s \mid T > s\} = \Pr\{T > t\}$ for s, $t > 0$. The exponential distribution is the only continuous distribution with lack-of-memory, while the geometric distribution is the only discrete distribution with lack-of-memory. See **Exponential arrivals**; **Markov processes**; **Markov property**.

MENU PLANNING

A diet problem in which the variables represent complete menu items such as appetizers and entrees, instead of individual foods. The problem is formulated as an integer-programming problem in which the integer binary variables represent the decision of selecting or not selecting a complete menu item.

METAGAME ANALYSIS

A problem structuring method which addresses situations of conflict and cooperation between independent actors. Based on game theoretic concepts, it identifies explicit and implicit threats and promises between the actors to analyse the stability of alternative scenarios.

METAMODELING

Israel Pressman and
Linda Weiser Friedman

Baruch College of the City University
of New York

INTRODUCTION: The term *metamodel* refers to any auxiliary model that is used to aid in the interpretation of a more detailed model. Metamodeling is frequently part of a simulation study. The simulation model, although simpler than the real-world system, is still a very complex way of relating input to output. A simpler model, however, may be used as an auxiliary to the simulation model in order to better understand the more complex model and to provide a framework for testing hypotheses about it. This auxiliary model is the metamodel.

Several authors have pointed out the need for an analytic auxiliary model to aid in interpretation of the more detailed model: Geoffrion (1976), whose concern was with mathematical programming models; Blanning (1974, 1975a,b), who proposed the use of metamodels for all kinds of management science models; Lawless *et al.* (1971) and Rose and Harmsen (1978), who made explicit use of metamodels for sensitivity analysis; Kleijnen (1975, 1979, 1981, 1982; Kleijnen *et al.*, 1979), who introduced the metamodel concept to simulation analysis; Friedman (1986, 1989) who used a multivariate metamodel to explore relationships among factors and response measurements in queueing simulations.

One metamodel frequently postulated in simulation analysis is the general linear model or, regression model, of experimental design. The procedure begins with a valid simulation program that is run to generate data in the region of interest. This simulation-generated data is then used as input in developing a model relating the performance characteristic of interest to one or more independent, predictor variables.

The use of a metamodel is often implicit in the statistical procedures used to analyze simulation output. After all, many of these procedures assume an underlying model relating factors and measurements; often, this is the linear additive model of experimental design. The explicit use of a metamodel in post-simulation analysis has many benefits. Aside from the obvious advantage that working with a simple mathematical function has over running and rerunning costly simulation programs, and aside from the pleasing elegance of a solution obtained by the union of numerical and analytic techniques, the simulation metamodel has been lauded for its many other uses. Among these are the following: model simplification; enhanced exploration, optimization, and interpretation of the model; the unraveling of a model's dynamics in order to gain a better understanding of the system's behavior; generalization to models of other systems of the same type; the ability to test many hypotheses regarding the system without performing additional runs; efficient sensitivity analysis; ease in answering inverse questions, for example, given a particular value for a response variable, what input value (factor level) is needed?

THE METAMODEL IN SYSTEM SIMULATION: As Figure 1 indicates, the true real-world relationships targeted by a simulation study may be represented by:

$$\mu = g(x_1, x_2, \ldots, x_q) \tag{1}$$

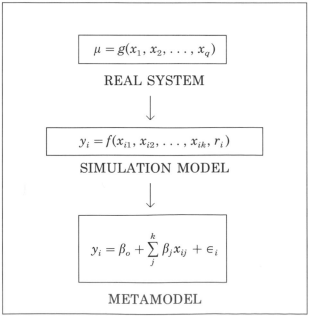

Fig. 1. The metamodel in system simulation.

where is the system response, a measure which in some way characterizes a time path of the real system; the q different x's are the factors, either controllable or environmental, which determine the value of the system response; g represents the (unknown) relationship governing the effect of the factors on the system response, that is, the way in which the factors determine the value of the system response. The objective of simulation modeling is to approximate this unknown relationship as closely as possible (at least with regard to the variables that are important to the model user) in order to study the system in ways that would be impossible or impractical in the physical world. If we represent the simulation model (flowchart, computer program, etc.) by a function, f, then

$$y_i = f(x_{i1}, x_{i2}, \ldots, x_{ik}, r_i), \quad i = 1, \ldots, n \quad (2)$$

where y_i is the value of the response variable in the ith factor combination; n represents the number of combinations; k (probably less than q) is the number of input variables (factors); x_{ij} is the value of the jth input variable in the ith combination; r_i is the set of random number streams upon which the simulation depends and serves to model the effect of all $q - k$ factors in the real system that have been excluded from the simulation.

Figure 1 is a pictorial representation of the three levels of explanation of the dynamics of a system simulation (Friedman and Friedman, 1984). The first level, the real system itself, is "unapproachable" by the researcher, who can never hope to understand it completely. The system analysis and data collection functions take place here. At the second level, although the simulation model is "leaner" than the real system, it does attempt to replicate the real system at least with regard to variables that are important to the goals of the simulationist. The simulation model building, verification, and validation functions take place here.

At the third level, the metamodel attempts to approximate and aid in the interpretation of the simulation model and, ultimately, of the real system itself. The experimental design and analysis function takes place here, and the general linear metamodel can often be used as a generalization of the various types of analyses performed on simulation output data. One simple metamodel favored by some simulation researchers is the additive model of experimental design, often generalized as a "regression" model (Kleijnen 1979, 1982; Kleijnen *et al.*, 1979). Often the linear regression metamodel can provide additional information regarding the relative contribution of

each input factor to the response variable of interest.

THE GENERAL LINEAR METAMODEL: The general linear metamodel (GLM) would be:

$$y_i = \beta_0 + \sum_j \beta_j x_{ij} + e_i \quad i = 1, \ldots, n; j = 1, \ldots, k \quad (3)$$

where e_i is the experimental error in the ith replication.

In the multiple-response simulation study, the multivariate general linear metamodel is a set of regression-type equations, each representing the contributions of the factors to the value of a specific response measure (Friedman 1986, 1987, 1989). This metamodel may be tested for overall multivariate significance via the mutivariate general linear hypothesis outlined in Morrison (1976) and automated in Friedman and Friedman (1985a).

It can be shown that the linear additive models underlying many multivariate and univariate statistical procedures, including those of experimental design, are specific cases of the general linear model (Morrison, 1976). Thus, depending on the experimental layout, whether the factors are quantitative or qualitative, and the aim of the study, the general linear metamodel reduces to the specific models of regression analysis, analysis of variance, t-test, paired t-test, etc. In fact, the design of a simulation experiment that will be analyzed via one of these statistical tools implies just such a general linear metamodel in one of its forms.

Simulationists may be legitimately concerned that metamodeling, being a data-dependent technique, may produce different results with different sets of data. Indeed, it is probably a good idea to randomly split the data in half and construct two different metamodels on the two resulting sets of data in order to see if they are relatively consistent (Friedman and Friedman, 1985b). On the whole, studies indicate that the simulation metamodel tends to be relatively stable. In an analysis of three different simulated systems, the simulation metamodel was fairly consistent from one set of data to the next (Friedman and Pressman, 1988). As a technique, it is probably just as good as the simulation upon which it is based.

See **Mathematical model; Simulation of discrete-event stochastic systems; Verification, validation, and testing of models.**

References

[1] Blanning, R.W. (1974). "The Sources and Uses of Sensitivity Information," *Interfaces*, 4, 32–38.

[2] Blanning, R.W. (1975a). "Response to Michel Kleijnen, and Permut," *Interfaces*, 5, 24–25.

[3] Blanning, R.W. (1975b). "The Construction and Implementation of Metamodels," *Simulation*, 24, 177–184.

[4] Friedman, L.W. (1986). "Exploring Relationships in Multiple-Response Simulation Experiments," *Omega*, 14, 498–501.

[5] Friedman, L.W. (1987). "Design and Analysis of Multivariate Response Simulations, The State of the Art," *Behavioral Science*, 32, 138–148.

[6] Friedman, L.W. (1989). "The Multivariate Metamodel in Queueing System Simulation," *Computers & Industrial Engineering*, 16, 329–337.

[7] Friedman, L.W. and H.H. Friedman (1984). "Statistical Considerations in Simulation, The State of the Art," *Jl. Statistical Computation and Simulation*, 19, 237–263.

[8] Friedman, L.W. and H.H. Friedman (1985a). "MULTIVREG, A SAS Program," *Jl. Marketing Research* (Computer Abstracts), 22, 216–217.

[9] Friedman, L.W. and H.H. Friedman (1985b). "Validating the Simulation Metamodel, Some Practical Approaches," *Simulation*, 25, 144–146.

[10] Friedman, L.W. and I. Pressman (1988). "The Metamodel in Simulation Analysis, Can It Be Trusted?" *Jl. Operational Research Society*, 39, 939–948.

[11] Geoffrion, A.M. (1976). "The Purpose of Mathematical Programming is Insight, not Numbers," *Interfaces*, 7(1), 81–92.

[12] Kleijnen, J.P.C. (1975). "A Comment on Blanning's 'Metamodel for Sensitivity Analysis': The Regression Metamodel in Simulation," *Interfaces*, 5, 21–23.

[13] Kleijnen, J.P.C. (1979). "Regression Metamodels for Generalizing Simulation Results," *IEEE Transactions Systems, Man, and Cybernetics*, SMC-9, 93–96.

[14] Kleijnen, J.P.C (1981). "Regression Analysis for Simulation Practitioners," *Jl. Operational Research Society*, 32, 35–43.

[15] Kleijnen, J.P.C. (1982). "Regression Metamodel Summarization of Model Behavior." In *Encyclopedia of Systems and Control* (M.G. Singh, ed.), New York, Pergamon Press.

[16] Kleijnen, J.P.C., A.J. Van den Burg, and R. Th. van der Ham (1979). "Generalization of Simulation Results," *European Jl. Operational Research*, 3, 50–64.

[17] Lawless, R.W., L.H. Williams, and C.G. Richie (1971). "A Sensitivity Analysis Tool for Simulation with Application to Disaster Planning," *Simulation*, 17, 217–223.

[18] Morrison, D.F. (1976). *Multivariate Statistical Methods*, McGraw-Hill, New York.

[19] Rose, M.R. and R. Harmsen (1978). "Using Sensitivity Analysis to Simplify Ecosystem Models: A Case Study," *Simulation*, 31, 15–26.

METHOD OF STAGES

An analysis method that extends the birth and death type analysis to queueing systems with Erlangian service or interarrival times. Since an Erlangian variable is the sum of independent and identically distributed exponential random variables, the method of stages increases the state space to coincide with the underlying exponential random variables and solves the resulting system of equations using generating functions. See **Queueing theory**.

MILITARY OPERATIONS RESEARCH

Clayton J. Thomas

Air Force Studies & Analyses Agency, Washington, DC

INTRODUCTION: To say that Military Operations Research (MOR) is the application to *military* operations of the methods of operations research, is strictly correct, but gives only one clue to understanding the subject. The military operations research accomplishments in World War II, sketched below, pioneered and greatly influenced the early development and institutionalization of operations research generally. Also, they led to the continuation of military operations research after the war, in the governments of World War II participants, in academia, in industry, in not-for-profit "think tanks," and its adoption in similar institutions of other nations. The emphasis in this article is on practice and trends in the United States.

The general methods of operations research apply in particular to many aspects of military applications. Such differences as exist pertain mainly to the needs of military security and classification procedures, the nature of military operations and equipment, and the concerns of strategy, operational art, and tactics that relate to the use of military forces as instruments of national policy.

To follow current developments in the field, one may obtain the quarterly *Bulletin of Military Operations Research* from the Military OR Society (MORS). MORS and the Military Applications Section (MAS) of INFORMS publish the *Bulletin*. Also, MORS conducts annual classified symposia, as well as smaller mini-symposia and workshops (some unclassified), and MAS sponsors unclassified military operations research papers at National meetings of INFORMS. Both MORS and MAS have published proceedings of meetings and unclassified monographs. From these societies, one gets an idea of the overall MOR community. MAS has a membership of several hundred, while MORS has about three thousand members.

WORLD WAR II MOR ACCOMPLISHMENTS:

Although there were individual contributions to the scientific study of military operations, ranging from Archimedes to the work of Thomas A. Edison in World War I, it was in World War II that MOR became widespread and institutionalized. Solandt (1955) recalled that MOR as we think of it now began in the services in England as "operational research" in the early days of the war.

The British work centered about different subjects depending on the service: in the Air Force it was the problem of how to use radar, in the Navy it was the problem of anti-submarine warfare, and in the Army it was first limited to antiaircraft problems and again centered around radar. Professor Blackett is sometimes said to have started the work in all three services, and his account in Blackett (1962) drew on earlier papers to describe both results and methods.

Morse and Kimball (1946) drew on the work of many early MOR analysts of the Operations Research Group, U.S. Navy, to give results and methods. That work, once it was declassified and slightly modified, was re-published in 1951 and was very influential, not only in introducing MOR to future analysts, but also in introducing the potential applications of operations research generally to a wider audience.

The above work quotes a letter from Admiral King that enumerated helpful MOR applications (suggestive also of the work in other services):

(a) The evaluation of new equipment to meet military requirements.

(b) The evaluation of specific phases of operations (e.g., gun support, AA fire) from studies of action reports.

(c) The evaluation and analysis of tactical problems to measure the operational behavior of new material.

(d) The development of new tactical doctrine to meet specific requirements . . .

(e) The technical aspects of strategic planning.

(f) The liaison for the fleet with the development and research laboratories, naval and extra-naval.

Morse and Kimball also gave some reasons for the emergence in World War II of the practical value of the methods of MOR. As opposed to earlier wars there were:

more repetitive operations susceptible to analysis – strategic bombing, submarine attacks on shipping, landing operations, etc.

the increased mechanization of warfare, in that ". . . a men-plus-machines operation can

be studied statistically, experimented with, analyzed, *and predicted* by the use of known scientific techniques just as a machine operation can be"

". . . the increasing tempo of obsolescence in military equipment . . . . When we can no longer have the time to learn by lengthy trial and error on the battlefield, the advantages of quantitative appraisal and planning become more apparent."

Another useful account of World War II MOR, centering about the United States Air Force, is Brothers (1954). In addition to illuminating examples such as aerial bombing accuracy improvement, it gave valuable guidance on the organization of military operations research groups and operating procedures. In World War II, most of the MOR practitioners were civilians (though sometimes in uniform), and they had to earn the trust of military operators over time through useful work. This, of course, is by no means unique to MOR in World War II.

POSTWAR MOR DEVELOPMENTS:

After World War II ended, a majority of the MOR practitioners returned to non-military pursuits: universities, laboratories, industry, etc. The military services wondered how much MOR would be needed in peacetime. Each did decide to institutionalize its use of MOR. An early chapter of Tidman (1984) gave an interesting account of how the Navy chose to continue MOR after World War II. Each service had a different choice or mix of civil service groups, not-for-profit groups, use of industry, etc., and their emphases varied over time. Fairly soon, as the cold war emerged, there was general recognition that it would be necessary to increase the use of MOR. The chapter of Tidman mentioned above is appropriately entitled "A Period of Consolidation and Growth."

Some of the postwar applications of MOR resembled wartime MOR, with combat operations replaced by tests or exercises. However, some of the operations research (or "operations analysis" or "operations evaluation," as it was often termed) was devoted to operations of supply, logistics, recruiting, and training. Moreover, much of the effort went to thinking through the implications of new weapons for new types of combat operations.

World War II had seen the introduction of radar, atomic weapons, cruise missiles, and ballistic missiles, but each type was still improving rapidly at war's end. Their implications for, and fuller integration into, military forces needed more thought. The cold war climate also provided a sense of urgency, and MOR offices took on these problems as important foci of effort.

THE EMERGENCE OF SYSTEMS ANALYSIS:
MOR also took on problems at a level higher than that of individual weapon systems or engagements between two opposing weapon systems. Even in a cold war climate, there were significant limits on national expenditures for armed forces. It was necessary for government to decide "how much is enough," and MOR sought to aid this decision.

Applications of operations research at this high level, often termed "systems analysis," face difficulties far greater than the difficulties of World War II MOR, significant as the latter were. Wartime combat analysis, sometimes without recognizing it, had already faced criterion problems of sub-optimization, as Hitch (1953) pointed out. These become still more significant when one is structuring forces for the future, seeking to be prepared to deal with contingencies still beset with great uncertainty.

Hitch (1955) gave an understanding of the relative difficulty of systems analysis by comparing the World War II problem of improving bomber accuracy with the postwar problems of weapon system development and force composition. In the former problem, difficult as it seemed at the time, we knew the types of aircraft involved, how many there were, much about their characteristics, the kind of bombs available, and much about enemy targets and their defenses. These become variables when we prepare for an uncertain future that may sometimes hold a multiplicity of potential opposing forces.

The difficulties are in the problems, as Hitch went on to point out. Despite these difficulties, governments must make decisions, and systems analysis, with all of its limitations, has much to offer. MOR analysts developed judgment in "cutting problems down to size," and Quade (1954) collected some of the helpful approaches in an influential volume. Quade and Boucher (1968) and Miser and Quade (1988) gave refinements and extensions to non-defense analysis.

THE INSTITUTIONALIZATION AND IMPACT OF SYSTEMS ANALYSIS: Hitch and McKean (1960)
did much to introduce cost-effectiveness studies as instruments of defense systems analysis. In the new Kennedy administration in 1961, Secretary of Defense McNamara brought Hitch into the Office of the Secretary of Defense (OSD) as Comptroller to install a system of planning-programming-budgeting (PPB), and Enthoven, as Hitch's assistant, started an office of systems

analysis. Although the titles and organizational placement have changed over the years, OSD has continued both PPB and systems analysis.

These new offices had great impact. The government sought to create other similar offices in other departments, Bureau of the Budget (1965). Within the Department of Defense, the new OSD offices played an important role in departmental decisions. As its emphasis on, and requests for, quantitative analysis increased, the military services organized and enlarged their MOR offices to meet the demand.

The above developments came at a time when computer capabilities were rapidly increasing. Many MOR offices sought to use the new capabilities in producing cost-effectiveness studies required for systems analysis. Computer simulation models began to proliferate in the effort to understand what new or proposed weapon systems would contribute to the future battlefields. Because this effort contributed to studies with great impact on weapon systems acquisition, it has continued to grow.

WARTIME COMBAT OPERATIONS RESEARCH IN KOREA AND VIETNAM: Although its successes in World War II had led to its postwar continuation, MOR efforts had come to emphasize future weapon system acquisition as described above. In Chapter I of Hughes (1989), Thomas observed that combat operations research both in Korea and later in Vietnam was very similar to that of World War II. Despite the postwar increase in modeling and computer capabilities, it did not make nearly as much contribution in Korea or Vietnam as might have been expected. "Though the menu of available techniques increased with time, much that had been learned in World War II was forgotten and relearned in later conflicts."

MOR LESSONS FROM DESERT SHIELD/DESERT STORM: The new computer and modeling capabilities seemed to have more impact in MOR for the Gulf War combat of 1991. Vandiver *et al.* (1992) concluded that while some of its analytic lessons were reminiscent of World War II, and some lessons were probably peculiar to wars "like" the Gulf War, there were trends indicative of future combat analysis:

- Computer influence on analysis is increasingly varied and pervasive.
- Software analytical tools are increasingly available to all—including "non-analysts."
- The demand for good data bases is growing more rapidly than the supply.

– There is growing need for coalition and joint service analysis.
– There is increasing analytical interest in operational art and campaign focus.
– There is less danger of "central" misuse of field analysis and data than formerly.

MOR ISSUES: Although MOR has been a flourishing enterprise with an expanding technological menu, there are still issues to resolve, some of long standing. A significant fraction of the issues relate to modeling and simulation, or are frequently so characterized. Some of the more serious concerns address:

– Scientific foundations (including verification and validation),
– DoD organization (including that for MOR),
– Management,
– Filling a perceived need, and
– Taking suitable advantage of technological opportunities.

Although most of the following illustrative critiques deal with more than one issue, a majority of the titles suggest one of the above. For example:

Scientific Foundations

GAO (1980), *Models, Data, and War: A Critique of the Foundation for Defense Analyses*
Davis and Blumenthal (1991), *The Base of Sand Problem: A White Paper on the State of Military Combat Modeling*

DoD Organization

BLUE RIBBON DEFENSE PANEL (1970), *Report to the President and the Secretary of Defense on the Department of Defense*

Management

DoDIG (1993), *Audit Report on Duplication/Proliferation of Weapon Systems' Modeling and Simulation Efforts Within DoD*

Filling a Perceived Need

Davis and Hillestad (1992), *Proceedings of Conference on Variable Resolution Modeling*
Bankes (1993), *Exploratory Modeling for Policy Analysis*

Taking Advantage of Technological Opportunities

Defense Science Board (1993), *Impact of Advanced Distributed Simulation on Readiness, Training and Prototyping*

There have been in DoD both organizational and procedural changes that relate to the design, management, and use of models and simulations, some undertaken to address the kind of concerns exemplified in reports of the first three and the last of the categories above. Although the changes

should result in continuing improvements in realizing the potential of models and simulations, their effect will not be immediate. It may take even longer to meet some of the perceived needs. For example, two consensus conclusions of the Conference on Variable-Resolution Modeling were:

– Aggregated models are essential, but they often need to be stochastic; and
– The "configuration problem" is severe.

Each of the conclusions leads to a significant and demanding research program, and many problems will long remain. As observed long ago in Hitch (1960), there are important "Uncertainties in Operations Research."

See **Air Force operations analysis; Center for Naval Analyses; Cost analysis; Cost effectiveness; Exploratory modeling; Operations Research Office and Research Analysis Corporation; RAND; Systems analysis.**

References

[1] Bankes, S. (1993). "Exploratory Modeling for Policy Analysis." *Opns. Res.* **41**, 435–449.
[2] Blackett, P.M.S. (1962). *Studies of War*. Hill and Wang, New York.
[3] Blue Ribbon Defense Panel (1970). *Report to the President and the Secretary of Defense on the Department of Defense*. U.S. Govt. Printing Office, Washington, D.C.
[4] Brothers, L.A. (1954). "Operations Analysis in the United States Air Force." *Opns. Res.* **2**, 1–16.
[5] Bureau of the Budget (1965). *Planning-Programming-Budgeting*, Bulletin 65–5.
[6] Davis, P.K. and D. Blumenthal (1991). "The Base of Sand Problem: A White Paper on the State of Military Combat Modeling," N-3148-OSD/DARPA, RAND, Santa Monica, Calif.
[7] Davis, P.K. and R. Hillestad (eds.) (1992). *Proceedings of Conference on Variable-Resolution Modeling*, Washington, D.C., CF-103-DARPA, RAND, Santa Monica, Calif.
[8] Defense Science Board (1993). *Impact of Advanced Distributed Simulation on Reatiness, Training and Prototyping*. DDR&E, Washington, D.C.
[9] DoD, Office of the Inspector General (1993). *Duplication/Proliferation of Weapon Systems' Modeling and Simulation Efforts Within DOD*. Rpt 93–060, Arlington, Virginia.
[10] Government Accounting Office (1980). *Models, Data, And War: A Critique of the Foundation For Defense Analyses*. PAD-80-21, Comptroller General, Washington, D.C.
[11] Hitch, C.J. (1953). "Sub-optimization in Operations Problems." *Opns. Res.* **1**, 87–99.
[12] Hitch, C.J. (1955). "An Appreciation of Systems Analysis." *Opns. Res.* **3**, 466–481.
[13] Hitch, C.J. (1960). "Uncertainties in Operations Research." *Opns. Res.* **4**, 437–445.

[14] Hitch, C.J. and R.N. McKean (1960). *The Economics of Defense in the Nuclear Age.* R-348, RAND, Santa Monica, Calif.

[15] Hughes, W.P. Jr. (ed.) (1989). *Military Modeling* (2nd ed.). MORS, Alexandria, VA.

[16] Miser, H.J. and E.S. Quade (eds.) (1988). *Handbook of Systems Analysis: Craft Issues and Procedural Choices.* North-Holland, New York.

[17] Morse, P.M. and G.E. Kimball (1946). *Methods of Operations Research.* OEG Rpt. 54, Office of the Chief of Naval Operations, Navy Dept., Washington, D.C.

[18] Quade, E.S. (ed.) (1954). *Analysis for Military Decisions.* R-387-PR, RAND, Santa Monica, Calif.

[19] Quade, E.S. and W.I. Boucher (eds.) (1968). *Systems Analysis and Policy Planning*: *Applications in Defense.* R-439-PR, RAND, Santa Monica, Calif.

[20] Solandt, O. (1955). "Observation, Experiment, and Measurement in Operations Research." *Opns. Res.* **3**, 1–14.

[21] Tidman, K.R. (1984). *The Operations Evaluation Group, A History of Naval Operations Analysis.* Naval Institute Press, Annapolis, Maryland.

[22] Vandiver, E.B. *et al.* (1992). "Lessons are Learned from Desert Shield/Desert Storm." *PHALANX.* **25**, 1, 6–87.

MIMD

Acronym for Multiple Instruction, Multiple Data. A class of parallel computer architectures in which each processing element fetches and decodes its own stream of instructions, possibly different from the instruction streams for other processors.

MINIMUM

A real-valued function $f(x)$ is said to have a minimum on a set S when the greatest lower bound of $f(x)$ on S is assumed by $f(x)$ for some x^0 in S. Thus, $f(x^0) \leq f(x)$ for all x in S. See **Global maximum (minimum)**.

MINIMUM (MAXIMUM) FEASIBLE SOLUTION

In a mathematical-programming problem, the solution that both satisfies the constraints of the problem and minimizes (maximizes) the objective function is a minimum (maximum) feasible solution. Such solutions may not be unique.

MINIMUM-COST NETWORK-FLOW PROBLEM

In a directed, capacitated network with supply and demand nodes, the problem is to determine the flows of a single, homogeneous commodity from the supply nodes to the demand nodes that minimize a linear cost function. In its general form, when the network contains transshipment or intermediate nodes, that is, nodes that are neither supply or demand nodes, the problem is called the transshipment problem. Conservation of flow through each node is assumed. Due to its special mathematical structure, this problem has a solution in integer flows, given that the data that define the network are integers. It is a linear-programming problem whose major constraints form a node-arc incidence matrix. See **Conservation of flow**; **Maximum-flow network problem**; **Network optimization**.

MINIMUM SPANNING TREE PROBLEM

Given a connected network with n nodes and individual costs associated with all edges, the problem is to find the least-cost spanning tree. See **Network optimization**; **Spanning tree**.

MINOR

See **Matrices and matrix algebra**.

MIP

See **Combinatorial and integer optimization**; **Mixed-integer programming problem**.

MIS

Management information systems. See **Information systems and database design**.

MIXED-INTEGER PROGRAMMING PROBLEM (MIP)

A mathematical-programming problem in which the constraints and objective function are linear, but some of the variables are constrained to be integer valued. The integer variables can either be binary or take on general integer values. See **Binary variable**; **Linear programming**; **Mathematical programming**.

MIXED NETWORK

A queueing network in which some customers can enter and leave the network while others neither enter nor leave but cycle through the nodes endlessly. A queueing network in which the routing process contains at least one closed set of states for some types of customers but not others.

MODEL

A model is an idealized representation –an abstract and simplified description – of a real world situation that is to be studied and/or analyzed. Models can be classified in many ways. A mental model is an individual's conceptual, unstated, view of the situation under review; a verbal or written model is a description (a model) of one's mental model; an iconic model looks like what it is supposed to represent (e.g., an architectural model of a building); an analogue model relates the properties of the entity being studied with other properties that are both descriptive and meaningful (e.g., the concept of time as described by the hands and markings of a clock); symbolic or mathematical model represents a symbolic representation of the process under investigation (e.g., Einstein's equation $e = mc^2$). See **Decision problem; Descriptive model; Deterministic model; Mathematical model; Normative model; Predictive model; Prescriptive model; Stochastic model.**

MODEL ACCREDITATION

Saul I. Gass

University of Maryland, College Park

Model accreditation is an official determination that a model is acceptable for a specific purpose (Williams and Sikora, 1991; Ritchie, 1992). Accreditation certifies that the element being accredited meets given standards. For a model, accreditation must be done with respect to the model's explicit specifications and the demonstration that the computer-based model does or does not meet the specifications. This demonstration is the responsibility of the model developers who must show that their work passes agreed-to user and developer acceptance tests. If the modeling process was done properly and was accompanied by appropriate documentation, accreditation of the model for its specified uses should follow.

Accreditation of a model must rely on a review and evaluation of its available documentation. Such an evaluation, usually done by an independent third-party, is made against various criteria to determine the levels of accomplishment of the criteria, in particular those of verification and validation. The review is made with a specific user and uses in mind. The review should produce a report that gives guidance to the user on whether or not the model in question can be used with confidence for the designated uses, that is, the model is or is not accredited for specific uses (Gass, 1993).

Definitive guidelines as to how one accredits a model have not been promulgated. However, the ideas, if not the general process behind model accreditation, are becoming accepted by modeling agencies within the U.S. Department of Defense.

See **Model evaluation; Model management; Practice of OR/MS; Validation; Verification.**

References

[1] Williams, M.K. and J. Sikora (1991). "SIMVAL Minisymposium – A Report," *Phalanx, Bulletin of the Military Operations Research Society*, 24, 2.

[2] Ritchie, A.E., ed. (1992). *Simulation Validation Workshop Proceedings* (SIMVAL II), Military Operations Research Society, Alexandria, Virginia.

[3] Gass, S.I. (1993). "Model Accreditation: A Rationale and Process for Determining a Numerical Rating," *European Jl. Operational Research*, 66, 2, 250–258.

MODEL BUILDER'S RISK

Probability of rejecting the credibility of a model when in fact the model is sufficiently credible. See **Verification, validation and testing of models.**

MODEL EVALUATION

Saul I. Gass

University of Maryland, College Park

Model evaluation or assessment is a process by which interested parties, who were not involved in a model's origins, development and implementation, can assess the model's results in terms of its structure and data inputs so as to determine, with some level of confidence, whether or not the results can be used in decision making (Gass, 1977). Model evaluation encompasses: (1) verification, validation, and quality control of the usability of the model and its readiness for use, and (2) investigations into the assumptions and limitations of the model, its appropriate uses, and why it produces the results it does (Greenberger and Richels, 1979).

There are three reasons for advocating evaluation of models: (1) for many models, the ultimate decision maker is far removed from the modeling process and a basis for accepting the model's results by such a decision maker needs to be established; (2) for complex models, it is difficult to assess and to comprehend fully the interactions and impact of a model's assumptions, data availability, and other elements on the model structure and results without a formal, independent evaluation; and (3) users of a complex model that was developed for others must be able to obtain a clear statement of the applicability of the model to the new user problem area (Gass, 1977a).

All procedures for evaluating a model are basically information gathering activities, with the detail and level of information being a function of the purposes of the assessment and the skills of the assessors. Specific evaluation approaches are given in Gass (1977a, 1977b), Wood (1980), U.S. GAO (1979), with an evaluation case study given in Fossett, Harrison, Weintrob and Gass (1991).

A model evaluation procedure and its objectives should be tailored to the scope and purposes of the model and will vary with the model, model developers, assessors, users, and available resources. Model assessment is an expensive and involved undertaking; all models need not be assessed. Model developers and users should recognize that by applying proper modeling management procedures, the burdens that evaluators of models have to contend with are alleviated greatly (Gass, 1987).

See **Model accreditation; Model management; Practice of operations research and management science; Project management; Verification, validation and testing of models.**

References

[1] Fossett, C., Harrison, D., Weintrob, H., and Gass, S.I. (1991). "An Assessment Procedure for Simulations Models: A Case Study," *Operations Research*, 39, 710–723.

[2] Gass, S.I. (1977a.) "Evaluation of Complex Models," *Computers and Operations Research*, 4, 27–35.

[3] Gass, S.I. (1977b). "A Procedure for the Evaluation of Complex Models," *Proceedings of the First International Conference in Mathematical Modeling*, 247–258.

[4] Gass, S.I., ed. (1980). *Validation and Assessment Issues of Energy Models*, National Bureau of Standards Special Publication 569, U.S. GPO Stock No. 033-003-02155-5, Washington, D.C.

[5] Gass, S.I., ed. (1981). *Validation and Assessment of Energy Models*, National Bureau of Standards Special Publication 616. U.S. Government Printing Office, Washington, D.C.

[6] Gass, S.I. (1983). "Decision-Aiding Models: Validation, Assessment, and Related Issues for Policy Analysis," *Operations Research*, 31, 603–631.

[7] Gass, S.I. (1987). "Managing the Modeling Process: A Personal Perspective," *European Jl. Operational Research*, 31, 1–8.

[8] Wood, D.O. (1981). "Energy Model Evaluation and Analysis: Current Practice," in S.I. Gass, ed., *Validation and Assessment of Energy Models*, National Bureau of Standards Special Publication 616. U.S. Government Printing Office, Washington, D.C.

[9] U.S. GAO (1979). *Guidelines for Model Evaluation*, GAO/PAD-79-17, Washington, D.C.

MODEL MANAGEMENT

Ramayya Krishnan

Carnegie Mellon University,
Pittsburgh, Pennsylvania

INTRODUCTION: The term model management was coined in the mid-1970s in the context of work on decision support systems (DSS) (Will, 1975; Sprague and Watson, 1975). An important objective of the DSS concept was to provide an environment in which decision makers could gain materially useful insights by interactively exercising OR/MS models. However, developing such an environment required principled solutions to problems of specifying, representing and interacting with models. This focus on models, and in turn on modeling, led to the study of model management – defined broadly to encompass the study of model representation, the set of operations facilitated by such representation, and of computer-based environments that facilitate modeling.

Following is a brief review of work in two areas that have been actively studied in model management. First, we review work on languages to specify models, and on the development of techniques to facilitate operations that support modelers in both the pre-solution and post-solution phases. Next, we review work on the representation of a *collection of models* (e.g., a model library) and the development of techniques to enable model selection and configuration.

MODEL MANAGEMENT-I: *Modeling languages –* The need to represent a model in a notation that is easy to validate, verify, debug, maintain and communicate motivated the development of modeling languages (Fourer, 1983). Prior to their development, the only computer-executable representation of a model was in an arcane format optimized for efficient solution (e.g., the MPS format). Current modeling languages provide a high-level symbolic notation to specify models. Solution operations can also be declared and all the required details of binding the model instance to the data structures required by solver is done transparently. This has greatly increased the productivity of model-based work.

Four principles have been articulated as essential to modeling language design (Fourer, 1983; Geoffrion, 1992a, 1992b; Bhargava and Kimbrough, 1993; Krishnan, 1993). These are:

- *Model data independence:* requires the mathematical structure of the model to be independent of the data used. This permits model data to be modified in format, dimension, units or values without any modification to the model representation.
- *Model solver independence:* requires the model representation to be independent of the representation required by the solver. This permits more than one solver to be used with a given model. Further, it recognizes the fundamental differences in the requirements placed on model representations and representations required by the solver.
- *Model paradigm independence:* requires that the modeling language allow the representation of models drawn from different paradigms (e.g., mathematical programming and discrete event simulation).
- *Meta level representation and reasoning:* requires that the modeling language represent information *about* model components and models in addition to their mathematical structure in order to enable semantic consistency checking.

Modeling languages incorporate these principles to varying degrees. Examples of modeling languages include spreadsheet-based languages such as IFPS (Gray, 1987), algebraic modeling languages such as GAMS (Bischop and Meeraus, 1982), AMPL (Fourer *et al.*, 1990), and MODLER (Greenberg, 1992), relational modeling languages such as SQLMP (Choobineh, 1991), graphical modeling languages such as NETWORKS (Jones, 1991) and typed modeling languages such as ASCEND (Piela *et al.*, 1992).

Two developments have had a significant impact on modeling languages. One is the seminal work on Structured Modeling (SM) (Geoffrion, 1987). While previous work on modeling languages had sought to provide a computer executable representation of the notation tradiitionally used by modelers, SM defines a theory which treats models as hierarchical collections of definitional dependencies. This permits structured modeling languages to satisfy all of the four design principles discussed above. While several languages have implemented SM, the most completely developed of these is SML (Geoffrion, 1992a, 1992b). The other important development is the embedded languages technique which can be used to define an architecture of considerable generality for modeling environments. The technique is used to specify modeling languages as well as information about the terms and expressions stated in these languages. The TEFA modeling environment has been implemented using the technique (Bhargava and Kimbrough, 1993).

Operations – The early work on model management focussed on model solution. The objective was to transparently bind solution algorithms to model instances. As noted above, modeling languages have realized this objective. Model management research has since focussed on operations required to support both pre-solution and post-solution phases of the modeling life cycle. In the following, we review research related to a pre-solution phase, model formulation, and a post-solution phase, model interpretation.

Model Formulation – Model formulation is the task of converting a precise problem description into a mathematical model (Krishnan, 1993). It is a complex task requiring diverse types of knowledge. The appropriateness of a model depends on a variety of factors such as accuracy, tractability, availability of relevant data and understandability. Model formulation research has primarily focussed on the development of theory, tools and techniques to support the formulation of mathematical programming models.

Using protocol analysis, detailed studies of the expert modeling process have been conducted and process models have been developed (Orlikowski and Dhar, 1986; Sklar and Pick, 1990; Raghunathan *et al.*, 1993; Krishnan *et al.*, 1992). Domain-independent and domain-specific model formulation strategies have been implemented in model formulation support systems (Ma, Murphy and Stohr, 1989; Binbasioglu and Jarke, 1986; Krishnan, 1990; Raghunathan *et al.*, 1993) and a variety of representation and (deductive) reasoning schemes have been investigated. Work by Liang and Konsynski (1993) has also investigated alternative approaches such as analogical reasoning and case-based reasoning to implement model formulation systems. A survey of the state of the art of this research is in Bhargava and Krishnan (1993).

Model Interpretation – Model interpretation consists of a variety of techniques to help a modeler comprehend a model. These include parametric analysis, structural analysis, and structure inspection.

Parametric analysis has long been supported in model management systems. Spreadsheets routinely support *what if* analysis and goal seeking. Modeling languages for mathematical programming implement the theory of sensitivity analysis.

The pioneering work on structural analysis is due to Greenberg on the ANALYZE system (Greenberg, 1987). ANALYZE extracts model structures that cause exceptions such as redundancy and infeasibility in linear programming models. Kimbrough and Oliver (1994) examined the issue of post-solution analysis for models other than linear programs and have attempted to fashion a solution along the lines of ANALYZE. An important feature of their approach is the analysis of the impact on the solution to a model when changes are made to the parameters of a *surrogate* model.

Piela *et al.* (1992) described the use of a browser to inspect the structure of a model, and Dhar and Jarke (1993) and Raghunathan *et al.* (1993) examined the usefulness of recording the rationale underlying a model. The documented rationale is used to aid comprehension as well as to correctly and consistently propagate the changes made to the structure of a model.

MODEL MANAGEMENT-II: *Model libraries* – In contrast to the work reviewed in the previous section, the focus here assumes the existence of a library of debugged and validated models. This has led to the study of issues such as the representation of model libraries, and of operations such as model selection and configuration.

Model representation – Predominantly, models are abstractly represented as black boxes, that is, as a set of named inputs and outputs. This is in contrast to the detailed representation of the structure of the model in the previous section. A variety of representations, including virtual relations (Blanning, 1982) and predicate logic (Bonczek, Holsapple, and Whinston, 1978) have been used to represent models. Additional structure has been imposed on these representations. Mannino *et al.* (1990) proposed the use of categories such as model type, model template, and model instance to organize the collection of models in a library. A model type is a general description of a model class such as linear programming. A model template is a refinement of a model type such as a production planning LP model, and a model instance is an instance of a model template

in which the source of values for each parameter has been declared.

Model Selection – Model selection leverages the existence of previously developed models to create a model for a new problem. In addition to the set of inputs and outputs associated with a model, additional information such as model assumptions need to be represented. Mannino *et al.* (1990) described model selection operators which match, either exactly or fuzzily, the assumptions associated with a model and those that are part of a problem statement. Banerjee and Basu (1993) adopted the same framework as Mannino *et al.* (1990) but their work differed in its use of structuring technique called the Box Structure method (Mills *et al.*, 1986), which was borrowed from the domain of systems analysis and design to develop its taxonomy of model types.

Model Configuration – Model configuration leverages previously developed models by either linking them together (referred to as model composition) or by integrating them (referred to as model integration).

Model composition links together independent models such that the output of one model becomes an input to another. Model composition is often used in conjunction with model selection when no one model meets the requirements of a problem. An example of model composition is the linking together of a demand forecasting model and a production scheduling model.

While the early work only permitted links between variables with the same name, the recent work of Muhanna (1992) and Krishnan *et al.* (1993) permitted linkages between objects (variables, arrays, instances of types, etc.) as long as certain semantic constraints are met. Muhanna (1992) also proposed methods which determine the order in which a collection of linked models should be solved. Representation methods and algorithms which can determine the set of models that need to be composed in order to obtain a set of outputs from a given set of inputs have been a major focus of model composition research. Here, we have work based on virtual relations (Blanning, 1982), on predicate logic (Bonczek, Holsapple and Whinston, 1978), and on a construct called metagraphs (Basu and Blanning, 1994).

Model integration differs from model composition in allowing modifications to be made to the models being integrated. Model integration involves both schema integration and solver integration (Dolk and Kotteman, 1993). Schema integration is the task of merging the internal structure of two or more models to create a new model. Process integration is the task of interweaving associated solution processes in order to solve the integrated model.

Support for conflict resolution is a major focus of research in schema integration. This has involved the development of a variety of typing schemes that seek to integrate data typing (Muhanna, 1992), and concepts such as quiddity and dimensions (Bhargava *et al.*, 1991).

Detailed procedures for integrating models specified in the Structured Modeling Language have been proposed (Geoffrion, 1989, 1992a, 1992b). The method uses the ability in structured modeling to trace the effects of changes and the formal definition of what constitutes a structured model to advantage.

The pioneering work on solver integration was the work of Dolk and Kotteman (1993). They used the theory of communicating sequential processes (Hoare, 1985) to address the problem of solver integration. A simplified version of the problem was addressed by Muhanna (1992) in the SYMMS system.

Additional discussion and surveys on model management are given in Stohr and Konsynski (1992), Blanning (1993), Shetty (1993), and Krishnan (1993).

See **Decision support systems; Model evaluation; Modeling languages; Structured modeling; Verification, validation and testing of models.**

References

[1] Banerjee, S. and A. Basu (1993), "Model Type Selection in an Integrated DSS Environment," *Decision Support Systems*, 75–89.

[2] Basu, A. and R. Blanning (1994), "Metagraphs: A Tool for Modeling Decision Support Systems," *Management Science*, 40, 1579–1600.

[3] Bhargava, H.K. and R. Krishnan (1993), "Computer-aided Model Construction," *Decision Support Systems*, 9, 91–111.

[4] Bhargava, H.K., S. Kimbrough and R. Krishnan (1991), "Unique Names Violations: A Problem for Model Integration," *ORSA Jl. Computing*, 3, 107–120.

[5] Bhargava, H.K. and S.O. Kimbrough (1993), "Model Management: An Embedded Languages Approach," *Decision Support Systems*, 10, 277–300.

[6] Binbasioglu, M. and M. Jarke (1986), "Domain Specific DSS Tools for Knowledge-based Model Building," *Decision Support Systems*, 2, 213–223.

[7] Bischop, J. and A. Meeraus (1982), "On the Development of a General Algebraic Modeling System in a Strategic Planning Environment," *Mathematical Programming Study*, 20, 1–29.

[8] Blanning, R. (1982), "A Relational Framework for Model Management," *DSS-82 Transaction*, 16–28.

[9] Blanning, R. (1993), *Decision Support Systems: Special Issue on Model Management*, in Blanning, R., C. Holsapple, A. Whinston, eds., Elsevier.

[10] Bonczek, R., C. Holsapple, and A. Whinston (1978), "Mathematical Programming within the Context of a Generalized Data Base Management System," *R.A.I.R.O. Recherche Operationalle*, 12, 117–139.

[11] Choobineh, J. (1991), "SQLMP: A data sublanguage for the representation and formulation of linear mathematical models," *ORSA Jl. Computing*, 3, 358–375.

[12] Dhar, V. and M. Jarke (1993), "On Modeling Processes," *Decision Support Systems*, 9, 39–49.

[13] Dolk, D.K. and J.E. Kotteman (1993), "Model integration and a theory of models," *Decision Support Systems*, 9, 51–63.

[14] Fourer, R. (1983), "Modeling Languages versus Matrix Generators for Linear Programming," *ACM Transactions on Mathematical Software*, 2, 143–183.

[15] Fourer, R., D. Gay, and B.W. Kernighan, "A Mathematical Programming Language," *Management Science*, 36, 519–554.

[16] Geoffrion, A.M. (1987), "An Introduction to Structured Modeling," *Management Science*, 33, 547–588.

[17] Geoffrion, A.M. (1989), "Reusing Structured Models via Model Integration," J. F. Nunamaker, ed., *Proceedings Twenty-Second Annual Hawaii International Conference on the System Sciences*, III, 601–611, IEEE Press, Los Alamitos, California.

[18] Geoffrion, A.M. (1992a), "The SML Language for Structured Modeling: Levels 1 and 2," *Operations Research*, 40, 38–57.

[19] Geoffrion, A.M. (1992b), "The SML Language for Structured Modeling: Levels 3 and 4," *Operations Research*, 40, 58–75.

[20] Gray, P. (1987), *Guide to IFPS*, McGraw-Hill, New York.

[21] Greenberg, H.J. (1987), "ANALYZE: A computer-assisted analysis system for linear programming models," *Operations Research Letters*, 6, 249–255

[22] Greenberg, H.J. (1992), "MODLER: Modeling by object-driven linear elemental relations," *Annals Operations Research*, 38, 239–280.

[23] Hoare, C.A.R. (1992), *Communicating Sequential Processes*, Prentice-Hall, Englewood Cliffs, New Jersey.

[24] Jones, C.V. (1991), "An Introduction to Graph Based Modeling Systems, Part II: Graph Grammars and the Implementation," *ORSA Jl. Computing*, 3, 180–206.

[25] Kimbrough, S. and J. Oliver (1994), "On Automating Candle Lighting Analysis: Insight from Search with Genetic Algorithms and Approximate Models," J. F. Nunamaker, ed., *Proceedings*

Twenty-Seventh Hawaii International Conference on the System Sciences, III, 536–544, IEEE Press, Los Alamitos, California.

[26] Krishnan, R. (1993), "Model Management: Survey, Future Research Directions and a Bibliography," *ORSA CSTS Newsletter*, 14, 1.

[27] Krishnan, R. Piela, and A. Westerberg (1993), "Reusing Mathematical Models in ASCEND," in *Advances in Decision Support Systems*, C. Holsapple and A. Whinston, eds., 275–294, Springer-Verlag, Munich.

[28] Krishnan, R., X. Li, and D. Steier (1992), "Development of a Knowledge-based Model Formulation System," *Communications ACM*, 35, 138–146.

[29] Krishnan, R. (1990), "A Logic Modeling Language for Model Construction," *Decision Support Systems*, 6, 123–152.

[30] Liang, T.P. and B.R. Konsynski (1993), "Modeling by Analogy: Use of Analogical Reasoning in Model Management Systems," *Decision Support Systems*, 9, 113–125.

[31] Ma, P.-C., F. Murphy, and E. Stohr (1989), "A Graphics Interface for Linear Programming," *Communications ACM*, 32, 996–1012.

[32] Mannino, M.V., B.S. Greenberg, and S. N. Hong (1990), "Model Libraries: Knowledge Representation and Reasoning," *ORSA Jl. Computing*, 2, 287–301.

[33] Mills, H., R. Linger and A. Hevner (1986), *Principles of Information Systems Analysis and Design*, Academic Press, Orlando.

[34] Muhanna, W. (1992), "On the Organization of Large Shared of Model Bases", *Annals Operations Research*, 38, 359–396.

[35] Orlikowski, W. and V. Dhar (1986), "Imposing Structure on Linear Programming Problems: An Empirical Investigation of Expert and Vice Modelers," *Proceedings National Conference on Artificial Intelligence*, Philadelphia.

[36] Piela, R. McKelvey, and A. Westerberg (1992), "An Introduction to ASCEND: Its Language and Interactive Environment," in J.F. Nunamaker Jr., ed., *Proceedings Twenty-Fifth Annual Hawaii International Conference on System Sciences*, Vol. III, 449–461, IEEE Press, Los Alamitos, California.

[37] Raghunathan, S., R. Krishnan and J. May (1994), "MODFORM: A Knowledge Tool to Support the Modeling Process," *Information Systems Research*, 4, 331–358.

[38] Shetty, B. (1993), *Annals of Operations Research: Special Issue on Model Management*, Shetty, B., ed., J.C. Baltzer Scientific Publishing, Amsterdam.

[39] Sklar, M.M., R.A. Pick, G.B. Vesprani, and J.R. Evans (1990), "Eliciting Knowledge Representation Schema for Linear Programming," D.E. Brown and C.C. White, eds., *Operations Research and Artificial Intelligence: The Integration of Problem Solving Strategies*, 279–316, Kluwer, Amsterdam.

[40] Sprague, R.H. and H.J. Watson (1975), "Model Management in MIS," *Proceedings Seventeenth National AIDS Conference*, 213–215.

[41] Stohr, E. and B. Konsynski (1992), *Information Systems and Decision Processes*, IEEE Press, Los Altimos, California.

[42] Will, H.J. "Model Management Systems" in *Information Systems and Organization Structure*, Edwin Grochia and Norbert Szyperski, eds., 468–482, Walter de Gruyter, Berlin.

MODEL TESTING

Investigating whether inaccuracies or errors exist in a model. See **Validation**; **Verification**.

MODEL USER'S RISK

Probability of accepting the credibility of a model when in fact the model is not sufficiently credible. See **Verification, validation and testing of models**.

MODEL VALIDATION

See **Verification; Verification, validation and testing of models**.

MODEL VERIFICATION

See **Validation; Verification, validation and testing of models**.

MODI

Modified Distribution Method. A procedure for organizing the hand computations when solving a transportation problem using the transportation simplex method. See **Transportation simplex method.**

MOIP

Multi-objective integer programming. See **Multiple criteria decision making**.

MOLP

Multi-objective linear programming. See **Multiobjective programming**.

MONTE CARLO SAMPLING AND VARIANCE REDUCTION

Jack P.C. Kleijnen

Tilburg University, The Netherlands

Reuven Y. Rubinstein

Technion, Israel

INTRODUCTION: Let $\ell(v)$ be the expected performance of a discrete-event system (DES):

$$\ell(v) = E_v \varphi\{L(\mathbf{Y})\} \qquad (1)$$

where $L(\mathbf{Y})$ is the sample performance function (simulation model) driven by an input vector $\mathbf{Y}$ with probability distribution function (pdf) $f(\mathbf{y}, v)$; the subscript v in E_v means that the expectation is taken with respect to $f(\mathbf{y}, v)$; φ is a real-valued function. To estimate $\ell(v)$ through simulation, one generates a random sample $\mathbf{Y}_i$ with $i = 1, \ldots, N$ from $f(\mathbf{y}, v)$, computes the sample function $L(\mathbf{Y}_i)$, and the simple sample-average estimator

$$\tilde{\ell}_N = \frac{1}{N} \sum_{i=1}^{N} \varphi[L(\mathbf{Y}_i)]. \qquad (2)$$

This is called *crude Monte Carlo (CMC)* sampling. As N gets large, laws of large numbers may be invoked (assuming simple conditions) to verify that the sample-average estimator stochastically converges to the actual expectation.

To illustrate, consider a slight reframing of the problem to require the evaluation of

$$\theta = \int \varphi(y) f(y) dy,$$

where y shall be assumed to be one-dimensional for simplicity. Monte-Carlo integration is an especially good way to estimate the value of the integral when the dimension is much higher than one, but the concept is still the same. For the Monte-Carlo estimate, we sample $Y_1, \ldots, Y_N$ independently from f and calculate

$$\hat{\theta} = \frac{1}{N} \sum_{i=1}^{N} \varphi(Y_i).$$

Thus $\hat{\theta}$ estimates θ and the precision of the estimator is proportional to $1/\sqrt{N}$.

A further, simple example might be the estimation of the probability of winning a computer-based game. The game is replayed repeatedly on the computer and a win on attempt i is given a value of 1 and a loss a value of 0. Then the estimator of the Bernoulli expectation is

$$\tilde{\ell}_N = \frac{1}{N} \sum_{i=1}^{N} w_i = \frac{r}{N},$$

where r is the number of wins out of N tries. Since the trials are assumed independent and identically distributed, it follows that the estimator converges almost surely to the actual success probability $E(W) = p$. The variance of such an estimator is easily computed as $p(1-p)/N$ since the variance of a single Bernoulli trial is $p(1-p)$. Given this, we can derive "good" approximate interval estimators for the actual value of p using the central limit theorem since N is likely to be quite large.

However, the desired level of "precision" in the estimator is problem specific and may require, for example, sample sizes that go well beyond available resources. Hence, people have looked for ways to reduce the variance of the estimator as much as possible for the same amount of sampling resources. Thus the development of *variance reduction techniques (VRTs)*.

VRTs transform an underlying simulation model into a related one. Typically, the more one knows about the system, the more effective VRTs are. Well-known VRTs are antithetic and common random variates, control variates, and importance sampling. Other VRTs are presented in Glynn and Iglehart (1988), Kleijnen (1974), and Wilson (1984).

There are good reasons for applying VRTs. First, there may be negligible extra costs in terms of computer time and human effort. Examples are common and antithetic variates. The performance of the VRT may represent the probability of a "rare event"; for example, the failure probability of a highly reliable computer system may be 10^{-25}. Then the only practical alternative is importance sampling, which may make reliable estimation of the rare event probability feasible.

ANTITHETIC AND COMMON RANDOM VARIATES: Consider a simple example where X and Y are random variates (RVs) with known and fixed cumulative distribution functions (CDFs) F_1 and F_2. We seek a minimum variance estimator of $E(X - Y)$. Since

$$Var(X - Y) = Var(X) + Var(Y) - 2Cov(X, Y) \qquad (3)$$

it follows that $Cov(X, Y)$ should be maximized to minimize the variance of the difference between the variables. Assume that both X and Y are generated by the inverse transformation method:

$$\begin{cases} X = F_1^{-1}(U_1) = inf\{x | F_1(x) \geq U_1\} \\ Y = F_2^{-1}(U_2) = inf\{y | F_2(y) \geq U_2\} \end{cases} \qquad (4)$$

where U_1 and U_2 are uniformly distributed on $(0,1)$. We say that *common random variates (CRVs)* are used if $U_1 = U_2 = U$. We say that *antithetic random variates (ARV)* are used if $U_2 = 1 - U_1$.

Since both F_1^{-1} and F_2^{-2} are monotonic nondecreasing functions of U, it is readily seen that the use of CRVs implies a nonnegative covariance in (3). It can be proved that $Cov(X, Y)$ is *maximized*, so that $Var(X - Y)$ is *minimized*. Similarly, $Var(X + Y)$ is minimized when ARV are used (Glasserman and Yao, 1992).

Consider now a more complete case, namely, the minimum variance estimation of the expected value of two sample performance functions, $E[L_1(X) - L_2(Y)]$, for the n-dimensional random vectors X and Y, where L_1 and L_2 are real-valued monotone functions in each component of X and Y. In practice, L_1 and L_2 may correspond to two comparable queueing systems; if there are (say) three servers in series, then n equals four (one U per customer arrival plus one U per service time). Rubinstein, Samorodnitsky and Shaked (1985) proved that if L_1 and L_2 are monotonic in the same direction in each component of the vectors X and Y, respectively, and the dependence is permitted only between the components $X^{(i)}$ and $Y^{(j)}$ having the same indices $(i = j)$, then the variance of $L_1(X) - L_2(Y)$ is minimized when $U_1 = U_2 = U$, componentwise.

The use of antithetic random variates (ARVs) means that pairs of *negatively* correlated samples are generated in (2): Y_{2i-1} and Y_{2i} with $i = 1, \ldots, N/2$, using $U_{1,2i-1}$ and $U_{2,2i} = 1 - U_{1,2i-1}$, respectively.

When comparing two or more systems $(L_1, L_2, \ldots, L_Q$ with $Q > 1)$, then both common and antithetic variates can be applied. Their optimal combination (in the context of metamodeling, using the blocking concept taken from experimental design theory), is discussed in Schruben and Margolin (1978), and Donohue, Houck, and Myers (1992).

Applications of common random variables are abundant in practice, since simulationists find it the natural way to run their experiments: compare alternative systems under "the same circumstances" (same sampled traffic rate). Their analysis is often crude: a few runs with no formal statistical analysis. If the analysis is not neglected, then the only extra work involves the estimation of the correlations among the sample performances $L(Y)$. Applications of antithetics are rare, even though their implementation is very simple (Kleijnen and Van Groenendaal, 1992).

CONTROL RANDOM VARIABLES: Suppose X is an unbiased estimator of μ. A random variable C is called a *control variate* for X if it is correlated with X and its expectation γ is known. The *linear*

control random variable $X(\alpha)$ is defined as

$$X(\alpha) = X - \alpha(C - \gamma), \qquad (5)$$

where α is a scalar parameter. The variance of $X(\alpha)$ is minimized by

$$\alpha^* = Cov\{X, C\}/Var\{C\}. \qquad (6)$$

The resulting minimal variance is

$$Var\{X(\alpha^*)\} = (1 - \rho_{XC}^2)Var\{X\} \qquad (7)$$

where ρ_{XC} denotes the correlation coefficient between X and C. Because $Cov(X, C)$ is unknown, the optimal control coefficient α^* must be estimated from the simulation. Estimating both $Cov(X, C)$ and $Var(C)$ means that linear regression analysis is applied to estimate α^*. Estimation of α^* implies that the variance reduction becomes smaller than (7) suggests, and that the estimator may become biased. This VRT can be easily extended to multiple control variables (simulation input variables) and multiple response variables (simulation outputs) (Kleijnen and Van Groenendaal, 1992, pp. 200–201; Lavenberg, Moeller, and Welch, 1982; Rubinstein and Marcus, 1985; Wilson, 1984).

IMPORTANCE SAMPLING: The idea of importance sampling is related to weighted and stratified sampling ideas from classical statistical sampling theory. Suppose we wish to estimate $\theta = E\varphi(Y)$ for a random observation Y. But we know that there are some outcomes of Y which may be more *important* than others in estimating θ, so we wish to select such values more often than others from the outcome space. As a result, suppose we change the pdf of Y to (say) g instead of the actual f. Then it follows that

$$\hat{\theta}_0 = \frac{1}{N} \sum_{i=1}^{N} \frac{\varphi(Y_i)f(Y_i)}{g(Y_i)}, \quad \text{with } Y_i \sim g$$

is an unbiased estimator of θ. The key point is that $\hat{\theta}_0$ is a weighted mean of the $\varphi(Y_i)$ values with weights inversely proportional to the "selection factor" g/f and the resultant variance of the estimator can be smaller than it was before the change provided g is chosen to make $\varphi f/g$ nearly constant. The choice is very much problem dependent, however, and unfortunately, it is difficult to prevent the gross misspecification of the function g, particularly in multiple dimensions. Such possibilities are well documented in Bratley, Fox and Schrage (1983).

There are a number of other methods for reducing estimator variance, a few of which are extremely complicated but well suited for some specific problems, particularly when the sequence of observations is no longer independent and identically distributed. Recommended references on this include Kleijnen and Van Groenendaal (1992), Kriman and Rubinstein (1993), Rubinstein and Shapiro (1993), Asmussen, Rubinstein, and Wang (1994).

See **Random number generators; Simulation of discrete-event stochastic systems.**

References

[1] Asmussen, S., R.Y. Rubinstein, and C.-L. Wang (1994). "Regenerative rare events simulation via likelihood ratios," *Jl. Applied Probability*, **31**, 797-815.

[2] Bratley, P., B. L. Fox and L.E. Schrage (1983). *A Guide to Simulation*, Springer-Verlag, New York.

[3] Bucklew, J.A., P. Ney, and J.S. Sadowsky (1991). "Monte Carlo simulation and large deviations theory for uniformly recurrent Markov chains," *Jl. Applied Probability*, **27**, 44-59.

[4] Donohue, J.M., E.C. Houck, and R.H. Myers (1992). "Simulation designs for quadratic response surface models in the presence of model misspecification," *Management Science*, **38**, 1765-1791.

[5] Heidelberger, P. (1993). "Fast simulation of rare events in queueing and reliability models," Manuscript, IBM Research Center, Yorktown Heights, New York.

[6] Glasserman, P. and D.D. Yao (1992). "Some guidelines and guarantees for common random numbers," *Management Science*, **38**, 884-908.

[7] Glynn, P.W. and D.L. Iglehart (1988). "Simulation method for queues: an overview," *Queueing Systems*, **3**, 221-256.

[8] Kleijnen, J.P.C. (1974). *Statistical Techniques in Simulation*, Part I, Marcel Dekker, New York.

[9] Kleijnen, J.P.C. and W. Van Groenendaal (1992). *Simulation: a Statistical Perspective*, John Wiley, Chichester, UK.

[10] Kriman, V. and R. Rubinstein (1993). "The complexity of Monte Carlo estimators with applications to rare events," Manuscript, Technion, Haifa, Israel.

[11] Lavenberg, S.S., T.L. Moeller and P.D. Welch (1982). "Statistical results on control variables with application to queueing network simulation," *Operations Research*, **30**, 182-202.

[12] Rubinstein, R.Y. and R. Marcus (1985). "Efficiency of multivariate control variates in Monte Carlo simulation," *Operations Research*, **33**, 661-667.

[13] Rubinstein, R.Y., G. Samorodnitsky, and M. Shaked (1985). "Antithetic variates, multivariate dependence and simulation of complex stochastic systems," *Management Science*, **31**, 66-77.

[14] Rubinstein, R.Y. and A. Shapiro (1993). *Discrete Event Systems: Sensitivity Analysis and Stochastic Optimization via the Score Function Method*, John Wiley, New York.

[15] Schruben, L.W. and B.H. Margolin (1978). "Pseudorandom number assignment in statistically designed simulation and distribution sampling experiments," *Jl. American Statistical Association*, **73**, 504-525.

[16] Wilson, J.R. (1984). "Variance reduction techniques for digital simulation," *American Jl. Mathematical Management Science*, **4**, 277-312.

MOR

See **Military operations research**.

MORS

Military Operations Research Society. See **Military operations research**.

MRP

See **Material requirements planning**.

MS

Management science.

MSE

Mean square error.

MULTI-ATTRIBUTE UTILITY THEORY

Rakesh K. Sarin

University of California, Los Angeles

Consider a decision problem such as selection of a job, choice of an automobile, or resource allocation in a public program (education, health, criminal justice, etc.). These problems share a common feature—decision alternatives impact multiple attributes. The attractiveness of an alternative therefore depends on how well it scores on each attribute of interest and the relative importance of these attributes. Multi-attribute utility theory (MAUT) is useful in quantifying relative attractiveness of multi-attribute alternatives.

The following notations are useful:

X_i The set of outcomes (scores, consequences) on the ith attribute

x_i A specific outcome in X_i

X $X_1 \times X_2 \times \cdots \times X_n$ (Cartesian product)

u_i A single attribute utility function $u_i : X_i \to \mathbb{R}$

u The overall utility function, $u : X \to \mathbb{R}$

$\gtrsim$ "is preferred or indifferent to"

A decision maker uses the overall utility function, u, to choose among available alternatives. The major emphasis of the work on multi-attribute utility theory has been on questions involving u: on conditions for its decomposition into simple polynomials, on methods for its assessment, and on methods for obtaining sufficient information regarding u so that the evaluation can proceed without its explicit identification with full precision.

The primitive in the theory is the preference relation $\gtrsim$ defined over X. Luce *et al.* (1965) and Fishburn (1968) provide conditions on a decision maker's preferences that guarantee the existence of a utility function u such that

$$(x_1, \ldots, x_n) \gtrsim (y_1, \ldots, y_n),$$

$$x_i, y_i \in X_i, i = 1, \ldots, n$$

if and only if

$$u(x_1, \ldots, x_n) \geq u(y_1, \ldots, y_n) \tag{1}$$

Additional conditions are needed to decompose the multi-attribute utility function u into simple parts. The most common approach for evaluating multi-attribute alternatives is to use an additive representation. For simplicity, we will assume that there exist the most preferred outcome x_i^* and the least preferred outcome x_i^0 on each attribute $i = 1$ to n. In the additive representation, a real value u is assigned to each outcome $(x_1, \ldots, x_n)$ by

$$u(x_1, \ldots, x_n) = \sum_{i=1}^{n} w_i u_i(x_i) \tag{2}$$

where the $\{u_i\}$ are single attribute utility functions over X_i that are scaled from zero to one ($u_i(x_i^*) = 1$, $u_i(x_i^0) = 0$ for $i = 1$ to n) and the $\{w_i\}$ are positive scaling constants reflecting relative importance of the attributes with $\sum_{i=1}^{n} w_i = 1$.

If our interest is in simply rank-ordering the available alternatives, then the key condition for the additive form in (2) is *mutual preferential independence*. The resulting utility function is called an *ordinal value function*. Attributes X_i and X_j are preferentially independent if the trade-offs (substitution rates) between X_i and X_j are independent of all other attributes. Mutual preferential independence requires that preference independence holds for all pairs X_i and X_j. Essentially, mutual preferential independence implies that the indifference curves for any pair of attributes are unaffected by the fixed levels of the remaining attributes. Debreu (1962), Luce and Tukey (1964), and Gorman (1968) provide axiom systems and analysis for the additive form (2).

If, in addition to rank order, one is also interested in the strength of preference between pairs of alternatives, then additional conditions are needed. The resulting utility function is called a *measurable value function* and it may be used to order the preference differences between the alternatives.

The key condition for an additive measurable value function is *difference independence* (see Dyer and Sarin, 1979). This condition asserts that the preference difference between two alternatives that differ only in terms of one attribute does not depend on the common outcomes on the other $n - 1$ attributes.

Finally, perhaps the most researched topic is the case of decisions under risk where the outcome of an alternative is characterized by a probability distribution over X. We denote $\tilde{X}$ as the set of all simple probability distributions over X. We shall assume that for any $p \in \tilde{X}$ there exists an alternative that can be identified with p, and thus p could be termed as a risky alternative. The outcome of an alternative $p \in \tilde{X}$ might be represented by the lottery which assigns probabilities $p_1, \ldots, p_l$, $\Sigma_{j=1}^{l} p_j = 1$, to the outcomes $x^l, \ldots, x^l \in X$, respectively. For the choice among risky alternatives $p, q \in \tilde{X}$, von Neumann and Morgenstern (1947) specified conditions on the decision maker's preference relation $\gtrsim$ over $\tilde{X}$, that imply:

$$p \gtrsim q$$

if and only if

$$\sum_{x \in X} p(x)u(x) \geq \sum_{x \in X} q(x)u(x). \tag{3}$$

Notice that we have abused the notation by using the same symbol u to denote ordinal value function, measurable value function, and now the von Neumann-Morgenstern utility function. The context, however, makes the interpretation clear.

A majority of the applied work in multi-attribute utility theory deals with the case when the von Neumann-Morgenstern utility function is decomposed into the additive form (2). Fishburn (1965) has derived necessary and sufficient conditions for a utility function to be additive. The key condition for additivity is the marginality condition which states that the preferences for any lottery $p \in \tilde{X}$ should depend only on the marginal probability distributions over X_i and not on their joint probability distribution. Thus, for additivity to hold, the two lotteries below must be indifferent:

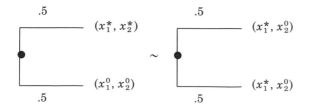

Notice that in either lottery, the marginal probability of receiving the most preferred outcome or the least preferred outcome on each attribute is identical. A decision maker may however prefer the right-hand side lottery over the left-hand side lottery if he/she wishes to avoid a 0.5 chance of the poor outcome (x_1^0, x_2^0) on both attributes.

The assessment of single attribute utility functions $\{u_i\}$ in (2) will require different methods depending on whether the overall utility represents an ordinal value function, a measurable value function, or a von Neumann-Morgenstern utility function. Keeney and Raiffa (1976) discuss methods for assessing multi-attribute ordinal value function and multi-attribute von Neumann-Morgenstern utility function. Dyer and Sarin (1979) and von Winterfeldt and Edwards (1986) discuss assessment of multi-attribute measurable value function.

Besides the additive form (2), a *multiplicative* form for the overall utility function has also found applications in a wide variety of contexts. In the multiplicative representation, a real value u is assigned to each outcome $(x_1, \ldots, x_n)$ by

$$1 + ku(x_1, \ldots, x_n) = \left[\prod_{i=1}^{n} [1 + kk_i u_i(x_i)] \right] \qquad (4)$$

where the $\{u_i\}$ are single attribute utility functions over X_i that are scaled from zero to one, the $\{k_i\}$ are positive scaling constants, k is an additional scaling constant satisfying $k > -1$, and

$$1 + k = \prod_{i=1}^{n} [(1 + kk_i)].$$

If u is a measurable value function then *weak difference independence* along with mutual preference independence provides the desired result (4). An attribute is weak difference independent of the other attributes if preference difference between pairs of levels of that attribute do not depend on fixed levels of any of the other attributes. Thus, for $x_i, y_i, w_i, z_i \in X_i$ the ordering of preference difference between x_i and y_i, and w_i and z_i remains unchanged whether one fixes the other attributes at their most preferred levels or at their least preferred levels.

If the overall utility function is used for ranking lotteries as in (3), then a *utility independence* condition, first introduced by Keeney (1969), is needed to provide the multiplicative representation (4). An attribute is said to be utility independent of the other attributes if the decision maker's preferences for lotteries over this attribute do not depend on the fixed levels of the remaining attributes. Mutual preferential independence and one attribute being utility independent of the others are sufficient to guarantee either the multiplicative form (4) or the additive form (2). The additive form results if in (4) $k = 0$ or $\Sigma_{i=1}^{n} k_i = 1$. Keeney and Raiffa (1976) discuss methods for calibrating the additive and multiplicative forms for the utility function. In the literature, other independence conditions have been identified that lead to more complex nonadditive decompositions of the utility function. These general conditions are reviewed in Farquhar (1977).

If utilities, importance weights, and probabilities are incompletely specified, then the approaches of Fishburn (1964) and Sarin (1975) can be used to obtain a partial ranking of alternatives.

The key feature of multi-attribute utility theory is to specify verifiable conditions on a decision maker's preferences. If these conditions are satisfied, then the multi-attribute utility function can be decomposed into simple parts. This approach of breaking the complex value problem (objective function) into manageable parts has found significant applications in decision and policy analysis. In broad terms, multi-attribute utility theory facilitates measurement of preferences or values. The axioms of the theory have been found to be useful in suggesting approaches for measurement of values. In physical measurements (e.g., length), the methods for measurement have been known for a long time and the theory of measurement has added little to suggesting new methods. In the measurement of values, however, several new methods have been developed as a direct result of the theory. In this regard multi-attribute utility theory has truly been a remarkable achievement.

See **Decision analysis; Preference theory; Utility theory.**

References

[1] Debreu, G. (1960). "Topological Methods in Cardinal Utility Theory," in *Mathematical Methods in the Social Sciences.* Stanford University Press, Stanford, California, 16–26.

[2] Dyer, J.S. and R.K. Sarin (1979). "Measurable Multiattribute Value Functions," *Operations Research*, 27, 810–822.

[3] Farquhar, P.H. (1977). "A Survey of Multiattribute Utility Theory and Applications," *TIMS Stud. Management Science*, 6, 59–89.

[4] Fishburn, P.C. (1964). *Decision and Value Theory.* Wiley, New York.

[5] Fishburn, P.C. (1965a). "Independence in Utility Theory with Whole Product Sets," *Operations Research*, 13, 28–45.

[6] Fishburn, P.C. (1965b). "Utility Theory," *Management Science*, 14, 335–378.

[7] Gorman, W.M. (1968). "Symposium on Aggregation: The Structure of Utility Functions," *Rev. Econ. Stud.*, 35, 367–390.

[8] Keeney, R.L. (1969). *Multidimensional Utility Functions: Theory, Assessment, and Applications.* Technical Report No. 43, Operations Research Center, M.I.T., Cambridge, Massachusetts.

[9] Keeney, R.L. and H. Raiffa (1976). *Decisions with Multiple Objectives: Preferences and Value Tradeoffs.* Wiley, New York.

[10] Luce, R.D and J.W. Tukey (1964). "Simultaneous Conjoint Measurement: A New Type of Fundamental Measurement," *J. Math Psychol.*, 1, 1–27.

[11] Luce, R.D., R.R. Bush and E. Galantor (1965). *Handbook of Mathematical Psychology*, Vol. 3. Wiley, New York.

[12] Sarin, R.K. (1975). "Interactive Procedures for Evaluation of Multi-Attributed Alternatives." Working Paper 232, Western Management Science Institute, University of California, Los Angeles.

[13] von Neumann, J. and O. Morgenstern (1947). *Theory of Games and Economic Behavior*, Princeton University Press, New Jersey.

[14] Edwards, W. and D. von Winterfeldt (1986). *Decision Analysis and Behavioral Research.* Cambridge University Press, Cambridge, England.

MULTICOMMODITY NETWORK-FLOW PROBLEM

This is a minimum-cost network flow problem in which more than one commodity simultaneously flows from the supply nodes to the demand nodes. Unlike the single commodity problem, an optimal solution is not guaranteed to have integer flows. The problem takes on the block-angular matrix form that is suitable for solution by Dantzig-Wolfe decomposition. Applications areas include communications, traffic and logistics. See **Dantzig-Wolfe decomposition**; **Minimum-cost network-flow problem**; **Network optimization**.

MULTICOMMODITY NETWORK FLOWS

Bala Shetty

Texas A&M University, College Station

INTRODUCTION: The multicommodity minimal cost network flow problem may be described in terms of a distribution problem over a network $[V, E]$, where V is the node set with order n and E is the arc set with order m. The decision variable x^{jk} denotes the flow of commodity k through arc j, and the vector of all flows of commodity k is denoted by $\boldsymbol{x}^k = [x^{1k}, \ldots, x^{mk}]$. The unit cost of flow of commodity k through arc j is denoted by c^{jk} and the corresponding vector of costs by $\boldsymbol{c}^k = [c^{1k}, \ldots, c^{mk}]$. The total capacity of arc j is denoted by b^j with corresponding vector $\boldsymbol{b} = [b^1, \ldots, b^m]$. Mathematically, the multicommodity minimal cost network flow problem may be defined as follows:

minimize $\sum_k \boldsymbol{c}^k \boldsymbol{x}^k$

s.t.

$$\boldsymbol{A}\boldsymbol{x}^k = \boldsymbol{r}^k, \quad k = 1, \ldots, K$$

$$\sum_k \boldsymbol{x}^k \leq \boldsymbol{b}$$

$$0 \leq \boldsymbol{x}^k \leq \boldsymbol{u}^k, \text{ for all } k,$$

where K denotes the number of commodities, $\boldsymbol{A}$ is a node-arc incidence matrix for $[V, E]$, $\boldsymbol{r}^k$ is the requirements vector for commodity k, and $\boldsymbol{u}^k$ is the vector of upper bounds for decision variable $\boldsymbol{x}^k$.

Multicommodity network flow problems are extensively studied because of their numerous applications and because of the intriguing network structure exhibited by these problems (Ahuja *et al.*, 1993; Ali *et al.*, 1984; Assad, 1978; Kennington, 1978). Multicommodity models have been proposed for planning studies involving urban traffic systems (Chen and Meyer, 1988; LeBlanc, 1973; Potts and Oliver, 1972) and communications systems (LeBlanc, 1973; Naniwada, 1969). Models for solving scheduling and routing problems have been proposed by Bellmore, Bellington, and Lubore (1971) and by Swoveland (1971). A multicommodity model for assigning students to achieve a desired ethnic composition was suggested by Clark and Surkis (1968). Multicommodity models have also been used for casualty evacuation of wartime casualties, grain transportation, and aircraft routing for the USAF. A discussion of these applications can be found in Ali *et al.* (1984).

SOLUTION TECHNIQUES: There are two basic approaches which have been employed to develop specialized techniques for multicommodity network flow problems: *decomposition and partitioning.* Decomposition approaches may be further characterized as price-directive or resource directive. A price-directive decomposition procedure directs the coordination between a master program and each of several subprograms by changing the objective functions (prices) of the subprograms. The objective is to obtain a set of prices (dual variables) such that the combined solution for all subproblems yields an optimum for the original problem. A resource-directive decomposition procedure (Held *et al.*, 1974 and Kennington and Shalaby, 1977), when applied to a multicommodity problem having K commodities, is to distribute the arc capacity among the individual commodities in such a way that solving K subprograms yields an optimal flow for the coupled problem. At each iteration, an allocation is made and K single commodity flow problems are solved. The sum of capacities allocated to an arc over all commodities is equal to the arc capacity in the original problem. Hence, the combined flow from the solutions of the subproblems provides a feasible flow for the original problem. Optimality is tested and the procedure either terminates or a new arc capacity allocation is developed. Partitioning approaches are specializations of the simplex method where the current basis is partitioned to exploit its special structure. These techniques are specializations of primal, dual, or primal-dual simplex method. The papers of Hartman and Lasdon (1972), and Graves and McBride (1976) are primal techniques, while the work of Grigoriadis and White (1972) is a dual technique. An extensive discussion of these techniques can be found in Ahuja *et al.* (1993), and Kennington and Helgason (1980).

Several researchers have recently suggested new algorithms for the multicommodity flow problem: Gersht and Shulman (1987), Barnhart (1993), Farvolden and Powell (1990), and Farvolden *et al.* (1993) all present various new approaches for the multicommodity model. Parallel optimization has also been applied for the solution of multicommodity networks. Pinar and Zenios (1990) present a parallel decomposition algorithm for the multicommodity model using the penalty functions. Shetty and Muthukrishnan (1990) develop a parallel projection which can be applied to resource-directive decomposition. Chen and Meyer (1988) decompose a nonlinear multicommodity problem arising in traffic assignment into single commodity network components that are independent by commodity. The difficulty of solving a multicommodity problem explodes when the decision variables are restricted to be integers. Very little work is available in the literature for the integer problem (Evans, 1978 and Evans and Jarvis, 1978).

Several computational studies involving multi-commodity models have been reported in the literature. Ali *et al.* (1980) present a computational experience using the price-directive decomposition procedure (PPD), the resource directive-decomposition procedure (RDD), and the primal partitioning procedure (PP). They find the primal partitioning and price directive decomposition methods take approximately the same amount of computing time, while the resource directive decomposition runs in approximately one-half the time of the other two methods. Convergence to the optimal solution is guaranteed for PPD and PP, whereas RDD may experience convergence problems. Ali *et al.* (1984) present a comparison of the primal partitioning algorithm for solving the multicommodity model with a general purpose LP code. On a set of test problems, they find that the primal partitioning technique runs in approximately one-half the time required by the LP code. Farvolden *et al.* (1993) report very promising computational results for a class of multicommodity network problems using a primal partitioning code (PPLP). On these problems, they find PPLP to be two orders of magnitude faster than MINOS and about 50 times faster than OB1.

Linear, nonlinear, and integer multicommodity models have numerous important applications in scheduling, routing, transportation, and communications. Real-world multicommodity models tend to be very large and there is a need for faster and more efficient algorithms for solving these models.

See **Large-scale systems; Linear programming; Logistics; Minimum cost network flow problem; Networks; Transportation problem.**

References

[1] Ahuja, R.K., T.L. Magnanti, and J.B. Orlin (1993), *Network Flows: Theory, Algorithms, and Applications*, Prentice Hall, New Jersey.

[2] Ali, A., R. Helgason., J. Kennington, and H. Lall (1980), "Computational Comparison Among Three Multicommodity Network Flow Algorithms," *Operations Research*, 28, 995–1000.

[3] Ali, A., D. Barnett, K. Farhangian, J. Kennington, B. McCarl, B. Patty, B. Shetty, and P. Wong (1984), "Multicommodity Network Flow Problems: Applications and Computations," *IIE Transactions*, 16, 127–134.

[4] Assad, A.A. (1978), "Multicommodity Network Flows – A Survey," *Networks*, 8, 37–91.

[5] Barnhart, C. (1993), "Dual Ascent Methods for Large-Scale Multicommodity Flow Problems," *Naval Research Logistics*, 40, 305–324.

[6] Bellmore, M., G. Bennington, and S. Lubore (1971), "A Multivehicle Tanker Scheduling Problem," *Trans. Sci.*, 5, 36–47.

[7] Chen, R. and R. Meyer (1988), "Parallel Optimization for Traffic Assignment," *Math. Programming*, 42, 347–345.

[8] S. Clark and J. Surkis (1968), "An Operations Research Approach to Racial Desegregation of School Systems," *Socio-Econ. Plan. Sci.*, 1, 259–272.

[9] Evans, J. (1978), "The Simplex Method for Integral Multicommodity Networks," *Naval Research Logistics*, 25, 31–38.

[10] Evans, J. and J. Jarvis (1978), "Network Topology and Integral Multicommodity Flow Problems," *Networks*, 8, 107–120.

[11] Farvolden, J.M. and W.B. Powell (1990), "A Primal Partitioning Solution for Multicommodity Network Flow Problems," Working Paper 90–04, Department of Industrial Engineering, University of Toronto, Toronto, Canada.

[12] Farvolden, J.M., W.B. Powell, and I.J. Lustig (1993), "A Primal Partitioning Solution for the Arc-Chain Formulation of a Multicommodity Network Flow Problem," *Operations Research*, 41, 669–693 (1993).

[13] Gersht, A. and A. Shulman (1987), "A New Algorithm for the Solution of the Minimum Cost Multicommodity Flow Problem," Proceedings of the IEEE Conference on Decision and Control, 26, 748–758.

[14] Graves, G.W. and R.D. McBride (1976), "The Factorization Approach to Large Scale Linear Programming," *Math. Programming*, 10, 91–110.

[15] Grigoriadis, M.D. and W.W. White (1972), "A Partitioning Algorithm for the Multi-commodity Network Flow Problem," *Math. Programming*, 3, 157–177.

[16] Hartman, J.K. and L.S. Lasdon (1972), "A Generalized Upper Bounding Algorithm for Multicommodity Network Flow Problems," *Networks*, 1, 331–354.

[17] Held, M., P. Wolfe, and H. Crowder (1974), "Validation of Subgradient Optimization," *Math. Programming*, 6, 62–88.

[18] Kennington, J. and M. Shalaby (1977), "An Effective Subgradient Procedure for Minimal Cost Multicommodity Flow Problems," *Management Science*, 23, 994–1004.

[19] Kennington, J.L. (1978), "A Survey of Linear Cost Multicommodity Network Flows," *Operations Research*, 26, 209–236.

[20] Kennington, J.L. and R. Helgason (1980), *Algorithms for Network Programming*, John Wiley, Inc., New York.

[21] LeBlanc, L.J. (1973), "Mathematical Programming Algorithms for Large Scale Network Equilibrium and Network Design Problems," Unpublished Dissertation, Industrial Engineering and Management Sciences Department, Northwestern University.

[22] Naniwada, M. (1969), "Multicommodity Flows in a Communications Network," *Electronics and Communications in Japan*, 52-A, 34–41.

[23] Pilar, M.C. and S.A. Zenios (1990), "Parallel Decomposition of Multicommodity Network Flows Using Smooth Penalty Functions," Technical Report 90-12-06, Department of Decision Sciences, Wharton School, University of Pennsylvania, Philadelphia.

[24] Potts, R.B. and R.M. Oliver (1972), *Flows in Transportation Networks*, Academic Press, New York.

[25] Shetty, B. and R. Muthukrishnan (1990), "A Parallel Projection for the Multicommodity Network Model," *Jl. Operational Research*, 41, 837–842.

[26] Swoveland, C. (1971), "Decomposition Algorithms for the Multi-Commodity Distribution Problem," Working Paper, No. 184, Western Management Science Institute, University of California, Los Angeles.

MULTI-CRITERIA DECISION MAKING (MCDM)

See **Multiple criteria decision making**.

MULTIDIMENSIONAL TRANSPORTATION PROBLEM

This problem is usually a transportation problem with a third index that refers to a product type. Here there are i origins, j destinations, and k types of products available at the origins and demanded at the destinations. The variables x_{ijk} represent the amount of the kth product shipped from the ith origin to the jth destination. The constraint set is a set of linear balance equations, with the usual linear cost objective function. It is also a special form of the multicommodity network-flow problem. Unlike the transportation problem, its optimal solution may not be integer-valued even if the network data are given as integers. The problem can also be defined with more than three indices. See **Transportation problem**.

MULTI-ECHELON INVENTORY SYSTEMS

Inventory systems comprised of multiple stages of individual inventory problems. See **Inventory modeling**.

MULTI-ECHELON LOGISTICS SYSTEMS

Logistics systems comprised of several layers of individual logistics problems. See **Logistics**.

MULTI-OBJECTIVE LINEAR-PROGRAMMING PROBLEM

This problem has the usual set of linear-programming constraints ($Ax = b$, $x \geq 0$) but requires the simultaneous optimization of more than one linear objective function, say p of them. It can be written as "Maximize" Cx subject to $Ax = b$, $x \geq 0$, where C is a $p \times n$ matrix whose rows are the coefficients defined by the p objectives. Here "Maximize" represents the fact that it is usually impossible to find a solution to $Ax = b$, $x \geq 0$, that simultaneously optimizes all the objectives. If there is such an (extreme) point, the problem is thus readily solved. Special multiobjective computational procedures are required to select a solution that is in effect a compromise solution between the extreme point solutions that optimize individual objective functions. The possible compromise solutions are taken from the set of efficient (nondominated) solutions. This problem is also called the vector optimization problem. See **Efficient solution**; **Multi-objective optimization**; **Multi-objective programming**; **Pareto-optimal solution**.

MULTIOBJECTIVE PROGRAMMING

Ralph E. Steuer

University of Georgia, Athens

INTRODUCTION: Related to linear, integer and nonlinear programming, multiobjective programming is concerned with the extensions to theory and practice that enable us to address mathematical programming problems with more than one objective function.

In single objective programming, we often settle on a single objective such as to maximize profit or to minimize cost. However, in many if not most real world problems, we may find that we are in an environment of multiple conflicting criteria. To illustrate problems that may be more adequately modeled with multiple objectives, we have:

Oil Refinery Scheduling
   min {cost}
   min {imported crude}
   min {high sulfur crude}
   min {deviations from demand slate}
Production Planning
   max {total net revenue}
   max {minimum net revenue in any period}
   min {backorders}
   min {overtime}
   min {finished goods inventory}
Forest Management
   max {timber production}
   max {visitor days of recreation}
   max {wildlife habitat}
   min {overdeviations from budget}

Emerging as a new topic in the early 1970s, multiobjective programming has grown to the extent that a chapter on programming with multiple objectives can be found in almost all new operations research/management science survey texts (Zeleny, 1982; Goicoechea, Hanson and Duckstein, 1982; Yu, 1985).

TERMINOLOGY: In multiobjective programming, we have

$$\text{maximize } \{f_1(x) = z_1\}$$
$$\vdots$$
$$\text{maximize } \{f_k(x) = z_k\}$$
$$\text{subject to}$$
$$x \in S$$

where k is the number of objectives, the z_i are *criterion values*, and S is the feasible region in *decision space*. Let $Z \subset R^k$ be the feasible region in *criterion space* where $z \in Z$ if and only if there exists an $x \in S$ such that $z = (f_1(x), \ldots, f_k(x))$. Let $K = \{1, \ldots, k\}$. *Criterion vector* $\bar{z} \in Z$ is *nondominated* if and only if there does not exist another $z \in Z$ such that $z_i \geq \bar{z}_i$ for all $i \in K$ and $z_i > \bar{z}_i$ for at least one $i \in K$. The set of all nondominated criterion vectors is designated N and is called the *nondominated set*. A point $\bar{x} \in S$ is *efficient* if and only if its criterion vector $\bar{z} = (f_1(\bar{x}), \ldots, f_k(\bar{x}))$ is nondominated. The set of all efficient points is designated E and is called the *efficient set*.

Let $U: R^k \to R$ be the *utility function* of the decision maker (DM). A $z^\circ \in Z$ that maximizes U over Z is an *optimal criterion vector* and any $x^\circ \in S$ such that $(f_1(x^\circ), \ldots, f_k(x^\circ)) = z^\circ$ is an *optimal solution* of the multiobjective program. Our interest in the efficient set E and the nondominated set N stems from the fact that if U is *coordinate-wise increasing* (i.e., more is always better than less of each objective), $x^\circ \in E$ and

$z^\circ \in N$. In this way, a multiobjective program can be solved by finding the most preferred criterion vector in N.

One might think that the best way to solve a multiobjective program would be to assess the DM's utility function and then solve

maximize $\{U(z_1, \ldots, z_k)\}$

subject to

$$f_i(\boldsymbol{x}) = z_i \qquad i \in K$$

$$\boldsymbol{x} \in S$$

because any solution that solves this program is an optimal solution of the multiobjective program. However, multiobjective programs are usually not solved in this way because (1) of the difficulty in assessing an accurate enough U, (2) U would almost certainly be nonlinear, and (3) the DM would not see other candidate solutions to learn from during the solution process.

Consequently, multiobjective programming employs mostly interactive procedures that only require *implicit*, as opposed to *explicit*, knowledge about the DM's utility function. In interactive procedures, the goal is to search the nondominated set for the DM's most preferred criterion vector. Unfortunately, because of the size of N, finding the best criterion vector in N is not a trivial task. As a result, an interactive procedure is concluded with what is called a *final solution*, a solution that is either optimal or close enough to being optimal to satisfactorily terminate the decision process.

BACKGROUND CONCEPTS: Along with the basics of conventional mathematical programming, multiobjective programming requires additional concepts not widely employed elsewhere in operations research. The key ones are as follows.

(1) *Decision Space vs. Criterion Space.* Whereas single objective programming is typically studied in decision space, multiobjective programming is mostly studied in criterion space. To illustrate, consider

maximize $\{x_1 - 1/2 x_2 = z_1\}$

maximize $\{\qquad x_2 = z_2\}$

subject to

$$\boldsymbol{x} \in S$$

where S in decision space is given in Figure 1, and Z in criterion space is given in Figure 2. For instance, $\boldsymbol{z}^4$, which is the *image* of $x^4 = (3, 4)$, is obtained by evaluating the point $(3, 4)$ in the objective functions to generate $\boldsymbol{z}^4 = (1, 4)$. In Fig-

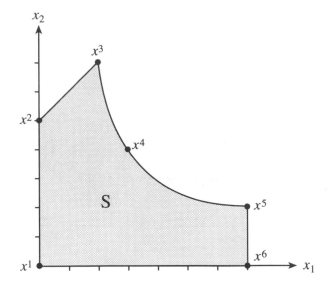

Fig. 1. Representation in decision space.

ure 2, the nondominated set N is the set of boundary criterion vectors $\boldsymbol{z}^3$ through $\boldsymbol{z}^4$ to $\boldsymbol{z}^5$ to $\boldsymbol{z}^6$. In Figure 1, the efficient set E is the set of *inverse images* of the criterion vectors in N, namely the set of boundary points $\boldsymbol{x}^3$ through $\boldsymbol{x}^4$ to $\boldsymbol{x}^5$ to $\boldsymbol{x}^6$. Note that Z is not necessarily confined to the nonnegative orthant.

(2) *Unsupported Nondominated Criterion Vectors.* A $\bar{z} \in N$ is *unsupported* if and only if it is possible to dominate $\bar{z}$ by a convex combination of other nondominated criterion vectors. In Figure 2, the set of unsupported nondominated criterion vectors is the set of criterion vectors from $\boldsymbol{z}^3$ through $\boldsymbol{z}^4$ to $\boldsymbol{z}^5$, exclusive of $\boldsymbol{z}^3$ and $\boldsymbol{z}^5$. The set of *supported* nondominated criterion vectors is the set that consists of $\boldsymbol{z}^3$ plus the line segment $\boldsymbol{z}^5$ to $\boldsymbol{z}^6$, inclusive. Unsupported nondominated criterion vectors can only occur in non-convex feasible regions; hence, they can only occur in integer and nonlinear multiobjective programs.

(3) *Identifying Nondominated Criterion Vectors.* To graphically determine whether a $\bar{z} \in Z$ is nondominated or not, visualize the nonnegative orthant in R^k translated so that its origin is now at $\bar{z}$. Note that, apart from $\bar{z}$, a vector dominates $\bar{z}$ if and only if the vector is in the nonnegative orthant translated to $\bar{z}$. Thus, $\bar{z}$ is nondominated if and only if the translated nonnegative orthant is empty of feasible criterion vectors other than for $\bar{z}$. Visualizing in Figure 2 the nonnegative orthant translated to $\boldsymbol{z}^4$, we see that $\boldsymbol{z}^4$ is nondominated. Visualizing the nonnegative orthant translated to $\boldsymbol{z}^2$, we see that $\boldsymbol{z}^2$ is dominated.

(4) *Payoff Tables.* Assuming that each objective is bounded over the feasible region, a payoff table is of the form

	z_1	z_2		z_k
z^1	z_1^*	z_{12}		z_{1k}
z^2	z_{21}	z_2^*		z_{2k}
			$\cdots$	
z^k	z_{k1}	z_{k2}		z_k^*

where the rows are criterion vectors resulting from individually maximizing the objectives. The z_i^* entries along the main diagonal of the payoff table are the maximum criterion values of the different objectives over the nondominated set. The minimum value in the ith column of the payoff table is an estimate of the minimum criterion value of the ith objective over N. Often these column minima are used in place of the minimum criterion values over N because the minimum criterion values over N are often difficult to obtain (Isermann and Steuer, 1988).

(5) $\boldsymbol{z}^{**}$ *Reference Criterion Vectors.* A $\boldsymbol{z}^{**} \in R^k$ *reference* criterion vector is a criterion vector that is suspended above the nondominated set. Its components are given by

$$z_i^{**} = z_i^* + \varepsilon_i$$

where the ε_i are moderately small positive values. An ε_i value that raises z_i^{**} to the smallest integer greater than z_i^* is normally sufficient.

(6) *Weighting Vector Space.* Without loss of generality, let

$$\Lambda = \left\{ \lambda \in R^k \big| \lambda_i \in (0,1), \sum_{i \in K} \lambda_i = 1 \right\}$$

be the weighting vector space. In an interactive environment, subsets of Λ called *interval defined subsets* are of the form

$$\Lambda^{(h)} = \left\{ \lambda \in R^k \big| \lambda_i \in (\ell_i^{(h)}, \mu_i^{(h)}), \sum_{i \in k} \lambda_i = 1 \right\}$$

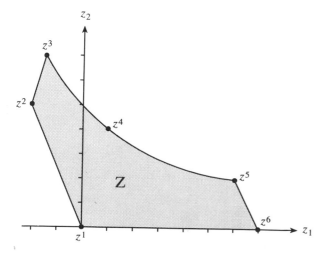

Fig. 2. Representation in criterion space.

where h is the iteration number and

$$0 < \ell_i^{(h)} < \mu_i^{(h)} < 1 \qquad i \in K$$
$$\mu_i^{(h)} - \ell_i^{(h)} = \mu_i^{(h)} - \ell_j^{(h)} \qquad \text{for all } i \neq j$$

Sequences of successively smaller interval defined subsets can be defined by reducing the $(\mu_i^{(h)} - \ell_i^{(h)})$ interval widths at each iteration.

(7) *Sampling Programs.* The *weighted-sums* program

$$\max \left\{ \sum_{i \in k} \lambda_i f_i(x) \big| x \in S \right\}$$

can be used to sample the nondominated set because, as long as $\lambda \in \Lambda$, the program returns an efficient point. A disadvantage of the weighted-sums program is that it cannot generate unsupported points.

To make "downward" probes of the nondominated set from a $\boldsymbol{z}^{**}$ as required in many of the interactive procedures of multiobjective programming, we employ the *augmented Tchebycheff* program

$$\text{minimize} \left\{ \alpha - \rho \sum_{i \in K} z_i \right\}$$

subject to

$$\alpha \geq \lambda_i(z_i^{**} - z_i) \qquad i \in K$$
$$f_i(\boldsymbol{x}) = z_i \qquad i \in K$$
$$\boldsymbol{x} \in S$$
$$\boldsymbol{z} \in R^k \text{ unrestricted}$$

where $\alpha \in R$, $\lambda \in \Lambda$, and ρ is a small, computationally significant positive scalar. A disadvantage of the augmented Tchebycheff program is that, regardless of the value of ρ, there may be unsupported members of the nondominated set that the program is unable to compute (Steuer, 1986).

A program that has better mathematical properties, but is more difficult to implement, is the *lexicographic Tchebycheff* program

$$\text{lex min} \left\{ \alpha, - \sum_{i \in k} z_i \right\}$$

subject to

$$\alpha \geq \lambda_i(z_i^{**} - z_i) \qquad i \in K$$
$$f_i(x) = z_i \qquad i \in K$$
$$\boldsymbol{x} \in S$$
$$\boldsymbol{z} \in \boldsymbol{R^k} \text{ unrestricted}$$

where $\lambda \in \Lambda$. At the first lexicographic level we solve to minimize α. Then, at the second lexicographic level, subject to only those solutions that minimize α, we minimize $-\Sigma_{i \in K} z_i$. Not only does the lexicographic Tchebycheff program always return a nondominated criterion vector, but if $\bar{z}$ is nondominated, there then exists a $\bar{\lambda} \in \Lambda$ such that $\bar{z}$ uniquely solves the program (Steuer, 1986).

(8) *Aspiration Criterion Vectors.* An aspiration criterion vector $q \in R^k$ is a criterion vector specified by a DM to reflect his or her hopes or expectations from the problem. An aspiration criterion vector, when specified, is typically *projected* onto N by an augmented or lexicographic Tchebycheff program in order to find the nondominated criterion vector closest to the aspiration criterion vector.

(9) *T-vertex λ-vector Defined by q and z^{**}.* The T-vertex (Tchebycheff-vertex) λ-vector defined by q and z^{**} is the $\lambda \in \Lambda$ whose components are given by

$$\lambda_i = \frac{1}{(z_i^{**} - q_i)} \left[\sum_{j \in K} \frac{1}{(z_j^{**} - q_j)} \right]^{-1}$$

The T-vertex λ-vector, when installed in an augmented or lexicographic Tchebycheff program, causes the program to probe the nondominated set along a line that goes through both z^{**} and q in the direction

$$-\left(\frac{1}{\lambda_1}, \dots, \frac{1}{\lambda_k} \right)$$

VECTOR-MAXIMUM ALGORITHMS: In the linear case, a multiple objective linear program (MOLP) is sometimes written in *vector-maximum* form

"max" $\{ Cx = z \,|\, x \in S \}$

where C is the $k \times n$ matrix whose rows are the coefficient vectors of the k objectives. A point is a solution to a vector-maximum problem if and only if it is efficient. Algorithms for characterizing the efficient set E of an MOLP are called vector-maximum algorithms. In the 1970s, considerable effort was spent on the development of vector-maximum codes to compute all efficient extreme points. The thought was that, by reviewing the list of nondominated criterion vectors associated with the efficient extreme points, a DM would be able to identify his or her efficient extreme point of greatest utility in hopes of satisfactorily terminating the decision process.

Unfortunately, in using vector-maximum codes such as ADBASE to explore the nondominated set, it was quickly found out, as shown in Table 1 (sample size of 10 for each $k \times m \times n$ size), that MOLPs possessed more efficient extreme points than anyone had imagined (Steuer, 1986). Whereas the number of variables and the number of constraints play a roll, the factor most dramatically affecting the number of efficient extreme points is the size of the *criterion cone*, the convex cone generated by the gradients of the k objective functions.

Table 1. Average number of MOLP efficient extreme points

MOLP size $k \times m \times n$	Efficient extreme points	C.P.U. time (seconds)
$3 \times 50 \times 25$	313	11.9
$3 \times 50 \times 50$	543	45.2
$3 \times 50 \times 100$	794	193.7
$4 \times 10 \times 50$	41	3.2
$4 \times 20 \times 50$	279	25.4
$4 \times 40 \times 50$	1702	203.2
$3 \times 50 \times 25$	313	11.9
$4 \times 50 \times 25$	1506	81.5
$5 \times 50 \times 25$	8905	939.7

The CPU seconds reported in Table 1 represent times on an IBM ES9000 Model 720 computer. With nondominated sets of sizes indicated by the numbers in Table 1, other methods in the form of *interactive procedures* have moved to the forefront in multiobjective programming.

INTERACTIVE PROCEDURES: In interactive multiobjective programming, we conduct an exploration over the feasible region for the best point in the nondominated set. Interactive procedures are characterized by phases of decision making alternating with phases of computation. We generally establish a pattern and keep repeating it until termination. At each iteration, a solution, or a group of solutions, is generated for examination. As a result of the examination, the DM inputs updated information about preferences to the solution procedure in the form of values of the *controlling parameters* (preference weights, aspiration criterion vectors, λ-vector interval widths, criterion vector components to be increased/decreased/held fixed, criterion vector lower bounds, etc., depending upon the particular interactive procedure).

While many interactive procedures have been proposed, virtually all of them more or less follow the same *general algorithmic outline*. As portrayed in Figure 3, the general algorithmic outline includes:

- an initial setting of the controlling parameters
- optimization of one or more mathematical programming problems to probe (i.e., sample) the nondominated set
- examination of the criterion vector results
- resetting of the controlling parameters for the next iteration in the light of what was learned on the current iteration.

With the consensus being that a range of interactive procedures is necessary because the most appropriate one to use is often application and user decision-making style dependent, ten of the most prominent interactive procedures are as follows:

1. ECON: e-Constraint Method
2. STEM: Benayoun, de Montgolfier, Tergny and Larichev (1971)
3. GDF: Geoffrion-Dyer-Feinberg Procedure (1972)
4. ZW: Zionts-Wallenius Procedure (1976)
5. ISWT Interactive Surrogate Worth Trade-off Method (Chankong and Haimes, 1978)
6. IGP: Interactive Goal Programming (Franz and Lee, 1981; Spronk, 1981)
7. WIERZ: Wierzbicki's Aspiration Criterion Vector Method (1982, 1986)
8. TCH: Tchebycheff Method of Steuer and Choo (1983)
9. SATIS: Satisficing Tradeoff Method of Nakayama and Sawaragi (1984)
10. RACE: Pareto Race (Korhonen and Laakso, 1986; Korhonen and Wallenius, 1988)

Other interactive multiobjective programming procedures include those by Sobol and Statnikov (1981), Gabbani and Magazine (1986), Climaco and Antunes (1987), Sakawa and Yano (1990), and Jaszkiewicz and Slowinski (1992).

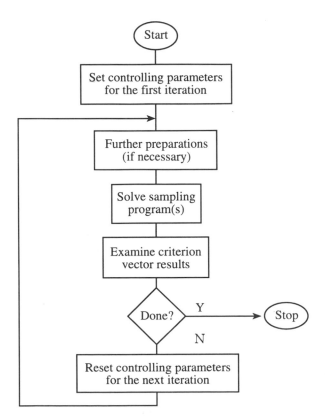

Fig. 3. General algorithmic outline.

SELECTED INTERACTIVE PROCEDURES: The *Aspiration Criterion Vector Method* (WIERZ) begins by asking the DM to specify an aspiration criterion vector $q^{(1)} < z^{**}$. Using the T-vertex λ-vector defined by $q^{(1)}$ and z^{**}, the augmented Tchebycheff program is solved, thus projecting $q^{(1)}$ onto N in order to produce $z^{(1)}$. In the light of $z^{(1)}$, the DM specifies a new aspiration criterion vector $q^{(2)}$. Using the T-vertex λ-vector defined by $q^{(2)}$ and z^{**}, the augmented Tchebycheff program is solved thus projecting $q^{(2)}$ onto N in order to produce $z^{(2)}$. In the light of $z^{(2)}$, the DM specifies a third aspiration criterion vector $q^{(3)}$, and so forth. Algorithmically, we have:

Step 1. $h = 0$. Construct a payoff table, form a z^{**} reference criterion vector, and specify $\rho > 0$ for use in the augmented Tchebycheff program. The DM specifies aspiration criterion vector $q^{(1)}$.

Step 2. $h = h + 1$. Compute T-vertex λ-vector defined by $q^{(h)}$ and z^{**}.

Step 3. Using the T-vertex λ-vector, solve the augmented Tchebycheff program for $z^{(h)}$.

Step 4. With $q^{(h)}$ and z^{**} as benchmarks, the DM contemplates $z^{(h)}$.

Step 5. If the DM wishes to cease iterating, stop with $(z^{(h)}, x^{(h)})$ as the final solution.

Step 6. The DM specifies aspiration criterion vector $q^{(h+1)}$. Go to Step 2.

Consider Figure 4 in which N is the set of boundary criterion vectors z^1 through $z^{(h)}$ to z^2. *In the figure, we see the way aspiration criterion vector $q^{(h)}$ is projected* onto the nondominated set by means of the augmented Tchebycheff program. Note that the direction of the arrow emanating from z^{**} and going through $q^{(h)}$ is given by

$$-\left(\frac{1}{\lambda_1}, \ldots, \frac{1}{\lambda_k}\right)$$

where the λ_i are the components of the T-vertex λ-vector defined by $q^{(h)}$ and z^{**}. Thus by changing $q^{(h)}$, we can change the $z^{(h)}$ generated by the sampling program.

Instead of generating only one solution at each iteration, the *Tchebycheff Method* (TCH) generates groups of solutions by making multiple probes of each subset in a sequence of progressively smaller subsets in N. Letting P be the number of solutions to be presented to the DM at each iteration, TCH begins by generating P well-spaced λ-vectors from $\Lambda^{(1)} = \Lambda$. Then the augmented Tchebycheff program is solved for each of the λ-vectors. From the P resulting nondominated criterion vectors, the DM selects his or her most preferred, designating it $z^{(1)}$. At this point, the interval widths of $\Lambda^{(1)}$ are reduced and cen-

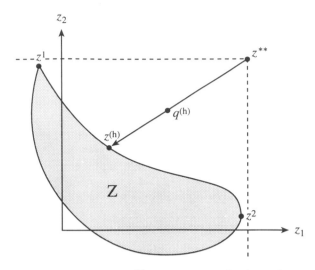

Fig. 4. Projection of $q^{(h)}$ onto the nondominated set.

tered about the T-vertex -vector defined by $z^{(1)}$ and z^{**} to form an interval defined subset $\Lambda^{(2)}$. Then P well-spaced λ-vectors are generated from $\Lambda^{(2)}$ and the augmented Tchebycheff program is solved for each of these λ-vectors. From the P resulting nondominated criterion vectors, the DM selects the most preferred, designating it $z^{(2)}$. Now, the interval widths of $\Lambda^{(2)}$ are reduced and centered about the T-vertex λ-vector defined by $z^{(2)}$ and z^{**} to form an interval defined subset $\Lambda^{(3)}$. Then P well-spaced λ-vectors are generated from $\Lambda^{(3)}$ and the augmented Tchebycheff program is solved for each of them, and so forth.

Another procedure that also generates multiple solutions at each iteration, but employs the weighted-sums program, is the *Geoffrion-Dyer-Feinberg* (GDF) procedure. GDF begins with the specification of an initial feasible criterion vector $z^{(0)}$. Then the DM specifies a λ-vector that is to be reflective of the local marginal trade-offs at $z^{(0)}$. Using this λ-vector, the weighted-sums program is solved for criterion vector $y^{(1)}$. Then the line through the feasible region in criterion space Z that starts at $z^{(0)}$ and ends at $y^{(1)}$ is divided into segments so as to create P equally spaced criterion vectors. The most preferred of the equally spaced criterion vectors becomes $z^{(1)}$. Then the DM specifies a new λ-vector that is to be reflective of the local marginal tradeoffs at $z^{(1)}$. Using this λ-vector, the weighted-sums program is solved for criterion vector $y^{(2)}$. Then the line segment through Z that starts at $z^{(1)}$ and ends at $y^{(2)}$ is divided into segments so as to create P new equally spaced criterion vectors. The most preferred of the new equally spaced criterion vectors becomes $z^{(2)}$, and so forth.

The *Satisficing Tradeoff Method* (SATIS) begins with the specification of a z^{**} reference

criterion vector and an initial aspiration criterion vector $q^{(1)}$. Then the augmented Tchebycheff program is solved using the T-vertex λ-vector defined by $q^{(1)}$ and z^{**} to produce $z^{(1)}$. The DM then specifies the components of $z^{(1)}$ that are to be increased, the amounts of each increase, the components that are to be relaxed, and the amounts of each relaxation in order to form a second aspiration criterion vector $q^{(2)}$. Using the T-vertex λ-vector defined by $q^{(2)}$ and z^{**}, the augmented Tchebycheff program is solved to produce $z^{(2)}$. The DM then specifies which components of $z^{(2)}$ are to be increased, the amounts of each increase, the components that are to be relaxed, and the amounts of each relaxation in order to form $q^{(3)}$. Using the T-vertex λ-vector defined by $q^{(3)}$ and z^{**}, the augmented Tchebycheff program is solved to produce $z^{(3)}$, and so forth.

Descriptions of the other interactive procedures given above can be found in their noted source references.

Conclusion: Because the weighted-sums, augmented Tchebycheff, and other variants of these programs that are used to sample the nondominated set are single criterion optimization problems, conventional mathematical programming software can in most cases be employed (Gardiner and Steuer, 1994). In this way, the interactive procedures can address multiobjective programming problems with as many constraints and variables as in single objective programming. Unfortunately, in multiobjective programming, there are limitations with regard to the number of objectives. Problems with up to about five objectives can generally be accommodated, but above this number, difficulties can easily arise because of the rate at which the nondominated set grows as the number of objectives increases. However, with further research progress anticipated, the future looks bright, particularly if the procedures of multiobjective programming can be implemented in such a way that they would subsume conventional single criterion mathematical programming as a special case when the number of objectives is one.

See **Decision analysis; Linear programming; Multiple criteria decision making; Utility theory.**

References

[1] Benayoun, R., J. de Montgolfier, J. Tergny, and O. Larichev (1971). "Linear Programming with Multiple Objective Functions: Step Method (STEM)," *Mathematical Programming* 1, 366–375.

[2] Chankong, V. and Y.Y. Haimes (1978). "The Interactive Surrogate Worth Trade-off (ISWT) Method for Multiobjective Decision Making," in S. Zionts, ed., *Multiple Criteria Problem Solving, Lecture Notes in Economics and Mathematical Systems* 155, 42–67.

[3] Climaco, J.C.N. and C.H. Antunes (1987). "TRIMAP – An Interactive Tricriteria Linear Programming Package," *Foundations Control Engineering* 12(3), 101–120.

[4] Franz, L.S. and S.M. Lee (1981). "A Goal Programming Based Interactive Decision Support System," *Lecture Notes in Economics and Mathematical Systems* 190, 110–115.

[5] Gabbani, D. and M. Magazine (1986). "An Interactive Heuristic Approach for Multi-Objective Integer-Programming Problems," *Jl. Operational Research Society* 37, 285–291.

[6] Gardiner, L.R. and R.E. Steuer (1994). "Unified Interactive Multiple Objective Programming," *European Jl. Operational Research* 74, 391–406.

[7] Geoffrion, A.M., J.S. Dyer, and A. Feinberg (1972). "An Interactive Approach for Multicriterion Optimization, with an Application to the Operation of an Academic Department," *Management Science* 19, 357–368.

[8] Goicoechea, A., D.R. Hanson, and L. Duckstein (1982). *Multiobjective Decision Analysis with Engineering and Business Applications*, John Wiley, New York.

[9] Isermann, H. and R.E. Steuer (1988). "Computational Experience Concerning Payoff Tables and Minimum Criterion Values over the Efficient Set," *European Jl. Operational Research* 33, 99–97.

[10] Jaszkiewicz, A. and R. Slowinski (1992). "Cone Contraction Method with Visual Interaction for Multi-Objective Non-Linear Programmes," *Jl. Multi-Criteria Decision Analysis* 1, 29–46.

[11] Korhonen, P.J. and J. Laakso (1986). "A Visual Interactive Method for Solving the Multiple Criteria Problem," *European Jl. Operational Research* 24, 277-287.

[12] Korhonen, P. J. and J. Wallenius (1988). "A Pareto Race," *Naval Research Logistics* 35, 615–623.

[13] Nakayama, H. and Y. Sawaragi (1984). "Satisficing Trade-off Method for Multiobjective Programming," *Lecture Notes in Economics and Mathematical Systems* 229, 113–122.

[14] Sakawa, M. and H. Yano (1990). "An Interactive Fuzzy Satisficing Method for Generalized Multiobjective Programming Problems with Fuzzy Parameters," *Fuzzy Sets and Systems* 35, 125–142.

[15] Sobol, I.M. and R.B. Statnikov (1981). *Optimal Parameters Choice in Multicriteria Problems*, Nauka, Moscow.

[16] Spronk, J. (1981). *Interactive Multiple Goal Programming*, Martinus Nijhoff Publishing, Boston.

[17] Steuer, R. E. (1986). *Multiple Criteria Optimization: Theory, Computation, and Application*, John Wiley, New York.

[18] Steuer, R.E. (1995). "ADBASE Multiple Objective Linear Programming Package," Faculty of Management Science, University of Georgia, Athens, Georgia.

[19] Steuer, R.E. and E.-U. Choo (1983). "An Interactive Weighted Tchebycheff Procedure for Multiple Objective Programming," *Mathematical Programming* 26, 326–344.

[20] Wierzbicki, A.P. (1982). "A Mathematical Basis for Satisficing Decision Making," *Mathematical Modelling* 3, 391–405.

[21] Wierzbicki, A.P. (1986). "On the Completeness and Constructiveness of Parametric Characterizations to Vector Optimization Problems," *OR Spektrum* 8, 73–87.

[22] Yu, P.L. (1985). *Multiple-Criteria Decision Making: Concepts, Techniques and Extensions*, Plenum Press, New York.

[23] Zeleny, M. (1982). *Multiple Criteria Decision Making*, McGraw-Hill, New York.

[24] Zionts, S. and J. Wallenius (1976). "An Interactive Programming Method for Solving the Multiple Criteria Problem," *Management Science* 22, 652–663.

MULTIPLE CRITERIA DECISION MAKING

R. Ramesh and Stanley Zionts

State University of New York at Buffalo

INTRODUCTION: Multiple Criteria Decision Making (MCDM) refers to making decisions in the presence of multiple, usually conflicting, objectives. Multiple criteria decision problems pervade almost all decision situations ranging from common household decisions to complex strategic and policy level decisions in corporations and governments. Prior to the development of MCDM as a discipline, such problems have been traditionally addressed as single-criterion optimization problems by (i) deriving a composite measure of the objectives and optimizing it, or (ii) by choosing one of the objectives as the main decision objective for optimization and solving the problem by requiring an acceptable level of achievement in each of the other objectives. The emergence of MCDM as a discipline has been founded on two key concepts of human behavior, introduced and explored in detail by Herbert Simon in the 1950s: *satisficing* and *bounded rationality*. The two are intertwined because satisficing involves finding solutions that satisfy constraints rather than optimizing, while bounded rationality involves setting the constraints and then searching for solutions satisfying the constraints, adjusting the constraints, and then continuing the process until a satisfactory solution is found. Some of the MCDM tools developed in recent years are based on these concepts.

We overview some of the important aspects of MCDM. These include basic concepts, modeling techniques, and algorithms. We outline the basic concepts in the next section and provide a taxonomy of MCDM methods in the following section. The algorithmic approaches are explored after that, and then some comments on trends and a few concluding remarks are provided in the closing section.

BASIC CONCEPTS: An MCDM problem can be broadly described as follows. Let $D = \{d_1, \ldots, d_n\}$ denote the decision space, comprising the set of possible decision alternatives to a problem. Let $C = \{C_1, \ldots, C_p\}$ denote the objective space, comprising of a set of p mutually conflicting objectives. Without loss of generality, we assume all objectives are to be maximized. Let $E: D \to C$ be a mapping of the decision space on to the objective space, where $E(d_i)$ is the vector $(C_1^i, \ldots, C_p^i)$. Each element of this vector is an assessment, or the value of the corresponding objective provided by the decision alternative d_i. We define a fundamental concept in MCDM as follows.

Definition 1 (*Dominance*): A decision alternative d_i said to be dominated by another alternative d_j if $C_k^i \le C_k^j$, $k = 1, \ldots, p$ with at least one strict inequality.

In the above definition, if all the inequalities hold as strict inequalities, then the dominance is said to be strong. Otherwise, it is called weak. The following concept is a logical extension of the dominance concept.

Definition 2 (*Convex Dominance*): An alternative d_i is said to be convex dominated by a subset $\hat{D} \subset D$ if it is dominated by a convex combination of the alternatives in $\hat{D}$.

The above definitions lead to a central theme of all MCDM techniques as follows.

Definition 3 (*Efficiency*): An alternative d_j is said to be 'efficient' or 'nondominated' in D if there is no other alternative in D that dominates it, even weakly.

The concept of efficiency can be extended to convex dominance as well. In this case, an efficient alternative is known as "convex-efficient" or "convex-nondominated." The following theorem of Geoffrion (1968) shows how the efficiency of an alternative can be determined.

Zionts and Wallenius (1980) introduced a different but equivalent methodology that solves a number of problems including that one.

Theorem 1: Consider any decision alternative d_i and its mapping on the objective space $(C_1^i, \ldots, C_p^i)$. The decision d_i is efficient if only if the following linear program is unbounded:

$$\text{Maximize} \sum_{j=1}^{p} w_j C_j^i$$

$$\text{subject to} \sum_{j=1}^{p} w_j C_j^k \le 0, \quad k = 1, \ldots, n, \quad k \ne i$$

$$w_j \ge 0, \quad j = 1, \ldots, p.$$

A TAXONOMY OF MCDM METHODS: The MCDM methods proposed in the literature cover a wide spectrum, and there are several alternative ways of organizing them into a taxonomy. Our taxonomy is based on Chankong *et al.* (1984), which is one of the interpretations of the world of MCDM models. At the outset, MCDM methods can be classified into two broad classes: vector optimization methods and utility optimization methods. Vector optimization is primarily concerned with the generation of all efficient decision alternatives. These methods do not require intervention of a decision maker. These methods do generate a subset of nondominated solutions. Some of the well-known vector optimization methods include those of Geoffrion (1968), Villarreal and Karwan (1981), and Yu and Zeleny (1976).

The utility optimization methods can be broadly organized according to the following dimensions:

(1) Nature of decision space: Explicit or Implicit
(2) Nature of decision outcomes: Stochastic or Deterministic

In an explicit decision space, decision alternatives are stated explicitly. A classical example is the home buying problem, where a decision maker is faced with a set of possible homes to consider. For an implicit decision, alternatives are stated using a set of constraints, such as in linear or nonlinear programming where a feasible alternative must satisfy the constraints. An implicit decision situation can be further categorized as continuous or discrete. The decision outcomes are stochastic or deterministic depending on whether the mapping function $E: D \to C$ is stochastic or deterministic. Table I classifies MCDM methods broadly along the two dimensions. There are many approaches in the various segments of this classification. Hence, we restrict our discussion to some of the best-known methods.

Table I. A taxonomy of MCDM approaches

Decision outcomes	Decision space	
	explicit	implicit
Deterministic	Deterministic multiattribute Decision analysis	Deterministic multiobjective Mathematical programming
Stochastic	Stochastic multiattribute Decision analysis	Stochastic multiobjective Mathematical programming

METHODOLOGICAL APPROACHES: We present the key ideas from selected methods in each category in the following discussion.

Deterministic Decision Analysis – Deterministic decision analysis is concerned with finding the *most preferred* alternative in decision space by constructing a value function representing a decision maker's preference structure, and then using the value function to identify the most preferred solution. A value function $v(C_1, C_2, \ldots, C_p)$ is a scalar-valued function defined with the property that $v(C_1, C_2, \ldots, C_p) > v(C_1^i, C_2^i, \ldots, C_p^i)$ if and only if $(C_1, C_2, \ldots, C_p)$ is at least as preferred as $(C_1^i, C_2^i, \ldots, C_p^i)$ (Keeney and Raiffa, 1976). The construction of the value function involves choice decisions made by the decision maker. Generating value functions is simplified if certain conditions hold, in which case it is possible to decompose the above functions into partial value functions $v_k(C_k)$ for each value of k.

The decomposition and certain simplifications of the value function may be carried out if certain underlying assumptions on the decision maker's preference structure hold. One of these is preferential independence, which is stated as follows: Consider a subset of objectives denoted as $\hat{C}$. If the decision maker's preferences in the space $C - \hat{C}$ are the same for any set of arbitrarily fixed levels of the objectives $\hat{C}$, then $\hat{C}$ is said to be preferentially independent of $C - \hat{C}$. The set C is said to be mutually preferentially independent if every subset of C is preferentially independent of its complement with respect to C. When mutual preferential independence holds, an additive value function of the form

$$v(d_i) = \sum_{k=1}^{p} \lambda_k v_k(C_k^i) \text{ where } \lambda_k \text{ is a scalar constant}$$

is appropriate. There are other nonlinear forms that can be used as well. Of course, an additive value function, if appropriate, is highly desirable. Once the value function has been determined, it can be used to evaluate and rank the alternatives.

Stochastic Decision Analysis – Stochastic decision analysis is similar to the deterministic case, except that the outcomes are stochastic, and utility functions are constructed instead of value functions. The ideas are similar. There is an analogous condition to that described for the discrete case above. It involves utility independence. A subset of objectives $\hat{C}$ is utility independent of its complement if the conditional preference order for 'lotteries' involving changes in $\hat{C}$ does not depend on the levels at which the objectives are fixed. Since utility independence refers to lotteries and preferential independence refers to deterministic outcomes, utility independence implies preferential independence, but not vice versa. Analogous to mutual preferential independence, the set C is said to be mutually utility independent if every subset of C is utility independent of its complement with respect to C. Keeney and Raiffa (1976) show that if C is mutually utility independent, then a multiplicative utility function is appropriate. This function is of the form

$$u(d_i) = \prod_{k=1}^{p} \mu_k u_k(C_k^i),$$

where $u(d_i)$ is the overall utility of the decision alternative d_i, $u_k(C_k^i)$ is the utility of its kth objective component, and μ_k is a scalar constant. A more stringent set of assumptions must hold in order that the utility function be additive. In the stochastic case, not only must a utility function be estimated, but probabilities of various outcomes must also be estimated by the decision maker.

Multiobjective Mathematical Programming – Considerable work has been done in the multiobjective mathematical programming area. These include *Multiobjective Linear Programming* (MOLP) and *Multiobjective Integer Programming* (MOIP). Goal programming (Lee, 1972), the method of Zionts and Wallenius (1976, 1983), the *Step Method* of Benayoun et al. (1971), and the method of Steuer (1976) are some of the better-known MOLP methods. We outline the approaches of goal programming and the method of Zionts and Wallenius in this section.

Goal programming is an extension of linear programming and was proposed by Charnes and Cooper in 1961. A description of this technique is as follows. Consider the following MOLP problem:

Maximize $\boldsymbol{Cx}$ (MOLP)

subject to $\boldsymbol{Ax} \leq \boldsymbol{b}$

          $\boldsymbol{x} \geq \boldsymbol{0}$

where $C = (c_{kj})$ is a $(p \times n)$ matrix, A is a $(m \times n)$ matrix and x is a $(n \times 1)$ vector. Let $(\alpha_1, \ldots, \alpha_p)$ denote the goals with respect to the desired levels of attainment in the objectives specified by a decision maker. Introduce over and under attainment variables y_k^+ and y_k^- for each objective and add the following constraints, where c_k is the kth row of C:

$$c_k x - y_k^+ + y_k^- = \alpha_k, \quad k = 1, \ldots, p$$

Let w_k denote the penalty for the net deviation from the goal of objective $k = 1, \ldots, p$. Then the goal programming problem is formulated as follows:

$$\text{Minimize} \sum_{j=1}^{p} w_k(y_k^+ + y_k^-) \qquad \text{(GP)}$$

$$\text{subject to} \sum_{j=1}^{n} c_{kj} x_j - y_k^+ + y_k^- = \alpha_k, \quad k = 1, \ldots, p$$

$$Ax \leq b$$

$$x, y \geq 0$$

The above problem minimizes a weighted sum of deviations from the desired goals, where weights are required from the decision maker. The goal programming formulation is an attempt to find a solution that is closest to the decision maker's desired goals, while also responding to his differential emphasis on the nonattainment of the various goals.

The Zionts and Wallenius method follows an interactive approach using pairwise evaluations of decision alternatives by a decision maker to solve problem MOLP. The method starts by choosing an initial set of weights $\lambda \in R^p$, and maximizing a linear composite objective λCx. This generates a corner point of $\{Ax \leq b, x \geq 0\}$ that is efficient. Call this solution x^0. Next, the adjacent corner points of x^0 that are also efficient (and whose edges leading to them are *also* efficient) are determined. Call this set S^0. The decision maker is asked to choose between x^0 and a solution from S^0 until: (i) either he or she prefers x^0 to all the points in S^0, or (ii) prefers some solution in S^0 to x^0. If x^0 is preferred to all the points in S^0, then the method stops with x^0 as a "locally" best preferred corner-point solution. Otherwise, if some solution in S^0 is preferred to x^0, then it is devoted as x'. Linear constraints of the form $\lambda(Cx' - Cx'') \leq -\varepsilon$ where x' is preferred to x'' and ε is a small positive quantity are generated from the decision maker's pairwise preferences. A new set of weights that satisfy these constraints are then obtained. If these constraints are in conflict, then some of them are dropped in determining the new weights. Call the new set λ''. Maximiz-

ing the composite objective $\lambda'' Cx$, a new efficient corner point is generated, and the above steps are repeated until a corner point that is preferred to all its adjacent efficient corner points is obtained.

Compared to MOLP, research on MOIP is rather limited. Some of the earlier works on MOIP have been in the domain of vector optimization. Bitran and Rivera (1982) provided an implicit enumeration algorithm for determining the efficient set of zero-one MOIP problems. Pasternak and Passy (1973) studied the vector optimization problem for two objectives. Klein and Hannan (1982) extended Pasternak and Passy's work to more than two objectives. Villarreal and Karwan (1981) generalized the classical dynamic programming recursions to a multicriteria framework. Ramesh *et al.* (1989) followed the utility optimization approach to find the most preferred solution to an MOIP problem.

The method of Ramesh *et al.* (1989) follows a branch-and-bound search strategy using the Zionts and Wallenius method for bounding. The decision maker's preference structure is assessed using pairwise evaluations and an internal representation of the preference structure is successively built during the course of the branch-and-bound search. This representation is used to deduce the decision maker's preferences wherever possible so that the cognitive load arising out of the pairwise judgments can be minimized. The internal representation is based on the concept of convex cones as described below (Korhonen *et al.*, 1984).

Consider a two-dimensional objective space as shown in Figure 1. Let $\bar{C}$ and $\hat{C}$ be two points in this space such that $\hat{C}$ is preferred to $\bar{C}$. Assuming a quasiconcave and nondecreasing utility function for the decision maker, it follows that every point falling on the ray $\{C | C = \hat{C} + \mu(\bar{C} - \hat{C}), \mu \geq 0\}$ is less preferred than $\hat{C}$ and no more preferred than $\bar{C}$. Consequently, every point in this ray and those dominated by it can be eliminated from consideration. This ray is called a convex cone, and is illustrated in Figure 1. Every pairwise judgment of a decision maker yields a convex cone and the cones are ordered into a tree structured to eliminate search regions efficiently and minimize the need for the decision maker's pairwise evaluations throughout the search procedure.

Other Explicit Decision Space Methods – Several methods have been proposed for finding the most preferred alternative from an explicitly stated decision space without estimating a value function. These techniques are methods of deterministic decision analysis, and there is substan-

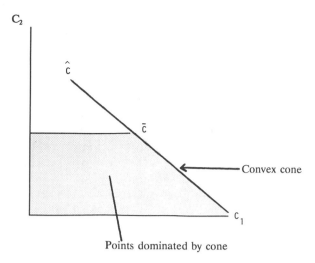

Fig. 1. Illustration of convex cones.

tial interest in these problems. Three important methods in this category are the *Analytic Hierarchy Process* (AHP – Saaty, 1980), the method of Korhonen *et al.* (1984), and the AIM method (Lotfi *et al.*, 1992).

The idea of AHP is that one can structure a problem hierarchically, and then make judgments regarding the relative importance of various aspects of the problem. As a result of these judgments, a ranking is produced. A simple decision problem would have a hierarchy that consists of three levels, from the top down: 1) the goal; 2) the criteria involved; and 3) the alternatives. The number of levels depends on the nature of the problem involved. In general, suppose that we have an n-alternative, p-criteria problem. Then the decision maker is asked to fill in entries in $p + 1$ reciprocal matrices as follows:

1. One $(p \times p)$ matrix relating each criterion to all others; and

2. $p(n \times n)$ matrices, each relating one criterion to all alternatives.

Each reciprocal matrix has all diagonal elements one, and off-diagonal elements reciprocal, that is, $a_{ij} = 1/a_{ji}$. Accordingly, the decision maker need only provide just less than half the entries, more specifically, the $[p(p - 1)/2] + p[n(n - 1)/2]$ off-diagonal (lower or upper) entries in the matrix. Though the amount can be reduced to as few as $(p - 1) + p(n - 1)$ entries (having no redundancy), the reduction in information required increases the cognitive load on the decision maker to provide entries, and does not provide the redundancy and cross checking that furnishing the complete input provides.

In filling in the matrices, the decision maker is asked to provide numbers between 1/9 and 9 reflecting the relative importance between the aspects involved. One of the matrices reflects the comparison among criteria and the p other matrices reflect evaluations of alternatives with respect to each criterion. AHP next solves for the right eigenvector, or characteristic vector, of each matrix. An eigenvector of a matrix may be estimated by taking the geometric mean of the elements of each row of the matrix (for a $p \times p$ matrix, the pth root of the product of the p elements of a row), and then normalizing the resulting vector so that the sum of the elements is unity. The consistency of the matrix (as differentiated from a matrix generated at random) may be tested using a calculation on the matrix. By the user furnishing fewer than all $p(p - 1)/2$ entries required in the matrix, the test on consistency is compromised. The scaled eigenvectors are then used to score and rank each alternative.

Korhonen *et al.* (1984) presented an interactive method employing pairwise comparisons for solving the discrete, deterministic MCDM problem. Assuming a quasiconcave and nondecreasing utility function, they introduce the concept of convex cones. Choosing an arbitrary set of positive weights w_i, $i = 1, \ldots, p$, a composite linear utility function is initially generated. Using the composite as a proxy for the true utility function, the decision alternative maximizing the composite is generated. Call this solution d^0. Using the mapping $E: D \to C$, all adjacent efficient decision alternatives to d^0 (as in the Zionts-Wallenius method) are determined. This is done for the region that consists of all convex combinations of feasible solutions. Call the set of such solutions S^0. The decision maker is asked to choose between d^0 and some solution from S^0. Based on the response, a constraint on the weights is generated, as in the Zionts and Wallenius method for MOLP, and a convex cone is derived. Any solution in the set S^0 dominated by the cone is removed from S^0, and the above step is repeated until either d^0 is preferred to all solutions in S^0 or some solution in S^0 is preferred to d^0. The constraints on the weights and the convex cones generated at each iteration of this step are accumulated. The set of cones is used to deduce the decision maker's preferences wherever possible. This reduces the search space, while also minimizing the number of pairwise comparisons the decision maker has to perform.

Every solution in S^0 that is less preferred than d^0 is dropped from consideration. If d^0 is preferred to all the solutions in S^0, then it is denoted as $\bar{d}$. If some solution in S^0 is preferred to d^0, then the preferred solution is denoted as $\bar{d}$. If $\bar{d}$ is the only efficient solution remaining in the decision space, then the procedure stops with as the most preferred decision. Otherwise, choosing a set of weights consistent with the weight constraints (after dropping any conflicting constraints), a new composite linear utility function is generated. Denoting the decision alternative maximizing this composite function as $\hat{d}$, the decisionmaker chooses between $\bar{d}$ and $\hat{d}$. Denoting the preferred solution as d^0, the above steps are repeated.

Lotfi, Stewart, and Zionts (1992) develop an eclectic method called the *Aspiration-Level Interactive Method* (AIM) for MCDM. It involves a philosophy that aspiration levels and feedback regarding the relative feasibility of the aspiration levels provide a powerful tool for decision making. The method is embodied in a computer program called AIM for an IBM-compatible PC. The method provides the decision maker with various kinds of feedback as he or she explores the solutions. Several different kinds of objectives may be included: objectives to be maximized; objectives to be minimized; target objectives; any of the above kinds of objectives with thresholds, or levels beyond which the user is indifferent to further gains in the objective; and qualitative objectives. To further explain the idea of thresholds, suppose that in the purchase of a house, the age of the house is an attribute to be minimized. Suppose further that the buyer treats as equivalent, however, any houses ten years or less in age. In this case, there is a threshold of ten years, so that an eight-year-old house is considered to be no better than a ten-year-old house with respect to age.

To begin with, the decision maker has the following basic information:

1. A current goal or aspiration level for each objective, initially set to the median, together with the proportion of alternatives having values of the objective at least as good as that value.
2. Two other aspiration levels, the next better and the next worse than the current goal occurring in the data base.
3. The ideal and nadir solutions to the problem.
4. The proportion of alternatives that simultaneously satisfy aspiration levels given in 1 and 2.
5. A "nearest nondominated solution" to the current goal. The nearest solution is found by mapping the current goal to a solution on the efficient frontier or in the set of nondominated solutions.

The current goal may be (and should be) changed by the user, component by component, to any desired realizable level of any objective. The intention, however, is to keep the current goal "near" the efficient frontier and therefore "nearly" achievable. As the user changes the current goal, all but item(s) 3 above change.

The user can invoke various options to help in decision making. He or she can see which solutions, if any, satisfy his current goal. Second, he or she can obtain a ranking of solutions based on a function resulting from his choice of a current goal. Third, he or she can use a simplified version of a concept called outranking to identify "neighbor" solutions that are similar to his nearest solution. The decision maker may also review the "weights" implied by the current goal, see a quartile distribution of the problem by objective, and identify and possibly delete dominated solutions.

WHERE DO WE GO FROM HERE? The field of multiple criteria decision making has been an active one for over twenty years. Many interesting approaches have been developed, explored, and implemented in solving problems. Current trends in MCDM include multiple criteria decision support systems (MCDSS) and negotiations, which may be regarded as multiple criteria problems involving multiple decision makers.

MCDSS integrate the multiple criteria approaches in user-friendly microcomputer systems. Several of the currently available systems include the VIG/VIMDA system of Korhonen and Laakso (1986), the Expert Choice software that implements AHP, and the AIM package of Lotfi *et al.* (1992). An objective of most of the MCDSS is to ultimately provide inexpensive stand-alone software that is sufficiently easy to use that it becomes widely used.

Negotiations or multiperson MCDM is a natural extension of MCDM. Many decisions are made by groups, and negotiation theory involves using some of the MCDM concepts to simplify and assist negotiations.

In addition to the journals devoted to management science and operations research and behavioral science, there are two journals that contain articles more exclusively in this area. They are *Multi-Criteria Decision Analysis* published by Wiley and *Group Decision and Negotiation* published by Kluwer.

See **Analytic hierarchy process; Decision analysis; Decision problem; Multi-attribute utility analysis; Multiobjective programming; Utility theory; Value function.**

References

[1] Belton, B. (1986), "A Comparison of the Analytic Hierarchy Process and a Simple Multi-Attribute Value Function," *European Jl. Operational Research*, 26, 7–21.

[2] Belton, B., and Gear, A.E. (1983), "On a Shortcoming of Saaty's Method of Analytic Hierarchies," *Omega*, 11, 227–230.

[3] Benayoun, R., De Montgolfier, J., Tergny, J. and Larichev, O. (1971), "Linear Programming with Multiple Objective Functions: Step Method (STEM)," *Mathematical Programming*, 1, 366–375.

[4] Bitran, G.R. and Rivera, J.M. (1982), "A Combined Approach to Solving Binary Multicriteria Problems," *Naval Research Logistics*, 29, 181–201.

[5] Chankong, V. Haimes, Y.Y., Thadathil, J. and Zionts, S. (1984), "Multiple Criteria Optimization: A State of the Art Review," in *Decision Making with Multiple Objectives*, Springer-Verlag, Berlin, 36–90.

[6] Geoffrion, A.M. (1968), "Proper Efficiency and the Theory of Vector Maximization," *Jl. Mathematical Analysis and Applications*, 22, 618–630.

[7] Keeney, R.L. and Raiffa, H. (1976), *Decisions with Multiple Objectives: Preferences and Value Tradeoffs*, John Wiley, New York.

[8] Klein, D. and Hannan, E. (1982), "An Algorithm for the Multiple Objective Integer Linear Programming Problem," *European Jl. Operational Research*, 9, 378–385.

[9] Korhonen, P., and Laakso, J. (1986), "A Visual Interactive Method for Solving the Multiple Criteria Problem," *European Jl. Operational Research*, 24, 277–287.

[10] Korhonen, P., Wallenius, J. and Zionts, S. (1984), "Solving the Discrete Multiple Criteria Problem Using Convex Cones," *Management Science*, 30, 1336–1345.

[11] Lee, S.M. (1972), *Goal Programming for Decision Analysis*, Auerbach Publishers, Philadelphia.

[12] Lotfi, V., Stewart, T.J., and Zionts, S. (1992), "An Aspiration-Level Interactive Model for Multiple Criteria Decision Making," *Computers and Operations Research*, 19, 671–681.

[13] Pasternak, H. and Passy, V. (1973), "Bicriterion Mathematical Programs with Boolean Variables," in *Multiple Criteria Decision Making*, University of South Carolina Press, Columbia.

[14] Ramesh, R., Karwan, M.H. and Zionts, S. (1989), "Preference Structure Representation Using Convex Cones in Multicriteria Integer Programming," *Management Science*, 35, 1092–1105.

[15] Saaty, Thomas L. (1980), *The Analytic Hierarchy Process*, McGraw-Hill, New York.

[16] Simon, H. (1957), *Administrative Behavior*, The Free Press, New York.

[17] Steuer, R.E. (1976), "Multiple Objective Linear Programming with Interval Criterion Weights," *Management Science*, 23, 305–316.

[18] Villarreal, B. and Karwan, M.H. (1981), "Multicriteria Integer Programming: A (Hybrid) Dynamic Programming Recursive Approach," *Mathematical Programming*, 21, 204–223.

[19] Yu, P.L. and Zeleny, M. (1976), "Linear Multiparametric Programming by Multicriteria Simplex Method," *Management Science*, 23, 159–170.

[20] Zionts, S. and Wallenius, J. (1976), "An Interactive Programming Method for Solving the Multiple Criteria Problem," *Management Science*, 22, 652–663.

[21] Zionts, S. and Wallenius, J. (1980), "Identifying Efficient Vectors: Some Theory and Computational Results," *Operations Research*, 28, 788–793.

[22] Zionts, S. and Wallenius, J. (1983), "An Interactive Multiple Objective Linear Programming Method for a Class of Underlying Nonlinear Utility Functions," *Management Science*, 29, 519–529.

MULTIPLE OPTIMAL SOLUTIONS

In an optimization problem, when different feasible solutions yield the same optimal value for the objective function, the problem has multiple optimal solutions. If a linear-programming problem has multiple optimal solutions, then such solutions correspond to extreme point solutions and their convex combinations. See **Unique solution**.

MULTIPLE PRICING

When solving a linear-programming problem using the simplex method, it is computationally efficient to select a small number, say 5, possible candidate vectors from which one would be chosen to enter the basis. The candidate set consists of columns with large (most negative or most positive) reduced costs, and the vector in this set that yields the largest change in the objective function is selected. Succeeding iterations only consider candidate basis vectors from the vectors that remain in the set that have properly signed reduced costs. When all vectors in the set are chosen or none can serve to change the objective function in the proper direction, a new set is determined. See **Partial pricing**; **Simplex method**.

MULTIPLIER VECTOR

For a given feasible basis B to a linear-programming problem, let the row vector c_B be the ordered set of cost coefficients for the vectors in B. The multiplier vector is defined as $\pi = c_B B^{-1}$. If B is an optimal basis, then the components of π are the dual variables associated with the corresponding primal constraints. The vector π is also called the simplex multiplier vector, with the components of π being the simplex multipliers. See **Simplex method**.

NASH SADDLE-POINT

See **Game theory**.

NATURAL RESOURCES

Andrés Weintraub

University of Chile, Santiago, Chile

INTRODUCTION: The field of natural resources covers various related areas: agriculture, fishing, forestry, mining. Though usually viewed separately, they share common problems, such as ecological concerns, use of scarce resources, and sustainability. There is also a common thread in what has happened in the last decade. On the one hand, driven in part by population growth and economic development, many natural resources are beginning to reach or exceed sustainable levels of exploitation, or in the case of non-renewable resources there are limits on known reserves. A second main issue is the new awareness of the need to preserve natural habitats, protect endangered species, provide water and air quality, and promote biodiversity. This has often led to serious conflicts between production goals and ecological impacts, with increased public participation in decision processes. A third basic issue is the emergence of global, competitive markets with the need to derive efficient production processes. In this context, operations research and management science have played a significant role in managing natural resources. We must distinguish between methodological proposals through case studies and actual applications. This is an issue of importance. Typically, these problems are often complex, involve uncertainty and consider multiple objectives. Also natural resources problems are often of large size and scope, with reliable data difficult to obtain. This partly explains the important gap that exists in some areas between modeling proposals shown through representative examples and actual use in planning and production processes. The introduction of personal computers, new information gathering systems, improved data processing, and algorithmic software plays a vital role in supporting the use of OR/MS. A wide range of problems has been approached using typical OR/MS techniques in each area. In some cases, their solution has led to algorithmic developments. It is interesting to analyze the nature of the main problems in each area and the techniques proposed for their solution, with linear programming and simulation being the techniques most commonly used across the areas. Forestry is a good example of how OR/MS has influenced decision making with a number of successful applications, as shown in Garca (1990), Kallio *et al.* (1986), Bare *et al.* (1984), and Dykstra (1984).

FORESTRY: Most of the forestry issues are related to the management of forests. Native forests, often publicly owned, are viewed as multiple use entities, considering timber and range production, recreation, wildlife habitat preservation, water and soil quality. Plantations, such as pine or eucalyptus, are usually privately owned, sometimes integrated with pulp and sawmill processing plants, with timber production as their main objective within legal preservation regulations. Decisions in forestry management go from long range planning to short range operations.

Long range planning, which, depending on the species under consideration, can go from 40 to more than 200 years to include up to two tree rotations, reflects basic silvicultural and economic options and in the case of plantations can include decisions on high level investments in plants. Mathematical tools have been used successfully in this area. For the purpose of predicting tree growth under different management alternatives, simulation models based on regression techniques and sampling plots have proved reasonably accurate. Decision making has been supported mostly by linear programming models. FORPLAN, a multiple output linear programming model (Kent *et al.*, 1991) developed by the US Forest Service is one of the better known. Other LP models have been developed by private enterprises oriented towards managing plantations and are used in the USA, Canada, Europe, New Zealand, Chile, Brazil, and Australia.

Medium range management decisions must consider spatial decisions, such as road building to access areas to be harvested. One major spatial issue that has emerged during the last decade is that of spatial location of activities. To favor wildlife habitat or scenic beauty, adjacent lots should be harvested in different periods to allow for new growth to establish itself, as some animals will graze only near cover provided by mature trees. This problem adds considerable combinatorial complexity to the planning problems, particularly when combined with road building. One form of solving this is to include the adjacency and road building issues explicitly into the models (Kirby *et al.*, 1986). Given the integer variables involved, this approach has only been applied to small cases or solved using heuristic approximations. New approaches based on column generation techniques adding lifting constraints Barahona *et al.* (1992) or applying Lagrangian relaxation have been proposed lately. In many applications, heuristic techniques have been used. Random search techniques, which basically consist of randomly choosing a large set of feasible solutions and then selecting the best among them have been widely proposed and used for the adjacency problem (Nelson and Brodie, 1990). Random search has been combined with heuristic rules and shortest path algorithms to solve jointly the adjacency and road building problem (Sessions and Sessions, 1991). Simulated annealing and tabu search have been also proposed to handle the adjacency problem (Lockwood and Moore, 1993).

Short term operations involve problems such as selection of units to cut, volume to harvest, selection of bucking patterns, log allocation, selection of harvesting equipment and transportation scheduling. A variety of models and algorithms have been proposed. Most used are linear programming based models for harvesting and allocation. For the more complex stem bucking problem, dynamic programming and heuristic algorithms have been proposed though its use has been mostly for training purposes. Scheduling of harvesting equipment such as towers, skidders or helicopters, has been carried out interacting with digital terrain models or geographic information systems in simulation models.

Given the multiple uses of forests, it is only natural to view forest management as a *multiple objective* problem, considering diverse issues such as timber and range production, recreation, scenic beauty, preservation of endangered species, wildlife habitat, water quality, costs, income and social impacts. The most common approach to handle this problem has been through goal programming or multiple objective linear programming (Bare and Mendoza, 1988). However, these developments have seldom been adopted by practitioners, mostly due to the difficulties in implementation.

The explicit treatment of *risk and uncertainty* is receiving increased attention of forest planners. The main issues related to uncertainty are in future timber markets, timber growth and yield projections and the possibility of catastrophes such as large fires or pests. Basic approaches proposed to handle uncertainty are: (a) parametric or scenario analysis; (b) probability-based models such as stochastic dynamic programming, portfolio theory, chance constrained linear programming and simulation (Hof *et al.*, 1992); and (c) fuzzy models, in which a certain ambiguity is assumed for restrictions or parameters (Bare and Mendoza, 1994). These efforts are mostly at a developmental stage with few applications reported.

Hierarchical Planning. Forestry problems range from decisions involving spatial concerns over 20 acres to entire forests of 2,000,000 acres, from short term horizons of a few months to long range planning over 150 or 200 years. Decision levels go from high level management to operations on the ground. At first, large scale monolithic models were proposed to solve global models. Given the difficulties in running and analyzing these models, several hierarchical decomposition approaches of global problems have been proposed to handle in a separate but linked way problems at different decision levels (Martell *et al.*, 1993).

Other OR/MS forestry applications are pest management, global trade analysis, and forest fire management and control, where OR/MS has been used in prevention of fires, fuel management, detection of fires, resources acquisition, initial attack dispatching, extended attacks management and training (Martell, 1982).

WILDLIFE MANAGEMENT: While it is very difficult to quantify wildlife dynamics, models have been developed to evaluate behavior of different species, based on aspects such as predator-prey relationships or habitat, climate and competition effects on food availability. Linear programming, decision analysis and mainly simulation are used for example to analyze control of lion populations, develop protection plans for antelopes, or to evaluate hunting policies for moose or deer (Starfield and Bleloch, 1986; Golden and Wasil, 1993).

AGRICULTURE: Mathematical models have been proposed extensively to deal with agricultural problems (Glen, 1987; Hazel and Norton, 1986; Kennedy, 1986). The main areas where quantitative approaches have been proposed are:

Crop Production Problems at Farm Level – These include the determination of cropping patterns, planning of harvesting operations, design of harvesting equipment, and control of pests and diseases. These interelated decisions include planting design, use of fertilizers, irrigation schemes and capital investment. The main techniques proposed to handle these problems are mainly linear programming and simulation, and also mixed integer programming, dynamic programming, and decision theory.

Uncertainty in crop yields and prices has also been introduced via portfolio theory, stochastic dominance, stochastic dynamic programming, and games against nature. Another important issue is that of multiple criteria, relevant in most areas of agricultural decisions, handled mostly using goal programming and compromise programming (Romero and Rehman, 1989).

These proposed models have not been applied intensively. One main reason is the farming tradition of using judgment based on experience, rather than looking for technical optimality which is also constrained by the lack of accurate information (Simon, 1982).

Regional Plannning Problems – These are oriented toward centralized decisions such as the evaluation of development projects, determination of tax or price support policies, or to analyze environmental impacts (Kutchner and Norton, 1982). Spatial market equilibrium models serve for analysis of domestic or international trade. These approaches, however, have mostly had indirect influence on practice or are of research interest only.

Livestock Production – In this area, mathematical models have been widely and successfully used (Glen, 1987). In the classical diet and ration formulation problems, a variety of models, mostly linear programming and also quadratic programming, have been proposed for different animal stocks. Simulation has been used for modeling pasture-based livestocks systems. The problem of livestock breeding and replacement has been approached through simulation, linear programming, deterministic and stochastic dynamic programming. Most applications are in the area of evaluation of replacement policies, particularly in large-scale dairy, egg production and poultry.

MINING: Quantitative techniques have played a significant role in the mining industry, accelerated in the last decade (Elbrond 1988; Lane 1988). Major decision problems in mining are:
- *The optimal design of open-pit mines*, to determine the feasibility of operations and the con-

tours in mining extraction processes, where extraction is viewed as a series of nested blocks in three dimensions, so as to maximize the difference between sale value and extraction and processing costs within geological and mining restrictions. Graph theory, linear programming, and heuristic methods have been used for this problem.
- *The geological modeling of ore bodies*, combining the classical geological approach with geostatistics, which is derived from the theory of spatial variation.
- *Mine operations*, both for underground and especially for the more complex open pit mines, which include decisions on capacity planning, machine scheduling, production planning, and transportation. Simulation, network flows, linear and nonlinear programming have been used for this problem.
- *Economic evaluation and optimization* of the rate of production and the cut-off grade.
- *Plant operation*, where problems of smelting, blending, mineral processing, and internal transport have been solved through simulation and optimal control.

New techniques are being introduced such as expert systems for ore reserve valuation and scheduling of production and maintenance, computer aided design and computer generated animation to simulate mine development under different alternatives and to support the introduction of robotics.

FISHING: Fisheries Management presents a different perspective from other natural resources in that since the resource is of free access, production is usually shared among different enterprises, and its allocation process is difficult and fuzzy. The long term conservation of fish stocks is a high priority issue and a set of regulations to protect them has been developed worldwide. Fisheries systems are concerned mainly with two basic issues (Lane, 1989): biological analysis of fish stock behavior and the allocation and exploitation of the resources.

Biological issues include all aspects of population dynamics to understand how fish stock evolves (growth and mortality rates, reproductive properties) and fish stock assessment, given environmental impacts (pollution, warming or cooling trends), stock interactions, and exploitation.

The problem of resource allocation involves assigning and regulating fishing rights (quotas, licenses, capture taxes, area closures). Exploitation or management decisions include: fleet design and harvesting operations, determination of catching effort (the response of fishing captures to fishing effort provides important information for stock assessment), design of fish plants.

Quantitative approaches have been widely proposed for all these problems: Descriptive mathematical modeling (in particular for the biological aspects), mathematical programming methods such as linear programming, nonlinear programming, optimal control and dynamic programming, statistical estimation and simulation. While the range of methodological proposals is wide, actual applications lag behind, mainly due to the lack of reliable data. Most applications are in the areas of exploitation and allocation (Golden and Wasil, 1993; and Guimaraes, 1990).

OR/MS methods have also been used to approach other areas of natural resources, including energy sources such as coal and oil. An extensive literature covers the field of water resources, with management problems in operating reservoirs, generating hydroelectrical energy, and designing irrigation shemes (Golden and Wasil, 1993; and Yeh, 1985).

See **Agriculture and food industry; Environmental systems analysis; Fuzzy sets; Global model; Goal programming; Linear programming; Multi-objective programming; Simulated annealing; Tabu search.**

References

[1] Barahona F., A. Weintraub and R. Epstein (1992). "Habitat Dispersion in Forest Planning and the Stable Set Problem," *Operations Research*, 40(1), S14–S21.

[2] Bare B.B., D.G. Briggs, J.P. Roise and G.F. Schreuder (1984). "A Survey of Systems Analysis Models in Forestry and the Forest Products Industries," *European Jl. Operational Research*, 18, 1–18.

[3] Bare B.B. and G.A. Mendoza (1988). "Multiple Objective Forest Land Management Planning: An Illustration," *European Jl. Operational Research*, 34, 44–55.

[4] Bare B.B. and G.A. Mendoza (1994). "A Fuzzy Approach to Natural Resource Management from a Regional Perspective," *International Transactions in Operational Research*, 1, 51–58.

[5] Dykstra D. (1984). *Mathematical Programming for Natural Resource Management*, McGraw-Hill, New York.

[6] Elbrond J. (1988). "Evolution of Operations Research Techniques in the Mining Industry" in *Computer Applications in the Mineral Industry* (K. Fytas, J.L. Collins, and R. Singhal, eds.). A.A. Balkema, Rotterdam.

[7] Garcia O. (1990). "Linear Programming and Related Approaches in Forest Planning," *New Zealand Jl. Forest Science*, 20, 307–331.

[8] Guimaraes, Rodrigues A., ed. (1990). *Operations Research and Management in Fishing*, Kluwer Academic Publishers, Dordrecht, The Netherlands.

[9] Glen J.J. (1987). "Mathematical Models in Farm Planning: A Survey," *Operations Research*, 35, 641–666.

[10] Golden B. and E. Wasil (1993). "Managing Fish, Forests, Wildlife and Water Successful Applications of Management Science and Operations Research Models to Natural Resource Decision Problems," in *Operations Research and Public Systems* (S. Pollock, A. Barnett and M. Rothkopf, eds.), Handbook in Operations Research and Management Science, 7.

[11] Hof J.G., B.M. Kent and J.B. Pickens (1992). "Chance Constraints and Chance Maximization with Random Yield Coefficients in Renewable Resource Optimization," *Forest Science*, 38, 305–323.

[12] Hazell P.B.R. and R.D. Norton (1986). *Mathematical Programming for Economic Analysis in Agriculture*, Macmillan, New York.

[13] Kallio M., A.E. Andersson, R. Seppala and A. Morgan, eds. (1986), "Systems Analysis in Forestry and Forest Industries," *TIMS Studies in the Management Sciences*, 21.

[14] Kennedy J.O.S. (1986). *Dynamic Programming: Applications to Agriculture and Natural Resources*, Elsevier Applied Science Publishers, London, England.

[15] Kent B., B.B. Bare, R.C. Field and G.A. Bradley (1991). "Natural Resource Land Management Planning Using Large-Scale Linear Programs: The USDA Forest Service Experience with FORPLAN," *Operations Research*, 39, 13–27.

[16] Kirby M.W., W.A. Hager and P. Wong (1986). "Simultaneous Planning of Wildland Management and Transportation Alternatives," *TIMS Studies in the Management Sciences*, 21, 371–387.

[17] Kutcher G.P. and R.D. Norton (1982). "Operations Research Methods in Agricultural Policy Analysis," *European Jl. Operational Research*, 10, 333–345.

[18] Lane D.E. (1989). "Operational Research and Fisheries Management," *European Jl. Operational Research*, 42, 229–242.

[19] Lane K.F. (1988). *The Economic Definition of Ore*, Mining Journal Books Ltd.

[20] Lockwood C. and T. Moore (1993). "Harvest Scheduling with Spatial Constraints: A Simulated Annealing Approach," *Canadian Jl. Forest Research*, 23, 468–478.

[21] Martell D., Davis L. and A. Weintraub, eds. (1993). *Proceedings of the Workshop on Hierarchical Approaches to Forest Management in Public and Private Organizations.* University of Toronto.

[22] Martell D.L. (1982). "A Review of Operational Research Studies in Forest Fire Management," *Canadian Jl. Forest Research*, 12, 119–140.

[23] Nelson J. and D. Brodie (1990). "Comparison of a Random Search Algorithm and Mixed Integer Programming for Solving Area-Based Forest Plans," *Canadian Jl. Forest Research*, 20, 934–942.

[24] Romero C. and T. Rehman (1989). *Multiple Criteria Analysis for Agricultural Decisions*. Elsevier, Amsterdam.

[25] Sessions J. and J.B. Sessions (1991). "Tactical Harvest Planning," in *Proceedings of 1991 Society of American Foresters National Convention*, SAF Publication 91-05, 362–368, Bethesda, Maryland.

[26] Simon H. (1982). *Models of Bounded Rationality: Economic Analysis and Public Policy*. The MIT Press, Cambridge, Massachusetts.

[27] Starfield A.M. and A.L. Bleloch (1986). *Building Models for Conservation and Wildlife Management*, Macmillan, New York.

[28] Yeh W.G. (1985). "Reservoir Management and Operations Models: A State of the Art Review," *Water Resources Research*, 21&22, 1797–1818.

NEAR-OPTIMAL SOLUTION

For an optimization problem, a near-optimal solution is a feasible solution with an objective function value within a specified range from the (usually unknown) optimal objective function value.

NEIGHBORING EXTREME POINT

In a convex set of solutions to a linear-programming problem, two extreme points are neighbors if they are connected by an edge of the convex set. The path of solutions determined by the simplex method is one that moves from one neighboring extreme point to another. See **Simplex method**.

NETWORK

A network is a pair of sets (N,A), where N is a set of nodes (points, vertices) and A is a set of arcs (edges, lines, links). If i and j are nodes, then the arc joining them is denoted by the ordered pair (i, j). An arc may have a cost c_{ij} that denotes the cost per unit flow across that arc, and an upper bound flow capacity denoted by u_{ij}. For some applications, a node may be a supply node in which goods enter the network, a demand node in which goods leave the network, or a transshipment node through which goods are shipped without a gain or a loss. In most network applications, it is assumed that the flow of goods that enter a node is equal to the flow that leaves the node. This is the conservation of flow assumption. However, in some applications, the amount of goods that enter a node can be more than the amount that leaves the node (e.g., due to the expansion of a liquid) or can be less than the amount that leaves a node (e.g., due to a leak or pilferage). These latter situations are termed networks with gains or losses, respectively. In most instances, network problems are special forms of linear-programming problems. See **Network optimization**.

NETWORK DESIGN

A decision problem concerning the configuration (the nodes and links to be included/excluded) of a logistics network. See **Network optimization; Network planning**.

NETWORK OPTIMIZATION

Thomas L. Magnanti

Massachusetts Institute of Technology, Cambridge

INTRODUCTION: Networks are familiar to all of us, as the highways, telephone lines, railways, electric power systems, airline route maps, and more recently, computer and cable television networks that we use constantly in our everyday lives. Networks also arise in other, perhaps less visible settings: for example, manufacturing or distribution networks determine the flow of products through plants or between plants, warehouses, and retail outlets; and networks of interconnected components in integrated semiconductor chips and printed circuit boards provide electronic processing capabilities in thousands of commercial products.

In these settings, we typically would like to resolve two sets of network optimization issues:

(1) *Operational Planning* – How do we use a given (distribution, telecommunication, or manufacturing) network as efficiently as possible? In this setting, the underlying network structure (topology and facilities) is known and we need to find the best way to route flow on it. For this reason, the set of optimization models for supporting these decisions have become known as *network flow problems*.

(2) *System Design* – What is the best design of a network, one that will offer cost efficient and yet effective service to its users? In this setting, we need to simultaneously create the network structure and route flow on it. These models have become generally known as *network design problems*.

Throughout the last five decades, the OR/MS community has developed a rich array of network models and solution methods for operational planning and system design, applying these techniques in thousands of applications. Indeed, network optimization has served as one of the most active and fertile application, modeling, and theoretical domains within the fields of applied mathematics, computer science, engineering, and OR/MS.

NETWORK MODELS: Figure 1 illustrates a typical application that contains the basic ingredients of network optimization. In this application context, which is typical of the automotive, computer and many other industries, a company produces product components in several plants/countries and assembles the products in other plants/countries. For convenience, we will refer to any product component as a commodity. Rather than shipping all commodities directly from each component plant to each assembly plant, to achieve economies of scale, the firm uses a set of intermediate distribution centers (or warehouses). The distribution centers could also hold inventory and thereby permit the company to meet fluctuating demand requirements in the assembly plants.

To formulate this problem mathematically, we first define an underlying network. In general, a network is (i) a set N of nodes, together with (ii) a set E of directed edges (i, j) that connect certain pairs i and j of the nodes. The application in Figure 1 has a special network structure with one node corresponding to an "input" and to an "output" for each plant and distribution center; the edges are of two types: those that connect plants and distribution centers and those that connect the input and the output node of each plant or distribution center. Each edge (i, j) has an associated per unit flow cost c_{ij}^k for each commodity $k = 1, 2, \ldots, K$, a flow capacity u_{ij}^k imposed upon commodity k, and flow capacity u_{ij} imposed upon the total flow of all commodities. For edges connecting plants and distribution centers, these quantities model the flow of commodities between the facilities. For edges joining the input and output nodes of each distribution center, these quantities model the throughput costs and capacities of the distribution center

(similarly, for the plants). The use of two nodes to model the throughput at any node is a common modeling device for representing node costs and capacities as edges costs and capacities.

The arrows in Figure 1 directed into the component plant input nodes specify the supplies of the commodities at the component plants. This model assumes that we have already determined the production of each component in each component plant. To use the network optimization model to allocate the production of each commodity among the component plants, we could introduce an additional component supply node s^k for each commodity k with the total supply of that commodity as the node's input. The flow on edges (s^k, q) connecting this node to the plants q would allocate the total supply of that component to the available plants (see Insert A in Figure 1). The introduction of additional nodes and edges like this is another modeling device used frequently in practice.

To model general network optimization problems (and thus various versions of the production and distribution planning problem), we let f_{ij}^k denote the flow of commodity k from node i to node j (i.e., the flow on edge (i, j) in the direction i to j). We also let b_i^k denote the net supply of commodity k at node i; this quantity is positive at the input nodes of the network (component plants in our example), is negative (to model demand) at the output nodes of the network (the assembly plants in our example), and is zero at all the other nodes. The model has the following general form:

$$\text{minimize} \quad \sum_{k=1}^{K} \sum_{(i,j) \in E} c_{ij}^k f_{ij}^k + \sum_{(i,j) \in E} F_{ij} y_{ij} \qquad (1)$$

$$\text{subject to} \quad \sum_{j:(i,j) \in E} f_{ij}^k - \sum_{j:(j,i) \in E} f_{ji}^k = b_i^k$$

$$\text{for all } i \in N \text{ and } k = 1, \ldots, K \qquad (2)$$

$$\sum_{k=1}^{K} f_{ij}^k \le u_{ij} y_{ij} \quad \text{for all } (i,j) \in E \qquad (3)$$

$$f_{ij}^k \le u_{ij}^k y_{ij} \quad \text{for all } (i,j) \in E \quad \text{and } k = 1, \ldots, K \qquad (4)$$

$$f_{ij}^k \ge 0 \quad \text{for all } (i,j) \in E \quad \text{and } k = 1, \ldots, K \qquad (5)$$

$$0 \le y_{ij} \le 1 \quad \text{and } y_{ij} \text{ integer for all } (i,j) \in E. \qquad (6)$$

This model has the following interpretation. The *flow conservation equation* (2) for node i states the total flow out of that node minus the total flow into that node must equal the node's supply. The binary (0 or 1) decision variables y_{ij} model the network design decision, "do we include edge (i, j) in the network ($y_{ij} = 1$) or not ($y_{ij} = 0$)." (In the application in Figure 1, these variables model two types of decisions: (i)

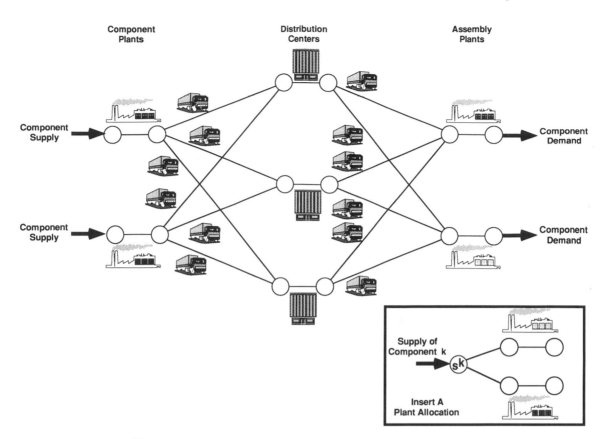

Fig. 1. Network model of a production/distribution system.

whether or not we locate a plant or a distribution center at one of the available locations and whether the network contains the corresponding throughput edge; and (ii) whether or not we use a particular transportation lane joining a plant to distribution center combination or distribution center to plant combination.) The fixed cost F_{ij} associated with edge (i, j) is the cost for constructing/renting/operating that edge (independent of its flow). The "forcing constraints" (3) and (4) force the flow on edges (i, j) for each commodity k to be zero if the network does not contain that edge ($y_{ij} = 0$). If $y_{ij} = 1$, constraint (3) states that the total flow on edge (i, j) cannot exceed the installed capacity u_{ij} of that edge and constraint (4) states that the flow of commodity k on edge (i, j) cannot exceed the flow capacity u_{ij}^k for that commodity.

This model assumes that the edges are directed – that is, we can flow on any edge in only one direction. But, in some applications, the edges will be undirected. To model these situations, we impose the condition that $y_{ij} = y_{ji}$ and replace f_{ij}^k in constraints (3) and (4) with $f_{ij}^k + f_{ji}^k$, the total flow in both directions on edge (i, j).

Note that we can use the network optimization model (1)–(6) to make facility location decisions, as we have in our example. We can also use it to make routing decisions: the choice of transporta-tion lanes. Moreover, we can use the model in telecommunication and other applications to design a physical network. For example, we can use the edge decision variables to determine where to locate fiber optic cables in a telecommunication system.

The model (1)–(6) is a special type of mixed integer programming problem (i.e., it contains both continuous and integer/binary variables). In practice, solving this network optimization problem is a challenging (on the surface almost daunting) task. The model has a flow conservation equation for each node and commodity. Since networks with thousands of nodes arise frequently in practice, even for situations with a single commodity, the model often has thousands of equations. Many telecommunications and transportation applications require the flow of a commodity (message or freight) between every pair of nodes in the network. Therefore, with as few as one hundred nodes, the problem will have $100*99 \approx 10,000$ commodities and $10,000*100 = 1$ million flow conservation equations (one for each combination of commodity and node).

The problems become even more difficult when they have design variables. With as few as twenty nodes and a binary design variable y_{ij} for each of the $20(19)/2 = 190$ possible edges connecting these nodes, the model has 2^{190} different design alternatives (since any design can include or exclude each of the 190 edges). This number is as large as the number of grains of sand needed to fill the solar system! Therefore, solving these problems requires considerable ingenuity; since it is impossible to enumerate all possible solutions, the methods must consider them only implicitly.

TYPES OF MODELS: The network optimization model (1)–(6) has many specializations and variants, each generating a considerable literature on its own (applications, solution methods, and underlying theory). Tables I and II show some of these models and indicate typical solution times for solving them (on a modern computer workstation).

The tables separate the models into two categories:

(1) *Network flow models* – For these models, each binary variable y_{ij} is fixed at value 0 or 1 (and so the network topology is fixed) and the problem becomes a linear program. Notice that the problem has a very special structure since each flow variable f_{ij}^k appears in exactly two flow conservation equations: as an output of node i and an input of node j. Researchers have been able to use this feature to develop special purpose algorithms that solve the problems much more efficiently than solving them using general purpose linear program software.

(2) *Network design problems* – In these models, both design decisions and flow decisions are relevant. For some of these models, the flow costs are zero and so the problems become that of finding a least cost network configuration that meets the required flow requirements.

Figure 2 gives examples of *minimum spanning tree* and *Steiner tree problems*, assuming that the cost of each edge is proportional to its length. The underlying network in these examples is typical of those in printer circuit board applications that have East-West and North-South "channels" for making wiring connections (therefore, all the edges are in a rectangular pattern). Note that the minimum spanning tree needs to connect all the nodes and the Steiner tree needs to connect only a subset of the nodes (so called terminal nodes), but can optionally use some of the other nodes (so called Steiner nodes). In both cases, we wish to find the least cost network configuration, as measured by the total cost of the chosen edges (flow costs are irrelevant). The other network design problems also wish to find optimal network configurations, but might include flow costs as well.

Our discussion has shown that the network optimization problem (1)–(6) has a wide range of applications and has shown how to introduce

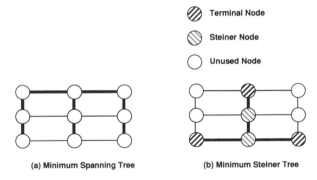

Fig. 2. Minimum spanning tree and Steiner tree problems.

Table I. Network flow models (each y_{ij} is fixed at value zero or one).

Model type	Problem description	Solution methods	Computational experience. Number of nodes: Solution time
Multicommodity flows	General flow model (1)–(6) with multiple commodities	Linear programming, decomposition methods	Hundreds: minutes
Minimum cost flows	Single commodity ($K = 1$)	Specialized path flow methods	Thousands: seconds
Maximum flows	Single commodity, no flow cost; send maximum flow between single source and destination node pair	Specialized node labeling (sequential search) methods	Thousands: seconds
Shortest paths	Single commodity, single origin, no flow capacities	Specialized node labeling (sequential search) methods	Tens of thousands: seconds

Table II. Network design models.

Model type	Problem description	Solution methods	Computational experience. Number of nodes: Solution time
Fixed cost network design	General model (1)–(6)	Integer programming, heuristics	Tens: minutes or hours
Network loading	No flow costs, load network to meet required point to point demands (y_{ij} as integers, not binary)	Integer programming, heuristics	Tens: minutes or hours
Network connectivity	Find prescribed number of edge disjoint paths between various node pairs	Integer programming, heuristics, and linear programming dual ascent methods	Hundreds: minutes
Network synthesis	Given flow requirements, determine capacities on edges (at minimum cost) so that the network has the capability to meet prescribed demands between various node pairs	Minimum spanning tree if capacity on every edge costs the same	Thousands: seconds if all capacities have same cost Tens to hundreds: minutes (in general)
Minimum spanning trees	No flow costs, no capacities; find a network that connects all nodes	Specialized one pass "greedy" algorithms	Thousands: seconds
Steiner trees	No flow costs, no capacities; find a network that connects prescribed set of nodes (and possibly others)	Heuristics and linear programming dual ascent methods	Thousands: seconds

Note: For all network design problems, except the minimum spanning tree problem, the methods generally produce approximately optimal, not optimal solutions.

additional nodes and edges to enhance the model's ability to capture varied application features (e.g., to allocate production at the component plants or to represent the throughput of a node). Many other such modeling techniques have proven to be useful in practice. As indicated in Table I, computer software is available for solving large scale network flow problems very quickly. Except for spanning tree and Steiner trees, capabilities for solving network design problems is much more limited.

SOLUTION METHODS: Solving network optimization problems requires considerable ingenuity, in developing solution methods, implementing them efficiently on a computer, and analyzing them to determine their efficiency (in theory or practice). To illustrate these issues, we consider one of the easiest network flow problems, *the shortest path problem*. After describing

a basic algorithm for solving this problem, we show how to organize the computations to implement the algorithm more efficiently and then how to improve on it even further when we know more about the underlying data (the edge lengths).

Suppose we are given a network with nonnegative lengths d_{ij} on the edges (i, j) and wish to find the shortest path between two designated nodes, a source node s and a terminal node t. To solve the problem, we could use the following algorithm (solution method). If node j is the node closest to the source node, then the shortest path distance $d(j)$ from the source node to this node is the direct path on the edge (s, j) whose distance is d_{sj}. Next consider the node k that is closest to node s either along the direct edge (s, k) or on the shortest path through node j, that is, with the distance $d_{sj} + d_{jk} - d(j) + d_{jk}$. To choose the best of these two alternatives, we compute $d(r) = \min\{d_{sr}, d(j) + d_{jr}\}$ for each node $r \neq s$ or j and select as the next node k, a node r with the smallest value of $d(r)$. It is easy to see that this choice gives the shortest distance along any path from node s to node k. In general, suppose that after several steps, we have found the shortest path distances $d(j), d(k), \ldots, d(p)$ from the source node s to each of the nodes $j, k, \ldots, p$. Then to find the shortest path distance to the next node q, for all nodes $r \neq s, j, k, \ldots p$, we compute

$$d(r) = \min\{d_{sr}, d(j) + d_{jr}, d(k) + d_{kr}, \ldots, d(p) + d_{pr}\} \tag{7}$$

We choose node q to be a node r with the smallest of the values $d(r)$. Once we have chosen node t in any of these steps, we have solved the problem, that is, found the shortest path distance from node s to node t. (See Ahuja *et al.*, 1993, for a proof.)

This algorithm computes the shortest path distance to one more node at each step. If the network contains a total of n nodes and we have found the shortest path distance to v of them, then the computation (7) requires v additions and comparisons for each of $n - v$ nodes and so $v(n - v)$ computations. Therefore, to find the shortest path distance to all nodes, the algorithm requires $1(n-1) + 2(n-2) + 3(n-3) + \cdots + (n-1)(1) = n^2(n-1)/6$ computations. Can we do better? We can if we notice that this algorithm performs many redundant computations. For example, after the first step, for each node r not yet chosen in one of the previous steps, the algorithm computes the quantity $d(j) + d_{jr}$. Note that after we have chosen node q, the computation (7) becomes

$$d^{\text{new}}(r) = \min\{d_{sr}, d(j) + d_{jr}, d(k) + d_{kr}, \ldots,$$
$$d(p) + d_{pr}, d(q) + d_{qr}\}. \tag{8}$$

Comparing (7) and (8) shows that $d^{\text{new}}(r) \leftarrow \min\{d(r), d(q) + d_{qr}\}$. Therefore, if we store the values of $d(r)$ from one step to another and use the computation

$$d(r) \leftarrow \min\{d(r), d(q) + d_{qr}\}, \tag{9}$$

then at step v when we have $n - v$ nodes that are candidates to choose next, we make $v - r$ computations and so overall the algorithm now requires only $(n-1) + (n-2) + \cdots + 1 = n(n-1)/2$ computations. As this simple example shows, by organizing the computations intelligently, we can often considerably reduce the computational requirements of an algorithm. Much of the literature of network flow algorithms involves the use of similar ideas for designing and analyzing

algorithms (though, in general, the ideas are much more complex).

To illustrate further how researchers have used problem structure to design efficient algorithms, suppose that the cost structure for our shortest path problem were even simpler, such that each distance d_{ij} has a limited value; for concreteness assume it has value 1 or 2. Notice that in this case, the shortest path distance from the source node s to any node k is one of the integers $1, 2, \ldots, 2(n - 1)$. To obtain an improved algorithm, we will use this fact and implement the computations (9) in a more streamlined fashion. We maintain a collection of $2(n - 1)$ "buckets," storing all the nodes r whose distance $d(r)$ is k in the kth bucket. Then we choose the buckets one at a time from smallest to largest, starting with bucket number 0. If the bucket is nonempty, we select a node q from it and then any edge (q, r) incident to node r. We then use the expression (9) to update the distance $d(r)$ of node r and if the distance of node r decreases, we move it to a new bucket. Note that this algorithm considers each edge (i, j) only once and must search at most $2(n - 1)$ buckets to see if they are empty or not and to extract their contents. Therefore, for a network with m edges, this implementation of the algorithm requires $m + 2(n - 1)$ computations, since m is often far less than its maximum possible value of n^2, this algorithm is typically much faster than the implementation embodied by the previous implementation of the computations (8). When the edge lengths are limited within some range $0 \leq d_{ij} \leq C$ for some constant C, a similar type of bucket implementation can be very efficient and produces some of the most effective algorithms for solving shortest path problems.

The design and analysis of network optimization algorithms has an enormous literature. This brief introduction to the topic has illustrated several important aspects of this field:

- Network algorithms often use simple computations (like those invoked in expressions (7) and (9)) for solving problems rather than the more sophisticated methods needed to solve other optimization problems such as general linear programs. Indeed, software based upon specialized methods like these are able to solve shortest path problems with thousands of nodes in just a few seconds of computational time (see Table I) even though these problems are linear programs with thousands of constraints (one conservation equation in the model (1)-(6) for each node).

- In solving a particular problem, it is often just as efficient to solve a broader class of problems (the algorithm we have described finds the shortest paths from the source node to all other nodes, not just the terminal node).
- Organizing computations carefully can improve an algorithm's efficiency. In our shortest path example, we reduced the number of computations from $n^2(n-1)/6$ to $n(n-1)/2$ by merely avoiding redundant computations [see (8) and (9)].
- The creative use of data structures (for our example, buckets) often leads to more efficient algorithms [$m + 2(n-1)$ instead of $n(n-1)/2$ computations for our example].
- Often algorithms can be designed to exploit the nature of the data and not just the type of problem being solved. We have seen two illustrations of this fact: (i) the algorithm we have described might not solve a shortest path problem when some the edge lengths are negative, and so it has exploited the fact that all the edge lengths are nonnegative; and (ii) when the data have restricted ranges (e.g., the edge costs are between 0 and C in our discussion), we frequently can devise more efficient algorithms.

FURTHER READINGS: Several books amplify on the topics we have considered and introduce many other applied and theoretical aspects of network optimization. Ford and Fulkerson (1962) provides a seminal account of early developments in this field. Ahuja, Magnanti and Orlin (1993) offer a modern treatment of this subject covering both theory and applications. Glover, Klingman and Phillips (1992) provide valuable insight into network modeling and applications. The handbooks edited by Ball, Magnanti, Monma, and Nemhauser (1995) contain comprehensive reviews by many leading researchers in network optimization. Lawler (1976) draws valuable connections between network flows and a related topic in combinatorial optimization known as matroids.

See **Combinatorial and integer optimization; Facility location; Linear programming; Location analysis; Maximum-flow problem; Minimum-cost network flow problem; Multicommodity network-flow problem; Shortest-path problem; Steiner tree.**

References

[1] Ahuja R.K., T.L. Magnanti and J.B. Orlin (1993), *Network Flows: Theory, Algorithms and Applications*, Prentice Hall, Englewood Cliffs, New Jersey.

[2] Ball M., T.L. Magnanti, C. Monma and G.L. Nemhauser (1995), *Network Models*, vol. 7, Handbooks of Operations Research and Management Science, Elsevier, New York

[3] Ball M., T.L. Magnanti, C. Monma and G.L. Nemhauser (1995), *Network Routing*, vol. 8, Handbooks of Operations Research and Management Science, Elsevier, New York.

[4] Ford L.R. and D.R. Fulkerson (1962), *Flows in Networks*, Princeton University Press, New Jersey.

[5] Glover F., D. Klingman and N. Phillips (1992), *Network Models in Optimization and Their Applications in Practice*, John Wiley, New York.

[6] Lawler E.L. (1976), *Combinatorial Optimization: Networks and Matroids*, Holt, Rinehart and Whinston, New York.

NETWORK PLANNING

Graham K. Rand

Lancaster University, United Kingdom

Network planning (or analysis) is a generic name for techniques that present information to assist in planning and controlling projects. These techniques, such as the *Critical Path Method* (*CPM*) and the *Program Evaluation and Review Technique* (*PERT*), require the determination of each of the activities that are involved in the project, the sequence in which these activities must be performed and the activities that can be performed concurrently with other activities. Network planning can deal with problems involving time minimization, resource limitations and cost minimization.

HISTORY: The need to improve planning techniques to help control major projects was recognized in the 1950s. CPM arose from a jointly sponsored venture of E.I. du Pont de Nemours and Company and the Sperry-Rand Corporation. By September 1957, an actual application was conducted on a pilot system using the UNIVAC I computer, and from this CPM evolved (Kelley and Walker, 1959; Kelley, 1961). At the same time, the U.S. Navy was developing a system to plan and coordinate the Polaris missile program. From this PERT evolved, and was credited with helping to advance Polaris by at least two years (Malcolm, Roseboom *et al.*, 1959). Further details of these early developments can be found in Moder, Phillips *et al.* (1983) and Sculli (1989).

Since these early days, variations on these methods have been developed (*VERT*, *GERT* and *SCERT*), and research in this field is still active. In 1986, the EURO Working Group on Project Management and Scheduling was established to assess the state of the art and to stimulate progress in theory and practice. As a result workshops have been held every two years, and three special issues of the *European Journal of Operational Research* (volumes 49(1), 1990; 64(2), 1993; and 78(2), 1994) and Slowinski and Weglarz (1989) have been published.

TIME MINIMIZATION: *Construction of the Network* – The network, or *arrow diagram* (also, and more accurately, called activity-on-arc diagram) is designed to represent the entire work content of the project. It consists of all the activities comprising the project, identifies all the *events* associated with various degrees of project completion, that is, completion of one or more activities, and indicates the inter-relationships and inter-dependencies of activities and events.

Activities are usually represented by arrows, the direction of which indicates the flow of time (but the length of the arrow is not related to a time scale), and the relationship between the immediately preceding event (i) and the succeeding event (j). The activity is then known as activity (i, j). The events are represented by *nodes*.

Dummy activities are sometimes required to maintain network logic or to maintain a convention that no two activities may span the same pair of nodes (they are represented by a dashed arrow). They consume no time, require no resources, and require no expenditure of effort.

The procedure for construction of a network is:

1. Produce an activity list.
2. Determine the sequence in which the activities are to be performed and the inter-relationships between them. For each activity ask –
   a. which activities must precede it,
   b. which activities cannot start till it is complete.
3. Draw the network so that these relationships, and only these, are implied. The network is drawn so that there is only one initial event and one terminal event. Events in general should be numbered in such a manner that any preceding event has a lower event number than any succeeding event number. (This requirement is mandatory for some computer solutions.)

The procedure for the analysis of the network is:

1. Determine the durations of each activity.
2. Determine the earliest start times for each activity.
3. Determine the latest start times for each activity.
4. Calculate the float for each activity. The total float of an activity can be classified into free float, independent float, and interfering float. An activity which has free float can more readily be arranged to be performed at different times than one which is subject only to interfering float, as the effect on succeeding activities need not be considered.
5. The critical path consists of those activities for which the total float is zero, in the situation when a date for the scheduled finish has not been specified. The activities which comprise this path are critical activities.

Activity on Node – Although arrow diagrams are still the most widely used system for network planning an alternative system was independently developed by Roy (1964), and called by him the "method of potentials," and Fondahl (1961). The essence of the system is that activities are represented by nodes, and logical links are represented by arrows. Variants of this system, now more commonly called "activity on node" or "precedence diagramming" systems are gaining ground on arrow diagrams and it is claimed that they possess a number of advantages; for example,

1. they are easier to learn: (e.g., dummies are not required);
2. they are more compact;
3. it is easier to make changes to the network; and
4. they are unique (whereas there may be alternative ways to represent networks using the activity on arc mode of presentation, because of the presence of dummy activities).

Project control – Once the project has started, control is maintained by a system of status reporting. Some activities will take longer than estimated and some shorter. Sometimes estimates for activities not yet completed require revision. At regular intervals the network must be updated, re-analyzed and new schedules prepared, taking into account all this new information.

Uncertain Durations – It is sometimes impossible to estimate durations (e.g. in projects of an exploratory nature, as in R&D work). It is sometimes inexpedient to estimate durations (because a "contingency allowance" or "fiddle factor" would be included). The method for overcoming this problem adopted by PERT is the use of three time estimates: an optimistic estimate (t_0), an estimate of the most likely duration (t_m), and a pessimistic estimate (t_p). An approximation to the expected time can be found by taking a weighted average of the three figures (in the ratio 1:4:1) which has a standard deviation

$$\sigma_t = (t_p - t_0)/6.$$

The assumption, made by PERT, is that the expected time has a beta distribution, so that

$$f(t) = k(t - t_0)^\alpha (t_p - t)^\beta \quad (t_0 < t < t_p),$$

where α and β are "shape" parameters.

The PERT approach has been widely criticized on theoretical grounds. In the PERT system, the mean value for the project duration is taken as the sum of the mean values of the durations of the activities on the critical path. This assumption is only correct for a project which consists of a single chain of activities, but progressively underestimates the mean project duration as the complexity of the network increases. In the PERT system the variance for the project duration is taken as the sum of the variances of the actitivies on the critical path. Again, this assumption is only correct for a project which consists of a single chain of activities, but progressively underestimates the variance as the complexity of the network increases. The formula for expected time gives the true mean only for particular values of α and β. There have been serious doubts expressed as to the appropriateness of the beta distribution. The PERT approach does not take into account the probability of completion of sub-critical paths. This inability to take account of sub-critical paths is the most serious criticism of PERT. Because of this PERT typically underestimates the "true statistical project mean duration" by up to 30% and also seriously underestimates the probability of meeting a deadline. The only satisfactory way of generating these measures is by simulation and program packages are available for this purpose. Elmaghraby (1977) and Golenko-Ginzburg (1989) provide further discussion on these points.

Paradoxically, PERT works very well in practice and the reason for this is not hard to find. The power of network techniques is their ability to detect deviations from plan. Their use ensures that better control of projects is maintained. Regular monitoring of ongoing projects quickly reveals deviations from plan and enables appropriate action to be taken to bring the project back to its planned completion time. Thus the "true statistical project mean duration" is a meaningless concept. There is no point in simu-

lating the effects of random influences, like the weather, on project preformance unless we also include in our simulation the control measures which can be adopted in response to these random actions. Project planning using PERT/CPM can be highly successful (as also proved in practice) even in situations involving uncertainty.

COST OPTIMIZATION: While it may be argued that the critical objective for important projects in industrialized countries during the relatively affluent 1950s and 1960s was the minimization of their duration, more recently project management has been carried out in a time of tight resources and stringent financial constraints. As Tavares (1987) has pointed out, the objective for the project management team may well be to maximize "the net present value of a project in terms of its schedule under eventual restrictions concerning its total (or partial) duration."

Some, or all, of the activities in a network may be carried out at various alternative levels of cost with correspondingly different durations. It is usually assumed that the cost/duration relationship is linear between a normal and a *crash duration* and that any intervening duration may be attained. The objective is, by selection of activity durations and their corresponding costs, to minimize total activity costs for a given project duration. A related problem is concerned with detecting the shortest project duration available within a given budget.

The solution to the problem of reducing the project duration (by some specified amount) at the lowest possible cost is quite elementary whenever the desired project duration is equal to or greater than the second longest path. All that is required in this case is to rank the activities on the critical path in order of ascending *cost slope*, shorten the lowest cost slope activities as much as possible, and continue to shorten the progressively higher cost slope activities until the critical path has been shortened by the required amount.

However, when the desired project duration is less than the second longest network path, it is necessary to shorten not only the critical path, but one or more sub-critical paths as well. Now the problem solution may well be considerably more difficult and a systematic procedure is required to achieve the goal of cost optimization. The problem may be formulated as a linear program providing the cost/duration relationship is assumed to be linear, and solved by the simplex method, but a more efficient, network flow algorithm has been developed by Fulkerson (1961). The algorithm, although straightforward and very suitable for computer applications is rather lengthy for hand operations. Other approaches are described by Ritchie (1985).

RESOURCE ALLOCATION: Frequently, *resource* requirements are estimated for each activity. Resources are usually labor, machinery of various types, and money, particularly important when "cash flow" is of concern to the project management. Each activity may require several units of one or more types of resource. A summary of the literature on this subject can be found in Boctor (1990).

Analysis when resource restrictions are operative means using a computer for all but trivial networks. All types of resource analysis start by carrying out the event and activity time calculations. The simplest type of analysis then assumes all activities are scheduled according to some common rule, for example, at their earliest start times, and accumulates the total units of each resource required in each period; this is called *resource aggregation* or accumulation.

More complex methods are used to schedule activities taking into account their resource requirements. These methods are used either to smooth out the resource requirements as far as possible while maintaining the project duration at its minimum or to keep the project duration as short as possible while not allowing resources to rise above predetermined levels. The first approach is called time-limited scheduling or *resource smoothing* and the second resource-limited scheduling or *resource levelling*. It is often helpful to represent the results of such scheduling on a *Gantt* or *bar chart*.

There are two main approaches that have been considered for resource constrained problems: mathematical programming (Pritsker, Walters, *et al.*, 1969; Patterson and Huber, 1974; Talbot and Patterson, 1978), and heuristic methods (e.g., the SPAR-1 model of Weist, 1967, and Davies, 1973). It is still the case that realistic problems of this nature cannot be solved easily, despite recent advances in discrete optimization (Patterson, 1984), and, therefore, several simplifying assumptions are usually made. These reduce the flexibility of the models, but do not always increase their usefulness, as can be seen by consideration of the absence of procedures for scheduling optimization from commercial software (Tavares, 1989).

COMPUTERS: Computers are widely used for network analysis. The calculations involved are not complex but for projects of any significant size the amount of data processing required can be large. Microcomputers are increasingly being used for network analysis, and many software packages are available for doing routine time analysis and presenting the results. Such packages enable planners to update their plans quickly and effectively. Many packages are relatively unsophisticated, for example in dealing with resource constrained problems, but progress is being made. Surveys have been published by Wasil and Assad (1988) and de Wit and Herroelen (1990), but these will become rapidly out of date because of the rapid rate of developments of these packages. Increasingly, attention is being given to the application of expert systems in network analysis (Probst and Worlitzer, 1988).

See **Critical path method; Gantt charts; Networks; Program evaluation and review technique; Project management; Scheduling and sequencing.**

References

[1] Boctor F.F. (1990). "Some efficient multi-heuristic procedures for resource-constrained project scheduling." *European Journal of Operational Research* 49, 3–13.

[2] Davies E.M. (1973). "An experimental investigation of resource allocation in multiactivity projects." *Operational Research Quarterly* 24, 587–591.

[3] de Wit J. and W. Herroelen (1990). "An evaluation of microcomputer-based software packages for project management." *European Jl. Operational Research* 49, 102–139.

[4] Elmaghraby S.E. (1977). *Activity Networks: Project Planning and Control by Network Models.* Wiley, New York.

[5] Fondahl J.W. (1961). *A noncomputer approach to the Critical Path Method for the construction industry.* Dept. of Civil Engineering, Stanford University, Stanford, California.

[6] Fulkerson D.R. (1961). "A network flow computation for project cost curves." *Management Science* 7, 167–178.

[7] Golenko-Ginzburg D. (1989). "PERT assumptions revisited." *Omega* 17, 393–396.

[8] Kelley J.E. (1961). "Critical-path planning and scheduling: mathematical basis." *Operations Research* 9, 296–320.

[9] Kelley J.E. and M.R. Walker, eds. (1959). "Critical-path planning and scheduling." In *Proceedings of Eastern Joint Computer Conference,* Boston, December 1–3, 1959, 160–173.

[10] Malcolm D.G., J.H. Roseboom, C.E. Clark and W. Fazar (1959). "Application of a technique for research and development program evaluation." *Operations Research* 7, 646–669.

[11] Moder J.J., C.R. Phillips and E.W. Davis (1983). *Project Management with CPM, PERT and Precedence Diagramming*. Van Nostrand, New York.

[12] Patterson J.H. (1984). "A comparison of exact approaches for solving the multiple constrained resource, project scheduling problem." *Management Science* 30, 854–867.

[13] Patterson J.H. and W.D. Huber (1974). "A horizon-varying, zero-one approach to project scheduling." *Management Science* 20, 990–998.

[14] Pritsker A.A.B., L.J. Walters and P.M. Wolfe (1969). "Multi-project scheduling with limited resources: a zero-one programming approach." *Management Science* 16, 93–108.

[15] Probst A.R. and J. Worlitzer (1988). "Project management and expert systems." *International Jl. Project Management* 6, 11–17.

[16] Ritchie E. (1985). "Network based planning techniques: a critical review of published developments." In *Further Developments in Operational Research* (G.K. Rand and R.W. Eglese, eds.), 34–56. Pergamon, Oxford.

[17] Roy B. (1964). "Contribution de la theorie des graphes a l'etude des problems d'ordonnancement." In *Les problems d'ordonnancement: applications et methodes* (B. Roy, ed.), 109–125. Paris, Dunod.

[18] Sculli D. (1989). "A historical note on PERT times." *Omega* 17, 195–196.

[19] Slowinski R. and J. Weglarz (1989). *Advances in Project Scheduling*. Studies in Production and Engineering Economics. North Holland, Amsterdam.

[20] Talbot F.B. and J.H. Patterson (1978). "An efficient integer programming algorithm with network cuts for solving resource-constrained scheduling problems." *Management Science* 24, 1163–1174.

[21] Tavares L.V. (1987). "Optimal resource profiles for program scheduling." *European Jl. Operational Research* 29, 83–90.

[22] Tavares L.V. (1989). "A multi-stage model for project scheduling under resource constraints." In *Advances in Project Scheduling* (R. Slowinski and J. Weglarz, eds.), 315–326. Amsterdam, The Netherlands, Elsevier Science Publishers.

[23] Wasil E.A. and A.A. Assad (1988). "Project management on the PC: software, applications, and trends." *Interfaces* 18(2), 75–84.

[24] Weist J.D. (1967). "A heuristic model for scheduling large projects with limited resources." *Management Science* 13, B369–B377.

NETWORK SIMPLEX ALGORITHM

For the minimum-cost network-flow problem, a special adaptation of the simplex method that takes advantage of the mathematical structure of the network constraints to produce a computationally fast and efficient solution algorithm.

The main idea behind this algorithm is the recognition that a basic feasible solution to the network problem, treated as a linear-programming problem, corresponds to a spanning tree of the defining network. See **Minimum-cost network-flow problem**; **Network optimization**; **Simplex method**.

NETWORKS OF QUEUES

Ralph L. Disney

Texas A&M University, College Station

INTRODUCTION: Studies of queueing networks are about as old as any studies of queueing. The field has progressed steadily and is well developed in some aspects, notably the queue length problem. However, there are still unresolved issues, as in the study of the sojourn time problem.

There is a large and growing literature about network problems using an extensive collection of methods to study them. Computer simulation is probably the most often used, but one finds analytical studies using martingales or Petri nets. Approximations of many kinds have been suggested including diffusion, as well as numerical estimates.

We concentrate here on the queue length processes and the quasi-reversible queues which have product-form solutions. To help explain what is occurring in these networks, we discuss the sojourn time problem and traffic processes. A short review of single server networks is also given.

Basic Notation – Many studies of queueing networks are based on steady-state Markov process theory. We let $X(t)$ be a Markov process; states for such processes are vectors $x = (x_1, x_2, \ldots x_J)$ where J is the number of nodes in the network. The x_k are integers which in the simpler models represent the number of customers at node k. In more complicated models, one needs a more general state space to retain the Markov property. In such cases we shall use x generically. In all cases, the generator for the Markov process will be Q with elements $q(x, x')$. We deal only with steady-state results, which are assumed to exist. If π is the vector of steady-state probabilities for $X(t)$, $\pi Q = 0$ gives the usual steady-state equations called the *global balance equations*.

JACKSON NETWORKS: A *Jackson network* is one of the simplest cases of a queueing network and forms the basis on which many subsequent models are based. We discuss it here to illustrate some concepts which are generalized later.

Arrivals and Departures – We assume that the network consists of a finite number, J, of single server nodes. At each node j, service times are sequences of independent, identically distributed (i.i.d.) exponential random variables with parameter $\mu_j \phi_j(n_j)$ that depends on the node, j, and possibly the queue length n_j at that node. (Multiserver nodes in more general Jackson networks also have such parameters.) The service time processes are independent of each other. Each node has an infinite capacity queue and service disciplines are all FCFS.

External arrivals may occur to node j and if they do, they form a Poisson process with a parameter v_j that depends only on node j. The arrival processes are independent of each other and of the service processes.

States of the Process – The states of the Jackson networks are vectors $\boldsymbol{n} = (n_1, n_2, \ldots, n_J)$ where n_k for $k = 1, 2, \ldots, J$, is the queue length at node k where, as usual, this queue length includes any customers at the server. If the network is *open*, we can append a node 0 to the set of states and call this the outside of the network. For closed networks we do not need an outside. The state space, E, of the Markov process is then the set of vectors $\boldsymbol{n}$ with 0 appended if necessary.

Routing – Customers move from node to node and, perhaps, into and eventually out of the network. Several schemes to do this are possible. *Markov routing* is probably assumed most often. Others will be discussed in later sections.

Define a matrix P' with elements $p(j, k)$, $j, k = 1, 2, \ldots, J$ where, given a customer is exiting node j, the customer proceeds next to node k with probability $p(j, k)$. Let P' be augmented with $p(j, 0) \geq 0$ and $p(0, k) \geq 0$ representing the probability that a customer leaves the network from node j and a customer enters the network through node k, respectively. Call the augmented matrix P. [This is the same as having an overall Poisson arrival process to the network with parameter and having the arrival stream to node k be a Poisson process with parameter $v_k = vp(0, k)$.]

Let $\boldsymbol{R}(t)$ be an irreducible, positive recurrent, and aperiodic Markov chain on the nodes of the network with one-step transition matrix P. Then $\boldsymbol{R}(t)$ is called a *routing process* and the network is said to be an *open network*. A diagram representing a possible open network configuration appears as Figure 1.

When for all j and k, $p(j, 0)$ and $p(0, k)$ are 0, and if the routing process $\boldsymbol{R}(t)$ is an irreducible,

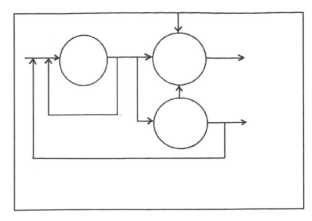

Fig. 1. Possible open network configuration.

positive recurrent, Markov chain on the nodes of the network, the network is *closed*. For such networks, there is a fixed number of customers endlessly circulating in the network.

There may be cases where the network is open for some customers but closed for others (i.e., the Markov chain is not irreducible). These are *mixed networks*.

Steady-State Distribution of the Process – The common measure of effectiveness in these Jackson networks is the steady-state probability vector of queue lengths. Let $\pi(\boldsymbol{n})$ be the steady-state vector of the queue lengths. Then it follows for single-server, open networks that

$$\pi(\boldsymbol{n}) = \prod_{j=1}^{J} (1 - \rho_j)\rho_j^{n_j} \tag{1}$$

where $\rho_j = \lambda_j / \mu_j < 1$ for $j \in E$, is the traffic intensity for node j. Here, the $\{\lambda_j\}$ satisfy the *traffic equations*

$$\lambda_j = v_j + \sum_{k=1}^{J} \lambda_k p(k, j) \quad (j = 1, 2, \ldots, J), \tag{2}$$

since the total flow into a node from all sources (right-hand side) must equal the flow out (left-hand side). This $J \times J$ linear system is guaranteed to have full rank for any legitimate routing matrix.

There are several observations about Jackson networks that are useful more generally. We list them here and refer to them as needed.

- Defining a vector $\boldsymbol{n} = (n_1, n_2, \ldots, n_J)$, where n_j is the number of customers at node j, as a *state*, the open Jackson network $\boldsymbol{X}(t) = [X_1(t), X_2(t), \ldots, X_J(t)]$ is a Markov process on the set of all such vectors. For a closed network, the state space is the set of vectors for which $n_1 + n_2 + \cdots + n_J = N$, the total number in the network at all times.

- Equation (1) is of the form

$$\pi(\boldsymbol{n}) = B\pi_1(n_1)\pi_2(n_2) \cdots \pi_J(n_J)$$

called a *product-form solution*, with B its normalizing constant. The open Jackson networks have $B = 1$ and the $\pi_k(n_k)$, when they exist, are distributions on $0, 1, 2, \ldots$, the number of customers at node k.

In closed networks, the normalizing function B depends on the number of customers in the network. A consequence is that one does not have the independence implied by the product form of the Jackson networks noted below. Nonetheless, the closed networks have the product form of the steady-state distribution for the network queue length. Finding the normalizing constant B is, however, a nontrivial task in closed networks.

- Equation (1) implies that the open network queue lengths at the nodes are independent in the steady state and each appears to be generated by a M/M/1 node.

- Notice that the traffic equations (2) are "balance equations" in that one can interpret them as the balance of probability "flux" out of node j (the left-hand side) no matter where that flux goes next. The right-hand side is then the probability flux into node j no matter where that flux is coming from. In steady-state problems, these (and other) fluxes must be balanced in this sense. In more general networks, there may be several kinds of "balance equations." We discuss some later.

- We can view the Jackson networks in at least two useful ways. We can take the network as given and analyze it, or we can start with a collection of single nodes having certain, known properties and construct a network with them. In building up a network, individual nodes may lose some or all of their desirable properties, but they might gain some new properties from combining with other nodes. (Such a construction is discussed below.)

- Due to Burke's Theorem (described below), nodes in an open Jackson network have the property that, in isolation, if the input process to a node is a Poisson process, one obtains a Poisson output process. Each node in such a network is said to be *quasi-reversible*. If all the nodes in the network are quasi-reversible, then the network is said to be a *quasi-reversible network*.

AN EXTENSION: These results can be extended in several ways. We may relax the Markov structure of the routing process and at the same time include types or classes of customers, perhaps each with its own route through the network. There is a cost to be paid for this specification, however. The state space now is not the queue

length process at the nodes, and the queue length process by itself is not a Markov process. Rather, one needs to calculate the steady-state queue length probabilities from a more general and complicated Markov process.

Types of Customers – Suppose there are T types of customers arriving to the network with the ith arrival type being generated by a Poisson process with parameter v_i. The several arrival processes are independent of each other and of the service processes. Types may be identified by their route through the network. Thus, let $r(i, s)$ be the node being visited by customers of type i when they are in the sth stage of completing their journey through the network. Then a *route*, $r(i, 1), r(i, 2), \ldots, r[i, L(i)]$, is a sequence of nodes for type i customers where r(i, 1) is the first node entered by customers of type i and $r[i, L(i)]$ is the last node entered by these types before leaving the network.

Dynamics of the System – Suppose that

(1) each type of customer requires an exponentially distributed amount of service where, without loss of generality, we take 1 as its parameter.

(2) at node j, the server supplies an effort $\phi_j(n_j)$ when the queue length is n_j. We need $\phi_j(n_j) > 0$ when $n_j > 0$ and $\phi_j(0) = 0$, and we say that the node is *work conserving*.

(3) a proportion, $\gamma_j(\ell, n_j)$, of the $\phi_j(n_j)$ is allocated to the customer in position ℓ at node j when there are n_j at that node. We require for each node j that

$$\sum_{\ell=1}^{n_j} \gamma_j(\ell, n_j) = 1.$$

For example, in a first come, first served queue, $\gamma_j(1, n_j) = 1$ and $\gamma_j(\ell, n_j) = 0$ for any other describes the first-served part of the description (FS).

(4) when a customer enters node j, the customer enters into position ℓ with probability $\delta_j(\ell, n_j + 1)$ when there are $n_j(\geq 0)$ customers already at the node. We require that for each node j that

$$\sum_{\ell=1}^{n_j} \delta_j(\ell, n_j) = 1.$$

The FCFS discipline has $\delta_j(n_j + 1, n_j + 1) = 1$ and $\delta_j(\ell, n_j + 1) = 0$ otherwise describes the first come part of the description (FC).

(5) when a customer at position ℓ finishes at node j, the customers at positions $\ell + 1, \ell + 2, \ldots, n_j$ move to positions $\ell, \ell + 1, \ldots, n_j - 1$. When a customer enters node j and moves into position ℓ, the customers formerly at positions

$\ell, \ell+1, \ldots, n_j$ move to positions $\ell+1$, $\ell+2, \ldots, n_j+1$.

The States – With these generalizations, the queue length process is no longer a Markov process as it was in the Jackson network case. However, a Markov process can be obtained on the following states. Let $c_j(\ell) = c_j[t_j(\ell), s_j(\ell)]$ be the *class* of position ℓ at node j where $t_j(\ell)$ is the type of customer at position ℓ and $s_j(\ell)$ is its stage of progress through the network. Then, let $c_j = [c_j(1), c_j(2), \ldots, c_j(n_j)]$ be the state of node j. Finally, let $C = (c_1, c_2, \ldots, c_J)$ be the state of the network. Then, $X(t)$ is a Markov process on the set of network states.

Transition Intensities – There are three kinds of transitions possible in these Markov networks. Let $T_{jm}C$ be the state that arises when a customer at node j in position leaves and moves into position m at the next node on its route. Let $T_{j\ell}C$ be the state that arises when a customer at node j in position ℓ leaves the network and $T^{im}C$ be the state that arises when a customer of type i enters the network at position m at node $j = r(i, 1)$. It then follows that the transition intensities for the Markov process $X(t)$ are given as

$$q(C, T_{j\ell}C) = \sum_k \phi_j(n_j)\gamma_j(k, n_j).$$

The right-hand side is the rate of service completions at node j for position k, where node j is this customer's last node before leaving the network, that is, $j = r[., L(i)]$ whatever the type. The summation is over all types that could cause such a transition. If, for example, all customers at node j are of the same type and this node is the last they will visit before leaving the network, then the sum is over all customers at node j.

Another transition is caused by having a customer of type i arrive to the network and join position m at node $j = r(i, 1)$. Once again the sum is over all types that could cause such a state change at position m at node j, that is, such that $T^{im}C = T^{ih}C$:

$$q(C, T^{im}C) = \sum_h v_i \delta_j(h, n_j+1).$$

Finally, we have the possibility of a customer leaving node j from position ℓ and moving to position m at node $k = r[t_j(\ell), s_j(\ell)+1]$:

$$q(C, T_{j\ell m}C) = \sum_g \sum_h \phi_j(n_j)\gamma_j(g, n_j)\delta_k(h, n_k+1).$$

The sums here are over all those states $T_{jgh}C$ giving rise to the same terminal state as illustrated above.

The Steady-State Probabilities – In this network, one finds the product form of solution as

$$\pi(C) = \prod_{j=1}^J \pi_j(c_j)$$

where

$$\pi_j(c_j) = b_j \prod_{\ell=1}^{n_j} \frac{\alpha_j[t_j(\ell), s_j(\ell)]}{\phi_j(\ell)}.$$

Here $b_j > 0$, otherwise a steady state does not exist and $\alpha_j(., .)$ is the total arrival rate of type i customers to node j.

From this, it follows that the probability that there are n customers at node j is given by

$$\Pr\{N_j = n\} = b_j \frac{a_j^n}{\prod_{\ell=1}^n \phi_j(\ell)},$$

with $a_j = \Sigma_i \Sigma_s \alpha_j(i, s)$, the total arrival rate into node j.

Summary – The open networks of this section have two important properties.

- The reverse of a network as described here is another such network (see below).
- In the steady state, the departure process of type i customers is a Poisson process and the individual departure processes are independent of each other and of the state of the process subsequent to a departure epoch.

REVERSING OF QUEUEING NETWORKS: We have modeled the networks in the previous sections as Markov processes on a vector state space. Such models have proven to have quite useful applications. However, there is more information in some of these models that we have not examined. To pursue this, we need the concepts of *reversing* and *reversibility* of a Markov process.

Reverse of a Process – For a stationary Markov process $X(t)$ with countable state space E, steady-state vector π, and generator Q, there is another Markov process $X'(t)$ with generator Q' such that

$$\pi(x)q'(x, x') = \pi(x')q(x', x) \quad \text{for } x, x' \in E \tag{3}$$

called the *reverse* of $X(t)$ and π is the steady-state vector for both processes. This concept is quite useful. In many otherwise complicated problems, one might make a shrewd guess as to what the reverse of a process should be and from (3) determine the steady-state elements from it rather than from the global balance equations. But the reverse process has useful properties in its own right.

REVERSIBILITY AND THE DETAILED BALANCE EQUATIONS: The equations

$$\pi(x)q(x, x') = \pi(x')q(x', x)x, \quad x' \in E$$

are called *detailed balance equations*. They require that the probability flux from state x to state x' be equal to that from state x' to state x. This is a rather stringent requirement and is not often found. As shown below, however, when it is available, the consequences are important. For example, see the insensitivity property discussed later.

The detailed balance equations imply that $Q = Q'$ which in turn implies that in the steady state the forward process (unprimed process) and the reverse process have the same joint finite dimensional distributions meaning that they are probabilistically the same process whether considered looking forward or backward in time. On the other hand, if one knows that a queue is reversible, by summing over all nodes, any solution to the detailed balance equations also solves the global balance equations. Thus, the detailed balance equations offer a system of equations that can be used to obtain the elements of π without resorting to the global balance equations for that purpose. Once again, these equations imply more about queueing behavior than simply a way to solve complicated equations.

Birth-death queues are reversible: the departure process from a M/M/s queue is a Poisson process with the same parameter as the arrival process, an argument used in one of the early proofs of what is known as Burke's Theorem (see later).

Using a death process for the service time process at the nodes in a network to model multiple server nodes, one obtains the original Jackson network alluded to earlier and, because of reversibility, still retains the properties discussed there.

QUASI-REVERSIBILITY AND PARTIAL BALANCE:

Let us now consider a model, based on the preceding results, as follows. We will still have customers whose class may depend on their path through the J-node network. For simplicity we assume there is only a finite number of classes and a customer does not change class while in a node but can change when moving from one node to the next. We assume there is a Markov process $X(t) = [X_1(t), X_2(t), \ldots, X_J(t)]$ where a state is a vector $x = (x_1, x_2, \ldots, x_J)$ such that each element x_j may be a vector describing the state of node j.

Quasi-reversibility – The nodes in isolation will now have the property that for $t > 0$ the arrival process following t and the departure process preceding t are independent of the state of the system at t. Then the queue is *quasi-re-*

versible. From this, it follows that the arrival processes and the departure processes are independent Poisson processes. Furthermore, one defines the set of states, say $S(c, x)$ such that any state in S has one more class c customer than state x and has the same number of customers of other classes. From the quasi-reversible assumption, the arrival process of class c customers can only depend on c and not on x.

Then let

$$\alpha(c) = \sum_{x' \in S(c, x)} q(x, x').$$

be the arrival rate of these class c customers. By reversing the queue, the arrival process to the reverse process will have the same rate as this arrival process. That is

$$\alpha(c) = \sum_{x' \in S(c, x)} q'(x, x').$$

Therefore one has

$$\pi(x) \sum_{x' \in S(c, x)} q(x, x') = \sum_{x' \in S(c, x)} \pi(x')q'(x', x).$$

We can interpret this as a *partial balance equation* saying that the probability flux out of a state due to a customer of class c being put into the state is equal to the flux into that state due to the output of a class c customer.

If a network of these quasi-reversible nodes is built, then the network is said to be a *quasi-reversible network*. Note carefully that this is not intended to mean that in the network these nodes have Poisson arrivals and Poisson departure processes. As we shall see subsequently, such Poisson processes are usually rare in queueing networks. What is intended is that we have taken a collection of quasi-reversible nodes and constructed a network.

Transitions in the Network – As earlier, there are three kinds of transitions here: arrivals of class c customers to a node from either the outside or from other nodes in the network and departures of class c customers from a node. We will allow customers to change types as they move from node to node but not while they are at a node.

Consequences of Quasi-reversibility – Consider the customers at node j whose class is designated by the pair (i, s) for i its type and s its stage of completion. Let $\pi_j(x_j)$ be the steady-state probability of node j being in state x_j for those customers at node $j = r(i, s)$. Then

- a quasi-reversible network has a product form for its steady-state distribution.
- if the network is reversed, the reversed process is another network of the same type.

- the network itself is quasi-reversible in that the departure processes are independent Poisson processes and each is independent of the state of the system subsequent to the departure epoch.
- the closed network, as discussed earlier, is a quasi-reversible network and thus has a product-form solution.

SYMMETRIC NETWORKS AND INSENSITIVITY: If a network is one of quasi-reversible nodes and if $\gamma = \delta$, then the nodes are said to be symmetric and if all of the nodes are symmetric, one has a *symmetric network*. In such a network, the steady-state probabilities at the nodes depend on the service time distribution only through its mean value. Such nodes are said to be *insensitive*. Except for its expected value, the service-time distribution is irrelevant. This is quite a useful result, for when such a network occurs, one can replace all service times with exponentially distributed service times which exhibit the important "memoryless" property.

There are few insensitive nodes because of the severe restrictions $\gamma = \delta$. FCFS disciplines at the nodes, for example, are not insensitive. Those nodes that are insensitive include processor sharing, last come-first served with a preemptive resume discipline and infinite-server queues.

Erlang appears to have originally conjectured that his famous "lost call formula," that is, that the probability that a customer arriving to a finite-capacity M/M/c node (total system capacity $= c$) will not gain access (i.e., overflows the node) depends on the service-time distribution only through its mean value. The formula is said to be insensitive to the form of the service-time distribution. Other insensitivities have been shown in other cases.

TRAFFIC PROCESSES: Some of the foregoing results require interpretation. They seem to imply that the traffic processes of customers in these networks are Poisson processes. The result is not correct except in special cases and even then one needs to be careful how to interpret the results. We will give three examples here to illustrate these remarks which make use of the preceding results.

Burke's Theorem – Burke's Theorem says that the departure process from an M/M/c queue is a Poisson process with the same parameter (usually denoted by λ) as the arrival process. But notice that a departure interval depends on the queue length at the start of the interval. If the queue is not empty, the time until the next departure is one service time. If the queue is empty

following a departure, the time until the next departure is the total of the time awaiting the next arrival (exponentially distributed) and the service time of that next arrival (also exponentially distributed and independent of the arrival interval). Thus, the departure intervals must surely depend on the queue length process. It does.

The Burke result concludes that a departure interval does depend on the state of the queue (empty or not) at the start of the interval but not on the state at the end of the interval. This dependence is sufficient for the departure intervals to be *marginally* independent, a result more generally true of Markov renewal processes.

Notice the close connection between quasi-reversibility and the Burke result. In both cases, we have the departures independent of the state of the system. The important consideration from either view is that a departure interval does not depend on the state of the system at some future time although it certainly depends on the state of the system in the past. In FCFS queues of M/G/1 type with an infinite queue capacity, it is only the M/M/1 queues that have Poisson departure processes. But by quasi-reversibility, those queues in isolation will have Poisson departure processes.

The Queue with Feedback – The simplest example which explains some of the peculiarities of traffic behavior in these networks is provided by the FCFS M/M/1 queue with instantaneous Bernoulli feedback with feedback customers going to the tail of the line. It can be pictured as in Figure 2 and described as follows.

Consider the M/M/1 queue where a customer leaving the server can, with probability p, return to the tail of whatever waiting line exists to eventually receive another service. With probability $(1-p)$, the customer leaves the system at the conclusion of any service completions. As a result, there are five traffic streams of customers in this queue.

- The *arrival process* is by assumption a Poisson process.
- The *departure process* (the process of customers leaving the system) is a Poisson process with the same parameter as the arrival process by a simple extension of Burke's Theorem.
- The *output process* (the process of customers leaving the server some of which might feedback for more service) is never a Poisson process (except trivially when there is no feedback).
- The *input process* (the process entering the server which includes arrivals from outside as

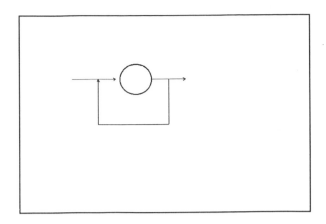

Fig. 2. Schematic of simple feedback queue.

well as those feeding back) is the reverse random process of the output process and hence is never a Poisson process (unless the output process is).

- The *feedback process* (the stream of customers feeding back) is its own reverse so it is reversible but is never a Poisson process.

The implication, illustrated by this example, is that in these queueing networks one cannot expect Poisson traffic processes in spite of the seeming appearance of such in various product-form networks especially the Jackson networks. What is correct is that if the graph of the network is a tree, then the traffic processes on the nodes are Poisson processes (but then see the discussion of sojourn times to follow to see that even here there can be dependent processes e.g., sojourn times at the nodes). If there is a loop in the network, the feedback queue being a simple example, the internal traffic streams will not be Poisson processes on loops even though traffic going to the outside may well be Poisson processes.

Sub-networks – From earlier results it would appear that each node in the Jackson network is a simple M/M/1 queue and, consequently, one could decompose the network to study it node-by-node. Let us consider this further.

Let $X(t)$ be a Jackson network. Partition the nodes of the network into three sets, say, A, B, C and call these sets *sub-networks*. Assume that all arrivals occur to sub-network B, proceed directly to sub-network A and then to sub-network C. One would like to consider A as a sub-Jackson network to reduce the number of nodes to be considered at one time, perhaps.

The question is, "Is sub-network A a Jackson queueing network?" The result is probably *no*. The problem is this. Suppose there are paths from C to B. Then, the network has at least one loop, a path from A back to A. What is known is

that the flow of customers on these loops is not a Poisson process (the feedback queue discussed above provides an especially simple case of this). Hence, the input process to sub-network A is not a Poisson process in this instance and A cannot be a Jackson sub-network.

The general property here is that output processes from nodes are Poisson processes only in special cases. The only FCFS queues producing Poisson departure processes in the M/G/1 class are MM1 queues. Other exceptions occur in the quasi-reversible queueing cases.

Care does have to be exercised here (as we tried to do earlier). One must make a distinction between the arrival process to a queue and the input process, the customers actually entering the service facility. These are not the same in, for example, finite capacity queues. Likewise, one must distinguish between the output process of a queue, those customers actually leaving the server after having received service, and the departure process, customers leaving the service facility whether or not they have received service. The Burke theorem, for example, addresses the arrival and departure processes not the input and output processes. The input and output processes are reverse processes for finite FCFS MM1 queues, for example, and are neither Poisson processes nor renewal processes when the queue capacity is greater than two.

SOJOURN TIMES: One consequence of the results of the previous section is that the *sojourn time* process (the sequence of total times to traverse the network) can be quite complicated and except for special cases there is little known about it in queueing networks. What little is known is for the sojourn time for a specific customer. The sojourn time process is largely unstudied at present.

The Feedback Queue Again – If one considers the sojourn time of a specific customer in the feedback queue to be made up of several passes of the customer through the queue, then it is clear that at the first pass a customer will spend whatever time it would normally spend going through a M/M/1 queue, the M/M/1 sojourn time.

When the customer starts the second pass, ahead will be a number of other customers, some of whom were originally ahead during the first pass, while others may have arrived after the subject customer was in the first pass. How many customers of this latter kind there will be depends on how long the customer spent going through the queue on the first pass. Thus the customer's time to get through the queue the second time will depend on the total number of customers it finds ahead of it upon returning to the queue and this number will depend on how long the customer was in the queue on the first pass. Thus, the sojourn times on these two passes are dependent.

What can be shown is that the two-tuple process of the number of customers faced by the customer entering another pass coupled with the time to get through that pass is a Markov renewal process. The total time through the network is a first passage time in that process.

A Three-Node Network – Consider the Jackson network consisting of three nodes each with a FCFS discipline. There is one Poisson arrival process to node 1. Outputs from node 1 choose with probability p to proceed to node 2 and with probability $(1-p)$ choose to proceed directly to node 3 from where all departures from the network occur. The queue length process here has the product form noted earlier.

Let S_j be the time spent at node j and let $S = S_1 + S_2 + S_3$ be the total time a customer using node 2 takes to traverse this network. Then S_1 and S_2 are independent random variables as are S_2 and S_3. But S_1 and S_3 are dependent.

An Explanation: Overtaking – The result in both examples above is caused by what is called *overtaking*. The following largely explains what is occurring not only with the overtaking phenomenon but with other seeming anomalies in some network problems.

Choose two times, t_0 and $t_0 + \tau$. Since $X(t)$ is a Markov process, $X(t_0)$ and $X(t_0 + \tau)$ are dependent for any t_0 and τ, steady state or not. However, as in the Jackson networks, the elements within either of these vectors may be independent. This, unfortunately, does not mean that $X_j(t_0)$ and $X_j(t_0 + \tau)$ or any nodes have independent state processes at different times, nor does it imply that any two elements within the vectors at different times are independent. There may be these types of independence. In series networks of single server nodes, for example, it is well known that for a given customer, sojourn times at the nodes are independent so that the total time to traverse such a network is simply the sum of times to traverse each node. But such independence is not guaranteed.

The feedback queue noted earlier illustrates that at the same node at different times, the queue length processes will be dependent, even in the steady state. This results in the sojourn times being dependent at that node.

In the three-node sojourn problem the sojourn time for a customer at node 1 depends on the queue length as seen by that arrival at, say, t_0,

while the customer's sojourn time at node 3 depends on the queue length at node 3 when the customer arrives there, at, say, $t_0 + \tau$. The two vectors of queue lengths are not independent. Even though the elements in each are independent under steady-state conditions by the Jackson results, the elements in $X(t_0)$ and those in $X(t_0 + \tau)$ may be dependent. It is this dependency that is largely unstudied at present in queueing network theory. Therefore there are many things unknown about sojourn times in these steady-state product-form networks.

SINGLE-SERVER NETWORKS: Recent interest has turned to networks with a single server serving many queues. Vacation queues and polling systems are examples of such systems.

Vacation Models – There are many variations of these models but essentially they consist of a single node served by a single server. At the conclusion of service of some customer, the server vacations, that is, leaves the node. After some random time, called the vacation period, the server returns to the node to continue service or immediately begin another vacation.

The simplest rules governing the times to take a vacation assume the server leaves the node at each service completion and immediately takes another vacation if, upon returning to the node, the server finds it empty. Various other assumptions allow the server to empty the queue before vacationing or serve some but not all customers before vacationing. Vacation times may be independent of the network state or dependent on some aspect of it, for example, the queue length.

Polling Models – In these models, there are several nodes and a single server servicing all of them. Under some rule of movement, the server visits a node and services customers there according to disciplines akin to those for the vacation models. The server then moves to the next node on the path, eventually returning to the start node, thereby completing a cycle.

A common model assumes the server visits the nodes cyclically, serves a single customer then moves to the next node. Such models have been used to study token ring networks. An older example is called a *patrolling repairman* model. In this scenario, a repairman patrols among a collection of machines repairing those in need and by-passing those not in need.

FURTHER READING: For one starting into a study of queueing networks, the 1960 Syski book is still a favorite for a detailed overview of the early history of Markov queueing theory, including networks. The Kelly (1979), van Dijk

(1993), Walrand (1988), and Whittle (1986) books are all excellent introductions to queueing networks and product forms. Each has its own approach to the field and each contains a variety of applications and examples in diverse fields. Van Dijk (1993), for example, is a nice informal introduction to the interplay between the several balance equations, product forms, and reversibility.

Walrand (1988) included material on both the control aspects and what is known about the sojourn problem, while Whittle (1986) provided more details on the insensitivity property. For the study of traffic processes in queueing networks, there is the book by Disney and Kiessler (1987). Vacation models are thoroughly discussed in Takagi (1991) and polling models are found in Takagi (1986). For a more up-to-date picture of the scope of the field and its approaches, *QUESTA* (1993) is a good place to start.

See **Markov chains; Markov processes; Queueing theory.**

References

[1] Disney R.L. and Kiessler P.C. (1987), *Traffic Processes in Queueing Networks: A Markov Renewal Approach*, Johns Hopkins University Press, Baltimore.

[2] Kelly F.P. (1979), *Reversibility and Stochastic Networks*, John Wiley, New York.

[3] *QUESTA* (1993), *Queueing Systems: Theory and Applications*, 13, May.

[4] Syski R. (1960), *An Introduction to Congestion Theory in Telephone Systems*, Oliver and Boyd, London.

[5] Takagi H. (1991), *Queueing Analysis: A Foundation for Performance Analysis, Vacation and Priority Systems*, part 1, North-Holland, Amsterdam.

[6] Takagi H. (1986), *Analysis of Polling Systems*, MIT Press, Cambridge, Massachusetts.

[7] van Dijk N.M. (1993), *Queueing Networks and Product Forms: A Systems Approach*, John Wiley, New York.

[8] Walrand J. (1988), *An Introduction to Queueing Networks*, Prentice-Hall, Englewood Cliffs, New Jersey.

[9] Whittle P. (1986), *Systems in Stochastic Equilibrium*, John Wiley, New York.

NEURAL NETWORKS

James P. Ignizio

University of Virginia, Charlottesville

Laura I. Burke

Lehigh University, Bethlehem, Pennsylvania

INTRODUCTION: While the early research in neural network models of intelligence surfaced several decades ago, the field nearly died in the 1960s due to obstacles which seemed insurmountable. Critical advances in the late 1970s and early 1980s, however, led to a resurgence of interest and activity in the area (Hopfield, 1984; Rumelhart, Hinton and Williams, 1986). The classical model of human cellular activity, proposed by McCulloch and Pitts (1943), still forms the foundation of much of the work being conducted in current neural network research.

A real neural network is that interconnection of elements within the mammalian brain, and those activities that go on within and between these elements, that evidently serve to carry out the decision making process (e.g., memory, recognition, prediction, planning, problem solving). Artificial neural networks represent mankind's rather feeble attempts to replicate such biological processes by means of algorithms in conjunction with either physical or computer simulated neural network abstractions (e.g., electronic components, digital simulations). There are widely divergent philosophies as to just how rigorously such replication should be followed. The neural network purists might insist that their neural networks have an almost one-for-one analogy between its architecture and performance and that of the brain. The more pragmatic advocates are likely to simply seek networks and associated algorithms that serve to solve the problems at hand, regardless as to just how well it does, or does not, follow the theory of the operation of the brain.

From an operations researcher's perspective, neural networks may be thought of as algorithms with certain characteristics and restrictions that serve to facilitate parallelism in, ultimately, hardware implementation. More precisely, computation proceeds through a system of relatively simple, nonlinear, processing elements linked by weighted connections. Such processing elements compute from local information stored and transmitted via connections. The key result is that parallel computations occur and that all storage is restricted to the information (weights) on connections (Burke and Ignizio, 1992). A graphical illustration of a so-called feedforward neural network is shown in Figure 1.

WHY USE NEURAL NETWORKS: Neural networks may be used, in most instances as a strictly heuristic approach, to provide (acceptable) solutions for problems of prediction/forecasting, or problems of pattern classification/pattern recognition, or problems of optimization

Network Response

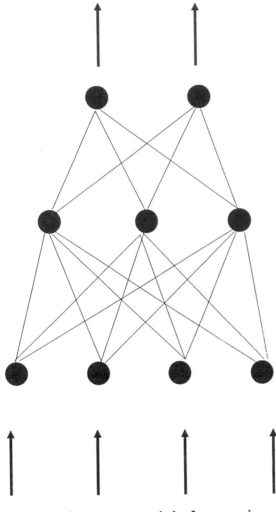

Environmental Information

Fig. 1. Artificial neural structure.

(involving either continuous or discrete-valued variables). As such, it should be noted that alternative methods for producing solutions to such problems already exist (i.e., statistical tools for prediction and pattern recognition, mathematical programming algorithms for problems of optimization). In fact, in the case of optimization, existing mathematical programming algorithms may well be able to provide optimal solutions to problems which neural networks can only solve heuristically. Thus, neural networks do not really offer us the ability to solve some new type of problem type or to solve problems previously unsolvable by any existing methods. They do, however, provide us with some exceptional advantages, including:

- they are nonparametric (i.e., independent of any assumptions with regard to the data) and

thus more robust than statistical methods employing parametric assumptions,
- they provide one with the opportunity to use parallel processing in the problem solving procedure and thus offer the potential for faster processing, more fault-tolerant processing, and ultimate hardware replication (i.e., replication on a silicon chip)

It is then for these reasons that one might consider the employment of a neural network approach (Burke, 1991; Lippmann, 1987).

STRUCTURE OF NEURAL NETWORKS: One may define a given neural network on the basis of three attributes: the network architecture, or topology; the characteristics of the nodes; and the training or learning process employed to develop the weights on the branches between node pairs. Thus, the neural network designer must first determine an appropriate network architecture for the problem at hand (i.e., number of levels, number of nodes at each level, and the interconnections of the nodes), select the type of (nonlinear) processing to be carried out in each node, and then train the network to accomplish its purpose in either a supervised or unsupervised mode. Supervised training results in networks that are capable of prediction (e.g., given a set of values for the attributes of interest, associate that set with a particular outcome) or classification (e.g., given a set of values for the attributes of interest, associate that set with a specific class, or group).

However, before one can perform a classification it is first necessary to determine the specific set of classes to which objects are to be assigned. Networks employing unsupervised learning are typically used to develop these groups. In other words, they may be used to perform an operation known as cluster analysis.

The architecture of the neural network in conjunction with the weights on the connections between nodes serves as a representation of the knowledge replicated by that specific network. This knowledge may then be used to either predict (or associate) an outcome on the basis of values of the input signals, or to classify an object on the basis of these input signals, or to even solve a problem of continuous or combinatorial optimization on the basis of achieving network stability. As just one illustration of the use of neural networks, consider the network depicted in Figure 2.

Note that, in a neural network of any number of nodes and architecture, node j receives the weighted summation of all its inputs; that is,

$$\text{input}(j) = \sum_i w_{ij} v_i \quad i = 1, \ldots, N$$

where N is the number of nodes in the network, $w_{ij} = 0$ if no connection exists between nodes i and j, and v_i is the state of node i. Here, we have four nodes in total. Nodes one through three, however, are input nodes and thus simply transmit signals applied to them. Node four, on the other hand, functions as a processing element (PE). Thus the three weighted inputs to node four are summed, and the PE then processes the resultant sum. One type of nonlinear processing is known as hard-limiting. Specifically, if the weighted sum of the input signals exceeds a certain level then the output of the node is activated. Otherwise, the node output is not activated. In this manner, one might use the above network for the purpose of pattern classification and, in such an event, the three weights serve to actually define the plane separating one class from another.

Most classification problems will require nonlinear separating surfaces and these cannot be developed via the single PE, single layer network shown. Fortunately, we can produce nonlinear separating surfaces by means of multi-layer neural networks. Consequently, the bulk of neural networks used for prediction and classification involve three layers (e.g., as depicted earlier in the three layer feedforward network of Figure 1). Such an architecture, coupled with a means to *train* the networks to recognize classes, or predict outcomes, produces the branch weights on the final network structure. A method for training such networks requires differentiable processing functions (rather than the hard limiting type) and its discovery was a key contribution to the renewed interest in neural networks.

One frustrating aspect of neural network research has been the apparent "arbitrariness" associated with choice of architecture and training strategies for successful applications. While some prefer to believe that neural networks will replace statistical methods, the truth is that appropriate transference of statistical concepts to the neural network realm will aid in understanding how neural nets can augment statistical techniques. For example, a neural network trained to classify patterns, such as attributes of patients used in diagnosis of dermatological problems, will perform quite differently than a similar network trained on different examples. Determining an effective training set represents as important a task in neural network applications as it does in statistical approaches.

Neural networks for solving problems of optimization represent a somewhat more complex topic. However, in essence, one typically establishes an energy function for which we seek a local minima solution. Such a network is provided with an initial input and iteratively adjusts its internal branch weights until a stable state is realized; where such a state coincides with a local minima. Hopfield and Tank (1985) have demonstrated that one may use such networks to solve various optimization problems, including linear programming. However, unlike the simplex method of linear programming (which moves along the constraint set boundaries), the neural network is an interior point solution process converging, hopefully, at or near the global optimal solution.

In both kinds of applications (viz., pattern recognition and optimization), an important side benefit of neural network research has been a fresh perspective on problem solving. In many cases, a statistical approach could as well be applied to a problem as a neural network. However, the neural network approach tends to expand the creativity of the researcher, and often leads to dramatically new insights into old problems.

See **Artificial intelligence; Control theory; Cybernetics.**

References

[1] Burke Laura I. and James P. Ignizio (1992). "Neural Networks and Operations Research: An Overview," *Computers and Operations Research*, 19(3/4), 179–189.

[2] Burke Laura I. (1991). "Introduction to Artificial Neural Systems for Pattern Recognition," *Computers and Operations Research*, 18(2), 211–220.

[3] Lippmann Richard P. (1987). "An Introduction to Computing with Neural Networks," *IEEE ASSP Magazine*, April, 4–22.

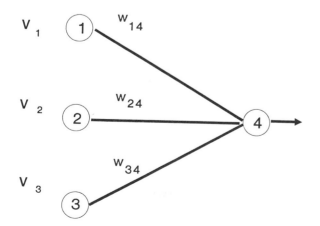

Fig. 2. Four node, single PE network.

[4] McCulloch Warren S. and Walter Pitts (1943). "A logical calculus of the ideas immanent in nervous activity." *Bulletin of Mathematical Biophysics*, 5, 115–133.

[5] Rumelhart David E., Geoffrey E. Hinton and Ronald J. Williams (1986). "Learning representations by back-propagating errors." *Nature*, 323, 533–536.

[6] Hopfield J.J. (1984). "Neurons with graded response have collective computational properties like those of two-state neurons." *Proceedings of the National Academy of Sciences*, 81, 3088–3092.

[7] Hopfield J.J. and D.W. Tank (1985). "Neural Computation of Decisions in Optimization Problems." *Biological Cybernetics*, 52, 141–152.

NEWSBOY PROBLEM

An item, here newspapers, has to be procured at the beginning of a time period and is discarded (or sold at a discounted price) at the end of the time period. The demand is assumed to be a random variable with known distribution. The problem is to determine how many items to stock at the beginning of the time period to minimize expected cost. This leads to a closed-form, single-period inventory model with stochastic demand. The problem statement also applies to items such as Christmas trees, time-dependent fashions, and items that can be stored until the next season like snow tires and Chanukah candles. See **Inventory modeling**.

NEWSVENDOR PROBLEM

See **Inventory modeling**; **Newsboy problem**.

NEWTON'S METHOD

See **Nonlinear programming**; **Numerical analysis**; **Quadratic programming**.

NLP

See **Nonlinear programming**.

NODE

(1) A simple queueing subsystem in a queueing network consisting of a server center with one or more servers, a queueing capacity infinite or finite, and a queue discipline. (2) An element of a graph or network, pairs of which are connected by arcs or edges. Nodes are sometimes referred to as points or vertices. See **Graph theory**; **Net-**work optimization. (3) Show graphically in a project network plan as a circle depicting the beginning or end of an activity. A node represents an instantaneous point in time at the junction of arrows. See **Network planning**.

NODE-ARC INCIDENCE MATRIX

For the minimum-cost network-flow problem, this is a matrix in which the rows i correspond to the nodes and the columns j correspond to the arcs. For an arc (i,j), with its flow directed from i to j, the entry in matrix location (i, j) is $a + 1$ and the entry in location (j, i) is $a - 1$. All other entries are zero. Thus, every column has only two nonzero entries. See **Minimum-cost network-flow problem**; **Multicommodity network flows**; **Network optimization**.

NONACTIVE (NONBINDING) CONSTRAINT

An inactive constraint. See **Active constraint**; **Inactive constraint**.

NON-ARCHIMEDEAN NUMBER

A number that does not satisfy the Archimedean axiom. Such numbers arise in setting preemptive (lexicographic) priorities in goal programming, the "Big-M" for finding a feasible basis to a linear-programming problem, and in selecting an infinitesimal in data envelopment analysis. See **Archimedean axiom**; **Big-M**; **Data envelopment analysis**; **Goal programming**.

NONBASIC VARIABLE

Given a feasible basis to a linear-programming problem, a variable is nonbasic if it does not correspond to one of the vectors in the basis. See **Basic variables**.

NON-COMPENSATORY CHOICE STRATEGIES

Not employing trade-offs between the dimensions of choice alternatives but, rather, using thresholds (or cut-offs) that need to be achieved for choice of an alternative. See **Choice theory**.

NONDEGENERATE BASIC FEASIBLE SOLUTION

A feasible basis to a linear-programming problem is nondegenerate if all basic variables are strictly positive. See **Basic feasible solution**; **Degeneracy**; **Degenerate solution**.

NONDOMINATED SOLUTION

See **Efficient solution**.

NONLINEAR GOAL PROGRAMMING

A goal programming methodology used to solve goal programming problems that have nonlinear elements in their model formulation. See **Goal programming**.

NONLINEAR PROGRAMMING

Anthony V. Fiacco

The George Washington University, Washington, DC

INTRODUCTION AND PERSPECTIVE: *Nonlinear programming* [a term coined by Kuhn and Tucker (Kuhn, 1991)] has come to mean that collection of methodologies associated with any optimization problem where nonlinear relationships may be present in the objective function or the constraints. Since maximization and minimization are mathematically equivalent, without loss of generality, we shall assume that the nonlinear programming problem is the problem of finding a solution point or optimal value of

minimize $f(\boldsymbol{x})$ subject to $g_i(\boldsymbol{x}) \geq \boldsymbol{0}$ $(i = 1, \ldots, m)$

and $h_j(\boldsymbol{x}) = \boldsymbol{0}$ $(j = 1, \ldots, p)$ (P)

where all problem functions are real valued. The underlying space can be more general, but here we shall assume $\boldsymbol{x} \in E^n$, the well known Euclidian n-space. In this terminology and context, problem (P) is a linear program (LP) if f, g_i and h_j are linear (actually linear-affine, i.e., linear plus a constant) for all i and j.

Another very important instance of problem (P) is where the constraints g_i and h_j are not present or where every point in the domain of f is feasible (i.e., satisfies the constraints). This is called an "unconstrained" problem and goes back to the very early days of mathematics.

SIMPLE EXAMPLES: Finding the highest point of a pyramid may be viewed as a linear programming problem. Assume that we can find equations of the planes that contain the sides and base of the pyramid. Then, the pyramid is essentially the feasible region, the set of points in the volume contained by these planes. Our problem is to find the point in the region that yields the greatest height.

A somewhat analogous example is that of finding the deepest point in a lake. We may view the shoreline as the constraint, the feasible region as the surface of the water, and the objective function as the depth. This example would generally be highly nonlinear.

Since we are all accustomed to using maps, another readily understandable example is that of finding a point where the maximum or minimum altitude is attained in a given area, for example, find the highest point in the state of Virginia. The constraints are determined by the state boundaries. Lines of equal altitude are often displayed on a map and correspond to iso-value contours or "level curves" of our objective function (altitude), in mathematical programming terminology. Thus, we are looking for the latitude and longitude of a location in Virginia (the feasible region) on a level curve of maximum value, a problem that is (logically) equivalent to a realization of a nonlinear problem of the form of problem (P). There are many local maxima (i.e., hills and peaks) in this problem that are not global, a formidable challenge to solving (P).

MATHEMATICAL EXAMPLES: Obtaining a solution to a system of equations $h_1(\boldsymbol{x}) = 0, \ldots, h_q(\boldsymbol{x}) = 0$ where $\boldsymbol{x} \in E^n$ may be posed as the unconstrained NLP, P: $\min \Sigma_{j=1}^{q} h_j(\boldsymbol{x})^2$ s.t. $\boldsymbol{x} \in E^n$, or equivalently, P: $\min \|\boldsymbol{h}(\boldsymbol{x})\|^2$, where $\boldsymbol{h} = (h_1, \ldots, h_q)^\top$ and the norm is the usual Euclidian norm. Alternatively, we could choose to solve $\min \|\boldsymbol{h}(\boldsymbol{x})\|^2$ or $\min \|\boldsymbol{h}(\boldsymbol{x})\|$ for any suitable choice of norm. There may be no solution to the system of equations, but the indicated NLP problems can still be addressed and their solutions would yield points that minimize the "residual error," that is, the deviation of $\boldsymbol{h}(\boldsymbol{x})$ from $\boldsymbol{0}$, in the sense of the given norm. The choice Σh_j^2 leads to a so-called "least-squares" solution and is undoubtedly the most popular, yielding a smooth (i.e., differentiable) problem if $\boldsymbol{h}$ is differentiable (the other measures usually being nonsmooth). If the solutions are underdetermined, one could use the degrees of freedom to seek to determine a solution having a desired quality, for example, minimum norm, that is, a solution of the constrained NLP, $\min \|x\|$ s.t. $h(x) = 0$.

The minimum norm-residual idea turns out to be extremely fruitful and is a driving mechanism for several important classes of NLP problems, including regression, also known as parameter estimation or data-fitting and essentially a type of curve-fitting, minimum distance problems, and eigenvalue problems. The idea of regression is to assume a functional form, say $y = F(\alpha, x) + \varepsilon$ that relates observation y to the input data vector x, parameter vector α, and a random experimental error ε. If y_i $(i = 1, \ldots, r)$ is observed, respectively, when x^i $(i = 1, \ldots, r)$ occurs, then, under suitable assumptions, a least-squares regression problem can be formulated. We may seek to find a parameter vector $\bar{\alpha}$ that solves the unconstrained NLP, $\min_\alpha \Sigma_{i=1}^r [y_i - F(\alpha, x^i)]^2$. If F is linear, then this is called "linear regression" and is well known, heavily used, and has a rich statistical basis and interpretation. Other norms could be used. A similar approach, so called "curve-fitting," could be used to "fit" a function F to a given set of points or to approximate another function, possibly introducing constraints on F, for example, requiring bounds on F or its derivatives, without necessarily being supported by the attendant statistical rationale and utilizing a variety of norms.

The problem of finding the minimum eigenvalue and associated eigenvector of a real symmetric $n \times n$ matrix A can be posed as that of determining the optimal value and solution vector, respectively, of the constrained NLP problem, $\min x^\top A x$ s.t. $\|x\|^2 = 1$, again a minimum-norm-type problem. The problem of finding the shortest distance between one point and another, or a point and a line, or a point and a set or, more generally, between one set S_1 and another set S_2 takes on the rather natural NLP-constrained form, $\min \|x - y\|^2$ s.t. $x \in S_1$ and $y \in S_2$. Many extensions and ramifications of this idea can be envisioned.

PRACTICAL APPLICATIONS: Hancock (1960, p. 151) stated that "by means of Gauss's principle all problems of mechanics may be reduced to problems of maxima and minima." Principles in optics, wave mechanics, quantum physics, astronomy, chemistry, biology, etc., can usually be formulated in terms of "extremal" (i.e., maximum or minimum) principles, for example, a path of least resistance, minimum energy, maximum entropy, etc.

The practical applications of nonlinear programming are incredibly vast. Regression and curve-fitting applications abound in mathemat-

ics and physics, the natural and applied sciences, econometrics, and engineering statistics. Generalizations include possible constraints on the parameters and extensions to higher dimensions (surface fitting), with applications in pattern recognition, geography, agriculture and quantum physics, for example (Hobson and Weinkam, 1979). As early as 1980, Hillier and Lieberman (1980, Ch. 1) reported that the most widely used operations research techniques were statistical techniques (mainly those involving regression analysis), simulation, and linear programming. They noted that the most important applications of mathematical programming were those in production management (e.g., in allocation of resources to maximize some measure of profit, quality, efficiency, effectiveness, etc.), followed next by financial and investment planning, and they reported that about 25% of all scientific computation on computers was devoted to linear programming and related techniques. It seems clear that these trends have sustained. Winston (1991, p. 51) noted that 85% of the respondents of a survey of Fortune 500 firms report use of LP, and that about 40% of his book is devoted to related optimization techniques.

Practically every research and textbook in NLP discusses important current applications. Fletcher (1987, p. 4) noted important applications in structural design, scheduling, and blending, as well as numerical analysis and differential equations. McCormick (1983) analyzed problems in chemical equilibrium, inventory control, engineering design and water pollution control. Bazaraa and Shetty (1979, Ch. 1) discussed problems in discrete and continuous optimal control, mechanical and structural design, electrical networks, and location of facilities.

A methodology for NLP apparently all started coming together in about 1960. This was largely motivated by applications, for example, to petroleum refinery problems which inspired algorithmic work by Rosen in 1960–61 and a paper-pulp manufacturing process that led to a technique proposed by Carrol in 1959 and 1961. A collection of case studies in such diverse areas as bid evaluation, stratified sampling, launch vehicle design and alkylation process optimization was given by Bracken and McCormick (1968). The algorithms used were based on Carrol's interior barrier function and Courant's exterior quadratic penalty function proposed in 1943, developed and extended by Fiacco and McCormick in 1963, and implemented via the SUMT computer program by McCormick, Mylander and Fiacco in 1965 (Fiacco and McCormick, 1968, 1990).

BASIC THEORY: We first indicate some of the significantly different problem types: (i) one dimensional or many dimensional, (ii) finite-dimensional (e.g., in E^n) or infinite-dimensional (e.g., as in variational calculus and optimal control), (iii) a finite number of constraints or an infinite number (as in "semi-infinite programming"), (iv) unconstrained or constrained, (v) involving real numbers (standard NLP) or integers (integer programming), (vi) convex or nonconvex, (vii) smooth (i.e., differentiable) or nonsmooth (nondifferentiable), and (viii) deterministic or stochastic.

A local minimizer of (P) is a feasible point $\bar{x}$ such that $f(\bar{x}) \leq f(x)$ for all x in a feasible neighborhood of $\bar{x}$. If $f(\bar{x}) \leq f(x)$ for *all* feasible x, then $\bar{x}$ is called a global minimizer. If $\bar{x}$ is a local minimizer and $f(\bar{x}) < f(x)$ for all $x \neq \bar{x}$ in a feasible neighborhood of $\bar{x}$, then $\bar{x}$ is called a strict local minimizer. If $\bar{x}$ is the only local minimizer in some feasible neighborhood of $\bar{x}$, it is called an isolated local minimizer.

A fundamental result of great importance is the fact that a feasible global minimizer of a continuous function f exists in the feasible region R if R is nonempty and compact, a result attributed to Weierstrass. If f is once continuously differentiable and $\bar{x}$ is a local unconstrained minimizer, then the gradient, $\nabla f(\bar{x}) = 0$. If f is twice continuously differentiable, then $\nabla f(\bar{x}) = 0$ and the Hessian (matrix of second partial derivatives) $\nabla^2 f(\bar{x})$ is positive-semi-definite (p.s.d.) at a local minimizer $\bar{x}$. If $\nabla f(\bar{x}) = 0$ and $\nabla^2 f(\bar{x})$ is positive definite (p.d.), then $\bar{x}$ is an isolated (hence, also strict) local minimizer.

The usual "Lagrangian" of problem P is defined as

$$L(x, u, w) = f(x) - \sum_{i=1}^{m} u_i g_i(x) + \sum_{j=1}^{p} w_j h_j(x),$$

where the $\{u_i\}$ and $\{w_j\}$ are the Lagrange multipliers. John in 1948, Karush in 1939, and Kuhn and Tucker in 1951 (Fiacco and McCormick, 1968, 1990) independently generalized and extended the classical "Lagrange multiplier rule" (Lagrange, 1762) for equalities to include inequalities, arriving at the following first-order conditions that we simply call the Karush-Kuhn-Tucker conditions and abbreviate as KKT($\bar{x}, \bar{u}, \bar{w}$): there exists $\bar{u}_i \geq 0$ $(i = 1, \ldots, m)$ and $\bar{w}_j$ $(j = 1, \ldots, p)$ such that $\nabla_x L(\bar{x}, \bar{u}, \bar{w}) = 0$ and $\bar{u}_i g_i(\bar{x}) = 0$ $(i = 1, \ldots, m)$, for $\bar{x}$ feasible. If a suitable constraint qualification (CQ) holds at a local minimizer $\bar{x}$, then KKT($\bar{x}, \bar{u}, \bar{w}$) holds. Though a more general CQ was given in Kuhn and Tucker (1951), it turns out that this holds if the "binding" constraint gradients are linearly independent, that is, if $\{\nabla g_i(\bar{x}), i \in B(\bar{x}); \nabla h_j(\bar{x}), j = 1, \ldots, p\}$ are linearly independent, where $B = \{i : g_i(\bar{x}) = 0\}$. We denote this CQ by LI($\bar{x}$). The KKT($\bar{x}, \bar{u}, \bar{w}$) are sufficient for $\bar{x}$ to be a minimizer if (P) is a convex program, that is, if f is convex, the $\{g_i\}$ concave and the $\{h_j\}$ affine. Convex programs have additional attributes: local solutions are global, they have associated with them a rich duality theory, and they are among the easiest to analyze and solve. Second-order optimality conditions are now also well known and heavily used.

When P is convex, a dual problem is the following:

$$\max_{(x, u, w)} L(x, u, w) \quad \text{s.t.} \quad \nabla_x L(x, u, w) = 0, \quad u \geq 0, \tag{D}$$

where $u = (u_1, \ldots, u_m)$ and $u \geq 0$ means that $u_i \geq 0$ for all $i = 1, \ldots, m$. This simple but remarkably useful formulation was first proposed and developed by Wolfe in 1961 (Fiacco and McCormick, 1968, 1990). It turns out that the optimal value of **P** is bounded below by the optimal value of **D**. Further, if LI($\bar{x}$) holds, or one of several other well known CQs, then if $\bar{x}$ solves **P**, it follows that KKT($\bar{x}, \bar{u}, \bar{w}$) holds, ($\bar{x}, \bar{u}, \bar{w}$) solves the dual **D**, and $f(\bar{x}) = L(\bar{x}, \bar{u}, \bar{w})$. Duality has significant computational applications; for example, algorithms that generate dual-feasible points also yield lower bounds on the primal optimal value.

Other duals have been developed, most notably the Fenchel Dual, significantly extended and utilized by Rockafellar (1970) and others. Not surprisingly, a rich and finely tuned duality theory has been developed for LP.

ALGORITHMS: An algorithm is essentially a recipe, in our context a numerical procedure, for starting with given initial conditions and calculating a sequence of steps or iterations until some stopping rule is satisfied. Up until the recent past, the uncontested winner for LP has been some version of Dantzig's simplex method, a technique based on the idea of moving from one vertex of the feasible region to an adjacent vertex while reducing the objective function with each move. The elegance of the mathematics, industrialization and economic planning needs, and the advent of the electronic digital computer in the 1940s, and a host of important practical applications that followed, all conspired to make LP widely accepted and heavily utilized.

The same forces that stimulated the development of LP in the latter 1940s were encouraging research on theory and algorithms for NLP. The

1930s and 1940s saw a flurry of theoretical activity in variational calculus and optimization at the University of Chicago and other mathematical centers by mathematicians like Valentine, Reid, McShane, Karush, Bliss, Graves, Hestenes, Courant, John and others. The early 1950s brought a sharper focus to first and second order optimality conditions for inequality constrained NLP by Kuhn and Tucker in 1951 and Pennisi in 1953, respectively, and others. As early as 1951, for example, a paper by Arrow in 1951 on a gradient method for solving constrained saddle-point problems, there is evidence of serious algorithmic work. Two key results during this period were the "conjugate direction" method, an iterative procedure for solving a system of linear equations, by Hestenes and Stiefel in 1952, and a "variable metric" method (a "quasi-Newton" algorithm wherein the required Hessian inverse is calculated iteratively) by Davidon in 1959. Such developments significantly enhanced the steepest-descent and Newton tools for solving equations, the heart of solving NLP problems, and were followed by an intense period of activity in the 1960s, with a rapid solidification of the theory and computational and methodological breakthroughs such as cutting plane algorithms by Kelley in 1960, methods of feasible directions by Zoutendijk in 1960, gradient projection methods by Rosen in 1960–1961, and SUMT by Fiacco and McCormick in 1963 (Fiacco and McCormick, 1968, 1990).

Many algorithms have subsequently been developed for both unconstrained and constrained problems. The most popular prototype starts with a "merit function" (e.g., the objective function) and initial conditions, determines a search direction vector, then calculates a step in the given direction based on a "line search," some one-dimensional curve fitting scheme aimed at both reducing the value of the merit function and maintaining feasibility. The process is repeated until some convergence criteria are satisfied. An algorithm for a well posed problem generally attempts to satisfy first-order necessary optimality conditions for a local minimizer. More sophisticated algorithms satisfy second-order necessary conditions and others even seek out global minimizers, though techniques for the latter continue to be under intense development. Algorithms may be deterministic or stochastic, continuous or discrete-step, accumulate information or not, etc. A host of special-purpose algorithms have been developed for one-dimensional optimization, for example, variations of successive bisection, Newton's method, the secant method, false position, Fi-

bonacci search and golden section (McCormick, 1983).

Some of the most effective contemporary algorithms for smooth unconstrained problems are generally some variant or mixture of a "quasi-Newton" (approximate Newton) or conjugate direction algorithm. The survivors in the competition must fare well overall in meeting several demanding and sometimes opposing criteria: computational effort, speed of convergence, accuracy, robustness, ease of implementation, accessibility, and so on. Efforts to develop rigorous computational and theoretical standards for measuring these attributes are ongoing, for example, a rate of convergence theory is in place that tells us that steepest descent converges at least at a "linear rate" (as in a geometric series) and Newton's method at a "quadratic rate" (exponentially, at least quadratic), under rather ideal circumstances. Hybrid methods, conjugate directions and variable metric methods, are thought to perform adequately when they converge "superlinearly" (more or less, the best linear rate possible, a compromise between linear and quadratic). Another important criterion is that an unconstrained algorithm be able to calculate the minimizer of a positive definite quadratic form in n variables, in at most n iterations. A key driving principle is exploitation of problem structure.

Some important algorithms for constrained problems are sequential linear programming (SLP), for example, separating or cutting plane algorithms; sequential quadratic programming (SQP), for example, constrained Newton approaches; generalized reduced gradient (GRG) methods, essentially, variable elimination simplex-type algorithms; feasible direction (constrained steepest descent) methods; projected gradient methods; and auxiliary function methods, for example, "augmented Lagrangian function" (i.e., Lagrangian plus penalty term) techniques, "penalty function" (objective function plus constraint violation cost) and "barrier function" (objective function plus feasibility enforcing) methods. Algorithms and software are given in the references.

Additional important topics and suggested references are the following: global optimization, (Kan and Timmer, 1989); parametric programming, sensitivity and stability analysis (Fiacco, 1983; Fiacco, ed., 1990), which we discuss next; stochastic programming (Wets, 1989); semi-infinite programming (SIP) (Fiacco and Kortanek, eds., 1983); multi-objective programming (Sawaragi, Nakayama and Tanino, 1985); multilevel programming (Anandalingam, ed., 1992);

control theory (Hocking, 1991); numerical methods and implementation (Gill, Murray and Wright, 1981); software, evaluation and comparison of algorithms (Waren, Hung and Lasdon, 1987) and (Mor and Wright, 1993); parallel and large-scale programming (Rosen, 1990); integer programming (Schrijver, 1986); basic barrier and penalty function methodology (Fiacco and McCormick, 1968, 1990); and nonsmooth optimization (Neittaanmaki, 1992).

The last few years have seen intense activity in devising polynomial-complexity *interior point methods* for LP and NLP, sparked by a theoretical breakthrough with Khachian's ellipsoid method (Khachian, 1979) and a theoretical-computational breakthrough with Karmakar's potential method (Karmakar, 1984). The literature is already incredibly vast and rapidly growing, along with computational advances. The reader is referred to the excellent surveys by Gonzaga (1992) and Wright (1992) for a good introduction to this important development and for many good references, and to the books of Megiddo (1989) and Nesterov and Nemirovski (1993) for technical advances.

Software exists to implement variations of all the methods described here. As to computational capability, we settle for a general assessment from Rosen (1990, Preface): "Based on the computational results described, it now seems possible to solve a wide variety of optimization problems with many thousands of variables and constraints in real times of a few minutes." Problems in a hundred or so variables can now be solved on a PC. Large problems may require parallel processors. However, a meaningful measure of computational difficulty is elusive in NLP. For example, consider that a high degree polynomial in one variable may have many local minima and may be much more difficult to solve globally than a large convex program with hundreds of variables and constraints.

SENSITIVITY ANALYSIS: The question motivating this topic can be raised in connection with almost any method of inquiry that results in a conclusion: How does the answer change when the assumptions change? The assumptions can be any conditions or data that are given and the changes can be qualitative or quantitative, controlled or uncontrolled, deterministic or stochastic, small or large, known or estimated, immediate or staged over time. The issue is somewhat inevitable and universal, since we live in an imperfect world and must maintain a certain flexibility in dealing with ever-present errors and ranges in approximation and interpretation,

whether it be in carrying on a conversation, steering a car, hitting a tennis ball, or calculating expected return on investment. In business, industry, recreation or mathematics, we are seldom interested in a single answer to one scenario.

In the early days of mathematics and physics, a related issue was apparently frequently raised: When is a problem well-posed, that is, when does a solution change continuously with continuous changes in the problem data? Variations on this theme must have quickly followed: When are the solution changes well-behaved in some sense, for example, when are they finite or bounded or smooth, and when are they not? Can we calculate the solution in closed form as a function of the changes in the data, or can we at least calculate bounds on the changes or rate of growth of the changes? Can we identify or measure any useful properties of perturbed solutions, for example, whether a unique solution remains unique, whether a solution function or set is convex as a function of the changes, if a solution trajectory is differentiable, whether an assumption persists under given perturbations, etc.? At a slightly more sophisticated level, can we calculate some of these properties from information available at a solution . . . without resolving the problem with new data?

We briefly summarize a collection of such results in the context of nonlinear programming (NLP) where there are parameters (e.g., data) present that are subject to perturbations. Here, we confine most of our attention to characterizations involving small specified changes in the parameters, the study of which we call "sensitivity analysis."

Very simple problems can be stable or unstable, for different parameter values. Consider the linear problem, minimize x_1 s.t. $x_1 \geq -1$, $x_2 \leq \varepsilon x_1$, and $x_2 \geq 0$, where $x \in E^2$ and $\varepsilon \geq 0$. If $\varepsilon > 0$, then the solution is $x(\varepsilon) = (0, 0)$ and does not change with small enough changes in ε. However, if $\varepsilon = 0$, then the solution is $x(0) = (-1, 0)$ and this changes to $x(\varepsilon) = (0, 0)$ for arbitrarily small positive changes of ε, an extremely erratic change. Our desire is to understand the causes and implications associated with such stability or instability.

Preliminaries – The parametric NLP is defined as

minimize $f(x, \varepsilon)$ subject to

$$\{x \in E^n: g_i(x, \varepsilon) \geq 0, i = 1, \ldots, m;$$

$$h_j(x, \varepsilon) = 0, j = (1, \ldots, p)\} \qquad \text{P}(\varepsilon)$$

where $x \in E^n$ and ε is a perturbation parameter in T, a nonempty subset of E^k. If ε is held constant, then problem P(ε) is simply a realization of a standard NLP problem of the form P that we discussed at the outset. The Lagrangian associated with P(ε) is defined as $L(x, u, w, \varepsilon) = f(x, \varepsilon) - \sum_{i=1}^{m} u_i g_i(x, \varepsilon) + \sum_{j=1}^{p} w_j h_j(x, \varepsilon)$. The optimal-value function f^* and the optimal-solution map S of P(ε) are defined as

$$f^*(\varepsilon) = \begin{cases} \inf_{R(\varepsilon)} f(x, \varepsilon) & (\text{if } R(\varepsilon) \neq \emptyset) \\ +\infty & (\text{if } R(\varepsilon) = \emptyset) \end{cases}$$

and $S(\varepsilon) = \{x \in R(\varepsilon) : f(x, \varepsilon) = f^*(\varepsilon)\}$. The set of optimal Lagrange multipliers for a given solution $x \in S(\varepsilon)$ is the set $\{(u, w): \text{KKT}(x, u, w) \text{ holds}\}$.

The directional derivative of the function f^* at the point ε in the direction z is defined as

$$D_z f^*(\varepsilon) = \lim_{\alpha \to 0^+} \frac{f^*(\varepsilon + \alpha z) - f^*(\varepsilon)}{\alpha}$$

if the limit exists.

The problem P(ε) is said to be convex in x if f and the $-g_i$ are convex in x and the h_j are affine in x for each fixed $\varepsilon \in T$, and jointly convex if these functions have the respective properties in (x, ε) and T is a convex set. We assume that the functions defining problem P(ε) are continuous jointly in (x, ε) in the sequel.

Some Basic Theoretical Results – We make use of the following conditions that may hold at a feasible point x for some parameter value ε. Differentiability is assumed as needed, at (x, ε).

(a) The Karush-Kuhn-Tucker conditions, as before, designated KKT(x, u, w): there exist $u_i \geq 0$ $(i = 1, \ldots, m)$ and $w_j (j = 1, \ldots, p)$ such that $\nabla_x L(x, u, w, \varepsilon) = 0$, $u_i g_i(x, \varepsilon) = 0$ $(i = 1, \ldots, m)$, and $h_j(x, \varepsilon) = 0$ $(j = 1, \ldots, p)$;

(b) Linear independence, as before, designated LI(x): $\nabla_x g_i(x, \varepsilon)$ $(i \in B(x, \varepsilon))$, $\nabla_x h_j(x, \varepsilon)$ $(j = 1, \ldots, p)$ are linearly independent, where $B(x, \varepsilon) = \{i: g_i(x, \varepsilon) = 0\}$;

(c) Strict Complementary Slackness, designated SCS(x): $u_i > 0$ $(i \in B(x, \varepsilon))$;

(d) The Mangasarian-Fromovitz Constraint Qualification, designated MFCQ(x): (i) $\nabla_x h_j(x, \varepsilon)$ $(j = 1, \ldots, p)$ are linearly independent and there exists z such that $\nabla_x g_i(x, \varepsilon)z > 0$ $(i \in B(x,))$ and $\nabla_x h_j(x, \varepsilon)z = 0$ $(j = 1, \ldots, p)$;

(e) The Second-Order Sufficient Condition, designated SOSC(x, u, w): $z^\top \nabla_x^2 L(x, u, w, \varepsilon)z > 0$ for all $z \neq 0$ such that $\nabla_x g_i(x, \varepsilon)z \geq 0$ $(i \in B(x, \varepsilon))$, $\nabla_x g_i(x, \varepsilon)z = 0$ $(i \in D(x, \varepsilon) = \{i \in B(x, \varepsilon): u_i > 0\})$, and $\nabla_x h_j(x, \varepsilon)z = 0$ $(j = 1, \ldots, p)$, for some (u, w) such that KKT(x, u, w) holds;

Known facts relevant to this brief overview are the following.

The SOSC $(\bar{x}, \bar{u}, \bar{w})$ implies that $\bar{x}$ is a strict local minimizer (i.e., the unique global minimizer in some feasible neighborhood of $\bar{x}$) of P with optimal Lagrange multipliers $(\bar{u}, \bar{w})$. The condition MFCQ($\bar{x}$) holds at a local solution $\bar{x}$ if and only if the set of (u, w) satisfying KKT($\bar{x}, u, w$) is nonempty, compact and convex. If LI($\bar{x}$) holds at a local solution $\bar{x}$, then there exists a unique $(\bar{u}, \bar{w})$ satisfying KKT($\bar{x}, \bar{u}, \bar{w}$).

Using these and other well-known facts, we indicate some of the important results that hold for problem P(ε):

(i) If all f, g_i, h_j are once differentiable in (x, ε), $R(\bar{\varepsilon}) \neq \emptyset$, $R(\varepsilon)$ is contained in a compact set for ε near $\bar{\varepsilon}$, and MFCQ($\bar{x}$) holds for some $\bar{x} \in S(\bar{\varepsilon})$, then $f^* \in C$ at $\varepsilon = \bar{\varepsilon}$.

(ii) If problem P(ε) is jointly convex, then f^* is convex on T. If f is concave in ε and R does not depend on ε, then f^* is concave on T. Global parametric optimal value bounds can readily be calculated at a solution point when f^* is convex or concave and optimal Lagrange multipliers exist and are known. Also, when f^* is convex or concave, it follows from well-known results that f^* is continuous in the interior of T.

(iii) If T is a convex set and $R(\varepsilon) \neq \emptyset$ and is compact and independent of ε and f and $\nabla_\varepsilon f$ are jointly continuous in (x, ε), then $D_z f^*(\varepsilon) = \min \nabla_\varepsilon f(x, \varepsilon)z$ s.t. $x \in S(\varepsilon)$.

(iv) If f, g_i, h_j are twice continuously differentiable in (x, ε) and KKT($\bar{x}, \bar{u}, \bar{w}$), SOSC($\bar{x}, \bar{u}, \bar{w}$), LI($\bar{x}$) and SCS($\bar{x}$) hold at $\varepsilon = \bar{\varepsilon}$, then (x, u, w) is locally unique and once continuously differentiable as a function of, such that the assumptions persist to hold at $(x(\varepsilon), u(\varepsilon), w(\varepsilon))$, near $\bar{\varepsilon}$, f^* is twice continuously differentiable where $f^*(\varepsilon) = f[x(\varepsilon), \varepsilon]$ and $\nabla_\varepsilon f^*(\varepsilon) = \nabla_\varepsilon L[x(\varepsilon), u(\varepsilon), w(\varepsilon), \varepsilon]$ near $\varepsilon = \bar{\varepsilon}$. Thus, $x(\varepsilon)$ is an isolated (i.e., locally unique) and hence also strict local minimizer and $[u(\varepsilon), w(\varepsilon)]$ is unique. Strengthening SOSC by relaxing the restriction that $\nabla_x g_i(x, \varepsilon)z \geq 0$ $(i \in B(x, \varepsilon))$ and dropping SCS, we again find that $x(\varepsilon)$ is an isolated local minimizer with unique $[u(\varepsilon), w(\varepsilon)]$ since the assumptions again persist near $\varepsilon = \bar{\varepsilon}$ at $[x(\varepsilon), u(\varepsilon), w(\varepsilon)]$ locally unique, but now (x, u, w) is not once continuously differentiable in ε but only directionally differentiable, with f^* once continuously differentiable and $\nabla_\varepsilon f^* = \nabla_\varepsilon L$ as before. Relaxing LI to MFCQ and further strengthening SOSC as above and assuming this holds for all (u, w) in the set of optimal multipliers, the assumptions again locally persist, although now the Lagrange multipliers are not unique but are known to form a nonempty compact convex set, $x(\varepsilon)$ is a locally isolated minimizer as before and is known to be at least continuous and f^* is only directionally differentiable. It may also interest the reader to know that KKT, SOSC, LI and SCS are satisfied at the (unique vertex) solution of a nondegenerate linear programming problem.

Extensions and Future Research – With additional problem structure, more analytic results follow. For example, a fairly highly developed post-optimality sensitivity analysis is known and extensively used in linear programming, including parametric expressions for local solution changes and error bounds. Likewise, more can be said about unconstrained minimization, right-hand-side perturbations in the constraints, separable programs, geometric programs, etc. Closed form formulas or detailed characterizations have been given for optimal value, solution point and Lagrange multiplier parameter derivatives or directional derivatives when these exist, in addition to those noted in the last section.

Extensions of the kind of results indicated have been developed for problems in more general spaces or with less structure, for example, utilizing weaker constraint qualifications, involving an infinite number of variables or constraints such as in control theory and semi-infinite programming, multiobjective optimization, and integer programming, and stochastic programming. Further generalization of structure leads to variational inequalities, equilibrium problems and, at a more abstract level, generalized equations, for which a sophisticated parameter perturbation theory exists, specializations of which yield deep results for NLP. Qualitative extensions include significant additional more general convexity and concavity characterizations of the optimal value function for generalized convexity or concavity assumptions on the problem functions, more general optimal value derivative measures such as the Clarke generalized derivative, and other solution continuity concepts such as Holder continuity. Other significant extensions are those involving other (than parametric) classes of perturbations, for example, functional perturbations or abstract set-theoretic perturbations. A considerable literature exists on variations of all these ideas.

Two more research directions must be mentioned: the approximation of sensitivity information from information available as an algorithm makes progress towards a solution; and measurement of the effect of perturbations on the convergence and rate of convergence of solution algorithms. A solid basis for algorithmic approximation, the first topic, has been developed for barrier and penalty methods, but little else. Some results on the latter topic are known for a few standard algorithms.

Applications – All the results mentioned have significant theoretical and practical applications.

Perhaps one of the most obvious is in extrapolating from a solution with given data to a solution with perturbed data. Another is the approximation of the change in the optimal value resulting from perturbations of the constraints, a measure directly related to the associated optimal Lagrange multipliers ("shadow prices" in linear programming) which in turn are involved in duality relationships. Applications exist in decomposition, min-max problems, bilevel and multilevel programming, semi-infinite programming, implicit function optimization and other areas where optimal value functions of subproblems are encountered and certain variables are viewed as a function of others that are treated as parameters during a given iteration. Sensitivity analysis results provide valuable inputs to "parametric programming," where one endeavors to approximate a solution over a finite range or given set of parameter values.

Post-optimality sensitivity analysis for linear programming is a standard option in many commercial packages and is heavily used in practice, as noted. The potential applications of NLP sensitivity analysis are even more vast, though far from being realized as yet. NLP computational implementations on practical problems have been extremely limited, sporadic and *ad hoc*, largely experimental, and applied only to a few well-structured models. A number of experiments have been conducted on geometric programs, for example. Some of the other models and parameters for which a variety of sensitivity results have been generated are stream water pollution with maximum allowable dissolved oxygen deficit and on the order of 70 other parameters perturbed, a continuous review multi-item inventory model with several parameters such as item unit cost and the standard deviation of the lead-time demand, the structural design of a vertically corrugated transverse bulkhead of an oil tanker with many design parameters, portfolio analysis with parameters affecting risk and expected return on investments, and a power system energy model requiring the development of a turbine exhaust annulus and condenser system design with objective and constraint function parameter changes. Sensitivity information was calculated by SENSUMT, a computer code developed in 1973 by Fiacco, Armacost and Mylander (see Fiacco, 1983), using barrier function approximations. SENSUMT is apparently the first code to offer sensitivity analysis for NLP as a user option.

Notes and Literature – Most of the theoretical results given on sensitivity analysis can be found in Fiacco (1983), particularly in the survey given in Chapter 2 of that work. Much has been done elsewhere and since 1983, but we have confined our attention to a nucleus of early basic results that provides a good profile of the variety of qualitative and quantitative sensitivity measurements. Other directly relevant surveys are Fiacco and Hutzler (1982), Fiacco and Kyparisis (1992), and Fiacco and Ishizuka (1990). For a recent compendium on the state of the art of sensitivity and stability analysis in variational inequalities, and stochastic, semi-infinite, integer, nonlinear, geometric, linear and multi-objective programming with parameters, including results on continuity, differentiability, bounds, algorithmic perturbation results and continuation and parametric methods, the reader is referred to the collection of tutorials edited by Fiacco (1990). Hundreds of references are given to significant current work, including numerous references to other important areas such as generalized equations, curve-following techniques, multi-level programming and other topics mentioned in this article and beyond. Recent books in sensitivity analysis and related topics are those by Jongen, Jonker and Twilt (1986) on parametric results; Brosowski (1982) on semi-infinite optimization; Brosowski and Deutsch (1985) on approximation; Guddat, Jongen, Kummer and Nozicka (1987) on parametric optimization; Fiacco (1984) on a wide variety of topics; Bank, Guddat, Klatte, Kummer, and Tammer (1982) on continuity results in particular and nonlinear parametric optimization in general; Dontchev and Zolezzi on well-posed optimization (1993); and Levitin (1993) for a unified general perturbation theory.

See **Barrier and distance functions; Calculus of variations; Combinatorial and integer optimization; Linear programming; Parametric programming; Regression; Unconstrained optimization.**

References

[1] Anandalingam G., ed. (1992), *Hierarchical Optimization*, special issue of *Annals of Operations Research* 34, J.C. Baltzer, Basel, Switzerland.

[2] Bank B., J. Guddat, D. Klatte, B. Kummer and K. Tammer (1982), *Nonlinear Parametric Optimization*, Akademie-Verlag, Berlin.

[3] Bazaraa M.S. and C.M. Shetty (1979), *Nonlinear Programming, Theory and Algorithms*, Wiley, New York.

[4] Bracken J. and G.P. McCormick (1968), *Selected Applications of Nonlinear Programming*, Wiley, New York.

[5] Brosowski B. (1982), *Parametric Semi-Infinite Optimization*, Verlag Peter Lang, Frankfurt am Main.

[6] Brosowski B. and F. Deutsch, eds. (1985), *Parametric Optimization and Approximation*, Birkhauser.

[7] Dontchev A.L. and T. Zolezzi (1993), *Well-Posed Optimization Problems*, Springer-Verlag, Berlin Heidelberg.

[8] Fiacco A.V. (1983), *Introduction to Sensitivity and Stability Analysis in Nonlinear Programming*, Academic Press, New York.

[9] Fiacco A.V., ed., (1984), *Sensitivity, Stability and Parametric Analysis*, Mathematical Programming Study 21, Elsevier Science Publishers B.V., North-Holland.

[10] Fiacco A.V., ed. (1990), *Optimization with Data Perturbations*, special issue of *Annals of Operations Research* 27, J. C. Baltzer, Basel, Switzerland.

[11] Fiacco A.V. and W.P. Hutzler (1982), "Basic Results in the Development of Sensitivity and Stability Analysis in Nonlinear Programming," in *Mathematical Programming with Parameters and Multi-level Constraints*, J.E. Falk and A.V. Fiacco, eds., *Special Issue on Computers and Operations Research* 9, Pergamon, New York, 9–28.

[12] Fiacco A.V. and Y. Ishizuka (1990), "Sensitivity and Stability Analysis for Nonlinear Programming," *Annals of Operations Research* 27, 215–235.

[13] Fiacco A.V. and K.O. Kortanek, eds. (1983), *Semi-Infinite Programming and Applications*, Lecture Notes in Economics and Mathematical Systems, Number 215, Springer, Berlin.

[14] Fiacco A.V. and J. Kyparisis (1992), "A Tutorial on Parametric Nonlinear Programming Sensitivity and Stability Analysis," in *Systems and Management Science by Extremal Methods*: *Research Honoring Abraham Charnes at Age* 70, F. Y. Phillips and J. J. Rousseau, eds., Kluwer Academic Publishers, Boston, 205–223.

[15] Fiacco A.V. and G.P. McCormick (1968), *Nonlinear Programming, Sequential Unconstrained Minimization Techniques*, Wiley, New York. An unabridged corrected version was published by SIAM in the series *Classics in Applied Mathematics* (1990).

[16] Fletcher R. (1987), *Practical Methods of Optimization*, Wiley, New York.

[17] Gill P.E., W. Murray and M.H. Wright (1981), *Practical Optimization*, Academic Press, London.

[18] Gonzaga C.C. (1992), "Path Following Methods for Linear Programming," *SIAM Review* 34, 167–224.

[19] Guddat J., H. Th. Jongen, B. Kummer and F. Nozicka, eds. (1987), *Parametric Optimization and Related Topics*, Akademie-Verlag, Berlin.

[20] Hancock H. (1960), *Theory of Maxima and Minima*, Dover, New York.

[21] Hillier F.S. and G.J. Lieberman (1980), *Introduction to Operations Research*, Holden-Day, Oakland, California.

[22] Hobson R.F. and J.J. Weinkam (1979), "Curve Fitting," in *Operations Research Support Methodology* (A.G. Holzman, ed.), Marcel-Dekker, New York, 335–362.

[23] Hocking L.M. (1991), *Optimal Control: An Introduction to the Theory with Applications*. Oxford University Press, New York.

[24] Jongen H. Th., P. Jonker and F. Twilt (1986), *Nonlinear Optimization in R^n: II. Transversality, Flows, Parametric Aspects*, Verlag Peter Lang, New York.

[25] Kan A.H., G. Rinnooy and G.T. Timmer (1989), "Global Optimization," Ch. IX in *Handbooks in OR&MS*, 1, G. L. Nemhauser *et al.*, eds., North-Holland, Amsterdam.

[26] Karmakar N. (1984), "A New Polynomial-time Algorithm for Linear Programming," *Combinatoria* 4, 373–395.

[27] Khachian L.G. (1979), "A Polynomial Algorithm in Linear Programming," *Soviet Mathematics Doklady* 20, 191–94.

[28] Kuhn H.W. (1991), "Nonlinear Programming: A Historical Note," in *History of Mathematical Programming* (J. K. Lenstra, A.H.G. Rinnooy Kan and A. Schrijver, eds.), North-Holland, Amsterdam, 82–96.

[29] Kuhn H.W. and A.W. Tucker (1951), "Nonlinear Programming," *Proceedings of the Second Berkeley Symposium on Mathematical Statistics and Probability* (J. Neyman, ed.), University of California Press, Berkeley, 481–493.

[30] Lagrange J.L. (1762), "Essai sur une Nouvelle Methode pour Determiner les Maxima et Minima des Formules Integrales Indefinies," in *Miscellanea Taurinensia* II, 173–195.

[31] Levitin E.S. (1993), *Perturbation Theory in Mathematical Programming*, Wiley-Interscience, New York.

[32] McCormick G.P. (1983), *Nonlinear Programming: Theory, Algorithms and Applications*, Wiley, New York.

[33] Megiddo N., ed. (1989), *Progress in Mathematical Programming – Interior Point and Related Methods*, Springer, New York.

[34] Mor J.J. and S.J. Wright (1993), *Optimization Software Guide*, Frontiers in Applied Mathematics 14, SIAM.

[35] Neittaanmaki M. (1992), *Nonsmooth Optimization*, World Scientific Publishing, London.

[36] Nesterov Y. and A. Nemirovski (1993), *Interior-Point Polynomial Algorithms in Convex Programming*, Studies in Applied Mathematics 13, SIAM.

[37] Rockafellar R.T. (1970), *Convex Analysis*, Princeton University Press, Princeton, NJ.

[38] Rosen J.B., ed. (1990), *Supercomputers and Large-Scale Optimization: Algorithms, Software, Applications*, special issue of *Annals of Operations Research* 22(1)–(4), J.C. Baltzer, Basel, Switzerland.

[39] Sawaragi Y., H. Nakayama and T. Tanino (1985), *Theory of Multiobjective Optimization*, Academic Press, New York.

[40] Schrijver A. (1986), *Theory of Linear and Integer Programming*, Wiley, New York.

[41] Waren A.D., M.S. Hung and L.S. Lasdon (1987), "The Status of Nonlinear Programming Software: An Update," *Operations Research* 35, 489-503.

[42] Wets Roger J.-B. (1989), "Stochastic Programming," in *Handbooks in OR&MS*, 1, Elsevier/North-Holland, Amsterdam.

[43] Winston W.L. (1991), *Operations Research: Applications and Algorithms*, PWI-Kent Publishing, Boston.

[44] Wright M.H. (1992), "Interior Methods for Constrained Optimization," in *Acta Numerica 1* (A. Iserles, ed.), Cambridge University Press, New York, 341–407.

NONNEGATIVE SOLUTION

A solution to a problem in which all variables $x_j \geq 0$.

NONNEGATIVITY CONDITIONS

A restriction that limits a variable or a set of variables to be either zero or positive. The set of conditions $x_j \geq 0$ $(j = 1, \ldots, n)$ are the usual nonnegativity conditions that apply to the variables of a linear-programming problem.

NON-PREEMPTIVE

(1) In queueing models, when customers in service are not replaced when a customer with higher priority arrives. See **Queueing theory**. (2) See **Goal programming**.

NONSINGULAR MATRIX

A square matrix that has an inverse. A nonsingular matrix has a nonzero value for its determinant. See **Matrices and matrix algebra**.

NONTRIVIAL SOLUTION

For the set of homogeneous linear equations $Ax = 0$, a solution $x \neq 0$. See **Null space; Trivial solution**.

NONZERO-SUM GAME

A game in which the payoffs p_i to the players do not sum to zero. Here the payoff to player i is positive if it is a win and negative if it is a loss. See **Payoff matrix**; **Game theory**; **Zero-sum game**.

NORMATIVE MODEL

A model that attempts to describe standards of behavior of a man/machine system; the "what ought to be." Normative models identify feasible and desirable configurations of the system to serve as goals or norms. For a decision problem, such a model specifies logically consistent decision procedures that indicate how an individual should decide. Normative models are often based on an axiomatic foundation. See **Decision problem**; **Descriptive model**; **Mathematical model**; **Prescriptive model**.

NORTHWEST-CORNER SOLUTION

A procedure for finding a basic feasible solution to a transportation problem. For a problem with m origins and n destinations, the approach is to form an array with m rows and n columns, where a cell (i, j) of the array represents the shipment of goods from origin i to destination j. The algorithm starts with all shipments zero and first assigns the maximum shipment possible to the most northwest cell ($i = 1$, $j = 1$). Each time an allocation is made, either a row or column of the array is crossed out. The algorithm continues to make the maximum possible shipments in the northwest corners of the reduced arrays, until the shipment is made in cell $i = m$ and $j = n$. The resulting shipments form a basic feasible solution to be the underlying linear-programming problem. A degeneracy avoiding procedure may have to used in determining whether a row or column is to be crossed out in the intermediate steps.

NP, NP-COMPLETE, NP-HARD

See **Computational complexity**.

NULL MATRIX

A matrix with all entries equal to zero. See **Matrices and matrix algebra**.

NULL SPACE

The set of solutions to the equations $Ax = 0$ is called the nullspace of A. See **Trivial solution**.

NUMERICAL ANALYSIS

Stephen G. Nash

George Mason University, Fairfax, Virginia

INTRODUCTION: Numerical analysis uses computation as a tool to investigate mathematical models. At its most basic, this might mean computing an answer, such as the optimal value of a linear program. Beyond this, one might want error estimates (how accurate is the optimal value computed by the algorithm) or sensitivity information (how sensitive is the optimal value to changes in the data). One might even wish to analyze the effects of randomness in the data. Numerical analysis can also be used as an experimental tool (perhaps combined with computer graphics) to reveal properties of models that may be inaccessible by analytic means.

The techniques of numerical analysis have been widely adopted. It is rare for someone to solve a linear program by hand – except perhaps in a classroom. Large-scale simulations would be all but impossible without the aid of a computer. For many people, numerical techniques have superseded analytic techniques as a tool for solving mathematical problems. There are many cases (such as when evaluating integrals or solving differential equations) where no closed form analytic solution exists, but where a numerical solution is easy to compute. There are also cases where, even when an analytic solution is available, it is preferable to use a numerical method because it can compute the solution more efficiently and more accurately. In many areas of application, numerical analysis offers a routine, reliable, and often automated way of solving mathematical problems.

THE IMPACT OF COMPUTERS: It makes sense to speak of numerical analysis together with the computer. Numerical analysis only developed as a separate discipline after the invention of the computer. Although computation was an important subject at earlier times, it is only with the invention of the computer that the full range of numerical analysis techniques becomes necessary. Pencil and paper calculations tend to be small scale, and are carefully supervised. There is less opportunity for accumulation of error. In addition, the precision of the calculations can be adjusted during a calculation, if that becomes necessary. On a computer, however, it is easy to perform a sequence of millions of calculations. These calculations will normally be performed at a fixed precision, without supervision. Further, algorithms that are satisfactory for small problems, may not scale well to larger problems. Automatic computation carries with it both opportunities and risks. The techniques of numerical analysis attempt to exploit these opportunities while understanding and minimizing the risks involved.

There are some central questions in the study of numerical analysis. Is there an efficient algorithm to solve the given mathematical problem? How sensitive is the solution of the problem to errors in the data? How accurate is the computed solution? Can the algorithm provide an error estimate?

The most important and immediate question is whether there exists any algorithm to solve a particular problem. Currently a wide variety of numerical software is available, so for many classes of problems good methods are available. (Some sources are listed in the references.) These methods are capable of solving a great many problems that lack closed-form solutions. Even when closed-form solutions exist, the methods used in the software may be unrelated, for reasons of efficiency and accuracy. For example, the eigenvalues of a matrix will generally be calculated without forming the characteristic polynomial.

Good numerical methods, together with powerful modern computers, have made possible the routine solution of many large and difficult computational problems. Linear programs with thousands of variables pose no great challenge, for example. Although there still exist problems that strain the fastest supercomputers (such as three-dimensional fluid flows), off-the-shelf software and desk-top computers are capable of serving many people's needs.

LINEAR EQUATIONS: The ideas of numerical analysis are perhaps most clearly expressed in the context of solving systems of linear equations. We can write such a system as $Ax = b$, where A is an $n \times n$ invertible matrix and b is the vector of right-hand side coefficients.

The most commonly used technique for solving linear equations is Gaussian elimination. Gaussian elimination requires about n^3 arithmetic operations to solve a linear system, where n is the number of variables. On many current computers, a linear system with a few hundred vari-

ables can be solved in several seconds, and supercomputers can solve problems with a thousand variables in well under a second. If the number of variables doubles, the number of arithmetic operations increases by a factor of eight.

Gaussian elimination does not compute A^{-1}, and in fact there are a number of reasons why computing A^{-1} is undesirable in many circumstances (Golub and Van Loan, 1989). This is especially true for large sparse problems, problems where many of the entries in the matrix A are zero. Gaussian elimination can take advantage of the presence of these zeros. Often the number of arithmetic operations required to solve such a system will be proportional to the number of nonzeros in the matrix, which in turn will often be proportional to the number of variables n. In contrast, A^{-1} may have virtually no zero entries, even when A is sparse, and computing and applying the inverse will require between $O(n^2)$ and $O(n^3)$ operations. This is one case, among many, where the "mathematical" solution $x = A^{-1}b$ and the computer solution are calculated in different ways.

Gaussian elimination is not the only algorithm available for solving linear equations. There exist algorithms with costs proportional to n^{α} with $\alpha < 3$, but these are not widely used. There are also techniques (called "iterative" methods) that are especially effective on large sparse problems (Golub and Van Loan, 1989).

ERROR ANALYSIS: Suppose that one of these algorithms is applied to a system of linear equations. How accurately can the solution x be computed? We will phrase this question in another way: How sensitive is the solution x to errors in the data A and b? We will be concerned if small errors in the data are magnified into large errors in the solution. Such magnification can occur for two reasons. Either we have a "bad" problem (the solution is poorly determined by the data) or we have a "bad" algorithm (an algorithm that magnifies errors in the data). If the problem is bad, we will say that it is "ill conditioned," If the algorithm is bad, we will say that it is "unstable."

If the data (either the matrix A or the right-hand side b) are subject to errors of order ε, then the relative errors in the solution x will in general be proportional to $cond(A)\varepsilon$, where $cond(A)$, the condition number of A, is a measure of how close A is to being singular. (These errors in the solution are due solely to the errors in the data; for now we are assuming that the system of equations is solved exactly.) If A is singular, then $cond(A) = \infty$; otherwise,

$$cond(A) = \|A\| \cdot \|A^{-1}\|$$

in terms of some matrix norm $\|\cdot\|$. If, say, the Euclidean norm is used, then $cond(A) \geq 1$ for all matrices A. To illustrate this result, suppose that the data were accurate to $\varepsilon = 10^{-6}$ and that $cond(A) = 10^4$; then we would expect the relative errors in x to be proportional to 10^{-2}, that is, x would be accurate to two decimal digits. Thus, the condition number can be used as a quantitative tool to predict the accuracy of the solution to a linear system.

Even though A^{-1} is not computed by Gaussian elimination, it is still possible to *estimate* $cond(A)$ as a byproduct of the algorithm. Some additional calculations are required (about n^2 arithmetic operations), but this is much less than the n^3 operations required to solve the linear system. Thus, not only can the solution be computed, but an error estimate can be provided as well.

For most computational problems it will be possible to determine the sensitivity of the solution to errors in the data. This sensitivity can be considered as a "condition number" for that problem. If this condition number is large, it can be expected that the errors in the solution will be large, regardless of what algorithm is used to solve the problem. Of course, it will be desirable to use an algorithm that does not further magnify errors.

Computers only store numbers to a finite number of digits, often using binary arithmetic, so that just storing numbers in a computer can introduce errors in the data. For example, $1/3 = 0.333\ldots$ cannot be represented exactly with a finite number of binary digits. The precision of computer arithmetic – referred to as "machine epsilon" ε_{mach} or "unit roundoff" – limits the accuracy of computer calculations. Even if the data in a linear system are otherwise known exactly, when they are stored in the computer and computer arithmetic is used, the solution of the linear system can be expected to have errors proportional to ε_{mach} times $cond(A)$. If $cond(A) \approx 1/\varepsilon_{mach}$ then, from the point of view of computer arithmetic, the matrix might as well be singular.

Mathematically a matrix is either singular or nonsingular, and there are sharp differences between the two cases. Computationally we refer to the condition number of a matrix, and use this to measure how close a matrix is to being singular. Whether a matrix is "sufficiently nonsingular" to be useful will depend on the accuracy of the data and the desired accuracy of the solution. For many sorts of computational problems, the meaning of singularity or degeneracy will be blurred, with the accuracy of the solution deteriorating as the problem becomes closer to being degenerate.

So far, we have only considered errors arising from the data in the problem. The algorithm used to solve the linear system will also introduce errors. Let us assume that Gaussian elimination is used to solve the linear system. Gaussian elimination is unstable in its raw form, and can fail even when A is nonsingular. With minor modifications (such as the use of "partial pivoting") it becomes a stable algorithm that can be applied to any nonsingular system. It can be proved that Gaussian elimination with partial pivoting computes the *exact* solution to a perturbed system of the form $(A + E)x = b$, where $\|E\|$ is proportional to machine epsilon times $\|A\|$. Thus $(A + E)$ can be interpreted as a perturbation of A where the relative errors are proportional to machine epsilon. As has been mentioned, just storing A in the computer can introduce relative errors of this magnitude. Thus the errors introduced by Gaussian elimination are comparable to the errors introduced by storing the problem on the computer. Thus Gaussian elimination is considered to be a benign algorithm.

When we say that the computed solution from Gaussian elimination is the exact solution to a perturbed problem $(A + E)x = b$, we are adopting a distinctive point of view. For many, it will be more common to ask about the error in the computed solution. Instead we indicate by how much Gaussian elimination distorts the original problem. This is a property of the algorithm. The error in the solution (or the amount by which this distortion is magnified in the solution) is a property of the data, and depends on the condition number of the matrix. This point of view allows us to isolate the effect of the algorithm on the accuracy of the solution. In this case, it shows that Gaussian elimination computes the exact solution to a "nearby" problem. This point of view is referred to as a "backward" error analysis.

The error analysis for linear systems is particularly elegant. For other computational problems, the error analysis may not be so favorable (the computed solution may exactly solve a perturbed problem where the perturbations are large), or a backward error analysis may not be possible. In the latter case, other techniques must be used to assess the stability of an algorithm.

These comments illustrate some of the major questions of numerical analysis. In some settings additional questions arise. For example, the linear system might be obtained by discretizing a differential equation, that is, by approximating a continuous function using its values at finitely many points. Then it is natural to ask how accurately the solution of the linear system approximates the solution of the original continuous problem. In addition, we would like that the discrete solution converge to the continuous solution as the size of the finite-dimensional problem increases.

There are a further set of issues that must be faced when implementing an algorithm in software. The ultimate goal is to try to produce software that can efficiently compute a solution to full accuracy whenever the data and the solution consist of numbers that can be stored on the computer, and to design the software so that it works reliably on as large a collection of computers as possible. This goal can be difficult to achieve. Even seemingly innocuous tasks, such as computing the Euclidean norm of a vector, can require great care when the components of the vector are pathologically large or small, near the limits of computer arithmetic.

Numerical analysts continually try to solve ever larger and more difficult computational problems. This has often meant turning to vector and parallel computers, computers capable of carrying on multiple computations simultaneously. This has led to further questions. Can an efficient parallel algorithm be found to solve the problem? Is the algorithm "scalable," that is, does it continue to perform well as the problem size and the number of processors increase? How does the parallel algorithm compare to the best "scalar," or non-parallel, algorithm? Because of the variety of parallel and vector computers

available, the answers to these questions can vary from machine to machine, making it ever more difficult to design effective algorithms and software.

There is a vast literature on numerical analysis. General introductions to the topic can be found in Atkinson (1992), Kahaner, Moler, and Nash (1989), Kincaid and Cheney (1991), and Press *et al.* (1992). An extensive discussion of numerical linear algebra is given in Golub and Van Loan (1989), while a large online collection of software is described in Grosse (1994). The issues involved in developing software for parallel computers are mentioned in Anderson *et al.* (1992).

See **Gaussian elimination; Matrices and matrix algebra.**

References

[1] E. Anderson *et al.* (1992), *LAPACK Users' Guide*, SIAM, Philadelphia.

[2] K. Atkinson (1985), *Elementary Numerical Analysis*, Wiley, New York.

[3] G.H. Golub and C.F. Van Loan (1989), *Matrix Computations* (second edition), The Johns Hopkins University Press, Baltimore, Maryland.

[4] E. Grosse (1994), "Netlib Joins the World Wide Web," *SIAM News*, 27(5), 1–3.

[5] D.K. Kahaner, C.B. Moler and S.G. Nash (1989), *Numerical Methods and Software*, Prentice-Hall, Englewood Cliffs, New Jersey.

[6] D.R. Kincaid and E.W. Cheney (1991), *Numerical Analysis: Mathematics of Scientific Computing*, Brooks/Cole, Pacific Grove, California.

[7] W.H. Press, S.A. Teukolsky, W.T. Vetterling and B.P. Flannery (1992), *Numerical Recipes in FORTRAN: The Art of Scientific Computing* (second edition), Cambridge University Press, England.

O, o NOTATION

O means "of order of" and o means "of lower order than." If $\{u_n\}$ and $\{v_n\}$ are two sequences such that $|u_n/v_n| < K$ for every n greater than some fixed value n_0, where K is a constant independent of n, we write $u_n = O(v_n)$; for example, $(2n - 1)/(n^2 + 1) = O(1/n)$. The symbol O (colloquially called "big O") is also extended to the case of functions of a continuous variable; for example, $(x + 1) = O(x)$. We denote by $O(1)$ any function x which is defined for all values of x sufficiently large, and which either has a finite limit as x tends to infinity, or at least for all sufficiently large values of x remains less in absolute value than some fixed bound; for example, $\sin x = O(1)$.

If the limit of $u_n/o(v_n) = 0$, we write instead that $u_n = o(v_n)$ (colloquially called "little o"). Thus, $u_n < v_n$ and $u_n = o(v_n)$ are two different ways of expressing the same relation; for example, $\sin x - x = o(x)$. The notation is also extended to functions of a continuous variable. Furthermore, $u_n = o(1)$ as n tends to infinity means that u_n tends to 0 as n tends to infinity. In probability modeling (e.g., Markov chains and queueing theory), it is common to see $o(\Delta t)$ used to represent functions going to 0 faster than the small increment of time Δt, so that the $\lim_{\Delta t \to 0}[o(\Delta t)/\Delta t] = 0$.

OBJECTIVE FUNCTION

The mathematical expression that is to be optimized (maximized or minimized) in an optimization problem. See **Measure of effectiveness; Optimality criteria**.

OBJECT-ORIENTED DATABASE

See **Information systems and database design**.

OEG

Operations Evaluation Group. See **Center for Naval Analyses**.

OFFERED LOAD

The ratio of mean service time to mean interarrival time; the rate at which work is brought to a queueing system. See **Erlang; Queueing theory**.

OPEN NETWORK

A queueing network in which all customers enter and eventually leave the network, that is, routing process contains no closed subsets of states for any type of customer. See **Networks of queues; Queueing theory**.

OPERATIONS EVALUATION GROUP (OEG)

See **Center for Naval Analyses**.

OPERATIONS MANAGEMENT

Mark A. Vonderembse and
William G. Marchal

The University of Toledo, Ohio

INTRODUCTION: Organizations exist to meet the needs of society that people working alone cannot. Operations are one part of an organization, and they are responsible for producing the tremendous array of products in the quantities consumed each day. *Operations* are the processes which transform inputs (labor, capital, materials, and energy) into outputs (services and goods) consumed by the public. Operations employ people, build facilities, purchase equipment in order to change materials into finished goods such as computer hardware and/or to provide services such as computer software development.

Services are intangible products, and goods are physical products. According to the classification scheme used by the U.S. Department of Commerce and Labor, services include transportation, utilities, lodging, entertainment, health care, legal services, education, communications, wholesale and retail trade, banking and finance, public administration, insurance, real estate, and other miscellaneous services. Goods are described as articles of trade, merchandise, or wares. Manufacturing is a specific term referring to the production of goods. In this description, the term "product" is used to refer to both goods and services.

Whether an organization is producing services or goods as part of the for-profit private sector or the not-for-profit public sector, the output from its operations should be worth more to customers than the total cost of its inputs. As a result, an organization creates wealth for society through decisions and actions made in the management of operations.

Operations management is a multi-disciplinary sub-field of the study of management with particular focus on the production or "operations" function of the firm. Its scope includes decision making involving the design, planning, and management of the many factors that affect operations. Decisions include: what products to produce, how large a facility to build, how many people to hire, and what methods to use to increase product quality. Operations managers apply ideas and technologies to increase productivity and reduce costs, improve flexibility to meet rapidly changing customer needs, enhance quality, and improve customer service.

Organizations can use operations as an important way to gain an advantage on the competition. By linking operations with the overall strategy of the organization (including engineering, financial, marketing, and information system planning) synergy can result. Operations become a positive factor when facilities, equipment, and employee training are viewed as a means to achieve organizational rather than suboptimal departmental objectives.

Increasing demand for product variety and shorter product life cycles are forcing operations to respond more frequently and more quickly to customer needs. Competition is no longer based only on price or price and quality. Competition is becoming time-based with customers expecting high-quality, low-cost products that are designed and produced quickly to meet specific customer requirements. When flexibility is designed into operations, an organization is able to rapidly and inexpensively respond to changing customer needs. Organizations can use computers and information technology to become more flexible. Improvements in productivity and product quality provide the basis for competing in global markets.

To be successful, an organization should consider issues related to: designing a system that will be capable of producing the appropriate services and goods in the needed quantities; planning how to use the system effectively; and managing key elements of the operations. Each of these topics are described briefly in the following sections.

DESIGNING THE SYSTEM: Designing the system includes all the decisions necessary to determine the characteristics and features of the goods and services to be produced. It also establishes the facilities and information systems required to produce them. When designing a system which is capable of producing services and/or goods several questions arise.

What products will the organization produce (product development and design)?

What equipment and/or methods will be used (process design)?

How much capacity will an organization acquire?

Where will the facility be located?

How will the facility be laid out?

How will individual jobs and tasks be designed?

Product Development is a process for (1) assessing customer needs, (2) describing how products (both services and goods) can be designed to meet those needs, (3) determining how processes can be designed to make quality products efficiently and reliably, and (4) developing marketing, financial, and operating plans to successfully launch those products. Product development is a cross-functional decision process that requires teamwork. It is a key factor for success because it shapes how the organization competes.

Product Design is the determination of the characteristics and features of the product, that is, how the product functions. Product design determines a product's cost and quality as well as its features and performance, and these are the primary criteria on which customers make the purchase decision. Techniques such as Design for Manufacturing and Assembly (DFMA) are being implemented in many companies with very successful results. The objective is to improve product quality and lower product costs by focusing on manufacturing issues during product design. DFMA is implemented through computer software that points designers towards designs that would be easy to build by focusing on the economic and quality implications of design decisions. This is often critical because even though design is a small part of the overall cost of a product, the design decision may fix 80 to 90 percent of the manufacturing costs. Quality

Functional Deployment is also being used. It is a set of planning and communication routines that focuses and coordinates actions. The foundation is the belief that a product should be designed to reflect customers' desires and tastes.

Process Design is describing how the product will be made. The process design decision has two major components; a technical or engineering component and a scale economy or business component. The technical side requires that decisions be made regarding the technology to be used. For example, a fast food restaurant should decide whether its hamburgers will be flame broiled or fried. Decisions must also be made about the sequence of operations. For example, should a car rental agency inspect a car that has been returned by the customer first or send it to be cleaned and washed by maintenance? Decisions need to be made regarding the type of equipment to be used in making the good or providing the service. In addition, the methods and procedures used in performing the operations must also be determined.

The scale economy or business component involves applying the proper amount of mechanization (tools and equipment) to leverage the organization's work force to make it more productive. This involves determining if the demand for a product is large enough to justify mass production; if there is sufficient variety in customer demand so that flexible production systems are required; or if demand is so small that it cannot support a dedicated production facility.

Capacity is a measure of an organization's ability to provide the demanded services or goods in the amount requested and in a timely manner. More specifically, capacity is the maximum rate of production that can be sustained over a long period of time. Capacity planning involves estimating demand, determining the capacity of facilities, and deciding how to change the organization's capacity to respond to demand.

Facility Location is the placement of a facility with respect to its customers, suppliers, and other facilities with which it interacts. Normally, facility location is a strategic decision because it is a long-term commitment of resources which cannot easily or inexpensively be changed. When evaluating a location, management should consider: customer convenience, initial investment for land and facilities, operating costs, transportation costs, and government incentives. In addition, qualitative factors, such as availability of financial service, cultural activities for employees, and university research pro-

grams that relate to the firm's needs, should be considered.

Facility Layout is the arrangement of the work space within a facility. At it highest level, it considers which departments or work areas should be adjacent so that the flow of product, information, and people can move quickly and efficiently through the production system. Next, within the department or work area, where should people be located with respect to equipment and storage? How large should the department be? Finally, how should each work area within a department be arranged?

Job Design specifies the tasks, responsibilities, and methods used in performing the job. For example, the job design for x-ray technicians would describe what equipment they would need and would explain the standard operating procedures including the safety requirement that should be followed.

PLANNING THE SYSTEM: A plan is a list of actions management expects to take to deal with opportunities and problems present in the environment. Production planning is how management expects to utilize the resource base created when the production system was designed. One of the outcomes may be to change the design such as increasing or decreasing capacity and rearranging layout to enhance efficiency.

Production planning decisions depend upon the planning time horizon. Long-range decisions include how many facilities to add to match capacity with forecasted demand and how technological change might affect techniques used to manufacture goods and provide services. The time horizon for long-term planning varies with the industry and depends on how long it would take an organization to build new facilities. For example, in electric power generation it often takes ten or more years to build a new plant. So electrical utilities must plan at least that far into the future.

In medium-range production planning, which is normally about one year, organizations find it difficult to make substantial changes in facilities. In this case, production planning involves determining work force size and developing training programs, working with suppliers to improve product quality and improve delivery, and determining how much material to order on an aggregate basis.

Scheduling has the shortest planning horizon. As production planning proceeds from long-range planning to short-range scheduling, the decisions become more detailed. In scheduling, management must decide what product or products will be made; who will do the work; what equipment will be used; which materials will be consumed; when the work will begin; and what will happen to the product when the work is complete. All aspects of production come together to make the product a reality.

Some techniques used in production planning require special mention. Aggregate planning, material requirements planning, just-in-time, and the critical path method are important techniques that can be useful in production planning.

MANAGING THE SYSTEM: The impact of people, information, materials, and quality on operations is growing. As a result, managing these areas is a key factor for organizational success. Participative management and teamwork are becoming an essential part of successful operations. Motivation, leadership, and training are receiving new impetus.

Information systems are mechanisms for gathering, classifying, organizing, storing, analyzing, and disseminating information. Information requirements in some operations are extensive. From product development through job design and from long-range planning to scheduling, timely information is required to make better decisions.

Material management includes decisions regarding the procurement, control, handling, storage, and distribution of materials. Materials and material management are becoming more and more important because in many operations purchased material costs are over 50 percent of the total product cost. How much material should be ordered, when should it be ordered, and which supplier should it be ordered from are some of the important questions.

Producing quality products is a minimum requirement for a customer to consider an organization's product. Quality has progressed from an era of inspection to one of building quality at the source. Quality is increasingly becoming customer-driven with emphasis put on obtaining a product design that builds quality into the product. Then, the process is designed to transform the product design into a quality product and the employees are trained to execute it. The role of inspection is not to enhance quality but to determine if the designs are effective.

Over time, operations management has grown in scope. It has elements that are strategic; it relies on behavioral and engineering concepts; and it utilizes management science/operations research tools and techniques for systematic decision making and problem solving. As operations

management continues to develop, it will increasingly interact with other functional specialties such as research and development, marketing, engineering, and finance to develop integrated answers to complex interdisciplinary problems.

See **Facility location; Flexible manufacturing systems; Information systems/database design; Inventory modeling; Job shop scheduling; Organization; Production management; Quality control; Scheduling and sequencing; Total quality management.**

References

[1] Blackstone, J.H. (1989), *Capacity Management*, South-Western Publishing Co., Cincinnati, Ohio.

[2] Chase, R.B. and D.A. Garvin (1989), "The Service Factory," *Harvard Business Review*, **67, 4,** 61–69.

[3] Clark, K.B. and T. Fujimoto (1991), *Product Development Performance*, Harvard Business School Press, Boston, Massachusetts.

[4] Doll, W.J. and M.A. Vonderembse (1991), "The Evolution of Manufacturing Systems: Towards the Post-Industrial Enterprise," *OMEGA Int. Jl. Mgmt. Sci.* **19, 5,** 401–411.

[5] Garvin, David A. (1988), *Managing Quality*, The Free Press, New York.

[6] Skinner, W. (1969), "Manufacturing-Missing Link in Corporate Strategy," *Harvard Business Review* **52, 3,** 136–145.

[7] Sule, D.R. (1988), *Manufacturing Facilities: Location, Planning, and Design*, PWS-Kent Publishing Company, Boston, Massachusetts.

[8] Tersine, R.J. (1982), *Principles of Inventory and Materials Management*, Elsevier Science Publishing Co., New York.

[9] Umble, M.M. and M.L. Srikanth (1990), *Synchronous Manufacturing*, South-Western Publishing Co., Cincinnati, Ohio.

[10] Utterback, J.M. and W.J. Abernathy (1975), "A Dynamic Model of Process and Product Innovation," *OMEGA Int. Jl. Mgmt. Sci.* **3, 6,** 639–656.

[11] Vonderembse M.A. and G.P. White (1988). *Operations Management: Concepts, Methods, Strategies*, West Publishing Company, St. Paul, Minnesota.

OPERATIONS RESEARCH OFFICE AND RESEARCH ANALYSIS CORPORATION

Eugene P. Visco

Office of Deputy Under Secretary of the Army (Operations Research) Washington, DC

Carl M. Harris

George Mason University, Fairfax, Virginia

EARLY MILITARY OR: In discussing the early days of operations research in the United Kingdom, Blackett (1962) notes: "The Armed Services have for many decades made use of civilian scientists for the production of new weapons and vehicles of war, whereas the tactical and strategical use of these weapons and vehicles has been until recently almost exclusively a matter for the uniformed Service personnel. During the first years of the Second World War circumstances arose in which it was found that civilian scientists could sometimes play an important role in the study of tactics and strategy. The essential feature of these new circumstances was the very rapid introduction of new weapons and devices, pre-eminently radar, into the Services at a time both of great military difficulty and of such rapid expansion that the specialist officers of the Armed Services, who in less strenuous times can and do adequately compete with the problems raised, found themselves often quite unable to do so. I will attempt to describe . . . . how it was that civilian scientists, with initially little or no detailed knowledge of tactics or strategy, came to play a sometimes vital role in these affairs, and how there grew up a virtually new branch of military science – later to be dignified in the United Kingdom by the name 'Operational Research,' or 'Operations Analysis' in the United States. By the end of the war, all three Services had operational research groups of mainly civilian scientists either at headquarters or attached to the major independent Commands. These groups were, in varying degrees, in close touch with all the main activities of the Service operational staffs and were thus in a position to study the facts of operations in progress, to analyse them scientifically, and, when opportunities arose, to advise the staffs on how to improve the operational direction of the war . . . ."

While British military operational analysis was in place in all three uniformed services during World War II, U.S. military operations research during the war was carried out primarily in the Army Air Corps (later the Army Air Force) and the Navy. There was no single U.S. Army group comparable to the British Army Operational Research Group. There was a scattering of small groups doing operational analyses in various parts of the Army. The Signal Corps set up an Operational Research Division to prepare instruction manuals for radio communications by using operational experience data. The Office of Field Service, a major subdivision of the Office of Scientific Research and Development, provided civilian scientists, initially to conduct operational analyses, to Army units in the Pacific

Theater. However, the scientists were often called upon to carry out work other than operational analysis. Only the Navy and Army Air Force groups were dedicated to operational analyses. By war's end, the U.S. Army Air Force had 26 Operations Analysis sections assigned to the numbered Air Forces, Commands, Areas, Wings, Boards, and Schools. Approximately 250 analysts served in those sections. A wide range of professions were involved: typically 50 engineers, 40 educators and trainers, 35 mathematicians, 25 lawyers, and 21 physicists. Other professions represented included architects, meteorologists, physiologists, historian (one), agriculturists, investment analysts and stock brokers(!), astronomer (one), biologists, and many others – true adherence to the mixed team concept introduced by the British founders. Some analyses conducted at the Army's Aberdeen Proving Ground in its recently formed Ballistic Research Laboratory (BRL) can certainly be considered Army operations analysis, even though those words were not recognized there. A variety of survivability and vulnerability studies, particularly on Army aircraft, were carried out at BRL, as were many weapons effectiveness and bombing pattern analyses. However, there was no central overall Army operations research group, so identified, other than those working with the Air Forces, during World War II.

POST-WORLD WAR II ACTIVITIES: After the war, the British and United States wartime groups, in one form or another, continued to conduct operations research and analysis for their respective services. In the U.S., it was clear, due to the demonstrated relevance and importance of military operations research, that operations analysis organizations were needed in all the services. Thus, within a few years after the end of World War II, we find in place the Navy's Operations Evaluation Group (OEG) [later to become the Center for Naval Analyses (CNA)], the Air Force's Operations Analysis Division and Project RAND, and the Army's Operations Research Office (ORO). Each has played an important role in the history and development of OR/MS. The Operations Research Office of The Johns Hopkins University (JHU) was founded in 1948 by the U.S. Army to serve as the Army's civilian run organization for operations research analysis and studies, with offices in the Washington, DC area. ORO had the major goal of providing independent, objective, and scientifically sound studies of national security and defense issues.

THE ORO DIRECTOR: The history of ORO is one with the history of Ellis A. Johnson, its founder and only Director. After earning his M.S. and D.Sc. degrees at the Massachusetts Institute of Technology, Johnson went to Washington in 1934 to work on magnetic instruments at the U.S. Coast and Geodetic Survey. In 1935, he joined the Department of Terrestrial Magnetism, Carnegie Institution as a geophysicist. Early in 1940 he moved to the Naval Ordnance Laboratory (NOL), first as a consultant, then as Associate Director of Research, where he worked on degaussing as a countermine tactic, among other things. He quickly became interested in the operational offensive use of mines and countermeasures to mines. Even during the early days of the mine-countermine analysis, Johnson believed that analysts and researchers had to maintain a close association with those who had the ultimate responsibility for military operations – the very essence of operations research. Thus, from the outset, the ORO reflected Johnson's wartime experiences and the philosophy of analysis. [Much of what follows here, particularly in relation to Johnson, draws heavily on the tribute to him published by the Operations Research Society of America following his death (Page *et al.*, 1974).] With respect to the start-up of ORO, Page *et al.* (1974) notes:

"Thus, as ORO began its work, there was a working assumption that something called operations research was in being, and the Army anticipated its value enough to be willing to try to use it. But for the Army, this did not mean that it was clearly defined. Ground warfare was recognized as a more difficult field for operations research than air and sea warfare; on the one hand, ground warfare could not be affected so much by one new technical factor as air warfare was by radar, while, on the other, the analysis of the convenient geometry of the open space in the air or on the sea was quite inapplicable for troop movement on terrain. So, if OR was to play a significant role in support of Army planning, it would have to learn how to structure the problems, identify the elements amenable to analysis, and find methods of analysis by adaptation or invention. There were almost no direct precedents as to what could be expected . . . ." (Page *et al.*, 1974).

ORO ACTIVITIES AND PROJECTS: The organizational principles that quickly evolved included: a wide breadth of study topics; control and management of analysis in the hands of the researchers conducting the analyses; and close involvement with the operational elements of the Army, including access to real and often raw operational data representing performance of organizations and systems. Research leaders at ORO were also expected to conduct research themselves, to maintain a connection with the reality of research management.

The first two years of ORO included assignments from the Army covering a major study of military aid to other nations, a study of the causation of artillery firing errors, and armored force operations. During this time the staff was brought to the level of about 40 full time analysts, and a pool of more than 100 consultants was established, with ORO linkages to a number of research and analytic companies to provide additional, on-call support. Arrangements were concluded with the Army to establish a broad program of continuing research on nuclear weapons, tactics, logistics, military costing, psychological warfare, guerrilla warfare, and air defense. A core set of 15 projects was authorized and funded, thus providing a formative and formidable base from which to proceed. When the Korean War broke out in June 1950, the ORO was a functioning institution with a developing reputation for sound and practical analysis on behalf of the operational Army. Johnson quickly recognized a need and an opportunity in the war. He made an early visit to Korea to establish a *modus operandus* for field analysis teams in the theater of operations. By the fall, ORO had 40 analysts in the field (as many as the full staff only a few months earlier). At the end of the war, over 50% of the professional staff had spent time in the combat theater. Many hundreds of reports were written, with considerable impact on military operations. The ORO influence was felt in the UK and Canada, and operations analysts from those two UN participating countries joined their respective countries' military units operating in Korea.

A small ORO field team was organized at the Continental Army Command Headquarters, Fortress (now Fort) Monroe, Virginia. At that time, CONARC, as the Command was known, was responsible for development of operational doctrine for the Army and for the training related to that doctrine. It was Johnson's view that operations research could make important contributions to the development of doctrine, particularly considering the need for combat formations to adapt to the new considerations of ground combat under conditions of the potential use of atomic (later nuclear) weapons on the battlefield. ORO help design formations, assisting in the structure and doctrine for the Pentomic Division and the Pentagonal (for five combat commands) Division. Other studies looked at the vulnerability of armored formations to tactical nuclear weapons and at the potential for the offensive use of low yield nuclear weapons. Much attention was paid to tactical operations and logistics in the early days. Later, there were studies related to strategic matters, the most demanding and significant was a large study devoted to defense of the U.S. mainland from manned bomber attack involving nuclear weapons.

A field office of the ORO was also established at the headquarters of the U.S. Army Europe in Heidelberg, Federal Republic of Germany, where major contributions to the defense of Europe and NATO operations were made. Heavy use was made of war gaming and exercises for European operations at the Heidelberg office.

ORO was a continuing and positive force in advancing military OR and OR in general. It conducted a series of conferences designed to evaluate the Army's proposed research and development budget to help the Army understand the potential effects of R&D investments and improve the allocation of funds to the many R&D projects competing for support. The PISGAH (named for the mountain from which Moses saw the promised land) conferences brought uniformed officers, operations analysts, industrial scientists, and academics together to examine the Army's future needs. Seminars and colloquia were regular weekly events; the former related to planned or on-going research or outside speakers of note, while the latter focused on more abstruse mathematical analysis topics. ORO conducted experiments to test the capability of bright high school students to conduct (relatively) independent analyses, under the guidance of senior ORO analysts. Through the years, studies by teams of students were done on a wide range of topics, for example, the characteristics of effective air raid warning systems for civilians, and deep-thrust armored operations in difficult terrain. During the five years of the program, 75 students spent at least one summer at ORO; a number joined the regular staff after completing college.

During the 13 years of ORO activity, a full range of Army study topics were addressed: air operations and air defense; guerrilla, urban and unconventional warfare; tactical, intratheater and strategic mobility and logistics; weapons systems; civil defense; intelligence, psychological warfare and civil affairs; and, overall, Army readiness for operations in a complex national security world (Operations Research Office, 1961). Two examples are cited next to demonstrate the wide-ranging impact of ORO studies.

In the arena of tactical operations, ORO examined ways to improve the casualty producing capability of small arms fire. Two unique ideas were introduced and assessed. One was a salvo concept, developed from a patent taken out in the 19th century by an Army officer serving with the cavalry on the western frontier. The concept consisted of a system of two projectiles of rifle ammunition, one nested behind the other, with a single cartridge casing and propellant. With one trigger pull, the two rounds came out of the weapon in tandem. ORO analysts predicted, using probability theory, that the natural spread of the two projectiles would greatly increase the hit probability on a man-sized target at operational combat ranges. An ORO analyst, returning to an earlier principle of operations analysis as an experimental science, cast a few bullets in the salvo mode, loaded them with his hand-loading equipment, and fired them on his backyard range. The crude experimental results confirmed the statistical analysis. The Army accepted the results and standardized a two-round salvo projectile for the M14 .30 caliber rifle. The second concept concerning improved effectiveness of small arms fire focussed on infantry rifle training. The ORO analysts developed and tested a simulated infantry battlefield target array as an alternative to the known-distance range traditionally used to train infantry. Sets of man-sized targets were scattered over the battlefield and linked with electronic controls that caused the targets to pop-up to vertical positions simulating enemy shooters. The concept was adopted by the Army as the TRAINFIRE system.

The second example is a study the ORO did on the use of black soldiers in Korea and the extension of the study to the broader issue of full integration of black troops throughout the Army (Hausrath, 1954). From the Revolutionary War to the Korean War, it was traditional for the U.S. military services not to integrate its forces. President Truman's 1948 Presidential Executive Orders, directing equal opportunity in the Executive Branch and the Armed Forces, plus the growing post-World War II economy and major demographic changes, gave impetus for the study requested of ORO by the Army. The study used a wide range of tools: demographic analysis, opinion and attitude surveys, content analysis, critical incident technique, statistical analysis, and community surveys.

The study's summary notes the following:

"...this study provided policy-makers in the U.S. Army with objective evidence in support of integrated units of Negro and white soldiers. This evidence indicated: first, that integrated units allow more effective use of the manpower available through a more even distribution of aptitudes than is possible in segregated units; second, that performance of integrated units is satisfactory; and, third, that the resistance to integration is greatly reduced as experience is gained. The limit, if any, on the level of integration was shown to be above 20 per cent Negroes, and difficulties in extending integration to all parts of the Army were identified and arranged in a sequential order so that a program leading to Army-wide integration could be formulated." (Hausrath, 1954)

The ORO findings, conclusions and recommendations supported the Army process and success in integration during the 1950s.

END OF THE ORO: In 1961, The Johns Hopkins University, following a disagreement with the Army over management, withdrew from the contractual relationship. At midnight on August 31 1961, the Johns Hopkins University Operations Research Office ceased to exist. Its activities were transferred to the newly formed Research Analysis Corporation (RAC), a Federal Contract Research Center.

We close this brief history of ORO with a statement from the late Ellis A. Johnson, written in the summer of 1961 (Operations Research Office, 1961):

> "During the last 13 years ORO's accomplishments have indeed been noteworthy. ORO published 648 studies containing thousands of conclusions and recommendations. A majority of these have been adopted and acted on. This survey was written to summarize ORO accomplishments so that these could be considered in perspective and with satisfaction by those responsible for the accomplishments – the entire ORO staff: research staff, support staff, and administrative staff.
>
> "We can all be proud of this record."

TRANSITION TO RAC: Although the great bulk of RAC's work would be done for defense agencies, RAC sought to diversify its capabilities and its clients. As a result, RAC's activities were expanded to include the White House and the National Security Council; the Department of Defense; nine other governmental agencies with national security interests; some forty other governmental agencies of all levels; and private foundations whose primary interests lie outside the field of national security. This variety of clients, and their varying interests and requirements for research support, broadened RAC's resources of data and knowledge, analytic capability and interpretative skills. Work done on foreign aid and development further enriched RAC's capabilities to deal with the Army's problems, as did its work for urban clients on problems of crime and delinquency.

Throughout its existence, RAC viewed its major mission to be service to the public interest, chiefly by providing the Army with services, studies, research, and counsel based on operations research and systems analysis. RAC's Army work concentrated on force structure analysis and planning, logistics, military manpower, resource analysis, cost studies, and military gaming. These studies encompassed problems of operations, planning, intelligence, and research and development. In addition, RAC made studies of the nature and purposes of insurgency, counterinsurgency, and operations undertaken to stabilize societies under "threat." RAC examined the politico-military aspects of these regions where US land forces were already operating or providing advice and training, or might be called on to do so. These studies considered both current and projected environments. As ORO did, RAC set up field offices to study problems on the scene and to provide analytic support in direct conjunction with local tests and local operations.

RAC's PROJECT PORTFOLIO: Some of the largest projects RAC undertook involved very complex, worldwide logistical and transportation systems. RAC's studies of air mobility for army forces (in both wartime and peacetime) led to specific evaluations later on, after such systems were built, deployed, and put to work, thus creating bodies of hard data suitable for operations analysis.

There was also the continuation of ORO's emphasis on the assessment of weapons requirements and of the comparative effectiveness of competing weapons systems. RAC engineers and scientists also sought ways to improve the management of military research and development, seeking more efficient ways of allocating uncommitted resources to research and development projects. RAC analysts also dealt with communications, proposing new ways to allocate radio frequencies to military users and to improve the dependability of communication nets, early forerunners of today's highly sophisticated command, control and communications systems.

RAC inherited an especially strong program from ORO of basic research into quantitative methods for analyzing a wide variety of OR problems, particularly in the areas of mathematical programming and decision analysis. The term "think tank" was most appropriately applied to RAC at the time.

RAC's work also included economic, political and social science studies of problems arising outside formal military institutions. Most prominent were studies of public safety problems, the administration of justice and control of crime and delinquency, and economic and social development at home and abroad. These grew logically out of RAC's work for defense clients.

Also prominent in RAC's work on military subjects were manpower and personnel. Problems in these areas took on new dimensions in the 1960s, under the impact of the Vietnam war, the draft, and the later shift to an all-volunteer army.

ORO had pioneered the study of military costs and cost analysis, war gaming and simulations, and strategic and limited war. RAC continued these efforts, improving methods and exploring further the possibilities for applying more sophisticated and powerful analytic procedures to the unfolding problems of the 1960s. RAC also conducted studies in arms control and disarmament. It inaugurated a broad range of politico-military analyses relevant to the needs not only of military planners, but also of those concerned with broader questions of national security.

When RAC took over from ORO, a program of advanced research studies was well under way and maintained momentum throughout most of RAC's existence. This continuation was not without conflict, both within the Army and within RAC, since there were serious differences of opinion about how much effort (if any) should be devoted to basic research, as opposed to applications of existing techniques that would be directly useful to the Army in the short run. At the outset, top Army officials decided that such a program was needed and suggested that it form about 10% of the total effort under the RAC-Army contract. They felt it was necessary to explore ways in which advances in basic methodology might be used to deal with short-run problems. RAC's Advanced Research Goup devoted considerable attention to applications, while it continued to make advances primarily in mathematical programming and decision analysis. Fiacco and McCormick's 1968 seminal book on nonlinear programming was a product of the Advanced Research Group.

RAC also conducted a number of studies in the areas of decision, utility, and cognitive theory. In total, these studies were aimed at a compre-hensive understanding and theory of decision making at its various levels. In addition to client research applications, this work provided guidance for the establishment of a problem solving rationale within RAC. In 1972, the General Research Corporation (GRC), a for-profit organization, bought RAC and partly took over its staff, physical assets, and contract relationships with the Army and other RAC clients.

CONCLUDING REMARKS: ORO and RAC were important elements of OR/MS history. It would probably be fair to say that their combined contributions played a major role in establishing operations research as a paradigm for rational decision making.

See **Air Force operations analysis; Battle modeling; Center for Naval Analyses; Military operations research; RAND Corporation**.

References

[1] Blackett, P.M.S. (1962). *Studies of War, Nuclear and Conventional*, Hill and Wang, New York.
[2] Fiacco, A.V. and G.P. McCormick (1968). *Nonlinear Programming: Sequential Unconstrained Minimization Techniques*. John Wiley, New York.
[3] Harris, C.M., ed. (1993). "Roots of OR, I: The Research Analysis Corporation, The World War II years," *OR/MS Today*, 20(6), 30–36.
[4] Harris, C.M., ed. (1994). "Roots of OR, II: The History of the Research Analysis Corporation Continues with the Vietnam Era," *OR/MS Today*, 21(3), 46–49.
[5] Hausrath, A.H. (1954). "Utilization of Negro Manpower in the Army," *Jl. Operations Research Soc. Amer.*, 2, 17–30.
[6] Moore, L. (1968). "20th Anniversary of ORO/RAC," *The Raconteur*, 4(15).
[7] Operations Research Office (1961). *A Survey of ORO Accomplishments*, The Johns Hopkins University, Baltimore, Maryland.
[8] Page, T., G.S. Pettee and W.A. Wallace, (1974). "Ellis A. Johnson, 1906–1973," *Jl. Operations Research Soc. Amer.*, 22, 1141–1155.
[9] Parker, F.A. (1967). *An Introduction to the Research Analysis Corporation*, RAC, McLean, Virginia.

OPERATIONS RESEARCH SOCIETY OF AMERICA (ORSA)

Founded in 1952, the Operations Research Society of America was the major U. S. society for operations researchers. It was merged with The Institute of Management Sciences (TIMS) into the Institute for Operations Research and the Management Sciences (INFORMS) effective January 1, 1995. The purposes of ORSA were (1) the advancement of operations research through the exchange of information, (2) the establishment and maintenance of professional standards of competence for work known as operations research, (3) the improvement of the methods and techniques of operations research, (4) the encouragement and development of students of operations research, and (5) the useful applications of operations research. During the period, ORSA published the journal *Operations Research* (in 42 volumes), as well as other journals (some jointly with TIMS). In addition, ORSA sponsored national meetings (jointly with TIMS), and other meetings organized by its technical sections and geographic chapters. It was the U.S. representative to International Federation of Operational Research Societies (IFORS). See **Institute for Operations Research and the Management Sciences**; **The Institute of Management Sciences**.

OPPORTUNITY COST

The cost associated with forgoing an opportunity; the money or other value sacrificed by choosing a nonoptimal course of action. In linear programming, the opportunity cost is the reduced cost of a variable not in the optimal basic solution. If a unit of a nonbasic variable is introduced into the solution, the optimal value of the objective function would decrease by an amount equal to the associated reduced cost. See **Simplex method**.

OPTIMAL FEASIBLE SOLUTION

For an optimization problem, an optimal feasible solution is a solution that satisfies all the constraints of the problem and optimizes the objective function.

OPTIMALITY CRITERIA

For many optimization problems there are mathematical formulas by which one can test whether or not a given feasible solution is also optimal. For some nonlinear-programming problems, we have the Karush-Kuhn-Tucker conditions; for linear-programming problems, we have the simplex algorithm test applied to the reduced costs of the nonbasic variables. See **Linear programming**; **Nonlinear programming**; **Simplex method**.

OPTIMAL SOLUTION

See **Optimal feasible solution**.

OPTIMAL VALUE

The best value that can be realized or attained; for a mathematical programming problem, the minimum or maximum value of the objective function over the feasible region.

OPTIMAL VALUE FUNCTION

The optimal value of a mathematical programming problem as a function of problem parameters, such as objective function coefficients.

OPTIMIZATION

The process of searching for the best value that can be realized or attained. In mathematical programming, this is the minimum or maximum value of the objective function over the feasible region. See **Unconstrained optimization**.

OPTIMIZATION OF QUEUES

The process of determining the optimal setting of a particular queueing system parameter. The optimization refers to the minimization or maximization of a cost function where the parameter (or parameters) of interest appear as variables. See **Queueing theory**.

OR

Operations research or operational research.

ORGANIZATION

Richard M. Burton

Duke University, Durham, North Carolina

Børge Obel

Odense University, Denmark

INTRODUCTION: Organization studies encompass two areas: *organization theory* is a positive science to explain and understand the structure, behavior and effectiveness of an organization; and *organizational design* is a normative science to recommend better designs for increased effectiveness and efficiency. Organization theory attempts to understand and explain; organizational design creates and constructs an organization.

Organizing behavior is evident in history of even the earliest of recorded time. Ancient China was a highly organized society; a meritocracy with labor specialization. The Roman Empire, and in particular the Roman army, were efficiently designed. The modern organization is part and parcel to our civilization and its understanding fundamental to modern life. Not only is organization both timely and timeless, its study is basic in management science, political science, economics, sociology, business, and military science, to name a few. Organization study is interdisciplinary and central to all of social science.

The great insight that management science brought to our understanding of organization is that its basic work is information processing. In rough analogy, the nerve system, which channels information even more than the blood (energy carrier) or the skeleton (structure), provides the fundamental basis for understanding organization in modern life. Despite centuries of organization study, the organization as an information processor is an insight of the twentieth century—even the latter half of the twentieth century. To study organization without information is analogous to studying the human body ignoring the nervous system; it can be done, but so much is lost, or ignored.

We focus here on the contribution of management science and operations research to the study of organization. We give a more formal description and definition of an organization. Then, we consider a number of management science theories, models, and methods which help illuminate our ability to understand and design organizations. Finally, we briefly mention some alternative approaches to organization and what the future holds. Throughout, we give reference to the management science literature which will provide a beginning point to pursue the issue in greater detail.

ORGANIZATION AS INFORMATION PROCESSOR: What is an organization? Definitions abound. All have certain elements in common. An organization is a created social entity which is composed of individuals (and machines) who must be coordinated to achieve its purpose. March and Simon (1959), in an early and perhaps the most influential book in modern organization studies, wrote (p. 4):

A biological analogy is apt here, if we do not take it literally or too seriously. Organizations are assemblages of interacting human beings and they are the largest assemblages in our society that have anything resembling a central coordinative system. Let us grant that these coordinative systems are not developed nearly to the extent of the central nervous system in higher biological organism – that organizations are more earthworm than ape. Nevertheless, the high specificity of structure and coordination within organizations – as contrasted with the diffuse and variable rations among organizations and among unorganized individuals – marks off the individual organization as a sociological unit comparable in significance to the individual organism in biology.

March and Simon's organization examined attitudes, values, and goals and developed propositions about decision makers and problem solvers – a new organizational vocabulary to replace authority, responsibility and span of control as organizing principles. They provided a new basis for thinking about organizing – the principle of bounded rationality (p. 140-1):

Most human decision-making whether individual or organizational, is concerned with the discovery and selection of satisfactory alternatives; only in exceptional cases it is concerned with the discovery and selection of optimal alternatives.

This is in contrast to the rational economic man who makes optimal decisions in well defined environments. The information processing model is a powerful metaphor of an organization which processes information to obtain coordination:

- reads information, or observes the world,
- stores information, or remembers facts and programs,
- transmits information, or communicates among the members,
- transposes information, or makes decisions.

These are the work tasks of the organization. Information processing includes choosing, decision making and problem solving. At a very base level, the work of an organization is symbol manipulation. Whether human or machine, the organization is rational only in a bounded sense, reaching less than optimal decisions with the less than perfect information available to it.

Coordination of decisions and their implementation is the fundamental problem. Team theory models (Marschak and Radner, 1972) of organization were explicit mathematical models of multi-person organizations who had to make multiple decisions in the face of uncertainty – both uncertainty about the true state of nature, and about the information and decisions of other team members. Better prediction and communications and its use in deriving decision rules reduce the level of uncertainty, and obtain more nearly optimal decisions. The team theory models explicitly incorporated information: reading, storing, communicating, and calculating. The best information scheme, or organizational design balances the returns from nearly optimal coordinated actions and the costs of organizing.

The ship builder's problem is a deceptively simple, yet fundamental problem (p. 132):

Let a firm have two sales managers, each specializing in a different market for its product. Let it have two production facilities, one producing at low cost and another, more costly, to be used as a standby. This second facility can be visualized as a separate plant or as the use of the same plant at "overtime" periods, which involves higher wages. A conveniently simple case is offered by a shipyard firm with two docks (a new one and an old, less efficient one) and two markets ("East" and "West"). Each sales manager is offered a price for a ship to be delivered in his market. The prices offered in each of the two markets are the two state variables. (That is, the market prices have a priori *known probabilities of high or low.) There are two decision variables, each of them taking one of two values: either accept or reject the order.*

There are nine possible organizational designs about reading and communicating. For each case, there are decision rules which maximize the expected returns. Here are four of the possible designs:

1) No market information is gathered or communicated.
2) Both market prices are observed and communicated to a central headquarters.
3) Each market price is observed and a decision is made to accept the offer, or not, and,
4) The market price is obtained in one market and sent to a central headquarters, but not in the other.

The best design depends upon the returns from better information and cost of observing, communicating and choosing. The more costly the observations, the fewer the observations. The more costly communications are, the more decentralized the organization. The decision rules depend upon the information available.

Information processing is a core issue. March and Simon begin with the boundedly rational individual and build implications for the organization. Marschak and Radner take a number of perfectly rational individuals who are bounded rationally as a team, because of the limited information available, that is, imperfect information at the right time for the right person.

Focusing directly on organizational design, Galbraith (1974) assertively conjectured that the principal managerial task is to reduce uncertainty by processing information: "A basic proposition is that the greater the uncertainty of the task, the greater the amount of information that has to be processed between decision makers during the execution of the task" (p. 28).

Building upon March and Simon, Galbraith (p. 29) offered three mechanisms to obtain greater coordination among the decision makers:

1) *coordination by rules or programs.* Operational contingent rules can be stated in "if-then" terms, for example, if the inventory stock is less than 4, then re-order 10 items. Programs are compositions of large numbers of rules.
2) *hierarchy.* With greater uncertainty and no rules, exceptions and new situations are referred up the hierarchy for resolution. (This is a rule itself; *if* there is great uncertainty and no rule about what to do, *then* refer the issue up the hierarchy.)
3) *coordination by targets or goals.* Here the rules may be largely unspecified but the desired ends or goals can be stated. Subgoals are developed to obtain coordination among the units.

Adding to these organizational design alternatives, Galbraith (p. 30) then offered four information processing strategies. The first two reduce the need for information processing by creating slack resources, for example, excess personnel to complete a task and secondly, the creation of self-contained units, that is, small quasi-independent units. Alternatively, increased information processing capacity can be obtained by investing in vertical information system, for example, MIS, or creating lateral relations, that is, the genesis of the matrix organization.

Baligh (1990) develops a process of design that is systematic if it uses the algebra of decision rules. He defines an organization structure as a set of people connected by decision rules. These are mappings of which an element is of the form (If A, choose (do) one of the elements of B), where A represents a circumstance, and B a set of possible decisions. Each rule has a set of makers and a set of users. Different rules and different sets of rules describe many different structures, including all those mentioned above.

Decision rules identify their own information needs and design decisions must consider the returns to decision rules and the costs of the information they require.

Each organizational design alternative is developed and its appropriateness rationalized on the need for information to coordinate activities in the face of uncertainty.

OTHER MANAGEMENT SCIENCE CONTRIBUTIONS: The key feature of the management science approach has been the explicit organizational modeling of who does what, who talks to whom, and who takes what action under what conditions, that is, reading, storing, communicating and calculating. Management science approaches include both the positive science which explains the world and the normative science which identifies organizations which are efficient. All are explicit information models of organization. In this section, we provide a brief overview of these management science approaches.

A mathematical organization theory by R.F. Drenick (1986) was motivated by his "state of exasperation caused by organizational practices." How could a firm be so badly managed? Can we obtain better understanding? His motivation is fundamental to operations research: begin with practice and bring understanding through the rigor of mathematics and logical reasoning. He begins with Simon's succinct bounded rationality for organizing "it is only because human beings are limited in knowledge, foresight, skill and time, that organizations are useful instruments for the achievement of human purpose." Drenick incorporates individual characteristics of (in)efficiency, fallibility, and resistance to change and loyalty with mathematical precision. Motivated by his experience, Drenick focusses on alternative strategies to reduce the workload of the executive, that is, cope with the limited time to process information. Decentralization is precisely defined and shown to reduce the executive's workload, to obtain greater performance reliability, and to utilize private information effectively. Drenick concludes with a discussion on how to keep the design up to date, that is, organizational design.

Agency theory is a different mathematical approach to the ubiquitous principal-agent problem. An agent acts for, or on behalf of the principal. Agent-principal examples abound: most employee-employer contracts, investor-portfolio manager, auditor-stockholder, client-lawyer, and the sales agents in the shipyard problem above, to name a few. A principal may engage an agent for his/her effort, expertise, knowledge and special information. It is an asymmetric situation where the agent has information not available to the principal. Usually, the agent is assumed to be risk averse and the principal risk neutral; that is, the agent will take $1 rather than a 50/50 bet on $0 or $2, where the principal is indifferent between the two choices. The main question is how to write a risk sharing contract between the principal and the agent such that the agent acts faithfully on the principal's behalf where the principal cannot monitor the agent's special information, or even know the agent's risk preference. For our earlier example, should a salesperson refuse a ship order even when a sale can be made? It may maximize the firm's profits to refuse a possible sale but not the salesperson's commission. Should the ship salesperson share in the firm's profits rather than be given a sales commission? What are incentives for employees to work hard when the work task is uncertain, incentives for the portfolio manager to maximize the investor's return, etc. The analytical modeling is quite sophisticated: maximize the principal's utility constrained by the agent's utility under conditions of uncertainty. One organizational design problem is how to encourage innovation by risk averse employees for a risk neutral firm. Eisenhardt (1989) provides an excellent review and explores in more depth the application to organization studies.

Simulation offers a complementary approach to the study of organization. Simulation models are frequently complicated but mathematically ill-formulated, and hence, do not lend themselves to analytical closed form solutions. As such simulation is very powerful approach to investigate complex phenomenon without the need for inappropriately simplifying assumptions. Simulation models are usually explicit in their modeling of organization as an information processing task.

Cohen, March and Olsen (1972) developed a garbage can model of organizational choice – an intriguing metaphor, to say the least. They, too, began with observing how organizations (here, educational institutions) choose, or make decisions. Their discovery was a process in marked contrast to the normal scientific method which gathers data, defines the problem, lists the alternatives, chooses the best one and then implements it. Rather, the organization was a garbage can or unordered set of choices looking for problems, issues and feelings seeking forums for airing, solutions looking for issues and decision makers looking for work. They translated these observations into an explicit computer simulation model and were able to verify and explain a number of observations. Can such an organization or super bounded rationality ever accomplish anything? Perhaps surprisingly, yes! One conclusion is that "important problems are more likely to be solved than unimportant ones" (p. 10). This much is reassuring: such organizations can and do function and can be quite effective. Information is used in very complex ways. Their simulation model was devised to explain and understand these very complex organizational processes.

Burton and Obel (1984) formally modeled a hierarchical, decentralized organization using a decomposed linear program. Divisional units pass up local planning information to the headquarters unit which evaluates these plans and sends revised guidance on limited resource costs to the units. To replicate actual planning systems, only a very few iterations were permitted, prior to implementation. Research questions focussed on which organizational design, that is, information and decision making system, would yield the best performance in the face of uncertainty. The empirical results verify Williamson's M-form hypothesis that a divisional organization yields better performance than a functional organization. Other simulations investigated the best design for loosely coupled and tightly coupled technologies and the value of historical information for planning. They found that tightly coupled systems required more information processing. They also found that the value of historical information for future planning depends both on the uncertainty of the tasks and the choice of planning process.

These mathematical computer simulations were modified for laboratory experiences (Burton and Obel, 1988) to investigate the importance of opportunism, that is, will individuals give misleading information to better their own situation at the expense of others and the organization as a whole. Indeed, some will behave opportunistically, but, contrary to agency theory assumptions, some were altruistic. These simulation and laboratory studies were controlled experiments to investigate basic hypotheses in organizational design.

THE FUTURE: The management science information perspective described above remains a base for the future – after all, information processing is fundamental to what an organization does. Huber and McDaniel (1986) provide a complementary view. Nonetheless, there are other approaches or lenses to view the world and frameworks for description. Scott's (1992) is an extraordinarily comprehensive lucid integrating review of the sociological approach to organization theory. It is a positive science review and considers organizational design only implicitly.

Daft and Lewin (1990), in launching a new journal *Organization Science*, begin with a provocative question: "Is the field of organization studies irrelevant?" Where is the audience in business and government? They go on to argue that organization studies must break out of the normal science straightjacket. Knowledge from diverse sources and multiple perspectives should be encouraged, given the complexity of organization. No one method or approach can reveal true understanding – it takes multiple views. Organizational design includes the organization processes of culture, decision making, information processing, CEO values and style, at least. They conclude with a challenge for a new era in organization studies which uphold the rigor of scientific inquiry and embrace multiple perspectives and approaches. In a very short life, *Organization Science* has answered the challenge.

Other novel and interesting approaches include MacKenzie's (1991) description of organization utilizing a hologram. Carley and Prietula (1993) elevate the simulation approach to a more comprehensive computational organization theory, as a complement to deductive analytical modeling and field based empirical study. Computational organization theory is again an information processing perspective. Carley, *et al.* (1992) generalize the SOAR model. SOAR models simulate goal driven search behavior through problem spaces. SOAR acts through a series of decision cycles which include working memory, permanent memory of if-then production rules and a preference memory. The task is the retrieval of requested warehouse items involving multiple agents. Baligh, Burton and Obel (1990) develop an expert system which incorporates an organization theory knowledge base for organizational diagnostics and design. They, too, use an information processing rationale to create the knowledge base.

We have focussed on the information processing view of organization studies, that is, the nerve system and brain functions of organization. This view remains fundamental; yet, organization studies require a diversity of perspectives which you will find in *Management Science* and *Organization Science*, among others. Ever an evolving science, Daft and Lewin (1993) call for new theories to help us understand and manage organizational forms in a modern society.

See **Decision making; Economics; Hierarchical production planning; Information systems/database design; Marketing.**

References

[1] Baligh, H.H. (1990), "Decision Rule Theory and Its Use in the Analysis of the Organization's Performance," *Organization Science*, 1, 360–374.

[2] Baligh, H.H., R.M. Burton and B. Obel (1990), "Devising Expert Systems in Organization Theory: The Organizational Consultant," in *Organization, Management, and Expert Systems: Models of Automated Reasoning*, M. Mascuch, ed., Walter de Gruyter, Berlin.

[3] Burton, R.M. and B. Obel (1984), *Designing Efficient Organizations: Modelling and Experimentation*, North-Holland, Amsterdam.

[4] Burton, R.M. and B. Obel (1988), "Opportunism, Incentives and the M-form Hypothesis," *Jl. Economic Behavior and Organization*, 10, 99–119.

[5] Carley, K., J. Kjaer-Hasen, A. Newell and M. Prietula (1992), "Plural-Soar: A Proglegomenon to Artificial Agents and Organization Behavior," in *Artificial Intelligence in Organization and Management Theory*, M. Masuch and M. Waglien, eds., North-Holland, New York.

[6] Carley, K.M. and M.J. Prietula, eds. (1993), *Computational Organization Theory*, Lawrence Erlbaum Associates, Hinsdale, New Jersey.

[7] Cohen, M.D., J.G. March and P.J. Olsen (1972), "A Garbage Can Model of Organizational Choice," *Administrative Science Qtly.*, 17, 1–25.

[8] Daft, R.L. and A.Y. Lewin (1990), "Can Organization Studies Begin to Break out of the Normal Science Straitjacket? An Editorial Essay," *Organization Science*, 1, 1–9.

[9] Daft, R.L. and A.Y. Lewin (1993), "Where Are the Theories for the 'New' Organizational Forms? An Editorial Essay," *Organization Science*, 4.

[10] Drenick, R.F. (1986), *A Mathematical Organization Theory*, North-Holland, New York.

[11] Eisenhardt, K.M. (1989), "Agency Theory: An Assessment and Review," *Academy Management Review*, 14(1), 57–74.

[12] Galbraith, J.R. (1974), "Organizational Design: An Information Processing View," *Interfaces*, 4, 28–36.

[13] Huber, G.P. and R.R. McDaniel (1986), "The Decision-making Paradigm of Organizational Design," *Management Science*, 32, 572–589.

[14] MacKenzie, K.D. (1991), *The Organizational Hologram: The Effective Management of Organizational Change*, Kluwer Academic, Norwell, Massachusetts.

[15] March, J.G. and H.A. Simon (1958), *Organizations*, John Wiley, New York.

[16] Marschak, J. and R. Radner (1972), *Economic Theory of Teams*, Yale University Press, New Haven, Connecticut.

[17] Scott, W.R. (1992), *Organizations: Rational, Natural, and Open Systems*, Prentice-Hall, Englewood Cliffs, New Jersey.

ORIGIN NODE

A node in a network through which goods can enter the network. It is sometimes useful to define a special origin node through which all goods enter the network.

OR/MS

Operations research and management science.

ORO

See **Operations Research Office and Research Analysis Corporation.**

ORSA

See **Operations Research Society of America**.

OUT-OF-KILTER ALGORITHM

A special primal-dual algorithm for solving minimum cost network flow problems. See **Minimal-cost network-flow problem**.

OUTPUT PROCESS

The stochastic point or marked point process whereby the marks represent some aspect of the queueing customers or the state of the service stage or node and the points represent the times of customers leaving the server. This is not to be confused with the departure process, which requires that the customers leave the entire queueing system for good. For example, in queues with feedback, the output process (X°, T°) is the superposition of the departure and feedback processes. See **Departure process; Networks of queues; Queueing theory**.

OUTSIDE OBSERVER DISTRIBUTION

The probability distribution of the state of a queueing system at an arbitrarily chosen point in time, as opposed to what it would be at arrival or service-completion epochs. For queueing systems with a Poisson arrival process, all these steady-state distributions are the same. See **Queueing theory**.

OVERACHIEVEMENT VARIABLE

A nonnegative variable in a goal-programming problem constraint that measures how much the left-hand side of the constraint is greater than the right-hand side. See **Goal programming**.

OVERFLOW PROCESS

The stochastic marked point or point process of customers arriving to a queueing service center or node but not receiving service there. For example, the arrival process is composed of two stochastic processes, those gaining access to the server (i.e., the input process) and the overflow process of those not gaining access to the server. These distinctions are needed to model finite capacity-nodes. See **Arrival process**; **Input process**; **Queueing theory**.

OVERTAKING

In queueing networks with alternate paths, the ability of customers leaving a node to use arcs to move ahead of customers not taking those arcs so as to arrive at a subsequent node ahead of customers they were previously behind. See **Networks of queues**.

P

P⁴

Partitioned preassigned pivot procedure. A procedure for arranging the basis matrix of a linear-programming problem into as near a lower triangular form as possible. Such an arrangement helps in maintaining a sparse inverse, given that the original data set for the associated linear-programming problem is sparse. See **Linear programming**; **Revised simplex method**.

PACKING PROBLEM

The integer-programming problem defined as follows:

Maximize cx

subject to

$$Ex \le e$$

where the components of E are either one or zero, the components of the column vector e are all ones, and the variables are restricted to be either zero or one. The idea of the problem is to choose among items or combinations of items that can be packed into a container and to do so in the most effective way. See **Bin packing**; **Set covering problem**; **Set partitioning problem**.

PALM MEASURE

See **Markovian arrival process**.

PARALLEL COMPUTING

Jonathan Eckstein

Thinking Machines Corporation, Cambridge, Massachusetts

To the applications-oriented user, *parallel computing* is the use of a computer system that contains multiple, replicated arithmetic-logical units (ALUs), programmable to cooperate concurrently on a single task. This definition does not include other kinds of concurrency not typically visible to the applications programmer, such as the overlapping of floating point and integer operations, or launching of multiple concurrent instructions on superscalar microprocessor chips.

KINDS OF PARALLEL COMPUTERS: The taxonomy of Flynn (1972) classifies parallel computers as either "SIMD" or "MIMD." In a SIMD (**S**ingle **I**nstruction, **M**ultiple **D**ata) architecture, a single instruction stream controls all the ALUs in a synchronous manner: if one processor performs an add instruction, so do all others. However, local register values, memory contents, and processor state bits may vary between processing units. MIMD (**M**ultiple **I**nstruction, **M**ultiple **D**ata) architectures replicate instruction decoding hardware as well as ALU circuitry, and are typically built out of multiple standard microprocessor chips. There are a number of different, concurrent, and asynchronous streams of instructions, typically one for each ALU.

Another distinction is between local and shared memory. In pure local-memory architectures, each processor has its own memory bank, and information can be moved between different processors' banks only by messages passed between processors along a communication network. On the other end of the spectrum are pure shared-memory machines in which there is a single bank of global memory that is equally accessible to all processors. Such designs are difficult to scale to large numbers of processors due to memory contention problems.

Although the SIMD/MIMD and local/shared memory distinctions are useful general principles, many real systems do not fall neatly into these categories: for example, some systems have some local memory attached to each processor, and some global system memory. Others may look like MIMD collections of smaller SIMD systems, or provide an illusion of fully shared memory or a global address space, even though physical memory units are closely tied to particular processors.

In a large-scale system, it is not generally practical to provide a dedicated connection between every pair of processors. Popular interconnection patterns include rings, grids, meshes, toroids, butterflies, and hypercubes. One wants to minimize the cost and complexity of the communications hardware, but at the same time to keep the diameter of the network low. The topology yielding the best trade-off depends on the relative speed of communication versus computation, and the intended applications of the system. Some systems have "compound" networks, such as small rings connected by larger rings, or small dense graphs connected by a larger hypercube. In some architectures, such as the "fat tree" class (Leighton, 1991), processors occupy only a subset of the nodes of the communications network; the remaining nodes contain only communication switches. Chapter 1 of Bertsekas and Tsitsiklis (1989) also analyzes some common interconnection architectures.

"Distributed" systems, or "workstation farms" combine the background or off-hours capacity of collections of desktop computers to solve a single problem. This type of system can be inexpensive if the hardware is already available. However, the local-area networks that usually connect such systems impose severe restrictions on communication: at a given point in time, only one processor may be sending, and one receiving, on any branch of a typical network.

PROGRAMMING MODELS: Parallel computing is still evolving, so there is a tendency for each system to have its own unique programming environment, designed for the hardware at hand. However, a move towards standardization is under way, and certain basic principles are beginning to emerge. The first is the distinction between *data-parallel* and *control-parallel* specification of concurrency. In the data-parallel model, the program consists a single thread of control, but individual statements may manipulate large arrays of data in an implicitly parallel way. For example, if A, B, and C are arrays of the same size of shape, the statement $A = B + C$ might replace each element of A by the sum of the corresponding elements of B and C. Responsibility for portions of each array is typically partitioned between multiple processors, so they divide the work and perform it concurrently. Communication in data-parallel programs is typically invoked through certain standard intrinsic functions. For instance, the expression $SUM(A)$ might represent the sum, across all processors, of all A's elements, computed by whatever algorithm is optimal for the current hardware.

Data-parallel languages were initially developed for SIMD architectures, but "data-parallel" and "SIMD" are not synonymous. MIMD systems can be and are programmed in a data-parallel manner when it suits the application at hand. In fact, data-parallel programming styles may sometimes be appropriate for *sequential* computers. The most common basis for today's data-parallel languages is FORTRAN 90 (Metcalf and Reid, 1990).

In control-parallel programming models, restricted to MIMD architectures, there is a separate thread of control for each processor. If shared memory is available, threads communicate via memory, using mechanisms called "locks," "critical sections," or "monitors" to prevent simultaneous or inconsistent writes to the same location. Otherwise, processors communicate by sending and receiving messages, a style called *message passing*. Control parallel programs are typically written in standard sequential programming languages, handling messages and memory interlocks via special subroutine libraries.

It is generally accepted that control-parallel programs are harder to analyze, understand, develop, and debug than data-parallel programs, due to complicated "race" and "deadlock" conditions that can easily develop between threads. On the other hand, the data-parallel programmer must sacrifice some flexibility. An ideal compromise would be to invoke control parallelism only in the parts of a program where it is truly essential, and only to the extent that it is needed. "Global-local" programming models that would support such an approach are being developed.

SPEEDUP, EFFICIENCY AND SCALABILITY: If T_p is the time to solve a given problem using p processors, and T_1 is the time to solve the same problem with a single processor (using the best sequential algorithm, if it can be defined), then a key concept is *speedup*, defined to be $S_p = T_1/T_p$. *Efficiency* is then defined to be S_p/p, or, roughly speaking, the effectively used fraction of the raw computing power available. The main goal of parallel algorithm designers is to obtain *linear* speedups that grow roughly as p, or, equivalently, efficiencies that do not approach 0 as p increases.

A key motivation for using parallel computing is to solve ever-larger problems. Thus, rather than worrying about obtaining very large speedups for a fixed-size problem, it may be more important to study the effect on total solution time as the problem data and number of processors grow in some proportional or related way. This concept is called *scalability* (Kumar and Gupta, 1994).

APPLICATIONS IN OPERATIONS RESEARCH: Despite over a decade of active research, parallel computing has not had nearly the effect on practice in operations research as it has, for example, in computational fluid dynamics. This is due largely to the lack of efficient parallel methods for factoring and related operations on unstructured sparse matrices. Such operations are essential to the sparse active set and Newton methods that form the core of operations research's numerical optimization algorithms. However, successes have been reported for specially structured problems amenable to decomposition methods, including stochastic programming, and on dense problems. Parallelism has also proved very useful in branch-and-bound and related search algorithms, and in randomized algorithms for global optimization of complicated, non-convex functions. Simulation applications with many independent trials or scenarios are also natural applications for parallel computing. A general principle seems to be that one should take advantage of problem structure to localize troublesome operations, most typically sparse matrix factorization, onto individual processors. Another approach is to try radically new algorithms that avoid such operations completely, and are highly parallelizable. One should remember, however, that parallelism is not a panacea that can frequently make inappropriate or "brute force" methods competitive. The relationships between parallel computing and OR/MS are described and discussed in Barr and Hickman (1993), Eckstein (1993) and Zenios (1989).

See **Combinatorial and integer optimization; Simulation and discrete-event stochastic systems; Stochastic programming.**

References

[1] R.S. Barr and B.L. Hickman (1993). Reporting Computational Experiments with Parallel Algorithms: Issues, Measures and Experts' Opinions. *ORSA Jl. Computing* **5**, 2–18.

[2] D.P. Bertsekas and J. Tsitsiklis (1989). *Parallel and Distributed Computation: Numerical Methods.* Prentice-Hall, Englewood Cliffs, New Jersey.

[3] J. Eckstein (1993). Large-Scale Parallel Computing, Optimization, and Operations Research: A Survey. *ORSA Computer Science Technical Section Newsletter* **14(2)**, 1, 8–12.

[4] M.J. Flynn (1972). Some Computer Organizations and their Effectiveness. *IEEE Transactions Computers* **C-21**, 948–960.

[5] G.A.P. Kindervater and J.K. Lenstra (1988). Parallel Computing in Combinatorial Optimization. *Annals Operations Research* **14**, 245–289.

[6] V. Kumar and A. Gupta (1994). Analyzing Scalability of Parallel Algorithms and Architectures. *Jl. Parallel and Distributed Computing.*

[7] F.T. Leighton (1991). *Introduction to Parallel Algorithms and Architectures: Arrays, Trees, and Hypercubes.* Morgan Kaufmann, San Mateo, California.

[8] M. Metcalf and J. Reid (1990). *Fortran 90 Explained.* Oxford University Press, Oxford, United Kingdom.

[9] S.A. Zenios (1989). Parallel Numerical Optimization: Current Status and an Annotated Bibliography. *ORSA Jl. Computing* **1**, 20–43.

[10] S.A. Zenios (1994). Parallel and Supercomputing in the Practice of Management Science. *Interfaces*, **24**, 122–140.

PARAMETER

A quantity appearing in a mathematical model that is subject to controls beyond those affecting the decision variables.

PARAMETER-HOMOGENEOUS STOCHASTIC PROCESS

A stochastic process in which distribution properties between two index parameter points t_1 and t_2, $t_1 \leq t_2$, depend only on the difference $t_2 - t_1$, and not on the specific values of t_1 and t_2.

PARAMETRIC BOUND

An optimal value function or solution point bound as a function of problem parameters.

PARAMETRIC LINEAR PROGRAMMING

In the general linear-programming problem of

Minimize cx

subject to

$$Ax = b$$

$$x \geq 0$$

it is often appropriate to study how the optimal solution changes when some of the data are functions of a single parameter λ. Most mathematical programming systems allow parametric analysis of the cost coefficients (PARAOBJ), the right-hand-side elements (PARARHS), joint analysis of the objective function and right-hand-side elements (PARARIM), and the parametric analysis of the data in a row (PARAROW). See **Parametric programming.**

PARAMETRIC PROGRAMMING

Tomas Gal

FernUniversitaet, Hagen, Germany

INTRODUCTION: The meaning of a parameter as used here is best explained by a simple example. Recall that a parabola can be expressed as follows: $y = ax^2$, $a \neq 0$. Setting $a = 1$, we obtain a parabola that has a different shape from the parabola when we set $a = 5$. In both cases, however, we have parabolas that obey specific relationships; only the shapes are different. Hence, the parabola $y = ax^2$ describes a family of parabolas and the parameter a specifies the shape.

Consider the general mathematical-programming model:

$$\text{Max } z = f(\boldsymbol{x}) \tag{1}$$

subject to:

$$g(\boldsymbol{x}) \leq 0 \tag{2}$$

If we introduce one or more parameters, the model stays the same, but for each value of the parameter(s) we obtain a specific problem.

In setting up a mathematical optimization model, one of the first tasks is to collect data. The collected data might, however, be inaccurate, be of a stochastic character, be uncertain or be deficient in other ways. Therefore, it is appropriate to introduce parameters that enable us to analyze the influence of specific data elements on the optimal solution. This can be done by:

1) Introducing the parameter(s) at the beginning when setting up the model, or
2) Introducing the parameter(s) after an optimal solution has been found.

The latter case is called postoptimal analysis (POA) and is applied much more frequently than the first case.

Postoptimal analysis is a very important tool that should be used in the framework of a good report generator (Gal, 1993). The corresponding decision maker (DM) would then have information with which the DM can select a "firm-optimum." POA consists of several analyses, the most important of which is sensitivity analysis (SA). A sort of extended SA is parametric programming (PP). In nonlinear programming, SA corresponds to perturbation analysis, in which, after having found an optimal solution, some of the initial data are perturbed and the influence of the perturbation on the outcome is analyzed.

HISTORICAL SKETCH: Advanced methods for SA and PP for linear programming have been developed. In the 1950s, Orchard-Hays (in his master's thesis), Manne (1953), Saaty and Gass (1954), Gass and Saaty (1955) published the first works on parametric programming. By the end of the 1960s, the first monograph on parametric programming appeared (Dinkelbach, 1969), followed by the monograph and book by Gal (1973, 1979). In 1979, the first Symposium on "Data Perturbation and Parametric Programming" was organized by A.V. Fiacco in Washington, D.C., with such a symposium being held every year since. Several monographs (Bank *et al.*, 1982; Guddat *et al.*, 1991) and special journal issues have been published in the 1970s and 1980s. More details on the history of PP are given in Gal (1980, 1983). A bibliography with almost 1,000 items is given in Gal (1994a).

POSTOPTIMAL ANALYSIS: Let us now assume that the mathematical optimization model under consideration is a linear program of the form:

$$\text{Max } z = \boldsymbol{c}^T \boldsymbol{x} \tag{3}$$

subject to

$$\boldsymbol{A}\boldsymbol{x} = \boldsymbol{b}, \ \boldsymbol{x} \geq \boldsymbol{0} \tag{4}$$

where $\boldsymbol{c}$ is an n-vector of objective function coefficients (OFC) c_j, $\boldsymbol{x}$ is an n-vector of the variables, $\boldsymbol{A}$ is an $m \times n$ matrix of the technological coefficients a_{ij}, $m < n$, $\boldsymbol{b}$ is an m vector of the right-hand-side (RHS) elements b_i. All vectors are column vectors.

Suppose that the problem defined by (3) and (4) has an optimal basic feasible solution $\boldsymbol{x}_{\mathrm{B}} = \boldsymbol{B}^{-1}\boldsymbol{b}$, where $\boldsymbol{B}^{-1}$ is the inverse of the $m \times m$ basic matrix $\boldsymbol{B}$ (the basis) consisting of m linearly independent columns of $\boldsymbol{A}$. Here, $\boldsymbol{x}$ is an m-dimensional solution vector. This means that we have determined the following solution elements and simplex method elements:

1) The maximal value of the objective function (OF), $z_{\max}$,
2) The values of the basic variables x_i, $i = 1, \ldots, m$, and
3) The reduced costs $d_j = z_j - c_j$, $j = 1, \ldots, n$.

In the framework of POA, an evaluation of the above solution elements is to be performed. This means, that we provide the DM with information about the meaning of the values of the basic variables, tell the DM which resources are used and are critical (values of slack variables), and interpret the values of the opportunity costs and shadow prices. We also might carry out a suboptimal analysis, that is, show the DM what happens if one or several nonbasic variables were introduced into the solution at a positive level.

SENSITIVITY ANALYSIS: The POA would continue by performing a SA with respect to the OF and the RHS. This analysis is usually a part of the solution output for just about all linear-programming software. It is called OFC-ranging and RHS-ranging. Behind such analyses is the introduction of a scalar parameter, t or λ, in the form

$$c_j(t) = c_j + t \text{ , } j \text{ fixed} \tag{5}$$

or

$$b_i(\lambda) = b_i + \lambda, \text{ } i \text{ fixed} \tag{6}$$

SA finds a critical interval T_j or Λ_i, such that for all $t \in T_j$ or $\lambda \in \Lambda_i$, respectively, the (found) optimal basis B remains the same. The critical values, that is, the upper and lower bounds of the critical interval can be easily determined by certain formulas (Gass, 1985). A change in a RHS element b_i causes the values of the basic variables and the value of $z_{\max}$ to change, while a change in an OFC c_j causes the values of the reduced costs and the value of $z_{\max}$ to change. Such information is of great value to the DM. An assumption of this type of SA is that we are investigating how the optimal solution would vary with respect to a change in one data element, while holding all other data fixed. Analysis of multiple changes can be done in a limited manner by the techniques of the hundred percent rule (Bradley, Hax, and Magnanti, 1977) and tolerance analysis (Wendell, 1985).

PARAMETRIC ANALYSIS: For an element b_i of the RHS, we ask the question: for what range of values of the parameter λ in (6) does there exist an optimal solution to (3) and (4)? Given such values, we can move from the original optimal basis and generate a sequence of optimal bases, with each basis associated with a critical interval of the parameter. Such an analysis provides the DM with a full range of possible solutions from which a subset of optimal solutions appropriate for the given problem can be selected. The DM then chooses a certain value of the parameter and, thus, a corresponding optimal solution for the parametric range of b_i.

Note that a similar analysis can be performed with respect to the parametric OFC, as given by (5). Moreover, taking into account the possibility that a parameter introduced in the RHS may influence some (or several) OFC or vice versa, it is possible to perform a RIM parametric analysis, that is, find a sequence of optimal bases to each of which a critical interval for the RHS- and OFC-parameters are associated simultaneously. Standard RHS, OFC and RIM parametric analysis procedures are usually included in linear-programming software.

It is also possible to perform a sensitivity or parametric analysis with respect to the elements a_{ij} of the matrix A. The corresponding procedures are, unfortunately, not incorporated into linear-programming software as the underlying formulas are a bit too complex. However, some software enables one to compute a series of linear programs in each of which slightly changed values of the $\{a_{ij}\}$ are chosen.

Up to now, we have discussed the simplest parametric case having one parameter with a coefficient equal to 1. The above cases can, however, also be carried out when:

   (i) a scalar parameter is introduced into several elements of the RHS and/or OFC with coefficients which differ from 1, and

   (ii) a parameter-vector (vector of parameters) is introduced into several elements of the RHS and/or OFC with their respective coefficients different from 1.

As far as case (i) is concerned, to each optimal basis a critical interval is associated. In case (ii), each optimal basis is associated with a higher dimensional convex polyhedral set of parameters. In the RIM case, each optimal basis is associated with a higher dimensional interval, a box, provided that the parameters in the RHS and OFC are independent from each other. The larger the number of parameters in the parameter-vector, the more difficult it is to interpret the results and for the DM to find an appropriate optimal basis. In such cases, an interactive approach is recommended in which the "parametric specialist" helps the DM to select an appropriate solution.

APPLICATIONS: There are two kinds of uses of PP:

   1) introducing parameters into various classes of mathematical programming problems for solving these problems via parametrization; and

   2) practical applications.

As to 1), the introduction of parameters helps to solve problems from the areas of nonconcave-mathematical programming, decomposition, approximation, and integer programming. Also, note that by replacing the OFC in (3) and (4) with a matrix C times a parameter-vector t we obtain the problem

$$\text{Max } z = (C^T t)x,$$

subject to

$$Ax = b, \ x \geq 0$$

which is a scalarized version of a linear multi-objective-programming problem (Steuer, 1986). Methods for solving the corresponding homogeneous multiparameter-programming problem provide a procedure to determine the set of all efficient solutions of the corresponding multiobjective problem (Gal, 1994b).

As to 2), SA and/or PP has been used in the pipeline industry, in capital budgeting, for farm decision making, refinery operations, for return maximization in an enterprise, and a number of other applications (Gal, 1994b).

SA AND PP IN OTHER FIELDS: Theoretical and methodological works have been published about SA and/or PP in linear and nonlinear complementarity problems, control of dynamic systems, fractional programming, geometric programming, integer and quadratic programming problems, transportation problems. A more detailed survey with corresponding references is given in Gal (1994b).

DEGENERACY: Recall that a basic feasible solution to a linear-programming problem is called primal degenerate when at least one element of this solution equals zero. The corresponding extreme point of the feasible set, that is, of the convex polyhedron, is then also called degenerate. Degeneracy causes various kinds of efficiency and convergence problems and special precautions must be taken when performing SA for a degenerate extreme point. Degeneracy influences even POA, especially the determination of opportunity costs and shadow prices. Performing SA, the main rule, that is, "determine the critical interval such that the original optimal basis does not change," is not valid any more because, for a degenerate solution, many bases are associated with it. A theoretical discussion of this problem is given in Kruse (1986). Note that standard software analysis for RHS- or OFC-ranging yield false results when degeneracy is involved.

CONCLUSION: For linear programming and related mathematical areas, SA and PP have become important tools for analyzing variations in initial data, for obtaining better insight into and gaining more information about the related mathematical model, for improving our understanding of model building in general, and as aids in solving a wide range of mathematical problems.

See **Linear programming; Multiobjective programming; Nonlinear programming; Sensitivity analysis.**

References

[1] Bank, B., J. Guddat, D. Klatte, B. Kummer, and T. Tammer (1982). *Non-linear parametric optimization.* Akademie-Verlag, Berlin.

[2] Bradley, S.P., A.C. Hax, and T.L. Magnanti (1977). *Applied Mathematical Programming.* Addison-Wesley, Reading, Massachusetts.

[3] Dinkelbach, W. (1969). *Sensitivitsänalysen und parametrische Programmierung.* Springer Verlag, Berlin.

[4] Gal, T. (1973). *Betriebliche Entscheidungsprobleme, Sensitivitsänalyse und parametrische Programmierung.* W. de Gruyter, Berlin.

[5] Gal, T. (1979). *Postoptimal analyses, parametric programming and related topics.* McGraw Hill, New York.

[6] Gal, T. (1980). "A 'historiogramme' of parametric programming." *Jl. Operational Research Society,* 31, 449–451.

[7] Gal, T. (1983). "A note on the history of parametric programming". *Jl. Operational Research Society,* 34, 162–163.

[8] Gal, T. (1992). "Putting the LP survey into perspective." *OR/MS Today,* December, 93.

[9] Gal, T. (1994a). "Selected Bibliography on Degeneracy." *Annals Operations Research.*

[10] Gal, T. (1994b). *Postoptimal analyses and parametric programming.* Revised and updated edition. W. de Gruyter, Berlin.

[11] Gass, S.I. (1985). *Linear Programming,* 5th ed. McGraw-Hill, New York.

[12] Gass, S.I. and T. L. Saaty (1955). "The parametric objective function." *Naval Research Logistics Quarterly,* 2, 39–45.

[13] Guddat, J., F. Guerra Vazquez, H. Th. Jongen (1991). *Parametric optimization: singularities, pathfollowing and jumps.* B.G. Teubner, Stuttgart, and John Wiley, New York.

[14] Kruse, H.-J. (1986). *Degeneracy graphs and the neighborhood problem.* Lecture Notes in economics and mathematical systems No. 260. Springer Verlag, Berlin.

[15] Manne, A.S. (1953). "Notes on Parametric Linear Programming." RAND Report P-468. The Rand Corporation, Santa Monica, California.

[16] Saaty, T.L., and S.I. Gass (1954). "The Parametric Objective Function, Part I." *Operations Research,* 2, 316–319.

[17] Steuer, R.E. (1986). *Multiple criteria optimization: Theory, computation, and application.* John Wiley, New York.

[18] Wendell, R.E. (1985). "The Tolerance Approach to Sensitivity Analysis in Linear Programming." *Management Science,* 31, 564–578.

PARAMETRIC SOLUTION

A solution expressed as a function of problem parameters.

PARETO-OPTIMAL SOLUTION

If a feasible deviation from a solution to a multi-objective problem causes one of the objectives to improve while some other objective degrades, the solution is termed a Pareto-optimal. Such a solution is also called an efficient or nondominated solution. See **Efficient solution**.

PARTIAL BALANCE EQUATIONS

In queueing problems like networks, a subset of the global balance equations that may be satisfied at a node, that is, a balance of probability flux. See **Networks of queues**; **Queueing theory**.

PARTIAL PRICING

When determining a new variable to enter the basis by the simplex method, it is somewhat computationally inefficient to price out all nonbasic columns, as is the way of the standard simplex algorithm or its multiple pricing refinement. The scheme of partial pricing starts by searching the nonbasic variables in index order until a set, say 5, candidate vectors has been found. These vectors are then used as possible vectors to enter the basis, as is done in multiple pricing. After the candidate set is depleted, another set is found by searching the nonbasic vectors from the point where the first set stopped its search. The process continues in this manner by searching and selecting candidate sets until the optimal solution is found. Although the total number of iterations to solve a problem usually increases, computational time is saved by this type of pricing strategy. See **Simplex method**.

PASTA

Means Poisson Arrivals See Time Averages. If arrivals to a queue are Poisson, the (limiting) fraction of arrivals who find the process in some state is equal to the overall fraction of time the process is in that state. See **Queueing theory**.

PATH

A path in a network is a sequence of nodes and arcs that connect a designated initial node to a designated terminal node. See **Chain**; **Cycle**.

PAYOFF FUNCTION

An expression that defines the gains and losses for the players of a game in terms of the strategies they use is called a payoff function. See **Game theory**.

PAYOFF MATRIX

For a zero-sum, two-person game, the payoff matrix is an $m \times n$ matrix of real numbers with the entry a_{ij} representing the payoff to the maximizing player if the maximizing player plays strategy i and the minimizing player plays strategy j. See **Game theory**.

PDA

Parametric decomposition approach. See **Production management**.

PDF

Probability density function.

PDSA

Plan, do, study, act. See **Total quality management**.

PERIODIC REVIEW

A type of inventory control policy in which the inventory position is assessed at the end of each of a prescribed number of discrete time periods. It is to be contrasted with the continuous-review format. See **Inventory modeling**.

PERT

An event-oriented, project-network diagramming technique used for planning and scheduling. See **Network planning**; **Program evaluation and review technique**; **Project management**; **Research and development**.

PERTURBATION

A change in a parameter, function or set.

PERTURBATION METHODS

(1) Procedures that modify the constraints of a linear-programming problem so that all basic feasible solutions will be nondegenerate, thus removing the possibility of cycling in the simplex method. The modification can be either explicitly done by adding small quantities to the right-hand-sides or implicitly by using lexicographic procedures. See **Cycling**; **Degeneracy**; **Lexicographic ordering**. (2) A form of sensitivity analysis for discrete event systems. See **Score functions**.

PETRO-CHEMICAL INDUSTRY

Thomas E. Baker

Chesapeake Decision Sciences, Inc., New Providence, New Jersey

INTRODUCTION: Almost from its inception, the petro-chemical industry has been dominated by very large, fully integrated, multi-national companies competing in world markets. This competitive environment has led to the application of OR/MS tools in literally all facets of the business from search and estimation techniques in exploration to production and processing optimization to transportation and vehicle routing techniques in product delivery. In fact, it is difficult to identify an OR/MS tool or approach that has not been successfully applied in the petro-chemical industry.

During the 1960s and 1970s, most large companies in the industry maintained sizable OR/MS groups or departments with concentrations of expertise in linear programming, simulation and statistical analysis. These groups stretched the limits of the available problem-solving technologies and provided the impetus for many advancements in the OR/MS field during that era. Today, most of these central OR/MS groups have been disbanded; however the application of OR/MS tools continues as before since the tools themselves have long been embedded in the business functions that use them.

The embedding of OR/MS techniques in various business functions has the unfortunate side effect of making the applications literature harder to navigate. Applications in the petro-chemical industry are as likely to appear in the *Oil and Gas Journal* or in *Chemical Engineering* as they are in *Management Science*. In addition, most petroleum companies have gone through periods of publishing bans and the technological leaders in the chemical industry have had a tradition of keeping their good work to themselves. Fortunately, these policies are changing and more applications are finding their way into the literature today.

The exploration and production side of the industry has applied the widest variety of OR/MS tools. Simulation has been a common approach for oil exploration (Higgins, 1993) and gas exploration (Power, 1992). Hansen (1992) used mixed-integer programming and tabu search for exploration planning. Equilibrium modeling and game theory have been used in the lease bidding arena. Findlay (1989) described stochastic dynamic programming applied to oil production. Several companies have applied hybrid dynamic programming/linear programming approaches to long term planning in gas production.

Simulation has been the mainstay in oil and gas transportation facilities design. During the 1970s, most oil companies were engaged in developing discrete event simulation models for marine terminal design and for offshore platform design. Bammi (1990) describes the application of simulation for pipeline planning at Northern Plains Natural Gas Co.

USE OF LINEAR PROGRAMMING: The refining industry began using linear programming shortly after its invention (Bodington and Baker, 1990). In the early 1950s, many major oil companies began using LP-based product blending models (Charnes and Cooper, 1952) which severely tested the available computational capabilities of that time. As computer capabilities expanded, so did the scope of LP models, encompassing whole refineries (Symonds, 1955) and the US refining industry (Manne, 1958).

The simplest form of a quality-based, multi-period refinery model, ignoring capacities, is described below.

Indices:	$i \in$ OPR operations
	$j \in$ STR all refinery streams
	$k \in$ PRO blended products
	$m \in$ QUA qualities
	$n \in$ TIM time periods
Variables:	f_{jkn} – flow from stream j to product k in period n
	p_{in} – operation i in period n
	h_{jn} – inventory for stream j at end of period n
	q_{jmn} – stream j quality m at end of period n

Data: Y_{ij} – yield of stream j from operation i
         Q_{ijm} – stream j quality m from operation i
         D_{kn} – demand for product k in period n

Equations:

$$h_{jn} = h_{jn-1} + \sum_{i \in \text{OPR}} Y_{ij} p_{in} - \sum_{k \in \text{PRO}} f_{jkn}$$
$$\text{for all } j \in \text{STR}, \, n \in \text{TIM} \quad (1)$$

$$q_{jmn} h_{jn} = q_{jmn-1} h_{jn-1} + \sum_{i \in \text{OPR}} Y_{ij} Q_{ijm} p_{in}$$
$$- \sum_{k \in \text{PRO}} q_{jmn} f_{jkn}$$
$$\text{for all } j \in \text{STR}, \, m \in \text{QUA}, \, n \in \text{TIM} \quad (2)$$

$$h_{kn} = h_{kn-1} + \sum_{j \in \text{STR}} f_{jkn} - D_{kn}$$
$$\text{for all } k \in \text{PRO}, \, n \in \text{TIM} \quad (3)$$

$$q_{kmn} h_{kn} = q_{kmn-1} h_{kn-1} + \sum_{j \in \text{STR}} q_{jmn} f_{jkn} - D_{kn} q_{kmn}$$
$$\text{for all } k \in \text{PRO}, \, m \in \text{QUA}, \, n \in \text{TIM} \quad (4)$$

Equation (1) represents the material balance row for streams, while Equation (2) represents the quality (e.g., quality-volume) balance row for stream qualities. Equations (3) and (4) represent the material and quality balance rows for blended products. In this formulation qualities are variables governed by equations (2) and (4). Even with constant yields, Y_{ij}, and constant yield qualities, Q_{ijm}, the problem is nonlinear due to the inventory-quality and the flow-quality terms in Equations (2) and (4). In the literature, this problem is commonly referred to as the "pooling problem" due to that fact that inventory carry-overs and blends of blended pools create the nonlinearities.

The nonlinear nature of petro-chemical processes was first incorporated by Shell Oil via successive linear programming (SLP), a straightforward technique based on the iterative solution of linearized models (Griffith and Stewart, 1961). SLP, which was mistakenly dubbed "recursion" in the petro-chemical industry, was applied by most major companies in the 1960s (Baker and Lasdon, 1985). Some of the literature on the subject will be found under the name of "distributed recursion," a specific form of SLP dealing with the distribution of nonlinear error terms across blended pools.

Literally, every other form of nonlinear optimization has been applied in the industry. Lasdon and Waren (1980) provide a comprehensive survey of applications. Edgar and Himmelblau (1988) present techniques specific to the design and operation of chemical processes. Some approaches to refinery optimization are found in Ciriani and Leachman (1993).

As the scope of the optimization applications increased, so did the burden of model management. Many major companies developed their own systems. Beale (1978) describes British Petroleum's approach. At one point, Exxon's model management system handled over 100 mathematical programming applications which were run on a routine basis (Palmer, 1984). Other companies' efforts are referenced in Bodington and Baker (1990).

The area of production planning and scheduling has seen a wide variety of hybrid approaches combining mathematical programming, expert systems, decision support systems, forecasting techniques and simulation. Klingman (1987) describes the integrated logistics system developed at Citgo. A combination of network flow algorithms, mixed-integer programming, decision support were applied to ship scheduling at Ethyl Corporation (Miller, 1987). Brown (1987) reports on a vehicle loading and routing system developed for Mobil Oil. Some of DuPont's efforts in developing a modular system for planning and scheduling are described by Miller (1994). The design and development of integrated systems for planning and scheduling is an area of active interest both in academic and industrial settings (Baker, 1994).

Few people outside the petro-chemical industry realize the extent to which mathematical programming concepts and techniques have permeated all aspects of the business from strategic planning to process control. Shell's use of scenario optimization influences decision making at the strategic level (de Geus, 1988). Price-directed and resource-directed decomposition techniques are routinely applied to coordinate distributed logistics systems. Most of the advanced process control applications are derived from concepts developed by the OR/MS community. Bodington (1995) describes the current state-of-the-art in the integration of planning, scheduling and control.

See **Combinatorial and integer optimization; Decomposition; Linear programming; Nonlinear programming; Simulation.**

References

[1] Baker, T.E. and L.S. Lasdon (1985). "Successive Linear Programming at Exxon," *Management Science*, 31, 264–274.

[2] Baker, T.E. (1994). "An Integrated Approach to Planning and Scheduling," *Foundations of Computer-Aided Process Operations*, D.W.T. Rippin, ed., CACHE, Austin, Texas, 237–251.

[3] Bammi, D. (1990). "Northern Border Pipeline Logistics Simulation," *Interfaces*, 20(3), 1–13.

[4] Beale, E.M.L. (1978). "Nonlinear Programming Using a General Mathematical Programming System," in *Design and Implementation of Optimization Software*, H.J. Greenberg, ed., Sijthoff and Noordhoff, The Netherlands, 259–279.

[5] Bodington, C.E. and T.E. Baker (1990). "A History of Mathematical Programming in the Petroleum Industry," *Interfaces*, 20(4), 117–127.

[6] Bodington, C.E. (1995). *Planning, Scheduling and Control Integration in the Process Industries*, McGraw-Hill, New York.

[7] Brown, G.G. *et al.* (1987). "Real-Time, Wide Area Dispatch of Mobil Tank Trucks," *Interfaces*, 17(1), 107–120.

[8] Charnes, A., W.W. Cooper and B. Mellon (1952). "Blending Aviation Gasoline – A Study in Programming Interdependent Activities in an Integrated Oil Company," *Econometrica*, 20(2), 135–139.

[9] Ciriani, T.A. and R.C. Leachman (1993). *Optimization in Industry*, John Wiley, New York.

[10] de Geus, A.P. (1988). "Planning As Learning," *Harvard Business Review*, 88(2), 70–77.

[11] Edgar, T.F. and D.M. Himmelblau (1988). *Optimization of Chemical Processes*, McGraw-Hill, New York.

[12] Findlay, P.L. *et al.* (1989). "Optimization of the Daily Production Rates for an Offshore Oilfield," *Jl. Operational Research Society*, 40, 1079–1088.

[13] Griffith, R.E. and R.A. Stewart (1961). "A Nonlinear Programming Technique for the Optimization of Continuous Processing Systems," *Management Science*, 7, 379–392.

[14] Hansen, P. *et al.* (1992). "Location and Sizing of Offshore Platforms for Oil Exploration," *European Journal Operational Research*, 58(2), 202–214.

[15] Higgins, J.G. (1993). "Planning for Risk and Uncertainty in Oil Exploration," *Long Range Planning*, 26(1), 111–122.

[16] Klingman, D. *et al.* (1987). "The Successful Deployment of Management Science Throughout Citgo Petroleum Corporation," *Interfaces*, 17(1), 4–25.

[17] Lasdon, L.S. and A.D. Waren (1980). "A Survey of Nonlinear Programming Applications," *Operations Research*, 28, 102–1073.

[18] Manne, A. (1958). "A Linear Programming Model of the US Petroleum Refining Industry," *Econometrica*, 26(1), 67–106.

[19] Miller, D. *et al.* (1994). "A Modular System for Scheduling Chemical Plant Production," *Foundations of Computer-Aided Process Operations*, D.W.T. Rippin ed., CACHE, Austin, Texas, 355–372.

[20] Miller, D. (1987). "An Interactive, Computer-Aided Ship Scheduling System," *European Jl. Operational Research*, 32(3), 363–379.

[21] Palmer, K.H. *et al.* (1984). *A Model-Management Framework for Mathematical Programming*, John Wiley, New York.

[22] Power, M. (1992). "Simulating Natural Gas Discoveries," *Interfaces*, 22(2), 38–51.

[23] Symonds, G.H. (1955). *Linear Programming – The Solution of Refinery Problems*, Esso Standard Oil Company, New York.

PFI

See **Product form of the inverse**.

PHASE I PROCEDURE

That part of the simplex method directed towards finding a first basic feasible solution. See **Artificial basis**; **Linear programming**; **Phase II procedure**; **Simplex method**.

PHASE II PROCEDURE

The part of the simplex algorithm that finds an optimal basic feasible solution, starting with a Phase I basic feasible solution or an initial basic feasible solution. See **Linear programming**; **Phase I procedure**; **Simplex method**.

PHASE-TYPE DISTRIBUTION

Since the Erlang random variable is the sum of exponential random variables, one can consider going through a series of phases (one for each exponential random variable) before realizing the complete Erlangian random variable. Thus, the Erlang distribution is sometimes called a phase-type distribution. A more general formulation of the phase-type idea allows the movement through the phases to be governed by a Markov chain that permits movement back and forth between the interior phases, with the final stage being an absorbing barrier. See **Erlang distribution**; **Queueing theory**.

PHASE-TYPE PROBABILITY DISTRIBUTIONS

Marcel F. Neuts

The University of Arizona, Tucson

The probability distributions of phase-type, or *PH*-distributions, form a useful general class for the representation of nonnegative random variables. A comprehensive discussion of their basic properties is given in Neuts (1981). There are parallel definitions and properties of *discrete* and *continuous PH*-distributions. This discussion emphasizes the continuous case.

A probability distribution $F(\cdot)$ on $[0,\infty)$ is *of phase-type* if it can arise as the absorption time distribution of an $m+1$-state Markov chain with m transient states $1, \ldots, m$ and an absorbing state 0. The generator Q of such a Markov chain is written as

$$Q = \begin{vmatrix} T & T^\circ \\ 0 & 0 \end{vmatrix},$$

where T is a nonsingular $m \times m$ matrix with negative diagonal elements and nonnegative off-diagonal elements. If e denotes a column vector with all components equal to one, then the vector T° satisfies $T^\circ = -Te$. The initial probability vector of the Markov chain is specified as (α, α_0). Without loss of generality, it may be assumed that the generator, $Q^* = T + (1-\alpha_0)^{-1}T^\circ\alpha$, is irreducible.

The general formula for the *PH*-distribution $F(\cdot)$ is then

$$F(x) = 1 - \alpha \exp(Tx)e, \text{ for } x \geq 0.$$

The pair (α, T) is called a *representation* of $F(\cdot)$. The *PH*-distribution $F(\cdot)$ has a point mass α_0 at 0 and a density $F'(x) = -\alpha \exp(Tx)Te = \alpha \exp(Tx)T^\circ$, on $(0,\infty)$. The Laplace-Stieltjes transform $f(s)$ of $F(\cdot)$ is

$$f(s) = \alpha_{m+1} + \alpha(sI - T)^{-1}T^\circ, \text{ for Re } s \geq 0.$$

Its moments λ'_ν, $\nu \geq 1$, are all finite and given by $\lambda'_\nu = (-1)^\nu \nu! \alpha T^{-\nu} e$. Some special classes of *PH*-distributions are the *hyperexponential* distributions

$$F(x) = \sum_{\nu=1}^{m} \alpha_\nu (1 - e^{-\lambda_\nu x}),$$

which may be represented by $\alpha = (\alpha_1, \ldots, \alpha_m)$, $\alpha_{m+1} = 0$, and $T = -\text{diag}(\lambda_1, \ldots, \lambda_m)$, and the (mixed) *Erlang* distributions

$$F(x) = \sum_{\nu=1}^{m} p_\nu E_\nu(\lambda; x),$$

which are represented by $\alpha = (p_m, p_{m-1}, \ldots, p_1)$, $\alpha_{m+1} = 0$, and

$$T = \begin{vmatrix} -\lambda & \lambda & 0 & \ldots & 0 & 0 & 0 \\ 0 & -\lambda & \lambda & \ldots & 0 & 0 & 0 \\ & \ldots & & & & \ldots & \\ 0 & 0 & 0 & \ldots & 0 & -\lambda & \lambda \\ 0 & 0 & 0 & \ldots & 0 & 0 & -\lambda \end{vmatrix}.$$

USES OF PHASE-TYPE DISTRIBUTIONS: The utility of *PH*-distributions is due, in the first place, to their *closure properties*. These allow standard operations such as convolution and mixing to be represented by matrix operations. Many classical simplifying properties of the exponential distribution have analogues in the matrix formalism for *PH*-distributions. In the analysis of probability models, *PH*-distributions often lead to tractable results without the severe restriction of exponential assumptions. *Integrals* involving *PH*-distributions also can usually be evaluated by stable recurrence relations or differential equations. Moreover, the phase-type distributions form a *dense subset* of the probability distributions on $[0,\infty)$. Any such distribution can, in principle, be uniformly approximated by a sequence of *PH*-distributions.

Examples of closure properties are:

(a) If $F(\cdot)$ is a *PH*-distribution with representation (α, T) and mean λ'_1, the corresponding *delay distribution* $F^*(\cdot)$ with density $(\lambda'_1)^{-1}[1 - F(x)]$ is *PH* with representation (π, T) where $\pi = (\lambda'_1)^{-1}\alpha(-T)^{-1}$.

(b) If $F(\cdot)$ (with $\alpha_0 = 0$) is the service time distribution of a stable $M/G/1$ queue with arrival rate θ and service time distribution $H(\cdot)$ of mean μ'_1, such that $\rho = \theta\mu'_1 < 1$, the (steady-state) distribution $W(\cdot)$ of the waiting time is *PH*. Its representation is given by (γ, L), where $\gamma = \rho\pi$, $L = T + \rho T^\circ\pi$. For the $M/PH/1$ queue, the distribution $W(\cdot)$ may therefore be computed by integrating a system of linear differential equations, rather than by solving the Pollaczek-Khinchin integral equation.

The fact that any probability distribution on $[0,\infty)$ can be approximated by *PH*-distributions is of somewhat limited practical application, although very good *PH*-approximations to classes such as the Weibull distributions have been obtained. Because of the following general result, that denseness property is, however, of considerable theoretical utility.

Suppose that a stochastic model involves one or more general probability distributions $F_j(\cdot)$, $1 \leq j \leq N$, on $[0,\infty)$ and that we wish to evaluate a *continuous* functional $\Phi[F_1(\cdot), \ldots, F_N(\cdot)]$ of these probability distributions. If an expression for $\Phi(\cdot)$ can be found for the case where $F_1(\cdot), \ldots, F_N(\cdot)$ are *PH*-distributions and if that expression does not explicitly depend on the formalism of *PH*-distributions, then it is also valid for arbitrary distributions $F_1(\cdot), \ldots, F_N(\cdot)$. This result has been used to establish various moment and other formulas in the theory of queues.

There is now an extensive literature on phase-type distributions and their applications. Subjects of current research interest are the structural geometric properties of families of *PH*-distributions; the approximation of other families of distributions by those of phase-type, and the fitting of *PH*-distributions to data. An important characterization of *PH*-distributions was proved in O'Cinneide (1990). Procedures for the approximation by *PH*-distributions are discussed in Asmussen, Haggström, and Nerman (1992), Johnson (1993) and Schmickler (1992). The appearance of phase-type distributions in some unexpected places in queueing theory was noted in Asmussen (1992).

See **Markov chains; Markov processes; Queueing theory.**

References

[1] Asmussen, S. (1992), "Phase-type representations in random walk and queueing problems," *Annals Probability*, 20, 772–789.

[2] Asmussen, S., Haggström, O., and Nerman, O. (1992), "EMPHT – A program for fitting phase-type distributions," in *Studies in Statistical Quality Control and Reliability*, Mathematical Statistics, Chalmers University and University of Gteborg, Sweden.

[3] O'Cinneide, C.A. (1990), "Characterization of phase-type distributions," *Stochastic Models*, 6, 1–57.

[4] Johnson, M.A. (1993), "Selecting parameters of phase distributions: Combining nonlinear programming, heuristics, and Erlang distributions," *ORSA Jl. Computing*, 5, 69–83.

[5] Johnson, M.A. (1993), "An empirical study of queueing approximations based on phase-type distributions," *Stochastic Models*, 9, 531–561.

[6] Neuts, M.F. (1981), *Matrix-Geometric Solutions in Stochastic Models: An Algorithmic Approach*. The Johns Hopkins University Press, Baltimore. Reprinted by Dover Publications, 1994.

[7] Pagano, M.E. and Neuts, M.F. (1981), "Generating Random Variates from a Distribution of Phase Type," 1981 *Winter Simulation Conference Proceedings*, T.I. Oren, C.M. Delfosse, C.M. Shub (eds.), 381–387.

[8] Schmickler, L. (1992), "MEDA: Mixed Erlang distributions as phase-type representations of empirical distribution functions," *Stochastic Models*, 8, 131–156.

PIECEWISE LINEAR FUNCTION

A function that is formed by linear segments or one that approximates a nonlinear function by linear segments.

PIVOT COLUMN

The column vector of coefficients associated with the entering basis variable in a simplex method iteration. Also, more generally, the column that contains the pivot element of a Gaussian elimination step or similar process. See **Eta vector; Gaussian elimination; Matrices and matrix algebra; Pivot element; Pivot row; Simplex method.**

PIVOT ELEMENT

In the simplex method, the coefficient of the pivot column whose row index corresponds to the basic variable that is to be dropped from the basis. Also, the element of the pivot column in a Gaussian elimination step that is selected to be on the diagonal of the associated upper triangular matrix. See **Eta vector; Gaussian elimination; Matrices and matrix algebra; Pivot column; Pivot row; Simplex method.**

PIVOT ROW

The row corresponding to the position of the basic variable that is to be dropped from the basis in a simplex method iteration. In general, the row corresponding to the row position of a pivot element in a Gaussian elimination step. See **Eta vector; Gaussian elimination; Matrices and matrix algebra; Pivot column; Pivot element; Simplex method.**

PIVOT-SELECTION RULES

In the simplex method, the pivot selection rules determine which variable is to enter the basic solution and which variable is to be dropped. Depending on the solution at hand, the rules are designed to preserve feasibility (nonnegativity) of the solution (primal-simplex method), or to preserve the optimality conditions (dual-simplex method). In either case, the rules attempt to select an entering variable that would cause an improvement in the objective function. These rules are often augmented with antidegeneracy or anticycling rules, and procedures for maintaining sparsity and numerical accuracy. See **Bland's anticycling rules; Density; Devex pricing; Linear programming; Matrices and matrix algebra; Perturbation methods; Simplex method.**

PO

See **Postoptimal analysis**.

POINT STOCHASTIC PROCESSES

Igor Ushakov

SOTAS, Rockville, Maryland

INTRODUCTION: A *point process* is a stochastic process $\{N(t) = $ number of occurrences by time $t\}$ which describes the appearance of a sequence of instant random events in time. Usually (though not always) intervals between two neighboring events are considered to be independently distributed. A process of this type is called a *point process with restricted memory*. If times between occurrences (e.g., called interarrival times in queueing theory, etc.) are a sequence of independent and identically distributed (i.i.d.) random variables, the point process is called a *renewal* or *recurrent point process*. The Poisson process represents a particular case of a renewal process in that the intervals between occurrences are identically exponentially distributed (Cox and Isham, 1980; Franken *et al.*, 1981).

Point processes of a special type can be formed by two random variables which alternate with the sequence $X_1, Y_1, X_2, Y_2, \ldots$ and so on. Such a process is called *alternating point process* – more specifically, an alternating renewal process when the X and Y subsequences are themselves ordinary renewal processes.

THINNING OF A POINT PROCESS: In practice, one often meets cases when, by some reasons, events are excluded from the point process with a specified probability. For instance, a unit failure leads to a system failure only if several additional random circumstances happen. This exclusion of events is called a *thinning procedure*. If in the result of such a thinning procedure the (normalized) probability of the event exclusion goes to 1, the resulting point process converges to a Poisson process. This statement is reflected in strong terms in Renyi's Limit Theorem and in its generalization made by Yu.K. Belyaev (see Gnedenko *et al.*, 1969).

For practical purposes, it means that if the mean time between neighboring events in the initial recurrent process equals T, and each event is excluded from this process with the probability p close to 1, the resulting process will be a Poisson one with parameter

$$\lambda = \frac{1-p}{T}.$$

THE SUPERPOSITION OF POINT PROCESSES: The next important statement concerns the so-called superposition of point processes, which is formulated in the Khinchine-Ososkov Limit Theorem (Khinchine, 1960; Ososkov, 1956) and later generalized in the Grigelionis-Pogozhev Limit Theorem (Grigelionis, 1964; Pogozhev, 1964). On a qualitative level, the theorem states that a limiting point process, which is formed by the superposition of independent "infinitesimally rare" point processes, converges to a Poisson process. For instance, if a piece of equipment consists of a *large* number of blocks and modules, the flow of its failures may well be considered to form a Poisson process. The parameter of this resulting process is expressed as a sum of the parameters of the initial processes, that is, if there are n recurrent processes ($n \gg 1$), each of them with mean T_i, then the resulting process will be close to a Poisson process with parameter

$$\lambda = \sum_{1 \le i \le n} \frac{1}{T_i}.$$

As a consequence of these results, the Poisson process plays a role in the theory of stochastic processes which is analogous to that of the normal distribution in general probability and statistical theory.

See **Poisson process; Renewal processes; Queueing theory; Stochastic model**.

References

[1] Cox, D.R. and V. Isham (1980). *Point Processes*, Chapman and Hall, New York.

[2] Franken, P., D. Knig, U. Arndt, and V. Schmidt (1981). *Queues and Point Processes*, Akademie-Verlag, Berlin.

[3] Gnedenko, B.V., Yu.K. Belyaev, and A.D. Solovyev (1969). *Mathematical Methods of Reliability Theory*, Academic Press, New York.

[4] Grigelionis, B.I. (1964). "Limit Theorems for Sums of Renewal Processes," in *Cybernetics in the Service of Communism, vol. 2: Reliability Theory and Queueing Theory*, A.I. Berg, N.G. Bruevich, and B.V. Gnedenko, eds. Energiya, Moscow, 246–266.

[5] Khintchine, A.Ya. (1960). *Mathematical Methods in the Theory of Queueing*, Charles Griffin, London.

[6] Osokov, G.A. (1956). "A Limit Theorem for Flows of Similar Events," *Theory Probability & Its Applics*. 1, 246–255.

POINT-TO-SET MAP

A function that maps a point of one space into a subset of another.

POISSON ARRIVALS

When interarrival times between successive customers coming to a queueing system are independent and identical exponential random variables, it follows that the number of arrivals to the system in a given period of time has a Poisson distribution (i.e., is a Poisson process). As a result, we would say that this queueing system has Poisson arrivals. See **Exponential arrivals**; **Poisson process**; **Queueing theory**.

POISSON PROCESS

A stochastic, renewal-counting point process beginning from time $t = 0$ with $N(0) = 0$ that satisfies the following assumptions is called a Poisson process with rate λ: (1) the probability of one event happening in the interval $(t, t + h]$ is $\lambda h + o(h)$, where $o(h)$ is a function which goes to zero faster than h; (2) the probability of more than one event happening in $(t, t + h]$ is $o(h)$; and (3) events happening in non-overlapping intervals are statistically independent. (Either (1) or (2) can be replaced by: the probability of no event happening in the interval $(t, t + h]$ is $1 - \lambda h + o(h)$.) The number of arrivals to a M/G/1 queueing system is a Poisson process. If the arrival process to a queueing system is a Poisson process with rate λ, then the interarrival times are independent and identically, exponentially distributed with mean $1/\lambda$. See **Markov chains**; **Markov processes**; **Queueing theory**.

POLITICS

Sidney W. Hess and Carlos G. Wong-Martinez

Drexel University, Philadelphia, Pennsylvania

We consider here applications of OR/MS to the representation and electoral processes. We follow the narrower definition of politics denoting the theory and practice of managing political affairs in a party sense (Webster's, 1951).

In particular we will consider applications to:
– apportionment
– districting
– voting methods and logistics, and
– promotion of candidates.

APPORTIONMENT: This is the process of equitably assigning a fixed number of legislators to a lesser number of political subdivisions. In the United States, 435 congressional districts must be apportioned to 50 states with each state receiving at least one district. The method of rounding to an integer solution will influence the political result.

Balinski and Young (1982) have provided an exceptional mathematical analysis of the issue along with a historical, nontechnical exposition. In 1791, following the first U.S. census, Jefferson and Hamilton proposed alternate methods for apportionment, the "method of greatest divisors" and the "method of greatest remainders." Washington exercised the first presidential veto when he disagreed with Congress' support of Hamilton's method.

Most methods are biased; for example Jefferson's favors the more populated states while the "method of equal proportions" (also known as the Hill or Huntington method and used since 1941) discriminates against them. Other methods exhibit the paradox of a state's apportioned number of seats declining as the total number of representatives increases even when all states' populations are unchanged!

Balinski and Young (1982) conclude that there can be no perfect method. However, Senator Daniel Webster promoted a method called "major fractions" (frequently used between 1842 and 1932) which has been felt by many to be preferable. It is simple and exhibits neither bias nor the population paradox. Furthermore, Webster's method is more likely than the other methods to give each state its proportional number of seats, either rounded up or rounded down (Ernst, 1994).

REDISTRICTING: This is the process of defining geographic boundaries for each representative that has been apportioned to the larger area. Historically, the party controlling the legislature drew districting maps to protect incumbents and increase their party's chances of maintaining control.

In 1962, the Supreme Court required population equality among districts demanding more careful mapping than the usual prior political process (*Baker v. Carr*, 369 U.S. 186, 1962). A variety of techniques to "computerize" the mapping process appeared. Most approaches incorporated population equality with the additional criteria that each district be:
– contiguous, a single land parcel,
– compact, consolidated rather than spread out, and
– designed without political consideration.

The districting problem is analogous to the warehouse-location problem, i.e. locate a specified number of warehouses (district centers) and assign equal demand (population) to each.

Hess, Weaver, Seigfeldt, Whelan, and Zitlau (1965) solved a sequence of transportation linear programs. In each LP, equal population was allocated to trial district centers to minimize total "cost." The measure of cost was compactness defined as the second moment of population about its district center. Centroids of the resultant districts became new centers for repeating the linear program. Successive solution of the transportation problems trended to more compactness while maintaining near population equality. Their heuristic handled problems as large as 350 population units by 19 districts. Larger problems were "apportioned" into smaller ones. This Ford Foundation supported program was used for districting in at least seven states.

Garfinkel and Nemhauser (1969) developed a tree search algorithm that would minimize compactness while constraining maximum allowable population deviation. Their measure of district compactness was the diameter squared divided by area. Computation speed and capacity limited the problem size to about 50 population units by 7 districts.

Nygreen (1988) redistricted Wales by three different solution methods: integer programming, set partitioning (a variant of Garfinkel and Nemhauser's technique) and implicit enumeration. Although his example was small, he concluded that the integer programming technique was inferior. He felt problems to about 500 population units by 60 districts could be solved efficiently by set partitioning. Twenty years of computer improvement permit a tenfold larger problem!

All these redistricting techniques require apportioning a problem too large for solution into many smaller and solvable ones. Apportioning first has added benefits: small political subdivisions are more likely to remain intact and district boundaries will more often coincide with political boundaries. Hess (1971) showed how first apportioning New York legislative seats to groups of counties minimizes the number of counties that must be in more than one district.

Meanwhile, the courts and legislatures have been slow to articulate permissible or required criteria for districting. In the United States "one man one vote" is still the law of the land. The 1982 Voting Rights Act requires states with histories of racial discrimination to provide a rea-sonable chance of minority elections (Van Biema, 1993).

While the courts scrutinize the results of districting, they have not yet challenged the process (Browdy, 1990). As a result, political parties have been using proprietary software to generate districting plans which would make Governor Gerry blush. Computer services generated over one thousand plans for Florida alone, making it difficult for the press and public to criticize gerrymandering (Miniter, 1992). Should the courts order an "open" districting process or bipartisanship necessitate, OR/MS algorithms could again provide an acceptable way to redraw representative boundaries (Browdy, 1990).

VOTING METHODS AND LOGISTICS: The application of approval voting was pioneered in the election processes of The Institute of Management Sciences (Fishburn and Little, 1988). In this, a voter checks off (approves) any number of the candidates on a ballot, from a single one to potentially every one, with the winner determined as the one with the most checks.

Savas, Lipton, and Burkholz (1972) reduced the number of New York City election districts by locating multiple voting machines at polling places. The City achieved significant cost savings and increased the probability voters would find functioning machines, without a significant increase in voter distance to the polls.

PROMOTION OF CANDIDATES: Barkan and Bruno (1972) used allocation techniques and statistical analysis to aid the 1970 California election campaign of Senator Tunney. Their analyses identified precincts where voter registration and get-out-the-vote effort should be targeted. The key to their success was the ability to identify swing precincts by estimating party loyalty.

See **Combinatorial and integer optimization; Linear programming; Location analysis; Transportation problem.**

References

[1] *Baker v. Carr*, 369 U.S. 186 (1962).
[2] Balinski, M.L. and H.P. Young (1982), *Fair Representation. Meeting the Ideal of One Man, One Vote*, Yale University Press, New Haven, Connecticut.
[3] Barkan, J.D. and J.E. Bruno (1972), "Operations Research in Planning Political Campaign Strategies," *Operations Research*, 20, 925–941.
[4] Browdy, M.H. (1990), "Computer Models and Post-*Bandemer* Redistricting," *Yale Law Journal*, 99, 1379–1398.

[5] Ernst, L.R. (1994), "Apportionment Methods for the House of Representatives and the Court Challenges," *Management Science*, 40, 1207–1227.

[6] Fishburn, P.C. and J.D.C. Little (1988), "An Experiment in Approval Voting," *Management Science*, 34, 555–568.

[7] Garfinkel, R.S. and G.L. Nemhauser (1969), "Optimal Political Districting by Implicit Enumeration Techniques," *Management Science*, 16, B495–B508.

[8] Hess, S.W. (1971), "One-Man One-Vote and County Political Integrity: Apportion to Satisfy Both," *Jurimetrics Journal*, 11, 123–141.

[9] Hess, S.W., J.B. Weaver, H.J. Seigfeldt, J.N. Whelan, and P.A. Zitlau (1965), "Nonpartisan Political Redistricting by Computer," *Operations Research*, 13, 998–1006.

[10] Miniter, R. (1992), "Running Against the Computer; Stephen Solarz and the Technician-Designed Congressional District," *The Washington Post*, September 20, C5.

[11] Nygreen, B. (1988), "European Assembly Constituencies for Wales – Comparing of Methods for Solving a Political Districting Problem," *Mathematical Programming*, 42, 159–169.

[12] Savas, E.S., H. Lipton, and L. Burkholz (1972), "Implementation of an OR Approach for Forming Efficient Districts," *Operations Research*, 20, 46–48.

[13] Van Biema, D. (1993), "Snakes or Ladders," *Time*, July 12, 30–33.

[14] *Webster's New Collegiate Dictionary*, (1951). "Politics," p. 654, Mirriam, New York.

POLLACZEK-KHINTCHINE FORMULA

For the M/G/1 queueing system, with L defined as the steady-state expected number of customers in the system, λ the customer arrival rate, $1/\mu$ the expected service time and σ^2 the variance of the service distribution then

$$L = \rho + (\rho^2 + \lambda^2\sigma^2)/[2(1-\rho)]$$

where $\rho = \lambda/\mu$. This equation is known as the Pollaczek-Khintchine (P-K) formula. Sometimes, the formulas for mean queue size, L_q, mean line delay, W_q, and mean system waiting time, W, which can be easily derived from L using Little's formula, are also called P-K formulas. See **Queueing theory**.

POLLING SYSTEM

Where a single server visits each group of customers (queue) in cyclic order and then polls to see if there is anyone present. If yes, the service facility serves those customers under such rules as gated (serve only those present when polled) or exhaustive (serve until no customers are left at the location). See **Networks of queues**; **Queueing theory**.

POLYHEDRON

The solution space defined by the intersection of a finite number of linear constraints, an example of which is the solution space of a linear-programming problem. Such a space is convex. See **Convex set**; **Linear programming**.

POLYNOMIAL HIERARCHY

A general term used to refer to all of the various computational complexity classes (see **Computational complexity**).

POLYNOMIALLY BOUNDED (-TIME) ALGORITHM (POLYNOMIAL ALGORITHM)

An algorithm for which it can be shown that the number of steps required to find a solution to a problem is bounded by a polynomial function of the problem's data. See **Computational complexity**; **Exponential-bounded algorithm**.

POLYNOMIAL-TIME

See **Computational complexity**.

POLYNOMIAL-TIME REDUCTIONS AND TRANSFORMATIONS

See **Computational complexity**.

PORTFOLIO ANALYSIS

See **Mean-value portfolio analysis**.

PORTFOLIO THEORY: MEAN-VARIANCE

John L.G. Board
London School of Economics

William T. Ziemba
University of British Columbia, Vancouver

PORTFOLIO SELECTION PROBLEM: The heart of the portfolio problem is the selection of an optimal set of investment assets by rational economic agents. The particular feature of this problem is that the attributes of a portfolio may be different from the attributes of any of its constituents. Although elements of portfolio problems were discussed in the 1930s and 1950s by Allais, De Finetti, Hicks, Marschak and others, the first formal specification of such a selection model was by Markowitz (1952, 1959) who defined a mean-variance model for calculating optimal portfolios. Following Tobin (1958, 1965), Sharpe (1970) and Roll (1977), this portfolio selection model may be stated as:

$$\text{Minimize} \quad x'Vx$$
$$\text{subject to} \quad x'r = r_p \tag{1}$$
$$x'e = 1$$

where x is a column vector of investment proportions, V is a variance-covariance matrix of asset returns (which must be at least positive semi-definite), r is a column vector of expected asset returns, r_p is the investor's target rate of return and e is a column unit vector. An explicit solution for the problem can be solved using the procedures described in Merton (1972), Ziemba and Vickson (1975), or Roll (1977).

Restrictions on short selling can be modeled by augmenting (1) by the series of constraints:

$$x \geq 0 \tag{2}$$

where 0 is a column vector of zeros. The problem now becomes a classic example of quadratic mathematical programming; indeed, the development of the portfolio problem coincided with early developments in nonlinear programming. Formal investigations of the properties of both formulations, and variants, appear in Szeg (1980), Huang and Litzenberger (1988), and the references above.

THE USE OF MEAN AND VARIANCE: The economic justification for this model is based on the von Neumann-Morgenstern expected utility results, discussed in this context by Markowitz (1959). The model can also be conveniently viewed in terms of consumer choice theory together with the the characteristics model developed by Lancaster (1971). His argument is that goods purchased by consumers seldom yield a single, well defined service; instead, each good may be viewed as a collection of attributes each of which gives the consumer some benefit (or dis-benefit). Thus preference is defined over those characteristics embodied in a good rather than over the good itself.

The analysis focusses attention on the attributes of assets rather than on the assets *per se*. This requires the assumption that utility depends only on the characteristics. With k characteristics, C_k, we need

$$U = f(W) = g(C_1, \ldots, C_k)$$

where U and W represent utility and wealth. Modeling too few characteristics will yield apparently false empirical results. Clearly, the benefits of this approach increase as the number of assets rises relative to the number of characteristics. The objects of choice are the characteristics $C_1, \ldots, C_k$. In portfolio theory, these are taken to be payoff (return) and risk. At Markowitz's suggestion, when dealing with choice among risky assets, payoff is measured as the expected return of the distribution of returns and risk is measured by the standard deviation of returns. Apart from minor exceptions, like Ziembda and Vickson (1975), this pair of characteristics form a complete descriptions of assets which is consistent with expected utility theory in only two cases: assets have normal distributions, or investors have quadratic utility of wealth functions. The adequacy of these assumptions has been investigated by a number of authors (e.g., Borch, 1969; Feldstein, 1969; Tsiang, 1972). Although returns have been found to be non-normal and the quadratic utility has a number of objectionable features (not least diminishing marginal utility of wealth for high wealth), several authors demonstrate approximation results which are sufficient for mean variance analysis (Samuelson, 1970; Ohlson, 1975; Levy and Markowitz, 1979).

A number of authors, including Markowitz (1959), consider alternatives to the variance and suggest the use of the semi-variance. This suggestion has been extended into workable portfolio selection rules. Fama (1971) and Tsiang (1973) have argued the usefulness of the semi-interquartile range as a measure of risk. Kraus and Litzenberger (1974) and others have examined the effect of preferences defined in terms of the third moment which allows investor choice in terms of skewness. Kallberg and Ziemba (1979, 1983) show that risk aversion preferences are sufficient to determine optimal portfolio choice if assets have normally distributed returns whatever the form of the assumed, concave, utility function.

SOLUTION OF PORTFOLIO SELECTION MODEL: In the absence of short sales restrictions, (1) can be rewritten as

$$\text{Minimize} \quad L = \tfrac{1}{2}x'Vx - \lambda_1(x'r - r_p) - \lambda_2(x'e - 1) \tag{3}$$

The first order conditions are

$$Vx = \lambda_1 r + \lambda_2 e$$

which shows that, for any efficient x, there is a linear relation between expected returns r and their covariances, Vx.

Solving for x:

$$x = \lambda_1 V^{-1}r + \lambda_2 V^{-1}e = V^{-1}[r \quad e]A^{-1}[r_p \quad 1]' \quad (4)$$

where

$$A = \begin{bmatrix} a & b \\ b & c \end{bmatrix} = \begin{bmatrix} r'V^{-1}r & r'V^{-1}e \\ r'V^{-1}e & e'V^{-1}e \end{bmatrix}$$

Substituting (4) into the definition of portfolio variance, $x'Vx$, yields

$$V_p = [r_p \quad 1]A^{-1}[r_p \quad 1]'$$

$$S_p = \left[\frac{cr_p^2 - 2br_p + a}{ac - b^2} \right]^{1/2} \quad (5)$$

where V_p and S_p represent portfolio variance and standard deviation, respectively. This defines the efficient set, which is a hyperbola in mean/standard-deviation space (or a parabola in mean/variance space). The minimum risk is at $S_{\min} = c^{1/2}$, and $r_{\min} = b/c$ (both strictly positive). Rational risk averse investors will hold portfolios lying on this boundary with $r \geq r_{\min}$. Equation (5) shows a two fund separation theorem, that linear combinations of only two portfolios are sufficient to describe the entire efficient set.

Each efficient portfolio, p, has an orthogonal portfolio z (i.e., such that $\text{Cov}(r_p, r_z) = 0$) with return

$$r_z = (a - br_p)/(b - cr_p).$$

Using this, the efficient set degenerates into the straight line tangent to the hyperbola at p which has intercept r_z. This is the Capital Market Line:

$$r = r_z + \lambda s$$

where r and s represent vectors of the expected return and risks of efficient portfolios, and $\lambda = (r_p - r_z)/S_p$ can be interpreted as the aditional expected return per unit of risk. This is known as the Sharpe ratio (Sharpe, 1966, 1994).

Under the additional assumptions of homogeneous beliefs (so that all investors perceive the same parameters) and equilibrium, the Security Market Line (i.e., the Capital Asset Pricing Model – CAPM) can be derived by premultiplying (4) by V:

$$r = r_z e + (r_p - r_z)\beta$$

where $\beta = \dfrac{Vx}{V_p}$. $\quad (6)$

If it exists, the risk-free rate of interest may be substituted for r_z (definitionally, the risk-free return will be uncorrelated with the return on all risky assets). Equation (6) then becomes the original CAPM in which expected return is calculated as the risk-free rate plus a risk premium (measured in terms of an asset's covariance with the market portfolio). The CAPM forms one of the cornerstones of modern finance theory and is not appropriately addressed here. However, discussions can be found in Huang and Litzenberger (1988) and Ferson (1994), as well as Ziemba (1994) for systematic fundamental and seasonal violations of the theory.

SHORT SELLING: The assumption that assets may be sold short (i.e., $x_i < 0$) is justified when the model is used to derive analytical results for the portfolio problem. Also, when considering equilibrium (e.g., the CAPM), none of the short selling constraints should be binding (because in aggregate, short selling must net out to zero). However, significant short selling restrictions do face investors in most real markets. These restrictions may be in the form of absolute prohibition imposed by markets, extra costs of deposits to back short selling or self imposed controls designed to limit potential losses. For example, the NYSE imposes the 'uptick rule' under which short sales are allowed only if the price of the immediately preceding trade was higher than or equal to the trade preceding it (as short selling is profitable in a falling market, this rule substantially limits its attractiveness).

In contrast, most models based on portfolio theory, in particular the CAPM, ignore short selling constraints (Markowitz, 1983, 1987). This change is consistent with the development of equilibrium models for which institutional restrictions are inappropriate (and if imposed would not be binding), although applied work usually finds that up to 50% of assets are sold short, often in large amounts and sometimes in amounts exceeding the initial value of the investment portfolio. Indeed, this is the main activity of 'short seller' funds.

ESTIMATION PROBLEMS: The model (1) requires estimates of r and V for the period during which the portfolio is to be held. This estimation problem has been given relatively little attention, and many authors, both practitioners and academics, have used historical values as if they were precise estimates of future values. However, Hodges and Brealey (1973), among others, demonstrate the benefits obtained even from slight improvements on historical data.

Estimation risk can be allowed for either by using different methods to forecast asset returns, variances and covariances, which are then used in place of the historical values in the Markowitz model, or by using the historical values in a modified portfolio selection technique (Bawa *et al.*, 1979). Since the Markowitz approach takes these estimates as parametric, there is no theoretical guidance on the estimation method and a variety of methods have been proposed to provide the estimates. The *single index market* model of Sharpe (1963) has been widely applied in the literature to forecast the covariance matrix. Originally proposed to reduce the computation required by the full model, it assumes a linear relation between stock returns and some measure of the market, $\mathbf{r} = \alpha + \beta'\mathbf{m} + \varepsilon$ (for market index $\mathbf{m}$ and residuals ε). This uses historical estimates of the means and variances; however, the implied covariance matrix is $\mathbf{V}_\mathrm{I} = v_m \beta\beta' + \mathbf{V}$, where v_m is the variance of the index, β is a column vector of slope coefficients from regressing each asset on the market index and $\mathbf{V}$ is a diagonal matrix of the variances of the residuals from each of these regressions. A number of studies have found that models based on the single index model outperform ones based on the full historical method.

The *overall mean* method, first proposed by Elton and Gruber (1973), is based on the finding that, although historical estimates of means are satisfactory, data are typically not stable enough to allow accurate estimation of the $N(N-1)/2$ covariance terms. The crudest solution is to assume that the correlations between all pairs of assets expected in the next period are equal to the mean of all the historic correlations. An estimate of $\mathbf{V}$ can then be derived from this. Elton, Gruber and Urich (1978) compared the overall mean method of forecasting the covariance matrix with forecasts made using historical values, and four alternative versions of the single index model. They concluded that the overall mean model was clearly superior. A simplified procedure for estimating the overall mean correlation appears in Aneja, Chandra and Gunay (1989).

In recent years, statisticians have shown increasing interest in *Bayesian* methods (Hodges, 1976) and particularly *James-Stein* estimators (Efron and Morris, 1975, 1977; Judge and Bock, 1978; Morris, 1983). The intuition behind this approach is that returns which are far from the norm have a higher chance of containing measurement error than those close to it. This is analogous to arguments about mean reversion in stock prices. Thus, estimates of returns, based on individual share data, are cross-sectionally 'shrunk' towards a global estimate of expected returns which is based on all the data. Although these estimators have unusual properties, they are generally expected to perform well in large samples.

Jorion (1985, 1986) examined the performance of Bayes-Stein estimation using both simulated and small real data sets and concluded that the Bayes-Stein approach outperformed the use of historical estimates of returns and the covariance matrix. However, Jorion (1991) found that the index model outperformed Stein and historical models. Board and Sutcliffe (1994) applied these and other methods to large real data sets. They found that, in contrast to earlier studies, the relative performance of Bayes-Stein was mixed. While it produced reasonable estimates of the mean returns vector, there were superior methods (e.g., use of the overall mean) for estimating the covariance matrix when short sales were permitted. They also found that, when short sales were prohibited, actual portfolio performance was clearly improved, although there was little to choose between the various estimation methods.

An alternative approach is to try to control for errors induced by the parameter estimates by imposing additional constraints on (1). Clearly, *ex-ante* the solution to such a model cannot dominate (1), however, *ex-post*, dominance might emerge (i.e., what seems, in advance, to be an inferior portfolio might actually perform better than others). The argument is that adding constraints to (1) to impose lower bounds (i.e., prohibiting short sales) and/or upper bounds (forcing diversification) can be used as an *ad hoc* method of avoiding the worst effects of estimation risk. As a result, although *ex-ante* performance will be inferior, performance over a holding period may improve. Of course, extreme, but possibly desirable, corner solutions will also be excluded by this technique. Cohen and Pogue (1967) imposed upper bounds of 2.5% on any asset. Board and Sutcliffe (1988) studied the effects of placing upper bounds on the investment proportions, which may be interpreted as a response to estimation risk. Using historical forecasts of returns and the covariance matrix, and with short sales excluded, they found that forcing diversification leads to improved actual performance over the unconstrained model.

Chopra and Ziemba (1993), following the work of Kallberg and Ziemba (1984), showed that errors in the mean values have a much greater effect than errors in the variances, which are in turn more important than errors in the covariances. Their simulations show errors of the order of 20 to 2 to 1. This quantifies the earlier findings and stresses the importance of having good estimates of the asset means.

Another approach is to use fundamental analysis to provide external information to modify the estimates (Hodges and Brealey, 1973). Clearly, among the simplest external data to add are the seasonal (e.g., turn of the year, and month and weekend effects) which have been found in most stock markets around the world. Incorporation of these into the parameter estimates can substantially improve the performance of the model. Ziemba (1994) demonstrated the benefits of factor models to estimate the mean returns.

CONCLUSION: We have considered only the single period mean-variance portfolio theory model. Although recent developments have focussed on extending the model to multiple periods, most of these models which assume the frictionless capital markets require the solution of a sequence of instantaneous mean-variance models in which the existence of transactions costs adds enormously to the complexity of the problem. Surveys covering dynamic portfolio theory appear in Constantinides and Malliaris (1994), Ziemba and Vickson (1975), Huang and Litzenberger (1988), and Ingersoll (1987).

See **Linear programming; Nonlinear programming.**

References

[1] Aneja Y.P., Chandra R. and Gunay E. (1989), "A Portfolio Approach to Estimating the Average Correlation Coefficient for the Constant Correlation Model," *Jl. Finance*, 44, 1435–1438.

[2] Bawa, V.S., Brown S.J., and Klein R.W. (1979), *Estimation Risk and Optimal Portfolio Choice*, North Holland, Amsterdam.

[3] Board, J.L.G. and Sutcliffe C.M.S. (1988), "Forced Diversification," *Quarterly Review Economics and Business*, 28(3), 43–52.

[4] Board, J.L.G. and Sutcliffe C.M.S. (1994), "Estimation Methods in Portfolio Selection and the Effectiveness of Short Sales Restrictions: UK Evidence," *Management Science*, 40, 516–534.

[5] Borch, K. (1969), "A Note on Uncertainty and Indifference Curves," *Review Economic Studies*, 36, 1–4.

[6] Chopra V.K. and Ziemba W.T. (1993), "The Effect of Errors in Means Variances and Covariances on Optimal Portfolio Choice," *Jl. Portfolio Management*, 19, No. 2, 6–13.

[7] Cohen K.J. and Pogue J.A. (1967), "An Empirical Evaluation of Alternative Portfolio Selection Models," *Jl. Business*, 40, 166–193.

[8] Constantinides G. and Malliaris G. (1995), "Portfolio Theory," in *Handbook of Finance*, Jarrow, Maksimovic and Ziemba, eds., North-Holland, Amsterdam.

[9] Efron B. and Morris C. (1975), "Data Analysis Using Stein's Estimator and its Generalizations," *Jl. American Statistical Assoc.*, 70, 311–319.

[10] Efron B. and Morris C. (1977), "Stein's Paradox in Statistics," *Scientific American*, 236, No. 5, 119–127.

[11] Elton E.J. and Gruber M.J. (1973), "Estimating the Dependence Structure of Share Prices Implications for Portfolio Selection," *Jl. Finance*, 28, 1203–1232.

[12] Elton, E.J., Gruber M.J., and Urich T.J. (1978), "Are Betas Best?," *Jl. Finance*, 33, 1375–1384.

[13] Fama E. (1971), "Risk, Return and Equilibrium," *Jl. Political Economy*, 79, 30–55.

[14] Fama E.F. (1976), *Foundations of Finance*, Basil Blackwell, Oxford.

[15] Feldstein M. (1969), "Mean Variance Analysis in the Theory of Liquidity Preference and Portfolio Selection," *Review Economic Studies*, 36, 5–12.

[16] Ferson W. (1995), "Theory and Testing of Asset Pricing Models," in *Handbook of Finance*, Jarrow, Maksimovic, and Ziemba, eds., North-Holland, Amsterdam.

[17] Hodges S.D. (1976), "Problems in the Application of Portfolio Selection," *Omega*, 4, 699–709.

[18] Hodges S.D. and Brealey R.A. (1973), "Portfolio Selection in a Dynamic and Uncertain World," *Financial Analysts Jl.*, 29, March, 50–65.

[19] Huang C.F. and Litzenberger R.H. (1988), *Foundations for Financial Economics*, North-Holland, Amsterdam.

[20] Ingersoll J. (1987), *Theory of Financial Decision Making*, Rowman & Littlefield.

[21] Jarrow R., Maksimovic V., and Ziemba W.T., eds. (1995), *Finance*, North-Holland, Amsterdam.

[22] Jobson J.D. and Korkie B. (1981), "Putting Markowitz Theory to Work," *Jl. Portfolio Management*, Summer, 70–74.

[23] Jobson, J.D., Korkie B., and Ratti V. (1979), "Improved Estimation for Markowitz Portfolios Using James-Stein Type Estimators," *Proceedings Business Economics Statistics Section*, American Statistical Association, 279–284.

[24] Jorion P. (1985), "International Portfolio Diversification with Estimation Error," *Jl. Business*, 58, 259–278.

[25] Jorion P. (1986), "Bayes-Stein Estimation for Portfolio Analysis," *Jl. Financial and Quantitative Analysis*, 21, 279–292.

[26] Jorion P. (1991), "Bayesian and CAPM Estimators of the Means: Implications for Portfolio Selection," *Jl. Banking and Finance*, 15, 717–727.

[27] Judge G.G. and Bock M.E. (1978), *The Statistical Implications of Pre-Test and Stein-Rule Estimators in Econometrics*, North-Holland.

[28] Kallberg J.G. and Ziemba W.T. (1979), "On the Robustness of the Arrow-Pratt Risk Aversion Measure," *Economics Letters*, 2, 21–26.

[29] Kallberg J.G. and Ziemba W.T. (1983), "Comparison of Alternative Utility Functions in Portfolio Selection," *Management Science*, 29, 1257–1276.

[30] Kallberg J.G. and Ziemba W.T. (1984), "Misspecification in Portfolio Selection Problems," in G. Bamberg and K. Spremann (eds), *Risk and Capital*: *Lecture Notes in Economics and Mathematical Systems*, Springer-Verlag, New York.

[31] Kraus A. and Litzenburger R.F. (1976), "Skewness Preference and the Valuation of Risk Assets," *Jl. Finance*, 31, 1085–1100.

[32] Lancaster K. (1971), *Consumer Demand*: *A New Approach*, Columbia University Press.

[33] Levy H. (1969), "A Utility Function Depending on the First Three Moments," *Jl. Finance*, 24, 715–719.

[34] Levy H. and Markowitz H (1979), "Approximating Exected Utility by a Function of Mean and Variance," *American Economic Review*, 69, 308–317.

[35] Markowitz H.M. (1952), "Portfolio Selection," *Jl. Finance*, 7, No 1, March, 77–91.

[36] Markowitz H.M. (1959), *Portfolio Selection*: *Efficient Diversification of Investments*, Yale University Press.

[37] Markowitz H.M. (1983), "Nonnegative or Not Nonnegative: a Question about CAPMs," *Jl. Finance*, 38, 283–295.

[38] Markowitz H.M. (1987), *Mean-Variance in Portfolio Choice and Capital Markets*, Blackwell.

[39] Merton R.C. (1972), "An Analytic Derivation of the Efficient Portfolio Frontier," *Jl. Financial and Quantitative Analysis*, 7, 1851–1872.

[40] Morris C. (1983), "Parametric Empirical Bayes Inference: Theory and Applications," *Jl. American Statistical Assoc.*, 78, 47–55.

[41] Ohlson J. (1975), "Asymptotic Validity of Quadratic Utility as the Trading Interval Approaches Zero," in *Ziemba and Vickson, op. cit.*

[42] Roll R. (1977), "A Critique of the Asset Pricing Theory's Tests," *Jl. Financial Economics*, 4, 129–176.

[43] Samuelson P. (1970), "The Fundamental Approximation Theorem of Portfolio Analysis in Terms of Means Variances and Higher Moments," *Review Economic Studies*, 37, 537–542.

[44] Sharpe, W.F. (1963), "A Simplified Model for Portfolio Analysis," *Management Science*, 9, 277–293.

[45] Sharpe, W.F. (1966), "Mutual Fund Performance," *Jl. Business*, 39, 119–138.

[46] Sharpe, W.F. (1970), *Portfolio Theory and Capital Markets*, McGraw-Hill, New York.

[47] Sharpe, W.F. (1994), *The Sharpe Ratio*, Technical Report, Stanford University, California.

[48] Szegö, G.P. (1980), *Portfolio Theory, with Application to Bank Asset Management*, Academic Press, New York.

[49] Tobin, J. (1958), "Liquidity Preference as Behaviour Towards Risk," *Review Economic Studies*, 26, 65–86.

[50] Tobin, J. (1965), "The Theory of Portfolio Selection," in *The Theory of Interest Rates*, F. Brechling, ed.

[51] Tsiang, S. (1972), "The Rationale of the Mean Standard Deviation Analysis, Skewness Preference and the Demand for Money," *American Economic Review*, 62, 354–371.

[52] Tsiang, S. (1973), "Risk, Return and Portfolio Analysis: Comment," *Jl. Political Economy*, 81, 748–751.

[53] Ziemba, W.T. (1994), "World Wide Security Markey Regularities," *European Jl. Operational Research*, 74(2).

[54] Ziemba, W.T. and R.G. Vickson, eds. (1975), *Stochastic Optimization Models in Finance*, Academic Press, New York.

POS

Point of sale. See **Retailing**.

POSTOPTIMAL ANALYSIS

The study of how a solution changes with respect to (usually) small changes in the problem's data. In particular, this term is applied to the sensitivity analysis and parametric analysis of a solution to a linear-programming problem. See **Linear programming**; **Parametric programming**; **Sensitivity analysis**.

POSYNOMIAL PROGRAMMING

See **Geometric programming**.

POWER MODEL

See **Learning curves**.

PP

See **Parametric programming**.

PPB(S)

Planning-programming-budgeting (system). See **Cost analysis**; **Military operations research**.

PRACTICE OF OPERATIONS RESEARCH AND MANAGEMENT SCIENCE

Hugh J. Miser

Farmington, Connecticut

By the practice of OR/MS, we mean using the appropriate models, tools, techniques, and craft skills of these sciences to understand the problems of man/machine/nature systems with a view toward ameliorating these problems, possibly by new understandings, new decisions, new procedures, new structures, or new policies. Such practice calls for a suitable form of professionalism in dealing not only with the phenomena of the problem situation but also with the persons with relevant responsibilities, as well as other parties at interest.

OR/MS AS A SCIENCE: Following Ravetz (1971), we may describe science in general as "craft work operating on intellectually constructed objects," each object defining a class. Scientific work is thus aimed at establishing new properties of these objects and verifying that they reflect the reality of the classes of phenomena that they represent (Miser, 1993). This description has four implications:

1. The intellectual objects – that OR/MS workers usually call models – are created by the imagination, informed by earlier knowledge of the phenomena and objects that have described them successfully, as well as innovative ideas or new evidence from reality.
2. There is a continuing reference to the phenomena of reality.
3. Scientific inquiry then becomes the search for new properties of the classes both by manipulating the objects and seeking new evidence from reality as a basis for revising them.
4. The new properties deduced from the objects – or models – must then be compared with the appropriate aspects of the phenomena of reality.

It is essential to observe that the different sciences – such as physics, biology, or OR/MS – are distinguished, not by their methods, techniques, or models (many of which are widely shared among the sciences), but by the portions of reality that they are undertaking to understand, explain, and solve problems in Kemeny (1959).

Within the framework established by this conception, it is convenient to distinguish three classes of problems, depending on their goals: to paraphrase Ravetz (1971), scientific problems (where the goal of the work is to establish new properties of the objects of inquiry, and the ultimate function is to achieve knowledge in its field); technical problems (those where the function to be performed specifies the problem); and practical problems (where the goal of the task is to serve or achieve some human purpose and the problem is brought into being by recognizing a problem situation in which some aspect of human welfare should be improved).

Against this background, we may recognize practice as the activity centered on practical problems, even while noting that to solve a practical problem often involves solving technical problems, and, when the basic phenomena underlying a problem situation are not understood, solving scientific problems in order to have the models needed for understanding the practical problem. It is also important to note that this view of science includes work on all three classes of problems within the conception of science as a whole. (For a more extended summary of Ravetz's view of science, see Miser and Quade, 1988.)

THE CONTEXT OF OR/MS: Since sciences are distinguished by their fields of inquiry, it is important to describe this context for OR/MS if it is to be differentiated from other sciences. In this endeavor the OR/MS community has not reached any sort of brief consensus, so what is said here must be regarded as a personal view, based in part on the literature and in part on personal experience.

While we have noted that OR/MS deals with systems involving men, elements of nature, and machines (where this last term is intended to include not only man's technical artifacts but also his laws, standard procedures, common behaviors, and social structures and customs), attempts to take the concept of system beyond this primitive statement as the basis for describing the context of OR/MS have, however, not proved fruitful.

The concept of an action program (Boothroyd, 1978) is more useful: a function, operation, or response that is related to and given coherence by a human objective, need, or problem, together with the system of people, equipment, portion of nature, organizational elements, and management or social structure involved.

It is easy to see that an element in an action program may also have membership in other action programs; for example, an executive in one may also play a role in many others, as may also be the case for a major facility or organization, such as a large corporation or a government. Too, an action program may produce effects on other action programs, both through the cross memberships of elements and by the direct impacts of what it does. (For a more extended summary of Boothroyd's concept, see Miser and Quade, 1988.)

We may then describe the practice of OR/MS as the activity that brings the knowledge and skills of the science of OR/MS to bear on the problems of action programs. While this brief description will suffice as a basis for the argument here, the reader should be aware of the facts that, while it is quite general and covers most of what OR/MS does in practice now, it not only may not cover all of today's activities of practice but also may become even more incomplete with the passage of time.

THE SITUATIONS OF PRACTICE: While each situation in practice may properly be seen as unique, it is nevertheless possible to describe one that contains elements central to most – if not all – of practice, as follows.

An OR/MS analyst is often consulted when someone with a suitable responsibility in an action program discerns a problem situation that needs improvement. While this responsible person may have diagnosed the problem and even may have a notion about a possible solution, it is commonly the case that the forces actually yielding the source of dissatisfaction lie buried deeply enough to make such a diagnosis questionable, and the preconceived fix inappropriate. Thus, typically it is best for the analyst – or the team of analysts if the problem situation is complex – to approach it with an open mind, and aim to explore it thoroughly before deducing its properties and using them to devise a scheme for ameliorating its undesirable properties.

The analysts may be drawn from two sources:

1. There may be an analysis group inside the organization or action program with which the responsible problem-situation identifier – or client – is associated.
2. Analysts may have to be drawn from outside this organization or action program.

In either case, there is abundant experience to support the conclusion that a successful outcome of the practice engagement calls for creating a constructive partnership between the analysis team and the parties at interest in the problem situation, as will be discussed in more detail later.

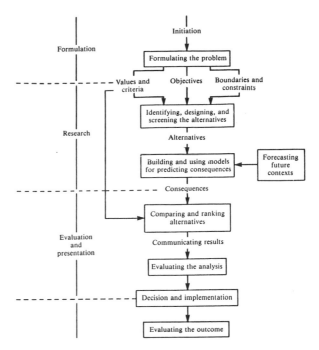

Fig. 1. Important elements in an OR/MS practice engagement that runs from problem formulation through research and implementation to evaluating the outcome. [Source: Miser and Quade (1988), p. 23; reproduced by permission.]

THE PROCESSES OF PRACTICE: Figure 1 offers a synoptic view of the elements that may be included in a practice engagement that proceeds from the general unease of a problem situation to the implementation of some policy or course of action and evaluates its effects. Since each situation has its own unique properties, few OR/MS practice engagements follow such a procedure exactly, but it is a common experience for many – if not most – of these elements to occur at some stage of the work.

FORMULATION: The work begins with a thorough exploration of the problem situation in which the client and his/her action program cooperate. The purpose is to formulate the problem to be addressed, which commonly is quite different from the one originally conceived by the client. Once this is done, and the client has agreed with the analysis team on the problem, it is possible to plan the work to be done. This early work also identifies the values and criteria that should inform the choice of what eventually will be done to ameliorate the client's concerns, sets up the objectives to be sought by the solution, and agrees with the client on the boundaries and constraints that must be observed in devising it.

RESEARCH: This stage extends the information- and data-gathering that began in the formulation stage. The findings that emerge from processing these results allow the analysis team to identify, design, and screen possible alternatives that may help with the problem. Against this background, the analysis team can build models capable of deducing the consequences of adopting each of the alternatives chosen for further investigation within the contexts of possible future conditions.

EVALUATION AND PRESENTATION: With estimates of the consequences in hand, the analysts may compare – and possibly rank – the alternatives against the criteria chosen earlier in the analysis, plus any new ones that may have emerged during the work. These findings must then be presented to the client and other parties at interest in a way that enables them not only to appreciate the results but also have at least a broad overview of the logic that produced them. These understandings may then enable the client to adopt a suitable policy or course of action.

Although the client, and not the analysts, must decide on what to do and how to carry ·it out effectively, experience shows that it is very important for the analysis team, or at least analysts who understand and appreciate what was done, to work cooperatively throughout the implementation stage.

VARIATIONS: While it is possible to specify a core diagram of the principal elements of OR/MS practice, it must be admitted immediately that few, if any, such engagements follow this outline exactly. Rather, since each problem situation is different, the analysis activity must be adapted to it. Thus, in studying a series of cases, one sees variations like these:

- Instead of proceeding linearly from the top to the bottom of Figure 1, the work cycles from intermediate stages back to earlier ones as the progress brings new insights and fresh intermediate results that may prompt reconsideration of the beginning foundations of the work.
- Some work may be aimed more at fleshing out the client's understanding of his situation than prompting him/her to change it significantly, so it may stop at one of the intermediate stages.
- The relative effort expended in the various stages may vary tremendously from case to case: one case may have to expend its major effort in just the information- and data-gathering stage, after which what needs to be done may be fairly apparent without much further analysis. Another case may proceed fairly ex-

peditiously through the outline of Figure 1 and then have a very long and complicated period of work to achieve what may appear to the outsider to be the implementation of a relatively simple set of proposals.
- In some cases an intermediate stage may dominate the work, owing to such factors as technical difficulty in devising proper models, major uncertainties in forecasting future contexts, complexities of the underlying situation, and so on.

In any case, the procedure specified here as the basis for discussion must be regarded as one that has stitched together the key elements that may enter OR/MS practice to varying extents depending on the peculiarities of the situation being studied.

IMPORTANCE OF FOLLOWING THROUGH: The interest of the OR/MS professional, particularly if academically oriented, may flag after the research stage is completed and its results obtained. However, experience shows strongly that to stop there is almost always to waste the earlier effort. Two essential steps must follow: effective communication of the results, and cooperative aid in the implementation process.

Communication. This process, which may not be as appealing to the analyst as the research that preceded it, is nevertheless equally important and deserves great care, since communicating the findings inadequately can vitiate their potential effect, and thus waste the earlier effort. In view of the importance of this step in the OR/MS process, it is surprising that there is no systematic literature describing the skills needed and setting forth how they are best used (for a brief exception see Miser, 1985). Here we must restrict the discussion to these points:

- Few clients will devote a large block of time to such communications, so it is very important to work very hard to condense the principal ideas and findings into as economical a space as possible, whether the form used is oral or written. For example, a top executive may want the key findings presented to him or her in a two-page memorandum or a 20-minute briefing. It is perhaps surprising to the uninitiated to see how much important information can be condensed into so small a space, but only if great care is taken to make the best use of it. Graphs and charts accompanying the words can do much to aid this condensation.
- To communicate effectively, the client's vocabulary must be used, with as few technical terms introduced as possible.

– The whole must be focused on the interests of the client or the audience; after a major study many different groups may have to be addressed, and when this is the case the communication instruments must in each case be tailored to the group in view.

– The analysts must be prepared to stand behind their work and to discuss its implications, even those that may go beyond what was done as part of the analysis.

Implementation. No matter how thoroughly the client – or members of his or her staff who participated in the analysis – understand what was found and its prospective implementation, it is a common experience that the implementation process demands the continuing interest and co-operation of the analysis team, or at least some member of it who is able to follow through. The process of change invariably brings up new problems and issues that, wrongly handled, can vitiate the effects of what the original implementation set out to do. Too, these new problems may call for additional complementary analysis that must take account of what was done earlier (Tomlinson, Quade and Miser, 1985).

THE RELATION BETWEEN ANALYST AND CLIENT:

Emerging from a close scrutiny of the relations that should exist between analyst and client for effective cooperation, Schön (1983) advocates a "reflective contract" that works in this way: . . . "in a reflective contract between practitioner and client, the client does not agree to accept the practitioner's authority but to suspend disbelief in it. He agrees to join the practitioner in inquiring into the situation for which the client seeks help; to try to understand what he is experiencing and to make that understanding accessible to the practitioner; to confront the practitioner when he does not understand or agree; to test the practitioner's competence by observing his effectiveness and to make public his questions over what should be counted as effectiveness; to pay for services rendered and to appreciate competence demonstrated. The practitioner agrees to deliver competent performance to the limits of his capacity; to help the client understand the meaning of the professional's advice and the rationale for his actions, while at the same time he tries to learn the meanings his actions have for the client; and to reflect on his own tacit understanding when he needs to do so in order to play his part in fulfilling the contract."

Under this concept for OR/MS work, the client's obligation to share his experience and understanding of the problem situation is often discharged by assigning a member of his staff to work with the analysis team, an arrangement that has many benefits, among which these may be listed: it helps the analysis team identify and gather the information that it needs as a background and basis for its work; it helps the analysts avoid foolish mistakes related to the client's operations; and it acts to keep the client informed of what is emerging from the analysis, which often helps to pre-sell the findings that eventually emerge.

Since OR/MS practice may be viewed as a dialogue between analyst and client related to the problem situation and the problem from it that is eventually chosen for analysis, this arrangement serves as a useful continuing conduit for this dialogue, beyond what can be achieved with periodic progress meetings with the client (Miser, 1994).

Other practical arrangements between the analysis team and the client to implement Schön's concept of a reflective contract must, of necessity, be evolved in the light of the circumstances peculiar to each engagement. An in-house analysis group that has been able to achieve a reflective contract with the organization of which it is a part has a special opportunity: it can often identify problem situations that may not yet have been observed by executives in the organization, and thus set to work on them before they grow in size and importance.

HOW TO LEARN THE SKILLS OF PRACTICE:

The OR/MS community has, unfortunately, not evolved a comprehensive epistemology of practice and set it down in easily accessible literature that can be used widely in training courses. Some first steps in this direction for systems analysis, the large-scale efforts that can be thought of as part of OR/MS practice, are given in Miser and Quade (1985, 1988); much of what they say can apply equally to OR/MS as a whole. Thus, to learn the needed scientific and craft skills, someone aiming for an OR/MS career must pursue a tripartite program assembled from a variety of sources.

The intellectual basis. The foundation of effective OR/MS practice must be a thorough education in mathematics, with special attention to probability and statistics. Since by now certain models have become associated with OR/MS (as any introductory college textbook makes clear), these should be mastered as well. And a broad view of science with knowledge of other branches is also sure to be helpful.

Beyond a good mathematical and scientific education, however, the potential practitioner must not only be willing but also eager to learn from the problem situation, from the people in it, and from the representatives of other specialties, both practical and intellectual, that may have to be called on to help. As Schön's concept makes clear, to undertake an engagement in practice is to enter a multipartite partnership, and the flow of information must reflect this if the work is to be effective.

Since the action programs that OR/MS practice deals with contain people as essential elements, the analysts must know how to deal effectively and sympathetically with them, since they will enter the problem situation at many levels. In sum, interpersonal skills are an important requisite of good practice.

Familiarity with successful cases. There are by now a great many published accounts of successful cases of OR/MS practice. The journal *Interfaces* specializes in presenting them, and since 1975 has been a treasure-house of such accounts, as well as proven advice about the arts of practice. Assad, Wasil, and Lilien (1992) accompany a selection of these cases with valuable commentary. For a much wider view, one can consult the "Applications Oriented" section of the *International Abstracts in Operations Research*, the comprehensive abstract journal that has been published since 1961; it will not only exhibit the wide variety of practice being undertaken throughout the world but also identify the many journals and books in which cases appear.

Apprenticeship. Since the OR/MS community has yet to achieve a widely agreed and centrally documented view of its epistemology of practice, the best way for a person to observe and learn the myriad craft skills of practice is to work with an accomplished and skillful analysis team – in sum, to serve an apprenticeship. (Miser and Quade, 1985, 1988, offer a substantial body of additional information relating to the craft skills needed for effective OR/MS.)

EXAMPLES OF GOOD PRACTICE: Since 1975, *Interfaces* has published the finalist papers in the Franz Edelman competition for the best papers on practice each year; there are five or more finalists in each competition. These accounts are an excellent central source of examples of good practice; in recent years tapes of the finalist presentations have also been made available.

There are many other sources of such work – too many to list here; however, both *Operations Research* and the *Journal of the Operational Research Society* contain one or more examples of good practice in each issue, as do the sources mentioned earlier.

See **Decision making; Ethics; Implementation; Problem structuring methods; Systems analysis.**

References

[1] Assad, A.A., E.A. Wasil and G.L. Lilien (1992). *Excellence in Management Science Practice: A Readings Book.* Prentice Hall, New Jersey.
[2] Boothroyd, H. (1978). *Articulate Intervention.* Taylor and Francis, London.
[3] Kemeny, J.G. (1959). *A Philosopher Looks at Science.* Van Nostrand Reinhold, New York.
[4] Miser, H.J. (1985). "The Practice of Systems Analysis." In Miser and Quade (1985), 287–326.
[5] Miser, H.J. (1993). "A Foundational Concept of Science Appropriate for Validation in Operational Research." *European Jl. Operational Research* **66**, 204–215.
[6] Miser, H.J. (1994). "Systems Analysis as Dialogue: An Overview." *Technological Forecasting and Social Change* **45**, 299–306.
[7] Miser, H.J. and E.S. Quade, eds. (1985). *Handbook of Systems Analysis: Overview of Uses, Procedures, Applications, and Practice.* Wiley, Chichester, United Kingdom.
[8] Miser, H.J. and E.S. Quade, eds. (1988). *Handbook of Systems Analysis: Craft Issues and Procedural Choices.* Wiley, Chichester, United Kingdom.
[9] Ravetz, J.R. (1971). *Scientific Knowledge and its Social Problems.* Oxford University Press, Oxford.
[10] Schön, D.H. (1983). *The Reflective Practitioner: How Professionals Think in Action.* Basic Books, New York.
[11] Tomlinson, R., E.S. Quade and H.J. Miser (1985). "Implementation." In Miser and Quade (1985), 249–280.

PRECEDENCE DIAGRAMMING

A graphic analysis of a project plan in which the nodes are the work activities (or tasks) and are connected by arrows. Relationships among tasks are designated as start-to-start, start-to-finish, and finish-to-finish, which eliminates the use of dummy arrows. See **Network planning.**

PREDICTIVE MODEL

A model used to predict the future course of events and as an aid to decision making. See **Decision problem; Descriptive model; Mathematical model; Model; Normative model; Prescriptive model.**

PREEMPTION

(1) In a goal-programming problem, a statement that stipulates the ordering of the goals, so that a solution that satisfies the priority k goal is always to be preferred to solutions that satisfy the lower priority goals $k + 1, \ldots$. See **Goal programming**. (2) In queueing theory, this means that an arriving higher priority item pushes a lower one out of service because the newcomer has higher priority. Service of the preempted customer can either continue from the point of its interruption or totally start anew. See **Queueing theory**.

PREEMPTIVE PRIORITIES

See **Goal programming**; **Preemption**; **Queueing theory**.

PREFERENCE THEORY

James S. Dyer and Jianmin Jia

University of Texas at Austin

INTRODUCTION: Preference theory studies the fundamental aspects of individual choice behavior, such as how to identify and quantify an individual's preferences over a set of alternatives, and how to construct appropriate preference representation functions for decision making. An important feature of preference theory is that it is based on rigorous axioms which characterize an individual's choice behavior. These preference axioms are essential for establishing preference representation functions, and provide the rationale for the quantitative analysis of preference.

Preference theory provides the foundation for economics and the decision sciences. A basic topic of microeconomics is the study of consumer preferences and choices (Kreps, 1990). In decision analysis and operations research, knowledge about the decision maker's preference is necessary to establish objective (or preference) functions that are used for evaluating alternatives. Different decision makers usually have different preference structures, which may imply different objective functions for them. Preference studies can also provide insights into complex decision situations and guidance for simplifying decision problems.

The basic categories of preference studies can be divided into characterizations of preferences under conditions of certainty or risk, and over

alternatives described by a single attribute or by multiple attributes. In the following, we will begin with the introduction of basic preference relations, and then discuss preference representation under certainty and under risk. We shall refer to a preference representation function under certainty as a value function, and to a preference representation function under risk as a utility function.

BASIC PREFERENCE RELATIONS: Preference theory is primarily concerned with properties of a binary preference relation $>_p$ on a choice set X, where X could be a set of commodity bundles, decision alternatives, or monetary gambles. For example, we might present an individual with a pair of alternatives, say x and y (e.g., two cars), and ask how they compare (e.g., do you prefer x or y?). If the individual says that x is preferred to y, then we write $x >_p y$, where $>_p$ means strict preference. If the individual states that he or she is indifferent between x and y, then we represent this preference as $x \sim_p y$. Alternatively, we can define $\sim_p$ as the absence of strict preference; that is, not $x >_p y$ and not $y >_p x$. If it is not the case that $y >_p x$, then we write $x \geq_p y$, where $\geq_p$ represents a weak preference (or preference-indifference) relation. We can also define $\geq_p$ as the union of strict preference $>_p$ and indifference $\sim_p$; that is, both $x >_p y$ and $x \sim_p y$.

Preference studies begin with some basic assumptions (or axioms) of individual choice behavior. First, it seems reasonable to assume that an individual can state preference over a pair of alternatives without contradiction; that is, the individual cannot strictly prefer x to y and y to x simultaneously. This leads to the following definition for *preference asymmetry:* preference is asymmetric if there is no pair x and y in X such that $x >_p y$ and $y >_p x$.

Asymmetry can be viewed as a criterion of preference consistency. Furthermore, if an individual makes the judgment that x is preferred to y, then he or she should be able to place any other alternative z somewhere on the ordinal scale determined by the following: either better than y, or worse than x, or both. Formally, we define *negative transitivity* by saying that preferences are negatively transitive if given $x >_p y$ in X and any third element z in X, it follows that either $x >_p z$ or $z >_p y$, or both.

If the preference relation $>_p$ is asymmetric and negatively transitive, then it is called a *weak order*. The weak order assumption implies some desirable properties of a preference ordering, and is a basic assumption in many preference studies. If the preference relation $>_p$ is a weak order, then the associated indifference and weak preference relationships are well behaved. The following results summarize some of these.

If strict preference $>_p$ is a weak order, then

1) strict preference $>_p$ is *transitive* (if $x >_p y$ and $y >_p z$, then $x >_p z$);
2) indifference $\sim_p$ is transitive, *reflexive* ($x \sim_p x$ for all x), and symmetric ($x \sim_p y$ implies $y \sim_p x$);
3) exactly one of $x >_p y$, $y >_p x$, $x \sim_p y$ holds for each pair x and y; and
4) weak preference $\geq_p$ is transitive and *complete* (for a pair x and y, either $x \geq_p y$ or $y \geq_p x$).

Thus, an individual whose preferences can be represented by a weak order can rank all alternatives considered in a unique order. Further discussions of the properties of binary preference relations are presented in Fishburn (1970, Chapter 2) and Kreps (1990, Chapter 2).

PREFERENCE REPRESENTATION UNDER CERTAINTY:

If strict preference $>_p$ on X is a weak order, then there exists a numeric representation of preference, a real-valued function v on X such that

$$x >_p y \text{ if and only if } v(x) > v(y),$$

for all x and y in X (Fishburn, 1970). A preference representation function v under certainty is often called a *value function* (Keeney and Raiffa, 1976). A value function is said to be order-preserving since the numbers $v(x)$, $v(y)$, ... ordered by $>$ are consistent with the order of x, y, ... under $>_p$. Thus, any monotonic transformations of v will be order-preserving. As a result, the units of v have no particular meaning.

We may wish to consider a "strength of preference" notion that involves comparisons of preference differences between pairs of alternatives. To do so, we need more restrictive preference assumptions, including that of a weak order over preferences between exchanges of pairs of alternatives (Krantz *et al.*, 1971, Chapter 4). These axioms imply the existence of a real-valued function v on x such that, for all w, x, y, and z in X, the difference in the strength of preference between w and x exceeds the difference between y and z if and only if

$$v(w) - v(x) > v(y) - v(z).$$

Furthermore, v is unique up to a positive linear transformation; that is, if v' also satisfies the above difference inequality, then we must have $v'(x) = av(x) + b$, where a (> 0) and b are constants. This means that v provides an *interval scale of measurement*, such that v is often called a *measurable value function* in order to distinguish it from an order-preserving value function.

For multi-attribute decision problems, $X = X_1 X_2 \ldots X_n$, where n is the number of attributes and an element $x = (x_1, x_2, \ldots, x_n)$ in X represents an alternative. A multi-attribute value function can be written as $v(x_1, x_2, \ldots, x_n)$. Using some preference independence conditions, we can simplify the multi-attribute value model.

The subset Y of attributes in X is said to be *preferentially independent* of its complementary set $\bar{Y}$ if preferences for levels of these attributes Y do not depend on the fixed levels of the complementary attributes $\bar{Y}$. Attributes X_1, $X_2, \ldots, X_n$ are mutually preferentially independent if every subset of these attributes is preferentially independent of its complementary set.

A multi-attribute value function $v(x_1, x_2, \ldots, x_n)$, $n \geq 3$, has the following additive form

$$v(x_1, x_2, \ldots, x_n) = \sum_{i=1}^{n} v_i(x_i), \tag{1}$$

where v_i is a value function over X_i if and only if the attributes are mutually preferentially independent (Keeney and Raiffa, 1976; Krantz *et al.*, 1971). When v is bounded, it may be more convenient to scale v such that each of the single-attribute value functions ranges from zero to one. Thus, we will have the following form of the additive value function:

$$v(x_1, x_2, \ldots, x_n) = \sum_{i=1}^{n} w_i v_i(x_i), \tag{2}$$

where v and v_i are scaled from zero to one, and the w_i are positive scaling constants summing to one. The assessment of models (1) and (2) are discussed in Keeney and Raiffa (1976, Chapter 3).

Dyer and Sarin (1979) proposed multi-attribute measurable value functions based on the concept of preference differences between alternatives that are much easier to assess than the additive form based on preferential independence. In addition to preferential independence, they considered some additional conditions that, loosely speaking, require that the decision maker's comparisons of preference differences between pairs of alternatives that differ in the levels of only a subset of the attributes do not depend on the fixed levels of the other attributes. These conditions allow the decomposition of a multi-attribute value model into additive and multiplicative forms. This development also provides a link between the additive value function and the multi-attribute utility model.

PREFERENCE REPRESENTATION UNDER RISK: Now we turn to preference representation for risky options (i.e., lotteries or gambles). Perhaps the most significant contribution to this area was the formalization of *expected utility theory* by von Neumann and Morgenstern (1947). This development has been refined by a number of researchers, and is most commonly presented in terms of three basic axioms (Fishburn, 1970).

Let P be a convex set of simple probability distributions or lotteries $\{X, Y, Z, \ldots\}$ on a nonempty set X of outcomes. (We shall use X, Y and Z to refer to probability distributions and random variables interchangeably.) For lotteries X, Y, Z in P and all λ, $0 < \lambda < 1$, the expected utility axioms are:

A1. (*Ordering*) $>_p$ on P is a weak order;
A2. (*Independence*) If $X >_p Y$, then $\lambda X + (1-\lambda)Z >_p \lambda Y + (1-\lambda)Z$ for all Z in P;
A3. (*Continuity*) If $X >_p Y >_p Z$, then there exist some $0 < \alpha < 1$ and $0 < \beta < 1$ such that $\alpha X + (1-\alpha)Z >_p Y >_p \beta X + (1-\beta)Z$.

The von Neumann-Morgenstern expected utility theory asserts that the above axioms hold if and only if there exists a real-valued function u such that for all X, Y in P,

$$X >_p Y \text{ if and only if } E[u(X)] > E[u(Y)]$$

where the expectation is taken over the probability distribution of a lottery. Moreover, such a u is unique up to a positive linear transformation.

The expected utility model can also be used to characterize an individual's risk attitude (Keeney and Raiffa, 1976, Chapter 4). If an individual's utility function is concave, linear, or convex, then the individual is risk averse, risk neutral, or risk seeking, respectively.

The von Neumann-Morgenstern theory of risky choice presumes that the probabilities of the outcomes of lotteries are provided to the decision maker. Savage (1954) extended the theory of risk choice to allow for the simultaneous development of subjective probabilities for outcomes and for a utility function u defined over those outcomes.

As a normative theory, the expected utility model has played a major role in the prescriptive analysis of decision problems. However, for descriptive purposes, the assumptions of this theory have been challenged by empirical studies (Kahneman and Tversky, 1979). Some of these empirical studies demonstrate that subjects may choose alternatives that imply a violation of the independence axiom (A2). One implication of A2 is that the expected utility model is "linear in probabilities." A number of contributions have been made by relaxing the independence axiom and developing some *nonlinear utility models* to accommodate actual decision behavior (Fishburn, 1988).

When $X = X_1 \times X_2 \times \ldots \times X_n$ in a von Neumann-Morgenstern utility model, and the decision maker's preferences are consistent with some additional independence conditions, then $u(x_1, x_2, \ldots, x_n)$ can be decomposed into additive, multiplicative, and other well-structured forms that simplify assessment.

The attributes $X_1, X_2, \ldots, X_n$ are said to be *additive independent* if preferences over lotteries on $X_1, X_2, \ldots, X_n$ depend only on the marginal probabilities assigned to individual attribute levels, but not on the joint probabilities assigned to two or more attribute levels.

A multi-attribute utility function $u(x_1, x_2, \ldots, x_n)$, can be decomposed as

$$u(x_1, x_2, \ldots, x_n) = \sum_{i=1}^{n} w_i u_i(x_i) \qquad (3)$$

if and only if the additive independence condition holds, where u_i is a single-attribute function over X_i scaled from 0 to 1, and the w_i are positive scaling constants summing to one. The additive model (3) has been widely used in practice.

If the decision maker's preferences are not consistent with the additive independence condition, a weaker independence condition that leads to a multiplicative preference representation may be satisfied.

An attribute X_i is said to be *utility independent* of its complementary attributes if preferences over lotteries with different levels of X_i do not depend on the fixed levels of the remaining attributes. Attributes $X_1, X_2, \ldots, X_n$ are mutually utility independent if all proper subsets of these attributes are utility independent of their complementary subsets.

A multi-attribute utility function $u(x_1, x_2, \ldots, x_n)$ can have the multipicative form

$$1 + ku(x_1, x_2, \ldots, x_n) = \prod_{i=1}^{n} [1 + kk_i u_i(x_i)] \qquad (4)$$

if and only if the attributes $X_1, X_2, \ldots, X_n$ are mutually utility independent, where u_i is a single-attribute function over X_i scaled from 0 to 1, the k_i are positive scaling constants, and k is an additional scaling constant. For approaches to the assessment of model (4) and other extensions of multi-attribute utility theory, see Keeney and Raiffa (1976).

See **Decision analysis; Multi-attribute utility theory; Utility theory.**

References

[1] Dyer, J.S. and R.K. Sarin (1979). "Measurable Multi-attribute Value Functions," *Operations Research*, 27, 810–822.

[2] Fishburn, P.C. (1970). *Utility Theory for Decision Making*. Wiley, New York.

[3] Fishburn, P.C. (1988). *Nonlinear Preference and Utility Theory*. The Johns Hopkins University Press, Baltimore, Maryland.

[4] Kahneman, D.H. and Tversky, A. (1979). "Prospect Theory: An Analysis of Decision under Risk," *Econometrica*, 47, 263–290.

[5] Keeney, R.L. and H. Raiffa (1976). *Decisions with Multiple Objectives: Preferences and Value Tradeoffs*. Wiley, New York.

[6] Krantz, D.H., R.D. Luce, P. Suppes, and A. Tversky (1971). *Foundations of Measurement*. Academic Press, San Diego.

[7] Kreps, D.M. (1990). *A Course in Microeconomics Theory*. Princeton University Press, New Jersey.

[8] Savage, L.J. (1954). *The Foundations of Statistics*. Wiley, New York.

[9] von Neumann, J. and O. Morgenstern (1947). *Theory of Games and Economic Behavior*. Princeton University Press, New Jersey.

PRESCRIPTIVE MODEL

A model that attempts to describe the best or optimal solution of a man/machine system. For a decision problem, such a model is used as an aid in selecting the best alternative solution. See **Decision problem; Descriptive model; Mathematical model; Normative model; Prescriptive model.**

PRICES

In the simplex method, for a nonbasic variable x_j, the price is defined as $d_j = c_j - z_j$ or $d_j = z_j - c_j$, where c_j is the variable's original cost coefficient and $z_j = \pi A_j$, with A_j the variable's original column of coefficients and π the multiplier (pricing) vector of the current basis. The d_j is termed the reduced or relative cost. It is the difference between the direct cost c_j and indirect cost z_j. The d_j indicates how much the objective function would change per unit change in the value of x_j. The d_j for the variables in the basic feasible solution are equal to zero. See **Devex pricing; Opportunity cost; Simplex method.**

PRICING MULTIPLIERS

See **Multiplier vector.**

PRICING OUT

In the simplex method, the calculation of the prices associated with the current basic solution. See **Prices; Simplex method.**

PRICING VECTOR

See **Multiplier vector; Prices; Simplex method.**

PRIMAL-DUAL ALGORITHM

An adaptation of the simplex method that starts with a solution to the dual problem and systematically solves a restricted portion of the primal problem while improving the solution to the dual. At each step, a new restricted primal is defined and the process continues until solutions to the original primal and dual problems are obtained. See **Simplex method.**

PRIMAL-DUAL LINEAR-PROGRAMMING PROBLEMS

See **Dual linear-programming problem; Linear programming.**

PRIMAL PROBLEM

The primal problem is usually taken to be the original linear-programming problem under investigation. See **Dual-programming problem.**

PRIM'S ALGORITHM

A procedure for finding a minimum spanning tree in a network. The method starts from any node and connects it to the node nearest to it. Then, for those nodes that are now connected, the unconnected node that is closest to one of the nodes in the connected set is found and connected to these closest nodes. The process continues until all nodes are connected. Ties are broken arbitrarily. See **Greedy algorithm; Kruskal's algorithm; Minimal spanning tree.**

PRISONER'S DILEMMA GAME

This is a two-person game where neither player knows the other's decision until both are made.

Imagine a situation where two criminals are isolated from each other and the police interrogator offers each the following deal: if the prisoner confesses and the confession leads to the conviction of the other prisoner, he goes free and the other prisoner gets 10 years in prison. If both confess, they each get 5 years. If neither confesses, there is enough evidence to convict both on a lesser offense and they both get one year. If there is no trust, then both will confess. If there is complete trust, neither will. Since complete trust is rare, when the game is played one time, players almost always defect. When the game is played repeatedly and there is a chance for a long-term reward, wary cooperation with a willingness to punish defection is the best strategy. This game illustrates many social and business contracts and is important for understanding group behavior, both cheating and cooperation. See **Game theory**.

PROBABILISTIC ALGORITHM

An algorithm which leaves some of its decisions to chance (as opposed to a deterministic algorithm). See **Genetic algorithms**.

PROBABILISTIC PROGRAMMING

A mathematical programming problem in which some or all of the data are random variables. See **Chance-constrained programming**; **Stochastic programming**.

PROBABILITY DENSITY FUNCTION (PDF)

When the derivative $f(x)$ of accumulative probability distribution function $F(x)$ exists, it is called the density or probability density function.

PROBABILITY DISTRIBUTION

The term is used very loosely and typically refers to the function which describes the probabilistic behavior of a random variable. If the variable is discrete, then the probability mass function is often also called the probability distribution. However, for an absolutely continuous random variable, probability distribution may be meant to be the same as the cumulative distribution function or even its density function.

PROBABILITY DISTRIBUTION SELECTION

See **Distribution selection for stochastic modeling**.

PROBABILITY GENERATING FUNCTION

For any discrete random variable with integer outcomes (possibly including negatives) and probability function $p_j = \Pr\{r.v. = j\}$, its probability generating function is given by

$$P(s) = \sum_{j=-\infty}^{\infty} s^j p_j.$$

PROBABILITY INTEGRAL TRANSFORMATION METHOD

A major method for creating random variates for simulation experiments, using the fact that the cumulative distribution function of a random variable has a uniform (0,1) distribution when it is itself considered to be a random function. The basic procedure for deriving the variate is as follows. Given the pseudorandom number z on the unit interval and the cumulative distribution function (CDF) $F(x)$ of the random variable X, a random variate x can (theoretically) be generated by the formula

$$x = F^{-1}(z).$$

We say theoretically because the function F need not be (and often is not) easily invertible. As an example of a situation where the inverse transformation method works well, consider the standard negative exponential distribution, where $F(x) = 1 - e^{-\theta x}$. The random variates can be found here by

$$x = -\frac{1}{\theta} \ln (1 - z),$$

or equivalently,

$$x = -\frac{1}{\theta} \ln (z).$$

This procedure can be used for discrete as well as continuous probability distributions and even for empirical distributions, as long as the CDF is invertible. For those problems in which the CDF F will not be analytically or numerically invertible, other methods must be employed. See

Monte Carlo sampling and variance reduction; Random number generators; Random variates; Simulation of discrete-event stochastic systems.

PROBLEM SOLVING

The process of deciding on actions aimed at achieving a goal. Initially, the goal is defined to represent a solution to a problem. During the reasoning process, subgoals are formed, and problem-solving becomes recursive. See **Artificial intelligence; Expert systems; Decision analysis; Decision making; Decision support systems.**

PROBLEM STRUCTURING METHODS

Jonathan Rosenhead

The London School of Economics & Political Science

INTRODUCTION: Problem structuring methods (PSMs) are a broad group of problem handling approaches whose purpose is to assist in the structuring of problems rather than directly with their solution. They are participative and interactive in character, and in principle offer OR/MS access to a range of problem situations for which more classical OR techniques have limited applicability.

PSMs developed out of, or at least intertwined with, a critique of the restricted scope of traditional OR techniques. From the late 1960s there developed an active debate over claims for the objectivity of OR/MS models, and about the limitations imposed on OR/MS practice by its concentration on well-defined problems. Critics held that standard OR *techniques* assume that relevant factors, constraints, and objective function are both established in advance and consensual; commonly the function of the technique is to determine an optimal setting of the controllable variables. Consistently with this, standard formulations of OR *methodology* were seen to assume a single uncontested representation of the problematic situation under consideration.

Critics have recognized that OR's *practice* has been considerably more diverse than this, and in particular is far from dominated by considerations of optimality; however, the available tools were held to offer little appropriate assistance outside this area. The methodological framework on offer was equally seen as giving scant guidance to analysts confronting less well-behaved situations.

TAME VS. WICKED PROBLEMS: One aspect of the critique was an identification of two substantially different types of problem situation – tame vs. wicked problems (Rittel and Webber, 1973), problems vs. messes (Ackoff, 1981), technical vs. practical problems (Ravetz, 1971), and the moon-ghetto metaphor (Nelson, 1974). The distinction is well captured in Schon's (1987) extended metaphor contrasting the 'high ground' where problems are of great technical interest but of limited social importance, with the 'swamp' where messy, confusing problems defy technical solution.

The analytic methods which work well for the 'tame' problems of the high ground are of little use for the 'wicked' problems of the swamp. The traditional OR/MS approach works well only where there is an organization with a strong hierarchical chain of command and relatively low analytic sophistication, performing well-defined and repetitive tasks (conducive to the generation of volumes of reliable data) in performing a function for which there is a broad consensus on priorities (Greenberger, Crenson and Crissey, 1976). Elsewhere, it is liable to come to grief.

There are difficulties, therefore, in transferring the techniques and methodology of traditional OR/MS from tame to wicked problems. However, critics were unwilling to accept that only 'non-rigorous enquiry' (as assumed by Schon, 1987) was possible in swampy problem terrain. In practice alternative methods were developed piecemeal by individual innovators. (These methods also accumulated the collective designation of 'soft OR', an imprecise and unhelpful terminology.) Different PSMs have widely differing rationales, purposes, technical apparatus, etc. It will, however, be useful here to describe their common characteristics rather than concentrating on the detail of particular methods.

Rosenhead (1981) proposed an 'in principle' methodological framework for alternative approaches by inverting the characteristics of the conventional paradigm, as shown in Figure 1.

More firmly based (though similar) specifications can be derived from a consideration of the defining properties of 'swamp' conditions. Any decision is taken in a context of other decision-makers with a substantial degree of autonomy, their own agendas, and their own interests to advance or protect. Conflict is as relevant a feature as cooperation, and uncertainty (about

- trade-off onto single objective for optimization	- satisficing, separate dimensions, own units
- data greed, distortion	- analysis complementing judgment reduces data demand
- assumption of consensus, scientization of politics, opacity	- transparency to clarify conflict
- people as passive objects	- people as active subjects
- hierarchy, abstract objectives	- bottom up planning
- future certainty, so "predict and prepare"	- accept uncertainty, keep options open
Tame Problems	Wicked Problems

Fig. 1. Characteristics and assumptions of methodologies by problem category.

system behavior, priorities, the decisions of others) is endemic.

To engage with this problem environment imposes a range of social requirements on any decision-aiding methodology. It is more likely to be both used and helpful if it accommodates multiple alternative perspectives, facilitates the negotiation of a joint agenda, functions through interaction and iteration, generates partial rather than comprehensive solutions. These social requirements in turn have various technical implications. The representation of problem complexity by graphical means (rather than algebraically or in tables of numerical results) will assist participation. The existence of multiple perspectives invalidates the search for an optimum; the need is rather for systematic exploration of the solution space. Judgments are generally more meaningfully expressed between discrete alternatives rather than across continuous variables. Estimation of numerical probabilities will need to give way to identification of relevant possibilities. And alternative scenarios will substitute for future forecasts.

These outline specifications for a more appropriate decision-aiding technology eliminate much of the scope for advanced mathematics, probability theory, complex algorithms. They identify, rather, an alternative approach employing representation of relationships, symbolic manipulation, and limited quantification within a systematic framework. Such an approach can be seen as rigorous in an expanded sense of the word.

PROBLEM STRUCTURING: There is no definitive list of problem structuring methods. They can be distinguished from non-OR modes of working with groups, such as Organizational Development, by the core element of an explicit model of cause-effect relationships. They can also be demarcated from other OR approaches

which purport to tackle messy, ambitious problems (for example, Analytic Hierarchy Process) by PSMs' transparency of method, restricted mathematization, and focus on supporting judgment rather than representing it. These limits are imprecise and arguable; and there is scope for approaches developed for other or broader purposes (e.g., spreadsheet models) to be used in a similar spirit. Methods which have some degree of similarity to PSMs but which for coherence are best regarded as falling outside the category include decision analysis, decision conferencing, PROMETHEE, scenario planning, system dynamics and Viable System Diagnosis. Other parts of the perimeter are bordered by the focus group approach and by Rapid Rural Appraisal and other participative third world development approaches.

The best known and/or most widely used PSMs are (with their principal originators): Idealized Planning (Ackoff, 1981) – also known as interactive planning; Soft Systems Methodology (SSM) (Checkland and Scholes, 1990); Strategic Choice Approach (Friend and Hickling, 1987); and Strategic Options Development and Analysis (SODA) (Eden *et al.*, 1983). Other methods include Hypergame Analysis, Metagame Analysis, Robustness Analysis (Rosenhead, 1981, 1989) and Strategic Assumption Surfacing and Testing (SAST). Some PSMs have distinctive features suiting them to particular types of problem situation. Thus the two game theory-related methods focus on issues of cooperation and conflict; robustness analysis is concerned with the maintenance of flexibility, while Strategic Choice's emphasis is on the management of uncertainty and commitment; and SAST concentrates on generating agreement on key assumptions. Other methods are arguably more general purpose. SODA aims to identify factors in use by participants to make sense of a situation, so as to turn them into common property.

Idealized Planning aims to secure consensus around futuristic organizational re-designs. SSM aims to generate debate about alternative system modifications.

Many PSMs consist of a loosely articulated set of processes (part social, part technical), with considerable freedom to switch mode or re-cycle. They therefore lend themselves to creative re-assembly. Only some stages of a method may be carried through; or methods may be combined: for example, SODA and SSM, Strategic Choice and Robustness, Hypergames and SODA. There is also some limited experience of combining specific 'hard' and 'soft' methods. Clearly there is scope, at the least, for using PSMs for the initial problem formulation or problem framing phase of a conventional OR project.

Several established PSMs have associated software; examples include CONAN (for Metagame Analysis), STRAD (for Strategic Choice) and COPE (for SODA). These packages perform a variety of functions. They may display and re-organize concepts and their inter-relationships; identify a feasible range of options for action; elicit preferences using paired comparisons; and so on. They may also perform a variety of roles in the project, from 'back-room' technical assistance to the facilitator between group sessions, through enabling individual participants to pursue solo investigations, to the provision of an on-line Group Decision Support System.

Practical application of PSMs has been reported with a wide variety of client organizations. These include industry-government negotiations on environmental policies, land-use development planning by local government units, health service agencies, and a range of major industrial users. A developing field of practice in which PSMs have made a significant contribution is that of *community operational research*.

Practitioners who use PSMs have to manage not only the complexity of substantive subject matter but also the dynamics of interaction among participants. The dual roles of analyst and of facilitator of group process place heavy demands on the consultant, who is called upon to deploy a wider range of skills than in conventional operational research practice. Training should if possible include at least a brief experience of practical apprenticeship.

See **Practice of operations research and management science; Systems analysis.**

References

[1] Ackoff, R.L. (1981). "The art and science of mess management," *Interfaces*, 11, 20–26.

[2] Checkland, P. and Scholes, J. (1990). *Soft Systems Methodology in Practice*. Wiley, Chichester, UK.

[3] Eden, C., Jones, S. and Sims, D. (1983). *Messing About in Problems*. Pergamon, Oxford.

[4] Flood, R.L. and Jackson, M.C. (1991). *Creative Problem Solving: Total Systems Intervention*. Wiley, Chichester, UK.

[5] Friend, J.K. and Hickling, A. (1987). *Planning Under Pressure*. Pergamon, Oxford.

[6] Greenberger, M., Crenson, M.A. and Crissey, B.L. (1976). *Models in the Policy Process*. Russell Sage, New York.

[7] Nelson, R.R. (1974). "Intellectualizing about the moon-ghetto metaphor: a study of the current malaise of rational analysis of social problems," *Policy Science*, 5, 375–414.

[8] Ravetz, J.R. (1971). *Scientific Knowledge and Its Social Problems*. Oxford University Press, Oxford.

[9] Rittel, H.W.J. and Webber, M.M. (1973). "Dilemmas in a general theory of planning," *Policy Science*, 4, 155–169.

[10] Rosenhead, J. (1981). "Operational research in urban planning," *Omega*, 9, 345–364.

[11] Rosenhead, J., ed. (1989). *Rational Analysis for a Problematic World: Problem Structuring Methods for Complexity, Uncertainty and Conflict*. Wiley, Chichester, UK.

[12] Schon, D.A. (1987). *Educating the Reflective Practitioner: Toward a New Design for Teaching and Learning in the Professions*. Jossey-Bass, San Francisco.

PROCESSOR SHARING

A queueing discipline whereby the server shares its effort over all customers present. See **Queueing theory**.

PRODUCT FORM

See **Product-form solution**.

PRODUCT FORM OF THE INVERSE (PFI)

The inverse of a matrix expressed as the product of sequence of matrices. The matrices in the product are elementary elimination matrices. See **Eta file; Simplex method**.

PRODUCT-FORM SOLUTION

When the steady-state joint probability of the number of customers at each node in a queueing network is the product of the individual probabilities times a multiplicative constant, as in $\Pr\{N_1 = n_1, \quad N_2 = n_2, \ldots, N_J = n_J\} = B\pi(n_1)\pi(n_2)\ldots\pi(n_J)$, the network is said to have a product-form solution. Sometimes the designation of a product-form solution requires that the multiplicative constant also decompose into separate factors for each node, as it will in open Jackson networks. Variants of these sorts of solutions also occur in some non-network queues, such as those with vacations. See **Networks of queues; Queueing theory**.

PRODUCTION FUNCTION

See **Economics**.

PRODUCTION MANAGEMENT

Gabriel R. Bitran

Massachusetts Institute of Technology, Cambridge

Sriram Dasu

University of California at Los Angeles

INTRODUCTION: Some of the important objectives of a manufacturing system are to produce in a timely manner products that conform to specifications, while minimizing costs. Often hundreds of products are produced by a facility, and the entire production process may span several facilities that are geographically dispersed. In many industries the production network consists of plants that are located in different countries.

Production management entails many decisions that are made at all levels of the managerial hierarchy. Manufacturing processes involve a large number of people in many different departments and organizations, and utilize a variety of resources. In addition to the quality of human resources employed, operational efficiency depends upon the location and capacity of the plants, choice of technology, organization of the production system, and planning and control systems used for coordinating the day-to-day activities. Performance metrics that are most critical for today's production managers are: throughput, costs, inventories, quality and delivery performance. The complexity of the problems associated with effectively and efficiently utilizing all the resources – manpower, machines, materials – needed for producing goods often necessitates the development of mathematical models to aid decision making.

Manufacturing decisions can be classified into three categories: *strategic*, *tactical* and *operational*. Strategic decisions pertain to actions such as degree of vertical integration, items to be produced in-house, plant location, choice of technology, organization design, etc., that have long term consequences and can not be easily reversed. Tactical decisions have shorter horizons and include decisions such as aggregate production planning, facility layout, manpower planning, and incremental capacity expansion. Operational decisions pertaining to issues such as detailed production scheduling, maintenance routines, and inventory control rules, drive the day to day activities.

The nature of the problems faced by a production manager depends on the characteristics of the market that the facility is competing in. For this reason it is useful to distinguish between different types of manufacturing systems. The variety and volume of products produced are critical for determining the type of the manufacturing system. Manufacturing systems have been classified into job shops, batch shops, flow lines and continuous processes on the basis of the volume and variety of the product mix. Job shops produce many different products in small quantities, each with different processing requirements. Typically the products are customized and are made only after receiving an order. At the other end of the spectrum we have flow lines and continuous processes that produce a limited number of products in very high volumes. Demand is met from finished goods inventories. Batch shops lie in between these two extremes. The remainder of the paper describes some of the management science models that have been developed for each of these systems.

JOB SHOPS: Job shops specialize in producing customized products, and the production process has the flexibility to produce many different products. Due to the high variety the flows in job shops are jumbled, thus making it very difficult to predict and manage the completion times of jobs. Since most of the jobs are produced after receiving an order from a customer, very important managerial tasks are to accurately predict due dates, ensure that the quoted dates are not violated, and use resources effectively and efficiently.

Operational Problems – The challenge of managing day to day operations has given rise to a rich set of combinatorial optimization problems. The most basic operational problem is to determine a schedule that specifies when each job will be allocated different resources. Associated with each job are the arrival time, a due date and a set of operations. Each operation requires a set of resources for some duration, and there may be "precedence" constraints on the order in which the operations can be performed.

A variety of performance measures have been considered for evaluating alternative schedules. Common performance measures are the average or maximum time a set of jobs remains in the facility, number of jobs that are late, or the average or maximum tardiness for a set of jobs. Most of the problems of job shop schedule optimization problems, except for a small class, are computationally intractable (Lenstra *et al.*, 1977; French, 1982). Hence for most practical problems the emphasis has been on heuristics.

Researchers have successfully analyzed job shops with special structures. Many insights have been gained into the single machine and single stage, multiple machine scheduling problems. For multiple stage job shops, analysis has been possible, provided all the jobs follow the same route.

Job shop scheduling models can be classified into static and dynamic models. In static models the set of requirements including job arrival times and processing requirement are known in advance. In contrast, in dynamic job shop models new arrivals are permitted. The arrival times may be stochastic and the processing requirements may also vary dynamically.

Mathematical programming approaches have been employed to study static job shop problems. For performance measures that are non-decreasing in the completion time of the job, dynamic programming techniques have been employed to generate optimal solutions for problems of modest size. Dynamic programming based approaches have also been useful in identifying dominance criteria to reduce the number of schedules to be evaluated. Several heuristics have been developed that exploit dominance criteria. Integer programming formulations of scheduling problems have also been used to generate near optimal solutions. Typically some complicating constraints in the integer program are relaxed to yield tractable sub-problems.

While most of the theory focuses on static job shop models that assume deterministic requirements, most practical problems are dynamic and stochastic. For such complex environments analysis has largely been restricted to simulations of local dispatching rules. Each station employs a dispatching rule – for example, process jobs in increasing order of processing times – and the overall performance of the shop is evaluated via Monte Carlo simulations. Many dispatching rules have been discussed in the literature. Further details regarding scheduling algorithms are given in Conway, Maxwell and Miller (1967), Graves (1981), and O'hEigeartaigh, Lenstra and Rinnooy Kan (1985).

An important development in the area of scheduling dynamic shops has been to approximate the job shop scheduling problem by a Brownian control problem. Although the size of the networks analyzed is small, since the focus is on bottleneck stations the method is useful in many practical situations. The Brownian control problems have been useful in identifying near optimal scheduling policies for minimizing the average lead times (Wein, 1990).

Strategic and tactical problems – Since most of the operational problems of sequencing and scheduling jobs through a shop floor are computationally intractable, there is a need to design the job shops such that simple real time control rules are adequate to obtain good performance. The long term performance of the shop will depend on the types of jobs processed by the facility (product mix), the capacity and technology of different stations, and the rules employed to quote due dates and manage the flow through the shop floor. Tactical and strategic decisions regarding each of these variables require models that predict the medium to long term performance of the job shops.

One approach for assessing the long term performance is to employ Monte Carlo simulations. The strength of simulation models lies in their ability to incorporate many features such as (i) complex control rules – for example, local dispatching rules, control of input to the shop, etc.; (ii) complex arrival patterns – for example, correlated demands, non-stationary demand, etc.; and (iii) complex resource requirements and availability – for example, multiple resources, machine failures, etc. A broad range of performance measures can also be assessed through simulation models. These models, however, are time consuming and cannot identify optimal parameters for the policies being investigated.

Open queueing network models have been proposed to evaluate the long term performance of job shops. Good approximation procedures have been developed to estimate the average queue lengths in networks with features such as general processing and interarrival time distributions, multiple job classes, and class dependent deterministic routing through the network.

An approximation procedure that has been frequently employed is the parametric decomposition approach (PDA). Under the PDA, each node is treated as being stochastically independent and all the performance measures are estimated based on the first two moments of the interarrival and service time distributions at each node. Extensive testing has shown that PDA provides accurate estimates of the average queue length at each node in very general networks. Limitations of the approach are that all the measures are for steady state, only the average queue lengths are accurately predicted, and the analysis is based on the assumption that the jobs are processed on a first come first served basis. Nevertheless, the power of this approach lies in the ease with which complex networks can be analyzed, which in turn facilitates the design of networks.

The PDA has enabled the analysis of several optimal facility design problems. One such problem is:

Objective: Minimize total cost of equipment.
Decision Variables: Capacity of each station in the network, and technology.
Constraints: Upper bounds on the average lead time for different job classes.

This model addresses the relationship between average lead times and the choice of equipment. Since system design is based on multiple criteria, it is useful to develop curves that reflect the trade-off between lead times and cost of equipment. This can be done by parametrically varying the upper bound on the permissible lead times. Figure 1 illustrates the trade-off curves (Bitran and Tirupati, 1988). Details regarding the application of queueing models to job shops are given in Bitran and Dasu (1992).

BATCH SHOPS: The variety of jobs processed in a batch shop is less than that in job shops; furthermore, the set of products that are produced by the facility may be fixed. Nevertheless, the production volume of each product is such that several products may share the same equipment. Often the demand for final goods is met from finished goods inventory and production

plans are based on demand forecasts. A large number of discrete part manufacturing systems can be classified as batch shops.

Operational problems – The time and cost for switching machines from one product to the next poses one of the biggest problems in managing batch shops. Although job shops can also have significant set-ups, since each job is unique we can incorporate the set-up time in that job's total processing time. On the other hand, in batch shops, the same products are produced repeatedly and there is an opportunity to mitigate the effects of set-ups by combining or splitting orders. Consequently much attention has been paid to problems of determining batch quantity of and the sequence in which each item is produced. The primary trade-offs are between inventory carrying, shortage and set-up costs.

A classic lot sizing problem is the economic lot scheduling problem (ELSP). The ELSP seeks the optimal lot size at a single production stage when the demand rate for each item is fixed and deterministic (Panwalkar and Iskander, 1977). The objective of the analysis is to determine the frequency with which each item is to be produced so as to minimize the average set-up and holding costs without ever stocking out.

Many of the solution procedures for ELSP consist of three steps. First, ignoring the capacity constraint, the optimal production frequency for each item is determined. Next the frequencies are rounded off to an integer multiple of a base period. In the final step, a solution that specifies the sequence in which each item is produced is generated. Roundy (1986) showed that in the second step if the integer multiple is restricted to some power of 2, then a near optimal solution can be found. In recent years researchers have begun to extend the approaches developed for ELSP to multistage multimachine problems.

ELSP is a continuous time model. In practice production plans are made on a periodic basis, prompting several researchers to develop and analyze discrete time models of the lot-sizing problems. Below we formulate a single stage multi-item, multi-period, capacitated lot-sizing problem:

$$\text{Minimize} \sum_{t=1}^{T} \left\{ p_t(X_t) + h_t(I_t) + \sum_{i=1}^{I} s_{it}\delta(X_{it}) \right\}$$

$$s.t.\ I_{i,t-1} + X_{it} - I_{it} = d_{it} \quad t = 1, 2, \ldots, T; \ i = 1, 2, \ldots, I.$$

$$\sum_{i=1}^{I} X_{it} \leq C_t \quad t = 1, 2, \ldots, T.$$

$$\delta(X_{it}) = \begin{cases} 1 \text{ if } X_{it} = 0 \\ 0 \text{ otherwise} \end{cases}$$

$$I_{it},\ X_{it} \geq 0 \quad t = 1, 2, \ldots, T; \ i = 1, 2, \ldots, I,$$

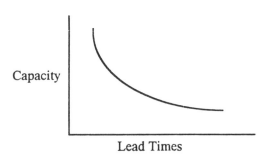

Fig. 1. Illustration of trade-off curves.

where X_{it}, I_{it}, C_t, d_{it} and s_t denote respectively for period t and product i, the production quantity, the ending inventory, the capacity, the demand, and the setup cost; X_{it} and I_{it} are the only decision variables; and X_t and I_t are vectors with elements $\{X_{it}\}$ and $\{I_{it}\}$, respectively. The functions $p_t(.)$ and $h_t(.)$ denote respectively the variable production and inventory holding costs.

Once again, except for a small class, the lot-sizing problems are NP-hard (Garey and Johnson, 1979; Bitran and Yanasse, 1982). The following two lot-sizing problems, however, can be solved in polynomial time and have been the basis of many approximation procedures: (a) the single item lot-sizing problem without capacity constraints, and concave variable production and inventory holding costs; and (b) single item problem, with constant capacity, and concave variable production and inventory holding costs.

Multistage systems producing multiple products with dynamic demands, usually require extensive information and considerable computational effort to find optimal solutions. For these reasons, hierarchical planning systems have been proposed. At the highest level in the hierarchy an aggregate plan with a horizon of several, usually twelve, months is developed. If the demand is seasonal, the horizon should cover the full demand cycle. Over such horizons it is impractical to obtain detailed information about demand for each item and the availability of every resource. Hence, it becomes necessary to aggregate the items into families, and the machines into machine centers, etc. The aggregate plan determines the time phased allocation of aggregate resources to different part families. The plan focuses on the primary trade-offs among the cost of varying production resources employed by the firm, the costs of carrying inventory (and possibly backordering demand), and major setup costs. The extended horizon enables the facility to respond to seasonality in demand.

The aggregate plan becomes the basis for determining the detailed production schedule for each item. The detailed resource allocation decisions are constrained by the decisions made at the aggregate planning level.

The number of hierarchical planning stages, the degree of aggregation at each level, and the planning horizon lengths affect the quality of the plan and must be carefully determined for each context. Many researchers have studied hierarchical planning systems. Bitran and Tirupati (1992), and Hax and Candea (1984) contain excellent discussions of this approach.

Once the plans have been disaggregated and the monthly requirements of each item are known, there are a number of approaches for scheduling and controlling the flow of the items through the shop. One approach is to time the release of the orders to the shop so that the required quantities of the items become available by the date specified by the hierarchical planning system. In this approach, also referred to as the push system, an estimate is made of production lead times and order releases are offset by the lead times. The scheduling decisions at each work station may be made on the basis of the queue in front of each work station. Scheduling models developed for job shops are also useful here.

An alternate approach for operating the shop is the pull system. Under this approach the work-in-process inventory level after a production stage determines the production decisions at that stage. The buffer inventories are maintained at planned levels and a production order is triggered if the inventory level drops below the threshold.

Since the push system operates on the basis of planned lead times, management science models have been developed to understand the relationship between release rules, capacity and lead times. The key decision variable in pull systems is the size of each buffer. Several researchers have examined the impact of buffer sizes on the shop performance (Conway *et al.*, 1988).

STRATEGIC AND TACTICAL PROBLEMS: An approach advocated for simplifying the operations of batch shops is to partition the facility into cells. Parts produced by the facility are grouped into families and each family is assigned to a cell. Ideally all operations required for a family of parts are performed in the same cell. The advantages of cellular manufacturing systems are simplified flows, and reduced lead times and setup costs. These benefits may be partially offset by the need for additional equipment. Many different criteria – such as part geometry, production volumes, setups, and route through the shop – have been proposed for forming part families. Researchers have also investigated several algorithms for identifying alternative partitions. Typically these algorithms begin with a product-process matrix. In this matrix rows correspond to parts and columns correspond to machines. An element ij in this matrix is one if a part i requires a machine j and zero otherwise. The columns and rows of the matrix are interchanged so as to produce a block diagonal matrix. Each block identifies a set of resources and jobs that does not interact with the remaining operations, and so corresponds to a cell.

As in the case of job shops, batch shop system design can be improved if the medium to long term performance of the shop can be assessed. Closed and open queueing network models and simulation based models are useful for assessing the long term performance of batch shops. The objective of these models is to determine the relationship among capacity of different cells, lot sizes, and lead times (Bitran and Dasu, 1992).

Queueing network models assume that the processing rate at each station is fixed. In practice the processing rate at each station may vary. Variations may be due either to the allocation of additional (human) resources to a stage or simply because the queue length has a motivational effect on the machine operator. Based on these observations, in recent years an alternative class of tactical models of the shop has been proposed (Graves, 1986). Here the production rates are assumed to vary as a function of the size of the queue length. The processing rates at each stage are allowed to vary so as to ensure that the time spent at a station is the same for every job. The model therefore enables managers to plan the lead times for each stage.

FLOW LINES AND CONTINUOUS OPERATIONS:

We include in this class all systems that are dedicated to the production of (one or few) items in large volumes. Examples of such systems include assembly lines, transfer lines, and continuous operations such as cement and oil derivatives manufacture. The demand is often met from finished goods inventory and thus the main focus tends to be on the management of the corresponding inventory levels and the supply chain. The operational problems are relatively simple and we omit their description.

Tactical problems – An important operational problem is to manage the trade-off between the cost of varying the production rate and the cost of finished goods inventory. The aggregate planning models discussed above are applicable here. Typically, all the stages of the production system have equal capacity, hence, managing the flow through the facility does not pose a significant problem. In assembly lines, the balance is achieved by carefully assigning tasks to different work stations – a complex combinatorial optimization problem. Several algorithms have been developed for assembly line balancing.

Strategic problems – High volume production systems frequently compete on the basis of low costs and supply large geographically dispersed markets. It is therefore not uncommon to have many plants that cater to different markets. OR/MS models have been developed to aid in the design of the multiplant networks and the distribution systems (Erlenkotter, 1978; Cohen *et al.*, 1987; Federgruen and Zipkin, 1984). Here we restrict the discussion to the plant location problems.

The number of plants, their capacity and location have a big effect on production and distribution costs. Models have been developed to analyze the trade-off between the fixed costs of setting up plants and the variable (transportation and production) costs of operating the plants. The models assume that a set of markets with known demands have to be supplied and the decision variables are the number of plants, their location and capacities (Erlenkotter, 1978).

CONCLUSIONS: Production management involves many complex trade-offs. As a result many mathematical models have been developed to aid decision makers. We have described some of these models. This is certainly not an exhaustive list and excludes many important problem areas such as inventory management, preventive maintenance, capacity expansion, and quality control. Our focus has been on models that are concerned with the flow of goods through a manufacturing system. Even within this domain, in order to provide a broad overview, we did not discuss many important models that deal with specialized systems such as flexible manufacturing systems.

We described the problems arising in each type of production system as if each plant operated in isolation. In practice, a production system is likely to consist of a network of plants. While some plants may be batch or job shops others are likely to be assembly or continuous processes. We have not discussed the problems of coordinating these networks.

Most of the management science models focus on managing the trade-offs among setup costs, inventory carrying costs and cost of varying production rates. On the other hand, many gains in productivity are due to the elimination (or mitigation of) the factors that give rise to these trade-offs. For example reduction in set-up costs and times reduces lead times, increases the ability of the system to produce a wider mix of products, diminishes the role of inventories and simplifies the management of batch shops. Researchers have begun to develop models that quantify the benefits of and guide such process improvement efforts (Porteus, 1985; Silver, 1993).

See **Combinatorial and integer programming; Dynamic programming; Facility location; Flexible manufacturing systems; Hierarchical production planning; Inventory modeling; Job shop scheduling; Location analysis; Networks of queues; Operations management; Queueing theory.**

References

[1] Bitran, G.R. and S. Dasu (1992), "A Review of Open Queueing Network Models of Manufacturing Systems," *Queueing Systems*, 12, 95–134.

[2] Bitran, G.R. and D. Tirupati (1989), "Trade-off Curves, Targeting and Balancing in Manufacturing Networks," *Oper. Res.*, 37, 547–564.

[3] Bitran, G.R. and D. Tirupati (1993), "Hierarchical Production Planning," in *Logistics of Production and Inventory*, Handbooks in O.R. and M.S, Vol 4, Edited by S.C. Graves, A.H.G. Rinnooy Kan and P. Zipkin, Elsevier Science Publishers, Amsterdam.

[4] Bitran G.R. and H.H. Yanasse (1982), "Computational Complexity of Capacitated Lot Sizing Problem," *Mgmt. Sci.*, 28, 1174–1186.

[5] Burbridge, J.L. (1979), *Group Technology in the Engineering Industry*, Mechanical Engineering Publications, London.

[6] Conway, R.W., W.L. Maxwell, and L.W. Miller (1967), *Theory of Scheduling*, Addison-Wesley, Reading, Massachusetts.

[7] Conway, R.W., W. Maxwell, J.O. McClain, and L. J. Thomas (1988), "The Role of Work-in-process Inventory in Serial Production Lines," *Oper. Res.*, 36, 229–241.

[8] Erlenkotter, D. (1978), "A Dual-based Procedure for Uncapacitated Facility Location," *Oper. Res.*, 26, 992–1005.

[9] Federgruen, A. and P. Zipkin (1984), "Approximation of Dynamic Multi-location Production and Inventory Problems," *Mgmt. Sci.*, 30, 69–84.

[10] French, S. (1985), *Sequencing and Scheduling, An Introduction to the Mathematics of the Job-Shop*, John Wiley and Sons, New York.

[11] Garey, M.R. and D.S. Johnson (1979), *Computers and Intractability: A Guide to the Theory of N.P. Completeness*, Freeman, San Francisco.

[12] Graves, S.C. (1981), "A Review of Production Scheduling," *Oper. Res.*, 29, 646–675.

[13] Graves, S.C. (1986), "A Tactical Planning Model for a Job Shop," *Oper. Res.*, 34, 522–533.

[14] Hax, A.C. and D. Candea (1984), *Production and Inventory Management*, Prentice-Hall, New Jersey.

[15] Lenstra J.K., A.H.G. Rinnooy Kan and P. Brucker (1977), "Complexity of Machine Scheduling Problems," *Ann. Discr. Math.*, 1, 343–362.

[16] O'hEigeartaigh, M., J.K. Lenstra, and A.H.G. Rinnooy Kan (1985), *Combinatorial Optimization – Annotated Bibliographies*, John Wiley, New York.

[17] Panwalker, S.S. and W. Iskander (1977), "A Survey of Scheduling Rules," *Oper. Res.*, 25, 45–61.

[18] Porteus, E.L. (1985), "Investing in Reduced Setups in the EOQ Model," *Mgmt. Sci.*, 31, 998–1010.

[19] Roundy, R. (1986), "A 98% Effective Lot-sizing Rule for a Multi-product, Multi-stage Production/Inventory System," *Math. Oper. Res.*, 11, 699–727.

[20] Silver, E.A. (1993), "Modeling in Support of Continuous Improvements Towards Achieving World Class Operations," in *Perspectives in Operations Management: Essays in Honor of Elwood S. Buffa*, R. Sarin, ed., Kluwer, Boston.

[21] Wein, L.M. (1990), "Optimal Control of a Two-station Brownian Network," *Math. Oper. Res.*, 15, 215–242.

PRODUCTION RULE

A mapping from a state space to an action space, generally used in modular knowledge representation. With roots in syntax-directed parsing of language, production rules comprise a basic reasoning mechanism, particularly in heuristic search. See **Artificial intelligence**; **Expert systems**.

PRODUCT-MIX PROBLEM

See **Activity-analysis problem**; **Blending problem**.

PROGRAM EVALUATION

Edward H. Kaplan and Todd Strauss

Yale School of Management, New Haven, Connecticut

Program evaluation is not about mathematical programming, but about assessing the performance of social programs and policies. Does capital punishment deter homicide? Which job training programs are worthy of government support? How can emergency medical services be delivered more effectively? What are the social benefits of energy conservation programs? These are the types of questions considered in program evaluation.

Notable evaluations include the Westinghouse evaluation of the Head Start early childhood program (Cicarelli, 1969), the Housing Allowance experiment (Struyk and Bendick, 1981), the Kansas City preventive patrol experiment (Kelling *et al.*, 1974), and evaluation of the New Haven needle exchange program for preventing HIV transmission among injecting drug users (Kaplan and O'Keefe, 1993). As these examples suggest, questions and issues deserving serious evaluation often are in the forefront of social policy debates in areas such as public housing, health services, education, welfare, and criminal justice.

Closely related to program evaluation are the activities of cost-benefit and cost-effectiveness analysis. These resource allocation methods help decision makers decide which social programs are worth sponsoring, and how much money should be invested in competing interventions. Program evaluation may be construed as an attempt to understand and estimate the benefits associated with the social program under study. While some evaluations attempt to relate these benefits to the costs of program activities, most program evaluations are viewed as attempts to measure benefits alone.

Program evaluation is often conducted by social scientists at the behest of organizations with some interest in the program, either as participants, administrators, legislators, managers, program funders, or program advocates. In such a charged atmosphere, how can OR/MS be useful? Program evaluation contributes to policy making chiefly by informing policy debate. Evaluation can be construed as an activity that produces important information for decision makers in the policy process (Larson and Kaplan, 1981). Evaluation is also useful for framing issues, and for identifying and choosing among policy options. Evaluation is crucial to program administrators concerned with improving service delivery. These tasks are about gathering, analyzing, and using information. It is the orientation toward decision making that renders OR/MS particularly useful in the evaluation of public programs.

PROGRAM COMPONENTS AND THE SCOPE OF EVALUATION: In the language of systems analysis, the components of social programs can be classified as inputs, processes, and outputs (Rossi and Freeman, 1993). Inputs are resources devoted to the program, while outputs are products of the program. In this framework, program evaluation is usually about assessing a program's effects on outputs. Such evaluation is often called outcome or impact evaluation. Typically, the result of outcome evaluation is the answer to the question: did the program achieve its goals?

In contrast to outcome evaluation, process evaluation is often referred to, perhaps pejoratively, as program monitoring. As the myriad details of real programs are classified simply as processes in monitoring studies, programs become black boxes. Such a framework is anti-operational. On the other hand, an OR/MS approach to process evaluation focuses on program operations, often with the assistance of appropriate mathematical models. Typical program evaluations too often lead to simplistic conclusions regarding which programs work. Focusing on program operations often results in understanding why some programs are successful and other programs fail. As an example, consider Larson's analysis of the Kansas City Preventive Patrol Experiment (Larson, 1975). This experiment attempted to discern the impact of routine preventive patrol on important outcomes such as crime rates and citizen satisfaction, in addition to important intermediate outcomes such as response time and patrol visibility. The empirical results of this experiment resulted in several findings of "no difference" between patrol areas with supposedly low, regular, and high intensities of police preventive patrol. In contrast, Larson's application of back-of-the-envelope probabilistic models to this experiment showed that one should have expected such results due to the nature of the experimental design. He showed, for example, that one should not have expected large differences in police response times given the peculiarities of patrol assignments and call-for-service workloads evident in the experiment. The same models suggested that different experimental conditions, better reflecting police operations in other large American cities, could lead to different results.

An advantage of an OR/MS approach to program evaluation is that goals and objectives are stated as explicitly as possible. What is the purpose of the program under study, and how does one characterize good versus poor program performance? While the importance of such questions may be self-evident to OR/MS practitioners, most actors on the policy stage are not accustomed to such explicitness. The act of asking such questions is often, by itself, a contribution to policy debate.

A defining feature of the OR/MS approach to problem solving is the association of one or more performance measures with program objectives. A performance measure quantifies how well a system functions. Performance measures should be measurable (computable if not actually observable), understandable, valid and reliable, and responsive to changes in program operations. Operational modeling of public programs can even yield performance measures not apparent *a priori*. For example, the evaluation of the New Haven needle exchange program involved a mathematical model of HIV transmission among drug injectors as modified by the operations of needle exchange (Kaplan and O'Keefe, 1993). The model revealed needle circulation time, that is, the amount of time a needle is available for use by drug injectors, as a critical performance measure. Reducing needle circulation time reduces opportunities for needle sharing on a per needle basis. This reduces both the chance that a needle becomes infected, and the chance that an injection with a used needle transmits infection. Needle exchange adjusts the distribution of needle circulation times. The model uncovered a direct link between the exchange of needles and the probability of HIV transmission.

METHODOLOGIES: Much of program evaluation is qualitative in nature. Social science methods relying on field observation, case histories, and the like are often used. However, such qualitative data often fail to satisfy critics of particular social programs. In addition, qualitative data generally allow only coarse judgments about program effectiveness. While no panacea, quantitative assessment methods have become standard in evaluating social programs and policies. Assessments of program effects are often made by statistically comparing a group participating in the program to a control group. The randomized experiment is the archetype for this kind of comparison. Since true randomized experiments may be difficult to execute under real program settings, quasi-experimental designs are often used instead. Rather than randomly assigning participants to program and control groups, quasi-experimental methods attempt to find natural or statistical controls. Multiple regression, analysis of variance, or other statistical techniques are often used; Cook and Campbell (1979) is a classic reference on quasi-experimental methods.

The model-based techniques of OR/MS are also applicable to program evaluation. Decision analysis is obviously useful in prospectively selecting among policy options. Queueing theory may be used to analyze the delivery of a wide range of programs, including public housing assignments, 911 hotlines, and dial-a-ride van services for the elderly and disabled. Applied probability

models are generally useful, while statistical methods are widely valued. Techniques for multi-criteria optimization, data envelopment analysis, and the analytical hierarchy process may be useful in identifying tradeoffs among multiple objectives.

While it seems that a solid understanding of OR/MS modeling is useful in conducting program evaluation, OR/MS has been underutilized. For example, basic optimization techniques such as linear programming have not been widely applied, perhaps because formulating a consensus objective function is usually very difficult. Training in OR/MS is less common than training in statistics and other social sciences. Few of those who have been trained in OR/MS have chosen to concentrate their efforts in the evaluation of public programs. Nonetheless, we believe that social program evaluation remains an important and fertile area for further development and application of OR/MS methods.

PROFESSIONAL OPPORTUNITIES AND ORGANIZATIONS: Departments and agencies of federal, state, and municipal government and international organizations typically have offices that perform evaluation activities. Examples include the U.S. Environmental Protection Agency's Office of Policy Planning and Evaluation, the New York City Public School's Office of Research, Evaluation, and Assessment, and the World Bank's Operations Evaluations Unit. A few large private or non-profit organizations undertake many program evaluations. Among such organizations are The Urban Institute, Abt Associates, RAND Corporation, Mathematica Policy Research, and Westat. Much program evaluation is done by academics, largely social scientists. There are opportunities for OR/MS practitioners to get involved. One outlet is the INFORMS College on Public Programs and Processes. The American Evaluation Association is an interdisciplinary group of several thousand practitioners and academics. The journal *Evaluation Review* publishes examples of quality evaluations.

See **Cost analysis; Cost effectiveness analysis; Emergency services; Problem structuring methods; Practice of OR/MS; Public policy analysis; Urban systems; Systems analysis.**

References

[1] Cicarelli, V.G. *et al.* (1969). *The Impact of Head Start*. Westinghouse Learning Corporation and Ohio University, Athens, Ohio.

[2] Cook, T.D. and D.T. Campbell (1979). *Quasi-Experimentation: Design and Analysis Issues for Field Settings*. Houghton Mifflin, Boston.

[3] Kaplan, E.H. and E. O'Keefe (1993). "Let the Needles Do the Talking! Evaluating the New Haven Needle Exchange," *Interfaces* 23, 7–26.

[4] Kelling, G.L. *et al.* (1974). *The Kansas City Preventive Patrol Experiment: Summary Report*. The Police Foundation, Washington, D.C.

[5] Larson, R.C. (1975). "What happened to patrol operations in Kansas City? A review of the Kansas City Preventive Patrol Experiment," *Jl. Criminal Justice* 3, 267–297.

[6] Larson, R.C. and E.H. Kaplan (1981). "Decision-Oriented Approaches to Program Evaluation," *New Directions for Program Evaluation* 10, 49–68.

[7] Rossi, P.H. and H.E. Freeman (1993). *Evaluation: A Systematic Approach*, 5th ed. Sage Publications, Newbury Park, California.

[9] Struyk, R.J. and M. Bendick, Jr. (1981). *Housing Vouchers for the Poor: Lessons from a National Experiment*. Urban Institute, Washington, D.C.

PROGRAM EVALUATION AND REVIEW TECHNIQUE (PERT)

A method for planning and scheduling a project which models uncertainties in activity by using optimistic, likely and pessimistic time estimates for each activity. See **Critical path method; Network planning; Project management**.

PROJECT MANAGEMENT

Graham K. Rand

Lancaster University, United Kingdom

Project management means different things to different people. In OR/MS, it has traditionally been almost exclusively concerned with technical aspects using PERT/CPM where the objective is to complete the project on time with an efficient use of resources. Outside of OR/MS, there has been much more of a concern with human aspects: the management of a group of people who are engaged in a project. In recent years, it would seem that this has increasingly become a concern of operations researchers, and particularly managers of OR/MS groups.

Organizational structure plays an important role in project management. The increasing need for fast response has motivated both the trend to flatter organizational structures and much of the recent thinking about project management (Handy, 1993). OR/MS expertise is becoming more dispersed within organizations: for instance, project teams may include an OR/MS person, but less frequently do exclusively OR/MS project teams exist.

The management of a project team involves skills such as leadership, coaching, supporting, decision-making (Boddy and Buchanan, 1992). The individual members of the project team require skills such as team-working (Belbin, 1981); time management (Adair, 1988; but see Eilon, 1993, for a more conservative view of the efficacy of the approaches recommended); report writing and presenting (Woolcott and Unwin, 1983); networking and negotiating (Kennedy, 1982); and interviewing.

Certification in project management is available through The Association of Project Managers (UK), 85 Oxford Road, High Wycombe, Bucks HP11 2DX, UK. Graduate courses are offered by The Project Management Institute, PO Box 189, Webster, NC 28788, USA, who also publish *Project Management Journal*. The International Project Management Association also issues a journal, *The International Journal of Project Management*.

See **CPM; Network planning; Organization; PERT; Practice of operations research and management science.**

References

[1] Adair, J. (1988). *Effective Time Management*. Pan Books, New York.

[2] Belbin, R.M. (1981). *Management Teams: Why They Succeed or Fail*. Heinemann, London.

[3] Boddy, D. and D. Buchanan (1992). *Take the Lead: Interpersonal Skills for Project Managers*. Prentice Hall, Hemel Hempstead, UK.

[4] Eilon, S. (1993). "Time management." *Omega* 21, 255–259.

[5] Handy, C. (1993). *Understanding Organizations*. Penguin, London.

[6] Kennedy, G. (1982). *Everything Is Negotiable*. Business Books, London.

[7] Woolcott, L.A. and W.R. Unwin (1983). *Mastering Business Communication*. Macmillan, Basingstoke, UK.

PROJECTION MATRIX

For a given matrix A, its associated projection matrix is defined as $P = A(A^{\mathrm{T}}A)^{-1}A^{\mathrm{T}}$. The matrix P projects any vector b onto the column space of A. See **Matrices and matrix algebra**.

PROJECT SCOOP

SCOOP stands for the Scientific Computation of Optimal Programs. Project SCOOP was a research program of the U.S. Air Force whose main objective was to study and solve Air Force programming and scheduling problems. It was while working on Project SCOOP problems that George B. Dantzig formulated the linear-programming model and developed the simplex method for solving such problems. Project SCOOP functioned from the late 1940s to early 1950s.

PROPER COLORING

An assignment of colors to nodes in a graph in which adjacent nodes are colored differently. See **Graph theory**.

PROSPECT THEORY

A descriptive theory of choice that explains certain deviations from expected utility theory. See **Choice theory**.

PROTOCOLS

The elicitation of an expert's procedure by asking the expert to describe aloud how he or she is solving a problem, such as making a forecast or a decision. See **Artificial intelligence**; **Expert systems**.

PSEUDOCONCAVE FUNCTION

Given a differentiable function $f(\cdot)$ on an open convex set X, we say that the function is pseudoconcave if $f(y) > f(x)$ implies that $(y - x)^T \nabla f(x) > 0$ for all $x, y \in X$ where $x \neq y$. See **Concave function**; **Quasi-concave function**.

PSEUDOCONVEX FUNCTION

Given a differentiable function $f(\cdot)$ on an open convex set X, we say that the function is pseudoconvex if $-f(\cdot)$ is pseudoconcave. See **Convex function**; **Pseudoconcave function**; **Quasi-convex function**.

PSEUDOINVERSE

See **Matrices and matrix algebra**.

PSEUDO-POLYNOMIAL-TIME ALGORITHM

An algorithm whose running time is technically not polynomial because it depends on the magnitudes of the numbers involved, rather than their logarithms. See **Computational complexity**.

PSEUDORANDOM NUMBERS

See **Random number generators**.

PUBLIC POLICY ANALYSIS

Warren E. Walker and Gene H. Fisher

The RAND Corporation, Santa Monica, California

INTRODUCTION: Public policy analysis is a systematic approach to making policy choices in the public sector. It is a process that generates information on the consequences that might be expected to follow the adoption of various policies. Its purpose is to *assist* policymakers in choosing a preferred course of action from among *complex* alternatives under *uncertain* conditions.

The word "assist" emphasizes that policy analysis is used by policymakers as a decision aid, just as check lists, advisors, and horoscopes can be used as decision aids. Policy analysis is not meant to replace the judgment of the policymakers (any more than an X-ray or a blood test is meant to replace the judgment of medical doctors). Rather, the goal is to provide a better basis for the exercise of that judgment by helping to clarify the problem, presenting the alternatives, and comparing their consequences in terms of the relevant costs and benefits.

The word "complex" means that the system being studied contains so many variables, feedback loops, and interactions that it is difficult to project the consequences of a policy change. Also, the alternatives are often numerous, involving mixtures of different technologies and management policies, and producing multiple consequences that are difficult to anticipate, let alone predict.

The word "uncertain" emphasizes that the choices must be made on the basis of incomplete knowledge about alternatives that do not yet physically exist, and whose projected consequences will occur – if at all – only in an unknown future. Alternatives must be compared not only by their expected consequences, but also by the risks of being wrong.

Policy analysis is performed in government, at all levels; in independent policy research institutions, both for-profit and not-for-profit; and in various consulting firms. It is not a way of solving a specific problem, but is a general approach to problem solving. It is not a specific methodology, but it makes use of a variety of methodologies in the context of a generic framework.

THE POLICY ANALYSIS STEPS: The policy analysis process generally involves performing the same set of logical steps, not always in the same order (Walker, Chaiken and Ignall, 1979, p. 70; Miser and Quade, 1985, p. 123). The steps, summarized in Figure 1, are:

1. *Identify the problem.* This step sets the boundaries for what follows. It involves identifying the questions or issues involved, fixing the context within which the issues are to be analyzed and the policies will have to function, clarifying constraints on possible courses of action, identifying the people who will be affected by the policy decision, discovering the major operative factors, and deciding on the initial approach.

2. *Identify the objectives of the new policy.* Loosely speaking, a policy is a set of actions taken to solve a problem. The policymaker has certain objectives that, if met, would "solve" the problem. In this step, the policy objectives are determined. (Most public policy problems involve multiple objectives, some of which conflict with others.)

3. *Decide on criteria (measures of performance and cost) with which to evaluate alternative policies.* Determining the degree to which a policy meets an objective involves measurement. This step involves identifying consequences of a policy that can be measured and that are directly related to the objectives. It also involves identifying the costs (negative benefits) that would be produced by a policy, and how they are to be measured.

4. *Select the alternative policies to be evaluated.* This step specifies the policies whose consequences are to be estimated. It is important to include as many as stand any chance of being worthwhile. If a policy is not included in this step, it will never be examined, so there is no way of knowing how good it may be. The current policy should be included as the "base case" in order to determine how much of an improvement can be expected from the other alternatives.

5. *Analyze each alternative.* This means determining the consequences that are likely to follow if the alternative is actually implemented, where the consequences are measured in terms of the criteria chosen in Step 3. This step usually involves using a model or models of the system.

6. *Compare the alternatives in terms of projected costs and benefits.* This step involves ranking the alternatives in order of desirability and choosing the one preferred. If none of the alternatives examined so far is good enough to be implemented (or if new aspects of the problem have been found, or the analysis has led to new alternatives), return to Step 4.

7. *Implement the chosen alternative.* This step involves obtaining acceptance of the new procedures (both within and outside the government), training people to use them, and performing other tasks to put the policy into effect.

8. *Monitor and evaluate the results.* This step is necessary to make sure that the policy is actually accomplishing its intended objectives. If it is not, the policy may have to be modified or a new study performed.

The individual steps in the process are described in detail by Miser and Quade (1985, Chapter 4) and by Quade (1989, Chapter 4).

OR/MS AND PUBLIC POLICY: Policy analysis is closely related to operations research; in fact, in many respects it grew out of operations research as it was being applied at the RAND Corporation and other applied research organizations in the 1960s and 1970s. (Miser, 1980; Majone, 1985, describe this evolution.) In the beginning, operations research techniques had been applied primarily to problems in which there

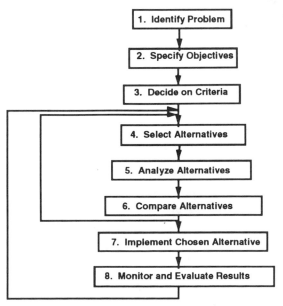

SOURCE: Warren Walker, et al. 1979.

Fig. 1. Steps in a policy analysis study.

were few parameters and a clearly defined single objective function to be optimized (e.g., aircraft design and placement of radar installations). Gradually, the problems being analyzed became broader and the contexts more complex. Health, housing, transportation, and criminal justice policies were being analyzed. Single objectives (e.g., cost minimization or single variable performance maximization) were replaced by the need to consider multiple (and conflicting) objectives (e.g., the impacts on health, the economy, and the environment and the distributional impacts on different social or economic groups). Non-quantifiable and subjective considerations had to be considered in the analysis. Schlesinger (1967) provided an early discussion of this issue. Optimization was replaced by satisficing.

Simon (1969, pp. 64–65) defined *satisficing* to mean finding an acceptable or satisfactory solution to a problem instead of an optimal solution. He said that satisficing was necessary because "in the real world we usually do not have a choice between satisfactory and optimal solutions, for we only rarely have a method of finding the optimum."

Operations research techniques are among the many tools in the policy analyst's tool kit. The analyses and comparisons of alternative policies are usually carried out with the help of mathematical and statistical models. Simulation, mathematical programming, and queueing theory are among the many tools that are used in a policy analysis study. But modeling is just one part of the process; all of the steps are important.

The policy analysis process has been applied to a wide variety of problems. Miser and Quade (1985, Chapter 3) provided examples of some of these, including:

- Improving blood availability and utilization;
- Improving fire protection;
- Protecting an estuary from flooding; and
- Providing energy for the future.

More generally, the policy analysis approach has been used in the formulation of policies at the national level, including national security policies, transportation policies, and water management policies. Other examples that illustrate the approach can be found in a variety of publications, including Drake, Keeney, and Morse (1972), House (1982), and Mood (1983).

See **Choice theory; Cost analysis; Cost effectiveness analysis; Decision analysis; Decision making; Multi-attribute utility theory; Practice of operations research and management science; RAND Corporation; Systems analysis.**

References

[1] Drake, A.W., R.L. Keeney, and P.M. Morse, eds. (1972), *Analysis of Public Systems*, MIT Press, Cambridge, Massachusetts.
[2] Findeisen, W. and E.S. Quade (1985), "The Methodology of Systems Analysis: An Introduction and Overview," Chapter 4 in H.J. Miser and E.S. Quade, eds., *Handbook of Systems Analysis: Overview of Uses, Procedures, Applications, and Practice*, Elsevier, New York.
[3] House, P.W. (1982), *The Art of Public Policy Analysis*, Sage Library of Social Research Vol. 135, Sage Publications, Beverly Hills, California.
[4] Majone, G. (1985), "Systems Analysis: A Genetic Approach," Chapter 2 in H.J. Miser and E.S. Quade, eds., *Handbook of Systems Analysis: Overview of Uses, Procedures, Applications, and Practice*, Elsevier, New York.
[5] Miser, H.J. (1980), "Operations Research and Systems Analysis," *Science*, 209, 4 July, 139–146.
[6] Miser, H.J. and E.S. Quade, eds. (1985), *Handbook of Systems Analysis: Overview of Uses, Procedures, Applications, and Practice*, Elsevier, New York.
[7] Mood, A.M. (1983), *Introduction to Policy Analysis*, North-Holland, New York.
[8] Quade, E.S. (1989), *Analysis for Public Decisions*, Elsevier, New York.
[9] Schlesinger, J.R. (1967), *On Relating Non-Technical Elements to System Studies*, P-3545, The RAND Corporation, Santa Monica, California.
[10] Simon, H.A. (1969), *The Sciences of the Artificial*, MIT Press, Cambridge, Massachusetts.
[11] Walker, W.E. (1988), "Generating and Screening Alternatives," Chapter 6 in Miser, H.J. and E.S. Quade, eds., *Handbook of Systems Analysis: Craft Issues and Procedural Choices*, Elsevier, New York.
[12] Walker, W.E., J.M. Chaiken, and E.J. Ignall, eds. (1979), *Fire Department Deployment Analysis: A Public Policy Analysis Case Study*, Elsevier North Holland, New York.

PULL SYSTEM

See **Production management**.

PURE-INTEGER PROGRAMMING PROBLEM

A mathematical programming problem in which all variables are restricted to be integer. Usually refers to a problem in which the constraints and the objective function are linear. See **Mixed-integer programming problem**.

PUSH SYSTEM

See **Production Management**.

QC

Quality control; quality circles. See **Statistical quality control**.

Q-GERT

Queue graphical evaluation and review technique. See **GERT; Network planning; Project management; Research and development**.

QP

See **Quadratic programming**.

QUADRATIC ASSIGNMENT PROBLEM

An assignment problem with a quadratic form as part of the objective function. The problem is usually one that is concerned with the assignment of facilities to sites and the minimization of the cost of the flow of materials between the facilities. See **Assignment problem; Facilities layout; Facility location**.

QUADRATIC FORM

A function that can be written as $x^T C x$, where the $n \times n$ matrix C is a matrix of known coefficients and x is a column vector. Matrix C is usually assumed to be symmetric or can be transformed into a symmetric matrix. The form is said to be positive definite if $x^T C x > 0$ for $x \neq 0$. The form is positive semidefinite if $x^T C x$ 0 for all x. Negative definite and negative semidefinite forms are defined by appropriate reversal of the inequality signs in the preceding definitions. See **Matrices and matrix algebra**.

QUADRATIC-INTEGER PROGRAMMING

A mathematical program involving the minimization of a quadratic function subject to linear constraints and integer variables. See **Quadratic-programming problem**.

QUADRATIC PROGRAMMING

Katta G. Murty

University of Michigan, Ann Arbor

INTRODUCTION: Quadratic programming (QP) deals with a special class of mathematical programs in which a quadratic function of the decision variables is required to be optimized (i.e., either minimized or maximized) subject to linear equality and/or inequality constraints.

Let $x = (x_1, \ldots, x_n)^T$ denote the column vector of decision variables. In mathematical programming it is standard practice to handle a problem requiring the maximization of a function $f(x)$ subject to some constraints, by minimizing $-f(x)$ subject to the same constraints. Both problems have the same set of optimum solutions. Because of this, we restrict our discussion to minimization problems.

A quadratic function of decision variables x is a function of the form

$$Q(x) = \sum_{i=1}^{n} \sum_{j=i}^{n} q_{ij} x_i x_j + \sum_{j=1}^{n} c_j x_j + c_0.$$

Define $c = (c_1, \ldots, c_n)$, and a square symmetric matrix $D = (d_{ij})$ of order n, where

$$d_{ii} = 2q_{ii} \quad \text{for all } i = 1, \ldots, n$$
$$d_{ij} = d_{ji} = q_{ij} \text{ for } j > i$$

Then in matrix notation, $Q(x) = (1/2)x^T D x + cx + c_0$. Here, D is the Hessian matrix (i.e., the matrix of second-order partial derivatives) of $Q(x)$.

As an example, consider $n = 3$, $x = (x_1, x_2, x_3)^T$, and $h(x) = 81x_1^2 - 7x_2^2 + 5x_1 x_2 - 6x_1 x_3 + 18x_2 x_3$. This quadratic function $h(x) = (1/2)x^T D x$ where

$$D = \begin{pmatrix} 162 & 5 & -6 \\ 5 & -14 & 18 \\ -6 & 18 & 0 \end{pmatrix}.$$

A quadratic function is the simplest nonlinear function, and hence, it has always served as a model function for approximating general nonlinear functions by local models. A square matrix D of order n is said to be

Positive semidefinite (PSD) if $x^T D x \geq 0$ for all $x \in R^n$;

Positive definite (PD) if $x^T D x > 0$ for all nonzero $x \in R^n$.

These matrix theoretic concepts are important in the study of QP because the quadratic function $Q(x) = (1/2)x^T Dx + cx + c_0$ is a convex function over R^n iff the matrix D is PSD. Given a square matrix of order n, there are efficient algorithms for checking whether it is PD or PSD, requiring at most n Gaussian pivot steps along its main diagonal (Murty, 1988, 1994).

CLASSIFICATION OF QUADRATIC PROGRAMS:

QPs can be classified into the following types.

Unconstrained quadratic minimization problems require the minimization of a quadratic function $Q(x)$ over the whole space.

Equality constrained quadratic minimization problems require the minimization of a quadratic function $Q(x)$ subject to linear equality constraints on the variables, $Ax = b$. These equations can be used to eliminate some variables by expressing them in terms of the others, and thereby transforming the problem into an unconstrained one in the remaining variables. Thus these problems are mathematically equivalent to (and can be solved by techniques similar to those of) unconstrained quadratic minimization problems.

Inequality constrained quadratic minimization problems require the minimization of a quadratic function $Q(x)$ subject to linear inequality constraints $Bx \geq d$, and possibly bounds on individual variables $\ell_i \leq x_i \leq u_i$, and maybe some equality constraints $Ax = b$.

Quadratic network optimization problems are quadratic programs in which the constraints are flow conservation constraints on a pure or generalized network.

Bound constrained quadratic minimization problems require the minimization of a quadratic function subject only to bounds (lower and/or upper) on the variables.

Convex quadratic programs (CQPs) are any of the above problems in which the objective function to be minimized, $Q(x)$, is convex.

Nonconvex quadratic programs are any of the above problems in which the objective function to be minimized, $Q(x)$, is nonconvex.

Linear complementarity problem (LCP) is a special problem dealing with a system of equations in nonnegative variables in which the variables are formed into various pairs called *complementary pairs*. A feasible solution in which at least one variable in each pair is zero is desired. There is no objective function to be minimized in this problem. The first-order necessary optimality conditions for a QP are in the form of an LCP. And in turn, every LCP can be posed as a QP.

UNCONSTRAINED QUADRATIC MINIMIZATION IN CLASSICAL MATHEMATICS:

Historically, quadratic functions became prominent because they provide simple local models for general nonlinear functions. A quadratic function is the simplest nonlinear function, and when used as a local approximation for a general nonlinear function, it can capture the important curvature information that a linear approximation cannot.

The use of quadratic approximations to handle general nonlinear functions goes back a very long time. We discuss some important instances of this.

1. *Newton's method* – Newton used it when he developed the celebrated Newton's method for finding an unconstrained minimum of a twice continuously differentiable function, $f(x)$ say. This method constructs the local model for $f(x^r + y)$ at the current point x^r to be the quadratic function $Q(y) = f(x^r) + \nabla f(x^r)y + (1/2)y^T H(f(x^r))y$, where $\nabla f(x^r)$ is the row vector of the first-order partial derivatives of $f(x)$ at x^r, and $H(f(x^r))$ is the Hessian matrix of $f(x)$ at x^r. The term $Q(y)$ is the second-order Taylor series approximation for $f(x)$ at x^r. The method takes the minimizer of the model function $Q(y)$, y^r (assuming that $H(f(x^r))$ is PD, we have $y^r = -(H(f(x^r)))^{-1}(\nabla f(x^r))^T)$, and the next point to be $x^{r+1} = x^r + y^r$.

Thus Newton's method solves an unconstrained quadratic minimization problem in each step. Starting from an initial point x^0, it generates the sequence $\{x^r\}$, which under certain conditions can be shown to converge to the minimum of the original function $f(x)$.

To treat the case where the Hessian $H(f(x^r))$ may not be PD, several modified Newton methods based on quadratic models different from the second-order Taylor series approximation at x^r have been developed. Also, the mathematical theory of quasi-Newton methods for unconstrained minimization has also been developed through the study of quadratic models (Dennis and Schnabel, 1983; Fletcher, 1987; Bazaraa, Sherali, and Shetty, 1993).

Conjugate gradient method – There is the very efficient Gaussian elimination method for solving a square nonsingular system of linear equations, $Ax = b$ say, of order n. However, when n is very large, this method becomes unwieldy and difficult to implement. The least squares formulation of this system of equations is the unconstrained quadratic minimization problem

$$\text{Minimize } (Ax - b)^T(Ax - b)$$

and Hestenes and Stiefel (1952) developed the conjugate gradient method for solving this problem. Subsequently, through the study of the quadratic model, several researchers have extended this method directly into a variety of conjugate gradient methods for the unconstrained minimization of general nonlinear functions.

Linear least squares – Suppose we have a large system of linear equations (typically overdetermined, that is, where the number of equations exceeds the number of variables), say $Ax = b$, which has no exact solution. A common approach for handling such a system is to look for a least squares solution, that is, an optimum solution of the unconstrained quadratic minimization problem

$$\text{Minimize } (Ax - b)^{\mathrm{T}}(Ax - b).$$

This is known as the *linear least squares problem*. Powerful numerical linear algebra techniques such as singular value decomposition (SVD) have been developed to solve large scale versions of this special class of QPs (Dennis and Schnabel, 1983). Statisticians have been using the linear least squares model for computing the estimates of the coefficients in a linear regression model for a long time.

TYPES OF SOLUTIONS: In linear programming (LP) we talk about optimum solutions, but not about different types of optima such as local and global optima. That is because every local optimum, and every point satisfying the first-order necessary optimality conditions for an LP, is also a global optimum. Unfortunately, this is not the case in general QPs.

For a QP, or any mathematical program in which an objective function $\theta(x)$ is required to be minimized, a

local minimum is a feasible solution $\bar{x}$ for which there exists an $\varepsilon > 0$ such that $\theta(x) \geq \theta(\bar{x})$ for all feasible solutions within a Euclidean distance of ε from $\bar{x}$;

global minimum is a feasible solution $\hat{x}$ satisfying $\theta(x) \geq \theta(\hat{x})$ for all feasible solutions x;

stationary or KKT point is a feasible solution satisfying the first-order necessary optimality conditions, also called the KKT (Karush, Kuhn, Tucker) optimality conditions, for the problem.

In a convex QP, every stationary point (KKT point), or local minimum, is a global minimum, and hence, all these concepts converge in a convex QP. The same may not be true in nonconvex QPs, that is, there may be local minima which are not global minima, and stationary points which are neither global nor local minima. Also, the problem may have some local minima, even

when the objective function is unbounded below on the set of feasible solutions. We will refer to the first order (KKT) necessary optimality conditions for a QP as its *KKT system*.

WHAT TYPES OF SOLUTIONS CAN BE COMPUTED EFFICIENTLY? Like LPs, QPs have the property that when the set of feasible solutions is nonempty, either a global minimum exists, or the objective function is unbounded below on the set of feasible solutions. And for both convex and nonconvex QPs, there exist finite algorithms for checking whether the objective function is unbounded below on the set of feasible solutions, and for computing a global optimum solution when one exists.

For convex QPs, there are very efficient algorithms for computing a global minimum when it exists, and very high quality software implementing these algorithms is available commercially. For nonconvex QPs, even though finite algorithms for computing a global minimum are available, they are impractical because the computational effort needed by them grows exponentially with the size of the problem being solved. Nonconvex QP is NP-hard, and so far there is no algorithm known that is guaranteed to find a global minimum for it within a reasonable time.

Can we at least compute a local minimum for a nonconvex QP efficiently? Unfortunately, even the problem of checking whether a given feasible solution is a local minimum for a nonconvex QP may be a hard problem. In Murty and Kabadi (1987), it has been shown that the problem of checking whether 0 is a local minimum in the simple QP

$$\text{Minimize } \quad x^{\mathrm{T}}Dx$$

$$\text{subject to } \quad x \geq 0$$

is a co-NP-complete problem when D is not PSD. In this paper, it has been explained that when dealing with a nonconvex QP, a reasonable goal is to look for an algorithm that produces a descent sequence (i.e., a sequence of feasible points along which the objective value strictly decreases) converging to a KKT point. Some of the algorithms discussed below have this property.

SOME IMPORTANT APPLICATIONS OF QP: *Finance* – Analysis using QP models is an established part of selecting optimum investment strategies. Perhaps Markowitz (1959) is the first published book in this area. The Markowitz model employs the variation in return as measured by the quadratic function $x^{\mathrm{T}}Dx$, where D

is the variance/covariance matrix, and x is the vector of stock investments, used for measuring the risk. This risk is the objective function to be minimized. Constraints in the model guarantee conservation on the flow of funds, and a lower bound on the expected returns from the portfolio. There may also be bounds placed on the investments in particular sectors of the economy (such as pharmaceuticals, utilities, etc.) to make sure that the model does not put too many eggs in any basket, thus achieving diversification. Many other practical aspects of investing can easily be included by either adding appropriate constraints or modifying the objective function by including quadratic penalty terms. Many authors have designed similar multiperiod quadratic generalized network flow models in which interest, dividends, and loans are modeled by means of arc multipliers (Crum and Nye, 1981; Mulvey, 1987).

Taxation – QP models play a very important role these days in the analysis of tax policies. Political leaders at the national and state levels are relying more and more on such analyses to forecast growth rates in tax revenues, and to set various taxes at levels that are likely to ensure growth at desired rates. White (1983) gives a detailed description of such an analysis carried out for the state of Georgia. National and state government taxes such as sales tax, motor fuels tax, alcoholic beverages tax, personal income tax, etc. are all set at levels to ensure a healthy economic growth. Government finance is based on the assumption of predictable and steady growth of each tax over time.

If s is the tax rate for a particular tax and S_t the expected tax revenue for this tax in year t, then a typical regression equation used to predict S_t as a function of s and t is: $\log_e S_t = a + bt + cs$ where a, b, c are parameters to be estimated from past data to give the closest fit by the least squares method, a QP technique. The annual growth rate in this tax revenue is then the regression coefficient b multiplied by 100 to convert it to percent.

The decision variables in the model are: s_j = the tax rate for tax j in the base year (0th year) as a fraction. From the known tax base for tax j in the 0th year, the revenues from tax j in this year can be obtained as: s_j (tax base for tax j) = x_j. The instability or variability in this revenue is measured by the quadratic function $Q(x) = x^T V x$ where V is the variance/covariance matrix estimated from past data and $Q(x)$ is to be minimized. The constraints in the model consist of bounds on the x_j, and a condition that $\Sigma x_j = T$, the total expected tax revenue in the 0th year.

And there is an equation that the overall growth rate which can be measured by the weighted average of the growth rates of the various taxes j, $\Sigma(x_j b_j)/T$ should be equal to the desired growth rate λ. Any other linear constraints that the decision variables are required to satisfy can also be included. In fact, λ can be treated as a parameter, and the whole model solved as a parametric QP model. Exploring the optimum solution for different values of λ in the reasonable range yields information for the political decision makers to determine good values for the various tax rates that are consistent with expected growth in tax revenues.

Equilibrium models – Economists use equilibrium models to analyze expected changes in economic conditions, predict prices, inflation rates, etc. These models often involve QPs. As an example, in Glassey (1978), a simple equilibrium model of interregional trade in a single commodity is described. He considers N regions, and the following data elements and variables:

Data:	$a_i > 0$	the equilibrium price in the ith region in the absence of imports and exports;
	$b_i > 0$	the elasticity of supply and demand in the ith region;
	c_{ij}	the cost/unit to ship from i to j.
Variables:	p_i	equilibrium price in the ith region;
	y_i	net imports into the ith region (may be >0, 0, or <0);
	x_{ij}	actual exports from region i to region j.

If $p_i > a_i$, supply locally exceeds demand in the ith region, the difference being available for export. From this we have $p_i = a_i - b_i y_i$. Also, the y_i and x_{ij} are linked through flow conservation equations. The interregional trade equilibrium conditions are

$$\begin{cases} p_i + c_{ij} \geq p_j & \text{for all } i, j \\ (p_i + c_{ij} - p_j)x_{ij} = 0 & \text{for all } i, j \end{cases}$$

If the first condition above does not hold, exports from i to j will increase until the elasticity effects in markets i and j rise, and prices will adjust so that additional profit from export no longer exists. Also, if $x_{ij} > 0$, we must have $p_i + c_{ij} - p_j = 0$. It can be verified that these conditions are the first-order necessary optimality conditions for a quadratic network flow problem in which the quadratic objective function can be interpreted as a net social payoff function. Using this observation, Glassey (1978) described a procedure for computing the equilibrium prices and flows based on solving the QP.

In the same way, traffic engineers use traffic equilibrium models solved by quadratic network flow algorithms for road and communication network planning. These traffic equilibrium models typically have hundreds of thousands of variables and constraints, and are probably the largest QP models solved on a regular basis.

Electrical networks – During the physicist J.C. Maxwell's time in the second half of the 19th century, it was recognized that the equilibrium conditions of an electrical or a hydraulic network are attained at the point where the total energy loss is minimized. Dennis (1959) has formally shown that the sum of the energy losses in the resistors and at the voltage sources in an electrical network is a quadratic function of the branch currents, if all devices in the network are of a linear (i.e., ohmic) nature. Using this, he formulated the problem of determining the branch currents at equilibrium in an electrical network connecting various devices, voltage sources, diodes, and resistors, as a QP. He then showed that the optimality conditions for this QP are precisely the Kirchhoff laws governing the equilibrium conditions of the network, with the Lagrange multipliers representing node potentials. In the distribution of electrical power, this QP model is used to solve the load flow problem concerned with the flow of power through the transmission network to meet a given demand.

Power system scheduling – The economic dispatch problem in an electrical power system operation deals with the problem of allocating the demand for power – or system load – among the generating units in operation at any point of time. The optimal allocation of load among the units to achieve a least cost allocation, depends on the relative efficiencies of the units; and can be modeled as a QP (Wood, 1984). In power system operation, this model is usually solved many times during the day with appropriate load adjustments.

Application to solving general nonlinear programs – One of the most popular algorithms for solving general nonlinear programming problems is the SQP (sequential or recursive quadratic programming) method. It is an iterative method which in each iteration solves a convex QP to find a search direction, and a line search problem (one dimensional minimization problem for a merit function) in that direction. The original concepts of this method are outlined in Wilson (1963), Han (1976), and Powell (1978); but it has been developed into a successful approach through the work of many researchers, including Eldersveld (1991); Bazaraa, Sherali, and Shetty (1993); and Murty (1988).

The success of these methods has made QP a very important topic in mathematical programming. A nice software package for nonlinear programs based on this approach is FSQP (Zhou and Tits, 1992).

RECENT ALGORITHMIC DEVELOPMENTS:
Frank-Wolfe method – One of the first methods for QP developed in recent times is that of Frank and Wolfe (1956). It is an iterative method which in each iteration solves an LP to find a search direction, and a line search problem in that direction. It produces a descent sequence such that every limit point of this sequence is a KKT point. However, the method has slow convergence, and is not popular except on problems with special structure that make it possible to solve the LP in each iteration by an extremely fast special method taking advantage of the structure.

Reduced gradient methods – The simplex method for LP has been extended to solve problems involving the minimization of a quadratic (or in general a smooth nonlinear) function subject to linear constraints. The method is called the *reduced gradient method* and was discussed in Wolfe (1959). The name, reduced gradient method, refers to any method which uses the equality constraints to eliminate some variables (called the dependent or basic variables) from the problem, and treats the remaining problem in the space of the independent (or nonbasic variables) only, either explicitly or implicitly. The reduced gradient is the gradient of the objective function in the space of independent variables. The method is quite popular. The OSL software package uses this method for solving QPs. The MINOS 5.4 software package uses this method for minimizing a smooth nonlinear function subject to equality constraints. This method has been generalized directly into the GRG (generalized reduced gradient) method for solving nonlinear programs involving nonlinear constraints (Abadie and Carpentier, 1969). The GRG is a popular method on which several successful nonlinear programming software packages are based.

Methods based on the LCP – In the 1950s and 60s, several researchers proposed schemes for solving the QP by solving its KKT system. Lemke (1965) formulated the KKT system for a QP as an LCP and developed an algorithm for it called the *complementary pivot algorithm*. The data for an LCP of order n consists of a square matrix M of order n, and a column vector $q \in R^n$, and we need to find a $w = \{w_j\} \in R^n$ and a $z = \{z_j\} \in R^n$ satisfying

$$w - Mz = q$$

$$w, z \geq 0$$

$$w_j z_j = 0 \quad \text{for all } j.$$

Checking whether the general LCP has a solution is an NP-complete problem, and there are no efficient algorithms known for it. But the complementary pivot algorithm is a finite path following method for finding a solution, when one exists, to a class of LCPs which includes the KKT systems corresponding to convex QP.

However, the complementary pivot method, and several other methods developed for the LCP are not popular for solving even convex QPs, because a QP involving m inequality constraints in n nonnegative variables leads to an LCP of order $m + n$, blowing up the size. For tackling nonconvex QPs, the complementary pivot approach is clearly unsuitable, as it focusses attention purely on the KKT system, and never even computes the objective value; and if it leads to a KKT point at termination, that point may not even be a local minimum.

But the theoretical contribution of the formulation of the LCP and the complementary pivot method for it is great. The LCP has a fascinating geometrical interpretation. The study of the geometry of LCP was initiated in Murty (1968). The mathematical principle behind the complementary pivot method has been used to develop simplicial methods (which are also called complementary pivot methods) to solve systems of nonlinear equations and fixed point problems (Murty, 1988; Cottle, Pang, and Stone, 1992).

Active set methods – A popular method for solving QP is based on a combinatorial approach to determine iteratively the set of active constraints at the optimum. This type of strategy for handling inequality constrained optimization problems is called the *active set strategy*. The method solves a sequence of equality constrained QPs by treating some of the inequality constraints as equations (the active set) and temporarily ignoring the others. Several rules are employed to modify the active set from one iteration to the next, to guarantee finite convergence of the procedure (Theil and van de Panne, 1961). Several researchers have extended this method to minimize a smooth nonlinear function subject to linear equality and inequality constraints.

Interior point methods – Since the development of a very successful interior point method for LP by Karmarkar in 1984, a variety of interior point methods have been developed for convex QPs and the LCPs associated with them. These methods are polynomially bounded, and some versions of them give excellent computational performance on large sparse problems. The monograph by Kojima, Megiddo, Noma, and Yoshise (1991) established the theoretical foundations for primal-dual interior point methods for LP and LCP. Some other references on these methods are Ye (1991), and Fang and Puthenpura (1993).

(f) *Methods for nonconvex QP* – Vavasis (1992) described efficient polynomially bounded algorithms that are guaranteed to find a local minimum for some special classes of nonconvex QP.

SOFTWARE: There are two commercially available software packages for solving QPs. One is MINOS 5.4, available from Stanford Business Software or from The Scientific Press as part of either of the algebraic modeling systems AMPL or GAMS. The other is OSL that is available from IBM or from The Scientific Press as part of AMPL.

AMPL is a modeling language for mathematical programming which provides a natural form of input for linear, integer, and nonlinear mathematical models besides QP (Fourer, Gay, and Kernighan, 1993). The book is accompanied by a PC student version of AMPL and representative solvers, enough to easily handle problems of a few hundred variables and constraints. Versions that support much larger problems are available from the publisher. AMPL uses either the MINOS 5.4 solver or the OSL solver for solving QP models.

GAMS (Brooke, Kendrick, and Meeraus, 1988) is a high-level language that is designed to make the construction and solution of large and complex mathematical programming models straightforward for programmers, and more comprehensible to users of models. It uses the MINOS solver for solving QPs and also has solvers for linear, integer, and nonlinear programming problems. A student version and a professional version are available.

IBM's OSL is a collection of high-performance mathematical subroutines for solving linear, integer and quadratic programming models. MINOS 5.4 (Murtagh, and Saunders, 1987) is a Fortran-based computer system designed to solve large-scale linear, quadratic, and nonlinear models.

See **Banking; Complementarity problem; Computational complexity; Electrical power;**

Interior point methods; Linear programming; Nonlinear programming; Optimization; Portfolio theory; Wolfe's quadratic programming-problem algorithm.

References

[1] J. Abadie and J. Carpentier (1969), "Generalization of the Wolfe Reduced Gradient Method to the Case of Nonlinear Constraints," in *Optimization*, R. Fletcher, ed., Academic Press.

[2] M.S. Bazaraa, H.D. Sherali and C.M. Shetty (1993), *Nonlinear Programming Theory and Algorithms*, Wiley-Interscience, New York.

[3] A. Brooke, D. Kendrick and A. Meeraus (1988), *GAMS: A User's Guide*, Scientific Press.

[4] R.W. Cottle, J.S. Pang and R.E. Stone (1992), *The Linear Complementarity Problem*, Academic Press.

[5] R.L. Crum and D.L. Nye (1981), "A Network Model of Insurance Company Cash Flow Management," *Mathematical Programming Study*, 15, 86–101.

[6] J.B. Dennis (1959), *Mathematical Programming and Electrical Networks*, Wiley, New York.

[7] J.E. Dennis, Jr. and R.B. Schnabel (1983), *Numerical Methods for Unconstrained Optimization and Nonlinear Equations*, Prentice Hall, New Jersey.

[8] S.K. Eldersveld (1991), "Large Scale Sequential Quadratic Programming," SOL91, Dept. OR, Stanford University.

[9] S.C. Fang and S. Puthenpura (1993), *Linear Optimization and Extensions Theory and Algorithms*, Prentice Hall, New Jersey.

[10] R. Fletcher (1987), *Practical Methods of Optimization* (2nd ed.), Wiley, New York.

[11] R. Fourer, D.M. Gay and B.W. Kernighan (1993), *AMPL A Modeling Language for Mathematical Programming*, The Scientific Press, San Francisco.

[12] M. Frank and P. Wolfe (1956), "An Algorithm for Quadratic Programming," *Naval Research Logistics Quarterly*, 3, 95–110.

[13] C.R. Glassey (1978), "A Quadratic Network Optimization Model for Equilibrium Single Commodity Trade Flows," *Mathematical Programming*, 14, 98–107.

[14] S.P. Han (1976), "Superlinearly Convergent Variable Metric Algorithms for General Nonlinear Programming Problems," *Mathematical Programming*, 11, 263–282.

[15] M.R. Hestenes and E. Stiefel (1952), "Method of Conjugate Gradients for Solving Linear Systems," *J. Res. N. B. S.*, 49, 409–436.

[16] IBM (1990), *OSL – Optimization Subroutine Library Guide and Reference*, IBM Corp., New York.

[17] M. Kojima, N. Megiddo, T. Noma and A. Yoshise (1991), *A Unified Approach to Interior Point Algorithms for Linear Complementarity Problems*, Lecture Notes in Computer Science 538, Springer-Verlag, New York.

[18] C.E. Lemke (1965), "Bimatrix Equilibrium Points and Mathematical Programming," *Management Science*, 11, 681–689.

[19] H.M. Markowitz (1959), *Portfolio Selection: Efficient Diversification of Investments*, Wiley, New York.

[20] J.M. Mulvey (1987), "Nonlinear Network Models in Finance," *Adv. Math. Program. Finan. Plann.* 1, 253–271.

[21] B.A. Murtagh and M.A. Saunders (1987), "MINOS 5.4 User's Guide," SOL 83-20R, Dept. OR, Stanford University.

[22] K.G. Murty (1968), "On the Number of Solutions to the Complementary Quadratic Programming Problem," Ph. D. dissertation, University of California, Berkeley.

[23] K.G. Murty (1988), *Linear Complementarity, Linear and Nonlinear Programming*, Heldermann Verlag, Berlin.

[24] K.G. Murty (1994), *Operations Research: Deterministic Optimization Models*, Prentice Hall, New Jersey.

[25] K.G. Murty and S.N. Kabadi (1987), "Some NP-complete Problems in Quadratic and Nonlinear Programming," *Mathematical Programming*, 39, 117–129.

[26] M.J.D. Powell (1978), "Algorithms for Nonlinear Constraints That Use Lagrangian Functions," *Mathematical Programming*, 14, 224–248.

[27] H. Theil and C. van de Panne (1961), "Quadratic Programming as an Extension of Conventional Quadratic Maximization," *Management Science*, 7, 1–20.

[28] S.A. Vavasis (1992), "Local Minima for Indefinite Quadratic Knapsack Problems," *Mathematical Programming*, 54, 127–153.

[29] F.C. White (1983), "Trade-off in Growth and Stability in State Taxes," *National Tax Journal*, 36, 103–114.

[30] R.B. Wilson (1963), "A Simplicial Algorithm for Convex Programming," Ph.D. dissertation, School of Business Administration, Harvard, Cambridge, Massachusetts.

[31] P. Wolfe (1959), "The Simplex Method for Quadratic Programming," *Econometrica*, 27, 382–398.

[32] A.J. Wood (1984), *Power Generation, Operation, and Control*, Wiley, New York.

[33] Y. Ye (1991), "Interior Point Algorithms for Quadratic Programming," 237–261 in *Recent Developments in Mathematical Programming*, S. Kumar, ed., Gordon and Breach, Philadelphia.

[34] J.L. Zhou and A.L. Tits (1992), "User's Guide to FSQP Version 3.1," SRC TR-92-107r2, Institute for Systems Research, University of Maryland, College Park.

QUALITY CONTROL

Frank Alt and Kamlesh Jain

University of Maryland, College Park

INTRODUCTION: While interest in quality is as old as industry itself, quality control as a technical and managerial discipline started to become accepted and widely practiced only in the 1940s and 1950s. Statistical methods of quality control, though developed in the United States and Britain, found their most ardent followers among Japanese businessmen and managers in the post-War decades. Statistical quality control (SQC) consultants such as Deming became household names in Japan while they were scarcely known in their own countries. During the last decade or so, however, there has been a renewed interest in quality control in the West, spurred no doubt by the globalization of competition and increasing customer awareness of quality. SQC procedures are often a key element in total quality management (TQM) and continuous improvement programs in industry. Such programs have made it possible for customers to demand and get higher quality products at lower costs.

While basic *statistical process control* (SPC) techniques have remained unchanged for over fifty years, exciting developments are taking place which utilize the concepts of cumulative sum, moving averages, multivariate statistics, and Bayesian decision theory. After a brief introduction, we discuss the basic concepts of control charts and classical SPC methods (Shewhart control charts). Cumulative sum and moving average quality control procedures are discussed next; multivariate and Bayesian procedures are briefly reviewed; and we end with a review of some other recent developments in the management of quality in organizations.

USES OF SQC PROCEDURES: Over the years, SQC techniques have found literally thousands of applications in manufacturing and, more recently, service organizations. The basic SQC techniques are relatively simple to use and are often applied by shop floor personnel with little training in statistical methods. SQC techniques can be broadly divided into two categories, "statistical process control" and "acceptance sampling."

Statistical process control (SPC) typically involves the use of "control charts" for monitoring a manufacturing process at regular intervals of time to detect any problems that may develop in the process and to take corrective action if necessary. Control charts provide a clear and visual representation of the status of the process. Often, problems are identified before they become serious – thus minimizing economic losses. When a problem is indicated, the pattern of measurements on the control chart can usually indicate the source of the problem so that corrective action may be taken.

Montgomery (1991) lists five main uses of control charts. Control charts are a proven technique for improving productivity, preventing defects and unnecessary process adjustments, and providing diagnostic and process capability information.

The quality of inputs into a process has a significant bearing on the quality of its output. Raw materials from a supplier or semifinished output from a work station may be the input for a particular process. In such a situation, one may receive a batch of products/materials from a source external to the process and one has to then decide whether to accept or reject the batch based on the quality of a representative sample taken from it. This type of quality control where decision to accept or reject a batch is based on inspection of a sample of incoming/outgoing goods is termed *acceptance sampling*. This is a well developed branch of SQC, though its use has declined over the years because it is based on the false premises that quality can be inspected into products and that a certain number of nonconforming items in the lot may be acceptable. Also, as pointed out by Montgomery (1991), sorting out good units from bad ones after the fact is never cheaper than making them right to start with. Quality control is most effective when it is preventive in nature rather than curative.

The techniques of statistical process control and acceptance sampling have been around since the 1920s. Walter A. Shewhart of the Bell Telephone Laboratories developed statistical quality control charts in 1924. In late 1920s, Harold F. Dodge and Harold G. Romig developed the concept of acceptance sampling, again at the Bell Telephone Laboratories. The development and use of SQC techniques grew rather slowly initially. The exigencies created by the defense requirements of World War II provided an impetus to the use of SQC techniques in industry. The late 1940s and 1950s were a period characterized by the consolidation of technical gains in SQC methodology achieved during the war. While the use of SQC spread in the industrialized and industrializing nations, Japan embraced these techniques with a missionary zeal and showed their potential to the world.

The 1950s saw the use of experimental designs for making sequential product and process improvements. Box (1957) suggested an innovative industrial application of (statistical) design of experiments and termed it Evolutionary Operation (EVOP). EVOP involves introducing small, deliberate variations in a process according to a systematic plan (i.e., according to a designed experiment). After a certain number of trials, sufficient information becomes available to guide future trials (in an evolutionary manner) in order to improve productivity, reduce costs, or both. The *experiment* is carried out under the direct supervision of the production supervisor and typically does not require any additional resources, the key idea being that process improvement can be effected right in the plant during regular production runs rather than in a research lab. The concept of designed experiments, however, gained popularity only in late 1970s and early 1980s after it was found that the Japanese had been using the technique and benefiting from it. Taguchi's (1976, 1977) volumes on experimental design and Taguchi and Wu's (1979) book on offline quality control have been instrumental in the revival and recent popularity of the concept of experimental design for product and process improvement. Taguchi's rationale was that quality has to be built into the product. According to him, any deviation from the target leads to quality losses that can and should be minimized. Books edited by Bendell, Disney, and Pridmore (1989) and by Dehnad (1989) contain many case studies.

The concept of total quality control (TQC) was proposed by Feigenbaum in the 1950s. His idea was to unify all phases of quality control, namely, (1) design and development, (2) manufacturing, (3) inspection of incoming/outgoing product, and (4) feedback of information to design/development, manufacturing, and sales to facilitate corrective action, if and when necessary (Feigenbaum, 1951). Focusing on quality assurance, TQC was intended to shift management's attention from a simple assessment of quality to the prevention of unsatisfactory conditions in the manufacturing organization.

The latest developments furthering the ideas of total quality are embodied in TQM and continuous improvement programs. TQM programs are organization-wide efforts designed to maximize customer satisfaction. According to Roberts and Sergesketter (1993), "TQM is a people-focussed management system that aims at continual increase of customer satisfaction at continually lower real cost." TQM programs have both a technical component and a managerial component, the latter including such aspects as teamwork, leadership practices, quality assurance systems, measures of customer satisfaction, and so on. The technical components of TQM programs in manufacturing organizations are typically based on SPC techniques. Over the last few years, such programs have become very popular in the United States and Western Europe and have become institutionalized in organizations through the Malcolm Baldridge National Quality Award, the Deming Award, and ISO 9000.

TERMINOLOGY: The term *quality* means different things to different people. The traditional definition of quality stipulates conformance to specifications. In a manufacturing environment, however, the term quality has two distinct meanings: Quality of design and quality of conformance.

Quality of design refers to the fact that different products for the same end use may have different design specifications. For instance, both a Rolls Royce and a Toyota Tercel provide transportation but have very different design specifications. The two cars are aimed at different market segments, a decision made by the managements of the two companies before the cars were even at the drawing board stage. Once an organization has decided on the product it wants to manufacture for targeting a specific market segment, it articulates design specifications and tolerances for the quality characteristics of the raw materials, semifinished components, and the finished product. Furthermore, in terms of process design, management must make sure that the process adopted is capable of producing a product that conforms to the design specifications. While there are statistical methods which can be used for improving the quality of design, we will not be concerned with the quality of design in this chapter.

Quality of conformance refers to the ability of a (manufacturing) process to hold to the specified quality of design, that is, to produce the output that meets design specifications. It's here that SQC procedures can be most effectively applied. In particular, SPC techniques offer systematic approaches for monitoring a process to ensure that the quality characteristics of its output conform to the specifications laid down by the product design people. These techniques are the focus of this article.

Statistical quality control professionals and researchers have suggested other definitions of quality. According to the Japanese engineer Taguchi (1979), quality is the loss impaired to society from the time a product is shipped. Loss to the society is measured using a loss function which assigns measurable penalties that are proportional to the distance a measurable characteristic is away from its target value.

Montgomery (1991) defines quality as "fitness for use," where the fitness for use is jointly determined by quality characteristics. For example, for a drilling process, the characteristics of interest may be the width and the depth of bores. One could monitor the drilling process using univariate process control charts – one for each of the quality characteristics being monitored – or one multivariate quality control chart for the several correlated quality characteristics of interest. Even for a single quality characteristic, one may employ two control charts, one for monitoring the process mean and the other for monitoring the process variability.

Specification Limits – For any measurable quality characteristic, design engineers typically specify a target value (the product specification) as well as the minimum and/or maximum allowable values (tolerances). The maximum allowable value for a quality characteristic is referred to as the upper specification limit and the minimum allowable value is the lower specification limit.

The terms *specification limits* and *tolerance limits* are often used interchangeably. The definition of the two terms in the ASQC Glossary (1983) is the same. The terms are defined as "the conformance boundaries for an individual unit of a manufacturing or service operation." However, the Glossary suggests that *tolerance limits* are generally preferred in evaluating the manufacturing or service environment, whereas *specification limits* are more appropriate for categorizing materials, products, or services in terms of their stated requirements. According to Banks (1989), *tolerances* refer to physical measurements only, whereas *specifications* refer to all characteristics, and thus specification limits include tolerance limits.

SPECIFICATION LIMITS AND PROCESS CAPABILITY:
Once the specification limits for a quality characteristic have been established, one needs to make sure that the process is capable of meeting these requirements. A commonly used measure of process capability relative to specification limits is called the Process Capability Index, C_p. It is defined as

$$C_p = (U - L)/6s,$$

where U is the upper specification limit, L the lower specification limit, and s is the standard deviation of the distribution of the quality characteristic. A value of C_p less than 1 indicates that the process is not capable of delivering the quality required by the specification limits. A process with a C_p of at least 1.33 is considered to be quite good. For such a process the standard deviation will have to be $(1/8)$th of $(U - L)$. For example, Motorola wants a C_p of 2 for many of its processes. To achieve a C_p of 2, the process standard deviation needs to be $(1/12)$th of $(U - L)$. Whereas C_p gives an indication of a process being capable of meeting specification limits, it does not indicate if the process is also centered at the target mean.

A commonly used measure which gives an indication of how far the process average varies from the target value, T, is K:

$$K = (T - \bar{X})/[(U - L)/2],$$

where $\bar{X}$ is an estimate of the current process average. When the process mean equals the target mean, $K = 0$; $K = -1$ when the process mean shifts to the upper specification limit; and $K = 1$ when the process mean equals the lower specification limit. A measure of process capability based on the measure K is

$$C_{pk} = C_p(1 - K).$$

When $\bar{X}$ equals the target mean T, $C_{pk} = C_p$; $C_{pk} = 0$ when the process mean shifts to the lower specification limit; and $C_{pk} = 2C_p$ when the process mean equals the upper specification limit.

Conformity; Nonconforming Units – For years, quality control professionals used terms such as "defects" and "defectives." Based on inspection, a product/raw material could be classified as defective or nondefective. This led to the development of control charts for "number of defects per unit," "number of defectives," and "percent defective." Defects could be further classified as critical, major, or minor. The recent trend has been towards the use of the terms *nonconformity* for a defect and *nonconforming unit* for a defective (ASQC, 1983).

Thus, nonconformity refers to the departure of a quality characteristic from its intended level or state that occurs with a severity sufficient to cause an associated product not to meet a specification requirement. A nonconforming unit is one which contains at least one nonconformity. While nonconformity refers to a quality characteristic, nonconforming unit refers to a complete unit. A unit having several nonconformities may or may not be classified as nonconforming. For example, the presence of a few paint bubbles on a car's surface is not a severe nonconformity; the car may still be classified as a conforming unit.

Process – A manufacturing process may be visualized as consisting of people, machines, materials, operating procedures, inspection equipment, tools, dies, jigs, fixtures, settings, and environmental conditions. Such a visualization could later help in identifying possible sources of problems, which is the basic purpose of SPC.

BASIC SPC CONCEPTS: In this section, we present some of the key ideas based on which SPC procedures (control charts) were developed.

Sources of Variation – A key concept is the realization that two items manufactured on the same machine under practically identical conditions, one after the other, may nevertheless be quite different in terms of their quality characteristics. This is to say, variability is inherent in all production processes and that it is impossible to eliminate all variability in manufactured products irrespective of the precision of the process used. Some processes may exhibit lesser variability than others, but variability is present nonetheless.

In an automatic or semi-automatic manufacturing process, there are two broad sources of variation: "Chance" variation, and variation produced by "assignable causes." Chance (random) variation is the sum total of the effects of numerous factors impacting on a process, each of which has too small an impact to be identified individually. Chance variation is the inherent variation in a process and is taken into account by design engineers through the specification of tolerances (specification limits). When a process is operating as planned, it will produce products with the desired quality characteristics within specification limits. In such a situation, the process is said to be *in control*.

The other source of variation in manufacturing processes is that due to one or more assignable causes. When an assignable cause is present in a manufacturing process (such as wear and tear of a tool, a displaced setting, temperature change, introduction of poor quality raw materials, or even an inspection gauge needing recalibration), the process is said to be *out of control*. Assignable causes can be identified and eliminated through the use of control charts, thus bringing the process back in control.

When a process is in control (i.e., when only chance variation is present), experience has shown that its measurements tend to follow a pattern which may be taken as the model of satisfactory process behavior. A normal distribution is frequently used as a model of satisfactory process behavior. Thus, a process in control has predictable behavior, that is, its measurements follow the normal distribution. If measurements indicate that they do not follow the predictable pattern, the implication is that there must be an assignable cause leading up to an out-of-control status. Recall our definition of the process. Looking at the pattern of measurements on a control chart, it is often possible to narrow down the assignable causes to one or more elements in the process. The purpose of SPC is to detect an assignable cause as early as possible so that corrective action may be taken – often even before defectives are produced.

Control Charts – A process is monitored by regularly taking small samples of size n from the output, taking some measurements on the selected items, computing a relevant statistic (such as the mean, standard deviation, range, proportion nonconforming, etc.), and plotting the summary statistic on a chart. This is the Shewhart (1931) quality control chart and has a center line (CL), an upper control limit (UCL), and a lower control limit (LCL). It is used for monitoring a single quality characteristic; the UCL is typically plotted three standard deviations above the center line, and the LCL three standard deviations below the center line. The chart can be based on individual or grouped observations, and the plotted statistic could be a sample mean, range, standard deviation, proportion nonconforming, number nonconforming, etc. Figure 1 shows a typical univariate Shewhart chart.

If the sample statistic falls between the UCL and the LCL, it is concluded that the process is

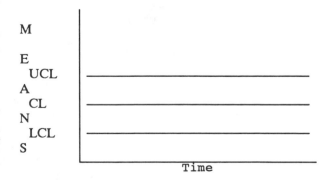

Fig. 1. A typical univariate Shewhart chart for means.

in control, that is, no special causes are present. On the other hand, if the statistic falls outside the UCL and LCL, or if the statistic shows some kind of a trend, the possibility of the existence of an assignable cause is entertained. The process is possibly stopped and efforts made to detect any special cause(s) that may be present. Only after confirming that either no special cause is present or that an assignable cause has been eliminated is the process restarted.

Determination of Control Limits – For the Shewhart charts, the specification of the probability of false alarms (Type I error) associated with an in-control process determines the control limits. A false alarm occurs when a statistic plots outside the control limits when the process is actually in control. The forms of the control limits are presented later for a variety of control charts.

Sampling Considerations – The relevant sampling issues in the use of control charts include the determination of sample size, frequency of sampling, and sampling technique.

In terms of the sample size, the larger the sample, the higher the probability of detecting a shift in the process mean. In practice, however, small samples ($n = 4$ or 5) are used when the quality characteristic of interest can be measured. And, in terms of the frequency of sampling, samples are taken at time intervals depending on how fast a process change could occur. If a process could change rapidly, samples of output are taken more frequently. Taking smaller samples more frequently is better than taking large samples infrequently. When the characteristic of interest cannot be measured, the sample sizes required are even larger.

Since we are generally interested in monitoring a process over time to detect any process shifts, it makes sense to use the time order of production as a logical basis for drawing samples. Shewhart suggested that samples (subgroups) should be so drawn that, if assignable causes are present, the chance for within group differences is minimized while the chance for between group differences is maximized. He also suggested the use of "rational subgroups" in sample selection. The most commonly used approach for rational subgrouping is to include items produced around the same time in a sample. Another approach to rational subgrouping is to take a random sample from all the units produced since the last sample was taken. This approach is useful when the entire production interval is of interest. The concept of rational subgroups is more relevant when the control chart is going to be based on grouped observations. The proper selection of samples is absolutely essential for gaining as much useful information as possible from control chart analysis. Outputs from different machines, work centers, shifts, or operators should not be pooled together to form a rational subgroup because that would make it impossible to pinpoint assignable causes should a problem arise.

Judging Control Chart Performance – The performance of a control chart is judged in terms of its average run length (ARL). The ARL of a procedure is the average number of samples taken before an out-of-control signal is given. An out-of-control signal for an in-control process is nothing but a false alarm. It can be shown that the in-control ARL for a Shewhart chart equals $1/\alpha$, where α is the probability of Type I error associated with the control chart.

Types of Control Charts – Shewhart charts, being based only on the latest sample, are effective at detecting large shifts in the process but are insensitive to small and medium shifts as well as to incremental shifts over time. Other quality control charts, specifically the cumulative sum (CUSUM) chart and the exponentially weighted moving average (EWMA) chart are, however, useful for detecting small to moderate shifts in the process though they are less sensitive for detecting large shifts.

The univariate CUSUM chart is based on a repeated application of the sequential probability ratio test while the EWMA chart is based on exponentially weighted moving averages of all the past observations. The CUSUM procedure gives equal weight to all of the observations. The EWMA procedure, on the other hand, gives the most weight to the latest observation and exponentially decreasing weights to prior observations. The EWMA charts are often considered to be easier to implement and interpret than the CUSUM charts, though both EWMA and CUSUM charts are relatively more difficult to design and use than the Shewhart charts.

Much of the research on CUSUM techniques, since their inception in 1954 by Page, has been on computing the ARL (Goldsmith and Whitfield, 1961; Kemp, 1961; Van Dobben de Bruyn, 1968; Goel and Wu, 1971; Brook and Evans, 1972; Reynolds, 1975; Khan, 1978; Zacks, 1981; Lucas, 1976, 1982; Lucas and Crosier, 1982; Woodall, 1983, 1984; Pollack and Siegmund, 1986; Waldman, 1986). Reynolds (1975) found that good ARL approximations can be determined through simulation. Nomograms of Goel and Wu (1971), the charts of Gan (1991), or the tables of Van Dobben de Bruyn (1968) can be used for determining the design parameters of the CUSUM chart based on factors like the acceptable quality level (μ_0), rejectable quality level (μ_1), and the desired in-control and out-of control ARLs for detecting the rejectable quality level.

Hunter (1986) provided a detailed exposition of the EWMA procedure. Crowder (1987) suggested a simple method for studying the run length distributions of EWMA charts. By providing plots of EWMA chart parameters, Crowder (1989) simplified the design of optimal EWMA charts. An optimal EWMA chart is defined as an EWMA chart with a fixed in-control ARL which has the smallest out-of-control ARL for a specified shift in the process mean. Lucas and Saccucci (1990) suggested some enhancements such as the fast initial response feature. They also provided a table of chart parameters for designing optimal EWMA charts.

Setting Up a Statistical Process Control Scheme – Finally, a word about how the control charts are implemented. If the use of the control chart is to detect deviation from a target mean (μ_0) and standard deviation (σ_0), the design of control chart procedures is relatively straightforward. If μ_0 and σ_0 are unknown, they are estimated by sampling a process *under stable conditions of production*. Data from 20 or more samples (rational subgroups) are used for estimating the target mean and standard deviation.

Using these estimates, trial limits are computed and we retrospectively test whether the process was in control when the samples were taken. If necessary, the trial limits are revised by excluding samples that resulted in out-of-control signals using the trial limits. Then, we continue monitoring the process in future using the revised limits, which are based on revised estimates of the process mean and standard deviation.

SHEWHART CHARTS: Control charts can be classified into two groups, *univariate* and *multivariate*, depending on the quality characteristics being monitored. Each of these groups contains three main types of control charts, namely, Shewhart, CUSUM, and EWMA charts. In this section, we present the univariate Shewhart charts. Univariate CUSUM and EWMA charts are discussed in the following section. Immediately after that, we present a brief discussion of the multivariate control charts.

If the quality characteristic of interest is a continuous variable, control charts based on measurements of that characteristic are referred to as "control charts for variables." If the data being collected comprise a discrete variable, the relevant control charts are termed "control charts for attributes."

Control Charts for Variables – For a continuous variable with a probability distribution (assumed to be normal in this discussion), we generally want to monitor both the central tendency and the variability of the process. The central tendency can be monitored using either individual observations (X chart) or sample means ($\bar{X}$ chart) based on grouped observations. Measures typically used for monitoring variability are the sample range (R chart) or sample standard deviation (S chart). Typically, control charts are used in pairs such as $\bar{X}$ and R charts (or $\bar{X}$ and S charts). It is recommended that process variability be monitored while monitoring the process average.

Control Charts for Attributes – Monitoring some quality characteristics may involve the classification of each inspected item as conforming or nonconforming to the specifications, such as can be determined by using a go/no-go gauge. Quality characteristics of this type are referred to as attributes, and the control chart is either based on proportion nonconforming (p chart) or number nonconforming (np chart). The basis for these charts is the binomial distribution. It is assumed that the probability (p) of getting a nonconforming unit is constant.

Another set of attribute control charts is based on counting nonconformities in a unit. If each sample unit, say, a square inch of a cloth, or a roll of wire of given length, can have a number of different nonconformities, we may be interested in a control chart based on the number (or average number) of nonconformities per unit. These charts are relevant if the probability of an occurrence of a nonconformity is constant and very low relative to the opportunities for its occurrence. These charts are based on the Poisson distribution.

Control Charts for Variables (With Known Target Mean and Variance) – Control Chart for Individual Observations (X Chart): The process mean can be monitored using either individual observations or means of samples of observations from rational subgroups. The rationale behind these charts is that the measurements from a process constitute a random sample from a normal distribution with mean μ_0 and standard deviation σ_0. If this is the case, then 99.73% of the individual observations will lie in the interval $\mu_0 \pm 3\sigma_0$. In general, $100(1 - \alpha)\%$ of the observations will be in the interval $\mu_0 \pm z_{\alpha/2}\sigma_0$. When $(1 - \alpha) = 0.9973$, this implies that $\alpha/2 = 0.00135$ and $z_{0.00135} = 3.0$.

A control chart using individual observations for monitoring the process mean is designed as follows:

$$UCL = \mu_0 + z_{\alpha/2}\sigma_0, \quad CL = \mu_0, \quad LCL = \mu_0 - z_{\alpha/2}\sigma_0.$$

The commonly used value of $z_{\alpha/2}$ for designing control charts in the U.S. is 3.0, and these control limits are referred to as 3σ control limits. Thus, for the control chart based on individual observations, the UCL is $\mu_0 + 3\sigma_0$ and the LCL is $\mu_0 - 3\sigma_0$.

Control Chart for Sample Means ($\bar{X}$ Chart) – The control limits for a chart based on sample means are developed from the result that if observations from a process are normally distributed with mean μ_0 and standard deviation σ_0, then the sample mean based on n observations from the same process is normally distributed with mean μ_0 and standard deviation $\sigma_0/\sqrt{n}$. The control limits for a control chart based on sample means are as follows:

$$UCL = \mu_0 + z_{\alpha/2}\sigma_0/\sqrt{n} = \mu_0 + A\sigma_0, \quad CL = \mu_0,$$

$$LCL = \mu_0 - z_{\alpha/2}\sigma_0/\sqrt{n} = \mu_0 - A\sigma_0$$

where $z_{\alpha/2} = 3.0$ and the values of A for $n = 2, 3, \ldots, 25$ can be found in Duncan (1986).

For successive random samples of size n, this control chart can be viewed as repeated tests of significance of the form $H_0: \mu = \mu_0$ vs. $H_1: \mu \neq \mu_0$ where α is the Type I error associated with the test.

Control Charts for Measures of Variability – If rational samples of size n are taken at regular intervals, the dispersion of the process could also be monitored using either a control chart based on the range (the R chart) or the standard deviation (the S chart).

The R Chart: The range is given by $R = X_{max} - X_{min}$. To find the mean and standard deviation of R, let us define the relative range as $W = R/\sigma_0$, where W is a function of the sample size n. Its mean, $E(W)$, is given by d_2 and its standard deviation by d_3. Therefore, the mean of the range R, $E(R)$, equals $d_2\sigma_0$ and the standard deviation of R equals $d_3\sigma_0$. The values of d_2 and d_3 for $n = 2, \ldots, 25$ are given in Duncan (1986).

The control limits for the R chart can now be computed:

$$UCL = E(R) + 3\sigma_R = d_2\sigma_0 + 3d_3\sigma_0 = (d_2 + 3d_3)\sigma_0 = D_2\sigma_0$$

$$CL = E(R) = d_2\sigma_0$$

$$LCL = E(R) - 3\sigma_R = d_2\sigma_0 - 3d_3\sigma_0 = (d_2 - 3d_3)\sigma_0 = D_1\sigma_0$$

where the values of D_1, D_2, and d_2 for $n = 2, \ldots, 25$ are given in Duncan (1986).

The S Chart: Because the computation of the sample standard deviation is more difficult than the computation of the range, R charts have been used more often than the S charts. However, now with the increasing availability of computing power, the use of S charts should grow.

The control limits for the S chart are as follows:

$$UCL = B_6\sigma_0, \quad CL = c_4\sigma_0, \quad LCL = B_5\sigma_0.$$

The values of c_4, B_5, and B_6 for $n = 2, \ldots, 25$ can be found in Duncan (1986). Both R and S charts are based on the notion that most of the probability distribution of each statistic lies within three standard deviations of the mean of the statistic.

Control Charts for Variables (Target Mean and Variance Unknown) – As indicated earlier, we first estimate the target mean and the variance based on data from, say, m rational subgroups; we then compute trial limits and retrospectively check whether the process was in control when the data for those m subgroups were collected. Then, by excluding the data from rational subgroups that resulted in out-of-control signals, the estimates of target mean and variance are revised. Finally, revised control limits are calculated for monitoring the process in future with subsequent updating as required.

A common practice is to use unbiased estimates of the process mean and standard deviation. Such estimates are found as follows. Using data from m subgroups, we compute the mean ($\bar{X}$) and standard deviation (S_i) for each subgroup and then compute their overall mean ($\bar{\bar{X}}$) and average of the standard deviations ($\bar{S}$). If range is used instead of standard deviation, then $\bar{R}$ is calculated where $\bar{R}$ is the average of the ranges from the subgroups. Note that $\bar{S}$ and $\bar{R}$ are, however, not unbiased estimators of the process standard deviation. Unbiased estimators are obtained by using $'c$ or d.

Control Chart for $\bar{X}$ – The control limits for this chart are similar to those presented in earlier except that estimates of μ_0 and σ_0 must now be used. Here, the $\bar{X}$ control limits when $\bar{R}$ is used to measure variability are:

$$UCL = \bar{\bar{X}} + A_2\bar{R}, \quad CL = \bar{\bar{X}}, \quad LCL = \bar{\bar{X}} - A_2\bar{R}.$$

The control limits when S is used are:

$$UCL = \bar{\bar{X}} + A_3\bar{S}, \quad CL = \bar{\bar{X}}, \quad LCL = \bar{\bar{X}} - A_3\bar{S}.$$

The values of A_2 and A_3 for $n = 2, \ldots, 25$ are given in Duncan (1986).

Control Charts for R and S – Both these charts are also similar to those presented earlier except that estimates of σ_0 need to be used. Here, the 3σ control limits for R and S charts should not be thought of as probability limits since their distributions are asymmetric. Details about the use of probability limits for R and S charts as well as the design of a chart based on the maximum likelihood estimator of the process deviation may be found in Ryan (1989).

Control Charts for Attributes – As mentioned earlier, these charts are usually based on the binomial or the Poisson distribution. Charts for proportion nonconforming (p charts) and number nonconforming (np charts) are based on the binomial distribution, and charts for number of nonconformities per unit (c charts) or average number of nonconformities per unit (u charts) are based on the Poisson distribution.

p Chart for Proportion Nonconforming Units – Suppose that a process is operating in a stable manner and that the probability of producing a nonconforming unit is p. If this process is monitored by taking samples of size n at regular intervals of time and counting the number (x) of nonconforming units in each sample, then the random variable (the number of nonconforming units) follows a binomial distribution with parameters n and p. The mean of this random variable is np and its variance is $np(1-p)$. If we are interested in using the statistic fraction nonconforming (x/n), its mean is p and its variance is $p(1-p)/n$.

If p is known, the control limits for the p chart are:

$$\text{UCL} = p + 3[p(1-p)/n]^{1/2}, \quad \text{CL} = p,$$
$$\text{LCL} = p - 3[p(1-p)/n]^{1/2}.$$

If the LCL is negative, it is set equal to 0.

If p is unknown, or not given, it is estimated by using data from, say, 20 to 30 rational subgroups. The sample fraction nonconforming from each of these subgroups is computed and the $\bar{R}$ average of these sample fractions ($\bar{p}$) is used as an estimate of p. Replacing p by $\bar{p}$ in the above control limits results in trial control limits for the desired p chart. If necessary, the trial limits can be revised as more information becomes available.

The control limits given above apply only if the sample size n is constant. If the sample size changes from subgroup to subgroup, then the upper and lower control limits will change from subgroup to subgroup by using the relevant sample size in place of n in the control limits formulae. However, if the changes in the sample sizes from one subgroup to another are fairly small, average sample size could be used for n in the computation of the control limits.

As mentioned earlier, sample sizes required for fraction nonconforming charts are much larger than the sample sizes used in control charts for variables. A rule of thumb is that the sample size should be such that the probability of detecting a specified shift, d, on the next sample is 0.5 (Duncan, 1986). Using the normal approximation to the binomial distribution, based on 3σ limits, the sample size, n, is given by:

$$n = 9p(1-p)/d^2.$$

np Chart for Number of Nonconforming Units – The control limits of this chart are based on the fact that for a process with probability (p) of producing a nonconforming unit, the mean number of nonconforming units in a sample of size n is np and the variance of the number of nonconforming units is $np(1-p)$. The 3σ control limits are:

$$\text{UCL} = np + 3[np(1-p)]^{1/2}, \quad \text{CL} = np,$$
$$\text{LCL} = np - 3[np(1-p)]^{1/2}.$$

If p is unknown, $\bar{p}$ is used in its place. Furthermore, in the interest of continuous improvement, it is desirable to reduce the level of nonconforming units.

c Chart for Nonconformities (Constant Sample Size) – In this case, the sample unit may be a single product (e.g., a square inch of cloth) or a group of units (e.g., 50 DRAM chips), but the sample size is constant. We also assume that the probability of occurrence of a nonconformity (p) relative to the number of opportunities (n) for its occurrence is extremely small and constant. Then, the random variable (number of nonconformities) follows a Poisson distribution which is a limiting form of the binomial distribution. If n is very large and p approaches 0 such that $np = c$ remains constant, then c is the parameter of the Poisson distribution and both the mean and variance of the number of nonconformities would equal c.

Thus, 3σ control limits for the c chart are:

$$\text{UCL} = c + 3c^{1/2}, \quad \text{CL} = c, \quad \text{LCL} = c - 3c^{1/2}.$$

If LCL is negative, it is set equal to 0. Again, every effort should be made to reduce the level of c. If no standard is given, c is estimated by the average number of nonconformities ($\bar{c}$) in a preliminary sample of size n and trial limits are computed using $\bar{c}$ instead of c.

u Chart for Average Number of Conformities per Unit – The u chart can be used in place of the c chart when the sample unit contains more than one inspection unit and we want to chart the number of conformities per unit, where $u = c/n$ and n is the number of inspection units from which c is obtained.

If the sample size (or the area of opportunity) varies from sample to sample, then a u chart should be used instead of a c chart. For example, if we are interested in counting the number of paint blemishes on a car door, the area of opportunity would vary depending on the car model. One could find the average number of blemishes per square meter and use that information in a u chart.

The control limits for a u chart when the sample size (n) is constant are:

$$\text{UCL} = \bar{u} + 3\sqrt{\bar{u}/n}, \quad \text{CL} = \bar{u}, \quad \text{LCL} = \bar{u} - 3\sqrt{\bar{u}/n},$$

where $\bar{u}$ is the total number of nonconformities in all samples divided by the total number of units in all samples. If the sample size varies from sample to sample, one could either use variable control limits by replacing n with n_i or use control limits obtained by replacing n by the average sample size ($\bar{n}$).

Interpreting Control Chart Patterns – If a process is operating only under chance or natural causes of variation, the points on the control chart should be randomly scattered around CL. Approximately half the points should be below the CL and half above the CL. Nonrandom control chart patterns of any kind signal the possibility of assignable causes being present. So, in addition to the 3σ limits, one should use other rules to identify nonrandom patterns on the control chart. Grant and Leavenworth (1988) give the following rules for identifying possible assignable causes:

- 7 successive points on the control chart are on the same side of the CL.
- Out of 11 successive points, at least 10 are on the same side of the CL;
- Out of 14 successive points, at least 12 are on the same side of the CL;
- Out of 17 successive points, at least 14 are on the same side of the CL;
- Out of 20 successive points, at least 16 are on the same side of the CL.

The *Statistical Quality Control Handbook* (1958) also gives some rules to identify nonrandom patterns. The rules given in the handbook are such that the probability of occurrence of those patterns is approximately equal to the probability of a point falling outside the 3σ limits. These rules are: Two out of three successive points outside 2σ limits, or four out of five successive points outside 1σ limits, or eight successive points on the same side of the CL. The occurrence of any of these patterns indicates the possible presence of an assignable cause.

Duncan (1986) gives the probabilities of the occurrence of different types and lengths of runs.

If five successive points on a control chart fall below the mean, we have a run below the mean of length 5. Similarly, one could have a run above the mean. If, starting with a point on the control chart, the next 6 observations are decreasing in value, we have a run down of length 6. Nelson (1984) provides a list of rules where the probabilities of out-of-control signals are about equal.

If one uses more than one rule for monitoring a process, the probability of detection of special causes increases but the probability of false alarms increases too. For example, if two different rules have false alarm rates of α_1 and α_2, respectively, then the false alarm rate associated with using both the rules simultaneously is approximately $1 - (1 - \alpha_1)(1 - \alpha_2)$, which is greater than both α_1 and α_2.

Some other patterns to watch for are: cycles, trends, sudden shifts, etc. More information on the interpretation of patterns is available on pages 161–180 of the *Statistical Quality Control Handbook* (1958).

CUSUM AND EWMA CHARTS: *Cumulative Sum (CUSUM) Charts* – Univariate CUSUM procedures were first proposed by Page (1954) and have since been studied by many researchers. Gibra (1975), Goel (1981), Vance (1983), and Woodall (1986) all provide references and reviews of the research done on this topic.

We will discuss only the design of cumulative sum charts for sample means. Cumulative sum procedures based on other sample statistics, such as fraction nonconforming, number nonconforming, number of nonconformities per unit, range, standard deviation, etc., can also be designed. For information on the design of these CUSUM procedures, refer to Johnson and Leone (1962), Page (1961), and Lucas (1985). CUSUM procedures for monitoring multivariate processes have also been designed.

A CUSUM procedure combines information from both previous samples and the current sample to detect process shifts. If we want to control a single quality characteristic, CUSUM charts for controlling the mean and variance of the process can be designed based on repeated applications of the sequential probability ratio test proposed by Wald in 1947.

Since, in the case of a univariate process, the shift in the process mean may be an increase or a decrease in the target mean, we design a two-sided CUSUM procedure. A two-sided procedure can be operated either using a mobile V-mask or an equivalent procedure with a decision interval. The procedure discussed in this article is based on a decision interval.

It is assumed that the sample means are independent and normally distributed with mean μ_0 and variance σ_0^2/n. Instead of working with the sample mean $\bar{X}$, we transform it into the standard normal variable Z_t, where $Z_t = \sqrt{n}(\bar{X}_t - \mu_0)/\sigma_0$. Then, at time t, we compute two cusums, T_t and S_t, T_t for detecting a downward shift and S_t for detecting an upward shift in the process mean:

$$T_t = \min\{0, T_{t-1} + Z_t + k\} \quad \text{and}$$

$$S_t = \max\{0, S_{t-1} + Z_t - k\}$$

where $T_0 = S_0 = 0$; $k \geq 0$ is referred to as the reference value. This reference value is chosen halfway between the standardized target value of the process mean and the standardized rejectable process mean. Usually, k is taken as 0.5. An out-of-control signal is given at the first t for which either $T_t < -h$ or $S_t > h$, where h is another chart parameter called the decision interval. Typically, $h = 4$ or 5. A value $T_t < -h$ signals a downward shift in the process mean whereas $S_t > h$ signals an upward shift in the process mean.

The design parameters, h and k, of the CUSUM procedure depend on the in-control ARL, acceptable and rejectable quality levels, and the out-of-control ARL for the specified rejectable quality level. Nomograms of Goel and Wu (1971), tables of Van Dobben de Bruyn (1968), or charts of Gan (1991) can be used for finding the design parameters.

CUSUM procedures are more effective than Shewhart charts in detecting small to moderate shifts, say of the order of 0.5 to 2 standard deviations, in the process mean. These charts, however, are very slow in detecting large shifts. Lucas (1982) suggests the use of a combined CUSUM-Shewhart procedure for improving the ability of the CUSUM chart to detect large shifts.

Lucas and Crosier (1982) also suggested a procedure with fast initial response or head start to improve the sensitivity of the procedure at process start-up. The procedure recommends putting T_0 and S_0 equal to $h/2$.

Exponentially Weighted Moving Average (EWMA) Charts – The EWMA chart was introduced by Roberts (1959). It can be used with individual observations or sample means from grouped data. We will only discuss the case when the target mean μ_0 and variance σ_0 are known, and we will design the EWMA chart for

sample means. We assume that the sample means $\bar{X}$ are independent and normally distributed with mean μ_0 and variance σ_0^2/n. At time t, we compute the EWMA statistic z_t as:

$$z_t = r(\bar{X}_t) + (1 - r)z_{t-1}$$

where $0 < r \leq 1$ and $z_0 = \mu_0$. It can be shown that z_t is a weighted average of all previous observations where the weights add to 1. Also, the weights decrease geometrically with the age of the sample observation. The variance of z_t is given by:

$$\text{Var}(z_t) = (\sigma_0^2)(r)[1 - (1 - r)^{2t}]/[n(2 - r)].$$

As t increases, the variance of z_t approaches $r\sigma_0^2/[n(2 - r)]$. Therefore, the control limits for the EWMA charts are:

$$\text{UCL} = \mu_0 + 3\sigma_0\{r/[n(2 - r)]\}^{1/2}, \quad \text{CL} = \mu_0,$$

$$\text{LCL} = \mu_0 - 3\sigma_0\{r/[n(2 - r)]\}^{1/2}$$

for moderately large t. For small t, control limits should be based on the exact variance of z_t.

If the target mean and variance are unknown, the target mean μ_0 can be replaced by $\bar{X}$ and the target variance can be replaced by $\bar{R}/d_2$ or $\bar{S}/c_4$ to get trial control limits. The EWMA chart can be extended to other sample statistics such as the number of conformities c, the number nonconforming np, etc.

The EWMA chart is very effective in detecting small shifts in the process mean but is not as effective as the Shewhart chart for detecting large shifts in the process mean. Crowder (1989) and Lucas and Saccucci (1990) have provided tables for designing optimal EWMA procedures. An optimal EWMA chart is defined as an EWMA chart with a fixed in-control ARL which has the smallest out-of-control ARL for a specified process shift.

Comparison of the Shewhart, CUSUM, and EWMA Procedures – The performance of a control chart is judged in terms of its average run length. The ARL of a procedure is the average number of samples taken before an out-of-control signal is given. The ARLs are computed for different shifts in the process mean by assuming that the magnitude of the shift is equal to $k\sigma$. When $k = 0$, the process is in control. It can be shown that the in-control ARL for a Shewhart chart equals $1/\alpha$, where α is the probability of Type I error associated with the control chart. The ARL for a Shewhart chart for an out-of-control process with a shift of $k\sigma$ equals $1/(1 - \beta)$, where β is the probability of not detecting the shift. Table 1 shows the average run lengths for the three charts mentioned. They were de-

Table 1. ARLs of Shewhart, optimal CUSUM, and EWMA charts

Shift	Shewhart $\alpha = 0.004$	CUSUM $h = 4.39$ $k = 0.50$	EWMA $h = 0.76$ $r = 0.15$
0.0	250.00	250.00	250.00
0.5	110.33	30.80	26.90
1.0	33.08	9.16	8.75
1.5	11.89	5.14	5.05
2.0	5.27	3.60	3.59
2.5	2.84	2.81	2.82
3.0	1.82	2.34	2.36

SOURCE: Gan (1991)

signed so that all three have an in-control ARL of 250. For small shifts, the EWMA performs best; for moderate shifts (1.5σ to 2.0σ), CUSUM and EWMA are comparable; there is little difference in the performance of all three charts when the shift is 2.5σ; and the Shewhart chart performs the best only for large shifts (3.0σ or larger).

MULTIVARIATE CONTROL CHARTS: Manufacturing processes typically involve the control of several quality characteristics simultaneously. This is often achieved by using a univariate control chart for each of the characteristics. However, in view of possible relationships between variables, the use of bivariate (or multivariate) data on the same chart can provide much richer interpretations than individual univariate charts. In addition, Jackson (1956) showed that the use of simultaneous univariate control charts could give misleading results. The need for designing multivariate quality control procedures was, thus, recognized early in the development of the field. Hotelling (1947) was the first statistician to recognize this need, and several researchers have since studied the bivariate case extensively as well as other multivariate cases to some extent. Alt (1985) and Jain (1993) provide discussions of these developments over the last five decades.

TQM, ISO AND OTHER DEVELOPMENTS: The concepts of total quality control, quality circles, and zero defects surfaced in the 1960s. The rationale for total quality control is that all departments, not only the quality control department, must have some role in the outgoing quality level. Quality circles motivate workers by involving them on improvement projects using simple tools like check sheets, histograms, Pareto charts, graphs, and Ishikawa cause-and-effect diagrams. These improvement projects are typically in related areas such as quality, productivity, safety, costs, etc. The basic idea behind zero defects is that once a defect is identified, its source has to be isolated and rectified to ensure that the defect does not occur again. Zero defect programs aim to motivate employees to reduce their own errors as well as to strive to reduce systematic controllable errors.

The modern version of such programs can be found in the TQM and continuous improvement efforts of organizations which focus as much (and perhaps more) on the managerial aspects of quality as on its technical aspects. Much of the improvement in quality demanded today is, however, beyond the classical tools of statistical process control and sampling inspection (Feigenbaum, 1983).

Since the 1980s, considerable emphasis has been placed on the human and managerial aspects of quality management. The domain in which quality is currently viewed in the United States is much broader than what it was even a few years ago. The full spectrum of quality management spans the product and service life cycles – from product conception to customer service – and is no longer restricted to the shop floor. The TQM programs are no doubt a response to the globalization of competition and changing customer needs. Customers know that, in a global economy, they can get better quality products at lower costs, and companies have no choice but to satisfy customer needs merely to survive in an increasingly complex and uncertain world. Those who want to succeed and grow must not only satisfy customers but attempt to "delight" them by offering them more than value for money.

The adoption of TQM and continuous improvement programs in the United States has been encouraged in both service and manufacturing organizations during the last few years through the Malcolm Baldridge National Quality Award, patterned somewhat after the Deming Prize of Japan. (The Deming Prize is Japan's highly coveted award that recognizes successful quality effort in terms of statistical quality control.) Up to six Baldridge Awards are given each year in three categories: manufacturing, service, and small business. The Award has had a great impact on bringing quality on top managements' agendas throughout the nation.

With a view to developing a common and consistent set of guidelines for suppliers to meet the growing need for better quality products needed by the European Union (formerly, the European Community), the International Organization for Standards (ISO) developed and issued a set of international standards called ISO 9000. The ISO 9000 Standard Series was published in 1987 and had five components containing extensive guidelines for quality measurement and management. The Series has been expanded since then. An organization wanting to obtain the registration and approval of its quality system needs to go through extensive documentation and assessment, and periodic audits, by a third party. Quality system registration, however, does not imply product conformity to any given set of specifications. An organization's quality system is registered and not any of its products. Over 50 countries have already adopted the ISO 9000 standards as their national standards, and the ISO 9000 certification has now become quite essential for companies interested in doing business globally, especially with the European Union.

See **Total quality management.**

References

[1] Alt F.B. (1973). *Aspects of Multivariate Control Charts.* Unpublished MS Thesis. School of Industrial and Systems Engineering, Georgia Institute of Technology, Atlanta, Georgia.

[2] Alt F.B. (1982). "Multivariate Quality Control: State of the Art," *ASQC Quality Congress Transactions-Detroit*, 886–893.

[3] Alt F.B. (1985). "Multivariate Quality Control." In Kotz, S. and N.L. Johnson, Eds., *Encyclopedia of Statistical Sciences*, Volume 6: 110–122. John Wiley, New York.

[4] Alt F.B., S.J. Deutsch and J.W. Walker (1977). "Control Charts for Multivariate, Correlated Observations," *ASQC Quality Congress Transactions*, Philadelphia: 360–369.

[5] ASQC (1983). *Glossary and Tables for Statistical Quality Control*, 2nd ed. Milwaukee, WI: American Society for Quality Control.

[6] Banks J. (1989). *Principles of Quality Control.* John Wiley, New York.

[7] Bendell A., J. Disney and W.A. Pridmore, Eds. (1989). *Taguchi Methods: Applications in World Industry.* Springer-Verlag, New York.

[8] Box G.E.P. (1957). "Evolutionary Operation: A Method of Increasing Industrial Productivity," *Applied Statistics*, 6(2), 81–101.

[9] Calvin T.W. (1990). "Bayesian Analysis." In Harrison M. Wadsworth, Ed., *Handbook of Statistical Methods for Engineers and Scientists.* McGraw-Hill, New York.

[10] Brook D. and D.A. Evans (1972). "An Approach to the Probability Distribution of Cusum Run Length," *Biometrica*, 59, 539–549.

[11] Crosier R.B. (1988). "Multivariate Generalizations of Cumulative Sum Quality Control Scheme." *Technometrics*, 30, 291–303.

[12] Crowder S.V. (1987). "A Simple Method for Studying Run Length Distributions of Exponentially Weighted Moving Average Charts," *Technometrics*, 29, 401–407.

[13] Crowder S.V. (1989). "Design of Exponentially Weighted Moving Average Schemes." *Technometrics*, 31, 156–162.

[14] Dehnad K., Ed. (1989). *Quality Control, Robust Design, and the Taguchi Method.* Wadsworth and Brooks, Pacific Grove, California.

[15] Duncan A.J. (1986). *Quality Control and Industrial Statistics*, 5th ed., Irwin, Homewood, Illinois.

[16] Feigenbaum A.V. (1951). *Quality Control: Principles, Practice and Administration: An Industrial Management Tool for Improving Product Quality and Design and for Reducing Operating Costs and Losses*, 1st ed., McGraw-Hill, New York.

[17] Feigenbaum A.V. (1983). *Total Quality Control-Engineering and Management*, 3rd ed. McGraw-Hill, New York

[18] Gan F.F. (1991). "An Optimal Design of CUSUM Quality Control Charts," *Jl. Quality Technology*, 23, 279–286.

[19] Ghare P.M. and P.E. Torgersen (1968). "The Multicharacteristic Control Chart," *Jl. Industrial Engineering*, 19, 269–272.

[20] Gibra I.N. (1975). "Recent Developments in Control Chart Techniques," *Jl. Quality Technology*, 7, 183–192.

[21] Goel A.L. and S.M. Wu (1971). "Determination of ARL and a Contour Nomogram for Cusum Charts to Control Normal Mean," *Technometrics*, 13, 221–230.

[22] Goel A.L. (1981). "Cumulative Sum Control Charts," In S. Kotz and N.L. Johnson, Eds., *Encyclopaedia of Statistical Sciences*. John Wiley, New York.

[23] Goldsmith P.L. and H. Whitfield (1961). "Average Run Length Cumulative Chart Control Schemes," *Technometrics*, 3, 11–20.

[24] Grant E.L. and R.S. Leavenworth (1988). *Statistical Quality Control*, 6th ed., McGraw-Hill, New York.

[25] Hotelling H. (1947). "Multivariate Quality Control-Illustrated by the Air Testing of Bombsights." In C. Eisenhart, M.W. Hastay and W.A. Willis, Eds., *Techniques of Statistical Analysis*. McGraw-Hill, New York.

[26] Hunter J.S. (1986). "The Exponentially Weighted Moving Average," *Jl. Quality Technology*, 18, 203–210.

[27] Jackson J.E. (1956). "Quality Control Methods for Two Related Variables," *Industrial Quality Control*, 12, 2–6.

[28] Jackson J.E. and R.A. Bradley (1966). "Sequential Multivariate Procedures for Means with Quality Control Applications." In P.R. Krishnaiah, Ed., *Multivariate Analysis* I, 507–519. Academic Press, New York.

[29] Jain K. (1993). *A Bayesian Approach to Multivariate Quality Control.* Unpublished doctoral dissertation, University of Maryland at College Park.

[30] Johnson N.L. and F.C. Leone (1962). "Cumulative Sum Control Charts, Mathematical Principles Appplied to Their Construction and Use," Part I, II, & III. *Industrial Quality Control*, June 1962, 15–22.

[31] Joseph J. and V. Bowen (1991). "A Cumulative Bayesian Technique for Use in Quality Control Schemes," *Proceedings of the American Statistical Association.*

[32] Kemp K.W. (1961). "The Average Run Length of the Cumulative Sum Chart When a V-Mask is Used," *Jl. Royal Statistical Society, Series B*, 23, 149–153.

[33] Khan R.A. (1978). "Wald's Approximations to the Average Run Length in Cusum Procedures," *Jl. Statistical Planning and Inference*, 2, 63–77.

[34] Lowry C.A., W.H. Woodall, C.W. Champ and S.E. Rigdon (1992). "A Multivariate Exponentially Weighted Moving Average Control Chart," *Technometrics*, 34, 1992.

[35] Lucas J.M. (1976). "The Design and Use of V-Mask Control Schemes," *Jl. Quality Technology*, 8, 1–12.

[36] Lucas J.M. (1982). "Combined Shewhart-Cusum Quality Control Schemes," *Jl. Quality Technology*, 14, 51–59.

[37] Lucas J.M. (1985). "Counted Data CUSUM's," *Technometrics*, 27, 129–144.

[38] Lucas J.M. and R.B. Crosier (1982). "Fast Initial Response for Cusum Quality Control Schemes," *Technometrics*, 24, 199–205.

[39] Lucas J.M. and M.S. Saccucci (1990). "Exponentially Weighted Moving Average Control Schemes, Properties and Enhancements," *Technometrics*, 32, 1–12.

[40] Montgomery D.C. (1991). *Introduction to Statistical Quality Control*, 2nd ed. John Wiley, New York.

[41] Nelson L.S. (1984). "The Shewhart Control Chart-Tests for Special Cases," *Jl. Quality Technology*, 16, 237–239.

[42] Page E.S. (1954). "Continuous Inspection Schemes," *Biometrika*, 41, 100–115.

[43] Page E.S. (1961). "Cumulative Sum Charts," *Technometrics*, 3, 1–9.

[44] Pignatiello J.J., Jr. and M.D. Kasunic (1985). "Development of Multivariate Cusum Chart," *Proceedings of the* 1985 *ASME International Computers and Engineering Conference and Exhibition*, 2, 427–432.

[45] Pignatiello J.J., Jr., G.C. Runger and K.S. Korpela (1986). Truly Multivariate Cusum Charts,. Working Paper, School of Systems & Industrial Engg., Univ. of Arizona, Tucson.

[46] Pignatiello J.J. Jr. and G.C. Runger (1990). "Comparisons of Multivariate CUSUM Charts," *Jl. Quality Technology*, 22, 173–186.

[47] Radharaman R. (1986). "Bicharacteristic Quality Control in Manufacturing," in *Proceedings of the 8th Annual Conference on Computers and Industrial Engineering*, 209–214.

[48] Pollack, M. and D. Siegmund (1986). *Approximations to the Average Run Length of Cusum Tests.* Technical Report 37. Department of Statistics, Stanford University, California.

[49] Reynolds M.R. Jr. (1975). "Approximations to the Average Run Length in Cumulative Sum Control Charts," *Technometrics*, 17, 65–71.

[50] Roberts S.W. (1959). "Control Chart Based on Geometric Moving Averages," *Technometrics*, 1, 239–250.

[51] Roberts H.V. and B.E. Sergesketter (1993). *Quality is Personal, A Foundation for Total Quality Management.* The Free Press, New York.

[52] Ryan T.R. (1989). *Statistical Methods for Quality Improvement.* John Wiley, New York.

[53] Shewhart W.A. (1931). *Economic Control of Quality of Manufactured Product.* Van Nostrand, New York.

[54] Smith N.D. (1987). *Multivariate Cumulative Sum Control Charts.* Ph.D. Dissertation, University of Maryland at College Park.

[55] Taguchi G. (1976). *Experimental Design*, Vol. 1, 3rd ed., Maruzen, Tokyo.

[56] Taguchi G. (1977). *Experimental Design*, Vol. 2, 3rd ed., Maruzen, Tokyo.

[57] Taguchi G. and Y. Wu (1979). *Introduction to Off-line Quality Control.* Central Japan Quality Control Association, Nagoya.

[58] Van Dobben de Bruyn, C.S. (1968). *Cumulative Sum Tests-Theory and Practice.* Hafner Publishing, New York.

[59] Vance L.C. (1983). "A Bibliography of Statistical Quality Control Chart Techniques, 1970–1980," *Jl. Quality Technology*, 15, 59–62.

[60] Wald A. (1947). *Sequential Analysis.* John Wiley, New York.

[61] Waldman K.H. (1986). "Bounds for the Distribution of the Run Length of One-Sided and Two-Sided CUSUM Quality Control Schemes," *Technometrics*, 28, 61–67.

[62] Woodall W.H. (1983). "The Distribution of the Run Length of One-Sided Cusum Procedures for Continuous Random Variables," *Technometrics*, 25, 295–301.

[63] Woodall W.H. (1984). "On the Markov Chain Approach to the Two-Sided Cusum Procedure," *Technometrics*, 26, 41–46.

[64] Woodall W.H. (1986). "The Design of CUSUM Quality Control Charts," *Jl. Quality Technology*, 18, 99–102.

[65] Woodall W.H. and M.M. Ncube (1985). "Multivariate CUSUM Quality Control Procedures," *Technometrics*, 27, 285–292.

[66] Zacks S. (1981). "The Probability Distribution and the Expected Value of a Stopping Variable Associated with One-Sided Cusum Procedures for Non-Negative Integer Valued Random Variables," *Communications in Statistics-Theory and Methods*, A10, 2245–2258.

[67] Western Electric Company (1958), *Statistical Quality Control Handbook*, 2nd ed., Indianapolis.

QUASI-CONCAVE FUNCTION

Given a function $f(\cdot)$ and points x, $y \in X$, with $x \neq y$ and X convex, if $f(y) \geq f(x)$ implies that $f[\lambda x + (1 - \lambda)y] \geq f(x)$ for all $0 < \lambda < 1$, then we say that f is a quasi-concave function. See **Concave function**.

QUASI-CONVEX FUNCTION

Given a function $f(\cdot)$ and points x, $y \in X$, with $x \neq y$ and X convex, if $-f(y) \geq -f(x)$ implies that $-f[\lambda x + (1 - \lambda)y] \geq -f(x)$ for all $0 < \lambda < 1$, then we say that f is a quasi-convex function. See **Concave function**; **Convex function**.

QUASI-REVERSIBILITY

A property of a node in a queueing network where the state of the system at t_0, the departure prior to t_0, and the arrival process subsequent to t_0 are independent. See **Networks of queues**.

QUEUE INFERENCE ENGINE

Richard C. Larson

Massachusetts Institute of Technology, Cambridge

Imagine receiving your monthly bank statement and with it is your *personal probability distribution* of the times you spend waiting in bank queues. The queues could include both those involving human tellers and automatic teller machines (ATMs). With the technology of the Queue Inference Engine (QIE) such an innovation is now well within the realm of possibility.

BACKGROUND, MOTIVATION AND OVERVIEW: The QIE was born in the late 1980s as a result of M.I.T.-based queueing research for Bay-Banks, an eastern Massachusetts bank, under the auspices of a grant from the National Science Foundation. BayBanks had provided a large sample of transactional data from three of their ATM sites. Their question was, "Which, if any, of these sites is 'too congested' from a queueing point of view, thereby requiring additional ATM capacity at the site?" The transactional data consisted of the times of each ATM transaction by each customer over a period of up to a month.

The first approach to this problem was traditional: estimate arrival rates and service times from the data and then apply well known (steady state) queueing models, such as Erlang's results, or the M/G/1 model, etc. Examining the data set, it was realized that a substantial portion of the "sample path" of the queue had been preserved in the data set. That is, the data set contained a large subset of the information one would have if one "tracked" the actual queue with "clipboard and stopwatch." For instance, one could identify which customers had been delayed in queue (rather than enter service immediately) by noting the "signature" of a queued customer: a back to back service completion and service initiation at the same ATM, during a time when all N ATMs are busy with customers. The customer entering service in such a back-to-back situation was, with probability near one, delayed in queue. Moreover, by following this signature over time-adjacent customers, one could identify the entire set of customers who were delayed in queue during a single *congestion period*, a continuous period of time during which all N servers are continuously busy (excepting the small intervals during which a customer whose service is completed departs and the new [queued] customer enters service). We explored further the information content of the data set to see if it contained additional queue-related information.

Surprisingly, the partial information in the data set allowed a wide variety of queueing measures for each congestion period to be computed efficiently. Assuming Poisson arrivals, these measures include mean queue delay, mean queue length, probability distribution of the queue length and even the transient mean queue length over the course of the congestion period. Later research extended these first results in a number of important directions.

In this article, the focus is four-fold: (1) to illustrate the types of physical situations in which the QIE can be applied; (2) to describe one of three alternative analytical approaches to obtaining QIE results; (3) to guide the reader through the emerging literature in this new and exciting field; and (4) to discuss briefly several implementation experiences.

ILLUSTRATIVE QUEUE INFERENCING PROBLEMS: *Retail Sales* – With most "human server retail service systems," one has to collect the transactional data either from a modern POS (Point Of Sale) computer system that does the time marking or from some type of customer sensing device (e.g., pressure sensitive mats, infrared or ultrasonic sensors). For an ATM, the transactional data are recorded automatically, by time marking the moment that a customer inserts a bank card (corresponding to service initiation) and the moment that the ATM ejects the card (corresponding to service completion). The queue statistics generated by the QIE for ATMs may be used by bank managers to monitor the use of ATM sites, thereby providing an accurate method of identifying those sites requiring additional (or fewer) machines. With human servers in retail sales, at banks, post offices, fast food restaurants, etc., the manager would most likely use the results to (1) monitor service levels throughout the day and week, to assure that queue delays are within prescribed quality limits, and (2) to schedule servers optimally over the course of a day and week.

Invisible Queues in Telecommunication Systems – Many finite capacity telecommunications systems have during periods of congestion invisible queues of customers outside the system, continuously trying to gain access to it. One example is a k-channel land mobile radio system. Whenever all k channels are simultaneously in use, potential users having a message to transmit (often in the field, in vehicles) continuously monitor channel use and attempt to acquire a channel as soon as any one of the current k communications is completed. If at any given time t there are $n(t)$ such potential users awaiting a channel, they constitute a spatially dispersed invisible queue, a queue in which one of the waiting customers enters service very shortly after another customer completes service. This queue can grow in size due to the Poisson arrivals of new potential users desiring channel access. The user entering service next is the one who successfully "locks in" the channel very shortly after termination of a previous message. Service discipline is most likely not first come first served (FCFS). Within the context of the QIE the customer transaction times are the moments of gaining channel access (service initiation) and message termination (service completion). These times can be routinely monitored and recorded by technology, and thus the QIE can be used to deduce queueing behavior. The same argument, perhaps with minor modifications, can be applied to other telecommunications systems, including phone systems from airplanes, mobile cellular telephone systems, standard telephone systems and various digital communications networks.

USING ORDER STATISTICS TO DERIVE QIE PERFORMANCE MEASURES: The analysis of

the queue inferencing problem is rooted in order statistics. Suppose we consider a homogeneous Poisson process with rate parameter $\lambda > 0$. Over a fixed time interval $[0,T]$, we are told that precisely N Poisson events (e.g., "queue system arrivals") occur. The N *ordered* arrival times are $0 \leq X_{(1)} \leq X_{(2)} \leq \cdots \leq X_{(N)} \leq T$ (by implication $X_{(N+1)} > T$). The N *unordered* arrival times are $X_1, X_2, \ldots, X_N$, $0 \leq X_i \leq T$ ($i = 1, 2, \ldots, N$). Since the Poisson process is time homogeneous, it is well known that the $\{X_i\}$ are independent and uniformly distributed over $[0, T]$. If the Poisson process is non-homogeneous, that is, having time varying rate parameter $\lambda(t)$, then the N unordered arrival times are independent identically distributed (iid) over $[0, T]$, with a pdf (probability density function) proportional to $\lambda(t)$. For simplicity in this discussion, we focus on homogeneous processes.

A Pedestrian Queueing Example – To illustrate queue inference, consider a signalized pedestrian crosswalk having fixed cycle time T. Poisson-arriving pedestrians queue at curbside waiting to cross the street during a time interval of length T, and all such queuers are served in "bulk" fashion when the light changes at time T allowing them safely to cross the street. The number N of queued arrivals in any particular light cycle is Poisson distributed with mean λT. Given N, $X_{(i)}$ is the arrival time during $[0, T]$ of the ith queued pedestrian. Here X_i could be viewed as the arrival time of a *random* queued pedestrian, selected from, say, a photograph of all queued pedestrians taken just before the light changed at time T. The Poisson arrival assumption is usually thought of as evolving sequentially over time, with customer interarrival times selected in an iid manner from a negative exponential pdf with mean λ^{-1}.

An equivalent way to conduct the pedestrian crosswalk experiment is to first select N from the Poisson distribution, and then for each of the N queuers to select the arrival time over $[0, T]$ independently from a uniform pdf; this experiment is probabilistically identical to the sequential Poisson arrival realization of the experiment. Suppose now at some intermediate time T we focus on the total number of queued pedestrians $N(t)$, defined as the total number of arrivals (at curbside) during the interval $[0, t]$. The following results, derived from the second model of the process, are well known for $N(t)$:

$$\begin{cases} E[N(t)] = (t/T)T \\ \text{var}[N(t)]A \equiv \sigma^2 N(t) = [N(t)/T]y[(T-t)/T] \\ \Pr\{N(t) = k\} = \binom{N}{k}\left(\frac{t}{T}\right)^k\left(\frac{T-t}{T}\right)^{N-k}. \end{cases} \quad (1)$$

Here the *transactional* data are N, the total number of queuers, and T, the time until bulk service. From these data we have found transient values of conditional mean, variance and probability distribution of the queue length. Similar logic can be applied to find other performance measures, such as mean delay in queue, which in this case is trivially equal to $T/2$. This is one of the simplest examples of queue inferencing.

Queue Inference in More General Queues – In most queues, customers usually leave one-at-a-time. Their service completion times within a congestion period, recorded as part of the transactional data set, impose a set of inequality constraints on the arrival times of customers who waited in queue. It is this set of inequality constraints that produces precise conditioning information within the general context of order statistics, conditioning information that we use to deduce queue behavior.

Suppose for a M/D/1 system, we examine a congestion period having precisely $N = 2$ queued customers. For simplicity the service time is one minute per customer and the server's congestion period starts at time zero. Then since $N = 2$, we know that precisely 2 customers queued during this congestion period and after their service the server was again idle. The busy period for the server is 3 minutes in length, the time to serve 3 customers, the two who queued and the first arrival who initiated the congestion period. From the transactional data, we know that zero customers arrived during service of the last customer, the third in the congestion period and the second to queue (assuming FCFS queueing). We know that at least one of the 2 queued customers must have arrived in [0,1], else there would be no queued customer to select for service commencement at time $T = 1^+$. Similarly, the second queued customer must have arrived by time $T = 2$.

Without the ordering information, the conditional arrival times for the two queued customers are independent uniformly distributed over [0,2]. In the joint sample space of r.v.'s X_1 and X_2, this corresponds to X_1 and X_2 uniformly distributed over the square of size 2 in the positive quadrant. We can split the sample space into four equal subsquares, (1) $0 \leq X_1 \leq 1$, $0 \leq X_2 \leq 1$; (2) $1 \leq X_1 \leq 2$, $0 \leq X_2 \leq 1$; (3) $0 \leq X_1 \leq 1$, $1 \leq X_2 \leq 2$; (4) $1 \leq X_1 \leq 2$, $1 \leq X_2 \leq 2$. Without the additional conditioning information regarding service completion times, the outcome of the experiment is equally likely to be within each of the four subsquares, and conditional on being in a subsquare the r.v.'s X_1 and X_2 are conditionally uniformly distributed over that subsquare.

But the additional conditioning information from the transactional data imposes the constraints: $X_{(1)} \leq 1$, $X_{(2)} \leq 2$, thereby eliminating subsquare (4). The a priori probability of this event, called the *master probability*, is 3/4. *For any number of queued customers N, the master probability is the a priori probability that the order statistics will obey the ordered inequalities imposed by the transactional data.* Once we can efficiently calculate the master probability, most other quantities of interest are easy to compute.

Continuing with the $N = 2$ example, if we know that X_1 and X_2 fall in subsquare (1), then these two arrival times are uniform identically distributed over [0,1]. If one falls in [0,1] and the other falls in [1,2], i.e., subsquare (2) or (3), then the minimum is uniformly distributed over [0,1] and the maximum is uniformly independently distributed over [1,2]. This property generalizes: *once we know that n_1 of N arrival times are contained in subinterval $[t_k, t_{k+1})$, where t_k and t_{k+1} are the entry into service times of queued customers k and k+1, respectively, during the congestion period, then the n_1 arrival times are conditionally uniform and independently distributed over $[t_k, t_{k+1})$* (Larson, 1990). These facts allow us to obtain many useful performance characteristics of the queueing system, conditioned on the transactional data.

A simple application of the above observation yields for the pdf of the arrival time A of a randomly queued customer the step-wise decreasing pdf shown in Figure 1. The form of this pdf generalizes to arbitrary N: *the marginal pdf for the arrival time of a random queued customer has a step-wise decreasing pdf over the duration of the congestion period, with each step occurring at an end-of-service time t_i* (Hall, 1992; Larson, 1990).

As a second illustration, the conditional arrival time $X_{(1)}$ of the first queued customer is either the minimum of two uniform independent r.v.'s over [0,1] or simply uniform over [0,1], with the former situation applying only if the experimental outcome is within subsquare (1). Likewise the conditional arrival time $X_{(2)}$ of the second queued customer is either the maximum of two uniform independent r.v.'s over [0,1] or simply uniform over [1,2], the former applying again only within subsquare (1). Recalling that such minimum and maximum r.v.'s have triangular pdf's and combining results appropriately, we immediately obtain the pdf's for the arrival times of the two respective customers, as shown in Figure 2. Finally, assuming a FCFS queueing discipline, the queueing delay for the first queued customer is $1 - X_{(1)}$ and the queueing delay of the second is $2 - X_{(2)}$. The corresponding

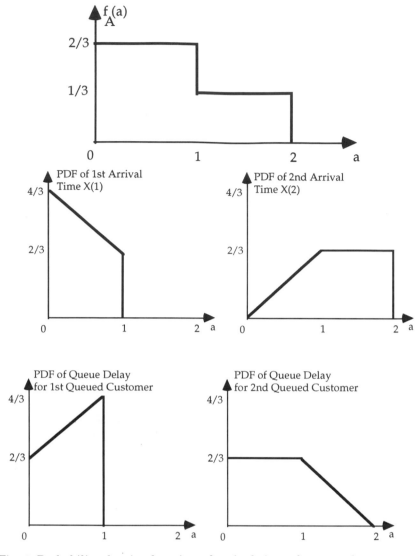

Fig. 1. Probability density function of arrival time of a queued customer.

queue delay pdf's are inverted forms of those in Figure 2, as shown in Figure 3. If a bank knows that you were the second customer in this congestion period, it then has the required information to begin to build your personal pdf for bank queueing; to obtain the monthly pdf, the bank simply has to add together such conditional pdf's for each banking service session that you had during the month.

A General Result in Order Statistics and Application to Queue Inference – Suppose that the service end/start time transactional data are given by the vector $t = \{t_i : i = 1, \ldots, N\}$. In a queue inferencing setting, t_i has two definitions: (1) it is the observed time of departure of the ith departing customer to leave the system during the congestion period; (2) it is also the observed time for the ith customer from the queue to enter service, not necessarily in a FCFS manner. The two sets of individuals comprising the set of arriving customers and the set of departing customers during a congestion period are never identical, and may be disjoint. The number of servers M does not enter into the analysis, nor do any distributional properties of the service times (e.g., there is no requirement for *iid* service times). We do assume that service times are independent of arrival times. For any given congestion period, the QIE computations may occur any time after completion of the congestion period.

Let $X_1, X_2, \ldots, X_{N(1)}$ be an *iid* sequence of r.v.'s with values in [0, 1], where the sequence length $N(1)$ is an independent random integer. We seek a computationally efficient algorithm to calculate the probability of an order statistics vector lying in a given N-rectangle,

$$\Gamma(\underline{s}, \underline{t}) \equiv Pr\{s_1 < X_{(1)} \le t_1, s_2 < X_{(2)} \le t_2, \ldots, s_N < X_{(N)}$$

$$\le t_N | N(1) = N\}, \tag{2}$$

where $\underline{s} \equiv (s_1, s_2, \ldots, s_N)$, $\underline{t} \equiv (t_1, t_2, \ldots, t_N)$ and without loss of generality the sequences $\{s_i\}$ and $\{t_i\}$ are increasing. Using the fact that the N unordered Poisson arrival times during any fixed time interval $(0, T]$ are *iid* and now (for convenience) scaling the congestion period to $(0, 1]$, then in our notation, $\Lambda(0, \underline{t})$ is the a priori probability that the (unobserved) arrival times $X_{(1)}$, $X_{(2)}, \ldots, X_{(N)}$, obey the inequalities $X_{(i)} \leq t_i$ for all $i = 1, 2, \ldots, N$, that is, it is the "master probability" discussed above. That is, "$X_{(i)} \leq t_i$" simply says that the ith arriving queued customer must arrive (and enter the queue) before completion of service of the ith departing customer from service.

If the Poisson arrival process is homogeneous, then the unordered arrival times are *iid* uniform and the rate parameter of the process *does not* enter the analysis. If the arrival process is nonhomogeneous, then the time-dependent arrival rate parameter $\lambda(t)$ must be known up to a positive multiplicative constant for use in computing the cdf $F(x)$, that is,

$$F(x) = \frac{\int_0^x \lambda(t)\, dt}{\int_0^1 \lambda(t)\, dt}, \quad 0 \leq x \leq 1.$$

For simplicity we assume that $F(x)$ is strictly monotone nondecreasing continuous. Jones and Larson (1995) have derived an $O(N^3)$ algorithm for finding $(\underline{s}, \underline{t})$. We next discuss briefly several of its queue inference applications.

The Maximum Experienced Queue Delay – Assume we have a FCFS queue. Suppose we consider a congestion period having N customers with observed departure time vector $\underline{t}$, and we are interested in the maximum time that any of the N customers was delayed in queue, given $\underline{t}$. More precisely, we are interested in the cdf of the maximum of N nonindependent r.v.'s, the in-queue waiting times of the N queued customers, given $\underline{t}$.

Define $D(\tau|\underline{t})$ as the conditional probability that none of the N customers waited τ or more time units, given the observed departure time data. Set $s_i = \max\{t_i - \tau, 0\}$ for all $i = 1, 2, \ldots, N$. Then $\Gamma(\underline{t} - \tau, \underline{t})$ is the a priori probability that the observed departure time inequalities will be obeyed and that no arrival waits τ or more time units in queue. Clearly,

$$D(\tau|\underline{t}) = \Gamma(\underline{t} - \tau, \underline{t})/\Gamma(0, \underline{t}). \quad (3)$$

Maximum Queue Length – Without any assumption regarding queue discipline, suppose we define $\underline{s} = \underline{s}^{*K}$ such that $s_i^{*K} = t_{(i-K)}$ for all $i = 1, 2, \ldots, N; K = 1, 2, \ldots, N$, where a non-positive subscript on $\underline{t}$ implies a value of zero. These

values for $\underline{s}$ imply that each arriving customer i has to arrive after the departure time of departing customer $i - K$ during the congestion period. Now we can compute the conditional probability that the queue length did not exceed K during the congestion period, given $\underline{t}$:

$$\Pr\{Q \leq K|\underline{t}\} = \Pr\{\text{queue length did not exceed } K$$
$$\text{during the congestion period } |$$
$$\text{observed departure time data}\}, \text{ or}$$

$$P(Q \leq K|\underline{t}) = \Gamma(\underline{s}^{*K}, \underline{t})/\Gamma(0, \underline{t}). \quad (4)$$

Probability Distribution of Queue Length – Following the same arguments as in Larson (1990), we can use the $O(N^3)$ computational algorithm to determine for any queue discipline the probability distribution of queue length at departure epochs, and by a balance of flow argument, this distribution is also the queue length distribution experienced by arriving customers.

The Cumulative Distribution of Queue Delay – The algorithm allows computation of points on the conditional in-queue waiting time distribution, given the observed departure data. Again assume we have a FCFS queue. Suppose we define $\beta_j(\tau|\underline{t})$ $\Pr\{j$th customer to arrive during the congestion period waited less than τ time units $|$ observed departure time data}. Then if we set $\underline{s} = \underline{s}^j$, defined so that

$$\begin{cases} s_i^j = 0 & i = 1, 2, \ldots, j-1 \\ s_i^j = \text{Max}\{t_j - \tau, 0\} & i = j, j+1, \ldots, N \end{cases}$$

we can write

$$\beta_j(\tau|\underline{t}) = \Gamma(\underline{s}^j, \underline{t})/\Gamma(0, \underline{t}). \quad (5)$$

This result allows us to determine for any congestion period the probability that a random customer waited less than τ time units, given the observed departure data. We simply compute Eq. (5) for each value of j and average the results. Jones and Larson (1995) have developed a separate algorithm that allows $O(N^3)$ computation of this average probability of queue delay exceeding some threshold.

RESEARCH LITERATURE: Research in queue inferencing is rather extensive. For $O(N^3)$ algorithms for queue performance estimation, see Bertsimas and Servi (1992), Larson (1990), Daley and Sevi (1992, 1993); for personnel queue delay pdf, see Hall (1002); for balking, see Larson (1990), Daley and Servi (1993), Jones (1994). Applications of QIE are discussed in Gawlick (1990), and Chandrs and Jones (1994). QIE concepts have been incorporated into a commercial software product Queue Management System (QMS) and has been used by banks, an airline, and the United States Postal Service.

See **Queueing theory; Retailing.**

References

[1] Bertsimas, D.J. and L.D. Servi (1992), "Deducing Queues from Transactional Data: The Queue Inference Engine Revisited," *Operations Research*, 40(S2), 217–228.

[2] Chandrs, K. and L.K. Jones (1994), "Transactional Data Inference for Telecommunication Models," presentation at First Annual Technical Conference on Telecommunications R & D in Massachusetts, University of Massachusetts, Lowell, Massachusetts.

[3] Daley, D.J. and L.D. Servi (1992), "Exploiting Markov Chains to Infer Queue-Length from Transactional Data," *Jl. Applied Probability*, 29, 713–732.

[4] Daley, D.J. and L.D. Servi (1993), "A Two-Point Markov Chain Boundary-Value Problem," *Adv. Applied Probability*, 25, 607–630.

[5] Gawlick, R. (1990), "Estimating Disperse Network Queues: The Queue Inference Engine," *Computer Communication Review*, 20, 111–118.

[6] Hall, S.A. (1992), "New Directions in Queue Inference for Management Implementations," Ph.D. dissertation in Operations Research, Massachusetts Institute of Technology, available as Technical Report No. 200, Operations Research Center, M.I.T., Cambridge.

[7] Jones, L.K. (1995), "Inferring Balking Behavior and Queue Performance From Transactional Data," submitted to *Management Science*.

[8] Jones, L.K. and Richard C. Larson (1995), "Efficient Computation of Probabilities of Events Described by Order Statistics and Applications to Queue Inference," *INFORMS Journal on Computing*, 7, 89–100.

[9] Larson, R.C. (1990), "The Queue Inference Engine: Deducing Queue Statistics From Transactional Data," *Management Science*, 36, 586-601. Addendum, 37, 1062, 1991.

QUEUEING DISCIPLINE

The rules used to select a next customer to be served. The FCFS discipline chooses the head of the line customer, LIFO chooses the tail of the line, random order chooses the next customer at random, usually from a uniform distribution, etc.

QUEUEING NETWORKS

See **Networks of queues.**

QUEUEING THEORY

Daniel P. Heyman

Bell Communications Research, New Jersey

HISTORY: Queueing theory is the study of service systems with substantial statistical fluctuations in either the arrival or service rates. Other names for the subject are *stochastic service systems* and the *theory of mass storage*. An example of a stochastic service system from everyday life is a line for bank tellers (human or automatic); customers arrive at random, and the transaction lengths will vary depending on the services requested. An example from the world of technology is a computer system; jobs arrive randomly and require different amounts of system resources. An all-too-common source of service-rate variability is a hardware or software crash, which probably occurs randomly even though it might appear that they happen just when you want to use the computer. Looking inside the computer system reveals some more stochastic-service systems. The components (e.g., disk drives, I/O devices, the CPU) have randomly arriving tasks, and the time required to execute a task may be subject to significant statistical fluctuations.

Queueing theory started with the work of A. K. Erlang in 1905. He was designing automatic telephone exchanges and needed to know how many calls might be carried simultaneously. Since the calls start at random times and have random durations, the number of calls in progress fluctuates as a stochastic process. Erlang developed several concepts (e.g., birth-and-death processes and statistical equilibrium) about stochastic processes before the formal mathematical theory of stochastic processes was developed. Most of the first thirty years of queueing theory was done in the context of telephony, and telephony continues to be a major consumer of queueing theory. The creation of operations research during World War II led to other applications, such as capacity evaluation of toll booths and port facilities, the order to assign "stacked" airplanes to runways, and scheduling patients in hospital clinics. Two areas of extensive current activity are the analysis of production systems and computer/communication systems.

BASIC NOTIONS: The paradigm of a queueing model is that there is a facility consisting of some servers, and customers arrive at the facility to receive some sort of service. Upon arrival, the customers will go to a server if one is available;

if all the servers are busy, the customer will either join the *queue* (also called the *waiting room* or *buffer*) or leave. There are two typical reasons that a customer will leave before obtaining service. The queue may be full (in this case we say that the customer is *blocked*), or the customer may be adverse to waiting on line (in this case we say that the customer has *balked*). If the customer leaves, it may depart forever, or retry after some time. Customers who join the queue may wait until a server is free (the alternative is to bolt from the queue, which is called *reneging*), and then one of them enters service; the rule that selects the lucky customer is called the *queue discipline*. Some queue disciplines allow newly arrived customers to displace a customer that is in service, which is called *preemption*. Once at the server, the customer receives the desired service, and then departs. When there is a single facility, we do not keep track of where a departing customer goes. When there are multiple facilities, the model is called a *queueing network*, and we need to specify a *routing rule* that determines where a departing customer goes.

A statistical description of the arrival and service times is almost always given. The objective of the theory is to describe some performance measures, which include the following. The *delay* of a customer is the time spent in the queue waiting to start service; sometimes it is called the queueing time. The *sojourn time* of a customer is the total time spent in the facility; it is sometimes called the *total waiting time* or *waiting time*. The *queue length* is the number of customers in the queue. The *number in the system* is the queue length plus the number of customers in service. A *busy period* is a time interval that starts when all the servers become busy, and ends when at least one server is free. Other performance measures include the number of busy servers, the proportion of blocked customers, and the proportion of non-blocked customers who have a positive delay.

TAXONOMY: D.G. Kendall introduced a compact notation for describing queueing models. A model is described by five parameters, written $A/S/c/K/Q$: A describes the distribution of the times between arrivals, S describes the service time distribution, c is the number of parallel servers, K is the maximum number of customers that can be in queue or in service, and Q is the queue discipline. It is required that $K \geq c$; when $K = \infty$, it is often omitted.

The following symbols are used for both A and S: M for exponential (Markov), D for deterministic, E_k for Erlang k, H_k for hyperexponential of order k, and PH for phase-type. When the service-times have a general distribution, the letter G is used. The symbol G is used for inter-arrival times when they are not necessarily independent; independence is emphasized by using the pair of letters GI.

A common queue discipline is FIFO (First-In First-Out), which is usually taken to be the same as FCFS (First-Come, First-Served). They are identical when there is a single server, but when there are multiple servers, FIFO is stronger than FCFS. This is the default queue discipline, so it is frequently omitted. Two other rules are LIFO (Last-In, First-Out) and SIRO (Service In Random Order). Customers may be partitioned into priority classes, so that more important customers get favored treatment. There are several priority disciplines that have been examined.

There are tacit assumptions that service times are independent and identically distributed (i.i.d.), that service times are independent of interarrival times, that (except for the G case) interarrival times are i.i.d., and that arrivals and services occur one at a time. *Ad hoc* notation is used to represent bulk arrivals or departures, dependencies, and other features.

It is common to denote the mean interarrival time by $1/\lambda$ (λ is the arrival *rate*) and the mean service time by $1/\mu$. Then $a = \lambda/\mu$ is the rate at which work is brought to the system; it is called the *offered load*. The offered load is dimensionless, but it is often expressed in erlangs to honor the contributions of A.K. Erlang. When there are c servers, the load on each server is a/c which is usually denoted by ρ and is called the *traffic intensity*.

GENERAL THEOREMS: Most results in queueing theory are formulas for operating characteristics in particular models. There are some theorems that apply to many queueing models, and two of them will be described here. Before doing so, we need to introduce the notion of *statistical equilibrium*, also called the *steady state*.

Let zero be the time that a queueing system starts operating; for example, for a computer system, it is the time that the installation procedures are completed. Let t be the current time, and let $X(t)$ be the operating characteristic we are modeling at time t, for example, the number in the system t time units after the system started. The initial conditions at time zero usually affect $X(t)$. If there was a backlog of work at time zero, $X(t)$ would be larger than if there were no backlog. The effects of the initial conditions usually decrease as t increases; statistical equilibrium is reached when the effects of the initial conditions have faded away. The mathematical description of this idea starts with defining $p_{ij}(t) = \Pr\{X(t) = j, \text{ given } X(0) = i\}$, and then showing that $p_j = \lim_{t \to \infty} p_{ij}(t)$ exists and is independent of i. Another interpretation of the steady state is that probabilities are not changing with time, that is, that the derivatives of $p_{ij}(t)$ with respect to t are zero.

The steady-state solution $\{p_j\}$ is typically much easier to obtain than the *transient* solution $\{p_{ij}(t)\}$. We would like to interpret p_j as the long-run proportion of time that X is in state j. To express this formally, let $T_j(t)$ be the amount of time $X(s)$ equals j during $(0, t]$. We want to say that

$$p_j = \lim_{t \to \infty} \left[\frac{1}{t} \int_0^t T_j(s) \, ds \right]. \tag{1}$$

Results of this type are called *ergodic theorems*, and some conditions on the model are needed for (1) to be valid. The theories of stationary and regenerative processes often are used to prove ergodic theorems.

Among the general theorems, these two are used most frequently.

Little's Theorem – For any queueing system, or part of a queueing system, let λ be the arrival rate, L be the steady-state mean number of customers present, and W be the steady-state mean waiting time. If λ and W are well-defined, then so is L and $L = \lambda W$.

The use of this theorem is clearly to obviate the need to compute separately both L and W. Three subtler uses of the theorem are the following. It is important to know, and difficult to measure, the average time to get a telephone dial tone (W). It is not so difficult to measure the arrival rate of calls (λ) and the average number of calls waiting to receive dial tone (L). Little's theorem gives an indirect way to estimate the dial tone delay.

In a model with homogeneous servers where all arriving customers are served and the steady-state queue length is finite, suppose we want to calculate the mean number of busy servers; this can be very intricate if done in a straightforward way. By considering the servers as "the system," the arrival rate is the arrival rate of customers λ, the "waiting time" of a customer in the system is the service time, with mean $1/\mu$ say, and the "number in the system" is the number of busy servers. Little's theorem shows that our answer is simply λ/μ, which is the offered load. This result shows that at least λ/μ servers are needed. When the queue discipline is such that all servers are equally used, the traffic intensity is the propor-

tion of time a server is busy. The third application concerns comparisons among queue disciplines. Disciplines that produce the same L as FIFO (LIFO and SIRO are examples) must produce the same value of W. Some disciplines use information about service times and reduce L (compared to FIFO); these must also reduce W.

PASTA is an acronym for Poisson Arrivals See Time Averages. Eq. (1) shows that p_i can be interpreted as a time average, that is, as the proportion of time i customers are present. We say that a customer *sees* the stochastic process $X(t)$ in state i if $X = i$ just before the customer arrives. Let t_n be the arrival epoch of the nth customer. The state seen upon arrival by the nth customer is $X(t_n^-)$, and

$$\pi_i = \lim_{N \to \infty} \left[\frac{1}{N} \sum_{n=1}^{N} X(t_n^-) \right] \tag{2}$$

is the *customer average* for state i. A simple example where $\pi_i \neq p_i$ is a D/D/1 queue where the times between arrivals are one and all the service times are 1/2. Here, $\pi_0 = 1$ yet $p_0 = 1/2$. The PASTA theorem asserts that when the arrivals occur according to a Poisson process, if either π_i or p_i exist, then the other one exists and is equal to it. There is a technical proviso that roughly states that at any time, the future of the arrival process is independent of the past of the X-process. This theorem will be invoked several times in the sequel.

BIRTH-AND-DEATH QUEUES: The simplest stochastic queueing model has exponentially distributed service and interarrival times. Let $X(t)$ be the number of customers present at time t. When $X(t) = i$, the probability that an arrival will occur in $(t, t + h]$ is $\lambda_i h$, and the probability that a service will be completed in the interval is $\mu_i h$. It is implicit that h is small, and terms of order h^2 can be (and are) ignored. The probability of an arrival and a service occurring, or of more than one arrival or service occurring in $(t, t + h]$, is of order h^2. Implicit in this description is the *memoryless property* of the exponential distribution. The λ's are called *birth rates* and the μ's are called *death rates* because of the interpretation of this model as the size of a population.

The parameter μ_0 will have no role in the analysis; it is convenient to set it equal to zero. We insist that $\mu_i > 0$ for $i > 0$ so that the population does not have an *a priori* lower bound. When $\lambda_n = 0$, a population of size $n + 1$ will not occur after the population drops below $n + 1$, so this is a device to model a finite capacity system.

We want to obtain $p_i(t) = \Pr\{X(t) = i\}$, where the $\{p_i(0)\}$ are given initial conditions.

The flow of probability argument: Think of probability as a fluid flowing between buckets numbered $0, 1, 2, \ldots$, and $p_i(t)$ as the amount of probability in bucket i at time t. Think of λ_i as the rate at which each molecule of probability in bucket i flows to bucket $i + 1$, and μ_i is the rate at which each molecule of probability flows to bucket $i - 1$. Then the rate at which probability flows from bucket i to bucket $i + 1$ is $\lambda_i p_i(t)$, and the rate of flow in the reverse direction is $\mu_{i+1} p_{i+1}(t)$. Since the rate-of-change of the contents of a bucket is the inward flow-rate minus the outward flow-rate, we have for $i = 0, 1, 2, \ldots$ that

$$\frac{dp_i(t)}{dt} = \lambda_{i-1} p_{i-1}(t) + \mu_{i+1} p_{i+1}(t) - (\lambda_i + \mu_i) p_i(t) \quad (3)$$

ignoring the $p_{i-1}(t)$ term when $i = 0$. These are called the *backward Kolmogorov equations* of the birth-and-death process.

In the steady-state, the derivatives in (3) equal zero and the probabilities on the right-side equal their steady-state limits, and we obtain the *steady-state balance equations*

$$(\lambda_i + \mu_i) p_i = \mu_{i+1} p_{i+1} + \lambda_{i-1} p_{i-1}, \quad i = 0, 1, 2, \ldots \quad (4)$$

These are currently written as second-order difference equations (because they have $i - 1$, i, $i + 1$); the flow-of-probability argument can be used to make them first-order difference equations. In the steady-state, the flow rate of probability into a state must equal the flow rate out of that state, so that $\lambda_i p_i = \mu_{i+1} p_{i+1}$, $i = 0, 1, 2, \ldots$ The solution of (4) is obtained by iteration; it is

$$p_i = p_0 \frac{\lambda_0 \lambda_1 \ldots \lambda_{i-1}}{\mu_1 \mu_2 \ldots \mu_i}, \quad i = 0, 1, 2, \ldots, \quad (5)$$

where p_0 is chosen to make the sum of all the probabilities equal to one. This can only be done if the sum of the product terms in (5) converges, so some restrictions on the birth and death parameters apply.

The M/M/1 queue – In the M/M/1 queue, the memoryless property of the exponential distribution implies that $\lambda_i = \lambda$ for all i, and $\mu_i = \mu$ for all i, so (5) yields $p_i = p_0 \rho^i$, where $\rho = \lambda/\mu$. The summability condition requires that $\rho < 1$, and then $p_0 = 1 - \rho$ is obtained. (No calculation is required because we proved this via Little's theorem.) The mean of this distribution is $\rho/(1 - \rho)$, and Little's theorem yields $W = 1/[\mu(1 - \rho)]$. The probability that more than k customers are present is ρ^k. The formulas exhibit some of the qualitative features of all stochastic service systems. The operator of the system typically wants to keep the server busy, so the closer ρ is to one the better. However, keeping ρ close to one will produce very long waiting times, which tends to make customers complain.

To obtain the delay distribution, we use the memoryless property of the exponential distribution and PASTA to argue that at arrival epochs, the remaining service time of the customer in service (if any) is exponential. Thus, a customer who arrives to find i customers in the system has a delay that is distributed as the sum of i independent and identically distributed exponential random variables, which is a gamma distribution with shape parameter i. We use PASTA again to interpret p_i as the probability that an arriving customer sees i other customers. Hence, with probability $1 - \rho$ the delay is zero, and with probability ρ the delay is exponentially distributed with mean $1/(\mu - \lambda)$.

The M/M/1/N queue – When at most N customers can be present, set $\lambda_i = 0$ for $i \geq N$. Then (5) yields $p_i = p_0 \rho^i$ and the normalizing condition produces $p_0 = (1 - \rho)/(1 - \rho_{N+1})$ for $\rho \neq 1$. When $\rho = 1$, $p_i = 1/(N + 1)$. The condition $\rho < 1$ is not needed here because the normalizing sum has finitely many terms.

The M/M/c queue – Here $\lambda_i = \lambda$ for all i, $\mu_i = i\mu$ for $i \leq c$ and $\mu_i = c\mu$ for $i > c$. This is called the *Erlang delay* or *Erlang C model*. From (5) we obtain

$$p_i = \begin{cases} \dfrac{p_0 a^i}{i!} & \text{if } 1 \leq i \leq c \\[2mm] \dfrac{p_0 a^i}{c^{i-c} c!} & \text{if } i \geq c \end{cases}$$

where

$$p_0 = \left[\sum_{j=0}^{c-1} \frac{a^j}{j!} + \frac{ca^c}{c!(c-a)} \right]^{-1}, \quad a < c.$$

The probability that all servers are busy is given by

$$C(c, a) = \frac{p_0 ca^c}{c!(c-a)}$$

which is called the Erlang-C formula.

The M/M/c/c queue – This is the same as the M/M/c queue except that $\lambda_i = 0$ for $i \geq c$. It is called *Erlang's loss model* or sometimes the *Erlang B model*. Eq. (5) yields

$$p_i = \frac{a^c/c!}{\displaystyle\sum_{k=0}^{c} a^k/k!}$$

which is a truncation of the Poisson distribution. From PASTA, p_c is the probability that a customer is blocked; it is called *Erlang's loss (or sometimes B) formula* and denoted by $B(c, a)$. A remarkable feature of this formula is that it is valid for any service-time distribution with mean

$1/\mu$. This is an example of an *insensitivity* theorem.

The $M/M/\infty$ queue – This is the previous model with $c = \infty$. It may be an appropriate model for a self-service system. The steady-state probabilities are given by the Poisson distribution with mean a.

The machine-repair (finite-source) queue – This is a model where there are m machines attended by r mechanics. When the times between machine failures are iid and exponential, and so are the repair times, then the number of inoperative machines is a birth-and-death process with $\lambda_i = (m - i)\lambda$ and $\mu_i = \min(i, r)\mu$. Eq. (5) can be used to calculate the steady-state probabilities, which will be denoted by $p_i(m)$ to emphasize the dependence on m. Since the machine failures do not constitute a Poisson process, we cannot conclude that $\Sigma_{i \geq r} p_i(m)$ is the probability that a failed machine has to wait for repair to begin. A surprising feature of this model is that $p_i(m - 1)$ *is the probability that* i other machines are down at a failure epoch.

Balking and reneging – Balking and reneging can be incorporated into the models above by adjusting the birth-and-death rates. Suppose that customers will balk at a queue of length i with probability b_i. To describe this, we replace λ_i with $\lambda_i b_i$. Suppose that when i customers are present, the probability that one of them will renege in a short time interval of length h is $r_i h$. To describe this, we replace μ_i with $\mu_i + r_i$.

Output theorem – Let $\Delta(t)$ be the number of departures in an interval of length t in the steady-state. When $\lambda_i \equiv \lambda$, $\Delta(t)$ is a Poisson process. This result is also known as *Burke's theorem*.

Additional details on these fundamental models are presented in the classic texts of Morse (1958), Cox and Smith (1961), and Prabhu (1965).

MARKOV CHAIN MODELS: The birth-and-death process is the special case of a continuous-time Markov chain in which all transitions are to neighboring states. The added flexibility of the continuous-time Markov chain permits analysis of bulk service and arrival, and some forms of non-exponential service and interarrival times.

A continuous-time Markov chain is described by its rate matrix $Q = (q_{ij})$ where q_{ij} is the rate of making transitions from state i to state j, $i \neq j$. We set $q_{ii} = -\Sigma_{i \neq j} q_{ij}$; it is the rate of making transitions out of state i. The flow of probability argument is valid for continuous-time Markov chains, and the generalization of (5) is that the row vector of steady-state probabilities, $p = (p_0, p_1, \ldots)$ satisfies the matrix equation $pQ = 0$, with the elements of p summing to one.

Erlang distributions – Erlang devised the following way to use exponential distribution arguments for some non-exponentially distributed random variables. For a random variable with mean $1/\lambda$, imagine that it is constructed by adding k iid exponential random variables called *stages*, each having mean $1/k\lambda$. The resulting distribution is called an *Erlang* distribution of order k; it is a gamma distribution with an integer shape parameter, and the density function at t is $k\lambda(k\lambda t)^{k-1} e^{-k\lambda t}/(k-1)!$. The standard deviation is $1/(\lambda\sqrt{k})$, which is less than $1/\lambda$, the standard deviation of the exponential distribution with the same mean.

For the $M/E_k/1$ model we take as our state the number of customers present and the stage of the customer in service (if any). A customer in stage $j < k$ that completes a service stage moves to the next larger numbered stage. A customer that completes stage k actually leaves the server. This is called the method of stages. The balance equations for this model are more intricate than for the $M/M/1$ queue, and solving them requires more work. One result is that the expected delay, $E(D)$ say, is given by

$$E(D) = \frac{k+1}{2k} \frac{\rho}{1-\rho} \frac{1}{\mu},$$

which is the expected delay of the $M/M/1$ model multiplied by $(k+1)/2k$ and turns out to be less than one for $k > 1$.

Extended Erlang family of distributions – Extensions of the method of stages are based on the following distributions. A *hyperexponential* random variable is formed by selecting from among k different exponential distributions according to a probability distribution. Let a_j be the probability of choosing distribution j and $1/\lambda_j$ be the mean of distribution j. Then the density function at t is $\Sigma_1^k a_j\lambda_j e^{-\lambda_j t}$. This produces a larger standard deviation than an exponential distribution with the same mean. This distribution can be used in queueing models in the same way as the Erlang distributions.

Erlang distributions can be pictured as exponential stages in series, and hyperexponential distributions can be pictured as exponential stages in parallel. Replacing these exponential stages by Erlang or hyperexponential distributions yields a broader class of distributions. Repeating this procedure as many times as desired produces the family of *general Erlangian distributions*. General Erlangian distributions can be pictured as directed graphs. The time required to traverse an edge is an exponential random variable. At each node, an edge is traversed; the choice of which edge to take is determined by a chance event that is independent of how the node was reached. The time to go from the source node to the sink node has a generalized Erlang distribution.

The generalized Erlang distribution is the time to absorption in some continuous-time Markov chain where the initial state is fixed. A *phase-type distribution* is the time to absorption of a finite continuous-time Markov chain with a single absorbing state, where the initial state can be chosen at random. This representation expands and unifies the extensions of the exponential distribution described above. The family of PH distributions has properties that can be exploited to obtain algorithms for solving the balance equations when they are used for either the interarrival or service times.

Bulk queues – Heretofore we have assumed that arrivals and services occur one at a time; this need not be valid. A busload of customers may arrive at a ticket counter, or a bus may serve several customers waiting at a bus stop. The batch sizes may be random variables; for example, partially filled buses. In the bus example, it is natural to assume that the number of customers served will be the smaller of the spaces available and the number of waiting customers. This may not always be valid; when there are large setup costs, a minimum number of customers may be required. Similarly, if there is not enough queueing space for an entire arriving batch, sometimes the entire batch is blocked (some communication systems will not accept part of a message) and sometimes part of the batch is blocked.

In the $M/M/1$ queue with batch arrivals, let c_k be the probability that a batch consists of k customers, and let $\bar{c}$ be the mean batch size. The arrival rate is $\lambda \bar{c}$, so $\rho = \lambda \bar{c}/\mu < 1$ is required for the steady-state to exist, and Little's theorem yields $p_0 = 1 - \rho$. The number in system goes from i to $i + k$ at rate λc_k and from i to $i - 1$ at rate μ, so the balance equations are

$$\lambda p_0 = \mu p_1, \quad (\lambda + \mu)p_i = \mu p_{i+1} + \sum_{k=1}^{i} p_{i-k}c_k,$$

$$i = 1, 2, \ldots$$

These equations can be solved for the probability generating function of the $\{p_n\}$ in terms of the probability generating function of the $\{c_k\}$. When $c_k = (1-\alpha)\alpha^{k-1}$, $0 < \alpha < 1$, an explicit solution is

$$p_i = (1-\rho)[\alpha + (1-\alpha)\rho]^{i-1}(1-\alpha)\rho, \quad i > 0.$$

The mean of this probability function is $\rho[(1-\rho)(1-\alpha)]^{-1}$, which is the mean of the $M/M/1$ queue with the same traffic intensity multiplied by the mean batch size. A reason why the performance is worse with batches than without is that batches make the arrival process 'more bursty.' That is, in any interval of time the batch process will tend to have less epochs where arrivals occur, but those epochs will have several customers appearing at once and the arrivals cluster.

NON-MARKOVIAN QUEUES: The exponential interarrival and service times render the queue length processes Markovian; when these conditions are not valid, we look elsewhere for a Markov chain. The models below have *embedded Markov chains* instead, which are obtained by restricting attention to selected times.

The $M/G/1$ queue – Here, the service times have a known distribution $G(\cdot)$ with mean $v_g = 1/\mu$, second moment v_{2g}, variance σ_g^2, and Laplace-Stieltjes transform $G^*(\cdot)$. The mean delay can be obtained without detailed calculations using some general theorems. In the steady-state, let D be the mean delay and Q be the mean queue length. Let R be the mean of the remaining service time of the customer in service (if any) when a customer arrives in the steady-state. Then $D = Qv_g + R$, and Little's theorem asserts that $Q = \lambda D$; solving simultaneously yields $D = R/(1 - \rho)$, where $\rho = \lambda v_g$ is the traffic intensity and is the probability that the server is busy. The next two statements are justified by PASTA. When the server is idle, $R = 0$. When the server is busy, renewal theory can be applied to argue that R is the mean of the *forward-recurrence time* associated with $G(\cdot)$, which is $v_{2g}/2v_g$. Hence, $R = \rho v_{2g}/2v_g$ and $D = \lambda v_{2g}/2(1 - \rho)$; this is the *Pollaczek-Khintchine formula* for the mean delay. It is instructive to write this formula in terms of the squared coefficient of variation $c^2 = \sigma_g^2/v_g^2$,

$$D = \frac{c^2 + 1}{2} \frac{\rho}{1 - \rho} \frac{1}{\mu}. \qquad (6)$$

From this equation it is easily seen that constant service times produce one-half the mean delay of exponential service times.

The waiting-time distribution is obtained by looking at the number in the system only at service-completion epochs. In any queue where arrivals and services occur one at a time, the number of customers that see state i upon arrival differs from the number of customers that leave state i upon departure by at most one, so the steady-state distributions at arrival and departure epochs are equal. The PASTA theorem allows us to conclude that looking at only departure epochs will yield time-average probabilities.

Let X_n be the number present just after the nth departure, and let A_n be the number of arrivals during the nth service time. To make matters easy, assume that the zeroth customer arrives at time zero and sees an empty system. Then $X_{n+1} = \max(0, X_n - 1) + A_n$, which shows that the X_ns form a Markov chain. When $\rho < 1$, this Markov chain has a limiting distribution, $\{\pi_i\}$ say, with probability generating function $\hat{\pi}(\cdot)$. It is not hard to show that

$$\hat{\pi}(z) = \frac{(1-\rho)(1-z)\tilde{G}(\lambda - \lambda z)}{\tilde{G}(\lambda - \lambda z) - z}$$

which is another equation associated with Pollaczek and Khintchine. Let $W(\cdot)$ be the waiting-time distribution when the customers are served FIFO, and $\tilde{W}(\cdot)$ be the Laplace-Stieltjes transform of $W(\cdot)$. Then π_i is the probability that i customers arrive in an interval of time whose length is distributed as $W(\cdot)$, so $\pi_i = \int_0^\infty e^{-\lambda t}[(\lambda t)^i/i!]\,dW(t)$; whence $\hat{\pi}(z) = \tilde{W}(\lambda - \lambda z)$ or

$$\tilde{W}(s) = \frac{(1-\rho)s\tilde{G}(s)}{s - \lambda[1 - \tilde{G}(s)]}.$$

An $M/G/1/Priority$ queue – Let the customers be partitioned into K priority classes. Class i has priority over class j if $i < j$. At a service completion epoch, the next customer to enter service is a member of the class with the lowest priority number among those present. Priority is *non-preemptive*; the customer in service is not replaced when a customer with more priority arrives. The notation is the same as above with the script k denoting class k. The service-time distribution for an arbitrary customer is a mixture of the service-time distributions of the classes, and has mean $1/\mu$ and coefficient of variation c^2. The mean delay of a class j customer is

$$D_j = \frac{c^2 + 1}{2}\frac{\rho}{\left(1 - \sum_{i=1}^{j-1}\rho_i\right)\left(1 - \sum_{i=1}^{j}\rho_i\right)}\frac{1}{\mu},$$

$j = 1, 2, \ldots, K$.

Comparison with the FIFO formula (6) shows that class 1 does better with priorities than without, and class K does worse.

Suppose that the cost of keeping a member of class j waiting in queue per unit time is C_j, and that we can reorder the priorities anyway we like. The way to minimize the waiting costs is to assign priorities in increasing order of C_j/μ_j; this is called the *$C\mu$-rule*. Taking $C_j \equiv 1$ shows that the overall mean delay is minimized when priorities are assigned in increasing order of mean service time. Letting the number of priority classes become infinite, so that customer i has priority over customer j if its service time is shorter, shows that "serve the shortest job first" is the optimum non-preemptive priority rule. When preemption is allowed, the optimum rule is "serve the job with the shortest remaining- processing-time first."

$GI/M/c$ and $GI/G/1$ queues – The $GI/M/c$ queue is analyzed similarly to the $M/G/1$ queue. There is an embedded Markov chain at arrival epochs, and the steady-state probabilities and line delay distribution are found in terms of the real root in $(0, 1)$ of a single, potentially transcendental equation.

For the $GI/G/1/FIFO$ queue, for the nth arriving customer, let D_n be the delay, T_n be the arrival time, $U_n = T_{n+1} - T_n$, and S_n be the service time. The departure time is $T_n + D_n + S_n$, so $D_{n+1} =$ $\max(0, D_n + S_n - U_n)$. When $\rho < 1$, the D_n-process has a proper limit; let $D(\cdot)$ be the limiting distribution. The S's and U's are mutually independent and individually iid, so let $F(t) = P\{S_1 - U_1 \le t\}$. Then $D(t) = \int_{-\infty}^t D(t - x)dF(x)$, which is called *Lindley's equation*. A tractable general solution to this equation is not known; but the equation has proved useful for obtaining qualitative information. For example, it has been shown that the mean delay is no larger than $\lambda(\sigma_U^2 + \sigma_S^2)/[2(1 - \rho)]$, where the σ^2 values are the variances of U_1 and S_1, respectively. When ρ approaches one from below, $D(\cdot)$ approaches an exponential distribution whose mean is the upperbound multiplied by ρ. This is an example of a *heavy-traffic approximation*.

QUEUEING NETWORKS: In the models above, a customer gets service from one server and then departs. In a queueing network, the departures may join another queue. This may be described as a network, where the nodes represent service centers (a queue and some parallel servers) and a directed arc connects service centers i and j if departures from node i can join the queue at node j. New issues that arise include specifying the routing rule and the disposition of customers that attempt to go to a service center where all the waiting positions are occupied.

Feedback queues – The simplest network is a single node where some departures rejoin the queue. This is called a *feedback queue* and it can be used to model rework in a manufacturing context. Placing the items to be reworked at the head of the queue is equivalent to using an expanded service time, but placing them at the tail of the queue resembles an increase in the arrival rate.

Tandem Networks – When the arrivals first appear at node one, then go to nodes $2, 3, \ldots$, and N in order, and then depart, we have a *tandem network*. This is useful for describing repair or assembly operations. When the arrivals are Poisson, the service times are exponential and independent from node to node, and every arrival to a node will be granted a waiting space, Burke's theorem shows that in the steady-state, node i is a birth-and-death queue. Let $p_n(i)$ be the steady-state probability that i customers are present at node n, and let $p(\boldsymbol{i})$, where $\boldsymbol{i} = i_1, i_2, \ldots, i_n$ be the probability that i_n customers are at node $n, n = 1, 2, \ldots, N$; then $p(\boldsymbol{i}) = \Pi_n p_n(i_n)$. This is a *product-form solution*, which greatly simplifies computing the joint distribution and shows that in the steady-state, at each point in time the queue lengths at the various nodes are independent.

Jackson networks – Let r_{ij} be the probability that a customer will go from node i to node j, and assume that this probability is independent of all other routings of this and all other customers. This is called

Markovian routing. The probability that a departure from node i leaves the network is $1 - \Sigma_j r_{ij}$. A network with birth-and-death assumptions for arrival and service times, and Markovian routing, is called a *Jackson network*.

Let α_j be the arrival rate to node j from outside the network, and λ_j be the arrival rate including arrivals from other nodes. The arrival rates are related by the *traffic equations*

$$\lambda_j = \alpha_j + \sum_i \lambda_i r_{ij}.$$

A network is called *open* if some $\alpha_j > 0$. Open networks have been used to model flexible manufacturing systems and communication networks. When the routing matrix (r_{ij}) is irreducible, the traffic equation has a unique solution for open networks when some row sum of (r_{ij}) is less than one. Open Jackson networks have a product-form solution based on birth-and-death queues for the steady-state probability $p(i)$, with arrival rate λ_j used at node j.

A network is called *closed* if every $\alpha_j = 0$ and all of the row sums of (r_{ij}) equal one and a fixed number of customers, say M, circulate among the nodes of the network. Closed networks have been used to model time-shared computer systems. When the system is almost always heavily loaded, the number of jobs is essentially constant, and they sequentially require work from different components (processors, disks, etc.). The traffic equation has infinitely many positive solutions for a closed network; if λ is a solution, then so is $C\lambda$ for any $C > 0$. There is a product-form solution, $p(i) = C\Pi_n p_n(i_n)$, where the $p_n(\cdot)$ are computed from birth-and-death formulas with λ_n taken from some particular solution of the traffic equation. The normalizing constant C must be chosen to make the probabilities sum to one, and it can be a computationally burdensome task. There are $_{N+M-1}C_M$ ways to place M customers in N service centers, which is roughly 4.25×10^{12} for $M = 100$ and $N = 10$.

There have been many texts written on queueing theory over the years. We note the following ones: Cohen (1969), Cooper (1984), Cox and Smith (1961), Gross and Harris (1985), Heyman and Sobel (1982), Kelly (1979), Kleinrock (1975), Mehdi (1991), Morse (1958), Neuts (1981), Prabhu (1965), Takács (1962), Walrand (1988), and Wolff (1989).

RESEARCH TOPICS: Modern ideas in stochastic processes such as marked-point processes, martingale calculus, and Palm probabilities are being used to derive general theorems concerning the existence of limits and relations among averages, in the spirit of Little's theorem and PASTA. Advances in numerical Laplace transform inversion and algorithms for solving the linear equations in Markov chain analyses have been inspired by queueing problems, and have provided numerical solutions to a wide variety of models. Bounds and approximations for currently intractable problems are being explored, particularly for large networks of queues. Many issues in the design and control of queues are under investigation; high-speed telecommunications are a major area of inspiration and application.

See **Birth-death process**; **Little's law**; **Markov chains**; **Markov processes**; **Networks of queues**.

References

[1] Cohen J.W. (1969). *The Single Server Queue*. John Wiley, New York.

[2] Cooper R.B. (1984). *Introduction to Queueing Theory*, 2nd ed., North-Holland, New York.

[3] Cox D.R. and W.L. Smith (1961). *Queues*, Methuen, London.

[4] Gross D. and C.M. Harris (1985). *Fundamentals of Queueing Theory*, 2nd ed., John Wiley, New York.

[5] Heyman D.P. and M.J. Sobel (1982). *Stochastic Models in Operations Research*, vol. 1. McGraw-Hill, New York.

[6] Kelly F.P. (1979). *Reversibility and Stochastic Networks*. John Wiley, New York.

[7] Kleinrock L. (1975). *Queueing Systems, vols. 1 and 2*. John Wiley, New York.

[8] Medhi J. (1991). *Stochastic Models in Queueing Theory*. Academic Press, Boston.

[9] Morse P.M. (1958). *Queues, Inventories and Maintenance*. John Wiley, New York.

[10] Neuts M.F. (1981). *Matrix-Geometric Solutions in Stochastic Models*. Johns Hopkins University Press, Baltimore.

[11] Prabhu N.U. (1965). *Queues and Inventories*. John Wiley, New York.

[12] Takács L. (1962). *Introduction to the Theory of Queues*. Oxford University Press, New York.

[13] Walrand J. (1988). *An Introduction to Queueing Networks*. Prentice Hall, Englewood Cliffs.

[14] Wolff R.W. (1989). *Stochastic Modeling and the Theory of Queues*. Prentice Hall, Englewood Cliffs, New Jersey.

RAC

See **Operations Research Office and Research Analysis Corporation.**

RAIL FREIGHT OPERATIONS

Carl D. Martland

Massachusetts Institute of Technology, Cambridge

In North America, the railroad industry is predominantly privately owned and overwhelmingly oriented toward freight rather than passenger operations. An emphasis on profitability and practical problems has been a characteristic of OR/MS applications in the rail industry. We address freight car utilization, operations planning and control, and line dispatching, the three rail areas that have had the longest history of successful OR/MS applications.

FREIGHT CAR UTILIZATION: There are three main issues in freight car utilization: fleet sizing, allocation of equipment to specific services, and distribution of empty equipment. Each of these issues is complicated by the fact that cars are interchanged among the North American railroads. A complex set of rules has been developed for the use of "foreign" cars owned by another railroad or "private" cars owned by a shipper or a car-supply company. The Freight Car Utilization Research/Demonstration Program (1980), jointly established in 1975 by the Association of American Railroads and the Federal Railroad Administration, conducted a series of studies addressing all facets of freight car utilization. Each study was supervised by an industry task force in order to promote consideration of the most important issues, to provide access to data, and to enhance implementation. Task Force I-2 (1977), for example, developed an integrated set of performance measures for car utilization and showed how these measures could be used in both fleet sizing and fleet allocation decisions.

The process of moving cars from an unloading point to a location where they can be reloaded is called empty car distribution. Railroads typically try to hold enough empty cars at each major yard to cover the expected demand in the surrounding region; extra cars can be sent to locations where additional, or more profitable, loads are required. Southern Railway was a leader in the application of linear programming to car distribution (AAR, 1976). They divided the railroad into 37 zones and created monthly supply and demand estimates for each of 13 car types. They then used a linear program to define "flow rules" that determined where local car distributors should send any extra cars. After Southern merged with Norfolk & Western to form the Norfolk Southern Railroad, this approach to car distribution was expanded to the entire merged system. The program has gone through several revisions and is now run weekly, but the underlying logic is similar to what was originally installed in the late 1960s (Gohring *et al.* 1993). In 1993, the Norfolk Southern used 70 distribution areas and 15 car types; the revised program addressed shortages and surpluses more realistically, and it also provided more flexibility for forecasting supply and demand.

When Philip (1980) surveyed car distribution models, Southern Railway had the most successful intra-road application in the industry. The review by Dejax and Crainic (1987) cited 151 separate studies of the "empty vehicle distribution problem," many of which involved railroads, but the Southern Railway's model and two discontinued models were the only ones identified as having been implemented by a railroad.

Two other LP models have been used to overcome problems in the car service rules, which govern car distribution among railroads. The car service rules generally favor the use of system as opposed to foreign cars, which tends to increase fleet sizes and empty mileage. The "Clearinghouse" encouraged member railroads to use each other's cars as if they were system cars, thereby reducing the flow of empty cars (Task Force I-5 1978). A linear program was run on a weekly basis to determine how best to balance the supply of empties among the member railroads. A study conducted in 1977 showed that the percentage of cars reloaded had risen from 55 to 62%, while the percentage of loaded car days increased from 50.7 to 56.2% and the percentage of loaded miles increased from 60 to 64%. The success of the Clearinghouse led to changes in the car service rules that produced similar benefits for the entire industry.

The Multilevel Reload program (I.C.C. Finance Docket 29653, 1981) targeted a particularly expensive and poorly utilized portion of the fleet, namely the two- or three-decked equipment used to transport new automobiles from assembly plants to distribution centers. Historically, separate fleets were assigned to movements between each assembly plant and distribution center, so that the empty mileage equalled the loaded mileage. In 1979, the Multilevel Reload Program combined all the fleets serving General Motors assembly plants (later expanded to the other major manufacturers) and created an industry group that used an LP to minimize empty movements. This led to an immediate, significant reduction in empty mileage of multilevel cars. By 1981, more than 9,000 cars were managed under this program, and the ratio of empty to loaded miles had dropped from 0.95 at the outset to 0.55 for the GM fleet and 0.84 for the Ford fleet. The program was still in operation at the end of 1993. It is noteworthy that the analytic application in both of these very successful programs was only a small part of major institutional changes.

As suggested by the number of articles cited in the Crainic review, there has been a great deal of methodological research on car distribution modeling over the last 20 years. Much of this research focussed on achieving a better optimization by considering such things as a longer forecast period, variability in demand, or unreliability in travel times (Turnquist and Jordan, 1983). However, as Turnquist and Markowicz (1989) point out, the methodological gains were possible only by making the unrealistic assumption that all vehicles were either identical or completely substitutable. In implementing a car distribution system for CSX Transportation, they therefore used a less advanced network linear program and devoted more attention to the practical aspects of supply and demand.

BLOCKING AND SCHEDULING: The consolidation of individual cars into blocks and the movement of blocks on trains is the essence of railroading. A block is a group of cars that move together from one location to another; a block can be carried by one train or by two or more trains. Blocks can be defined based upon the type of traffic, the destination, the priority of the customer, and many other factors. The operating plan specifies how and where cars are classified into blocks, which trains can carry a block, and which blocks are carried by each train. Unfortunately, it has proven impossible so far to define

an optimization technique that can solve simultaneously for blocking and scheduling for realistic networks. Successful OR applications have therefore focussed on specific aspects of operations planning and paid close attention to the institutional and organizational contexts.

Algorithms have been developed for blocking policy, for assigning blocks to trains, and for scheduling trains. The Automatic Blocking Model and the Train Scheduling System have been widely used in the North American rail industry to determine yard work loads under alternative operating plans (Van Dyke and Davis, 1990).

The operating plan implies a trip plan for cars, where a trip plan is the sequence of train movements required to move the car from its origin, through a series of yards, to its destination. For a typical boxcar movement, the shipment will depart on the first available local train after the car is made available by the shipper. The local train is scheduled to bring the car to a nearby yard. The car is next scheduled to move in a particular block that could move on various outbound trains; the car can be scheduled to the first outbound train whose cut-off is later than the car's scheduled arrival time. This process is repeated until the car has a scheduled arrival time at its destination. The first computerized freight car scheduling system was developed by the Missouri Pacific Railroad (1976). In 1991, the rail industry established a plan to implement inter-line car scheduling as a major element of interline service measurement (Ad Hoc Committee, 1991).

Predicting the time required for a train connection is the most difficult portion of car scheduling and it is also a critical problem in establishing standards for terminal control systems. Many railroads have terminal control systems that include connection standards, which are usually based upon cutoffs. Problems arise with these systems because it is difficult to maintain a coherent system of cutoffs or connection standards. There is also a conceptual problem. Since operating conditions are variable, better predictions of yard times and connection reliability can be obtained by considering the probability of making a connection, which can be called "PMAKE." It is possible to calibrate PMAKE functions that express the probability of making a connection as a function of the time available, the priority of the car, the priority of the inbound and outbound trains, the time of arrival, and other factors (Martland, 1982); PMAKE functions can also be calibrated by convolving yard processing time distributions.

The Service Planning Model uses PMAKE analysis to predict trip times and reliability for a given operating plan and traffic flows (McCarren and Martland, 1980). The SPM has been used by the rail industry to set standards for trip time reliability, to evaluate alternative operating plans, and to evaluate merger possibilities.

Network models have also been widely used in railroad rationalization studies. These models are more concerned with traffic flows and line capacity than with the details of operating plans. In many cases, shortest path algorithms are used to route traffic over various proposed networks, and the results are displayed graphically (Hornung and Kornhauser, 1979).

LINE DISPATCHING: Effective control systems are essential to rail systems. Dispatching is the process of giving trains authority to move along a route, maintaining a safe distance between trains, deciding which sidings to use for meets between trains traveling in opposite directions on a single track railroad, and allowing faster trains to overtake slower trains. In systems where trains routinely run on schedule, train meets and passes are worked out carefully as part of the development of the operating plan. In complex environments, such as is the case in systems with high density passenger operations, it may take a year or more to develop a workable schedule, and various algorithms have been developed to assist in these processes. In these cases, dispatching involves enforcing the plan and responding to emergencies.

In North American operations, train schedules are seldom planned at this level of detail, and departure times vary considerably from one day to the next. As a result, meets and passes are continually different, and the dispatching function is very critical to train performance. Several approaches have been taken to provide support for dispatching. Sauder and Westerman (1983) formulated the dispatching problem as an integer program, which they solved using a branch and bound solution technique on Southern Railway. Their procedure identifies the optimal set of meets and passes for the upcoming 4 hours, which is updated continually and presented as information to the dispatcher. Other systems have used less complex algorithms to plan meets and passes, but implement these plans automatically unless overridden by the dispatcher. It appears that the savings from making routine decisions in a timely manner (e.g., avoiding delays while the dispatcher is on the phone) are at least as large as the savings from making "optimal" decisions.

Models have been used to study line capacity, line scheduling, and dispatching, many of them building upon the work of Petersen and Taylor (1982). Jovanovich and Harker (1991) developed the SCAN system (1991), which was used by the Burlington Northern Railway to evaluate the potential improvements from advanced line control systems (Smith, Patel, Resor, and Kondapalli, 1990).

See **Linear programming; Networks; Scheduling and sequencing.**

References

[1] Ad Hoc Committee to Develop ETA and Trip Plan Capabilities Among Railroads (1991), *A Proposal for Systems to Support Interline Service Management*, Association of American Railroads Report R-776, Washington, DC.

[2] Association of American Railroads (1976), *Manual of Car Utilization Practices and Procedures*, Association of American Railroads Report R-234, Washington, DC.

[3] Dejax P.J. and T.G. Crainic (1987), "A Review of Empty Flows and Fleet Management Models in Freight Transportation," *Transportation Science*, 21, 227–246.

[4] Freight Car Utilization Program (1980), *Catalog of Projects and Publications*, 2nd ed., Association of American Railroads Report R-453, Washington, DC.

[5] Gohring K.W., T.W. Spraker, P.M. Lefstead and A.E. Rarvey (1993), "Norfolk Southern's Empty Freight Car Distribution System Using Goal Programming," presented to ORSA/TIMS Annual Spring Meeting, Chicago, Illinois.

[6] Hornung M.A. and A.L. Kornhauser (1979), "An Analytic Model for Railroad Network Restructuring," Report 70-TR-11, Princeton University, New Jersey.

[7] I.C.C. Finance Docket 20653 (1981), "Application of Certain Common Carriers by Railroad Under 49 U.S.C. Paragraph 11342 for Approval of an Agreement for the Pooling of Car Service" (see especially verified statements of H.H. Bradley, W.E. Leavers, and J.M. Slivka).

[8] Jovanovic D. and P.T. Harker (1991), "Tactical Scheduling of Rail Operations: The SCAN I System," *Transportation Science*, 25, 46–64.

[9] Martland C.D. (1982), "PMAKE Analysis: Predicting Rail Yard Time Distributions Using Probabilistic Train Connection Standards," *Transportation Science*, 16, 476–506.

[10] McCarren J.R. and C.D. Martland (1980), "The MIT Service Planning Model," *MIT Studies in Railroad Operations and Economics*, vol. 31.

[11] Missouri Pacific Railroad (1976), *Missouri Pacific's Computerized Freight Car Scheduling System*, Federal Railroad Administration Report No. FRA-OPPD-76-5, Washington, D.C.

[12] Petersen E.R. and A.J. Taylor (1982), "A Structural Model for Rail Line Simulation and Optimization," *Transportation Science*, 16, 192–205.

[13] Philip C.E. (1980), "Improving Freight Car Distribution Organization Support Systems: A Planned Change Approach," *MIT Studies in Railroad Operations and Economics*, vol. 34.

[14] Sauder R.L. and W.M. Westerman (1983), "Computer Aided Train Dispatching: Decision Support Through Optimization," Norfolk Southern Corporation/Southern Railway Corporation, Atlanta, Georgia.

[15] Smith M., P.K. Patel, R.R. Resor and S. Kondapalli (1989), "Benefits of the Meet/Pass Planning and Energy Management Subsystems of the Advanced Railroad Electronics System (ARES)," *Jl. Transportation Research Forum*, 301–309.

[16] Task Force I-2 (1977), *Freight Car Utilization: Definition, Evaluation and Control*, Association of American Railroads Report R-298, Washington, DC.

[17] Task Force I-5 (1978), *Freight Car Clearinghouse Experiment – Evaluation of the Expanded Clearinghouse*, Association of American Railroads Report R-293, Washington, DC.

[18] Van Dyke C. and L. Davis (1990), "Software Tools for Railway Operations/Service Planning: the Service Planning Model Family of Software," Comprail Conference, Rome, Italy.

RAND CORPORATION

Gene H. Fisher and Warren E. Walker

RAND, Santa Monica, California

BACKGROUND: As World War II was ending, a number of individuals both inside and outside the U.S. government saw the need for retaining the services of scientists for government and military activities after the war's end. They would assist in military planning, with due attention to research and development. Accordingly, Project RAND was established in December 1945 under contract to the Douglas Aircraft Company. The first RAND report was published in May 1946. It dealt with the potential design, performance, and use of man-made satellites. In February 1948, the Chief of Staff of the Air Force approved the evolution of RAND into a nonprofit corporation, independent of the Douglas Company.

On November 1, 1948, the Project RAND contract was formally transferred from the Douglas Company to the RAND Corporation. The Articles of Incorporation set forth RAND's purpose: "To further and promote scientific, educational, and charitable purposes, all for the public welfare and security of the United States of America."

It accomplishes this purpose by performing both classified and unclassified research in programs treating defense, international, and domestic issues. The staff numbers approximately 600 researchers and 500 support persons, with about 36% of the researchers being operations researchers, mathematicians, physical scientists, engineers, and statisticians. For most of its history, RAND's research departments have been discipline based (for example, mathematics, economics, physics, etc.). However, recently the departments were re-constituted around five broad policy areas: defense and technology planning, human capital, international policy, resource management, and social policy.

This article focuses on RAND's contributions to the theory and practice of operations research. However, RAND has also made major theoretical and practical contributions in other areas, including engineering, physics, political science, and the social and behavioral sciences.

THE FIRST TEN YEARS (1948–1957): The first decade saw RAND accomplishments ranging from the beginning of the development of systems analysis, which evolved from the earlier more specific and more narrowly focused operations analyses, to the creation of new methodological concepts and techniques to deal with problems involving many variables and multiple objectives.

Systems analysis may be defined as the systematic examination and comparison of alternative future courses of action in terms of their expected costs, benefits, and risks. The main purpose of system analysis is to provide information to decision makers that will sharpen their intuition and judgment and provide the basis for more informed choices. From the beginning, it was evident that to be successful, systems analysis would require the conception and development of a wide range of methodological tools and techniques. One of the most important sources of these tools and techniques was the emerging discipline of *operations research*.

In the early 1950s, Edwin Paxson led the project that produced a report entitled *Strategic Bombing Systems Analysis*, which is generally regarded as the first major application of the concept of systems analysis, as well as the source of the name for the new methodology. Among other things, the report advocated the use of decoys to help mask bombers from enemy defenders. This study was a catalyst that stimulated the development and rise of a number of analytical methods and techniques. Some of the more important examples are:

- *Game theory*: Mathematics and game theory were prominent subjects in the early research agendas of Project RAND. Lloyd Shapley, J.C.C. McKinsey, Melvin Dresher, Merill Flood, Oliver Gross, Irving Glicksberg, Rufus Isaacs, and Richard Bellman were among the numerous early RAND contributors to this area. John von Neumann, who is often cited as the father of game theory, and Oskar Morgenstern, who linked game theory to economic behavior, were active RAND consultants, as were many others with connections to major universities. Dresher and Flood developed the prisoner's dilemma game, and A.W. Tucker, RAND consultant, gave it its name.

- *Enhanced computer capabilities*: The Paxson project required computer capabilities beyond those available at that time. This stimulated developments that lead to the JOHNNIAC digital computer in 1954. Based on a design by John von Neumann, it was one of the six "Princeton class" stored programming machines and the first operational computer with core memory in the world. In the first on-line time-shared computer system (1960), RAND built the JOHNNIAC Open Shop System (JOSS), one of the first interactive programming languages for individual users.

- *Dynamic Programming*: The Paxson project also demanded the examination of the dynamic programming of key strategic bomber components (for example decoys) in the context of an overall enhanced strategic capability. This, along with the demands of other projects in the early 1950s, provided a significant part of the motivation for the development of the mathematical theory of dynamic programming. Richard Bellman, together with a few collaborators, almost exclusively pioneered the development of this theory. The first RAND report on dynamic programming was published in 1953. Bellman's well-known book (*Dynamic Programming*) followed in 1956, and his book with Stuart Dreyfus (*Applied Dynamic Programming*) was published in 1962.

A second large systems analysis study of this period was a study led by Albert Wohlstetter on the selection and use of strategic air bases. It developed basing and operational options for improving the survivability of SAC forces and helped shift the focus of strategic thinking in the United States toward deterrence based on a secure second-strike force.

Another major effort beginning in the 1950s that led to the development of operations research tools was research on logistics policy issues. RAND's involvement with Air Force logistics stressed the demand for spare parts and the need for logistics policies that could cope with demand uncertainty. Major players in this effort were Stephen Enke, Murray Geisler, James Peterson, Chauncey Bell, Charles Zwick, and Robert Paulson. The key analytical issue here was the examination of alternative policy issues under conditions of strategic uncertainty. Early research used "expected value" analysis. Later, RAND researchers developed and used more sophisticated methods, such as:

- The use of sensitivity analysis to determine what areas of uncertainty really matter in final outcomes;

- Iteration of the analysis across several relevant future scenarios to seek problem solutions that are robust for several of the possible (uncertain) scenarios;

- Given the outcomes of the above, design R&D activities that will (1) reduce key areas of uncertainty, (2) provide hedges against key uncertainties, (3) preserve options for several possible courses of action, any one of which might be used when the future environment becomes less uncertain.

Finally, the first decade witnessed the development of a number of methods and techniques that were useful across a range of RAND projects and elsewhere.

Some important examples are:

- *Problem Solving with the Monte Carlo Techniques*: Although not invented at RAND, the powerful mathematical technique known as the Monte Carlo method received much of its early development at RAND in the course of research on a variety of Air Force and atomic weapon problems. RAND researchers pioneered the use of the method as a component of a digital system simulation.

- *A Million Random Digits with* 100,000 *Normal Deviates*: The tables of random numbers in this 1955 report have become a standard reference in engineering and econometrics textbooks and have been widely used in gaming and simulations that employ Monte Carlo trials. It is RAND's best selling book (RAND, 1955).

- *Approximations for Digital Computers*: This book, by Cecil Hastings and J.P. Wong, Jr., contained function approximations for use in digital computations of all sorts.

- *Systems Development Laboratory*: This laboratory was set up under the leadership of John

Kennedy to help examine how groups of human beings and machines work under stress. The work ultimately led to the formation of the System Development Corporation.

THE SECOND TEN YEARS (1958–1967): This period in RAND's history witnessed the beginning of the evolution of systems analysis into policy analysis. It also witnessed the beginning of research on domestic policy issues.

One of the most important dimensions of change as systems analysis evolved into policy analysis was the *context* of the problem being analyzed. Contexts became broader and richer over time. What was taken as "given" (exogenous to the analysis) before became a variable (endogenous to the analysis) later. For example in the typical systems analysis of the 1950s and early 1960s, many considerations were not taken into account very well, often not at all: for example, political, sociological, psychological, organizational, and distributional effects. Thus, as systems analysis evolved into policy analysis, the boundaries of the problem space expanded. This had important implications for changes in concepts and methods of analysis. For example with respect to models, the demands of the expanded boundaries of the problem space could not be met by merely trying to make models used in policy analysis bigger and more complex. Of equal importance, was the development of sophisticated strategies for the development and use of models.

While the evolution of systems analysis into policy analysis did not progress very far during this period, there are several areas of RAND research that were conducted in broader contexts than were typical of the 1950s. These included Ed Barlow's Strategic Offense Forces Study (SOFS), Bernard Brodie's work on the development of a strategy for deterrence in the "new" age of abundant nuclear weapons and ballistic missiles, Herman Kahn's analysis of civil defense in the event of a nuclear war, and Charles Hitch and Roland McKean's book *Economics of Defense in the Nuclear Age*, which espoused the view that the economic use of scarce resources should be a critical aspect of defense planning. This view was adopted by Secretary of Defense Robert McNamara and led to RAND's involvement in the development of the defense Planning, Programming, and Budgeting System (PPBS).

In addition to policy strides like those discussed above, the second ten years witnessed further development of methodological tools for quantitative analysis – primarily operations research tools. Major advances were made at RAND in the areas of mathematical programming, queueing theory, computer simulation, stochastic processes, and operational gaming.

Linear programming was probably RAND's most important and most extensive contribution to the theory and practice of operations research as well as to economic decision making. Between 1947 and 1952, George Dantzig and others who worked in the Pentagon on the Air Force's Project SCOOP developed the simplex method and other basic features of LP. Dantzig moved to RAND in 1952. During the following decade, RAND was the world's center of LP developments. In addition to methodological developments by Dantzig and other RAND employees and consultants (e.g., the dual simplex algorithm), there was seminal work on classic problems like production planning and the traveling salesman problem. In addition, most of the pioneering programming of LP algorithms (e.g., the first code for the revised simplex method) was carried out by William Orchard-Hays and others at RAND. Much of the work of this period is captured in Dantzig's book, *Linear Programming and Extensions*, published in 1963 (which includes an extensive bibliography).

Seminal work in other areas of mathematical programming also took place at RAND during the 1950s and 1960s. Ralph Gomory developed the first integer programming algorithms; Philip Wolfe, George Dantzig, and Harry Markowitz initiated work on quadratic programming; and George Dantzig, and Albert Madansky initiated work on stochastic programming.

Three other examples of RAND work in the "tools" area during this period are worthy of note:

- *Simulation.* In the early 1960s, after doing complex simulation modeling "the hard way," Harry Markowitz and Herb Karr developed SIMSCRIPT, a programming language for implementing discrete event simulation models. This work led in 1968 to SIMSCRIPT II, which is still widely-used.

- *Artificial intelligence (AI).* The man-machine partnerships explored in the Systems Research Laboratory gained new impetus as Allen Newell, Herb Simon, and Cliff Shaw began to construct a general problem solving language that employed symbolic (non-numerical) processes to simulate human thinking on a computer. One of their initial efforts to carry out a "theory of thinking" involved programming computers to play chess. On a broader scale, this research resulted in several information processing languages (e.g., IPL V), which were similar to LISP and were used in some of the early AI computer work.

- *Flows in networks.* In 1962, Lester Ford, Jr. and Delbert R. Fulkerson, RAND mathematicians, published the first unified treatment of methods for dealing with a variety of problems that have formulation in terms of single commodity flows in capacity-constrained networks. Their book, entitled *Flows in Networks*, introduced concepts (e.g., "max-flow/min-cut") and algorithms (e.g., "out-of-kilter") that have been used to treat network problems ever since.

The year 1964 saw the publication of the first of several RAND books by mathematician Edward Quade, who played a major role in developing and disseminating the methodology of systems analysis and (later) policy analysis. *Analysis for Military Decisions* documents an intensive five-day course that RAND offered to military officers and civilian decisionmakers in 1955 and 1959.

THE THIRD TEN YEARS (1968–1977): This period in RAND's history saw an acceleration of many of the trends begun in the previous ten years. One of these trends involved the development of improved procedures for the use of expert judgment as an aid to military decision making. The Delphi procedures grew out of this effort. These procedures incorporate anonymous response, iteration and controlled feedback, and statistical group response to elicit and refine group judgments where exact knowledge is unabilable.

Other trends involved the continued evolution of systems analysis into policy analysis and an increasing emphasis on analyzing major domestic research issues. Important in the last two trends was the establishment of the New York City-RAND Institute (NYCRI). Important RAND research efforts during 1968–1977 include the work performed at the NYCRI and policy analysis studies for the government of the Netherlands.

The New York City-RAND Institute – In 1968, RAND began a long-term relationship with the City of New York to tackle problems in welfare, health services, housing, fire protection, law enforcement, and water resources. The NYCRI was formally established in 1969. The research staff evaluated job training programs, suggested solutions to shortages of nurses in municipal hospitals, helped change rent control, altered fire department management policies, reallocated police manpower, and helped improve Jamaica Bay's water quality.

The most successful of the NYCRI's projects was the one devoted to improving the operations and deployment of the Fire Department of New York. In 1968, the major problem facing the Department was the rising alarm rate. Its increasing workload was not significantly relieved by adding more men and equipment; nor were traditional methods of fire company allocation, dispatching, and relocation working. The Institute's studies altered the way the Department managed and deployed its men and equipment and operated its dispatching system. An integral part of the research involved creation of a wide variety of computer models to analyze and evaluate deployment, which led to the formulation of new policies. Warren Walker and Peter Kolesar were awarded ORSA's 1974 Lanchester Prize for a paper that described how mathematical programming methods were applied to the problem of relocating available fire companies to firehouses vacated temporarily by companies fighting fires. The entire body of work from this project is documented in Walker, Chaiken, and Ignall (1979).

Policy Analysis Studies for the Dutch Government – Reflecting an increasing interest in doing policy analysis studies in international contexts, RAND started working for the Dutch government in the 1970s. One important study was concerned with protecting an estuary from floods. In April 1975, RAND began a joint research venture with the Dutch government to compare the consequences of three alternative approaches for protecting the Oosterschelde, the largest Dutch estuary, from flooding. Seven categories of consequences were considered for each alternative: financial costs, ecology, fishing, shipping, recreation, national economy, and regional effects. Within each category, several types of consequences were considered. The study required the development of sophisticated computer models of estuaries and coastal seas. In June 1976, the Dutch Parliament adopted one of the alternatives based in large part on the results of the RAND study: to build a 10 km, multi-billion dollar storm surge barrier with large movable gates across the mouth of the estuary.

A second study was focused on improving water management in the Netherlands. Begun in April 1977, the Policy Analysis for the Water Management of the Netherlands (PAWN) project was conducted jointly by RAND, the Dutch Government, and the Delft Hydraulics Laboratory. It analyzed the entire Dutch water management system and provided a basis for a new national water management policy for the country. It developed a methodology for assessing the multiple consequences of possible policies, and applied it to generate alternative policies and to assess and compare their consequences. Considering both research and documentation, it directly involved over 125 person-years of effort. The project won a Franz Edelman Award for Management Science Achievement in 1984.

THE FOURTH TEN YEARS (1978–1988): This period witnessed a number of major institutional milestones. Some examples:

- In 1982 the joint RAND/UCLA Center for Health Policy Study was funded by the Pew Memorial Trust. A year later RAND and UCLA established a joint Center for the Study of Soviet International Behavior.
- In 1984 a new Federally Funded Research and Development Center – the National Defense Research Institute (NDRI) – was established, funded by the Office of the Secretary of Defense.
- The Arroyo Center, the Army's Federally Funded Research and Development Center for studies and analysis, was established at RAND in 1984.
- The Center for Policy Research in Health Care Financing, sponsored by the Department of Health and Human Services, was created in 1984.

These institutional developments helped RAND to enhance its work in existing areas of the research program; for example, Health Policy – and to stimulate work in new areas – for example, analysis of Army policy issues.

During this period, RAND's research program increased substantially in size and diversity. Many of the trends of the past continued; for example, the increase in efforts devoted to domestic policy research and the tendency to conduct research in broader contexts. Several new trends began to emerge; for example, an increase in emphasis on research done in international contexts other than the (then) USSR. The development of analytical concepts, methods, and techniques also continued. Some of the more important of these were:

- *RAND Strategy Assessment System (RSAS):* Because of perceived limitations in methods of strategic analysis, in 1982 RAND began to develop methods for strategic analysis that combined classical gaming, systems analysis methods and techniques, artificial intelligence, and advanced computer technology. The RSAS provides a structure and tools for analyzing strategic decisions at the national command level as well as decisions at the

operational level. It also provides great flexibility in choosing which roles are to be played by people and which by machines.

- *Dyna-METRIC:* The Dyna-METRIC logistics support model provided a major new tool for relating the availability of spare parts to wartime aircraft sortie-generation capability. The model, which was developed by Richard Hillestad and Irving Cohen, combines elements of queueing theory, inventory theory, and simulation. It is now an integral part of the Air Force logistics and readiness management system.
- *CLOUT (Coupling Logistics to Operations to Meet Uncertainties and the Threat)* is a RAND-developed set of initiatives for improving the ability of the Air Force logistics system to cope with uncertainties and disruptions of a conventional war overseas. The CLOUT initiatives are intended to offset the substantial variability expected in the demand for spare parts, maintenance, and other support activities, as well as the consequences of damage to theater air bases.
- *The Enlisted Force Management System (EFMS):* The EFMS project is notable for the scope and complexity of the decision support system that it developed, and for demonstrating how the tools of operations research can be married with emerging information technologies to provide real-time decision support throughout an organization. Warren Walker led a large RAND team that worked with Air Force counterparts beginning in the early 1980s. Together, they produced an organizational decision support system (ODSS) to help make decisions about the grade structure of the enlisted force, enlisted promotion policies, and the recruitment, assignment, training, compensation, separation, and retirement of Air Force enlisted personnel. Since 1990, the EFMS has been the primary analytical tool used to support major policy decisions affecting the enlisted force. The success of the system motivated the publication of Carter *et al.* (1992).

CURRENT SITUATION: After the first 40 years, most of the main trends outlined above continue to play themselves out; for example, domestic research represents nearly 30% of RAND's $100 million research budget, and methodological enhancements driven by the practical needs of the research continue to have high priority. In 1992, RAND established the European-American Center for Policy Analysis (EAC) in the Netherlands. Major research efforts so far have included studies of the safety of Schiphol Airport, ways of improving river dikes in the Netherlands while preserving the environment, and a systematic examination of alternative strategies for reducing the negative effects of road freight transport in the Netherlands.

See **Artificial intelligence; Cost analysis; Delphi method; Dynamic programming; Emergency services; Game theory; Gaming; Inventory modeling; Linear programming; Logistics; Military operations research; Networks; Nonlinear programming; Optimization; Public policy analysis; Simulation; Systems analysis; Traveling salesman problem.**

References

[1] Bellman E. Richard (1957), *Dynamic Programming*, Princeton University Press, Princeton, New Jersey.

[2] Bellman E. Richard and Stuart E. Dreyfus (1962), *Applied Dynamic Programming*, Princeton University Press, Princeton, New Jersey.

[3] Carter Grace M., Michael P. Murray, Robert G. Walker and Warren E. Walker (1992), *Building Organizational Decision Support Systems*, Academic Press, Inc., San Diego.

[4] Dalkey N.C., D.L. Rourke, R.J. Lewis and D. Snyder (1972), *Studies in the Quality of Life: Delphi and Decision-making*, D.C. Heath and Company, Lexington, Massachusetts.

[5] Dantzig George (1963), *Linear Programming and Extensions*, Princeton University Press, Princeton, New Jersey.

[6] Ford Lester R. Jr. and Donald R. Fulkerson (1962), *Flows in Networks*, Princeton University Press, Princeton, New Jersey.

[7] Goeller Bruce F., *et al.* (1983), *Policy Analysis of Water Management for the Netherlands: Vol. 1, Summary Report*, R-2500/1-NETH, The RAND Corporation, Santa Monica, California.

[8] Hitch Charles J. and Roland McKean (1960), *The Economics of Defense in the Nuclear Age*, Harvard University Press, Cambridge, Massachusetts.

[9] Kahn Herman (1960), *On Thermonuclear War*, Princeton University Press, Princeton, New Jersey.

[10] Markowitz Harry M., Bernard Hausner, and Herbert W. Karr (1963), *SIMSCRIPT: A Simulation Programming Language*, Prentice Hall, Inc., Englewood Cliffs, New Jersey.

[11] Newell Allen ed. (1961), *Information Processing Language-V Manual*, Prentice-Hall, Inc., Englewood Cliffs, New Jersey.

[12] Paxson Edwin (1950), *Strategic Bombing Systems Analysis*.

[13] Quade E.S. (1964), *Analysis for Military Decisions*, Rand McNally & Co., Chicago.

[14] RAND (1955), *A Million Random Digits with 100,000 Normal Deviates*, The Free Press, Glencoe, Illinois.

[15] Shaw J.C. (1964), *JOSS: A Designer's View of an Experimental On-line Computing System*, P-2922, The RAND Corporation, Santa Monica, California.

[16] Sherbrooke Craig C. (1966), *METRIC: A Multi-Echelon Technique for Recoverable Item Control*, RM-5078-PR, The RAND Corporation, Santa Monica, California.

[17] Walker Warren E., Jan M. Chaiken and Edward J. Ignall, eds. (1979), *Fire Department Deployment Analysis: A Public Policy Analysis Case Study*, Elsevier North Holland, Inc., New York.

[18] Williams John D. (1954), *The Compleat Strategyst: Being a Primer on the Theory of Games of Strategy*, McGraw-Hill Book Company, Inc., New York.

[19] Wohlstetter A.J., F.S. Hoffman, R.J. Lutz and H.S. Rowen (1954), *Selection and Use of Strategic Air Bases*, R-266, The RAND Corporation, Santa Monica, California.

RANDOM FIELD

A stochastic process with a multi-dimensional index set; for example, $\{R(x,y), -\infty < x,y < \infty\}$, where $R(x, y)$ equals the amount of rain falling during a given day at location (x,y).

RANDOM NUMBER GENERATORS

Pierre L'Ecuyer

Université de Montréal, Québec, Canada

INTRODUCTION: Several algorithms and heuristics in operations research require a source of random numbers. Such numbers are needed, for example, for Monte Carlo integration, stochastic discrete-event simulation, and probabilistic algorithms (like genetic algorithms or simulated annealing). In practice, typically, the so-called "random numbers" are produced by a deterministic computer program, and are therefore not random at all. The aim of such a program, called a *random number generator*, is to produce a sequence of values which "look" as if they were a typical sample of i.i.d. (independent and identically distributed) random variables, say from the U(0, 1) distribution (the uniform distribution between 0 and 1). Some generators may also produce random integers, or random bits, etc. Since the sequence produced is really deterministic, it is often called a *pseudorandom sequence* and the generator producing it is sometimes called a *pseudorandom number generator*. Here, we adopt the well-accepted practice of just using the word *random*. Of course, random variables are often required from non-uniform probability distributions, like the normal, exponential, Poisson, and so on. In practice, such random variables are typically generated by transforming uniform random numbers in some appropriate way (Bratley *et al.*, 1987; Devroye, 1986), as we discuss below.

Physical devices for producing random numbers (e.g., picking balls from a box or using specialized "noisy" electrical circuits) are practically never used in combination with computers, mainly because they are not convenient, and because it is not clear that the numbers produced by such methods are independent and really follow the uniform distribution.

There is a well developed body of theory concerning the construction and analysis of (pseudo)random number generators (Knuth, 1981; L'Ecuyer, 1990; L'Ecuyer, 1994; Niederreiter, 1992). But, unfortunately, unreliable and "dangerous" generators still abound in the scientific literature and on computer systems. For many practical applications, mediocre generators may nevertheless produce useful results, but care must be taken, because when things start to go wrong, they can quickly become disastrous. Examples of specific well-tested and recommended generators, with computer codes, can be found in L'Ecuyer *et al.* (1993), L'Ecuyer *et al.* (1991), and Tezuka and L'Ecuyer (1991).

A *random number generator* (RNG) can be defined as a structure $\mathscr{F} = (S, s_0, T, U, G)$, where S is a finite set of *states*, $s_0 \in S$ is the *initial state*, $T: S \to S$ is the *transition function*, U is a finite set of *output* symbols, and $G: S \to U$ is the *output function*. The generator starts from state s_0 (called the *seed*), evolves according to $s_i := T(s_{i-1})$, and at each step i, outputs the *observation* $u_i := G(s_i)$. One would expect the observations to behave from the outside as if they were the values of i.i.d. random variables, uniformly distributed over U. The set U is often a set of integers of the form $\{0, \ldots, m-1\}$, or a finite set of values between 0 and 1 to approximate the $U(0, 1)$ distribution. In what follows, we will assume the latter. Since S is finite, the sequence of states is ultimately periodic. The *period* is the smallest positive integer ρ such that for some integer $\tau \geq 0$ and for all $n \geq \tau$, $s_{\rho+n} = s_n$. The smallest τ with this property is called the *transient*. When $\tau = 0$, the sequence is said to be *purely periodic*.

To introduce some real randomness, one can choose s_0 randomly, say by drawing balls from a box. Generating a truly random seed is much less work and is more reasonable than generating a long sequence of truly random numbers. An RNG with a random seed can be viewed as an *extensor* of randomness, whose purpose is to save "coin tosses." It stretches a short truly random seed into a long sequence of values that is supposed to appear and behave like a true random sequence.

Constructing a RNG should not be done at random; it requires a lot of care and a certain amount of theoretical support. Knowing the period length is not enough, even if it is astronomical. One must also analyze theoretically how the points (vectors of successive or non-successive values) produced by the generator are distributed in multidimensional spaces. For example, one can look at the set of all t-dimensional vectors of successive observations

$$\Omega_t = \{\mathbf{u}_n = (u_n, \ldots, u_{n+t-1}); \quad \tau \leq n \leq \tau + \rho - 1\},$$

over the full period of the generator, for a given t. Common practice is to demand that these points are very evenly distributed in the t-dimensional unit hypercube $[0, 1]^t$. Of course, points that are too uniformly distributed do not look random and fail to imitate i.i.d. uniform variates as well as points whose distribution is too far from even. So, why demand that the set Ω_t have such a "superuniform" distribution?

The rationale is that Ω_t should be viewed as a sample space, from which points are taken at random by the generator, *without replacement*. One can imagine a large box containing all the points of Ω_t. If those points are very evenly distributed over $[0, 1]^t$, then a good way to generate an i.i.d. sample from (approximately) the uniform distribution over $[0, 1]$ would be to pick up points randomly from this box, *with replacement*, and use all the components of each vector. If the generator's period is so long (i.e., the cardinality of Ω_t is so huge) that only a tiny (negligible) fraction of it can be used in practice, then we can further assume (or hope) that selecting points from Ω_t with or without replacement should make no practical difference. When the set of points produced by the generator is viewed as a random sample taken from Ω_t, then it makes sense to have Ω_t as evenly distributed as possible over $[0, 1]^t$. This argument also suggests that RNGs should have *huge* periods, many orders of magnitude larger than whatever can be exhausted in practice.

Good equidistribution over the whole period improves our confidence in good statistical behavior over the fraction of the period that we use. That is usually complemented by additional empirical statistical tests. However, those empirical tests do not always discriminate easily between good and mediocre generators. So, the right approach is to select a generator first on the basis of its theoretical properties, and then submit it to appropriate empirical tests. There is no limit on the number of tests that can be designed and implemented. Several "standard" tests are described in Knuth (1981) and Marsaglia (1985). Ideally, the tests should be selected in relation with the target application. So, before using a "general purpose" generator, provided by a software package or library, it is wise to submit it to additional "specialized" empirical testing. Of course, for any RNG whose output sequence is periodic, it is possible to build a statistical test that the generator will fail miserably, if enough time is allowed. So, the idea of applying statistical tests to such a RNG may look ridiculous. However, from a pragmatic point of view, people usually feel good if the RNG passes a certain set of statistical tests which can be run in "reasonable" time.

Other relevant criteria for the choice of a general purpose RNG include speed, memory requirement, repeatability, portability, ease of implementation, and availability of jumping ahead and splitting facilities. There are simulation applications (e.g., in particle physics) which require billions of random numbers and for which the generator's speed will always remain a critical factor, regardless of the available computing power. Memory utilization could become important when many "virtual" generators (i.e., many substreams) must be maintained in parallel. This is required, for example, for proper implementation of certain variance reduction techniques (Bratley *et al.*, 1987; L'Ecuyer and Côté, 1991; L'Ecuyer, 1994). *Portability* means that the generator can be implemented efficiently in a standard high-level language, to produce exactly the same sequence (at least up to machine accuracy) with all "standard" compilers and on all "reasonable" computers. Being able to reproduce the same sequence of random numbers on a given computer or on different computers (called *repeatability*) is important for program verification and for variance reduction purposes (Bratley *et al.*, 1987; Ripley, 1990). Repeatability is a major advantage of pseudorandom sequences with respect to sequences generated by physical devices. Of course, for the latter, one could store an extremely long sequence on a large disk or tape, and reuse it as needed thereafter. But this is not as convenient as a good pseudorandom number generator which stands in a few lines of code. *Jumping ahead* means the ability to quickly compute, given the current state s_n, the state s_{n+v} for any large v (without generating all the intermediate states). This is useful for breaking up the sequence into long disjoint substreams and jumping ahead quickly from one substream to the other. The package given in L'Ecuyer and Côté (1991) implements "virtual" parallel generators using such facilities.

LINEAR RECURRING SEQUENCES: Most of the currently used random number generators are based on linear recurrences of the form

$$x_n = (a_1 x_{n-1} + \cdots + a_k x_{n-k}) \bmod m, \tag{1}$$

where m is a positive integer called the *modulus*, $a_1, \ldots, a_k$ are integers between $-m$ and m called the *multipliers* (with $a_k \neq 0$), and k is the order of the recurrence. We can define the *state* of the recurrence at step n as $s_n = (x_n, \ldots, x_{n+k-1}) \in \mathbf{Z}_m^k$. Taking into account that the all-zero state is absorbing and must be avoided, one can see that the maximal possible period length is $m^k - 1$. That maximal (or full) period is in fact achieved when m is prime and the characteristic polynomial of the recurrence,

$$P(z) = z^k - a_1 z^{k-1} - \cdots - a_k$$

is a primitive polynomial modulo m. Such primitive polynomials are easy to find once the factorizations of $m-1$ and $(m^k-1)/(m-1)$ are available (Knuth, 1981; L'Ecuyer, 1990; L'Ecuyer *et al.*, 1993). For a full-period recurrence, in any subsequence of ρ consecutive values of s_n, each element of $\{0, \ldots, m-1\}^k$ appears once and only once, except for $s_n = (0, \ldots, 0)$.

One way of producing the output sequence is to transform s_n into an observation u_n at each step of the recurrence. For example, a value in $[0, 1)$ can be obtained by taking $u_n = x_n/m$, in which case the generator is called a *multiple recursive generator* (MRG) (L'Ecuyer *et al.*, 1993; L'Ecuyer, 1994; Niederreiter, 1992). A special case is when $k = 1$, where we obtain the well-known and popular *multiplicative linear congruential generator* (MLCG) (Bratley *et al.*, 1987; Knuth, 1981).

In terms of practical implementation considerations, certain non-prime moduli are much more attractive than the prime ones. Some can be implemented efficiently via combined generators (L'Ecuyer and Tezuka, 1991; Tezuka and L'Ecuyer, 1991; Wang and Compagner, 1993), while for the important special case where m is a power of two, the "mod m" operation can be performed trivially by just "chopping-off" the high-order bits, without caring for overflow. However, there is a price to pay in terms of period length and statistical robustness. For example, for a MLCG with power-of-two modulus, the largest possible period is only $m/4$, reached, for example, when $a \bmod 8 = 5$ and x_0 is odd (Knuth, 1981). It is in fact possible to attain a period of length m by using the more general form:

$$x_n = (ax_{n-1} + c) \bmod m, \tag{2}$$

where c is a constant (Knuth, 1981). But in any case, power-of-two moduli have many other drawbacks and should probably be discarded altogether. For example, if $m = 2^e$, the period of the ith least significant bit of x_n is at most 2^i and the pairs (x_n, x_{n+i}), for $i = 2^{e-d}$, lie in at most $\max(2, 2^{d-1})$ parallel lines. Techniques for implementing (1) for prime m are discussed in Bratley *et al.* (1987), L'Ecuyer *et al.* (1993), and L'Ecuyer and Côté (1991).

A slightly more general way of producing the output is to use, say, s terms of the recurrence (1) at each stage:

$$u_n = \sum_{j=1}^{L} x_{ns+j-1} m^{-j}, \tag{3}$$

where s and $L \le k$ are positive integers. In that case, we redefine the generator's state at step n as $s_n = (x_{ns}, \ldots, x_{ns+k-1})$, and the sequence $\{u_n\}$ is called a *digital multistep sequence* (L'Ecuyer, 1994; Neiderreiter, 1992). If (1) has period $\rho = m^k - 1$ and s is coprime with ρ, then the sequence (3) also has period $\rho = m^k - 1$. Using such a digital expansion yields a better resolution than just using $u_n = x_n/m$, and permits one to take smaller values of m. As explained in Couture *et al.* (1993), L'Ecuyer (1994), and Tezuka and L'Ecuyer (1991), the sequence (3) can be obtained equivalently from an MLCG in a space of formal Laurent series. This is useful for analyzing some of the structural properties of the sequence when m is small.

An important special case of (3) is when $m = 2$: each u_n is constructed by taking blocks of L successive bits from the binary sequence (1), with spacings of $s - L \ge 0$ bits between the blocks. This is known as a *Tausworthe* generator (Knuth, 1981; Neiderreiter, 1992; Tezuka and L'Ecuyer, 1991). Implementation issues are discussed in Bratley *et al.* (1987), L'Ecuyer (1994), and Tezuka and L'Ecuyer (1991).

One may want to consider an MLCG in matrix form:

$$X_n = (A X_{n-1}) \bmod m \tag{4}$$

where A is a $k \times k$ matrix and

$$X_n = \begin{pmatrix} x_{n,1} \\ \vdots \\ x_{n,k} \end{pmatrix}$$

is a k-dimensional column vector. Then, it turns out that $\{X_n\}$ follows the recurrence

$$X_n = (a_1 X_{n-1} + \cdots + a_k X_{n-k}) \bmod m. \tag{5}$$

This gives k copies of the same recurrence (1) evolving in parallel, perhaps with different lags between the copies. Implementing (5) directly instead of (4) takes k times as much memory, but may lead to much quicker implementations (the generator's state becomes $s_n = (X_n, \ldots, X_{n+k-1})$. We may call (5) the *parallel MRG* implementation of the matrix generator (4). If we define $y_n = \sum_{j=1}^{k} b_j x_{n,j} \bmod m$, where $b_1, \ldots, b_k$ are any integers, then the sequence $\{y_n\}$ again obeys the recurrence (1), which means that no further generality is achieved by (4). However, other ways of combining the elements of X_n might lead to a different recurrence than (1). One of them is the so-called *matrix MLCG*, which uses each component X_n to produce k uniform variates per step; another one is the *digital matrix MLCG*, which uses the digital method on each X_n to produce one uniform variate per step (L'Ecuyer, 1994). For the latter, if we assume that the lags between the k successive copies of the MRG in (5) are all the same (say, equal to d) and that $\gcd(d, \rho) = 1$, we have:

$$u_n = \sum_{j=1}^{L} x_{n,j} m^{-j} = \sum_{j=1}^{L} x_{ns+j-1} m^{-j}, \tag{6}$$

where $L \le k$, $x_{n,j} = x_{n+(j-1)d,1}$, $x_i \stackrel{\text{def}}{=} x_{id,1}$, and s is the inverse of d modulo $\rho = m^k - 1$. So, we are back to (3), with the proviso that the coefficients in (1) are changed in accordance with the redefinition of x_i. In other words, a digital matrix MLCG gives an alternative way of implementing (3). In that case, only the first L components of the vectors X_n in (5) need to be computed and maintained in memory.

An important special case of the digital matrix MLCG is the Generalized Feedback Shift Register (GFSR) generator, with $m = 2$ (Fushimi and Tezuka, 1983; Neiderreiter, 1992). Each X_n is then a vector of bits, obtained by making a bitwise exclusive-or (XOR) of the X_{n-j}'s for which $a_j \ne 0$. From the discussion in the preceding paragraph, it follows that the GFSR (with uniformly spaced lags) can be viewed as a (speed-wise) efficient way of implementing a Tausworthe generator. Several instances of GFSR generators based on a primitive trinomial of the form $P(z) = z^k - z^{k-r} - 1$ have been proposed in the literature, typically with $L = 32$

(Fushimi and Tezuka, 1983). In that case, (5) becomes

$$X_n = X_{n-r} \text{ XOR } X_{n-k}, \qquad (7)$$

which really gives a speedy generator. However, the excessive simplicity of this recurrence leads to major statistical defects (L'Ecuyer, 1994; Wang and Compagner, 1993). As argued in Compagner (1991), Tezuka and L'Ecuyer (1991), and Wang and Compagner (1993), it seems that the right way to go is to have a characteristic polynomial with *many* (e.g., approximately $k/2$) non-zero coefficients. This can be implemented efficiently by combining Tausworthe generators. The XOR operation in (7) can also be replaced by some other operation like $+$, $-$, or modulo m, where m is typically a power of two. This more general form is often called a *lagged-Fibonacci* generator (Marsaglia, 1985).

Recently, some efforts have been deployed to design generators with extremely long periods, while keeping a fast and very simple generating rule. Notorious examples include the *twisted GFSR* and *add-with-carry/subtract-with-borrow* generators, of which subsequent analyses have unveiled major defects (L'Ecuyer, 1994; Tezuka *et al.*, 1993). In general, keeping a very simple generating rule appears to conflict with statistical robustness.

LATTICE STRUCTURE: An important characteristic of MRGs is their lattice structure, which means that all t-dimensional vectors $\boldsymbol{u}_n = (u_n, \ldots, u_{n+t-1})$, for $n \geq 0$, lie in a relatively small number of equidistant parallel hyperplanes (Knuth, 1981; L'Ecuyer, 1990). The shorter the distance d_t between those hyperplanes, the better, because this means thinner empty slices of space, and so a more evenly distributed set Ω_t. Computer programs now exist for computing d_t in reasonably large dimensions, up to around 50 or more. Good generators should have a small value of d_t for all t up to some large constant. Specific examples are given in Knuth (1981), L'Ecuyer (1990), L'Ecuyer *et al.* (1993), and L'Ecuyer and Tezuka (1991).

Tausworthe and GFSR generators also have a lattice structure in the space of formal series, and that structure can also be used to analyze the geometrical distribution of the points Ω_t. Let $m = 2$ and partition the unit hypercube $[0, 1]^t$ into $2^{t\ell}$ cubic cells of equal size. If each cell contains the same number of points of Ω_t (with the possible exception of the zero vector), we say that the sequence is (t, ℓ)-equidistributed. Since $\Omega_t \cup \{0\}$ has cardinality 2^k, this is possible only for $\ell \leq \lfloor k/t \rfloor$. When the sequence is $(t, \lfloor k/t \rfloor)$-equidistributed for $t = 1, \ldots, k$, we say that it is *maximally equidistributed*, or *asymptotically random*. The equidistribution properties of Tausworthe generators can be analyzed via their lattice structure (Couture *et al.*, 1993; Tezuka and L'Ecuyer, 1991). Generators which are almost maximally equidistributed are suggested in Tezuka and L'Ecuyer (1991).

DISCREPANCY AND LOW-DISCREPANCY SEQUENCES: Consider a set of Nt-dimensional points $\boldsymbol{u}_n = (u_n, \ldots, u_{n+t-1})$, $0 \leq n \leq N-1$. For any hyper-rectangular box of the form $R = \Pi_{j=1}^t [\alpha_j, \beta_j)$, with $0 \leq \alpha_j < \beta_j \leq 1$, let $I(R)$ be the number of points $\boldsymbol{u}_n$ falling into R, and $V(R) = \Pi_{j=1}^t (\beta_j - \alpha_j)$ be the volume of R. If $\mathbb{R}$ is the set of all such regions R, then

$$D_N^{(t)} = \max_{R \in \mathbb{R}} |V(R) - I(R)/N|$$

is called the t-dimensional (*extreme*) *discrepancy* for the set of points $\boldsymbol{u}_0, \ldots, \boldsymbol{u}_{N-1}$. If we impose $\alpha_j = 0$ for all j, then we obtain a variant called the *star discrepancy*.

According to the law of the iterated logarithm, the discrepancy of a "truly random" sequence should be approximately in $O(N^{-1/2})$ (Neiderreiter, 1992). So, ideally, if true randomness is to be imitated, the set of points that are used during a simulation should have a discrepancy approximately in that order. A too high discrepancy indicates a non-uniform distribution, while a too low discrepancy indicates that the points are too evenly distributed. As we previously indicated, we might seek a very low discrepancy for N equal to the period length of the generator, and use only a negligible fraction of the period during the simulation, hoping that the discrepancy will be in the right order over that portion. Neiderreiter (1992) gave general discrepancy bounds for several classes of generators, mostly for $N = \rho$. However, no efficient algorithm is available for computing the discrepancy exactly, except for a few special cases.

The notion of discrepancy is also useful to obtain error bounds in Monte Carlo numerical integration. In that context, the smaller the discrepancy, the better, because bounds on the numerical error are given in terms of the discrepancy and the aim is to minimize that error, not really to imitate true randomness. Specific *low-discrepancy* sequences, also called *quasi-random* sequences, have been designed for that purpose. They are infinite sequences for which the discrepancy is low for all N (Neiderreiter, 1992).

COMBINED GENERATORS: Combining different generators has been often advocated for increasing the period and (hopefully) improving the statistical properties. Indeed, most of the fast and simple generators, like MLCGs and Tausworthe generators based on primitive trinomials, do not have enough statistical robustness for the computationally intensive simulations which are performed on modern computers. The structures of the combined generators are not always well understood, but empirical results tend to support combination (L'Ecuyer, 1994; Marsaglia, 1985). Furthermore, some classes of combined generators turn out to be equivalent, or approximately equivalent, to MLCGs with large non-prime moduli or to Tausworthe generators with large-degree reducible characteristic polynomials. This is what happens if one adds the states of two or more MLCGs with different prime moduli, running in parallel, modulo some integer m_1, or if one adds the output values of those MLCGs modulo 1 (L'Ecuyer and Tezuka, 1991). Similarly, performing a bitwise exclusive-or of the output values of two or more Tausworthe generators is equivalent to using a Tausworthe generator whose characteristic polynomial is the product of the characteristic polynomials of the individual generators (Tezuka and L'Ecuyer, 1991; Wang and Compagner, 1993). As a result, these combined generators can be viewed as efficient ways of implementing MLCGs with large moduli or Tausworthe generators based on characteristic polynomials with many non-zero coefficients. Currently, those methods appear to give the most reliable sources of pseudorandom numbers among the methods that are fast enough for most current simulation applications.

NONLINEAR GENERATORS:

Many believe that the structure of linear sequences is too regular and that the right way to go is *nonlinear* (Eichenauer-Herrmann, 1992; Neiderreiter, 1992). One can introduce nonlinearity by either (a) using a linear-type generator but transforming the state nonlinearly to produce the output, or (b) constructing a generator with a nonlinear transition function T. One example of (a) is the inversive generator, which uses (1) but then takes the inverse of x_n modulo m (discarding the zeros) before dividing by m to produce the output (Eichenauer-Herrmann, 1992). Another instance of nonlinear generator is the BBS generator, proposed by Blum, Blum, and Shub (1986) for cryptographic applications. It evolves according to the recurrence $x_n = x_{n-1}^2 \bmod m$, where m is the product of two (distinct) k-bit primes, both congruent to 3 modulo 4, and $\gcd(x_0, m) = 1$. At each step, the generator outputs the last v bits of x_n, where v is in the order of $\log(k)$. Under the reasonable assumption that factoring is hard, and that m and x_0 are chosen somewhat "randomly," it has been proved that no polynomial-time (in k) statistical test can distinguish (in some specific sense) a BBS generator from a truly random one. This means that for large enough k, the generator should behave very nicely from a statistical point of view.

A common property of nonlinear generators is that they do not produce a lattice structure like the linear ones. They also behave very much like truly random sequences with respect to discrepancy (Neiderreiter, 1992). However, specific well-tested parameter values with fast implementations are currently not available. Software implementations of the BBS and inversive generators are too slow for general simulation applications.

NON-UNIFORM RANDOM NUMBERS:

The standard approach for generating random variables from non-uniform distributions is to apply further transformation to the output u_n of the uniform generator. This is easily done for some distributions, but not so for others. The Devroye (1986) and Bratley *et al.* (1987) books give in-depth coverage of the most useful methods. For some distributions, compromises must be made between simplicity of the algorithm, quality of the approximation, robustness (with respect to parameter changes), and efficiency (speed and memory requirements). Generally speaking, simplicity should not be sacrificed for small speed gains.

Inversion: Conceptually, the simplest way of generating a random variable X from distribution F is to use *inversion*: let $X = F^{-1}(U) = \min\{x | F(x) \geq U\}$, where U follows the uniform distribution between 0 and 1. With that definition of X, one has $\Pr\{X \leq x\} = \Pr\{F^{-1}(U) \leq x\} = \Pr\{U \leq F(x)\} = F(x)$, and so X has distribution F. The use of this method requires that F^{-1} (or a good approximation of it) be available.

For a specific example, if X has the Weibull distribution with parameters α and β, then F has the form $F(x) = 1 - \exp[-(x/\beta)^\alpha]$ for $x > 0$, and $F^{-1}(U)\beta[-\ln(1-U)]^{1/\alpha}$, so X is easy to generate. As another example, let X be geometric with parameter p. Then $F(x) = 1 - (1-p)^x$ for $x = 1, 2, \ldots$ and $F^{-1}(U) = 1 + \lfloor \ln(1-U)/\ln(1-p) \rfloor$.

For some distributions, however, F^{-1} cannot be written in closed form, but good numerical approximations of it are often available. For example, functional approximations of F^{-1} for the normal, Student's t and chi-squared distributions (among others), with FORTRAN codes, are given in Bratley *et al.* (1987). For general discrete distributions with finite support, the distribution function can be stored in a table and inversion performed by either linear search, binary search, or index search with "buckets." When the support is infinite (e.g., the Poisson distribution), then one can store a truncated table and compute the remaining values only as needed (which would typically be rare).

In most simulation applications, inversion should be the method of choice, because it gives a *monotone* transformation of U into X, which makes it most compatible with major variance reduction techniques like antithetic variates and common random numbers (Bratley *et al.*, 1987). But there also exist situations where *speed* is the real issue and where monotonicity is no real concern. Then, non-inversion methods might be appropriate.

Non-inversion methods: Suppose that X has a discrete distribution over the finite set $\{x_1, \ldots, x_n\}$, with $p_i = \Pr\{X = x_i\}$ for each i. Then, a very fast way of generating X is given by the following *alias method*. This method is not monotone, but generates random variables in time $O(1)$ per call, after a table setup which takes time in $O(N)$. Consider the histogram of the distribution, where each index i has an associated rectangle of width 1 and height p_i. Roughly, the idea is to "level" the histogram so that each index i has probability $1/N$, by cutting-off rectangle pieces and transfering them to different indices. This is done in such a way that in the new histogram, each rectangle i contains one piece of size q_i (say) from the original rectangle i and one piece of size $1/N - q_i$ from some other rectangle whose index j, denoted $A(i)$, is called the *alias* value of i. The setup procedure initializes two tables A and R where $A(i)$ is the alias value of i and $R(i) = (i-1)/N + q_i$. Then, to generate X, take a $U(0, 1)$ variate U, let $i = \lceil N \cdot U \rceil$, and return $X = x_i$ if $U < R(i)$; $X = x_{A(i)}$ otherwise. There also exists a version of the alias method for continuous distributions, which is called the *acceptance-complement* method (Bratley *et al.*, 1987; Devroye, 1986).

Suppose now that you want to generate X from a complicated density f, and that f can be majorized by some simpler function t (i.e., $f(x) \le t(x)$ for all x), so that generating variables from the density r defined by $r(x) = t(x)/a$, where $a = \int_{-\infty}^{\infty} t(s) \, ds$, is easy (note that r is just the rescaling of t into a density). The random variable X can be generated by repeating: generate Y from the density r and an independent $U(0, 1)$ variate U, until $U \le f(Y)/t(Y)$. Then, return $X = Y$. This is called the *acceptance/rejection*

method (Bratley *et al.*, 1987; Devroye, 1986). The number of turns into the "repeat" loop is random; it is actually a geometric random variable with parameter $1/a$. So, the expected number of turns into the loop is $1/a$, which can be minimized by having a as close to 1 as possible. This means that the dominating density t should squeeze f as closely as possible. However, in practice, there is usually a compromise between bringing a close to 1 and keeping t simple.

A variant of the acceptance/rejection method, called *thinning*, is often used for generating events from a non-homogeneous Poisson process. Suppose that the process has rate $\lambda(t)$ at time t, with $\lambda(t) \le \bar{\lambda}$ for all t, where $\bar{\lambda}$ is a finite constant. One can generate Poisson pseudo-arrivals at constant rate $\bar{\lambda}$ by generating inter-arrival times as i.i.d. exponentials of mean $1/\bar{\lambda}$. Then, any pseudo-arrival at time t is accepted (becomes an arrival) with probability $\lambda(t)/\bar{\lambda}$ (i.e., if $U \le \lambda(t)/\bar{\lambda}$, *where* U is an independent $U(0, 1)$), and rejected with probability $1 - \lambda(t)/\bar{\lambda}$ Note that non-homogeneous Poisson processes can also be generated by inversion (Bratley *et al.*, 1987).

Suppose that F is a convex combination of several distributions: $F(x) = \Sigma_j p_j F_j(x)$, or more generally $F(x) = \int F_y(x) \, dH(y)$. To generate from F, one can generate $J = j$ with probability p_j (or Y from H), then generate X from F_j (or F_Y). This is called the *composition algorithm*. This method is useful for generating from "compound" distributions like the hyperexponential or from compound Poisson processes. It is also often used, as follows, to design specialized algorithms to generate from complicated densities (like the normal density). The idea is to partition the area under the complicated density into pieces, where piece j has surface p_j. To generate, first select a piece (choose piece j with probability p_j, using perhaps the alias method), then draw a random point uniformly over the surface of that piece. The trick is to make the partition in such a way that large pieces will be fast and easy to generate from. Then, X will be returned very quickly most of the time.

A dual method to composition is the *convolution method*, which can be used when X can be written as $X = Y_1 + Y_2 + \cdots + Y_n$, where the Y_i are independent with specified distributions. With that method, one just generates the Y_i and sums up. This requires at least n uniforms. Examples of random variables which can be expressed as sums like that include the hypoexponential, Erlang, and binomial distributions, as well as many others. Besides the general approaches that we have briefly discussed, several specialized and fancy techniques have been designed for different commonly used distributions like the Poisson, normal, and so on. Further details are given in Bratley *et al.* (1987) and Devroye (1986).

See **Monte Carlo sampling and variance reduction; Simulation of discrete-event stochastic systems.**

References

[1] L. Blum, M. Blum and M. Shub (1986), "A Simple Unpredictable Pseudo-Random Number Generator," *SIAM Jl. Comput.* **15**, 364–383.

[2] P. Bratley, B.L. Fox and L.E. Schrage (1987), *A Guide to Simulation*, second edition. Springer-Verlag, New York.

[3] A. Compagner (1991), "Definitions of Randomness." *Amer. Jl. Physics* **59**, 700–705.

[4] R. Couture, P. L'Ecuyer and S. Tezuka (1993), "On the Distribution of k-Dimensional Vectors for Simple and Combined Tausworthe Sequences," *Math. of Computation* **60**, 749–761 & S11–S16.

[5] L. Devroye (1986), *Non-Uniform Random Variate Generation*, Springer-Verlag, New York.

[6] J. Eichenauer-Herrmann (1992), "Inversive Congruential Pseudorandom Numbers: a Tutorial," *International Statist. Revs.* **60,** 167–176.

[7] M. Fushimi and S. Tezuka (1983), "The k-Distribution of Generalized Feedback Shift Register Pseudorandom Numbers," *Communications of the ACM* **26**, 516–523.

[8] D.E. Knuth (1981), *The Art of Computer Programming: Seminumerical Algorithms*, vol. 2, second edition. Addison-Wesley, Reading, Massachusetts.

[9] P. L'Ecuyer (1990), "Random Numbers for Simulation," *Communications of the ACM* **33**, 85–97.

[10] P. L'Ecuyer (1994), "Uniform Random Number Generation," to appear in *Annals of Operations Research*.

[11] P. L'Ecuyer, F. Blouin and R. Couture (1993), "A Search for Good Multiple Recursive Random Number Generators," *ACM Trans. Modeling and Computer Simulation* **3**, 87–98.

[12] P. L'Ecuyer and S. Côté (1991), "Implementing a Random Number Package with Splitting Facilities," *ACM Trans. Math. Software* **17**, 98–111.

[13] P. L'Ecuyer and S. Tezuka (1991), "Structural Properties for Two Classes of Combined Random Number Generators," *Math. of Computation* **57**, 735–746.

[14] G. Marsaglia (1985), "A Current View of Random Number Generation," *Computer Science and Statistics, Proceedings of the Sixteenth Symposium on the Interface*, Elsevier/ North Holland, 3–10.

[15] H. Niederreiter (1992), *Random Number Generation and Quasi-Monte Carlo Methods*, SIAM CBMS-NSF Regional Conference Series in Applied Mathematics, vol. 63, SIAM, Philadelphia.

[16] B.D. Ripley (1990), "Thoughts on Pseudorandom Number Generators," *Jl. Computational and Applied Math.* **31**, 153–163.

[17] S. Tezuka and P. L'Ecuyer (1991), "Efficient and Portable Combined Tausworthe Random Number Generators." *ACM Trans. Modeling and Computer Simulation* **1**, 99–112.

[18] S. Tezuka, P. L'Ecuyer, and R. Couture (1993), "On the Lattice Structure of the Add-with-Carry and Subtract-with-Borrow Random Number Generators," *ACM Trans. Modeling and Computer Simulation*, to appear.

[19] D. Wang and A. Compagner (1993), "On the Use of Reducible Polynomials as Random Number Generators," *Math. of Computation* **60**, 363–374.

RANDOM VARIATES

Values created algebraically, typically on a computer using uniform-(0,1) pseudo random numbers, to appear as if they were random observations following a particular probability distribution. A basic procedure for doing this is the inverse transformation method or probability integral transformation. Given the pseudorandom number z on the unit interval and the cumulative distribution function (CDF) $F(x)$ of the random variable X, a random variate x can (theoretically) be generated by the inverse transformation method as

$$x = F^{-1}(z).$$

As long as the CDF is invertible, this approach can be used for discrete as well as continuous probability distributions and even for empirical distributions. For those problems in which the CDF F will not be analytically or numerically invertible, methods other than the inverse transformation are employed, including acceptance-rejection, convolution, and composition. See **Monte Carlo sampling and variance reduction; Probability integral transformation method; Random number generators; Simulation of discrete-event stochastic systems.**

RANDOM WALK

If $S_n = X_1 + X_2 + \ldots + X_n$, then S_n is a special discrete-time Markov process called a random walk if $S_0 = 0$ and the random variables $\{X_i\}$ are independent and identically distributed. The most common form of the random walk is the discrete one in which $X_i = -1$ or $+1$. See **Markov chains; Markov processes**.

RANGING

A term equivalent to a full sensitivity analysis of an optimal solution to a linear programming problem. Ranging refers to how much the cost coefficients and the right-hand-side elements of the linear program can vary before the optimal feasible basis is no longer optimal or feasible. A ranging analysis could also include variations of a technological coefficient, but it is not standard practice. A full cost and right-hand-side ranging analysis is part of computer-based simplex method solutions. See **Linear programming; Sensitivity analysis; Simplex method**.

RANK

The rank of an $m \times n$ matrix A is the maximum number of linearly independent columns in A. The rank of A equals the rank of its transpose A^T, with the rank not greater than m or n. See **Matrices and matrix algebra**.

RATE MATRIX

The matrix of infinitesimal rates that determine how transitions occur in a birth and death stochastic process. See **Birth-death process; Markov chains; Markov processes**.

RAY

A ray is a collection of points $(x_0 + \lambda d)$, where d is a nonzero vector and $\lambda \geq 0$. The vector d is called the direction of the ray and x_0 the origin of the ray.

R-CHART

A quality control chart that shows the variations in the ranges of samples. See **Quality control; $\bar{X}$-bar chart**.

R&D

See **Research and Development**.

READINESS

See **Availability**.

REASONING

A problem-solving process. Two paradigms are logical and analogical reasoning. Logical reasoning includes deductive and inductive. Deductive reasoning is arriving at a conclusion from premises and rules of inference. Inductive reasoning is forming a general conclusion that explains multiple observations. Analogical reasoning uses analogy of a current situation to familiar ones from previous experiences. One paradigm for analogical reasoning is a neural network. See **Artificial intelligence; Expert systems; Neural networks**.

REASONING KNOWLEDGE

Knowledge about what circumstances allow particular conclusions to be considered to be valid. See **Artificial intelligence; Expert systems**.

RECOGNITION PROBLEM

A computational problem whose answer is "yes" or "no." For example, "given a graph G, is there an Euler tour?" See **Computational complexity**.

RECOURSE LINEAR PROGRAM

See **Stochastic programming**.

REDUCED COSTS

See **Prices**.

REDUCED GRADIENT METHODS

See **Nonlinear programming; Quadratic programming**.

REDUNDANCY

Igor Ushakov

SOTAS, Rockville, Maryland

Redundancy is an engineering method of improving system and equipment reliability. Mainly, redundancy consists in using extra units (subsystems, modules and/or additional elements) within the system. Redundant units are usually considered in two main regimes: (1) loaded ones where a redundant unit has the same failure rate, $\lambda(t)$, as a corresponding main unit of the system; (2) unloaded (spare) ones where redundant units cannot fail at all, so that $\lambda(t) = 0$ for all t before switching on. There also exists an intermediate case where redundant units are in an underloaded regime where $\lambda(t)$ differs from 0 but is lower than for a main unit. Usually, their relative loading factor is not known with much certainty.

Redundancy might be individual or common. In the first case, a group of n redundant units supports a single main unit. In the second case, a group of n redundant units is predestined to support a group of k main units. The latter case is typical for spare units. The possibility of renewal or repair of main and redundant units plays a dominant role in reliability improvement.

Though it improves system reliability, redundancy requires extra resources and money. Cost-effectiveness analysis of redundancy is considered as the problem of *optimal redundancy*.

Consider a system consisting of n independent units. For simplicity, assume that the considered system is series, that is, failure of any main unit of the system leads to total system failure. To increase the reliability of the system, one uses redundant units in the following way. Let unit i of the system have x_i redundant units and write the probability of successful operation (PSO) of this group as $R_i(x_i)$.

For independent groups of units, the system PSO can be written as

$$R(\boldsymbol{X}) = \prod_{i=1}^{n} R_i(x_i).$$

where $\boldsymbol{X} = (x_1, \ldots, x_n)$. At the same time, introducing x_i redundant units leads to the expenditure of $C(x_i)$ cost units. Usually, one assumes that $C(x_i) = c_i x_i$. In this case, the system total cost equals

$$C(\boldsymbol{X}) = \sum_{i=1}^{n} c_i x_i$$

Then the optimal redundancy problem consists of solving one of the following problems: find the vector solution $\boldsymbol{X}^0 = (x_1^0, \ldots, x_n^0)$ which delivers either

$$\max_{\boldsymbol{X}} \left\{ \prod_{i=1}^{n} R_i(x_i) \middle| \sum_{i=1}^{n} c_i x_i \leq C^0 \right\} \qquad (1)$$

or

$$\min_{\boldsymbol{X}} \left\{ \sum_{i=1}^{n} c_i x_i \middle| \prod_{i=1}^{n} R_i(x_i) \geq R^0 \right\}. \qquad (2)$$

These problems are conditional discrete optimization ones which can be solved by means of some standard tools as steepest descent, branch and bound, dynamic programming, and integer programming. Notice that both goal functions, $R(X)$ and $C(X)$, are concave. One of the best ways to solve is by use of Kettelle's Algorithm, which represents a convenient computational modification of dynamic programming (Kettelle, 1962).

Cases of multi-constraint versions of problem (1) and its solution are considered in Barlow and Proschan (1981) and Ushakov (1994). The optimal redundancy problem for multi-functional systems, an important extension of problem (2), is solved in Ushakov (1994) as well.

See **Cost analysis; Cost effectiveness analysis; Reliability of systems.**

References

[1] Barlow R.E. and F. Proschan (1981). *Statistical Theory of Reliability and Life Testing: Probability Models*, 2nd ed. To Begin With, Silver Spring, Maryland.

[2] Kettelle J.D. Jr. (1962). "Least-Cost Investment of Reliability Investment," *Operations Research*, 10, 249–265.

[3] Kozlov B.A. and I.A. Ushakov (1970). *Reliability Handbook*. Holt, Rinehart and Winston, New York.

[4] Ushakov I.A. ed. (1994). *Handbook of Reliability Engineering*. John Wiley, New York.

REDUNDANT CONSTRAINT

An inequality or equation of a mathematical programming problem that does not define part of the solution space. An equivalent problem can be formed by removing redundant constraints.

REGENERATION POINTS

Suppose that there exists a time epoch T_1 in a stochastic process such that the continuation of the process beyond T_1 is a probabilistic replica of the process starting at time 0. The existence of subsequent times $T_2, T_3, \ldots$ having the same properties follows by repetition, and the set $\{T_1, T_2, T_3, \ldots\}$ are said to be regeneration points of the process. See **Renewal processes.**

REGRESSION ANALYSIS

Irwin Greenberg

George Mason University, Fairfax, Virginia

INTRODUCTION: In almost all fields of study, the researcher is frequently faced with the problem of trying to describe the relation between a response variable and a set of one or more input variables. Given data on input (predictor, independent) variables labeled $x_1, x_2, \ldots, x_p$ and the associated response (output, dependent) variable y, the objective is to determine an equation relating output to input. The reasons for developing such an equation include the following:

1. to predict the response from a given set of inputs;
2. to determine the effect of an input on the response; and
3. to confirm, refute, or suggest theoretical or empirical relations.

To illustrate, the simplest situation is that of a single input for which a linear relation is assumed. Thus, if the relation is exact, it is given for appropriate values of β_0 and β_1 by

$$y = \beta_0 + \beta_1 x. \qquad (1)$$

The determination of β_0 and β_1 in this case is easy, requiring only two distinct pairs of observations (x_1, y_1) and (x_2, y_2).

In general, the problem is more complex in that the response is not given exactly by (1). This may be true because, although the relation is theoretically given by (1), the observations are not measured without error. Alternatively, there may be no theoretical justification for an exact linear relation but it is used as an approximation.

A model, commonly used in both cases, is

$$y = \beta_0 + \beta_1 x + e. \qquad (2)$$

Here e denotes the measurement error or other random fluctuations in y which cause the response to depart from (1); it is assumed that the input variables are either specified by the user or measured without error.

The appropriate analysis of (2) is dictated by the assumptions made on the distribution of errors. Typically, it is assumed that the errors have mean zero and variance σ^2 and that the errors associated with distinct observations are uncorrelated. That is, if a very large number of pairs (x_i, y_i) were observed for a situation modeled by (2), then (a) the errors

$$e_i = y_i - \beta_0 - \beta_1 x_1; \qquad (3)$$

would average to zero; (b) the error associated with one observation would in no way influence any other error; and (c) the mean of the squares of the errors would be σ^2.

Based on n pairs of observations (x_i, y_i), $i = 1, \ldots, n$, the objective of the analyst is to estimate β_0, β_1, and σ^2 and to make inferences about these parameters. In addition, it may be desirable to indicate the precision of a prediction obtained for a given input when the estimates b_0 and b_1 of β_0 and β_1 are used in (1). These inferences require further specification of the distribution of the errors. The classical results are developed assuming a Gaussian (or normal) distribution.

A generalization of this simple model is the multiple, linear regression model

$$y = \beta_0 + \beta_1 x_1 + \beta_2 x_2 + \cdots + \beta_p x_p + e. \qquad (4)$$

Here the assumption on the errors is the same as given above and the analysis is to be based on n $(p+1)$-tuples $(x_{1i}, x_{2i}, \ldots, x_{pi}, y_i)$, $i = 1, \ldots, n$. The sense in which (4) is a linear model must be emphasized. As written, the average response in (2) is a linear function of x and in (4) is a linear (planar) function of $x_1, \ldots, x_p$, but this is not the essential linearity. The critical feature is that the average response is a linear function of the coefficients $\beta_0, \beta_1, \ldots, \beta_p$. The variables indicated by y and x_i, $i = 1, \ldots, p$ may represent *functions* of the variables which are actually observed as long as these functions do not depend on unknown parameters. For example,

$$\log z = \beta_0 + \beta_1/w + e; \qquad (5)$$

the model does not represent z as a linear function of w, but by letting $y = \log z$ and $x = 1/w$ this model is seen to be equivalent to (2). Similarly, the polynomial model

$$y = \beta_0 + \beta_1 x + \beta_2 x^2 + e \qquad (6)$$

is a special case of (4) with $x_1 = x$ and $x_2 = x^2$.

CLASSICAL LEAST-SQUARES ANALYSIS: The estimation of the unknown parameters in the general linear regression model is most frequently achieved by the method of *least squares*. Given n observations (or cases) $(x_{1i}, x_{2i}, \ldots, x_{pi}, y_i)$, $i = 1, \ldots, n$, let the ith residual be

$$e_i = y_i - \beta_0 - \sum_{j=1}^{p} \beta_j x_{ji}.$$

The method of least squares determines values, b_j, as estimates of β_j so as to minimize the sum of squared residuals. The estimated regression function (predicted value) for the ith set of inputs, $\hat{y}_i$, and the estimated residual r_i are given by

$$\hat{y}_i = b_0 + \sum_{j=1}^{p} b_j x_{ji}, \quad r_i = y_i - \hat{y}_i. \tag{7}$$

There are essentially two major advantages of this method. The first is computational, since the method only requires the solution of a system of linear equations. The second is statistical in that the estimates possess desirable small sample properties. In particular, the b_j are unbiased estimates of the β_j which have minimum variance in the class of estimators which are unbiased. Further, the assumption of normality allows for simple inferences on the β_j. The estimate of σ^2 is also unbiased and minimum variance.

Note that these properties refer to the $(x_1, x_2, \ldots, x_p, y)$ relationship and not to the underlying variables. For example, the b_0 and b_1 values derived to estimate the β_0 and β_1 of Equation (5) provide unbiased estimators of log z, not of z, and minimize the sum of the squares of the deviations from the linear plot of log z vs. $1/w$, not from the curvilinear plot of z vs. w.

With the advent of high-speed computing, the computational advantage is less compelling than in the past. This has encouraged a study of alternatives to least squares, some of which will be described subsequently.

DEPARTURES FROM THE CLASSICAL ASSUMPTIONS: The standard analysis assumes that the model is correct and that the data are good. In practice, this is rarely the case and it is essential that the violations be detected and evaluated. We now provide a brief discussion of several of the main problems:

1. Incorrect functional form for the regression function. Additional variables and/or different functions of the variables may be required.
2. Violations of the assumptions of independence, constant variance, and normality of errors.
3. Outliers and extreme points. The former are observations in which the response is abnormally large or small and the latter are cases in which the inputs are different from the rest of the data.
4. Multicollinearity among the input variables, that is, nearly exact linear relations among subsets of the input variables. This includes the case where one of the inputs is nearly constant.

One or more of these problems may completely invalidate the analysis and, unfortunately, the standard statistics provide little indication of their presence. Several additional indicators have been proposed to address these possibilities. Unfortunately, there are no guaranteed solutions to any of the problems cited. The following remedies are typical but must be used with caution.

1. Nonuniform residual plots may suggest nonlinear functions. Individual points which are outstanding may suggest other variables that could be included, especially categorical variables defining subgroups of cases.
2. The most common cause of variance inhomogeneity is that the variance is proportional to one of the inputs. Division of the equation by this variable, or some power of it, will help. Normality may be achieved by transformations.
3. Outliers and extreme points may be deleted from the analysis but care must be taken, as these may be valid, informative observations. Alternatively, one of the robust procedures might be used.
4. An eigenvector analysis may identify the multicollinearity, but the action to be taken depends on the cause. If the linear relation is inherent in the system being modeled and the relation is strong, it may be appropriate to eliminate one of the variables in the relation. If the apparent linear relation is due to the peculiarities of the particular sample, then, if possible, additional data should be taken which are more uniformly spread over the sample space. Alternatively, one might simulate this by using *ridge regression* or a related method.

ALTERNATIVES TO CLASSICAL LEAST SQUARES: Since least-squares analysis is vulnerable to departures from the basic assumptions, several alternatives have been suggested.

One of best known of these alternatives is *robust regression*, where the basic idea is that observations with large residuals are given less weight and hence become less influential.

When multicollinearities are present, least squares estimates of the coefficients may be abnormally large or even have the wrong sign. *Ridge regression* is the method that effectively adjoins fictitious data.

One of the oldest modifications of least squares is that of eliminating variables. This has been a confusing and controversial topic primarily because it has often applied indiscriminately to data that have not been subjected to proper diagnostics. Variable elimination only should be applied after the data have been examined for extremes, outliers, and multicollinearities, and appropriate action has been taken. Variables which are then not contributing to the description of the response may be eliminated.

An alternative to eliminating variables, implemented in most of the popular statistical software packages, is *stepwise regression*. The most significant of the x_i is determined and the parameters of Equation (1) are estimated. New x variables are added to the equation until the resulting decrease in the portion of the variance of errors not explained by the regression becomes statistically insignificant.

See **Exponential smoothing; Time series analysis**.

References

[1] Belsley D.A., Kuh E. and Welsch R.E. (1980). *Regression Diagnostics*. Wiley, New York.
[2] Daniel C. and Woods F.S. (1971). *Fitting Equations to Data*. Wiley, New York.
[3] Draper N.R. and Smith H. (1966). *Applied Regression Analysis*. Wiley, New York.
[4] Gunst R.F. and Mason R.L. (1980). *Regression Analysis and Its Applications*. Marcel Dekker, New York.
[5] Neter J. and Wasserman W. (1974). *Applied Linear Statistical Models*. Richard D. Irwin. Homewood, Illinois.

RELATIONAL DATABASE

See **Information systems and database design**.

RELATIVE COSTS

See **Prices**.

RELAXED PROBLEM

The term given to a constrained optimization problem in which some of the constraints have been weakened or relaxed. In particular, it is applied to an integer-programming problem in which the variables are no longer restricted to be integer. The objective function of the relaxed problem serves as a bound for the original problem. See **Integer-programming problem**.

RELIABILITY

The ability of a component or system to be operable when called upon to do its intended job.

Reliability is most often quantified as the probability that the device or system has not failed (is alive) at a particular time: $R(t) = \Pr\{\text{lifetime} > t\} = 1 - F(t)$, where F is the cumulative distribution function of the lifetimes. This function is often also called the survival function. See **Failure-rate function; Reliability function; System reliability**.

RELIABILITY FUNCTION

The reliability at time t, $R(t)$, is defined as $\Pr\{\text{lifetime} > t\} = 1 - F(t)$, where F is the cumulative distribution function of the lifetimes. See **Reliability; System reliability**.

RELIABILITY OF SYSTEMS

Donald Gross

The George Washington University, Washington, DC

INTRODUCTION: The coming decades may well be known for the popularization of the Q word – *quality*. Phrases like *quality circles*, *total quality management*, and *total quality integration* practically became household words, even though old standbys like *quality control* and *quality assurance* have been the subject of study of industrial engineers and statisticians for many of the preceding decades. Intricately related to *quality*, in fact, a necessary ingredient, is *reliability*, loosely defined as the probability that a system, subject to random failures, will perform properly over some time span of interest. This definition shall be made more precise in the following. One might be able to have reliability without quality, but one can never have quality without reliability.

The major issue here is a consideration of the probability structure of systems made up of individual components, each with a known lifetime density, say, $f_i(t)$. The two basic combinations of system design are the series and parallel systems, with more complex structures built up from these. By series, we mean the arrangement whereby any single item's failure leads to total system failure. For the parallel case, we require all component devices to fail for total system failure.

There are a number of important variations on the parallel theme. They differ in the manner in which the set of devices are permitted to operate simultaneously and if not, what form of switch is necessary to call upon any alternative. When all are going together, we say that such a system is parallel redundant. When items are not in use but waiting to be switched to use if needed, and the items not in use do not deteriorate with age, the structure is said to be a cold standby system. The hybrid combination which finds the standby elements possibly aging at a slower pace than if they were in use is called a warm standby system. If, items not in use age at the same rate as they do when in use, then the system is often referred to as a hot standby system and is equivalent to a parallel redundant system as long as the switching mechanism which brings the standby item on line when the operating item fails is itself 100% reliable (zero probability of failing).

LIFETIME PROBABILITIES: The direct application of the basic laws of probability permits the easy derivation of the lifetime probabilities associated with each of these fundamental structures. The cumulative distribution function (CDF) for the simple series system without maintenance is

$$F(t) = 1 - \prod_{i=1}^{n} [1 - F_i(t)]$$

where $F_i(t)$ is the lifetime CDF of the ith component. This result is made slightly more compact by defining a *reliability function* $R(t)$ as the complementary CDF, $1 - F(t)$, namely, the probability of a lifetime longer than t. Then we see that the system reliability may be written in terms of the component device reliabilities as

$$R(t) = \prod_{i=1}^{n} R_i(t).$$

In the special case where each component's life follows the exponential distribution, we have

$$R(t) = \exp\left(-\sum_{i=1}^{n} \lambda_i t \right).$$

For the parallel redundant (or hot standby with a 100% reliable switch) case, the system lifetime CDF is found as

$$F(t) = \prod_{i=1}^{n} F_i(t)$$

and thus, its reliability function is

$$R(t) = 1 - \prod_{i=1}^{n} F_i(t).$$

In the event that the devices are independent and identically distributed exponential distributions with parameter λ, then

$$F(t) = [1 - \exp(-\lambda t)]^n.$$

In the special case where $n = 2$, the exponential

system has reliability function

$$R(t) = \exp(-\lambda t)[2 - \exp(-\lambda t)].$$

It is interesting then to compare this result to that for the two-unit exponential parallel cold standby 100% reliable switch system. The latter can be derived as the sum of two probabilities: the probability that the original component lives past time t plus the probability that the original component fails in some time v, $0 \le v \le t$, and the standby component lives longer than $t - v$, integrated over v from 0 to t. This cold-standby reliability, which, for the case of identical exponential components, turns out to be $R(t) = (1 + \lambda t)\exp(-\lambda t)$, is greater than that of the redundant structure for all values of λ. We can, of course, extend this result to all sizes n. We can also build in a probability of switch failure for the standby case and observe the effects of an unreliable switch on the relative merits of standby versus redundant systems. If we have a probability p that the switch will work, the reliability is adjusted to $R(t) = (1 + p\lambda t)\exp(-\lambda t)$. The figures that follow show plots of $R(t)$ versus t for $p = 0$, 0.5 and 1, respectively. When the probability of the switch working is zero, the graph shows that the parallel redundant case is superior. When the probability of the switch working is one, the cold standby case is superior, as we saw above. But, when the switch probability is between 0 and 1 as it is for the 0.5 case, there is a point of time where the reliabilities of the parallel redundant and the cold standby cases cross.

N-OUT-OF-K SYSTEMS: We generalize the parallel redundant system to define failure if more than $n - k$ of the n components are not working (less than k working). So $R(t) = \Pr\{$at least k of n are up$\}$. Let $p =$ probability of a component working until $t = \exp(-\lambda t)$. Then, from the binomial probability law we see that

$$R(t) = \sum_{i=k}^{n} \binom{n}{i} \exp(-\lambda t i)[1 - \exp(-\lambda t)]^{n-i}.$$

MAINTAINED SYSTEMS: It should be realized that in the typical reliability application, failed units are often put into repair. As a first illustration of such a maintained system, consider a single device with time-to-failure exponential, mean $1/\lambda$, and the time to repair exponential, mean $1/\mu$. It then turns out in this simple single component system that the probabilities that the system is operating or is down at time t, respectively, are

$$p_0(t) = \frac{\mu + \lambda \exp[-(\lambda + \mu)t]}{\lambda + \mu}$$

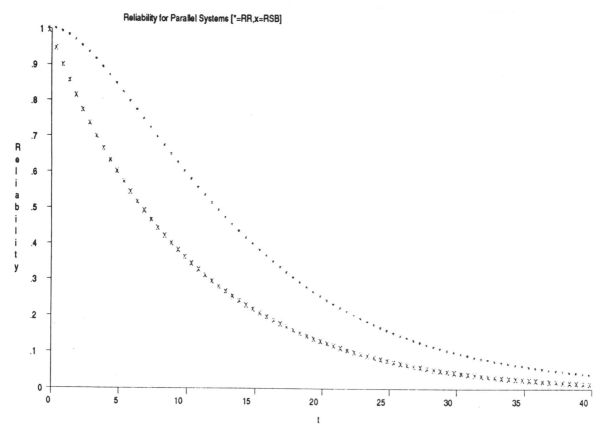

Fig. 1. Reliability function for redundant & standby parallel systems (switch probability = 0).

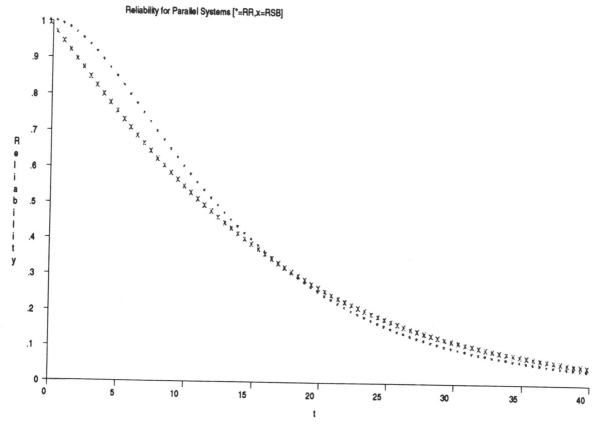

Fig. 2. Reliability function for redundant & standby parallel systems (switch probability = 0.5).

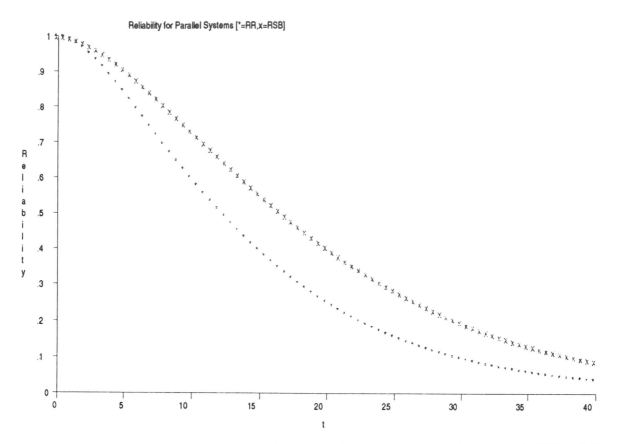

Fig. 3. Reliability functions for redundant & standby parallel systems (switch probability $= 1.0$).

and

$$p_1(t) = \frac{\lambda\{1 - \exp[-(\lambda + \mu)t]\}}{\lambda + \mu}.$$

We often call $p_0(t)$ the system availability (written as $A(t)$ since it is the probability that the system is "available" at time t). The long-run average availability is computed from $A(t)$ as $t \to \infty$ to be $A = \mu/(\lambda + \mu)$.

Next, we look at a two-item series system with identical exponential failure distributions and one exponential repair facility. The time-dependent probabilities are found as the solution to a 3×3 system of difference/differential equations. To avoid details, we instead offer the limiting probabilities. The steady-state availability is the limiting fraction of time no devices are "down" and is given by $A = \mu^2/(2\lambda^2 + 2\lambda\mu + \mu^2)$. The limiting probabilities that one and two units are down are respectively given as

$$\begin{cases} p_1 = \dfrac{2\lambda\mu}{2\lambda^2 + 2\lambda\mu + \mu^2} \\ p_2 = 1 - p_1 - A \end{cases}$$

The final maintained system we discuss is the simple two-item exponential parallel redundant structure with repair. Here

$$A = \frac{\mu^2 + 2\lambda\mu}{\mu^2 + 2\lambda\mu + 2\lambda^2},$$

which is clearly larger than that just presented for the series system.

THE STRUCTURE FUNCTION: For more complicated systems of components, we establish what is commonly called the structure function as a convenient vehicle for characterizing the reliability. First, for any component i, we define a binary indicator random variable X_i as 1 if the device is operating and 0 otherwise. The structure function of a system of n components is then written as $\phi(X_1, \ldots, X_n)$ and will likewise be 1 if the system is operating and 0 otherwise. If all the components are queried at time t, then the system reliability $R(t)$ at that point is the probability that $\phi = 1$.

For a full series system then, all X_i must be 1 for ϕ to be 1, so that

$$\phi(X_1, \ldots, X_n) = \prod_{i=1}^{n} X_i.$$

In the pure parallel case, $\phi = 1$ if any $X_i = 1$, so that

$$\phi(X_1, \ldots, X_n) = 1 - \prod_{i=1}^{n} (1 - X_i).$$

The beauty of the structure function is its ability to model the most complex of systems in a fairly natural Boolean way. For example, consider a structure of five components with 1 and 4 in series, parallel with 2 and 5 in series, and also allowing the combination 1, 3, 5 for operation (together called a bridge structure). The system operates as long as at least one of these three combinations is up. Thus we see that

$$\phi(X_1, \ldots, X_n) = 1 - (1 - X_1 X_4)(1 - X_2 X_5)(1 - X_1 X_3 X_5).$$

As a general rule. we limit our attention to structures that make sense (the technical term is coherent). A system is coherent if

(a) its structure function ϕ is increasing in each argument (that is, ϕ improves as X goes to 1 from 0); and

(b) each component is relevant (that is, its reliability affects system performance). A fairly complete theory has been developed and is given in Barlow and Proschan (1975).

There is quite an extensive literature on systems reliability and related problems. Barlow and Proschan (1975) is a key reference, and further material of special importance on systems problems may be found in Kaufmann *et al.* (1977). Introductory material on systems reliability may also be found in Chapter 21 of Hillier and Lieberman (1990), with a more advanced treatment in Chapter 9 of Crowder *et al.* (1991).

See **Distribution selection for stochastic modeling; Markov chains; Markov processes; Quality control; Queueing theory; Total quality management.**

References

[1] Barlow R.E. and Proschan F. (1975). *Statistical Theory of Reliability and Life Testing.* Holt, Rinehart and Winston, New York.

[2] Crowder M.J., Kimber A.C., Smith R.L. and Sweeting T.J. (1991). *Statistical Analysis of Reliability Data.* Chapman and Hall, London.

[3] Hillier F.S. and Lieberman G.J. (1990). *Introduction to Operations Research*, Ch. 21. McGraw Hill, New York.

[4] Kaufmann A., Grouchko D., and Cruon R. (1977). *Mathematical Models for the Study of the Reliability of Systems.* Academic Press, New York.

RENEGING DISCIPLINE

When queueing customers get impatient and leave their queue before their service is begun, or even before their service ends. See **Queueing theory.**

RENEWAL EQUATION

See **Point stochastic processes; Renewal processes.**

RENEWAL PROCESSES

Igor Ushakov

SOTAS, Rockville, Maryland

A *renewal process* is a stochastic point process $\{N(t) = \text{number of occurrences by time } t\}$ which describes the appearance of a sequence of instant random events where the times between occurrences (e.g., called interarrival times in queueing theory) are a sequence of independent and identically distributed (i.i.d.) random variables. It is common to write the inter-occurrence distribution function as $F(t)$ and its density (if it exists) as $f(t)$, with expected value $1/\mu$. The Poisson process represents a particularly important renewal process in which the intervals between occurrences are identically exponentially distributed (Cox, 1960; Cox and Isham, 1980; Feller, 1966; Smith, 1955).

The so-called *renewal equation* for the process expectation (or renewal function) $H(t) = E[N(t)]$, plays a fundamental role in all renewal problems:

$$H(t) = F(t) + \int_0^t H(t - x) dF(x).$$

The derivative of $H(t)$, $h(t) = dH/dt$, is often called the *intensity function* and has a simple interpretation: $h(t)dt$ is the approximate probability of an occurrence within the time interval $[t, t + dt]$.

One can write an equation for the intensity function similar to the one above:

$$h(t) = f(t) + \int_0^t h(t - x) dF(x).$$

With t increasing, it follows that

$$\lim_{t \to \infty} \frac{H(t)}{t} = \frac{1}{\mu}.$$

In a physical sense, this means that, over a large interval of size t, the mean number of events is inversely proportional to the expected interarrival time.

Very close to the previous statement is the following. If the renewal process is formed by continuous random variables, then

$$\lim_{t \to \infty} h(t) = \frac{1}{\mu}.$$

This reflects the fact that with increasing t, the renewal process becomes stationary and its intensity becomes independent of the current time.

A further generalization comes from Blackwell's Theorem: for continuous renewal-time random variables and an arbitrary interval width τ (Feller, 1966):

$$\lim_{t \to \infty} [H(t + \tau) - H(t)] = \frac{\tau}{\mu}.$$

The next important result is contained in Smith's Theorem (1955). If the renewal-time random variables are continuous and $V(t)$ is a monotone non-increasing function, integrable on $(0, \infty)$, then

$$\lim_{t \to \infty} \int_0^t V(t - x)dH(t) = \frac{1}{\mu} \int_0^\infty V(t)dt.$$

The actual choice of the function $V(t)$ depends on the particular problem of concern.

Point processes of a special type can be formed by two random variables which alternate, where a realization of such a process has the sequence $X_1, Y_1, X_2, Y_2, \ldots$ and so on. Such a process is called an *alternating renewal process* when the X and Y subsequences are themselves ordinary renewal processes. This sort of thing is very convenient for describing equipment failure and repair.

References

[1] Cox D.R. (1960). *Renewal Theory*, Methuen, New York.

[2] Cox D.R. and V. Isham (1980). *Point Processes*, Chapman and Hall, New York.

[3] Feller W. (1966). *Introduction to Probability Theory and Its Applications*, vol. II, John Wiley, New York.

[4] Smith W.L. (1955). "Regenerative Stochastic Processes," *Proc. Royal Society, Ser. A,* 232, 6–31.

REPRESENTATION THEOREM FOR POLYHEDRAL SET

Given a nonempty polyhedral set S, then a point X is in S if and only if X can be expressed as a convex combination of the set's extreme points plus a nonnegative combination of its extreme directions.

RESEARCH ANALYSIS CORPORATION (RAC)

See **Operations Research Office and Research Analysis Corporation.**

RESEARCH AND DEVELOPMENT

John C. Papageorgiou

University of Massachusetts at Boston

INTRODUCTION: Products and services have a finite life cycle, and the speed with which they go through their life cycle stages has been continuously increasing. Through the functions in an organization called research and development (R&D), new products and services are developed, existing ones are improved, and the respective transformation processes are improved to increase efficiency and minimize cost.

The United States spent 2.6% of its Gross Domestic Product (GDP) in 1991 on R&D. The other major industrialized countries Japan, Germany (unified), France, and UK spent 3.0, 2.6, 2.4, and 2.1%, respectively. Part of this R&D expenditure was aimed at pure research, that is, research meant for pursuit of knowledge. This research takes place mainly in research laboratories of universities, research centers, government agencies, and major corporations. The other part of R&D expenditure was made on applied research, in which existing knowledge is used to design new products, services and processes, as well as on development, that is, the conversion of the results of applied research into the actual transformation systems that will produce the new products and services.

In industry, ideas for R&D projects originate primarily in a firm's R&D department. However, other departments such as marketing, production and engineering are frequent contributors, as is top management. In some cases, suppliers, clients/customers, and government departments are sources of ideas. R&D project management is often difficult, due to the high degree of uncertainty involved, and OR/MS has developed several approaches to help R&D managers. OR/MS has addressed mainly two major problems in R&D management: 1) Project evaluation, selection and resource allocation; and 2) project planning and control.

R&D PROJECT SELECTION: The R&D project evaluation, selection and resource allocation problem deals with the evaluation of candidate R&D projects, and the selection of a subset of such projects to which available R&D resources (manpower, funds, equipment and facilities) will be allocated. Because of the investment commitment, the uncertainties involved, and the impact of the decisions upon the future of the organization, project selection is a very important and difficult problem. As a result, hundreds of papers have been published discussing the problem and suggesting various approaches for its solution. They have been reviewed in several literature survey papers (Augood, 1973; Baker and Pound, 1964; Baker, 1974; Baker and Freeland, 1975; Liberatore and Titus, 1983; Souder, 1972; Souder and Mandakovic, 1986). Only a small number of all the approaches that have been proposed can be discussed here, indicating the evolution of this area of OR/MS.

Before World War II, the project selection problem was non-existent. Companies were relatively small and competition was limited, which resulted in a limited need to develop new products. There was no distinction between "research" and "development," and the relevant function was not viewed as important. Usually, the chief technical officer would come up with a production related project and pursue it through its implementation. It was after World War II that the business environment changed, increased competition resulted in increased demand for new products/services and improved processes, and R&D project selection became a problem.

Project evaluation and selection methods started appearing in the mid-1950s. The first methods used, called *checklists or profile charts*, are based on a checklist of criteria. The checklist consists of factors considered to be important to the success or failure of the project, and is used as the basis on which each project is subjectively rated by one or more individuals. The checklist may include both economic and non-economic factors, such as social impacts and environmental concerns. The degree of favorableness of each criterion is checked for each project, with the objective to derive an overall pattern for each project and determine its degree of favorableness.

Because this method does not differentiate among the importance of different criteria and is based on qualitative judgment, *scoring models* were later developed. These models use weights assigned to both the different criteria and the degree of favorableness of each of them for each project. As a result, a weighted score is computed for each project. For the project scores to be comparable, similar criteria have to be used for all the projects. Different methods have been proposed for deriving the set of weights representative of the preference function of the particular decision-maker, such as having the decision-maker rank order the criteria or make comparisons of different pairs of projects.

Since these methods give dimensionless results, *benefit-cost ratio* approaches were developed. The different costs and benefits associated with a project, including non-economic costs and benefits, are expressed in terms of a common measure and their present value computed and expressed as a ratio. Risk factors can also be included in terms of probabilities of research, development and market success. For a project to be considered, its benefit-cost ratio should be greater than one.

The above methods, usually called *classical methods*, have been used extensively in R&D project evaluation and selection due to their simplicity and ease of use. They can prove useful in preliminary project analysis and screening. However, they cannot solve the project selection problem because they ignore several key aspects. For example, that projects are selected in sequences rather than individually, where the outcome of one affects that of another. Other ignored aspects include the dynamic nature of the R&D environment in terms of project funding at different levels, in which case the value and preferability of a project is a function of its funding level; dynamic resource constraints; and that the set of candidate projects continuously changes over time.

Decision trees were then introduced to deal with sequences of interrelated projects. A decision tree consists of decision nodes and event nodes from which alternative decisions and events, respectively, are branched out. Economic consequences expressed in monetary or utility values for decisions, and probabilities for events are added and, starting at the end of the branches and working backwards, the expected payoff for a sequence of decisions is computed. The optimum sequence of decisions (optimum sequence of projects/sub-projects) can thus be identified. Since the number of events that can be branched out of an event node is limited, *stochastic decision trees* were developed, where the event nodes are represented by a probability distribution. However, resource constraints cannot be included in a decision tree, which is a serious drawback, amongst others.

With the advent of OR/MS and the wide availability of computers, different *portfolio models* were developed to overcome the shortcomings of the above methods. Several types of mathematical programming (linear, integer, mixed integer, zero-one, nonlinear, dynamic, goal, multi-objective, and stochastic) have been used to select a subset of projects which maximizes a particular objective without violating a set of constraints. The objective is usually to maximize the expected net present value of the subset of projects. In addition to providing the optimum portfolio of

R&D projects and allocating the budget among them, mathematical programming presents the additional advantage of making sensitivity analysis easy, suggesting ranges of solutions, and answering what-if types of questions. However, the optimality of the portfolio is a function of the assumptions associated with each particular type of mathematical programming and the estimates used in the input data. This, combined with the difficulty on the part of decision-makers in understanding the mathematical aspects of the models, has resulted in limited reported successes of the portfolio models in terms of actual implementation.

During the 1970s, issues regarding the usefulness of the existing methods for solving real life R&D project selection problems and their acceptability by R&D managers were widely discussed. Two studies by Baker (1974) and Baker and Freeland (1975) identified several limitations of the methods that had been proposed up until that time.

Such limitations included the inadequate treatment of uncertainty and risk with respect to benefit contribution and parameter estimation; project and parameter relationships with respect to both benefit contribution and resource utilization; multiple, interrelated decision criteria; the time variant property of data and criteria; and the problems associated with the continuity in the research program and the research staff.

Further limitations were the lack of explicit recognition of the experience and knowledge of the R&D manager; the non-monetary aspects such as establishing and maintaining a balance between basic and applied research, product and process development, in-house and contracted projects, improvement and breakthrough work, and different levels of risk-payoff opportunities; the perceptions held by R&D managers that the models are difficult to understand and use; and the importance of certain individuals in the R&D organization.

Other limitations include the failure to treat the problem as an intermittent stream of investment alternatives and as a hierarchical diffuse decision process; to include in the model the timing of decisions, the generation of additional alternatives, and project recycling by gathering new information, reformulating criteria, variables and constraints, and defining new alternatives; and to recognize the diversity of projects from basic research to engineering.

Several OR/MS researchers tried to develop approaches which do not have these limitations. As a result, different models have been proposed which take into consideration a particular aspect of the shortcomings of the existing methods. Among the approaches which have been developed, the emphasis has been on *multi-objective mathematical programming* methods. In this respect, *goal programming methods* have been developed, where several goals are considered and expressed as constraints, with deviational variables used to express under-achievement or overachievement of the goals. The objective function minimizes these deviations. The goals can be prioritized, so that their achievement is considered according to their priority sequence.

As difficulties arise in setting the aspiration levels of the goals and in including trade-offs among goals, *multi-objective linear programming methods* were used, including *multi-attribute utility theory* (Ringuest and Graves, 1989; Mehrez *et al.*, 1982). In applying multi-attribute utility theory, utility values are assigned to each possible subset of projects for each of the goals and, using integer linear programming, a list of all nondominated solutions is generated, consisting of solutions in which the performance in one goal cannot be improved without sacrifice in one or more other goals. One drawback is that the list of nondominated solutions can be very large in a real life situation, creating a complex selection problem for the decision-maker. Screening methods which have been developed, could provide some help in selecting one of the non-dominated solutions.

Recent emphasis, however, has been on a different philosophy in approaching the problem of R&D project selection and resource allocation, by including in the decision making process the people at every level of the organization who would influence the project selection process. As a result, *Behavioral Decision Aids* (BDA) have been proposed, which use the output of project selection models not as a solution to the problem, but as aids to communication and interaction among the parties involved, in order to achieve a consensus.

One such approach is *Q-Sorting*, where each individual is given a stack of cards, each bearing the title or number of one project. Through a series of sorting operations and the use of a specific criterion, the projects are sorted into five piles ranging from very high level of the criterion to very low. Another BDA approach is the *Nominal Interactive Decision Process*, used in combination with various other methods depending on the type of the project, where consensus is built using a modified Delphi approach.

The *Analytic Hierarchy Process* (AHP) has also been used in what is called *Decentralized*

Hierarchical Modeling (DHM), where the involved parties communicate electronically until they reach a consensus on the project portfolio. The dialogue takes place among the different hierarchical levels. Top management initiates the process through budgetary guidelines sent to the divisional managers; the divisional managers then send the guidelines, maybe modified, together with suggested prioritized program areas to the R&D managers; and the R&D managers and their staff propose an R&D portfolio and send it up the hierarchy. The R&D people, in coming up with the portfolio, may use any of the available OR/MS techniques. This process may be repeated several times. Back and forth communication among the different hierarchical levels may take place at any stage of the process.

PROJECT PLANNING AND CONTROL: The second major area of R&D management to which OR/MS has made a significant contribution is project planning and control. Several approaches have been developed in this area as well.

One of the first approaches is *Program Evaluation and Review Technique (PERT)*, developed in 1958 for planning and controlling the Polaris Fleet Ballistic Missile project by the Navy Special Projects Office and Lockheed Aircraft Corporation, in cooperation with Booz, Allen and Hamilton. In using PERT, the project is represented by a network, consisting of events (nodes) standing for specific accomplishments at a point in time, or milestones, and activities (arrows) representing the actual performance of tasks. The events and activities follow one another in their proper technological and logical sequence, and the PERT network, also called the precedence relationships network, has a beginning event and an ending event. Activities consume time and resources such as manpower, materials, equipment, funds, and so on, and each activity is represented by the beginning and ending nodes. This means that only one activity can connect two nodes and that the network cannot have a loop.

Activity times are assumed to follow the beta distribution and three time estimates are given for each activity: an optimistic standing for the practically minimum time, a most likely standing for the best estimate of time, and a pessimistic standing for the practically maximum time. On the basis of these estimates, the mean and variance for each activity time is computed, the longest time path is determined (critical path), and the probability of reaching an event or of completing the critical path by a certain sched-

uled time is computed. The latter is usually taken as the probability of completing the project. Other important information derived from PERT is the earliest start and finish time, the latest start and finish time, and the slack time for each activity. Activities with zero slack time are considered critical activities that require special monitoring to avoid delays in the completion of a project.

At about the same time as the PERT development, the *Critical Path Method (CPM)* was developed in 1957 at the du Pont company, in consultation with Remington Rand, in scheduling maintenance shutdowns of chemical processing plants. PERT is a probabilistic approach whereas CPM is deterministic. CPM uses normal and crash time estimates for each activity, and their associated costs. The normal time estimate would be equivalent to PERT's expected time, and the crash time would be the minimum possible time needed to complete the activity irrespective of cost increases. Aside from these differences, CPM is almost identical to PERT, and for this reason the two techniques are referred to as the PERT/CPM technique. PERT and CPM have generated several variations, modifications, and new techniques with added capabilities and wider applications. Although it is not possible to review all of them here, a brief synopsis can be found in (Wiest, 1985).

PERT and CPM have been used in a variety of project planning and control situations, including R&D project management. However, some of their underlying assumptions are not always valid. For example, the originally developed network may become irrelevant in the future because of the changed content of a project. Precedence relationships cannot always be specified as they sometimes depend on the outcome of previous activities. Project completion time is not always determined by the longest time path, as a delay in a non-critical activity may result in a longer completion time of the project. The beta distribution is not the only distribution that could be used, the formulae used to estimate its mean and variance may give erroneous estimates compared with the original beta distribution formulae, and the three time estimates may include a high degree of subjectivity (Chase and Aquilano, 1989).

In addition to the criticisms of PERT and CPM, their use in R&D project planning and control includes some additional limitations (Clayton and Moore, 1972; Pritsker, Sigal, and Hammesfahr, 1989). One of them is that branching from the nodes is deterministic, that is, each activity must be completed before the project is completed. In R&D projects, however, branching is usually probabilistic, for example, successful test and performance of the next stage, failure and abandonment of that part of the project, inconclusive results and repeat of the test. All the activities leading to a node must be realized before the relevant event can be realized. In R&D projects, given the probabilistic nature of branching, not all activities leading to a node can be realized. Looping is not allowed, though an activity in an R&D project may have to be repeated, for example, a test. Activity times are assumed to be solely described by a beta distribution while R&D activities may follow different other distributions. One terminal node is allowed (completion of the project) while in R&D one of several end events can be realized, for example, successful completion, failure and abandonment, redesign of the project.

It is obvious that these limitations render PERT inflexible in modeling complex R&D projects. To overcome PERT's limitations, the *Graphical Evaluation and Review Technique (GERT)* was developed which has the following characteristics: An activity has an associated probability of being selected, ranging from zero to one. As a result, the nodes are constructed differently to denote their nature as deterministic or probabilistic. The realization of a node may be specified to occur upon the realization of one or more of the activities leading to it, it may be realized one or more times, and the first time it is realized the number of activities to be completed may be different from subsequent repeats. Looping in simple or complex forms is allowed. The network can have more than one source node and/or sink node. Modifications of the network following the completion of certain activities can be incorporated. Several types of probability distributions can be used to represent activity times. Cost can be assigned to each activity in terms of a fixed part and a variable per unit time component. Statistics on time, cost, and activity counts for specified activities can be collected for the sink as well as other designated nodes.

GERT is a network-simulation approach that has been further improved into a more powerful version, *Q-GERT*, that allows the inclusion and the simulation of more than one project, can model queues at nodes and route projects through teams based on user established decision rules (Taylor and Moore, 1980; Pritsker, Sigal, and Hammesfahr, 1989).

Another network-simulation based technique that was developed after GERT is *Venture Eval-*

uation and Review Technique (VERT) (Moeller and Digman, 1981). VERT, like GERT, has been developed as a technique for analyzing potential outcomes of projects, expected values of various project parameters, and criticality indices rather than for scheduling projects. It is used in assessing the risks involved in undertaking new ventures and in resource planning, control monitoring, and overall evaluation of on-going projects with respect to time, cost, and performance. It is considered to be more powerful than GERT due to the fact that performance enters the network in numerical terms. It can be modeled in terms of any unit of measurement or a dimensionless index. VERT introduced six new types of node logics and the capability of establishing a mathematical relationship between an arrow's parameter values (time, cost, performance) and any other arrow or node's parameter values, as well as mathematical relationships between the time, cost, and performance variables of a given arrow.

Finally, there have been a few approaches that have been proposed for large-scale R&D *program planning.* Such programs involve multiple interdependent technologies, are initially defined in terms of broad, qualitative policy directives, serve broad constituencies of sponsors, and R&D is performed at separate external organizations or at remote sites within an organization. Decisions regarding resource allocation among projects, establishment of objectives, assignment of projects to program sub-divisions, and scheduling of the projects need to be made. For a decision support approach and pertinent bibliography see (Mathieu and Gibson, 1993).

CONCLUSION: In summary, there are thousands of OR/MS models that have been proposed for R&D management. The evolution of the field continues through newly proposed models that appear in the bibliography which try to improve upon existing ones by including additional aspects of the problem situation. The usage of these approaches is expected to increase in the future due to the wider exposure of R&D managers to OR/MS approaches, the wider availability of microcomputers and user friendly software, and the emphasis on using them "as a laboratory for testing policies, sharing opinions, asking what-if types of questions and simulating interdepartmental interactions throughout the organization" (Souder and Mandakovic, 1986).

See **Analytic hierarchy process; Decision trees; Goal programming; Linear program-**

ming; **Multi-attribute utility theory; Multi-objective programming; Portfolio theory; Project management.**

References

[1] Augood D. (1973). "A Review of R&D Evaluation Methods," *IEEE Trans. Engineering Management*, EM-20, 114–120.

[2] Baker N.R. and W.H. Pound (1964). "R&D Project Selection: Where We Stand," *IEEE Trans. Engineering Management*, EM-11, 124–134.

[3] Baker N.R. (1974). "R&D Project Selection Models: An Assessment," *IEEE Trans. Engineering Management*, EM-21, 165–171.

[4] Baker N.R. and J. Freeland (1975). "Recent Advances in R&D Benefit Measurement and Project Selection Methods," *Management Science*, 21, 1164–1175.

[5] Chase R.B. and N.J. Aquilano (1989). *Production and Operations Management: A Life Cycle Approach*, Irwin, Homewood, Illinois, pp. 501–505.

[6] Clayton E.R. and L.J. Moore (1972). "PERT vs. GERT," *Journal of Systems Management*, 22, 11–19.

[7] Liberatore M.J. and G.J. Titus (1983). "The Practice of Management Science in R&D Project Management," *Management Science*, 29, 962–974.

[8] Mathieu R.G. and J.E. Gibson (1983). "A Methodology for Large-Scale R&D Planning Based on Cluster Analysis," *IEEE Trans. Engineering Management*, 40, 283–292.

[9] Mehrez A.S., S. Mossery and Z. Sinuany-Stern (1982). "Project Selection in a Small R&D Laboratory," *R&D Management*, 12, 169–174.

[10] Pritsker A.A.B., C.E. Sigal and R.D.F. Hammeswahr (1989). *SLAM II – Network Models for Decision Support*, Prentice Hall, Englewood Cliffs, New Jersey.

[11] Ringuest J.L. and S.B. Graves (1989). "The Linear Multi-Objective R&D Project Selection Problem," *IEEE Trans. Engineering Management*, 36, 54–57.

[12] Schroder H.H. (1971). "R&D Project Evaluation and Selection Models for Development: A Survey of the State of the Art," *Socio-economic Planning Sciences*, 5, 25–39.

[13] Souder W.E. (1972). "A Comparative Analysis of R&D Investment Models," *AIIE Trans.*, 4, 57–64.

[14] Souder, W.E. and T. Mandakovic (1986). "R&D Project Selection Models," *Research Management*, 29, 36–42.

[15] Taylor B.W. and L.J. Moore (1980). "R&D Project Planning with Q-GERT Network Modeling and Simulation," *Management Science*, 26, 44–59.

[16] Wiest J.D. (1985). "Gene-Splicing PERT and CPM: The Engineering of Project Network Models" in B.V. Dean (Ed.), *Project Management: Methods and Studies*, Elsevier (North-Holland), Amsterdam, pp. 67–94.

RESOURCE AGGREGATION

In a project network, a method of scheduling activities within their available float times according to a specific rule, for example, at their earliest start times, and determining the consequent total units of each resource required in each time period. See **Network planning**.

RESOURCE LEVELING

A method of scheduling activities of a project to meet a limit in the amount of a resource that is available. This may mean that the project completion date is allowed to slip. See **Network planning; Project management**.

RESOURCE SMOOTHING

A method of scheduling activities of a project within their available float times to minimize fluctuations in day-to-day resource requirements. This approach would be used when the project completion time is not allowed to slip. See **Network planning; Project management**.

RESPONSE TIME

Often used to describe the time between the arrival of a new queueing customer (e.g., as in the receipt of a call by an emergency dispatcher) and the initiation of service (as in the arrival of the emergency unit at the scene of the call), thus equal to the queueing delay.

RESTRICTED-BASIS ENTRY RULE

In the adaptation of the simplex algorithm for solving separable-programming problems in which variables are approximated by a set of grid variables, the restricted basis entry rule only allows, for each original variable, no more than two such neighboring grid variables to be in a solution. Such a rule is also used in solving quadratic-programming problems for certain complementarity conditions to hold. See **Separable programming; Special ordered sets of type 2; Wolfe's quadratic programming-problem algorithm**.

RETAILING

Kaoru Tone

Saitama University, Japan

Retailing can be seen as the third phase in the flow of goods, following production and logistics. OR/MS has been used in retailing for some time, typical examples being the "newsboy problem" (a classical inventory problem) and the "traveling salesman problem," one of the origins of combinatorial optimization.

Retail shops can be classified into several categories; independent stores, department stores, supermarkets, discount stores and convenience stores. In recent years, stores in the last three categories have outperformed the first two in profit efficiency, reflecting a change in consumer behavior. It is therefore crucial for independent and department stores to restructure their methods of retailing. Moreover, the net profit for some goods can be as low as 1–3%, the life cycle of goods is becoming shorter and the diversity of products wider. Consequently, the application of scientific management methods has assumed even greater importance.

OR/MS can be applied in many aspects of retailing, such as store policymaking, location problem, marketing/merchandising, inventory control and advertising/sales promotion. Higgins (1981) presents an excellent survey of management science in retailing up to 1980.

In the mid-1980s, the popularity of the "Point of Sale" (POS) information system increased. This system uses a photo-electronic cash register to scan the bar code on the commodities, transforming the information on retail sales into the host computer on a real-time base. Thus, a store has access to a massive amount of consumer information from its database. This system has two kinds of benefits. Firstly, there are the so-called *hard benefits*, e.g. the speeding-up of cash register operations, the increased accuracy in accounting and manpower reduction. Such benefits help to offset the costs of the system to some extent. The second type of benefits can be termed *soft benefits*. These are gained from the processing and analysis of the POS data using appropriate software methodologies. Using such software efficiently can result in large difference in profits and the role of OR/MS in this regard is described below.

PREDICTION OF THE NUMBER OF CUSTOMERS VISITING A STORE: Information on the number of customers visiting a store has significant influence over store management strategies. This information relates closely to sales volume and staff scheduling. If management can predict with a fair degree of accuracy the future number of customers, be it one day, a week or a month in advance, it then becomes possible to map out an operation program efficiently. Regular and part-time employees can be scheduled effectively in order to reduce operating costs and/or improve customer service. On the product side, stock orders can be gauged more accurately, reducing the risk of lost sales due to shortages and wastage due to an overstocking of goods.

The causals apparent in calculating customer numbers are usually given as the day of the week, weather, temperature and sales campaigns. Based upon the POS data, multiple regression analysis (including categorical data) can be used to arrive at a formula which explains the volume and distribution of customers. See Chatterjee and Price (1991) for details of statistical methods. By using new data, the formula can be updated on a daily or weekly basis, or when the difference between the predicted number and the actual number falls out of a predetermined range.

CLASSIFICATION OF GOODS BY SALES VOLUME: In an average size supermarket, several thousands of goods are displayed. We can divide them into three classes according to sales (or gross sales). This is known as *ABC* analysis or the Pareto chart. Class *A* goods, while accounting for around 50% of sales, constitute only a small portion of the entire range of goods, usually about 10%. Class *B* comprise approximately 40% of both the range of goods and sales value. Class *C* goods comprise about half of the goods available but only around 10% of the sales.

Empirical studies have shown that the sales of Class *C* goods reflect a Poisson-type distribution, while that of Class *B* goods is represented by a normal or log-normal distribution. The sales of Class *A* goods are well explained by regression analysis using causals (including price and sales campaigns).

Retailers, wholesalers and producers will be able to get valuable information by classifying goods in this way and extend their understanding of their characteristics as represented by the parameters set. This information can then be used to make appropriate choices for items in the store. For Class *A* goods, the effects of price and sales campaigns in the regression formula provide useful information for strategic management. In addition, as will be discussed below, goods procurement and inventories will be more efficiently controlled.

INVENTORY CONTROL: Based on the results of the type of data analysis described above, we can estimate future sales with a certain degree of accuracy. The estimated values relate to the buying-in amount and the inventory of goods. It is especially important to predict and control the inventories of goods delivered daily, such as fresh foods and dairy products.

A basic inventory policy is described as follows. Let the remaining amount of a commodity at the store's closing time be Z, of which a part, D, has to be scrapped due to expiration. Then, the stock at the end of the day is $U = Z - D$. If the predicted sales for the following day are Y, and the leadtime for buying-in is one night, then the order volume, P, is determined by $P = \max\{Y - U + \alpha, 0\}$, where α corresponds to the safety stock, the slack which prevents opportunity loss. The safety stock α relates directly to the trade-off between the chance loss and the stock cost. An inventory simulation such as this can be applied using past data to estimate a good α.

ANALYSIS OF MOVEMENT OF CUSTOMERS: Although a POS record of a customer tells us what kinds of goods were bought, his or her movement through the store is not clear from the record alone. However, by comparing the layout of the store and the POS record, we can deduce the route taken via a "traveling salesman" scenario. If we then superimpose the solutions for a given number of customers, we can estimate the congestion likely in each pathway of the store. In addition, by changing the distance table to correspond with the new assignment, it becomes much easier to analyse the effects of display changes on congestion. Traditional methods, such as the use of video and first-hand observations, are less efficient in terms of cost and precision. Using analysis such as that described above will result in less *dead* corners and fewer *busy* corners.

CONCLUSION: The last two decades have seen an increase in competition and innovation in retailing. The hard benefits of POS are readily apparent – real-time information and the instant processing of data into a variety of forms. The more significant benefits however, can be found in the software which enables managers to improve their decision-making processes. OR/MS can contribute to the more effective use of database, thereby increasing retailing efficiency.

This review has touched on only a few of the myriad of potential OR/MS applications.

See **Marketing; Regression analysis.**

References

[1] Chatterjee S. and B. Price 1991. *Regression Analysis by Example*, 2nd ed., John Wiley & Sons, New York.

[2] Eliashberg J. and G.L. Lilien (eds.). 1993. *Marketing*, Handbook in Operations Research and Management Science. Vol. 5. North-Holland.

[3] Higgins J.C. 1981. "Management Science in Retailing," *Eur. J. Opnl. Res.* 7, 317–331.

[4] Mason J.B. and M.L. Mayer 1980. "Retail Merchandise Information Systems for the 1980s," *J. Retailing*, 56 (Spring), 56–76.

[5] Shugan S.M. 1987. "Estimating Brand Positioning Map Using Supermarket Scanning Data," *J. Marketing Research*, 24, 1–18.

[6] Sinkula J.M. 1986. "Status of Company Usage of Scanner Based Research," *J. Academy of Marketing Science*, 14, 63–71.

REVENUE EQUIVALENCE THEOREM

Revenue equivalence theorems in bidding theory establish conditions under which the expected revenue from various auction types (e.g. standard sealed bidding sales and progressive oral auctions) is the same. See **Bidding models**.

REVENUE NEUTRALITY THEOREM

See **Revenue equivalence theorem.**

REVERSIBLE MARKOV PROCESS

A stationary Markov process whose generator has elements given by

$$q(k, j) = \frac{\pi_j \, q(j,k)}{\pi_k} \text{ for } j,k \in E;$$

where π_j is the steady-state probability that the chain is in state j and $q(j,k)$ is the rate at which the chain goes from state j to k. That is, the mean flow rates or probability flux satisfies detailed balance equations for every pair of nodes. See **Markov chains; Markov processes; Networks of queues; Queueing theory.**

REVISED SIMPLEX METHOD

A version of the simplex method that uses an explicit or implicit expression of the inverse of the current basis to calculate the simplex multipliers (prices) and related information. See **Product form of the inverse**; **Simplex method**; **Simplex tableau**.

RHS

See **Right-hand-side**.

RIGHT-HAND-SIDE

The column vector of coefficients b in a general system of linear constraints $Ax = b$.

RIGHT-HAND-SIDE RANGING

See **Sensitivity analysis**.

RISK

The risk of a decision d is the expected value of the loss incurred using d taken over all possible states of nature. A risk averse person is one who prefers to behave conservatively. A decision maker (DM) is said to be risk averse if the DM prefers the expected consequence of a nondegenerate lottery to that lottery (a nondegenerate lottery is one where no single consequence has a probability of one of occurring). A DM is risk averse if and only if the DM's utility function is concave. In contrast, a risk prone person is one who does not prefer to behave conservatively. A DM is said to be risk prone if the DM prefers any nondegenerate lottery to the expected consequences of that lottery. A DM is risk prone if and only if the DM's utility function is convex. Finally, a DM is risk neutral, if and only if the DM's utility function is linear. See **Lottery**; **Risk assessment**; **Risk management**; **Utility theory**.

RISK ASSESSMENT

Clyde Chittister

Carnegie Mellon University, Pittsburgh, Pennsylvania

Yacov Y. Haimes

University of Virginia, Charlottesville

Carl M. Harris

George Mason University, Fairfax, Virginia

INTRODUCTION: To the layman, risk is often quantified in terms of probabilities, whereby it might be said that gambling on any event with a "low probability" of occurring is a risky proposition. Or, the mere existence of a "catastrophic" event with non-zero probability exposes those involved to risk. The risk of nuclear plant failure, global warming, or the depletion of the ozone layer are examples. We might say that the risk of not carrying an automobile insurance policy is not worth the risk. One result of these simple considerations is that the standard components of risk are the chance of a loss, the possible magnitudes of the loss, and the exposure to that loss.

People take different views of risk. There are *risk-neutral* individuals for whom per unit change in the "usefulness of wealth" (i.e., their utility function) per unit change in wealth (+ or −) is constant. Thus we might say that they are willing to tolerate the possibility of "bad" outcomes as long as there is compensating probability of "good" results. To these decision makers, the value of an extra unit of wealth is the same no matter how much wealth they hold.

On the other hand, there are those who are *risk-seekers* or *risk-prone* in that the per unit change in their utility function per unit change in wealth is always more when the wealth increment is positive than when it is negative. Thus such people have convex utility functions – the incremental value of an extra unit of wealth increases as the wealth of the individual increases. These decision makers might ignore bad outcomes as long as there are good ones which have non-zero probabilities, particularly when the favorable outcomes might have very high ultimate payoffs even at low probabilities.

Then there are the *risk-averse* individuals whose utility functions are such that the incremental value of an extra unit of wealth decreases as the wealth of the individual increases. For them, the per unit change in utility per unit change in wealth is always more when the wealth increment is negative than when it is positive, and thus their utility functions are concave.

These concepts are integral parts of utility theory, and examples of possible utility functions for the three types of risk-related behavior are presented in Figure 1. Again, note that pure risk-seeking behavior implies a convex utility function, while pure risk aversion would lead to

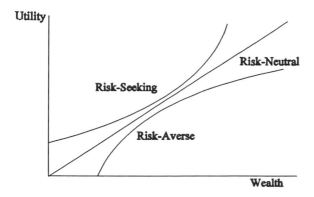

Fig. 1. Possible shapes for utility functions.

a concave representation. However, real life is not so simple, and risk aversion and risk seeking are quite often mixed together in an individual decision maker.

In the more formal context of statistical decision theory, risk is defined in terms of decision rules and the state of nature as follows. Suppose that X is the information available to a decision maker, who must choose a strategy telling what decision to make for each possible value that X may take – let us denote this dependence by $d(x)$. The decision maker is of course interested in choosing a function d that would be optimal in some sense. Now, because the action is a function of the random variable X, then clearly $d(x)$ is a random variable as well, and the loss associated with the decision maker's action also therefore depends on X and the random state θ of nature. We write that loss function therefore as $\ell[d(X, \theta)]$. A common measure of the total result of this process is thus the expected value of the loss function with respect to X. This expectation is typically known as the *risk function* and may be written in terms of the state of nature as the function

$$R(d, \theta) = \mathrm{E}\{\ell[d(X, \theta)]\}.$$

A decision maker's loss function will typically include the cost of experimentation, and the probabilities of the states of nature are revised *a posteriori* as a result of the experimental outcomes.

The loss function used by a decision maker can take on many different forms, including, for example, the impact of a decision on mortality and morbidity, the effect on the profitability of a company, or the change in enrollments at a university. Optimality is hard to reach here, since this would require an action that minimizes the risk for all possible values of the state of nature θ. Unfortunately, such decision rules do not often exist in real problems. For example, a strategy that may be work well in an increasing stock

market may likely be very poor in a declining one.

The modeling of decision maker's risk is often covered under the topics of decision analysis, utility and choice theories. But since the ultimate issue in a risk assessment is the tradeoff between the risk perceived and the possible benefit or return, risk is often included as part of a multi-criterion decision problem. It is often useful, however, to examine risk modeling all by itself.

ILLUSTRATIONS: As a first example, consider the classical single-period inventory model subject to exponential demand with (unknown) mean θ (= state of nature). Let the unit cost of purchasing or producing the item be u, the holding cost per unit be h, and the per unit shortage cost be s. Then the expected cost of stocking amount y is found to be

$$C(y) = h(y - \theta) + uy + \theta(s + h)e^{-y/\theta} .$$

It follows that the optimum value of the order size is $y^* = -\theta \ln[(u + h)/(s + h)]$, with minimum cost of

$$C(y^*) = \theta[u - (h + u)\ln[(u + h)/(s + h)].$$

Suppose that t is an estimate of θ, giving us, in turn, an estimated optimal order size of

$$\hat{y}^* = -t \ln[(u + h)/(s + h)] = (t/\theta)y^*,$$

with corresponding estimated optimal value of the cost function

$$C(\hat{y}^*) = h(\hat{y}^* - \theta) + u\hat{y}^* + \theta(s + h)e^{-\hat{y}^*/\theta}.$$

Now, define the loss function as

$$\ell(t, \theta) = C(\hat{y}^*) - C(y^*)$$
$$= \theta(s + h)[(u + h)/(s + h)]^{t/\theta} - (u + h)/(s + h)]$$
$$+ (u + h)(t - \theta)\ln[(u + h)/(s + h)].$$

The risk function would then be found as the expected value (integral) of $\ell(t, \theta)$ with respect to the conditional probability distribution of t, given θ. If we assume for this problem that θ is estimated as the average of n observed demands, then this conditional distribution is an Erlang with shape parameter n and scale parameter θ/n.

For a second illustration, consider the estimation of a statistical parameter θ by a statistic t. Define the loss function $\ell(t, \theta)$ as the "cost" of estimating θ by t and the sampling distribution of t as $f(t|\theta)$. It follows that the risk function may be written as

$$R(d, \theta) = \int_{-\infty}^{\infty} \ell(t, \theta)f(t|\theta)dt.$$

A common form for the loss function is the quadratic function

$$\ell(t, \theta) = A(t - \theta)^2,$$

because it is proportional to the variance of $t|\theta$. How does this relate to the ordinary hypothesis testing problem $H_0: \theta = \theta_0$ versus $H_1: \theta \neq \theta_0$, for which the decision rule is to accept H_0 if $\theta_{LB} \leq t \leq \theta_{UB}$, and to accept H_1 otherwise? Combining the hypothesis test with the risk framework tells us that we are indifferent with loss λ_0 between all points t, $\theta_{LB} \leq t \leq \theta_{UB}$, while all values of t outside that interval are equally bad with loss equal $\lambda_1 > \lambda_0$.

See **Choice theory; Decision analysis; Risk management; Utility theory.**

References

[1] Clemen R.T. (1991), *Making Hard Decisions*, PWS-Kent, Boston.

[2] Kaplan S. and B.J. Garrick (1981),"On the Quantitative Definition of Risk," *Risk Analysis*, 1, 11–27.

[3] Lowrence W.W. (1976), *Of Acceptable Risk: Science and Determination of Safety*. William Kaufmann, Los Altos, California.

[4] MacCrimmon K.R. and D.A. Wehrung (1986), *Taking Risks: The Management of Uncertainty*. Free Press, New York.

RISK MANAGEMENT

Clyde Chittister

Carnegie Mellon University, Pittsburgh, Pennsylvania

Yacov Y. Haimes

University of Virginia, Charlottesville, Virginia

INTRODUCTION: Most, if not all, engineering systems are conceived, designed, constructed, marketed, and maintained under great unknowns and immense uncertainties. This lack of knowledge is not limited to technological issues such as strength of material, functionality, performance, accuracy, and quality of the components and the total product, but in fact spans a diversity of non-technical areas as well, such as predictions of customer, competitor, and market behaviors, or the anticipation of the product's impact on the organization that manufactures it. Thus, there are risks (measured by the probability and severity of adverse effects) associated with engineering systems; these risks must be managed. Managing risk has been an integral part of engineering since time immemorial; however, what distinguishes the management of risk as practiced two or three decades ago from the management of risk as practiced today is the systemic and methodical approach that must be followed by current management. And, we must emphasize the critical role that computer software plays in the operation of engineering systems. Because of the pervasive nature of computers in society, plus the integrality of software to the business of OR/MS, we direct our discussion in this article of risk management to the software aspect of computing.

Because software has its foundation in mathematics and logic and not in physical laws, the ability of a software engineer to introduce uncertainty into a software system is greater than in any other field. *Only through very stringent management can those uncertainties introduced during the software developement cycle be effectively controlled.* To date these uncertainties have not been effectively controlled; this is because there has not been a well-defined risk assessment and management process for software development.

The increased influence of software in decisionmaking has introduced a new dimension to the way business is done in engineering quarters: many of the used-to-be-engineering decisions have been or soon will be transferred and transformed, albeit in a limited and controlled manner, to the *software* function. This powershift in software functionality, the explicit responsibility and accountability of software engineers, and the expertise required on the job of technical professionals has interesting *manifestations*, *implications*, and *challenges* for the software engineers to adapt to new realities and to *change* – all of which affect the assessment and management of risk associated with software development (Chittister and Haimes, 1994).

Perhaps one of the most striking manifestations of this powershift relates to real-time control systems. Quality control in the manufacture of an engineering component, for example, is no longer primarily the responsibility of the operator; instead, the software controlling the process also controls the quality. Thus, in many respects, the software, which is designed and developed by software engineers, actually controls the process, not the engineers who originally designed the product. This implies that a shift has taken place from a strictly hardware-engineering perspective to a hardware- and software-engineering perspective. Software now fundamentally influences the design of the system. For example, the C-17 transport aircraft has been called the most computerized, software-intensive transport aircraft ever built (General Accounting Office, 1992). Similarly, the Space Station has "[o]n-board computers . . . critical to space craft safety and mission" (General Accounting Office, 1989). Likewise, neither the A320 Airbus nor a number of high technology civilian and military systems can perform their functions without software. Such examples illustrate the difference between software as a manufacturing or implementation mechanism and software as a system-design component.

For example, the decision to update or change operating parameters or entire algorithms based on real-time sensor data received from other sources is now embedded in the software system design. As another example, the data selected to be displayed on one system may be based on information received from other systems. Indeed, the types of changes or updates being implemented by software today would have, in the past, required either system hardware modification or a fundamental redesign of the system. In spite of these examples and others like them, software risk assessment and management, as a specialized entity, with all its importance and implications on other engineering systems and humans, remains an emerging rather than a well-understood activity.

Since software development, in the majority of cases, is an *ad hoc* process (Humphrey, 1990), it is not surprising that the risk identification and management process has been by and large *ad hoc* also. That process, however, can be made systematic and structured even if the software development process is not. The advances in hardware technology and reliability and the seemingly unlimited capabilities of computers render the reliability of most systems to be more heavily dependent on the integrity of the software used. Thus, software failure must be scrutinized with respect to its contribution to overall system failure and with the same diligence and tenacity that have been devoted to hardware failure.

SOFTWARE RISK: In the software development process, the following three basic questions must be posed and answered at each stage regarding risk: (1) What can go wrong? (2) What is the likelihood that it will go wrong? (3) What are the consequences? (Kaplan and Garrick, 1981). Only after these questions have been answered can the final question be asked: What can be done? Determining what can be done entails developing alternative design options; evaluating trade-offs; selecting one or more acceptable options (in terms of cost, reliabil-

ity, performance, total quality, and safety); and evaluating the impact of current policies on future options. To answer the first three questions in the risk assessment process, however, one may benefit from knowledge of the four major sources of failure of systems in general, as well as in software development (Haimes, 1991):

1. Hardware failure.
2. Software failure (which includes software used in the development of software).
3. Organizational failure.
4. Human failure.

The evolving role of the software engineer in decisionmaking has created and continues to create enormous new challenges. The risk of not meeting specified product quality has also shifted. What was once solely the responsibility of traditional engineers who had technical know-how, expertise, and experience is now responsibility shared with software engineers, who design and develop the controlling software.

Although all engineering managers practice risk management in one way or another, only a minority follow this systemic process by looking for sources of failure across the entire system. The intricacy and complexity of the risk assessment and management process (when applied to complex engineering systems) and the need for quantitative analysis (which requires knowledge in probability and statistics, and frequently other content knowledge) have contributed to the emergence of the subspecialization of risk management in engineering. Thus, seeds for two seemingly distinct groups – *engineers as managers of risk*, and *risk experts as managers of engineering systems* – have been sown. In a parallel way, one may trace the distinction between (a) the "engineer" as a technical expert, one primarily concerned with the technical aspects of a project and to a lesser degree with managerial issues; and (b) the "manager," one primarily concerned with management (in the broader and more encompassing sense of the term) and to a lesser degree with technical aspects.

Here again, the *"engineer"* (*as a local manager*) and the *"manager"* (*as a more global manager with a broader vision and perspective*) share responsibilities, tools, and methodologies, yet at the same time, each performs distinct functions, matures in different professional cultures, often uses a different jargon, and communicates with a different language. Understanding this emerging paradigm surrounding the three entities – software engineering, management, and risk analysis – is at the heart of understanding the emergence of software technical risk management.

Furthermore, to appreciate the connectedness among the three elements of this paradigm, one must also understand the hierarchical managerial structure and the consequences of its divisions:

1. *Upper management*: This group views risk almost exclusively in terms of profitability, schedule, and quality. Risk is also viewed in terms of the organization as a whole, and the effect on multiple projects or a product line.

2. *Program management*: Although this group is concerned with profitability, it concentrates more on cost, schedules, product specificity, quality, and performance, usually for a specific program or project.

3. *Technical staff* (software engineers, hardware engineers, etc.): This group of professionals concerns itself primarily with technical details of components, subassemblies, and products for one or more projects.

Clearly, differences among the risk managers at each level of this hierarchical decisionmaking structure are caused by numerous factors, including the scope and level of responsibilities, time horizon, functionality, as well as requirements of skill, knowledge, experience, and expertise. Consequently, these differences determine, to a large extent, the tools and methodologies employed by risk managers at various levels. The management of risk associated with the development of software is governed by the same hierarchical decisionmaking structure and by the same interconnected engineering-management-risk subspecialization paradigm.

TECHNICAL VS. NON-TECHNICAL RISK: The increase in the influence and dominance of software on the system necessarily accompanies an increase in the elements of risk and uncertainty. Although no single classification of risk associated with software development has been developed, a dichotomous model of *software technical risk* vs. *software non-technical risk* is adopted here (Chittister and Haimes, 1994).

This dichotomy between software technical and non-technical risk is introduced not for the purpose of distinguishing between two types of software products; rather, this classification distinguishes various functions in the developmental process of software, and thus, is concerned with the expertise required to deliver each function. Clearly, software technical and non-technical risks are dependent on and influence one another. For example, during a systems integration phase, the developed software may not meet some performance criteria or requirements. In this case, management has several options, including fixing the product and thus delaying the delivery time (and possibly exceeding the budgeted cost) or shipping the product as-is on time. In either case, however, the sources of software technical risk have not changed: only the consequences have been altered.

Software technical risk is defined as *a measure of the probability and severity of adverse effects inherent in the development of software that does not meet its intended functions and performance requirements*. Thus, software technical risk connotes the risk associated with those aspects in the software developmental process that are concerned with the quality, precision, accuracy, and performance over time of the developed software. In other words, software technical risk connotes the risk associated with building a software product that meets intended functions and performance.

On the other hand, software non-technical risk connotes the risk associated with the programmatic aspects in the developmental process of software that are concerned with general management, that is, with personnel, contractor selection, scheduling, budget, and marketing.

Software non-technical risk is defined as *a measure of the probability and severity of adverse effects that are inherent in the development of software and are associated with the programmatic aspects in the development process of software*. Although each type of risk may have an impact on the other, this distinction is still useful because it improves the process of risk assessment and management by establishing causality. Indeed, the distinction between software technical risk (e.g., noncompliance with expected product quality) and software non-technical risk (e.g., cost overruns and delays in scheduled delivery of the product) is helpful in four ways. Indeed, cognizance of the differences between the types of risks should improve their assessment and management, not serve as a detriment to dealing with them. In other words, while the distinction among the multifarious sources and types of risks is important only to the extent that the totality of these sources and type can be accounted for through their inherent differences, *the successful management of risk can be achieved only through an integrated and holistically-based approach*. Since software development is an intellectual, labor-intensive activity, the role of humans and human factors must be carefully understood to properly assess and manage software technical risk.

The sources of risk associated with software development are many and varied. Indeed, at each stage of the software life-cycle (design, development, testing, installation, integration into a larger system, and its ultimate use), one can identify numerous sources of risk.

The road to building a software system is full of surprises. Software often goes through numerous changes, upgrades, fixes, recompiles, and system builds, etc., to address problems; nevertheless, new problems invariably arise. These changes take place because requirements change, people make mistakes, hardware manufacturers make changes to the system in response to marketing information, engineers introduce improvements, and software vendors upgrade their tools. In addition, often there is a break in communication, all these changes necessarily introduce uncertainties into the software development process and into the road to building software.

The ability to predict software problems beforehand has three major components:

1. Identifying and anticipating problems before they happen.
2. Determining the magnitude of potential or existing problems or risks.
3. Communicating the problems or risks to the appropriate people (people who cause, fix, are affected by, or are responsible for the problems).

THE ROLE OF SOFTWARE IN A LARGER SYSTEM:
To understand what software development risk is, and to contribute to its assessment and management through the transfer of knowledge from the hardware engineering field to the software engineering field, one must (a) recognize the salient features and differences between the development processes of hardware engineering and software engineering; (b) understand the role of software engineering within the entire system; (c) appreciate, in the context of design and development, the uniqueness of software failure as juxtaposed against hardware failure, recognizing the importance of all four sources of system failure – hardware, software, organizational, and human; and (d) be familiar with the process of risk assessment and management from a total systems viewpoint.

There would be, in general, a finite and unambiguous number of fundamentally different paths or design options for hardware development to meet a given set of design specifications. Indeed, the extensive use of fault-tree analysis builds on this premise of finiteness. This is not so in the case of the architectural design of software; the number of significantly distinguishable paths or design options of software, for any given specifications, is significantly larger, more ambiguous, and broader. This inherently large number of degrees of freedom in design defies attempts to rely on historical statistics in predicting potential defects, faults, and errors in the development of software.

The design and development of software do not typically follow well-established protocol and commonly accepted procedures. Indeed, in most cases, each software development is envisioned as a unique and distinct product. This lack of a well-developed and acceptable protocol has major implications on several dimensions for the assessment of software development risk.

Hardware has been increasingly taking the *component* role, whereas software has been forcefully assuming the overall *systems* role. Clearly, however, this is not the case in all organizations. This seemingly pivotal development has significant implications for the evolving influence of software engineers on important and critical decisions concerning product design, development, and marketing. The coordination of myriads of components in one system can often be accomplished more cost effectively and with higher reliability through software; this is a marked departure from past practices. This fact has also brought the role of systems engineers to greater prominence in formulating policy affecting product design and development. Furthermore, the software engineers' evolving role in implementation, more than the knowledge that they bring into the project, constitutes another important force in the powershift from hardware to software engineers. Indeed, the last step in the development of a system always involves software engineers, a fact that carries with it more responsibility and implied authority in final product development. It is important to recognize, however, that both hardware and software engineers play an equally important role in the overall system design. In this sense, in the evolution of the powershift from hardware to software one must keep in mind that software development is an intellectually intensive activity where human factors are central. In his chapter on cognitive ergonomics, Sage (1992) makes a forceful argument about the centrality of human interaction with various aspects of the system throughout all phases of a systems engineering life cycle. Therefore, to assess and manage software technical and non-technical risks, one must explicitly address the human element. Sage argues that "a systematic study of human error and approaches to ameliorate the effects of human error in systems and in organizations" is essential in this regard.

As the role that software is assuming in meeting system requirements grows, the impact of software on system risk grows. To be effective and meaningful, risk management must be an integral part of overall system management. This is particularly important in the management of technological systems, especially software-intensive systems, where the failure of a system can be caused by the failure of hardware, software, the organization, or its people.

HIERARCHICAL HOLOGRAPHIC MODELING FOR RISK ASSESSMENT: In the quest to develop an analytical framework for risk management of software engineering it is important to focus on the *sources and causes* of these problems, attempt to group them into a meaningful, yet manageable, number of categories, and then develop a comprehensive framework for dealing with the causes rather than the symptoms. To streamline the discussion and add order to it, a hierarchical structure will be adopted.

Indeed, it is impossible to do justice to a comprehensive framework for the risk assessment and management of software development by boxing it into one planar structure (model). By allowing cross-representations and overlapping models of the various facets and dimensions of the process, hierarchical holographic modeling (HHM) alleviates some of the limitations of a single schema or a single vision of the complex system (Haimes, 1981; Haimes *et al.*, 1990).

Fundamentally, HHM is grounded on the premise that large-scale and complex systems, such as software development, should be studied and modeled by more than a single representation, vision, or schema. And, because such complexities cannot be adequately modeled or represented through a planar or a single vision, overlapping among these visions is not only unavoidable, but can be helpful in a holistic appreciation of the interconnectedness among the various components, aspects, objectives, and decision makers associated with such systems.

The stratagem presented here for risk identification evolves around three hierarchical levels (Chittister and Haimes, 1994). The three major decompositions, visions, or perspectives include: the functional perspective, the source-based perspective, and the temporal perspective.

From a functional perspective, the software development process may be decomposed into the following seven subsystems: requirement, product, process, people, management, environment, and the development system (Figure 1). These terms may be defined as follows:

1. *Requirement*: The highest-level definition of what the product is supposed to do: what needs it must meet, how it should behave, and how the customer will use it. It corresponds to the production perspective.

2. *Product*: The output of the project that will be delivered to the customer. It includes the complete system: hardware, software, and documentation.

3. *Process*: The way by which the contractor proposes to satisfy the customer's requirement. The process is the sequence of steps – their inputs, outputs, actions, validation criteria, and monitoring activities – that leads from the initial requirement to the final delivered product. It includes such phases as requirements analysis, product definition, product creation, testing, and

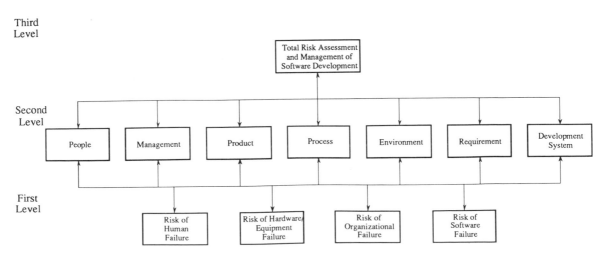

Fig. 1. Total risk assessment and management of software engineering: functional-based hierarchical holographic structure.

delivery. It includes both general management processes, such as costing, schedule tracking, and personnel assignment, and project-specific processes, such as feasibility studies, design reviews, or regression testing.

4. *People*: All those who will be associated with the technical work on the project and all the support staff. It also includes the technical advisers, overseers, and experts, whether in the chain of command or matrixed.

5. *Management*: The line managers at every level who have authority over the project, including those responsible for budget, schedule, personnel, facilities, and customer relations.

6. *Environment*: The "externals" of the project: the factors that are outside the control of the project but can still have major effects on its success or be sources of substantial risk.

7. *Development system*: The methods, tools, and supporting equipment that will be used in the product development. This includes, for instance, CASE tools, simulators, design methodologies, compilers, and host computer systems.

Another vision of the HHM can be obtained through the four sources of system failure discussed earlier (Figure 2):

1. Hardware failure
2. Software failure (software used in the development of software)
3. Organizational failure
4. Human failure

These four sources of failure are not necessarily independent of each other. Just as the distinction between software and hardware is not always straightforward, neither is the separation between human and organizational failure. Nevertheless, these four categories of sources of failure provide a meaningful foundation upon which

to build the decisionmaking hierarchy for the proposed framework. Note that software development is an intellectual, labor-intensive activity that must be streamlined through a well-managed organizational infrastructure and nurtured by an organizational culture and vision that are conducive to and driven by a continuous improvement philosophy.

The third vision of the HHM relates to the evolution of software development over time. Each of the various stages of software development, although often not sharply distinguishable, overlapping, and iterative, constitutes a subsystem in the temporal decomposition (Figure 3). For purposes of this section, the temporal stages are identified as in Humphrey (1990) as: (1) system requirements; (2) software requirements; (3) analysis; (4) program design; (5) coding; (6) testing; and (7) operations.

Each stage (subsystem) in the temporal decomposition can be viewed as one frame in a fixed time (e.g., testing) during the software development process. It is at this fixed-time frame that risks associated with the functional decomposition (e.g., requirement) and with the source-based decomposition (e.g., organizational failure) are identified and articulated. As another example, consider the following four risks that are common during each stage of software development: cost overrun, time delay, not meeting requirements, and not meeting technical quality specifications. The temporal domain has significance far beyond the schedule of the project; it articulates how risks change and evolve over time.

Each of the three hierarchical holographic (HH) submodels developed here contributes to the identification of risk associated with soft-

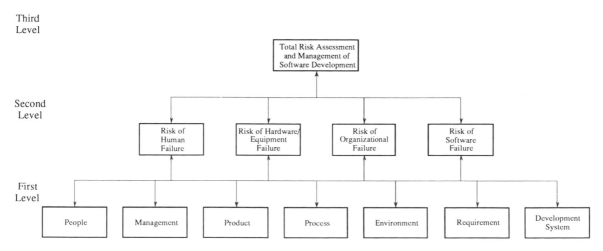

Fig. 2. Total risk assessment and management of software engineering: source-based hierarchical holographic structure.

ware development. The overlappings among these HH submodels mimic the fuzziness that characterizes the real world of software development with respect to the inability to make a clear distinction among the various causes of failures. Figure 1, which is an inverted image of Figure 2, presents an entirely different perspective in answering the set of triple questions: What can go wrong? What is the likelihood that it will go wrong? And what would the consequences be? Since a central objective of risk assessment is to identify, to the extent possible, everything that can go wrong, then a hierarchical holographic modeling structure is superior to a planar single model in this respect. Note, for example, that in Figure 1, the four sources of risk (human, hardware, organizational, and software) are investigated for each subsystem of the func-

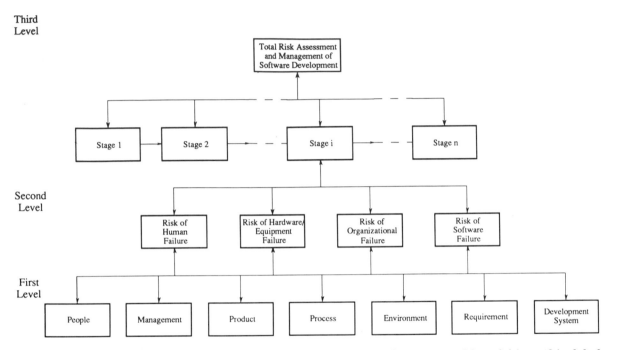

Fig. 3a. Total risk assessment and management of software engineering: temporal-based hierarchical holographic structure.

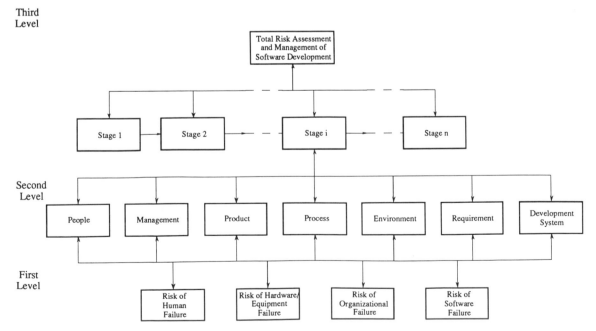

Fig. 3b. Total risk assessment and management of software engineering: temporal-based hierarchical holographic structure.

tional decomposition (people, management, product, process, environment, requirement, and development systems). On the other hand, Figure 2 depicts a different perspective; namely, the seven functional decompositions are investigated for each of the four sources of risk of failure. Figure 3 incorporates the temporal decomposition that captures the stagewise evolutionary process of software development, and thus the risk associated with each stage and for each subsystem of the functional and source-based decompositions. In particular, Figures 3a and 3b extend Figures 1 and 2 by incorporating the temporal domain into the HHM.

SUMMARY AND CONCLUSIONS: Software will continue to grow in size, complexity, and importance as it assumes more functionality in large, complex systems. If engineers and managers working in the community do not embrace a risk management ethic for software development, then software problems will continue to grow as well. Although practicing risk management does not guarantee fewer problems, it does provide a structure with which to make better decisions about the uncertainty and impact of future events. If risks can be measured, then contingency strategies can be provided; however, if risks are unknown, then surprise is likely when it is least convenient.

Although software engineering is different from other engineering disciplines, the management of risk in the developmental process is critical for all engineering disciplines. The framework for identifying and assessing risk in the software development process is grounded on the premise that software development is an intellectual, labor-intensive activity, thus making the human factor central to the assessment and management of risk.

As systems become larger and more complex, the assessment and management of risk must be a team effort. The team has to include, among others, the system developers, support staff from the organization, and management. Risk management is neither just the program manager's job nor just a technical issue. Financial and quality risks are as important as software technical risk.

The more diversified the team, the more important it is to have a common and agreed-upon risk assessment and management process. The members of the team will have their own technical jargon and their own frames of reference. If each subgroup identifies and manages risks differently, there will be no common ground for com-

munication or measurement. A systematic and structured process that is used by everyone will provide a foundation for discussion and for mitigation strategies. This process will also greatly reduce confusion caused by misunderstanding, which is itself a source of risk in large complex systems.

Indeed, modeling and managing software technical risk must be an activity that recognizes the intricacy of the internal and external environment within which software development is practiced. Depending on the forces exerted and on the software development practice itself, two types of risks are likely to emerge – software technical and non-technical risks. Indigenous to these forces is the powershift from hardware to software; consequently, such change must be recognized and managed. In *Changed Agents*, London (1990) summarizes his views on organizational change:

Incremental change merges the new with the old. It requires a willingness to be open to new ideas and to continuously refine and possibly extend the goals of the organization. Frame-breaking change is dramatic and often sudden. Though resistance is likely, the organization's survival depends on re-creating the organization's mission, structure, staff, and modes of operation.

Indeed, the risks of not meeting product quality and performance, cost, and schedule can be successfully identified, quantified and measured, evaluated, and managed only when a systemic and holistic process of assessment and management is employed. Such a process is an organizational recipe for long-term sustainable development. Toffler (1990) best articulated the imperative of properly coping with technological change. The software engineering community is traveling in unexplored terrain; the success of its journey depends, to a large extent, on its ability to bring the larger systems community into the realization that acknowledging and responding to the powershift in the software area is a first and critical step in a successful management of software technical and non-technical risk.

No one would argue that the best way to manage problems is to keep them from happening. A risk ethic that is embraced and practiced by an entire organization will significantly reduce the chaos created by unknown risks and crisis situations.

See **Quality control; Risk assessment.**

References

[1] C. Chittister and Y.Y. Haimes (1993), "Risk Associated with Software Development: A Holistic Framework for Assessment and Management," *IEEE Trans. Systems, Man, & Cybernetics*, 23, 710–723.

[2] C. Chittister and Y.Y. Haimes, "Assessment and Management of Software Technical Risk," *IEEE Trans. Systems, Man, & Cybernetics*, 24, 187–202.

[3] General Accounting Office (1989), "Automated information system," Washington, D.C.: Government Printing Office.

[4] General Accounting Office (1992), "Embedded computer systems: Significant software problems on C-17 must be addressed," Washington, D.C.: CAO/IMTEC-92-48, Government Printing Office.

[5] Y.Y. Haimes (1981), "Hierarchical holographic modeling,"*IEEE Transactions on Systems, Man, and Cybernetics*, SMC-11, 606–617.

[6] Y.Y. Haimes (1991),"Total risk management," editorial in *Risk Analysis*, 11, 169–171.

[7] Y.Y. Haimes, K. Tarvainen, T. Shima and J. Thadathil (1990), *Hierarchical Multiobjective Analysis of Large Scale Systems*. Hemisphere Publishing, New York.

[8] W.S. Humphrey (1990), *Managing the Software Process*, Addison-Wesley, Reading, Massachusetts.

[9] S. Kaplan and B.J. Garrick (1981), "On the Quantitative Definition of Risk," *Risk Analysis*, 1, 11–27.

[10] M. London (1990), *Change Agents*: *New Roles and Innovation Strategies for Human Resource Professionals*, Jossey-Bass, San Francisco.

[11] W.W. Lowrence (1976), *Of Acceptable Risk*: *Science and Determination of Safety*. William Kaufmann, Los Altos, California.

[12] A.P. Sage (1992), *Systems Engineering*, John Wiley, New York.

[13] A. Toffler (1990), *Powershift*, Batman Books, New York.

RITTER'S PARTITIONING METHOD

A procedure for decomposing and solving a linear-programming problem that has both coupling constraints and coupling variables. See **Block-angular system**.

ROBUSTNESS ANALYSIS

(1) A problem structuring method for sequential decision-making under uncertainty. It provides a criterion for assessing the useful flexibility maintained by initial decision commitments. Single- and multiple-scenario methodologies exist, as do versions which incorporate the extent to which unwelcome future configurations remain accessible. See **Problem structuring methods**. (2) More generally, examining whether the solution to a problem is stable through minor alterations of the model inputs. See **Sensitivity analysis**.

ROSEN'S PARTITIONING METHOD

A procedure for decomposing and solving a linear-programming problem that is a block-angular system with either coupling constraints or coupling variables. See **Block-angular system**.

ROUNDOFF ERROR

The computational error due to the significant-digit arithmetic inherent in digital calculations.

ROUTE CONSTRUCTION HEURISTIC

A vehicle-routing heuristic that builds a feasible solution by inserting at every iteration unrouted customers into a current partial vehicle route. See **Vehicle routing**.

ROUTE IMPROVEMENT HEURISTIC

A local improvement heuristic for vehicle routing. See **Vehicle routing**.

ROW VECTOR

One row of a matrix or a matrix consisting of a single row.

RULE

A named fragment of reasoning knowledge consisting of a premise and a conclusion. In addition, a rule may have other attributes such as a priority, a cost, a preaction sequence, a premise testing strategy, a textual description, and an internal comment. See **Artificial intelligence**; **Expert systems**.

RULE SET

A named collection of rules that represent reasoning knowledge about some problem area. A rule set is used by an inference engine to solve specific problems in that area. In addition to rules, a rule set may also contain an initialization sequence, a completion sequence, and variable descriptions. See **Artificial intelligence**; **Expert systems**; **Inference engine**.

RUNNING TIME OF AN ALGORITHM

See **Computational complexity**.

S

SA

See (1) **Sensitivity analysis**; (2) See **Simulated annealing.**

SADDLE-POINT OF A FUNCTION

For an arbitrary payoff function $F(x, y)$, the point (x^0, y^0) is a saddle point if $F(x^0, y) \leq F(x^0, y^0) \leq F(x, y^0)$. See **Saddlepoint problem.**

SADDLE-POINT OF A GAME

For a zero-sum, two-person game, if an element a_{ij} of the payoff matrix is the minimum of its row and maximum of its column, it is a saddle point. The value of the game is equal to the value of the saddle point, with the maximizing player's optimal strategy being the pure strategy i and the minimizing player's optimal strategy being the pure strategy j. See **Game theory; Saddle-point of a function.**

SADDLEPOINT PROBLEM

For the mathematical-programming problem: Minimize $f(x)$, subject to $\{g_i(x) \leq b_i\}$, the saddle-point problem is to find vectors x^0 and y^0 such that $F(x^0, y) \leq F(x^0, y^0) \leq F(x, y^0)$, where $F(x, y)$ is the associated Lagrangian function, $y \geq 0$. See **Saddle-point of a function.**

ST. PETERSBURG PARADOX

See **Bayesian decision theory, subjective probability and utility.**

SAFETY

Igor Ushakov

SOTAS, Rockville, Maryland

Safety is a property of a system which permits the system to operate without dangerous consequences for people (including serving personnel) and the environment. For many systems (as aircraft, submarines, chemical plants, nuclear power stations, etc.), some kinds of failures can lead to catastrophic results. In these cases, the safety indices coincide with reliability indices after the choice of the appropriate criteria for defining failure. These might be the (complementary) probability of successful operation without accident, the mean time to accident appearance, and so on.

Sometimes, the safety of systems (as for dams of hydro power stations, constructions in seismic zones, etc.) is considered to be only under the influence of nature. In this case, probabilistic measures may be insufficient and one should instead consider conditional safety under some specified levels of external influence.

But many systems may be harmful even under ideal conditions, without accidents. Examples are various chemical and metallurgical technological processes, power stations, and other objects polluting the environment with various toxic substances.

To quantify, begin by letting $f(t)$ be the poisonous emission function in time. One useful index of safety of such a system is the condition that, for some specified time interval of width Δ,

$$\int_t^{t+\Delta} f(t) dt \leq f^o$$

where the threshold f^o is given.

If there is some reduction process $\phi(t)$ which lowers the harmful consequences described by $f(t)$, then an appropriate index could be

$$\int_t^{t+\Delta} [f(t) - \phi(t)]_+ dt \leq f^o$$

where $[\cdot]_+$ denotes the positive part of the number in brackets.

Many harmful processes (radioactive emission, dioxide pollution, etc.) exponentially calm down with time. In this case, a good safety criterion might be

$$\int_t^{t+\Delta} f(t) e^{-\alpha t} dt \leq f^o$$

where α is the intensity of the "recovery."

Rudenko and Ushakov (1989) and Ushakov (1994) have provided more complete discussions of the issues outlined here, particularly the relationship of classical reliability modeling to the analysis of safety.

See **Redundancy; Reliability of systems.**

References

[1] Rudenko Yu. N. and I.A. Ushakov (1989). *Reliability of Power Systems* (in Russian), 2nd edition, edited by B.V. Gnedenko. Nauka, Novosibirsk.

[2] Ushakov I.A. ed. (1994). *Handbook of Reliability Engineering*. John Wiley, New York.

SAND TABLE BATTLE MODEL

A model or game with a physical representation of the geography and units. The classical sand table used sand because it could be molded into a model of the terrain's relief. Tin soldiers were used to represent the troops. See **Battle modeling.**

SATISFICING

In a decision problem, the selection by the decision maker (DM) of a satisfactory alternative as opposed to the selection of an "optimal" alternative. Here, the DM sets aspiration levels or acceptable levels on the outcomes and chooses the (first) alternative that satisfies these levels. This compromise selection is due to the DM's inability to encompass all the complexities of the decision problem and/or lack of a method that can determine an optimal solution. See **Bounded rationality; Choice theory; Decision analysis; Decision maker; Decision problem; Goal programming.**

SCALING

The pre-solution transformation of the data of a problem that attempts to make the magnitudes of all the data as close as possible. Such scaling is important for mathematical- and linear-programming problems as it helps to reduce roundoff error. Most mathematical-programming systems have a SCALE command that automatically adjusts the magnitudes of the data in the rows and columns. This can be done by multiplying the technological coefficient matrix A by suitable row and column transformation matrices. A frequently used scaling algorithm is to divide each row by the largest absolute element in it, and then divide each resulting column by the largest absolute element in it. This ensures that the largest absolute value in the matrix is 1.0 and that each column and row has at least one element equal to 1.0.

SCENARIO

The set of conditions and characteristics that define the situation or environment under which a system or policy has to perform. There is often a baseline scenario (what will happen if trends continue) and an ideal scenario (what future one would like to have). See **Battle modeling; Sensitivity analysis.**

SCENARIO ANALYSIS

See **Stochastic programming.**

SCERT

Synergistic Contingency Evaluation and Response Technique, which uses a systematic approach for the identification and articulation of the risks to which a project is subject and the uncertainties and contingencies which might significantly affect the outcome of the project. See **Network planning.**

SCHEDULING AND SEQUENCING

Michael Magazine

University of Waterloo, Ontario, Canada

INTRODUCTION: The term scheduling represents the assignment of resources over time to perform some tasks, jobs or activities. We concentrate here on applications to machine scheduling in factory and computer systems. Other uses of the term scheduling include:

(a) Production scheduling – the determination of which and how many products to produce over time;

(b) Workforce scheduling – the determination of the number of workers and their duty cycles to meet certain labor restrictions; and

(c) Timetabling – the determination of the matching of participants with each other and resources, such as sports scheduling or student/room exam assignments.

Scheduling has been done informally for centuries. The Gantt chart, developed in World War I for logistics purposes, is a graphical representation of tasks and resources over time, and was the first formal model used for scheduling purposes. Critical Path Methods (CPM) followed and are still widely used. The 1950s saw the growth of models used to analyze scheduling problems; their proliferation continues.

The reason for this continuing activity is the importance and amount of time that is spent by organizations on the scheduling function. This function is typically at the operational level and assumes the planning phases of which tasks are to be done and what resources are available to do them have been completed.

In spite of much research in this area, there are few results that we can point to as having a major impact in applications and very little available as a unifying theory (McKay *et al.*, 1988). This is because of the very difficult nature of these problems and the myriad of special constraints and circumstances of the applications. Nevertheless, there are some results which have had a major impact. These, along with some of the basic tools of scheduling will be described, after we give some common definitions.

For those interested in reviews of the field there are some excellent texts and surveys: Baker (1992); Conway, Maxwell and Miller (1967); French (1982); Lawler *et al.* (1993); and Morton and Pentico (1993).

PRELIMINARIES: Scheduling involves the determination of two types of decisions:

- allocation decisions, namely, which resources (here called machines) will be assigned to perform each of a given set of tasks (referred to as jobs); and
- sequencing decisions, namely, when do we perform each of these jobs.

Even though many applications involve the determination of a schedule that is simply feasible, given the scarce resources, most models use some economic measure of performance enabling us to compare different schedules. These objectives typically represent some surrogate of throughput, customer satisfaction or costs. In general, the problems focus on a single objective at a time, although there are some results on multicriteria problems (Daniels, 1990).

Some of the constraints or considerations that must be included in these models are precedence relations amongst jobs, job priorities, setup times of machines, and preemption capabilities. McKay *et al.* (1988) identified over 600 types of constraints present in manufacturing environments. The machine environment also determines our ability to solve these problems. We consider single machine problems, several machines in parallel, *flow shops* which involve several machines in series, and *job shops* which involve several machines without the series structure.

Also influencing the difficulty of solving these problems and the techniques that can be used is the preciseness of the knowledge of the problem data. Most models and results assume the data are deterministic, although there exist some results for stochastic or probabilistic environments (Pinedo and Schrage, 1982).

SOLUTION TECHNIQUES: Since most scheduling problems have economic criteria and constraints, it is natural to use optimization methods to solve these problems. The number of feasible schedules grows quickly with problem size, making almost all such problems extremely difficult to solve. Consider the simplest problem with n jobs on a single machine, with no additional constraints. There are $n!$ sequences which must be considered, and even for relatively small problems of size, say $n = 100$, the enumeration task is virtually impossible. In the mid 1970s, all but a handful of scheduling problems were classified by complexity theory as hard, or more formally, NP-complete (or NP-hard). In some sense, the work by Dempster, Lenstra and Rinnooy Kan (1982) and others cited therein represents the only unified theory in the area.

Nevertheless, these hard problems need to be solved. Traditional optimization techniques such as mathematical programming, enumeration methods, and dynamic programming have had limited success. Quite often, it is necessary to resort to heuristic techniques which do not guarantee an optimal solution. These techniques have been of two types: limit the space of search by only considering schedules that meet some specified criteria or search in a limited neighborhood of some known feasible schedule. These heuristics have shown some, but, unfortunately, most of this work considers only their worst-case performance, which is not necessarily a true measure of the goodness of a heuristic. Advances in heuristics, including genetic algorithms, tabu search and simulated annealing open new possibilities in our ability to find good solutions.

Surprisingly, it is a simple heuristic that has given us the most significant scheduling results. The principle used is pairwise interchange. This neighborhood search technique starts with a (typically) feasible schedule to the problem and interchanges the sequence of two of the scheduled tasks according to some rule. In some classes of problems, this procedure will always result in an optimal solution.

In problems with stochastic form, some of these same combinatorial techniques can be used. Some simple problems can take advantage of queueing theory results, but, by and large, simulation becomes the technique of choice. Often, the complicated environment or simply the highly combinatorial nature of the problem is enough to make simulation the only viable alternative.

SCHEDULING RESULTS: We next describe some results which are key building blocks to scheduling theory and applications. These arise from problems with a single, continuously available machine; with independent, one-operation, and deterministic jobs; and with machine set-up times that are negligible or can be ignored, or are included in the processing time of the jobs.

Single-Machine Models – We assume there are n jobs with characteristics:

p_j = processing time required by job j;

r_j = ready time of job j, that is, the earliest time the job can commence processing;

d_j = the time job j is due to have processing completed.

Once a schedule has been determined additional characteristics emerge:

C_j = completion time of job j;

F_j = flowtime of job $j = C_j - r_j$;

L_j = lateness of job $j = C_j - d_j$;

T_j = tardiness of job $j = \max(0, L_j)$.

The simplest models assume criteria which are non-decreasing in the completion time of jobs. These are called regular measures and include:

$$\text{Total Flowtime} = F = \sum_{j=1}^{n} F_j;$$

$$\text{Total Lateness} = L = \sum_{j=1}^{n} L_j,$$

$$\text{Maximum Lateness} = L_{\max};$$

$$\text{Total Tardiness} = T = \sum_{j=1}^{n} T_j,$$

$$\text{Maximum Tardiness} = T_{\max}.$$

In these problems, a permutation of the integers $1, 2, \ldots n$, called a *permutation schedule*, is always optimal. That is, optimal schedules exist in which job preemption and machine idle time do not occur.

The following is an important result of scheduling theory:

Shortest Processing Time (SPT) sequence, that is, sequencing the jobs from shortest to longest processing time, minimizes total flowtime when $r_j = 0$ (Smith, 1956).

This is proved by interchange arguments and can be extended to include jobs of different importance or weights.

For customer service measures, we include due date information. It may be surprising but SPT minimizes total lateness (L). The most important result using due date measures is:

Earliest Due Date (EDD), i.e., sequence jobs from earliest to latest due date, minimizes both $L_{\max}$ and $T_{\max}$.

In many other problems with due-date measures, EDD provides basic information in reaching an optimal solution.

There are several other single machine results. Having $r_j > 0$ causes problems with due date measures unless jobs can be preempted without penalty. If this is not the case, it is necessary to look ahead to jobs not yet in the system, forcing simple static sequencing rules, such as SPT or EDD, to yield suboptimal solutions.

Another condition imposed on jobs may be precedence relations. They represent a partial ordering of the jobs which is imposed for technological reasons. These relationships (which can be depicted by graphs where job i immediately precedes job j if there is an arc from i to j), may, under certain conditions, still yield relatively simple algorithms. When the graph has strings or is series-parallel, many problems can be solved optimally, whereas, general relationships between jobs will not yield optimal solutions (Monma, 1981).

As noted, these simple models ignore setup times. When setup times are present and sequence-dependent, many simple problems become very difficult to solve. For example, minimizing *makespan*, that is, $C_{\max}$ = completion time of last job, is as hard to solve as the traveling salesman problem.

We have thus far considered only regular measures of performance. One non-regular measure, that represents another attribute of customer service, is a job earliness penalty. That is, a job's earliness is $E_j = \max(0, d_j - C_j)$. When there are many jobs and a common due date, minimizing a combination of earliness and tardiness penalties is still possible (Baker and Scudder, 1990). When the due dates are distinct for each job, the problems are very difficult to solve.

Multiple-Machine Models – Results for several machines are scarce. The existence of several machines requires not only sequencing decisions, but also allocation decisions. The first of these models assumes there are n jobs to be processed and m machines available for processing. The simplest of such models assumes the machines are identical and that all jobs have the same ready time.

The most basic model attempts to minimize makespan, that is, the time to complete all n jobs. Since changing the sequence of jobs allocated to a particular machine does not affect the makespan, the only decision is allocating jobs to machines. If preemption is permitted, Mc-Naughton (1959) shows that there is a simple algorithm to do this allocation. When preemption is not permitted, the problem is NP-hard even if $m = 2$.

Here, a reasonable heuristic seems to be to list the jobs in some prespecified order, placing the next job in the list onto the first machine that becomes available. By cleverly ordering the jobs, list scheduling heuristics have acceptable performance. For example, a simple rule such as longest processing time first (LPT) guarantees a solution within 33% of the optimal makespan (Graham, 1969). Graham also finds several anomalies in list scheduling, such as increasing the number of machines or reducing the processing time of the jobs can increase makespan. Dobson (1984) finds similar results on performance guarantees, even when the machines have different speeds. Naturally, when precedence relationships exist amongst jobs, the problem is more difficult, although, under certain instances, we can find performance guarantees.

When our objective shifts to minimizing total flow time on m identical machines, an SPT list scheduling algorithm gives an optimal solution. Almost all other problems that can be generalized from the single machine case, unfortunately, do not yield optimal solutions in polynomial time.

The next important class of problems are called *flow-shop models*. The simplest instance assumes there are m machines and each job requires processing on each machine, that is, each job has m operations. In addition, processing moves from machine to machine in a prespecified order that is the same for each job. These problems are much more difficult than single machine problems. One additional consideration is that inserted idle time may be desirable. This was not the case for single machines with regular performance measures. In addition, different permutations of the jobs must be considered for each machine, giving rise to $(n!)^m$ possible sequences.

When $m = 2$, however, Johnson's rule gives rise to an efficient algorithm for minimizing makespan (Johnson, 1954). When there are more than two machines the makespan problem becomes hard. There are several special cases, however, when $m = 3$ that use variations of Johnson's algorithm to guarantee optimal solutions.

We state Johnson's theorem (algorithm) using his notation (Johnson, 1954). Here,

A_i = the processing-time (including setup, if any) of the first operation of ith job;

B_i = the processing-time (including setup, if any) of the second operation of ith job;

F_i = the time at which the ith job is completed.

Johnson's algorithm is applied to the following (Johnson's) problem: Sequence an arbitrary number of jobs in a two-machine flow shop to minimize the maximum flow time. We assume that all jobs are simultaneously available.

Johnson's Theorem: An optimal schedule for a two-machine flow shop that minimizes maximum flow time will be obtained when job j precedes job $j + 1$ if

$$\min(A_j, B_{j+1}) < \min(A_{j+1}, B_j).$$

Under certain conditions, the processing of an operation of job j can start before the completion of the previous operation. This overlapping of operations certainly can improve the makespan. Baker (1992) calls these problems lot streaming.

One variation of the flow shop model requires that processing a job cannot be delayed once processing of that job starts. A survey of these no-wait models is given in Hall and Sriskandarajah (1994).

Another extension of the flow shop model allows the job to pass through the machines in any order. This is referred to as an open shop and, except for the case of two machines with makespan criterion, the problems are essentially all very hard.

Job shop scheduling encompasses a significant portion of all factory scheduling problems. In the job shop, the number and order of operations for each job may differ. These problems are virtually intractable as permutation schedules are not available. The notorious 10-job, 10-machine problem of Fisher and Thompson (1963) provides the benchmark for the computational difficulty of these problems. It took 25 years of research for its optimal solution to be verified (Carlier and Pinson, 1988). Heuristic procedures attempt to consider dispatching rules. These algorithms choose, according to some rule, the next job from those available to start on a machine only when that machine becomes idle. Many such rules are tested using simulation studies.

OTHER RESULTS: *Generalized Resources* – Scheduling problems often involve resources that are not continuously available, get depleted by use, or are subject to failure or deterioration. Scheduling in this environment is called resource-constrained project scheduling. Many results can be found in Blazewicz, Lenstra and Rinnooy Kan (1983).

Stochastic Machine Scheduling – In many situations, job parameters are known with certainty. Processing times may, however, be subject to fluctuation or job arrivals may not be known in advance. At times the deterministic problem has a direct stochastic counterpart. There are several cases where the algorithm which is used as a heuristic in the deterministic version of a problem can guarantee an optimal solution in the stochastic version. In most cases, however, stochastic problems are hard, not only in computational effort, but require a different set of techniques such as semi-Markov decision theory.

Scheduling Families or Groups – Modern manufacturing facilities contain flexible machines which can produce or assemble a variety of products. When products are similar, switching between them requires no set-up time. These groups of products are called product *families*. Switching between different product families is possible but requires a set-up. The difficulty in scheduling these environments is the trade-offs that occur when large batches (a batch represents the set of items produced in a single set-up) occur – causing delays for jobs in other families – and small batches which will incur many setups. The issues, typically, involve sequencing items within families, the determination of the batch sizes and the sequencing of the batches from different families (Santos and Magazine, 1985).

See **Combinatorial and integer optimization; Combinatorics; Flexible manufacturing; Genetic algorithms; Inventory modeling; Job shop scheduling; Operations management; Production management; Simulated annealing; Tabu search.**

References

[1] Baker K. (1992), "Elements of Sequencing and Scheduling," Technical Report, Dartmouth College, Hanover, New Hampshire.

[2] Baker K. and Scudder G. (1990), "Sequencing with Earliness and Tardiness Penalties: A Review," *Opns. Res.*, 38, 22–36.

[3] Blazewicz J., Lenstra J.K. and Rinnooy Kan A. (1983), "Scheduling Subject to Resource Constraints," *DAM*, 5, 11–24.

[4] Carlier J. and Pinson E. (1988), "An Algorithm for Solving the Job-Shop Problem," *Mgmt. Sci.*, 35, 164–176.

[5] Conway R., Maxwell W. and Miller L. (1967), *Theory of Scheduling*, Addison-Wesley, Reading, Massachusetts.

[6] Daniels R. (1990), "A Multi-Objective Approach to Resource Allocation in Single Machine Scheduling," *Euro. Jl. Opns. Res.*, 48, 226–241.

[7] Dempster M., Lenstra J. and Rinnooy Kan A. (1982), *Deterministic and Stochastic Scheduling*, Reidel, Dordrecht.

[8] Dobson G. (1984), "Scheduling Independent Tasks on Uniform Processors," *SIAM Jl. on Comp.*, 13, 705–716.

[9] Fisher H. and Thompson G. (1963), "Probabilistic Learning Combinations of Local Job-Shop Scheduling Rules" in J. Muth and G. Thompson (eds.), *Industrial Scheduling*, 225–251, Prentice-Hall, Englewood Cliffs, New Jersey.

[10] French S. (1982), *Sequencing and Scheduling*, Ellis Horwood, U.K.

[11] Graham R. (1969), "Bounds on Multiprocessor Timing Anomalies," *SIAM Jl. of App. Math.*, 17, 416–425.

[12] Hall N. and Sriskandarajah C. (1995), "A Survey of Machine Scheduling Problems with Blocking and No-Wait in Process," to appear in *Opns. Res.*

[13] Johnson S. (1954), "Optimal Two and Three Stage Production Schedules with Setup Times Included," *Naval Res. Logistics Qtrly.*, 1, 61–68.

[14] Lawler E., Lenstra J., Rinnooy Kan A. and Shmoys D. (1993), "Sequencing and Scheduling: Algorithms and Complexity," in *Handbooks in Operations Research and Management Science*, Vol. 4, Logistics of Production and Inventory, Graves, S., Rinnooy Kan, A. and Zipkin, P., Editors.

[15] McKay K., Safayeni F. and Buzacott J. (1988), "Job-Shop Scheduling Theory: What is Relevant," *Interfaces*, 18(4), 84–90.

[16] McNaughton R. (1959), "Scheduling with Deadlines and Loss Functions," *Mgmt. Sci.*, 6, 1–12.

[17] Monma C. (1981), "Sequencing with General Precedence Constraints," *Math. Opns. Res.*, 4, 215–224.

[18] Morton T. and Pentico D. (1993), *Heuristic Scheduling Systems*, John Wiley, New York.

[19] Pinedo M. and Schrage L. (1982), "Stochastic Shop Scheduling: A Survey" in M. Dempster, *et al.*, *Deterministic and Stochastic Scheduling*, Reidel, Dordrecht, 181–196.

[20] Santos C. and Magazine M. (1985), "Batching in Single Operation Manufacturing Systems," *Opns. Res. Letters*, 4, 99–103.

[21] Smith W. (1956), "Various Optimizers for Single Stage Production," *Naval Res. Logistics Qtrly.*, 3, 59–66.

SCORE FUNCTIONS

Reuven Y. Rubinstein

Technion – Israel Institute of Technology, Haifa

Alexander Shapiro

Georgia Institute of Technology, Atlanta

Stanislav Uryasev

Brookhaven National Laboratory, Upton, New York

Many complex real world systems can be modeled as discrete-event systems (DES). Examples are computer-communication networks, flexible manufacturing systems, probabilistic fracture mechanics models, PERT-project networks and flow networks. In view of the complex interaction of such DES they are typically studied via stochastic simulation.

In designing and analyzing such complex DES, we are often interested not only in performance evaluation but in *sensitivity analysis* and *optimization* as well. Consider for example *manufacturing systems*. Here (i) the performance measure may be the average waiting time of an item to be processed at several work stations (robots) according to a given schedule and route; (ii) the sensitivity and decision parameters may be the average rate at which the workstations (robots) process the item. In such a system we might be interested in minimizing the average make-span consisting of the processing time and delay time with allowance for some constraints (e.g., cost).

Alternatively, consider *failure probability models for mechanical passive components* (like pipes, vessels). Here (i) the performance measure may be cumulative failure or leakage probability of a passive component over some time; (ii) the sensitivity and decision parameters may be the geometry of the component (thickness of the walls), fracture toughness of the material and stress intensity factors. In such a system failure probability is a function of the sizes of defects (cracks) and their time dependent stochastic development.

In the last decade, two new methods for sensitivity analysis and optimization of DES have been developed. They are called *infinitesimal perturbation analysis* (IPA) (Ho and Cao, 1991; Glasserman, 1991) and the *score function* (SF) (also called *likelihood ratio* (LR)) methods (Rubinstein, 1976; Reiman and Weiss, 1989; Glynn, 1990; L'Ecuyer, 1990; Rubinstein and Shapiro, 1993). The SF method allows one to evaluate, *simultaneously* from a *single* sample path (simulation experiment) not only the performance and all its sensitivities (gradient, Hessian, etc.), but to solve an entire optimization problem as well.

ESTIMATION AND SENSITIVITY ANALYSIS OF DISCRETE-EVENT STATIC SYSTEMS:

Let $\ell(\theta)$ be a real-valued function representable in the form $\ell(\theta) = E_\theta[L(Y)]$. Here Y is a random vector whose cumulative distribution function (CDF) $F(y, \theta)$ depends on the parameter vector $\theta \in \Theta$ with Θ being a subset of a finite dimen-

sional vector space, say $\Theta \subset \mathbb{R}^n$. The function $L(Y)$ can be viewed as a sample performance driven by the input vector Y. The notation E_θ stands for the expectation with respect to the CDF $F(y, \theta)$.

In order to estimate the expected value $\ell(\theta)$ by simulation (Monte Carlo) techniques, one can proceed as follows. Generate a sample $Y_1, \ldots, Y_N$ from the CDF $F(y, \theta)$ and set the sample mean $N^{-1}\Sigma_{i=1}^N L(Y_i)$ as the corresponding estimator. Of course, this procedure requires generation of a new sample every time the expectation $\ell(\theta)$ should be estimated for a different value of the parameter vector $\theta \in \Theta$. In order to overcome this difficulty, let us consider the following procedure based on a change of probability measure techniques. Suppose that the random vector Y has a probability density function (PDF) $f(y, \theta)$ corresponding to the CDF $F(y, \theta)$. Let us choose a PDF $g(y)$. Then, provided the outcome space of $f(\cdot, \theta)$ lies in the outcome space of $g(\cdot)$, we can write:

$$\ell(\theta) = \int L(z)[f(z, \theta)/g(z)]g(z)d\mathbf{z} = E_g[L(Z)W(Z, \theta)],$$

(1)

where $W(z, \theta) = f(z, \theta)/g(z)$. Note that the integration variable y was replaced by z when the involved densities were changed from $f(\cdot, \theta)$ to $g(\cdot)$, and the integrals are over the outcome spaces of the corresponding density functions. Formula (1) suggests the following way for estimating $\ell(\theta)$. Generate a sample $Z_1, \ldots, Z_N$, from the PDF $g(z)$ and estimate $\ell(\theta)$ by

$$\hat{\ell}(\theta) = N^{-1} \sum_{i=1}^N L(Z_i)W(Z_i, \theta).$$

(2)

The function $W(z, \theta)$ is called the *likelihood ratio* (LR) function and $g(y)$ is the *dominating density*. Typically one takes $g(z) = f(z, \theta_0)$ for a particular value $\theta_0 \in \Theta$ of the parameter vector. We refer to the chosen θ_0 as the *reference value* of the parameter vector.

The LR function is given explicitly through the corresponding density functions and is typically smooth (differentiable) in θ. As soon as the sample $Z_1, \ldots, Z_N$ is generated, the obtained LR estimator $\ell_N(\theta)$ becomes an analytical function of θ and provides an estimate of the entire function (response surface) $\ell(\theta)$. Moreover, under mild regularity conditions ensuring interchangeability of the integration and differentiation operators, it follows that derivatives of the expected performance $\ell(\theta)$ can be taken inside the expected value representation given in the right-hand side of equation (1). Consequently, the gradient $\nabla\hat{\ell}_N(\theta)$ of the sample estimate, as defined in Equation (2), provides an unbiased estimator of the corresponding gradient of $\ell(\theta)$. The Hessian matrix can be similarly estimated (Rubinstein and Shapiro, 1993).

In particular, for $g(\cdot) = f(\cdot, \theta)$, the gradient of the LR function $W(y, \theta)$ is called the *score function* (SF) and can be written in the form $S(y, \theta) = \nabla \log f(y, \theta)$. Then, given a random sample $Y_1, \ldots, Y_N$ from $f(y, \theta)$, the gradient $\nabla \ell(\theta)$ can be estimated by:

$$\bar{\nabla}\ell_N(\theta) = N^{-1} \sum_{i=1}^{N} L(Y_i) S(Y_i, \theta) \ . \tag{3}$$

High-order derivatives can be handled in a similar way. Note that gradient $\nabla W(z, \theta)$ of the LR function can be written in the form $W(z, \theta) S(z, \theta)$ and is called the *generalized score function*.

A word of caution is due. Although the generalized SF estimators typically are unbiased and consistent, their accuracy is determined by the corresponding variances and can be quite sensitive to the choice of the dominating PDF $g(y)$ (reference value θ_0). The problem of an optimal choice of $g(y)$ is closely related to the *importance sampling* method in simulation. A detailed discussion of this problem and relevant variance reduction techniques may be found in Rubinstein and Shapiro (1993).

EXAMPLE 1 (SYSTEM RELIABILITY): Consider the sample performance function:

$$L(Y) = \max_{1 \le k \le p} \min_{j \in \mathfrak{I}_k} Y_j,$$

where $\mathfrak{I}_1, \ldots, \mathfrak{I}_p$ are the complete paths from a source to a sink and Y_j are durations (lifetimes) of the components in the system. Suppose that the random variables $Y_1, \ldots, Y_m$ are independent and each distributed as a gamma. We find that, given the (vector) random sample $Y_1, \ldots, Y_N$, the gradient of the corresponding expected performance $\ell(\lambda)$ can be estimated by the SF estimator

$$\bar{\nabla}\ell_N(\lambda) = N^{-1} \sum_{i=1}^{N} L(Y_i)(\beta\lambda^{-1} - Y_i) \ . \tag{4}$$

ESTIMATION AND SENSITIVITY ANALYSIS OF DISCRETE EVENT DYNAMIC SYSTEMS: The SF approach presented in the previous section can be extended to dynamic systems as well. Consider a Discrete Event Dynamic System (DEDS) driven by an input sequence of iid random vectors $Y_1, Y_2, \ldots$ generated from a PDF $f(y, \theta)$ depending on the parameter vector $\theta \in \Theta$. Let $L_1, L_2, \ldots$ be an output process driven by this input sequence. That is, $L_t = L_t(Y_t)$, $t = 1, \ldots$, where vector $Y_t = (Y_1, \ldots, Y_t)$ represents a history of the input

process up to time t, and $L_t(\cdot)$ is a sequence of real valued functions.

Suppose that $\{L_t\}$ is a discrete-time regenerative process with the regenerative cycle of length τ. For example, consider a GI/G/1 queue with FIFO discipline. In that case, the input sequence is represented by the two-dimensional vector $Y_t = (Y_{1t}, Y_{2t})$, with Y_{1t} being the service time of the tth customer, Y_{2t} being the interarrival time between the $(t-1)$st and tth customers, and τ is the number of customers served during the busy period. The output process L_t can be, for example, the system waiting time of the tth customer.

Consider the expected long-run average $\ell(\theta)$ of the process L_t. It is well known in the theory of regenerative processes that $\ell(\theta)$ is equal to the expected steady-state performance of L_t and can be represented as the ratio $\ell(\theta) = \ell_1(\theta)/\ell_2(\theta)$ of the expectations $\ell_1(\theta) = E_\theta[\sum_{t=1}^{\tau} L_t]$ and $\ell_2(\theta) = E_\theta[\tau]$, respectively (Asmussen, 1987). Note that the above expectations and hence the expected performance $\ell(\theta)$ are functions of the parameter vector $\theta \in \Theta$.

In order to derive an estimate of the expected performance $\ell(\theta)$ and its sensitivities $\nabla \ell(\theta)$ for different values of θ, we can again use the SF approach. That is, for a chosen dominating PDF $g(y)$, the expected value functions $\ell_1(\theta)$ and $\ell_2(\theta)$ can be written in the form:

$$\ell_1(\theta) = E_g\left[\sum_{t=1}^{\tau} L_t(Z_t)\tilde{w}_t(Z_t, \theta)\right] \quad \text{and}$$

$$\ell_2(\theta) = E_g\left[\sum_{t=1}^{\tau} \tilde{w}_t(Z_t, \theta)\right], \tag{5}$$

where $\tilde{w}_t(Z_t, \theta) = f_t(Z_t, \theta)/g_t(Z_t)$ with $f_t(z_t, \theta) = \Pi_{i=1}^{t} f(z_i, \theta)$ and $g_t(z_t) = \Pi_{i=1}^{t} g(z_i)$. The latter term is the density function of the random vector $Z_t = (Z_1, \ldots, Z_t)$, with the random vectors $Z_1, \ldots, Z_t$ drawn according to the PDF $g(\cdot)$.

Under standard regularity conditions, the derivatives of $\ell_1(\theta)$ and $\ell_2(\theta)$ can be taken inside the expectation. Consequently, by generating a random sample of N regenerative cycles based on the PDF $g(\cdot)$, one can estimate the above expectations by the corresponding sample means (averages), and hence can estimate $\ell(\theta)$ and $\nabla \ell(\theta)$.

EXAMPLE 2 (QUEUEING DELAYS): Let L_t be the system waiting time of tth customer in a GI/G/1 queue driven by the input sequences of the service times Y_{1t} and the interarrival times Y_{2t} with respective density functions $f_1(y_1, \theta_1)$ and $f_2(y_2, \theta_2)$. By Lindley's equation we have:

$$L_t = Y_{1t} + [L_{t-1} - Y_{2t}]_+, \quad t = 1, 2, \ldots$$

Let τ be the number of customers served in the first busy period. Then the expected long run average waiting time of a customer can be written as:

$$\ell(\boldsymbol{\theta}) = \frac{E\left[\sum_{t=1}^{\tau} L_t\right]}{E_0[\tau]} = \frac{E\left[\sum_{t=1}^{\tau}\sum_{j=1}^{\tau} Y_{1j} - \sum_{t=2}^{\tau}\sum_{j=2}^{\tau} Y_{2j}\right]}{E_0[\tau]},$$

where $\boldsymbol{\theta} = (\theta_1, \theta_2)$ and $f(\boldsymbol{y}, \boldsymbol{\theta}) = f_1(y_1, \theta_1)f_2(y_2, \theta_2)$. By generating N regenerative cycles (busy periods) of the service and interarrival times according to the PDF $f(z, \boldsymbol{\theta}_0)$ for a chosen value $\boldsymbol{\theta}_0$ of the parameter vector, one can estimate $\ell(\boldsymbol{\theta})$ and $\nabla\ell(\boldsymbol{\theta})$ for various values of $\boldsymbol{\theta}$.

OPTIMIZATION: Consider the following (unconstrained) optimization problem involving the expected performance function of a static or dynamic system

(**P₀**)   minimize $\ell(\boldsymbol{\theta})$, $\boldsymbol{\theta} \in \Theta$.

Let $\boldsymbol{\theta}^*$ be an optimal solution of the program (**P₀**). In order to estimate the optimal solution $\boldsymbol{\theta}^*$ from a single simulation we can apply the SF approach. That is, let $\hat{\ell}_N(\boldsymbol{\theta})$ be the LR estimator of $\ell(\boldsymbol{\theta})$ calculated via the corresponding LR function (process). Consider the optimization problem:

(**P̂**$_N$)   minimize $\hat{\ell}_N(\boldsymbol{\theta})$, $\boldsymbol{\theta} \in \Theta$.

We refer to the program (**P̂**$_N$) as the *stochastic counterpart* of the program (**P₀**).

The LR estimator $\hat{\ell}_N(\boldsymbol{\theta})$ and, hence, the program (**P̂**$_N$) depend on the generated random sample and in that way are stochastic. However, as soon as the sample is generated, the function $\hat{\ell}(\boldsymbol{\theta})$ and its derivatives are given explicitly through the corresponding density functions and can be calculated for various values of $\boldsymbol{\theta}$. Consequently, (**P̂**$_N$) becomes a deterministic optimization program and can be solved by standard methods of mathematical programming. Rubinstein and Shapiro (1993) showed that, under mild regularity conditions: (i) The optimal solution $\hat{\boldsymbol{\theta}}_N$ of the program (**P̂**$_N$) converges with probability one as $N \to \infty$ to its "true" counterpart $\boldsymbol{\theta}^*$. That is, $\hat{\boldsymbol{\theta}}_N$ is a consistent esimator of $\boldsymbol{\theta}^*$; (ii) $N^{1/2}(\hat{\boldsymbol{\theta}}_N - \boldsymbol{\theta}^*)$ converges in distribution to multivariate normal with zero mean vector and covariance matrix $B^{-1}\Sigma B^{-1}$, where $B = \nabla^2\ell(\boldsymbol{\theta}^*)$ and Σ is the asymptotic covariance matrix of $N^{1/2}\nabla\hat{\ell}_N(\boldsymbol{\theta}^*)$. That is, $\hat{\boldsymbol{\theta}}_N$ is asymptotically normal $N(\boldsymbol{\theta}^*, N^{-1}B^{-1}\Sigma B^{-1})$. Extensive simulation studies with the SF approach, as well as extension of the above simulation based approach to constraints optimization, can be found in Rubinstein and Shapiro (1993).

See **Monte Carlo sampling and variance reduction; Nonlinear programming; Optimization; Parametric programming; Simulation of discrete-event stochastic systems.**

References

[1] Asmussen S. (1987). *Applied Probability and Queues*, Wiley, New York.

[2] Glasserman P. (1991). *Gradient Estimation via Perturbation Analysis*, Kluwer Academic Publishers, Norwell, Massachusetts.

[3] Glynn P.W. (1990). "Likelihood ratio gradient estimation for stochastic systems," *Communications of the ACM*, 33, 75–84.

[4] Ho Y.C. and Cao X.R. (1991). *Perturbation Analysis of Discrete Event Dynamic Systems*, Kluwer Academic Publishers, Boston.

[5] L'Ecuyer P.L. (1990). "A unified version of the IPA, SF, and LR gradient estimation techniques," *Management Science*, 36, 1364–1383.

[6] Reiman M.I. and Weiss A. (1989). "Sensitivity analysis for simulations via likelihood rations," *Operations Research*, 37, 830–844.

[7] Rubinstein R.Y. (1976). "A Monte Carlo method for estimating the gradient in a stochastic network," technical report, Technion, Haifa, Israel.

[8] Rubinstein R.Y. and Shapiro A. (1993). *Discrete Event Systems: Sensitivity Analysis and Stochastic Optimization via the Score Function Method*, Wiley, Chichester.

SCORING MODEL

See **Research and development**.

SCRIPTED BATTLE MODEL

A model or game in which some (or all) major events are predetermined to ensure that particular points are addressed. The list of predetermined events (with time and description) is the script. See **battle modeling**.

SEARCH THEORY

Lawrence D. Stone

Metron Inc., Reston, Virginia

INTRODUCTION: The terms search and search theory often include a wide variety of topics such as search through data bases, search for the maximum of a function, and search for a job. For this article, we limit our discussion to topics that are based on the classical search problem where there is one target that a "searcher" wishes to detect in an efficient manner. The searcher's knowledge about the target's location is represented by a probability distribution. There is a detection sensor whose performance is characterized by a function that relates search effort placed in a region to the probability of detecting the target given it is in that region. The searcher has a limited amount of effort and wishes to allocate this effort to maximize the probability of detecting the target.

In mathematical terms, the problem is formulated as follows: Let

X = the search space (typically n dimensional Euclidean space);

$p(x)$ = the prior probability (density) of the target being located at x for $x \in X$;

$f(x)$ = the amount of search effort (density) allocated to x for $x \in X$;

$b(x, z)$ = the probability of detecting the target with z effort (density) placed at x given the target is located at x for $x \in X$ and $z \geq 0$; and

$c(x, z)$ = the cost (density) of placing z effort (density) at x for $x \in X$.

In the above definitions, we are using the same notation for continuous and discrete search spaces. The plane is an example of a continuous search space, and a finite set of cells is an example of a discrete one. When X is continuous, p becomes a probability density and f an effort density. We call f an allocation function and b a detection function. All functions are assumed to be Borel measurable.

Let $F(X)$ = the set of allocation functions defined on X, and define

$$P[f] = \int_X p(x)b(x,(x))dx,$$

$$C[f] = \int_X c(x,(x))dx \quad \text{for} \quad f \in F(X)\}$$

Then $P[f]$ is the probability of detecting the target with allocation f and $C[f]$ is the cost of that allocation. Suppose that the searcher's budget constraint is K. Then the optimal search problem is to find an allocation $f^* \in F(X)$ such that

$$C[f^*] \leq K, \quad P[f^*] = \max\{P[f]: f \in F(X), C[f] \leq K\}$$

The allocation f^* maximizes the probability of detecting the target subject to the constraint K on the cost of the search. This is called the *optimal detection search problem*.

HISTORY: Search theory had its beginnings in the work performed by the U. S. Navy's Antisubmarine Warfare Operations Research Group (ASWORG) in 1942 to counter the German submarine threat in the Atlantic. The ASWORG became ORG and later the Operations Evaluation Group (OEG).

The first memorandum published by the AS-WORG was on search. George Kimball suggested that Bernard Koopman, James Dobbie, and a few others be given the job of assembling the existing results on search into a coherent theory. Philip Morse credits Koopman with providing the basic probabilistic foundation of the subject and finding the first results on optimal allocation of search effort, namely, the optimal allocation of a fixed amount of search effort to detect a stationary target with a bivariate normal target location distribution and exponential detection function. Koopman defined the elements of the basic problem of optimal search: a prior distribution on target location, a function relating search effort and detection probability, a constrained amount of search effort, and the optimization criterion of maximizing probability of detection subject to a constraint on effort.

The resulting synthesis was published in *Search and Screening* (Koopman 1946). It defined many of the basic search concepts: sweep width, sweep rate, detection function, kinematic enhancement. It presented models for radar and visual search. This and its updated version, Koopman (1980), are still the classic references on basic search theory.

SEARCH FOR A STATIONARY TARGET: Koopman (1946) solved the following problem. Let the search space, X, be the plane and

$$\begin{cases} p(x_1, x_2) = (2\pi\sigma)^{-1} \exp\left(-\frac{x_1^2 + x_2^2}{2\sigma^2}\right) & \text{for } (x_1, x_2) \in X \\ b(x,z) = 1 - e^{-Wz} & \text{for } x \in X, z \geq 0 \\ c(x,z) = z & \text{for } x \in X, \quad z \geq 0. \end{cases}$$

The target location distribution is circular normal, the detection function is exponential, and effort is measured in time. The parameter W is called the sweep rate which has dimensions of area/time. Effort density, z, has dimensions of time/area. Sweep rate is a measure of the efficiency of the search sensor, for example, how many square miles per hour it can sweep. Suppose that there is an amount of time T for search. What is the allocation, f^*, of search effort density that maximizes probability of detection? Koopman found the answer to be

$$f^*(x_1, x_2)$$

$$= \begin{cases} \left(\frac{WT}{\pi\sigma^2}\right)^{1/2} - \frac{x_2^1 + x_2^2}{2\sigma^2} & \text{for } \frac{x_2^1 + x_2^2}{2\sigma^2} \leq \left(\frac{WT}{\pi\sigma^2}\right)^{1/2} \\ 0 & \text{for } \frac{x_2^1 + x_2^2}{2\sigma^2} > \left(\frac{WT}{\pi\sigma^2}\right)^{1/2}. \end{cases}$$

The resulting probability of detection is

$$P[f^*] = 1 - \left[1 + \left(\frac{WT}{\pi\sigma^2}\right)^{1/2}\right]\exp\left[-\left(\frac{WT}{\pi\sigma^2}\right)^{1/2}\right].$$

Subsequent work by researchers in the United States and elsewhere has brought the theory of search for a stationary target to a mature state of development. Stone (1989) summarizes this development.

UNIFORM AND INCREMENTAL OPTIMALITY:

Let $b'(x, \cdot)$ denote the derivative of $b(x, \cdot)$. If for all $x \in X$, $b'(x, \cdot)$ is positive, continuous, and strictly decreasing, then b is a *regular* detection function. If the detection function is regular, then Theorems 2.2.2 and 2.2.3 of Stone (1989) give formulas (which may be computed numerically) for f^*, the optimal allocation of effort K for any $K > 0$. (Stone's Example 2.2.8 gives an algorithm for finding optimal plans for discrete search spaces and exponential detection functions.) Theorems 2.2.4 and 2.2.5 of Stone (1989) show that for a regular detection function, one can compute an optimal allocation of effort in space and time so that the effort expended by time t is optimally allocated for all $t \geq 0$. Such an allocation is called *uniformly optimal* since it is optimal at all times.

Optimal search plans for regular detection functions have another interesting property, called *total* optimality. Suppose that a searcher implements a plan that is optimal for K_1 effort, but the plan fails to find the target. Suppose he is given another increment of effort, K_2, and implements a plan that is optimal for this increment of effort given the failure of the first search. Consider the allocation of the total $K_1 + K_2$ effort that results from both searches. Is this allocation optimal for the total effort? For regular detection functions, the answer is yes. Incrementally optimal search plans produce totally optimal plans.

SEARCH FOR A MOVING TARGET:

For one-sided search problems, the target's position and motion through the search space X are specified by the stochastic process $Y = \{Y_t, t \geq 0\}$ where $Y_t \in X$ gives the target's position at time t. We specify a time horizon $[0, T]$ and seek to maximize the probability of detecting the target by time T. A search plan ψ specifies the allocation of search effort (density) in space and time, that is,

$\psi(x, t) =$ effort (density) placed at x at time t for $x \in X$ and $0 \leq t \leq T$.

Search effort is constrained by the rate at which effort can be applied. Specifically, $m(t)$ (rate of) effort may be expended at time t for $0 \leq t \leq T$. A search plan ψ must satisfy

$$\int_X \psi(x, t)dx \leq m(t) \quad \text{for } 0 \leq t \leq T$$

$$\psi(x, t) \geq 0 \quad \text{for } x \in X, \quad 0 \leq t \leq T.$$

Let Ψ be the set of plans satisfying these conditions. For each sample path ω of the process Y, the probability of detecting the target by time T, given that it follows that path, is a function of the weighted total effort density,

$$\zeta(\psi, \omega, t) = \int_0^T W(Y_S(\omega), s)\psi(Y_S(\omega), s)ds$$

that accumulates by time T on the target over the course of the path. Let E denote expectation over the sample paths of Y. Then the probability of detection by time T using plan ψ is

$$P_T[\psi] = E[b(\zeta(\psi, \cdot, T))].$$

The optimal detection problem for a moving target is to find $\psi^* \in \Psi$ such that $P_T[\psi^*] \geq P_T[\psi]$ for all $\psi \in \Psi$. Such a plan is called *T-optimal*.

BROWN'S ALGORITHM:

When the detection function is exponential and the target motion process, Y, is a discrete-time-and-space Markov chain, optimal plans may be computed quickly and efficiently. Brown (1980) applied the Karush-Kuhn-Tucker conditions to find an algorithm for computing T-optimal plans for this case. By writing these conditions in a suitable form, he observed that the optimal plan for this moving target problem has the following property. Select a time t and condition on failure of the optimal plan to detect at all times other than t (both before and after t). Let g_t be the posterior target location distribution given failure to detect both before and after t. The optimal plan allocates the $m(t)$ effort for time t to maximize the detection probability for the *stationary* target problem with target distribution g_t. Since there are efficient methods for finding optimal plans for stationary targets when the detection function is exponential, Brown was able to devise an iterative and efficient algorithm that produces a T-optimal search plan. Washburn (1983) extended this algorithm to other payoff functions.

OPTIMAL SEARCHER PATH PROBLEM:

Consider a discrete time and space problem in which the searcher moves among the cells according to constraints, for example, in one time period the searcher can move only to adjacent cells. The optimal searcher path problem is to find a path that satisfies the constraints and maximizes detection probability by time T. Eagle and Yee (1990) have developed an algorithm for solving this problem.

TWO SIDED SEARCH PROBLEMS: In this class of problems, the target is trying to avoid detection. In the case of a stationary target, the target's objective is to choose its location to make the search as difficult as possible. This problem is usually modeled as a two-person game with the target wishing to maximize the mean time to detection and the searcher wishing to minimize it. The solutions are typically mixed strategies for searcher and target. When the target moves, the two-sided game becomes much more complex (see Gal 1980 and Thomas and Washburn 1991).

APPLICATIONS: Search theory has been applied to finding the U.S. nuclear submarine, *Scorpion*, lost near the Azores in 1968 (Richardson and Stone 1971) and to finding the steamship *S.S. Central America* which was lost in a hurricane in 1857 while carrying three tons of gold from California to New York (Stone 1992). Search theory was used to develop a computer-assisted, search-and-rescue planning system for the U.S. Coast Guard (Richardson and Discenza 1980).

References

[1] Brown S.S. 1980. "Optimal Search for a Moving Target in Discrete Time and Space." *Opns. Res.* **28**, 1275–1289.

[2] Eagle J.N. and Yee J.R. 1990. "An Optimal Branch–and–Bound Procedure for the Constrained Path, Moving Target Search Problem." *Opns. Res.* **38**, 110–114.

[3] Gal S. 1980. *Search Games.* Academic Press, New York.

[4] Koopman B.O. 1946. *Search and Screening.* Operations Evaluation Group Report No. 56. Center for Naval Analyses, Alexandria, Virginia.

[5] Koopman B.O. 1980. *Search and Screening: General Principles with Historical Applications.* Pergamon Press, New York.

[6] Richardson H.R. and Discenza J.H. 1980. "The United States Coast Guard computer-assisted search planning system (CASP)." *Naval Research Logistics Quarterly* **27**, 659–680.

[7] Richardson H.R. and Stone L.D. 1971. "Operations Analysis During the Underwater Search for Scorpion," *Naval Research Logistics Quarterly* **18**, 141–157.

[8] Stone L.D. 1989. *Theory of Optimal Search.* Second edition, Operations Research Society of America, Baltimore, MD.

[9] Stone L.D. 1992. "Search for the SS Central America: Mathematical Treasure Hunting." *Interfaces* **22**, 32–54.

[10] Thomas L.C. and Washburn A.R. 1991. "Dynamic Search Games." *Opns. Res.* **39**, 415–432.

[11] Washburn A.R. 1983. "Search for a Moving Target: The FAB Algorithm." *Opns. Res.* **31**, 739–751.

SECOND-ORDER CONDITIONS

Conditions involving second derivatives.

SELF-DUAL PARAMETRIC ALGORITHM

A variation of the simplex method and parametric programming in which the given linear-programming problem is adjusted so that the same parameter is added to each cost coefficient and each right-hand-side element. By using a sequence of primal and dual simplex transformations, the problem will be optimal for some value of the parameter, with the process continuing until a solution with a zero value of the parameter is found. See **Simplex method**.

SEMI-MARKOV PROCESS

A discrete-valued stochastic process which evolves via a Markov chain, but the time spent on each visit to a state before making a transition is a random variable that may be a function of the source and destination states. It is another basis for analyzing queueing and related systems. See **Markov processes**.

SEMI-STRICTLY QUASI-CONCAVE FUNCTION

Given a function $f(\cdot)$ and points $x, y \in X$, with $x \neq y$ and X convex, if $f(y) > f(x)$ implies that $f(\lambda x + (1 - \lambda)y) > f(x)$ for all $0 < \lambda < 1$, then we say that f is a semi-strictly quasi-concave function. See **Concave function**; **Quasi-concave function**; **Strictly quasi-concave function**.

SEMI-STRICTLY QUASI-CONVEX FUNCTION

Given a function $f(\cdot)$ and points $x, y \in X$, with $x \neq y$ and X convex, if $-f(y) > -f(x)$ implies that $-f(\lambda x + (1 - \lambda)y) > -f(x)$ for all $0 < \lambda < 1$, then we say that f is a semi-strictly quasi-convex function. See **Concave function**; **Convex function**; **Quasi-concave function**; **Quasi-convex function**; **Strictly quasi-concave function**; **Strictly quasi-convex function**.

SENSITIVITY ANALYSIS

The study of how an optimal solution or, in general, the output of a mathematical model changes with (usually) small changes in the given data. In linear programming, a sensitivity study is a basic part of the problem solution. It shows how much a cost coefficient or a right-hand-side element can vary, holding all other data constant, before the optimal feasible basis is either no longer optimal or feasible. See **Hundred percent rule; Linear programming; Nonlinear programming; Parametric-linear programming; Ranging; Robustness; Tolerance analysis.**

SEPARABLE FUNCTION

A function $f(x_1, \ldots, x_n)$ is a separable function if $f(x_1, \ldots, x_n) = f_1(x_1) + \ldots + f_n(x_n)$. Certain nonlinear-programming problems that contain separable functions can be suitably represented by a linear approximation and solved by a variation of the simplex method. See **Separable-programming problem; Simplex method.**

SEPARABLE-PROGRAMMING PROBLEM

A nonlinear-programming problem in which some or all of the constraints and the objective function are separable functions of one variable. Using linear approximations to the separable functions, the problem can be approximated and solved by a variation of the simplex algorithm that uses a restricted-basis entry rule. See **Separable function; Special-ordered sets.**

SEPARATING HYPERPLANE THEOREM

Let C_1 and C_2 be two nonempty disjoint convex sets in n-dimensional space. Then there exists an n-dimensional hyperplane $ax = b$, $a \neq 0$, that separates them. That is, for x in C_1, $ax \leq b$, and for x in C_2, $ax \geq b$. See **Hyperplane.**

SERIES QUEUES

A network of queueing systems with serial routing. They are also called tandem queues. See **Network of queues.**

SERVICE SYSTEMS

Queueing systems. See **Queueing theory.**

SET-COVERING PROBLEM

The set-covering problem is an integer-programming problem defined as follows:

Minimize cx

subject to

$$Ex \geq e$$

where the components of E are either one or zero, the components of the column vector e are all ones, and the variables are restricted to be either one or zero. The idea of the problem is to find the minimum cost set of columns from E such that the one's in vector e are "covered" by at least one of the ones in the selected set of columns. Note that multiple coverage is allowed. See **Bin packing; Packing problem; Set-partitioning problem.**

SET-PARTITIONING PROBLEM

The set-partitioning problem is an integer-programming problem defined as follows:

Minimize cx

subject to

$$Ex = e$$

where the components of E are either one or zero, the components of the column vector e are all ones, and the variables are restricted to be either one or zero. It is similar to a set-covering problem except multiple coverage is not allowed. See **Packing problem; Set-covering problem.**

SEU

Subjective expected utility. See **Decision analysis.**

SHADOW PRICES

The term shadow prices is used to describe the optimal dual variables (marginal values) to a linear-programming problem. For an activity-analysis and similar problems, the shadow price associated with a constraint can be interpreted as the change in the value of the objective function per unit change of the constraint's right-hand-side (resource). See **Marginal value.**

SHAPLEY VALUE

See **Group decision**.

SHELL

An expert system development tool providing a prefabricated inference engine. See **Expert systems**; **Inference engine**.

SHEWHART CHART

See **Statistical quality control**.

SHORTEST-ROUTE PROBLEM

A network problem in which one wants to find the shortest route from a home node to a destination node, or the shortest routes from a home node to all other nodes. This is a linear-programming problem and can be solved by the simplex method, but special shortest route algorithms exist that are computationally more efficient. See **Dijkstra's algorithm**; **Network optimization**; **Simplex method**.

SIGNOMIAL PROGRAMMING

See **Geometric programming**.

SIMD

An acronym for Single Instruction, Multiple Data. A class of parallel computer architectures in which a single stream of instructions controls multiple processing elements. Processors synchronously perform the same computations on differing data.

SIMPLE UPPER-BOUNDED PROBLEM (SUB)

A linear-programming problem in which some or all of the variables x_j are constrained by upper-bound conditions of the form $x_j \leq u_j$, where u_j is a given finite bound. It can be solved by a special adaptation of the simplex method in which the upper-bounded constraints are considered implicitly. See **Simplex method**.

SIMPLEX

A polyhedron of the form $x_1 + \ldots, + x_n \leq 1$, $x_j \geq 0$. Also, a simplex is the convex hull of $n + 1$ points in general position in Euclidean n-space.

SIMPLEX METHOD (ALGORITHM)

A computational procedure for solving linear-programming problems of the form: Minimize (maximize) cx, subject to $Ax = b$, $x \geq 0$, where A is an $m \times n$ matrix $(m < n)$, c is an n-dimensional row vector, b is an m-dimensional column vector, and x is an n-dimensional variable vector. The simplex method was developed by George B. Dantzig in the late 1940s. The method starts with a known basic feasible solution or an artificial basic solution, and, given that the problem is feasible, finds a sequence of basic feasible solutions (extreme-point solutions) such that the value of the objective function improves or does not degrade. Under a nondegeneracy assumption, the simplex algorithm will converge in a finite number of steps, as there are only a finite number of extreme points and extreme directions of the underlying convex set of solutions. At most m variables can be in the solution at a positive level. In each step (iteration) of the simplex method, a new basis is found and developed by applying Gaussian elimination in a manner that preserves the nonnegativity of the solution. The elimination step replaces a variable in the current solution with a new one. The inverse of the basis (revised-simplex method) is used to develop a pricing vector for "pricing out" the variables not in the current basic solution and to select one to enter the solution if the current solution is not optimal. The optimal solution to the corresponding dual problem is also generated by the simplex method as part of the solution to the original, primal problem. The simplex method has been implemented on just about all major computer systems and a wide range of simplex-based software is available for personal computers and in spreadsheets. See **Artificial basis**; **Dual-programming problem**; **Linear programming**; **Prices**; **Primal problem**; **Revised simplex method**.

SIMPLEX MULTIPLIERS

See **Multiplier vector**.

SIMPLEX TABLEAU

A schematic, numerical representation which displays the transformed data set associated with a basic solution to a linear-programming problem. For the problem: Minimize cx, subject to $Ax = b$, $x \geq 0$, if we let the $m \times m$ matrix B be a feasible basis and the $m \times 1$ row vector c_0 be the ordered cost coefficients for the variables in the basis, then the simplex tableau displays the following information in an $(m + 1) \times (n + 1)$ rectangular array

$$\begin{bmatrix} B^{-1}A & | & B^{-1}b \\ \pi A - c & | & \pi b \end{bmatrix}$$

where π, the pricing vector, is equal to $c_0 B^{-1}$. The last or $(m + 1)$st, row of the tableau contains the reduced costs and the current value of the objective function, respectively. If the simplex method is in Phase I, then an additional row is added to the tableau which contains the reduced costs associated with the artificial basis. In some arrangements of the tableau, the rows associated with the reduced costs are given at the top of the tableau; also, a reduced tableau is obtained by leaving out the columns that correspond to the columns in the basis as they transform to unit columns with reduced costs of zero. The simplex tableau is useful when solving small problems by hand and as an instructional tool. Computer-based simplex method software do not use the tableau as it is computationally inefficient. Instead, they use some form of the revised simplex method. See **Basis**; **Phase I**; **Phase II**; **Revised simplex method**; **Simplex method**.

SIMULATED ANNEALING

G. Anandalingam

University of Pennsylvania, Philadelphia

INTRODUCTION: Simulated annealing (SA) is motivated by the behavior of mechanical systems with a very large number of degrees of freedom. According to the general principles of physics, any such system can be coaxed into a minimum energy condition or state by a slow annealing process. For example, atoms of a molten metal, when cooled to a solidifying temperature will tend to assume relative positions in a lattice which minimizes the potential energy associated with their mutual forces. The reduction of temperature confines the system (or metal) to a smaller and smaller region of the phase space; if the reduction is carried out slowly enough, it allows the system to pass out of metastable local energy minima, and move towards the global minimum. Simulated annealing is the algorithmic counterpart to the physical annealing process. "Annealing" is carried out by the decline of a parameter called "temperature." The "simulation" is performed by using the well known Metropolis algorithm in order to reach "equilibrium" within each temperature state (Metropolis *et al.*, 1953). Kirkpatrick, Gelatt and Vecchi (1983) and Cerny (1985) independently showed that a fruitful connection existed between simulated annealing, and combinatorial optimization which involves a search for the global extremum of a function that has many local extrema.

In generating the global extremum of a function, potential solutions are generated in a pseudo-random manner, and then tested for optimality, or more correctly, goodness. As in pure random search and traditional local optimization techniques, simulated annealing accepts all new solutions that are better than the previous ones. (In the case where a function is being minimized, "better" solutions are those that give a lower functional value than the previous ones.) SA differs from pure random search and local optimization techniques in a significant way: new solutions are accepted with nonzero probability even if they are worse. The key to simulated annealing is the interaction between the probability acceptance parameter, which is analoguous to temperature, which is lowered over time, and the search. At high temperature, many of the worse solutions are accepted. Lowering the temperature of the acceptance distribution causes the search to accept only a few worse solutions, for example, only slightly worse ones, thereby rejecting the really bad solutions. In analogy to annealing in physics, the search is run at each temperature using simulation until an "equilibrium" is reached as measured by the number of consecutive rejected and/or accepted solutions. Then the temperature is set to a lower value, the starting solution is set to the best or last (depending on the researcher) solution obtained at the previous temperature, and the process repeats until the final temperature setting has been run.

In optimization, simulated annealing works as follows: At relatively high temperatures, many solutions are accepted, even if they are bad. This allows the search to jump from mode (peak) to mode in the search space; if only improved solutions were accepted, the search would seek a local optima, but in general would be precluded from seeking the global optimum. The intent of the relatively high temperature phase of the search is to discover the gross features of the search space. Successively lower temperatures identify more and more detail while solutions become more and more localized. Ultimately, simulated annealing yields a very good solution on a very good mode of the search space.

A number of books have appeared on simulated annealing: Aarts and Korst (1989) and van Laarhoven and Korst (1987). An extensive bibibliography appears in the paper by Collins *et al.* (1988).

A MATHEMATICAL EXPOSITION: We are concerned with the minimization of a function $Z: X \to R$, where $X \subset I^N$ is a vector of integer variables of dimension N. Assume that for each $x \in X$, there is a set $N(x)$ which we call the neighborhood set of x. Assume that there is a transition probability matrix P such that $P(x', x) > 0$ if and only if $x' \in N(x)$. We will denote $Z = Z(x)$ and $Z' = Z(x)$. Also let $T_1, T_2, \ldots$ be a sequence of strictly positive numbers (the temperature), such that

$$T_1 \geq T_2 \geq \cdots \tag{1}$$

and

$$\lim_{k \to \infty} T_k = 0. \tag{2}$$

The simulated annealing algorithm, with current solution x, works as follows:

0. Let the current solution be x; set the best solution $x^* = x$; and set $T_k = T_0$.
1. Generate $x' \in N(x)$.
2. Set x^* to x' with probability p given by

$$p = \text{minimum } [1, \exp((Z - Z')/T_k)], \tag{3}$$

   and leave it unchanged with probability $(1 - p)$.
3. Replace x with x', and repeat Steps 1 to 3, until "equilibrium" is reached at temperature T_k.
4. Replace T_k with T_{k+1} according to a "cooling schedule," and repeat Steps 1 to 4 until $T \approx 0$.

Note that in Step 2, for most values of T_k, if $Z' < Z$, p given by equation (3) equals unity, and x' is chosen to be the next value of x^* for sure. Conversely, if $Z' > Z$, $p < 1$, for most values of T_k. The exception is the case when $T_k \gg Z - Z'$, which gives $\exp(Z - Z')/T_k \approx 1$, even if $Z' > Z$. In this case, x' would be chosen to be the next value of x^*, even though it might give a worse value of Z; for example, initially, nearly all x are candidates for the best solution. As the temperature decreases according to a cooling schedule, the worse solutions get selected with lower probability.

CONVERGENCE, COOLING SCHEDULE AND EQUILIBRIUM: There are a number of rigorous mathematical justifications for the simulated annealing approach. Well-known papers that provide mathematical models of simulated annealing include Geman and Geman (1984), Gidas (1985), Lundy and Mees (1986), Mitra *et al.* (1986) and Hayek (1988). Most of these papers have formulated Markov models to analyze simulated annealing and consider "equilibrium" in SA to be directly associated with the equilibrium distribution of a Markov chain. One of the main thrusts of the analyses is to show that there are cooling schedules that yield limiting equilibrium distributions, over the space of all solutions, in which, essentially, all the probability is concentrated on the optimal solutions.

All the theoretical papers prove that the simulated annealing algorithm converges asymptotically. Unfortunately, as Sasaki and Hajek (1988) have found, convergence times are exponential, even for simple problems, if the assumptions of SA are adhered to very strictly. Thus, researchers in the field have had to try a number of approximations. For instance, Fetterolf (1990) tried a number of cooling schedules and found that the simple exponential decay function

$$T_k = T_0 e^{-\alpha k}, \quad k = 0, 1, \ldots \tag{4}$$

provided the best results, although it did not satisfy the conditions of the theoretical papers exactly. While not that practical, the mathematical papers do provide the important intuition that the slower the cooling rates (and hence, longer computational times), the closer the solutions would be to the global optimum (optima).

In practice, one also has to decide the value of T_0, the initial temperature. Here, there is a trade-off between choosing a value of T_0 so high that all $x' \in N(x)$ would be candidates for x in Step 2, thus considerably slowing down the convergence of the algorithm, versus having a T_0 so low that only x for which $Z(x') < Z(x)$ are chosen to be the next value of x, in which case the gross features of the solution space are missed leading to local rather than global optima. The latter case also amounts to reducing simulated annealing to a straightforward steepest descent algorithm.

Finally, although $\lim_{k \to \infty} T_k = 0$ guarantees convergence, it may take exponentially long to obtain a result. Thus the choice of T_f, the final temperature determines the final error in the solution. If the total number of possible solutions to the combinatorial optimization problem is N (only one of which is the global optimum), the sensitivity of optimal solution (i.e. the allowed error bound) is δ, and the error probability is λ, then we need $\lambda/(1 - \lambda) \geq (N - 1)e^{-\delta/T_f}$. This gives us:

$$T_f \leq \delta/(\log(N - 1) - \log \lambda). \tag{5}$$

One can use equation (5) to set the final temperature for the simulation. The usual practice, however, is to set the final temperature small, for example, close to zero.

Equilibrium is reached by using the Metropolis algorithm in most cases (Metropolis *et al.*, 1953). A number of variations are used to stop the computations at each temperature level. The most popular one is to stop when a certain number of pseudo randomly generated solutions are rejected consecutively. Often, "equilibrium" is considered to be reached when the accept to reject ratio becomes very low.

APPLICATIONS: Simulated annealing has been applied to a variety of problems, in areas as diverse as graph partitioning and graph coloring (Johnson *et al.*, 1989 and 1991), traveling salesman problem (Bonomi and Lutton, 1984), telecommunications network design (Fetterolf and Anandalingam, 1992), VLSI design (Kirkpatrick *et al.*, 1983), pattern recognition (Geman and Geman, 1984), code generation (El Gamal *et al.*, 1987). It should be noted, of course, that the above is not an exhaustive list.

PERFORMANCE: Many authors have found simulated annealing to be relatively slow for solving many problems (Collins *et al.*, 1988, for instance). Successful practical applications of simulated annealing have been in complicated problem domains where, either there were no previous algorithms, or the derived algorithms performed poorly. The usual practice in operations research is to simplify problem structure by making assumptions about the structure or about the solution space, and then applying algorithms derived from existing theory. The main power of simulated annealing is that it allows analysts to keep complicated, but realistic, problem structures.

In order to increase the speed of the simulated annealing algorithm in practice, most researchers use hybrid techniques. A popular method is to only use simulated annealing for integer valued variables, and then use linear/nonlinear programming to solve the problem every time the integer variables are generated pseudo randomly (for instance, Vernekar *et al.*, 1989). Another popular method is to incorporate SA within a penalty function approach (Johnson *et al.*, 1991). One can get a number of helpful insights for improving the performance of SA by examining

the paper by Johnson *et al.* (1989) which reports on extensive experiments on the application of SA to the graph partitioning problem. There, the authors examine many issues such as optimizing parameter settings, using different cooling schedules including adaptive ones, using cutoffs to remove unneeded trials, using better-than-random starting solutions, and using other approximations.

See **Combinatorial and integer optimization; Combinatorics; Markov chains; Unconstrained optimization.**

References

[1] Aarts E.H.L. and J.H.M. Korst (1989), *Simulated Annealing and Boltzmann Machines*, John Wiley, Chichester, United Kingdom.

[2] Cerny V. (1985), "A thermodynamical approach to the travelling salesman problem: an efficient simulation algorithm," *Jl. Optimization Theory and Applications*, 45, 41–51.

[3] Bonomi E. and J.L. Lutton (1984), "The N-city Travelling Salesman Problem: Statistical Mechanics and the Metropolis Algorithm," *SIAM Review*, 26, 551–568.

[4] Collins W.E., R.W. Eglese and B.L. Golden (1988), "Simulated Annealing: An Annotated Bibiliography," *Amer. Jl. Mathematical Management Science*, 8, 205–307.

[5] El Gamal A.A., L.A. Hemachandra, I. Shperling and V.K. Wei (1987), "Using Simulated Annealing to Design Good Codes," *IEEE Trans. Information Theory*, 33, 116–123.

[6] Fetterolf P. (1990), "Optimal Design of Heterogeneous Networks with Transparent Bridges," Ph.D. dissertation, University of Pennsylvania, Department of Systems.

[7] Fetterolf P. and G. Anandalingam (1992), "Optimal Design of LAN-WAN Internetworks: An Approach Using Simulated Annealing," *Annals Operations Research*, 36, 275–298.

[8] Geman S. and D. Geman (1984), "Stochastic Relaxation, Gibbs Distribution, and Bayesian Restoration of Images," *IEEE Trans. Pattern Analysis and Machine Intelligence*, PAMI-6, 721–741.

[9] Gidas B. (1985), "Non-stationary Markov Chains and Convergence of the Annealing Algorithm," *Jl. Statistical Physics*, 39, 73–131.

[10] Hajek B. (1988), "Cooling Schedules for Optimal Annealing," *Math. Operations Research*, 13, 311–321.

[11] Johnson D.S., C.R. Aragon, L.A. McGeoch and C. Schevon (1989), "Optimization by Simulated Annealing: An Experimental Evaluation; Part I, Graph Partitioning," *Operations Research*, 37, 865–892.

[12] Johnson D.S., C.R. Aragon, L.A. McGeoch and C. Schevon (1991), "Optimization by Simulated

Annealing: An Experimental Evaluation; Part II, Graph Coloring and Number Partitioning," *Operations Research*, 39, 378–395.

[13] Kirkpatrick S., C.D. Gelatt and M.P. Vecchi (1983) "Optimization By Simulated Annealing," *Science*, 220, 671–680.

[14] Lundy M. and A. Mees (1986), "Convergence of an Annealing Algorithm," *Math. Programming*, 34, 111–124.

[15] Metropolis W., A. Rosenbluth, M. Rosenbluth, A. Teller and E. Teller (1953), "Equation of State Calculations by Fast Computing Machines," *Jl. Chem. Physics*, 21, 1087–1092.

[16] Mitra D., F. Romeo and A. Sangiovanni-Vincentelli (1986), "Convergence and Finite-time Behavior of Simulated Annealing," *Advances Applied Probability*, 18, 747–771.

[17] Sasaki G.H. and B. Hajek (1988), "The Time Complexity of Maximum Matching by Simulated Annealing," *Jl. Assoc. Computing Machinery*, 35, 387–403.

[18] van Laarhoven P.J.M. and E.H.L. Aarts (1987), *Simulated Annealing*: *Theory and Practice*, Kluwer Academic Publishers, Dordrecht, The Netherlands.

[19] Vernekar A., G. Anandalingam and C.N. Dorny (1990), "Optimization of Resource Location in Hierarchical Computer Networks," *Computers and Operations Research*, 17, 375–388.

SIMULATION OF DISCRETE-EVENT STOCHASTIC SYSTEMS

Donald Gross

The George Washington University, Washington, DC

One of the most powerful modeling tools in the operations research analyst's toolbox is simulation. Simulation provides the ability to study complex systems in great detail. But, simulation is not without its drawbacks, and generally, when analytical results are available, that is the preferred path to follow. Nevertheless, with the advances in simulation methodology, it is becoming increasingly competitive with analytic modeling in some situations, and in many situations, simulation is the *only* way to proceed.

Simulation has found major uses in modeling manufacturing systems and communication systems. Such systems are usually stochastic in nature, with a variety of random processes interacting in complex ways. Without quite simplifying assumptions about the nature of the randomness, analytical modeling is usually not an option. The following presents an overview of what we believe to be the key points in discrete event stochastic simulation (Law and Kelton, 1991; Hoover and Perry, 1989; Fishman, 1978).

ELEMENTS OF A SIMULATION MODEL: One can look at simulation modeling as being comprised of three major elements: input generation, bookkeeping and output analysis. Simulation most often involves modeling stochastic systems and thus, it is necessary to generate (usually via the computer), the appropriate stochastic phenomena. For example, a manufacturing system may consist of a network of queues with a variety of different interarrival time and service time distributions. Random variates from these different distributions must be generated so that the system can be observed "in action." Once these distributions are chosen and random variates generated, the bookkeeping phase keeps track of transactions moving around the system and keeps counters on ongoing processes in order to calculate appropriate performance measures. Output analysis has to do with computing measures of system effectiveness and employing the appropriate statistical techniques required to make valid statements concerning system performance.

The following simple hypothetical example illustrates the three basic elements described above.

A small manufacturer of specialty items has signed a contract with a very prestigious customer for twenty orders of its premiere product. Management is concerned with current capacity and wishes to analyze the situation using discrete-event simulation. The customer will place orders at random times and of course would like them filled as soon as possible. Orders are placed only at the beginning of a month and could come as frequently as two months apart or as infrequently as seven months apart, or anything in between, all with equal probability. Currently, the production capability for this product is such that orders are shipped only at the end of a month and order filling time is equally likely between one and six months, inclusive. Only one order at a time can be processed, so that if a second order comes in while one is being prepared, it must wait until the order ahead of it is completed. For this capability, management would like to get an idea of the average number of orders in the system, the average time an order spends in the system, the maximum time an order spends in the system and the percentage

Table 1. Input data

Time between orders:	$-$,7,2,6,7,6,7,2,5,4,5,3,2,6,2,4,2,6,5,5
Service times:	1,3,2,3,6,5,4,5,1,1,3,1,3,2,2,6,5,1,3,5

Using the above input data, we can construct the following abbreviated bookkeeping table:

Table 2. Bookkeeping
key: $[n],t = $ [transaction number n],time of occurrence

Master clock time	Next events		Transaction in queue	Transaction in service
	arrival	departure		
0	[2],7	[1],1		-> [1]
1	[2],7	[2],10		[1]->
7	[3],9	[2],10		-> [2]
9	[4],15	[2],10	-> [3]	[2]
10	[4],15	[3],12	[3]->	-> [3] [2]->
●			●	●
●			●	●
●			●	●
81	[20],86	[19],84		-> [19]
84	[20],86			[19]->
86		[20],91		-> [20]
91				[20]->

of time the system is idle. The date of the first order is known and the production line will be set up just in time to receive the first order. The production line will be taken down after the last (20*th*) order is completed.

The first element, that of generating the random input data, can be easily done by the roll of a fair die. Times between placement of orders are uniform (2,7) and service times are uniform (1,6). Thus, for generating the interarrival times, we simply roll the die 19 times and add one to each value to get the times between successive orders after the first one. For the service times, the value of the roll itself suffices and we simply need to roll the die 20 more times. The following table gives the results of using a fair die to generate the input data.

Table 1 was developed from interarrival and service-time data. At clock 0, the first order comes into the system, has a service time of 1 month and is due to depart at clock 1. At clock 1, the next arrival, order number 2, is due in at clock $0 + 7 = 7$, and since no order is in the system, will depart at its arrival time plus service time, that is, $7 + 3 = 10$. The clock is advanced to time 7, and the next arrival (order 3) scheduled at $7 + 2 = 9$. Since order 3 arrives before order 2 leaves, the clock is advanced to time 9, the arriving order 3 enters the queue and order number two is still in service, but due to

depart at time 10. Order 4 is due in at $9 + 6 = 15$. The clock is then advanced to time 10, where order 2 leaves the system, order 3 enters service and is scheduled to depart at $10 + 2 = 12$. Order 4 is next to arrive and it is due in at 15, so the clock advances to 12. The bookkeeping continues on in this fashion until the 20th order is processed.

Such a bookkeeping table (which can get very complicated for realistic, complex systems), allows one to make the performance measure calculations. It is relatively straightforward to use Table 2 and obtain the queue wait and total time in system for each order. For example, order 1 came into the system at time 0, went right into service and left at time 1, spending zero time in queue and one month in the system. Order 2 arrived at time 7, also went directly into processing and left at time 10, spending 3 months in the system. Order 3, however, arriving at time 9, had to enter the queue since 2 was still in process. It left the queue for processing at time 10 and exited the system at time 12 (not shown in the abbreviated table) spending 1 month in queue waiting for processing and 3 months total time in the system. Average waiting times and maximum waiting times can then be easily calculated.

Queue and system size values, as well as idle periods, can also be obtained from Table 2, although it is a little more work getting average figures since the sizes must be weighted by the amount of time the queue and system stayed at their various sizes.

For the above example, the maximum number of orders in the queue was one, the maximum number of orders in the system, 2, the maximum time an order spent in the system waiting to be processed was 4 months (order number 17) and the maximum time an order spent in the system in toto was nine months (also order number 17). The average queue size was 0.13, the average system size, 0.81, the average percent of the time the system was empty and idle was 32% and the average waiting times in queue and system respectively were 0.6 and 3.7 months.

INPUT DISTRIBUTION SELECTION AND RANDOM VARIATE GENERATION:

Keeping in mind the old acronym, *GIGO* (Garbage In, Garbage Out), care must be given to choosing the distributions that best describe the environment being modeled. This involves knowing as much about the modeling environment as possible (the "physics" of the situation, if you will) and valid data analyses. There are some cases where data are not available nor can they be collected (e.g., design of a new system) and for these, only environmental knowledge can be used. The actual sensitivity of output performance measures to specific input distributions is still an area of active research.

Once the appropriate distributions are chosen, it is necessary to be able to generate representative samples from these distributions for running the simulation. Much study has been done in this area as discussed in Fishman (1978), Law and Kelton (1991), and Hoover and Perry (1989).

The basis for generating random variates from a desired probability distribution lies in being able to generate uniform random numbers in an interval (0, 1), possibly closed on either end. This is done via a *pseudorandom number generator*, which, using a mathematical formula, generates over some large range, integers that act as if they are random. These integers can then be normalized to fall in (0, 1). In simplest form, this is accomplished with a linear congruential generator, represented by

$$Y_i = (kY_{i-1} + a) \bmod m,$$

where k is a multiplier, a is an increment, and mod m refers to modulo arithmetic (the quantity in the parentheses is divided by m and only the remainder kept). The relation is recursive and thus a starting point (or seed), Y_0, is required. The numbers generated from this recursion will be in the interval $[0, m-1)$ and thus dividing by

m normalizes the values to [0, 1). A great deal of thought must go into choosing the multiplier, the increment and the seed.

From these pseudorandom numbers, random variates from virtually any probability distribution (including empirical data) can be generated. A basic procedure for doing this is via inverse transformation. Let the pseudorandom numbers Y_i after normalization on [0, 1) be denoted by Z_i, and the cumulative distribution function (cdf) of a random variable X be denoted by $F(x)$. To generate random variates, X_i from the distribution $F(x)$ by the inverse transformation method, we employ the relationship

$$X_i = F^{-1}(Z_i).$$

As an example, suppose we wish to generate random variates from the exponential probability distribution, $f(x) = \theta e^{-\theta x}$. Here, $F(x) = 1 - e^{-\theta x}$, and the random variates X_i are found by solving

$$Z_i = 1 - e^{-\theta X_i}$$

which yields

$$X_i = -\frac{1}{\theta} \ln(1 - Z_i).$$

This procedure can be used for discrete as well as continuous probability distributions and even for empirical distributions, as long as the cdf is invertible. This is essentially equivalent to entering the Z_i value as the ordinate on the plot of F, projecting over to the cdf curve, and then projecting down to the abscissa to read off the corresponding X_i value. However, for most continuous problems, the cdf F will not be analytically invertible, though a simple numerical procedure can often be used. But for many other cases, methods other than inverse transformation may be more efficient. These include acceptance-rejection, convolution, and composition.

SIMULATION PROGRAMMING LANGUAGES:

The simulation modeler has a large variety of languages and packages from which to choose. These can be categorized into three main types: general purpose languages, simulation languages, and simulators. General purpose languages such as FORTRAN, C, PASCAL, BASIC, etc., allow the most flexibility in modeling but require the most effort to program. One can get a feel from the earlier very simple example of what might be involved in generating variates from the input probability distributions, programming the bookkeeping tasks, and programming the statistical calculations needed for obtaining output measures of performance.

Requirements for bookkeeping, random variate generation, and data collection necessary for statistical analyses of output are similar for large classes of simulation models and this situation has given rise to the development of a variety of simulation language packages. These simulation language packages (e.g., GPSS/H, SIMAN, SIMSCRIPT II.5 and SLAM, to mention a few) make programming a simulation model much, much, easier. However, some flexibility in modeling is sacrificed since the model must fit into the general language environment. For most cases, this is not a problem. As a general rule, one can expect that the easier the programming becomes, the less flexibility there is in deviating from the language environment.

Most of the popular simulation languages use a next event approach to bookkeeping (as opposed to fixed time increments), in that the master clock is advanced, as in the example above, to the next event scheduled to occur, rather than the clock being advanced in fixed increments of time, where, for many of the fixed increments, nothing might have happened. Further, most of the event oriented routines employ a transaction process technique which keeps track of the entire experience of a transaction as it proceeds along its way in the system. Even easier to use than simulation languages are packages called *simulators*. These require very little, if any, programming, and the model can be built usually by choosing among icons, filling out tables, etc. However, these offer the least flexibility in modeling in that the modeling environment is quite fixed, for example, a manufacturing plant, or a communications network. Some examples of simulators are SIMFACTORY, ProModel, and WITNESS for manufacturing environments and NETWORK and OPNET for communications environments.

OUTPUT ANALYSIS: Making reliable conclusions from simulation output requires a great deal of thought and care. When simulating stochastic systems, a single run yields output values which are statistical in nature so that sound experimental design and sound statistical analyses are required for valid conclusions. Unlike sampling from a population in the classic sense where great effort is made to have random samples with independent observations, we often purposely induce correlation in simulation modeling as a variance reduction technique so that classical, off-the-shelf statistical techniques for analyzing sample data are often not appropriate. We present next some basic procedures for analyzing simulation output.

There are two major types of simulation models: terminating and continuing (non-terminating). A terminating model has a natural start and stop time, e.g., a bank opens its doors at 9:00 a.m. and closes its doors at 3:00 p.m. On the other hand, a continuing model does not have a start and stop time, e.g., a manufacturing process where, at the beginning of a shift, things are picked up exactly as they were left at the end of the previous shift, so that, in a sense, the process runs continually. In these latter cases, steady-state results are usually of interest, and in simulating such a system, a determination must be made as to when the initial transients are "damped out" and the simulation is in steady state.

Considering first a terminating simulation such as the earlier example, from a single run, we cannot make any statistical statements. For example, the maximum waiting time is a single observation, that is, a sample of one. What can be done is to *replicate* (repeated runs) the experiment, using different random number streams for the order arrival and processing times for each run, and thus generate a sample of independent observations to which classical statistics can be applied. Assuming we replicate n times, we then will have n values for the maximum waiting times, say, $w_1, w_2, \ldots, w_n$. Assuming n is large enough to employ the central limit theorem, we can get a $100(1-\alpha)\%$ confidence interval (CI) by first calculating the mean and sample standard deviation of the maximum waiting time by

$$\bar{w} = \sum_{i=1}^{n} w_i/n$$

and

$$s_w = \sqrt{\sum_{i=1}^{n} (w_i - \bar{w})^2/(n-1)},$$

and then obtaining the CI as

$$[\bar{w} - t(n-1, 1-\alpha/2)s_w/\sqrt{n},$$
$$\bar{w} + t(n-1, 1-\alpha/2)s_w/\sqrt{n}],$$

where $t(n-1, 1-\alpha/2)$ is the upper $1-\alpha/2$ critical value for the t distribution with $n-1$ degrees of freedom.

For continuing simulations for which we are interested in steady-state results, we know from the ergodic theory of stochastic processes that

$$\lim_{T \to \infty} \frac{1}{T} \int_0^T X^n(t)dt = E[X^n]$$

so that if we run long enough, we would get close to the limiting average value. But we never know precisely how long is long enough, and we often wish to be able to obtain a CI statement. Thus, we have two additional problems: determining when we reach steady state and deciding when to terminate the simulation run. Assuming for the moment that these problems are solved and we decide to run the simulation for n transactions after reaching steady state and measure the time a customer spends waiting in a particular queue for service, we can obtain n queue wait values, which we shall again denote by w_i (now these are actual waits, not maximum waits). It might be tempting to calculate the average and standard deviation of these n values and proceed as above to form a CI. However, these w_i are correlated and using the above formula for s_w greatly underestimates the true variance. There are procedures for estimating the required correlations and obtaining an estimate of the standard deviation from this correlated data, but this requires a great deal of estimation from this single data set which has its drawbacks in statistical precision.

To get around the correlation problem, we may again replicate the runs m times, using a different random number seed each time as we did in the terminating case. For each run, we still calculate the mean of the w_i, denoting the mean for the jth replication by $\bar{w}_j$, that is,

$$\bar{w}_j = \sum_{i=1}^{n} w_{ij}/n$$

where w_{ij} is the waiting time for transaction i on replication j, $i = 1, 2, \ldots, n$ and $j = 1, 2, \ldots, m$. The $\bar{w}_j$ are now independent and in an analogous fashion to that of the terminating simulation, we can form a $100(1 - \alpha)$ percent CI by calculating

$$\bar{w} = \sum_{j=1}^{m} \bar{w}_j/m$$

and

$$s_{\bar{w}_j} = \sqrt{\sum_{j=1}^{m} (\bar{w}_j - \bar{w})^2/(m-1)}$$

and then the CI becomes

$$[\bar{w} - t(m-1, 1-\alpha/2)s_{\bar{w}_j}/\sqrt{m},$$
$$\bar{w} + t(m-1, 1-\alpha/2)\, s_{\bar{w}_j}/\sqrt{m}]$$

Returning to the previous two problems mentioned for continuing simulations, namely those of when steady state is reached and when to stop each run, we first discuss the latter and then the former. The run length (n) and the number of replications (m) will both influence the size of the standard error ($s_{\bar{w}_j}/\sqrt{m}$) required for making the CI above. The smaller the standard error, the more precise is the CI (narrower limits) for a

given confidence $(1 - \alpha)$. We know that the standard error goes down by the square root of m, so that the more replications made, the more precise the CI. Also, we might expect that as run length n is increased, the computed value of $s_{\bar{w}_j}$ itself will be smaller for a given number of replications, so that longer run lengths will also increase the precision of the CI. Thus for a fixed amount of computer running time, we can trade off size of n versus size of m. Setting n and m is still somewhat of an art and trial runs can be made to evaluate the trade-offs.

The warm-up period (the initial amount of time required to bring the process near steady-state conditions) is also not an easy thing to determine. The basic idea is that if we could compute the time or number of transactions required so that the process is near steady-state, we could simply not start recording data for calculating output measures until that master clock passes that point. A variety of procedures have been developed and the reader is referred to the basic simulation texts referenced previously.

One particular method that seems to work fairly well (again, the warm-up period analysis is still somewhat of an art rather than a science) is that due to Welch (1981, 1983). The procedure involves choosing one of the output performance measures (e.g., average waiting time for service), calculating means of this measure over the replications for *each* transaction (i.e., if we have m replications of run length n, we average the m values we obtain for transaction i from each replication, for $i = 1, 2, \ldots, n$), taking moving averages of neighboring values of these transaction averages, plotting these and visually determining when the graph appears to be stabilizing. It is recommended to try various moving average windows (the number of adjacent points in the moving average). The size of the moving average window and the point at which the graph "settles down" are again judgment calls.

Another approach using the transaction averages calculated as described above, is a regression approach suggested by Kelton and Law (1983). The transaction average data stream of n values is segmented into b batches. A regression line is fit for the last batch, and if the slope of the line is not significantly different from zero, steady state is assumed. Then the next to last batch of data is included in the regression, the test made again, and if not significantly different from zero, the last $2b$ values are said to be in steady state. This procedure is continued, and when the test for slope becomes significantly different from zero, that batch and all preceding ones are considered in the transient region. This approach is predicated on the assumption that the performance measure is monotonic in time, which should be the case when the initial conditions assume the system starts empty and idle. Again, decisions on the values for n, m, and b must be made subjectively.

One approach which differs from the two previous approaches in that rather than attempting to find a point at which the process enters steady state, the bias of the transient effects is estimated in order to determine if the data do show an initial conditions bias (Schruben, 1982). This procedure could be used in conjunction with one of the two above to check whether the warm-up period chosen was adequate to remove the initial conditions bias.

In order to avoid "throwing away" the initial warm-up period for each of the m replications in a non-terminating simulation experiment, the procedure of *batchmeans* has been suggested (Law, 1977; Schmeiser, 1982). Rather than replicating, a single long run (say mn transactions) is made, and the run "broken up" into m segments (batches) of n each. The performance measures for each segment are assumed to be approximately independent (if the segments are long enough, the correlation between segments should be small), so that the classical estimate of the standard deviation can be employed, that is, the segments act as if they are independent replications. The methodology for determining CIs is identical to that for m independent replications. But now, one only has to discard a single warm-up period instead of m as before. Other methods have been suggested, such as the regenerative method (Crane and Iglehart, 1975; Fishman, 1973), and time series analyses (Fishman, 1971; Schruben, 1983).

In comparing two alternative system designs, the technique most commonly used is a paired t CI on the difference of a given performance measure for each design. For example, if we are simulating a queueing system, and one design has two servers serving at a particular rate at a service station in the system and a competing design replaces the two servers with automatic machines, we may be interested in the average holding time of a transaction. We make a run for design one, calculating the average holding time, make a run for design two, calculating the average holding time, and then compute the difference between the two average holding times. Then we replicate the pair of runs m times, obtaining m differences, d_i, $i = 1, 2, \ldots, m$. The mean and standard deviation of the d_i are calculated and the t distribution used to form a $100(1 - \alpha)$ CI on the mean difference in a manner

analogous to that described above, yielding

$$[\bar{d} - t(m - 1, 1 - \alpha/2)s_{d_i}/\sqrt{m},$$
$$\bar{d} + t(m - 1, 1 - \alpha/2)s_{d_i}/\sqrt{m}].$$

Whenever possible, the same random number stream(s) should be used for each design *within* a replication, so that the difference observed depends only on the design parameter change and not the variation due to the randomness of the random variates generated. Of course, different random number streams are used between the replications. This is a variance reduction technique called *common random numbers* (discussed in more detail below) and is quite effective in narrowing the CI limits.

Often, we wish to compare more than two designs which necessitates using multiple comparison techniques. There are a variety of procedures which can be of help. All systems could be compared in a pairwise fashion using the methodology for comparing only two systems. However, if the confidence level of a CI for a single pair is $1 - \alpha$, and we have k pairs, the confidence associated with a statement concerning all the pairs simultaneously drops to $1 - k\alpha$ (Bonferroni inequality). Therefore, if we desire the overall confidence to be $1 - \alpha$, then it is necessary to have confidence for each pair to be $1 - \alpha/k$.

Also available are a variety of ranking and selection procedures, such as selecting the best of k systems, selecting a subset of size r containing the best of the k systems, or selecting the r best of k systems. These are treated in some detail in Law and Kelton (1991).

VARIANCE REDUCTION TECHNIQUES: Unlike sampling from the real world, the simulation modeler has control over the randomness generated in the system. Often, purposely introducing correlation among certain of the random variates in a simulation run can reduce variance and provide narrower CIs. One example of this was shown above in forming a paired t CI by using *common random numbers* within a replication, which introduces positive correlation between the two performance measures within a replication, yielding a smaller variance of the mean difference over the replications.

Another technique, called *antithetic variates*, introduces negative correlation between two successive replications of a given design with the idea that a large random value in one of the pairs will be offset by a small random value in the other. The performance measures for the pairs are averaged to give a single "observed" performance measure. Hence, if m replications are run, only $m/2$ independent values end up being averaged for the CI calculation, but the variance of these values should be considerably lower than m independent observations.

Among other variance reduction techniques are *indirect estimation*, *conditioning*, *importance sampling*, and *control variates*, and again we refer the reader to one of the basic simulation texts.

MODEL VALIDATION: Model validation is a very important step in a simulation study which often is "glossed over" by modelers. Prior to embarking in developing a simulation model, it behooves the simulation analyst to become very familiar with the system being studied, to involve the managers and operating personnel of the system, and thus to agree on the level of detail required to achieve the goal of the study. The appropriate level of detail is always the "coarsest" that can still provide the answers required. One problem with simulation modeling is that since any level of detail can be modeled, models are often developed in more detail than necessary and this can be very inefficient and counterproductive.

Validity is closely associated with *verification* and *credibility*. Verification has to do with program debugging to make sure the computer program does what is intended. This is generally the most straightforward of the "triumvirate" to accomplish as there are well known and established methods for debugging computer programs.

Validation deals with how accurate a representation of reality the model provides and credibility deals with how believable the model is to the users. To establish validity and credibility, users must be involved in the study early and often. Goals of the study, appropriate system performance measures and level of detail must be agreed upon and kept as simple as possible. A log book of assumptions should be kept, updated frequently, and signed off periodically by the model builders and users.

When possible, simulation model output should be checked against actual system performance, if the system being modeled is in operation. If the model can duplicate (in a statistical sense) actual data, both validity and credibility are advanced. If no system currently exists, then if the model can be run under conditions where theoretical results are known (e.g., if studying a queueing system, compare the simulation results to known queueing theoretic results), and if the simulation results duplicate theoretical results, then validity is promoted. The model can be run under a variety of conditions and results examined by the users for plausibility. Most simulation texts have at least one chapter devoted to this important topic. Other references include Carson (1986), Gass and Thompson (1980), Sargent (1988) and Schruben (1980).

See **Distribution selection for stochastic modeling; Mathematical model; Model validation, verification and testing; Networks of queues; Queueing theory; Random number generation.**

References

[1] Carson J.S. (1986). "Convincing Users of Model's Validity Is Challenging Aspect of Modeler's Job," *Industrial Engineering*, **18**, 74–85.

[2] Crane M.A. and Iglehart D.L. (1975). "Simulating Stable Stochastic Systems, III: Regenerative Processes and Discrete Event Simulation," *Operations Research*, **23**, 33–45.

[3] Fishman G.S. (1971). "Estimating Sample Size in Simulation Experiments," *Management Science*, **18**, 21–38.

[4] Fishman G.S. (1973). "Statistical Analysis for Queueing Simulations," Management Science, **20**, 363–369.

[5] Fishman G.S. (1978). *Principles of Discrete Event Simulation*. Wiley, New York.

[6] Gass S.I. and Thompson B.W. (1980). "Guidelines for Model Evaluation: An Abridged Version of the U.S. General Accounting Office Exposure Draft," *Operations Research*, **28**, 431–439.

[7] Hoover S.V. and Perry R.F. (1989). *Simulation: A Problem-Solving Approach*. Addison-Wesley, Reading, Massachusetts.

[8] Kelton W.D. and Law A.M. (1983). "A New Approach for Dealing with the Startup Problem in Discrete Event Simulation," *Naval Research Logistics Quarterly*, **30**, 641–658.

[9] Law A.M. (1977). "Confidence Intervals in Discrete Event Simulation: A Comparison of Replication and Batch Means," *Naval Research Logistics Quarterly*, **27**, 667–678.

[10] Law A.M. and Kelton W.D. (1991). *Simulation Modeling*. McGraw-Hill, New York.

[11] Sargent R.G. (1988). "A Tutorial on Validation and Verification of Simulation Models," *Proceedings of the 1988 Winter Simulation Conference*, 33–39.

[12] Schmeiser B.W. (1982). "Batch Size Effects in the Analysis of Simulation Output," *Operations Research*, **30**, 556–568.

[13] Schruben L.W. (1980). "Establishing the Credibility of Simulations," *Simulation*, **34**, 101–105.

[14] Schruben L.W. (1982). "Detecting Initialization Bias in Simulation Output," *Operations Research*, **30**, 569–590.

[15] Welch P.D. (1981). "On the Problem of the Initial Transient in Steady-State Simulation." IBM Watson Research Center, Yorktown Heights, New York.

[16] Welch P.D. (1983). "The Statistical Analyses of Simulation Results," *The Computer Performance Modeling Handbook*, S. S. Lavenberg, ed. Academic Press, New York.

SIMULATOR

(1) A machine that mimics the real world (e.g., a vehicle or weapon system) by generating cues for the operator(s), accepts inputs by the operator(s), and simulates the proper responses. The cues may include visual, auditory, and tactile sensations. The operator inputs consist of manipulation of control systems similar to those of the simulated equipment. Examples include vehicle and flight simulators. See **Control theory**. (2) The computer code of a simulation model (e.g., Monte Carlo or discrete-event type). See **Simulation of discrete-event stochastic systems.**

SINGLE-SERVER NETWORK

A queueing network with one or more nodes and one server servicing all of them. Examples are a token-ring system with (rotating) possession of the token giving a particular node the right to access the server, and polling systems whereby a rule governs the movement of the server around a sequence or loop of queueing stations. Practical applications include local area computing networks, robots working on a production line, sequential physical service operations, etc. See **Networks of queues; Queueing theory.**

SINGULAR MATRIX

A square matrix whose determinant value is zero. If this is the case, then the columns of the matrix form a linearly dependent set. See **Matrices and matrix algebra**.

SINK NODE

A node in a network through which all (or some) of the flow in the network leaves the network.

SIRO

Service In Random Order; a queueing service scheme in which customers are served in random order unrelated to their times of arrival. See **Queueing theory.**

SKEW-SYMMETRIC MATRIX

A square matrix $A = (a_{ij})$ is skew-symmetric if $a_{ij} = -a_{ji}$. Thus, all diagonal elements are zero. See **Matrices and matrix algebra; Symmetric zero-sum two-person game.**

SLACK VARIABLE

A nonnegative variable that is added to a linear inequality of the form $\Sigma_j a_{ij} x_j \leq b_i$ to transform the inequality into an equation. The slack variable measures the difference between the right- and left-hand-sides of the inequality. See **Logical variable; Slack vector; Surplus variable.**

SLACK VECTOR

The column representation of a slack variable in a linear-programming problem. See **Slack variable**.

SLP

Successive linear programming. See **Petrochemical industry**.

S-MODEL

See **Learning curve.**

SMOOTH PATTERNS OF PRODUCTION

In a production-planning problem with many production cycles, one usually attempts to have the amounts produced in the sequence of cycles to be the same or as similar as possible. Such a smooth production pattern (one with small fluctuations) tends to be more cost efficient than one that has large fluctuations. Many such production problems can be formulated as linear-programming problems. See **Inventory modeling; Operations management; Production management**.

SOFT SYSTEMS METHODOLOGY (SSM)

A problem structuring method which uses systems concepts to promote debate about organizational change among concerned participants. Parsimonious descriptions ("root definitions") are developed of a number of systems, each of which is relevant, from a particular world view ("Weltanschauung"), to the area of concern. For each root definition, a "conceptual model" is constructed of the minimum activities (and their inter-relationships) necessary for the system described in the root definition to function. Differences between conceptual models and participants' perceptions of the existing situation are used as a systematic basis for debate aimed at agreeing on desirable and culturally feasible change. See **Community operations research**; **Problem structuring methods**.

SOJOURN TIME

(1) The total time spent in a queueing facility, determined by the sum of the delay and service times; sometimes called the total waiting time or just waiting time. See **Queueing theory**. (2) Often used as the time spent in a visit to a state of a random process, such as a Markov chain. See **Markov processes**.

SOLUTION

A set of values for the variables of a problem that satisfy all the constraints of the problem. See **Feasible solution**.

SOLUTION SPACE

For a constrained problem, the solution space is a portion of Euclidean space defined by all the constraints of the problem. For a linear-programming problem, the solution space is defined by the intersection of the nonnegative portion of Euclidean n-space and the constraints of the problem.

SOS

See **Special-ordered sets**.

SOURCE NODE

A node in a network from which all (or some) of the flow in the network enters the network.

SPACE

Gerald W. Evans

University of Louisville, Kentucky

INTRODUCTION: Man's venture into outer space began on October 4, 1957 with the successful launch by the U.S.S.R. of the first artificial Earth satellite, Sputnik I. This was soon followed on November 3 by the "dog (Laika)-manned" Sputnik II. The first American satellite, Explorer I, was launched on January 31, 1958. From this period on, both the U.S.A. and the U.S.S.R. mounted extensive research and development activities for putting manned space vehicles into Earth orbit. In particular, the U.S. established the National Aeronautics and Space Administration (NASA) on October 1, 1958 with the mission to "... to achieve at the earliest practicable date orbital flight and successful recovery of a manned satellite, and to investigate the capabilities of man in this environment," (Swenson *et al.* 1966, p. 111). From this beginning, we can trace the U.S. man-in-space program from Project Mercury (Glenn in Earth orbit on February 20, 1992), to Project Gemini (two-manned space capsule in Earth orbit), to Project Apollo (the moon-landing on July 20, 1969), and to the Space Shuttle. However, the Soviet effort succeeded in sending the first manned satellite into Earth orbit with the launch of Gagarin on April 12, 1961. Throughout this time, OR/MS techniques were used by NASA and the space industry in the management and analysis of their space activities. Applications include project management, forecasting, scheduling, cost estimating, optimization, simulation, and multi-objective decision analysis. We discuss some of these applications next.

APPLICATIONS: A long range planning problem encountered by NASA is discussed in Evans and Fairbairn (1989). This research addressed the problem of determining which missions, out of dozens of possibilities, NASA should undertake during the next decades. A 0–1 integer programming model with linear constraints was formulated in which the decision variables determine whether or not to include a particular mission in

NASA's long range plan. The model allows for the consideration of several criteria relating to benefits derived in various areas (for example, intellectual, humanistic, and utilitarian), as well as cost. In addition, the model implicitly considers the dependence among the various missions in the plan by specifying appropriate constraints. An example of dependence would be the fact that a manned mission to Mars would require the undertaking of several precursor missions.

Each flight of a Space Shuttle requires the scheduling of thousands of activities which have certain precedence relationships, and require the use of various types of scarce resources. Scheduling these activities in order to meet criteria relating to time, cost, and quality is a complex process. To develop feasible schedules requires the use of sophisticated project management techniques, including project networks, heuristic scheduling rules, and simulation (Deale, 1992; Barth and Schafer, 1992).

Paté-Cornell and Fischbeck (1994) used a probabilistic risk analysis procedure to set priorities for the maintenance of the heat shield tiles of the space shuttle orbiter. They showed that implementation of the policy suggested by their procedure would result in a 70 percent reduction in the probability of a shuttle accident attributable to tile failure.

Muscettola *et al.* (1992) described the Heuristic Scheduling Testbed System (HSTS) used for generating observation schedules for the Hubble Space Telescope (HST). The HST is a $1.4 billion observatory, with an expected operational lifetime of 15 years. It was placed into earth orbit in 1990. In any period of time, there are many different observational requests for the telescope's resources. Scheduling these requests is a difficult process for several reasons. A particular observation may require that several different operations be performed with six different scientific instruments that make up the HST. Several observations may be grouped together within a particular "window of opportunity," depending on the locations of the telescope and the space objects to be observed. Parallel observations may be made with different viewing instruments; yet, not all six instruments may be turned on simultaneously because of energy constraints. The HSTS employs artificial intelligence procedures that provide a flexible approach to scheduling HST operations. This approach allows the "effective balancing of conflicting scheduling constraints and objectives."

Quirk *et al.* (1989) addressed the problem of selecting one of two energy module alternatives (photovoltaic or solar dynamic) for the space station. A chance constrained programming model was developed to select the system that minimized the expected cost, subject to the constraint that the probability that the net output would be less than or equal to some given net output would be no more than some prespecified value. The structure underlying the model is a stochastic Leontief system. Inputs to the model include subjective probability distributions of energy requirements associated with various activities, as given by Johnson Space Center engineers. Hence, the model accounted for the inherent uncertainties associated with each alternative.

Discrete-event simulation has been employed in several instances at NASA. For example, Morris and White (1987) employed a SLAM II simulation model (Pritsker, 1986) to analyze the operational support requirements of the space shuttle under a delivery payload scenario from the earth to the proposed space station. The model consists of three major modules: one each for ground-base operations, space station operations, and orbital operations. The main inputs to the model include the delivery requirements at the space station. The model can be used to help determine the delivery rate capability of the system, the support resources required, and the utilization of various system resources.

See **Chance-constrained programming; Combinatorial and integer programming; Leontief systems; Project management; Scheduling and sequencing; Simulation of discrete-event stochastic systems.**

References

[1] Barth T. and N. Schafer (1992). "Probabilistic Assessment of the Shuttle Processing Schedule Risks." Abstract published in the *TIMS/ORSA Joint Spring National Meeting Bulletin*, p. 59.

[2] Deale M.J. (1992). "Space Shuttle Ground Processing Scheduling." Abstract published in the *TIMS/ORSA Joint Spring National Meeting Bulletin*, p. 59.

[3] Emme E.M. ed. (1977). *Two Hundred Years of Flight in America: A Bicentennial Survey.* Univelt, Inc., San Diego, California.

[4] Evans G.W. and R. Fairbairn (1989). "Selection and Scheduling of Advanced Missions for NASA Using 0–1 Integer Linear Programming." *Jl. Opns. Res. Soc.* **40**, 971–982.

[5] Mark H. and A. Levine (1984). *The Management of Research Institutions: A Look at Government Laboratories*. U.S. Government Printing Office, Washington, D.C.

[6] Morris W.D. and N.H. White (1987). *Space Transportation System Operations Model*. NASA Technical Memorandum, NASA Langley Research Center, Hampton, Virginia.

[7] Muscettola N., S.F. Smith, A. Cesta and D. d'Aloisi (1992). "Coordinating Space Telescope Operations in an Integrated Planning and Scheduling Architecture." *IEEE Trans. Control Systems* 12, 28–37.

[8] NASA Office of External Relations, eds. (1986). *NASA Space Plans and Scenarios to 2000 and Beyond*. Noyes Publications, Park Ridge, New Jersey.

[9] Paine T.O. ed. (1991). "Leaving the Cradle: Human Exploration of Space in the 21st Century." *Science and Technology Series*, Vol. 28, Univelt, Inc., San Diego, California.

[10] Paté-Cornell H.M. and P.S. Fischbeck (1994). "Risk Management for the Tiles of the Space Shuttle," *Interfaces* 24, 64–86.

[11] Pritsker A.A.B. (1986). *Introduction to Simulation and SLAM II*. Systems Publishing Corporation, West Lafayette, Indiana.

[12] Quirk J., M. Olson, H. Habib-Agahi and G. Fox (1989). "Uncertainty and Leontief Systems: An Application to the Selection of Space Station System Designs." *Mgmt. Sci.* 35, 585–596.

[13] Roland A. (1985). *Model Research: The National Advisory Committee for Aeronautics*, 1915–1958. Volume 1, The NASA History Series, Scientific and Technical Information Branch, NASA, Washington, D.C.

[14] Swenson L.S. Jr., J.M. Grimwood and C.C. Alexander (1966). *This New Ocean: A History of Project Mercury*, National Aeronautics and Space Administration, Washington, D.C.

SPANNING TREE

A subnetwork (graph) of a given network that connects all the nodes of the network and which has the property that once a path travels through a node, it can not return to that node (the path has no cycles). A spanning tree is a tree of the network. If a network has n nodes, then the spanning tree has $n-1$ arcs. See **Kruskal's algorithm**; **Minimum spanning tree problem**; **Network optimization**; **Primal algorithm**; **Prim's algorithm**; **Tree**.

SPARSE MATRIX

A matrix whose elements are mostly zero. See **Density**; **Super sparsity**.

SPARSITY

See **Density**; **Large-scale systems**.

SPECIAL-ORDERED SETS (SOS)

Types of constraints in optimization models. SOS of type 1 require that only one variable in the set may be nonzero; SOS of type 2 require that only two variables in the set may be nonzero and they must be adjacent. SOS of type 1 are used in problems in which the variables in the set are binary and only one of them can be equal to one (e.g., assignment of personnel). SOS of type 2 occur when transforming a separable-programming problem into an equivalent linear structure. Special computational approaches are used to simplify the handling of both types of SOS problems. See **Separable-programming problem**.

SPLINES

Sharon A. Johnson

Worcester Polytechnic Institute, Massachusetts

Splines are an important class of mathematical functions used for approximation. A spline is a piecewise polynomial function that is commonly described as being "as smooth as it can be without reducing to a polynomial" (de Boor, 1978, p. 125). For example, the cubic spline shown as the solid line in Figure 1 is composed of individual cubic polynomials, each defined between two adjacent data points, such that the function values and first and second derivatives of adjoining polynomial pieces are the same. In general, a function defined on an interval $[a, b]$ is defined as a *polynomial spline of degree* k, having knots $t_1, \ldots, t_r$, if the following three conditions hold: (i) $a < t_1 < \cdots < t_r < b$, so the knots $t_1, \ldots, t_r$ partition the interval $[a, b]$ into $r+1$ smaller subintervals, (ii) on each subinterval $[t_i, t_{i+1}]$, the spline is given by a polynomial function of at most degree k, and (iii) the spline and its derivatives up to order $k-1$ are all continuous on $[a, b]$. The definition of splines is sometimes extended to allow the knots to be coincident (e.g., so that $t_i = t_{i+1}$), in which case the spline has less continuity at that knot (de Boor, 1978). Recognition in the early 1960s that splines could mathematically model the physical process of drawing a smooth curve with a mechanical spline resulted in further investigation of their

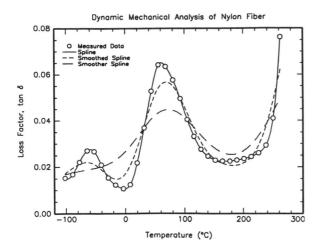

Fig. 1. Spline approximations to data collected from a dynamic mechanical analysis of nylon fiber (courtesy of Monsanto Chemical Company).

approximation and methods for computing them efficiently (Schumaker, 1981).

The benefits of splines as approximators are discussed below. Next, an application is described to illustrate when spline approximation can be effective. The B-spline representation of a spline is then introduced. Finally, multivariate approximation is briefly addressed.

SPLINES AS APPROXIMATING FUNCTIONS:

OR/MS scientists create mathematical models to describe a physical system or problem, then experiment with the model to draw conclusions about that system. Functions, relating decisions or independent variables to output or dependent variables, form the basis of these models. While these underlying functions are sometimes known explicitly, they are often created by collecting discrete data about the system, then constructing an approximation to the unknown underlying response function. Suppose that the value of a function is measured at points $x_1, \ldots, x_n$, yielding values $f(x_1), \ldots, f(x_n)$. In interpolation problems, the goal is to find an approximating function $s(x)$ which passes through the points $f(x_1), \ldots, f(x_n)$, so that $s(x_i) = f(x_i)$. When the measured values $f(x_1), \ldots, f(x_n)$ contain errors, an approximating function $s(x)$ is created to balance the desire to obtain an approximation with smooth behavior and the desire to fit the data closely enough (Dierckx, 1993). For example, the "smoothing splines" shown with dashed lines in Figure 1 are constructed with a smoothing factor that allows deviations from the data (de Boor, 1978); permitting larger deviations results in a cubic spline with a smaller second derivative. Broken curve regression can

also be used to fit data with errors (Seber and Wild, 1989).

The data fitting problems described above are one major category of valuable approximation (Schumaker, 1981). Other common approximation problems involve replacing a known function with one that is easy to compute, or estimating solutions to models involving differential equations, which can only be solved explicitly for simple cases.

Spline functions are effective approximators because they are relatively smooth, and splines of low degree usually provide an adequate fit with reasonable computational effort (Dierckx, 1993). Every continuous function on an interval can be approximated arbitrarily well by a polynomial spline of a particular order, provided a sufficient number of knots are allowed. For example, a linear spline (a piecewise linear function) will better fit a curve the more segments it has. Low order splines are flexible, and do not exhibit the oscillations usually associated with polynomials. The ease with which splines can be stored and evaluated on a computer make them powerful for a variety of applications. Spline fitting routines are included in most general software libraries (e.g., NAG 1988) or in specific spline packages (Dierckx, 1993).

AN APPLICATION: The cost-to-go function $f_t(x_t)$ and optimal solution for a finite horizon dynamic programming problem can be found by solving a functional equation

$$f_t(X_t) = \max_{R_t}\{B_t(X_t, R_t) + E_{q_t}[f_{t+1}(x_{t+1})]\} \quad (1)$$

which in period t represents the expected benefit from operating the system from period t to the end of the horizon, given the system begins period t in state x_t. Such problems arise in managing inventory or planning water reservoir operations, where the state x_t would represent the amount in inventory or the volume of water in a reservoir. In stochastic problems, the system is subject to random influences q_t, such as uncertain demand or the inflow to a reservoir. In each period t, the decision r_t is made to maximize both the benefit B_t occurring from operation in the current period as well as expected future benefits $E_q[f_{t+1}]$. For example, the decision r_t might correspond to how much to produce or how much to release. The status of the system x_{t+1} at the end of each period t is determined by a transition function $g(x_t, q_t, r_t)$; for example, in inventory planning, g is $x_t + r_t - q_t$.

When the state vector x_t is continuous, the cost-to-go function f_t and policy r_t are often found by discretizing x_t and recursively solving (1) backward, for $t = T$ (the last time period) to $t = 1$. Because the function $f_{t+1}(x_{t+1})$ is only known at a finite number of points, interpolation can be used to generate a value when it is needed at other points. Using cubic splines to approximate the cost-to-go function can significantly reduce the effort required to solve such dynamic programming problems, particularly when the state vector is of high dimension, because of their accuracy relative to piecewise linear approximations (so fewer knots are needed) and because their smoothness allows efficient optimization methods to be used to find the decisions r_t (Johnson *et al.*, 1993). Splines can also be applied to more general dynamic programming algorithms (Schweitzer and Seidmann, 1985).

THE B-SPLINE REPRESENTATION:

Any spline $s(x)$ of degree k can be written as a linear combination of B-splines $B_i(x)$:

$$S(x) = \sum_i a_i B_i(x),$$

where each B-spline $B_i(x)$ is a spline of degree k (de Boor, 1978; Dierckx, 1993). B-splines permit the efficient evaluation of a spline and its derivatives because they have local support; that is, outside of a small range, they take the value of zero.

A particular spline is selected as an approximating function by choosing the degree k of the spline, the number and position of the knots t_i, and the coefficients a_i (Dierckx, 1993). Choosing the number and/or the position of the knots is often a matter of trial and error. Theory may suggest points in the data where the underlying model changes (Smith, 1979). More knots should generally be placed in those regions where the underlying data change rapidly (de Boor, 1978; Dierckx, 1993). Algorithms have been developed for some spline problems where the knot locations are treated as parameters, and optimized. Because the problem is nonlinear, the computational effort in these algorithms increases significantly as the number of knots increases (Dierckx 1993).

The conditions that determine the coefficients a_i of the spline depend on how closely the spline is expected to fit the data, the desired smoothness, and specified boundary conditions (Dierckx, 1993). In problems where the function underlying the data are known to have certain properties, such as convexity or monotonicity, it may be desirable to develop an approximation with the same properties constraining the coefficients a_i. Such shape-preserving approximations may also be beneficial because they prevent undesirable oscillations.

MULTIVARIATE APPROXIMATION:

Tensor products are an efficient way to construct and evaluate multivariate approximations because they allow a multivariate problem to be solved as a series of one dimensional problems (de Boor, 1978). A bicubic spline constructed using tensor product methods would be a cubic spline in each coordinate direction. The major drawback of such splines is that they require the approximation domain to be a rectangle, or easily transformed to a rectangle (e.g., a sphere) (Dierckx, 1993). In addition, constructing tensor product approximations is appropriate only when it makes sense to have "preferred directions" in the approximant (de Boor, 1978). For example, a bicubic spline could efficiently approximate a peak that occurred along one axis. However, many knots would be required in each dimension if the peak occurred along a diagonal.

When tensor product methods are not appropriate, constructing multivariate spline approximations is much more complex, and thus computationally less attractive (Dierckx, 1993). First, an appropriate partition of the data must be chosen. Next an appropriate set of basis functions (similar to B-splines) must be defined that permit efficient evaluation of the approximating function. Dierckx (1993) describes two spline generalizations based on triangularizations of a surface, and also provides references.

References

[1] de Boor C. (1978). *A Practical Guide to Splines.* Springer-Verlag, New York.
[2] Dierckx P. (1993). *Curve and Surface Fitting with Splines.* Oxford University Press, New York.
[3] Johnson S.A., J.R. Stedinger, C.A. Shoemaker, Y. Li and J.A. Tejada-Guibert (1993). "Numerical Solution of Continuous-State Dynamic Programs Using Linear and Spline Interpolation," *Opns. Res.* **41**, 484–500.
[4] NAG 1988. *Fortran Library – Mark* 13, The Numerical Algorithms Group, Oxford.
[5] Schumaker L.L. (1981). *Spline Functions: Basic Theory.* John Wiley, New York.
[6] Schweitzer P.J. and A. Seidmann (1985). "Generalized Polynomial Approximations in Markovian Decision Processes," *J. Math. Anal. & Appl.* **110**, 568–582.
[7] Seber G.A.F. and C.J. Wild (1989). *Nonlinear Regression.* John Wiley, New York.
[8] Smith P.L. (1979). "Splines as a Useful and Convenient Statistical Tool," *The American Statistician,* **33**, 57–62.

SPORTS

Shaul P. Ladany

Ben-Gurion University of the Negev, Beer Sheva, Israel

INTRODUCTION: The history of the applications of quantitative methods and systems analysis to sports events is very much the history of systems analysis and its applications to many fields of human endeavor. For a thorough review of all the sports applications up to 1976 see Ladany and Machol (1977), while for the second half of the same two-pronged effort which culminated in invited research articles of the mid-1970s see Machol, Ladany and Morrison (1976). A further review, incorporating most of the recent applications, can be found in Gerchak (1994).

The first studies of sports were purely descriptive; the earliest such technical articles, on cricket, by Elderton (1909, 1927, 1945) and Wood (1941, 1945), are described in Pollard (1977). The application of sophisticated statistical analysis started with Mosteller (1952), who estimated the probability that the better team wins in the World Series competition. The next stage was Mottley's (1954) suggestion that operations research could be profitably applied to sports, specifying football and basketball examples. Bona fide optimization applications followed when Howard (1960) and Bellman (1964) applied dynamic programming to baseball.

In line with the beginning, most of the studies were applied to team sports, initially dealing with issues of individual teams, and more recently with organizational matters preoccupying leagues and associations. The applications to individual sports were much fewer.

TEAM SPORTS: *Baseball* – Most of the studies were applied to baseball, a sport particularly suitable for OR approaches because the action occurs in discrete events, and because the state of the game is simple to specify. The benefit of the strategy of intentional walk and base-stealing was thoroughly investigated by Lindsey (1959, 1961, 1963, 1977). The best batting order was analyzed by Cook and Garner (1964), Cook and Fink (1972), Freeze (1974) and Peterson (1977). A team's elimination from playoff consideration was explored by Robinson (1991).

Football – The value of field position was investigated by Carter (a former National Football League quarterback) and Machol (1971 and 1978), and Schaefer, Schaefer and Atkinson (1991). The value of a tie and extra-point strategy was analyzed by Porter (1967), Bierman (1968) and Hurley (1989), while Bisland (1977) developed a simulation model to evaluate offensive strategies.

Hockey – The problem when to pull the goalie was considered by Morrison (1976), Morrison and Wheat (1986), Erkut (1987), Nydic and Weiss (1989), and Washburn (1991). League playoff strategies were analyzed by Monahan and Berger (1977).

Soccer – Rivett (1975) modeled the attendance at soccer matches and suggested changes in the organization of clubs; Shikata (1977) tried to analyze the motions of ball and players in a four-dimensional space; Mehrez, Pliskin and Mercer (1987) evaluated a new point system for the soccer leagues, while Mehrez and Hu (1994) constructed predictors for outcomes in the league.

General League Issues – The problem of planning the schedule of the games for the entire season with the objective to minimize traveling distance and/or number of tours, under various constraints, was attacked – for various leagues – by Campbell and Chen (1976), Ball and Webster (1977), Cain (1977), Liittschwager and Haagenson (1978), Bean and Birge (1980), Schreuder (1980), Coppins and Pentico (1981), Tan and Magazine (1981), Lane (1981), de Werra (1982), Ostermann (1982), and Ferland and Fleurent (1991). The related counterpart problem of scheduling umpires was considered by Cain (1977) and by Evans (1988).

Draft issues prevailing in North American professional sports were the subject of investigations by Price and Rao (1976), Brams and Straffin (1979), and Gerchak and Mausser (1994), while the effect of a player's trade to a new team was considered by Bateman, Karvan and Kazee (1984).

The presence of streaks in baseball was analyzed and rejected by Tversky and Gilovich (1989), as well as by Albright (1992), while the advantage of splitting a league into smaller leagues to lead to more pennant races was proven by Winston and Soni (1982).

The problem of ranking teams (and applicable also to individuals) has drawn considerable attention by researchers. Leake (1976) used electrical network theory to rank football teams, Ushakov (1976) presented a methodology for ranking participants playing in a round robin tournament like in chess, while Sinuany-Stern (1988) applied the analytic hierarchy approach to predict the ranking of soccer teams.

INDIVIDUAL SPORTS: *Track and Field* – The derivation of optimal training plans for pentathlon (applicable also to decathlon, triathlon, biathlon, etc.) was pioneered by Ladany (1975b) using linear programming with physiological constraints. Whereas long jump was approached first by Brearley (1972) to analyze, using vector calculus, whether Bob Beamon's miracle jump in Mexico City was affected by the altitude, decision problems of aiming at take off attracted Ladany *et al.* (1975), Sphicas and Ladany (1976), Ladany and Singh (1978), and Mehrez and Ladany (1987). The tactical issues in pole-vaulting (which are similar to those prevailing in high-jump) for the selection of the optimal starting height were investigated by Ladany (1975a), and after change in the rules reinvestigated by Hersh and Ladany (1989) using dynamic programming. The sequential and competitive nature of several athletic events led to the coinage of the term "games of boldness" and to their analysis by Gerchak and Henig (1986), Henig and O'Neill (1992), and Gerchak and Kilgour (1992). Optimal assignments of runners (or swimmers) to relay teams, were put forward by Machol (1970), Heffley (1977), and Hannan and Chen (1987), advancing from the use of the simple deterministic assignment model to conditional and stochastic treatments. Optimization of the biomechanical aspects, like the optimal angle to release a shotput or a hammer, were discussed by Townend (1984).

Golf – The evaluation of the handicap system and its fairness occupied all researchers in the field, starting with Scheid (1972, 1977, 1990), and followed by Pollock (1974, 1977), Cochran (1990), and Freeze (1994). Handicapping was applied also to other sports events; Camm and Grogan (1988) applied it to road-running races using frontier analysis.

Tennis – Analysis of the most important points was performed by Morris (1977). Gale (1971) investigated the optimal serving strategies, and justified the greater risk taken on the first serve. Norman (1985) applied dynamic programming to determine when to use a fast serve. Blackman and Casey (1989) suggested a player rating system. The assignment of players to matches aspect of the sports was treated by Hannan (1979), and by Hamilton and Romano (1992).

Other Sports – Selection of teams for gymnastic competition was dealt with bivalent integer programming by Ellis and Corn (1984) and Eilon (1986). Optimal weight-lifting policies were derived by Lilien (1976). Favoritism and collusion in the judging of figure skating was detected by Klemm and Iglarsh (1984) using non-parametric statistical methods. Oar arrangements in rowing eights to prevent "fishtail" behavior were analyzed using the mechanical theory of moments by Brearley (1977). The unfairness of the existing scoring systems for jai-alai was evaluated using simulation by Hannan and Smith (1981) and Skiena (1988).

See **Analytic hierarchy process; Decision analysis; Dynamic programming; Combinatorial and integer optimization; Linear programming; Simulation; Systems analysis.**

References

[1] Albright C. (1992), "Streaks & Slumps," *OR/MS Today*, April, 94–95.

[2] Ball B.C. and D.B. Webster (1977), "Optimal Scheduling for Even-numbered Team Athletic Conferences," *AIIE Transactions*, 9, 161–167.

[3] Bateman T.S., K.R. Karwan and T.A. Kazee (1984), "Can a Trade to a New Team Affect a Baseball Player's Performance," working paper of the College of Business Administration, Texas A&M University, College Station.

[4] Bean J.C. and J.R. Birge (1980), "Reducing Travelling Costs and Player Fatigue in the National Basketball Association," *Interfaces*, 10, 98–102.

[5] Bellman R.E. (1964), "Dynamic Programming and Markovian Decision Processes with Particular Application to Baseball and Chess," Ch. 7 in *Applied Combinatorial Mathematics*, E. Beckenbach (ed.), Wiley, New York.

[6] Bierman H. (1968), A Letter to the Editor, *Management Science*, 14, B281–282.

[7] Bisland R.B. Jr. (1977), "The Football Strategy Simulator Model," paper presented at the ORSA/TIMS Joint National Meeting, Atlanta.

[8] Blackman S.S. and J.W. Casey (1980), "Developing of a Rating System for All Tennis Players," *Operations Research*, 28, 489–502.

[9] Brams S.J. and P.D. Straffin Jr. (1979), "Prisoner's Dilemma and Professional Sports Drafts," *The American Mathematical Monthly*, 86, 80–88.

[10] Brearly M.N. (1972), "The Long Jump Miracle of Mexico City," *Mathematics Magazine*, 45, 241–246.

[11] Brearley M.N. (1977), "Oar Arrangements in Rowing Eights," in *Optical Strategies in Sports*, S.P. Ladany and R.E. Machol, eds., North-Holland, 184–185.

[12] Cain W.O. Jr. (1977), "A Flexible Algorithm for Solving the Umpire-Scheduling Problem," paper presented at the ORSA/TIMS Joint National Meeting, Atlanta.

[13] Camm J.D. and T.J. Grogan (1988), "An Application of Frontier Analysis: Handicapping Running Races," *Interfaces*, 18, 52–60.

[14] Campbell R.T. and D.S. Chen (1976), "A Minimum Distance Basketball Scheduling Problem," in *Management Science in Sports*, R.E. Machol, S.P. Ladany and D.G. Morrison, eds., North-Holland/TIMS Studies in the Management Sciences, 4, 15–26.

[15] Carter V. and R.E. Machol (1971), "Operations Research in Football," *Operations Research*, 19, 541–544.

[16] Carter V. and R.E. Machol (1978), "Optimal Strategies on Fourth Down," *Management Science*, 24, 1758–1762.

[17] Cochran A.J. (ed.) (1990), *Science and Golf: Proceedings of the First World Scientific Congress of Golf*, St. Andrews, Scotland, Chapman and Hall.

[18] Cook E. and W.R. Garner (1964), *Percentage Baseball*, MIT Press, Cambridge, Massachusetts.

[19] Cook E. and D.L. Fink (1972), *Percentage Baseball and the Computer*, Waverly Press, Baltimore, Maryland.

[20] Coppins R.J.R. and D.W. Pentico (1981), "Scheduling the National Football League," paper presented at CORS/TIMS/ORSA Conference, Toronto.

[21] de Werra D. (1982), "Graphs and Sports Scheduling," O.R. Working Paper 82/2, Departement de Mathematiques, Ecole Polytechnique Federale de Lausanne.

[22] Eilon S. (1986), "Note: Further Gymnastics," *Interfaces*, 16, 69–71.

[23] Elderton W.P. and E.M. (1909), *Primer of Statistics*, Black; London.

[24] Elderton W.P. (1927), *Frequency Curves and Correlation*, 2nd ed., Layton; London.

[25] Elderton W.P. (1945), "Cricket Scores and Some Skew Correlation Distributions," *Jl. Royal Statist. Soc. A.*, 108, 1–11.

[26] Ellis P.M. and R.W. Corn (1984), "Using Bivalent Integer Programming to Select Teams for Intercollegiate Women's Gymnastics Competition," *Interfaces*, 14, 41–46.

[27] Erkut E. (1987), "More on Morrison and Wheat's 'Pulling the Goalie Revisited,'" *Interfaces*, 17, 121–123.

[28] Evans J.R. (1988), "A Microcomputer-Based Decision Support System for Scheduling Umpires in the American Baseball League," *Interfaces*, 18, 42–51.

[29] Ferland J.A. and C. Fleurent (1991), "Computer Aided Scheduling for a Sports League," *INFOR*, 29, 14–24.

[30] Freeze A.R. (1975), "Monte Carlo Analysis of Baseball Batting Order," in *Optimal Strategies in Sports*, S.P. Ladany and R.E. Machol, eds., North-Holland, 63–67.

[31] Freeze A.R. (1995), "A Simulation Analysis of the Golf Handicap System," paper under review.

[32] Gale D. (1971), "Optimal Strategy for Serving in Tennis," *Mathematics Magazine*, 44, 197–199.

[33] Gerchak Y. (1994), "Operations Research in Sports," in *Handbooks in OR & MS*, Vol. 6, S.M. Pollock *et al.*, eds., Elsevier Science, 507–527.

[34] Gerchak Y. and M. Henig (1986), "The Basketball Shootout: Strategy and Winning Probabilities," *Operations Research Letters*, 5, 241–244.

[35] Gerchak Y. and M. Kilgour (1993), "Sequential Competitions with Nondecreasing Levels of Difficulty," *Operations Research Letters*, 13, 49–58.

[36] Gerchak Y. and H.E. Mausser (1994), "The NBA Draft Lottery in the 90's: Back to Moral Hazard?," paper under review.

[37] Hamilton J. and R. Romano (1992), "Equilibrium Assignment of Players in Team Matches: Game Theory for Tennis Coaches," WP 1992–26, Graduate School of Industrial Administration, Carnegie Mellon University.

[38] Hannan E.L. (1979), "Assignment of Players to Matches in a High School or College Tennis Match," *Computers and Operations Research*, 6, 21–26.

[39] Hannan E. and C.D. Chen (1982), "Assignment of Swimmers to Events in a Swimming Meet," working paper, Institute of Administration and Management, Union College and University, Schenectady.

[40] Hannan E.L. and L.A. Smith (1981), "A Simulation of the Effects of Alternative Rule Systems for Jai Alai," *Decision Sciences*, 12, 75–84.

[41] Heffley D.R. (1977), "Assigning Runners to a Relay Team," in *Optimal Strategies in Sports*, S.P. Ladany and R.E. Machol, eds., North-Holland, 169–171.

[42] Henig M. and B. O'Neill (1992), "Games of Boldness, Where the Player Performing the Hardest Task Wins," *Operations Research*, 40, 76–87.

[43] Hersh M. and S.P. Ladany (1989), "Optimal Pole-Vaulting Strategy," *Operations Research*, 37, 172–175.

[44] Howard A. (1960), *Dynamic Programming and Markov Processes*, M.I.T. Press and John Wiley, Chapter 5.

[45] Hurley W. (1989), "Should We Go for the Win or Settle for a Tie?," unpublished paper.

[46] Klemm R. J. and H.J. Iglarsh (1983), "Skating on Thin Ice," working paper of the School of Business Administration, Georgetown University, Washington, D.C.

[47] Ladany S.P. (1975a), "Optimal Starting Height for Pole-Vaulting," *Operations Research*, 23, 968–978.

[48] Ladany S.P. (1975b), "Optimization of Pentathlon Training Plans," *Management Science*, 21, 10, 1144–1155.

[49] Ladany S.P., J.W. Humes and G.P. Sphicas (1975), "The Optimal Aiming Line," *Operational Research Quarterly*, 26, 3, 495–506.

[50] Ladany S.P. and Robert E. Machol eds. (1977), *Optimal Strategies in Sports*, North-Holland.

[51] Ladany S.P. and J. Singh (1978), "On Maximizing the Probability of Jumping Over a Ditch," *SIAM Review*, 20, 171–177.

[52] Lane D.E. (1981), "An Algorithm for the Construction of the Regular Season Schedule for the Canadian Football League," O.R. Working Paper

59, Departement de Mathematiques, Ecole Polytechnique Federale de Lausanne.

[53] Leake R.J. (1976), "A Method of Ranking Teams: With an Application to College Football," in *Management Science in Sports*, R.E. Machol, S.P. Ladany and D.G. Morrison, eds., North-Holland/TIMS Studies in the Management Sciences, 4, 27–46.

[54] Liittschwager J.M. and J.R. Haagenson (1978), "The Round Robin Athletic Scheduling Problem," presented at ORSA/TIMS joint meeting, Los Angeles.

[55] Lilien G.L. (1976), "Optimal Weightlifting," in *Management Science in Sports*, R.E. Machol, S.P. Ladany and D.G. Morrison, eds., North-Holland/TIMS Studies in the Management Sciences, 4, 101–112.

[56] Lindsey G.R. (1959), "Statistical Data Useful for the Operation of a Baseball Team," *Operations Research*, 7, 197–207.

[57] Lindsey G.R. (1961), "The Progress of the Score During a Baseball Game," *Jl. American Statistical Association*, 56, 703–728.

[58] Lindsey G.R. (1963), "An Investigation of Strategies in Baseball," *Operations Research*, 11, 477–501.

[59] Lindsey G.R. (1977), "A Scientific Approach to Strategy in Baseball," in *Optimal Strategies in Sports*, S.P. Ladany and R.E. Machol, eds., North-Holland, 169–171.

[60] Machol R.E. (1970), "An Application of the Assignment Problem," *Operations Research*, 18, 745–746.

[61] Machol R.E., S.P. Ladany and D. G. Morrison, eds. (1976), *Management Science in Sports*, North-Holland/TIMS Studies in the Management Sciences, 4.

[62] Mehrez A. and M.Y. Hu (1995), "Predictors of Outcomes on a Soccer Game – A Normative Analysis Illustrated for the Israeli Soccer League," *Zeitschrift Operations Research*.

[63] Mehrez A. and S.P. Ladany (1987), "The Utility Model for Evaluation of Optimal Behavior of a Long Jump Competitor," *Simulation & Games*, 18, 344–359.

[64] Mehrez A., J.S. Pliskin and A. Mercer (1987), "A New Point System for Soccer Leagues: Have Expectations Been Realized?," *European Jl. Operational Research*, 28, 154–157.

[65] Monahan J.P. and P.D. Berger (1977), "Playoff Structures in the National Hockey League," in *Optimal Strategies in Sports*, S.P. Ladany and R.E. Machol, eds., North-Holland, 123–128.

[66] Morris C. (1977), "The Most Important Points in Tennis," in *Optimal Strategies in Sports*, S.P. Ladany and R.E. Machol, eds., North-Holland, 131–140.

[67] Morrison D.G. (1976), "On the Optimal Time to Pull the Goalie: A Poisson Model Applied to a Common Strategy Used in Ice Hockey," in *Management Science in Sports*, R.E. Machol, S.P. Ladany and D.G. Morrison, eds., North-Holland/

TIMS Studies in the Management Sciences, 4, 137–144.

[68] Morrison D.G. and R.D. Wheat (1986), "Pulling the Goalie Revisited," *Interfaces*, 16, 28–34.

[69] Mosteller F. (1952), "The World Series Competition," *Jl. American Statistical Association*, 47, 259, 355–380.

[70] Mottley M. (1954), "The Application of Operations Research Methods to Athletic Games," *JORSA*, 2, 335–338.

[71] Norman J.M. (1985), "Dynamic Programming in Tennis: When to Use a Fast Serve," *Journal of the Operational Research Society*, 36, 75–77.

[72] Nydic R.L., Jr. and H.J. Weiss (1989), "More on Erkut's 'More on Morrison and Wheat's *Pulling the Goalie Revisited*," *Interfaces* 19, 45–48.

[73] Ostermann R. (1982), "Conversational Construction of a Sports Schedule," O.R. Working Paper, Département de Mathématiques, École Polytechnique Federale de Lausanne.

[74] Peterson A.V. Jr. (1977), "Comparing the Run-Scoring Abilities of Two Different Batting Orders: Results of a Simulation," in *Optimal Strategies in Sports*, S.P. Ladany and R.E. Machol, eds., North-Holland, 86–88.

[75] Pollard R. (1977), "Cricket and Statistics," in *Optimal Strategies in Sports*, S.P. Ladany and R.E. Machol, eds., North-Holland, 129–130.

[76] Pollock S.M. (1974), "A Model for Evaluating Golf Handicapping," *Operations Research*, 22, 1040–1050.

[77] Pollock S.M. (1977), "A Model of the USGA Handicap System and 'Fairness' of Medal and Match Play," in *Optimal Strategies in Sports*, S.P. Ladany and R.E. Machol, eds., North-Holland, 141–150.

[78] Porter R.C. (1967), "Extra-Point Strategy in Football," *The American Statistician*, 21, 14–15.

[79] Price, B. and A.G. Rao (1976), "Alternative Rules for Drafting in Professional Sports," in *Management Science in Sports*, R.E. Machol, S.P. Ladany and D.G. Morrison, eds., North-Holland/TIMS Studies in the Management Sciences, 4, 79–90.

[80] Rivett B.H. (1975), "The Structure of League Football," *Operational Research Quarterly*, 26, 801–812.

[81] Robinson L.W. (1991), "Baseball Playoff Eliminations: An Application of Linear Programming," *Operations Research Letters*, 10, 67–74.

[82] Schaefer M.K., E.J. Schaefer and W.M. Atkinson (1991), "Fourth Down Decisions in Football," Department of Mathematics, College of William and Mary.

[83] Scheid F. (1972), "A Least-squares Family of Cubic Curves with Application to Golf Handicapping," *SIAM Jl. Applied Mathematics*, 22, 77–83.

[84] Scheid F. (1977), "An Evaluation of the Handicap System of the United States Golf Association," in *Optimal Strategies in Sports*, S.P. Ladany and R.E. Machol, eds., North-Holland, 151–155.

[85] Scheid F. (1990), "On the Normality and Independence of Golf Scores, with Various Applications," in *Science and Golf*, A.J. Cochran, ed., Chapman and Hall, 147–152.

[86] Schreuder J.A.M. (1980), "Constructing Timetables for Sports Competitions," *Mathematical Programming Study*, 13, 58–67.

[87] Shikata M. (1977), "Information Theory in Soccer," *Journal of Humanities and Natural Sciences* (Tokyo College of Economics), 46, 35–94.

[88] Sinuany-Stern Z. (1988), "Ranking of Sport Teams via the AHP," *Jl. Operational Research Society*, 39, 661–667.

[89] Skiena S.S. (1988), "A Fairer Scoring System for Jai-Alais," *Interfaces*, 18, 35–41.

[90] Sphicas G.P. and S.P. Ladany (1976), "Dynamic Policies in the Long Jump," in *Management Science in Sports*, R.E. Machol, S.P. Ladany and D.G. Morrison, eds., North-Holland/TIMS Studies in the Management Sciences, 4, 113–124.

[91] Tan Y.Y. and M.J. Magazine (1981), "Solving a Sports Scheduling Problem with Preassignments," Department of Management Sciences, University of Waterloo.

[92] Townend M.S. (1984), *Mathematics in Sport*, Ellis Horwood, London.

[93] Tversky A. and T. Gilovich (1989), "The Cold Facts about the 'Hot Hand' in Basketball," *Chance*, 2, 16–21.

[94] Ushakov I.A. (1976), "The Problem of Choosing the Preferred Element: An Application to Sport Games," in *Management Science in Sports*, R.E. Machol, S.P. Ladany and D.G. Morrison, eds., North-Holland/TIMS Studies in the Management Sciences, 4, 153–162.

[95] Washburn A. (1991), "Still More on Pulling the Goalie," *Interfaces*, 21(2), 59–64.

[96] Winston W. and A. Soni (1982), "Does Division Play Lead to More Pennant Races?," *Management Science*, 28, 1432–1440.

[97] Wood G.H. (1945), "Cricket Scores and Geometrical Progression," *Jl. Royal Statist. Soc. A.*, 108, 12–22.

See **System analysis.**

SPREADSHEETS

Donald R. Plane

Rollins College, Winter Park, Florida

INTRODUCTION: The electronic spreadsheet is a computer application that displays on a computer screen a worksheet, which is a rectangular grid of numeric and text information. This row and column organization mimics many situations, such as an accountant's worksheet, a teacher's gradebook, an invoice, or a scientist's data journal. Through the use of the computer keyboard and pointing device (mouse), the user is able to manipulate the information using mathematical operations, logical operations, and text operations. The worksheet is the foundation for graphs and reports designed to communicate information in a user-friendly manner.

The computer spreadsheet, operating primarily on personal computers, is a serious business tool. It is used for myriad applications, including simple budgets, cash flow analysis, financial planning, optimizing investments, tracking production, forecasting, facilities analysis, and printing address labels.

VisiCalc, the first personal computer spreadsheet, was introduced in October, 1979. Until then, the personal computer had been viewed more as hobbyists' interests than as serious office equipment. Users soon began to realize that the electronic spreadsheet enabled them to change one number or numbers and immediately see the results of the changes in other parts of the worksheet. This capability was instrumental in causing the personal computer, with spreadsheet software, to become an important office tool. The 1982 introduction of Lotus Development Corporation's spreadsheet software called Lotus 1-2-3 marked the availability of a computer application that combined three functions (worksheet, graphics, and database) into software designed for the then new IBM personal computer. The electronic spreadsheet is a very important driving force in the development of the personal computer industry as we know it today. In 1994, the most widely used spreadsheets include Excel, Lotus 1-2-3, Quattro Pro, and Supercalc.

EXAMPLE: A monthly budget demonstrates the organization of a worksheet, and its basic capabilities. The columns of the worksheet show the monthly information, while the rows display expenditures by category. Totals are shown for each expenditure category and for each month. On a computer screen a very simple worksheet might appear, as shown in Table 1. The worksheet columns are identified by letters; the rows are identified by numbers. Cells are identified by the column letter(s) followed by the row number. In Table 1, cell B7 (column B, row 7) contains the budget total for January.

The user can change any monthly expenditure item, and the spreadsheet will automatically calculate the row and column totals. As the size and complexity of a worksheet grow, this ability to recalculate rapidly and automatically the entire spreadsheet replaces many hours of manual labor. The worksheet contains many more columns and rows than the screen can display at one

Table 1. Budget spreadsheet.

==

B7   **B2+B3+B4+B5**

	A	B	C	D	E
1		Jan	Feb	Mar	Total
2	Food	220	230	300	750
3	Housing	400	400	400	1200
4	Clothing	200	50	75	325
5	Entertainment	150	300	75	525
6					
7	Total	970	980	850	

==

time. The mouse or keystrokes can be used to show any part of the worksheet on the screen.

Prior to the development of spreadsheets, a user wishing to solve a problem using a computer had to write a program using a very precise programming language, with each command, or statement, fed into the computer in a definite order. Such programs often had to be run several times to discover programming and logical errors before they could be used to solve a problem or prepare an analysis. Spreadsheets, on the other hand, allow users to enter spreadsheet cell information in any order and location on the worksheet, even if that organization does not mimic familiar manual computation procedures.

BASIC OPERATIONS: Formulas, data, and text labels are entered from the keyboard. The formulas define the relationships and logic for each value calculated by the spreadsheet. Data values are used by the formulas. Text labels are important to users of the spreadsheet and users of reports generated by the spreadsheet. The formulas in a worksheet are the driving force behind a spreadsheet. In Table 1, cell B7 is selected to show its contents, which is the formula defining this cell:

$B2 + B3 + B4 + B5$

where the cells B2, B3, B4, and B5 contain values for food, housing, clothing, and entertainment expenditures for January. Similar formulas are used for other calculated cells in the worksheet.

There are many aspects of modern electronic spreadsheets that enhance their usefulness to people not trained in computer programming. Among the more important:

a. Formulas can be copied from one cell to another, without disturbing the logic of the formula. In the budget example, the formula for January total can be copied to corresponding cells for February and March.

b. The worksheet can be organized in a way meaningful to the user, rather than in a way required by computational procedures. A user may elect to put monthly totals at the top, and category totals in another part of the worksheet, without regard to the fact that these values depend upon cells below and to the right. This may be contrary to the way one would manually calculate the budget, which might be column by column or row by row. This *natural order of recalculation* is intelligence within the program that frees the user from the procedural steps followed in computer programming.

c. Reports can be printed by the spreadsheet. The spreadsheets available today provide substantial flexibility in formatting the report to meet the needs of the users. Some spreadsheets include spelling checkers to assist in report preparation. Spreadsheets also contain sophisticated formatting capabilities, allowing the user to change the appearance of the screen and report. The spreadsheet may permit the user to change many appearance items, including color, typeface, and character size. Reports can also include graphical images, lines, arrows, boxes, shading, and other visual enhancements often associated with desktop publishing.

d. Graphs can be created from the numbers calculated by the spreadsheet. The graphs are an integral part of the spreadsheet, so that changes in the numbers displayed by the graph also appear as changes in the graph. The graphs can be displayed on the screen as separate images, on paper, or as part of the spreadsheet. By including a graphical image as a part of the spreadsheet screen display, the graph may become a part of a report. Some spreadsheets include tools to assist in a "slide show" presentation of a sequence of screen images displaying tabes, graphs, and text.

CAPABILITIES: Spreadsheets can be used in many different ways by users with a wide variety of skills. A beginning spreadsheet user may view a worksheet as a way of saving time that would otherwise be spent using an adding machine or a calculator. This capability of spreadsheets is likely the initial reason for their popularity. As spreadsheets have progressed, their capabilities have grown immensely. Some of the more important of these enhanced capabilities are:

a. Spreadsheets may be linked to databases, so that data-intensive applications can be addressed in a spreadsheet environment. The entire database need not be a part of the spreadsheet; needed information can be selectively retrieved from the database.

b. Extensive tools for statistical analysis are included in newer spreadsheets. Techniques such as regression analysis, tests of significance, and analysis of variance are often a part of a spreadsheet's capabilities.

c. Mathematical capabilities required for engineering and scientific calculations are included in spreadsheets.

d. Matrix operations (multiply, transpose, invert) can be performed using the commands of a spreadsheet.

e. Extensive tools for financial analysis are included in spreadsheets.

f. Optimization algorithms are a part of some spreadsheets. These optimizers, sometimes called *solvers*, are capable of addressing linear and nonlinear constrained optimization problems with continuous or discrete decision variables. The method of communicating an optimization problem to a spreadsheet solver may be very different from the traditional methods used by OR/MS practitioners. Instead of formulating a problem as a set of equations and inequalities to be satisfied, the spreadsheet view of an optimization problem might be described by these steps:

1. Construct a model to evaluate or calculate the value of the objective, such as profit, for an arbitrary set of values for the decision variables. Include in the model the values that need to be checked to see if constraints have been exceeded. This includes limiting factors such as raw materials, production capacity, and human resources.

2. Identify to the spreadsheet solver the components of the optimization problem:
   - which cell is the objective to be maximized or minimized;
   - which cells are to be adjusted by the optimizer (the decision variables), and
   - which cells are constraints, and what are the limiting values.

3. Issue the appropriate "solve" command to the spreadsheet.

In this spreadsheet optimization environment, the spreadsheet is serving as a powerful problem generator, as an optimizing algorithm, and as a report generator to communicate the results of the optimization.

g. Spreadsheets may serve as the environment for developing sophisticated software. The capability to include *macros* or a set of procedures or steps to follow, gives the spreadsheet many of the structures of traditional programming languages, such as sequence, decision, loop, and case.

As spreadsheets continue to evolve, they are becoming the primary computer application software for many business managers and other professionals. For example, it is now possible to present virtually all of the "first course" management science topics using a spreadsheet approach. While this approach may ignore the algorithmic and mathematical aspects of OR/MS, it presents the basic tools of the discipline in a user-friendly manner, using spreadsheets as a language that is more comfortable than mathematics to many potential users of OR/MS. This provides both opportunities and dangers. As more end-users are aware of the spreadsheet OR/MS tools, these tools will be applied more widely. But as the use expands, those using the tools will be less familiar with mathematics and assumptions behind the tools, which may lead to difficult situations.

See **Information systems/database designs; Linear programming; Nonlinear programming; Verification, validation and testing of models; Visualization.**

References

[1] For a full description of contemporary spreadsheets, see the documentation provided with current versions of spreadsheet software. These publishers include:

Borland International, Inc., publishers of *Quattro Pro*.

Computer Associates, publishers of *Supercalc*.

Lotus Development Corporation, publishers of 1-2-3.

Microsoft Corporation, publishers of *Microsoft Excel*.

[2] Plane, Donald R. (1994). *Management Science: A Spreadsheet Approach*. Boyd & Fraser, Danvers, Massachusetts.

[3] Ragsdale, C.T. (1995). *Spreadsheet Modeling and Decision Analysis*, Course Technology, Inc., Cambridge Massachusetts.

[4] Saffo, Paul (1989). "Looking at VisiCalc 10 Years Later," *Personal Computing*, 13(11), 233–236.

SQC

See **Statistical quality control.**

SQUARE ROOT LAW

When a model result is proportional to the square root of input variables and/or parameters. One example is the formula that indicates that the average distance that an emergency unit must travel to a call scene is proportional to the square root of the area it services. Another example is the economic order quantity (EOQ) result from inventory modeling.

ST

Subject to, as in the linear-programming problem: Minimize cx st $Ax = b$, $x \geq 0$.

STAGES

(1) An artifice in which non-exponential service times are represented as a sum of exponential random variables, each of which is referred to as a stage. If the stages are independent and identically distributed, then we have an Erlang probability function as the distribution of the sum; if the stages are only independent, the resultant density is called a generalized Erlang. Further extensions of this sort of device lead to Coxian and phase-type distributions. See **Queueing theory.** (2) The subdivisions of a dynamic programming problem where a decision is required. Each stage has a number of states associated with it, and the decision at any stage describes how the state at the current stage moves to a state at the next stage. See **Dynamic programming.**

STAIRCASE STRUCTURE

A linear-programming problem in which the constraint set can be arranged into connecting blocks such that the first block is connected to the second block by a few variables, the second block is connected to the third block by a few variables, and so on. Staircase structures arise in production problems over time in which the connecting variables are inventories that carry over from one time period to the next. The matrix of coefficients defined by such structures is very sparse. See **Block angular systems; Large-scale systems; Super sparsity; Weakly-coupled systems.**

STANFORD-B MODEL

See **Learning curves.**

STATIONARY DISTRIBUTION

In a Markov chain, the state probability distribution resulting as the solution π to $\pi = \pi P$, where P is the single-step transition matrix. Mathematically, this is equivalent to finding the eigenvector associated with the eigenvalue 1 of the stochastic matrix P. See **Limiting distribution; Markov chains; Markov processes; Statistical equilibrium; Steady-state distribution.**

STATIONARY STOCHASTIC PROCESS

A stochastic process in which the state probability distributions are invariant over time.

STATIONARY TRANSITION PROBABILITIES

When the transition probabilities of a Markov process are time-invariant. That is, for times $s < t$ in the time domain T, and any state x and any set A in the state space, $\Pr\{X(t) \in A | X(s) = x\} = \Pr\{X(t - s) \in A | X(0) = x\}$. See **Markov chains; Markov processes.**

STATISTICAL EQUILIBRIUM

Let $p_{ij}(t)$ be the probability that a stochastic process takes on value j at "time" t (discrete or continuous), given that it began at time 0 from state i. If $p_{ij}(t)$ approaches a limit p_j independent of i at $t \to \infty$ for all j, we say that the process is in statistical equilibrium. If a Markov chain has a limiting distribution, then this distribution is identical to the stationary one found by solving

$\pi = \pi P$. See **Limiting distribution**; **Markov chains**; **Markov processes**; **Stationary distribution**; **Steady-state distribution**.

STATISTICAL PROCESS CONTROL

See **Statistical quality control**.

STEADY STATE

A stochastic process is said to be in its steady state if its state probabilities have (essentially) become independent of initial conditions. See **Statistical equilibrium**.

STEADY-STATE DISTRIBUTION

Let $p_{ij}(t)$ be the probability that a stochastic process takes on value j at "time" t (discrete or continuous), given that it began at time 0 from state i. If $p_{ij}(t)$ approaches a limit p_j independent of i at $t \to \infty$ for all j, the set $\boldsymbol{p} = \{p_j\}$ is called the limiting or stead-state distribution of the process. For Markov chains in discrete time, the existence of a limiting distribution implies that there is a stationary distribution found from $\pi = \pi P$ and that $\pi = \boldsymbol{p}$. For continuous-time chains, the steady-state distribution is the probability vector satisfying the global balance equations $\pi Q = 0$. See **Limiting distribution**; **Markov chains**; **Markov processes**; **Stationary distribution**; **Statistical equilibrium**.

STEEPEST DESCENT METHOD

A fundamental procedure for minimizing a differentiable function of several variables. Central to the method is that the direction of steepest descent, in moving from one intermediate solution point to another, is along the gradient of the function at the current intermediate solution point. See **Nonlinear programming**.

STEINER TREE PROBLEM

For a network with N nodes, we are given a subset S of its nodes. The problem is to determine a minimum length (cost) tree that contains all the nodes of S and, optionally, some other nodes from the set N. The Steiner tree problem is often defined on the Euclidean plane where the problem is to find the minimum length (distance) tree that spans a given set of S nodes. Here, the tree can contain nodes (points) other than those in S. See **Minimum spanning tree**.

STEPPING-STONE METHOD

A procedure for solving a transportation problem based on a simplification of the simplex method as applied to the constraint structure that defines a transportation problem. It starts with an initial basic feasible solution and then evaluates, for every nonbasic variable, whether an improved solution can be obtained by introducing one of the nonbasic variables into the basis. The problem is structured into an m-origin by n-destination rectangular array of cells in which the cell location (i,j) corresponds to the variable x_{ij} that represents the amount to be shipped from origin i to destination j. The evaluation process for a nonbasic variable x_{ij} starts in cell (i,j) and finds a path (steps) to current basic variable cells so that if x_{ij} does come into the basis, a new feasible solution is generated. Such a path always exists, although degeneracy procedures may be needed to define the path if the current basic solution is degenerate. Associated with the path is a cost that indicates whether or not the new feasible solution would improve (decrease) the value of the objective function. Although useful from a pedagogical point-of-view, the stepping-stone method is not efficient for hand or computer solution. Most computer-based procedures for solving the transportation problem use the transportation (primal-dual) simplex method or special network algorithms. See **Revised simplex method**; **Simplex method**; **Transportation problem**.

STIGLER'S DIET PROBLEM

A problem formulated by the economist George Stigler in the early 1940s that had as its goal the determination of a minimum cost diet for an adult that met, for a full year, the recommended daily allowances of nutrients and calories, using 77 foods and 1939 prices. It was one of the first problems solved by the simplex method. Stigler's nonoptimal solution cost $39.93, with a diet consisting of wheat flour, evaporated milk, cabbage, spinach, and dried navy beans. The optimal, linear-programming solution cost $39.67 and included wheat flour, evaporated milk, cabbage, spinach, corn meal, peanut butter, lard, beef liver, and potatoes. See **Diet problem**; **Linear programming**; **Simplex method**.

STOCHASTIC DUEL

A stochastic duel is a model of combat (originally between two individuals, expanded to two sides with finite numbers of individuals) which emphasizes the random nature of combat and finite attrition calculations. See **Battle modeling**.

STOCHASTIC MODEL

A mathematical model in which some data and parameters are random varaiables. See **Deterministic model**; **Mathematical model**.

STOCHASTIC PROCESS

A stochastic process is a family of random variables that are indexed by a parameter set. See **Inventory modeling**; **Markov chains**; **Markov processes**; **Point stochastic processes**; **Queueing theory**; **Reliability**; **Renewal processes**; **Simulation of discrete-event stochastic systems**.

STOCHASTIC PROGRAMMING

Alan J. King

IBM T. J. Watson Research Center, Yorktown Heights, New York

INTRODUCTION: There are few practical decision problems where the modeler is not faced with uncertainty about the value to assign to some of the parameters. The source of the uncertainty can be the lack of reliable data, measurement errors, or uncertainty about future, or unobserved, events; there may even be uncertainty about the structure of the problem itself.

In some instances no harm will come from ignoring these uncertainties. One may rely on "best estimates" and, if needed, follow up with post-optimality parametric analysis. But there are quite a number of situations when proceeding in this manner produces "solutions" whose implementation could lead to disaster! For example, designing a master production plan without taking into account the inherent uncertainty about future markets leaves the manufacturer exposed to large losses if the evolution of the market doesn't nearly match the predictions. A valid approach would account for a certain distribution of future sales, technological developments, commodity prices, etc.

There are a number of ways of handling uncertainty. One that has proved useful in a variety of situations is to assign to the uncertain parameters a probability distribution (based on statistical evidence or not), design "recourse" functions that model the risk if certain goals or targets are missed, and optimize the expected value of the recourse. This would cast the optimization model as a *stochastic programming problem*.

Stochastic programming is concerned with practical procedures for decision-making under uncertainty: modeling the uncertainty of events and the risk of decisions in a form suitable for optimization, and devising approximation and decomposition methods for computing solutions.

The subject matter of stochastic programming is shared by other fields. Statistical decision theory is concerned with the processing of sequential observations to make decisions (for example, whether or not to repair a certain machine). Stochastic dynamic programming, or stochastic control, is concerned with the computation of feedback, or control, laws that specify optimal actions based on the state of the system (for example, at what level of inventory should one place an order to restock). The scope of any of these fields can be generalized to cover every aspect of decision-making under uncertainty. But practical aspects of mathematical analysis and computation in this challenging subject lead to tangible differences between approaches. Articles illustrating the broad range of mathematical treatments of decision-making under uncertainty can be found in Dempster (1980) and Ziemba and Vickson (1975).

Models of uncertainty and risk that include enough detail to be useful in industrial, business, or government planning will generate problems that are impossible to solve because of the exponential explosion of states. For instance, a stochastic process with only four possible realizations in each of ten time periods generates over one million sample paths!

The approach employed by stochastic programming is to model and approximate the problem with a view to finding robust "first-stage decisions" that hedge against future uncertainty. When aspects of the uncertainty in the problem become known, "recourse decisions" responding to the new information may be made.

For example, in an investment problem, the uncertainty in asset prices might be modeled by diffusion processes. The investor might select a target performance level and formulate the risk in the problem as the total expected shortfall of the portfolio value below the target, net of taxes and transaction costs, over the next ten quarters. The first-stage decision is the initial allocation of available funds to assets, and the recourse decisions are the proportion of the portfolio bought or sold at each turn of the quarter.

A simplistic approach to the problem of modeling uncertainty that should be mentioned here is that of "scenario analysis." This practice is common in business and industry. One produces a number of simulations, say, of asset prices over the next ten quarters, then one views the impact of a decision policy, like a fixed allocation between long and short term bonds, by examining the outcome of the policy under each simulation. Through a process of exhaustive search one looks for a policy with a reasonable distribution of outcomes. While this procedure offers a simple way to incorporate uncertainty into a decision model, it is not operationally sound. It cannot find, unless through exceptional luck, a decision that hedges against uncertainty.

MATHEMATICAL STRUCTURES: The mathematical investigation of stochastic programming combines the subjects of probability theory, statistics, non-smooth analysis, and linear programming. To illustrate the mathematical structure, we outline the construction of a multistage stochastic linear program; for a more complete presentation, including nonlinear formulations, see the introduction to Ermoliev and Wets (1988). We first describe the dynamic structure of the problem as a multistage linear program, then we introduce the stochastics.

The first stage decision is a linear program (ignoring subsequent stages):

$$\text{minimize}_{x} \quad c_0 x$$
$$\text{subject to} \quad A_0 x \geq b_0 \qquad (1)$$

In each subsequent stage from time stage 1 until the ending time stage T, we make recourse decisions, also modeled as linear programs, which depend on decisions previously made. The recourse linear program is:

$$\text{minimize}_{y} \quad c_t y_t$$
$$\text{subject to} \quad A_{t0} x + A_{t1} y_1 + \cdots + A_{tt} y_t \geq b_t \qquad (2)$$

It is important to note that the inequalities in the recourse linear program include all interperiod dynamical relationships and intraperiod constraints.

In the general stochastic programming model, the stochastic process involves every coefficient of the objective, right-hand side, and matrix in the recourse linear programs; although in practice only a few such coefficients will be random. The distribution of coefficients in any time stage will in general depend on the history of the stochastic process up to that point. The subscript t appended to the expectation operator E will denote expectation conditioned on the history of the process up to (but not including) stage t. Now define, in recursive fashion, the value, or cost-to-go, functions. The value function depends on the stochastic process; we signal this dependence by including the Greek letter ω in the argument list.

$$f_{t-1}(x, y_{t-1}, \ldots, y_1; \omega)$$
$$= \text{minimize}_{y_t} \quad c_t y_t + E_t \{ f_t(x, y_t, \ldots, y_1; \omega) \}$$
$$\text{subject to} \quad A_{t0} x + A_{t1} y_1 + \cdots + A_{tt} y_t \geq b_t \qquad (3)$$

This definition begins at stage T (where the right side has no value-function term in the objective) and proceeds until stage 1 (where the left side has no dependence on any recourse decision). The stochastic programming problem can now be described in terms of the first stage decision variables alone:

$$\text{minimize}_{x} \quad c_0 x + E \{ f_0(x; \omega) \}$$
$$\text{subject to} \quad A_0 x \geq b_0 \qquad (4)$$

The mathematics of stochastic programming is devoted to understanding the formulation in Equation (4). Two noteworthy aspects distinguish it from other studies in nonlinear programming or classical probability and statistics. First, the objective function is defined by an integral. Second, the integrand is the value function of an optimization problem, which would therefore not generally possess a first derivative in the classical sense nor even be finite-valued. In general, it is hopeless to try to find a closed-form representation of the integral as a function of the first-stage decision (although in certain simple cases, like the Newsboy problem, it is possible to do so). Nevertheless, this formulation is amenable to analysis and computation. The basic tools are those of probability theory, and non-smooth analysis. This is an area of mathematics whose theoretical challenge is matched only by the practical importance of the applications themselves.

HISTORY, MODELS AND COMPUTATION: The history of stochastic programming closely follows that of the development of sophisticated optimization algorithms and ever more powerful computers. One of the earliest "stochastic programs'' was Markowitz's mean/variance formulation of the portfolio optimization problem: minimize variance of return subject to a constraint on expected return (Ziemba and Vickson, 1975). This amounts to selecting a target return and minimizing the expected value of a (quadratic) recourse function that penalizes the difference between portfolio return and target. The resulting objective is quadratic in the decision variables, and can be solved by a version of the simplex method. Chance constrained problems, where the probability of a bad event (e.g., bridge collapse) is constrained, are in wide use in engineering and power systems design. Prekopa showed that constraints specifying the probability that a random vector be coordinate-wise less than or equal to a given problem variable reduce to nonlinear convex constraints in the variable when the probability distributions involved are log-concave (Ermoliev and Wets, 1988). These two formulations are popular because components of the probability distribution (e.g., variance) can be incorporated directly into the optimization model, leading to low-dimensional problems with some hope of solution.

The more general subject of stochastic programming outlined above, or linear programming under uncertainty as it was called then, was independently introduced by Dantzig and Beale in 1955. Here, an explicit discrete description of the sample space is introduced either at the outset, or as part of an algorithmic procedure. This model is capable of representing a great variety of practical decision problems through various modeling devices of linear programming. The drawback of the general model of linear programming under uncertainty is the curse of dimensionality. Unless one can be clever, or lucky, one is faced with solving a problem with billions of variables and constraints.

The challenge of solving such problems has lead to many interesting computational and theoretical developments. Chief among these are the L-shaped method (Van Slyke and Wets, 1969) and its multistage extension (Birge, 1985) which partitions the stochastic program by decision stage, and the theory of epi-convergence pioneered by Wets and others which justifies sampling and other methods of approximating the integration (King and Wets, 1991). Two developments point to exciting prospects for the solution of these problems: the aggregation method which decomposes by information field (Rockafellar and Wets, 1991), and importance sampling in the L-shaped method (Dantzig and Glynn, 1991; Infanger, 1994). Decomposition permits very large problems to be solved on multiple processors. Sampling is used to represent the information in the uncertainty model with just a few data points. These two ideas, decomposition and approximation, are the keys to computational progress in stochastic programming.

Until recently, few stochastic programming applications could be formulated and successfully solved. Some exceptional recent efforts in two-stage stochastic programming are bank asset-liability management (Kusy and Ziemba, 1986), lake pollution management (Somlyody and Wets, 1988), and manufacturing capacity expansion (Eppen, Martin and Schrage, 1989). Due to the explosion in power and capability of computers and optimization algorithms, multistage stochastic programming formulations are emerging from academia and the top optimization laboratories into industrial applications. The major application areas include financial asset and liability management over multi-year horizons, multistage production planning models with uncertain demand, power systems management over multiple time periods, forest harvest management, and long-range energy-economic planning models.

See **Chance-constrained problem; Dynamic programming; Linear programming; Portfolio theory.**

References

[1] Birge J.R. (1985). "Decomposition and Partitioning Methods for Multi-stage Stochastic Linear Programs," *Operations Research*, 33, 989–1007.
[2] Dantzig G.B. and P.W. Glynn (1990). "Parallel Processors for Planning Under Uncertainty," *Annals Operations Research*, 22, 1–21.
[3] Dempster M.A.H. (1980). *Stochastic Programming*. Academic Press, New York.
[4] Eppen G.D., R.K. Martin and L.E. Schrage (1989). "A Scenario Approach to Capacity Planning," *Operations Research*, 37, 517–527.
[5] Ermoliev Y. and R.J.-B. Wets (1988). *Numerical Techniques for Stochastic Optimization*. Springer-Verlag, New York.
[6] Infanger G. (1994). *Planning Under Uncertainty*. Boyd and Fraser, Danvers, Massachusetts.
[7] King A.J. and R.J.-B. Wets (1991). "Epi-consistency of Convex Stochastic Programs," *Stochastics*, 34, 83–92.
[8] Kusy M.I. and W.T. Ziemba (1986). "A Bank Asset and Liability Management Model," *Operations Research*, 34, 356–376.
[9] Rockafellar R.T. and R.J.-B. Wets (1991). "Scenarios and Policy Aggregation in Optimization Under Uncertainty," *Math. Oper. Res.*, 16, 119–147.
[10] Somlyody L. and R.J.-B. Wets (1988). "Stochastic Optimization Models for Lake Eutrophication Management," *Operations Research*, 36, 660–681.

[11] VanSlyke R.M. and R.J.-B. Wets (1969). "L-shaped Linear Programs with Application to Optimal Control and Stochastic Programming," *SIAM Jl. Appl. Math.*, 17, 638–663.

[12] Ziemba W.T. and R.G. Vickson (1975). *Stochastic Optimization Models in Finance*. Academic Press, New York.

STRATEGIC ASSUMPTION SURFACING AND TESTING (SAST)

A problem structuring method for use in situations where decisive action is obstructed by internal disagreements. Coherent sub-groups are formed with the purpose of advocating differing strategies, and each identifies the significant assumptions on which its preferred strategy depends. The sub-groups are then reunited to debate the differences in assumptions, with the aim of achieving a compromise on assumptions so that a consensus strategy can be derived. See **Problem structuring methods**.

STRATEGIC CHOICE

A problem structuring method which assists in the strategic management of uncertainty and commitment. It is an interactive method for facilitating group decision-making. It provides techniques and processes for the agreement of a problem focus, the identification of feasible decision schemes, the establishment of a working shortlist, and paired comparisons designed to highlight relevant uncertainty. Outputs are structured into a commitment package of decisions, exploratory actions and deferred choices. See **Problem structuring methods**.

STRATEGIC OPTIONS DEVELOPMENT AND ANALYSIS (SODA)

A problem structuring method for group decision making. Individual cognitive maps are elicited for participants, and then merged into a strategic map which is used in workshop mode to facilitate discussion and commitment. See **Problem structuring methods**.

STRICTLY QUASI-CONCAVE FUNCTION

Given a function $f(\cdot)$ and points $x, y \in X$, with $x \neq y$ and X convex, if $f(y) \geq f(x)$ implies that $f(\lambda x + (1-\lambda)y) > f(x)$ for all $0 < \lambda < 1$, then we say that f is a strictly quasi-concave function. See **Concave function**; **Quasi-concave function**.

STRICTLY QUASI-CONVEX FUNCTION

Given a function $f(\cdot)$ and points $x, y \in X$, with $x \neq y$ and X convex, if $-f(y) \geq -f(x)$ implies that $-f(\lambda x + (1-\lambda)y) > -f(x)$ for all $0 < \lambda < 1$, then we say that f is a strictly quasi-convex function. See **Concave function**; **Convex function**; **Quasi-concave function**; **Quasi-convex function**.

STRONG DUALITY THEOREM

Consider the following primal linear-programming problem and its dual problem:

Primal

Minimize $c^{\mathrm{T}}x$

subject to

$$Ax \geq b$$
$$x \geq 0$$

Dual

Maximize $b^{\mathrm{T}}y$

subject to

$$A^{\mathrm{T}}y \leq c$$
$$y \geq 0$$

Some authors call the following basic duality theorem the strong duality theorem: If either the primal or the dual has a finite optimal solution, then the other problem has a finite optimal solution, and the optimal values of their objective functions are equal, that is, minimum-$c^{\mathrm{T}}x$ = maximum $b^{\mathrm{T}}y$. They give the name weak duality theorem to the theorem: If x is a feasible solution to the primal problem and y is a feasible solution to the dual problem, then $b^{\mathrm{T}}y \leq c^{\mathrm{T}}x$.

STRONGLY NP-COMPLETE (NP-HARD)

See **Computational complexity**.

STRONGLY POLYNOMIAL-TIME ALGORITHM

One whose running time is independent of the sizes of the numerical data of the instance. See **Computational complexity**.

STRUCTURAL VARIABLES

The original variables of a linear-programming problem as differentiated from slack, surplus and artificial variables. Structural variables are usually the variables of interest and have a physical interpretation such as production or shipments. They appear in the original defining inequalities or equations prior to the conversion of the problem to all equations. See **Linear inequality**; **Logical variable**; **Slack variable**; **Surplus variable**.

STRUCTURED MODELING

Arthur M. Geoffrion

Anderson Graduate School of Management, University of California, Los Angeles

INTRODUCTION: Structured modeling was developed as a comprehensive response to perceived shortcomings of modeling systems available in the 1980s. It is a systematic way of thinking about models and their implementations, based on the idea that every model can be viewed as a collection of distinct *elements*, each of which has a definition that is either primitive or based on the definition of other elements in the model. Elements are categorized into five types (so-called *primitive entity*, *compound entity*, *attribute*, *function*, and *test*), grouped by similarity into any number of classes called *genera*, and organized hierarchically as a rooted tree of *modules* to reflect the model's high-level structure. It is natural to diagram the definitional dependencies among elements as arcs in a directed acyclic graph. Moreover, this dependency graph can be computationally active because every function and test element has an associated mathematical expression for computing its value.

Using a model for any specific purpose involves subjective intentions. Structured modeling makes a sharp distinction between the resulting user-defined "problems" or "tasks" associated with a model, and the relatively objective model per se. A typical problem or task has to do with *ad hoc* query, drawing inferences, evaluating model behavior with specified inputs, determining a constrained solution, or optimization, and requires applying a computerized model manipulation tool ("solver"). For certain recurring kinds of problems and tasks, these tools are highly developed and readily available for incorporation into a structured modeling software system.

The theoretical foundation of structured modeling is formalized in Geoffrion (1989), which presents a rigorous semantic framework that deliberately avoids committing to a representational formalism. The framework is "semantic" because it casts every model as a system of definitions styled to capture semantic content. Ordinary mathematics, in contrast, typically leaves more of the meaning implicit. Twenty-eight definitions and eight propositions establish the notion of model structure at three levels of detail (so-called *elemental*, *generic*, and *modular* structure), the essential distinction between model *class* and model *instance*, certain related concepts and constructs, and basic theoretical properties. This framework has points in common with certain ideas found in the computer science literature on knowledge representation, programming language design, and semantic data modeling, but is designed specifically for modeling as practiced in OR/MS and related fields (Geoffrion, 1987, Sec. 4).

STRUCTURED MODELING LANGUAGES: An executable model description language called SML (Structured Modeling Language) fully supports structured modeling's semantic framework Geoffrion (1992). Other languages for structured modeling also exist, including ones that are graph-based, logic-based, SQL-oriented, subscript-free, or object-oriented. SML can be viewed in terms of four upwardly compatible levels of increasing expressive power. The first level encompasses simple definitional systems and directed graph models such as those found in Harary, Norman and Cartwright (1965). The second level covers more complex extensions of these, spreadsheet models, numeric formulas, and propositional calculus models. The third level encompasses mathematical programming and predicate calculus models with simple indexing over sets and Cartesian products. Finally, the

&NUT_DATA <u>NUTRIENT DATA</u>

NUTRi /pe/ There is a list of <u>NUTRIENTS</u>.

MIN (NUTRi) /a/ : Real+ For each NUTRIENT there is a <u>MINIMUM DAILY REQUIREMENT</u> (units per day per animal).

&MATERIALS <u>MATERIALS DATA</u>

MATERIALm /pe/ There is a list of <u>MATERIALS</u> that can be used for feed.

UCOST (MATERIALm) /a/ Each MATERIAL has a <u>UNIT COST</u> ($ per pound of material).

ANALYSIS (NUTRi, MATERIALm) /a/ : Real+ For each NUTRIENT-MATERIAL combination, there is an <u>ANALYSIS</u> (units of nutrient per pound of material).

Q (MATERIALm) /va/ : Real+ The <u>QUANTITY</u> (pounds per day per animal) of each MATERIAL is to be chosen.

NLEVEL (ANALYSISi., Q) /f/ ; @SUMm (ANALYSISim * Qm) Once the QUANTITIES are chosen, there is a <u>NUTRITION LEVEL</u> (units per day per animal) for each NUTRIENT calculable from the ANALYSIS.

T:NLEVEL (NLEVELi, MINi) /t/ ; NLEVELi > =MINi For each NUTRIENT there is a <u>NUTRITION TEST</u> to determine whether the NUTRITION LEVEL is at least as large as the <u>MINIMUM DAILY REQUIREMENT</u>.

TOTCOST (UCOST, Q) /f/ ; @SUMm (UCOSTm * Qm) There is a <u>TOTAL COST</u> (dollars per day per animal) associated with the chosen QUANTITIES.

SML schema for the classical feedmix model

fourth level covers sparse versions of the above plus relational and semantic database models.

The following exhibits from Geoffrion (1987) show an SML *schema* (third level) specifying the general structure of the classical feedmix model, and sample SML *elemental detail tables* specifying model elements. The latter, together with the schema, yield a specific feedmix model instance.

Space does not permit a proper description of SML's syntax, but a few hints are as follows. Schemas are organized as a tree of paragraphs whose leaves are the genera and whose interior nodes are the modules. The boldfaced part of each paragraph is the formal definition of the genus or module, as the case may be, and the rest consists of documentary comments about the formal part which are informal except for conventions about the use of underlining and upper case. The formal definition of a genus paragraph begins with the name of the genus, a parenthetical statement of definitional dependencies (if any), a slash-delimited statement of genus type, a colon-announced statement of data type if an attribute genus, and a semicolon-announced mathematical expression called a *generic rule* if a function or test genus. The formal definition of a module paragraph consists only of its name. Note that a schema is always specified indepen-

dently of any problem or task that might be posed on it. A common problem associated with the above schema is to find values for all Q elements such that all T:NLEVEL elements evaluate to true and the value of the TOTCOST element is minimal.

The structure and sequence of the elemental detail tables are determined procedurally from the schema. Each table is named, has column names that usually coincide with genus names, and has a row for each element of the corresponding genus.

The final figure shows the so-called *genus graph* associated with the above schema. It represents definitional dependencies at the level of genera.

STRUCTURED MODELING FUTURE: Owing to the design of the underlying semantic framework and of SML itself, SML-based modeling systems can have certain features often lacking in modeling systems of more conventional design, including:

* error-checking, of the formal specification of general model structure, that is exhaustive with respect to the underlying semantic framework;

* detailed semantic connections among model parts, a feature that facilitates maintaining,

NUTR		MATERIAL		
NUTR ‖ INTERP	MIN	MATERIAL ‖ INTERP	UCOST	
P ‖ Protein	16	std ‖ Standard Feed	1.20	
C ‖ Calcium	4	add ‖ Additive	3.00	

ANALYSIS			Q	
NUTR	MATERIAL ‖ ANALYSIS		MATERIAL ‖ Q	
P	std ‖ 4.00		std ‖ 2.00	
P	add ‖ 14.00		add ‖ 0.50	
C	std ‖ 2.00		‖	
C	add ‖ 1.00			

NLEVEL		TOTCOST
NUTR ‖ NLEVEL	T:NLEVEL	‖ TOTCOST
P ‖ 15.00	FALSE	‖ 3.90
C ‖ 4.50	TRUE	‖

Sample elemental detail tables for the feedmix schema

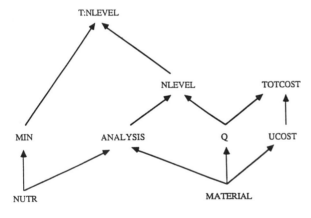

Genus graph for the feedmix schema.

enhancing, and integrating models, and that enables automatic generation of several kinds of model reference documents useful for communication, debugging, model maintenance and evolution, and other essential activities;

* the ability, owing to the generality of structured modeling's view of models as definitional systems, for a single modeling system to accommodate a wide variety of modeling paradigms, which leads to easier model integration and many of the benefits of standardization;

* browsable definitional dependency graphs at three levels of abstraction, constructs useful for visualizing and communicating the general structure of any model;

* the use of hierarchical organization as an approach to managing model complexity, and also as a visual device for model navigation;

* automatic generation of relational data table designs for model instance data, a feature that facilitates exploiting relational database tools for data management;

* partial consistency checking of SML's informal sublanguage for documenting the formal model specification, and also partial consistency and completeness checking of formal specifications by reference to this documentation;

* complete independence between the general structure of a class of models and instantiating data, a feature that promotes the reuse of each of these, conciseness, efficient communication, and dimensional flexibility;

* complete independence between models and solvers, a feature that promotes using multiple solvers with a single model, multiple models with a single solver, and conceptual clarity.

A research prototype implementation exhibiting the above features is described in Geoffrion (1991) and Neustadter *et al.* (1992). The first paper references several other research prototypes for structured modeling with different emphases, including: graph-based modeling, hybrid information/mathematical modeling systems, model management with a SQL database server in a networked environment, optimization-based applications, statistical analysis, and syntax-directed model editing. An ample foundation has been laid for the development of production prototypes.

Likely topics for future work include discrete event simulation (Lenard, 1993), graph-based modeling (Jones, 1992), integration with database systems (Dolk, 1988), language-directed editors (Vicuña, 1990), object-oriented systems (Muhanna, 1993), model integration, improved languages for model definition and manipulation, applications to early and late modeling life-cycle phases not supported by conventional modeling systems, and structured modeling-based enhancements and usage disciplines for other modeling approaches and systems. See Geoffrion (1994) for a survey of structured modeling and selected research directions, and a 50-item bibliography.

See **Mathematical model; Model management; Algebraic modeling languages.**

References

[1] Dolk D.R. (1988). "Model Management and Structured Modeling: The Role of an Information Resource Dictionary System," *Comm. ACM*, 31, 704–718.

[2] Geoffrion A.M. (1987). "An Introduction to Structured Modeling," *Management Science*, 33, 547–588.

[3] Geoffrion A.M. (1989). "The Formal Aspects of Structured Modeling," *Operations Research*, 37, 30–51.

[4] Geoffrion A.M. (1991). "FW/SM: A Prototype Structured Modeling Environment," *Management Science*, 37, 1513–1538.

[5] Geoffrion A.M. (1992). "The SML Language for Structured Modeling," *Operations Research*, 40, 38–75.

[6] Geoffrion A.M. (1994). "Structured Modeling: Survey and Future Research Directions," *ORSA CSTS Newsletter*, 15(1), 1&11–20.

[7] Harary F., R. Norman and D. Cartwright (1965). *Structural Models: An Introduction to the Theory of Directed Graphs*, Wiley, New York.

[8] Jones C.V. (1992). "Attributed Graphs, Graph-Grammars, and Structured Modeling," *Annals of Operations Research*, 38, 281–324. (Special volume on Model Management in Operations Research edited by B. Shetty, H. Bhargava, and R. Krishnan.)

[9] Lenard M.L. (1993). "A Prototype Implementation of a Model Management System for Discrete-Event Simulation Models," *Proceedings of the 1993 Winter Simulation Conference*, IEEE, Piscataway, New Jersey, 560–568.

[10] Muhanna W. (1993). "An Object-Oriented Framework for Model Management and DSS Development," *Decision Support Systems*, 9, 217–229.

[11] Neustadter L., A. Geoffrion, S. Maturana, Y. Tsai and F. Vicua (1992). "The Design and Implementation of a Prototype Structured Modeling Environment," *Annals Operations Research*, 38, 453–484. (Special volume on Model Management in Operations Research edited by B. Shetty, H. Bhargava, and R. Krishnan.)

[12] Vicuña F. (1990). *Semantic Formalization in Mathematical Modeling Languages*, Ph.D. Dissertation, Computer Science Department, University of California, Los Angeles.

SUBJECTIVE PROBABILITY

See **Bayesian decision theory, subjective probability and utility; Decision analysis.**

SUBOPTIMIZATION

The finding of a solution to an optimization problem by a procedure that does not guarantee that the solution will be optimal. The procedure usually includes heuristic rules that help eliminate the generation of poor solutions. See **Heuristic procedure**.

SUB PROBLEM

See **Simple upper-bounded problem**.

SUPER-SPARSITY

In most large-scale mathematical-programming problems, especially linear-programming problems, the number of nonzero elements in the problem matrix is quite small. Such problems are said to have a low density. Further, it has been noted that the number of distinct numerical values in the problem matrix is usually much smaller than the number of nonzero coefficients. This characteristic is known as super-sparsity. Computational savings in storage and processing time can be achieved by taking advantage of super-sparsity, as follows. Each distinct numerical value is recorded once in a value table stored in main memory. Each nonzero coefficient is recorded in an index array by means of a number triple: row index, column index, and a pointer. The pointer locates the coefficient's numerical value in the value table. See **Density; Large-scale systems; Sparse matrix.**

SUPPLEMENTAL VARIABLES

An analysis technique that introduces additional variables to allow non-Markovian systems to be made Markovian and thus analyzed more easily. See **Markov processes**; **Queueing theory**.

SURPLUS VARIABLE

A nonnegative variable that is added to a linear inequality of the form $\Sigma_j a_{ij} x_j \geq b_i$ to transform the inequality into an equation. The surplus variable measures the difference between the left- and right-hand-sides of the inequality. See **Logical variable**; **Slack variable**; **Surplus vector**.

SURPLUS VECTOR

The column representation of a surplus variable in a linear-programming problem. See **Surplus variable**.

SYMMETRIC MATRIX

A square matrix $A = (a_{ij})$ is symmetric if $a_{ij} = a_{ji}$. Thus, $A = A^{\mathrm{T}}$. See **Matrices and matrix algebra**.

SYMMETRIC NETWORK

A queueing network of quasi-reversible nodes with additional properties that make its major effectiveness measures insensitive to the service-time distribution. See **Insensitivity**; **Networks of queues**; **Queueing theory**.

SYMMETRIC PRIMAL-DUAL PROBLEMS

The two linear-programming problems with the following form:

Primal

Minimize $c^{\mathrm{T}} x$

subject to

$$Ax \geq b$$

$$x \geq 0$$

Dual

Maximize $b^{\mathrm{T}} y$

subject to

$$A^{\mathrm{T}} y \leq c$$

$$y \geq 0$$

See **Strong duality theorem**; **Unsymmetric primal-dual problems**.

SYMMETRIC ZERO-SUM TWO-PERSON GAME

A two-player game with a skew-symmetric payoff matrix. The amount lost by one player is the amount gained by the other player (zero-sum). Such a game has a value of zero and the optimal strategies of the two players are the same. See **Skew-symmetric matrix**; **Game theory**.

SYSTEM

A set of related elements organized to achieve a purpose. See **Systems analysis**.

SYSTEM DYNAMICS

George P. Richardson

State University of New York at Albany

INTRODUCTION: System dynamics is a computer-aided approach to policy analysis and design. It applies to dynamic problems arising in complex social, managerial, economic, or ecological systems – literally any dynamic systems characterized by interdependence, mutual interaction, information feedback, and circular causality.

The field developed initially from the work of Jay W. Forrester. His seminal book *Industrial Dynamics* (1961) is still a significant statement of philosophy and methodology in the field. Within ten years of its publication, the span of applications grew from corporate and industrial problems to include the management of research and development, urban stagnation and decay, commodity cycles, and the dynamics of growth in a finite world. It is now applied in economics, public policy, environmental studies, defense, theory-building in social science, and other areas, as well as its home field, management. The name industrial dynamics no longer does justice to the breadth of the field, so it has become generalized to system dynamics. The modern name suggests links to other systems methodologies, but the links are weak and misleading. System dynamics emerges out of servomechanisms engineering, not general systems theory or cybernetics (Richardson, 1991).

The system dynamics approach involves:

- Defining problems dynamically, in terms of graphs over time.
- Striving for an endogenous, behavioral view of the significant dynamics of a system, a focus inward on the characteristics of a system that themselves generate or exacerbate the perceived problem.
- Thinking of all concepts in the real system as continuous quantities interconnected in loops of information feedback and circular causality.
- Identifying independent stocks or accumulations (levels) in the system and their inflows and outflows (rates).
- Formulating a behavioral model capable of reproducing, by itself, the dynamic problem of concern. The model is usually a computer simulation model expressed in nonlinear equations, but is occasionally left unquantified as a diagram capturing the stock-and-flow/causal feedback structure of the system.
- Deriving understandings and applicable policy insights from the resulting model.
- Implementing changes resulting from model-based understandings and insights.

Mathematically, the basic structure of a formal system dynamics computer simulation model is a system of coupled, nonlinear, first-order differential (or integral) equations,

$$\frac{d}{dt}\boldsymbol{x}(t) = \boldsymbol{f}(\boldsymbol{x}, \boldsymbol{p}),$$

where $\boldsymbol{x}$ is a vector of levels (stocks or state variables), $\boldsymbol{p}$ is a set of parameters, and $\boldsymbol{f}$ is a nonlinear vector-valued function.

Simulation of such systems is easily accomplished by partitioning simulated time into discrete intervals of length dt and stepping the system through time one dt at a time. Each state variable is computed from its previous value and its net rate of change $x'(t)$: $x(t) = x(t - dt) + dt * x'(t - dt)$. In the earliest simulation language in the field (DYNAMO) this equation was written with time scripts K (the current moment), J (the previous moment), and JK (the interval between time J and K): $X.K = X.J + DT * XRATE.JK$ (Richardson and Pugh, 1981). The computation interval dt is selected small enough to have no discernible effect on the patterns of dynamic behavior exhibited by the model. In more recent simulation environments, more sophisticated integration schemes are available (although the equation written by the user may look like this simple Euler integration scheme), and time scripts may not be in evidence. Important simulation environments include the DYNAMO series on mainframes, workstations, and personal computers (Pugh, 1983, Pugh-Roberts Associates, 1986), DYSMAP2 (Dangerfield and Vapenikova, 1987), STELLA II and iThink (High Performance Systems, 1990), S**4 (Diehl, 1992), and Vensim (Ventana Systems, 1991).

Forrester's original work stressed a continuous approach, but increasingly modern applications of system dynamics contain a mix of discrete difference equations and continuous differential or integral equations. Some practitioners associated with the field of system dynamics work on the mathematics of such structures, including the theory and mechanics of computer simulation, analysis and simplification of dynamic systems, policy optimization, dynamical systems theory, and complex nonlinear dynamics and deterministic chaos.

The main applied work in the field, however, focuses on understanding the dynamics of complex systems for the purpose of policy analysis and design. The conceptual tools and concepts of the field – including feedback thinking, stocks and flows, the concept of feedback loop dominance, and an endogenous point of view – are as important to the field as its simulation methods. We next review the basic elements of systems dynamics; Richardson (1991a,b) provides detailed descriptions.

FEEDBACK THINKING: Conceptually, the feedback concept is at the heart of the system dynamics approach. Diagrams of loops of information feedback and circular causality are tools for conceptualizing the structure of a complex system and for communicating model-based insights. Intuitively, a feedback loops exists when information resulting from some action travels through a system and eventually returns in some form to its point of origin, potentially influencing future action. If the tendency in the loop is to reinforce the initial action, the loop is called a *positive* feedback loop; if the tendency is to oppose the initial action, the loop is called a *negative* feedback loop. The sign of the loop is called its *polarity*. Negative loops can be variously characterized as goal-seeking, equilibrating, or stabilizing processes. They can sometimes generate oscillations, as when a pendulum seeking its equilibrium goal gathers momentum and overshoots it. Positive loops are sources of growth or accelerating collapse; they are disequilibrating and destabilizing. Combined, positive and negative circular causal feedback loops can generate all manner of dynamic patterns.

Figure 1 illustrates the use of feedback loops. The positive loop on the left is a self-reinforcing corporate growth process; the negative loop on top constrains growth. The bold loop is an example of the generic goal-seeking, negative feedback loop structure underlying all purposeful behavior.

LOOP DOMINANCE AND NONLINEARITY: The loop concept underlying feedback and circular causality by itself is not enough, however. The explanatory power and insightfulness of feedback understandings also rest on the notions of active structure and loop dominance. Complex systems change over time. A crucial requirement for a powerful view of a dynamic system is the ability of a mental or formal model to change the strengths of influences as conditions change, that is to say, the ability to shift *active* or *dominant structure*.

In a system of equations, this ability to shift loop dominance comes about endogenously from nonlinearities in the system. For example, the S-shaped dynamic behavior of the classic logistic growth model ($dP/dt = aP - bP^2$) can be seen as the consequence of a shift in loop dominance from a positive, self-reinforcing feedback loop (aP) producing exponential-like growth to a negative feedback loop ($-bP^2$) that brings the system to its eventual goal. Only nonlinear models can endogenously alter their active or dominant structure and shift loop dominance. From a feed-back perspective, the ability of nonlinearities to generate shifts in loop dominance is the fundamental reason for advocating nonlinear models of social system behavior.

THE ENDOGENOUS POINT OF VIEW: The concept of endogenous change is fundamental to the system dynamics approach. It dictates aspects of model formulation: exogenous disturbances are seen at most as *triggers* of system behavior (like displacing a pendulum); the *causes* are contained within the structure of the system itself (like the interaction of a pendulum's position and momentum that produces oscillations). Corrective responses are also not modeled as functions of time, but are dependent on conditions within the system. Time by itself is not seen as a cause.

But more importantly, theory building and policy analysis are significantly affected by this endogenous perspective. Taking an endogenous view exposes the natural *compensating* tendencies in social systems that conspire to defeat many policy initiatives. Feedback and circular causality are delayed, devious, and deceptive. For understanding, system dynamics practitioners strive for an *endogenous point of view*. The effort is to uncover the sources of system behavior that exist within the structure of the system itself.

SYSTEM STRUCTURE: These ideas are captured in Forrester's (1969) organizing framework for system structure:

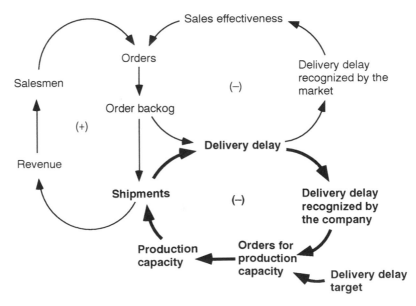

Fig. 1. Feedback loops for system dynamics study of inconsistent corporate growth (simplified from Forrester, 1968).

Closed boundary
  Feedback loops
    Levels
    Rates
      Goal
      Observed condition
      Discrepancy
      Desired action

The *closed boundary* signals the endogenous point of view. The word *closed* here does not refer to open and closed systems in the general system sense, but rather refers to the effort to view a system as *causally* closed. The modeler's goal is to assemble a formal structure that can, *by itself*, without exogenous explanations, reproduce the essential characteristics of a dynamic problem.

The causally closed system boundary at the head of this organizing framework identifies the endogenous point of view as the feedback view pressed to an extreme. Feedback thinking can be seen as a *consequence* of the effort to capture dynamics within a closed causal boundary. Without causal loops, all variables must trace the sources of their variation ultimately outside a system. Assuming instead that the causes of all significant behavior in the system are contained within some closed causal boundary forces causal influences to feed back upon themselves, forming causal loops. Feedback loops enable the endogenous point of view and give it structure.

LEVELS AND RATES: Stocks (levels) and the flows (rates) that affect them are essential components of system structure. A map of causal influences and feedback loops is not enough to determine the dynamic behavior of a system. A constant inflow yields a linearly rising stock; a linearly rising inflow yields a stock rising along a parabolic path, and so on. Stocks (accumulations, state variables) are the memory of a dynamic system and are the sources of its disequilibrium and dynamic behavior.

Forrester (1961) placed the operating policies of a system among its rates, many of which assume the classic structure of a negative feedback loop striving to take action to reduce the discrepancy between the observed condition of the system and a goal. The simplest such rate structure results in an equation of the form $RATE = (GOAL - LEVEL)/(ADJTIM)$, where $ADJTIM$ is the time over which the level adjusts to reach the goal.

BEHAVIOR IS A CONSEQUENCE OF SYSTEM STRUCTURE: The importance of levels and rates appears most clearly when one takes a *continu-ous* view of structure and dynamics. Although a discrete view, focusing on separate events and decisions, is entirely compatible with an endogenous feedback perspective, the system dynamics approach emphasizes a continuous view. The continuous view strives to look beyond events to see the dynamic patterns underlying them. Moreover, the continuous view focuses not on discrete decisions but on the *policy structure* underlying decisions. Events and decisions are seen as surface phenomena that ride on an underlying tide of system structure and behavior. It is that underlying tide of policy structure and continuous behavior that is the system dynamicist's focus.

There is thus a distancing inherent in the system dynamics approach – not so close as to be confused by discrete decisions and myriad operational details, but not so far away as to miss the critical elements of policy structure and behavior. Events are deliberately blurred into dynamic behavior. Decisions are deliberately blurred into perceived policy structures. Insights into the connections between system structure and dynamic behavior, which are the goal of the system dynamics approach, come from this particular distance of perspective.

DIRECTIONS FOR FURTHER READING: *The System Dynamics Review*, the journal of the System Dynamics Society, is the best source of activity in the field, including methodological advances and applications. Texts on the modeling process in system dynamics include Richardson and Pugh (1981), High Performance Systems (1990) (a software manual that is also an excellent introduction to simulation modeling in any language), and Wolstenholme (1990). Interesting collections of applications are contained in Roberts (1978) and Richardson (1996). A direction within the field is the use of model-based insights for organizational learning, represented most forcefully in Senge (1990). More about the system dynamics approach, including extensive bibliographies, is contained in Richardson (1991a,b).

See **Control theory; Global models; Public policy analysis; Simulation of discrete-event stochastic systems.**

References

[1] Dangerfield B.C. and O. Vapenikova (1987). *DYSMAP*2. Department of Business and Management Studies, University of Salford, Salford, United Kingdom.

[2] Diehl E.W. (1992). *S**4: The Strategy Support Simulation System*. MicroWorlds, Inc., Cambridge, Massachusetts.

[3] Forrester J.W. (1961). *Industrial Dynamics*. The MIT Press, Cambridge, Massachusetts.

[4] Forrester J.W. (1968). "Market Growth as Influenced by Capital Investment," *Industrial Management Review* (now *Sloan Management Review*), 9(2), 83–105.

[5] Forrester J.W. (1969). *Urban Dynamics*. The MIT Press, Cambridge, Massachusetts.

[6] High Performance Systems (1990). *STELLA II User's Guide* and *Ithink User's Guide*. High Performance Systems, Hanover, New Hampshire.

[7] Pugh A.L. (1983). *DYNAMO User's Manual*, 6th ed. The MIT Press, Cambridge, Massachusetts.

[8] Pugh-Roberts Associates (1986). *Professional DYNAMO Plus*. Pugh-Roberts Associates, Inc., Cambridge, Massachusetts.

[9] Richardson G.P. (1991a). "System Dynamics: Simulation for Policy Analysis from a Feedback Perspective." In P.A. Fishwick and P.A. Luker, eds., *Qualitative Simulation Modeling and Analysis*, Springer-Verlag, New York.

[10] Richardson G.P. (1991b). *Feedback Thought in Social Science and Systems Theory*. University of Pennsylvania Press, Philadelphia.

[11] Richardson G.P., ed. (1996). *Modeling for Management: Simulation in Support of Systems Thinking*. International Library of Management, Dartmouth Publ., Aldershot, England.

[12] Richardson G.P. and A.L. Pugh III (1981). *Introduction to System Dynamics Modeling with DYNAMO*. The MIT Press, Cambridge, Massachusetts.

[13] Roberts E.B. ed. (1978). *Managerial Applications of System Dynamics*. The MIT Press, Cambridge, Massachusetts.

[14] Senge P.M. *The Fifth Discipline: The Art and Practice of the Learning Organization*. Doubleday/Currency, New York.

[15] *System Dynamics Review* (1985-present). John Wiley, Chichester, United Kingdom.

[16] Ventana Systems (1991). *Vensim: Ventana Simulation Environment*. Ventana Systems, Inc., Belmont, Massachusetts.

[17] Wolstenholme E.F. (1990). *System Enquiry: A System Dynamics Approach*. John Wiley, Chichester, United Kingdom.

SYSTEM RELIABILITY

See **Reliability of systems**.

SYSTEMS ANALYSIS

Richard O. Mason & Sue A. Conger

Southern Methodist University, Dallas, Texas

HISTORY AND PHILOSOPHY: Systems analysis is a broad term applied to the study of real world processes. It requires breaking problems down into their component parts and formulating a conceptual definition of the situation. The purpose is to develop "an overall understanding of optimal solutions to executive type problems" (Churchman, *et al.*, 1957, p. 7). The resulting conceptual definition often is subsequently translated into a computer-based system or mathematical model. Systems analysis has been applied to understand complex, dynamic systems – both physical and social – such as business processes, governments, economics systems, weapons systems, mechanical and manufacturing systems, and computer software. The science is based on several key theories: systems, cybernetics, and modeling.

A system is a set of related elements organized to achieve a purpose. The elements form "an interconnected complex of functionally related components." (Churchman, *et al.*, 1957, p. 7) Each element has *inputs, processes, and outputs.* At the most detailed and fundamental level of analysis, the elements are generally treated as "black boxes." At a high level of abstraction, what goes into and out of each black box element is described but the activities within the element are not described. Each black box is analyzed in turn to define the *transformation* process on its inputs that generate its outputs. The concepts of *flow, relationship, message,* and *connection* are used to portray the *structure* of a system which has been analyzed. These terms describe the interrelationships among its elements. The transformation processes are described in terms including *transaction, process,* and *problem.*

A cybernetic connection integrates a feedback loop into a system, and provides communication about the system's outputs which are used, in turn, to make adjustments in either the inputs or the process necessary to achieve the system's purpose. This is called *control* (Weiner, 1948).

A mathematical model of a system is a "collection of mathematical relationships which characterize the feasible programs" for improving a system (Dantzig, 1963). Building a mathematical model provides insight into a system and its properties. The model can also be manipulated to derive conclusions about the system.

Mathematical models and other OR/MS techniques may be applied to the conceptual definition of a system and used to determine the best possible solution – the optimum decision, policy, or design for the system – for the problem the system represents. Churchman poses five necessary conditions for completing a systems analysis:

1. The total system objectives and, more specifically, the performance measures for the whole system;
2. The system's environment: the fixed constraints (which are 'outside' the system);
3. The resources of the system (which are capabilities found 'inside' the system and, therefore, can be reallocated);
4. The elements of the system, their activities, functions, goals, and measures of performance; and
5. The management of the system. That is, the process of allocating resources to the system's elements (Churchman, 1968).

Most systems contain recognizable sub-systems, sub-sub-systems, and so on, organized in a hierarchy. Their arrangement often involves a "Chinese box" form of nesting that permits them to be defined by recursion. A solution which is best for the system as a whole is called an 'optimum'; whereas, one which is best relative to the functioning of one or more elements is called a 'suboptimum.' One of the challenges of systems analysis is to improve the performance of a sub-system in terms of its own goals and purposes – sub-optimization – without detriment to the total system or, worse, defeating the system's overall purpose.

SYSTEMS ANALYSIS AND COMPUTER SYSTEMS:

Systems analysis is the first stage in presenting any large task to a computer (the other principal stages being design and implementation). It is performed by a systems analyst and consists of analyzing the whole task in its setting and deciding how to arrange it for processing by a computer. It includes an estimation of how much work is involved and hence how powerful a computer will be required. The problem is divided into a number of relatively independent parts which are specified, together with their interconnections, in sufficient detail for a programmer to take over.

Computer applications are developed through a series of translations. The first translation, as noted above, is from a real world situation to a conceptual definition of the situation. This conceptual model is then translated via a design activity to an implementation model that is still readable by humans and describes the conceptual model in a language related to the target computer environment. The implementation model is then translated into the specific coded language(s) of the target (machine) environment. These three translations define phases of activity that comprise an application's *development life cycle*. The translations relate to the

thinking processes involved and are called *analysis*, *design*, and *implementation*, respectively. Implementation can be divided into sub-phases for programming, testing, and cut-over.

Software development methodologies are used to guide the development processes through the life cycle. (Technically, *methodology is the study of tools and techniques, and methods* are tools and techniques. The common term for system development methods used as a package of tools is *methodologies* and it is used here.) Five different approaches are used currently: mathematical, process, data, object, and artificial intelligence. All use top-down strategies for problem-solving and progressively decompose a large problem into smaller, solvable problems for independent solution (Laszlo, 1972). The five approaches can be further divided into three classes – mathematical, transaction, and semantic – depending on the type of problem they attempt to solve (Figure 1). Mathematical methodologies solve selection and alternative analysis problems. Process, data, and object methodologies solve transaction processing problems. Semantic methodologies solve reasoning problems.

Mathematical Methodologies – Mathematical methodologies employ mathematical models of a system and focus on the logical relationships within the system. They are often formulated by interdisciplinary teams, who adapt scientific theories and methods to practical problems (Ackoff and Rivett, 1963). Operations research (OR), management science (MS), and cybernetic methods are applied. Classes of problems to which mathematical methods apply include inventory, allocation, sequencing, queueing, routing, replacement, competition, and search (Ackoff and Rivett, 1963, p. 34). The problems solved by OR techniques all deal with selection from many

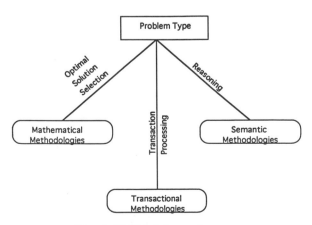

Fig. 1. Methodology classes.

alternatives and sensitivity analysis to develop alternative, contingent courses of action.

Mathematical, cybernetic systems seek optimal solutions based on unambiguous, but possibly incomplete, information (Churchman, 1957). The inputs to mathematical applications define the alternatives and resources available from which an optimal selection must be made. The tools and techniques used to develop mathematical models include linear, network, dynamic, and stochastic programming methods (Wagner, 1975). The results of these applications are usually presented in the form of suggested machine schedules, resource allocations, and so on.

Transactional Methodologies – Transactional methods focus on the flows of information between the elements of a system. Three different methodology classes have evolved to develop transactional, information retrieval, and data analysis types of applications: process, data, and object. No single methodology currently supports all three application types well. Further, as the demand for client/server systems and distributed systems evolve, improved methodologies will be required to support their development.

Process Methodologies – Process methods developed during the 1950s and 1960s to mirror von Neumann computer architecture which separates inputs and outputs from processes. Since computing was the difficult issue at the time, processing was the focus of process methods. In particular, the types of problems automated included accounting procedures, order entry, inventory processing and so on. These applications all deal with transactions that support the basic white collar operations of the organization.

The development techniques focus on data flowing between processes (DeMarco, 1979; Yourdon, 1979). The sample data flow diagram in Figure 2 shows the processes as circles connected via directed lines (i.e., data flows) to external entities or data stores. External entities are depicted on the diagram as squares and represent people, organizations, or other computer systems from which and to which information flows. Data stores, on the diagram as open-ended rectangles, indicate files of information that persist over time. The lines connecting the other icons indicate temporary data flowing through the system, hence the term data flow diagram.

Process methodologies and methods have undergone several iterations of refinement to support real-time systems development (Ward & Mellor, 1985, 1986). The lack of integration of data throughout analysis and design has led to an abandonment of process methods, *per se*, in favor of techniques which provide such integration.

Data Methodologies – Data methodologies developed as database technologies that came to the market in the 1960s and 1970s were found to require specific attention to data design. Data methods are based on theories of semantic modeling (Chen, 1981), relational database design (Codd, 1972), and data normalization (Kent, 1983). These theories are significant in business because they result in mathematically, provably correct processing of data, a key in mission critical applications. They are also significant because they signal the application of mathematical foundations for transaction processing which, before this, relied on analyst and programmer ingenuity and accuracy.

The essence of relational data design is that information should look to the user as if it were composed of rows and columns, similar to a spreadsheet (Figure 3). The physical implementation should be transparent, and entity and referential integrity must be maintained. *Entity integrity* refers to primary keys as a unique identifier of a relation and states that no component of a key may accept null values (Date, 1990). *Referential integrity* guarantees that no relation contains unmatched foreign key values. A foreign key is a primary key in one relation that appears as an attribute, or foreign key, in another relation (Date, 1990).

Early data methods focused only on data with the assumption that all primitive processing – create, retrieve, update, and delete – followed logically from the correct definition of data (Warnier, 1981). Demands to capture the complexities of the world led to significant extensions of process data flow analysis and the development of data modeling with integration

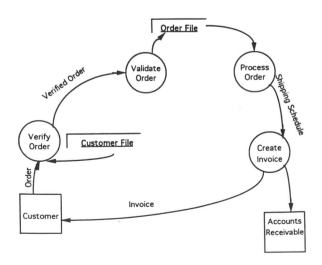

Fig. 2. Sample data flow diagram.

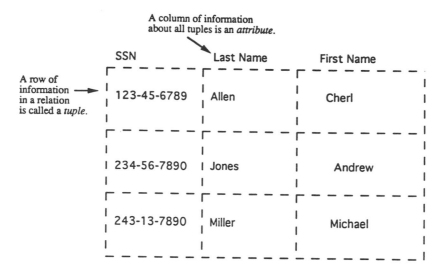

Fig. 3. Relational database view.

of data and process throughout the methodology. This is called information engineering – IE (Martin & Finkelstein, 1981). Data models in the form of entity-relationship diagrams, process models similar in form to data flow diagrams, and integration models that link data and process are all found at each stage of information engineering (Figure 4).

Information engineered applications are assumed to use traditional processing languages with integration of database technologies. Computer aided software engineering tools that support the development of IE applications also generate code that imbeds relational database code within it. Data methodologies assume on-line applications but can be used for batch processing as well. They are less adapted to real-time applications. Data methodologies are widely used in large, Fortune 500 organizations that rely on databases having millions of relations.

Object Methodologies –Object techniques were originally formalized at Xerox PARC in the 1970s with the development of Smalltalk™ and eventual commercialization of the Apple Lisa™. As on-line and real-time technologies migrated from the aerospace and defense industries to commerce, improved methods were needed to explain how elements in a system interacted with one another. Object-oriented analysis (OOA) involves development of three models: (1) an information model for describing elements in terms of objects and attributes, (2) a state model which describes the behavior of objects and relationships over time, and (3) a process model which specifies the actions in terms of elementary and reusable processes (Schlaer & Mellor, 1992).

The goal of object methods is complete integration of data and processes in *encapsulated* objects (Figure 5). Objects may be members of *classes* and exhibit *inheritance*, a property such that properties, data, and processes of related objects may be reused without redefinition, that is, inherited (Coad & Yourdon, 1990; Coad & Yourdon, 1991). Objects may have multiple inheritance from competing objects throughout a hierarchy (Figure 6). Objects may also exhibit *polymorphism*, that is, the ability to have the same process, using one public name, take different forms when associated with different objects (Booch, 1987; Booch, 1991). Client/server technology embodies the concepts of object orientation. *Server objects* perform a requested process; whereas, *client objects* request a process from a supplier.

Object orientation is based on the same theories as data methodologies, carrying normalization to the encapsulated (data + process) object units (Kent, 1983; Kim, 1989). The most visible example of object applications is MS Windows which uses windows, icons, menus, and pointers in an object-oriented human interface to personal computers. Object-oriented methodologies are currently adopted widely in the embedded systems and software markets (e.g., graphical user interfaces – GUIs such as MS Windows).

Object methods are in the experimental stage today and are running into obstacles that appear intractable in the short-run. The traditional von Neumann architecture separates data from the programs which process them, but object encapsulation assumes non-persistent data that are discarded upon use. Many applications, however, require data persistence. The problem re-

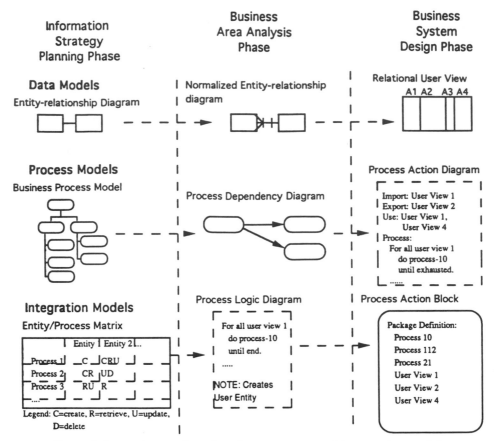

Fig. 4. Information engineering data, process, and integration models.

volves around the need for persistence without massive duplication of processes that occurs with strict encapsulation. Object-oriented databases have solved some of the design problems by imbedding references to procedures within the database schemas that define the data, making programmed procedures another element in the database. Speed and efficiency of processing for current object databases, plus the added requirements of auditability, reliability, and, eventually distributability have not been solved at this time (Kim & Lochovsky, 1989).

A further problem is that object-oriented methods do not have adequate standardized procedures or common definitions of terms. The fact that they are closely coupled to implementation languages requires that analysis and design methods be based on a desired target language. Some authors suggest C++ object-oriented analysis, others suggest Ada object-oriented analysis, while others suggest their favorites. These issues will preclude object-orientation from eclipsing other methods of application development for business applications in the near future.

Semantic Methodologies – Semantic refers to *meaning.* Semantic methods focus on the role of knowledge and meaning in a system. They imbed meaning in data and in reasoning rules used to process it by drawing on theories of cognitive development as it applies to computer reasoning and learning (Anderson, 1987; Kolodner, 1986). Semantic methodologies, such as Artificial Intelligence (AI), are used to design computer systems which understand languages, learn, reason, solve problems, and exhibit other characteristics associated with intelligence in human beings (Barr, 1981). They differ from mathematical and transactional methodologies because they produce a decision, a course of action or an answer to a question based on the application of qualitative knowledge and information. The more qualitative the systems situation being examined, therefore, the more advanced the techniques required. For instance, whether or not a person has toast for breakfast in the morning might depend on the previous night's sleep, one's dinner, recency of exercise, and so on. This simple example illustrates two of the problems AI applications must solve: making their reasoning sufficiently generic and identifying all of the relevant quantitative and qualitative relationships in the system.

A program called DENDRAL was an early result of AI research. It is a chemist's assistant that interprets data from a mass spectrograph

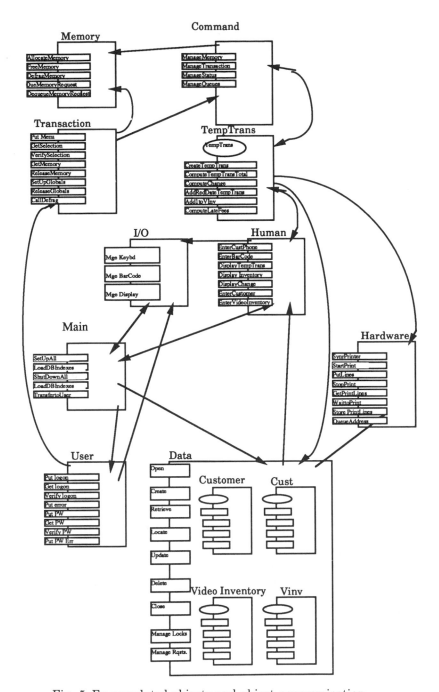

Fig. 5. Encapsulated objects and object communication.

and infers the chemical structure of an unknown organic compound. The program is based on an algorithm developed by J. Lederberg in 1964 which generates all possible acyclic graphs given the number of systems elements (the compound's chemical composition) and the number of links (relationships) pertaining to each element (the technical valences). The number of possibilities generated for any given compound is enormous. In order to avoid an exponential search, DENDRAL automates rules to apply heuristics and knowledge gained from practicing chemists to delimit radically the number of alternatives that must be evaluated to determine the compound's molecular structure. DENDRAL introduced the idea of using *rules* to represent expert knowledge, a concept that has prevailed in AI work since (Feigenbaum, 1971). Today, DENDRAL outperforms expert chemists on this task (Buchanan & Feigenbaum, 1981; Churchman, 1971; Feigenbaum, *et al.*, 1971; Smith, *et al.*, 1973). In another stream of ground-breaking re--

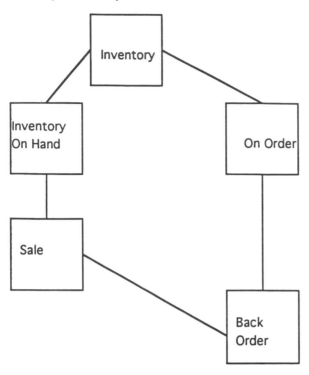

Fig. 6. Inheritance lattice of objects.

search, Minsky and Papert pioneered the application of parallel processing ideas to reasoning problems (Minsky, 1968; Papert, 1980).

Neural networks, the newest AI technique, takes the human brain as the system under analysis and attempts to model human intellectual activity on a broad scale by mirroring human brain functioning. A *neuron* is the smallest possible processing element and is related to other neurons via *synapses*. Objects called *dendrites* are message transmitters that flow between neurons over synaptic connections (Zahedi, 1993). Single neurons can have thousands of synapses. Inputs via dendrites can either excite (i.e., initiate) or inhibit action of a neuron. The number and frequency of messages to a neuron creates an activity level that can be triggered when some predefined threshold is reached. Each neuron has *axons* via which output signals are transmitted to the dendrite network (Zahedi, 1993). These terms have parallels in the other methodologies, but work slightly differently in neural nets. Neural networked problems, however, are different in kind from those solved by the other methodologies. AI and neural problems deal with incomplete information, probabilistic outcomes, and ambiguities in the reasoning and data to be used in the development of a solution. The techniques that comprise semantic methodologies are not really mature enough to be called methodologies. Rather they are individually applied, taught in an master-apprentice

relationship, based on practical experience with a given set of problems.

At this time, different types of reasoning problems require different types of methods and approaches to automating intelligence (Winograd, 1986). The problem types which are addressed by semantic methods include language understanding and translation, sensory understanding (i.e., sight, touch, etc.), memory recall and forgetting, and coordination and control of movement. The most common of these are *expert systems* which exhibit intelligence in selecting an action by reasoning through numerous, sometimes contradictory, rules. Expert systems have found acceptance in industry and government for applications such as surveillance of nuclear plant operations, selection of geological drilling sites, diagnosis of medical problems, and so on.

AI techniques and methodologies are in an emergent stage, experiencing continuous refinement and evolution. Like object-oriented methods, AI methods are also closely coupled to the target implementation language. For instance, some languages require data integration with reasoning rules while others require separation of data from reasoning rules. Most languages offer one reasoning approach which determines the nature of the reasoning process as forward, backward, depth-first, breadth-first, custom-defined, or other. The major business promise of AI is to augment existing applications to include reasoning about the processes and data they maintain. Neural nets are promising as generic reasoning systems that may coalesce the diversity of methods and techniques some time in the future.

EMERGING PROBLEM AREAS: As we move toward the new millennium, the problems companies seek to solve through automation include organizational reengineering, work flow design, and yield management. A few innovative companies also seek competitive advantage maximization through information technologies.

Organizational Reengineering – Organizational reengineering is a form of organizational design. First, processes and information are modeled and matched to the corporate goals and mission. Anything not contributing to the mission is eliminated. The analysis techniques for performing the modeling and matching are a natural extension of enterprise analysis techniques from information engineering (Conger, 1994).

Remaining work processes and information are next analyzed to design jobs that can maximize quality and quantity of work products. Job design theory of Hackman and Oldham (1980) to do whole processes, make meaningful decisions, exercise discretion, and contribute to the organizational mission are followed. Then, using dependency and coordination theories (Galbraith, 1967; Thompson, 1967), jobs are grouped to maximize group cohesion and focus and to minimize inter-group linkages. The resulting organization is analyzed to determine effective enterprise-wide potential for information technology to support the redesigned organization. The net result is identification of new and modified applications to support the reengineered organization. Thus, reengineering integrates systems analysis and organization theories in designing the new organization and its support systems.

Work Flow Redesign – Not all companies can cope with the radical change that accompanies a reengineering project. A scaled-down version of reengineering is work flow redesign. In work flow design, some particular work flow of an organization is analyzed for redesign. Paper flow through an insurance company is a good example. Policy and claims processing usually require the worker to have all policy related documents and information available to effect a change. In manual companies, this means a massive assembly, movement, and disassembly process for documents. The more people who touch a policy or claim, the more assembly, movement and disassembly. The process is error prone – lost papers are common and may not be known to be missing for years. The process is time-consuming – assembly movement and disassembly can take from four hours per policy/claim to indefinite time if problems in locating documents arise. The process is labor-intensive and necessarily sequential.

In work flow design, systems analysis techniques are used to identify the essential activities, skills, and information as in reengineering. Then, both the jobs and computer systems supporting the jobs are redesigned to increase quality of service and quantity of output. The resulting applications might include mathematical application components to support, for instance, actuarial work; might include expert systems components to guide service representatives through, for instance, a retirement counseling session; and might include traditional transaction processing capabilities to guide automated image storage, retrieval, movement, and tracking. Thus, work flow redesign combines and integrates the three problem types and methodologies into a new form of comprehensive application that supports parallel work of multiple people.

Yield Management – Yield management is undertaken to get the greatest return on a dollar of capital invested. It deals with optimal pricing of a limited number of products (e.g., rooms) when there are different product/price mixes (e.g., ski weekends, conventions, peak/low season), and when the demand for each product/price combination fluctuates and differs by location over time. Yield management, then, contains elements of mathematical and AI problems, with the solution based on data provided by transactional applications.

In the hotel industry, rooms wear out and must be refurbished on a fairly regular schedule, but rates charged for the room differ depending on season, lead time to rental, number of nights stay, promotions, and so on. Similarly, airline industry management involves pricing perishable airline seats at rates that will maximize revenues. Yield management in the manufacturing and hospitality industries is merging operations research methods with transaction and artificial intelligence methods to develop suggested rate changes throughout a given day or any other period of time.

The common element to all these emerging areas of systems analysis application is the need to integrate different theories and methodologies to develop solutions to these problems. This is a major change from the past 40 years during which the three problem types have been treated as distinct and separate problem-solving activities.

SUPPORT MECHANISMS: Several types of support mechanisms are useful to facilitating systems analysis in the development of applications. Organizational supports include data administration and joint application design. Automated support is provided by computer aided software engineering (CASE) tools.

Data Administration – Data administration is the management of data to support and foster data sharing across multiple divisions, and to facilitate the development of database applications (Conger, 1994). Data administration organizations develop and maintain a data architecture for the organization which depicts the structure and relationships of major data entities, such as customers. The data architecture identifies automated and nonautomated data and how they are used in the organization. Users, working with data administration, develop an "organizational" definition for each item. The architecture provides a framework for defining new applications and documents all data uses and responsibilities for existing applications. Also at the organization level, users and data administrators work to define data that is critical to the organization as a going concern. Critical data is subject to management, standards, audits, security, and recovery planning. Noncritical data, while useful, is not required in event of disaster and does not require the same managerial attention (Conger, 1994).

At a lower level of detail, data administrators develop, administer, and maintain policies and standards for data definition, sharing, acquisition, integrity, security, access and so on for the corporation's data resource. Data administration provides guidance to project teams on storage, access, use, disposition, and standardization of data (Conger, 1994).

Some benefits from data administration are improved communication and understanding of corporate data from formal recognition and agreement on entity definitions, business rules and relationships. Application development efficiencies also occur because normalization of data takes place across the organization and not just at the individual application level. Redundancy becomes planned, and, therefore, can be managed. Organizations can react to changing business conditions faster because fully specified data definitions exist and can be used in new applications easily making control of critical data possible.

Data administration should not be confused with database administration which is responsible for physical database design, disk space allocation, and day to day operations support for actual databases. Instead, data administration optimizes data redundancy, provides shared understanding of data definitions, and is a managed approach for planning future database environments.

Joint Application Development – Several techniques have been developed to describe intensive user-analyst sessions during which requirements, designs, or other application related work is accomplished. The most common names are joint application development (JAD), joint requirements planning (JRP), joint application requirements (JAR), and Fast-Track. (Note that JAD and JRP are design techniques of the IBM Corporation, while Fast-Track is a design technique of the Boeing Computer Services Company.) They are all similar in their goal of collaborative, user-analyst definition of application requirements. The planning and execution of joint sessions are also similar. The primary differences are the level of participants, subject matter, and level of detail of the discussions.

JAD is a team-based form of systems analysis which was designed to shorten the application development process and improve the quality of resulting application deliverable products. Some statistics show a 40% reduction in analysis time which leads to a faster implementation time. Other benefits of JADs are qualitative. Users and technicians develop a shared mental model and gain commitment to the joint work effort. Ideally, the spirit of teamwork continues into the other development stages.

A JAD team is composed of client representatives, a facilitator, systems representatives, and support personnel. The clients include decision makers at a level sufficient to resolve conflicts and make decisions that affect the scope and content of the application. They must also be at a low enough level to explain the daily functions and procedures of the organization. Clients from every functional area affected by the work should be represented. Systems representatives include the project manager, one or two senior analysts with technical expertise. The main role of the system representatives is to learn the problem area during the sessions to ensure accurate translation of the requirements in systems terms. In addition, systems representatives assess the feasibility and expected complexity of requirements for the target environment. The facilitator is a specially trained individual who runs the sessions, eliciting information, keeping the discussions on the topic, keeping the meeting moving, and identifying and resolving conflicts.

Both users and analysts prepare for the JAD sessions by attending training sessions in data and process modeling. Users are asked to prepare for the JAD session by gathering data examples and attempting to define the processes used in their own work. These user definitions become the basis of work the first day.

JAD sessions are three to five business days (or longer) at a site away from the normal work location. Daytime sessions are used to identify requirements, define terms, confirm business functions, identify constraints on processes and data, and so on. Evening sessions are used to document and circulate all work completed during the day for review the next morning.

The non-technical benefits of JAD are at least as important as the technical benefits. Users and analysts become friends and committed to a joint effort. Users and analysts develop a shared mental model that should be faithfully translated into the desired application. Staff time spent on analysis activities is reduced due to the intense, one-time session.

COMPUTER AIDED SOFTWARE ENGINEER-ING: Computer aided software engineering (CASE) tools automate aspects of the software engineering discipline. All CASE tools have a repository in which all design objects are defined and referenced. The more sophisticated the product, the more sophisticated the repository. The three distinguishing characteristics of CASE tools discussed here are level of the life cycle supported, level of intelligence imbedded in the software, and level of coupling to a methodology.

CASE products that automate analysis are referred to as *front-end CASE* or *Upper CASE*. CASE tools that automate program coding and testing are referred to as *back-end CASE* or *Lower CASE*. CASE products that are referred to as *I-CASE*, meaning integrated CASE, support multiple phases of work and integrate the work of one phase with that of succeeding phases.

In general, there are three levels of sophistication in CASE tools: dumb, semi-intelligent, and intelligent (Figure 7). A dumb CASE tool (e.g., Briefcase from the Briefcase Corporation) is essentially a paper and pencil substitute that accepts information for storage and outputs reports on its repository contents. A semi-intelligent CASE tool (e.g., Visual Analyst from the Visible Systems Corporation) is one in which intelligent consistency and completeness checking are performed within a diagram or set of like definitions, but not across sets. An intelligent CASE tool (e.g., IEF from Texas Instruments, Inc.) provides not only intra-diagram set consistency and completeness checking, but also inter-diagram set consistency and completeness. The more intelligent the tool, the more checking is performed. IEF, for instance, is an intelligent tool that also analyzes impacts of changes on process and data definitions, generates third normal form relational database schemas, analyzes implementation feasibility of definitions, and generates error-free COBOL code with imbedded DB2 database processing.

The third distinguishing characteristic of CASE is coupling to a methodology. Coupling and intelligence are linked concepts in that the more closely coupled a CASE tool is to a methodology, the more the methodology's rules about definitions of graphics and design elements can be defined and enforced by the CASE tool. Tight coupling also has drawbacks. Tightly coupled

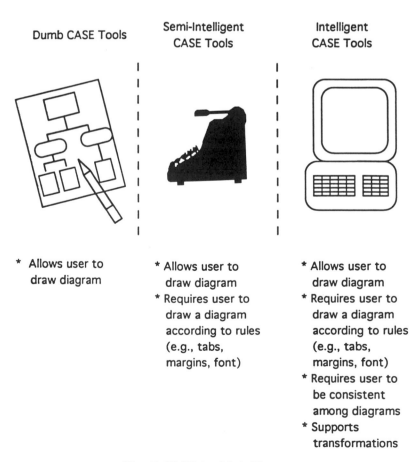

Dumb CASE Tools

Semi-Intelligent
CASE Tools

Intelligent
CASE Tools

* Allows user to
  draw diagram

* Allows user to
  draw diagram
* Requires user to
  draw a diagram
  according to rules
  (e.g., tabs,
  margins, font)

* Allows user to
  draw diagram
* Requires user to
  draw a diagram
  according to rules
  (e.g., tabs,
  margins, font)
* Requires user to
  be consistent
  among diagrams
* Supports
  transformations

Fig. 7. CASE tool intelligence.

CASE tools are inflexible to definitions that are not in the delivered product and sometimes require definition of graphics or items that are not required in a particular application. Both of these drawbacks can lead to extra work.

See **Artificial intelligence; Cybernetics; Experts systems; Mathematical model; Neural networks; Practice of operations research and management science; Yield management.**

References

[1] Ackoff Russell L. and Rivett Patrick (1963). *A Manager's Guide to Operations Research*. John Wiley & Sons, New York.

[2] Anderson J.R. ed. (1987). *Cognitive Skills and Their Acquisition*. Lawrence Erlbaum Associates, Hillsdale, New Jersey.

[3] Barr A. and Feigenbaum Edward (1981). *The Handbook of Artificial Intelligence*. William Kaufmann, Inc., Los Altos, California.

[4] Booch Grady (1987). *Software Engineering with Ada* (2nd ed.). Benjamin/Cummings, Menlo Park, California.

[5] Booch Grady (1991). *Object Oriented Design with Applications*. Benjamin/Cummings, Redwood City, California.

[6] Buchanan B. and Feigenbaum E. (1981). "DENDRAL and Meta DENDRAL: their application dimension." *Artificial Intelligence*, 11, 5–24.

[7] Chen Peter P.-S. (1981). "A Preliminary framework for entity-relationship models." In P. P.-S. Chen (Ed.), *Entity-Relationship Approach to Information Modeling and Analysis*. ER Institute, Saugus, California.

[8] Churchman C. West, Ackoff Russell L. and Arnoff E. Leonard (1957). *Introduction to Operations Research*. John Wiley, New York.

[9] Churchman C. West (1968). *The Systems Approach*. Delta (Dell) Publishing, New York.

[10] Churchman C. West (1971). *The Design of Inquiring Systems*. Basic Books, New York.

[11] Coad Peter and Yourdon Edward (1990). *Object Oriented Analysis* (2nd ed.). Prentice-Hall, Englewood Cliffs, New Jersey.

[12] Coad Peter and Yourdon Edward (1991). *Object Oriented Design*. Prentice-Hall, Englewood Cliffs, New Jersey.

[13] Codd Edgar F. (1972). "A relational model of data for large shared data banks." *Communications of the ACM*, 13, 377–387.

[14] Conger Sue (1994). *The New Software Engineering*. Wadsworth, Belmont, California.

[15] Dantzig G. (1963). *Linear Programming and Extensions*. Princeton University Press, New Jersey.

[16] Date Christopher J. (1990). *An Introduction to Database Systems* (5th ed.). Addison-Wesley, Reading, Massachusetts.

[17] DeMarco Tom (1979). *Structured Analysis*. Yourdon Press, New York.

[18] Feigenbaum Edward, Buchanan B. and Lederberg J. (1971). "On generality and problem solving." In B. Beltzer and Michie, D., eds., *Machine Intelligence*, 165–190. Elsevier, New York.

[19] Hackman J.R. and Oldham Gregory R. (1980). *Work Redesign*, Addison-Wesley, Reading, Massachusetts.

[20] Kent William (1983). "A simple guide to five normal forms in relational database theory." *Communications of the ACM*, 26, 120–125.

[21] Kim Won and Lochovsky Frederick H. eds. (1989). *Object-oriented Concepts, Databases, and Applications*. ACM Press, New York.

[22] Kolodner Janet L. and Riesbeck Christopher H. eds. (1986). *Experience, Memory and Reasoning*. Lawrence Erlbaum Associates, Hillsdale, New Jersey.

[23] Laszlo E. (1972). *The Systems View of the World*. John Wiley, New York.

[24] Martin James and Finkelstein Clive (1981). *Information Engineering*. Prentice-Hall, Englewood Cliffs, New Jersey.

[25] Minsky Marvin ed. (1968). *Semantic Information Processing*. MIT Press, Cambridge, Massachusetts.

[26] Papert Seymour (1980). *Mind-Storms: Children, Computers, and Powerful Ideas*. Basic Books, New York.

[27] Schlaer S. and Mellor Stephen J. (1992). *Object Lifecycles: Modeling the World in States*. Yourdon Press, Englewood Cliffs, New Jersey.

[28] Smith D., Buchanan B., Engelmore R., Adlercreutz J. and Djerassi C. (1973). "Application of artificial intelligence for chemical inference IX." *Journal of American Chemical Society*, 95, 6078.

T

TABLEAU

See **Simplex tableau**.

TABU SEARCH

Fred Glover

University of Colorado, Boulder

BACKGROUND: Tabu Search (TS) is a *meta-heuristic* that guides a local heuristic search procedure to explore the solution space beyond local optimality. Widespread successes in practical applications of optimization have spurred a rapid growth of tabu search in the past few years. New "records" have been set by TS, and by hybrids of TS with other heuristic and algorithmic procedures, in finding better solutions to problems in scheduling, sequencing, resource allocation, investment planning, telecommunications and many other areas. Some of the diversity of tabu search applications is shown in Table 1. (See also the survey of Glover and Laguna, 1993, and the volume edited by Glover, Laguna, Taillard and de Werra, 1993.)

Tabu search is based on the premise that problem solving, in order to qualify as intelligent, must incorporate *adaptive memory* and *responsive exploration*. The use of adaptive memory contrasts with "memoryless" designs, such as those inspired by metaphors of physics and biology, and with "rigid memory" designs, such as those exemplified by branch and bound and its AI-related cousins. The emphasis on responsive exploration (and hence purpose) in tabu search, whether in a deterministic or probabilistic implementation, derives from the supposition that a bad strategic choice can yield more information than a good random choice. (In a system that uses memory, a bad choice based on strategy can provide useful clues about how the strategy may profitably be changed. Even in a space with significant randomness – which fortunately is not pervasive enough to extinguish all remnants of order in most real world problems – a purposeful design can be more adept at uncovering the imprint of structure, and thereby at affording a chance to exploit the conditions where randomness is not all-encompassing.)

These basic elements of tabu search have several important features, summarized in Table 2.

Tabu search is concerned with finding new and more effective ways of taking advantage of the concepts embodied in Table 2, and with identifying associated principles that can expand the foundations of intelligent search. As this occurs, new strategic mixes of the basic ideas emerge, leading to improved solutions and better practical implementations. This makes TS a fertile area for research and empirical study.

TABU SEARCH FOUNDATIONS: The basis for tabu search may be described as follows. Given a function $f(x)$ to be optimized over a set X, TS begins in the same way as ordinary local search, proceeding iteratively from one point (solution) to another until a chosen termination criterion is satisfied. Each $x \in X$ has an associated *neighborhood* $N(x) \subset X$, and each solution $x' \in N(x)$ is reached from x by an operation called a *move*.

TS goes beyond local search by employing a strategy of modifying $N(x)$ as the search progresses, effectively replacing it by another neighborhood $N^*(x)$. As our previous discussion intimates, a key aspect of tabu search is the use of special memory structures which serve to determine $N^*(x)$, and hence to organize the way in which the space is explored.

The solutions admitted to $N^*(x)$ by these memory structures are determined in several ways. One of these, which gives tabu search its name, identifies solutions encountered over a specified horizon (and implicitly, additional related solutions), and forbids them to belong to $N^*(x)$ by classifying them *tabu*. (The tabu terminology is intended to convey a type of restraint that embodies a "cultural" connotation – i.e., one that is subject to the influence of history and context, and capable of being surmounted under appropriate conditions.)

The process by which solutions acquire a tabu status has several facets, designed to promote a judiciously aggressive examination of new points. A useful way of viewing and implementing this process is to conceive of replacing original evaluations of solutions by *tabu evaluations*, which introduce penalties to significantly discourage the choice of tabu solutions (i.e., those preferably to be excluded from $N^*(x)$, according

Table 1. Illustrative tabu search applications

Scheduling	**Telecommunications**
Flow-Time Cell Manufacturing	Call Routing
Heterogeneous Processor	Bandwidth Packing
Scheduling	Hub Facility Location
Workforce Planning	Path Assignment
Classroom Scheduling	Network Design for Services
Machine Scheduling	Customer Discount Planning
Flow Shop Scheduling	Failure Immune Architecture
Job Shop Scheduling	Synchronous Optical Networks
Sequencing and Batching	
	Production, Inventory and Investment
Design	Flexible Manufacturing
Computer-Aided Design	Just-in-Time Production
Fault Tolerant Networks	Capacitated MRP
Transport Network Design	Part Selection
Architectural Space Planning	Multi-item Inventory Planning
Diagram Coherency	Volume Discount Acquisition
Fixed Charge Network Design	Fixed Mix Investment
Irregular Cutting Problems	
Lay-Out Planning	**Routing**
	Vehicle Routing
Location and Allocation	Capacitated Routing
Multicommodity	Time Window Routing
Location/Allocation	Multi-Mode Routing
Quadratic Assignment	Mixed Fleet Routing
Quadratic Semi-Assignment	Traveling Salesman
Multilevel Generalized Assignment	Traveling Purchaser
	Convoy Scheduling
Logic and Artificial Intelligence	
Maximum Satisfiability	**Graph Optimization**
Probabilistic Logic	Graph Partitioning
Clustering	Graph Coloring
Pattern Recognition/Classification	Clique Partitioning
Data Integrity	Maximum Clique Problems
Neural Network Training	Maximum Planner Graphs
Neural Network Design	P-Median Problems
Technology	**General Combinational Optimization**
Seismic Inversion	Zero-One Programming
Electrical Power Distribution	Fixed Charge Optimization
Engineering Structural Design	Nonconvex Nonlinear Programming
Minimum Volume Ellipsoids	All-or-None Networks
Space Station Construction	Bilevel Programming
Circuit Cell Placement	General Mixed Integer Optimization
Off-Shore Oil Exploration	

to their dependence on the elements that compose tabu status). In addition, tabu evaluations also periodically include inducements to encourage the choice of other types of solutions, as a result of aspiration levels and longer term influences. The following subsections describe how tabu search takes advantage of memory (and hence learning processes) to carry out these functions.

Explicit and Attributive Memory – The memory used in TS is both explicit and attributive.

Explicit memory records complete solutions, typically consisting of elite solutions visited during the search (or highly attractive but unexplored neighbors of such solutions). These special solutions are introduced at strategic intervals to enlarge $N^*(x)$, and thereby provide useful options not in $N(x)$.

TS memory is also designed to exert a more subtle effect on the search through the use of attributive memory, which records information about solution attributes that change in moving

Table 2. Principal tabu search features

Adaptive Memory

    Selectivity (including strategic forgetting)

    Abstraction and decomposition (through explicit and attributive memory)

    Timing:
        recency of events
        frequency of events
        differentiation between short term and long term

    Quality and impact:
        relative attractiveness of alternative choices
        magnitude of changes in structure or constraining relationships

    Context:
        regional interdependence
        structural interdependence
        sequential interdependence

Responsive Exploration

    Strategically imposed restraints and inducements
        (*tabu conditions* and *aspiration levels*)

    Concentrated focus on good regions and good solution features
        (*intensification processes*)

    Characterizing and exploring promising new regions
        (*diversification processes*)

    Non-montonic search patterns
        (*strategic oscillation*)

    Integrating and extending solutions
        (*path relinking*)

---

from one solution to another. For example, in a graph or network setting, attributes can consist of nodes or arcs that are added, dropped or repositioned by the moves executed. In more abstract problem formulations, attributes may correspond to values of variables or functions. Sometimes attributes are also strategically combined to create other attributes, as by hashing procedures or by AI related chunking or "vocabulary building" methods (Hansen and Jaumard, 1990; Woodruff and Zemel, 1992; Battiti and Tecchioli, 1992a; Woodruff, 1993; Glover and Laguna, 1993).

Short-Term Memory and its Accompaniments – An important distinction in TS arises by differentiating between short-term memory and longer-term memory. Each type of memory is accompanied by its own special strategies. The most commonly used short-term memory keeps track of solution attributes that have changed during the recent past, and is called *recency-based* memory. To exploit this memory, selected attributes that occur in solutions recently visited are designated *tabu-active*, and solutions that contain tabu-active elements, or particular combinations of these attributes, are those that become tabu. This prevents certain solutions from the recent past from belonging to $N^*(x)$ and hence from being revisited. Other solutions that share such tabu-active attributes are also similarly prevented from being revisited. The use of tabu evaluations, with large penalties assigned to appropriate sets of tabu-active attributes, can allow tabu status to vary by degrees.

Managing Recency-Based Memory – The process is managed by creating one or several tabu lists, which record the tabu-active attributes and implicitly or explicitly identify their current status. The duration that an attribute remains tabu-active (measured in numbers of iterations) is called its *tabu tenure*. Tabu tenure can vary for different types or combinations of attributes, and can also vary over different intervals of time or stages of search. This varying tenure makes it possible to create different kinds of trade-offs between short-term and longer-term strategies. It also provides a dynamic and robust form of search. (See, e.g., Taillard, 1991, Dell'Amico and Trubian, 1993, Glover and Laguna, 1993.)

Aspiration Levels – An important element of flexibility in tabu search is introduced by means of aspiration criteria. The tabu status of a solution (or a move) can be overruled if certain conditions are met, expressed in the form of aspiration levels. In effect, these aspiration levels provide thresholds of attractiveness that govern whether the solutions may be considered admissible in spite of being classified tabu. Clearly a solution better than any previously seen deserves to be considered admissible. Similar criteria of solution quality provide aspiration criteria over subsets of solutions that belong to common regions or that share specified features (such as a particular functional value or level of infeasibility). Additional examples of aspiration criteria are provided later.

Candidate List Strategies – The aggressive aspect of TS is reinforced by seeking the best available move that can be determined with an appropriate amount of effort. It should be kept in mind that the meaning of best is not limited to the objective function evaluation. (As already noted, tabu evaluations are affected by penalties and inducements determined by the search history.) For situations where $N^*(x)$ is large or its elements are expensive to evaluate, candidate list strategies are used to restrict the number of solutions examined on a given iteration.

Because of the importance TS attaches to selecting elements judiciously, efficient rules for generating and evaluating good candidates are critical to the search process. Even where candidate list strategies are not used explicitly, memory structures to give efficient updates of move evaluations from one iteration to another, and to reduce the effort of finding best or near best moves, are often integral to TS implementations. Intelligent updating can appreciably reduce solution times, and the inclusion of explicit candidate list strategies, for problems that are large, can significantly magnify the resulting benefits.

The operation of these short term elements is illustrated in Diagram 1. The representation of penalties in Diagram 1 either as "large" or "very small" expresses a thresholding effect: either the tabu status yields a greatly deteriorated evaluation or else it chiefly serves to break ties among solutions with highest evaluations. Such an effect of course can be modulated to shift evaluations across levels other than these extremes. If all moves currently available lead to solutions that are tabu (with evaluations that normally would exclude them from being selected), the penalties result in choosing a "least tabu" solution.

The TS variant called *probabilistic tabu search* follows a corresponding design, with a short term component that can be represented by the same diagram. The approach additionally keeps track of tabu evaluations generated during the process that results in selecting a move. Based on this record, the move is chosen probabilistically from the pool of those evaluated (or from a subset of the best members of this pool), weighting the moves so that those with higher evaluations are especially favored. Fuller discussions of probabilistic tabu search are found in Glover (1989, 1993), Soriano and Gendreau (1993), and Crainic *et al.* (1993).

Longer-Term Memory – In some applications, the short-term TS memory components are sufficient to produce very high quality solutions. However, in general, TS becomes significantly stronger by including longer-term memory and its associated strategies.

Special types of *frequency-based* memory are fundamental to longer-term considerations. These operate by introducing penalties and inducements determined by the relative span of time that attributes have belonged to solutions visited by the search, allowing for regional differentiation.

Perhaps surprisingly, the use of longer-term memory does not require long solution runs before its benefits become visible. Often its improvements begin to be manifest in a relatively modest length of time, and can allow solution efforts to be terminated somewhat earlier than otherwise possible, due to finding very high quality solutions within an economical time span. The fastest methods for job shop and flow shop scheduling problems, for example, are based on including longer-term TS memory. On the other hand, it is also true that the chance of finding still better solutions as time grows – in the case where an optimal solution is not already found – is enhanced by using longer-term TS memory in addition to short term memory.

TABU EVALUATION

(Short Term Memory)

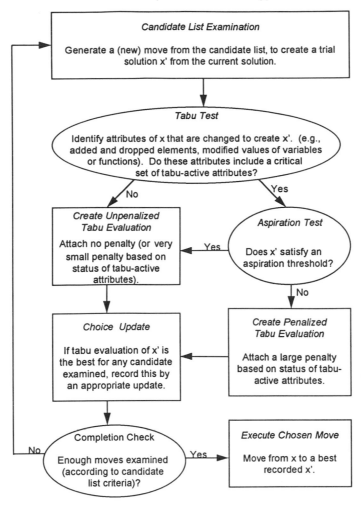

Diagram 1.

Intensification and Diversification – Two highly important longer-term components of tabu search are *intensification strategies* and *diversification strategies*. Intensification strategies are based on modifying choice rules to encourage move combinations and solution features historically found good. They may also initiate a return to attractive regions to search them more thoroughly. A simple instance of this second type of intensification strategy is shown in Diagram 2.

The strategy for selecting elite solutions is italicized in Diagram 2 due to its importance. Two variants have proved quite successful. One, due to Voss (1993), introduces a diversification measure to assure the solutions recorded differ from each other by a desired degree, and then erases all short term memory before resuming from the best of the recorded solutions. The other variant, due to Nowicki and Smutniki

(1993), keeps a bounded length sequential list that adds a new solution at the end only if it is better than any previously seen. The current last member of the list is always the one chosen (and removed) as a basis for resuming search. However, TS short-term memory that accompanied this solution also is saved, and the first move also forbids the move previously taken from this solution, so that a new solution path will be launched.

This second variant is related to a strategy that resumes the search from unvisited neighbors of solutions previously generated (Glover, 1990). Such a strategy keeps track of the quality of these neighbors to select an elite set, and restricts attention to specific types of solutions, such as neighbors of local optima or neighbors of solutions visited on steps immediately before reaching such local optima. This type of "unvisited neighbor" strategy has been little examined.

Simple TS Intensification Approach

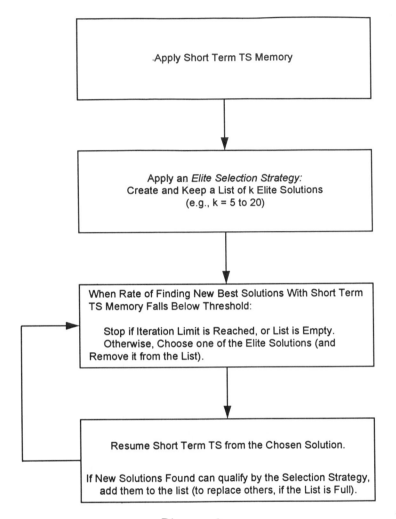

Diagram 2.

It is noteworthy, however, that the two variants previously indicated have provided solutions of remarkably high quality.

Diversification Strategies – TS diversification strategies, as their name suggests, are designed to drive the search into new regions. Often they are based on modifying choice rules to bring attributes into the solution that are infrequently used. Alternatively, they may introduce such attributes by partially or fully re-starting the solution process.

The same types of memories previously described are useful as a foundation for such procedures, although these memories are maintained over different (generally larger) subsets of solutions than those maintained by intensification strategies. A simple diversification approach that keeps a frequency-based memory over all solutions previously generated, and that has proved very successful for machine scheduling

problems, is shown in Diagram 3. Significant improvements over the application of short term TS memory have been achieved by this procedure.

Diversification strategies that create partial or full restarts are important for problems and neighborhood structures where a solution trajectory can become isolated from worthwhile new alternatives unless a radical change is introduced. Diversification strategies can also utilize a long term form of recency-based memory, which results by increasing the tabu tenure of solution attributes.

The two special TS strategies called *path relinking* and *strategic oscillation* embody aspects of both intensification and diversification and have proved highly effective in a variety of contexts (Glover and Laguna, 1993). The determination of effective ways to balance the concerns of intensification and diversification represents a

Simple TS Diversification Approach

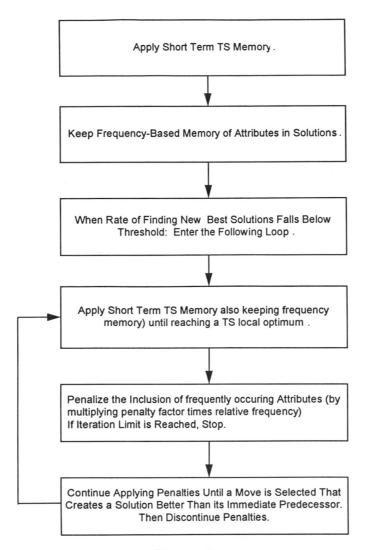

Diagram 3.

promising research area. These concerns also lie at the heart of effective parallel processing implementations. The goal from the TS perspective is to design patterns of communication and information sharing across subsets of processors in order to achieve the best trade-offs between intensification and diversification functions. General analyses and studies of parallel processing with tabu search are given in Taillard (1991, 1993), Battiti and Tecchioli (1992b), Chakrapani and Skorin-Kapov (1991, 1993), Crainic, Toulouse and Gendreau (1993a, 1993b), and Voss (1994).

CONCLUSION: Complementarities among the perspectives of tabu search and those favored by the artificial intelligence and neural network communities raise the possibility of creating systems that integrate their fundamental concerns.

Advances are already underway in this realm, with the creation of *tabu training and learning* models (de Werra and Hertz (1989), Beyer and Orgier (1991), Battiti and Tecchioli (1993), Gee and Prager (1994)), *tabu machines* (Chakrapani and Skorin-Kapov, 1993; Nemati and Sun, 1994) and *tabu design* procedures (Kelly and Gordon, 1994). The outcomes from this work have shown promising consequences for supplementing customary connectionist models – as by yielding levels of performance notably superior to that of models based on *Boltzmann machines*, and by yielding processes for modifying network linkages that give more reliable mappings of inputs to outputs.

The practical successes of tabu search have promoted useful research into ways to exploit its underlying ideas more fully. At the same time, many facets of these ideas remain to be explored.

The issues of identifying best combinations of short and long term memory and best balances of intensification and diversification strategies still contain many unexamined corners, and some of them undoubtedly harbor important discoveries for developing more powerful solution methods in the future.

See **Artificial intelligence; Heuristic procedure; Neural networks.**

References

[1] Battiti, R. and G. Tecchiolli (1992a). "The Reactive Tabu Search," IRST Technical Report 9303-13, to appear in *ORSA Journal on Computing*.

[2] Battiti, R. and G. Tecchiolli (1992b). "Parallel Biased Search for Combinatorial Optimization: Genetic Algorithms and TABU," *Microprocessors and Microsystems*, 16, 351–367.

[3] Battiti, R. and G. Tecchioli (1993). "Training Neural Nets with the Reactive Tabu Search," Technical Report UTM 421, Univ. of Trento, Italy, November.

[4] Beyer, D. and R. Ogier (1991). "Tabu Learning: A Neural Network Search Method for Solving Nonconvex Optimization Problems," *Proceedings of the International Joint Conference on Neural Networks*, IEEE and INNS, Singapore.

[5] Chakrapani, J. and Skorin-Kapov (1991). "Massively Parallel Tabu Search for the Quadratic Assignment Problem," SUNY at Stony Brook.

[6] Chakrapani, J. and Skorin-Kapov, J. (1993). "Connection Machine Implementation of a Tabu Search Algorithm for the Traveling Salesman Problem," *Journal of Computing and Information Technology*-CIT 1, 1, 29–36.

[7] Crainic, T.G., M. Gendreau, P. Soriano, and M. Toulouse (1993). "A Tabu Search Procedure for Multicommodity Location/Allocation with Balancing Requirements," *Annals of Operations Research*, 41(1-4): 359–383.

[8] Crainic, T.G., M. Toulouse, and M. Gendreau (1993a). "A Study of Synchronous Parallelization Strategies for Tabu Search." Publication 934, Centre de recherche sur les transports, Université de Montréal, 1993.

[9] Crainic, T.G., M. Toulouse, and M. Gendreau (1993b). "Appraisal of Asynchronous Parallelization Approaches for Tabu Search Algorithms." Publication 935, Centre de recherche sur les transports, Université de Montréal, 1993.

[10] Gee, A. H. and R.W. Prager (1994). "Polyhedral Combinatorics and Neural Networks," *Neural Computation*, 6, 161–180.

[11] Gendreau, M., P. Soriano, and L. Salvail (1993). "Solving the Maximum Clique Problem Using a Tabu Search Approach," *Annals of Operations Research*, 41, 385–404.

[12] Glover, F. (1989). "Tabu Search-Part I," *ORSA Journal on Computing*, 1, 190–206.

[13] Glover, F. (1990). "Tabu Search-Part II," *ORSA Journal on Computing*, 2, 4–32.

[14] Glover, F. (1993). "Tabu Thresholding: Improved Search by Nonmonotonic Trajectories," to appear in *ORSA Journal on Computing*.

[15] Glover, F. and M. Laguna (1993) "Tabu Search," *Modern Heuristic Techniques for Combinatorial Problems*, C. Reeves, ed., Blackwell Scientific Publishing, 70–141.

[16] Glover, F., M. Laguna, E. Taillard, and D. de Werra, eds. (1993) "*Tabu Search*," special issue of the *Annals of Operations Research*, Vol. 41, J. C. Baltzer.

[17] Hansen, P. and B. Jaumard (1990). "Algorithms for the Maximum Satisfiability Problem," *Computing*, 44, 279–303.

[18] Hertz, A. and D. de Werra (1991). "The tabu search metaheuristic: how we used it," *Annals of Mathematics and Artificial Intelligence*, 1, 111–121.

[19] Kelly, J. P. and K. Gordon (1994). "Predicting the Rescheduling of World Debt: A Neural Network-based Approach that Introduces New Construction and Evaluation Techniques." University of Colorado, Boulder, Colorado.

[20] Nemati, H. and M. Sun (1994). "A Tabu Machine for Connectionist Methods," Joint National ORSA/TIMS Meeting, Boston.

[21] Nowicki, E. and C. Smutnicki (1993). "A Fast Taboo Search Algorithm for the Job Shop Problem," Report 8/93, Institute of Engineering Cybernetics, Technical University of Wroclaw.

[22] Soriano, P. and M. Gendreau (1993). "Diversification Strategies in Tabu Search Algorithms for the Maximum Clique Problem," Publication # 940, Centre de Recherche sur les Transports, Université de Montréal.

[23] Taillard, E. (1991). "Parallel Tabu Search Technique for the Job Shop Scheduling Problem," Research Report ORWP 91/10, Departement de Mathematiques, Ecole Polytechnique Federale de Lausanne.

[24] Taillard, E. (1993). "Parallel Iterative Search Methods for Vehicle Routing Problems," *Networks*, 23, 661–673.

[25] Voss, S. (1994). "Concepts for Parallel Tabu Search," Technische Hochschule Dormstadt, Germany.

[26] de Werra, D. and A. Hertz (1989). "Tabu Search Techniques: A Tutorial and an Applications to Neural Networks," *OR Spectrum*, 11, 131–141.

[27] Woodruff, D.L. (1993). "Tabu Search and Chunking," working paper, University of California, Davis.

[28] Woodruff, D.L. and E. Zemel (1993). "Hashing Vectors for Tabu Search," *Annals of Operations Research*, 41, 123–138.

TAKUCHI LOSS FUNCTION

See **Total quality management**.

TANDEM QUEUES

Queues in series. See **Networks of queues**.

TECHNOLOGICAL COEFFICIENTS

The generic name given to the a_{ij} coefficients of the constraint set of a linear-programming problem.

TELECOMMUNICATION NETWORKS

See **Communication networks**; **Queueing theory**.

TERMINAL

A location used by a carrier for freight consolidation, break-bulk, interchange, and shipment and vehicle service.

THE INSTITUTE OF MANAGEMENT SCIENCES (TIMS)

Founded in 1953, The Institute of Management Sciences (TIMS) was an international organization for management science professionals and academics. It was merged with the Operations Research Society of America into the Institute for Operations Research and the Management Sciences (INFORMS) effective January 1, 1995. The objectives of TIMS were (1) to identify, extend and unify scientific knowledge contributing to the understanding and practice of management, (2) to promote the development of the management sciences and the free interchange of information about the practice of management among managers, scientists, scholars, students, and practitioners of the management sciences within private and public institutions, (3) to promote the dissemination of information on such topics to the general public, and (4) to encourage and develop educational programs in the management sciences. TIMS published the journal *Management Science* (in 40 volumes) and other publications (some jointly with ORSA). It

held national meetings (jointly with ORSA), sponsored meetings by its technical colleges and geographic sections, and held international meetings in various countries. See **Institute for Operations Research and the Management Sciences (INFORMS)**; **Operations Research Society of America (ORSA)**.

THEOREM OF ALTERNATIVES

Many such theorems exist, with a typical one being: either $Ax = b$ has a solution or $yA = 0$, $yb \neq 0$ has a solution. They can be shown to be equivalent to the strong duality theorem of linear programming. See **Farkas' lemma; Gordan's theorem; Strong duality theorem; Transposition theorems**.

THICKNESS

The minimum number of edge-disjoint planar subgraphs into which a graph can be decomposed. See **Graph theory**.

TIME/COST TRADE-OFFS

An approach to scheduling where the project duration is shortened with a minimum of added costs. See **Network planning**.

TIME SERIES ANALYSIS

Christina M. Mastrangelo

University of Virginia, Charlottesville

Douglas C. Montgomery

Arizona State University, Tempe

INTRODUCTION: A time series is an ordered sequence of observations. This ordering is usually through time, although other dimensions, such as spatial ordering, are sometimes encountered. A time series can be continuous, as when an electrical signal such as voltage is recorded. Typically, however, most industrial time series are observed and recorded at specific time intervals and are said to be discrete time series. If only one variable is observed, the time series is said to be univariate. However, some time series involve simultaneous observations on several variables. These are called multivariate time series.

There are three general objectives for studying time series: 1) understanding and modeling of the underlying mechanism that generates the time series, 2) prediction of future values, and 3) control of some system for which the time series is a performance measure. Examples of the third application occur frequently in industry. Almost all time series exhibit some structural dependency. That is, the successive observations are correlated over time, or autocorrelated. Special classes of statistical methods that take this auto-correlative structure into account are required.

Figure 1 shows examples of time series with distinctly different features. In Figure 1(a), the time series x_t appears to vary around a constant level. Such a time series is said to be stationary in the mean. In Figure 1(b), we see non-stationary behavior; that is, the time series x_t drifts with no obvious fixed level. Some nonstationary time series may exhibit trends, or the variance of the series may increase as the level of the time

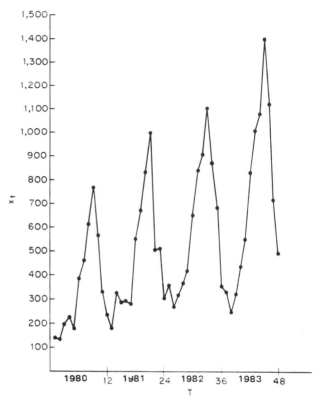

Fig. 1c. Monthly demand for a 48-oz soft drink in hundreds of cases.

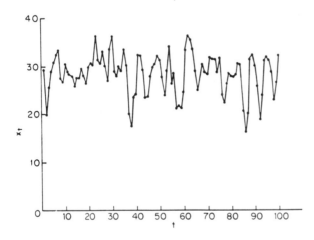

Fig. 1a. Viscosity of a chemical product.

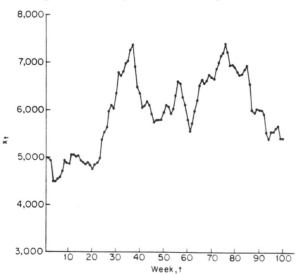

Fig. 1b. Demand for a plastic container.

series increases. Seasonal variation is illustrated in Figure 1(c).

The autocorrelation function is a very useful tool in characterizing time series behavior. The autocorrelation between x_t and x_{t+k} is defined as

$$\rho_k = \frac{\text{cov}(x_t, x_{t+k})}{\sqrt{V(x_t)V(x_{t+k})}} = \frac{\gamma_k}{\gamma_0}$$

where $\text{cov}(x_t, x_{t+k}) = \text{E}(x_t - m)(x_{t+k} - m)$. This is called the autocorrelation at lag k. The usual estimate of ρ_k, $k = 1, 2, \ldots, K$, is the sample autocorrelation function

$$r_k = \frac{\hat{\gamma}_k}{\hat{\gamma}_0} = \frac{\sum_{t=1}^{n-k} (x_t - \bar{x})(x_{t+k} - \bar{x})}{\sum_{t=1}^{n} (x_t - \bar{x})^2}.$$

Figure 2 shows the sample autocorrelation function for the time series in Figure 1(a). The dotted lines are two standard error limits. Notice that there is a large positive value or spike at lag 1 and the sample autocorrelation function decays as a damped sine wave from lag 1. The sample autocorrelation function is very useful in the identification of an appropriate time series model. In addition, the partial autocorrelation function, the inverse autocorrelation function, and the extended sample autocorrelation func-

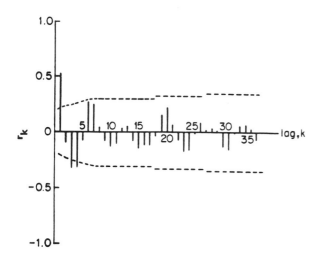

Fig. 2. Sample autocorrelation function.

tion are useful in time series model identification (Montgomery *et al.*, 1990; Cleveland, 1972).

TIME SERIES MODELING METHODS: There are several widely used approaches for modeling and analysis of time series data. Regression methods play a fundamental role. If y_t represents the time series of interest and x_{jt}, $j = 1, 2, \ldots, k$ are a collection of other time series thought to be related to y_t, then it is possible to fit a regression model of the form

$$y_t = \beta_0 + \sum_{j=1}^{k} \beta_j x_{jt} + \varepsilon_t, \quad t = 1, \ldots, n$$

using least squares or some suitable variation. Usually, however, the errors e_t are autocorrelated and more complex estimation schemes are needed. Several estimation methods are available which result in estimates similar to least squares estimates, but the standard errors may be very different. Yule-Walker estimation uses the Yule-Walker equations to estimate the autoregressive parameters of the errors and generalized least squares to estimate β. Harvey (1981) gives a full description of this and other methods.

Smoothing methods are frequently used in time series analysis. In particular, exponential smoothing is widely used for producing short term forecasts of many types of industrial time series. Much of the original working this area is by Brown (1962), Holt (1957), and Winters (1960). Exponential smoothing is often developed heuristically starting with a simple model such as $x_t = b + e_t$, where e_t are independent random variables and b is an unknown constant. Simple or first-order exponential smoothing is defined as

$$S_t = \alpha x_t + (1 - \alpha)S_{t-1}$$

where $0 \leq \alpha \leq 1$. The smoothed statistic S_t estimates the constant b, so the forecast for any future observation $x_{t+\tau}$ made at the end of period t is

$$\hat{x}_{t+\tau}(t) = S_t.$$

Extensions of this methodology to forecasting linear and quadratic trend and incorporating seasonal behavior are described in Montgomery *et al.* (1990).

The class of autoregressive integrated moving averages (ARIMA) models proposed by Box and Jenkins (1976) have been very successful for time series modeling and forecasting. The general form for this family of models is

$$(1 - \phi_1 B - \phi_2 B^2 - \cdots - \phi_p B^p)(1 - B)^d x_t$$
$$= \theta_0 + (1 - \theta_1 B - \theta_2 B^2 - \cdots - \theta_q B^q)\varepsilon_t$$

where ϕ_i are the autoregressive parameters, θ_j are the moving average parameters, B is a backshift operator defined such that $B^r x_t = x_{t-r}$, $(1 - B)^d = \nabla^d$ is the backward difference operator, and ε_t is an uncorrelated sequence of random disturbances with mean zero and variance σ^2. This model can also be extended to incorporate seasonal behavior (Box and Jenkins, 1976; Montgomery *et al.*, 1990). One chooses a model by specifying the integers p, d, and q, resulting in an ARIMA(p, d, q) model. This is usually done by examining the sample autocorrelation function and partial autocorrelation function. Then nonlinear regression methods are used to estimate the parameters ϕ_i and θ_j. Finally the residuals from the fitted model are studied to test model adequacy. Generally, one should examine the autocorrelation function of the residuals, for if the model is adequate, the residuals should be approximately uncorrelated. Residual plots, such as a plot of residuals versus the fitted x_t, and a normal probability plot of the residuals, are useful in detecting model inadequacy.

To illustrate, consider the viscosity data from Figure 1(b). It can be shown that an appropriate choice of p, d, and q is $p = 0$, $d = 1$, and $q = 1$, resulting in the ARIMA($0, 1, 1$) = IMA($1, 1$) model

$$(1 - B)x_t = (1 - \theta B)\varepsilon_t.$$

The least squares estimate of the parameter θ in this model is $\hat{\theta} = -0.70$. Therefore, the final model is

$$x_t = x_{t-1} + \varepsilon_t + 0.7\varepsilon_{t-1}.$$

This model is satisfactory with respect to the adequacy criteria cited above. We can also show that IMA($1, 1$) is equivalent to exponential smoothing, as described above.

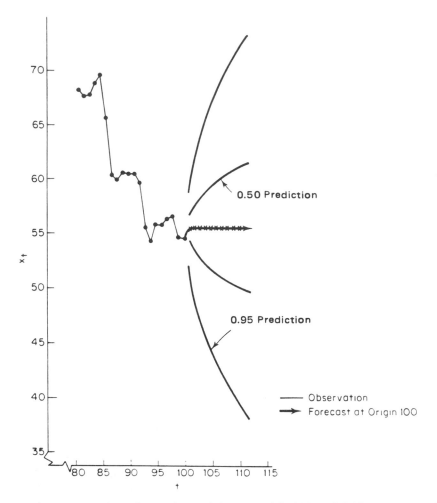

Fig. 3. Forecast of plastic container demand at origin 100, with 0.50 and 0.95 percent prediction limits.

FORECASTING: An important objective of any time series model is forecasting future values. The term forecasting is used in the time series analysis literature, although most results are based on the general theory of linear prediction developed by Kalman (1960), Box and Jenkins (1976) and many others. The objective is to produce minimum mean square error forecasts.

Minimum mean square error forecasts for ARIMA models are obtained by taking the conditional expectation $E(x_{t+\tau}|x_t, x_{t-1}, \ldots)$. For example, the minimum mean square error forecast for the ARIMA(0, 1, 1) = IMA(1, 1) model shown earlier for the viscosity data is

$$E(x_{t+\tau}|x_t, x_{t-1}, \ldots) \equiv \hat{x}_{t+\tau}(t) = x_t + 0.7\varepsilon_t(1)$$

where $e_t(1) = x_t - \hat{x}_t(t-1)$ is the one-step ahead forecast error. Figure 3 shows the forecasts obtained from this model. It is usually necessary to provide prediction intervals for forecasts as well as point estimates. Figure 3 shows the 50% and 95% prediction limits for the forecast of future viscosity values. The details of the construction of these limits are described in Box and Jenkins (1976) and Montgomery *et al.* (1990).

The form of the eventual forecast function for ARIMA models is also of interest, because it leads in some cases to efficient methods for forecast generation and updating. The form of the forecast function for several common ARIMA models is given in Box and Jenkins (1976).

TRANSFER FUNCTIONS AND RELATED TOP-ICS: If y_t and x_t are two stationary time series related through the mean filter

$$y_t = V(B)x_t + \varepsilon_t$$

then $V(B) = \Sigma_{j=-\infty}^{\infty} v_j B^j$ is called the transfer function of the filter and e_t is called the noise series of the system. Typically, x_t and e_t are assumed to follow ARMA = ARIMA(p, 0, q) models. It is customary to write

$$V(B) = \frac{\omega_s(B)B^b}{\delta_r(B)}$$

where $\omega_S(B) = \omega_0 - \omega_1 B - \omega_2 B^2 - \cdots - \omega_S B^S$, $\delta_r(B) = \delta_0 - \delta_1 B - \delta_2 B^2 - \cdots - \delta_r B^r$ and b is a delay representing the time before the input at time t produced an effect on the output. A transfer function model is identified by choosing appropriate values of s, r, b, and a model for the noise e_t. Usually s, r, and b will be no larger than 2. The cross-correlation function is useful in model identification.

Once a suitable transfer function model is identified, the parameters are estimated by nonlinear regression methods, and diagnostics checks are applied, much like in classical univariate ARIMA modeling. Minimum mean square error forecasts are generated using a similar approach, based on conditional expectation at time t of $y_{t+\tau}$. Examples of identification, estimation, diagnostic checking, and forecasting with transfer functions are given in Box and Jenkins (1976) and Montgomery *et al.* (1990).

An important special case of the transfer function occurs when the input series x_t is a sequence of indicator variables that represent the occurrence of identifiable, unique events that are thought to influence the output y_t. These events are called interventions, and the resulting models are called intervention models. An intervention model is often used to provide a statistical basis for concluding that the identifiable event has resulted in a change in the time series.

Box and Tiao (1975) developed the basic intervention analysis methodology and applied it to photo chemical pollution data from the Los Angeles basin. They showed that the opening of the Golden State Freeway and the adoption of a new law that reduced the proportion of reactive hydrocarbons in local gasoline, reduced ozone levels, and that required changes in automobile engines reduced ozone levels only in warm weather months. Intervention models are also useful in the study of time series outliers. Fox (1972) proposed two types of outliers, additive and innovational.

In some time series problems, we observe m different variables $x_{1t}, x_{2t}, \ldots, x_{mt}$ in a multivariate framework. Hannan (1970) and Granger and Newbold (1977) describe two ways to model multiple time series, a multivariate ARIMA model or state-space model.

See **Exponential smoothing; Forecasting; Quality control; Regression analysis.**

References

[1] Abraham B. and J. Ledolter (1983). *Statistical Methods for Forecasting*. John Wiley, New York.

[2] Box G.E.P. and G.M. Jenkins (1976). *Time Series Analysis, Forecasting and Control*, Revised Edition, Holden-Day, San Francisco.

[3] Box G.E.P. and G.C. Tiao (1975). "Intervention Analysis with Applications to Economic and Environmental Problems," *Jl. American Statistical Association*, 70, 70–79.

[4] Brown R.G. (1962). *Smoothing, Forecasting and Prediction of Discrete Time Series*, Prentice-Hall, Englewood Cliffs, New Jersey.

[5] Cleveland W.S. (1972). "The Inverse Autocorrelations of a Time Series and Their Applications," *Technometrics*, 14, 277–293.

[6] Fox A.J. (1972), "Outliers in Time Series," *Jl. Royal Statistical Society*, Ser. B, 43, 350–363.

[7] Granger G.W.C. and P. Newbold (1977). *Forecasting Economic Time Series*, Academic Press, New York.

[8] Hanan E.J. (1970). *Multiple Time Series*, John Wiley, New York.

[9] Harvey A.C. (1981). *The Econometric Analysis of Time Series*, John Wiley, New York.

[10] Holt C.C. (1957). "Forecasting Trends and Seasonal by Exponentially Weighted Moving Averages," *ONR Memorandum No. 52*, Carnegie Institute of Technology.

[11] Kalman R.E. (1960). "A New Approach to Linear Filtering and Prediction Problems," *ASME Jl. Basic Engineering for Industry*, Ser. D, 82, 35–45.

[12] Montgomery D.C., L.A. Johnson and J.S. Gardiner (1990). *Forecasting and Time Series Analysis*, 2nd ed., McGraw-Hill, New York.

[13] Winters P.R. (1960). "Forecasting Sales by Exponentially Weighted Moving Averages," *Operations Research*, 22, 858–867.

TIME-STEPPED SIMULATION

A computer model which models time by incrementing a simulated clock. Each appropriate function is recomputed after the clock is incremented in a cyclical manner. A model may be linearly coded and entirely time-stepped or an event-driven simulation may use time-stepping for some critical function with a cycle of sub-functions. See **Event-driven simulation; Simulation of discrete-event stochastic systems.**

TIME-TABLING

See **Scheduling and sequencing.**

TIMS

See **The Institute of Management Sciences.**

TOLERANCE ANALYSIS

A sensitivity analysis procedure applied to a linear-programming problem that allows for simultaneous changes of the objective function cost coefficients and/or right-hand-sides of the constraints. See **Hundred percent rule; Sensitivity analysis**.

TOTAL FLOAT

The amount of time a project work item can be delayed without affecting the duration of the project. Total float can be used in only one activity in a path. If no scheduled times are specified for starting and finishing the various activities, then the float is calculated as the difference between the latest start time and the earliest start time, or the difference between the latest finish time and the earliest finish time. Float can be positive, negative or zero. See **Network planning**.

TOTAL QUALITY MANAGEMENT

John S. Ramberg

University of Arizona, Tucson

INTRODUCTION: During the decade of the 1980s, U.S. corporations recognized the quality achievements of their Japanese counterparts and began to understand the messages being delivered by Deming, Juran and others on the importance of quality. They devised methods for obtaining, understanding and communicating customer needs and requirements within their organizations, developed strategies for improving their engineering design, development, manufacture and delivery processes, and created new corporate cultures that included the formation of self-directed working groups and encouragement of employee participation. Through this focus on quality and the development and adaptation of techniques for achieving customer satisfaction, some of these corporations have demonstrated improvement in achieving high quality, timely deliveries at low costs and ultimately improved their business performance. Many of these firms called this new management and operations philosophy *Total Quality Management* or simply TQM.

Other firms, frustrated by false starts and questionable implementations, began to question the value of total quality management, and some have given up, regarding it as just another fad (Senge, 1993). In many of these latter situations, quality efforts have been misdirected or unfocused. In some cases, quality improvement activities were simply knee jerk reactions to the customers who complained most vehemently to the highest level of the organization. Ramberg (1994) described some of the scurrilous characters who proclaim TQM, while delivering just another program, and raises the question, "TQM: Thought Revolution or Trojan Horse."

While TQM connotes much more than simply the three words *total*, *quality* and *management*, nevertheless, definitions of each of the three words seem an appropriate place to begin. A typical dictionary definition of *total* is: all or whole, that is constituting the whole; complete. The definition of *quality* is a bit more difficult to comprehend as U.S. firms have come to understand. A formal definition, as given by the American Society for Quality Control (ASQC) and the American National Standards Institute (ANSI) is "The totality of features and characteristics of a product or service that bear on its ability to satisfy stated or implied needs." Finally, *management* is the act, be it a science, an art or manner, of planning, directing, organizing and controlling of a firm's decisions and actions. As an aside, it is interesting to note that the phrase "to manage" originated as "to train (a horse) in his paces, or to cause to do the exercises of the manege!"

A PROFOUND UNDERSTANDING OF QUALITY: Quality is the pivotal word in TQM. A fundamental reason for the U.S. losing world leadership in manufacturing during the 1960s and 1970s was its lack of a profound understanding of the Q word. The gurus of quality, in the interest of developing a better understanding of the importance of quality, created shorter, more explicit, operationally oriented definitions such as "fitness for use" (Juran), "conformance to specifications" (Crosby), "long term loss to society" (Taguchi), and "a predictable degree of uniformity and dependability, at a low cost and suited to the market" (Deming, as paraphrased by Gitlow, Oppenheim, and Oppenheim, 1994).

Some have exploited the differences in these operational definitions of quality, and a few have concluded that even the quality gurus cannot agree on the definition of quality. We view them as being complementary, each definition emphasizing its definer's experience base in relation to the customer with whom he or she is communicating. "Fitness for use" is an appropriate operational definition of quality in the creation and marketing of a product or service on the production floor, where an employee may be far removed from the customer, the translation of quality performance measures into specific dimensions having specified targets and specification limits seems a necessity. Finally, if "loss to society" is thought of as "long-term business loss," then its relation to the other two operational definitions becomes clearer. Deming's definition exhibits his emphasis on variability and its reduction as a fundamental step in improving quality.

A first step in attaining a profound understanding of quality is the simple realization that it is customer-driven. It not only begins with the customer, in the end it is judged by the customer. Thus the "voice of the customer" is imperative. Equally important to understanding that quality is customer-driven is the recognition that the customer may not be able to articulate it fully. Even the most sophisticated customers are not likely to be able to envision all of the characteristics of a product that will satisfy and "delight" them. Expert panels can serve an important role, but they too have their limitations. Obtaining this information is a complex task. Based on this input, product creators, developers and deliverers must envision these dimensions of quality that will satisfy and delight their customer. Furthermore, they must maintain a dialog with the customer so that they will continue to understand and respond to this dynamic "voice of the customer."

Traditionally, managers have viewed quality, timeliness and cost as a zero-sum game. That is, any improvement in quality or timeliness will occur only at a substantial additional cost. The following quote from Vaughn (1990) illustrates this "conventional wisdom": "The trade off is between the effects of less emphasis on quality and the cost of more of it. . . ." Juran (1989) categorized quality associated costs (and estimates of the associated percentage for one industry) into four broad groups, those due to Internal Failures (30%), External Failures (40%), Appraisal (25%), and Detection and Prevention (5%). He also discussed how these percentages are dependent upon the maturity of the product line and effort expended on quality improvement. Juran's classic model for optimum quality levels also emphasized that there is a trade-off between quality and cost. He stated that "failure costs decline until they are overtaken by the increasing costs associated with appraisal and prevention. At this point total costs increase."

Cole (1992) made an excellent case for a fundamental paradigm shift regarding quality and costs and timeliness, based on the achievements of the Japanese. His conclusions are given in Table 1. Compare Cole's views with the old quality paradigm, "you get what you pay for." The truth is that high cost, alone, is not a guarantee

Table 1. Cole's underlying reasons for Japanese achievements in quality

"–realized that the costs of poor quality were far larger than had been recognized."

"–recognized that focusing on quality improvement as a firm-wide effort improved a wide range of performance measures."

"–established a system that moved toward quality improvement and toward low-cost solutions simultaneously."

"–focused on preventing error at the source, thereby dramatically reducing appraisal costs."

"–shifted the focus of quality improvement from product attributes to operational procedures."

"–evolved a dynamic model in which customer demands for quality rise (along with their willingness to pay for these improvements)."

that a product will be of high quality. Indeed, some times the contrary is true, resulting in what *Consumer Reports* refers to as "best buys." The six achievements of the Japanese, cited by Cole, have an important impact on conclusions drawn from quality cost models. Specifically, they indicate that the point at which it is no longer cost effective to improve quality is at a much lower defective rate than we previously thought.

Establishing, appraising or judging the quality of a product are far more difficult than simply defining it. In his highly acclaimed book *Management of Quality*, Garvin (1988) elaborated eight dimensions of quality, including performance, features, reliability, conformance, durability, serviceability, aesthetics, and perceived quality. Through his study on air conditioners, he illustrated the differences in the perception of quality of various constituencies, noting that customers, companies (as represented by first line supervisors), service personnel and *Consumer Reports* view quality quite differently and elaborated on the reasons for these different perceptions.

Quality is achieved by elaborating the important product characteristics, their targets and specifications. The ability of a product to meet these specifications depends upon its design, development and the processes employed in its manufacture. Product and process information is often gathered through capability studies, where measurements are obtained on important product characteristics, and control charts are employed to address the stability of the processes and the predictability of future performance. These product characteristics are frequently summarized by a statistical distribution, or even more succinctly, by the process capability, 6σ.

Process capability indices are typically employed to combine this voice of the customer with the voice of the product/process into a dimensionless measure. Pignatiello and Ramberg (1993) reviewed this approach, stressing the importance of an appropriate data collection scheme and the statistical analysis and summarization of results. These indices, which are dimensionless quantities, are then employed in quality improvement project selection.

While top-level management communicates in dollars, operations level personnel must be bilingual, communicating in both dollars and things, that is, in product units and performance measures. Taguchi popularized the use of loss functions to provide a link between these two languages. They provide a means for expressing the deviation of product characteristics from their targeted values in dollars. These loss functions can be determined through internal costs of a product at each stage of design, development, manufacture and delivery.

Juran (1994) has noted that many disagreements about achieving quality result from the fact that there are two fundamentally different quality issues, one income oriented and the other cost oriented. Features that produce customer satisfaction are income oriented. They are the key to attracting new customers and through satisfaction of retaining them. Cost oriented quality issues are the defects and failures that incur. They cause dissatisfaction and the loss of customers. As customers become aware of a product and indeed a producer's track record through publications such as *Consumer Reports*, they also impact the ability of attracting and retaining customers. Furthermore, they impact the profitability of the firm through the dollars lost internally in defectives and rework and externally through warranty costs and other required services.

TOTAL QUALITY: The term quality has traditionally been associated with manufacturing, and more explicitly with the products, processes, functions, and facilities associated with manufacturing. The modern total quality viewpoint extends this factory oriented view of quality to encompass all products, goods and services whether they are for sale or not.

Total quality proponents embrace training and education as universal, in direct contrast to the Taylor system, a system to which U. S. leadership in productivity has been attributed. Taylor made a strategic decision to separate planning and execution. This decision was based on his assessment that the then "immigrant" work force was uneducated and that it was not economically feasible to educate them in a timely manner. We now have a more highly educated work force and this work force represents an untapped resource for improving quality and productivity. Total quality proponents recognized this improvement in the education level, and the responsibility for not only utilizing this resource, but improving it through the continuing education of the work force. Furthermore, they recognized these workers as stakeholders and enhanced their role in productivity and quality their empowerment.

TOTAL QUALITY MANAGEMENT: While neither embraced the term *Total Quality Management*, its origins can be traced to the work of W. Edwards Deming and Joseph J. Juran and through the implementation of their quality philosophy, concepts and methods in Japanese industry. Kolesar (1995) discusses Deming's introduction of his quality philosophy to Japan. The importance of TQM became fully recognized in the U.S. only after its successful Japanese implementation became apparent through the domination of their products as a direct result of their outstanding quality. With this recognition, Deming and Juran gained the attention of enlightened U.S. corporate and government leaders, and Deming's name even became known to U.S. schoolchildren through public television programming.

Deming is perhaps best known for his *14 point manifesto*, which is fundamental to TQM philosophy, and the Shewhart/Deming PDCA cycle. The latter, now called PDSA, meaning "Plan, Do, Study, and Act," provides a fundamental structure for achieving quality. Gitlow, Oppenheim, and Oppenheim (1995) gave an excellent discussion of Deming's 14 points and employed the PDSA approach for achieving quality improvement. Scherkenbach (1986, 1991) has developed a balanced view of the key characteristics of the

Table 2. Key characteristics of the Deming philosophy, from W.W. Scherkenbach

* Reduce waste	* Add value
* Constancy of purpose	* Continual improvement
* Improvement	* Innovation
* Team	* Individual
* Long-term	* Short-term
* Inputs	* Outputs
* Synthesis	* Analysis
* Knowledge	* Action

philosophy of Deming given in Table 2. For example, one of Deming's 14 points is "reduce waste," which Scherkenbach has balanced with "add value."

Juran (1989) recognized the importance of including quality in the management game plan, as well as the need for developing managerial processes in managing quality. He noted financial management included three processes: financial planning producing the budget), financial control (assuring that the budget will be met), and financial improvement (ways of increasing income and decreasing costs). Translated to quality, these are known as the *Juran Trilogy:* Quality Planning, Quality Control and Quality Improvement. A major advantage facilitating the implementation of these ideas is that senior management already understands them in the financial arena. Juran also stated "universal sequences for accomplishing these processes, the quality planning road map, quality control and the quality improvement processes." Fundamental to his methodology is the recognition of the presence of "chronic quality wastes" resulting from "disconnected alarm systems."

Senge (1993) created a TQM paradigm that is based on the three cornerstones, Guiding Ideas, Infrastructure and Theory, Tools and Methods. He noted that guiding ideas are based on a vision. Without this vision, everything is mechanical and pedestrian. Leaders expressing this vision and these guiding ideas must practice them. When they make a decision differently, their colleagues and subordinates will know! However, these ideas and the behavior and actions of the leaders are not enough. An infrastructure is necessary for diffusing these ideas. Conflicts in goals must be resolved and this implies the importance of accountability and an appropriate reward structure. Finally, there is the theory, tools and methods cornerstone. Again, a necessary and important part of the structure, but certainly not sufficient on its own. OR/MS tends to be tool and methods oriented and we need to be aware of that limited view.

Table 3. Quality tools – the magnificent seven plus one.

Control Charts
Check Sheets
Histograms
Pareto Diagrams
Ishikawa Fishbone Diagrams
Scatter Plots
Flow Charts or Process Diagrams
Multi-Variate Charts

Table 4. Quality management – the seven tools

Affinity Diagram
Interrelationship Digraph
Tree Diagram
Prioritization Matrices
Matrix Diagram
Process Decision Program Chart
Activity Network Diagram

Tables 3 and 4 list these essential tools. These tools of TQM are communications enhancers that assist one in listening and talking to processes, products, systems and people.

TRANSFORMATION TO QUALITY ORGANIZATIONS: Implementation of total quality management in a firm requires a transformation of the organization, and any transformation of an organization is doomed to failure if it does not recognize the importance of the human aspect. Scherkenbach (1991) elaborated a theory of transformation that emphasizes this human aspect of quality. Scherkenbach notes how differently people view the world and thus why they are motivated by different means. Some, such as management scientists and operations researchers, live in the "logical world." They tend to proceed on the basis of logical actions. Others, including many top level managers and workers alike, live in a "physical world." This is the world of policies, procedures, standards and rewards punishments. They do it by the book. Still others, such as sales personnel, marketing specialists and artists live in the "emotional world," typified by the statement "The force is with you, Luke."

Scherkenbach's point is not to create stereotypes, but to enable a better understanding of why arguments made in one of these domains often do not have a substantive impact on people living in another domain. That is, when we are dealing with others, it is imperative to recognize that they may be motivated by different forces than we are. To make progress in our relationships with others, we need to be cognizant of their view and address them in an appropriate manner. As a point of exclamation to those who live in the logical world, Scherkenbach quotes Schopenhauer: "No one ever convinced anybody by logic; and even logicians use logic only as a source of income."

Scherkenbach goes on to describe transformation through three process relationships, one for each world view, and all given in terms of different mind states or attitudes: dependent, independent and interdependent. Many people function solely in either the dependent or independent mode. An important aspect of the quality transformation is to facilitate the move to the interdependent mode.

TQM AND PRINCIPLE BASED MANAGEMENT: Each of us holds an important key to any quality transformation process in which we are involved. Covey (1993) suggested that we begin the quality transformation by first taking action on ourselves; then proceed through the four steps of his inside-out principle based management. He described these four steps as self, interpersonal, managerial and organizational. At the self level, he stresses the need to carefully develop our vision, decide what our life is about and develop those principles that will serve as our guidelines in making all of our decisions in life. Next is the need to act on this vision in a consistent manner that builds an internal source of security. Immediate or complete success should not be expected since this is a learning process. Incorporating and practicing the Shewhart/Deming PDSA cycle in our own work is an important method for improving the quality of our own work. As we achieve some comfort with ourselves, and create a more positive opinion of ourselves, we will be able to move on to the interpersonal level. Covey stated that quality at the interpersonal level means that we live by the correct principles in our relationship with other people. Here Covey used the analogy of a bank account, that is, we make deposits to and withdrawals from an emotional bank account. He stated three important ground rules for achieving quality in interpersonal relationships. First, when we have a problem with a person, we should go directly to them and explain it. The second relates to the conduct of meetings. His ground rule is that no one is allowed to make a point in a meeting until they restate the point of their predecessor, and state it in a manner that is satisfactory to that person. He notes that this eliminates the majority of disagreements, since most of them are simply misunderstandings. Through this mechanism, potential misunderstandings can be quickly clarified, avoiding arguments, further miscommunications and withdrawals from the emotional bank account. Furthermore, having greatly reduced the number of misunderstandings, there is a better chance to disagree agreeably when new disagreements take place. An important question is "do we have the courage to practice this ground rule and continue to practice it even if the rest of group does not?"

Finally, when we do make mistakes, we need to have the courage to say that we were wrong. No excuses. We must apologize to the person; we must also apologize to the other people involved. At the managerial level, quality means that we attempt to empower people. In this way they become increasingly independent of us. They supervise themselves, and we become a source of help, rather than a micro-manager. Empowerment begins with self-control and self-inspection and extends to self-directing work teams. These teams plan processes, establish schedules, assign personnel and maintain discipline through peer pressure. They accomplish the work that was once limited to managers and specialists. Juran (1994) suggests that this system could be the successor to the Taylor system. It offers the opportunity to step off of the productivity and quality plateaus, which have been directly traced to the lack of involvement of the total work force, a result of not questioning the assumptions underlying Taylor's original separation of planning and execution. A craftsperson created a product from start to finish, and thus recognized the impact of each step on the following one. The production worker, as the execution of production was broken into individual components, had a smaller and decreasing opportunity to comprehend his role in achieving quality. As a result, inspection departments and later quality departments emerged and were used as policing units in the goal to achieve quality.

At the organizational level, the key is in the structures and the leadership styles. Are the leaders in harmony with the mission statement? Was everyone involved in the development of the mission statement?

TQM AND THE MALCOLM BALDRIGE AWARD: The Malcolm Baldrige Award framework provides an excellent road map for implementing TQM, as well as a method for evaluating a firm's progress (NIST, 1995). The framework emphasizes dynamic relationships between categories of core values and concepts: customer-driven quality, leadership, continuous improvement and learning, employee participation and development, fast response, design quality and prevention, long-range view of the future, management by fact, partnership development, corporate responsibility and citizenship and results orientation.

The stated goals are customer satisfaction, customer satisfaction relative to competitors, customer retention and market share gain as measured by product and service quality, productivity improvement, waste reduction/elimination, supplier performance and financial results. Leadership is viewed as the "driver" category of core values and concepts, driving the two categories – business results and customer focus and satisfaction through a system of processes. The system of processes consists of four "well-defined and well-designed processes" for achieving the firm's performance requirements and the firm's customer requirements. These four system categories are information and analysis, strategic planning, human resource development and management, and process management. The criteria, which are updated annually, are disseminated by the American Society for Quality Control and the National Institute of Standards and Technology.

HEALTH AND STATUS OF TQM: Senge began his 1993 ASQC Annual Conference keynote address on "The Health and Well Being of the TQM Movement" by posing the following questions about TQM: "Are fundamental breakthroughs being made? Are they being made in your organization?" Following this opening, he summarized surveys by the companies Arthur D. Little and McKinsey, and made the following conclusions. Out of 500 firms surveyed, less than a third were accomplishing anything! Two thirds of the TQM programs have ground to a halt!

Senge went on to diagnose TQM failures and "successes." Based on his case studies he concluded that there are only a few major reasons for failure – another example of the Juran/Pareto principle at work. He stated the three major reasons for failure as: conflict Between Time and Effort (it's the program versus my job); Wavering Goals (resulting from a host of threats); and Employee Perception (MY JOB IS AT RISK!).

Even where TQM has "succeeded," there are serious questions about the measures used to judge that success. That is, in many cases, even where the TQM indicators improved, the health of the company (e.g., as judged by its price) did not get any better, even over a reasonably long term. That is, TQM did not improve the health of the organization as judged by at least one of its customers, its stock holders. Reporting on the root cause of these problems, Senge concluded that a major reason was that most organizations viewed TQM as programmatic. Presented or implemented in this manner, TQM is certain to be D.O.A.

Comparative studies measuring the impact of TQM on a firm's business performance are available. Jarrell and Easton (1994) reported some evidence that long-term performance of firms adopting TQM is improved. This result is consistent across the accounting and stock price performance measures examined. Similar, but overall stronger results are found when the analysis is limited to a subsample of pilot firms identified as having more mature and well-integrated TQM systems.

The August 8, 1994 *Business Week* cover story on TQM emphasized the importance of institutionalizing the basics. These were: comparing the cost of quality initiatives against returns, determining the key factors for retaining customers as well as what drives them away, focusing quality efforts so as to improve customer satisfaction at a reasonable cost, establishing a link between each dollar spent on quality and its effect on customer retention, supporting promising programs and pruning those that are not.

See **Quality control; Reliability of systems.**

References

[1] Business Week (1994), "Quality: How to Make it Pay," August 8, 54–59.
[2] Cole, R.E. (1992), "The Quality Revolution," *Production and Operations Management*, 1, 118–120.
[3] Covey, S. (1993), "An Inside Out Approach to Change and Quality," audio tape, Covey Leadership Center, Provo, Utah.
[4] Crosby, P.B. (1987), "What Are Requirements?" *Quality Progress*, ASQC, August, 47.
[5] Deming, W. Edwards (1986), *Out of the Crisis*, MIT Press, Cambridge, Massachusetts.
[6] Easton, G.S. (1993), "The 1993 State of U.S. Total Quality Management: A Baldrige Examiner's Perspective," *California Management Review*, 35(3), 32–54.
[7] Garvin, D.A. (1988), *Managing Quality*, Free Press, New York.
[8] Gitlow, H., Oppenheim, A. and Oppenheim, R. (1994), *Tools and Methods for the Improvement of Quality*, Irwin, Homewood, Illinois.
[9] Jarrell, S.L. and Easton, G.S. (1994), "An Exploratory Empirical Investigation of the Effects of Total Quality Management on Corporate Performance," *The Practice of Quality Management*, P. Lederer, ed., Harvard Business School Press, Cambridge, Massachusetts.
[10] Juran, J.M. (1989), *Quality Control Handbook*, McGraw-Hill, New York.
[11] Kolesar, P.J. (1994), "What Deming Told the Japanese in 1950," *Quality Management Jl.*, Fall 1994, 9–24.
[12] NIST (National Institute of Standards and Technology) (1995), *The Malcolm Baldridge Award Application Package*, Gaithersburg, Maryland.
[13] Pignatiello, J.J., Jr. and J.S. Ramberg (1993), "Process Capability Studies, Just Say No!" *ASQC Annual Technical Conference Proceedings*.

[14] Ramberg, J.S. (1994), "TQM: Thought Revolution or Trojan Horse?" *OR/MS Today*, 21(4), 18–24.

[15] Scherkenbach, W.W. (1986), *The Deming Route to Quality and Productivity: Road Maps and Roadblocks*, ASQC Press and Washington CEE Press.

[16] Scherkenbach, W.W. (1991), *Deming's Road to Continual Improvement*, SPC Press, Knoxville, Tennessee.

[17] Scholtes, P.R. and Hacquebord, H. (1988), "Six Strategies for Beginning the Quality Transformation (Part III)," *Quality Progress*, July, 28–33.

[18] Scholtes, P.R. (1988), *The Team Handbook: How to Use Teams to Improve Quality*, Joiner and Associates, Madison, Wisconsin.

[19] Senge, P. (1990), *The Fifth Discipline: The Art and Practice of the Learning Organization.* Doubleday, New York.

[20] Senge, P. (1993), "Quality Management: Current State of the Practice," Keynote Speech at the American Quality Congress.

TQC

Total quality control. See **Statistical quality control**; **Total quality management**.

TQM

Total quality management. See **Statistical quality control**; **Total quality management**.

TRAFFIC ANALYSIS

Denos C. Gazis

IBM T.J. Watson Research Center, Yorktown Heights, New York

Traffic analysis has flourished over the last thirty-five years or so, stimulated from the need to address the ever growing traffic problems of our cities around the world. In true scientific tradition, it has yielded an understanding of the fundamental characteristics of automobile traffic, which in turn spawned significant contributions in the management and optimization of traffic facilities. In what follows, we shall outline some of the most important developments in one area of traffic analysis, that of traffic flow, including certain associated queueing phenomena. Aspects of control of traffic networks are outside the scope of this article. An encyclopedic review of such aspects has been given by Gazis (1992).

A KINEMATICAL THEORY OF TRAFFIC FLOW: One of the earliest, and most durable, contributions to the understanding of traffic flow

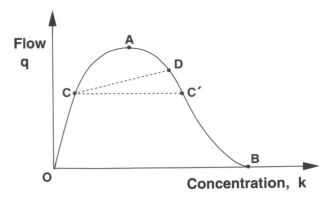

Fig. 1. Flow vs. concentration relationship.

was given by Lighthill and Whitham (1955). They viewed the traffic as a special fluid which obeys some basic laws consistent with the physical nature of traffic, such as its unidirectional influence of a vehicle only on the traffic behind it, the constraints on flow imposed by human limitations, etc. The Lighthill-Whitham theory is based on two basic postulates:

1) Traffic is conserved, in the sense that traffic units by and large are neither created nor annihilated.

2) There is a fundamental relationship between traffic flow and traffic density, resulting from the physical characteristics of the traffic system.

The first postulate is expressed in the relationship

$$\frac{\partial k}{\partial t} + \frac{\partial q}{\partial x} = 0 \tag{1}$$

where q is the traffic flow in vehicles per unit of time t, k is the density of traffic in vehicles per unit of distance x, and v is the (average) speed of the traffic fluid. The second postulate is expressed by the relationship

$$q = f(k) \tag{2}$$

between flow q and density k such as that shown in Figure 1. At zero density, there is zero flow. The flow is also zero at some "jam density," k_j, because traffic grinds to a halt as vehicles are packed bumper to bumper. Between these two extremes, traffic flow builds up to a maximum and then decreases down to zero.

A number of interesting properties of traffic can be described on the basis of these two postulates. They relate to observable phenomena such as "wave propagation," that is, the movement along the traffic stream of a transition point corresponding to a change in traffic characteristics, the "queueing" caused by an obstruction of the traffic movement, etc.

Wave Propagation – Traffic moving at a steady-state flow rate q_1 and density k_1 may shift to a different flow rate q_2, and a corresponding density k_2, by a change in roadway quality, obstruction, or other external influence. When this happens, vehicles situated in a transition region undergo maneuvers adjusting their speed and inter-vehicle spacing, and this transition region generally moves either forward or backward in space depending on the nature of the change. The adjustments of speed and spacing are gradual, but for the purpose of deriving the characteristics of wave propagation may be assumed abrupt as suggested by Lighthill and Whitham (1955). This assumption leads to the conclusion that a change from one steady-state flow condition to another is associated with a "shock wave," an expression now pervading traffic engineering literature.

The shock wave marks the transition from one speed to another, and moves always backwards with respect to the traffic stream, since vehicles exert an influence only on vehicles behind them. (The influence of an occasional tailgating vehicle "pushing" the vehicle in front is ignored as an unimportant aberration.) The speed of movement of the shock wave along a roadway may be obtained on the basis of Equation (1), and is given by

$$v = \frac{q_1 - q_2}{k_1 - k_2}. \tag{3}$$

It should be pointed out that the result given in Equation (3) depends only on the postulate of conservation of traffic, and is totally independent of any specific relationship between flow and concentration, or even on the existence of such a relationship. It results from kinematical considerations shown in Figure 2. The transition from one steady state flow situation to another results in a propagation of the change of the corresponding speed along the roadway. The "phase velocity" of this propagation depends only on the values of the initial and final pairs of flow, q, and concentration, k, and is given by Equation 3. If, in addition, we assume a relationship between flow and concentration (Figure 1), we can define different domains of traffic quality, and corresponding characteristics of wave propagation, as follows:

1) The range from zero flow at zero density to maximum flow (Section OA, Figure 1) corresponds to relatively uncongested traffic flow. A small increase in density in this domain moves forward along the roadway.

2) The range from maximum flow to zero flow at "jam density" (Section AB) corresponds

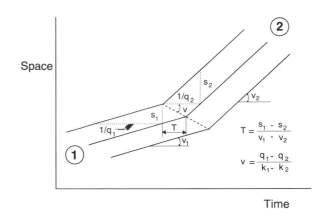

Fig. 2. Transition from one steady-state-flow situation to another.

to relatively congested, stop-and-go traffic. A small increase of density in this domain moves backwards along the roadway.

3) Any transition from one steady state flow to another (as from point C to point D in Figure 1) is associated with a wave propagation given by the slope of segment CD.

Queueing – Queueing may be caused by a reduction in roadway capacity at a fixed point on the roadway, or by an obstruction causing traffic to shift from the "uncongested" to the "congested" branches of the $q - k$ curve, even without reduction in flow rate (line CC′ in Figure 1). The rate of growth of the queue can be estimated using the same methodology described above. For example, a total obstruction of flow q and density k causes a queue formation, with the tail-end of the queue moving backwards along the roadway with speed equal to

$$v = \frac{q}{k_j - k}. \tag{4}$$

Additional results from the kinematic treatment of traffic – An extensive literature exists on applications of the Lighthill-Whitham model to various traffic phenomena. A word of caution is appropriate with regard to such applications. The Lighthill-Whitham model describes well only transitions from one steady-state to another. Any attempt to apply the model to a sequence of traffic maneuvers which do not allow enough relaxation time between changes of speeds violates the basic spirit of the model.

An interesting extension of the above kinematical treatment of traffic was applied by Gazis and Herman (1992) for the treatment of a moving obstruction such as that caused by a vehicle moving more slowly than the other vehicles in the traffic stream. The character of this "moving bottleneck" is different from that of a fixed bottleneck, and the Gazis-Herman treatment derives the characteristic queueing behavior associated with it. Gazis and Herman obtain description of the queueing caused by a slow vehicle on a two-lane highway. Both lanes are affected by such a vehicle, one by direct trapping of vehicles behind the slow one, and the other by interference from vehicles escaping from the queue behind this vehicle. The result is that queueing takes place in both lanes in the vicinity of the slow vehicle, with the affected vehicles moving at an average speed only marginally higher than that of the slow one, until they come abreast of this slow vehicle and are able to escape at their normal speed. Gazis and Herman also propose an explanation of the phenomenon of a "phantom bottleneck," the seemingly unexplainable regions of congestion that drivers often traverse. Some of them may be caused by a moving bottleneck caused by a vehicle that slows down temporarily and then resumes its normal speed; for example, a heavily loaded truck temporarily slowing down along an uphill portion of the roadway. The Gazis-Herman treatment provides a rational way of estimating the minimum allowable speed on a highway, which would not affect its throughput.

A BOLTZMANN-LIKE MODEL OF TRAFFIC FLOW:

In 1959, Prigogine suggested a model of traffic flow founded on statistical mechanics, analogous to the Boltzmann model of gases (Prigogine, 1961). The Prigogine model was subsequently developed extensively by Herman, Prigogine and their collaborators (Prigogine, Herman and Anderson, 1963, 1965; Herman and Prigogine, 1971). They considered a stream of traffic as an "ensemble" of units associated with certain statistical properties. In particular, a vehicle was associated with a "desired speed" which it would follow as long as it was not constrained by another vehicle in front with a lower desired speed.

Thus, traffic is described in terms of a probability density for the speed, v, of an individual car, $f(x,v,t)$. This density may vary as a function of time, t, and a coordinate x along the highway. The basic equation for this function f is assumed to be

$$\frac{\partial f}{\partial t} + v \frac{\partial f}{\partial x} = \left(\frac{\partial f}{\partial t} \right)_{relaxation} + \left(\frac{\partial f}{\partial t} \right)_{interaction} \tag{5}$$

The first term of the right-hand side of Equation 5 is a consequence of the fact that $f(x, v, t)$ differs from some desired speed distribution $f^0(v)$. A car tries to "relax" to its desired speed as soon as it finds an opportunity to do so. The second term of the right-hand side corresponds to the slowing

down of a fast vehicle by a slow one. True to his tradition as a leading expert in statistical mechanics, Prigogine frequently referred to this second term as the "collision" term – a rather unsettling choice of words in this context!

The form for these two terms was chosen for mathematical convenience and plausibility, leading to the equation

$$\frac{\partial f}{\partial t} + v \frac{\partial f}{\partial x} = \frac{f - f^0}{\tau} + (1 - p)k(V - v)f \tag{6}$$

where τ is a characteristic relaxation time, p is the probability of a car's passing another car, and V is the average speed of the stream of traffic. The second term of the right-hand side of Equation 6 corresponds to the interaction term, and tends to zero at very light traffic concentration when the probability of passing is close to unity, in which case the relaxation term is dominant. If, in addition, we assume a highway with constant properties along its length, then $\partial f / \partial x = 0$ and the solution of Equation 6 is

$$f(v,t) = f^0(v) + [f(v,0) - f^0(v)]e^{-t/\tau}. \tag{7}$$

If we are interested only in solutions of Equation 6 which are independent of time and space, then the left-hand side of this equation is zero. The equation may then be solved to yield an equation of state whose general form, for small values of the concentration, corresponds to an approximately linear increase of flow with concentration, for example:

$$q = V^0 k \tag{8}$$

where V^0 is the average of the desired speed. As k increases, the flow q falls below the straight line (Equation 8) due to the increasing influence of interactions.

In the range of high concentrations, q is independent of f^0 and depends only on τ and p, according to the equation

$$q = \frac{1}{\tau(1 - p)}. \tag{9}$$

The complete solution of Equation 6 for steady-state flow, independent of time and space, is shown in Figure 3. For any given f^0, the flow q rises with k, reaches a maximum, and then decreases until it intersects a curve corresponding to Equation 9. This curve may be viewed as a universal curve of "collective flow," characterized by high densities and very little passing. One very realistic feature of this theory is the fact that it predicts probable stoppage of some vehicles in the domain of collective flow, in agreement with the common experience of stop-and-go traffic at high concentrations.

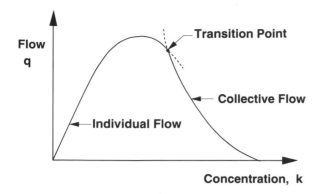

Fig. 3. Flow vs. concentration relationship according to the Boltzmann-like model of traffic flow.

Herman and Prigogine, together with several collaborators, went on to use the results of their model to develop a "two-fluid approach" to town traffic. This approach postulates that traffic in towns is a mixture of two fluids, one that moves and one that is stopped. Any individual vehicle traverses a network in a stop-and-go fashion, moving part of the time and being stopped part of the time. The quality of service in a particular urban network can be described in terms of two parameters which can be determined by circulating a test vehicle through the network and measuring the percentages of time during which the vehicle is moving, or is being stopped. Thus the two-fluid model yields a simple description of the system-wide traffic quality in congested urban networks. It allows comparison between different urban networks, and it offers the potential of identifying important elements of the network, related to its geometry or control features, which may be targeted for improvement of the service quality.

A CAR-FOLLOWING THEORY OF TRAFFIC FLOW:

Reuschel (1950) and Pipes (1953) proposed models to describe the detailed motion of cars proceeding close together in a single lane. This microscopic, car-following theory of traffic flow was extensively developed by Herman, Montroll *et al.* (1959). The theory is based on the fact that when drivers do not have the freedom to pass a vehicle in front, they follow it in a way that is controlled by the overriding need to avoid coinciding with the leader in space and time. In trying to achieve this reasonable objective, drivers react to a limited set of inputs. The postulate of the car-following theory, confirmed by experiments, was that drivers reacted mostly to the relative speed between their car and that of the one in front. Experiments showed a high correlation between the acceleration of a car and its speed relative to that of a leader, after a time-lag

of the order of 1 second. This led to the "linear car-following model"

$$\frac{d^2 x_n(t+T)}{dt^2} = \lambda \left[\frac{dx_{n-1}(t)}{dt} - \frac{dx_n(t)}{dt} \right], \qquad (10)$$

in which n denotes the position of a car in a line of cars (a platoon), λ is a constant "gain factor," T is the reaction time-lag, and x_n is the position of the nth car on the highway.

This model was used to investigate the *stability* of a traffic platoon when a perturbation in its movement is introduced. The movement of the platoon is said to be "locally stable" if the amplitude of a perturbation, for any given car in the platoon, decreases in time. It is "asymptotically stable" if the amplitude of the perturbation decreases as it propagates upstream. The value of the product λT is the determinant of stability or instability, local or asymptotic. When $\lambda T < 1/e$, a perturbation is damped exponentially as it is passed on to the following car, signifying a very stable situation. For λT between the values of $1/e$ and $\pi/2$, the perturbation produces oscillations of decreasing amplitude between pairs of cars, signifying still a locally stable situation. For $\lambda T > \pi/2$, a perturbation produces oscillations of increasing amplitude, signifying a locally unstable situation.

With regard to asymptotic stability, the dividing line is at $\lambda T = 1/2$. For values of λT below $1/2$, the amplitude of a perturbation decreases as it propagates backwards; for values of λT greater than $1/2$, it increases. This means that between $1/e$ (~ 0.368) and $1/2$ we can have a situation which is locally stable but asymptotically unstable. Any pair of cars in a platoon is able to absorb a perturbation, but it amplifies it as it passes it backwards, until the perturbation is so large that it causes a collision.

The linear car-following model may be satisfactory in describing fluctuations around a steady-state, constant speed situation. It cannot be expected to describe equally well transitions from one steady-state to another involving large changes of speed. For this reason, Gazis *et al.* (1961) proposed a nonlinear model in which the gain factor is not constant but depends on the speed of the follower and the relative spacing between leader and follower according to the relationship

$$\lambda = \frac{[v_n(t+T)]^l}{[x_{n+1}(t) - x_n(t)]^m} \qquad (11)$$

where c is a constant, $v = dx/dt$ is the speed, and (l,m) are integer exponents identifying particular nonlinear models.

Various values of pairs (l,m) were used to define car-following models and investigate their predictions concerning transitions between one steady-state flow situation and another. By integrating over time, Equation 10, with λ described by Equation 11, we obtain the functional relationship between changes of speed and concentration. Together with appropriate "boundary conditions," for example the condition of zero speed at "jam density," bumper-to-bumper concentration, we can then obtain a phenomenological relationship between flow and concentration such as that shown in Figure 1. Various pairs (l,m) have been used which yielded quite plausible relationships, consistent with observations.

The preceding discussion outlines most of the key contributions in the car-following treatment of traffic flow. Additional studies have been contributed by Gazis (1965), within the framework of control theory, in order to account for physical constraints on the system, such as limited acceleration or deceleration capability of cars.

CONCLUDING REMARKS: The analytical description of traffic flow has already had a profound influence on traffic engineering practice, and promises to have an even greater influence in the future. The advent of activities in the area of *Intelligent Vehicle-Highway Systems (IVHS)*, points to an increasing reliance on analytical investigations of traffic systems. One needs an improved understanding, and an improved analytical description of traffic phenomena such as the onset of congestion, queueing, and intervehicle signal propagation, in order to create the theoretical underpinning toward the use of high technology for the improvement of traffic systems, which is the central thrust of IVHS. Some improvement will come from direct application of analytical results. For example, the development of automatic highways will undoubtedly draw from knowledge based on car-following models. Other improvements may come from the improved understanding of traffic phenomena that traffic analysis provides, leading to improved heuristic schemes for the control and optimization of traffic systems.

See **Networks; Queueing theory.**

References

[1] Anderson, R.L., R. Herman, and I. Prigogine (1962), "On the Statistical Distribution Function Theory of Traffic Flow," *Operations Research*, 10, 180–196.

[2] Ardekani, S.A. and R. Herman (1985), "A Comparison of the Quality of Traffic Service in Downtown Networks of Various Cities around the World," *Traffic Engineering and Control*, 26, 574–581.

[3] Ardekani, S.A. and R. Herman (1987), "Urban Network–Wide Traffic Variables and Their Relations," *Transp. Science*, 21, 1–16.

[4] Bick, J.H. and Newell, G.F. (1960), "A Continuum Model for Two-directional Traffic Flow," *Quart. Appl. Math.*, 18, 191–204.

[5] Chandler, R.E., R. Herman, and E.W. Montroll (1958), "Traffic Dynamics: Studies in Car-Following," *Oper. Research*, 6, 165–184.

[6] Chang, M.-F. and R. Herman (1981), "Trip Time versus Stop Time and Fuel Consumption Characteristics in Cities," *Transportation Science*, 15, 183–209.

[7] Edie, L.C. and R.S. Foote (1960), "Effect of Shock Waves on Tunnel Traffic Flow," *Proc. Highway Research Board*, 39, 492–505.

[8] Edie, L.C., R. Herman, and T.N. Lam (1980), "Observed Multilane Speed Distribution and the Kinetic Theory of Vehicular Traffic," *Transportation Science*, 14, 55–76.

[9] Foster, J. (1962), "An Investigation of the Hydrodynamic Model for Traffic Flow with Particular Reference to the Effect of Various Speed-Density Relationships," *Proc. Australian Road Research Board*, 1, 229–257.

[10] Gazis, D.C. (1965), "Control Problems in Automobile Traffic," *Proc. IBM Scientific Symposium on Control Theory and Applications*, IBM Yorktown Heights, New York. 171–185.

[11] Gazis, D.C. (1992), "Traffic Modelling and Control: Store and Forward Approach," in *Concise Encyclopedia on Traffic and Transportation*, Markos Papageorgiou (ed.), Pergamon Press, New York, 278–284.

[12] Gazis, D.C. and R. Herman (1992), "The Moving and Phantom Bottlenecks," *Transp. Science*, 26, 223–229.

[13] Gazis, D.C., R. Herman, and R.B. Potts (1959), "Car-Following Theory of Steady-State Traffic Flow," *Oper. Research*, 7, 499–505.

[14] Gazis, D.C., R. Herman, and R.W. Rothery (1961), "Nonlinear Folow-the-Leader Models of Traffic Flow," *Oper. Research*, 9, 546–567.

[15] Greenberg, H. (1959), "An Analysis of Traffic Flow," *Operations Research*, 7, 79–85.

[16] Herman, R. and S.A. Ardekani (1984), "Characterizing Traffic Conditions in Urban Areas," *Transp. Science*, 18, 101–140.

[17] Herman, R. and R.B. Potts (1961), "Single-Lane Traffic Theory and Experiment," *Proc. 1st Intern. Symp. on the Theory of Traffic Flow*, R. Herman (ed.), Elsevier, 120–146.

[18] Herman, R. and I. Prigogine (1979), "A Two-Fluid Approach to Town Traffic," *Science*, 204, 148–151.

[19] Herman, R., E.W. Montroll, R.B. Potts, and R.W. Rothery (1959), "Traffic Dynamics: Analysis of Stability in Car Following," *Oper. Research*, 7, 86–106.

[20] Leutzbach, W. (1967), "Testing the Applicability of the Theory of Continuity on Traffic Flow at Bottlenecks," *Proc. 3rd Intern. Symposium on Theory of Traffic Flow*, L. C. Edie, R. Herman, and R.W. Rothery (eds.), Elsevier, 1–13.

[21] Lighthill, M.J. and G.B. Whitham (1955), "On Kinematic Waves: II. A Theory of Traffic Flow on Long Crowded Roads," *Proc. Royal Soc. (London)*, **A229**, 317–345.

[22] Makigami, Y., G.F. Newell, and R.W. Rothery (1971), "Three-dimensional Representations of Traffic Flow," *Transportation Science*, **5**, 302–313.

[23] Newell, G.F. (1965), "Instability in Dense Highway Traffic, a Review," *Proc. 2nd Intern. Symp. on Theory of Traffic Flow*, Joyce Almond (ed.), OECD, 73–83.

[24] Newell, G.F. (1991), "A Simplified Theory of Kinematic Waves," *Research Report UCB-ITS-RR-91-12*, University of California at Berkeley.

[25] Pipes, L.A. (1953), "An Operational Analysis of Traffic Dynamics," *Jl. Appl. Physics*, **24**, 274–281.

[26] Prigogine, I. (1961), "A Boltzmann-like Approach to the Statistical Theory of Traffic Flow," *Proc. 1st Intern. Symposium on the Theory of Traffic Flow*, R. Herman (ed.), Elsevier, 158–164.

[27] Prigogine, I. and F.C. Andrews (1960), "A Boltzmann-like Approach for Traffic Flow," *Operations Research*, **8**, 789–797.

[28] Prigogine, I. and R. Herman (1971), *Kinetic Theory of Vehicular Traffic*, American Elsevier, New York.

[29] Prigogine, I., R. Herman, and R.L. Anderson (1962), "On Individual and Collective Flow," *Acad. Roy. Belgique–Bull. de la Classe des Sciences*, **48**, 792–804.

[30] Prigogine, I., R. Herman, and R.L. Anderson (1965), "Further Developments in the Boltzmann-like Theory of Traffic Flow," *Proc. 2nd Intern. Symp. on the Theory of Traffic Flow*, Joyce Almond (ed.), OECD, 129–138.

[31] Prigogine, I., P. Resibois, R. Herman, and R.L. Anderson (1962), "On a Generalized Boltzmann-like Approach for Traffic Flow," *Acad. Roy. Belgique-Bull. de la Classe des Sciences*, **48**, 805–814.

[32] Reuschel, A. (1950), "Fahrzeugbewegungen in der Kolonne bei gleichfoermig beschleunigtem oder verzoegertem Leitfahrzeug," *Zeit. d. oesterreichischen Ing. u. Arch. Vereins*, **95**, 73–77.

[33] Underwood, R.T. (1962), "Some Aspects of the Theory of Traffic Flow," *Proc. Australian Road Research Board*, **1**, 35.

[34] Underwood, R.T. (1964), "Traffic Flow Models," *Traffic Eng. Control*, **5**, 699–701.

TRAFFIC EQUATIONS

In a queueing network, the balancing of flow into a node with the flow out. These are linear equations derived by recognizing that the total input seen at a node comes from the sum of fresh arrivals from outside the network and arrivals coming rerouted departures from all of the nodes within the network:

$$\lambda_i = \gamma_i + \sum_j \lambda_j r_{ij},$$

where λ_i is the total input flow rate seen at node i, γ_i is the external input rate to node i, and r_{ij} is the probability that a service completion at node i is routed to node j. See **Conservation of flow**; **Networks of queues**; **Queueing theory**.

TRAFFIC INTENSITY

The average load offered to each server in a queueing system. See **Offered load**; **Queueing theory**.

TRAFFIC PROCESS

A stochastic point or marked point process representing the flow of customers on the arcs of a queueing network. Marks represent some aspect of the customer or the state of the network and the points represent the epoch of the event. See **Arrival process**; **Departure process**; **Input process**; **Networks of queues**; **Output process**.

TRANSFER FUNCTION

See **Time series**.

TRANSIENT ANALYSIS

The time-dependent solution of a stochastic system (as in queueing theory). See **Queueing theory**.

TRANSITION FUNCTION

A function describing the transition probabilities of a Markov process $X(t)$ into a subset A of the state space as $p(s,x;t,A) = \Pr\{X(t) \in A | X(s) = x\}$. See **Markov chains**; **Markov processes**.

TRANSITION MATRIX

The matrix of (single-step) transition probabilities of a Markov chain $\{X_n\}$, $\mathbf{P} = (p_{i,j})$, where $p_{i,j} = \Pr\{X_{n+1} = j | X_n = i\}$ is the conditional probability that the chain moves to state j from state i in one step. See **Markov chains**; **Markov processes**.

TRANSITION PROBABILITIES

The conditional probabilities describing the movement from state to state of a Markov process $\{X(t), t \in T\}$. In general, the transition probabilities are written as $\Pr\{X(t) \in A | X(s) = x\}$ for times $s < t$ in the time domain T and state x and event A in the state space. For a discrete-time Markov chain (DTMC) $\{X_n, 0 \le n\}$ the transition probabilities are $\Pr\{X_{n+1} = j | X_n = i\} = p_{ij}$, for time n in the time domain and states i and j in the state space. See **Markov chains**; **Markov processes**.

TRANSPORTATION PROBLEM

A linear-programming problem of the following form is called a transportation problem:

Minimize $\sum_i \sum_j c_{ij} x_{ij}$

subject to

$$\sum_j x_{ij} = a_i, \; i = 1, \ldots, m \text{ (origins/supply)}$$

$$\sum_i x_{ij} = b_j, \; j = 1, \ldots, n \text{ (destinations/demand)}$$

$$x_{ij} \ge 0.$$

The variables x_{ij} represent a shipment of a homogeneous product from origin i to destination j, where the a_i are the amounts of the product to be shipped from the origins i, and the b_j are the amounts demanded by the destinations j. Here we restrict $\Sigma_i a_i = \Sigma_j b_j$. It can be shown that if the a_i and b_j are integers, then an optimal basic feasible solution exists that is all integer. The transportation problem is a special network problem whose network representation is called a bipartite graph. The special case with $m = n$ and all $\{a_i\}$ and $\{b_j\}$ equal to 1 is the assignment problem. A transportation problem can be solved by direct application of the simplex method, but due to its mathematical structure, the problem can be solved by an efficient modification of the simplex method called the transportation (primal-dual) simplex method. It can also be solved by specialized network algorithms. See **Assignment problem**; **Network optimization**; **Northwest corner solution**; **Transportation (primal-dual) simplex method**; **Unbalanced transportation problem**.

TRANSPORTATION PROBLEM PARADOX

Some transportation problems exhibit the paradox that an optimal solution can be improved if the total amount of units shipped is more than the total amount shipped by the optimal solution. That is, one can ship more for less.

TRANSPORTATION SIMPLEX (PRIMAL-DUAL) METHOD

The dual problem to the primal equation form of the transportation problem can be stated as follows:

Maximize $\sum_i a_i u_i + \sum_j b_j v_j$

subject to

$$u_i + v_j \le c_{ij} \text{ for all } (i, j).$$

Here the $(m + n)$ set of dual variables u_i and v_j are unrestricted (free) variables. Note that the primal has a redundant equation due to the equality of the total supply and demand. Thus, a feasible basis matrix to the transportation problem is of order $(m + n - 1) \times (m + n - 1)$. It can be shown that any feasible basis matrix can be arranged into a triangular form. For a given basis, the simplex method requires that the corresponding dual constraints must hold at equality, that is, we must have $u_i + v_j = c_{ij}$ for all variables x_{ij} in the basis. This $(m + n - 1) \times (m + n)$ set of dual equations can be reduced to an $(m + n - 1) \times (m + n - 1)$ system by arbitrarily setting one of the dual variables, say, $u_1 = 0$. This corresponds to removing, as a redundant constraint, the first equation of the transportation problem. The resulting dual square set of equations also has a triangular form that allows for the efficient calculation of the u_i and v_j that correspond to the current basic solution. Using these values of u_i and v_j, we calculate the $(u_i + v_j)$ terms for the nonbasic variables and, if each one is less than or equal to its corresponding c_{ij}, then by duality theory and complementary slackness, the current basis is optimal. If the latter condition does not hold, then the usual simplex criterion is used to select a variable to enter the basis and a new basic feasible solution is generated by simple adjustments to the flows in the network that describe the current basic feasible solution. This network is a tree that connects all origins and destinations, and the addition of the new variable (or arc to the tree) enables the new solution to be calculated readily. This primal-dual process is repeated until an optimal solution is found. Such a solution exists as the transportation problem always has feasible solutions and the solution set is bounded. See **Network optimization**; **Transportation problem**.

TRANSPOSITION THEOREMS

Transposition theorems deal with disjoint alternatives of solvability of linear systems. For example, the transposition theorem of Stiemke is: For a matrix $A \neq 0$, the following statements are equivalent: (1) $Ax = 0$, $x > 0$, has no solution, and (2) $uA \leq 0$, $uA \neq 0$ has a solution. See **Farkas' lemma; Gordan's theorem; Strong duality theorem; Theorem of alternatives**.

TRANSSHIPMENT PROBLEM

See **Minimum-cost network flow problem; Network optimization**.

TRAVELING SALESMAN PROBLEM

Karla L. Hoffman

George Mason University, Fairfax, Virginia

Manfred Padberg

New York University

INTRODUCTION: The *traveling salesman problem* (TSP) is one which has commanded much attention of mathematicians and computer scientists specifically because it is so easy to describe and so difficult to solve. The problem can simply be stated as: if a traveling salesman wishes to visit exactly once each of a list of m cities (where the cost of traveling from city i to city j is c_{ij}) and then return to the home city, what is the least costly route the traveling salesman can take? A complete historical development of this and related problems can be found in Hoffman and Wolfe (1985).

The importance of the TSP is that it is representative of a larger class of problems known as *combinatorial optimization problems*. The TSP problem belongs in the class of combinatorial optimization problems known as NP-complete. Specifically, if one can find an efficient (i.e., polynomial-time) algorithm for the traveling salesman problem, then efficient algorithms could be found for all other problems in the NP-complete class. To date, however, no one has found a polynomial-time algorithm for the TSP. Does that mean that it is impossible to solve *any* large instances of such problems? Many practical optimization problems of truly large scale are solved to optimality routinely. Recently a traveling salesman problem which models the production of printed circuit boards having 7,397 holes (cities) was solved to proven optimality, Applegate *et al.* (1994). Randomly generated problems having 10,000 cities have been solved using similar technology. So, although the question of what it is that makes a problem "difficult" may remain open, the computational record of specific instances of TSP problems coming from practical applications is optimistic.

How are such problems tackled today? Obviously, one cannot consider a brute force approach. In one example of a 16 city traveling salesman problem – the problem of Homer's Ulysses attempting to visit the cities described in *The Odyssey* exactly once – there are 653,837,184,000 distinct routes (Grötschel and Padberg, 1993)! Enumerating all such round-trips to find a shortest one took 92 hours on a powerful workstation. Rather than enumerating all possibilities, successful algorithms for solving the TSP problem have been capable of eliminating most of the round-trips without ever explicitly considering them.

FORMULATIONS: The first step to solving instances of large TSPs must be to find a good mathematical formulation of the problem. In the case of the traveling salesman problem, the mathematical structure is a graph where each city is denoted by a point (or node) and lines are drawn connecting every two nodes (called arcs or edges). Associated with every line is a distance (or cost). When the salesman can get from every city to every other city directly, then the graph is said to be *complete*. A round-trip of the cities corresponds to some subset of the lines, and is called a tour or a Hamiltonian cycle in graph theory. The length of a tour is the sum of the lengths of the lines in the round-trip.

Depending upon whether or not the direction in which an edge of the graph is traversed matters, one distinguishes the *asymmetric* from the *symmetric* traveling salesman problem. To formulate the *asymmetric* TSP on m cities, one introduces zero-one variables

$$x_{ij} = \begin{cases} 1 & \textit{if the edge } i \rightarrow j \textit{ is in the tour} \\ 0 & \textit{otherwise} \end{cases}$$

and given the fact that every node of the graph must have exactly one edge pointing towards it and one pointing away from it, one obtains the classic assignment problem. These constraints alone are not enough since this formulation would allow "subtours," that is, it would allow disjoint loops to occur. For this reason, a proper formulation of the asymmetric traveling salesman problem must remove these subtours from consideration by the addition of "subtour elimination" constraints. The problem then becomes

$$\min \sum_{j=1}^{m} \sum_{i=1}^{m} c_{ij} x_{ij}$$

$$s.t. \quad \sum_{j=1}^{m} x_{ij} = 1 \qquad \text{for } i = 1, \ldots, m$$

$$\sum_{i=1}^{m} x_{ij} = 1 \qquad \text{for } j = 1, \ldots, m$$

$$\sum_{i \in K} \sum_{j \in K} x_{ij} \leq |K| - 1 \quad \text{for all } K \subset \{1, \ldots, m\}$$

$$x_{ij} = 0 \text{ or } 1 \qquad \text{for all } i,j$$

where K is any nonempty proper subset of the cities $1, \ldots, m$. The cost c_{ij} is allowed to be different from the cost c_{ji}.

To formulate the *symmetric* traveling salesman problem, one notes that the direction traversed is immaterial, so that $c_{ij} = c_{ji}$. Since direction does not now matter, one can consider the graph where there is only one arc (undirected) between every two nodes. Thus, we let $x_j \in \{0,1\}$ be the decision variable where j runs through all edges E of the undirected graph and c_j is the cost of traveling that edge. To find a tour in this graph, one must select a subset of edges such that every node is contained in exactly two of the edges selected. Thus, the problem can be formulated as a 2-matching problem in the graph G. As in the asymmetric case, subtours must be eliminated through subtour elimination constraints. The problem can therefore be formulated as:

$$\min 1/2 \sum_{j=1}^{m} \sum_{k \in J(j)} c_k x_k$$

$$s.t. \quad \sum_{k \in J(j)} x_k = 2 \qquad \text{for all } j = 1, \ldots, m$$

$$\sum_{j \in E(K)} x_j \leq |K| - 1 \quad \text{for all } K \subset \{1, \ldots, m$$

$$x_j = 0 \text{ or } 1 \qquad \text{for all } j \in E,$$

where $J(j)$ is the set of all undirected edges connected to node j and $E(K)$ is the subset of all undirected edges connecting the cities in any proper, nonempty subset K of all cities. Of course, the symmetric problem is a special case of the asymmetric one, but practical experience has shown that algorithms for the asymmetric problem perform, in general, badly on symmetric problems. Thus, the latter need a special formulation and solution treatment.

ALGORITHMS: Exact approaches to solving such problems require algorithms that generate both a lower bound and an upper bound on the true minimum value of the problem instance. Any round-trip tour that goes through every city exactly once is a feasible solution with a given cost which cannot be smaller than the minimum cost tour. Algorithms that construct feasible solutions, and thus upper bounds for the optimum value, are called *heuristics*. These solution strategies produce answers but without any quality guarantee as to how far off they may be from the optimal answer. Heuristic algorithms that attempt to find feasible solutions in a single attempt are called *constructive heuristics*, while algorithms that iteratively modify and try to improve some given starting solution are called *improvement heuristics*. When the solution one obtains is dependent on the initial starting point of the algorithm, the same algorithm can be used multiple times from various (random) starting points. An excellent survey of *randomized improvement heuristics* is given in Jünger, Reinelt and Rinaldi (1994). Often, if one needs a solution quickly, one may settle for a well-designed heuristic algorithm that has been shown empirically to find "near-optimal" tours to many TSP problems. Research by Johnson (1990), and Jünger, Reinelt and Rinaldi (1994) describes algorithms that find solutions to extremely large TSPs (problems with tens of thousands, or even millions of variables) to within 2% of optimality in very reasonable times. Performance guarantees for heuristics are given in Johnson and Papadimitriou (1985); probabilistic analysis of heuristics are discussed in Karp and Steele (1985); and the development and empirical testing of heuristics is reported in Golden and Stewart (1985).

In order to know about the closeness of the upper bound to the optimum value, one must also know a lower bound on the optimum value. If the upper and lower bound coincide, a proof of optimality is achieved. If not, a conservative estimate of the true relative error of the upper bound is provided by the difference of the upper and the lower bound divided by the lower bound. Thus, one needs both upper and lower bounding techniques to find provably optimal solutions to hard combinatorial problems or even to obtain solutions meeting a quality guarantee.

So how does one obtain and improve the lower bound? A *relaxation* of an optimization problem is another optimization problem whose set of feasible solutions properly contains all feasible solution of the original problem and whose objective function value is less than or equal to the true objective function value for points feasible to the original problem. Thus, we replace the "true" problem by one with a larger feasible region that is more easily solvable. This relaxation is continually refined so as to tighten the feasible region so that it more closely represents the true problem. The standard technique for obtaining lower bounds on the TSP problem is to use a relaxation that is easier to solve than the original problem. These relaxations can have either discrete or continuous feasible sets. Several relaxations have been considered for the TSP. Among them are the *n-path relaxation, the assignment relaxation, the 2-matching relaxation, the 1-tree relaxation,* and the *linear programming relaxation.* For randomly generated asymmetric TSPs, problems having up to 7500 cities have been solved using an assignment relaxation which adds subtours within a branch and bound framework and which uses an upper bounding heuristic based on subtour patching (Miller and Pekny, 1991). For the symmetric TSP, the 1-tree relaxation and the 2-matching relaxations have been most successful. These relaxations have been embedded into a branch-and-cut framework.

The process of finding constraints that are violated by a given relaxation, is called a *cutting plane* technique and all successes for large TSP problems have used cutting planes to continuously tighten the formulation of the problem. It is important to stress that all successful computational approaches to the TSP utilize *facet-defining* inequalities as cutting planes. General-type cutting planes of the integer programming literature that use the simplex basis-representation to obtain cuts, such as Gomory or intersection cuts, have long been abandoned because of poor convergence properties. One of the simplest cuts that have been shown to define *facets* of the underlying TSP polytope are the subtour elimination cuts. Besides these constraints, comb inequalities, clique tree inequalities, path, wheelbarrow and bicycle inequalities, ladder inequalities and crowns have also been shown to define facets of this polytope. The underlying theory of facet generation for the symmetric traveling salesman problem is provided in Grötschel and Padberg (1985) and Jünger, Reinelt and Rinaldi (1994). The algorithmic descriptions of how these are used in cutting plane approaches is discussed in Padberg and Rinaldi (1992) and Jünger, Reinelt and Rinaldi (1994). Cutting plane procedures can then be embedded into a tree search referred to as *branch and cut.* Some of the largest TSP problems solved have used parallel processing to assist in the search for optimality. As our understanding of the underlying mathematical structure of the TSP problem improves, and with the continuing advancement in computer technology, it is likely that many difficult and important combinatorial optimization problems will be solved using a combination of cutting plane generation procedures, heuristics, variable fixing through logical implications and reduced costs, and tree search.

APPLICATIONS: One might ask, however, whether the TSP problem is important enough to have received all of the attention it has. Besides being a "polytope" of a difficult combinatorial optimization problem from a complexity theory point of view, there are important cases of practical problems that can be formulated as TSP problems and many other problems are generalizations of this problem. Besides the drilling of printed circuit boards described above, problems having the TSP structure occur in the analysis of the structure of crystals (Bland and Shallcross, 1987), the overhauling of gas turbine engines (Pante, Lowe and Chandrasekaran, 1987), in material handling in a warehouse (Ratliff and Rosenthal, 1981), in cutting stock problems (Garfinkel, 1977), the clustering of data arrays (Lenstra and Rinooy Kan, 1975), the sequencing of jobs on a single machine (Gilmore and Gomory, 1964) and the assignment of routes for planes of a specified fleet (Boland, Jones, and Nemhauser, 1994). Related variations on the traveling salesman problem include the resource constrained traveling salesman problem which has applications in scheduling with an aggregate deadline (Pekny and Miller, 1990). This paper also shows how the prize collecting traveling salesman problem (Balas, 1989) and orienteering problem (Golden, Levy and Vohra, 1987) are special cases of the resource constrained TSP. Most importantly, the traveling salesman problem often comes up as a subproblem in more complex combinatorial problems, the best known and important one of which is the vehicle routing problem. This is the problem of determining for a fleet of vehicles which customers should be served by each vehicle and in what order each vehicle should visit the customers assigned to it. For relevant surveys are given in Christofides (1985) and Fisher (1987).

Basic references are the seminal paper on the problem by Dantzig, Fulkerson and Johnson (1974) and the text *The TSP: A Guided Tour of Combinatorial Optimization,* edited by Lawler, Lenstra, Rinnooy Kan and Shmoys (1985), which summarizes most of the research up through 1984; and the excellent survey by Jünger, Reinelt and Rinaldi (1994). A problem library containing numerous test problems is available electronically (Reinelt, 1991) and is distributed by Rice University, Houston, Texas.

See **Assignment problem; Branch and bound; Chinese postman problem; Combinatorics; Combinatorial and integer optimization; Computational complexity; Graph theory; Heuristic procedure; Linear programming; Networks; Vehicle routing.**

References

[1] D. Applegate, R.E. Bixby, V. Chvatal, and W. Cook (1994). "Finding cuts in the TSP" a preliminary report distributed at The Mathematical Programming Symposium, Ann Arbor, Michigan, August 1994.

[2] E. Balas (1989). "The Prize Collecting Traveling Salesman Problem," *Networks* **19**, 621–636.

[3] R.E. Bland and D.F. Shallcross (1987). "Large Traveling Salesman Problem Arising from Experiments in X-ray Crystallography: a Preliminary Report on Computation," Technical Report No. 730, School of OR/IE, Cornell University, Ithaca, New York.

[4] N. Christofides (1985). "Vehicle Routing," in *The Traveling Salesman Problem*, Lawler, Lenstra, Rinooy Kan and Shmoys, eds., John Wiley, 431–448.

[5] G.B. Dantzig, D.R. Fulkerson and S.M. Johnson (1954). "Solution of a Large-scale Traveling Salesman Problem," *Operations Research* **2**, 393–410.

[6] M.L. Fisher (1988). "Lagrangian Optimization Algorithms for Vehicle Routing Problems," *Operational Research* '87, G.K. Rand, ed., 635–649.

[7] B.L. Golden, L. Levy and R. Vohra (1987). "The Orienteering Problem," *Naval Research Logistics* **34**, 307–318.

[8] B.L. Golden and W.R. Stewart (1985). "Empirical Analysis of Heuristics," in *The Traveling Salesman Problem*, Lawler, Lenstra, Rinooy Kan and Shmoys, eds., John Wiley, 207–250.

[9] M. Grötschel, C. Monma and M. Stoer (1991). "Polyhedral Approaches to Network Survivability," DIMACS Series in Discrete Mathematics and Theoretical Computer Science, Volume 5, Amer. Math. Soc., 121–141.

[10] M. Grötschel, M.W. Padberg (1985). "Polyhedral Theory," in *The Traveling Salesman Problem*, Lawler, Lenstra, Rinooy Kan and Shmoys, eds., John Wiley, 251–306.

[11] M. Grötschel and M. Padberg (1993). "Ulysses 2000: In Search of Optimal Solutions to Hard Combinatorial Problems," Technical Report, New York University Stern School of Business.

[11] A.J. Hoffman and P. Wolfe (1985), "History" in *The Traveling Salesman Problem*, Lawler, Lenstra, Rinooy Kan and Shmoys, eds., Wiley, 1–16.

[12] D.S. Johnson and C.H. Papadimitriou (1985). "Performance Guarantees for Heuristics," in *The Traveling Salesman Problem*, Lawler, Lenstra, Rinooy Kan and Shmoys, eds., John Wiley, 145–180.

[13] D.S. Johnson (1990). "Local Optimization and the Traveling Salesman Problem," *Proc. 17th Colloquium on Automata, Languages and Programming*, Springer Verlag, 446–461.

[14] M. Jünger, G. Reinelt and G. Rinaldi (1994). "The Traveling Salesman Problem," Technical Report 375, Instituto di Analisis dei Sistemi ed Informatica. Rome, Italy.

[15] R. Karp and J.M. Steele (1985). "Probabilistic Analysis of Heuristics," in *The Traveling Salesman Problem*, Lawler, Lenstra, Rinooy Kan and Shmoys, eds., John Wiley, 181–205.

[16] E.L. Lawler, J.K. Lenstra, A.H.G. Rinooy Kan, and D.B. Shmoys, eds. (1985). *The Traveling Salesman Problem*, John Wiley, Chichester.

[17] D. Miller and J. Pekny (1991). "Exact Solution of Large Asymmetric Traveling Salesman Problems," *Science* **251**, 754–761.

[18] M.W. Padberg and M. Grötschel (1985). "Polyhedral Computations," in *The Traveling Salesman Problem*, Lawler, Lenstra, Rinooy Kan and Shmoys, eds., John Wiley, 307–360.

[19] M.W. Padberg and G. Rinaldi (1991). "A Branch and Cut Algorithm for the Resolution of Large-scale Symmetric Traveling Salesmen Problems," *SIAM Review* **33**, 60–100.

[20] H.D. Ratliff and A.S. Rosenthal (1981). "Order-Picking in a Rectangular Warehouse: A Solvable Case for the Traveling Salesman Problem," PDRC Report Series No. 81-10. Georgia Institute of Technology, Atlanta, Georgia.

[21] G. Reinelt (1991) "TSPLIB-A traveling salesman library," *ORSA Journal on Computing* **3**, 376–384.

TREE

In a network, a tree is a subnetwork (graph) that has no cycles and connects all nodes of a subnetwork, that is, a unique path exists between each node. A tree that connects all n nodes of a network is called a spanning tree and has $(n-1)$ arcs. See **Minimal spanning tree; Network optimization.**

TRIANGULAR MATRIX

A square matrix $A = (a_{ij})$ such that either all the elements a_{ij} above the diagonal are zero or all the elements below the diagonal are zero. The former is called a lower triangular matrix and the latter an upper triangular matrix.

TRIM PROBLEM

The problem deals with determining how rolls or sheets of material should be cut to minimize the amount of wasted material (trim) while meeting the demand for different sizes of cuts. The problem originally arose in the context of cutting large rolls of newsprint into desired smaller sizes. The trim problem can be formulated and solved as a linear- or integer-programming problem. It was the problem that motivated column generation procedures. See **Column generation**.

TRIVIAL SOLUTION

For the homogeneous linear equations $Ax = 0$, the solution $x = 0$ is called a trivial solution. See **Nontrivial solution**; **Null space**.

TRUCK DISPATCHING

The dynamic assignment of trucks (drivers) to loads and/or customers. See **Logistics**; **Vehicle routing**.

TRUCKLOAD (TL) SHIPMENT

A shipment weighing at least the minimum weight to qualify for a TL-size rate reduction. See **Logistics**.

TS

See **Tabu search**.

TSP

See **Traveling salesman problem**.

TUCKER TABLEAU

A reduced simplex tableau of a linear-programming problem that considers the tableau as a representation of both the primal and dual problems.

TWO-PHASE SIMPLEX METHOD

Any version of the simplex method that requires the finding of a first basic feasible solution using artificial variables (Phase I) and then the finding of an optimal feasible solution (Phase II). See **Artificial variables**; **Phase I procedure**; **Phase II procedure**.

U

UNARY NP-COMPLETE (NP-HARD)

See **Computational complexity**.

UNBALANCED TRANSPORTATION PROBLEM

A transportation problem in which the total amount to be shipped is not equal to the total demand. The unbalanced problem can be stated as a standard transportation problem by the addition of a fictitious destination when the supply is greater than the demand, or by adding a fictitious origin if the demand is greater than the supply. In the first case, the demand at the fictitious destination is the difference between the total supply and total demand, while in the second case, the supply at the fictitious origin is the difference between the total demand and total supply. See **Transportation problem**.

UNBOUNDED OPTIMAL SOLUTION

A solution to a constrained optimization problem in which the objective function value can be shown to increase (or decrease) without bound on the feasible region. A real world problem whose mathematical model exhibits an unbounded optimal solution must have an incorrect formulation.

UNCONSTRAINED OPTIMIZATION

Ariela Sofer

George Mason University, Fairfax, Virginia

INTRODUCTION: Unconstrained optimization is concerned with finding the minimizing or maximizing points of a nonlinear function, where the variables are free to take on any value. Unconstrained optimization problems occur in a wide range of applications from the fields of engineering and science. A rich source of unconstrained optimization problems are data fitting problems, in which some model function with unknown parameters is fitted to data, using some criterion of "best fit." This criterion may be the minimum sum of squared errors, or the maximum of a likelihood or entropy function. Unconstrained problems also arise from constrained optimization problems, since these are often solved by solving a sequence of unconstrained problems.

In mathematical terms, an unconstrained minimization problem can be written in the form

minimize $f(x)$,

where x is a vector of n unrestricted variables. Ideally, one would like to find a *global minimizer* of the function, that is, a point x^* that yields the lowest value of f. Such a solution satisfies

$f(x^*) \leq f(x)$ for all x.

In many cases, however, finding a global minimizer is extremely difficult. For this reason, most algorithms attempt only to find a *local minimizer* of the function, that is, a point x^* that satisfies $f(x^*) \leq f(x)$ for all x in some neighborhood of x^*. If the objective f is a convex function, a local minimizer will also be a global minimizer; however, for nonconvex functions this property does not generally hold.

BACKGROUND: Much of the research in unconstrained optimization has focused on functions with continuous derivatives. Throughout this discussion we shall assume that the objective f is twice-continuously differentiable (that is, its second partial derivatives exist and are continuous). We shall denote the gradient of f at x (the vector of partial derivatives $\partial f(x)/\partial x_j$) by $\nabla f(x)$, and the Hessian of f at x (the matrix of second partial derivatives $\partial^2 f(x)/\partial x_i \partial x_j$) by $\nabla^2 f(x)$. When f is twice-continuously differentiable, the Hessian matrix is symmetric.

If there is a single fundamental tool in optimization of differentiable functions, it is the Taylor series, which provides an approximation to the function in a neighborhood of a point. The Taylor series is used in the derivation of the optimality conditions, in the development of solution methods and in analysis of their convergence.

Let $\bar{x}$ be a given point, and suppose that p is some direction. The first-order Taylor series expansion of f at $\bar{x}$ is
necessary conditions. The *first-order necessary*

$$f(\bar{x} + p) = f(\bar{x}) + \nabla f(\bar{x})^{\mathrm{T}} p + O(\|p\|^2),$$

where $O(q)$ indicates a term that goes to zero at least as fast as q does. By ignoring the last term in the expansion, we obtain a linear approximation to f in a neighborhood of $\bar{x}$; the error will be of order $\|p\|^2$. Similarly, the second-order Taylor series expansion of f is given by

$$f(\bar{x} + p) = f(\bar{x}) + \nabla f(\bar{x})^{\mathrm{T}} p$$
$$+ (1/2) p^{\mathrm{T}} \nabla^2 f(\bar{x}) p + O(\|p\|^3).$$

By ignoring the last term in this expansion we obtain a quadratic approximation to f, with an error of order $\|p\|^3$. We shall refer to this approximation as the *quadratic model*.

The quantity $\nabla f(\bar{x})^{\mathrm{T}} p$ is called the *directional derivative* of f along p at $\bar{x}$. If it is negative, then f is decreasing in the direction p, and p is termed a *direction of descent*. A small step $\varepsilon > 0$ taken in such a direction will lead to a point with a lower objective value: $f(\bar{x} + \varepsilon p) < f(\bar{x})$. The quantity $p^{\mathrm{T}} \nabla^2 f(\bar{x}) p$ is called the *curvature* of f along p. If the curvature is positive, the function is locally convex along the direction p.

Using the Taylor series approximation it is possible to derive conditions that must be satisfied by a local minimizer x^* of f. The conditions state that the function must have zero slope, and nonnegative curvature along any direction at x^*. They are summarized in the *necessary conditions*. The *first-order necessary condition* states that the gradient at x^* must vanish, so that $\nabla f(x^*) = 0$. The *second-order necessary condition* states that the Hessian must be positive semidefinite:

$$p^{\mathrm{T}} \nabla^2 f(x^*) p \geq 0 \quad \forall p.$$

(In the case of a local maximizer, the Hessian must be negative semidefinite.)

To illustrate these conditions consider the two-dimensional function $f(x) = x_1^2 + x_2^2$. The function attains its minimum at $x^* = (0, 0)^{\mathrm{T}}$. The gradient of f is $\nabla f(x) = (2x_1, 2x_2)^{\mathrm{T}}$, and indeed vanishes at x^*; the Hessian at x^* is twice the identity matrix, and hence is positive definite. Thus, the necessary conditions for a minimizer are satisfied at x^*.

A point at which the gradient is equal to zero is called a *stationary point*. Although such a point may be a local minimizer, it may also be a local maximizer, or neither of the above (in such case it is called a *saddle point*). As an example, $x^* = (0, 0)^{\mathrm{T}}$ is a stationary point of the functions $f_1(x) = -x_1^2 - x_2^2$ and: $f_2(x) = x_1^2 - x_2^2$; it is a local maximizer of f_1, and a saddle point for f_2.

It is possible to develop a condition that guarantees that a stationary point is a local minimizer. The *second-order sufficiency condition* states that if

$$\nabla f(x^*) = 0 \quad \text{and} \quad p^{\mathrm{T}} \nabla^2 f(x^*) p > 0 \quad \forall p \neq 0,$$

then x^* is a local minimizer of f.

METHODS: The vast majority of algorithms for unconstrained minimization are iterative descent methods. At each iteration, a direction of descent (called the *search direction*) is computed at the current solution estimate x_k; a step is then taken from x_k along the search direction, to obtain a new point a new point x_{k+1} such that $f(x_{k+1}) < f(x_k)$. The process is repeated till some test for convergence is satisfied.

The effectiveness of an algorithm is, of course, dramatically affected by the choice of the search direction. A key question, of course, is how to obtain a "good" search direction. The underlying idea of most methods, is to compute a direction that minimizes some local approximation to the function. Typically, this local model is obtained from the Taylor series. In Newton's method, the search direction is the vector p_k that minimizes the local quadratic model:

$$\text{minimize}_p \, f(x_k) + \nabla f(x_k)^{\mathrm{T}} p + (1/2) p^{\mathrm{T}} \nabla^2 f(x_k) p.$$

If the Hessian $\nabla^2 f(x_k)$ is positive definite, the minimizer of the quadratic model is the solution to the linear system of equations

$$\nabla^2 f(x_k) p = -\nabla f(x_k),$$

known as the *Newton equations*. The resulting iteration takes the form

$$x_{k+1} = x_k + p_k,$$

where p_k is the solution to the Newton equations.

If the initial point x_0 is sufficiently close to a local minimizer x^*, and if $\nabla^2 f(x^*)$ is positive definite, the iterates generated by Newton's method converge to x^*. Furthermore, the rate of convergence is quadratic. This means that for large k,

$$\|x_{k+1} - x^*\| \leq \gamma \|x_k - x^*\|^2$$

for some positive constant γ.

The rapid convergence of Newton's method near the solution makes it an extremely attractive method, and indeed, the method can be highly effective. However, the algorithm may fail if started from an initial point that is not sufficiently close to a minimizer. Why? First, if the Hessian $\nabla^2 f(x_k)$ is not positive definite the Newton direction may not be a descent direction, and if the Hessian is singular, the method is not even defined. Second, even if p_k is a descent direction, there is no guarantee that $f(x_{k+1})$ will actually be lower than $f(x_k)$. Thus, modifications to the basic Newton method are required to guarantee that the method will converge regardless of the starting point.

There are two major approaches to guarantee *global convergence* (convergence for any initial point): line search methods and trust-region method. Both approaches use the basic Newton method near the solution to exploit its rapid local convergence property. But they differ in the strategies they employ to guarantee convergence when far from the solution. Both approaches insist, however, on using a descent direction at each iteration.

Line search methods update the new estimate of the solution as $x_{k+1} = x_k + \alpha_k p_k$, where the steplength α_k is a positive scalar chosen so that $f(x_{k+1}) < f(x_k)$. Ideally, this steplength would be chosen to minimize $f(x_k + \alpha p_k)$ with respect to α. However, finding such a steplength is too time consuming. A more practical approach is to use a steplength that "approximately" minimizes f along p_k. One commonly used condition is to accept a trial step α_k *if*

$$|\nabla f(x + \alpha p_k)^\mathrm{T} p_k| \le \theta |\nabla f(x_k)^\mathrm{T} p_k| \text{ where } 0 \le \theta < 1,$$

that is, if a step of length α_k taken along p_k yields a "substantial decrease" in the magnitude of the directional derivative. This condition alone cannot guarantee convergence, since it does not guarantee decrease in the objective value. It is therefore common to impose an additional *sufficient decrease* condition on α_k:

$$f(x_k + \alpha_k p_k) \le f(x_k) + \eta \alpha_k \nabla f(x_k)^\mathrm{T} p_k$$

where $0 < \eta < 1$. If in addition, $\eta < \theta$, then under appropriate conditions, global convergence of the algorithm is guaranteed.

Line search-versions of Newton's method must also incorporate some strategy to handle the case when the Hessian is indefinite. One standard technique is to modify the Hessian matrix by a diagonal matrix, denoted $\mathbf{E}_k$, whose diagonal components are large enough to ensure that the modified Hessian is positive definite. The *modified Newton* direction is then computed as the solution to the system $(\nabla^2 f(x_k) + \mathbf{E}_k)p_k = -\nabla f(x_k)$. The approach generates descent directions and can overcome the numerical difficulties associated with near-singular Hessians.

Trust region methods differ from line search methods in that they determine *a priori* the maximum length of the search direction, say Δ. The direction is taken as the minimizer of the quadratic model, whose length does not exceed Δ. The motivation for this approach is that the quadratic model obtained from the Taylor series gives an adequate fit to the function for points that are close to x_k, but may not give an adequate fit for points far away. The length Δ is the radius of the *trust region*, the region in which we trust the quadratic model. It is adjusted from iteration to iteration, based on the agreement between the function and the quadratic model. It is increased if the agreement is considered to be good, and decreased if it is considered to be poor.

Modified Newton's methods (both line-search and trust region variants), are effective for solving small- or moderate-sized problems. As the number of variables increases, however, the cost of each iteration can become prohibitive. The solution of the $n \times n$ system of Newton equations is expensive, on the order of n^3 arithmetic operations. Furthermore, computation of the n^2 second partial derivatives can also be expensive and prone to errors. Thus, the benefits of fast local convergence are offset by the high costs of each iteration.

Quasi-Newton methods are a class of methods that are motivated by Newton's method but avoid the expense of computing second derivatives. The search direction is obtained by solving the system

$$\mathbf{B}_k p_k = -\nabla f(x_k),$$

where $\mathbf{B}_k$ is an approximation to the Hessian $\nabla^2 f(x_k)$. The matrix $\mathbf{B}_k$ is updated from $\mathbf{B}_{k-1}$ using gradient information from previous iterations. To make $\mathbf{B}_k$ resemble the Hessian, we require that it satisfy the *secant condition*

$$\mathbf{B}(x_k - x_{k-1}) = \nabla f(x_k) - \nabla f(x_{k+1}).$$

Quasi-Newton methods have been successful at solving a wide variety of practical problems, and are, perhaps, the most widely used methods for nonlinear optimization. However, their storage requirements and iteration costs can make them less suited for problems that have many variables. Limited-memory quasi-Newton methods are a modification to quasi-Newton methods that require much less storage and much lower arithmetic costs per iteration. Rather than store the matrix $\mathbf{B}_k$, they store a few vectors that provide the information to store a matrix close to $\mathbf{B}_k$.

Yet another class of methods suitable for large problems are truncated-Newton methods. These methods are a compromise on Newton's method. They obtain the search direction by finding an approximate solution to the Newton equations, using some iterative method such as the conjugate-gradient method. The iterative method is stopped before the exact solution has been found, hence the name of the method. The methods do not require explicit computation of the Hessian, and only require the storage of a few vectors. They have been used successfully to solve problems with large numbers of variables.

The theory and methods of unconstrained optimization are discussed in extensive detail in Dennis and Schnabel (1983) and Gill, Murray and Wright (1981). For a guide to software for numerical optimization, see Moré and Wright (1993). We have focused on methods for computing local optima of differentiable functions; a survey of methods for optimizing nondifferentiable functions is given in Lemarechal (1989). A survey of methods for global optimization is given in Rinnooy Kan and Timmer (1989).

See **Calculus of variations; Interior point methods; Linear programming; Nonlinear programming; Optimization.**

References

[1] J.E. Dennis and R.B. Schnabel (1983), *Numerical Methods for Unconstrained Optimization and Nonlinear Equations*, Prentice Hall, Englewood Cliffs, New Jersey.
[2] P.E. Gill, W. Murray, and M.H. Wright (1981), *Practical Optimization*, Academic Press, New York.
[3] C. Lemarechal (1989), "Nondifferentiable Optimization," in *Optimization*, G.L. Nemhauser, A.H.G. Rinnooy Kan, and M.J. Todd, eds., Elsevier, Amsterdam, 529–572.
[4] J.J. Moré and S.J. Wright (1993), *Optimization Software Guide*, SIAM, Philadelphia.
[5] A.H.G. Rinnooy Kan and G.T. Timmer (1989), "Global Optimization," in *Optimization*, G.L. Nemhauser, A.H.G. Rinnooy Kan, and M.J. Todd, eds., Elsevier, Amsterdam, 631–662.

UNCONSTRAINED SOLUTION

A solution that is independent or free of constraints.

UNCONTROLLABLE VARIABLES

In a decision problem, variables and other elements of a decision problem that are not under the control of the decision maker. See **Decision maker; Decision problem; Mathematical model.**

UNDERACHIEVEMENT VARIABLE

A nonnegative variable in a goal-programming problem constraint that measures how much the left-hand side of the constraint is less than the right-hand side. See **Goal programming**.

UNDERDETERMINED SYSTEM OF LINEAR EQUATIONS

An $m \times n$ system of linear equations $Ax = b$ in which $m < n$. Such systems may have an infinite number of solutions or are inconsistent. The equation form of a linear-programming problem is underdetermined.

UNDIRECTED ARC

In a network, an arc along which flow can go in either direction.

UNIMODULAR MATRIX

An $m \times n$ matrix A of rank r is said to be unimodular if every one of its square submatrices of order r has a determinant value of 0, $+1$, or -1.

UNIQUE SOLUTION

The optimal solution to an optimization problem that has one and only one optimal solution. See **Multiple optimal solutions**.

UNRESTRICTED VARIABLE

A variable that can take on any value. See **Free variable**.

UNSYMMETRIC PRIMAL-DUAL PROBLEMS

The two linear-programming problems with the following form:

Primal

   Minimize $c^T x$

   subject to

   $$Ax = b$$

   $$x \geq 0$$

Dual

   Maximize $b^T y$

   subject to

   $$A^T y \leq c$$

Note that the variables of the dual problem are unrestricted. See **Strong duality theorem**; **Symmetric primal-dual problems**.

UPPER-BOUNDED PROBLEMS

See **Simple upper-bounded problem**; **General upper-bounded problem**.

URBAN SERVICES

Kenneth Chelst

Wayne State University, Detroit, Michigan

INTRODUCTION: Urban services cover a broad range of activities. These include sanitation; street cleaning to remove trash, snow and ice; public housing; urban transportation systems; and other local government services. In this section we describe representative applications of operations research to improve the efficiency of these services (see, e.g., Larson and Odoni, 1981).

STREET CLEANING AND SANITATION: A planning function common to these services is the need to design efficient vehicle routes that minimize the cost of the service primarily through minimization of travel time. Garbage collectors travel up and down streets stopping in front of each house to pick up trash. Street sweepers move along curbsides sweeping up garbage while avoiding parked cars. Snow plows and/or salt trucks make their way through snow covered arteries preparing them for smooth safe traffic flow.

This class of routing problems that involve planning coverage of street segments is equivalent to the *Chinese Postman Problem*. In an urban context, route planning can involve a large number of vehicles and take into account one- and two-way streets and any opportunities for making U-turns. The garbage collection problem has an added complexity that involves the capacity of the trucks and randomness in the volume of trash. As routes are planned, it is not possible to predict exactly how many stops a truck will make before it reaches capacity. It then must leave its route and travel some distance to unload its garbage before returning to its route. Routing snow plows or salt trucks has its own added complexity. One pass though a street may not achieve the desired level of impact. In addition, as snow continues to fall or roads continue to freeze, streets will have to be revisited.

In New York City, operations research has had a broad impact on the sanitation department that goes beyond route planning. Workload forecasting models were developed to plan personnel needs, reduce overtime, shift vacations to off-peak seasons and plan for the hiring of new personnel. Analyses were carried out to assess the impact of increasing the capacity of trucks. A simulation model was developed to understand the impact of even one illegally parked car on the effectiveness of a street sweeper. This information was used to help coordinate efforts with the traffic enforcement division. One innovation involved the creation of Project Scorecard that was designed to sample 6,000 blocks each month to track how dirty the streets were and not just how much garbage was collected. The OR group also carried out a variety of studies to determine the impact of different regulations for separation of trash to facilitate recycling. In summary, operations research has fundamentally changed the way New York City makes decisions about street cleaning and the way the sanitation department manages its resources (Riccio, Miller and Litke, 1986)

PUBLIC HOUSING: In many urban environments, cities build and rent subsidized housing to the poor and the elderly. One of the first issues addressed by the Local Government Operational Unit of Reading, England was ranking applicants for the 100,000 housing units owned by the City of Manchester. They developed a housing points scheme that captured housing department officials' perspective on relative need. Through the use of paired comparisons, they were able to answer questions such as, "should an applicant with a medical problem be given more points than one living in over crowded conditions?"

In England, the Thurnscoe Housing Co-operative was the recipient of broader studies that are representative of the idea of Community OR. A team of operations researchers worked with the co-operative on a wide range of topics. These included financial planning and structured decision making. They were an integral part of all co-operative planning over several years. Both of these studies illustrate soft-OR, which does not necessarily use any of the classical OR techniques such as linear programming or queueing theory.

In the US, researchers have used queueing theory to evaluate two alternative tenant assignment policies, namely, first available unit vs. priority assignment. The measures of performance included mean waiting times and the impact on housing racial integration. In a second study, integer programming was used to plan the sequential relocation of housing tenants as part of the redevelopment of a housing project in East Boston (Kaplan and Berman, 1988).

URBAN TRANSPORTATION SERVICES: The issues surrounding the delivery of urban mass transit services have been studied from a broad range of disciplines. Economists have led the study of the relationship between fare structure (price) and demand. Urban and regional planners have researched the role of mass transit in urban and regional development. Statisticians have tackled the complex problem of estimating the origin-destination matrix that is critical in planning route specific demand. Civil engineers have made transportation planning, both road and mass transit, a major component of their discipline and have often used operations research models in their studies or teamed with operations researchers. The journal *Transportation Science* is a focal point for reporting the latest research in this and related fields. In this review, we limit the discussion primarily to the use of OR models.

Transportation services can be viewed from three perspectives, the passenger, the crew and the infrastructure needed to provide the service (e.g., vehicles, facilities). The passenger is interested in traveling from point A to point B in the most cost efficient way. The design of transit routes and the scheduled frequency of trains or buses (e.g., headway) are the key management decisions that influence passenger experiences. Probabilistic models, in general, and queueing models, in particular, have been developed to estimate passenger waiting times for both rail and bus services under a variety of operational strategies. Simulation was often the ideal flexible tool used for planning and evaluating the impact of a variety of polices on system performance.

The elderly and handicapped have difficulty using mass transit to meet their travel needs. Taxis are an expensive alternative. Dial-a-Ride mini-bus systems fill the gap by picking up upon request passengers from multiple points and delivering them to different locations. The *Traveling Salesman Problem* provides the basis for heuristic algorithms to efficiently manage the complex dispatching operation.

From a personnel perspective, the number one issue addressed by operations researchers has been manpower scheduling, which is a classic application of mathematical programming. Demand for transportation services varies significantly by time of day and personnel and vehicle schedules must adjust accordingly to be cost effective. HASTUS is a widely used tool for developing schedules for both personnel and vehicles. Random absences of scheduled personnel produce an added burden on managing an already complex system. Probabilistic models have been developed to help transportation managers plan area-wide pooled resources that are used to fill in unanticipated personnel shortages (Blais, Lamont and Rousseau, 1990).

Garages are an important element of any municipal bus system. Buses and drivers start and end their shifts at garages, and most maintenance occurs in these facilities. The decision as to the number and location of these garages has been analyzed by applying iteratively a minimum cost network flow model. A related question, common to all capital intensive systems, involves the maintenance of the capital equipment. This issue falls within the broad range of operations research methods that model reliability, optimal maintenance and replacement strategies. One statistical study of 2,000 buses in Montréal analyzed the relationship between inspection and breakdowns and suggested that the optimal inspection policy be changed from 5,000 kilometers to 8,000. Multi-criteria decision models have been used to develop component maintenance policies that focus not only on total maintenance cost but also transit vehicle availability and component reliability (Ball, Assad, Bodin, Golden, and Spielberg, 1984).

OTHER SERVICES: In an urban environment, one common problem that cuts across a broad range of both government and non-government services is how many facilities to build and where should they be located. Classic facility location models, both capacitated and uncapacitated, have been applied to address this decision in cities around the world. The urban setting often requires the organization of specialized delivery services that involve the scheduling and routing of multiple vehicles, usually with time constraints. One specific application area has been the delivery of meals. Traveling salesman based routing models were used to develop and maintain efficient routes for a Meal-on-Wheels program that provides regular service to home-bound elderly (Bartholdi, Platzman, Collins, and Warden, 1983; Ball and Beckett, 1991).

The role of operations researchers is not limited to model development; it also includes evaluation studies, discussion of performance evaluation and concerns over equity (Ball and Beckett, 1991). However, OR's overall impact on planning and managing urban services in cities world-wide has been extremely limited compared to its potential. The primary barriers to greater use are a) the lack of educated consumers of OR models in leadership positions within urban services, b) the limited availability of trained OR professionals in city government and c) no profit incentive to drive the search for continuous improvement.

See **Chinese postman problem; Community operations research; Crime and justice; Emergency services; Facility location; Location analysis; Manpower planning; Network optimization; Traveling salesman problem; Vehicle routing.**

References

[1] Ball, M., A. Assad, L. Bodin, B. Golden, and F. Spielberg (1984), "Garage Location for an Urban Mass Transit System," *Transportation Science*, 18, 56–75.

[2] Ball, R. and A. Beckett (1991), "Performance Evaluation in Local Government, The Case of a Social Work Department Meals-on-Wheels Service," *European Jl. Operations Research*, 51, 35–44.

[3] Bartholdi, J.J., L.K. Platzman, R.L. Collins, and W.H. Warden (1983), "A Minimal Technology Routing System for Meals on Wheels," *Interfaces*, 13(3), 1–8.

[4] Beltrami, E.M. and L. Bodin (1974), "Networks and Vehicle Routing for Municipal Waste Collection," *Networks*, 4, 65–94.

[5] Blais, J.Y., J. Lamont, and J.M. Rousseau (1990), "The HASTUS Vehicle and Manpower Scheduling System at the Société de Transport de la Communaute Urbaine de Montréal," *Interfaces*, 20(1), 26–42.

[6] Chang, S.K. and P. M. Schonfeld (1991), "Multiple Period Optimization of Bus Transit Systems," *Transportation Research-Applications*, 25B, 453–478.

[7] Goplalaswamy, V., J.A. Rice, and F.G. Miller (1993), "Transit Vehicle Component Maintenance Policy via Multiple Criteria Decision Making Methods," *Jl. Operational Research Society*, 44, 37–50.

[8] Kaplan, E. and O. Berman (1988), "OR Hits the Heights: Relocation Planning at the Orient Heights Housing Project," *Interfaces*, 18(6), 14–22.

[9] Nascimento, E.M. and J.E. Beasley (1993), "Locating Benefit Posts in Brazil," *Jl. Operational Research Society*, 44, 1063–1066.

[10] Larson, R.C. and A.R. Odoni (1981), *Urban Operations Research*, Prentice Hall, Englewood Cliffs, New Jersey.

[11] Lutin, J.M., M.A. Hornung and J. Beck (1988), "Staging Area Simulation Model for Seattle Metro Subway," *Transportation Research Record*, #1162, 58–66.

[12] Pinkus, C.E. and A. Dixson (1981), *Solving Local Government Problems*, George Allen & Unwin Ltd., London.

[13] Reinert, K.A., T.R. Miller, and H.G. Dickerson (1985), "A Location – Assignment Model for Urban Snow and Ice Control Operations," *Urban Analysis*, 8, 175–191.

[14] Riccio, L.J., J. Miller and A. Litke (1986), "Polishing the Big Apple, How Management Science Has Helped Make New York Streets Cleaner," *Interfaces*, 16(1), 83–88.

[15] Savas, E.S. (1978), "On Equity in Providing Public Service," *Management Science*, 24, 800–808.

[16] Stein, D.M. (1978), "Scheduling Dial-a-Ride Transportation Systems," *Transportation Science*, 12, 232–249.

UTILITY FUNCTION

See **Multi-objective programming; Utility theory.**

UTILITY THEORY

Peter Fishburn

AT&T Bell Laboratories, Murray Hill, New Jersey

Utility theory is the systematic study of preference structures and ways to represent preferences quantitatively. The objects on which preferences are defined could be potential outcomes of a decision, decision alternatives, individual or family consumption bundles in a fixed time period, time streams of net profits, investment portfolios, the entrees on a restaurant menu, or just about anything else. The preferences themselves are usually those of an individual, but are sometimes attributed to groups or organizations.

Let A denote the set of objects on which preferences are defined and let $\succsim$ be a binary relation on A, for example, a set of ordered pairs (x,y) of objects in A. When (x,y) is a member of $\succsim$, it is customary to write $x \succsim y$ and to say that *x is at least as preferred as y*. If $x \succsim y$ and not $(y \succsim x)$ then *x is* (strictly) *preferred to y*; if $x \succsim y$ and $y \succsim x$ then x and y are equally preferred, or are *indifferent*; if neither $x \succsim y$ nor $y \succsim x$ then x and y are preferentially *incomparable*. Strict preference and indifference are denoted by $x \succ y$ and $x \sim y$ respectively.

Utility theory typically regards the preference relation $\succsim$ on A as deterministic and interprets $x \succsim y$ as: if you have title to y you would be willing to trade it for title to x. There are also notions of uncertain or probabilistic preference that will not be described here. An excellent introduction to probabilistic preference and stochastic utility is provided by Luce and Suppes (1965).

Two book collections offer a broad overview of utility theory. Page (1968) contains historical essays, including an English translation of a 1738 paper by Daniel Bernoulli that introduced expected utility, a philosophical piece from 1823 by Jeremy Bentham that popularized the term *utility*, an excerpt from the game theory classic by John von Neumann and Oskar Morgenstern in 1944 that placed expected utility on a firm axiomatic foundation, and an economist's account by George Stigler of the development of utility theory from 1776 to 1915. The collection by Eatwell, Milgate and Newman (1990), covers many facets of utility theory, including several that are areas of contemporary research.

DISTINGUISHING FEATURES: There are numerous specific theories of utility. Each is distinguished by three features: (1) the structure of A; (2) the assumptions made about the properties of $\succsim$ on A; and (3) the quantitative representation that reflects $(A, \succsim)$ in a numerical structure.

Assumptions for feature 1 are *structural assumptions*, and those for feature 2 are *preference axioms*. Together they are used to deduce the *quantitative representation* of feature 3. The representation's numerical functions are often called *utility functions*. Other real-valued functions, including probability distributions and threshold functions, also occur in representations.

An important adjunct of a utility representation is a description of the class of all functions that satisfy the representation. This is the representation's *uniqueness* structure. Some representations have very demanding uniqueness structures; others allow great latitude for their utility functions.

Two examples illustrate these ideas. First, let $A = \{$beef, chicken, fish, lamb$\}$ in regard to entrees for dinner. Assume for feature 2 that $\succ$ on A is a *linear* (strict) *order* which, for all x, y and z in A, means that

$\succ$ is *irreflexive*: not$(x \succ x)$

$\succ$ is *complete*: $x \neq y \Rightarrow (x \succ y$ or $y \succ x)$

$\succ$ is *transitive*: $(x \succ y$ and $y \succ z) \Rightarrow x \succ z$.

One realization of $\succ$ on A is [beef $\succ$ lamb $\succ$ chicken $\succ$ fish]. This ordering is represented by a utility function u on A which assigns a number $u(x)$ to each x in A such that [u(beef) $>$ u(lamb) $> u$(chicken) $> u$(fish)]. There is great latitude for u. Every real-valued function on A whose $>$ ordering mirrors $\succ$ is a suitable utility function for the representation.

Second, let $A = [0, M]^3$, $0 < M$, the set of all triples (x_1, x_2, x_3) with $0 \leq x_i \leq M$ for each i. Interpret x_i as the income an individual earns in year i hence. One representation for feature 3 is the additive utility model

$$(x_1, x_2, x_3) \succsim (y_1, y_2, y_3)$$

$$\Rightarrow \sum_{i=1}^{3} u_i(x_i) \geq \sum_{i=1}^{3} u_i(y_i),$$

where each u_i is an increasing and continuous real-valued function on $[0, M]$. This requires that $\succsim$ on A be a *weak order* which, for all x, y and z in A, means that

$\succsim$ is *strongly connected*: $x \succsim y$ or $y \succsim x$

$\succsim$ is *transitive*: $(x \succsim y$ and $y \succsim z) \Rightarrow x \succsim z$.

Another axiom that concerns additivity says that if two triples have identical incomes in a given year then $\succsim$ between them remains unchanged if the identical income is changed, for example,

$$(x_1, x_2, x_3) \succsim (x_1, y_2, y_3)$$

$$\Rightarrow (y_1, x_2, x_3) \succsim (y_1, y_2, y_3).$$

Other axioms relate to monotonicity of utility in income and to continuity of each utility function.

The preceding model has a very tight uniqueness structure. In particular, when u_1, u_2 and u_3 satisfy the representation, then so do v_1, v_2 and v_3 in place of u_1, u_2 and u_3 respectively if, and only if, there are real numbers $\alpha > 0$ and β_1, β_2 and β_3 so that, for all m in $[0, M]$,

$$v_i(m) = \alpha u_i(m) + \beta_i, \quad i = 1, 2, 3.$$

Hence, except for an origin and unit, each u_i is unique.

The next few sections describe utility theories according to a three-part classification that mixes feature 1 with extra-mathematical interpretations:

certainty: there is no explicit use of chance or uncertainty;

chance: chance in the form of numerical probabilities appears in A, but unquantified uncertainty is excluded;

uncertainty: outcomes of decisions depend explicitly on uncertain events with not-yet-quantified probabilities.

Differences among classes can be illustrated by an object (m_1, m_2, m_3, m_4) in which each m_i is an amount of money. If the object describes a four-year income stream, the certainty designation applies. If the object is a gamble or risky prospect that pays off m_1, m_2, m_3 or m_4, each with probability 1/4, then chance applies. And if m_1 through m_4 are the amounts won for each dollar bet on your favorite horse in tomorrow's big race when the horse wins, places, shows and finishes out of the money, respectively, then uncertainty applies.

A differentiator for feature 2 is the extent to which preferences are transitive. The most restrictive case occurs when $\succsim$ is a weak order. Then each of $\succsim$, $\succ$ and $\sim$ is transitive. A more flexible case arises when $\succ$ but not $\sim$ is assumed transitive. *Intransitive indifference* is illustrated by a sequence of indifference comparisons $x_1 \sim x_2$, $x_2 \sim x_3, \ldots, x_{n-1} \sim x_n$ between similar objects the first of which is definitely preferred to the last $(x_1 \succ x_n)$. The most flexible case occurs when neither $\succ$ nor $\sim$ is assumed transitive. This allows preference cycles, such as $x \succ y \succ z \succ x$. Fishburn (1991) provides access to the nontransitive preference literature.

CERTAINTY: A basic theorem of utility theory for $(A, \succsim)$ says that a real number $u(x)$ can be assigned to each object in A so that, for all x and y in A,

$$x \succsim y \Leftrightarrow u(x) \geq u(y),$$

if and only if $\succsim$ on A is a weak order and there is a countable (finite or denumerable) subset B of A such that, whenever $x \succ y$, some z in B satisfies $x \succsim z \succsim y$. A relaxation of this *ordinal utility* representation that accommodates intransitive indifference and thresholds for preference is

$$x \succ y \Leftrightarrow > u(x) > u(y) + \sigma(y),$$

where $\sigma(y) \geq 0$ for each y. This representation assigns a *utility interval* $[u(x), u(x) + \sigma(x)]$ to each x and has x preferred to y if and only if the right end of y's interval is less than the left end of x's interval. One of its preference axioms is

$$(x \succ a \text{ and } y \succ b) \Rightarrow (x \succ b \text{ or } y \succ a).$$

A popular structure for preference theory formulates A as a subset of n-tuples $(x_1, \ldots, x_n)$, $(y_1, \ldots, y_n), \ldots$ in $X_1 \times X_2 \times X_n$. Index i for X_i could refer to an attribute of objects in A or a time period. This product structure gives rise to special forms for the utility function u of the preceding paragraph, including the *additive decomposition*

$$u(x_i, \ldots, x_n) = \sum_{i=1}^n u_i(x_i)$$

in which u_i is a marginal utility function for the ith attribute or time period. A generalization that does not presume transitivity but retains additivity is

$$(x_1, \ldots, x_n) \succsim (y_1, \ldots, y_n) \Leftrightarrow \sum_{i=1}^n \varphi_i(x_i, y_i) \geq 0,$$

where φ_i is defined on $X_i \times X_i$ and has $\phi_i(x_i, x_i) = 0$. Fishburn (1970,1991), Keeney and Raiffa (1976) and Wakker (1989) have extensive coverage of the preceding topics.

CHANCE: The primary structure for chance takes A as a set of probability distributions on an outcome set X. For p in A, $p(x)$ is the probability that risky prospect p will yield outcome x. It is usually assumed as part of feature 1 that A is *closed under convex combinations*: if p and q are in A and $0 < \lambda < 1$, then $\lambda p + (1 - \lambda)q$ is also in A.

Two common preference axioms for $(A, \succsim)$ are weak order and the *independence condition*

$$p \succ q \Rightarrow \lambda p + (1 - \lambda)r \succ \lambda q + (1 - \lambda)r$$

whenever p, q and r are in A and $0 < \lambda < 1$. When an Archimedean axiom is added to weak order and independence, the existence of a von Neumann-Morgenstern linear utility function u on A can be established. It has $p \succsim q \Leftrightarrow u(p) \geq u(q)$ along with the *linearity property*

$$u(\lambda p + (1 - \lambda)q) = \lambda u(p) + (1 - \lambda)u(q),$$

and is unique except for origin and unit, for example, unique up to transformations $\alpha u + \beta$ with $\alpha > 0$.

If A includes all distributions with finite support and $u(x)$ is defined as $u(p)$ when $p(x) = 1$, then linearity implies the *expected-utility form*

$$u(p) = \sum_x p(x)u(x)$$

for each finite-support distribution. Additional axioms are needed to obtain $u(p) = \int u(x)dp(x)$ for general probability measures.

Three variations on the expected-utility theme involve risk attitudes such as risk aversion when outcomes are monetary (Raiffa, 1968; Wakker, 1989), multiattribute expected utility when $X \times X_1 \times X_2 \times \ldots \times X_n$, including additive and multiplicative decompositions of $u(x_1, \ldots, x_n)$ (Fishburn 1970, Keeney and Raiffa 1976, Wakker 1989), and generalizations of expected utility that relax one or more of its axioms (Fishburn 1988). A representation that does not presume transitivity and substantially weakens the independence condition is $p \gtrsim q \Leftrightarrow \varphi(p,q) \geq 0$, where φ is *skew symmetric* $[\varphi(p,q) + \varphi(q,p) = 0]$ and linear separately in each argument.

UNCERTAINTY: The main structure for uncertainty (Savage 1954) takes A as the set of functions $f, g, \ldots$, called *acts* from a set S of *states* into an outcome set X. If "you" choose f and state s obtains, your outcome is $f(s)$. It is presumed that one and only one state will obtain, that you are uncertain which it will be, and that your chosen act will not affect its occurrence.

Savage's axioms (see also Fishburn 1970) for $(A, \gtrsim)$, which include weak order and independence assumptions, imply the existence of a bounded utility function u on X and a probability measure π on the set of all subsets of S such that, for all acts f and g,

$$f \gtrsim g \Leftrightarrow \int_S \varphi(f(s)d\pi(s) \geq \int_S ug(s)) \, d\pi(s) \geq 0.$$

Moreover, u is unique except for origin and unit, and π is unique.

Deduced probabilities in Savage's model are personal or *subjective probabilities*. The model itself is a *subjective expected utility representation*. The art of applying it to real-world problems is known as *decision analysis* (Raiffa 1968).

Multiattribute and/or time-stream outcomes occur in most applications.

Many other utility theories have been proposed for structures similar to Savage's. One strain relaxes his model by assuming monotonicity $[A \subseteq B \Rightarrow \pi(A) \leq \pi(B)]$ but not necessarily additivity $[\pi(A \cup B) = \pi(A) + \pi(B)$ when A and B are disjoint] for subjective probability. Another retains Savage's properties for π but relaxes transitivity to obtain

$$f \gtrsim g \Leftrightarrow \int_S \varphi(s), g(s)) \, d\pi(s) \geq 0,$$

with φ skew symmetric on $X \times X$. See Fishburn (1988) and Wakker (1989) for further details and references.

See **Choice theory; Decision analysis; Game theory; Preference theory.**

References

[1] Eatwell, J., M. Milgate and P. Newman (eds.). 1990. *The New Palgrave: Utility and Probability.* Macmillan, London.

[2] Fishburn, P.C. 1970. *Utility Theory for Decision Making.* Wiley, New York.

[3] Fishburn, P.C. 1988. *Nonlinear Preference and Utility Theory.* The Johns Hopkins University Press, Baltimore.

[4] Fishburn, P.C. 1991. Nontransitive Preferences in Decision Theory. *J. Risk & Uncertainty* 4, 113–134.

[5] Keeney, R.L., and H. Raiffa. 1976. *Decisions with Multiple Objectives: Preferences and Value Tradeoffs.* Wiley, New York.

[6] Luce, R.D., and P. Suppes. 1965. Preference, Utility and Subjective Probability. In *Handbook of Mathematical Psychology, III,* R.D. Luce, R.R. Bush and E. Galanter (eds.). Wiley, New York, pp. 249–410.

[7] Page, A.N. (ed.). 1968. *Utility Theory: A Book of Readings.* Wiley, New York.

[8] Raiffa, H. 1968. *Decision Analysis: Introductory Lectures on Choice under Uncertainty.* Addison-Wesley, Reading, Massachusetts.

[9] Savage, L.J. 1954. *The Foundations of Statistics.* Wiley, New York.

[10] Wakker, P.P. 1989. *Additive Representations of Preferences.* Kluwer, Dordrecht.

VACATION MODEL

A queueing model where the server at some point leaves the node and eventually returns. During a busy period before leaving, the server may serve one customer, all customers initially present, some customers, all customers present, or all who arrive. See **Cyclic service discipline**; **Queueing theory**; **Vacation time**.

VACATION TIME

In vacation models, the time starting when the server leaves the node and ending when the server returns to that node. See **Cyclic service discipline**; **Vacation model**.

VALIDATION

The process of determining how well the outputs of a mathematical model of a real-world problem conform to reality. Two key aspects of validity are face validity and predictive validity. Face validity is based on an examination of the assumptions and data going into the model for logical consistency and the review of the results by experts knowledgeable in the real world situation. Predictive validity is based on examining the model's predictions for events that were not used in building the model. See **Verification**; **Verification, validation and testing of models**.

VALUE FUNCTION

In a decision problem, let a be a feasible alternative from the set of all feasible alternatives A. Each alternative is measured against n attributes $(X_1, \ldots, X_n)$. The decision maker's (DM) problem is to choose an a in A that "maximizes" the payoff vector of scores $[X_1(a), \ldots, X_n(a)] = \mathbf{X}^a$. We define a real-valued, scalar function $v(.)$, the value function, as follows. The function has the property that $v(\mathbf{X}^a) > v(\mathbf{X}^b)$ if and only if the DM prefers alternative a to alternative b; and $v(\mathbf{X}^a) = v(\mathbf{X}^b)$ if and only if the DM is indifferent between alternative a and alternative b. The DM's decision problem is now the selection of an alternative that maximizes $v(\mathbf{X})$ over all alternatives. See **Choice theory**; **Decision analysis**; **Multiple criteria decision making**; **Preference theory**; **Utility theory**.

VAM

See **Vogel's approximation method**.

VARIANCE REDUCTION

See **Monte Carlo sampling and variance reduction**.

VECTOR MAXIMUM PROBLEM

See **Multi-objective programming**.

VECTOR OPTIMIZATION PROBLEM

See **Multi-objective optimization**.

VECTOR SPACE

A vector n-space is a set of vectors or points, each with n components, and rules for vector addition and multiplication by real numbers. Euclidean 3-space is a vector space.

VEHICLE ROUTING

Lawrence Bodin

University of Maryland, College Park

The traditional vehicle routing problem is to determine a set of minimum cost routes for a fleet of identical vehicles in order to service a collection of locations where each location has a known demand for service. If the number of vehicles in the fleet (*fleet size*) is known, then routes are formed so that the total travel time while performing nonproductive work (*deadhead travel time*) is minimized. If the fleet size is not known, then routes are formed in order to minimize the fleet size and the deadhead travel time.

Traditional constraints that can be present on route formation include the following:

1. Each route can be no longer than a specified length.
2. The volume on each route can be no larger than a specified amount, called the capacity of the vehicle.
3. Each route has to begin and end at the same location, called the depot. This is called a *single depot vehicle routing problem*.
4. If the service at location i must be carried out between L_i and U_i, then $[L_i, U_i]$ is called a *hard time window* at location i. If it is desired that the service at location i be carried out between L_i and U_i but L_i and U_i can be violated, then $[L_i, U_i]$ is called a *soft time window* at location i.

A vast literature on vehicle routing exists. Standard references include Bodin (1990) and Bodin *et al.* (1983), Golden and Assad (1986, 1988). These papers include extensive bibliographies.

PRACTICAL VEHICLE ROUTING PROBLEMS:

Practical vehicle routing problems include the delivery of goods from a depot to a set of locations, residential and containerized sanitation pickup, scheduling of meter readers, scheduling of field maintenance personnel, delivery of newspapers and telephone books, scheduling of fuel deliveries such as propane gas and gasoline, scheduling of paratransit vehicles, and scheduling of pickups and deliveries for courier services. The following conditions may be encountered in solving practical vehicle routing problems:

1. The length of each route must be between a prespecified lower and upper bound. Thus, the route length constraint becomes a goal rather than a constraint.
2. Each route has to begin and end at the same location but there can be several such locations or depots. This is called the *multiple depot vehicle routing and scheduling problem*.
3. Rather than all vehicles in the vehicle fleet being the same, there can be several types of vehicles in the fleet where the vehicles can be different in terms of capacity, size of crew, speed, etc. This problem is called the *multiple vehicle type routing and scheduling problem*.
4. If some of the locations can be serviced by some, but not necessarily all of the vehicle types and the specification of the vehicle types that can service a location can differ by location, then this problem is called the *vehicle/location or site dependency routing and scheduling problem*.
5. If demand at a location is not known in advance but can be stochastic, then this problem is called the *stochastic vehicle routing and scheduling problem*. A variant of the stochastic vehicle routing and scheduling problem, called the *inventory routing problem*, has occurred in the delivery of such items as propane gas. In the case of propane gas, the demand at each of the locations is forecasted using a factor such as degree-days.
6. In cases where the deliveries are so valuable that the hijacking of the vehicles is possible, the routes are broken down into pairs and the initial portion of each pair of routes has to coincide for security reasons. Upon completing the initial portion of the route, each vehicle "goes it separate way" for the remainder of the work day. One such application was the delivery of tobacco goods in a Latin American city. A crew over the initial portion of the day consisted of two delivery vehicles and a security car with an armed driver. The security car returned to the depot after completing the initial portion of the day.
7. In paratransit, courier delivery and shared cab ride problems, each customer demanding service has a specified pickup location and a specified delivery location. The pickup has to be scheduled on the route before the delivery is scheduled on the route. Some of these problems allow for transshipments; that is to say, packages are picked up and brought to a prespecified drop location where they are unloaded and another vehicle later picks up these packages and makes the deliveries.
8. In most vehicle routing and scheduling problems, the routes are developed assuming that the locations to be serviced are known in advance. However, in some paratransit, courier delivery, and shared cab ride problems, as well as other vehicle routing and scheduling problems, there is a real time aspect to the problem. In these problems, the assignment of locations to vehicles is decided upon as the "calls for service" are received by the dispatcher. Portions of the vehicle routes are scheduled in advance before the real time component of the routing is carried out.
9. In some vehicle routing problems, there are delivery locations and pickup locations. Each route has the restriction that all (or

most) of the deliveries are to be carried out before any of the pickups are to be carried out. In this way, the vehicle can be close to empty before it is filled up. This problem is called the *vehicle routing problem with backhauling*.

ALGORITHMS FOR SOLVING VEHICLE ROUTING PROBLEMS: Virtually all vehicle routing problems fall into the class of combinatorial optimization problems called NP-Hard. A problem is NP-Hard if the number of computations needed to solve this problem grows exponentially with a parameter of the problem, Garey and Johnson (1979), Karp (1975), Lenstra and Rinnooy Kan (1981) and Papadimitriou and Stieglitz (1982). Since finding the optimal solution for reasonable size problems is impossible, heuristic approaches are generally employed to find a close-to-optimal solution.

A standard heuristic approach for solving many node routing and scheduling problems consists of the following steps:

a. *Specify K, the fleet size.* Generally K is specified by the user.

b. *Tour Construction.* Aggregate the locations to be serviced into K clusters. Some of the approaches for tour construction use single location insertion heuristics and are sequential in nature (one location is assigned to a cluster on each iteration). Other approaches, such as the generalized assignment algorithm, are based on solving a mathematical program and are not sequential in nature. In some cases, routes and schedules are formed along with the clustering. If routes and schedules have not been formed while aggregating locations into clusters, then a route and schedule is found over the locations assigned to each of the K clusters, one cluster at a time. At the conclusion of this step, it is possible to have some locations unassigned to routes and/or the routes violating the upper bound on travel time.

c. *Tour Improvement:* In order to reduce the total travel time for all the routes and insert unassigned locations onto routes,

    i. Reorder the locations on each of the routes, one route at a time (if an optimal tour was not found in b),

    ii. Move locations between routes,

    iii. Insert unassigned locations onto routes using the insertion procedures described under Step b.

The tour improvement step continues until no more improvements are found or the time allocated to this process is exhausted. In tour improvement, a route is sometimes dissolved and the locations on this route inserted on other routes in order to reduce the size of the fleet.

When practical considerations are considered, K may be very difficult to estimate accurately. Therefore, the user may not know how many routes to form. Of course, one can always repeat the algorithm for different values of K and take the best solution. Therefore, this approach can be extremely time consuming and still not give reasonable results.

Furthermore, many of the tour construction approaches are sequential in nature in that one location is assigned to one of the K routes on each iteration. Therefore, a bad decision made at one step in the tour construction locks in the solution being generated, and can adversely affect the subsequent assignment of locations to routes. Having constructed a bad set of routes (for a given value of K), the tour improvement procedures (Step c) are either too time consuming or not powerful enough to derive a reasonable solution. This approach has been one of the "workhorse" procedures for solving routing and scheduling problems, both from a research standpoint and in commercially available software.

Other approaches for solving vehicle routing problems consist of mathematical programming, column generation, set partitioning and interactive optimization, Bodin (1990), Bodin *et al.* (1983), Golden and Assad (1988), and Karp (1975).

CLASSES OF VEHICLE ROUTING PROBLEMS: Vehicle routing problems are generally broken down into two classes: point-to-point or node routing and neighborhood or arc routing. In a point-to-point problem, the locations to be serviced are scattered over a region. In solving the point-to-point vehicle routing problem, the locations are partitioned into subsets and, within each subset, the locations are ordered to form a minimum travel time route. Traditionally, when solving point-to-point problems, the Euclidean distance between locations is used as the deadhead travel time metric.

The *traveling salesman problem* is the one vehicle point-to-point vehicle routing problem. More specifically, the traveling salesman problem requires the determination of a minimum deadhead time route that passes through each location demanding service exactly once. A traveling salesman solution is displayed in Figure 1. The route in Figure 1 represents a solution to a symmetric or undirected traveling salesman problem, since the travel time between each pair of

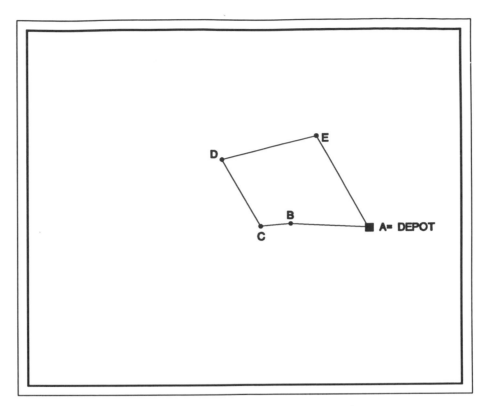

Fig. 1. Euclidean distance route over 5 points.

locations does not depend on the direction of travel. An asymmetric or directed traveling salesman problem occurs when the travel time between each pair of locations can be different.

The *capacitated arc routing problem* is an example of a neighborhood routing problem. In the undirected (directed) capacitated arc routing problem, an undirected (directed) network is given where each arc has demand $q_{ij} > 0$ and each vehicle has capacity Q. The problem is to break this network down into partitions where each arc is assigned to a partition, the demand in each partition is no greater than Q, a cycle can be found that traverses all of the arcs in each partition and the total travel time of the vehicles is minimized.

The *undirected (directed) Chinese postman problem* is the single vehicle arc routing problem where all of the arcs in the network require service, the arcs are undirected (directed) and the network is *connected*. A 1-match problem is solved when finding a minimum travel time route in the case where all of the arcs in the network are undirected. A transportation problem is solved when finding a minimum travel time route in the case where all of the arcs in the network are directed. Since the algorithms for optimally solving the 1-match problem and the transportation problems are polynomial, the undirected

and directed Chinese postman problems are not NP-Hard.

On the other hand, if the underlying network is not connected or if some of the arcs in the network are directed and other arcs in the network are undirected, then the problem of finding a minimum cost cycle that traverses all of the required arcs and other arcs (which are called deadhead arcs) becomes NP-Hard. The NP-Hard class of Chinese Postman problems are generally those encountered in practice.

STREET ROUTING AND SCHEDULING PROBLEMS: The Euclidean distance assumption may not be realistic when solving practical vehicle routing problems. In Figure 1, a five location traveling salesman solution, A-B-C-D-E-A, using Euclidean distances is displayed. In Figure 2, the same traveling salesman solution (A-B-C-D-E-A) is displayed but the locations to be serviced are superimposed on a street network. In this solution, the vehicle would drive past locations and not service these locations and then have to service these locations later in the route, increasing the total travel time of the route.

The solution in Figure 3 is generated by solving a traveling salesman problem where the shortest travel time path between each pair of locations is used rather than Euclidean dis-

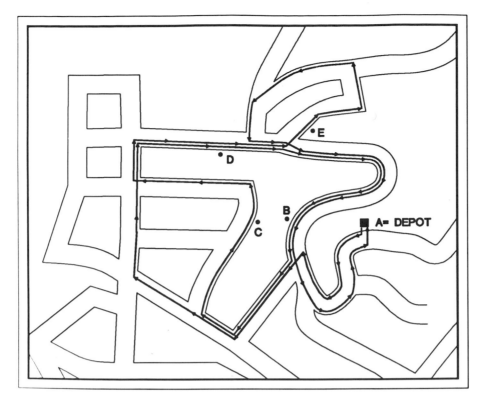

Fig. 2. Actual travel path for Euclidean distance route in Figure 1.

Notes on Figure 2

Order of servicing the locations on the route is maintained as in Figure 1.

No U-turns are allowed.

Service is carried out on the location's side of the street without making a left-hand turn.

To get from location A to location B without U-turns and ensuring that service is carried out on the location's side of the street, the route has to go around several blocks.

tances. In this solution, no U-turns are allowed and the vehicle is forced to traverse each street segment on the right-hand side. In this solution, the vehicle always services locations as it drives past these locations.

A special class of practical vehicle routing problems are called *street routing problems*. Street routing problems consist of point-to-point routing problems where locations to be serviced are dense and arc routing problems. In street routing problems, the locations to be serviced are located on a digital map and the deadhead travel times are computed as shortest paths rather than Euclidean distances. In arc routing problems, most street segments in the region are to be serviced.

GEOGRAPHIC INFORMATION SYSTEMS: To solve street routing and scheduling problems requires an accurate digital street network and a Geographic Information System (GIS). A *digital street network* is a street segment by street segment representation of a geographic region. A **GIS** is a system of computer hardware, software

and procedures designed to support the capture, management, manipulation, analysis, modeling and display of a digital street network. With an accurate GIS, the user is able to *address match* the locations to be serviced on the digital street network, to compute the travel times between locations as shortest travel-time paths and to give accurate street-by-street travel directions for each vehicle route. On the other hand, the Euclidean distance approach for solving vehicle routing problems only gives an ordering of the locations of a route.

CONCLUDING REMARKS: Vehicle routing systems are increasing in functionality and sophistication. These systems will have better graphics, user interfaces and allow for accurate and practical solutions to be found quickly. Moreover, the class of applications are increasing, leading to more new problems being formulated and new computational algorithms being developed. Computerized vehicle routing systems will become a necessary part of an organization's logistics/distribution system.

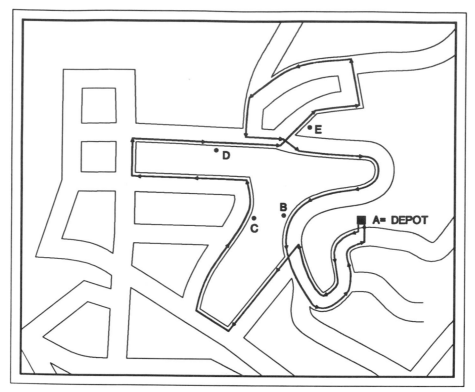

Fig. 3. Travel path when shortest paths are computed over the street network.

Notes on Figure 3

No U-turns are allowed.

Service is carried out on the location's side of the street without making a left-hand turn.

See **Chinese postman problem; Computational complexity; Geographic information systems; Graph theory; Logistics; Traveling salesman problem; Visualization.**

References

[1] Bodin, L. (1990), "Twenty Years of Routing and Scheduling," *Operations Research*, 38, 571–579.

[2] Bodin, L., B.L. Golden, A. Assad, and M. Ball (1983), "Routing and Scheduling of Vehicles and Crews: The State of the Art," *Computers & Operations Research*, 10(2), 63–211.

[3] Garey, M. and D. Johnson (1979), *Computer and Intractibility: A Guide to the Theory of NP-Completeness*, Freeman Press, San Francisco.

[4] Golden, B.L. and A. Assad, eds. (1986), *Special Issue on Time Windows, American Jl. Mathematical and Management Sciences*, 6 (3 and 4), 251–399.

[5] Golden, B.L. and A. Assad (1988), *Vehicle Routing: Methods and Studies.* North-Holland, Amsterdam.

[6] Karp, R. (1975), "On the Computational Complexity of Combinatorial Problems," *Networks*, 5, 45–68.

[7] Lenstra, J.K. and A. Rinnooy Kan (1981), "Complexity of Vehicle Routing and Scheduling Problems," *Networks*, 11, 221–227.

[8] Papadimitriou, C. and K. Stieglitz (1982), *Combinatorial Optimization*, Prentice Hall, Englewood Cliffs, New Jersey.

VEHICLE SCHEDULING

See **Vehicle routing**.

VERIFICATION

For a mathematical model, especially a computer-based one, verification is the process by which the computational procedure (computer software) is checked to determine if it is error free (debugged) and the determination that the model, as represented by the calculations or software, does what the analyst intended. A model is said to be verified if it (the computation) correctly executes the intended calculations. See **Validation; Verification, validation and testing of models.**

VERIFICATION, VALIDATION AND TESTING OF MODELS

Osman Balci

Virginia Polytechnic Institute & State University, Blacksburg

In operations research/management science (OR/MS) modeling studies, we work with a model of a problem rather than directly working with the problem itself. A model lacking a sufficiently accurate representation produces erroneous results which can be catastrophic when making critical decisions based on the model results. Concerned with the credibility of models used in the decision and policy-making functions of the Federal Government, the U.S. General Accounting Office (GAO) submitted a report to the Congress in 1976 on ways to improve management of federally funded computerized models (U.S. GAO, 1976). In 1979, the U.S. GAO published a report on guidelines for model evaluation (U.S. GAO, 1979). The National Bureau of Standards has produced several special publications (Gass, 1979, 1980, 1981) which advanced the state of the art in model assessment.

Our purpose here is to discuss the principles and techniques of verification, validation and testing (VV&T) for OR/MS models. After presenting some background information, the six principles of model VV&T are introduced. A taxonomical brief overview of 44 model VV&T techniques are given. The applicability of the techniques should be judged with respect to the model type (e.g., mathematical programming model, stochastic optimization model, simulation model).

BACKGROUND: A *model* is a representation and an abstraction of anything such as a system, concept, problem, or phenomena. It can have inputs, parameters, and outputs as illustrated in Figure 1. The term "system" is used to refer to whatever the model represents.

Model Verification is substantiating that the model is transformed from one form into another, as intended, with sufficient accuracy. Model verification deals with building the model *right*. The accuracy of transforming a problem formulation into a model specification or the accuracy of converting a model representation in micro flowchart into an executable computer program is evaluated in model verification.

Model Validation is substantiating that the model, within its domain of applicability, behaves with satisfactory accuracy consistent with the study objectives. Model validation deals with building the *right* model. It is conducted by executing/running the model under the "same" input conditions that drive the system and by comparing model behavior with the system behavior. (Note that a linear programming model is executed and a simulation model is run.) The comparison of model and system behaviors should not be made one output variable at a time, that is, O_1^m versus O_1^s, O_2^m versus O_2^s, etc. A multivariate comparison should be carried out to incorporate the correlations among the output variables.

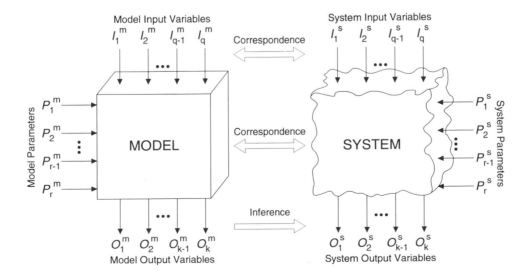

Fig. 1.

Model Testing is demonstrating that inaccuracies exist in the model or revealing the existence of errors in the model. In model testing, we subject the model to test data or test cases to see if it functions properly. "Test failed" implies the failure of the model, not the test. Testing is conducted to perform verification and validation. Some tests are intended to judge the accuracy of model transformation from one form into another (verification). Some tests are devised to evaluate the behavioral accuracy (i.e., validity) of the model. Therefore, we commonly refer to the whole process as model VV&T.

Model VV&T is conducted to prevent occurrences of three major types of errors in OR/MS modeling studies (Balci, 1990): *Type I Error* is the error of rejecting the model credibility when in fact the model is sufficiently credible; *Type II Error* is the error of accepting the model credibility when in fact the model is *not* sufficiently credible; and *Type III Error* is the error of solving the wrong problem. The probability of committing the Type I Error is called *Model Builder's Risk* and probability of committing the Type II Error is called Model User's Risk. Committing the Type I error increases the cost of model development. The consequences of committing the Type II and Type III errors may be catastrophic. Therefore, a cost-risk analysis should be conducted in those cases where data can be collected from the system under study (Balci and Sargent, 1981).

PRINCIPLES: Six principles of model VV&T are given below.

Principle 1: The outcome of model VV&T should not be considered as a binary variable where the model is absolutely correct or absolutely incorrect. Since a model is an abstraction of a system, perfect representation is never expected. A model is built for a specific purpose and its credibility is judged with respect to that purpose. The purpose dictates how representative the model should be. Sometimes, 60% representation accuracy may be sufficient; sometimes, 95% accuracy may be required.

Principle 2: The model VV&T must be conducted throughout the entire modeling life cycle starting with problem formulation and culminating with the presentation of model results. Errors should be detected as early as possible in the life cycle. Correcting errors detected in later phases of the life cycle is much more expensive. Some vital errors may not be detectable in later phases resulting in the occurrence of Type II or Type III error.

Principle 3: The model VV&T requires independence to prevent developer's bias. The organization which is contracted to conduct the modeling study is not qualified to perform the *final* model VV&T (acceptance testing). The sponsor of the modeling study should identify an independent third party to conduct the final model VV&T. To emphasize this principle, VV&T is called Independent VV&T (IVV&T) by many authors in the literature.

Principle 4: The model VV&T is difficult and requires creativity and insight. Knowledge of the problem domain, expertise in the modeling methodology, and prior modeling and VV&T experience are required. It is not possible for one person to fully understand all aspects of a large and complex model, especially if the model is a stochastic one containing hundreds of concurrent activities.

Principle 5: Complete model testing is not possible. Exhaustive (complete) testing requires testing the model under *all* possible inputs. Combinations of feasible values of model input variables can generate millions of logical paths in the model execution. Due to time and budgetary constraints, it is impossible to test the accuracy of millions of logical paths. Therefore, in model testing, the purpose is to increase our confidence in model credibility, as much as dictated by the study objectives, rather than trying to show 100% credibility.

Principle 6: The model VV&T must be planned and documented. Testing is not a phase or step in model development life cycle; it is a continuous activity *throughout* the entire life cycle. The tests should be identified, test data or cases should be prepared, tests should be scheduled, and the whole testing process should be documented. All test data and cases must be preserved for use in model maintenance.

TECHNIQUES: A taxonomy, presented in Figure 2, classifies 44 Model VV&T techniques into six categories. The level of mathematical formality of each category increases from very informal on the far left to very formal on the far right. Likewise, the complexity also increases as the category becomes more formal (Whitner and Balci, 1989).

Informal techniques are among the most commonly used ones. They are called informal because the tools and approaches used rely heavily on human reasoning and subjectivity without stringent mathematical formalism. The "informal" label does not imply any lack of structure and formal guidelines for the use of the techniques.

The *audit* is conducted by a single person to investigate the adequacy of model development process with respect to established practices, standards, and guidelines. *Desk Checking* consists of manually examining one's work to judge its accuracy. *Face Validation* is useful as a preliminary approach to validation. The modeling project team members, potential users of the model, people knowledgeable about the system under study, based on their estimates and intuition, subjectively compare model and system behaviors to judge whether the model and its results are reasonable. *Inspections* are formally conducted by a team, usually composed of moderator, designer, implementer, and tester, in six phases: planning, overview, preparation, inspection, rework, and follow-up. *Reviews* are formally conducted by a team of experts and seek to evaluate the model with respect to standards, guidelines, and specifications. The *Turing Test* is based upon the expert knowledge of people about the system under study. These people are presented with two sets of output data obtained, one from the model and one from the system, under the same input conditions. Without identifying which one is which, the people are asked to differentiate between the two. If they succeed, they are asked how they were able to do it. Their response provides valuable feedback for correcting model representation. If they cannot differentiate, our confidence in model validity is increased. *Walkthroughs* are conducted similar to inspections and reviews. In an organized manner, the team members walk through the details of the model design or source code to assess the completeness, consistency, and unambiguity of model representation.

Static techniques are intended to assess the quality of model structure without requiring model execution. *Consistency Checking* is concerned with: (a) justifying that model representation does not contain contradictions, (b) all specifications are unambiguous, (c) all model components fit together properly, and (d) all data elements are manipulated properly. *Data Flow Analysis* deals with justifying the accurate use of model variables. With the help of a data flow diagram, the definitions, referencing, dependencies, linkages, and transformations of data can be investigated. *Graph-Based Analysis* employs three types of diagnostics based on examination of graphs of model representation: analytical, comparative, and informative (Nance and Overstreet, 1987). *Semantic Analysis* is concerned with evaluating if the modeler's intentions are accurately translated into the model representation. *Structural Analysis* is used to examine the model structure and determine if it adheres to structured design and development principles. *Syntax Analysis* deals with checking if the mechanics of the executable modeling language are being applied correctly. This fundamental source code analysis is mostly done by the language compiler.

Dynamic techniques are employed to evaluate dynamic model characteristics and require model execution. *Black-Box Testing*, also called Functional Testing, is concerned with judging the accuracy of the input-output transformation of a model or submodel. *Bottom-Up Testing* is conducted submodel by submodel from base level submodels (the ones that are not decomposed further) all the way up to the model level. *Debugging* is an iterative process that (a) involves

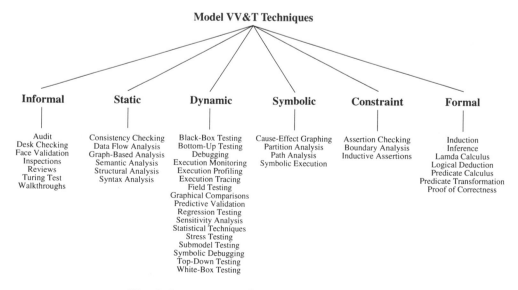

Fig. 2. A taxonomy of model VV&T techniques.

locating the source of error detected by testing, (b) finding a remedy to remove the error, (c) making the correction, and (d) retesting the model to ensure successful modification. *Execution Monitoring* produces information about events and activities that take place during model execution. This information is analyzed for the model verification and validation. *Execution Profiling* is similar to execution monitoring except that it constructs a model execution profile at a much higher level. *Execution Tracing* produces a low-level recording of model execution by way of instrumenting the model. The desired information is extracted from the trace data and is used in the verification and validation activities. *Field Testing* places the model in an operational situation for the purpose of collecting as much information as possible for model verification and validation. These tests are especially useful for validating models of military combat systems. Although it is usually difficult, expensive, and sometimes impossible to devise meaningful field tests for complex systems, their use wherever possible helps both the project team and decision makers to develop confidence in the model. *Graphical Comparisons* is a subjective, inelegant, and heuristic, yet quite practical technique especially useful as a preliminary approach. The graphs of values of model variables over time are compared with the graphs of values of system variables to investigate characteristics such as similarities in periodicities, skewness, number and location of inflection points, logarithmic rise and linearity, phase shift, trend lines, and exponential growth constants. *Predictive Validation* requires past data. The model is driven by past system input data and its forecasts are compared with the corresponding past system output data to test the predictive ability of the model. *Regression Testing* seeks to assure that model modifications do not cause errors and create adverse side-effects. *Sensitivity Analysis* is performed by systematically changing the values of model input variables and parameters over some range of interest and observing the effect upon model behavior. Unexpected effects may reveal invalidity. The input values can also be changed to induce errors to determine the sensitivity of model behavior to such errors. Sensitivity analysis can identify those input variables and parameters to the values of which model behavior is very sensitive. Then, model validity can be enhanced by assuring that those values are specified with sufficient accuracy. *Statistical Techniques* are used to conduct model validation by comparing model and system output data ob-

tained by running both model and system under the same input data. Some example statistical techniques for model validation include Confidence Intervals/Regions, Hotelling's T^2 Tests, Multivariate Analysis of Variance, Nonparametric Goodness-of-fit Tests, Nonparametric Tests of Means, and Time Series Analysis. *Stress Testing* is used to evaluate model accuracy under unexpectedly high levels of workload. *Submodel Testing* requires a top-down model decomposition in terms of submodels. The model is instrumented to collect data on all input and output variables of a submodel. The system is similarly instrumented (if possible) to collect similar data. Then, each submodel behavior is compared with corresponding subsystem behavior. *Symbolic Debugging* employs a debugging tool which enables a modeler to manipulate model execution while viewing the model at the source code level. *Top-Down Testing* is conducted concurrently with top-down model development. Testing starts with global model at the highest level and proceeds with lower level submodels. When testing submodels at a given level, calls to submodels at lower levels are simulated using dummy submodels. *White-Box Testing* is used to analyze the internal model structure and execution paths.

Symbolic techniques provide symbolic inputs to a model and produce expressions for the output which are derived from the transformation of the symbolic data along model execution paths. They are used to assess the model's input-output transformational accuracy. *Cause-Effect Graphing* seeks to determine the accuracy of transforming causes (input conditions) into effects (output conditions). *Partition Analysis* decomposes both model specification and model implementation into functional representations by using symbolic evaluation techniques which maintain algebraic expressions of model elements and show model execution paths. These functional representations are then compared to determine if they are sufficiently close to each other. *Path Analysis* attempts to establish model correctness on the basis of complete testing of all model execution paths. *Symbolic Execution* is performed by executing the model using symbolic values rather than actual data values for input. During model execution, the symbolic values are transformed along a model execution path and the resulting expressions are output for analysis.

Constraint techniques assess model accuracy based on comparisons between model assumptions and actual conditions arising during model execution. *Assertion Checking* examines what *is* happening during model execution against what the modeler *assumes* is happening. An *assertion* is a statement that should hold true as the model executes. Model assumptions are converted into assertions which are placed within the model at various logical points. The assertion statements are checked during model execution to make sure that all assumptions underlying the model are satisfied. *Boundary Analysis* is used to evaluate model correctness by using test cases generated just within, on top of, and just outside of the boundaries of input equivalence partitions. *Inductive Assertions* require the specification of input-to-output relations for all model variables in terms of assertion statements which are placed at the beginning and at the end of each model execution path. Model assessment is achieved by proving that for each path, if the assertion at the beginning of the path is true, and all statements along the path are executed, then the assertion at the end of the path is true.

Formal techniques are based on formal mathematical proof of correctness. If attainable, formal techniques provide the most effective means of model assessment. *Induction*, *Inference*, and *Logical Deduction* are acts of justifying conclusions on the basis of premises given. *Lamda Calculus* is a system of transforming the model representation into formal expressions for which mathematical proof techniques can be applied. *Predicate Calculus* provides rules for manipulating predicates (combinations of simple relations) which are derived from the model representation. *Predicate Transformation* is used to define the model semantics with a mapping which transforms model output states to all possible model input states. This definition provides the basis for proving whether or not the model is sufficiently correct. *Proof of Correctness* is employed to express the model in a precise notation and then mathematically proving that: (a) the executed model terminates and (b) it satisfies the requirements of its specification.

CONCLUDING REMARKS: In modeling studies, it is well to remember the dictum that "Nobody solves *the* problem. Rather, everybody solves the model that he [or she] has constructed of the problem" (Elmaghraby, 1968). This dictum clearly identifies the crucial importance of model credibility. If the model does not represent the problem with sufficient accuracy, the modeling study becomes useless.

The model VV&T principles and techniques presented here indicate that assessment of model credibility is an onerous task requiring multifaceted and interdisciplinary knowledge and experience. Out of the 44 techniques described, we try to apply as many of them as possible so as to increase our confidence in model credibility as much as needed. The amount of credibility required or when to stop testing is determined with respect to the study objectives.

See **Model accreditation; Battle modeling; Documentation; Model evaluation; Model management; Practice of OR/MS; Regression; Risk management; Sensitivity analysis; Simulation of discrete-event stochastic analysis; Structured modeling; Systems analysis; Validation; Verification.**

References

[1] Balci, O. (1990). "Guidelines for Successful Simulation Studies." *Proceedings 1990 Winter Simulation Conference*, IEEE, 25–32, Piscataway, New Jersey.

[2] Balci, O. and R.G. Sargent (1981). "A Methodology for Cost-Risk Analysis in the Statistical Validation of Simulation Models." *Communications ACM* 24, 190–197.

[3] Elmaghraby, S.E. (1968). "The Role of Modeling in I.E. Design." *Industrial Engineering* 19, 292–305.

[4] Gass, S.I., ed. (1979). "Utility and Use of Large-Scale Mathematical Models," Proceedings of a Workshop, NBS Special Publication 534, Washington, D.C.

[5] Gass, S.I., ed. (1980). "Validation and Assessment Issues of Energy Models," Proceedings of a Workshop, NBS Special Publication 569, Washington, D.C.

[6] Gass, S.I., ed. (1981). "Validation and Assessment of Energy Models," Proceedings of a Symposium, NBS Special Publication 616, Washington, D.C.

[7] Nance, R.E., and C.M. Overstreet (1987). "Diagnostic Assistance Using Digraph Representations of Discrete Event Simulation Model Specifications." *Transactions SCS* 4, 33–57.

[8] U.S. General Accounting Office (1976). "Ways to Improve Management of Federally Funded Computerized Models," LCD-75-111, U.S. GAO, Washington, D.C.

[9] U.S. General Accounting Office (1979). "Guidelines for Model Evaluation," PAD-79-17, U.S. GAO, Washington, D.C.

[10] Whitner, R.B. and O. Balci (1989). "Guidelines for Selecting and Using Simulation Model Verification Techniques." *Proceedings 1989 Winter Simulation Conference*, IEEE, 559–568, Piscataway, New Jersey.

VERT

Venture evaluation and review technique. A network simulation technique design for systematic assessment of the risks involved in undertaking a new venture and in resource planning, control monitoring and overall evaluation of ongoing projects, programs and systems. See **Network planning**; **Project management**; **Research and development**.

VERTEX

See **Extreme point**; **Node**.

VIRTUAL REALITY

An extension of the simulator concept in which the simulated external world is dynamic rather than static. Actions by the simulator operator(s) may change the simulated external world in the same way those actions would affect the real world through the use of the real equipment. The extent of the realism achieved varies with the system. See **Combat modeling**.

VISUALIZATION

Peter C. Bell

Western Business School

University of Western Ontario,
London, Canada

BACKGROUND: Operations research/management science (OR/MS) is increasingly involved in the development and use of *visualization techniques* at various stages of the problem solving cycle. Jones (1994) provides many examples including the use of natural language and informal diagrams at the problem conceptualization stage; spreadsheets, and block structured languages at the problem formulation stage; spreadsheets and relational databases during data collection; interactive optimization, and network flow graphics during problem solution; objective plots and matrix images at the solution analysis stage; and animation, hypertext, hypermedia, and presentation graphics for results presentation.

The use of visualization techniques such as these is not new: many of the earliest examples of OR/MS problem solving made use of visual concepts. Graphs were routinely used to summarize the results of modeling studies, flowcharts were used to sketch out the flow of an algorithm, and graphical techniques have long been a mainstay of teaching about the simplex method. However, visualization has moved from a position at the periphery of OR/MS to being an important driver of new developments in the field.

The emergence of visualization into the centre stage of OR/MS parallels developments in computing which have seen the industry's early emphases on "number-crunching" speed, and storage capacity replaced by those of "user-friendliness" and marketability. These developments have led to spreadsheet software with a visual presentation based on rows and columns, the "WIMP" (Windows/Icons/Mouse/Pull-down-menu) user interface now ubiquitous for personal computer operating systems, and a large variety of user-friendly software for the production of colorful, perhaps dynamic, computer generated pictures. Just as these developments have proved marketable for the computer industry, so have they proved marketable within OR/MS with the consequence that a large amount of OR/MS software which produces vivid visualizations is now being marketed.

Sophisticated computer-generated graphics represent the most elaborate extreme of the spectrum of visualization possibilities. Many other methods are employed which use many different spatial techniques to add information content to data. At the opposite end of this spectrum are some very simple tools where the picture elements are characters which are spatially arranged to have limited visual characteristics: for example, text, the arrangement of data in a table, a set of corporate accounts, and the block structure in a computer code all use a simple visual layout to improve understanding of the numbers and characters. Between these two extremes are a host of tools which have traditionally been characterized in two main groups: *presentation graphics*, and *iconic graphics*.

PRESENTATION GRAPHICS: Presentation graphics are pictures (for example, bar charts, line graphs, or pie charts) which are used to illustrate or summarize data. The use of these types of tools predates the computer era; from the earliest days of OR/MS, presentation graphics have been used to summarize data, to illustrate the results of OR/MS work, and to aid in communicating data or results to decision makers or management. Considerable research has been done that addresses issues such as when a graph is more useful than numbers, what type of graph is most useful in which situation, and when the use of color adds value to a presentation graphic (Desanctis, 1984). The results of this body of research suggest that the nature of the task is very important in determining the appropriateness of numerical display or various presentation graphic forms (Vessey, 1991).

ICONIC GRAPHS: Iconic graphics include picture elements which map to elements of the real world. A road map is an iconic graphic consisting of lines which are icons, representing roads, and blocks which represent urban areas. Other common iconic graphics include floor plans, PERT charts, and network flow diagrams. Again, iconic graphics have a long history within OR/MS, but research on the value of iconic formats is lacking. For many people, the value has been obvious: try driving from New York to San Francisco using only numeric data on road and town locations! Often, however, there are alternative iconic representations for a problem, but research has been slow to provide answers to resolve these choices. As a consequence, the market has been the determining factor in deciding which iconic formats survive and which die, with the result that the survivors are often high on color and "razzle-dazzle," but, perhaps, not those that are the most useful.

Iconic graphics can be categorized as *static* or *dynamic*. An important application area for static iconic graphics has been transportation systems routing and planning. Models which link mathematical programming models to computer-generated road or street maps have been used to solve truck routing and scheduling problems, mass transit system route planning and scheduling problems, and school bus routing problems (Florian *et al.*, 1987; Bodin and Levy, 1994).

Dynamic iconic graphics, or *animations*, were first applied to the study of operations problems by Hurrion (1980), and have proved to be a huge market success. The major application area for animation is simulation modeling, where animation is now routinely used to illustrate the progress of a simulation code. The use of animation seems to aid code debugging ("if it looks right and it moves right, then it probably is right"), model validation, and the presentation of the results of simulation studies to decision makers. *Visual interactive simulation* couples animation with interactive access to the running simulation model to produce *decision support systems* with visual *user interfaces* which provide useful tools to aid problem formulation and interactive problem solution (Bell, 1991). The use of animation and interaction with simulation models is now so pervasive that almost every major simulation software code now includes these facilities.

Animated sensitivity analysis uses dynamic graphics to illustrate the sensitivity of an optimal solution to changes in a parameter (Jones, 1992). As the parameter is changed, a visual screen display is updated 30 times/second to illustrate the response of the optimal solution to the change.

IMPACT OF NEW TECHNOLOGIES: The traditional view of visualization has been considerably expanded by new technologies. *Text* is a graphic format (the location of the characters has meaning) as is *hypertext*. These tools provide a host of visual formats, including choice of font, size, and layout, and the use of text and hypertext as a front-end for OR/MS models has developed into an important research area. Again, there exists a broad spectrum of possibilities from simple examples, such as the use of textual data on punched cards as input to mathematical programming software, to hypertext systems which provide the ability to navigate through a complex optimization problem (Kimbrough *et al.*, 1990).

The emergence of *multimedia* and *virtual reality* development tools at reasonable cost is driving new developments within OR/MS. As these technologies become more commonplace, we can expect to see many new kinds of OR/MS models which take advantage of the new delivery systems available for OR/MS work (Lembersky and Chi, 1984).

While the emergence of visualization as an important field within OR/MS appears to have been market driven, a body of research evidence is beginning to appear which supports a view that visualization helps decision makers solve problems. Surveys of model builders (Kirkpatrick and Bell, 1989) and of decision makers who have used visual and interactive models (Bell *et al.*, 1995) strongly support a view that model developers and decision makers believe that these types of tools lead to improved decision making, and explain the market success of software which provides animation capability for simulation models. Task based behavioral research comparing dynamic iconic graphic tools with non-visual tools has demonstrated the superiority of the graphic tools for some specific tasks (Bell and O'Keefe, 1995; Chau and Bell, 1995).

Finally, there is a growing body of evidence that suggests that the use of visualization and interaction in conjunction with OR/MS models and new information technology tools will have a revolutionary effect on OR/MS. These tools facilitate, or may even require, the use of innovative problem solving methodologies (Bell and O'Keefe, 1994), and the development of areas of new theory and new algorithms to support these methodologies (Bell, 1994). Jones (1994) is recommended for further reading.

See **Computational geometry; Computer science and operations research; Scheduling and sequencing; Simulation of discrete-event systems; Vehicle routing.**

References

[1] Bell, Peter C. (1991), "Visual Interactive Modelling: The Past, the Present, and the Prospects," *European Jl. of Operational Research*, 54, 274–286.

[2] Bell, Peter C. (1994), "Visualization and Optimization: The Future Lies Together," *ORSA Jl. on Computing*, 6, 258–260.

[3] Bell, Peter C., Elder, Mark and Staples, Sandy (1995), "Decision Makers' Perceptions of the Value and Impact of Visual Interactive Models," Technical Report, University of Western Ontario.

[4] Bell, Peter C. and O'Keefe, Robert (1994),"Visual Interactive Simulation: A Methodological Perspective," *Annals Operations Research*, **53**, Volume on Simulation and Modeling, Osman Balci, ed., 321–342.

[5] Bell, Peter C. and O'Keefe, Robert (1995), "An Experimental Investigation into the Efficacy of Visual Interactive Simulation," *Management Science* (forthcoming).

[6] Bodin, Lawrence and Levy, Laurence (1994), "Visualization in Vehicle Routing and Scheduling Problems," *ORSA Jl. on Computing*, 6, 261–269.

[7] Chau, Patrick and Bell, Peter C. (1995), "Designing Effective Simulation-Based Decision Support Systems: An Empirical Assessment of Three Types of Decision Support System," *Jl. of the Operational Research Society* (forthcoming).

[8] Desanctis, Gerardine (1984), "Computer Graphics as Decision Aids: Directions for Research," *Decision Sciences*, **15,** 463–487.

[9] Florian, M, Crainic, T. and Guelat, J. (1987), "FRET – An Interactive Graphic Method for Strategic Planning of Freight Flows," presented at the IFORS '87 Conference, Buenos Aires.

[10] Hurrion, Robert D. (1980), "An Interactive Visual Simulation System for Industrial Management," *European Jl. of Operational Research*, **5**, 86–93.

[11] Jones, Christopher V. (1992), "Animated Sensitivity Analysis," in *Computer Science and Operations Research: New Developments in Their Interface*, O. Balci, R. Sharda and S. A. Zenios, eds., Pergamon Press, Oxford, United Kingdom, 177–196.

[12] Jones, Christopher V. (1994), "Visualization and Optimization," *ORSA Jl. on Computing*, 6, 221–257.

[13] Kimbrough, Steven O., Pritchett, Clark W., Bieber, Michael P. and Bhargava, Hemant K. (1990), "The Coast Guard's KSS Project," *Interfaces*, 20, 5–16.

[14] Kirkpatrick, Paul, and Bell, Peter C. (1989), "Visual Interactive Modelling in Industry: Results from a Survey of Visual Interactive Model Builders," *Interfaces*, **19**, 5, 71–79.

[15] Lembersky, M.R. and Chi, U.H. (1984), "Decision Simulators Speed Implementation and Improve Operations," *Interfaces*, **14,** 4.

[16] Vessey, I. (1991), "Cognitive Fit: A Theory-Based Analysis of the Graphics versus Table Literature," *Decision Sciences*, 22, 219–241.

VOGEL'S APPROXIMATION METHOD (VAM)

A method for finding a first feasible solution to a transportation problem. The procedure begins by finding the two lowest cost cells for each row and column in the transportation problem array. Subtracting the smaller of these costs from the other produces a Vogel number for each row and column. Select the largest Vogel number and make the first assignment to the corresponding lowest cost cell, where the assignment is the maximum amount that can be sent from the corresponding origin to the corresponding destination. After each assignment, the Vogel numbers are recomputed based on the remaining rows and columns in the array. The procedure is repeated until all assignments (shipments) are made. Although VAM tends to find a good (low cost) first feasible solution, the extra computational work required has proven to be a detriment to its use in computer-based software for solving transportation problems. See **Northwest-corner rule; Transportation simplex method**.

VON NEUMANN-MORGENSTERN (EXPECTED) UTILITY THEORY

See **Decision analysis; Preference theory; Utility theory**.

VORONOI CONSTRUCTS

Isabel Beichl, Javier Bernal and Christoph Witzgall

National Institute of Standards & Technology, Gaithersburg, Maryland

Francis Sullivan

Supercomputing Research Center, Bowie, Maryland

INTRODUCTION: Given a finite set S of "sites" p_i located in Euclidean space $\mathscr{R}^d$, the *Voronoi polyhedron* $V(p_j)$ of site p_j is the set of all points $p \in \mathscr{R}^d$ which are at least as close to site p_j as to any other site p_i.

Such a Voronoi polyhedron (also called "Thiessen polygon" or "Wigner-Seitz cell") is convex, its facets determined by perpendicular bisectors – (hyper)planes or lines of equal Euclidean distance from two distinct sites. The Voronoi polyhedra $V(p_i)$, $p_i \in S$ cover the space $\mathscr{R}^d$ and define a polyhedral cell-complex known as a *Voronoi diagram* (Voronoi, 1908) or *Dirichlet tesselation* (Dirichlet, 1850). For a survey, consult Aurenhammer (1991), and the text by Okabe, Boots, and Sugihara (1992).

The cells of the dual complex are convex and, in general, simplicial. By partitioning nonsimplicial cells of the dual complex into simplices, the *Delaunay triangulation* results (Fig. 1). It provides a canonical scheme for triangulating the convex hull of an arbitrary set $S \subset \mathscr{R}^d$ of sites, with these sites as vertices. Under the assumption that sites are realizations of a homogeneous Poisson process, statistics for geometrical parameters of Voronoi diagrams and Delaunay triangulations have been derived (Miles, 1970; Stoyan, Kendall, and Mecke, 1987).

DELAUNAY TRIANGULATION: For each site $p_i \in S$, the Delaunay triangulation contains an edge from p_i to each of its nearest Euclidean neighbors $q \in S$. In particular, edges in that triangulation connect all pairs of points of minimum distance in S. The 1-skeleton of the Delaunay triangulation contains the *relative neighborhood graph* (Toussaint, 1979), which in turn contains a Euclidean *minimum spanning tree*. The Delaunay triangulation thus provides a convenient tool for solving various proximity problems (Shamos and Hoey, 1975). Delaunay triangulations avoid narrow triangles (see below) as much as possible, are essentially unique, and are readily determined. They are often the triangulations of choice for constructing piecewise-linear surfaces and for applications of "finite element" techniques in engineering.

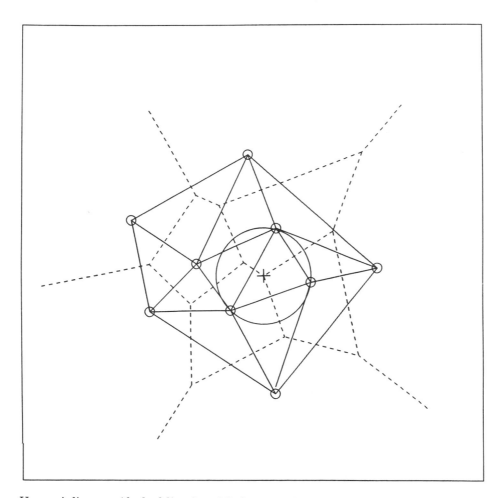

Fig. 1. Planar Voronoi diagram (dashed lines) and Delaunay triangulation of nine sites. The circle around one of the Delaunay triangles illustrates the "empty circle criterion."

Delaunay triangulations are characterized by the *empty sphere criterion*: the circumsphere of a simplex in a Delaunay triangulation does not contain any of the triangulation vertices in its interior (Delaunay, 1934). This criterion determines a triangulation uniquely in the absence of "degeneracy," that is, the occurrence of several simplices sharing a circumsphere.

In two dimensions, the empty circle criterion is equivalent to the requirement that the ascending sequence of angles, formed by selecting a smallest interior angle from each triangle in the triangulation, lexicographically maximizes the corresponding sequences for all triangulations of the same vertex set (*equiangularity*). The requirement that the sequence of *all* interior angles be lexicographically maximum is, in the presence of degeneracy, stronger, and can therefore serve in some instances as a tie-breaker in the presence of degeneracy.

The Delaunay triangulation of a set $S \subset \mathcal{R}^d$ of n sites can be obtained as a projection of the face lattice of the convex hull of n suitable points in $\mathcal{R}^{d+1}$. Those points can be chosen on a sphere – "stereographic projection" – or on a rotational paraboloid whose axis is perpendicular to the space of the triangulation. This implies that the Voronoi/Delaunay problem in d dimensions is computationally subsumed under the strong formulation of the convex hull problem in $d+1$ dimensions.

In order to check whether a given triangulation satisfies the empty sphere criterion, it is not necessary to verify that criterion for each simplex by scanning all sites which are not vertices of the simplex: only pairs of facet-adjacent simplices whose union is convex need to be examined as to whether any one of the two vertices not in the common facet might lie in the interior of its opposite circumsphere. This corresponds to establishing convexity of a (hyper)surface by examining the angles at which adjacent facets are joined. In two dimensions, the above criterion reduces to checking each strictly convex quadrangle formed by edge-adjacent triangles as to whether the correct diagonal of the quadrangle belongs to the triangulation (Lawson, 1977). Based on this observation, several simple and efficient methods such as the *insertion method* swap diagonals in quadrangles. Alternatively, "divide-and-conquer" as well as "plane sweep" techniques yield $O(n \log n)$ algorithms for planar Delaunay triangulation of n sites. Determination of Voronoi diagrams in linear expected time is discussed in Bentley, Weide, and Yao (1980), and Dwyer (1991).

In many applications, it is desirable to construct a planar triangulation with some prescribed edges while preserving the advantages – avoiding unnecessarily narrow triangles, essential uniqueness – of the Delaunay approach. In that case, the empty circle criterion can be generalized by testing for potential inclusion only those sites whose "visibility" from any point of the triangle is not blocked by a prescribed edge. This generalized empty circle criterion defines a *constrained Delaunay triangulation*, which is unique except for sites on the peripheries of empty circles (De Floriani and Puppo, 1988).

A second important generalization of the Voronoi diagram is the *power diagram* (Aurenhammer, 1987) or *radical Voronoi diagram* (Gellatly and Finney, 1982). Here sites may be enlarged to spheres of positive radius. The intersection, real or imaginary, of two spheres lies on and defines the "radical" (hyper)plane of that pair. These (hyper)planes then play the same role as the perpendicular bisectors in the classical Voronoi diagram. The radical Voronoi diagram of site spheres of radius $r_i \geq 0$ centered at locations $p_i \in \mathcal{R}^d$ respectively, can be obtained by intersecting the classical Voronoi diagram for the sites (p_i, r_i) in $d+1$ dimensions with the original d-dimensional space. Radical Voronoi diagrams are used in crystallography in order to account for differences in atomic radii.

The corresponding *radical Delaunay triangulation* in $\mathcal{R}^d$ satisfies a modified empty spheres criterion. For each simplex, consider the unique sphere K which is orthogonal to the $d+1$ site spheres. Then no other site sphere is "orthogonally interior" to sphere K, that is, is shifted towards the center of K from a position orthogonal to K.

There are numerous other generalizations of the Voronoi/Delaunay construct. Alternatives to the Euclidean norm, as well as general sets instead of single point sites, are considered. There are "order-k," "furthest point," "weighted," "discrete," and "abstract" Voronoi diagrams. Voronoi constructs based on the Euclidean metric are instances of cell-complexes derived from *arrangements* of hyperplanes. Data structures, algorithms and combinatorial results concerning such cell-complexes in general are presented by Edelsbrunner, O'Rourke, and Seidel (1986).

See **Computational geometry; Graph theory; Minimum spanning tree.**

References

[1] F. Aurenhammer (1987), "Power diagrams: properties, algorithms, and applications," *SIAM Jl. Comput.*, 16, 78–96.

[2] F. Aurenhammer (1991), "Voronoi Diagrams – a survey of a fundamental geometric data structure," *ACM Computing Surveys*, 23, 345–405.

[3] J.L. Bentley, B.W. Weide, and A.C. Yao (1980), "Optimal expected-time algorithms for closest point problems," *ACM Trans. Math. Software*, 6, 563–580.

[4] B. Delaunay (1934), "Sur la sphère vide," *Bull. Acad. Sci. USSR VII: Class. Sci. Mat. Nat.*, 793–800.

[5] G.L. Dirichlet (1850), "Uber die Reduction der positiven quadratischen Formen mit drei unbestimmten ganzen Zahlen," *Jl. Reine Angew. Math.*, 40, 209–227.

[6] L. De Floriani and E. Puppo (1988), "Constrained Delaunay triangulation for multiresolution surface description," *9th International Conference on Pattern Recognition*, Rome, Italy, 1, 566–569.

[7] R.A. Dwyer (1991), "Higher-dimensional Voronoi diagrams in linear expected time," *Discrete Comput. Geom.*, 6, 343–367.

[8] H. Edelsbrunner, J. O'Rourke, and R. Seidel (1986), "Constructing arrangements of lines and hyperplanes with applications," *SIAM Jl. Comput.*, 15, 341–363.

[9] B.J. Gellatly and J.L. Finney (1982), "Characterizations of models of multicomponent amorphous metals: the radical alternative to the Voronoi polyhedron," *Jl. of Non-Cryst-Solids*, 50, 313–329.

[10] C.L. Lawson (1977), "Software for C^1 surface interpolation," in *Mathematical Software III*, J.R. Rice (ed.), Academic Press, New York.

[11] R.E. Miles (1970), "On the homogeneous planar Poisson process," *Mathematical Biosciences*, 6, 85–127.

[12] A. Okabe, B. Boots, and K. Sugihara (1992), *Spatial Tesselations: Concepts and Applications of Voronoi Diagrams*, Wiley, New York.

[13] M.I. Shamos and D. Hoey (1975), "Closest-point problems," *Proc. 16th Annu. IEEE Sympos. Found. Comput. Sci.*, 151–162.

[14] D. Stoyan, W.S. Kendall, and J. Mecke (1987), *Stochastic Geometry and Its Applications*, Wiley, New York.

[15] G.T. Toussaint (1980), "The relative neighborhood graph of a finite planar set," *Pattern Recognition*, 12, 261–268.

[16] M.G. Voronoi (1908), "Nouvelles applications des paramètres continus à la théorie des formes quadratiques," *Jl. Reine Angew. Math.*, 134, 198–287.

VORONOI DIAGRAM

See **Computational geometry; Voronoi constructs.**

VV&A

Verification, validation and accreditation. See **Accreditation; Battle models; Model evaluation; Model management; Validation; Verification; Verification, validation and testing of models.**

VV&T

See **Verification, validation and testing of models.**

WAITING TIME

The time from entrance until leaving service in a queueing model or, sometimes, the time from entrance until the service is begun. These two are often differentiated by calling one the system waiting time and the other the line waiting time. See **Queueing theory**.

WAR GAME

A model whose object is military combat or some aspect of combat. War game is used to emphasize the competitive nature of the model, either through human interaction on one or more sides of the combat or automated, game-theoretic competition. See **Battle modeling**.

WAREHOUSE PROBLEM

A warehouse has a fixed capacity C and an initial stock s_0 of a certain product that is subject to known seasonal fluctuations in selling price and cost. The problem is to determine the optimal pattern of purchases, storage, and sales for the next n months. The problem can be formulated as a linear-programming problem. Its dual has an interesting form that enables the dual solution to be determined readily.

WATER RESOURCES

Roman Krzysztofowicz

University of Virginia, Charlottesville

Methodologies and techniques of operations research and management science have been applied to a vast array of water resource problems since the early 1960s. Conversely, water resource problems stimulated several methodological developments, notably in statistics of extremes, dynamic programming algorithms, and multi-objective optimization methods. Four classes of problems are discussed herein, techniques that have been employed are noted, and exemplary models are sketched. Fundamental to building operational models is the science of water transport processes.

HYDROLOGY AND HYDRAULICS: Variability of the quantity and quality of natural waters and their renewability in space and time are governed by the hydrologic cycle – a sequence of processes through which water is transported between the atmosphere, the land, and the ocean: precipitation, evaporation, transpiration, infiltration, groundwater flow, and river flow. Hydrology develops models of these natural processes; the models can be used to describe or predict the quantity and quality of water available in a given place and time (Young, 1993). This information, in turn, constitutes an input into models of water resource systems. Hydraulics develops models of flow in channels and lakes, and through constructed facilities such as spillways, sluices, fish ladders, turbines, pumps, pipelines, aqueducts, culverts, navigation locks, and floodways (Mays and Tung, 1992). These models serve as building blocks of control and management models.

PLANNING WATER RESOURCE DEVELOPMENT: The purpose of water resource development is to alter the natural hydrologic cycle so as to ensure the quantity and quality of water in places and times dictated by socio-economic objectives of human activities. Specific purposes are: (i) flood control, (ii) hydroelectric power generation, (iii) water supply for domestic, municipal, industrial and agricultural uses, and (iv) low-flow augmentation for navigation, recreation, water quality control (by diluting wastewater and contaminated runoff) and aquatic life maintenance (by increasing volume and decreasing temperature of rivers during summer).

For comprehensive planning, the natural boundary of a water resource system is a river basin (source of surface water) and the underlying acquifer (source of groundwater). Fig. 1 depicts an exemplary system. Planning involves tasks such as deciding the type, location and size of facilities, sequencing investments, and developing control policies; facilities may be operated individually or conjunctively (as two reservoirs in a cascade, or wells and reservoirs supplying irrigation water to the same district).

The planning process begins with identification of a time horizon (usually several decades) and objectives (usually multiple ones). Next, the available water resources are characterized:

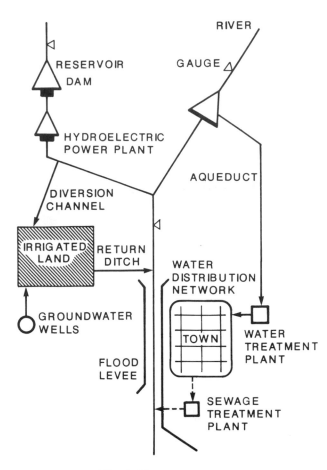

Fig. 1. Water resources.

groundwater supply and the rate of its recharge are estimated; river flows at gauge sites are modeled, for example, as time series (Hipel, 1985); extreme events such as floods (Potter, 1987) and droughts (SHH Special Issue, 1991) are modeled stochastically. Predictive models of water demands for various purposes are developed. Alternative system plans are designed, and operations of individual projects or sub-systems are described via simulation models or optimization models (in the form of integer, linear, nonlinear, chance-constrained, stochastic, or dynamic programming). Finally, all models are synthesized into a comprehensive river basin planning model which provides a decision support for a multi-objective analysis (WRB Special Issue, 1992). The purpose of the analysis is to screen a large number of alternative plans and to select a few which are Pareto-optimal. The choice of a plan for implementation is usually left to the political process (Loucks *et al.*, 1981).

OPERATION OF HYDROSYSTEMS: One of the most active and challenging research areas has been optimal control of reservoirs, aqueducts, irrigation systems, water distribution networks, urban drainage and sewage systems. Control

policies are almost always discrete-time (with the time interval of an hour, day, week, month or year), but otherwise they may be discrete- or continuous-state, finite- or infinite-horizon, stationary or nonstationary (e.g., periodic as the annual regime of river flows).

In a generic single reservoir control problem, the state x_n denotes the storage at the beginning of time interval n, the input ω_n represents the inflow during interval n, the control u_n is the release decided at the beginning of interval n, and the output y_n represents the outflow during interval n. With any finite $(n = 1, \ldots, N)$ trajectories $\mathbf{x} = \{x_n\}$ and $\mathbf{y} = \{y_n\}$, there is associated a performance measure $g(\mathbf{x}, \mathbf{y})$, whose form is dictated by reservoir purposes (e.g., generated hydropower, prevented flood damages). In a deterministic case, inflows $\{\omega_n\}$ are assumed to be known; hence $y_n = u_n$, and one wishes to find a policy $\boldsymbol{u}^* = \{u_n^*\}$ maximizing $g(\mathbf{x}, \mathbf{y})$ subject to the state dynamics, $x_{n+1} = x_n - u_n + \omega_n$, and constraints on storage and release. In a stochastic case, inflows follow a probabilistic law (usually of Markovian structure), and one wishes to find a strategy $\boldsymbol{\mu}^* = \{\mu_n^*\}$, a sequence of control rules, $u_n = \mu_n^*(x_n)$, that maximizes the expectation $\mathrm{E}[g(\mathbf{X}, \mathbf{Y})]$, subject to nonlinear state dynamics, output operators, and possibly probabilistic constraints.

The complexity of hydrosystems is reflected in the many control models described in the literature. They can be classified according to these features: (i) single reservoir vs. multi-reservoir, (ii) single purpose vs. multi-purpose, (iii) deterministic inflows vs. stochastic inflows, (iv) climatic statistics vs. hydrologic forecasts, (v) linear objective functions vs. nonlinear objective functions, (vi) separable objective functions vs. non-separable objective functions, (vii) single objective control vs. multi-objective control, (viii) short-term control (hourly, daily, weekly) vs. long-term control (monthly, yearly), (ix) one-level control vs. hierarchical control, (x) terminal condition vs. infinite horizon.

Deterministic control problems are often formulated as dynamic programs (DP) solved via discrete DP, successive approximation algorithms such as state incremental DP and differential DP (Yakowitz, 1982), or approximating linear-quadratic controllers (Protopapas and Georgakakos, 1990). Among other techniques one finds linear programming (Yeh *et al.*, 1980), and its chance-constrained variations, network flow algorithms, both linear and nonlinear (Rosenthal, 1981), and multi-objective optimization. Stochastic control problems are almost exclusively formulated as dynamic programs solved via discrete DP, policy iteration methods, or approximating linear-quadratic controllers. Various quasi-stochastic approaches have also been tried, such as sampling DP, simulation methods, combined simulation-optimization methods, and heuristic control strategies (Johnson *et al.*, 1991). Despite these advances, stochastic control of hydrosystems remains at the forefront of research – the challenges stemming from the dimensionality of the state space, spatial and temporal dependence of hydrologic inputs, nonlinear state dynamics, nonlinear and multiple objective functions.

MITIGATION OF FLOODS: Structural solutions, such as dams, diversion channels with retention basins, and levees, offer protection against floods up to a certain magnitude. Risk and benefit-cost analyses have guided decisions concerning the degree of protection and size of structures. Heuristic rules, simulation, and optimization methods have been employed to develop strategies for operation of reservoirs during floods.

Nonstructural solutions, such as floodplain zoning, flood insurance, and flood warning systems, aim at reducing the negative consequences of floods. Risk and decision analyses have been proposed for delineating land use zones, setting insurance rates, issuing flood warnings, and evaluating economic benefits of flood forecasts (Krzysztofowicz and Davis, 1984).

A decision-theoretic model of a flood warning system provides an example (Krzysztofowicz, 1993). Having received forecast (s,t) of $(H,\underline{\Lambda})$, the uncertain flood crest H and time to crest Λ at a river gauge, a manager must decide whether to issue ($w = 1$) or not to issue ($w = 0$) a warning for a zone of the floodplain above elevation y. Thereafter the zone is flooded ($\theta = 1$) or not ($\theta = 0$). Each decision-event vector (w,θ) leads to disutility

$$D_{w\theta}(s,t) = \int_y^\infty \int_0^\infty d_{w\theta}(h,\lambda)\phi(h,\lambda|s,t)d\lambda\,dh,$$

where $\phi(\cdot,\cdot|s,t)$ is the posterior density of (H,Λ), conditional on the forecast, and $d_{w\theta}(h,\lambda)$ is the disutility of all economic, social, and behavioral outcomes resulting from flood crest h occurring at time λ. The expected disutility associated with decision w, termed the risk function, is

$$R(s,t,w) = D_{w0}(s,t)\Pr\{\theta=0|s,t\} + D_{w1}(s,t)\Pr(\theta=1|s,t),$$

where Pr stands for probability. For each (s,t), the optimal warning rule W^* prescribes decision $w = W^*(s,t)$ which minimizes the risk $R(s,t,w)$.

MANAGEMENT OF WATER QUALITY: Water pollution comes from either point sources, which can be directly monitored (e.g., industrial wastewater discharges), or nonpoint sources, from which loadings can only be estimated (e.g., contaminated runoff from agricultural fields and urban areas). The preference of downstream users for clean water and the preference of upstream entities (such as municipalities, industries, and agricultural producers) for free discharging of contaminants create a societal conflict whose resolution requires legislative, economic, and institutional means.

Management models are typically formulated in support of planning by a regional authority faced with decisions such as locating and sizing wastewater treatment plants and effluent disposal fields, setting charges for release of wastewater, locating and operating monitoring networks, and devising enforcement policies. These decision problems are multi-objective and hierarchical in nature (Loucks *et al.*, 1981). At the upper level, the authority's objectives are (i) to minimize the total cost, (ii) to equitably allocate the cost to entities, and (iii) to improve the quality of waste-receiving waters. At the lower level, an entity's objectives are (i) to minimize its cost and (ii) to optimize its compliance with effluent standards and discharge regulations. Game-theoretic models are developed to predict the compliance behavior of entities, and thus the effectiveness of policies (WRB Special Issue, 1992). Water quality models – simulating the physical, chemical, and biological processes taking place in water bodies – are employed to predict impacts of alternative management plans on concentration of constituents (e.g., biochemical oxygen demand, dissolved oxygen deficit, nitrogen, phosphorus, metals, organics, bacteria) which collectively define water quality (Young, 1993).

See **Dynamic programming; Environmental system analysis; Global models; Linear programming; Multi-objective programming; Stochastic programming.**

References

[1] Hipel, K.W., ed. (1985). *Time Series Analysis in Water Resources*. American Water Resources Association, Bethesda, Maryland.
[2] Johnson, S.A., J.R. Stedinger and K. Staschus (1991). "Heuristic Operating Policies for Reservoir System Simulation." *Water Resources Research* 27, 673–685.
[3] Krzysztofowicz, R. (1993). "A Theory of Flood Warning Systems." *Water Resources Research* 29, 3981–3994.

[4] Krzysztofowicz, R. and D.R. Davis (1984). "Toward Improving Flood Forecast-Response Systems." *Interfaces* 14(3), 1–14.

[5] Loucks, D.P., J.R. Stedinger and D.A. Haith (1981). *Water Resource Systems Planning and Analysis*. Prentice-Hall, Englewood Cliffs, New Jersey.

[6] Mays, L.W. and Y-K. Tung (1992). *Hydrosystems Engineering and Management*. McGraw-Hill, New York.

[7] Potter, K.W. (1987). "Research on Flood Frequency Analysis: 1983–1986." *Reviews of Geophysics* 25, 113–118.

[8] Protopapas, A.L. and A.P. Georgakakos (1990). "An Optimal Control Method for Real-Time Irrigation Scheduling." *Water Resources Research* 26, 647–669.

[9] Rosenthal, R.E. (1981). "A Nonlinear Network Flow Algorithm for Maximization of Benefits in a Hydroelectric Power System." *Operations Research* 29, 763–786.

[10] SHH Special Issue (1991). "Drought Analysis." *Stochastic Hydrology and Hydraulics* 5, 253–322.

[11] WRB Special Issue (1992). "Multiple Objective Decision Making in Water Resources." *Water Resources Bulletin* 28, 1–231.

[12] Yakowitz, S. (1982). "Dynamic Programming Applications in Water Resources." *Water Resources Research* 18, 673–696.

[13] Yeh, W. W-G., L. Becker, D. Toy and A.L. Graves (1980). "Central Arizona Project: Operations Model." *Jl. Water Resources Planning and Management* 106, 521–540.

[14] Young, P.C., ed. (1993). *Concise Encyclopedia of Environmental Systems*. Pergamon Press, New York.

WEAK DUALITY THEOREM

See **Strong duality theorem**.

WEAKLY-COUPLED SYSTEMS

A linear-programming problem that has a few variables that connect (couple) the constraints or subsets of constraints. Such systems usually arise in time dimensioned large-scale problems that exhibit a block-angular structure. The dual of such systems are weakly-coupled in the sense of having a few constraints that tie the blocks together. Special adaptations of the simplex method exist that take advantage of such structures in their computation. See **Dualplex method**; **Large-scale systems**; **Rosen's partitioning method**.

WEBER PROBLEM

See **Location analysis**.

WILKINSON EQUIVALENT RANDOM TECHNIQUE

An approximation for the blocking probability that an overflow stream sees in an Erlang loss system. The method is used to analyze congestion in telecommunication networks. See **Queueing theory**.

WIMP

Window/Icons/Mouse/Pull-down-menu. See **Visualization**.

WOLFE'S QUADRATIC-PROGRAMMING PROBLEM ALGORITHM

An adaptation of the simplex method that solves quadratic-programming problems with positive definite or positive semidefinite quadratic forms. It is based on the simultaneous solution of the linear constraints of the problems and associated Karush-Kuhn-Tucker conditions. It uses a restricted basis entry for the solution of necessary complementarity conditions. See **Quadratic programming**.

WORK SCHEDULE

A schedule of hours and days to be worked. This issue is of special importance to emergency services which are usually provided 24 hours-a-day, 7 days-a-week, 12 hours aday. See **Emergency services**.

WORST-CASE ANALYSIS

For an algorithm and associated problem, the determination of an upper bound on the number of steps that the algorithm can take on any instance of the problem. For an optimization problem and an associated heuristic or nonoptimal algorithm, worst-case analysis may include a statement of how far the objective value returned by the algorithm will be from the true optimal solution.

$\bar{X}$-BAR CHART

A quality control chart that shows the variations in the averages of samples.

See **Quality control; R-chart.**

Y

YIELD MANAGEMENT

Richard O. Mason & Sue A. Conger

Southern Methodist University, Dallas, Texas

Yield management is undertaken to get the greatest return on a dollar of capital invested. It deals with optimal pricing of a limited number of products (e.g., rooms) when there are different product/price mixes (e.g., ski weekends, conventions, peak/low season), and when the demand for each product/price combination fluctuates and differs by location over time. Yield management, then, contains elements of mathematical and artificial intelligence problems, with the solution based on data provided by transactional applications.

In the hotel industry, rooms wear out and must be refurbished on a fairly regular schedule, but rates charged for the room differ depending on season, lead time to rental, number of nights' stay, promotions, and so on. Similarly, airline industry management involves pricing perishable airline seats at rates that will maximize revenues. Yield management in the manufacturing and hospitality industries is merging operations research methods with transaction and artificial intelligence methods to develop suggested rate changes throughout a given day or any other period of time (Gray, 1994).

See **Artificial intelligence; Inventory modeling; Systems analysis.**

References

[1] Gray, D.A. (1994), "Revenue-based capacity management, a survival technique developed by the airline industry, enters manufacturing sector," *OR/MS Today*, 21(5), 18–23.

Z

ZERO-ONE GOAL PROGRAMMING

A goal programming methodology that generates a solution for decision variables where the variables must be equal to one or zero.

ZERO-ONE VARIABLES

See: **Binary variables.**

ZERO-SUM

A competitive or economic situation is termed zero-sum when the total amount of money or comparable measure that is gained by some participants is exactly equal to the total amount of the measure that is lost by the remaining participants. The term is specifically associated with a game in which the sum of the payoffs lost or gained by the players is fixed.

See **Game theory**.

ZERO-SUM GAME

A game in which one side's gain is another side's loss. Outcomes in which both sides win are prohibited.

See **Game theory**.

ZERO-SUM TWO-PERSON GAME

See **Game theory**.

Index

Italic entries indicate figures.